UNITED STATES DEPARTMENT OF THE INTERIOR

BUREAU OF RECLAMATION

DESIGN OF SMALL DAMS

A Water Resources Technical Publication

Second Edition

1973

As the Nation's principal conservation agency, the Department of the Interior has basic responsibilities for water, fish, wildlife, mineral, land, park and recreational resources. Indian and Territorial affairs are other major concerns of America's "Department of Natural Resources."

The Department works to assure the wisest choice in managing all our resources so each will make its full contribution to a better United States—now and in the future.

UNITED STATES GOVERNMENT PRINTING OFFICE

WASHINGTON : 1973

For sale by the Superintendent of Documents, U.S. Government Printing Office, Washington, D.C. 20402, and the Bureau of Reclamation, Denver Federal Center, Denver, Colorado 80225, Attention: 922.
Price $12.65 domestic postpaid or $12.00 GPO Bookstore Stock Number 2403–0068

Shadow Mountain Dam and spillway on the Colorado River in Colorado. 7–400–192.

Preface to the Second Edition

THE FIRST edition of "Design of Small Dams" was published to serve primarily as a guide to safe practices for those concerned with the design of small dams in public works programs in the United States. Its publication was warmly received and, in the intervening years since then, it has been widely used in the United States, reprinted numerous times, and translated into many foreign languages, including Korean, Spanish, Japanese, and Chinese.

Since publication of the first edition of "Design of Small Dams," a large body of new literature has become available to dam designers, and many new design procedures used at the Bureau of Reclamation have been changed to reflect more modern techniques. As the number of changes in design techniques increased, it became apparent that their incorporation in a second edition would be beneficial to those individuals and agencies concerned with small dams.

The purpose of the second edition remains essentially the same as for the first edition. Many of the design procedures proposed in the first edition remain virtually unchanged. However, a number of new procedures have been developed by the Bureau and are currently in use. To make this new information available generally, it is included in the second edition. The increased concern of the Bureau of Reclamation with environmental problems is reflected by the inclusion of chapter II, "Ecological and Environmental Considerations." This chapter outlines some of the practical measures which may be taken to reduce the environmental and ecological impact of a project. Chapter III has been extensively revised to include current methods of design flood computation and to incorporate new graphical data.

Chapter V has been revised to reflect the availability of current information concerning foundation design and to include supplemental foundation investigation procedures. Chapter VI contains additional material on the design of earth dams, a discussion of the slurry trench method of cutoff construction, earthquake considerations, soil-cement design criteria, additional design details, and a more extensive reference list. The discussion of rockfill dam design has been considerably expanded in chapter VII to reflect the recent interest in rockfill dams and the growth of available information on this subject. Baffled spillway design procedures have been incorporated in chapter IX, and additional information on tunneling has been presented in chapter X. Appendix A includes new information on the estimation of rainfall runoff from soil cover data, and an expanded discussion of flow in natural channels is contained in appendix B. Appendix C includes new tables for the design of both reinforced concrete pressure pipe and cast-in-place conduits, and appendix E has a more complete discussion of the rapid method of compaction control. Appendix G has been expanded to include specifications concerning air and water pollution and each specification has been updated to reflect current Bureau requirements. A new appendix on reservoir sedimentation is presented in appendix H, which outlines current procedures used to estimate the rate of sedimentation and the period of time before sediment will interfere with the useful functions of the reservoir. A convenient list of conversion factors is presented in appendix I to facilitate the increased utilization of metric units. Many minor changes have been made throughout the text to reflect current design and construction techniques.

It is intended that this book will provide the designer with an important source of information. However, this text is not intended in any way to encourage the assumption of undue responsibility on the part of unqualified personnel, and the use or application of the methods and data contained herein is strictly the responsibility of the person utilizing the material. Designs should reflect the actual site conditions and should not merely be patterned after a successful design used at another location.

Periodically, the names of Bureau of Reclamation projects and features are changed by acts of Congress, Federal agencies, etc., and therefore there may be a few inconsistencies in the project and feature names referred to in the text.

Some recent changes include the following:

Cachuma Dam to Bradbury Dam

Wasco Reservoir to Clear Lake

Soap Park Reservoir to Milly K. Goodwin Lake

Missouri River Basin Project to Pick-Sloan Missouri Basin Program.

There are occasional references to proprietary materials or products in this publication. These must not be construed in any way as an endorsement since the Bureau cannot endorse proprietary products or processes of manufacturers or the services of commercial firms for advertising, publicity, sales, or other purposes.

The second edition was prepared by the engineers of the Bureau of Reclamation, U.S. Department of the Interior, in its Engineering and Research Center at Denver, Colo. A number of engineers and technicians participated in the preparation of the second edition and in its critical review, and the efforts of these persons are greatly appreciated. Special recognition is given to H. G. Arthur, Director of Design and Construction, for his overall guidance in preparation of the text and to Dr. J. W. Hilf, Chief of the Division of Design, for his technical advice.

The second edition of the text was coordinated, edited, and much supplemental technical information provided by L. W. Davidson, Civil Engineer, Earth Dams Section. Detailed editorial guidance, final review, and preparation of the manuscript for publication was performed by W. E. Foote of the Technical Services Branch.

The Bureau of Reclamation again expresses grateful appreciation to those organizations which have permitted the use of material from their publications, especially the National Oceanic and Atmospheric Administration, U.S. Department of Commerce, for material used in chapter III; the Soil Conservation Service, U.S. Department of Agriculture, whose material was used in appendix A; the U.S. Geological Survey of the Department of the Interior, who supplied material used in chapter V; and the Corps of Engineers, U.S. Department of the Army, whose report on slurry trench construction was used in the preparation of chapter VI. Acknowledgments to other organizations which furnished lesser amounts of material are given throughout the text.

Preface to the First Edition

THIS BOOK presents instructions, standards, and procedures for use in the design of small dams. It is intended to serve primarily as a guide to safe practices for those concerned with the design of small dams in public works programs in the United States. The book will serve this purpose in three ways: (1) It will provide engineers with information and data necessary for the proper design of small dams, (2) it will provide specialized and highly technical knowledge concerning the design of small dams in a form that can be used readily by engineers who do not specialize in this field, and (3) it will simplify design procedures for small earthfill dams.

An earlier publication, "Low Dams," which was prepared in 1938 by the National Resources Committee, presented much useful information on the design of small dams. In the 20 years that have elapsed since the printing of that book, however, there have been many technical advances in the design of dams, and the need for a new work incorporating the latest design techniques has become increasingly evident. It is believed that this book, "Design of Small Dams," will fill that need. The new book retains much of the format of "Low Dams" and some of the material from the earlier publication has been incorporated in the new one, but most of the text is wholly new.

Although this text is related almost exclusively to the design of small dams and appurtenant structures, it is important that the designer be familiar with the purposes of the project, the considerations influencing its justification, and the manner of arriving at the size and type of structure to be built. For these reasons, an outline discussion of a desirable project investigation has been included in chapter I.

Only the more common types of small dams now being constructed are discussed. These include concrete gravity, earthfill (rolled-type), and rockfill dams. Emphasis is placed on the design of rolled earthfill dams because they are the most common type. For the purpose of this book, small dams include those structures with heights above streambed not exceeding 50 feet except for concrete dams on pervious foundations. For the latter structures, the maximum height is further limited to dams whose maximum net heads (headwater to tailwater) do not exceed 20 feet. The text is not intended to cover dams of such large volumes that significant economies can be obtained by utilizing the more precise methods of design usually reserved for large dams. In recognition of the limited engineering costs justified for small dams, emphasis is placed on efficiency and relatively inexpensive procedures to determine the necessary design data. Simplified design methods are given to avoid the complex procedures and special investigations required for large dams or for unusual conditions. Adequate but not unduly conservative factors of safety are used in the simplified design methods.

Small dams are properly considered to be associated with small streams and drainage areas of limited extent. For these situations or for those in which spillway capacity is obtainable at relatively low cost, a sufficient approximation of the inflow design flood discharge may be determined by procedures given in this text. For important projects, particularly where the spillway cost is a major item of project cost and thus may have an important bearing on project feasibility, more exact and complex studies which are beyond the scope of this text may be justified.

This text is addressed to the designer of the structure and does not include in its scope the field of construction practices or methods. However, as the integrity of the design requires adherence to limiting specifications for materials and to the practice of good workmanship in construction, appendixes are included on "Construction of Embankments," "Concrete in Construction," and "Sample Specifications." More detailed specifications will be required to insure proper construction of any specific dam.

This text is not intended in any way to encourage assumption of undue responsibility on the part of unqualified personnel, but rather to point out the importance of specialized training and to stimulate wider use of technically trained and experienced consultants.

This text should be of service to all concerned with the planning of small water storage projects, but in no way does it relieve any agency or person using it of the responsibility for safe and adequate design. The stated limitations of the design procedures should be heeded.

This book was prepared by the engineers of the Bureau of Reclamation, U.S. Department of the Interior, at Denver, Colo., under the direction of Grant Bloodgood, Assistant Commissioner and Chief Engineer, and L. G. Puls, Chief Designing Engineer. More than 30 engineers and many technicians participated in the preparation of the book or in its critical review, and the efforts of all of these are gratefully acknowledged. Special recognition is given O. L. Rice, Chief of the Dams Branch, for his guidance and counsel, especially in determining the scope and treatment of the text.

The text was coordinated and edited by H. G. Arthur, Supervisor, Design Unit, Earth Dams Section, and final review and preparation of the manuscript for the printer was by E. H. Larson, Head, Manuals and Technical Records Section.

The Bureau of Reclamation expresses grateful appreciation to those organizations which have permitted the use of material from their publications, especially the Soil Conservation Service, U.S. Department of Agriculture, whose material was used in appendix A; and the Corps of Engineers, U.S. Department of the Army, whose Technical Manual TM 5-545 was freely used in the preparation of part D of chapter V. Acknowledgments to other organizations furnishing a lesser amount of material are given throughout the text.

Contents

Chapter II—Ecological and Environmental Considerations—Con.

D. DESIGN CONSIDERATIONS—Continued

Chapter III—Flood Studies

A. GENERAL

Chapter III—Flood Studies—Continued

B. PROCEDURES—Continued

Chapter IV—Selection of Type of Dam

A. CLASSIFICATION OF TYPES

Chapter IV—Selection of Type of Dam—Continued

B. PHYSICAL FACTORS GOVERNING SELECTION OF TYPE

C. LEGAL, ECONOMIC, AND ESTHETIC CONSIDERATIONS

Chapter V—Foundations and Construction Materials

A. SCOPE OF INVESTIGATIONS

B. SOURCES OF INFORMATION

C. SOIL CLASSIFICATION

Chapter V—Foundations and Construction Materials—Continued

C. SOIL CLASSIFICATION—Continued

D. ROCK CLASSIFICATION

E. SURFACE EXPLORATIONS

Chapter V—Foundations and Construction Materials—Continued

E. SURFACE EXPLORATIONS—Continued

F. SUBSURFACE EXPLORATORY METHODS

G. SAMPLING

H. LOGGING OF EXPLORATIONS

I. FIELD AND LABORATORY TESTS

Chapter V—Foundations and Construction Materials—Continued

I. FIELD AND LABORATORY TESTS—Continued

J. BIBLIOGRAPHY

Chapter VI—Earthfill Dams

A. INTRODUCTION

B. DESIGN PRINCIPLES

C. FOUNDATION DESIGN

Chapter VI—Earthfill Dams—Continued

C. FOUNDATION DESIGN—Continued

Chapter VI—Earthfill Dams—Continued

C. FOUNDATION DESIGN—Continued

D. EMBANKMENTS

E. EMBANKMENT DETAILS

Chapter VI—Earthfill Dams—Continued

F. DESIGN EXAMPLES OF SMALL EARTHFILL DAMS

G. BIBLIOGRAPHY

Chapter VII—Rockfill Dams

A. GENERAL

Chapter VII—Rockfill Dams—Continued

B. FOUNDATION DESIGN

C. EMBANKMENT DESIGN

D. MEMBRANE DESIGN

E. BIBLIOGRAPHY

Chapter VIII—Concrete Gravity Dams

A. INTRODUCTION

B. FORCES ACTING ON THE DAM

C. REQUIREMENTS FOR STABILITY

Chapter VIII—Concrete Gravity Dams—Continued

C. REQUIREMENTS FOR STABILITY—Continued

D. DAMS ON PERVIOUS (SOIL) FOUNDATIONS

E. DETAILS OF LAYOUT AND DESIGN

F. BIBLIOGRAPHY

Chapter IX—Spillways

A. GENERAL

B. DESCRIPTION OF SERVICE SPILLWAYS

Chapter IX—Spillways—Continued

B. DESCRIPTION OF SERVICE SPILLWAYS—Continued

C. HYDRAULICS OF CONTROL STRUCTURES

Chapter IX—Spillways—Continued

C. HYDRAULICS OF CONTROL STRUCTURES—Con.

Chapter IX—Spillways—Continued

F. HYDRAULICS OF SPILLWAYS—Continued

Chapter X—Outlet Works

A. GENERAL

Chapter X—Outlet Works—Continued

A. GENERAL—Continued

B. OUTLET WORKS COMPONENTS

C. HYDRAULIC DESIGN

Chapter X—Outlet Works—Continued

C. HYDRAULIC DESIGN—Continued

D. STRUCTURAL DESIGN DETAILS

E. BIBLIOGRAPHY

Chapter XI—Diversion during Construction

A. DIVERSION REQUIREMENTS

B. METHODS OF DIVERSION

C. SPECIFICATIONS REQUIREMENTS

Chapter XI—Diversion during Construction—Continued

D. BIBLIOGRAPHY

Chapter XII—Maintenance and Operation

A. GENERAL

B. INSPECTION AND MAINTENANCE OF DAMS

C. OPERATION

Appendix A—Estimating Rainfall Runoff from Soil and Cover Data

Appendix A—Estimating Rainfall Runoff from Soil and Cover Data—Con.

Appendix B—Hydraulic Computations

A. HYDRAULIC FORMULAS

Appendix B—Hydraulic Computations—Continued

A. HYDRAULIC FORMULAS—Continued

B. FLOW IN NATURAL CHANNELS

C. BIBLIOGRAPHY

Appendix C—Structural Design Data

Appendix C—Structural Design Data—Continued

Appendix D—Soil Mechanics Nomenclature

Appendix E—Construction of Embankments

Appendix F—Concrete in Construction

A. CONCRETE AND CONCRETING MATERIALS

Appendix F—Concrete in Construction—Continued

A. CONCRETE AND CONCRETING MATERIALS—Continued

Appendix G—Sample Specifications

Appendix G—Sample Specifications—Continued

A. ENVIRONMENTAL QUALITY PROTECTION—Continued

Appendix G—Sample Specifications—Continued

B. EXCAVATION—Continued

C. EMBANKMENT

Appendix G—Sample Specifications—Continued

C. EMBANKMENT—Continued

Appendix G—Sample Specifications—Continued

C. EMBANKMENT—Continued

D. TUNNEL AND SHAFT CONSTRUCTION

Appendix G—Sample Specifications—Continued

D. TUNNEL AND SHAFT CONSTRUCTION—Con.

E. PRESSURE GROUTING

Appendix G—Sample Specifications—Continued

E. PRESSURE GROUTING—Continued

F. CONCRETE SPECIFICATIONS

1. Concrete Specifications for Small Jobs

2. Concrete Specifications for Large Jobs

Appendix G—Sample Specifications—Continued

F. CONCRETE SPECIFICATIONS—Continued

Appendix G—Sample Specifications—Continued

F. CONCRETE SPECIFICATIONS—Continued

G. MISCELLANEOUS

Appendix G—Sample Specifications—Continued

G. MISCELLANEOUS—Continued

H. BIBLIOGRAPHY

Appendix H—Sedimentation

A. RESERVOIR SEDIMENTATION

Appendix H—Sedimentation—Con.

B. DOWNSTREAM CHANNEL EFFECTS

Appendix H—Sedimentation—Con.

C. BIBLIOGRAPHY

Appendix I—Conversion Factors

LIST OF TABLES

LIST OF TABLES IN APPENDICES

LIST OF FIGURES

LIST OF FIGURES IN APPENDICES

Chapter I

Project Planning

A. F. JOHNSON [1]

A. PURPOSE OF THE DEVELOPMENT

1. *General.*—In this text, the term "project" means a water resource development. It may be small or large, simple or complex, serving one purpose or several, but it should provide the facilities to accomplish the optimum development of related physical resources. Although this text is concerned with small dams as one type of facility, the investigations and studies made for these dams must be considered in relation to the function they perform in accomplishing the purposes of the project as a whole. The project's objectives, purposes, and scope determine what must be investigated with respect to dams.

In many cases, the project will be of a dual or multipurpose type. For this reason, the investigations may encompass a large number of matters, some or all of which will influence the selection of a dam site, the size of the dam, and the purposes it serves. Hence, the entire project must be investigated as a unit before the design requirements for a single feature, such as a dam, can be firmly established. Each project purpose and each increment of its size or scope must justify inclusion in the project by some appropriate measure of feasibility or justification which is usually related to the benefits it produces, the need it serves, or the investment it can repay with or without interest.

Feasibility studies of dams and reservoirs should always consider possible objections from a public health and nuisance standpoint, and a proper effort should be made to mitigate the damages involved. The exposed bed of a reservoir, as it is drawn down, is not only unattractive but may make access to the water difficult. As deposits of sediment dry, odors from decaying vegetation or windblown dust may produce a nuisance and actual damage to health and property. In some cases, sewage detention may augment the hazard. Impounded fresh water held at a constant level makes an ideal breeding place for mosquitoes, thereby creating a nuisance and the possibility of transmitting malaria and encephalitis.

The environmental and ecological aspects of the project should be studied and provisions made to minimize any deleterious effects.

Many reservoirs for which this text is intended will be in regions affected by drought and subject to flash floods. Under these climatic conditions, floodwater erosion of the watershed and streambanks will fill the streams with sediment to be caught in the reservoirs. The accumulations of sediment may soon reduce the usefulness of a reservoir and ultimately will completely eliminate its capacity. Loss of capacity and other damages due to silting of reservoirs and changes of the regimen of silt-laden streams as a result of reservoir operation should be considered for all proposed projects. The remaining paragraphs of this part of the chapter present pertinent aspects of common purposes, with particular emphasis upon design requirements for dams and reservoirs.

2. *Irrigation.*—The supply of water must be adequate for successful irrigation (considering occasional tolerable shortages) at an econom-

[1] Assistant to the Chief, Division of Project Investigations, Bureau of Reclamation (Retired).

ically reasonable cost per unit of area, both for capital investment and for operation, maintenance, and replacement. The quality of the water must be such that it will not be harmful to the crops or to the soils on which it will be used. If the distribution system is to depend on gravity flow, the reservoir must be high enough above the irrigated area to provide the necessary head for adequate delivery.

3. *Domestic and Municipal Purposes.*—The supply of water must be adequate to serve the requirements. The present demand and a surplus to care for the reasonably predictable increases in requirements are important considerations.

The quality of the water must be such that it can be rendered potable and usable for domestic and most industrial purposes by economical treatment methods. It should meet State public health standards as to bacterial purity. Standards with regard to taste, color, odor, and hardness may vary with different sections of the country. The degree to which objectionable characteristics can be corrected will depend on their nature and concentration in the raw supply, and on the cost of necessary remedial measures.

Control and protection of small watershed areas is desirable for municipal water supply reservoirs. Even though the entire watershed cannot be purchased or otherwise provided protection from contamination, an effort should be made to provide control through agreements or purchase of shoreline lands.

4. *Industrial Use.*—Although quality of water for municipal services is ordinarily satisfactory for industrial uses, certain industrial processes require more exacting standards with respect to freedom from chemicals injurious to equipment or to the product manufactured.

5. *Stock Water.*—The quality of the water for stock consumption must be suitable for the purpose. The pond should be situated where it is accessible to the stock, either directly or by the economical use of ditches or pipes.

6. *Power Development.*—Where power generation is included, the capacity of the power-generating equipment and the load demand are closely related to the quantity of water available and the amount of storage provided. The height of a power dam is usually dictated by these requirements. Special studies of this nature are outside the scope of this text.

7. *Flood Control.*—In the study and design of flood control projects and structures, the following factors should be considered:

(1) The relation of the cost of control to the benefits to be derived through the reduction of cumulative damage should be favorable as compared to alternative means of obtaining similar benefits, and in the light of public interest.

(2) The temporary storage must be sufficient to lower the major peak flows or to decrease the frequency of minor floods.

(3) So far as is practicable, the method of control should be automatic rather than manual.

(4) Any flood control must be effective. An implied downstream safety that does not exist is more dangerous than having no control at all.

8. *Recreation.*—The following factors should be considered in regard to development of recreational projects:

(1) There must be an adequate supply of water to provide for evaporation losses and to maintain the water level within those limitations assumed as a basis for the recreational and residential development of the shoreline.

(2) The water must be free of pollution within practical limits.

(3) If bathing is one of the purposes, there must be an adequate depth of water in the vicinity of a gently sloping shore.

(4) Where there is considerable recreational use of varied nature, the shoreline should be zoned to preserve privacy and to provide separation of conflicting uses, such as residential, camping, day picnicking, bathing, or boating. Facilities for public use and safety, such as access roads, parking, boat ramps or docks, fireplaces and tables, and sanitary facilities, are commonly provided. Other facilities, such as motels, stores, and commercial amusements, are generally provided under concession.

(5) The shoreline should have a relatively steep gradient where possible, so that a slight lowering of the water surface will expose a minimum of area. Also, the normal range of

the operating level should not include extensive areas of flat shoreline which will be unsightly when uncovered. Probable use of lands on shorelines should be considered in proposed plans and in estimates of required reservoir rights-of-way. Easements for maximum floods and floods of rare frequency are less costly than outright purchase and permit private use of the shorelines.

Chapter II discusses in greater depth the recreational development of a reservoir area.

9. *Wildlife.*—Projects to impound water for wildlife purposes should not be promoted without the advice of a biologist. An ability to catch fish or shoot game birds is not a guarantee of knowledge required for raising them. The following factors must be considered:

(1) There must be sufficient depth and supply of water to maintain livable conditions for wildlife throughout the dry season.

(2) Extensive fluctuations in water level are inimical to fish and other wildlife, as they prevent or destroy the development of aquatic vegetation required for food.

(3) Satisfactory quality of the water must be assured. Pollution may poison waterfowl or fish and reduce the oxygen content to the point where many game fish cannot survive. Excessively acid or highly alkaline water is harmful to many kinds of wildlife.

(4) The water and the basin must be suited to the production of the right kind of food supply and must provide adequate shelter and freedom from disturbance by man.

Additional considerations affecting the ecology and environment of fish and wildlife populations are given in chapter II.

10. *Water Storage for Streamflow Regulation.*—There is a need for projects of this type in those regions where streamflow either ceases entirely or is reduced to extremely low values during parts of the year. Where such natural streamflow is the principal source of water supply for one or more communities and where a dependable streamflow is necessary for the dilution of waste after proper economical treatment, water storage for streamflow regulation may be justified. For such projects, it must be determined that:

(1) The dependable streamflow, when properly regulated, is sufficient to produce the minimum values of regulated flow required for the purpose, after expected losses (including evaporation) have been deducted; and

(2) Storage for this purpose will not result in an objectionable alteration of the quality of the water.

11. *Miscellaneous Water Conservation Projects.*—Occasionally, projects are proposed to regulate the water level in shallow lakes, swamps, or bogs for purposes other than those heretofore enumerated. This classification also includes projects for the detention or diversion of streamflow to conserve it by transforming surface water to ground water through the process of infiltration.

It should be noted that natural shallow lakes, swamps, and bogs generally exist because of an underlying impervious subsoil and that additional storage for surface water at such locations will rarely be effective in increasing ground water, unless the stored water can be conveyed to and flooded over other areas where infiltration can take place.

Projects of the type under discussion will generally result in large increases in the evaporation and transpiration losses because of the enlargement of water surface, the increased time and exposure of the surface water to evaporation, or possibly an increase in the area of semiaquatic plants such as sedge grass and reeds. The net results on such projects, after deducting such losses, must be substantially beneficial. In many cases, there is a danger that projects of this type will result in the loss of water through evaporation that might otherwise be beneficially used elsewhere as streamflow.

For projects involving water spreading or the detention of surface water to increase infiltration and percolation opportunities, it must be determined that the soil characteristics will permit infiltration to occur in a quantity which justifies the project economically. The Groundwater Branch of the Geological Survey (Department of the Interior), the Bureau of Land Management (Department of the Interior), the Soil Conservation Service (Department of Agriculture), or appropriate State bureaus should be consulted with regard to projects of this nature.

B. PROJECT STUDIES

12. *Tests of Feasibility.*—The objective in project planning is the determination of the project's feasibility. This involves studies which will permit a sound analysis and conclusion with respect to the specific engineering-economic-environmental considerations. These are primarily:

(1) That the project is responsive to an urgent present or anticipated social or economic need;

(2) That the project as planned will adequately serve the intended purpose;

(3) That the services proposed to be performed through the project and the benefits it will produce will justify the cost; and

(4) That the project will cause minimal disturbance to the ecology and environment of the area.

The study should determine that the difficulties inherent in facility sites which affect economy, safety of construction, and quality of operation have been satisfactorily foreseen; and that the designs are technically sound and reasonably representative of the actual structures that may be expected to be built after more detailed investigation. The soundness of the conclusions regarding these matters will depend to a considerable degree on the completeness and accuracy of the investigation.

13. *Esthetic Values.*—Esthetic values may be of great importance to a project. In the location and design of dams and other major structures, they should be given appropriate consideration which should be recognized in the first studies made and carried out in later studies and in construction operations. However, esthetic values should not be allowed to outweigh the safety or adequacy of the structural design.

The application of the principles involved varies so greatly that the solution with respect to any particular dam requires individual study. When esthetic considerations are important, it is recommended that a qualified landscape architect be consulted; where warranted, an architect should collaborate in making the design.

14. *Extent of Studies.*—There is no simple rule for determining the extent of the investigation which is necessary in any particular case. For example, a dam that is to be constructed on a pervious or a weak foundation to impound a depth of water of 15 feet will require much more foundation investigation than a 50-foot-high dam to be built on solid, unfractured rock under a shallow soil mantle. Within the limits of height treated in this text, it may be said that the size of the physical structure has little relation to the extent of the investigation necessary. The maximum justifiable investigation cost is, however, limited by the magnitude of the project. The project is generally unjustified if the cost of the necessary investigation would offset a large portion of the constructed project's value. A cost reduction accomplished by the elimination of a portion of the fundamental investigation is rarely a saving. It generally results in unanticipated construction or functional costs.

15. *Stages of Investigation.*—Investigation, if carried to completion, is an expensive and time-consuming phase of project development. Moreover, it may indicate that the project is not economically or technically sound. Hence, an investigation should be planned and executed so that the probable soundness of the project will be determined as early and as inexpensively as possible. To accomplish this objective, the investigation may be divided into as many as three stages. The first, or reconnaissance stage, is designed primarily to support a decision on whether to proceed with more detailed investigations on the basis of rough data and shortcut studies. The second, or feasibility stage, determines the scope, magnitude, essential plan and feature, and the approximate benefits and costs of the project with sufficient dependability to support project authorization or approval for construction. The third, the specifications stage, supplements the feasibility stage to the degree needed for preparation of final plans and specifications after authorization or approval has been obtained and construction is imminent.

Many of the smaller projects will not require, at the specifications stage, any information in addition to that already obtained in the feasibility studies. The larger and more difficult projects will often require extensive additional surveys and investigations. However, project size is not the sole criterion with respect to the necessity for further detailed studies. This may rest on a question of complexity of the site, of the foundation conditions, and often of the hydrological factors.

The following sections discuss the elements of project studies within the scope of the various stages of investigation.

16. ***Related Projects and Studies of Need.***—The proposed project should be consistent with any long-range planning program which may have been adopted for the vicinity. The entire area to be served by the proposed project should be studied to determine whether there will be conflicts with other projects of a similar nature for use of land or water resources, power drops, etc., or whether economies are possible by securing optimum development of resources through joint use.

If a potential project conflicts with other similar facilities, either completed or contemplated, it is advisable to refer the matter to appropriate State or local authorities, especially where future needs or opportunities must be preserved for the good of the locality, State, or Nation. The probable existing demands for the services to be rendered by the project as well as a reasonable estimate of future demand should be carefully determined.

17. ***Development of the General Plan.***—The project plan normally originates with the desire to satisfy one sponsor's specific needs, objectives, or purposes. As support for project construction develops, those needs may increase, the objectives may broaden, and the purposes may multiply through the processes of project formulation until selections of the final magnitude and scope are reached.

At the outset of the reconnaissance study, considerable basic data are usually available in the form of maps, aerial photographs, streamflow records, regional geological reports, census statistics, crop yields, market statistics, power loads, previous investigation reports, etc. Data concerning the impact of the project upon the environment are not readily available and must be actively collected from appropriate sources. The investigator must evaluate these data, supplement them with rough additional data and conceive a workable basic plan that utilizes available resources to meet the needs. This basic plan may then be compared roughly with alternatives to accomplish the desired ends on progressively increased or diminished scope and scale. It will be possible, by judgment or cursory study, to eliminate many alternatives so that an approximate final plan emerges which includes purposes, approximate locations and heights of dams, capacities of reservoirs, spillways, outlets, canals, powerplants, and other features, and which minimizes the environmental impact of the project. This plan will be refined in the feasibility and specification design stages. The character and scope of required feasibility surveys, investigations, and studies may be visualized by judgment and experience from a consideration of the plan and knowledge of its components acquired during the reconnaissance stage study.

Governmental agencies have obtained a considerable amount of basic data of the nature mentioned above for many areas. The larger public libraries are generally designated as public document depositories and maintain more or less complete files of all United States Government publications and, of course, many State and local publications as well. U.S. Government publications in stock may be purchased from the Superintendent of Documents, Government Printing Office, Washington, D.C., and extensive lists of publications are available from that office and from issuing agencies. Individual agencies commonly provide sales rooms at their principal offices. Sponsors and engineers for small projects should consult with State and county planning officials, local offices of the Federal data-collecting agencies, and public libraries concerning the availability of pertinent data. Following is a source list of possible data:

(a) *Topographic Maps:*

(1) Quadrangle maps—U S. Department of the Interior, Geological Survey, Topographic Division; and U.S. Department of the Army, Army Topographic Command.

(2) River plans and profiles—U.S. Department of the Interior, Geological Survey, Conservation Division.
(3) National parks and monuments—U.S. Department of the Interior, National Park Service.
(4) Federal reclamation project maps—U.S. Department of the Interior, Bureau of Reclamation.
(5) Local areas—commercial aerial mapping firms.

The availability of topographic maps is discussed in chapter V.

(b) *Planimetric Maps:*
(1) Plats of public land surveys—U.S. Department of the Interior, Bureau of Land Management.
(2) National forest maps—U.S. Department of Agriculture, Forest Service.
(3) County maps—county surveyor or county engineer.
(4) City plats—city or county recorder.
(5) Federal reclamation project maps—U.S. Department of the Interior, Bureau of Reclamation.

(c) *Aerial Photographs:*
(1) The following agencies have aerial photographs of portions of the United States: U.S. Department of the Interior, Geological Survey, Topographic Division; U.S. Department of Agriculture, Commodity Stabilization Service, Soil Conservation Service, and Forest Service; U.S. Department of Commerce, National Oceanic and Atmospheric Administration, Coast and Geodetic Survey; U.S. Air Force; various State agencies; commercial aerial survey and mapping firms.

The availability of aerial photographs is discussed in chapter V.

(d) *Transportation Maps:*
(1) State transportation maps—U.S. Department of Transportation, Federal Highway Administration, Bureau of Public Roads.
(2) State and county highway maps—State highway departments.
(3) Sectional aeronautical charts—U.S. Department of Commerce, National Oceanic and Atmospheric Administration, Coast and Geodetic Survey.

(e) *Triangulation and Benchmarks:*
(1) U.S. Department of Commerce, National Oceanic and Atmospheric Administration, Coast and Geodetic Survey; and U.S. Department of the Interior, Geological Survey, Topographic Division.

(f) *Geology:*
(1) Geologic maps and reports—U.S. Department of the Interior, Geological Survey, Geologic Division; and State geological surveys or departments .

(*Note.*—Some regular quadrangle maps show geological data also.)

Geologic maps are discussed in chapter V.

(g) *Soils:*
(1) County soil survey reports—U.S. Department of Agriculture, Soil Conservation Service.
(2) Land use capability surveys—U.S. Department of Agriculture, Soil Conservation Service.
(3) Land classification reports—U.S. Department of the Interior, Bureau of Reclamation.

Agricultural soil maps are discussed in chapter V.

(h) *Climatological Data:*
(1) Daily weather reports—U.S. Department of Commerce, National Oceanic and Atmospheric Administration, National Weather Service.
(2) Climatological data (monthly and annual summaries) ; includes precipitation, temperature, evaporation, and wind velocity data—U.S. Department of Commerce, National Oceanic and Atmospheric Administration, National Weather Service.
(3) Daily synoptic weather maps—U.S. Department of Commerce, National Oceanic and Atmospheric

Administration, National Weather Service.

(4) Hydrologic bulletins—U.S. Department of Commerce, National Oceanic and Atmospheric Administration, National Weather Service.

(5) Technical papers—U.S. Department of Commerce, National Oceanic and Atmospheric Administration, National Weather Service.

(6) Hydrometeorological reports—U.S. Department of Commerce, and U.S. Department of the Army, Corps of Engineers.

(7) Cooperative study reports—U.S. Department of Commerce, National Oceanic and Atmospheric Administration, National Weather Service; and U.S. Department of the Interior, Bureau of Reclamation.

(8) Unofficial observers—such as banks, local businessmen, farmers, etc.

(i) *Streamflow Data:*

(1) Water supply papers—U.S. Department of the Interior, Geological Survey, Water Resources Division.

(2) Reports of State engineers.

(3) Annual reports—International Boundary and Water Commission, United States and Mexico.

(4) Annual reports—various interstate compact commissions.

(5) Water right filings, permits—State engineers, county recorders.

(6) Water right decrees—district courts.

(j) *Sedimentation:*

(1) Water supply papers—U.S. Department of the Interior, Geological Survey, Quality of Water Branch.

(2) Reports—U.S. Department of the Interior, Bureau of Reclamation; and U.S. Department of Agriculture, Soil Conservation Service.

(k) *Quality of Water:*

(1) Water supply papers—U.S. Department of the Interior, Geological Survey, Quality of Water Branch.

(2) Reports—U.S. Department of Health, Education, and Welfare, Public Health Service; and Environmental Protection Agency, Federal Water Quality Administration.

(3) Reports—State public health departments.

(l) *Irrigation and Drainage Data:*

(1) Agricultural census reports—U.S. Department of Commerce, Bureau of the Census.

(2) Agricultural statistics—U.S. Department of Agriculture, Agricultural Marketing Service.

(3) Annual crop summaries on Federal reclamation projects—U.S. Department of the Interior, Bureau of Reclamation.

(4) Reports of State departments of agriculture.

(5) Use of water—U.S. Department of Agriculture, Agricultural Research Service; and U.S. Department of the Interior, Bureau of Reclamation.

(m) *Power Data:*

(1) Directory of Electric Utilities—McGraw-Hill Publishing Co.

(2) Directory of Electric and Gas Utilities in the United States—Federal Power Commission.

(3) Power rates—Federal Power Commission, power companies, public utility districts, rural electric cooperatives, etc.

(4) Power markets—State universities, Bureau of Business Research.

(5) Reports—Federal Power Commission, various power companies, public utilities, State power commissions, etc.

(n) *Bulletins:*

(1) Water Resources Council—Regional Water Basin Commissions.

(2) Annotated bibliographies on hydrology.

(3) Annotated bibliographies on sedimentation.

(4) Population census—U.S. Department of Commerce, Bureau of the Census.

(o) *Basin and Project Reports and Special Reports:*

(1) U.S. Department of the Army, Corps of Engineers.

(2) U.S. Department of the Interior, Bureau of Land Management, Bureau of Mines, Bureau of Reclamation, Fish and Wildlife Service, and National Park Service.

(3) U.S. Department of Agriculture, Soil Conservation Service.

(4) U.S. Department of Health, Education, and Welfare, Public Health Service.

(5) Federal Power Commission.

(6) Water Resources Council—Regional Water Basin Commissions.

(7) State departments of water resources, departments of public works, power authorities, and planning commissions.

(p) *Consultation:*

(1) Sanitation and public health—U.S. Department of Health, Education, and Welfare, Public Health Service; Environmental Protection Agency, Federal Water Quality Administration; and State departments of public health.

(2) Fish and wildlife problems—U.S. Department of the Interior, Fish and Wildlife Service; State game and fish departments.

(3) Municipal and industrial water supplies—city water departments; State universities; Bureau of Business Research; State water conservation boards or State public works departments.

(4) Watershed management—U.S. Department of Agriculture, Soil Conservation Service, Forest Service; U.S. Department of the Interior, Bureau of Land Management, Bureau of Indian Affairs.

(5) Recreation—U.S. Department of the Interior, National Park Service, and Bureau of Outdoor Recreation; and State departments of parks and recreation.

(q) *Designs and Specifications, Water Rights:*

Consult State engineer.

(r) *Ecological and Environmental Information:*

(1) U.S. Department of the Interior, Bureau of Sport Fisheries and Wildlife, and Bureau of Reclamation.

(2) U.S. Department of Agriculture, Forest Service, and Soil Conservation Service.

(3) Environmental Protection Agency.

(4) State fish and game and conservation departments.

18. Outline of Investigations.—The outline given below provides a guide for the field engineer by indicating the items which should be considered for the investigations of dams and reservoirs. The outline includes items many of which may not be applicable in a specific instance. Most of the items in the outline are discussed in detail in the remainder of this chapter.

I. *General Data Required:*

A. Location and vicinity map:

1. Location of project.
2. Location of existing works affected by proposed development.
3. Existing location of highways, railroads, and other public utilities and proposed relocations.
4. Location of proposed construction headquarters, camp, or town, access roads, airports, etc.
5. Railroad shipping points.
6. Stream gaging and sampling stations, meteorological stations, etc.

B. Hydrologic data:

1. Streamflow records, including daily discharges, monthly volumes, and momentary peaks.
2. Streamflow and reservoir yield.
3. Project water requirements, including allowances for irrigation and power efficiencies, conveyance losses, reuse of return flows, and stream re-

leases for fish; and dead storage requirements for power, recreation, fish and wildlife, etc.

4. Flood studies, including inflow design floods and floods to be expected during periods of construction.
5. Sedimentation and quality studies, including sediment measurements, analysis of dissolved solids, etc.
6. Data on groundwater tables in the vicinity of the reservoir and damsite.
7. Water rights, including interstate compacts and international treaty effects, and contractual agreements with local districts, power companies, and individuals for subordination of rights, etc.

C. Climatic data:
1. Monthly temperatures and rainfall and storm intensities.
2. Evaporation rates.
3. Maximum, minimum, and mean temperatures.
4. Wind directions and velocities.
5. Ice thickness.

D. Geologic data:
1. Geological report by qualified geologist.
2. Discussion of geologic formations, particularly such as cavernous limestone or other soluble formations, exposed lava, exposed gravel, and glacial deposits of a permeable nature that might contribute to serious reservoir leakage.
3. Groundwater table observations.
4. Presence of deleterious mineral and salt deposits.
5. Photographs showing basin and character of lands.
6. Geologic cross sections where necessary.

E. Ecological and environmental data:[2]
1. Project's effect on fish and wildlife populations.
2. Provisions for fish and wildlife protection.
3. Provisions for public protection.
4. Type or types of fish to be stocked.
5. Current recreational development.
6. Future recreational development.
7. Discussion of environmental and ecological problems unique to this project.

II. *Reservoir Data:*

A. Reservoir map:
1. Topography.
2. Horizontal and vertical controls, preferably a triangulation survey system.
3. Coordinate system.
4. Cultural classification of reservoir lands, land ownership, and status in connection with acquisition of rights-of-way, easements, etc.
5. Ownership boundaries and names of owners.

B. Road and public utility surveys:
1. Relocation and reconstruction of railroads and highways.
2. Relocation and reconstruction of public utilities.
3. Preliminary report of joint reconnaissance made with the municipality or owner of public utilities with approximate costs of relocation, including the necessity for location surveys and who will perform the construction.

C. Miscellaneous data:
1. Estimate of probable life of reservoir—that is, considering loss of capacity due to silting.
2. Land evaluation surveys, including tabulations of areas and estimated costs for purchase and easements; discussion of necessity for appraisal survey; and discussion of possibility of securing easements for submergence of lands during maximum and infrequent floods.
3. Tabulation of areas to be cleared, with estimated cost.
4. Description of buildings, fences, and farm improvements which must be removed or salvaged, with estimates of cost.

[2] Refer to chapter II for a discussion of ecological and environmental considerations.

5. Description of lands adjacent to the proposed reservoir for public use, recreation, or other purposes.
6. Economic or physical limitations to maximum reservoir flow line.
7. Discussion of limitations to reservoir fluctuations.

III. *Data for Dams:*

A. Damsite map:
1. Topography of damsite and dike areas.
2. Horizontal and vertical controls, preferably by a triangulation system.
3. Coordinate system grid.
4. Location of rock outcrops and apparent geological features.
5. Location of manmade improvements in existing works at the site.
6. Location of drill holes, test pits, and other foundation explorations.

B. Foundation explorations:
1. Sufficient drill holes, auger holes, and/or test pits to determine character and depth of overburden for feasibility and specifications designs.
2. Description and logs of exploration, including ground elevation at the holes, location coordinates, and sufficiently detailed remarks for a clear interpretation of records.
3. Samples.
4. Sufficient explorations to determine character of bedrock or impervious foundation strata for feasibility and specifications designs.

C. Materials exploration:
1. Location and description of character of proposed material to be used in the construction of the dam, including earth, sand-gravel for aggregate and embankment, and rock for rockfill and riprap.
2. Map of borrow areas, showing location of test holes made for feasibility and specifications designs.
3. Logs of explorations.
4. Representative samples of materials in borrow areas.

D. Tailwater data:
1. Stage-discharge curves for streams, if available.
2. Cross sections of streams, with dated water surface elevations above and below dam.
3. Tailwater and backwater curves for stated high-water marks.

E. Local conditions controlling design of the dam:
1. Requirements for roadways.
2. Requirements for fishways or fish conservation measures.
3. Requirements for replacement or provisions for existing works.
4. Requirement of permanent building or quarters for operators.
5. Effect of local conditions as regards spillway and outlet gates, especially winter conditions, etc.
6. Availability of electric power for operation of mechanical equipment.
7. Capacities and elevations of required outlets as determined by local conditions.

F. Local conditions affecting construction:
1. Additional transportation facilities required for construction, including access roads, etc .
2. Location surveys for railroads, highways, or airports.
3. Requirement for improvements to existing transportation facilities.
4. Estimated cost for sufficient data for preparation of estimates for transportation facilities.
5. Hauling distance from nearest railroad shipping point and local trucking rates.
6. Availability of electric power for construction purposes.
7. Requirements for construction camp, with estimated population and quarters required for supervisory and construction employees; required water supply and sanitation facilities; local laws regarding sanitation, stream pollution, etc.

19. *Planning Programs for Surveys and Investigations.*—During the reconnaissance study, it is desirable to consider plans for detailed surveys and investigations, particularly if it is known that a feasibility study probably will be made. In planning such a program, consideration should be given to the following items: (1) the personnel required, (2) housing and subsistence for personnel, (3) location of field office if required, (4) transportation and other equipment, material, and supplies required for this work, (5) character of additional dam foundation investigation and equipment required, (6) availability of local labor and equipment, (7) arrangements with private landowners for entry to sites to avoid trespass of lands during surveys and explorations, (8) transportation of drilling equipment to isolated and inaccessible locations, (9) consideration of horizontal and vertical controls for surveys and available data from other surveys such as railroad and highway surveys, county engineers' surveys, etc., (10) location of stream gaging stations and available hydrologic data, (11) climatic conditions for work, (12) sanitary conditions if pollution will be a factor limiting the utilization of the project, and (13) estimates of time and funds required for work.

In the case of public works, investigation and construction funds generally must be budgeted and appropriated, and the time necessary for these processes must be anticipated. With local and private developments, similar time requirements usually are involved in public elections, bond issues, or other means of financing.

20. *Mapping.*—The project location should be shown on a general map, using a base of appropriate scale. U.S. Geological Survey quadrangle maps, particularly the 7½-minute quadrangles, 1 inch equal to 2,000 feet, are appropriate for preliminary purposes in locating land areas and principal features and in determining approximate reservoir capacities. If no topographic maps are available, rough reconnaissance maps, defined by a minimum of controlling points or sketched from cross sections, can be used for comparative studies. The general map should show governing elevations of water courses, canal routes, dams, and important cultural and occupational features such as woods, cultivated lands, swamps, roads, and railroads. Features pertinent to reservoirs and dams should be shown on separate maps of the reservoir and of the damsite.

Profiles at damsites may serve for reconnaissance purposes. During the feasibility stage, if practicable, and certainly during specifications investigations, reliable topographic maps of irrigable areas, reservoirs, dams, conduit locations, and other structural sites must be obtained. These should be to a convenient scale and contour interval laid out on a coordinate grid, preferably tied into State coordinate systems and public lands survey systems, with horizontal and vertical controls based on U.S. Coast and Geodetic Survey or U.S. Geological Survey nets and datum. Control points should be permanently monumented.

Topographic map scales should range from 1 inch equals 400 feet for small reservoirs to 1 inch equals 1,000 feet or 2,000 feet for very large reservoirs, with contour intervals ranging from 2 feet for small reservoirs created by low dams to 5 feet for reservoirs of greater depth. Damsite maps should be in greater detail—1 inch equals 20 feet for sites for the smallest dams, up to 1 inch equals 100 feet or 200 feet for sites in flat terrain, with contour intervals of 2 or 5 feet depending on the steepness of the topography. Coverage should extend beyond the limits of the reservoir or dam to include rights-of-way, utility relocations, camp and service areas, nearby borrow areas, etc. River sections should be obtained from the dam axis to 2,000 feet or more downstream as required for tailwater curves, degradation studies, channel changes, etc. Topography should be obtained at the heads of reservoirs where sediment deltas or backwater levels will be important. To an increasing degree, topographic surveys are utilizing photogrammetric methods and much of the work formerly done by project survey parties is done under contract by firms which specialize in photogrammetric surveys. Property boundaries and ownerships are desirable where rights-of-way must be purchased.

The location and elevations of all drill holes, test pits, trenches, etc., and of control monuments and temporary points should be shown on the detailed topographic maps of the damsite.

The principal difference between the surveys

and maps required for feasibility stage investigations and for specifications design is that, for the latter, certain features must be more complete and accurate so as to provide all of the information needed for the actual design of the structures. In many cases, an evaluation of the factors entering into the justification of the project may have required that surveys and investigations be carried out in considerable detail during the feasibility studies. If the feasibility stage surveys and maps are up to acceptable standards of accuracy, it will only be necessary to bring them up to date to reflect cultural changes and to add detailed data, such as supplemental drill holes. Where the feasibility studies have been based on reconnaissance surveys and sketch maps, extensive field surveys must be carried out in connection with the specifications stage investigations. Substandard surveys and maps should be brought up to standard or be replaced by new ones which have sufficient accuracy and detail for final design purposes.

21. *Hydrologic Investigations.*—Hydrologic investigations which may be required for project studies include the determination of the following: yield of streamflow, reservoir yield, water requirements for project purposes, sediment which will be deposited in the reservoir, flood flows, and groundwater conditions.

The most accurate estimate possible must be prepared of the portion of the streamflow yield that is surplus to senior water rights as the basis of the justifiable storage. Reservoir storage will supplement natural yield of streamflow during low-water periods. Safe reservoir yield will be the quantity of water which can be delivered on a firm basis through a critical low-water period with a given reservoir capacity. Reservoir capacities and safe reservoir yields may be prepared from mass curves of natural streamflow yield as related to fixed water demands or from detailed reservoir operation studies, depending upon the study detail which is justified. Reservoir evaporation and other incidental losses should be accounted for before computation of net reservoir yields. The critical low-water period may be 1 drought year or a series of dry years during the period of recorded water history. Water shortages should not be contemplated when considering municipal and industrial water use. For other uses, such as irrigation, it is usually permissible to assume tolerable water shortages during infrequent drought periods and thereby increase water use during normal periods with consequent greater project development. Tolerable irrigation water shortages will depend upon local conditions and crops to be irrigated. If the problem is complex, the consulting advice of an experienced hydrologist should be secured.

The annual rate at which sediment will be deposited in the reservoir should be ascertained to insure that sufficient sediment storage is provided in the reservoir so that the useful functions of the reservoir will not be impaired by sediment deposition within the useful life of the project or the period of economic analysis, say 50 to 100 years. The expected elevation of the sediment deposition may also influence the design of the outlet works, necessitating a type of design which will permit raising the intake of the outlet works as the sediment is deposited. Appendix H outlines the procedure for computation of reservoir sediment inflow, deposition, and the associated problems of stream channel aggradation and degradation.

Water requirements should be determined for all purposes contemplated in the project. For irrigation, consideration should be given to climatic conditions, soil types, type of crops, crop distribution, irrigation efficiency and conveyance losses, reuse of return flows, etc. For municipal and industrial water supplies, the anticipated growth of demand over the life of the project must be considered. For power, the factors to be considered are load requirements and anticipated load growth.

Knowledge of consumptive uses is important in the case of a large irrigation project, and especially for river systems as a whole. However, of equal and perhaps more importance in the case of an individual farm or project is the efficiency with which the water is conveyed, distributed, and applied. The losses incidental to application on the farm and the conveyance system losses and operational waste may in many instances exceed the water required by the growing crops. In actual operation, the amount of loss is to a great extent a matter of economics. In areas where water is not plentiful and high-value crops are grown, the use of pipe or lined conveyance systems and costly land preparation or sprinkler systems can be afforded to reduce losses to a minimum. A part of the losses may be consumed nonbeneficially by nonproductive areas adjacent to the irrigated land or in drainage channels. Usually most of the loss eventually returns to a surface stream or drain and is referred to as return flow.

In planning irrigation projects, two consumptive use values are developed. One, composed of monthly or seasonal values, is used with an adjustment for effective precipitation and anticipated losses mentioned above to determine the total water requirement for appraising the adequacy of the total water supply and determining reservoir storage requirements. The other, a peak use rate, is used for sizing the canal and lateral system.

Evapotranspiration, commonly called consumptive use, is defined as the sum of evaporation from plant and soil surfaces and transpiration from plants and is usually expressed in terms of depth (volume per unit area). Crop consumptive use is equal to evapotranspiration plus water required for plant tissue, but the two are usually considered the same. Predictions or estimates of evapotranspiration are basic parameters for the engineer or agronomist involved in planning and developing water resources. Estimates of evapotranspiration are also used in assessing the disposition of water in an irrigation project, evaluating the irrigation water-management efficiency, and projecting drainage requirements.

Reliable rational equations are available for estimating evapotranspiration when basic meteorological parameters such as net radiation, vapor pressure and temperature gradients, wind speed at a prescribed elevation above the crops or over a standard surface, and soil heat flux are available. When information on these parameters is not available, which is the usual case, recourse is made to empirical methods. Numerous equations, both empirical and partially based on theory, have been developed for estimating potential evapotranspiration. Estimates from these methods are generally accepted as being of suitable accuracy for planning and developing water resources. Probably the most widely used methods at this time are the Blaney-Criddle [1] [3] method and the Soil Conservation Service adaptation of the Blaney-Criddle method [2]. A more recent method, newly developed for general usage, is the Jensen-Haise solar radiation method [3]. In general terms, these methods utilize climatic data to estimate a climatic index. Then coefficients, reflecting the stage of growth of individual crops and their actual water requirement in relationship to the climatic index, are used to estimate the consumptive use requirements for selected crops.

Project studies must include estimates of floodflows, as these are essential to the determination of the spillway capacity. Consideration should also be given to annual minimum and mean discharges and to the magnitudes of relatively common floods of frequencies up to about 10-year recurrence intervals, as this knowledge is essential for construction purposes such as diverting the stream, providing cofferdam protection, and scheduling operations. Methods of arriving at estimates of floodflows are discussed in chapter III. If the feasibility studies are relatively complete, the flood determination may be sufficient for design purposes. If, however, floodflows have been computed for purposes of the feasibility study without making full use of all available data, these studies should be carefully reviewed and extended in detail before the actual design of the structure is undertaken. Frequently, new data on storms, floods, and droughts become available between the time the feasibility studies are made and construction starts. Where such changes are significant, the flood studies should be revised and brought up to date.

Project studies should also include a groundwater study, which may be limited largely to determining the effect of ground water on construction methods. However, some groundwater situations may have an important bearing on the choice of the type of dam to be constructed and on the estimates of the cost of

---

[3] Numbers in brackets refer to items in the bibliography, section 32.

foundations. Important groundwater information sometimes can be obtained in connection with subsurface investigations of foundation conditions.

As soon as a project appears to be feasible, steps should be taken in accordance with State water laws to initiate a project water right.

22. *Investigations of Foundations and Materials.* —On all dam projects the watertightness of the reservoir, the suitability of the foundations for the dam and appurtenant structures, and the construction materials sources, are important geological and engineering considerations. The extent of investigations for resolving these problems should be commensurate with the importance of the project and with the stage of the investigation. Obtaining highly detailed information on subsurface conditions may exceed justifiable costs on most small projects; hence, reliance will have to be placed on judgment and experience rather than on a large amount of subsurface exploration and laboratory testing. For example, considerations of economy require a minimum of exploration for alternative damsites prior to final selection. Once the site is established, however, the information to be obtained at each subsequent stage of the explorations should be of a quality suitable for use in specifications designs. Uneconomical duplication of effort and contractual difficulties are risked when incomplete data are obtained.

In the reconnaissance stage, sufficient information is obtained to select the damsite and to determine whether additional investigation is warranted. The first step is to search for and study all geological and soils data relating to the area, including maps, air photographs, and reports. Sources of such information are given in chapter V. It may be difficult to apply the data from these sources directly to the design of the dam; however, the information that can be gleaned from them is of considerable value in planning field investigations and, later, in interpreting their results. The second step in the reconnaissance stage is a field examination of the site and surrounding area by the engineer, who should be accompanied, if possible, by an engineering geologist. This examination should include construction materials sources and the reservoir and damsite geology that is visible on the ground.

It is important to record all of the data obtained in the reconnaissance stage. Even if economic factors at the time of the study preclude further investigations, changing economic conditions later may result in a reevaluation of the project; thus, costly duplication of earlier investigations can be avoided. Because reconnaissance data may be based primarily on limited surface information, large factors of safety should be used in estimates of potential quantities of construction materials, hauling distances, and depths for cutoffs.

The objective of the feasibility stage is to obtain data for a cost estimate. The estimate should be sufficiently accurate to determine whether the project is economically justified. The extent of the investigation depends on the data available from the reconnaissance stage. The accuracy of information required in the feasibility stage generally requires that subsurface explorations be performed. Detailed information on exploration methods, logging, and sampling procedures is given in chapter V.

Additional boreholes may be needed in the specifications stage to answer critical questions posed by the feasibility designs, such as, what should be the depth of cutoff trenches, or if borrow material is scarce, what quantity is available. Earlier explorations may also have raised critical geological questions requiring resolution such as, will a fault zone encountered in one of the bore holes be of sufficient extent to create a design problem or is further exploration necessary to outline definitely a buried channel filled with highly pervious materials that intersects the dam foundations.

In this final investigation stage, a limited number of laboratory index tests should be made on representative samples. The purpose of these tests is to verify soil classifications and to obtain the moisture-density characteristics of proposed borrow area soils. Permeability tests rarely will be required if adequate soil classifications are made. However, very sandy soils may be tested if the design requires specific permeability information. The properties of proposed sources of rock for riprap and rockfill and of concrete aggregate should be investigated, as described in chapter V.

23. *Sanitary Studies.*—The necessity for sanitary study is determined by the degree to which pollution will be a factor in limiting the utilization of the proposed project. All possible sources of pollution from human, animal, and industrial waste should be investigated and evaluated. If municipalities are situated on the catchment basin, their sewage disposal systems must be investigated. Water samples adequately covering the period of the year during which the proposed project will be in use should be taken from the water course below the municipality and analyzed, particularly if there is a sewer outfall emptying into the stream. If the disposal system includes a bypass to the stream around the treatment plant, its effect should be evaluated and definite provisions made whereby the authorities in charge of the reservoir project will be notified prior to its use. Often a municipality has an inadequate disposal system but has plans for eventual improvement. In such cases, it may be necessary to plan the proposed project for limited use, pending the correction.

24. *Recreation and Fish and Wildlife.*—A discussion of recreational and fish and wildlife considerations is given in chapter II.

25. *Design of Structures.*—Technical questions involved in the design of small dams of various types and in the designs and layout of the appurtenant structures, such as outlet works and spillways, are presented in the various chapters covering outlets, spillways, and different types of dams.

Except for very minor structures, most States require submittal of plans for review and inspection in the case of non-Federal works. Engineers or owners considering dam construction in any State should obtain concise information on what type of control is exercised by the State, as well as the name and address of the appropriate official and agency which exercises such control. Permits or licenses must be secured from the Federal Power Commission for power projects, and approval must be obtained from the Department of Defense if structures on navigable streams are involved.

Before proceeding with design of a dam, the designer should consult with the appropriate officials to determine the requirements of the supervising agency.

26. *Preparation of Cost Estimates.*—Rough overall estimates of project feature costs are commonly made during the reconnaissance stage for the purpose of comparing alternative sites and for determining the size and scope of development. More detailed estimates involving quantities and unit costs are necessary for inclusion in feasibility reports supporting authorization or approval for construction, after plan formulation studies have established the optimum scale for economic soundness of the development. Estimates for dams and reservoirs should include, in addition to the construction costs of the dam and appurtenant structures, the probable cost of lands, water rights (if existing rights must be purchased or subordinated), rights-of-way, and clearing the reservoir areas; costs of relocating public highways, railroads, buildings, and other properties; and engineering and administrative costs. Estimates will also be needed for annual costs for financing and for operation, maintenance, and replacement. The feasibility estimate may not be in full detail, but in overall amount it should represent a ceiling within which the project features can be built, barring significant advances in unit prices. The final estimate will be based on the subsequent detailed studies made in connection with the preparation of specifications and should be in sufficient detail to serve as a guide in securing bids and awarding a contract for construction.

27. *Quantity Estimates.*—After details of the design of the dam and auxiliary structures have been determined, the overall design can be completed and construction quantities computed. In preparing estimates for excavation and embankment work, allowances must be made for wasted and unsuitable materials, shrinkage from excavation to compacted fills, swell of rock excavation, and overbreak in tunnel and channel excavation. The contingency percentage usually added to an estimate is for the purpose of covering unforeseen difficulties, changes of plans, and detailed items of design that may be changed or possibly omitted because of limited funds. This contingency percentage does not cover overexcavation or excessive waste in construction.

In preparing estimates, correction factors or percentages should be applied to net computed quantities for certain classes of work. The major items likely to be in error are embankments, excavation, and concrete quantities. Soil for compacted earth embankments, if measured in borrow excavations, will usually shrink from 10 to 30 percent when compacted —in general, about 15 percent. An allowance of 20 to 40 percent, generally 25 percent, for the swell of rock from excavation to rockfill and riprap should be made. Adequate allowance for overbreak should be made for excavation and for concrete lining in tunnels. Overbreak will vary with the tunnel size; with the nature of the material as to its composition, blockiness, and laminar makeup; and with the construction methods employed. Overbreak in tunnels can amount to as much as 40 percent of the minimum bore quantities. Additional concrete lining quantity percentages in such instances will be considerably higher, since the ratio will be based on the area of the concrete ring rather than on the entire bore area. An allowance must be made also for cement and aggregate wastes in the concrete quantities.

The estimator must be generous in computing quantities and yet avoid unreasonable and excessively high estimates. Laboratory data available are of considerable importance to the estimator in making the proper allowances for the estimated quantities. This is especially true of embankment materials and concrete mixes. An important feature of estimating quantities is a general understanding of the definitions for the various items with respect to specifications or pay dimensions. In preparing a schedule of quantities it is generally advisable to list all items on a standard form and according to standard specifications so that the items may be carefully checked against quantities and so that omissions may be avoided.

28. Unit Costs.—Because of widely varying economic and labor conditions in different localities, it is not possible to furnish unit prices for use in estimating each class of work. Furthermore, unit costs fluctuate widely in response to economic conditions and other factors. The estimator should familiarize himself with local conditions, probable sources of materials and labor supply, cost of similar work in the locality, and probable changes in costs of materials and labor that may occur before actual construction due to economic adjustments. During periods of economic fluctuations, consideration should be given to construction cost indexes and trends, with an intelligent application to the particular project.

The costs of operations such as excavation, embankment, and mass concrete work have not risen in proportion to the increase in labor costs, because of improved methods and greater use of machinery and mechanical equipment. In fact, for some large-scale operations, unit costs have been reduced. On the other hand, the costs of construction operations involving a large percentage of labor (for example, reinforced concrete, formwork, and hand excavation) have greatly increased.

29. Finalizing the Project Plans.—Plan formulation is a continuous process of coordination, analysis, and extension of all the specific studies directed toward achieving optimum size and scope of project and maximum benefits. It involves trial studies of various combinations of purposes and sizes and designs of structures so that in the final plans the sponsor may know, for example, that the area to be irrigated has been properly selected in consideration of the available water supply from various sources; that the reservoir is of proper size to achieve the best regulation of that water supply in terms of the investment for construction, operation, and maintenance; and that there are economic balances between spillway capacities and reservoir surcharge capacities, and between height of diversion dam and length of canal.

It should be emphasized that incorrect conclusions from the study are more often arrived at through forced interpretation of the data in favor of the project than through a lack of ability to evaluate properly those data. Under no consideration should a factor be assumed as favorable until that assumption is supported by all available data.

In certain sections of the United States, water rights are commonly overlooked. Although no difficulties may have arisen in the past on a particular stream or in a particular area, water rights must be investigated for each proposed project.

The sanitary study will not be necessary for all projects. On the other hand, it is often incorrectly omitted in the study of projects where purity of water is a major consideration. Where such study is required, it is advisable to consider all potential sources of pollution in their most unfavorable light.

Where competent geologists are available, the geologic study will offer few difficulties. If such assistance is not available, the greatest caution should be exercised in interpreting geologic characteristics.

When thorough consideration has been given to each of the component studies for a particular project, they should each be briefed, listing: (1) the favorable, and (2) the unfavorable circumstances with regard to the project. Technical honesty should guide in the evaluation, and any unsoundness of the project should be recognized if soundness is not proved beyond a reasonable doubt.

30. ***Planning Reports.***—A reconnaissance report is generally prepared by the investigating engineer to make a record of the data available, to present a preliminary concept of the project plan with a rough economic and financial analysis, and to draw conclusions as to whether the project, based on data at hand, merits further study. If the recommendation is favorable, the report should outline the feasibility grade investigation to be made, the estimated costs and time required, and the requirements for personnel, equipment, etc.

The project feasibility report is generally prepared on completion of the feasibility investigations as a basis for advising the sponsor or owner and others who must approve or authorize the project of its merits. The report describes the project plans, features, costs, benefits, relationships to existing and future developments, problems, and financing. It should present definite recommendations, based upon probable accomplishments, regarding feasibility and acceptability under possible means of financing construction. The conclusions and recommendations should be adequately supported by the investigations, as documented in the report or its appendixes, in such form that, if necessary, the work may be readily reviewed by the proper authorities.

A feasibility report or a final report of a project involving the construction of a dam should include, as part of the report or as an appendix, a separate report on the design of the dam and its appurtenant structures. An outline for a report on the design of a dam is given in the following section.

31. ***Report on Design of Dam.***—A complete report should be prepared covering the investigation, engineering features, environmental considerations, and cost of the proposed dam and reservoir. It should contain a general description of the design, including the various factors involved, a copy of the detailed estimate, and a drawing showing the general plan and sections. Included on the drawing should be a location map and curves showing hydraulic capacities.

To insure a complete description and record of all essential data, calculations, and conclusions entering into the design, a uniform procedure is desirable. The following outline of the items which the report should cover is included as a guide. Obviously, all of the information listed in this outline is not necessary for any particular small dam, but the greater part of it will be required for the larger and more complex structures.

____________ PROJECT
FEASIBILITY (OR SPECIFICATIONS) DESIGN AND ESTIMATE FOR ____________ DAM

A. *Location and Purpose:*

1. Section, township, range, principal meridian, county, State, nearest city.
2. Location in respect to other features.
3. Accessibility.
4. Purpose:
   - (a) Amount of storage—live, dead.
   - (b) Type of storage—irrigation, flood, power, domestic, etc.
   - (c) Water surface elevations.
   - (d) Place where water will be used.
5. Alternate designs, if any.

B. *Summary of Design:*

1. Storage capacity	______	acre-feet.
2. Spillway capacity	______	second-feet at water surface elevation ______.
3. Outlet capacity	______	second-feet at water surface elevation ______.
4. Power outlet capacity	______	second-feet at water surface elevation ______.
5. Top of dam		elevation ______
6. Normal water surface		elevation ______
7. Maximum water surface		elevation ______ surcharge ______ acre-feet.
8. Minimum water surface		elevation ______
9. Freeboard above maximum high water.		______ feet.
10. Maximum height of dam above streambed.		______ feet.
11. Estimated cost of dam (or dam and reservoir).		$______.
12. Estimated cost per acre-foot of storage.		$______.
13. Total estimate, project		$______.
14. General plans and sections		Drawing No.______

C. *Design Data:*

1. Topography:
   (a) Scale.
   (b) Contour interval.
   (c) Planetable sheet numbers.
   (d) Surveyed by.
   (e) Date of survey.
2. Geological report—author and title.
3. Logs of test pits and drill holes.
4. Hydraulic data, capacities and requirements and by whom established:
   (a) Storage, irrigation, flood, power.
   (b) Spillway.
   (c) Outlet.
   (d) Diversion.
   (e) Area-storage capacity curves for various elevations of water surface.
5. Hydrologic data:
   (a) Hydrographs.
   (b) Maximum recorded flood.
   (c) Inflow design flood.
   (d) Mean annual runoff of drainage basin.
   (e) Tailwater curve.
   (f) Cross sections of streambed.
   (g) Design values.
   (h) Climatic conditions.
6. Borrow areas and aggregate deposits, location, and transportation facilities available:
   (a) Laboratory tests.
7. Right-of-way information.
8. Photographs.

D. *Reservoir Data:*

1. Proposed capacities with corresponding water-surface elevations.
2. General dimensions.
3. Existing structures affected.
4. Nature of land flooded and clearing required.
5. Relocations: Railroad, highway, telephone lines, oil lines, powerlines.
6. Limitations to maximum reservoir flow line.
7. Geology:
   (a) General formations.
   (b) Factors relating to reservoir losses.
   (c) Contributory springs.
   (d) Deleterious mineral and salt deposits.
8. Rights-of-way.

E. *Damsite Data:*

1. Geological features, formations:
   (a) Nature of streambed and abutments.
2. Interpretation of test pits and drill holes.
3. Percolation tests, groundwater.

F. *Dam Design:*

1. Number and types of estimates prepared.
2. Features governing design.
3. Drawing number.
4. Water-surface elevations, storage capacities, freeboard.
5. General dimensions:
   (a) Top width.
   (b) Description of section—slopes, height, zoning, etc.
   (c) Crest length; roadway.
   (d) Length of base at maximum section.
6. Percolation factor; sliding factor.
7. Cutoff trench and cutoff wall dimensions.
8. Grouting requirements.
9. Toe drains, drain holes.

10. Parapet and curbs.
11. Galleries.
12. Fishways, logways, etc.

G. *Outlet Works Design:*
1. Requirements:
   (a) Discharges and corresponding water surface elevations.
   (b) Diversion capacities and water surface elevations.
2. Factors affecting location.
3. Tunnel dimension—material encountered; liner plates.
4. Conduit dimensions.
5. Gate chamber:
   (a) Dimensions.
   (b) Location.
   (c) Accessibility.
6. Gates, valves, and pipes:
   (a) Dimensions.
   (b) Elevations.
7. Approaches, shafts, adits, plugs.
8. Location of controls.
9. Trashrack.
10. Stilling basin.

H. *Spillway Design:*
1. Requirements.
2. Factors governing design and location.
3. Type and description:
   (a) Controlled or uncontrolled.
   (b) Lining.
   (c) Dimensions.
   (d) Elevations.
4. Gates, gate structure:
   (a) Dimensions.
   (b) Operation.
5. Stilling basin:
   (a) General description.
   (b) Dimensions.
6. Approach and discharge channels.

I. *Construction Facilities:*
1. Estimated time to complete.
2. Power available.
3. Construction railroad, shipping points, hauls.
4. Construction camp.
5. Local conditions.

J. *Materials and Unit Prices:*
1. Location of borrow, hauls.
2. Aggregate deposits, hauls.
3. Cement, nearest mill, hauls.
4. Railroads, terminals.
5. Basis for unit prices.

K. *Environmental and Ecological Measures:*
1. Fish and wildlife protection.
2. Fish and wildlife development.
3. Recreational development plans.
4. Area beautification plans.
5. Environmental protection measures during construction.

C. BIBLIOGRAPHY

32. *Bibliography.*

[1] U.S. Department of Agriculture, Agricultural Research Service, "Determining Consumptive Use and Irrigation Water Requirements," Technical Bulletin No. 1275, December 1962.

[2] U.S. Department of Agriculture, Soil Conservation Service, "Irrigation Water Requirements," Technical Release No. 21, April 1967.

[3] Jensen, M. E., "Water Consumption by Agricultural Plants," Water Deficits and Plant Growth, vol. II, Academic Press, New York, 1968, pp. 1-22.

Ecological and Environmental Considerations

E. A. SEAMAN [1] and L. W. DAVIDSON [2]

"... we in this century have too casually and too long abused our natural environment. The time has come when we can wait no longer to repair the damage already done, and to establish new criteria to guide us in the future."

Richard M. Nixon.

A. INTRODUCTION

33. *The Basis for Environmental and Ecological Considerations.*—(a) *General Considerations.*— The rapid increase in the world's population and the increasing demands this population has made on the planet's natural resources have called into question the long-term effect of man upon his environment. The realization that man is an integral part of nature and that his interaction with the fragile ecological systems which surround him is of paramount importance to his continued survival, is prompting a reevaluation of the functional relationships that exist between the environment, its ecology, and man.

Increasingly of concern is the effect which man's structures have upon the ecosystems in which they are placed and especially on the fish, wildlife, and human inhabitants adjacent to these structures. The need to store water for use through periods of drought, to supply industry and agriculture with water for material goods and foodstuffs, to provide recreational water in ever-increasing amounts, and the skyrocketing electric power demand has required the use of dams and canals. These structures help man and yet at the same time cause problems in the environment and in the ecosystems into which they are placed. Many of these problems are exceedingly complex and few answers which encompass the total effect of a structure on its environment are readily available.

Included in the answers to these problems must be the development and protection of a quality environment which serves both the demands of nature for ecological balance and the demands of man for social and psychological balance. The present challenge is to develop and implement new methods of design and construction which minimize environmental disturbances while also creating esthetic and culturally pleasing conditions under which man can develop his most desirable potentialities. This challenge can only be answered by the reasoned, pragmatic approach of sensitive, knowledgeable human beings.

The purpose of this chapter is to provide practical solutions to some of the environmental and ecological problems which confront the designer. This discussion is not exhaustive and it is hoped that the reader will consult the references at the end of this chapter (and nu-

[1] Assistant to the Commissioner for Ecology, Bureau of Reclamation.

[2] Civil Engineer, Earth Dams Section, Bureau of Reclamation.

merous others available on this topic) for a more extensive coverage. The amount of scientific data concerning the environment and man's relation to it is expanding rapidly and new design methods should become available soon. The practical information presented here can provide a useful introduction to the designer and a basis for maximizing a project's benefits and minimizing its negative environmental and ecological effects.

The term environment is meant here to include the earth resources of land, water, air, and vegetation and manmade structures which surround or are directly related to the proposed structure. The term ecology is meant here to encompass the pattern of relationships that exist between organisms (plant, animal, and human) and their environment.

(b) *Planning Operations.*—One of the most important aspects of dealing correctly and completely with the ecological and environmental impact of any structure is proper planning. If possible, an environmental team should be formed of representatives from groups who will be affected by the structure and experts from various scientific fields who can contribute their ideas and experience. The team approach will help assure that environmental considerations are placed in proper perspective with other vital issues such as reliability, cost, and safety, and that the relative advantages and disadvantages of each proposal are carefully weighed. It should also assure that the project is compatible with the natural environment. A suggested list of participants is given below:

(1) Concerned local and community officials.
(2) Design personnel.
(3) Environment and ecology experts.
(4) Fish biologists and wildlife experts.
(5) Building architects.
(6) Landscape architects.
(7) Recreational consultants.

This team should be responsible for the submittal of an ecological and environmental report to the designers, with a list of criteria which the designs should encompass. Some of the topics which should be discussed in the report are covered briefly in this chapter. Since each site will present unique problems, only a general outline of the most important considerations is provided herein.

B. FISH AND WILDLIFE CONSIDERATIONS

34. *Fish and Wildlife Considerations.*—(a) *General.*—The placement of a dam or canal within the environment should be done with due consideration of the effects on the fish and wildlife populations of the specific area. These considerations often involve complex problems of feeding patterns and mobility and, where possible, an expert in this field should be consulted. The U.S. Bureau of Sport Fisheries and Wildlife, the U.S. Forest Service, and appropriate State agencies can supply considerable expertise on the environmental impact of a proposed structure. It should be remembered that dams and canals can be highly advantageous in that they provide a year-round supply of drinking water for wildlife, breeding grounds for waterfowl, and spawning areas for fish. At the time of design, as many benefits as possible to fish, wildlife, and waterfowl populations should be included and provisions should also be made for the future protection of these populations. The following sections discuss some of the items which affect fish and wildlife and outline what can be done to aid them.

(b) *Ecological and Environmental Considerations for Fish.*—Critically important to the survival of fish populations are: (1) water quality, (2) water temperature, and (3) mobility. Water quality is obviously important to the survival of fish and an effort should be made to see that the quantity of pollutants which enter the stream during construction and the reservoir after completion are kept to the minimum. Strict regulations concerning pollutants should be instituted and enforced. Quantities of degradable, soluble, or toxic pollutants should not be left within the reservoir area after construction. Heavy pesticide runoffs can cause fishkills and some method, such as a holding pond or contour ditches, should be used to reduce their presence and steps taken to eventually eliminate them. Substances that can cloud or darken the water interfere with the ability of sight-feeding fish to forage and should not be allowed to enter the water.

In mining areas where heavy erosion often occurs, careful consideration must be given to the effects of siltation which may rapidly reduce the reservoir capacity. Consideration must also be given to the acidic character of the water since it can cause fishkills. Control dams may be the solution and, in one case, the Bureau of Reclamation has constructed an off-reservoir dam, Spring Creek Dam in California, to reduce the rapid sedimentation of the main reservoir and to limit the amount of acidic inflow to an acceptable level. Further details of the factors surrounding the construction of this dam may be found in reference [2].[3]

The temperature of the water controls timing of migration, breeding, and hatching and affects the appetite, growth, rate of heartbeat, and oxygen requirements of all fish. Each specie of fish has an optimum temperature range within which it can survive and consideration must be given to the temperature range which will exist both in the reservoir and in the stream below the dam due to the reservoir releases. For example, if a reservoir which is constructed in a mountainous area is too shallow, the colder water entering the reservoir can be sun-warmed during the summer months to an undesirable extent. The warm temperatures of shallow areas within the lake and also of the downstream releases would then prevent the spawning of cold water species such as trout. To remedy this problem, care must be taken to provide sufficiently deep reservoir areas where cold water will remain and to use an outlet works which is capable of selectively withdrawing the colder water from the lower reservoir depths. Federal and State fish and wildlife agencies should be consulted as to the correct depth for a specific area. Figure 1 shows

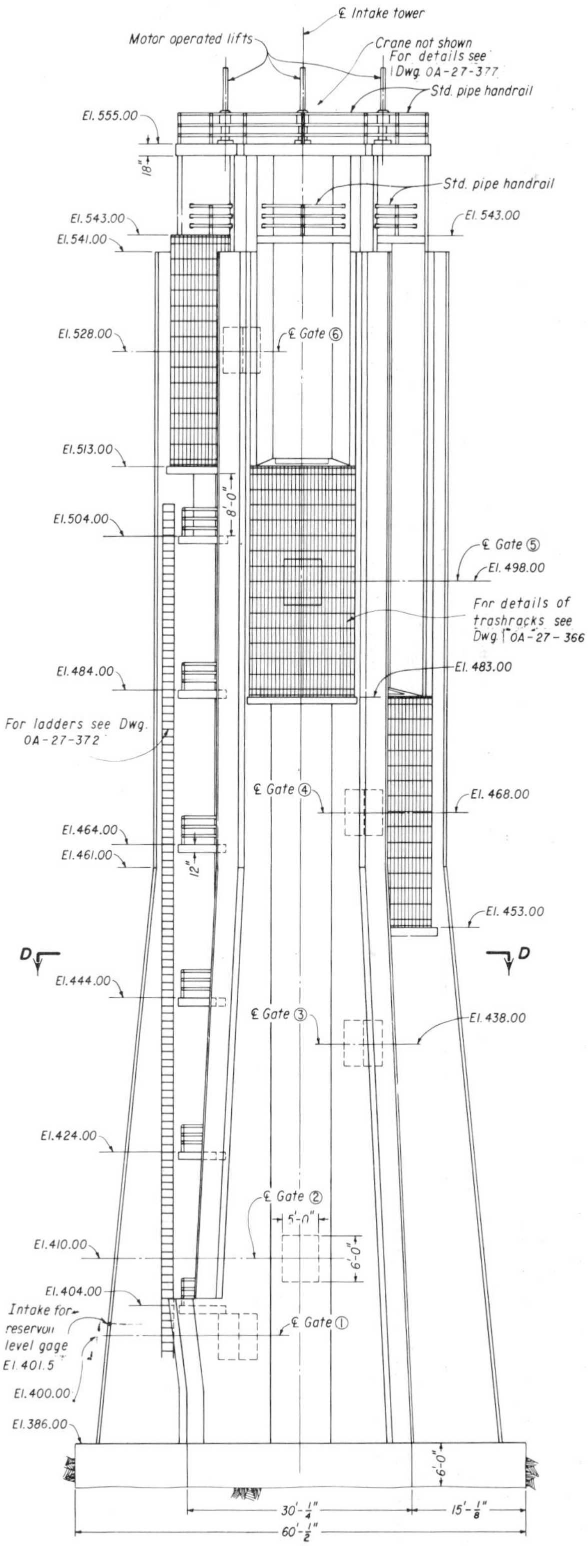

Figure 1. Selective withdrawal outlet works used at Toa Vaca Dam in Puerto Rico. OA–27–332.

[3] Numbers in brackets refer to items in the bibliography, section 37.

Figure 2. Fish ladder used on left abutment of Red Bluff Diversion Dam in California. P602–200–4543NA.

the selective withdrawal outlet works used at Toa Vaca Dam in Puerto Rico. Although this outlet works is considerably larger than those used for small dams, it illustrates the basic selective withdrawal principle. The Bureau of Reclamation has used selective withdrawal outlet works at several locations to create favorable temperatures for fish spawning downstream of the dam, and further information concerning these structures is available in references [17] and [18]. Reservoirs should also not be so shallow that fishkill occurs in the winter season. Conversely, shallow areas may be required to develop a warm water fishery or for waterfowl habitats.

Although salmon are commonly thought of as the only migrators, other species of fish such as shad, steelhead, and trout also require mobility considerations. The most common method for allowing fish to pass by a dam is the fish ladder. Figure 2 shows the fish ladder used by the Bureau of Reclamation on the Tehama-Colusa Canal in California. Specific design requirements for fish ladders may be obtained from the U.S. Forest Service, the U.S. Bureau of Sport Fisheries and Wildlife, or from appropriate State agencies. Fish should be prevented from entering spillways, outlet pipes, penstocks, and other restricted areas by the use of fish screens.

In flood control dams where water storage is temporary or in dams where the water flows are small during the dry season, a V-shaped notch (cunette) placed in the bottom of the outlet works pipe is used to facilitate fish migration. If the reservoir is to be drained at given intervals, it may be desirable to construct catch basins (kettles) below the outlet pipe so that fish may be transported to other areas or stored until repairs in the reservoir area have been made. Figure 3 shows the cunettes constructed in two outlet works conduits.

Where fish populations are concerned, care should be taken to avoid the destruction of

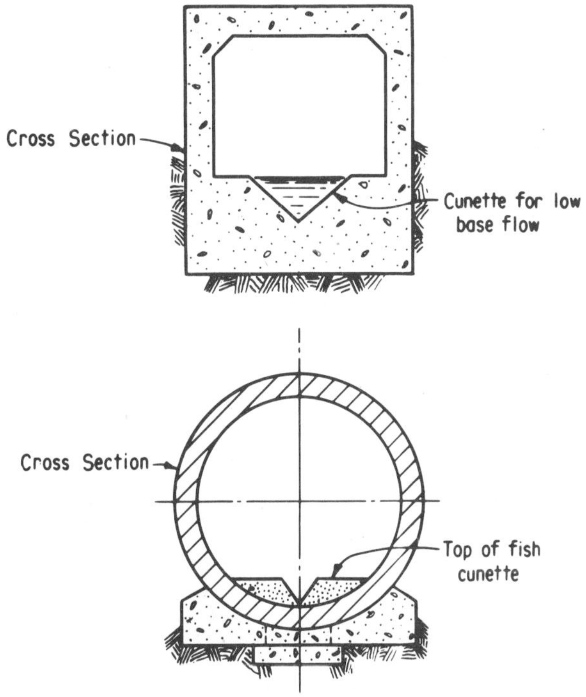

Figure 3. Cunettes used to facilitate fish passage at low water levels. 288–D–2867.

vegetation in the reservoir area since this becomes a food source after filling. Certain amounts of standing trees or tree debris left in the reservoir area can provide a habitat for several species of fish, as can brush piles which are staked down to prevent them from being washed away. Figure 4 shows a reservoir in which trees have been left standing to benefit the fish population.

Although certain aquatic plants are desirable for water birds, such as ducks, coots, wading birds, etc., they can be detrimental to fish production and should be controlled when necessary. Shallow shorelines in the inlet portions of the reservoir can be deepened to eliminate the growth of any plant life found not useful.

In newly constructed reservoirs, arrangements should be made for stocking with the appropriate type or types of fish. Consultation with a fisheries expert is recommended to determine the correct type of fish and the proper time for stocking them.

Figure 4. Aerial view of a small reservoir with trees left at the water's edge to help fish growth. 288–D–2869.

Figure 5. Fish hatchery below Lewiston Dam in California. 411–200–175A.

The oxygen content of some reservoirs can decrease with time and an examination of available reservoir reaeration devices may prove helpful. The oxygen content of the reservoir water may be increased during release from the reservoir by the use of reaeration devices such as the U-tube [19]. Reaeration is also aided by increasing the contact of the water released in the spillways and outlet works with air, and the use of baffled apron spillways (chapter IX) or other devices which cause extensive splashing may be helpful. A bibliography of reaeration devices, compiled by the Bureau of Reclamation, is contained in reference [20].

In some cases, fish hatcheries can be built in conjunction with the dam. Figure 5 shows the hatchery below Lewiston Dam in California. Canals also may provide spawning areas for fish although considerable cost and special equipment may be required. Figure 6 shows an artist's conception of the "gravel cleaner" which will be provided at a salmon spawning area in the Tehama-Colusa Canal in California. Special gravel and special gravel sizes were required in the canal bottom to facilitate spawning.

(c) *Environmental Considerations for Wildlife.*—Three common effects of reservoirs on wildlife involve: (1) removal of feeding areas, (2) loss of habitat, and (3) limitation of mo-

Figure 6. Artist's conception of gravel cleaner to be used at a salmon spawning area on the Tehama-Colusa Canal in California. CTP602–0–54534NA.

bility. The effects of each of these items can be significantly reduced.

When reservoirs inundate wildlife feeding areas, new areas should be planted to lessen the impact and, if possible, new types of grasses which are more suitable to the area and which provide more food per unit area should be planted. In addition, the new feeding areas can sometimes be irrigated with reservoir water to cause rapid, heavy growth. If the reservoir water is not immediately needed for irrigation, the water level can be left below the normal water surface to allow sufficient time for the feeding areas which are to be flooded to be replaced by equal areas of new growth.

Figure 7. Canal crossing on Charles Hansen Feeder Canal. Tracks show deer usage. P245–713–4111NA.

Where flooding of the homes of a large number of smaller animals, such as muskrat and beaver, will occur consideration should be given to adjusting the required excavation, reducing the reservoir water levels, or relocating the dam so that the number of animals affected is minimized. It may also be possible to provide special dikes and drainage conditions which lessen the effect. Problems in this area are difficult to solve and the advice of an expert should be sought.

Canals usually present the greatest problem to wildlife mobility. This problem has essentially been solved by requiring that sod-covered bridges be used at regular intervals over the canal. Figure 7 shows a deer crossing on a Bureau of Reclamation canal; deer tracks in the snow show usage. It is also probable that many smaller animals will use this bridge. Another problem with canals is that animals enter to drink and then are unable to get out. This problem can be solved by providing off-canal drinking areas and by requiring that steps be constructed on the canal sides at various locations. Another solution would be to flatten a portion of the canal and to roughen the surface so that the animals can get out. Figure 8 shows three possible methods of constructing canal steps. If heavy animal traffic along a certain stretch of canal is expected, the canal should be fenced.

Provisions for ducks, geese, and other waterfowl at small dams can be made by planting vegetation beneficial to nesting and by leaving areas of dense grass and weeds at the water's edge. If suitable areas already exist at the damsite, an effort should be made to selectively excavate the required material, leaving the habitat in place. Help concerning the appropriate reservoir treatment can be obtained from the Bureau of Sport Fisheries and Wildlife of the Department of the Interior, the Forest Service of the Department of Agriculture, and appropriate State agencies.

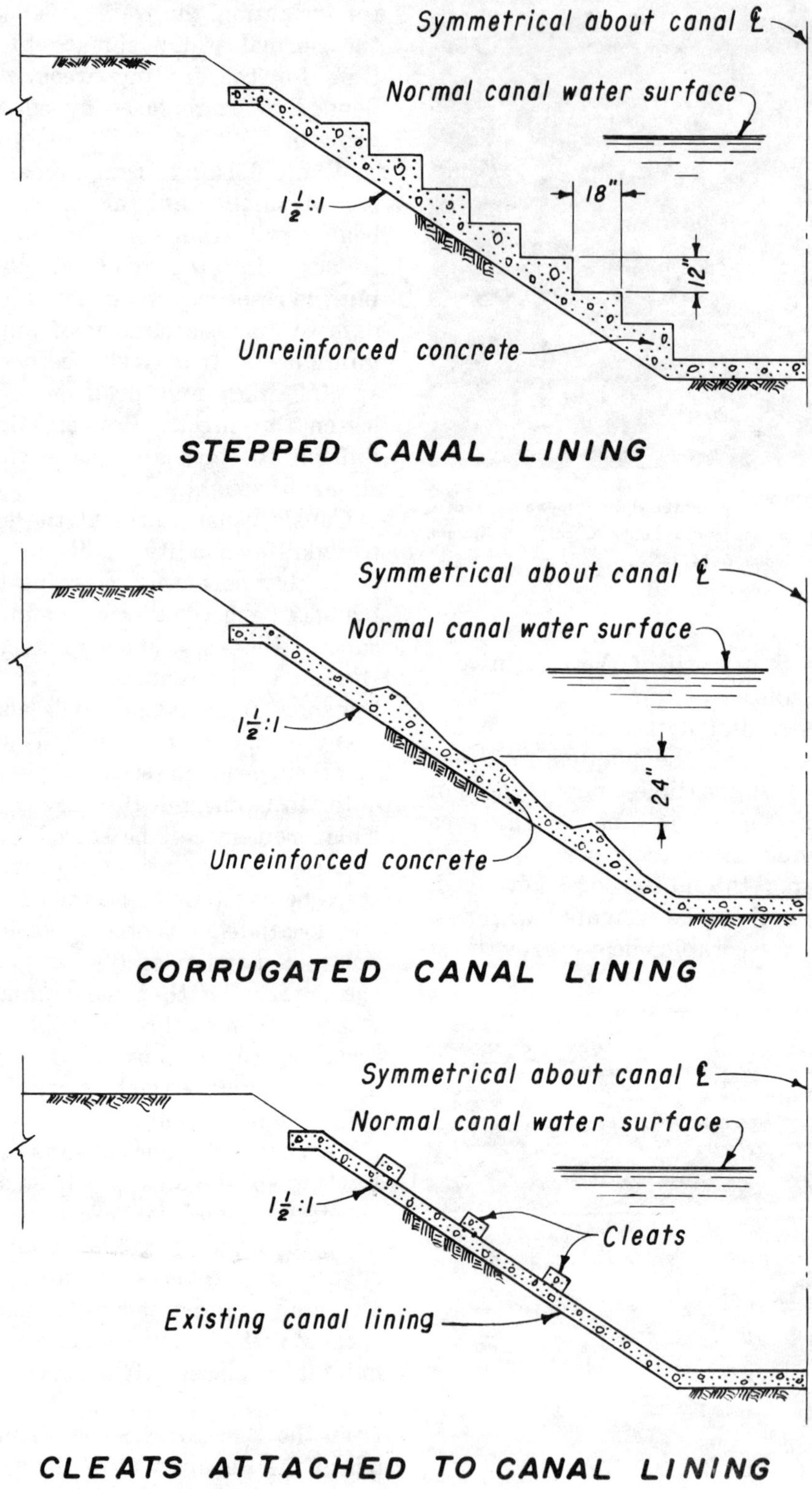

Figure 8. Three methods of providing canal steps (cleats) to aid humans and animals in removing themselves from the canal. 288–D–2868.

C. RECREATIONAL CONSIDERATIONS

35. *Recreational Considerations.*—(a) *General.*—The rapid increase in population, the decrease in working hours, and the great mobility of large numbers of people have caused a significant increase in the use of reservoirs for recreational activities. These activities include fishing, boating, water skiing, swimming, scuba diving, camping, picnicking and just simply enjoying the outdoor experience of the reservoir setting. Many of the reservoirs constructed in past years have become the recreation centers of the present and this will undoubtedly be repeated in the future. Provisions should be made to obtain the maximum recreational benefits from the completed reservoir and a future development plan should provide for area modifications as the recreation use increases.

(b) *Recreational Development.*—Recreation development considerations should start when project planning is begun and should be integrated into the total site plan. Areas of significant natural beauty should be left intact if possible, and recreational facilities should be developed around them. Boat ramps and boat-docking facilities are beneficial to most reservoir areas and should be constructed at the same time as the dam. Figure 9 shows the docking facilities at the Bureau of Reclamation's Canyon Ferry Reservoir in Montana. Camping facilities for truck campers, tenters, and picnicking areas can often be provided at

Figure 9. Boat docking facilities at Bureau of Reclamation's Canyon Ferry Reservoir in Montana. P269–600–949NA.

Figure 10. Viewing area at Bureau of Reclamation's Pactola Dam in South Dakota. P494–600–88A.

reasonably low costs.

Trash facilities should be provided at convenient locations to help in litter control, and the excessive use of signs and billboards near the reservoir area should be prohibited. The signs which are used should be blended with the surroundings. Toilet facilities should be available at all camping grounds and proper sewage disposal facilities should be installed, especially where the possibility of reservoir pollution exists.

If the reservoir is near a population center it may prove advantageous to provide bicycle paths, equestrian paths, and foot paths for public use. At the damsite or nearby, a reservoir viewing location and possibly a visitors center should be built. Exhibits showing the history of the project, local history, or other appropriate exhibits can enhance the visitor's enjoyment of the reservoir. These centers should be esthetically designed to fit the location. Figure 10 shows a simple, yet pleasant, viewing area at Pactola Dam in South Dakota. Where possible, sand swimming beaches should be constructed and maintained to provide maximum usage.

Buildings adjacent to the reservoir should be of low profile and blend with the reservoir surroundings; however, in some cases it may be desirable to contrast the buildings with their surroundings.

Fishing benefits can be maximized by stocking the reservoir with several types of fish and by replenishing these stocks yearly.

Proper maintenance requirements for the recreation areas should be instituted after completion of the dam and reservoir complex and should include repair of broken and damaged equipment, repainting, and rebuilding to meet expanded facility demands. Trash removal from the campgrounds and adjacent recreation facilities should be provided at regular intervals, and the possibility of recycling aluminum and metal products explored. Recreational areas which are overused should be rotated to prevent their deterioration, and single areas which receive exceptionally heavy use should be fenced off completely for short intervals to prevent ruination. Protection of the reservoir banks from sloughing may be required for steep slopes, and excessive erosion at any part of the site should be prevented.

D. DESIGN CONSIDERATIONS

36. *Environmental Design Considerations.*—(a) *General.*—Design requirements should be devoted to the accomplishment of three goals: (1) keeping the natural beauty of the surrounding area intact, (2) creating esthetically satisfying structures and landscapes, and (3) causing minimal disturbance to the area ecology. Designers should try to accomplish these goals in the most economical way. The following paragraphs should provide a partial list of items to be considered during design and will provide some practical suggestions for designers. Many of the items discussed here should be considered during the project planning stages and the critical decisions made at that time.

(b) *Structural Considerations.*—The main purpose of any structure is that it perform its function. Beyond this, thoughtful consideration should be given to blending the structure with the environment and to providing pleasing surroundings. If the structure is an earth dam, several things can be done which may prove beneficial to the project environment. A curved axis can be used in lieu of a straight axis to make the reservoir more nearly resemble a natural lake. The dam axis should always be convex in the upstream direction since this will help to provide a tighter bond between the embankment and its abutments when the reservoir loading is applied. The downstream slope of the embankment should be covered with topsoil and seeded to present a natural appearance; it should also be flat enough to allow mowing equipment to operate there. If it is not possible to reseed the downstream slope of the dam, cobbles and boulders which are indigenous to the area can provide a pleasing protective covering. The embankment design should utilize, to the maximum extent practicable, the materials found in the reservoir area and from adjacent structural excavations. Riprap material which blends with the area should be used if it meets the design criteria. Borrow areas in locations above the reservoir area should have their final slopes flattened to conform with the surrounding area and yet be flat enough for easy reseeding.

If it is necessary to excavate rock abutments above the crest of the dam, consideration should be given to the use of presplitting techniques since they leave clean, straight, esthetic surfaces. As discussed in part B of this chapter, a scenic overlook should be provided for viewing the dam and reservoir. The overlook should have adequate parking and, if possible, a visitors center.

The diversion schemes (see chapter XI) should be such that excessive siltation created during construction will not find its way into the downstream water. Materials from excavations which are not used in the embankment should be placed in the reservoir area and preferably along the upstream toe of the dam to prevent unsightly waste areas in the downstream approaches. In some cases, boat ramps, picnic areas, or view locations can be constructed with excavated material. Provided foundation conditions permit it, the use of drop-inlet spillway structures (chapter IX) will eliminate the required surface excavations on the dam abutments.

In the case of canals, spoil piles should be shaped to the natural landscape slopes along the canal length; this material can also be used to construct recreation areas where appropriate.

Buildings required for use at the site should be low profile structures and should blend with their surroundings. As an example, at the Elbert Forebay Dam in Colorado, the Bureau of Reclamation required that 14 stories of a 15-story powerplant be placed beneath the ground; the top story will be a visitors center. Pipelines should be buried, as should electrical wiring; where this is not possible the pipelines and electrical apparatus should be painted to blend with their background. Protective railings used on the dam crest and for bridges near the site should be low enough, consistent with safety, so that the reservoir may be seen from a passenger car.

(c) *Landscape Considerations.*—As much natural vegetation as possible should be left in place. If significant areas of natural beauty exist near the project, every effort should be made to utilize or preserve them and additional rights-of-way should be obtained to include any such adjacent areas. Borrow areas above the reservoir area should be revegetated with grass, trees, and shrubs soon after the job is completed. All slopes cut adjacent to the reservoir area should be reseeded and mulched; the topsoil removed from these slopes should always be stockpiled for reuse. Typical specifications for reseeding and borrow area beautification are given in appendix G. Figure 11 shows the mulching operations in a borrow area used at Pueblo Dam in Colorado.

Access roads to the damsite and roads used by the contractor during construction should be kept to a minimum, and those roads not planned for use after completion of the dam should be obliterated and reseeded. Access roads subject to excessive erosion should have protective surfacing. Roads which are required for maintenance of the dam or appurtenant works should be protectively fenced if excessive visitor usage will cause erosion.

Figure 11. Mulching operations for a reseeded borrow area at Pueblo Dam in Colorado. P382–706–8219NA.

Erosion control should be started at the beginning of the job. Roads, cutslopes, and borrow areas should be provided with terraces, berms, or other check structures if excessive erosion is likely. Exploratory trenches in borrow areas which are not used or adjacent to the damsite should be refilled and reseeded.

Quarry operations and rock excavations should be performed so that the minimum amount of material is removed, correct blasting techniques are used, large unsightly waste areas are not left, and final rock slopes for the completed excavation are shaped to have a pleasing appearance. Presplitting should again be considered for final slope cutting to permit a clean, pleasing view.

Road relocations near the dam provide an opportunity to eliminate deep cuts in hillsides, use scenic alinements, and provide reservoir viewing locations. Adequate road drainage should be used and slopes should be cut so that reseeding operations will be convenient. For projects where power transmission lines will be required, the publications, "Environmental Criteria for Electric Transmission Systems," [14] and "Environmental Considerations in Design of Transmission Lines" [26] will provide many helpful suggestions to lessen their environmental impact.

(d) *Protective Considerations.*—At locations where accidents can occur, protective devices and warning systems should be installed. The most dangerous locations at a damsite are near the spillway (especially if it is a chute-type), the outlet works intake tower, and the stilling basins of both the spillway and outlet works. Any portion of the spillway and outlet works stilling basins which might prove hazardous should be fenced off and marked by warning signs. Specifications for fencing materials are given in appendix G.

Canals often have steep side slopes which prevent climbing out and siphons which are extremely dangerous. Figure 8 shows three methods of providing steps in canals for humans and animals. These steps should be provided at specified intervals so that humans and animals which fall into the canal can get out easily. Safety racks on siphon entrances or stepped siphon entrances can prevent accidents, as can lifelines stretched across the canal upstream from siphon entrances and at potentially dangerous locations. Figure 12 shows the safety rack used on a canal siphon. Canals should be fenced at dangerous locations and warning signs clearly exhibited. Figure 13 shows the warning signs, protective fencing, and the lifeline near a siphon on the San Luis Canal in California. Considerable information concerning canal safety is contained in the Bureau's publication, "Reducing Hazards to People and Animals on Reclamation Canals." [15]

Figure 12. Safety rack installed in inlet transition of a siphon. P328–701–9467.

When the project encompasses the generation of electricity, the problems of providing adequate safety precautions are considerably increased and the advice of an expert should be sought.

(e) *Construction Considerations.*—The environmental and ecological design requirements provided in the specifications are converted from an abstraction into a reality by the build-

Figure 13. Protective equipment used at inlet to a siphon on the San Luis Canal in California. Note the warning signs, protective fencing, and lifeline. P805–236–14105NA.

er. The contractor and his personnel should be informed that this is an important step in the planning, design, and construction sequence. In this regard, a preconstruction conference may be invaluable for assuring an understanding of the job requirements by the builder and in enlisting his cooperation. The owner should insure compliance by having competent inspectors and by having specifications which clearly spell out the construction requirements. Excessive air, water, and dust pollution during construction should be prevented and specifications covering these items are included in appendix G; they should provide a framework for the inclusion of other important environmental provisions. The builder should also institute safety precautions during construction and the publication, "Construction Safety Standards" [13], may provide helpful information. The builder should be encouraged to bring forward any obvious defects in the environmental considerations which he encounters during construction and to improve on the considerations used.

Construction campsites should be placed within the reservoir area below normal water level. All trees, shrubs, and grassland areas which are to be protected should be staked or roped off. Any operations which would affect a large wildlife population should be performed in periods of low wildlife occurrence. Large volumes of water should not be taken from the stream if there are prior downstream commitments, and the water going downstream should be muddied as little as possible and be pollution free; siltation ponds may be needed in extreme cases. The builder should be required to remove or bury all trash and debris collected during the construction period and to remove all temporary buildings. Every opportunity should be taken to use the timber in the reservoir area for commercial operations. The burning of trees cleared within the reservoir area should be prevented if excessive air pollution will result or if State laws prevent it. At Pueblo Dam in Colorado, the Bureau is requiring that all brush and timber smaller than 7 inches in diameter be chipped into mulch and

Figure 14. Chipping operations at Pueblo Dam in Colorado. The chips will be stored for use as mulch on reseeded areas. P382–700–790NA.

stockpiled for future use on the reseeded areas. The chipping operation at Pueblo Dam is shown in figure 14. Slightly larger timber can be cut into firewood for use at camping and recreation areas and still larger timber can be channeled into commercial uses such as lumber, wallboard, boxes, etc.

A temporary viewing site for the project, having signs which show the completed project and explain its purpose, is helpful in promoting good community relations.

E. BIBLIOGRAPHY

37. *Bibliography.*

[1] Benson, N. G. (editor), "A Century of Fisheries in North America," American Fisheries Society, Washington, D.C., 1970.

[2] Prokopovich, N. P., "Siltation and Pollution Problems in Spring Creek, Shasta County, California," Journal of American Water Works Association, vol. 57, No. 8, August 1965, pp. 986–995.

[3] Dasmann, R. E., "Environmental Conservation," John Wiley & Sons, Inc., New York, 1968.

[4] "Transactions," American Fisheries Society, Washington, D.C.

[5] Seaman, E. A., "Small Fish Pond Problem-Management Chart," Technical Publication No. 2, West Virginia Conservation Commission, Charleston, W. Va.

[6] Reid, G. K., "Ecology of Inland Water and Estuaries," Reinhold, New York, 1961.

[7] Wing, L. W., "Practice of Wildlife Conservation," John Wiley & Sons, Inc., New York, 1951.

[8] U.S. Department of the Interior, "Quest for Quality," Government Printing Office, Washington, D.C., 1965.

[9] ———, "The Population Challenge—What It Means to America," Government Printing Office, Washington, D.C., 1966.

[10] ———, "The Third Wave," Government Printing Office, Washington, D.C., 1967.

[11] ———, "Man—An Endangered Species," Government Printing Office, Washington, D. C., 1968.

[12] ———, "River of Life, Water: The Environmental Challenge," Government Printing Office, Washington, D.C. 1970.

[13] Bureau of Reclamation, "Construction Safety Standards," June 1, 1968.

[14] U.S. Department of the Interior and U.S. Department of Agriculture, "Environmental Criteria for Electric Transmission Systems," Government Printing Office, Washington, D.C., 1970.

[15] Latham, H. S., and Verzuh, J. M., "Reducing Hazards to People and Animals on Reclamation Canals," REC-ERC-71-36, Bureau of Reclamation, January 1971.

[16] Bureau of Reclamation, "Environmental Quality—Preservation and Enhancement," Reclamation Instructions, series 350, pt. 376, 1969.

[17] "Register of Selective Withdrawal Works in United States," Task Committee on Outlet Works, Committee on Hydraulic Structures, Journal of the Hydraulics Division, ASCE, vol. 96, No. HY9, September 1970, pp. 1,841–1,872.

[18] Austin, G. H., Gray, D. A., and Swain, D. G., "Multilevel Outlet Works at Four Existing Reservoirs," Journal of the Hydraulics Division, ASCE, vol. 95, No. HY6, November 1970, pp. 1,793–1,808.

[19] Speece, R. E., and Orosco, R., "Design of U-Tube Aeration Systems," Journal of the Sanitary Engineering Division, ASCE, vol. 96, No. SA3, June 1970, pp. 715–725.

[20] King, D. L., "Reaeration of Streams and Reservoirs—Analysis and Bibliography," REC-OCE-70-55, Bureau of Reclamation, December 1970.

[21] "Report of the Committee on Water Quality Criteria," U.S. Department of the Interior, Federal Water Pollution Control Administration, April 1, 1968.

[22] Smith, G. (editor), "Conservation of Natural Resources," third edition, John Wiley & Sons, Inc., New York, 1965.

[23] Clawson, M., and Knetsch, J. L., "Economics of Outdoor Recreation," The John Hopkins Press, Baltimore, 1966.

[24] Watt, K. E., "Ecology and Resources Management; A Qualitative Approach," McGraw-Hill, New York, 1968.

[25] Flawn, P. T., "Environmental Geology: Conservation, Land-Use Planning, and Resources Management," Harper & Row, New York, 1970.

[26] Brenman, H., and Covington, D. A., "Environmental Considerations in Design of Transmission Lines," ASCE National Meeting on Transportation Engineering, Washington, D.C., July 1969.

[27] "Environmental Quality," First annual report of the Council on Environmental Quality, Government Printing Office, Washington, D.C., August 1970.

[28] Dober, R. P., "Environmental Design," Van Nostrand Reinhold Co., New York, 1969.

[29] Vernberg, J. F., and Vernberg, W. B., "The Animal and the Environment," Holt Reinhart & Winston, Inc., New York, 1970.

[30] McCullough, C. A., and Nicklen, R. R., "Control of Water Pollution during Dam Construction," Journal of the Sanitary Engineering Division, ASCE, vol. 97, No. SA1, February 1971, pp. 81–89.

Chapter III

Flood Studies

D. L. MILLER,[1] R. A. CLARK,[2] and S. SCHAMACH [3]

A. GENERAL

38. ***Scope***.—This chapter considers methods of determining flood flows to be expected from the drainage area tributary to the reservoir site, for which provision must be made in the design of a dam. The flood used for design against failure is termed the "inflow design flood." In most instances, particularly for structures impounding considerable storage, the inflow design flood is the maximum probable flood, which is defined as the largest flood that can reasonably be expected to occur on a given stream at a selected point. Adoption of an inflow design flood less than the maximum probable flood is a policy decision to be made by the owner, agency, or organization responsible for construction of the project. Factors to be considered in arriving at such a decision are discussed in chapter IX.

Determination of the maximum probable flood is based on rational consideration of the chances of simultaneous occurrence of the maximum of the several elements or conditions which contribute to the flood. A major consideration is the determination of the runoff that would result from an occurrence of a probable maximum storm based on meteorological factors. This hydrometeorological approach is necessary because streamflow records are of such relatively short duration in the United States that statistical analyses thereof do not provide reliable bases for estimates of maximum probable flood flows.

In addition to determination of inflow design floods, this chapter will indicate methods of determining the magnitude and frequency of floods as indicated by statistical analyses of streamflow records. These are primarily for use in connection with estimating diversion requirements during construction, establishing frequency of use of emergency spillways used in conjunction with outlets or small spillways, determining peak discharge estimates for diversion dams, or providing other information useful to the designers. Discussions in part A of this chapter are general; details of suggested procedures are given in part B.

39. ***Streamflow Data***.—The hydrologic data most directly useful in determining floodflows are actual streamflow records of considerable length at the location of the dam. Such records are rarely available. The engineer should obtain the streamflow records available for the general region in which the dam is to be situated. Locations of stream gaging stations and precipitation stations in the United States are shown on a series of maps entitled, "River Basin Maps Showing Hydrologic Stations," edition 1961 [4], prepared under the supervision of the National Weather Service. Such data collecting stations are subject to change in location, discontinuation, or initiation of new stations. These maps cannot be kept current, and information thereon must be supplemented by additional investigations in order to be sure of the location and operation of stations in a given area. The engineer should consult the water supply papers and indexes of the U.S.

---

[1] Engineer, Flood Hydrology Section, Bureau of Reclamation.
[2] Former Meteorologist, Bureau of Reclamation.
[3] Meteorologist, Bureau of Reclamation (retired).

Geological Survey [4] and, if possible, confer with the Survey's district engineer. He should also make a search of the records of other Federal agencies which may have collected information in the region, and the records of State water conservation agencies or State geological surveys; and he should determine whether any information may be available from other State departments, from county engineer offices, from municipalities in the vicinity, or from utility companies. Where streamflow records are not available, some agencies or inhabitants of the vicinity may have information about high-water marks caused by specific historic floods.

With respect to the character of the streamflow data available, flood flows at the damsite may be determined under one of the following conditions:

(1) *Streamflow record at or near the damsite.*—If such a record is available and covers a period of 20 years or more, the flood flows shown by the record may be analyzed to provide flood frequency values (sec. 59). Hydrographs of outstanding flood events can be analyzed to provide runoff factors for use in determining the maximum probable flood (secs. 51, 53, 55, and 56).

If such a record is available but covers only a few years, it may not include any flood of great magnitude within its limits and, if used alone, it would give false indication of flood potential. Analysis may, however, give some or all of the runoff factors needed to compute the maximum probable flood. Frequency values obtained from a short record should not be used without analysis of data from nearby watersheds of comparable runoff characteristics.

(2) *Streamflow record available on the stream itself, but at a considerable distance from the damsite.*—Such a record may be analyzed to provide unitgraph characteristics (secs. 52, 53, and 54), and frequency data which may be transferred to the damsite by appropriate area and basin-characteristic coefficients (secs. 54 and 59). This transfer can be made directly from one drainage area to another if the areas have comparable characteristics. Often damsites are located within the transition zone from mountains to plains and the stream gaging stations are located well out on the plains; in such instances, special care must be exercised when using the plains record for determination of flood flows at the damsite.

(3) *No adequate streamflow data available on the specific stream, but a satisfactory record for a drainage basin of similar characteristics in the same region.*—Such a record may be analyzed for unitgraph characteristics and frequency data, and these data transferred to the damsite by appropriate area and basin-characteristic coefficients.

(4) *Streamflow records in the region, but not satisfactorily useful for application and analysis under one of the above methods.*—These records may be assembled and analyzed as reference information on general runoff characteristics.

(5) *Use of high-water marks.*—High-water marks pointed out by inhabitants of the valley should be used with caution in estimating flood magnitudes. However, where there are a number of high-water marks in the vicinity of the project, and particularly if such marks are obtained from the records of public offices (such as State highway departments or county engineers), they may be used as the basis of a separate supplemental study. These records may be used to determine the water cross-sectional area and the water surface slope for the flood to which they refer, and from these data an estimate of that particular flood peak may be prepared using the slope-area method described in appendix B.

Whenever it appears that there will be one or more flood seasons between the selection of the damsite and construction of the dam, facilities for securing a streamflow record for the project should be set up as promptly as possible. This is of particular importance in order to obtain watershed data directly applicable to the computation of the inflow design flood for the dam, although a record usable for frequency computations cannot be secured. The facili-

[4] Published by the Government Printing Office and available in libraries designated as depositories of Government publications; most important libraries in the United States are so designated.

ties for obtaining such a record should be the best possible depending on the circumstances. A detailed discussion of these facilities, which may consist of either nonrecording or recording gages, is included in the following publications: "Equipment for Current-Meter Gaging Stations," U.S. Geological Survey Water Supply Paper 371; "Stream-Gaging Procedure," U.S. Geological Survey Water Supply Paper 888; and "Stream Flow," by Grover and Harrington, John Wiley & Sons, Inc., New York, 1943. The advice of Geological Survey engineers will be helpful in the site selection and installation, operation, and interpretation of records obtained.

Currently, a series of manuals, "Techniques of Water-Resource Investigations of the United States Geological Survey," is being published. Some parts of these manuals are presently available and some are being prepared; the published chapters may be obtained from the Superintendent of Documents, U.S. Government Printing Office, Washington, D.C. 20402.

The importance of utilizing records of runoff originating from the watershed above the damsite cannot be overemphasized. In the case of a damsite located on an ungaged stream, the establishment of measuring facilities as discussed above may produce basic data which would justify "eleventh hour" revision of the plans, thus improving the design of the dam.

40. *Precipitation Data.*—In each of the situations outlined in the preceding section, precipitation data are needed to evaluate factors for use in computing the maximum probable flood. The engineer should assemble the information with respect to precipitation during the greater storms in the region, and particularly for those storms for which runoff records are available. Such information can be obtained from publications of the National Weather Service [5] and Environmental Data Service. At present (1971), daily precipitation data for each month for each State are contained in the publication, "Climatological Data." Hourly data for each month for each State obtained by recording precipitation gages are contained in the publication, "Hourly Precipitation Data." [6] Often precipitation data obtained by the National Weather Service precipitation stations have been supplemented by "bucket survey" data, i.e., information on rainfall amounts of unusual storms obtained from residents within the storm area by personnel of the National Weather Service and other Government agencies.

Locations of precipitation stations as of 1961 are shown on the series of maps, "River Basin Maps showing Hydrologic Stations," previously referred to. The need may arise for utilizing frequency values of rainfall (sec. 47). National Weather Service publications presenting these data are included in the bibliography, section 60.

If plans are made to install streamflow measuring facilities as discussed in the preceding section, provision should also be made for obtaining precipitation records. An important item to consider is the selection of the location (or locations) of the precipitation gage, so that the catch will be a representative sample of average precipitation over the watershed. A comprehensive discussion of types of precipitation gages and observational procedures is contained in the National Weather Service publication, "Instructions for Climatological Observers," Circular B, 11th edition, January 1962.

41. *Use of Streamflow and Precipitation Records.*—From a flood hydrologist's point of view, an objective of analyzing streamflow and precipitation records is the development of procedures whereby the hydrograph (time versus distribution of runoff) that will result from a given amount of rainfall may be estimated. Another objective is the computation of a flood magnitude-frequency relationship based on experienced events. Certain procedures which have been developed for attaining these objectives have been selected for presentation in this text as applicable to problems encountered in the design of small dams. Engineers interested in a full discussion of the subject are referred to the texts included in the bibliography, section 60.

---

[5] Official designation: U.S. Department of Commerce, National Oceanic and Atmospheric Administration, National Weather Service.

[6] Subscription to these publications may be made through the Superintendent of Documents, U.S. Government Printing Office, Washington, D.C. 20402.

The primary use of recorded flows and precipitation records in the computation of an inflow design flood is the determination of a unitgraph and of retention loss characteristics of the watershed. Details regarding these analyses are presented in sections 51, 52, and 53. Where a streamflow record of sufficient length is available, a frequency curve may be computed by the procedure outlined in section 59.

42. *Watershed Data.*—All available information concerning watershed characteristics should be assembled. A map of the area above the damsite should be prepared showing the drainage system, contours if available, drainage boundaries, and locations of any precipitation stations and streamflow gaging stations. Available data on soil types, cover, and land usage provide valuable guides to judgment. Soil maps prepared by the Soil Conservation Service of the U.S. Department of Agriculture will prove helpful when the watershed lies within areas so mapped. The availability of this information is discussed in section 87.

The engineer preparing the flood study should make an inspection trip over the watershed to verify drainage area boundaries and soil and cover information, and to determine if any noncontributing areas are included within the drainage boundaries. The trip should also include visits to nearby watersheds if it is anticipated that records from nearby watersheds will be used in the study.

43. *Factors to be Considered in Estimating Flood Flows.*—(a) *General.*—Flood formulas primarily have been derived from and are directed toward peak discharge computations. Peak discharge values of major floods are more readily obtainable than volume values; hence, there are more data available for peak discharges. However, in most instances, the volume of runoff associated with the peak discharge and its time distribution is of vital concern to the designers, who usually need a hydrograph of the inflow design flood. For this reason, as well as for other reasons to be discussed later, flood formulas can be considered only as a guide to preliminary thinking and for making comparisons.

The detail with which hydrologic computations need be made in preparing a flood study depend on: (1) the character and applicability of the streamflow data available, and (2) the relationship of spillway cost to overall cost of the project. In regions of high rainfall potential, the cost of a spillway adequate to prevent failure sometimes makes construction of a dam exceedingly costly. For important projects, and particularly where the spillway cost is a major item of project cost and thus may have an important bearing on the feasibility of a project, the best possible use of streamflow data should be made. The hydrologic studies required in such instances are extremely complex, and consultation with an engineer or hydrologist experienced in this work is recommended. For small projects, or for those projects in which spillway capacity is obtainable at relatively low cost, a sufficient approximation of the inflow design flood discharge may be determined by procedures discussed in this chapter.

Statistical analyses of streamflow records do not provide reliable estimates of maximum probable flood discharges. Streamflow records provide the relationship between storm precipitation and runoff and information on runoff distribution. The determination of the maximum probable flood should be based on a study of storm potential, runoff potential, and runoff distribution as related to the physical characteristics of the watershed.

Each stream and each gaging station of a drainage area presents an individual problem because of the varying rainfall distribution and different runoff characteristics. It is not possible to express all factors affecting runoff in one simple formula for all watersheds. The simplicity of the Rational Method formula, $Q=CiA$, which is familiar to engineers encountering problems of runoff estimates, is deceptive in that a proper evaluation of the coefficient, C, and rainfall intensity, i, requires a detailed hydrologic study for each application. Engineers considering use of this formula should realize that the reliability of the results decreases with increase in size of drainage area.

For flood determinations, data concerning the following factors for each site need to be assembled and studied: geographical location, storm potential, drainage area, soil and cover, and runoff distribution. A general discussion of each of these factors is given in the remainder of this section, and a procedure is presented in section 44 which provides a means of progressively evaluating each of these factors and obtaining a maximum probable flood hydrograph.

It will be apparent from the following discussions that determination of a maximum probable flood hydrograph can become quite complex, and that an expert in this field should be consulted in such determinations for important projects. Enumeration of the complexities which may be encountered are not attempted; texts on hydrology are listed in the bibliography, section 60.

(b) *Geographical Location.*—Flood potential varies greatly between geographical subdivisions due to differences in geology, topography, and moisture sources. There is a close relationship between geographical location and storm characteristics. Depending on location, flood flows originate from rainfall, from snowmelt, or from a combination of varying amounts of rainfall and snowmelt. Certain topographic features exert significant influence on precipitation amounts, and certain geological formations are conducive to high flood discharges while others tend to reduce flood potential. An engineer preparing flood flow estimates should be familiar with the effect of these factors for the area in which he is interested.

(c) *Storm Potential.*—The first practical application of meteorology to studies of probable maximum precipitation dates from cooperative efforts by the U.S. Corps of Engineers and the U.S. Weather Bureau in 1937. Shortly afterward, the Hydrometeorological Section[7] of the Weather Bureau was established to perform such duties on a continuing basis. A Cooperative Studies Section[7] of the Weather Bureau made similar studies for the Bureau of Reclamation from 1946 to 1954. The several reports of these sections (listed in the bibliography, sec. 60) provide storm potential data for many parts of the United States.

[7] These sections are now designated as Hydrometeorological Branch and Special Studies Branch, Office of Hydrology, National Weather Service.

As the field of hydrometeorology is relatively new, a certain amount of confusion as to exact meaning of its terminology is to be expected, particularly when terms such as "maximum probable," "maximum possible," "probable maximum," etc., may or may not be intended to represent a difference in storm magnitude. In this text, the terminology used in regard to storm potential is in accordance with National Weather Service practice. The terms "probable maximum precipitation" and "probable maximum storm" represent two different things. Probable maximum precipitation for a particular area represents an envelopment of depth-duration-area rainfall relations for *all* storm types affecting that area adjusted meteorologically to maximum conditions, whereas probable maximum storm values consider such rainfall relations for only *similar* storm types. Probable maximum storm values are usually smaller than probable maximum precipitation values because different storm types are not combined. This subject is further discussed in section 50.

Generalized charts for estimating probable maximum precipitation east of the 105° meridian are published in Hydrometeorological Report No. 33 of the Hydrometeorological Branch, National Weather Service, Department of Commerce. Bureau of Reclamation hydrometeorologists have prepared some 300 probable maximum storm estimates for watersheds in the mountainous regions of western United States, in addition to numerous such estimates prepared by the Hydrometeorological and Special Studies Branch of the Office of Hydrology, National Weather Service. On the basis of these data, generalized maps of storm potential (sec. 50) have been prepared for this text to make possible the derivation of inflow design floods for areas throughout the United States by the hydrometeorological approach. It is recognized that continuing development of storm study techniques will lead to revision of storm potential estimates.

Generalized charts for estimating probable maximum precipitation west of the 105° meridian are contained in the National Weather Service Technical Paper No. 38. Special reports for limited portions of the West are also covered in the National Weather Service Hydrometeorological Reports Nos. 36 and 43.

(d) *Drainage Area*—An accurate map of the runoff contributing drainage area above a point of interest is essential to any flood study. This subject has been discussed in section 42.

(e) *Soil and Cover*.—The type of soil and vegetative cover of a watershed has a marked influence upon its runoff potential. It is difficult to express soil and cover types in numerical values which represent the difference between precipitation amounts and runoff amounts. If both precipitation and runoff data are available for several events, practical values can be obtained by direct analysis (sec. 51); if such data are not available, recourse must be made to comparisons with analyses for other watersheds. A method developed by the U.S. Soil Conservation Service for estimating runoff on the basis of soil type and cover as modified for use in this text is discussed in section 51 and is presented in detail in appendix A.

(f) *Runoff Distribution*.—The unitgraph has proved to be a very effective tool for hydrologic work. An engineer preparing flood estimates should be familiar with its basic principles. A brief outline of unitgraph derivation and application, including a dimensionless graph approach for derivation of a synthetic unitgraph for ungaged areas, is given in sections 52, 53, and 54. A concept of representing a unitgraph as a triangle and the relationships that can be derived therefrom is presented in section 55. Interested engineers are referred to the texts listed in the bibliography, section 60, for more detailed discussions of the unitgraph.

44. A Method of Computing Maximum Probable Flood Discharge.—The following discussion pertains in general to data contained in the "Hydrology Guide for Use in Watershed Planning."[8] published by the Soil Conservation Service. The introduction states: "The Hydrology Guide is intended for the use of Soil Conservation Service technicians and engineers. It presents material needed for fulfilling national service responsibilities in the field of soil and water conservation. Other technicians and engineers may find portions of the Guide useful in their own work, but they should recognize that the Guide attempts first to fill Service needs and requirements." A revised manual, "SCS National Engineering Handbook, Section 4, Hydrology, Part I—Watershed Planning," was issued in August 1964. Applicable data from this later manual have been included herein.

The Soil Conservation Service procedures are based on conclusions regarding applicable average values obtained from analyses of many natural flood events and are used to evaluate runoff for several purposes, including inflow design flood computations. The sequence for computing an inflow design flood as outlined by the SCS Guide has been adopted in this text, along with many of their conclusions regarding average applicable values for unitgraph determinations and retention rate estimates. However, the charts for obtaining design storm values (sec. 50) were prepared especially for this text to be more directly applicable to inflow design flood determinations, and are not the same as presented in the SCS Guide. Also, a minor modification of the SCS method of estimating direct runoff from design storm precipitation has been made, as discussed in section 58.

The method presented herein is applicable to watersheds where flow originates as direct runoff from precipitation in the form of rain (see sec. 51(a)). Those watersheds having runoff-delaying media such as lakes, swamps, heavy forest litter, and porous top soil require special study. Watersheds from which maximum runoff will likely include snowmelt also require special study. It is beyond the scope of this text to discuss these special studies in detail.

This procedure for computing a maximum probable flood is based on the hydrometeorological approach and requires estimates of storm potential and the amount and distribution of runoff. As in all generalized procedures, certain criteria have been adopted arbitrarily as applicable to the greatest number of cases. A step-by-step explanation of the procedure and computations for an example are given in section 58. A brief description of

[8] "Hydrology Guide for Use in Watershed Planning," *National Engineering Handbook*, Sec. 4, Hydrology, Supplement A, U.S. Department of Agriculture, Soil Conservation Service, 1956.

the procedure applicable to probable maximum general storm follows:

(1) A 6-hour point rainfall value is obtained for the geographical location from an appropriate chart.

(2) By means of graphs, this point rainfall value is adjusted to represent 6-hour average precipitation over the subject drainage area and also adjusted to give the cumulative rainfall for longer durations up to 48 hours.

(3) Charts and tables are used to determine the time-distribution of the rainfall.

(4) Soil and cover data are translated by means of tables into an appropriate runoff curve number (CN) from which the runoff volume is determined for increments of time D in equation (1).

(5) Based on the concept of representing hydrographs as triangles, a triangular hydrograph for 1 inch of runoff (unitgraph) from the subject watershed is computed from the following relationships:

$$T_p=\frac{D}{2}+0.6T_c \qquad (1)$$

$$T_b=2.67T_p \qquad (2)$$

$$q_p=\frac{484AQ}{T_p} \qquad (3)$$

where:

T_p=time to peak, hours,

D=rainfall excess period, hours,

T_c=time of concentration in hours, defined as travel time of water from hydraulically most distant point in the watershed to the point of interest,

T_b=time length of base of hydrograph, hours,

q_p=peak discharge in second-feet,

Q=volume of runoff, in inches, and

A=area of watershed in square miles.

Development of the above equations is presented on figure 29.

(6) The triangular hydrograph for each interval, D, is computed by direct ratio of the runoff for that interval to 1 inch of runoff and plotted.

(7) The total runoff hydrograph is obtained by graphical addition of the incremental hydrographs.

The above procedure is also used for computing inflow design floods from thunderstorm rainfall west of the 105° meridian, as discussed later in part B. However, a shorter time interval is used for the thunderstorm unitgraph as design thunderstorms presented in this text are of 1- or 3-hour duration.

This procedure might be termed a "generalized" unitgraph approach. An engineer using the above equations should take cognizance that the relationships of time-to-peak and base-time expressed by the equations represent average values for these relationships as adopted from numerous analyses of natural flood events made by Soil Conservation Service hydrologists.

If reliable streamflow hydrographs are available, analyses can be made to determine relationships expressed by the above equations that are applicable to the subject watershed, and these should be used. For ungaged watersheds, the above equations should give usable values for rain floods if the watershed has no unusual runoff characteristics. Representing the hydrograph as triangular instead of a natural curvilinear shape reduces computations. The error in shape thus introduced results in a slightly more severe hydrograph in that a triangle distributes a given amount of runoff in a shorter time interval than does a curvilinear hydrograph. (See fig. 32(A).)

45. *Inflow Design Flood.*—As stated in section 38, the inflow design flood is usually equivalent to the maximum probable flood, but in certain instances, as discussed in chapter IX, a flood of smaller magnitude than the maximum probable may be selected for use as an inflow design flood. Floods smaller than the maximum probable can be computed by using design precipitation values less than the probable maximum without changing runoff factors, or by using lesser design precipitation values in conjunction with varying degrees of runoff potential as determined from antecedent moisture conditions (AMC). The three antecedent moisture conditions, AMC-I, AMC-II, and AMC-III used to estimate runoff are described in appendix A. Procedures for computing floods less than the maximum probable are presented in section 58.

Before a decision is made as to the exact magnitude of the inflow flood to be adopted for design, a comparison should be made between computed synthetic values and records of floods that have occurred in the general area of interest. A method of making such comparisons is discussed in the following section.

46. Envelope Curves.—Peak discharge envelope curves and flood volume envelope curves can be prepared by drawing curves enveloping plotted points representing maximum recorded values for various drainage areas. The values plotted should represent similar type floods (rain floods or snowmelt floods) that have occurred within the broad geographical subdivision within which the subject watershed lies, and should not be limited to events of a single small river system. Preparation of envelope curves for a general area provides an engineer with valuable information on past flood history and an indication of the flood of record comparable to the subject area. However, they should not be relied upon as a means of estimating maximum probable flood values. Design flood values purporting to be the maximum probable should be higher than those obtained from envelope curves. Only in specific instances where a watershed has definitely lower flood potential than neighboring watersheds due to soil type, surface storage, etc., would it be good judgment to adopt an inflow design flood of smaller magnitude than that of a flood which has occurred nearby.

A simple method of preparation of envelope curves is to tabulate maximum peak discharges (or volumes of a selected duration) and respective drainage areas prior to plotting points. In most instances, the drainage area above a stream gaging station or the point of a large flood discharge measurement is given in the U.S. Geological Survey water supply paper listing the flood. When it is known that only a portion of the drainage area above a point of measurement contributed to a flood, the size of that contributing portion should be used in the envelope curve analysis. Discharges or volumes are plotted versus respective drainage areas using log-log paper. Data thus plotted usually indicate a curved line envelopment on log-log paper which may be approximated by a straight line for small ranges in areas. High discharges from local thunderstorms may suggest consideration of two curves—one for smaller areas subject to such occurrences and another for larger areas where maximum discharges originate from general storms.

47. Estimates of Frequency of Occurrence of Floods.—Estimates of the magnitude of floods which have frequencies of 1 in 5, 1 in 10, or 1 in 25 years are helpful in estimating requirements for stream diversion during construction. These floods are normally termed the "5-, 10-, or 25-year flood." The magnitude of more rare events such as the 50- or 100-year flood may be required for reasons such as to establish sill location of emergency spillways, to design diversion dams, etc. The usual term of expression, "x-year flood," should not lead to the wrong conclusion that the event indicated can happen only once in x years, and having occurred, will not happen again for another period of x years. It does mean that over a long span of years we can expect as many x-year floods (or larger) as there are x-year-long periods within that span. Floods occur randomly and may be bunched or spread out unevenly with respect to time. No predictions are possible for determining their distribution; the maximum probable flood *can* occur the first year after the project is built, though of course, the odds are heavily against it.

The frequency of a flood should be considered as the chances of occurrence of a flood of that size (or one larger) in any one year. Stated another way, the chances of the flood in any one year being equaled or exceeded by floods of the magnitudes indicated as the 5-, 10-, 25-, or 100-year floods have ratios of 20:100, 10:100, 4:100, and 1:100, respectively.

The hydrograph of a particular frequency flood is usually sketched to conventional shape using the peak discharge value and corresponding volume value obtained from computed frequency curves. In some instances, a peak discharge and associated volume of a recorded flood will correspond closely with a particular frequency value, in which case the recorded flood hydrograph is used.

If streamflow data for a period of 20 years or more are available for the subject watershed or comparable watersheds, frequency curve computations yield acceptable results for estimates up to the 25-year flood and may be extrapolated to indicate the 100-year flood with a fair assurance of obtaining acceptable values.

Many methods of flood frequency determinations based on streamflow data have been published. Excellent summaries of these methods, along with comments on factors affecting their accuracy and limitations, are contained in the papers entitled, "Review of Flood Frequency Methods"[9] and "Methods of Flow Frequency Analysis."[10] While the many methods of flood frequency determinations made from streamflow data are all based on acceptable statistical procedures, the difference in methodology can give appreciably different results when extensions are made beyond the range of adequate data. To provide for a uniformity in Federal water resources planning, the Water Resources Council has recommended that all Government agencies use the Log-Pearson Type III distribution as a base method. The method is described in the publication, "A Uniform Technique for Determining Flood Flow Frequencies".[11] Hazen's Method gives results that are comparable to those from the Log-Pearson Type III method and is easier to use when computations are made by hand with or without the aid of mechanical calculating machines. A procedural outline for such Hazen computations is presented in section 59.

For watersheds where runoff originates from rainfall and for which streamflow data are not available (usually small watersheds), an indication of flood frequencies can be obtained by estimating probable runoff from precipitation data of the desired frequency. Probable rainfall intensities for short durations can be obtained from National Weather Service publications (see bibliography, sec. 60), or in some instances by direct frequency analyses of records at nearby precipitation stations. These data provide means of obtaining probable "x-year" precipitation values for various time durations. A storm duration time is assumed equal to that of the time of concentration, T_c, estimated for the watershed. Triangular hydrographs representing the desired frequency flood are computed by the procedure presented in section 58. The uncertainties inherent in estimating the amount of runoff which will result from a given amount of rainfall make the results of this procedure less reliable than that of using streamflow data where such data are available.

48. *Special Cases.*—The discussions in this chapter have dealt primarily with flood flows resulting from runoff from rain falling on an unfrozen watershed. There are areas within the United States where large floods have originated from rapid snowmelt on a frozen watershed or from a combination of rain falling on a snow cover over a frozen watershed. The eastern and western coastal regions of the United States are subject to floods caused by rain falling on a snow cover over watersheds having usually unfrozen soil. These types of events require special study to ascertain the amount of water that becomes available for runoff. Once that is determined, distribution of the runoff by a unitgraph usually is feasible. Although such special cases may be rare within an area of interest, they may represent the maximum probable flood-producing conditions.

Instances in which snowmelt runoff provides the major portion of a maximum probable flood usually involve major streams and large drainage areas. Design of structures for such is beyond the scope of this text. However, problems may arise concerning small, mountainous watersheds where snowmelt runoff is a dominant characteristic. In these areas, the flow in the stream due to snowmelt runoff may be large and should be added to the computed rain flood. Also, when estimating runoff from rainfall in such areas, it should be recognized that part of the watershed will be covered by melting snow which is satisfying retention losses under the snow-covered area. The runoff from rain falling on the snow-covered area may

[9] "Review of Flood Frequency Methods," Final Report of the Subcommittee of the Joint Division Committee on Floods, Trans. ASCE, vol. 118, 1953, pp. 1,220-1,231.

[10] "Methods of Flow Frequency Analysis," Bulletin No. 13, Subcommittee on Hydrology, Inter-Agency Committee on Water Resources (now the Hydrology Committee, Water Resources Council), Washington, D.C., April 1966.

[11] "A Uniform Technique for Determining Flood Flow Frequencies," Bulletin No. 15, Hydrology Committee, Water Resources Council, Washington, D.C., December 1967.

be equal to the rainfall Therefore, the overall retention loss for rainfall on a partially snow-covered watershed during the melting season will be less than that for the same watershed when bare of snow.

It should be kept in mind that, although a usual sequence of events may produce floods, it is generally the unusual event or series of events that produce the great floods. The occurrence of two hurricane storms a few days apart following the same path over a large area in the northeastern States in August of 1955 is a prime example. Hasty conclusions as to flood potential should not be made on the basis of a long period of streamflow record. For example, although the recorded maximum peak discharge on the Pecos River near Comstock, Tex., was 116,000 second-feet for the period of years 1900 through 1953, a discharge of 948,000 second-feet occurred on June 28, 1954.

B. PROCEDURES

49. *Introduction.*—The selection of an appropriate inflow design flood is an essential part of the engineering studies for a project. The words "selection" and "appropriate" are used advisedly because a considerable amount of engineering judgment must be exercised in any hydrologic study of flood potential. It might be presumed that the problems of determining an inflow design flood would decrease in direct ratio to the size of the drainage areas involved and that such problems for drainage areas above small dams could be resolved quite easily. Such is not the case. In many instances, the hydrologic problems for small drainage areas are less easily resolved than those for large areas because relevant data for small natural watersheds are extremely meager.

It is believed that those using the material in this text most often will be concerned with projects for which little direct hydrologic data are available. Therefore, material and procedures presented herein have been selected with a view to assisting in solution of flood estimating problems for such projects. However, all available recorded streamflow and precipitation data should be utilized to the fullest extent possible, and outlines for methods of analyzing these data are included in this text. Discussions of procedural developments have been omitted or condensed, and discussions of applications of automatic data processing programs which have been developed for utilization in flood estimates are not included. The engineer so interested may obtain them from references cited in the bibliography, section 60. Discussions of the following subjects are presented:

50. *Estimating Storm Potential.*—(a) *General.*—An estimate of storm potential is an integral part of the hydrometeorological approach to computation of inflow design floods. The term "storm potential" is all-inclusive, embracing factors such as rainfall intensity, duration and areal extent. Meteorologists are able to establish estimates of maximum values for these factors which, judged by present knowledge, appear to be the limit of nature's capabilities. These maximum values differ throughout the United States (and the world). Knowledge of such limits, and the resulting probable maximum precipitation, provides the hydrologist with a good starting point for his estimate of a maximum probable flood as well as for floods less than the maximum probable The precipitation values adopted for computing the selected inflow design flood are usually referred to as design storm values.

(b) *Definitions.*—For the purposes of this text, the following terminology is used:

(1) *Probable maximum precipitation.*—Probable maximum precipitation values represent an envelopment of maximized intensity

duration values obtained from all types of storms. It is recognized that probable maximum precipitation values for all durations and for all areas will not occur from any one type of storm. For example, a maximized thunderstorm is very likely to provide probable maximum precipitation over an area of 50 square miles for a duration of 6 hours or less, but the controlling values for longer durations or for larger areas will almost invariably be obtained from general storms.

(2) *Probable maximum storm.*—The probable maximum storm values represent an envelopment of maximized intensity-duration values obtained from one type of storm only. Consideration is given to storm type and variations of precipitation with respect to location, areal coverage, and duration.

(c) *Probable Maximum Storm Considerations.*—Estimates of probable maximum storms are based on analyses which consist of three steps: (1) determining the areal and time distribution of the larger storms of record in the general area; (2) maximizing these observed storms by increasing their values to their physical upper limit as determined from a consideration of their observed moisture content in relation to the probable maximum moisture content that could be associated with a similar storm condition; and (3) considering transposition of these storms. The results of the first step will indicate which storms are best suited for further analysis and can also be used in the hydrograph analyses to estimate average retention loss rates and hydrograph lag times.

In the second step, the relation between maximum moisture potential and the moisture charge of the inflowing airmass is considered. Other factors that are effective in determining the efficiency of a storm in converting atmospheric moisture into precipitation have not been defined sufficiently at the present time to enable their use for making estimates of storm efficiency.

Storm transposition considered in the third step is based on the assumption that the location of a particular storm depended upon meteorological factors that could just as easily occur over other locations within given regions. Transposition, in the case of general-type storms, is limited to regions subject to similar types of storms and not separated by major orographic features. Thunderstorms are more widely transposed than general-type storms. Because the period of record for any particular drainage basin is generally quite short, the transposition of other storms within the same homogeneous meteorological and orographic area has the advantage of combining regional experience of a large number of storms.

(d) *Generalized Precipitation Charts.*—An engineer encountering a design flood estimating problem needs information regarding storm potential. Since such information pertains to magnitudes of storms which could occur from a more severe combination of meteorological events that has yet been observed, the engineer cannot make his estimate directly from recorded storm data. It is impossible to show all the refinements and variations that can influence the magnitude of design storm values for all individual locations within the United States on a generalized chart. However, broad areas do have like storm potential. Generalized charts have been prepared for this text to provide one means of rapidly determining design storm values for any specific area. The design storm values obtained from the generalized charts represent a reasonable upper limit and in most cases will exceed the values obtained for a specific watershed by a detailed hydrometeorological study. If such a study is desired because of the importance or size of the project, the services of a consulting hydrometeorologist should be secured.

(1) *Generalized chart for the United States east of 105° meridian.*—Figure 15 shows probable maximum 6-hour precipitation values for 10-square-mile areas of the United States east of the 105° meridian. This chart is based on one presented in Hydrometeorological Report No. 33, prepared by the Hydrometeorological Branch of the National Weather Service in collaboration with the U.S. Corps of Engineers (see bibliography, sec. 60). These 6-hour values for 10-square-mile areas can be modified for durations in excess of 6 hours and for larger areas up to 1,000 square miles by use of figure 16. No variation is assumed between point and 10-square-mile precipitation. For

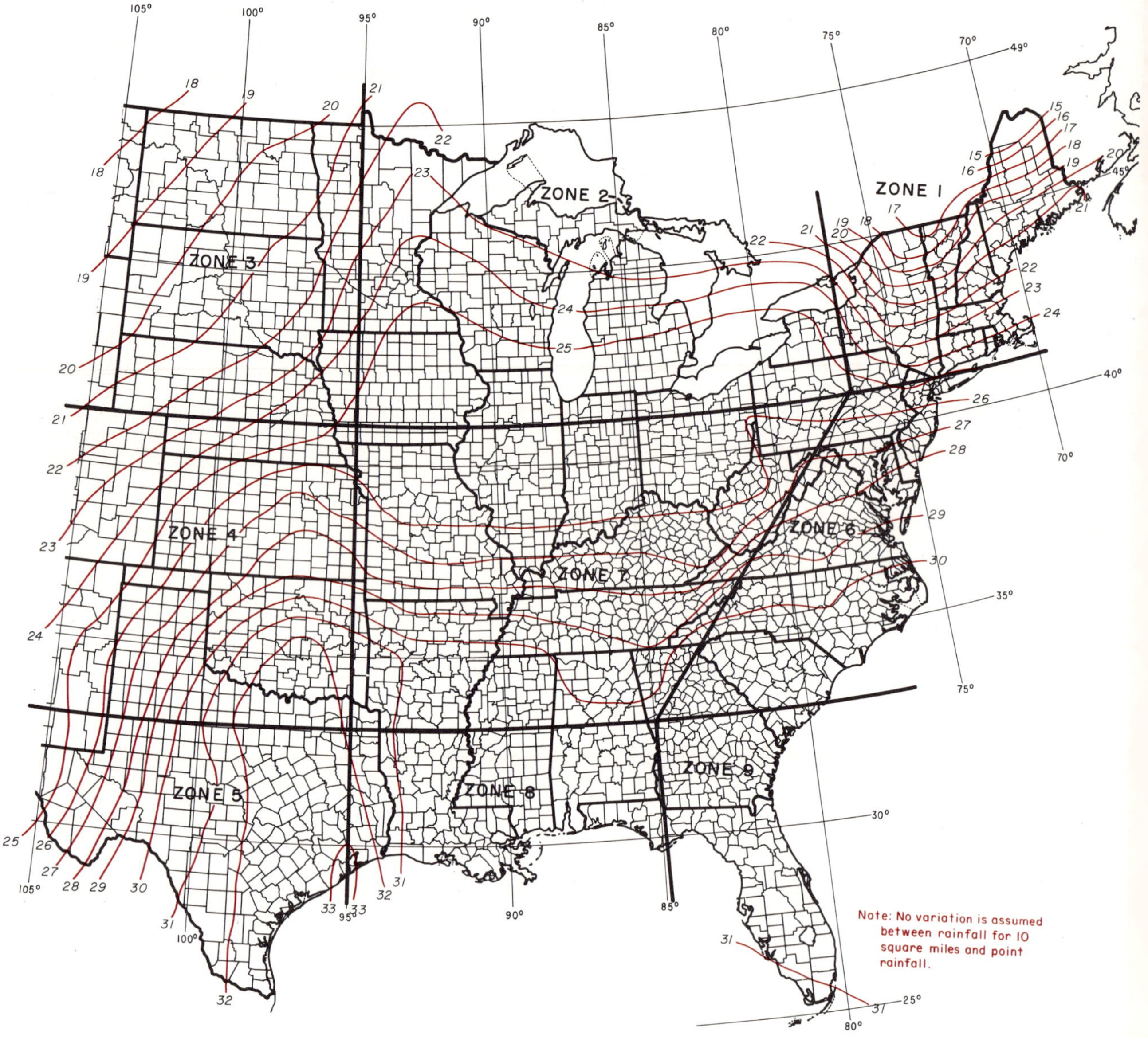

Figure 15. **Probable maximum precipitation (inches) east of the 105° meridian for an area of 10 square miles and 6 hours' duration. 288–D–2449A, 288–D–2754, 288–D–2755.**

durations shorter than 6 hours, the time distribution of precipitation can be obtained from curve C, figure 18. After the publication of Hydrometeorological Report No. 33, the Corps of Engineers recommended [12] that the following reductions be applied to the report values in order to provide for the imperfect "fit" of storm isohyetal patterns to the shape of a particular basin.

Drainage area (square miles)	*Reduction factor applicable to H.R. 33 rainfall values (percent)*
1,000	10.0
500	10.0
200	11.0
100	13.0
50	15.0
10	20.0

[12] Engineer Circular No. 1110-2-27, dated August 1, 1966, "Policies and Procedures Pertaining to Determination of Spillway Capacities and Freeboard Allowances for Dams."

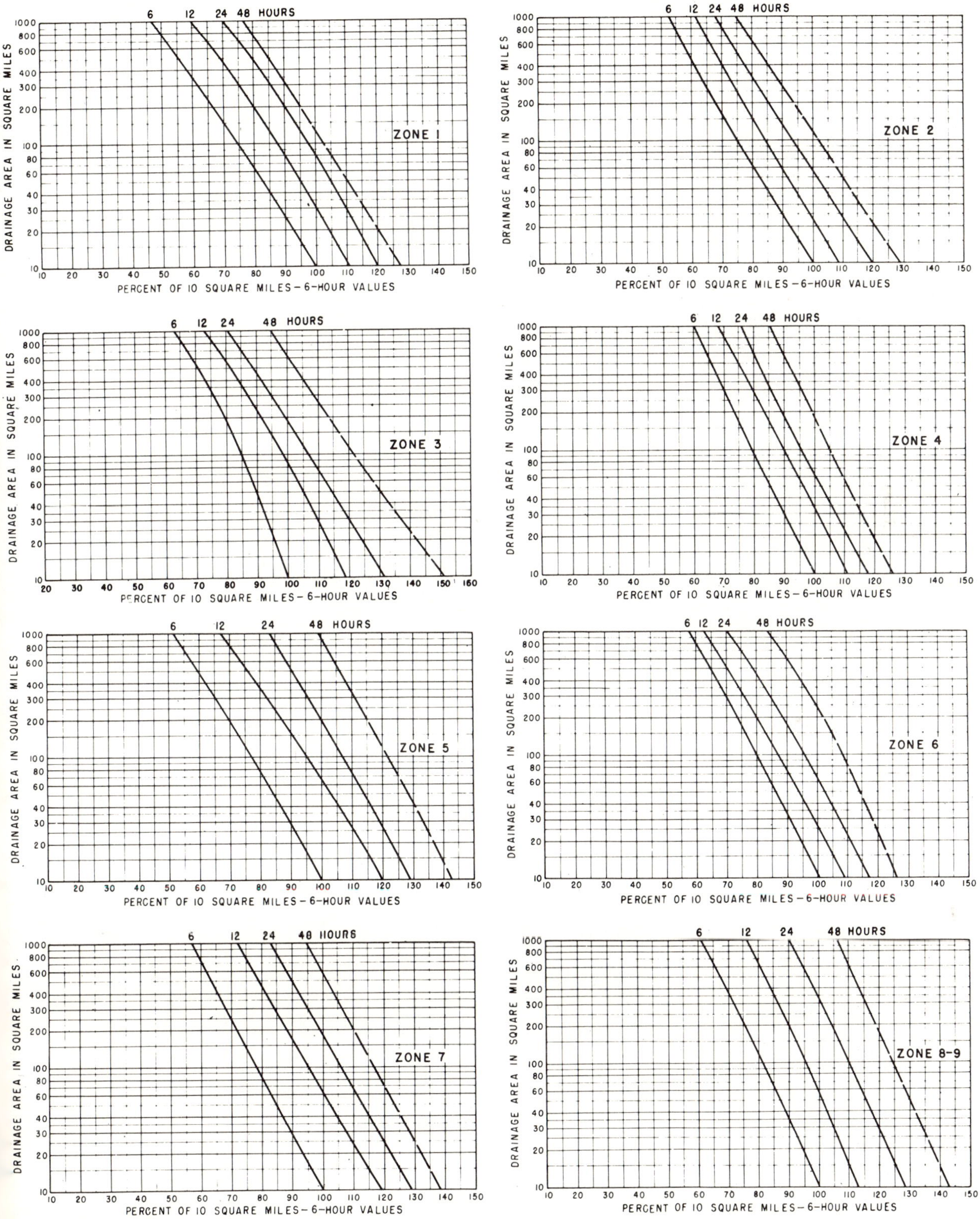

Figure 16. Depth-area-duration relationships. Percentage to be applied to 10 square miles, 6-hour probable maximum precipitation values. 288–D–2450.

Figure 17. **Probable maximum 6-hour point values in inches for general-type storms west of the 105° meridian. 288–D–2756, 288–D–2757.**

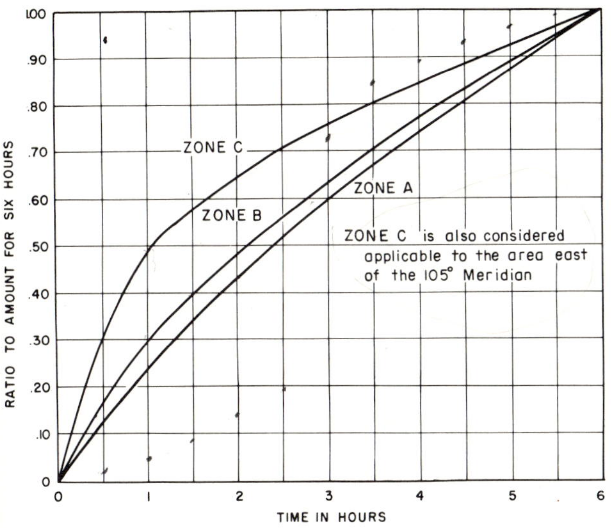

Figure 18. Distribution of 6-hour rainfall for area west of 105° meridian. See figure 17 for area included within each zone. 288–D–2758.

(2) *Generalized charts for the United States west of 105° meridian.*—Separate charts of probable maximum precipitation are presented for general-type storms which cover relatively large areas of rainfall, and for thunderstorm rainfall having limited areal coverage.

a. *General-type storm.*—Figure 17 shows probable maximum 6-hour point general-type storm values for areas of the United States west of the 105° meridian. This chart is based on the results of approximately 330 design storm analyses prepared by the Bureau of Reclamation for specific drainage basins west of the 105° meridian, as well as consideration of numerous design storm analyses made by the Special Studies and Hydrometeorological Branches of the National Weather Service. The variable topography of this part of the United States greatly influences the storm potential and permits only limited transposition of storms. These point storm values can be applied to areas up to 1,000 square miles by use of the curves presented on figure 19. The 6-hour general-type storm values can be extended for longer duration periods by multiplying the 6-hour value by the appropriate factor shown in table 1. Values for duration of less than 6 hours can be obtained from the appropriate curve of figure 18.

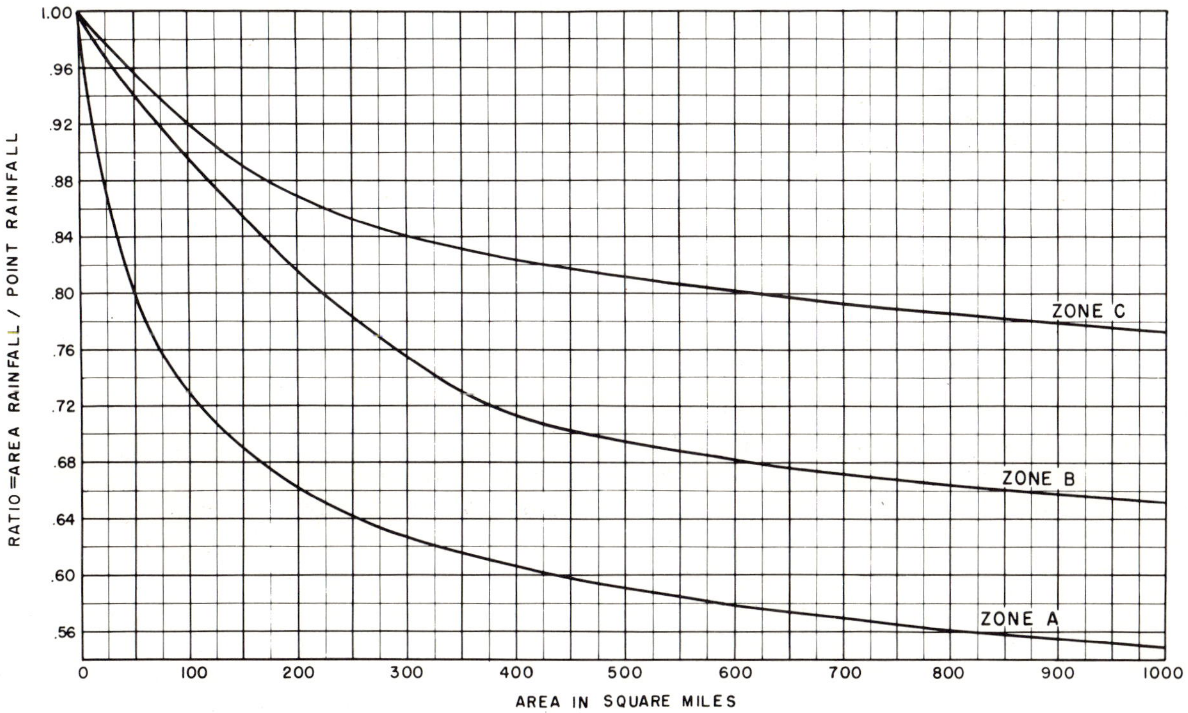

Figure 19. General storm: conversion ratio from 6-hour point rainfall to 6-hour area rainfall for area west of 105° meridian. 288–D–2759.

TABLE 1.—*Constants for extending 6-hour general-type design storm values west of 105° meridian to longer duration periods* [1]

Duration (hours) [2]	Constants		
	Zone A	Zone B	Zone C
8	1.20	1.18	1.14
10	1.39	1.36	1.26
12	1.58	1.53	1.36
14	1.76	1.66	1.43
16	1.93	1.77	1.50
18	2.10	1.87	1.57
20	2.26	1.95	1.64
22	2.42	2.03	1.71
24	2.57	2.10	1.78
30	2.95	2.28	1.97
36	3.26	2.38	2.15
42	3.55	2.40	2.25
48	3.79	2.41	2.28
60	4.14	...	...
72	4.34	...	...

[1] Multiply 6-hour point rainfall from figure 17 by indicated constant.

[2] For durations shorter than 6 hours, the time distribution of storm values can be obtained from the appropriate curve presented on figure 18.

TABLE 2.—*Probable maximum thunderstorm point rainfall in inches for various durations based on percentage of 1-hour rainfall values from figure 20*

Duration (minutes)	Zones I and III [1] (percent of 1-hour value)	Zone II [1] (percent of 1-hour value)
15	48	48
30	71	71
45	88	88
60	100	100
75	110	
90	117	
105	122	
120	126	
135	129	
150	131.5	
165	133	
180	134	

[1] Zones are outlined on figure 20.

b. *Thunderstorm.*—Estimates of probable maximum thunderstorm rainfall amounts for 1 hour over an extremely small area (point values) are shown on figure 20. Figure 20 was adapted with some changes from figure 6–3 of Technical Paper 38, prepared by the Cooperative Studies Branch of the National Weather Service in collaboration with the Soil Conservation Service (see bibliography, sec. 60). Table 2 presents a distribution of the figure 20 values into various durations for the three zones into which figure 20 is divided. To apply point values of figure 20 to various sizes of area, use the curves of figure 21. The curve for Zones I and II, figure 21, was adapted from the 1-hour curve of figure 5–13 in Hydrometeorological Report No. 43 (see bibliography, sec. 60). As indicated on figure 21, the area limits of probable maximum thunderstorm rainfall are estimated to be 100 square miles in Zone I, 50 square miles in Zone II, and 500 square miles in Zone III.

(3) *Use of generalized charts.*—For areas up to about 500 square miles west of the 105° meridian, inflow design floods should be prepared, using both the probable maximum thunderstorm rainfall and the probable maximum general-type storm rainfall. Probable maximum thunderstorm rainfall should be assumed to occur over the upstream area nearest the structure site for those drainage areas larger than 100 square miles in Zone I and 50 square miles in Zone II. The more critical of the two inflow design floods should be used in the design of the structure in all studies.

Design storm values obtained from figures 15 and 17 show considerable difference at their common boundary along the 105° meridian. This is due to the techniques used in determining the values shown on the charts. The values shown on figure 15 for the eastern United States are probable maximum precipitation based on envelopment of many types of storms, whereas the values shown on figure 17 for the western United States are based on studies of probable maximum general-type storms.

(e) *Design Storms for Low Hazard Areas.*—Some dams are located in areas where their

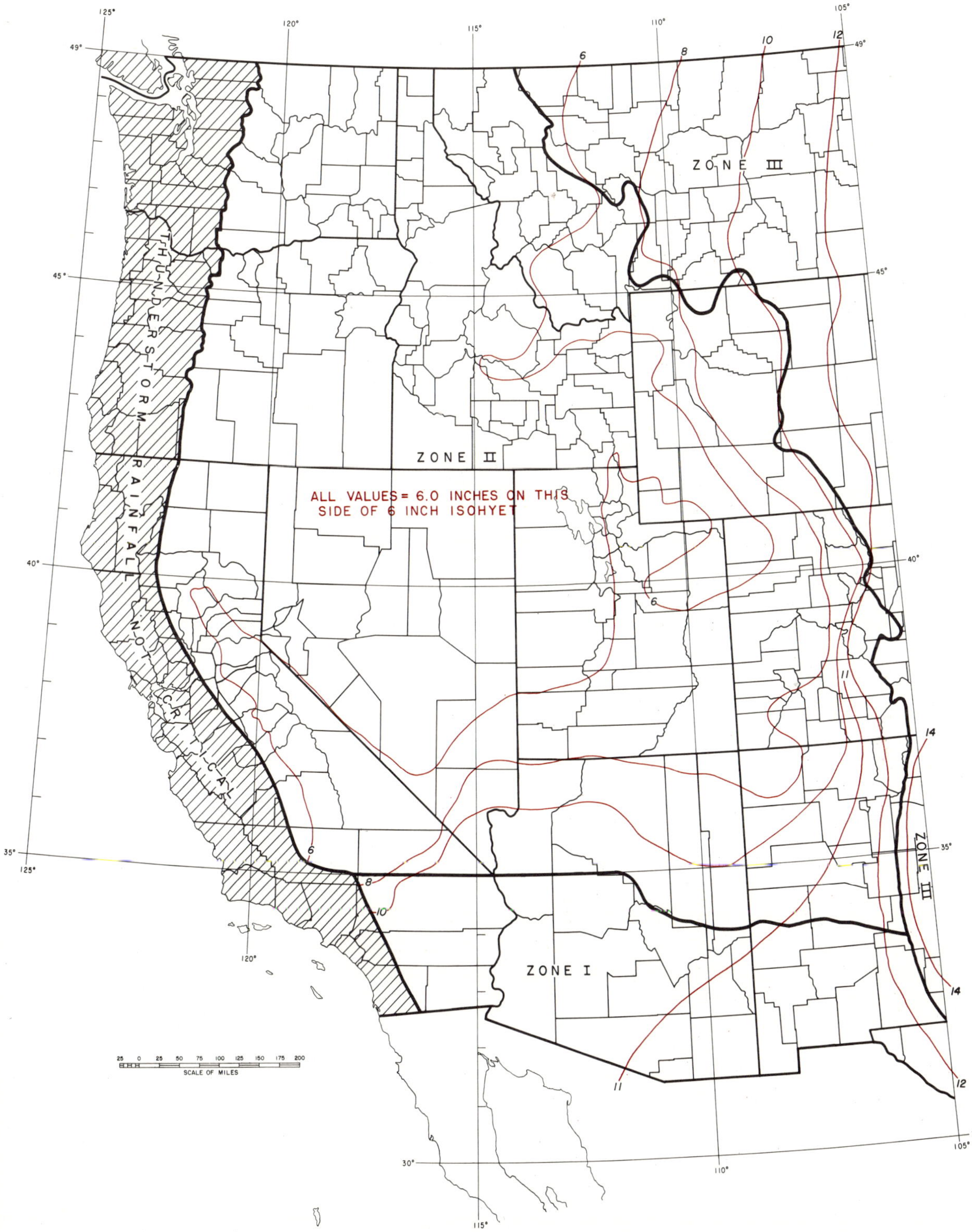

Figure 20. **Probable maximum thunderstorm 1-hour rainfall (point values in inches) for area west of 105° meridian. 288–D–2760, 288–D–2761.**

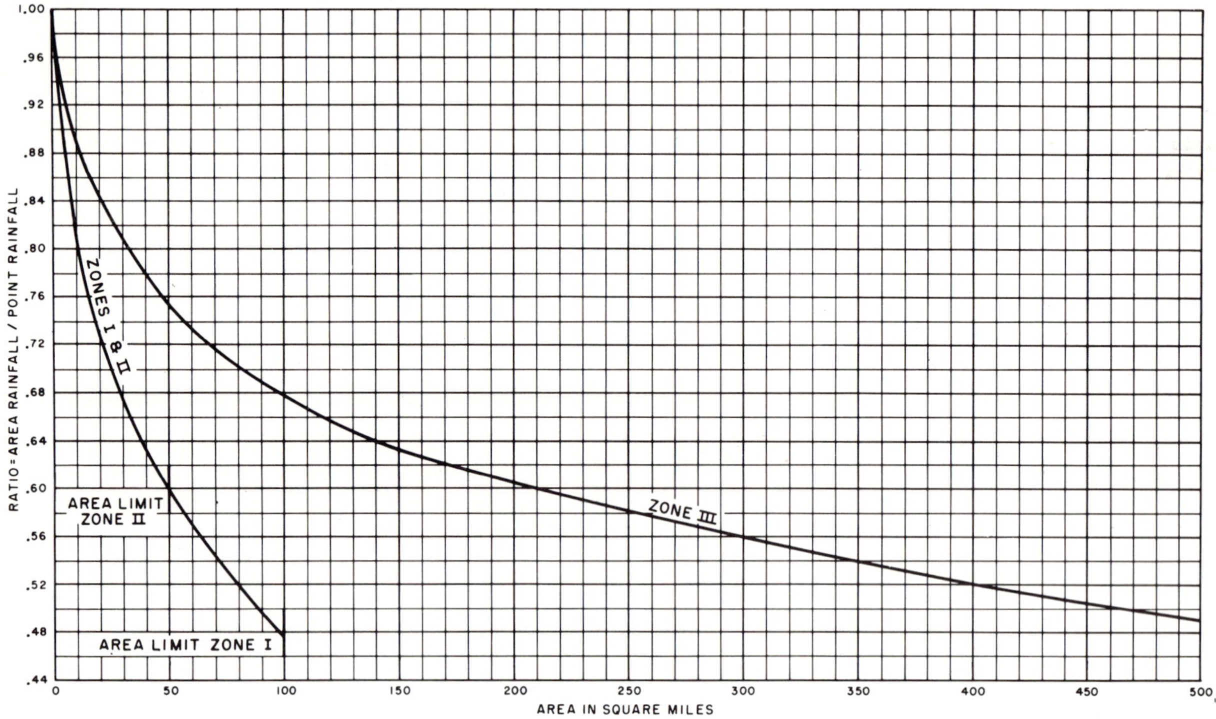

Figure 21. Thunderstorm: conversion ratio from 1-hour point rainfall to 1-hour area rainfall for area west of 105° meridian. 288–D–2762.

failure would not result in loss of life or extensive property damage. In such instances, the high cost of spillway construction may not warrant use of probable maximum precipitation in determination of the inflow design flood. Charts have been prepared indicating reduction factors to be applied to the probable maximum values for determination of a design storm for use in a low hazard area. Reduction factors for probable maximum values in figures 15, 17, and 20 are given in figures 22, 23, and 24, respectively. Values from figure 22 should be applied to the 6-hour probable maximum precipitation values from figure 15 without use of basin shape reduction percentages given in subsection 50(d)(1). For time distribution and area reduction, use the procedure described in subsection 50(d). Reductions produce values which have about a 1-percent chance of being equaled or exceeded in any given year.

51. *Estimating Runoff from Rainfall.*—(a) *General.*—The hydrometeorological approach to analyzing flood events and predicting the magnitude of floods requires a firm estimate of the difference between precipitation and resulting runoff. From a flood determination point of view, this difference is considered loss, i.e., loss from the precipitation falling over a given watershed. A simple solution to derive this loss value appears to be in finding the rate at which water will infiltrate into the soil. If this infiltration rate is known, along with the amount of precipitation, a simple subtraction should give the amount of runoff. However, there are other precipitation losses in addition to infiltration such as interception by vegetative cover, surface storage, and evaporation that may have material effect on runoff amounts.

Various types of apparatus have been devised to test the infiltration rates of soils, and studies have been made of interception and evaporation losses. Although maps to an extremely large scale could define most of the surface storage area, it is apparent that an accurate volumetric evaluation of all the loss factors can be made only for a highly instrumented, small plot of ground and that such an

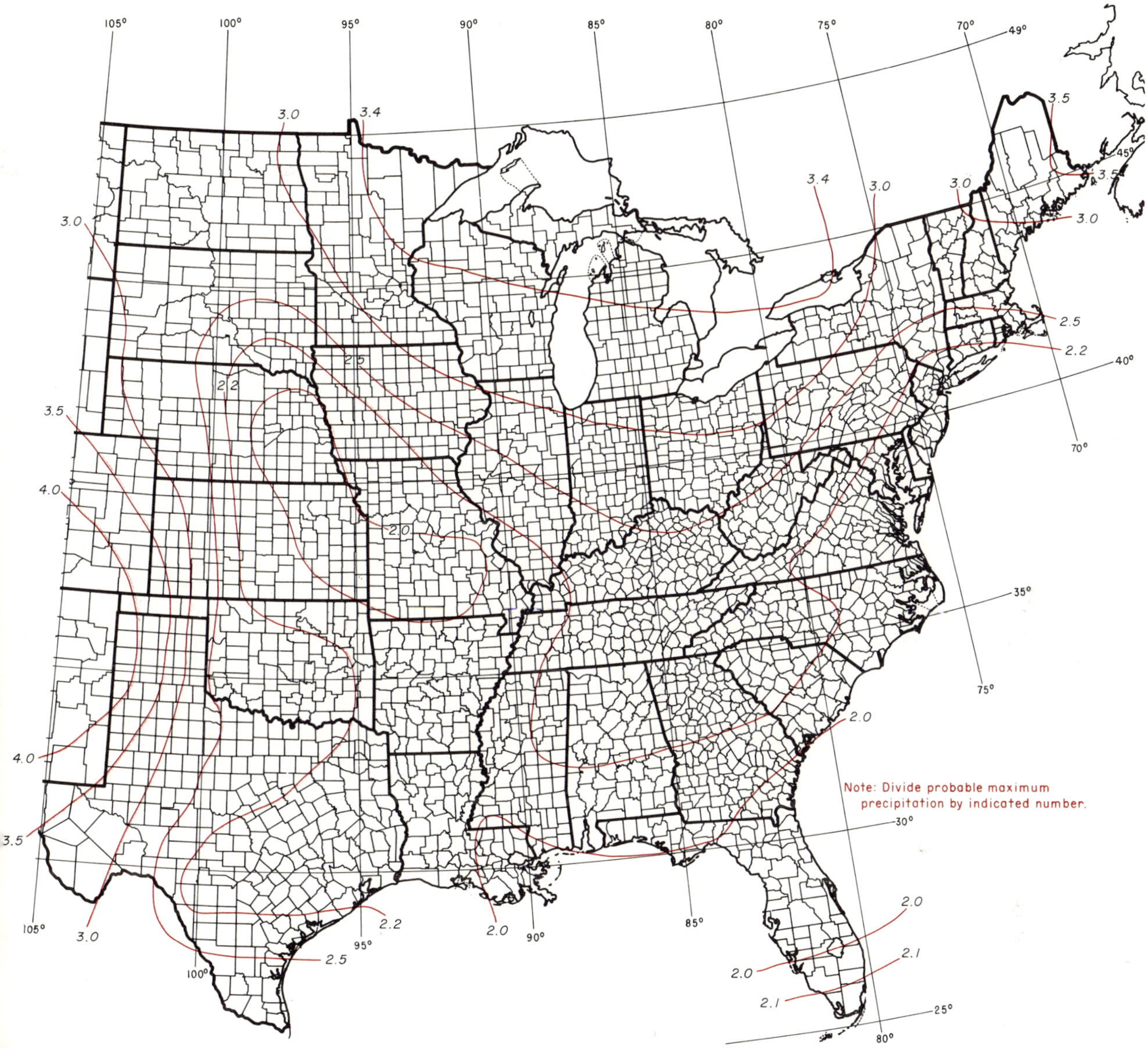

Figure 22. **Ratio for determining rainfall applicable for computing inflow design flood less than maximum probable for area east of 105° meridian. 288–D–2449A, 288 D 2763, 288 D 2764.**

evaluation is not practical for a natural watershed composed of many square miles of varying type soils, vegetative cover, and terrain features. For this reason, hydrologic literature contains arguments against the "infiltration rate approach" to determination of runoff amounts. However, the infiltration rate approach is applied on an empirical basis to obtain a practical solution to the problem of determining amounts of runoff, recognizing that the values used are of the nature of "index" values rather than "true" values. Natural events are studied and the difference between rainfall and runoff determined. Since this difference includes all the losses described above, it is usually called a retention loss or a retention rate. Such retention rates derived from available records may be adjusted to ungaged watersheds by analogy of soil type and cover.

An engineer making studies to compare rainfall with runoff must become familiar with the

Figure 23. Ratio for determining 6-hour general-type rainfall applicable for computing inflow design flood less than maximum probable for area west of 105° meridian. 288–D–2449A, 288–D–2765, 288—D2766.

units of measurement used and the factors for conversion to common units. These conversion factors are given in appendix B. In the United States, precipitation is measured in inches and runoff is measured in cubic feet per second (also referred to as second-feet or sec.-ft.). Also, the engineer must be cognizant that a hydrograph of storm runoff obtained at a streamflow gaging station represents one or more of the following types of runoff from the watershed: channel, surface, interflow, and base flow. Brief definitions of these types are:

Channel runoff: Caused by rain falling on the water surface of the stream. It begins with the start of precipitation and may be discernible from a slight rise of the hydrograph just after rainfall begins, but the quantity of channel runoff is so small it is ignored in hydrograph analyses.

Surface runoff: Occurs only when the rainfall rate is greater than the retention loss rate. This type of runoff causes most floods, and the computational procedures in this text consider this type of runoff dominant.

Interflow: Occurs when rainfall infiltrating the soil surface encounters an underground zone of lower permeability, travels above the zone to the surface downhill, and reappears to become surface runoff. This type of flow may also be called subsurface flow or "quick return flow."

Base flow: The fairly steady flow of a stream from natural storage, as shown by hydrographs during non-storm (or non-active snowmelt) periods.

In flood hydrology it is customary to deal separately with base flow and to combine all other types of flow into *direct runoff* in unknown proportions, as assumed in this text.

It is necessary to know the watershed area contributing the runoff at a given measuring point in order to express the runoff volume in inches of depth over the watershed for comparison with precipitation amounts. When making such comparisons, the amount of runoff, expressed as inches, is termed rainfall excess, and the difference between the rainfall excess and the total precipitation is considered retention loss.

The discussion which follows describes a method of making a rainfall-runoff analysis. The objectives of such analyses are: (1) the determination of a retention rate, and (2) the determination of the duration time interval of rainfall excess. A comparison of retention rates derived from several analyses leads to adoption of a rate for design flood computations. The determination of the duration of excess rainfall is necessary for the hydrograph analyses computations involving determinations of unitgraphs and lag times, which are discussed in sections 53, 54, 55, and 57. In all such analyses, the runoff volume which is compared with precipitation amounts is that which relates directly to the rainfall under study. Therefore, the base flow of the streamflow hydrograph must be subtracted out before comparisons are made (see sec. 53).

(b) *Analysis of Observed Rainfall Data.*—

(1) *Mass curves of rainfall.*—Mass curves of cumulative rainfall during the storm period should be plotted for all precipitation stations in and near the basin, as shown on figure 25(A). To show clearly the relation of rainfall to runoff, it is sometimes desirable to plot the mass curves to the same time scale as the discharge hydrograph of storm runoff. Usually, however, the curves should be given a more expanded time scale than it is desirable to use for the hydrograph analysis. When only one recording station is located nearby, and in the absence of better information, the mass curve of precipitation at a nonrecording station is usually considered to be proportional in shape to that of the recording station, except as otherwise defined by the observer's readings and notes (fig. 25(A)). The speed and direction of travel of the rainburst should be taken into account. Many rainfall observers enter the times of beginning and ending on the same line as the current daily reading. The notes may therefore refer to the previous day, especially when the gage is regularly read in the morning.

(2) *Isohyetal maps.*—The total amounts of rainfall occurring during the portion of the

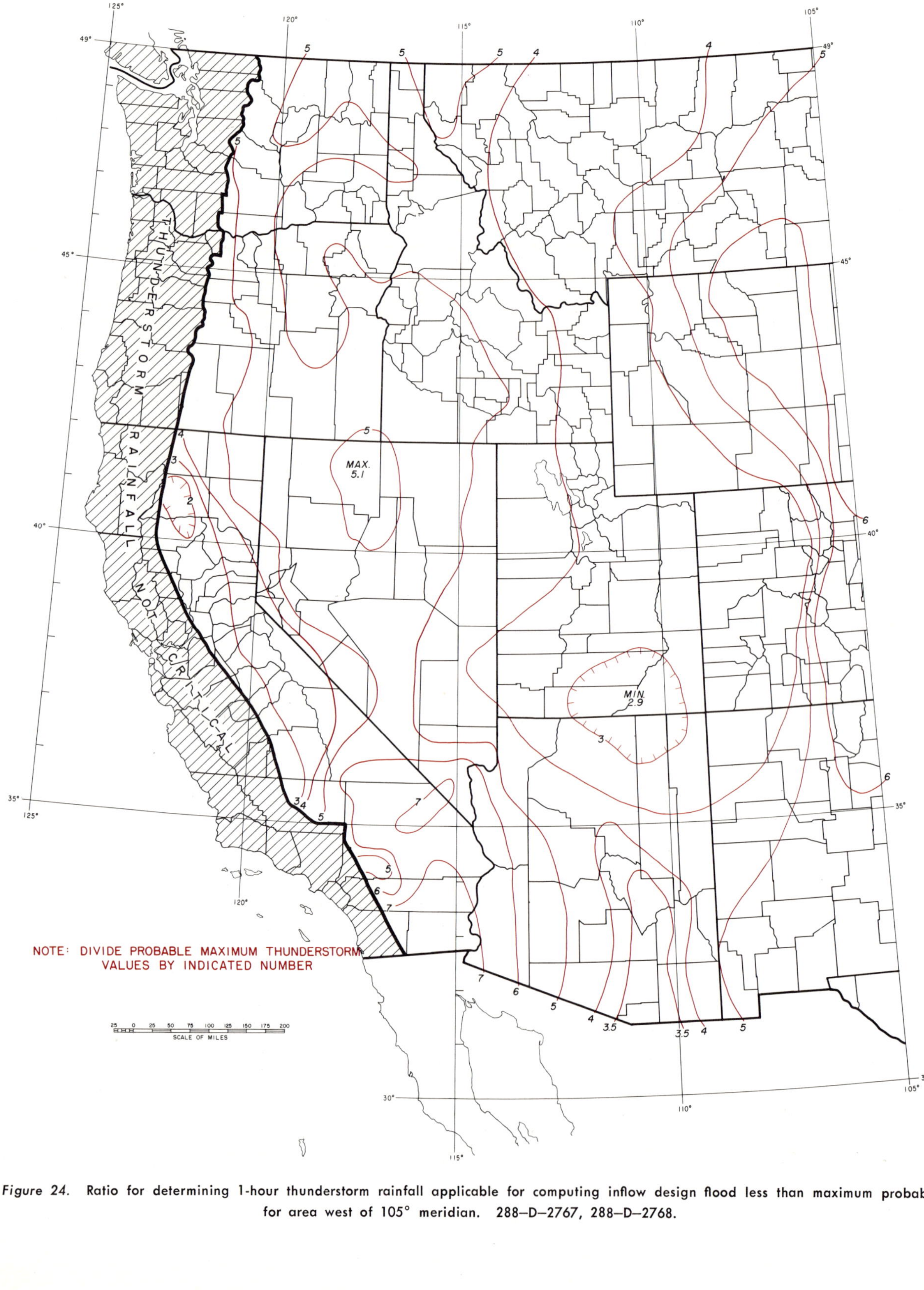

Figure 24. Ratio for determining 1-hour thunderstorm rainfall applicable for computing inflow design flood less than maximum probable for area west of 105° meridian. 288–D–2767, 288–D–2768.

storm that produced the flood hydrograph under study should be determined from the mass curves for each station in and near the drainage area. For a flood hydrograph consisting of a single event, this will be the total depth of precipitation occurring during the storm period. For a compound hydrograph, in which individual portions of the hydrograph are studied separately, temporary cessations of rainfall will usually be indicated in the mass curves, and from inspection it usually will be apparent which of the increments of rainfall caused the runoff event under study. The appropriate depths of rainfall are then used to draw an isohyetal map, using standard procedures. A typical isohyetal map is shown in figure 25(B).

Extreme caution should be used in drawing the isohyetal pattern in mountainous areas where the orographic effect is an important factor in the areal distribution of rainfall. For example, if there is a precipitation station in a valley on one side of a mountain range and another station in a valley on the opposite side of the range with no intervening station, it cannot be assumed that the rainfall during a storm would vary linearly between the two stations. It is likely that the rainfall would increase with increases in elevation on the windward side of the divide, whereas on the leeward side, precipitation would decrease rapidly with distance from the divide. This type of distribution can usually be verified in mountainous areas where there are sufficient precipitation stations to define the isohyetal pattern accurately.

After the preliminary hydrographs and the isohyetal maps have been drawn, the atypical flood events for unit hydrographs determination may readily be eliminated. *Those floods having a combination of large volume, uniform intensities, isolated periods of rainfall, and uniform areal distribution of rainfall should be chosen for further study.*

(3) *Average rainfall by Thiessen polygons.*—The average rainfall on a drainage area can be determined from precipitation station records by the Thiessen polygon method. A sample computation of average hourly rainfall from the mass curves in figure 25(A), using Thiessen polygons indicated in figure 25(B), is given in table 3.

The first step is to construct the Thiessen polygons, which are the areas bounded by the perpendicular bisectors of lines joining adjacent precipitation stations. The percentage of the drainage area controlled by each station's polygon is planimetered and entered in table 3. Next, the average depth of rainfall over each station's polygon is determined by planimetering areas between isohyets on figure 25(B). A factor to be used in weighing station rainfall values is obtained by multiplying the percentage of the drainage area controlled by each station's polygon by the ratio of the average depth of rainfall over each station's polygon to the observed rainfall at the station and dividing by 100.

Hourly incremental rainfall values are determined for each precipitation station from the mass curves of figure 25(A) and are multiplied by the appropriate weight factors, as shown in table 3, to obtain the total for the drainage area.

Additional information on determining average rainfall is given in "Cooperative Studies Technical Paper No. 1," published by the National Weather Service.

(4) *Determination of rainfall excess.*—Two methods may be used to determine rainfall excess: by assuming a constant average retention rate throughout the storm period, and by assuming a retention rate varying with time. The capacity rate of retention decreases progressively throughout the storm period until a constant minimum rate is reached if the rain is sufficiently prolonged. With dry antecedent conditions, the initial capacity rate will be greater and will decline faster. Because the use of a varying retention rate requires a complicated method of computation, it is often preferable to assume an average retention rate (sometimes referred to as infiltration index) with an estimate of initial loss being made if antecedent conditions are relatively dry.

The method of determining the period of rainfall excess, when an average retention rate is used, is a trial-and-error process in which a retention rate is assumed and subtracted from

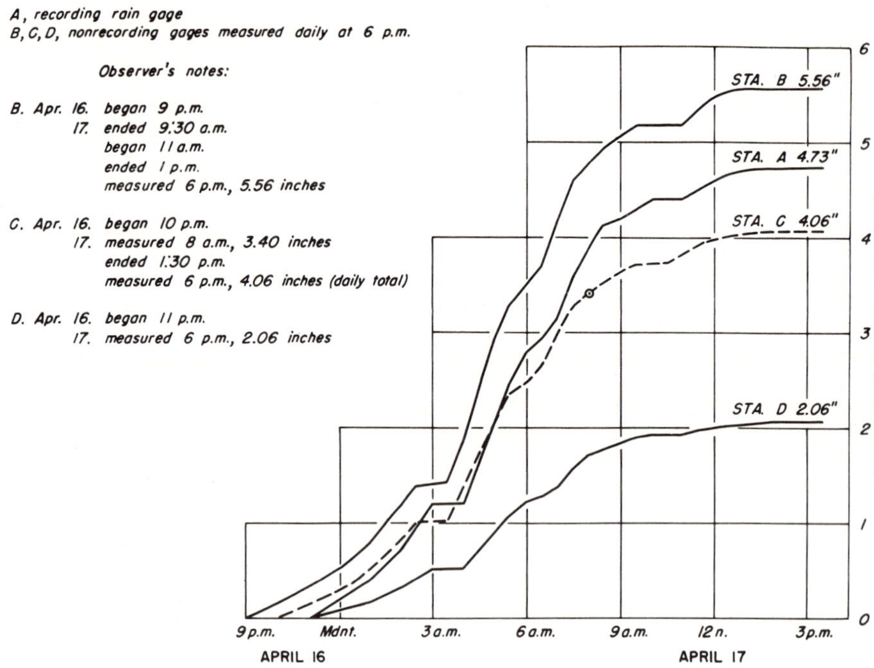

(A) MASS CURVES OF RAINFALL

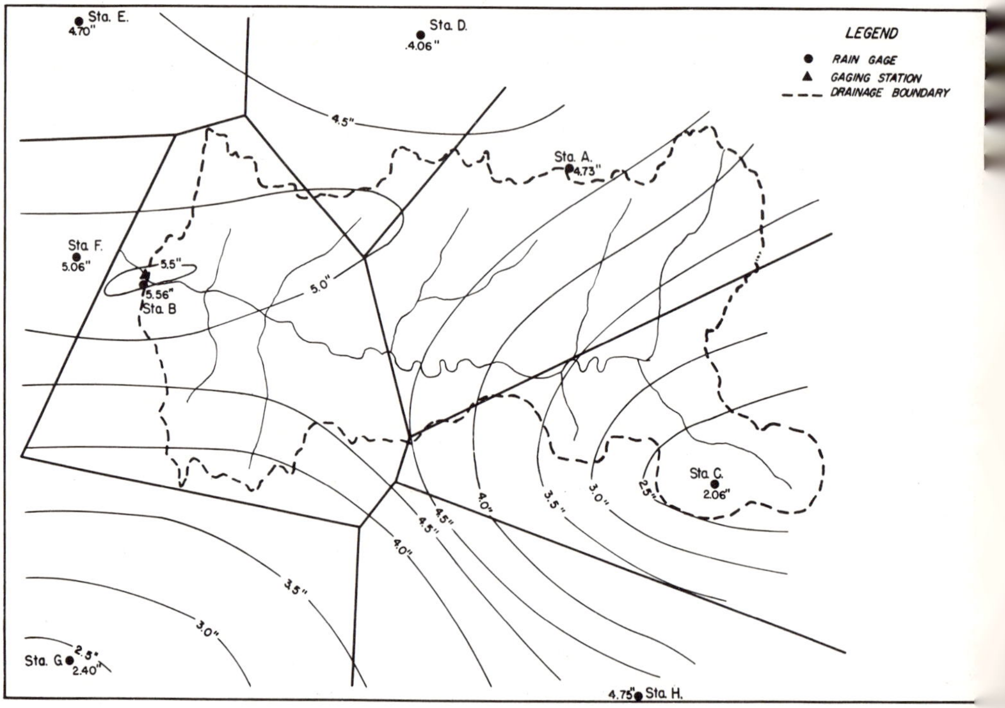

(B) ISOHYETS AND THIESSEN POLYGONS

Figure 25. Analysis of observed rainfall data. 288–D–2454.

TABLE 3.—*Computation of rainfall increments*

COMPUTATION OF STATION WEIGHTS

Station (1)	Rainfall over Thiessen polygon (2)	Percent of basin area (3)	Rainfall at station (4)	Weight, col. (2) x col. (3) / 100 x col. (4) (5)
A	4.3	38.9	4.73	0.35
B	4.6	37.0	5.56	.31
C	2.8	21.1	2.06	.29
D	5.0	3.0	4.06	.04

COMPUTATION OF WEIGHTED AVERAGE HOURLY RAINFALL OVER BASIN

Time, hours	Station A			Station B			Station C			Station D			Weighted average, sum of cols. (3)
	Mass rf. (1)	Δ rf. (2)	0.35xΔrf. (3)	Mass rf. (1)	Δ rf. (2)	0.31xΔrf. (3)	Mass rf. (1)	Δ rf. (2)	0.29xΔrf. (3)	Mass rf. (1)	Δ rf. (2)	0.04xΔrf. (3)	
0				0									
1				.17	0.17	0.053				0			0.053
2	0			.33	.16	.050	0			.15	0.15	0.006	.056
3	.20	0.20	0.070	.52	.19	.059	.09	0.09	0.026	.29	.14	.006	.161
4	.40	.20	.070	.80	.28	.087	.17	.08	.023	.52	.23	.009	.189
5	.73	.33	.116	1.20	.40	.124	.32	.15	.044	.84	.32	.013	.297
6	1.20	.47	.164	1.41	.21	.065	.52	.20	.058	1.01	.17	.007	.294
7	1.20	0	0	1.85	.44	.136	.52	0	0	1.34	.33	.013	.149
8	2.05	.85	.298	2.91	1.06	.329	.89	.37	.107	2.05	.71	.028	.762
9	2.80	.75	.262	3.49	.58	.180	1.22	.33	.096	2.47	.42	.017	.555
10	3.15	.35	.122	4.19	.70	.217	1.37	.15	.044	3.00	.53	.021	.404
11	3.90	.75	.262	4.79	.60	.186	1.70	.33	.096	3.40	.40	.016	.560
12	4.20	.30	.105	5.08	.29	.090	1.83	.13	.038	3.63	.23	.009	.242
13	4.40	.20	.070	5.18	.10	.031	1.92	.09	.026	3.73	.10	.004	.131
14	4.40	0	0	5.18	0	0	1.92	0	0	3.83	.10	.004	.004
15	4.59	.19	.066	5.49	.31	.096	2.00	.08	.023	3.97	.14	.006	.191
16	4.70	.11	.038	5.56	.07	.022	2.04	.04	.012	4.04	.07	.003	.075
17	4.73	.03	.010	5.56	0	0	2.06	.02	.006	4.06	.02	.001	.017
Total		4.73	1.653		5.56	1.725		2.06	.599		4.06	.163	4.140

hourly rainfall increments determined as the average over the basin. Various retention rates are assumed until the total of the computed rainfall excess equals the measured storm runoff. An example of this procedure is given in table 4. If the correct retention rate has not been assumed after two trials, a rainfall excess-retention curve will facilitate the solution. In the example of table 4, the curve could be drawn through the two points represented by the coordinates 0.25, 1.37, and 0.15, 2.15. The correct retention rate corresponding to a rainfall excess of 2.0 inches would then be taken from this curve.

The duration time of excess rainfall is that time during which rainfall increments exceed the average retention rate. In the third trial, table 4, the duration time is 9 hours.

(c) *Estimating Direct Runoff from Soil and Cover Data.*—The problem most often encountered in flood studies is the need for estimating runoff from a watershed for which there are no records of runoff or precipitation. An approach to solution of this problem is to compare runoff characteristics of the watershed under study with those of watersheds for which records are available. Basin characteristics which may be most readily compared for estimating the volume of runoff that will result from a given amount of rainfall are soil type and cover, which includes land usage. (The distribution of the runoff, i.e., the unitgraph, is compared on the basis of area, channel characteristics, etc.) Obviously, a great deal of judgment must be used in making such comparisons because of the great number of soil types and cover types and the many combinations thereof possible within a specific watershed. However, near minimum loss rates are associated with inflow design floods; therefore,

TABLE 4.—*Computation of rainfall excess*

Time, hours	Rainfall increment (basin average), inches	First trial: Assumed retention rate, inches per hour	First trial: Rainfall excess, inches	Second trial: Assumed retention rate, inches per hour	Second trial: Rainfall excess, inches	Third trial: Assumed retention rate, inches per hour	Third trial: Rainfall excess, inches
0							
1	0.05	0.25		0.15		0.17	
2	.06						
3	.16				0.01		
4	.19				.04		0.02
5	.30		0.05		.15		.13
6	.29		.04		.14		.12
7	.15		0		0		0
8	.76		.51		.61		.59
9	.56		.31		.41		.39
10	.40		.15		.25		.23
11	.56		.31		.41		.39
12	.24		0		.09		.07
13	.13				0		0
14	0				0		0
15	.19				.04		.02
16	.08						
17	.02	.25		.15		.17	
Total	4.14		1.37		2.15		1.96

Total rainfall, 4.14 inches; observed runoff, 2.0 inches; total retention in 17 hours, 2.1 inches. The average retention rate of 0.17 inches per hour assumed in the third trial gives the best agreement of computed rainfall excess with measured runoff.

only data on minimum loss rates for given soils and cover combinations need be determined.

Hydrologists of the Soil Conservation Service constantly encounter the problem of estimating direct runoff where no records are available for the specific watershed. As a result, a general procedure for estimating direct runoff has been developed to meet Soil Conservation Service needs. Discussions relating to this procedure taken from the "Hydrology Guide for Use in Watershed Planning," and the later publication, "Hydrology, Part I—Watershed Planning," published by the Soil Conservation Service (see bibliography, sec. 60) are given in appendix A. A feature of the Soil Conservation Service method that should prove particularly useful is the classification of about 4,000 soils of the United States into four hydrologic soil groups representing a progression in degree of runoff potential from high to low. Although the Soil Conservation Service procedure for estimating direct runoff has been selected for use in this text due to its simplicity, modifications have been introduced in adapting this procedure to inflow design flood computations.

The rainfall-runoff relationships shown by the curves of figure A-4 (appendix A) were developed by Soil Conservation Service hydrologists from analyses of rainfall and respective runoff records at numerous small area experimental watersheds. The relationships were developed for use of daily nonrecording rainfall data, which are more plentiful in the United States than are recording rainfall data. Data used in the development are totals for one or more storms occurring in a calendar day and nothing is known about their time distributions. The relationships developed therefore exclude time as an explicit variable, which means that rainfall intensity is ignored.

Rainfall intensity and variations of intensity during a storm are most important to individual flood hydrograph computations because of their relation to retention losses. A storm of low intensity for 24 hours may cause negligible runoff, whereas the same amount of rainfall in an hour may cause a damaging flood. It is beyond the scope of this text to discuss in detail rainfall-retention loss relationships, but the following characteristic should be remembered. A natural watershed of only a few square miles of area may contain different types of soil, cover and land usage, each having a different retention loss capability. When computing runoff from a specific storm, average retention loss relationships to rainfall are estimated and these losses subtracted from periods of rainfall which are also assumed to be average over the watershed for each rainfall period. Actually, rainfall is seldom uniform over a watershed so some portions receive rainfall that exceeds their respective retention losses, while other portions do not. As individual storm rainfall amounts of equal duration increase, so does the likelihood that all portions of a watershed receive rainfall greater than retention losses. Thus, an individual storm having rainfall intensity less than watershed average retention loss as determined from a higher intensity storm may cause runoff because of some portions of the watershed having soils with retention loss less than average.

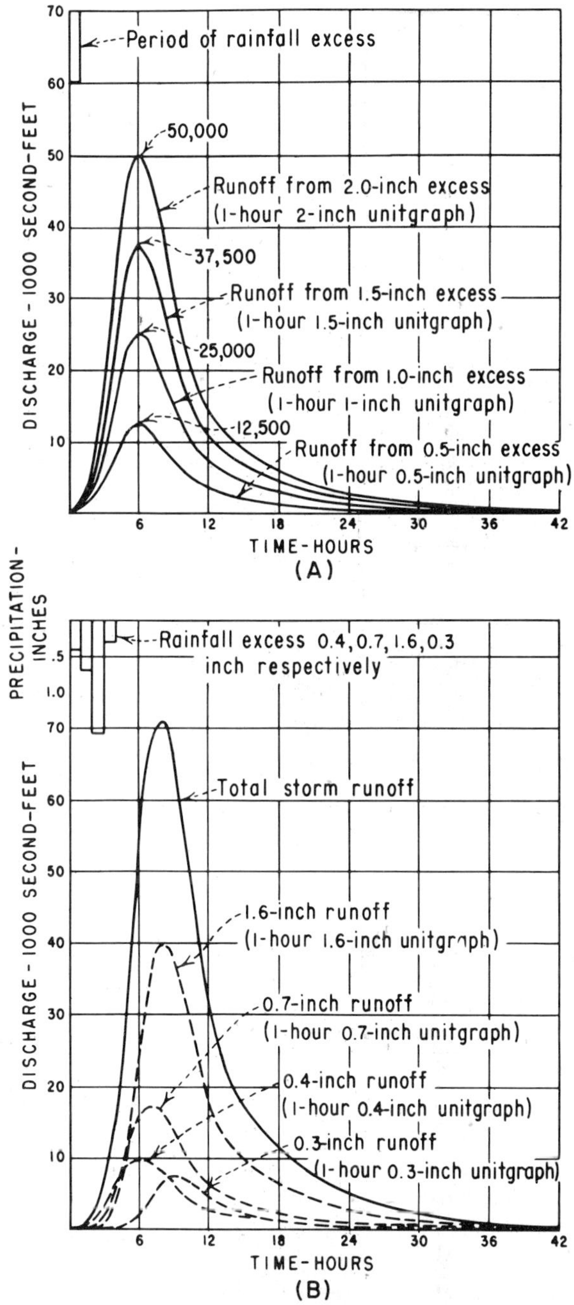

Definitions:

Unitgraph - A hydrograph of storm runoff at a given point that will result from an isolated event of rainfall excess occurring within a unit of time and spread in an average pattern over the contributing drainage area. Identified by the unit time and volume of the excess rainfall, that is 1-hour 1-inch unitgraph.

Rainfall excess - That portion of rainfall that enters a stream channel as storm runoff and produces the runoff hydrograph at the measuring point.

Basic Assumptions:

(1) The effects of all physical characteristics of a given drainage basin are reflected in the shape of the storm runoff hydrograph for that basin.

(2) At a given point on a stream, discharge ordinates of different unitgraphs of the same unit time of rainfall excess are mutually proportional to respective volumes. See (A) at left.

(3) A hydrograph of storm discharge that would result from a series of bursts of excess rain or from continuous excess rain of variable intensity may be constructed from a series of over lapping unitgraphs each resulting from a single increment of excess rain of unit duration. See (B) at left.

Practical Application:

For a given runoff contributing area, a unitgraph representing exactly one inch of runoff (rainfall excess) for a selected unit time interval is computed. Increments of rainfall excess for the same unit time interval are determined for a storm. A total hydrograph of runoff from the storm is then computed using assumptions (2) and (3) above. See graph (B) at left.

Figure 26. Unitgraph principles. 288–D–2455, 288–D–2456.

Strict adherence to use of the runoff curves in figure A-4 (appendix A) results in hourly runoff increments almost equal to hourly precipitation increments after a few hours for many of the design storm values obtained from data presented in section 50. Infiltrometer studies indicate that all but impervious clay soils have minimum constant infiltration rates after saturation that may range from 0.05 to greater than 1.00 inch per hour, depending on the type of soil. Therefore, to utilize the Soil Conservation Service data on rainfall-runoff relationships in the computational procedures given in this text, time-sequences of incremental rainfall for design storm are assumed and precipitation excesses are then computed using the runoff curves with the provision that hourly retention rates indicated by use of the runoff curves be tabulated for each hourly rainfall increment. Progressively through the arranged precipitation sequence, these hourly retention rates are compared with the following tabulated minimum retention rates assigned to the four hydrologic soil groups. When the retention rate given by use of a runoff curve becomes less than an assigned minimum retention rate, the minimum rate is then used to compute excesses thereafter for the balance of the storm.

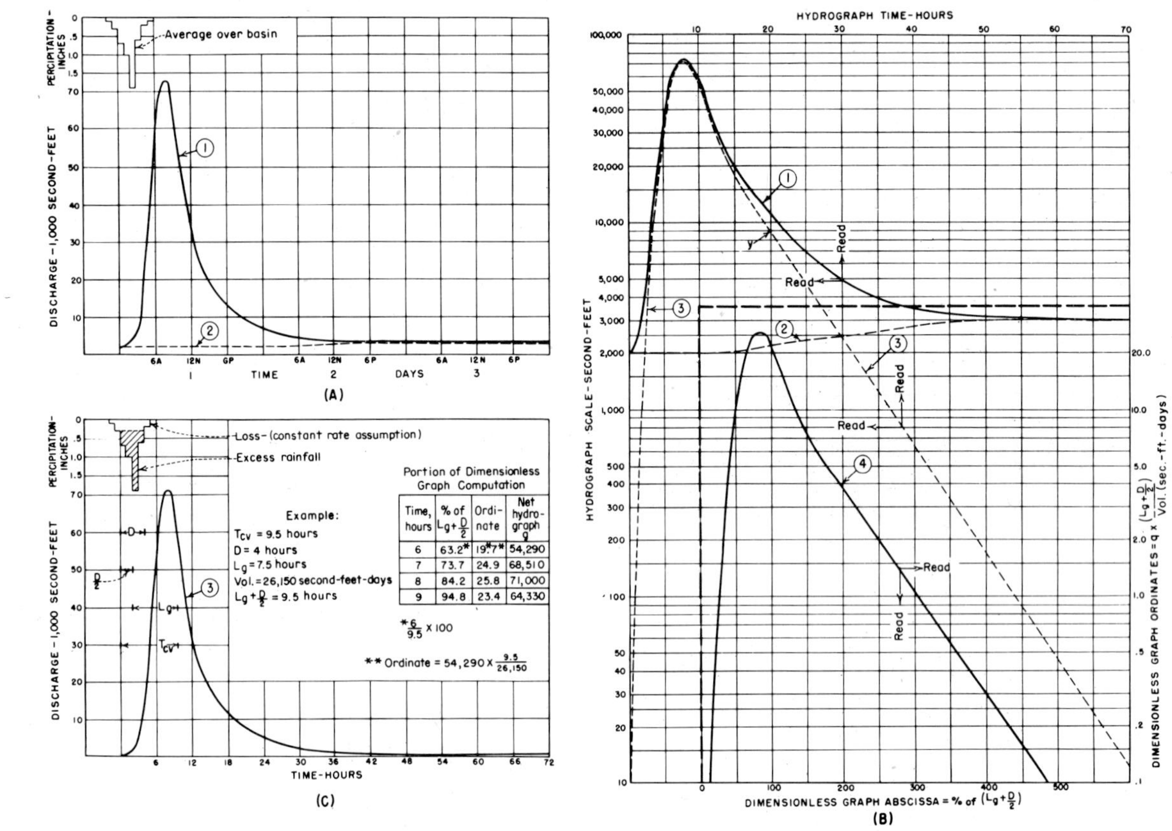

Figure 27. Hydrograph analysis. 288–D–2457.

Hydrologic soil group	Range of minimum retention rates, inch/hour	Recommended rate for use in general cases, inch/hour
A.	0.30—0.45	0.40
B.	0.15—0.30	0.24
C.	0.08—0.15	0.12
D.	0.02—0.08	0.04

An example of application of data in appendix A using the soil group minimum retention rate is given in section 58.

52. *Unitgraph Principles.*—The basic tool for hydrograph computation is the unitgraph. Its fundamental principles are presented in abbreviated form on figure 26.

53. *Hydrograph Analysis.*—A procedure of hydrograph analysis is presented on figure 27. Storm duration and distribution over a watershed affect the shape of the resulting unitgraphs. Direct averaging of unitgraphs of different storm durations gives erroneous results. However, such unitgraphs can be averaged by converting the unitgraphs to dimensionless form, as shown on figure 27 (B).

HYDROGRAPH ANALYSIS (Refer to Fig. 27)

Given: Recorded hydrograph at given point on a stream. Rainfall data may or may not be available.

Required: Factors for deriving unitgraph to be applied at point of derivation, at another point on stream if of comparable runoff characteristics, or to comparable ungaged watershed.

Procedures:

(a) Plot recorded hydrograph on cartesian coordinate paper and on semilog paper:

(1) on figure 27 (A), and
(1) on figure 27 (B).

(b) Estimate base flow, ② on figure 27(A) and (B), by trial and error. Subtract base flow from recorded hydrograph and plot net hydograph, ③ on figure 27(B). If the base flow has been estimated correctly, the descending limb of hydrograph ③ on figure 27(B) will be a straight line (exponential recession). (③ = ① minus ② on figure 27(B).

(c) Compute volume of net hydrograph ③ as follows:

(1) Add average hourly discharges (in sec.-ft.-hours) to a point such as y on the exponential recession, ③ on figure 27(B).

(2) Compute hourly recession constant, k_{hr}, from two points on exponential recession line by use of following equation:

$$k_{hr} = \sqrt[t]{\frac{q_t}{q_o}}$$

where:

q_o=discharge at first point,
q_t=discharge at second point, and
t=time interval, in hours, between points 1 and 2.

(3) Storage, or volume after y in (sec.-ft.-hours) equals:

$$\frac{-q_y}{\log_e k_{hr}}$$

where:

q_y=discharge in second-feet at point y, and
$\log_e k_{hr} = 2.3026\ (\log_{10} k_{hr})$.

(4) Total volume is sum of volume to y plus volume after y.

(d) For comparison with rainfall data, convert volume of ③ to inches of runoff:

Inches of runoff=

$$\frac{\text{volume in sec.-ft.-hours}}{(\text{area in sq. mi.}) \times 645.3}$$

(e) Analyze rainfall data, if available; determine period D of rainfall excess.

(f) Compute time of occurrence of one-half volume of hydrograph ③, figure 27 (C). The time to center of volume, T_{cv}, equals time from beginning of rise of net hydrograph to time one-half volume has passed measuring point.

(g) Find lag, L_g, time in hours from mid-point of excess rainfall period to time of occurrence of one-half volume.

(h) Compute dimensionless graph as follows and plot on semilog paper, ④ on figure 27(B).

(1) Abscissa—hours from beginning of excess rain expressed as percent of $(L_g+D/2)$. When rainfall data are not available $L_g+D/2$ may be taken equal to T_{cv}.

(2) Ordinates—discharge in second-feet of ③ (at respective abscissa) multiplied by $(L_g+D/2)$, all divided by net hydrograph volume expressed as sec.-ft.-days $\left(\frac{\text{sec-ft.-hours}}{24}\right)$.

54. Unitgraph Derivation for Ungaged Areas.—An example of the derivation of a unitgraph for an ungaged area is shown on figure 28. Since hydrographs recorded at a damsite are seldom available, a means of transferring unitgraphs to ungaged areas is necessary. The factor "lag time" is used for this purpose. Lag time is an index to the time of concentration of runoff from a basin. It may be determined from recorded hydrographs and empirically correlated with basin characteristics. As shown on figure 28, determination of a lag time for an ungaged watershed from basin measurements provides the key to unitgraph computation for that watershed if a representative dimensionless graph is known. There are several definitions of lag time, each pertaining to a specific form of unitgraph derivation; two different definitions are used in this text. Lag time is defined as (1) the time from the center of excess rainfall to the time of occurrence of one-half the volume of the hydrograph—used with the dimensionless graph procedure outlined on figure 27 and 28; and (2) the time from the center of excess rainfall to the time of peak discharge of the hydrograph—used with the triangular hydrograph concept, figure 29.

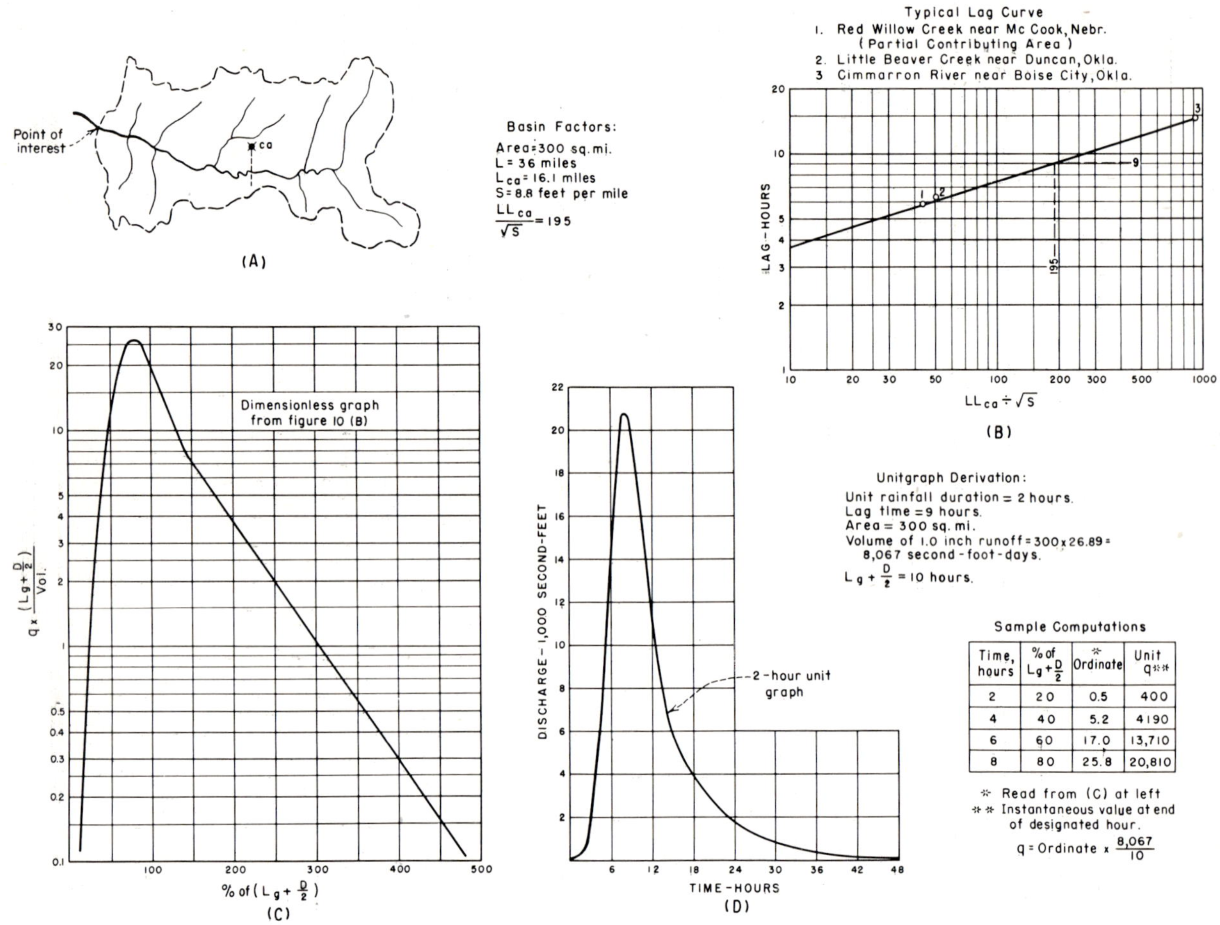

Time, hours	% of $L_g+\frac{D}{2}$	* Ordinate	Unit q**
2	20	0.5	400
4	40	5.2	4190
6	60	17.0	13,710
8	80	25.8	20,810

Figure 28. **Unitgraph derivation for ungaged area. 288–D–2458.**

UNITGRAPH DERIVATION FOR UNGAGED AREA (Refer to Fig. 28)

Given: Dimensionless graph (C) and lag relationship curve (B) for comparable area.

Required: Unitgraph for basin above given point.

Procedure:

(a) Outline drainage boundary, determine area (fig. 28(A)).

(b) Find:

L=length of longest watercourse from point of interest to watershed divide, measured in miles.

ca=centroid of basin—usually found by vertically suspending a cardboard cutout of basin shape successively from two or more points and finding intersection of plumb lines from each point.

L_{ca}=length of watercourse, in miles, from point of interest to intersection of perpendicular from ca to stream alinement.

S=overall slope in feet per mile of longest watercourse from point of interest to divide.

(c) Solve for factor:

$$\frac{LL_{ca}}{\sqrt{S}}$$

(d) Enter graph, figure 28(B), with $\frac{LL_{ca}}{\sqrt{S}}$ value and read the corresponding lag time. (Points establishing empirica curves such as figure 28(B) are constructed by plotting lag times obtained from hydrograph analyses (fig. 27) versus respective $\frac{LL_{ca}}{\sqrt{S}}$ factors for basins of similar runoff characteristics.)

(e) Select a dimensionless graph (fig. 28(C))—usually the mean dimensionless graph of a number of dimensionless graphs derived for the same stream or for streams of similar characteristics.

(f) Select unit rainfall duration time—should be one-fourth or less of lag time for basin.

(g) Compute unitgraph (fig. 28(D)) using:

(1) Basin area.

(2) Lag time plus one-half unit rainfall duration time.

(3) Dimensionless graph.

(h) Computed unitgraph may be used as shown on figure 26(B).

55. Triangular Hydrograph Analysis.—The practicability of representing a hydrograph as a triangle and the relationships that can be developed by such representation is presented on figure 29.

56. Estimating Time of Concentration.—As shown on figure 29, the general equation for computing the peak discharge resulting from a given amount of runoff is $q_p = \frac{484AQ}{T_p}$ in which $T_p = \frac{D}{2} + 0.6T_c$.

The term $0.6T_c$ is an empirical relationship adopted by hydrologists of the Soil Conservation Service as representative of L, lag time, which is defined as the time in hours from the midpoint of excess rainfall, D, to the time of peak discharge. The relationship of L to other hydrograph factors is shown on figure 29(B). Computation of hydrographs for ungaged watersheds by application of the equations given on figure 29 is dependent on an estimate of T_c, time of concentration of the watershed, defined as the travel time of water from the hydraulically most distant point of the watershed to the point of interest. Various methods of estimating T_c are given on figure 30. These data are admittedly generalized because the concentration of runoff is affected by storm distribution over the watershed and storm intensity, as well as by the watershed's hydraulic characteristics. If data are available, more than one method of estimating time of concentration should be used to select a representative value for a particular watershed.

Guides to assist in estimating T_c for obtaining a unitgraph for computation of flood inflow to a full reservoir are given in figure 31. Figure 31(A), (B), and (C) show three examples of stream lengths that control estimates for a full reservoir. Whenever it is possible to obtain a T_c value from streamflow records of the stream of interest or by detailed field inspection of stream channel and watershed characteristics, that value should be used. When it is necessary to estimate a T_c value, the following equation can be used:

$$T'_c = \left[\frac{11.9\,L^3}{H}\right]^{0.385}$$

In figure 30, an adjustment should be made using values from table 5 below.

TABLE 5.—*Adjustments for T_c for watersheds west of 105° meridian and mountainous timber-covered watersheds east of 105° meridian. (Interpolate for intermediate CN values.)*

CN	T'_c/T_c
80	1.0
70	1.4
60	1.8
50	2.2

Watershed cover and surface roughness cause a retardance of surface runoff. An estimate of this retardance effect has been incorporated in the runoff curve numbers CN 50 to CN 80 for AMC-II for watersheds west of the 105° meridian. This adjustment factor may also be used for timbered watersheds in mountainous regions east of the 105° meridian, but is not applicable to other type regions east of the 105° meridian.

57. Application of Triangular Hydrographs.—Examples of the application of the triangular hydrograph concept are presented on figure 32.

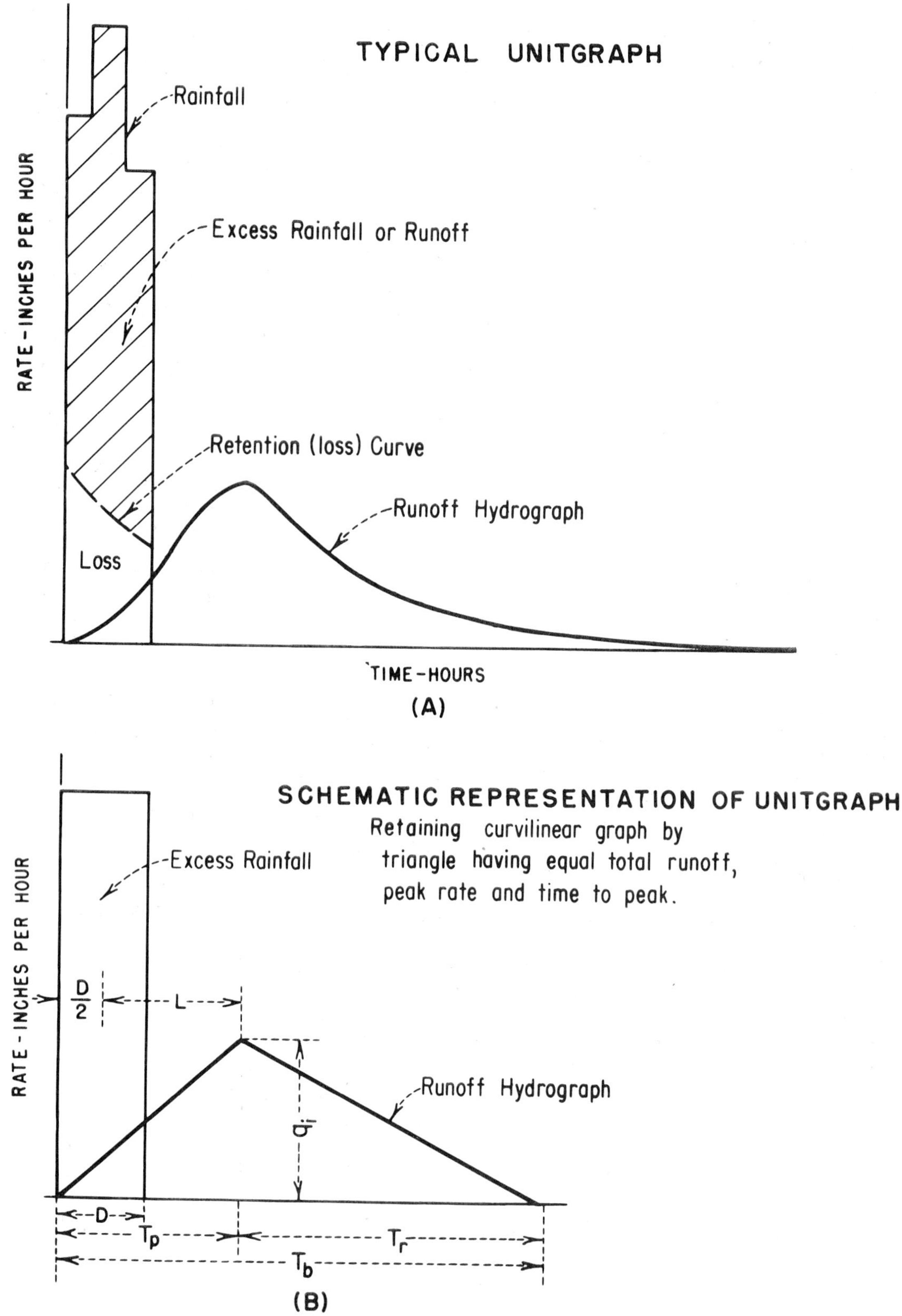

Figure 29. Triangular hydrograph analysis. (Sheet 1 of 2.) 288–D–2459.

PEAK EQUATION DEVELOPMENT

Using triangle from (B) at left

$$Q = \frac{q_i T_p}{2} + \frac{q_i T_r}{2}$$

$$q_i = \frac{2Q}{T_p + T_r}$$

Let $T_r = HT_p$, where H is a constant to be determined for a particular watershed.

$$q_i = \frac{2}{(1+H)} \frac{Q}{T_p}$$

Convert inches per hour to second-feet, introduce drainage area, A in square miles. (1 inch per hour = 645.3 second-feet per square mile.)

$$q_p = \frac{2(645.3)}{(1+H)} \frac{AQ}{T_p}$$

or

$$q_p = \frac{KAQ}{T_p}$$

where

$$K = \frac{1290.6}{1+H}$$

EXPLANATION

- Q = Total runoff in inches
- q_i = Peak rate, inches per hour
- T_p = Time in hours from start of rise to peak rate
- T_r = Time in hours from peak rate to end of triangle
- q_p = Peak rate in second-feet
- D = Rainfall excess period, hours
- L = Lag, time from center of excess rainfall to time of peak, hours
- T_c = Time of concentration-travel time of water from hydraulically most distant point to point of interest, hours
- T_b = Time base of hydrograph, hours

Empirical relationship for lag
$L = 0.6\,T_c$.

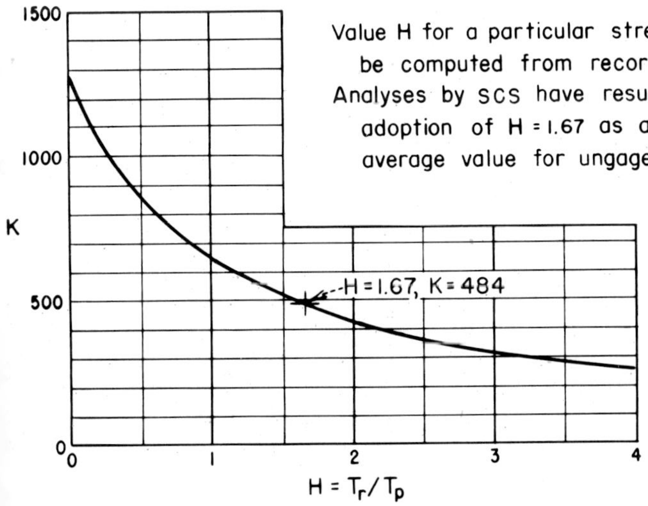

(C)

Value H for a particular stream may be computed from recorded hydrographs. Analyses by SCS have resulted in their adoption of H = 1.67 as a general average value for ungaged watersheds.

General Peak Equation:

$$q_p = \frac{484AQ}{T_p}$$

For H = 1.67
$T_b = 2.67\,T_p$

or using $L = 0.6\,T_c$

$$q_p = \frac{484AQ}{D/2 + 0.6\,T_c}$$

since

$$T_p = \frac{D}{2} + 0.6T_c$$

Figure 29. Triangular hydrograph analysis. (Sheet 2 of 2.) 288–D–2460.

Purpose: A time of concentration from which a lag time can be computed must be obtained for hydrograph construction representing runoff from a watershed. Various methods of estimating time of concentration, T_C, are as follows:

A. ESTIMATING T_c FROM STREAM HYDRAULICS (SCS GUIDE)

1. Obtain stream reaches and channel cross-sections from field surveys.
2. Find approximate channel bankfull discharge for each reach.
3. Compute average velocity for the bankfull discharge of each reach.
4. Use the average velocity and the valley length of the reach to compute travel time through each reach.
5. Add travel times of reaches to get T_C.

Note: Appendix B "Hydraulic Computations" presents methods of computing flows in natural channels.

B. ESTIMATING T_c FROM VELOCITY ESTIMATES AND WATERCOURSE LENGTHS

Velocity Estimate Guide

U.S. Navy - Technical Publication Navdocks TP-PW-5 Table 8B, March 1953

Average slope of channel from farthest point to outlet, in percent	Average velocity, feet per second
1 to 2	2.0
2 to 4	3.0
4 to 6	4.0
6 to 10	5.0

Texas Highway Department Rational Design of Culverts and Bridges, October 1946

Slope in percent	Average velocity, feet per second		
	Woodlands (upper portion watershed)	Pastures (upper portion watershed)	Natural channel not well defined
0 - 3	1.0	1.5	1.0
4 - 7	2.0	3.0	3.0
8 - 11	3.0	4.0	5.0
12 - 15	3.5	4.5	8.0

Figure 30. Time of concentration estimates. (Sheet 1 of 2.) 288–D–2461.

C. ESTIMATING T_c FROM LENGTHS AND SLOPES:

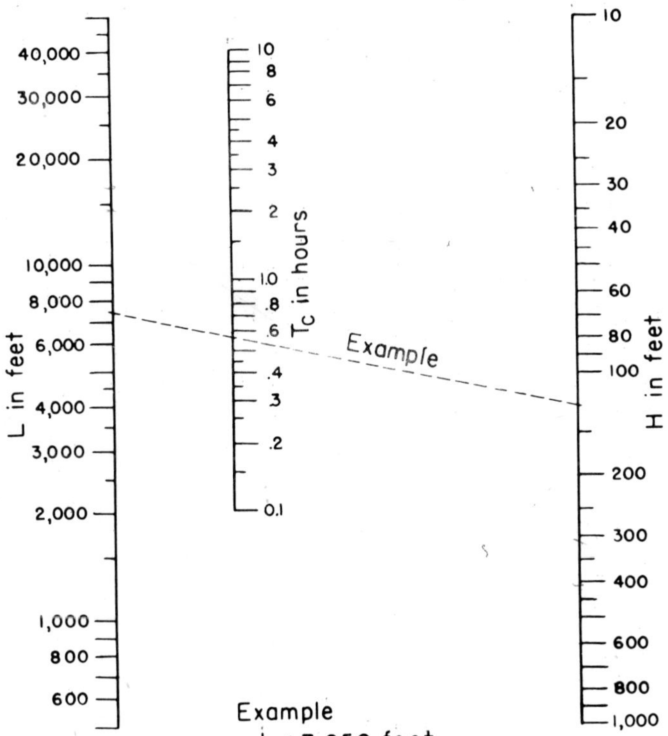

Example
L = 7,250 feet
H = 130 feet
then T_c = 0.57 hours

(a) Nomograph (SCS Guide)

L = length of longest watercourse in feet

H = difference in elevation in feet between outlet point and divide

(b.) Solution may be made by equation from California Culverts Practice, California Highways and Public Works, September 1942.

$$T = \left(\frac{11.9\,L^3}{H}\right)^{0.385}$$

T = T_c in hours

L = length of longest watercourse in miles

H = elevation difference in feet

Lag, L, (SCS Guide) may be estimated directly for a basin by subdividing into tributary drainage subareas and using the relationship:

$$L = \frac{\Sigma a_x T_x}{A}$$, where

L = lag in hours

a_x = the x-th increment of area in sq. mi.

T_x = travel time in hours from center of a_x to main basin outlet

A = total area of basin, sq. mi.

Figure 30. Time of concentration estimates. (Sheet 2 of 2.) 288–D–2462.

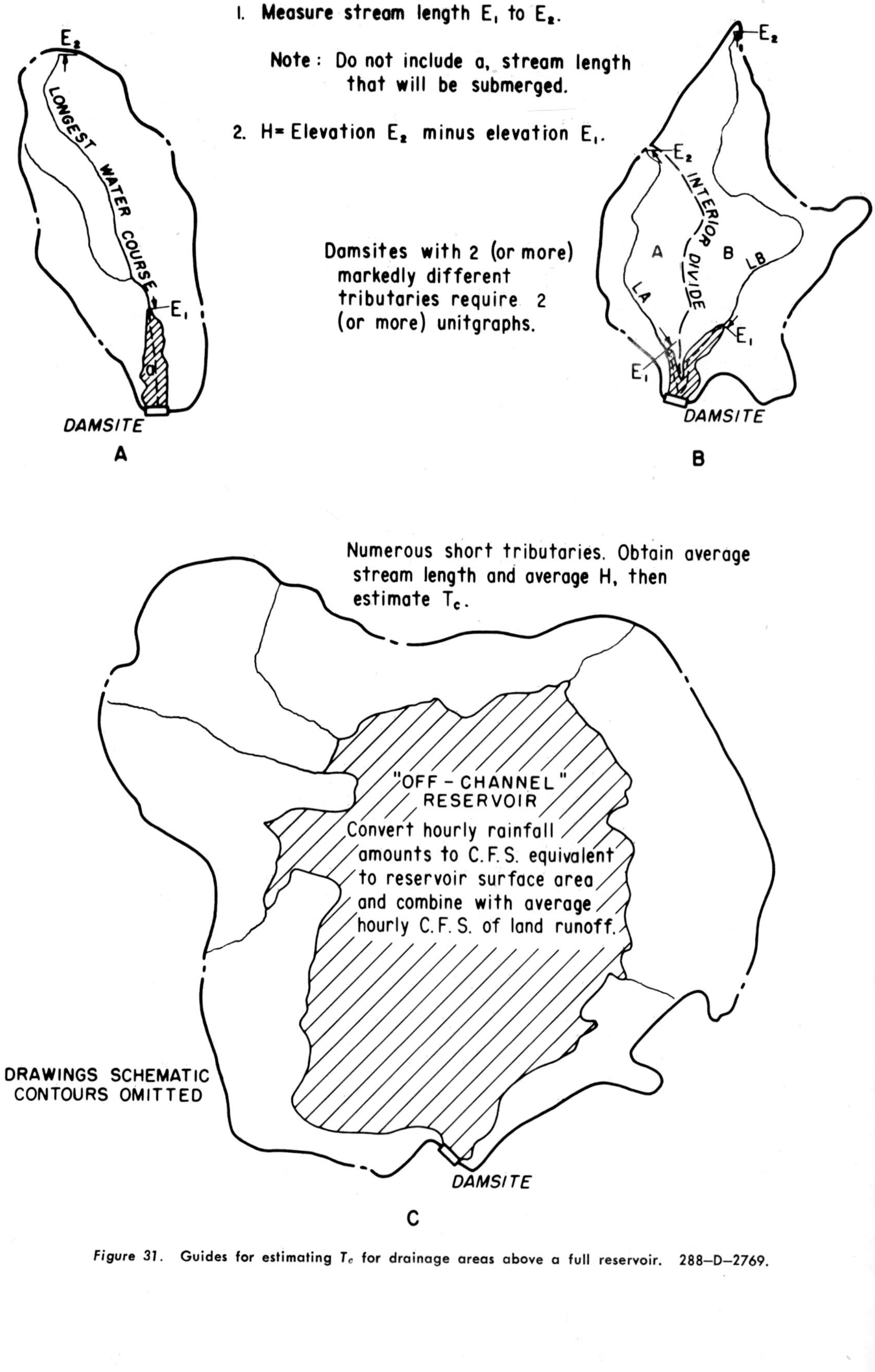

Figure 31. Guides for estimating T_c for drainage areas above a full reservoir. 288–D–2769.

58. *A Method of Computing an Inflow Design Flood.*—The following procedure is based to a large extent on that given in the Soil Conservation Service's "Hydrology Guide for Use in Watershed Planning," (see bibliography, sec. 60), but with modifications to meet the purpose of this text. The resulting inflow design flood hydrograph represents direct runoff from precipitation in the form of rain over a watershed having no unusual runoff characteristics. If a certain amount of flow is expected to be in the channel at the time an inflow design flood could occur, the amount of this base flow should be added to the inflow design flood hydrograph. The procedure outlined assumes an ungaged area. Computations for a gaged area will differ only in that data regarding unitgraph characteristics and retention losses obtained from analyses of recorded runoff should be used.

(a) *Watershed Data.*—The following watershed data are always needed:

(1) Geographical location.

(2) Map showing topography, streams, and runoff contributing drainage area.

a. The inflow design flood required is that which enters a reservoir full to top of conservation storage; therefore, the proposed reservoir surface contour for this elevation (normal water surface) should be drawn on the map. The proposed reservoir will have an influence, sometimes quite large, on the computation of watershed T_c because inflow into a reservoir will traverse the reservoir length much more rapidly than along a similar length of natural stream channel. Reservoir-length travel time is not included in T_c. Reservoir influence on measurements of stream length for T_c estimates are shown graphically on figure 31.

(3) Information about soils and vegetative cover and their distribution throughout the watershed.

(b) *Magnitude Criteria.*—It has been pointed out in section 45 that an inflow flood may not always be equivalent to the maximum probable flood. In those instances where it is permissible to design for less than the maximum probable flood, a reasonable reduction in the severity of the factors affecting flood potential can be made. In this text, conditions designated as assumption A and assumption B, to be used when considering design floods less than the maximum probable, are given to provide a method for achieving a consistent degree of magnitude as related to the risk of damage which would result from structure failure.

Design storm magnitude has been selected as basis for computing inflow design floods for assumptions A and B. Section 50 presents data for obtaining probable maximum precipitation values east and west of the 105° meridian and charts indicating reduction factors to be applied to probable maximum precipitation values to obtain design storm values for low hazard areas. Precipitation values computed by direct application of the reduction factors, figures 22, 23, and 24, are applicable as design storm values for computing inflow design floods for assumption B, low hazard areas, using the same retention losses as applicable for a maximum probable flood. Assumption A provides criteria for inflow design floods for structures where failure would not increase danger to human life and some risk of failure may be tolerated. Design storm values for assumption A are obtained by averaging design storm values applicable for maximum probable flood and for assumption B flood computations. Retention loss criteria are the same as for a maximum probable flood.

The following criteria for design floods are thus established:

If	*Then*
(1) Failure of structure would result in probable loss of human life.	(1) Inflow design flood is equivalent to the maximum probable flood.
(2) Failure would cause great damage to property and project operation, but loss of human life is not envisioned.	(2) Inflow design flood may be as much less than the maximum probable as that obtained by assumption A.
(3) Failure would cause only loss of structure with little additional damage to property and project operation.	(3) Inflow design flood may be as much less than the maximum probable as that obtained by assumption B.

Simple Triangular Hydrograph:

Given:

$A = 8.0$ sq. mi.

$T_c = 3.0$ hours

$D = 2.0$ hours

$Q = 1.0$ inch

Required: Triangular hydrograph (Needed: T_p, q_p, T_b)

Computations: (Ref.– Figure 29)

$$T_p = \frac{D}{2} + 0.6\,T_c = \frac{2}{2} + 0.6\,(3.0)$$

$$T_p = 2.8 \text{ hours}$$

$$q_p = \frac{484\,AQ}{T_p} = \frac{(484)(8.0)(1.0)}{2.8}$$

$$q_p = 1{,}380 \text{ second-feet}$$

$$T_b = 2.67\,T_p$$

$$T_b = 7.48 \text{ use } 7.5 \text{ hours}$$

Plot T_p, T_b. Connect origin and T_p, T_p and T_b by straight lines.

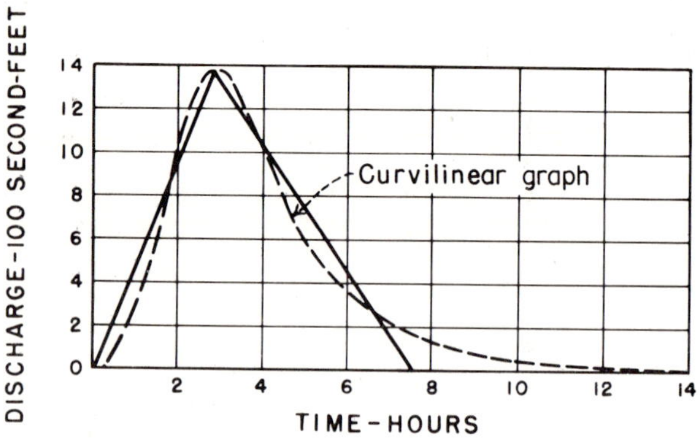

(A)

A curvilinear hydrograph may be constructed from values of q_p and T_p by using ratios tabulated below which were obtained by SCS analyses of many unitgraphs

Time ratio, T/T_p	Disch. ratio, q/q_p	Example–see (A) above	
		Time, hours	Discharge, second-feet
0	0	0	0
0.1	0.015	0.28	21
0.2	0.075	0.56	104
0.3	0.16	0.84	220
0.4	0.28	1.12	386
0.5	0.43	1.40	596
0.6	0.60	1.68	828
0.7	0.77	1.96	1,063
0.8	0.89	2.24	1,228
0.9	0.97	2.52	1,339
1.0	1.00	2.8	1,380
1.1	0.98	3.08	1,352
1.2	0.92	3.36	1,270
1.3	0.84	3.64	1,159
1.4	0.75	3.92	1,035

Time ratio, T/T_p	Disch. ratio, q/q_p	Example–see (A) above	
		Time, hours	Discharge, second-feet
1.5	0.66	4.20	911
1.6	0.56	4.48	773
1.8	0.42	5.04	580
2.0	0.32	5.60	442
2.2	0.24	6.16	331
2.4	0.18	6.72	248
2.6	0.13	7.28	179
2.8	0.098	7.84	135
3.0	0.075	8.40	104
3.5	0.036	9.80	50
4.0	0.018	11.20	25
4.5	0.009	12.60	12
5.0	0.004	14.00	5
Infinity	0		

APPLICATION OF TRIANGULAR HYDROGRAPH‡

Figure 32. Application of triangular hydrograph. (Soil Conservation Service.) (Sheet 1 of 2.) 288–D–2463.

Complex Hydrograph Due To Variable Rainfall Increments:

Given: Area = 100 sq. mi.

T_c = 10 hours

Storm of D = 6 hours, successive 2 – hour increments, ΔD, of 0.6, 1.4, 0.8 inches, rainfall excess.

Procedure: For variable intensity storms, total storm is subdivided into duration increments of time equal to 1/5 (or less) of T_c. In this example, ΔD equals 2 hours.

Computation: Compute simple hydrograph values for T_p, T_b and q_p for 1.0 inch of rainfall excess (runoff) occuring in a 2-hour period (1/5 T_c).

In example, T_p = 7.0 hours, T_b = 18.7 hours, q_p = 6,914 second-feet for 1.0-inch runoff, 2 hours. For ΔD = 2 hours, T_p and T_b remain constant, q_p varies as the ratio of Δ rainfall excess to 1.0-inch excess. Prepare plotting table, plot simple hydrographs for each ΔD, add ordinates of simple hydrographs to get total runoff.

PLOTTING TABLE

* Time, hours	Δ Rain excess	Δ Peak, second-feet (6914)(ΔQ)	Δ Hydrograph Begin time	T of peak	T of end
0					
	0.6	4,148	0	7	18.7
2					
	1.4	9,680	2	9	20.7
4					
	0.8	5,530	4	11	22.7
6					

* Time for total runoff.

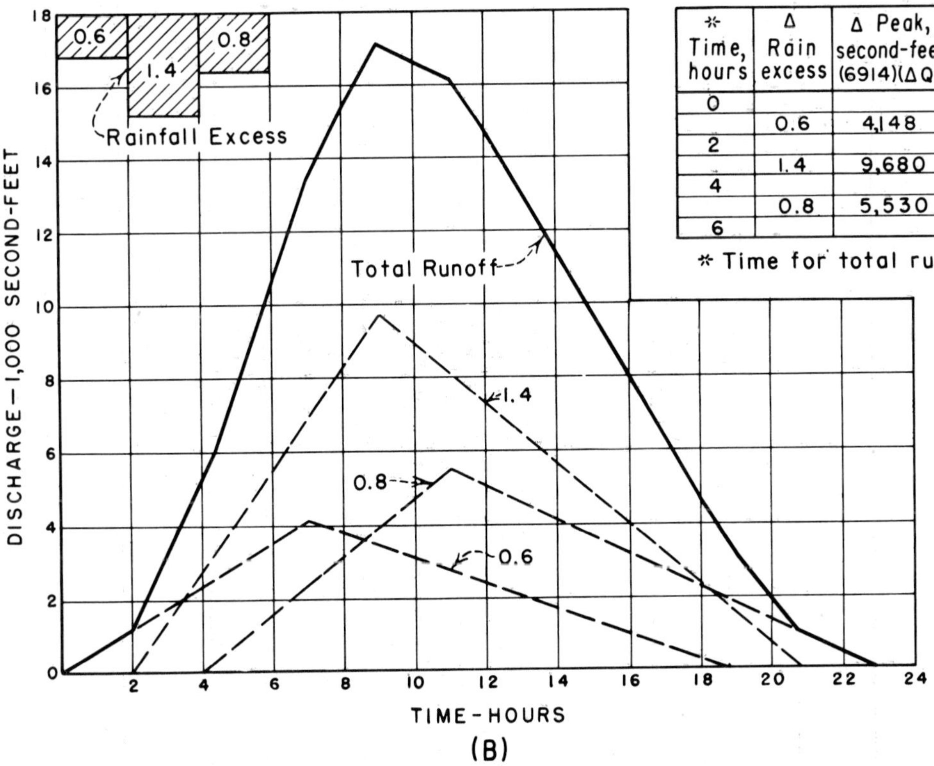

(B)

† Ref. Soil Conservation Service Hydrology Guide.

Figure 32. Application of triangular hydrograph. (Soil Conservation Service.) (Sheet 2 of 2.) 288–D–2464.

(c) *Computation of Inflow Design Floods—East of 105° Meridian.*—This subsection illustrates the procedure for computing the three floods for drainage areas east of the 105° meridian (examples 1, 2, and 3), and subsection (d) discusses the procedure for drainage areas west of the 105° meridian.

Example 1.—Maximum Probable Flood East of the 105° Meridian

General procedure	Specific application
1. Determine geographical location and size of drainage area.	1. For this example use a site in Peoria County, Ill. Drainage area is 61 square miles.
2. Determine zone number and 6-hour 10-square-mile probable maximum precipitation from fig. 15.	2. Zone 7: 6-hour 10-square-mile probable maximum precipitation is 25.5 inches.
3. Determine the probable maximum precipitation (PMP) increments as follows: (a) Adjust the 6-hour 10-square-mile PMP to values for the given drainage area size and to longer durations by use of fig. 16.	3. (a) Using fig. 16, obtain probable maximum precipitation (PMP) for area of 61 square miles for various durations:

Duration, hours	Percent of 10-square-mile 6-hour value	Total PMP, inches
0–6	83	21.2
0–12	100	25.5
0–24	111	28.3
0–48	121	30.9

General procedure	Specific application
(b) Determine hourly amounts of PMP within maximum (first) 6-hour period by percentage values of curve for zone C, fig. 18.	(b) Hourly PMP 6-hour period:

Time, hours	Percent, 6-hour PMP	Accumulative PMP, inches	Incremental PMP, inches
1	49	10.4	10.4
2	64	13.6	3.2
3	75	15.9	2.3
4	84	17.8	1.9
5	92	19.5	1.7
6	100	21.2	1.7

General procedure	Specific application
(c) Tabulate total 48-hour PMP sequence by incremental time periods showing incremental rainfall amounts and accumulative amounts. The incremental rainfall during the maximum 6-hour period should be rearranged from the descending order of magnitude obtained above to the following hourly magnitude sequence, 1 through 6 hours: 6, 4, 3, 1, 2, 5. This is a judicial arrangement that gives a computed flood greater than one based on the assumption that the greatest hourly increment of rain occurs during the first hour of a storm and a smaller flood than that computed by assuming the greatest hourly increment of rain occurs during the 6th hour of a storm. The hourly increments of a storm cannot be predicted.	(c) PMP design arrangement:

Time, ending hour	Incremental PMP, inches	Accumulative PMP, inches
1	1.7	1.7
2	1.9	3.6
3	2.3	5.9
4	10.4	16.3
5	3.2	19.5
6	1.7	21.2
12	4.3	25.5
24	2.8	28.3
48	2.6	30.9

Example 1.—Maximum Probable Flood East of the 105° Meridian—Continued.

General procedure	Specific application
4. Determine the hydrologic soil-cover complex number of the watershed (see appendix A). Availability of soil maps is discussed in sec. 87. If soil maps are not available, an estimate of the hydrologic soil-cover complex number can be made during the field inspection trip to the watershed.	4. Soil series indicated by maps are: Lindley, Clinton, Carrington, Grundy. Table A–1 (appendix A) shows the following hydrologic groups: Lindley—C, Clinton—B, Grundy—C, Carrington—B. Over 50 percent of the area is composed of either Clinton or Carrington soils; therefore, the hydrologic group for the watershed is B. Land usage is estimated as follows, and applicable runoff curve numbers obtained from table A–2 (appendix A):

Percent	Usage	Curve No.
45	Row crops, contoured, good	75
35	Legumes, contoured, good	69
20	Pasture, contoured, good	35

Hydrologic soil-cover complex number equals:

$$0.45\times75=33.75$$
$$0.35\times69=24.15$$
$$0.20\times35=\underline{\ 7.00}$$
$$64.90$$

Curve number equals 65, condition AMC-II (fig. A–4, appendix A).

Note.—The above example is based on general soils map and estimated land usage. These data should be verified by field inspection in actual practice.

Example 1.—Maximum Probable Flood East of the 105° Meridian—Continued.

General procedure	Specific application

General procedure

5. Estimate direct runoff.
   (a) Determine design rainfall values for computing the maximum probable flood (MPF) by applying the additional area reduction factors, given in section 50(d)(1) and shown below, to the accumulative PMP design arrangement from 3(c), above. Round design rainfall values to tenths of an inch.

   Area reduction factor from sec. 50(d)(1):

Drainage area, square miles [1]	Reduction factor, r_f, percent	$100-rf$
1,000	10	90
500	10	90
200	11	89
100	13	87
50	15	85
10	20	80

[1] Interpolate linearly for subject drainage area.

   (b) Compute direct runoff, using runoff curve for hydrologic soil-cover complex number for condition AMC-II (see appendix A) and design storm rainfall tabulated in step (a), above. Antecedent moisture condition II (AMC-II) is used for maximum probable flood computation east of 105° meridian. Condition AMC-II assumes soil moisture supply within the watershed to be similar to average conditions present before occurrence of the maximum annual flood. Runoff curves, fig. A-4 (appendix A), do not contain the dimension time.

Specific application

5. (a) Design rainfall, maximum probable flood (MPF)

Time, ending hour	Accumulative PMP, inches	Accumulative design rainfall, MPF [1], inches	Incremental design rainfall, MPF, inches
1	1.7	1.5	1.5
2	3.6	3.1	1.6
3	5.9	5.0	1.9
4	16.3	13.9	8.9
5	19.5	16.7	2.8
6	21.2	18.1	1.4
12	25.5	21.8	3.7
24	28.3	24.2	2.4
48	30.9	26.4	2.2

[1] $100-rf=85.4$.

(b) Computation of direct runoff
   (1) See table below. Accumulative runoff, Q, computed by equation figure A-4 (appendix A).
   (2) See table below.
   (3) Minimum retention rate range for hydrologic group B, 0.15-0.30 inch per hour, section 51(c). Minimum rate of 0.20 inch per hour selected for this example. In practice, minimum retention rate would be assigned on basis of field inspection of the watershed.

Example 1.—Maximum Probable Flood East of the 105° Meridian—Continued.

General procedure	Specific application

Because limiting loss values discussed in section 51(c) are rates with time as a factor, application of these curves to design flood computations requires some modifications as follows:

(1) Read appropriate curve, figure A-4 (appendix A), using accumulated rainfall amounts by progressive time increments, and determine accumulated direct runoff for respective progressive time increments.

(2) Tabulate incremental rainfall and respective incremental runoff; subtract incremental runoff from incremental rainfall to determine incremental loss. (Loss computations beyond 12 hours are seldom necessary—see (3) below.)

(3) Rainfall-runoff curves, figure A-4 (appendix A), give lower loss rates with increase in storm precipitation. When incremental loss rates determined by step (2), above, reach the limits stated in section 51(c), the runoff curves are no longer used. The incremental runoff is then computed by subtracting the limiting loss rate amounts from the incremental rainfall. Accumulative runoff, incremental runoff, and incremental loss values are usually computed to hundredths of an inch. The equation shown in figure A-4 (appendix A) may be used to compute accumulative runoff values, Q, to hundredths of an inch.

DIRECT RUNOFF INCREMENTS FOR COMPUTING MAXIMUM PROBABLE FLOOD

(Values in inches)

Time, ending hour	Incremental design rainfall	Accumulative design rainfall	Direct runoff		Incremental loss
			Accumulative [1]	Incremental	
1	1.5	1.5	0.03	0.03	1.47
2	1.6	3.1	0.55	0.52	1.08
3	1.9	5.0	1.65	1.10	0.80
4	8.9	13.9	9.03	7.38	1.52
5	2.8	16.7	11.62	2.59	0.21
6	1.4	18.1	12.82	1.20	0.20 [2]
12	3.7	21.8	15.32	2.50	1.20
24	2.4	24.2	15.32	0.00	2.40
48	2.2	26.4	15.32	0.00	4.8 [3]

[1] Using CN 65, AMC-II.

[2] Computation by CN 65 gives 0.09 inch per hour. Abandon curve and use 0.20 inch per hour for remaining storm increments.

[3] Potential loss, greater than incremental rainfall.

6. Determine time of concentration for the watershed (sec. 56, figs. 30 and 31).

6. From map study: L=18.5 miles. Elevation difference between headwater and site, 400 feet=H. Since these are the only data available, T_c is estimated by solution of equation given on figure 30.

$$T_c=\left(\frac{11.9\,L^3}{H}\right)^{0.385}$$

$$=\left[\frac{(11.9)(18.5)^3}{400}\right]^{0.385}$$

$$=7.5 \text{ hours}$$

Example 1.—Maximum Probable Flood East of the 105° Meridian—Continued.

General procedure

7. Compute triangular hydrograph for each increment of runoff (figs. 29 and 32) as follows:
   (a) Determine time increment, D, to be used. For the most intense period of the storm, the time increment, D, should be at least as small as one-fifth of the time of concentration.
      (1) For the first (most intense) 6 hours, D will usually be 1 hour. For rapid concentration times (T_c less than 3 hours), one-half hour is the minimum practical time increment, D, recommended for the most intense periods. For less rapid concentration times, D for the most intense periods may be longer. For T_c values of 10 to 15 hours, a D of 2 hours is recommended. For T_c values of 15 to 30 hours, a D of 3 hours is recommended.
      (2) The time period, D, may be lengthened in the later part of the storm to reduce computations. This will result in a poorer definition of the recession limb of the hydrograph, but it has little effect on the design.
      (3) Runoff from the 24–48-hour period, if obtained, has little effect on design and, therefore, may be neglected. Runoff hydrographs for only the first 24-hour period of design storm need be computed for watersheds having a T_c of 24 hours or less.
   (b) For each time interval, D, compute the time to peak, T_p, base time, T_b, and the peak discharge, q_p, for 1 inch of runoff. Use equations given on fig. 29.

Specific application

7. (a):
   (1) For first 6 hours, use $D=1$ hour.
   (2) For second 6 hours, use $D=6$ hours.
   For second 12 hours, use $D=12$ hours.*
   (3) Ignore runoff after first 24 hours as $T_c=7.5$ hours.

*Included for information; not applicable in this example.

(b) Given:

$$T_p=\frac{D}{2}+0.6T_c$$

$$T_b=2.67\ T_p$$

$$q_p=\frac{484\ A\ Q}{T_p}$$

$T_c=7.5$ hours

$A=61$ sq. mi.

$Q=1.00$ inch.

For $D=1$ hour:

$$T=\frac{1}{2}+0.6\ (7.5)$$
$$=5 \text{ hours}$$
$$T_b=2.67\ (5)$$
$$=13.4 \text{ hours}$$
$$q_p=\frac{484\ (61)\ (1.00)}{5}$$
$$=5{,}900 \text{ sec.-ft.}$$

For $D=6$ hours:

$$T_p=\frac{6}{2}+0.6\ (7.5)$$
$$=7.5 \text{ hours}$$
$$T_b=2.67\ (7.5)$$
$$=20 \text{ hours}$$
$$q_p=\frac{484\ (61)\ (1.00)}{7.5}$$
$$=3{,}940 \text{ sec.-ft.}$$

For $D=12$ hours:*

$$T_p=\frac{12}{2}+0.6\ (7.5)$$
$$=10.5 \text{ hours}$$
$$T_b=2.67\ (10.5)$$
$$=28.0 \text{ hours}$$
$$q_p=\frac{484\ (61)\ (1.00)}{10.5}$$
$$=2{,}810 \text{ sec.-ft.}$$

*Included for information; not applicable in this example.

Example 1.—Maximum Probable Flood East of the 105° Meridian—Continued.

General procedure	Specific application
(c) Compute the peaks of triangular hydrographs for each increment of runoff. These peaks are obtained by multiplying the peak discharge for 1 inch of runoff (computed in the previous step) by the increment of runoff. (d) Prepare a plotting table showing the peaks, starting time, time of peak, and ending time of each incremental hydrograph.	(c) and (d):

PLOTTING TABLE

Time, ending hour	Incremental runoff, inches [1]	q_p for 1.00-inch runoff, sec.-ft.	q_p for incremental runoff, sec.-ft.	Incremental hydrographs		
				Begin time [2]	Peak time [2]	End time [2]
1	0.03	5,900	180	0	5.0	13.4
2	0.52	5,900	3,070	1	6.0	14.4
3	1.10	5,900	6,490	2	7.0	15.4
4	7.38	5,900	43,540	3	8.0	16.4
5	2.59	5,900	15,280	4	9.0	17.4
6	1.20	5,900	7,080	5	10.0	18.4
12	2.50	3,940	9,850	6	13.5	26.0

[1] Step 5(b).
[2] Hour.

General procedure	Specific application
8. Plot maximum probable flood as follows: (a) Plot incremental triangular hydrographs on plain coordinate paper. (b) Add ordinates of plotted hydrographs to obtain ordinates of maximum probable flood hydrograph.	8. See fig. 33.

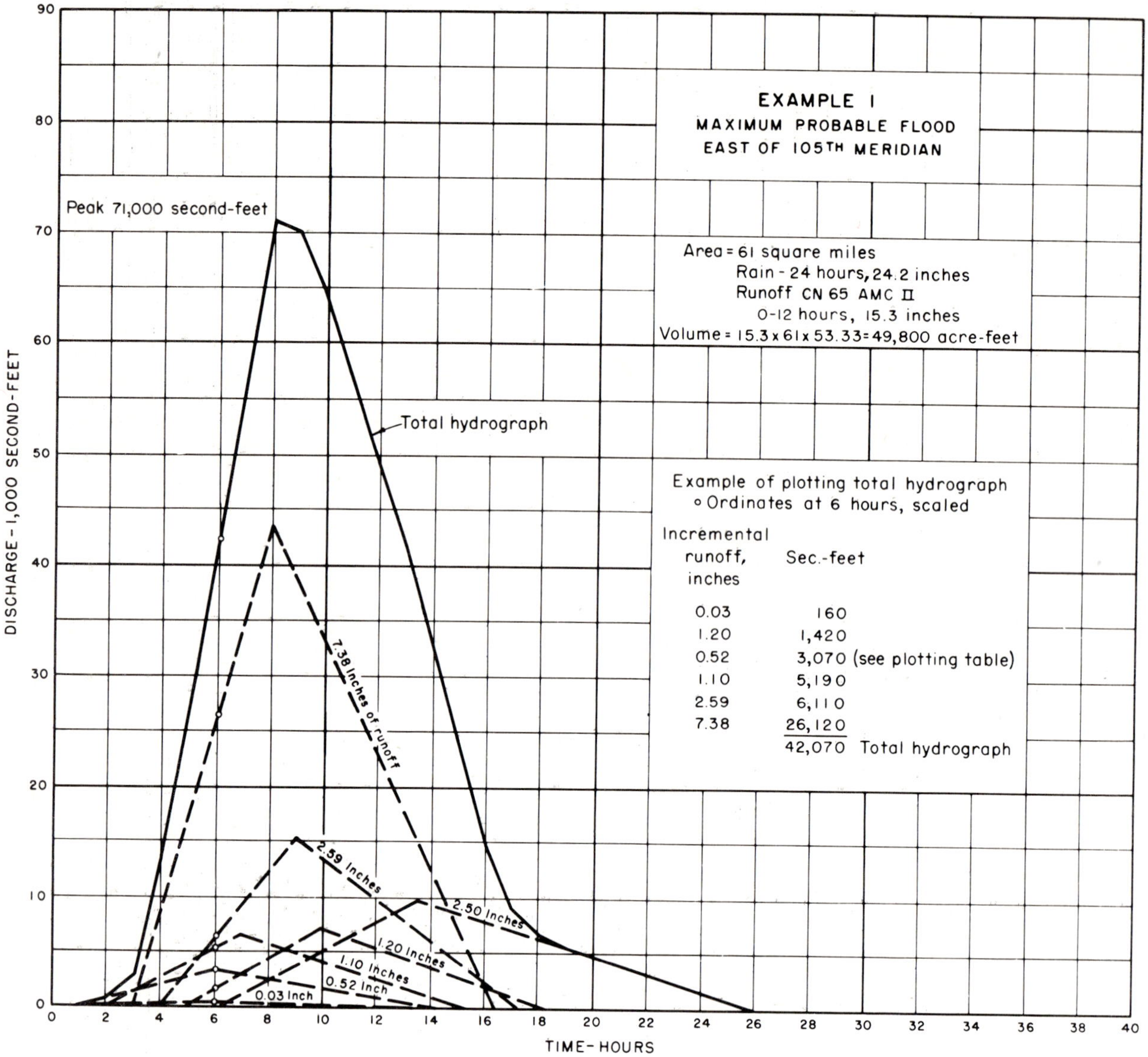

Figure 33. **Maximum probable flood east of 105° meridian (example 1.) 288–D–2770.**

Example 2.—Flood Less Than the Maximum Probable East of the 105° Meridian—Assumption A

General procedure	Specific application

General procedure

1. Design storm values less than the probable maximum provide the basis for this flood. A rainfall reduction factor is obtained for a geographical location from fig. 22. Apply this ratio directly to PMP values as arranged in step 3(c) of example 1; this gives design storm values applicable to computation of an inflow flood for assumption B, as shown by example 3. Design storm values applicable for computing an inflow flood meeting the hydrologic criteria for assumption A are obtained by averaging design storm values for the maximum probable flood step 5(a) of example 1 and for assumption B.

Specific application

1. For Peoria County, Ill., factor equals 2.6. Reduction =1/2.6, or 0.385.

DETERMINING DESIGN STORM—ASSUMPTION A

Time, ending hour	Accumulative PMP,[1] inches	Accumulative reduced rainfall,[2] inches	Accumulative design rainfall MPF,[3] inches	Assumption A Design Storm	
				Accumulative rainfall,[4] inches	Incremental rainfall, inches
1	1.7	0.7	1.5	1.1	1.1
2	3.6	1.4	3.1	2.2	1.1
3	5.9	2.3	5.0	3.6	1.4
4	16.3	6.3	13.9	10.1	6.5
5	19.5	7.5	16.7	12.1	2.0
6	21.2	8.2	18.1	13.2	1.1
12	25.5	9.8	21.8	15.8	2.6
24	28.3	10.9	24.2	17.6	1.8
48	30.9	11.9	26.4	19.2	1.6

[1] From example 1, step 3(c).
[2] Design arrangement accumulative PMP multiplied by 0.385.
[3] From example 1, step 5(a).
[4] Average of values, columns 3 and 4, to tenths of inch.

General procedure

2. The hydrologic soil-cover complex will have been established for computation of a maximum probable flood (step 4 of example 1). Direct runoff from the design storm for assumption A is computed by the same procedure as outlined in step 5(b) of example 1.

Specific application

2. Curve 65, condition AMC-II.

DIRECT RUNOFF—ASSUMPTION A
(Values in inches)

Time, ending hour	Incremental rainfall	Accumulative rainfall	Direct runoff		Incremental loss
			Accumulative	Incremental	
1	1.1	1.0	.0	.0	1.1
2	1.1	2.2	.19	.19	.91
3	1.4	3.6	.80	.61	.79
4	6.5	10.1	5.65	4.85	1.65
5	2.0	12.1	7.40	1.75	.25
6	1.1	13.2	8.30	.90	.20 [1]
12	2.6	15.8	9.70	1.40	1.20
24	1.8	17.6	9.70	.0	2.40 [2]
48	1.6	19.2	...	...	...

[1] Loss by CN 65 gives 0.11 inch. Use minimum loss rate of 0.20 inch per hour.
[2] Loss potential exceeds rainfall.

General procedure

3. The time of concentration for the watershed will be the same as that determined when computing the maximum probable flood (step 6 of example 1).

4. Compute triangular hydrograph for each increment of runoff. The triangular unitgraphs for the time intervals, *D*, used for computing the maximum probable flood are applicable (step 7 of example 1).

Specific application

3. T_c=7.5 hours.

4. PLOTTING TABLE—FLOOD ASSUMPTION A

Time, ending hour	Incremental runoff, inches	q_p for 1.00-inch runoff, sec.-ft.	q_p for incremental runoff, sec.-ft.	Incremental hydrographs		
				Begin time, hours	Peak time, hours	End time, hours
1	0.00	5,900	0	0	5	13.4
2	0.19	5,900	1,120	1	6	14.4
3	0.61	5,900	3,600	2	7	15.4
4	4.85	5,900	28,620	3	8	16.4
5	1.75	5,900	10,320	4	9	17.4
6	0.90	5,900	5,310	5	10	18.4
12	1.40	3,940	5,520	6	13.5	26.0

General procedure

5. Plot the total flood hydrograph as discussed in step 8 of example 1.

Specific application

5. The plotted hydrograph is shown on fig. 34.

Example 3.—Flood Less Than the Maximum Probable East of the 105° Meridian—Assumption B

General procedure

1. The computational procedure is the same as previously outlined in example 2. Design storm values are the reduced rainfall amounts obtained by using the reduction factor for geographical location given on figure 22 (see step 1, example 2). Direct runoff is computed assuming soil moisture of the watershed, before the design storm begins, to be represented by antecedent moisture condition II (AMC-II) CN 65 (see step 4, example 1).
2. Computation of direct runoff and plotting data can be combined into one tabulation as follows:

Time, ending hour	Incremental rain, inches [1]	Accumulative rain, inches [1]	Runoff, inches		Incremental loss, inches	q_p for 1.00-inch runoff, sec.-ft.	q_p for incremental runoff, sec.-ft.	Incremental hydrographs		
			Accumulative [2]	Incremental				Begin time, hours	Peak time, hours	End time, hours
0–1	0.7	0.7	0.0	0.0	0.7	....	0	0	0	0.0
1–2	0.7	1.4	0.02	0.02	0.68	5,900	120	1	6	14.4
2–3	0.9	2.3	0.22	0.20	0.70	5,900	1,200	2	7	15.4
3–4	4.0	6.3	2.58	2.36	1.64	5,900	13,920	3	8	16.4
4–5	1.2	7.5	3.50	0.92	0.28	5,900	5,430	4	9	17.4
5–6	0.7	8.2	4.00	0.50	[3] .20	5,900	2,950	5	10	18.4
6–12	1.6	9.8	4.40	0.40	1.20	3,940	1,580	6	13.5	26.0
12–24	1.1	10.9								

[1] From step 1 of example 2.

[2] From figure A-4 (appendix A) for Curve Number 65.

[3] Use 0.20 inch per hour; CN 65 gives 0.14 inch.

3. The plotted hydrograph is shown on figure 34.

The runoff hydrograph for 0.02-inch excess, hour 1-2, is so small in comparison with other values that it need not be plotted.

(d) *Computation of Inflow Design Floods—West of the 105° Meridian.*—Extreme diversity of terrain and precipitation characteristics westward from the 105° meridian to the Pacific Coast make doubtful the application of a simple general procedure for computing inflow design floods for the entire region. Therefore, two types of design precipitation have been developed for this text: (1) probable maximum general storm (PMGS), and (2) probable maximum thunderstorm (PMTS). Potential rainfall for each type of storm has been delineated for three different zones: PMGS for Zones A, B, and C (see fig. 17); PMTS for Zones I, II, and III (see fig. 20). An area along the West Coast has been delineated where potential thunderstorm rainfall is not considered critical for inflow design flood computations (see fig. 20). The charts, figures 17 and 20, present isohyetals of estimated probable maximum point rainfall, from which a total design storm for respective type rainfall may be developed by use of appropriate areal reduction curves, rainfall time-distribution curves, and tables, which express rainfall amounts for various time intervals as ratios to point rainfall.

Figures 23 and 24 have been prepared to provide estimates of reduced storm potential for computing floods less than probable maximum for low hazard areas west of the 105° meridian. Design storm data and the criteria for determining design storm values for assumption A floods thus provide precipitation amounts from which six inflow design floods may be computed for each watershed studied within regions subject to thunderstorm rainfall. However, it would be a rare circumstance that would require computation of all six flood types for a particular damsite. Site location will indicate whether consideration of floods less than probable maximum is appropriate. Drainage area size, damsite location, and proposed reservoir storage will usually in-

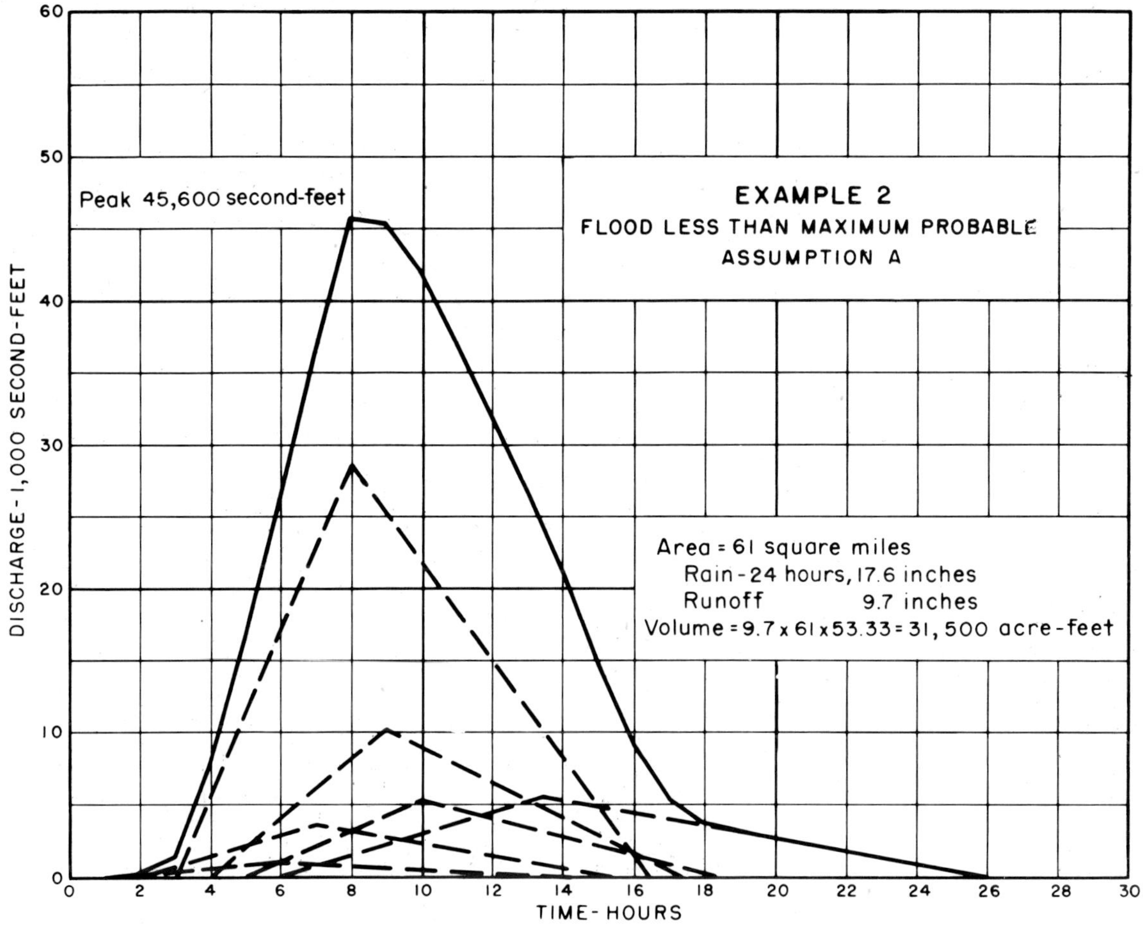

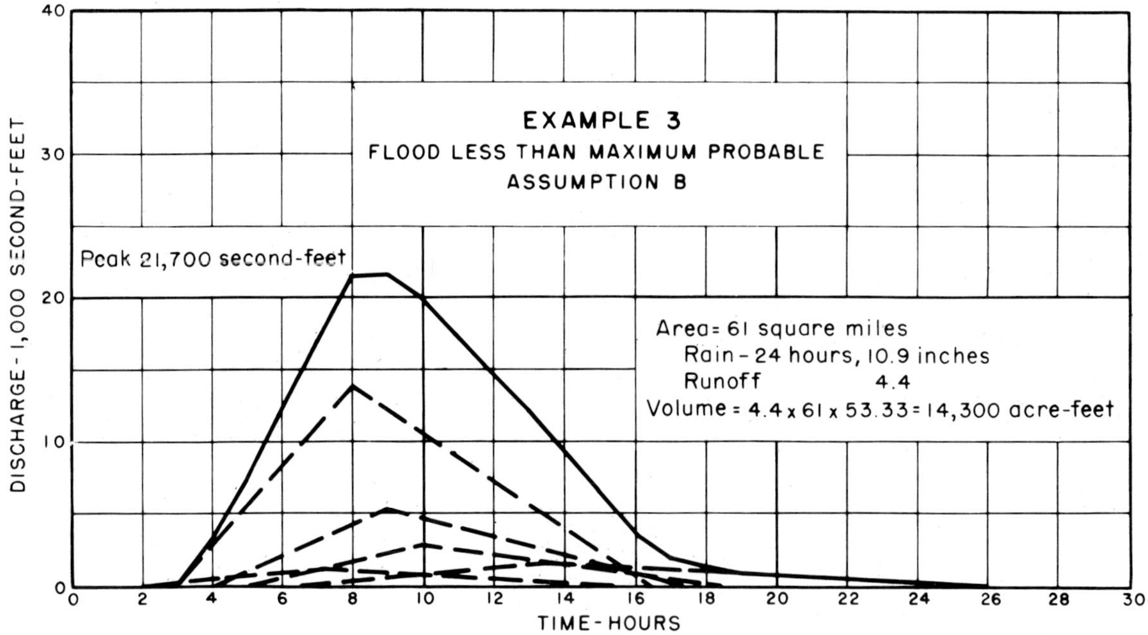

Figure 34. **Design flood less than maximum probable, Assumption A, east of 105° meridian (examples 2 and 3.) 288–D–2271.**

dicate whether peak discharge from a general storm or a thunderstorm or volume from a general storm will be critical for design. Snowmelt runoff potential in conjunction with design rainfall must always be considered for watersheds in the mountainous regions west of the 105° meridian. Computation of runoff from snowmelt has not been included in this text. When snowmelt runoff may become a significant part of an inflow design flood, an engineer should consult references containing methods of computing snowmelt runoff (see sec. 60) and an experienced hydrologist.

Criteria for computation of inflow design floods from design thunderstorms and general storms are presented in the following outline form, with references to preceding text discussions and mathematical procedures given in the three previous examples of computing inflow design floods east of the 105° meridian (sec. 58(c)). The outline is separated into two parts: preliminary steps which assemble data used in all the flood computations and from which an indication of critical type flood can be judged, and steps for specific type flood computation. The inflow design flood most critical for the particular structure proposed should be used for final design. Often it is necessary for designers to route inflow design floods from both thunderstorm rainfall and general storm rainfall through the proposed reservoir to determine which type flood is critical for design.

Outline of Computational Procedure for Inflow Design Flood as Determined for Design Thunderstorms and General Storms—West of the 105° Meridian.

Step description	References to preceding discussions or examples.
A. *Preliminary Steps*	
1. Determine geographical location and size of drainage area. Prepare accurate drainage map.	1. Example 1, step 1, sec. 58(a)(2).
2. By geographical location, determine zone for design thunderstorm rain and design general-type storm rain.	2. Figs. 17 and 20.
3. Determine soil group, and hydrologic soil-cover complex number for antecedent moisture condition II. CN, AMC-II	3. Example 1, step 4; appendix A.
4. Determine T_c, time of concentration for the watershed.	4. Sec. 56 and figs. 30 and 31, example 1, step 6.
(a) If CN, AMC-II, is 50 to 80, and equation, fig. 30, is used, compute T_c' from table 5 and use value of T_c' for triangular hydrograph computations.	(a) Sec. 56, table 5.
5. Select type of floods likely to be critical for design by study of: Purpose of structure. Site location. Drainage area size. Storm potential, snowmelt potential. Reservoir surface area.	
B. *Steps for Specific Type Floods*	B. Sec. 58(b)
1. *Floods from design thunderstorm rainfall*	(d) General.
(a) Inflow design flood, probable maximum thunderstorm rainfall.	
(1) Obtain probable maximum 1-hour point precipitation.	(1) Fig. 20.
(2) Apply area reduction ratio to otain basin 1-hour rain.	(2) Fig. 21.
(3) List accumulative basin rain at each 15-minute interval for duration of thunderstorm.	(3) Table 2. Percentages of 1-hour rainfall.

Outline of Computational Procedure for Inflow Design Flood as Determined for Design Thunderstorms and General Storms—West of the 105° Meridian—Continued.

Step description	References to preceding discussions or examples.
(4) Calculate 15-minute incremental amounts from accumulative basin rain.	(4) Step (3), above.
(5) Assign order of magnitude number to 15-minute increments, beginning with maximum 15-minute increment as number 1.	
(6) Arrange the 15-minute incremental rainfall amounts in the magnitude order sequence given in table A-8, appendix A.	(6) Table A-8, appendix A.

Time, hour	Order of magnitude of 15-minute incremental rainfall amounts	
	Zones	
	II	I & III
0	0	0
0.25	3	4
0.50	2	3
0.75	1	2
1.00	[1] 4	1
1.25		5
1.50		6
1.75		7
2.00		8
2.25		9
2.50		10
2.75		11
3.00		[1] 12

[1] Limiting duration of zone thunderstorm rainfall.

Step description	References to preceding discussions or examples.
(7) Determine thunderstorm runoff curve number and minimum retention loss rate.	(7) Using data assembled for preliminary step 3, assign specific thunderstorm cover-index. Obtain appropriate thunderstorm CN from table A-9 and thunderstorm minimum 15-minute retention loss rate for hydrologic soil group, table A-10, appendix A.
(8) Using fig. A-4 (appendix A), compute each 15-minute increment of direct runoff.	(8) Use CN and minimum retention rate, step (7), above. Computational procedure, example 1, step 5(b) (c).
(9) A time increment, *D*, of 15 minutes, 0.25 hour, is used for thunderstorm hydrograph computations regardless of watershed T_c because of the maximum 15-minute rainfall intensity. Compute T_p, T_b, and q_p for 1.00 inch 0.25 hour triangular hydrograph, using watershed T_c or T_c'	(9) Fig. 29. Preliminary step 4.
(10) Compute peak discharge for each 15-minute increment of excess precipitation. Prepare a hydrograph plotting table.	(10) Example 1, step 7 (c) (d).
(11) Plot the inflow design flood. Add applicable base flow to plotted hydrograph ordinates to complete inflow design flood.	(11) Example 1, step 8, fig. 33.

Outline of Computational Procedure for Inflow Design Flood as Determined for Design Thunderstorms and General Storms—West of the 105° Meridian—Continued.

Step description	References to preceding discussions or examples.
(b) Assumption A, inflow design flood, thunderstorm rainfall.	
(1) Obtain thunderstorm rainfall reduction factor for geographical location from fig. 24. Apply reduction factor to probable maximum thunderstorm rainfall arranged in given time sequence. Average the values of the maximum rainfall and the reduced rainfall to obtain rainfall amounts for assumption A.	(1) Example 2, step 1.
(2) Obtain thunderstorm runoff curve number, CN, from table A-9, appendix A.	
(3) Complete flood computations as outlined above, steps (8) through (11).	
(c) Assumption B, inflow design flood, thunderstorm rainfall.	
(1) The reduced rainfall amounts, step (1) assumption A above, are the thunderstorm rainfall values used for computing a flood for assumption B.	
(2) Obtain thunderstorm runoff curve number, CN, from table A-9, appendix A.	
(3) Complete flood computations as outlined above.	
2. *Floods from design general-type storms*	
(a) Inflow design flood, probable maximum general storm.	
(1) By geographical location, obtain probable maximum 6-hour point rainfall.	(1) Fig. 17.
(2) Apply area reduction ratio to obtain basin 6-hour rainfall.	(2) Fig. 19.
(3) Prepare a tabulation of accumulative maximum 18-hour probable maximum basin rainfall at 1-hour intervals. (Maximum 18-hour rainfall can be used to obtain design floods from general storms for most small dams west of 105° meridian. For damsites where runoff volume is critical for design, runoff from the total storm period should be computed.)	(3) Fig. 18, 1-6 hours, table 1, 6-18 hours. Use average of even-hour rain for intervening odd-numbered hour.
(4) Compute 6-hour increments of rainfall from above tabulation. Arrange 6-hour increments in magnitude order: 2, 1, 3.	
(5) Arrange the hourly increments of rainfall within each 6-hour period of rain in the following order of magnitude: 1-6 hours: ascending 6-12 hours: 6, 4, 3, 1, 2, 5 12-18 hours: descending	
(6) Convert watershed CN, AMC-II to CN, AMC-III.	(6) Preliminary step 3, table A-7, appendix A.

Outline of Computational Procedure for Inflow Design Flood as Determined for Design Thunderstorms and General Storms—West of the 105° Meridian—Continued.

Step description	References to preceding discussions or examples.
(7) Compute each hourly increment of direct runoff using CN, AMC-III and hydrologic soil group minimum retention loss rate.	(7) Example 1, step 5(b).
(8) A time increment, *D*, of 1-hour may be used for all areas greater than 5 square miles; a 0.5-hour for areas less than 5 square miles. Hourly excess rainfall increments are divided by 2 for obtaining 0.5-hour excesses. Compute T_p, T_b, and q_p 1.00 inch runoff for a triangular hydrograph of time increment *D*, using the watershed T_c or T_c'.	(8) Fig. 29, preliminary step 4.
(9) Compute triangular hydrograph for each increment of excess. Prepare plotting table. Plot hydrograph. Add base flow.	(9) Fig. 32; example 1, step 7(c) (d) and step 8.
(b) Assumption A, inflow design flood, general storm.	
(1) Obtain reduction factor for 6-hour point rainfall from fig. 23 for geographical location. Apply reduction factor to probable maximum rainfall values in design sequence arrangement. Average values of maximum probable and reduced storms to obtain general storm values for assumption A.	(1) Example 2, step 1.
(2) Compute hourly increments of excess using CN, AMC-III and soil group minimum retention loss rate.	(2) Example 1, step 5(b).
(3) Complete flood computation.	(3) Steps (a)(8) and (a)(9) just preceding.
(c) Assumption B, inflow design flood, general storm.	
(1) The reduced rainfall amounts, step (1) assumption A, above, are the storm values for computing a flood for assumption B.	
(2) Compute hourly increments of excess using CN, AMC-III, and soil group minimum retention loss rate.	(2) Example 1, step 5(b).
(3) Complete flood computation.	

Probability Data for
Annual Peak Discharge
Ashley Creek near Vernal, Utah

Computed by ________
Date July 1944

$$\text{Coefficient of Variation} = \sqrt{\frac{\Sigma \text{ Col. (8)}}{n-1}} = \sqrt{\frac{4.18}{23}} = 0.43$$

$$F = 1 + \frac{8.5}{n} = 1.35$$

$$\text{Coefficient of Skew} = \frac{\Sigma \text{ Col. (9)} \times F}{(n-1)\ (c.v.)^3} = \frac{(0.69)\ (1.35)}{(23)\ (.080)} = 0.51$$

① Year	② Items	③ Items in order of magnitude	④ Plotting position	⑤ Freq. in years	⑥ Ratio to mean	⑦ Col. ⑥ minus 1.0	⑧ Col. ⑦ squared	⑨ Col. ⑦ cubed	⑩ % of time	⑪ Skew factor CS=0.5	⑫ Col. ⑪ X C.V. (0.43)	⑬ Col. ⑫ plus 1.0	⑭ Col. ⑬ X mean
1920	1,360	2,050	2.08	48	2.05	1.05	1.10	1.16					
1921	2,050	1,700	6.25	16	1.70	.70	.49	.34	99	-1.99	-.86	.14	140
	⁄		⁄			⁄		⁄	95	-1.50	-.64	.36	361
									80	-.85	-.37	.63	631
1942	1,270	475	93.75	1.07	.47	-.53	.28	-.15	50	-.08	-.03	.97	972
1943	702	201	97.92	1.02	.20	-.80	.64	-.51	20	+.82	+.35	1.35	1,350
									5	+1.79	+.77	1.77	1,770
	Sum	24,038	Algebraic sum			+.03	4.18	+.69	1	+2.72	+1.17	2.17	2,170
Years of record n=24						Check							
	Mean	1,002				O.K.							

(A)

TABULATION AND COMPUTATIONS

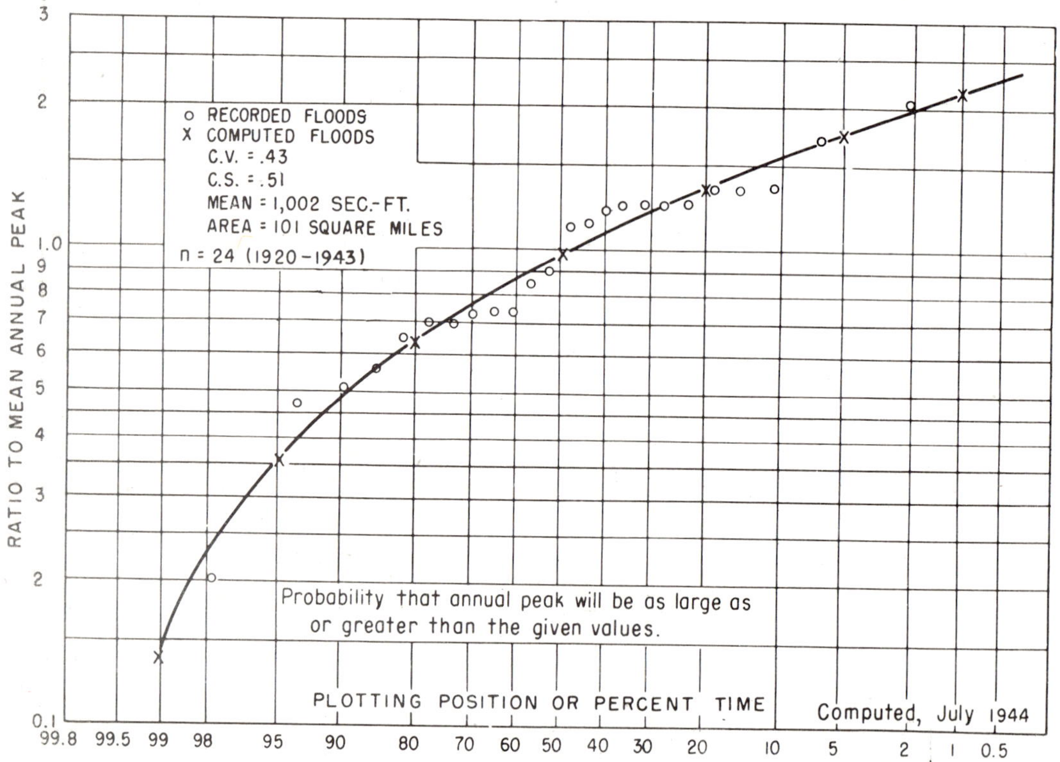

(B)

CUMULATIVE FREQUENCY CURVE OF ANNUAL PEAKS
ASHLEY CREEK NEAR VERNAL, UTAH

Figure 35. Frequency curve computations—Hazen's method. (Sheet 1 of 2.) 288–D–2467.

COMPUTATIONAL STEPS FOR FIGURE 35

1. Tabulate data in columns 1 and 2. If record is broken, try to fill gaps by correlation with nearby records. If no correlations are possible, use available data as continuous record. Consult other references for unusual cases such as one extreme event or incorporation of historical flows into current record.
2. Column 3.- List items from column 2 in descending order of magnitude.
3. Column 4.- See table 7.
4. Column 5.- 100 ÷ column 4 values.
5. Column 6.-Ratio of individual item in column 3 to arithmetical mean of items in column 3.
6. Complete columns 7, 8, 9 by indicated computations.
7. Compute Coefficient of Variation and Coefficient of Skew by equations shown in (A).
8. Fill in values column 11 from table 6.
9. Complete columns 12, 13, 14 by indicated computations.
10. Columns 13 and 14 give values for a curve which should fit the data in columns 6 and 3 respectively. Test of fit is obtained by plotting values column 4 vs column 6 and column 10 vs column 13 on logarithmic probability paper. See (B) on left. The same test of fit may be made by plotting on log-log paper the values in column 5 vs column 3, and drawing the curve through the points defined by values in column 10 divided into 100 vs column 14 values. (This plotting not shown.)

(C)

Figure 35. Frequency curve computations—Hazen's method. (Sheet 2 of 2.) 288–D–2468.

No.	Stream	Station	Years record	Record ending	Area, sq. mi.	Mean	c.v.	c.s.	n(c.s.)	n(c.v.)	Remarks
1.	Ashley Creek	Vernal, Utah	24	1943	101	1,002	.43	.50	12.00	10.32	
2.	Duchesne River	Tabiona, Utah	25	1943	352	1,357	.40	.20	5.00	10.00	
3.	Duchesne River	Duchesne, Utah	26	1943	660	2,752	.35	0	0	9.10	
4.	Uinta River	Neola, Utah	17	1943	181	1,119	.42	.50	8.50	7.14	
5.	Whiterocks River	Whiterocks, Utah	18	1943	115	1,162	.59	1.00	18.00	10.62	
			110						43.50	47.18	
		Avg. cs = $\frac{43.50}{110}$ = .396 use .40 Avg. cv = $\frac{47.18}{110}$ = .429 use .43									A.B.C. JULY 1944

(A) AVERAGE COEFFICIENTS OF VARIATION AND SKEW
(For Peak Discharges in Second-Feet)

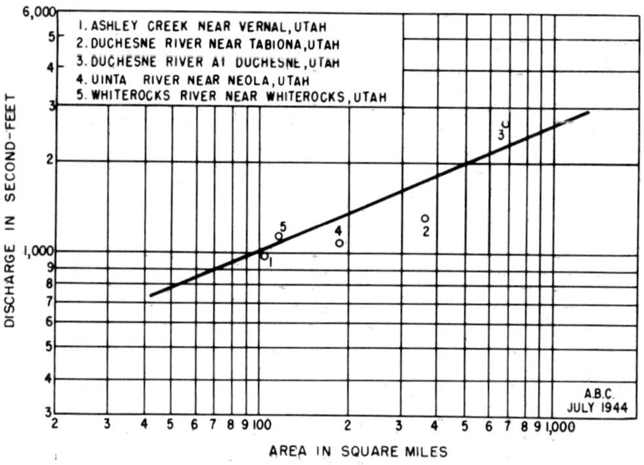

(B) MEAN ANNUAL PEAK DISCHARGE - VERSUS DRAINAGE AREA

EXPLANATION

Given: Area, ungaged water shed (500 sq. mi.).
Records of comparable streams in vicinity.
Required: Frequency curve of probable peak discharge for 500 sq. mi.
Procedure:
1. Compute mean, c.v. and c.s. for each record. See fig. 35(A).
2. Tabulate data as shown in (A) and determine weighted average c.v. and c.s. as shown.
3. Plot mean values vs respective drainage area on log-log paper as shown in (B).
4. Draw mean line through plotted points.
5. Enter plot (B) with ungaged area (500), read mean (2,000).
6. Compute curve for ungaged area using:
   a. Mean, 2,000 second-feet.
   b. Average c.s. and c.v. (see (A)).
   c. Procedure outlined by columns 10 through 14 of figure 35 (A).

Figure 36. Flood frequency study—ungaged basin. 288–D–2469.

59. *Frequency Curve Computations.*—Information shown on figures 35 and 36 gives the mathematical procedure for computing frequency curves by Hazen's method. Engineers basing decisions on results of flood frequency computations should keep in mind the points briefly discussed in section 47.

TABLE 6.—*Hazen's logarithmic skew curve factors* [1]

[Multiply coefficient of variation by these and add to or subtract from the mean]

Coefficient of skew	Terms over mean percent	Percent of term above limit									Corresponding coefficient variation
		99 (−)	95 (−)	80 (−)	50 (−)	20 (+)	5 (+)	1 (+)	0.1 (+)	0.01 (+)	
0	50.0	2.32	1.64	0.84	0	0.84	1.64	2.32	3.09	3.72	0
0.1	49.4	2.25	1.62	.85	.02	.84	1.67	2.40	3.24	3.96	.03
0.2	48.7	2.18	1.59	.85	.03	.83	1.71	2.48	3.39	4.20	.06
0.3	48.1	2.12	1.56	.85	.05	.83	1.74	2.56	3.55	4.45	.10
0.4	47.5	2.05	1.53	.85	.06	.82	1.76	2.64	3.72	4.72	.13
0.5	46.9	1.99	1.50	.85	.08	.82	1.79	2.72	3.90	5.00	.16
0.6	46.3	1.92	1.47	.85	.09	.81	1.81	2.80	4.08	5.30	.20
0.7	45.6	1.86	1.44	.85	.11	.80	1.84	2.89	4.28	5.64	.23
0.8	45.0	1.80	1.41	.85	.12	.79	1.86	2.97	4.48	6.00	.27
0.9	44.4	1.73	1.38	.85	.14	.77	1.88	3.06	4.69	6.37	.30
1.0	43.7	1.68	1.34	.84	.15	.76	1.90	3.15	4.92	6.77	.33
1.1	43.1	1.62	1.31	.84	.17	.75	1.92	3.24	5.16	7.23	.37
1.2	42.5	1.56	1.28	.83	.18	.74	1.94	3.33	5.40	7.66	.41
1.3	41.9	1.51	1.25	.83	.19	.72	1.96	3.41	5.64	8.16	.44
1.4	41.3	1.46	1.22	.82	.20	.71	1.98	3.50	5.91	8.66	.48
1.5	40.7	1.41	1.19	.81	.22	.69	1.99	3.59	6.18	9.16	.51
1.6	40.1	1.36	1.16	.81	.23	.67	2.01	3.69	6.48	9.79	.55
1.7	39.5	1.32	1.13	.80	.24	.66	2.02	3.78	6.77	10.40	.59
1.8	38.9	1.27	1.10	.79	.25	.64	2.03	3.88	7.09	11.07	.62
1.9	38.3	1.23	1.07	.78	.26	.62	2.04	3.98	7.42	11.83	.66
2.0	37.7	1.19	1.05	.77	.27	.61	2.05	4.07	7.78	12.60	.70
2.1	37.1	1.15	1.02	.76	.28	.59	2.06	4.17	8.13	13.35	.74
2.2	36.5	1.11	.99	.75	.29	.57	2.07	4.27	8.54	14.30	.78
2.3	35.9	1.07	.96	.74	.30	.55	2.07	4.37	8.95	15.25	.82
2.4	35.3	1.03	.94	.73	.31	.53	2.08	4.48	9.35	-----	.86
2.5	34.7	1.00	.91	.72	.31	.51	2.08	4.58	9.75	-----	.90
2.6	34.1	.97	.89	.71	.32	.49	2.09	4.68	10.15	-----	.94
2.7	33.5	.94	.86	.69	.33	.47	2.09	4.78	10.65	-----	.98
2.8	32.9	.91	.84	.68	.33	.45	2.09	4.89	11.20	-----	1.03
2.9	33.3	.87	.82	.67	.34	.43	2.09	5.01	11.75	-----	1.08
3.0	31.8	.84	.79	.66	.34	.41	2.08	5.11	12.30	-----	1.12
3.2	30.6	.78	.74	.64	.35	.37	2.06	5.35	13.50	-----	1.22
3.4	29.4	.73	.69	.61	.36	.32	2.04	5.58	-----	-----	1.33
3.6	28.2	.67	.65	.58	.36	.28	2.02	5.80	-----	-----	1.44
3.8	27.0	.62	.61	.55	.36	.23	1.98	6.10	-----	-----	1.57
4.0	25.7	.58	.56	.52	.36	.19	1.95	6.50	-----	-----	1.70
4.5	22.2	.48	.47	.45	.35	.10	1.79	7.30	-----	-----	2.10
5.0	19.2	.40	.40	.39	.34	0	1.60	8.20	-----	-----	2.50

$$\text{Adjusted skew} = \text{Computed skew}\left(1+\frac{8.5}{n}\right).$$

The figures in the last column show the value of the coefficient of variation that, in connection with the coefficient of skew shown in the first column will produce plotting point of a line that is straight on log. probability paper. The column "Terms over mean percent" represents the point of intersection of that straight line with the mean value of the variable.

[1] Reprinted by permission from A. Hazen's "Flood Flows."

TABLE 7.—*Hazen's plotting points for probability paper for series ranging from 13 to 40 terms* [1]

TERMS IN RECORD

13	14	15	16	17	18	19	20	21	22	23	24	25	26	27	28	29	30	31	32	33	34	35	36	37	38	39	40
3.85	3.57	3.33	3.13	2.94	2.78	2.63	2.50	2.38	2.27	2.17	2.08	2.00	1.92	1.85	1.79	1.72	1.67	1.61	1.56	1.52	1.47	1.43	1.39	1.35	1.32	1.28	1.25
11.54	10.71	10.00	9.38	8.82	8.33	7.89	7.50	7.14	6.82	6.52	6.25	6.00	5.77	5.56	5.36	5.17	5.00	4.84	4.69	4.55	4.41	4.29	4.17	4.05	3.95	3.85	3.75
19.23	17.86	16.67	15.62	14.71	13.89	13.16	12.50	11.90	11.36	10.87	10.42	10.00	9.62	9.26	8.93	8.62	8.33	8.06	7.81	7.58	7.35	7.14	6.94	6.76	6.58	6.41	6.25
26.92	25.00	23.33	21.87	20.59	19.44	18.42	17.50	16.67	15.91	15.22	14.58	14.00	13.46	12.96	12.50	12.07	11.67	11.29	10.94	10.61	10.29	10.00	9.72	9.46	9.21	8.97	8.75
34.62	32.14	30.00	28.12	26.47	25.00	23.68	22.50	21.43	20.45	19.57	18.75	18.00	17.31	16.67	16.07	15.52	15.00	14.52	14.06	13.64	13.24	12.86	12.50	12.16	11.84	11.54	11.25
42.31	39.29	36.67	34.37	32.35	30.56	28.95	27.50	26.19	25.00	23.91	22.92	22.00	21.15	20.37	19.64	18.97	18.33	17.74	17.19	16.67	16.18	15.71	15.28	14.86	14.47	14.10	13.75
50.00	46.43	43.33	40.62	38.24	36.11	34.21	32.50	30.95	29.55	28.26	27.08	26.00	25.00	24.07	23.21	22.41	21.67	20.97	20.31	19.70	19.12	18.57	18.06	17.57	17.11	16.67	16.25
57.69	53.57	50.00	46.87	44.12	41.67	39.47	37.50	35.71	34.09	32.61	31.25	30.00	28.85	27.78	26.79	25.86	25.00	24.19	23.44	22.73	22.06	21.43	20.83	20.27	19.74	19.23	18.75
65.38	60.71	56.67	53.12	50.00	47.22	44.74	42.50	40.48	38.64	36.96	35.42	34.00	32.69	31.48	30.36	29.31	28.33	27.42	26.56	25.76	25.00	24.29	23.61	22.97	22.37	21.79	21.25
73.08	67.86	63.33	59.37	55.88	52.78	50.00	47.50	45.24	43.18	41.30	39.58	38.00	36.54	35.19	33.93	32.76	31.67	30.65	29.69	28.79	27.94	27.14	26.39	25.68	25.00	24.36	23.75
80.77	75.00	70.00	65.62	61.76	58.33	55.26	52.50	50.00	47.73	45.65	43.75	42.00	40.38	38.89	37.50	36.21	35.00	33.87	32.81	31.82	30.88	30.00	29.17	28.38	27.63	26.92	26.25
88.46	82.14	76.67	71.87	67.65	63.89	60.53	57.50	54.76	52.27	50.00	47.92	46.00	44.23	42.59	41.07	39.66	38.33	37.10	35.94	34.85	33.82	32.86	31.94	30.08	30.26	29.49	28.75
96.15	89.29	83.33	78.12	73.53	69.44	65.79	62.50	59.52	56.82	54.35	52.08	50.00	48.08	46.30	44.64	43.10	41.67	40.32	39.06	37.88	36.76	35.71	34.72	33.78	32.89	32.05	31.25
	96.43	90.00	84.37	79.41	75.00	71.05	67.50	64.29	61.36	58.70	56.25	54.00	51.92	50.00	48.21	46.55	45.00	43.55	42.19	40.91	39.71	38.57	37.50	36.49	35.53	34.62	33.75
		96.67	90.62	85.29	80.56	76.32	72.50	69.05	65.91	63.04	60.42	58.00	55.77	53.70	51.79	50.00	48.33	46.77	45.31	43.94	42.65	41.43	40.28	39.19	38.16	37.18	36.25
			96.87	91.18	86.11	81.58	77.50	73.81	70.45	67.39	64.58	62.00	59.62	57.41	55.36	53.45	51.67	50.00	48.44	46.97	45.59	44.29	43.06	41.89	40.79	39.74	38.75
				97.06	91.67	86.84	82.50	78.57	75.00	71.74	68.75	66.00	63.46	61.11	58.93	56.90	55.00	53.23	51.56	50.00	48.53	47.14	45.83	44.59	43.42	42.31	41.25
					97.22	92.11	87.50	83.33	79.55	76.09	72.92	70.00	67.31	64.81	62.50	60.35	58.33	56.45	54.69	53.03	51.47	50.00	48.61	47.30	46.05	44.87	43.75
						97.37	92.50	88.10	84.09	80.43	77.08	74.00	71.15	68.52	66.07	63.79	61.67	59.68	57 81	56.06	54.41	52.86	51.39	50.00	48.68	47.44	46.25
							97.50	92.86	88.64	84.78	81.25	78.00	75.00	72.22	69.64	67.24	65.00	62.90	60.94	59.09	57.35	55.71	54.17	52.70	51.32	50.00	48.75
								97.62	93.18	89.13	85.42	82.00	78.85	75.93	73.21	70.69	68.33	66.13	64.06	62.12	60.29	58.57	56.94	55.41	53.95	52.56	51.25
									97.73	93.48	89.58	86.00	82.69	79.63	76.79	74.14	71.67	69.35	67.19	65.15	63.24	61.43	59.72	58.11	56.58	55.13	53.75
										97.83	93.75	90.00	86.54	83.33	80.36	77.59	75.00	72.58	70.31	68.18	66.18	64.29	62.50	60.81	59.21	57.69	56.25
											97.92	94.00	90.38	87.04	83.93	81.03	78.33	75.81	73.44	71.21	69.12	67.14	65.28	63.51	61.84	60.26	58.75
												98.00	94.23	90.74	87.50	84.48	81.67	79.03	76.56	74.24	72.06	70.00	68.06	66.22	64.47	62.82	61.25
													98.08	94.44	91.07	87.93	85.00	82.26	79.69	77.27	75.00	72.86	70.83	68.92	67.11	65.38	63.75
														98.15	94.64	91.38	88.33	85.48	82.81	80.30	77.94	75.71	73.61	71.62	69.74	67.95	66.25
															98.21	94.83	91.67	88.71	85.94	83.33	80.88	78.57	76.39	74.32	72.37	70.51	68.75
																98.28	95.00	91.94	89.06	86.36	83.82	81.43	79.17	77.03	75.00	73.08	71.25
																	98.33	95.16	92.19	89.39	86.76	84.29	81.94	79.73	77.63	75.64	73.75
																		98.39	95.31	92.42	89.71	87.14	84.72	82.43	80.26	78.21	76.25
																			98.44	95.45	92.65	90.00	87.50	85.14	82.89	80.77	78.75
																				98.48	95.59	92.86	90.28	87.84	85.53	83.33	81.25
																					98.53	95.71	93.06	90.54	88.16	85.90	83.75
																						98.57	95.83	93.24	90.79	88.46	86.25
																							98.61	95.95	93.42	91.03	88.75
																								98.65	96.05	93.59	91.25
																									98.68	96.15	93.75
																										98.72	96.25
																											98.75

Note:

$p=100\left(\frac{2m-1}{2n}\right)$, where

p=percent chance of occurrence in any year

m=order number, largest flood being number 1, second largest number 2, etc.

n=number of items (years)

[1] Reprinted by permission from A. Hazen's "Flood Flows."

C. SELECTED BIBLIOGRAPHY

60. *Bibliography.*

Creager, W. P., Justin, J. D., and Hinds, Julian, "Engineering for Dams," vol. I, John Wiley & Sons, Inc., New York, 1945.

Johnstone, Don, and Cross, W. P., "Elements of Applied Hydrology," The Ronald Press Co., New York, 1949.

Linsley, R. K. Jr., Kohler, M. A., and Paulhus, J. L. H., "Applied Hydrology," McGraw-Hill Book Co., Inc., New York, 1949.

"Snow Hydrology—A Summary Report of the Snow Investigations," North Pacific Division, Corps of Engineers, U.S. Army, Portland, Oreg., June 30, 1956.

"Unitgraph Procedures," Bureau of Reclamation, Division of Planning Coordination, Denver, Colo., 1952, Revised 1965.

"Hydrology Guide for Use in Watershed Planning," National Engineering Handbook, Sec. 4, Supplement A, U.S. Department of Agriculture, Soil Conservation Service, Central Technical Unit, Beltsville, Md.

Hazen, A., "Flood Flows," John Wiley & Sons, Inc., New York, 1930.

Corbett, D. M., et al., "Stream Gaging Procedure," U.S. Geological Survey Water Supply Paper 888, 1943.

"Soil and Water Conservation Engineering," second edition, The Ferguson Foundation Agricultural Engineering Series, January 1966.

"Manual of Engineering Practice, No. 28," Hydrology Handbook, American Society of Civil Engineers, 1949.

"Handbook of Applied Hydrology," Chow, Ven Te, editor-in-chief, McGraw-Hill, 1964.

"Hydrology, Part I—Watershed Planning," SCS National Engineering Handbook, sec. 4, U.S. Department of Agriculture, Soil Conservation Service, Washington, D.C., 1964.

"Soils and Men," Yearbook of Agriculture, 1938, U.S. Department of Agriculture, Washington, D.C.

Crawford, Norman H., and Linsley, Ray K., "Digital Simulation in Hydrology: Stanford Watershed Model IV," Technical Report No. 39, Department of Civil Engineering, Stanford University, Stanford, Calif., July 1966.

Bertle, F. A., "Effect of Snow Compaction on Runoff from Rain on Snow," Engineering Monograph No. 35, Bureau of Reclamation, 1966.

"Methods of Flow Frequency Analysis," Bulletin No. 13, Subcommittee on Hydrology, Inter-Agency Committee on Water Resources (now the Hydrology Committee, Water Resources Council), Washington, D.C., April 1966.

"A Uniform Technique for Determining Flood Flow Frequencies," Bulletin No. 15, Hydrology Committee, Water Resources Council, 1025 Vermont Ave., N.W., Washington, D.C., December 1967.

*National Oceanic and Atmospheric Administration—National Weather Service, Technical Papers**

*Current official designation is U.S. Department of Commerce, National Oceanic and Atmospheric Administration, National Weather Service. Prior to October 3, 1970, it was designated U.S. Department of Commerce, Environmental Science Services Administration, U.S. Weather Bureau.

Paper No. 2, "Maximum Recorded United States Point Rainfall for 5 minutes to 24 hours at 207 First-Order Stations," Washington D.C., 1963.

Paper No. 4, "Topographically Adjusted Normal Isohyetal Maps for Western Colorado," Russler, B. H., and Spreen, W. C. Washington, D.C., August 1947.

Paper No. 5, "Highest Persisting Dewpoints in Western United States," Washington, D.C., January 1948.

Paper No. 14, "Tables of Precipitable Water and Other Factors for a Saturated Pseudo-Adiabatic Atmosphere, Washington, D.C., 1951.

Paper No. 15, "Maximum Station Precipitation for 1, 2, 3, 6, 12, and 24 Hours." Part I: Utah; Part II: Idaho, 1951; Part III: Florida, 1952; Part IV: Maryland, Delaware, and District of Columbia; Part V: New Jersey, 1953; Part VI: New England, 1953; Part VII: South Carolina, 1953; Part VIII: Virginia, 1954; Part IX: Georgia, 1954; Part X: New York, 1954; Part XI: North Carolina, 1955; Part XII: Oregon, 1955; Part XIII: Kentucky, 1955; Part XIV: Louisiana, 1955; Part XV: Alabama, 1955; Part XVI: Pennsylvania, 1956; Part XVII: Mississippi, 1956; Part XVIII: West Virginia, 1956; Part XIX: Tennessee, 1956; Part XX: Indiana, 1956; Part XXI: Illinois, 1958; Part XXII: Ohio, 1958; Part XXIII: California, 1959; Part XXIV: Texas, 1959; Part XXV: Arkansas, 1960; Part XXVI: Oklahoma, 1961.

Paper No. 16, "Maximum 24-Hour Precipitation in the United States," Jennings, A. H., Washington, D.C., January 1952.

Paper No. 19, "Mean Number of Thunderstorm Days in the United States," Washington, D.C., December 1952.

Paper No. 24, "Rainfall Intensities for Local Drainage Design in the United States" (For Durations of 5 to 240 Minutes and 2-, 5-, and 10-Year Return Periods), Part I: West of 115th Meridian, Washington, D.C., August 1953; Part II: Between 105° W. and 115° W., Washington, D.C., August 1954.

Paper No. 25, "Rainfall Intensity-Duration-Frequency Curves" (For Selected Stations in the United States, Alaska, Hawaiian Islands, and Puerto Rico), Washington, D.C., December 1955.

Paper No. 28, "Rainfall Intensities for Local Drainage Design in Western United States" (For Durations of 20 Minutes to 24 Hours and 1- to 100-Year Return Periods), Washington, D.C., November 1956.

Paper No. 29, "Rainfall Intensity-Frequency Regime," Part 1: The Ohio Valley, 1957; Part 2: Southeastern United States, 1958; Part 3: The Middle Atlantic Region, 1958; Part 4: Northeastern United States, 1959; Part 5: Great Lakes Region, 1960.

Paper No. 38, "General Estimates of Probable Maximum Precipitation for the United States West of the 105th Meridian for Areas to 400 Square Miles and Durations to 24 Hours," Washington, D.C., 1969.

Paper No. 40, "Rainfall Frequency Atlas of the United States for Durations from 30 Minutes to 24 Hours and Return Periods from 1 to 100 Years," Washington, D.C., 1963.

Paper No. 42, "Generalized Estimates of Probable Maximum Precipitation and Rainfall-Frequency Data for Puerto Rico and Virgin Islands," Washington, D.C., 1961.

Paper No. 43, "Rainfall-Frequency Atlas of the Hawaiian Islands for Areas to 200 Square Miles, Durations to 24 Hours, and Return Periods from 1 to 100 Years," Washington, D.C., 1962.

Paper No. 47, "Probable Maximum Precipitation and Rainfall-Frequency Data for Alaska for Areas to 400 Square Miles, Durations to 24 Hours, and Return Periods from 1 to 100 Years," Washington, D.C., 1963.

Paper No. 49, "Two- to Ten-Day Precipitation for Return Periods of 2 to 100 Years in the Contiguous United States," Washington, D.C., 1964.

Paper No. 51, "Two- to Ten-Day Rainfall for Return Periods of 2 to 100 Years in the Hawaiian Islands," Washington, D.C., 1965.

Paper No. 52, "Two- to Ten-Day Precipitation for Return Periods of 2 to 100 Years in Alaska," Washington, D.C., 1965.

Paper No. 53, "Two- to Ten-Day Rainfall for Return Periods of 2 to 100 Years in Puerto Rico and Virgin Islands," Washington, D.C., 1965.

Cooperative Studies Technical Papers, Cooperative Studies Section, Division of Climatological and Hydrologic Services, National Weather Service, in cooperation with the Bureau of Reclamation

Paper No. 1, "Manual for Depth-Area-Duration Analysis of Storm Precipitation."

Cooperative Studies Reports, Cooperative Studies Section, Division of Climatological and Hydrologic Services, National Weather Service, in cooperation with the Bureau of Reclamation

Report No. 9, "Maximum Possible Flood-Producing Meteorological Conditions." (1) Colorado River Basin above Glen Canyon Damsite, (2) Colorado River Basin above Bridge Canyon Damsite, (3) San Juan River Basin above Bluff Damsite, (4) Little Colorado River Basin above Coconino Damsite, June 1949.

Report No. 11, "Critical Meteorological Conditions for Design Floods in the Snake River Basin," February 1953.

Report No. 12, "Probable Maximum Precipitation on Sierra Slopes of the Central Valley of California," Washington, D.C., March 1954.

National Oceanic and Atmospheric Administration, National Weather Service, Hydrometeorological Reports

Report No. 2, "Maximum Possible Precipitation over the Ohio River Basin above Pittsburgh, Pennsylvania," 1942.

Report No. 3, "Maximum Possible Precipitation over the Sacramento Basin of California," 1943.

Report No. 5, "Thunderstorm Rainfall," 1947.

Report No. 15, "A Report on Depth-Frequency Relations of Thunderstorm Rainfall on the Sevier Basin, Utah," 1941.

Report No. 18, "Tentative Estimates of Maximum Possible Flood-Producing Meteorological Conditions in the Columbia River Basin," 1945.

Report No. 20, "An Estimate of Maximum Possible Flood-Producing Meteorological Conditions in the Missouri River Basin Above Garrison Dam Site," 1945.

Report No. 21, "A Hydrometeorological Study of the Los Angeles Area," 1939.

Report No. 21A, "Preliminary Report on Maximum Possible Precipitation, Los Angeles Area, California," 1944.

Report No. 21B, "Revised Report on Maximum Possible Precipitation, Los Angeles Area, California," 1945.

Report No. 22, "An Estimate of Maximum Possible Flood-Producing Meteorological Conditions in the Missouri River Basin Between Garrison and Fort Randall," 1946.

Report No. 23, "Generalized Estimates of Maximum Possible Precipitation Over the United States East of the 105th Meridian, for Areas of 10, 200, and 500 Square Miles," 1947.

Report No. 24, "Maximum Possible Precipitation Over the San Joaquin Basin, California," 1947.

Report No. 25, "Representative 12-Hour Dewpoints in Major United States Storms East of the Continental Divide," 1947.

Report No. 25A, "Representative 12-Hour Dewpoints in Major United States Storms East of the Continental Divide," 2d edition, 1949.

Report No. 28, "Generalized Estimate of Maximum Possible Precipitation Over New England and New York," 1952.

Report No. 33, "Seasonal Variation of the Probable Maximum Precipitation East of the 105th Meridian for Areas from 10 to 1,000 Square Miles and Durations of 6, 12, 24, and 48 Hours," 1956.

Report No. 36, "Interim Report—Probable Maximum Precipitation in California," Washington, D.C., 1961.

Report No. 39, "Probable Maximum Precipitation in the Hawaiian Islands," Washington, D.C., 1963.

Report No. 40, "Probable Maximum Precipitation, Susquehanna River Drainage above Harrisburg, Pa.," Washington, D.C., 1965.

Report No. 41, "Probable Maximum and TVA Precipitation over the Tennessee River Basin above Chattanooga," Washington, D.C., 1965.

Report No. 42, "Meteorological Conditions for the Probable Maximum Flood on the Yukon River above Rampart, Alaska," Washington, D.C., 1966.

Report No. 43, "Probable Maximum Precipitation, Northwest States," Washington, D.C., 1966.

Report No. 44, "Probable Maximum Precipitation over the South Platte River, Colorado, and Minnesota River, Minnesota," Washington, D.C., 1969.

Chapter IV

Selection of Type of Dam

H. G. ARTHUR [1]

A. CLASSIFICATION OF TYPES

61. *General.*—Dams may be classified into a number of different categories, depending upon the purpose of the classification. For the purposes of this manual, it is convenient to consider three broad classifications according to: Use, hydraulic design, or materials comprising the structure.

62. *Classification According to Use.*—Dams may be classified according to the broad function which they are to serve, such as storage, diversion, or detention. Refinements of classification can also be made by considering specific functions involved.

Storage dams are constructed to impound water in periods of surplus supply for use in periods of deficient supply. These periods may be seasonal, annual, or longer. Many small dams impound the spring runoff for use in the summer dry season. Storage dams may be further classified according to the purpose of the storage, such as water supply, recreation, fish and wildlife, hydroelectric power generation, irrigation, etc. The specific purpose or purposes which are to be served by a storage dam often have an influence in the design of the structure, and may establish criteria such as the amount of reservoir fluctuation which may be expected and the amount of reservoir seepage which may be permitted. Figure 37 shows a small earthfill storage dam, and figure 38 shows a concrete-gravity structure serving both diversion and storage purposes.

Diversion dams are ordinarily constructed to provide head for carrying water into ditches, canals, or other conveyance systems to the place of use. They are used for irrigation developments, for diversion from a live stream to an off-channel-location storage reservoir, for municipal and industrial uses, or for any combination of the above. Figure 39 shows a typical small diversion dam.

Detention dams are constructed to retard flood runoff and minimize the effect of sudden floods. Detention dams fall into two main types. In one type, the water is temporarily stored, and released through an outlet structure at a rate which will not exceed the carrying capacity of the channel downstream. In the other type, the water is held as long as possible and allowed to seep into pervious banks or gravel strata in the foundation. The latter type is sometimes called a water-spreading dam or dike because its main purpose is to recharge the underground water supply. Detention dams are also constructed to trap sediment. These often are called debris dams.

Although this is not as common on small projects as on large developments, often dams are constructed to serve more than one purpose. Where multiple purposes are involved, a reservoir allocation is usually made to each of the separate uses. A common multipurpose project involving small dams combines storage, flood control, and recreational uses.

63. *Classification by Hydraulic Design.*—Dams may also be classified as overflow or nonoverflow dams.

Overflow dams are designed to carry dis-

[1] Formerly Engineer, Earth Dams Section, now Director of Design and Construction, Bureau of Reclamation.

Figure 37. Crescent Lake Dam, a small earthfill storage dam on Crescent Creek in Oregon. 806–126–92.

charge over their crests. They must be made of materials which will not be eroded by such discharges. Concrete, masonry, steel, and wood are required excepting for overflow structures only a few feet high.

Nonoverflow dams are those which are not designed to be overtopped. This type of design extends the choice of materials to include earthfill and rockfill dams.

Often the two types are combined to form a composite structure consisting of, for example, an overflow concrete gravity dam with dikes of earthfill construction. Figure 40 shows a composite structure built by the Bureau of Reclamation.

64. *Classification by Materials.*—The most common classification used for purposes of discussion of design procedures is based upon the materials comprising the structure. This classification also usually recognizes the basic type of design such as, for example, concrete *gravity* dam or concrete *arch* dam.

This text is limited in scope to consideration of the more common type of small dams which are constructed under present day conditions; namely, *earthfill, rockfill,* and *concrete gravity.* Other types of dams, including concrete arch, concrete buttress, and timber dams, are discussed briefly, with an explanation of why their designs are not contained in this text.

65. *Earthfill.*—Earthfill dams are the most common type of dam, principally because their construction involves utilization of materials in the natural state requiring a minimum of processing. Moreover, the foundation requirements for earthfill dams are less stringent than

Figure 38. Black Canyon Dam, a concrete-gravity storage and diversion structure on the Payette River in Idaho. Boise 2304.

for other types. It is likely that earthfill dams will continue to be more prevalent than other types for storage purposes, partly because the number of sites favorable for concrete structures is decreasing as a result of extensive water storage development, particularly in arid and semiarid regions where the conservation of water for irrigation is a fundamental necessity.

Although the earthfill classification includes several types, the development of modern excavating, hauling, and compacting equipment for earth materials has made the rolled-fill type so economical as to virtually replace the semihydraulic- and hydraulic-fill types of earthfill dams. This is especially true for the construction of small structures where the relatively small amount of material to be handled precludes the establishment of the large plant required for efficient hydraulic operations. For this reason, only the rolled-fill type of earthfill dam is treated in this text. Earthfill dams of the rolled-fill type are further classified as "homogeneous," "zoned," or "diaphragm," as described in chapter VI.

Earthfill dams require supplementary structures to serve as spillways. The principal disadvantage of an earthfill dam is that it will be damaged or even may be destroyed under the erosive action of water flowing over it if sufficient spillway capacity is not provided. It is also subject to serious damage or even failure due to burrowing of animals unless special precautions are taken. Unless the site is offstream, provision must be made for diversion of the stream during construction through the

Figure 39. Knight Diversion Dam, a small diversion structure on the Duchesne River near Duchesne, Utah. 66–400–3167.

damsite by means of a conduit, or around the damsite by means of a tunnel. Otherwise, special provisions, including the use of heavy rock sections, must be incorporated in the design to permit overflowing of the embankment during construction. This latter type of diversion should be attempted only by those experienced in this field.

66. *Rockfill.*—The rockfill dam uses rock of all sizes to provide stability and an impervious membrane to provide watertightness. The membrane may be an upstream facing of impervious soil, a concrete slab, asphaltic concrete paving, steel plates, or another similar device; or it may be an interior thin core of impervious soil.

Like the earth embankments, the rockfill dam is subject to damage or destruction by the overflow of water and so must be provided with a spillway of adequate capacity to prevent overtopping of the dam. An exception is the extremely low diversion dam where the rockfill facing is designed specifically to withstand overflows. Rockfill dams require foundations which will not be subject to settlement of magnitudes sufficient to rupture the watertight membrane. The only suitable foundations, therefore, are rock or compact sand and gravel.

The rockfill type is adapted to remote locations where the supply of good rock is ample, where suitable soil for an earthfill dam is not available, and where the construction of a concrete dam would be too costly.

67. *Concrete Gravity.*—The concrete gravity dam is adapted to sites where there is a reasonably sound rock foundation, although low

Figure 40. Olympus Dam, a combination earthfill and concrete-gravity structure on the Big Thompson River in Colorado. The concrete section contains the spillway and an outlet works to a canal. 245–704–3117.

structures may be founded on alluvial foundations if adequate cutoffs are provided. It is well adapted for use as an overflow spillway crest and, because of this advantage, it is often used for the spillway feature of earthfill and rockfill dams or as the overflow section of a diversion dam.

In the early 1900's, some gravity dams were constructed of stone. However, the amount of hand labor required in this operation has been responsible for the exclusive use of concrete in modern gravity dam construction.

Gravity dams may be either straight or curved in plan. The curved plan may offer some advantage in both cost and safety. Also, occasionally the upstream curvature will locate that part of the dam on higher bedrock foundation.

68. *Concrete Arch*.—A concrete arch dam is adaptable to sites where the ratio of width between abutments to height is not great and where the foundation at the abutments is solid rock capable of resisting arch thrust. Because the design of an arch dam is quite specialized, a detailed discussion of this type of design is not included in this text.

69. *Concrete Buttress*.—Buttress dams comprise flat deck and multiple arch structures. They require about 60 percent less concrete than solid gravity dams, but the increased formwork and reinforcement steel required usually offset the savings in concrete. A number of buttress dams were built in the 1930's when the ratio of labor cost to materials cost was comparatively low. This type of construction usually is not competitive with other types of dams when labor costs are high.

The design of buttress dams is based on the knowledge and judgment that comes only from specialized experience in that field. Because of this fact and because of the limited application for the construction of buttress dams under present day conditions, the design of this type of structure is not included in this text.

70. *Other Types.*—Dams of other types than those mentioned above have been built, but in most cases they meet some unusual local requirement or are of an experimental nature. In a few instances, structural steel has been used both for the deck and for the supporting framework of dams. Prior to 1920, a number of timber dams were constructed, particularly in the Northwest. The amount of labor involved in the timber dam, coupled with the short life of the structure, makes this type of structure uneconomical for modern construction. This and other uncommon types are not treated in this text.

B. PHYSICAL FACTORS GOVERNING SELECTION OF TYPE

71. *General.*—It is only in exceptional circumstances that an experienced designer can say that only one type of dam is suitable or most economical for a given damsite. Except in cases where the selection of type is obvious, it will be found that preliminary designs and estimates will be required for several types of dams before one can be shown to be most economical. It is, therefore, important to emphasize that the project is likely to be unduly expensive unless decisions regarding selection of type are based upon adequate study and after consultation with competent engineers.

In the selection of type for important structures, it is also usually advisable to secure the advice of an experienced engineering geologist in connection with the relative applicability of possible types to the foundation at the site.

In numerous cases, excessive cost of protection from spillway discharges, limitations of outlet works, and the problem of diverting the stream during construction have an important bearing on the selection of type. In certain cases, the selection of type may also depend upon the availability of labor and equipment. These may be particularly important considerations when the element of time is involved. Inaccessibility of site may also have an important bearing on the selection.

The selection of the best type of dam for a particular site calls for thorough consideration of the characteristics of each type, as related to the physical features of the site and the adaptation to the purposes the dam is supposed to serve, as well as economy, safety, and other pertinent limitations. The final choice of type of dam will generally be made after consideration of those factors. Usually, the greatest single factor determining the final choice of type of dam will be the cost of construction. The following paragraphs discuss important physical factors in the choice of type of dam.

72. *Topography.*—Topography, in large measure, dictates the first choice of type of dam. A narrow stream flowing between high, rocky walls would naturally suggest a concrete overflow dam. The low, rolling plains country would, with equal fitness, suggest an earthfill dam with a separate spillway. For intermediate conditions, other considerations take on more importance, but the general principle of satisfactory conformity to natural conditions is a safe primary guide.

The location of the spillway is an important item that will be governed very largely by the local topography and will, in turn, have a material bearing on the final selection of type of dam.

73. *Geology and Foundation Conditions.*—Foundation conditions depend upon the geological character and thickness of the strata which are to carry the weight of the dam, their inclination, permeability, and relation to underlying strata, existing faults, and fissures. The foundation will limit the choice of type to a certain extent, although such limitations will frequently be modified, considering the height of the proposed dam. The different foundations commonly encountered are discussed below:

(1) *Solid rock foundations,* because of relatively high bearing power and resistance to erosion and percolation, offer few restrictions as to the type of dam that can be built upon them. Economy of materials or overall cost will be the ruling factor. The removal of disintegrated rock together with the sealing of seams and fractures by grouting will frequently be necessary.

(2) *Gravel foundations,* if well compacted, are suitable for earthfill, rockfill, and low concrete gravity dams. As gravel foundations are frequently subject to water percolation at high rates, special precautions must be taken to provide effective water cutoffs or seals.

(3) *Silt or fine sand foundations* can be used for the support of low concrete gravity dams and earthfill dams if properly designed, but they are not suitable for rockfill dams. The main problems are settlement, the prevention of piping, excessive percolation losses, and protection of the foundation at the downstream toe from erosion.

(4) *Clay foundations* can be used for the support of earthfill dams but require special treatment. Since there may be considerable settlement of the dam if the clay is unconsolidated and the moisture content is high, clay foundations ordinarily are not suitable for the construction of concrete gravity dams, and should not be used for rockfill dams. Tests of the foundation material in its natural state are usually required to determine the consolidation characteristics of the material and its ability to support the superimposed load.

(5) *Nonuniform foundations.*—Occasionally, situations may occur where reasonably uniform foundations of any of the foregoing descriptions cannot be found and where a nonuniform foundation of rock and soft material must be used if the dam is to be built. Such unsatisfactory conditions can often be overcome by special design features. Each site, however, presents a problem for appropriate treatment by experienced engineers, and no attempt is made in this text to treat such unusual problems.

The details of the foundation treatments mentioned above are given in the appropriate chapters on the design of earthfill, rockfill, and concrete gravity dams.

74. *Materials Available.*—Materials for dams of various types, which may sometimes be available at or near the site, are:

(1) Soils for embankments.

(2) Rock for embankments and riprap.

(3) Concrete aggregate (sand, gravel, crushed stone).

Elimination or reduction of transportation expense for construction materials, particularly those which are used in great quantity, will effect a considerable reduction in the total cost of the project. The most economical type of dam will often be the one for which materials are to be found in sufficient quantity within a reasonable distance from the site.

The availability of suitable sand and gravel for concrete at a reasonable cost locally and perhaps even on property which is to be acquired for the project is a factor favorable to the use of a concrete structure. On the other hand, if suitable soils for an earthfill dam can be found in nearby borrow pits, an earthfill dam may prove to be the most economical. Advantage should be taken of every local resource to reduce the cost of the project without sacrificing the efficiency and quality of the final structure.

75. *Spillway Size and Location.*—The spillway is a vital appurtenance of a dam. Frequently its size and type and the natural restrictions in its location will be the controlling factors in the choice of the type of dam. Spillway requirements are dictated primarily by the runoff and streamflow characteristics, independent of site conditions or type or size of the dam. The selection of specific spillway types will be influenced by the magnitudes of the floods to be bypassed. Thus, it can be seen that, on streams with large flood potential, the spillway will become the dominant structure and the selection of the type of dam could become a secondary consideration.

The cost of constructing a large spillway is frequently a considerable portion of the total cost of the development. In such cases, combining the spillway and dam into one structure may be desirable, indicating the adoption of a concrete overflow dam. In certain instances, where excavated material from separate spillway channels can be utilized in dam embankment, an earthfill dam may prove to be advantageous. Small spillway requirements often favor the selection of earthfill or rockfill dams, even in narrow damsites.

The advisability or practice of building overflow concrete spillways on earth or rock embankments has generally been discouraged because of the more conservative design assumptions and added care needed to forestall failures. Inherent problems associated with such designs are: Unequal settlements of the structure due to differential consolidations of the embankment and foundation after the reservoir loads are applied; the need for special provisions to prevent cracking of the concrete or opening of joints which could permit leakage from the channel into the fill, with consequent piping or washing away of the surrounding material; and the construction delays necessitated by the requirement for having a fully completed and seasoned dam before spillway construction can be started. Consideration of the above factors, coupled with increased costs brought about by more conservative construction details such as arbitrarily increased lining thickness, increased reinforcement steel, cutoffs, joint treatment, drainage, and preloading, have generally led to selection of alternative solutions for the spillway design such as placing the structure over or through the natural material of the abutment or under the dam as a conduit.

One of the common spillway arrangements is the utilization of a channel excavated through one or both of the abutments outside the limits of the dam, or at some point removed from the dam. Where such a location is adopted, the dam can be of the nonoverflow type which extends the choice to include earthfill and rockfill structures. Conversely, failure to locate a site for the spillway away from the dam requires the selection of a type of dam which can include an overflow spillway. The spillway overflow can then be placed so as to occupy only a portion of the main river channel, in which case the remainder of the dam could be either of earth, rock, or concrete. (Olympus Dam (fig. 40) is an example.)

76. Earthquake.—If the dam lies in an area that is subject to earthquake shocks, the design must include provisions for the added loading and increased stresses. Earthquake design considerations for earthfill, rockfill, and concrete gravity dams are discussed in chapters VI, VII, and VIII, respectively. For earthquake areas, neither the selection of type nor the design of the dam should be undertaken by anyone who is not experienced in this type of work.

C. LEGAL, ECONOMIC, AND ESTHETIC CONSIDERATIONS

77. Statutory Restrictions.—Statutory restrictions exist with respect to control of the waters of navigable streams. Plans for diversion or control of waters in such streams are subject to approval by the Corps of Engineers, U.S. Department of the Army. There are numerous other Federal and State regulations relating to dam construction and operation which may be determining factors in the choice of type of structure. Almost every State has laws and regulations governing the design, construction, and operation of all dams and reservoirs of appreciable size. Engineers or owners considering dam construction in any of the 50 States should contact the proper State authorities before proceeding with detailed designs.

78. Purpose and Benefit-Cost Relation.—Consideration of the purpose which a dam is to serve will often suggest the type most suitable as, for example, whether its principal function is to furnish continuous and dependable storage of the water supply for irrigation, power, or domestic use; to control floods by detention; to regulate the flow of the streams; or to be a diversion dam or a weir without storage features.

Few sites exist where it would be impossible to build a dam that would be safe and serviceable, but in many instances conditions inherent in the site will result in a project cost in excess of the justifiable expenditure. The results of a search for desirable damsites often determine whether a project can be built at a cost which will be consistent with the benefits to be derived from it. Accepted procedures are available for evaluating the benefits for waterpower, irrigation, and water supply uses; these procedures are less well defined for flood control; and there is no satisfactory measure of value for recreational projects.

Justification for recreational development must be evaluated on the basis of a comparison of the population that will be benefited, the location of other projects of the same kind, and the trend of development in the district (appreciative and depreciative)—all as related to the cost of the project and the money available. In a case where the need is great but the number of people to be served is limited, the development of an expensive site may not be justified. In another case, the present need may be great but a tendency toward decline of population and property values must be considered. In both instances, the development should be made as inexpensive as possible, probably with a low dam of small storage capacity.

79. Appearance.—In general, every type of structure should have a finished, workmanlike appearance, compatible with its functional purpose. The alinement and texture of finished surfaces should be true to the design requirements and free from unsightly irregularities. Esthetic considerations may have an important bearing on the selection of type of structure, especially one which is designed primarily for recreational use.

Chapter V

Foundations and Construction Materials

DR. J. W. HILF[1]

A. SCOPE OF INVESTIGATIONS

80. *General.*—Information relating to foundation conditions and to the natural materials available for construction is essential for the design of small dams. The investigation is conducted in the office, in the field, and in the laboratory. For efficiency, the search for data must be properly planned. Subsurface explorations should not be started until all available geological and soils data have been evaluated. The investigator should know how to classify soils and rocks and should have an understanding of the geological and engineering characteristics of landforms. This background and a knowledge of the capabilities and limitations of the various methods of subsurface exploration will lead to selection of the most appropriate field methods, thereby avoiding time and effort lost through ineffectual procedures. The investigator should be familiar with logging and sampling methods and with the field and laboratory tests used for small dams.

The scope of investigation of foundations and of the various types of construction materials is given in this part of the chapter. Parts B through I provide information on the techniques and procedures for making these investigations.

81. *Foundations.*—Foundation investigations are necessary to ascertain whether a safe structure can be built at a selected site. The preferred type of structure can often be determined by inspection of the site, as described in chapter IV. The construction of a dam whose failure would result in a destructive flood involves a serious public responsibility. A large number of damaging floods have been caused by failures of small dams. Investigations have shown that many of these failures were due to insecure foundations. A considerable number of failures attributed to other causes probably originated in defective foundations. It is undoubtedly true that many failures could have been averted by thorough investigations, which would have led to the selection of safe and satisfactory sites or the adoption of provisions in design and construction necessary to overcome defects in the foundation.

The first and one of the most important steps in the investigation of a reservoir is a reconnaissance for the purpose of selecting, principally upon the basis of topography and areal geology, the most favorable of the potential damsites. Such a reconnaissance is a task for both the engineer and the engineering geologist and should be entrusted only to men experienced in work of this kind. The field work should be preceded by a study of all available data relating to the stream and to the area under consideration, including examination of maps, air photos, and reports, particularly those by the U.S. Geological Survey. Part B of this chapter discusses the various sources of information. A thorough reconnaissance leading to the selection of the most desirable site for the dam, or the elimination of as many of the potential damsites as possible where there is a choice of sites, may save many thousands of dollars in exploratory work.

Foundation conditions are often revealed by

[1] Formerly Engineer, Earth Dams Section, now Chief, Division of Design, Bureau of Reclamation.

or can be inferred from visual inspection of erosional features; rock outcrops; and manmade excavations such as highway or railroad cuts, building excavations, soil pits, and rock quarries. Some information on groundwater conditions often can be obtained from local wells. The results of the field studies should be sketched, preferably on a topographic map, although air photos can be used. At this stage of investigation, the sketch map often can show boundaries of soil deposits and rock outcrops, location of fault zones or other visible geologic irregularities, and the dip and strike of geologic features such as joints, bedding, and sheared zones.

The map should be accompanied by a report describing the various geologic features including rock and soil classifications, types of cementing materials that may occur in the rock and soil, and the origin and mode of deposition of the various soil deposits. The reconnaissance report should discuss the relationship of the geological conditions to the present and future permeability of the reservoir and dam foundation, and to the future stability and permanence of the dam, spillway, and other structures. Readily apparent geological problems requiring resolution by further investigations also should be discussed and a tentative program outlining the extent and character of more detailed explorations for the next stage of investigation should be recommended. The reservoir as well as the damsite area should be examined for the possible existence of old or potential landslides which might be activated by construction.

In the feasibility stage of investigation, subsurface exploration of the foundation is needed to determine definitely: (1) the depth to bedrock at the damsite, and (2) the character of both the rock and the soils under the dam and appurtenant structures. A line of holes usually is required at a damsite to determine the bedrock profile along the proposed axis. Since any axis selected in the field is necessarily tentative and subject to adjustment for design reasons, a few additional holes upstream and downstream from the axis are desirable. The number of holes required for foundation exploration of small dams is determined by the complexity of geological conditions, but the maximum spacing should not exceed 500 feet, and the depth of the holes should be at least equal to the height of the dam.

It is also necessary to develop the subsurface conditions at possible locations for the appurtenant structures, such as spillways, outlets, cutoff trenches, and tunnel portals, to clarify any unresolved geological problems. Holes in the areas of appurtenant structures usually should have a maximum spacing of 100 feet and should extend below the foundation at least 1½ times the base width of the structure. For the investigation of diversion dams, holes should be spaced a maximum of 100 feet apart in a pattern dictated by the complexity of the foundation. At least one hole should be located at each pier site.

Although there are many methods of penetrating overburden soils, only those that offer an opportunity for sampling and testing the foundation without excessive disturbance are recommended for exploring foundations for dams. Consequently, wash borings, for example, are not discussed in this text. Test pits, trenches, tunnels, and large-diameter auger borings, which permit visual examination of the foundation, are excellent methods of determining the character of the overburden and are recommended wherever they are practicable. The recommended boring methods for exploring the foundation of small dams are rotary drilling and drive sampling (standard penetration test). Density-in-place tests and determination of moisture content of soils above water table are desirable. Borings in rock require use of a rotary drill and diamond bits to obtain cores. Approximate values of the permeability of rock strata and soil overburden can be determined by water tests in bore holes. In all subsurface exploratory holes, it is important to measure and record the depths to water tables and the dates of these measurements.

The report prepared on completion of the feasibility stage should include a large-scale map showing surface geology, locations of all exploratory holes, and any geological sections measured in natural or manmade cuts. The map units used should be basically geologic,

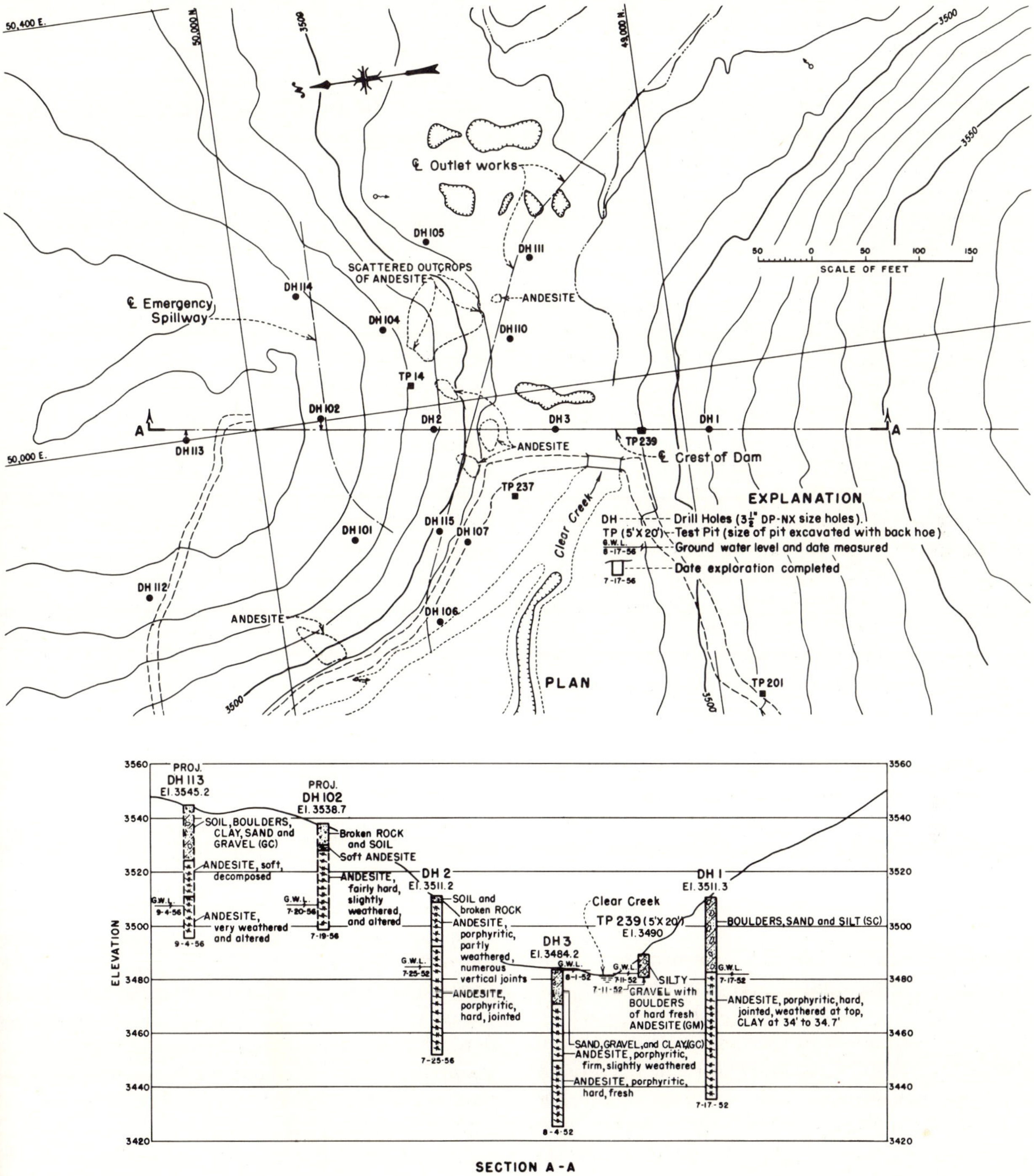

Figure 41. Geologic map and cross section of a damsite. 288–D–2470.

but modified or subdivided to show the distribution of materials with significantly different engineering or physical properties. Cross sections should be prepared showing the known and probable subsurface geologic features. Logs of all holes should be included. Figure

41 is an example of a geologic map of a damsite, and a section along the centerline of the proposed dam.

For preparation of specifications, additional boreholes may be required in the foundations to clarify critical geological questions or to pinpoint depth to rock for preparation of detailed construction drawings. Additional samples and laboratory tests may be required to establish a firm basis for design of soil foundations. During construction, the geologic map should be updated to show the conditions actually revealed in construction excavations. Such a map may prove to be an invaluable "as built" record in subsequent years of operation and maintenance if geologically oriented problems such as anomalies in foundation behavior or excessive seepage are recognized.

82. *Embankment Soils.*—Some damsites require considerable excavation to reach a competent foundation and, in many cases, the excavated material is satisfactory for use in portions of the embankment. Excavations for spillway and outlet works will also produce varying amounts of usable material. A major portion of embankment yardage for an earthfill dam, however, nearly always must be borrowed.

Investigation for embankment materials is a progressive procedure, ranging from a cursory inspection during the reconnaissance stage to extensive studies of all possible sources of material prior to undertaking the final design. A reconnaissance of the materials situation at each prospective location should be made before a damsite is selected. Careful examination of existing maps, soil surveys, and air photos will usually locate the areas which should be examined in the field. Highway and railroad cuts, arroyos, and banks along stream channels will provide valuable clues to the nature of the materials underlying an area, and should be examined. It will rarely be necessary to excavate test pits or auger holes in the reconnaissance stage. Quantity determinations can be made by consideration of topographic features and by a few rough measurements, either on the ground or on maps. The reconnaissance report should include a sketch map showing the locations of available borrow areas with respect to the damsite, the character of the materials in each area, and the probable quantity of each. Local factors which would affect the use of a deposit should be discussed in the report. In addition to the engineering properties of the soils, many other facets should be considered, including: proximity, accessibility, natural moisture content, and workability of the material; costs of rights-of-way and stripping; thickness of deposits; destruction of scenic features, and adverse topography.

A systematic plan for locating borrow areas should be followed during the feasibility stage of investigations subsequent to the selection of the damsite. To avoid overlooking nearby areas, the prospecting should start from the damsite and extend outward in all directions. All potential borrow areas within 1 mile of the dam should be investigated before more distant sources are considered. Holes should be dug at approximately 500-foot centers on a rough grid system in all practicable locations within the 1-mile limit. Earth augers should be used wherever possible, but test pits should be dug where boulders are encountered. Holes should extend 25 feet below ground surface except where bedrock or water table is encountered prior to reaching that depth. Holes should be sampled and logged in accordance with the procedures given in parts G and H of this chapter. Exploration at probable locations of cutoff trenches, for foundation stripping, and for spillways and outlet works should be given high priority in the investigation plan, and more detailed work in these areas is justified because of their possible early use as sources of embankment materials. When it becomes evident that a sufficient quantity of suitable material is not to be found within 1 mile of the damsite, more distant areas should be investigated according to the plan.

The ultimate purpose of a detailed borrow pit investigation is to determine the depth of shovel cut and the distribution of the materials in the embankment. This can be accomplished if a sufficient number of holes is dug to fix adequately the soil profiles in the borrow area. The plotting of profiles on at least 500-foot centers will indicate whether additional holes are needed. It is evident that the more homogeneous the soil is in a borrow area, the fewer holes will be required to establish the tentative profile. Figure 42 is an example of exploration for embankment materials for a dam. Soil classifications should be verified by laboratory tests on representative samples of the various materials encountered. A few density-in-place tests should be made in each borrow area to determine the shrinkage factor to be applied between borrow pit and compacted embankment yardages. Procedures for this test are given in section 119.

Because of changes in plan, errors in estimating and other contingencies, large safety factors should be used in estimating quantities

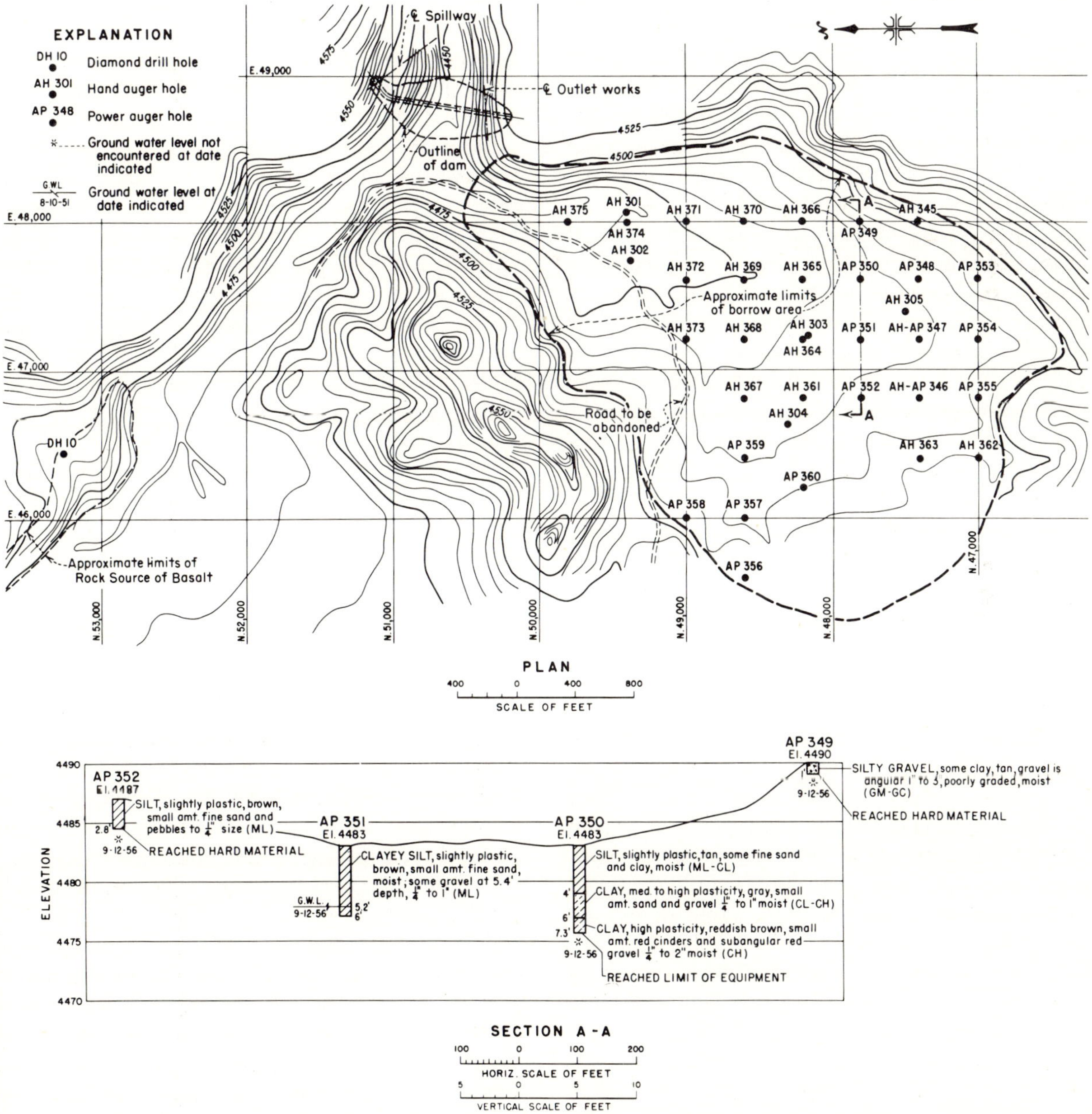

Figure 42. **Exploration for embankment materials—borrow area location map and typical cross section. 288–D–2471.**

available from borrow areas. The following criteria will assure adequate quantities: for a reconnaissance report, when it is estimated that less than 10,000 cubic yards of a type of material is needed, 10 times the estimated amount should be located; for requirements larger than 100,000 cubic yards, 5 times the estimated amount should be located. For feasibility estimates based on subsurface explorations, these safety factors can be cut in half. Even for well-explored borrow areas, at least 1½ times the required quantity customarily is outlined in the specifications in order to assure adequate quantities regardless of the contractor's choice of equipment and methods of excavation. Larger factors are often used when the existing information indicates the deposits are expected to be erratic.

83. *Riprap and Rockfill.*—Riprap is a layer of large, durable rock fragments. Its purpose is to preserve the shape of a surface slope or underlying structure by preventing erosion due to wave action or stream current. Rockfill is embankment constructed of rock fragments in portions of earthfill dams or in rockfill dams.

Search for suitable sources of riprap and rockfill is conducted in the same general sequence as is the search for earth embankment materials. Since riprap is almost essential for an earthfill dam, it is impractical to limit the extent of the area to be searched for it. Prospecting should extend radially outward from the site of the proposed work until a deposit of rock is located which is suitable in quality and sufficient in volume to fulfill the anticipated requirements. The best possible use should be made of existing data, such as geologic maps, air photos, topographic maps, and publications of State, Federal, or private agencies. From a study of these data, existing quarries, outcrops, and other promising areas can be marked on the map or photograph for later investigation in the field. It is often profitable to inquire about rock deposits or accumulations from residents of the locality, and from officials of local governments.

The primary criteria for riprap are quality and size of the rock fragments. Those who perform the investigations should attempt, by inspection, to evaluate the ability of the rock to resist wave action, freezing and thawing, and other disintegrating forces, and to determine whether the deposit will yield sufficient material of the required sizes. The most obvious place to begin exploration of the rock source is at any outcrop. Existing vertical faces cut back to unweathered material should be thoroughly examined for strain cracks, fracture patterns, bedding planes, and zones of unsound material. The joint and bedding plane systems are especially important because they indicate the sizes which probably will be produced in quarrying. Exploration required for determining the characteristics of the unexposed portions of a potential riprap source is usually accomplished with boreholes, test pits, or trenches. Core drilling is normally the most practicable and reliable method of determining the extent of the deposit, the depth of overburden, and the thickness of rock. The cores recovered provide information about the quality of the rock in portions of the deposit not otherwise exposed. In the example shown in figure 42, only one borehole was needed in the rock source to establish the depth of the extensive exposed lava flow.

Where bedrock formations are not of suitable quality for riprap, other sources must be investigated. There have been several cases where surface boulders were gathered and used for riprapping earthfill dams because quarry rock of suitable quality could not be found even within 100 miles of the site. The use of this type of riprap is normally feasible only when the boulders occur in fairly well concentrated accumulations, and when they cover areas sufficiently large to provide significant quantities. Nevertheless, the exploitation of several widely separated accumulations to yield the quantity required for one dam is not uncommon.

Occasionally, talus slopes are found which contain durable rock of the required sizes, and which are of sufficient extent to make quarrying from the source deposit unnecessary. Such slopes are especially desirable when they are more accessible than the source deposit, as is often the case. Little can be done in the way of exploration of talus material except for making a thorough examination and survey to determine the characteristics of the rock, the quantity available, and the range of sizes. Good photographs are part of the exploratory data for all riprap and rock sources, and they are especially valuable when talus slopes are being considered. Figure 43 shows a talus deposit of igneous rock suitable for riprap. of a superior rock at considerably greater expense. On the other hand, the use of lesser quantities of superior materials might offset their added cost. Information on the sampling of riprap sources is given in part G of this chapter. The suitability of rockfill mate-

Figure 43. Typical talus accumulation at base of basalt cliffs, Grant County, Washington. 3PPG—Aug—2—1969—1.

The availability of riprap or rockfill materials has a pronounced effect on the design of a structure; consequently, very careful studies of quantities must be made. It is occasionally possible to make use of a readily accessible material rather than to require the procurement rials is based on identification and field examination of the source.

84. *Concrete Aggregate.*—Field investigations for concrete materials prior to construction are confined chiefly to prospecting for aggregate and to explorations and sampling of available deposits. Those engaged in prospecting work should be familiar with the effects of grading, physical characteristics, and composition of aggregate on the properties of concrete. Judgment and thoroughness in conducting preliminary field investigations are usually reflected in durability and economy of the finished structures.

Most factors pertaining to the suitability of aggregate deposits are related to the geological history of the region. These factors include size, shape, and location of the deposit; thickness and character of the overburden; types and condition of the rock; grading, rounding, and degree of uniformity of the aggregate particles; and groundwater level. Aggregate may be obtained from deposits of natural sand and gravel, from talus, or from quarries in areas of bedrock outcroppings. Fine blending sand may sometimes be obtained from windblown deposits.

Stream deposits are the most common and generally most desirable because: (1) individual pieces are usually rounded; (2) streams exercise a sorting action which may improve grading; and (3) abrasion caused by stream transportation and deposition leads to a partial elimination of weaker materials. Alluvial fans are frequently used as sources of aggregate, but they often require more than usual processing. Glacial deposits provide sand and gravel, but they are generally restricted to northerly latitudes or high elevations. Those glacial deposits not influenced by fluvial agencies are usually too heterogeneous to be suitable as aggregate and at best are usable only after elaborate processing. Glacial deposits which have been affected by stream action frequently yield satisfactory aggregate materials.

When natural sand and gravel are not available it is necessary to produce concrete aggregate by quarrying and processing rock. Quarrying normally is done only when other materials of adequate quality and size cannot be obtained economically. Quarry deposits frequently contain stratified materials which make it difficult to obtain representative samples of the undeveloped source. Also, the presence of layers or zones of undesirable materials, such as clay or shale, in some instances necessitates selective quarrying and special processing.

The extent and justifiable expense of exploration for concrete aggregate are determined largely by the size of the job and the purpose of the structure. When searching for suitable aggregate it is important to remember that ideal materials are seldom found. Deficiencies or excesses of one or more sizes are very common. Objectionable rock types, coated and cemented particles, or particles of flat or slabby shape may occur in excessive amounts.

The most promising deposits should be explored and sampled by means of cased test holes, open test pits, or trenches, and the suitability of the aggregate should be determined. The methods of subsurface exploration, sampling, logging, and testing are given in parts F, G, H, and I, respectively, of this chapter. The quality and gradation requirements for aggregates are discussed in appendix F.

B. SOURCES OF INFORMATION

85. *Topographic Maps.*—A topographic map is indispensable in the design and construction of a dam. Topographic maps are of considerable value in the exploration of dam foundations and when exploring for construction materials for dams. The locations and elevations of exploratory holes, outcrops, and erosional features can be placed on the map, while the landforms portrayed by the contours will indicate to some degree the type of soil. Information on the origin and characteristics of some of the simpler landforms is given in part E of this chapter. Before undertaking the job of map making, a thorough search should be made for the existence of maps covering the areas of the reservoir, the damsite, and potential sources of construction materials. The U.S. Geological Survey (USGS) should be contacted for information on the availability of maps. This organization is making a series of standard topographic maps to cover the United States, Alaska, Hawaii, Puerto Rico, Virgin Islands, Guam, American Samoa, and the Trust Territory of the Pacific Islands.

The unit of survey for the USGS maps is a quadrangle bounded by parallels of latitude and meridians of longitude. Quadrangles covering 7.5 minutes of latitude and longitude are generally published at the scale of 1:24,000 (1 inch equals 2,000 feet). Quadrangles covering 15 minutes of latitude and longitude are published at the scale of 1:62,500 (1 inch equals approximately 1 mile). A series of topographic maps at the scale of 1:250,000 (1 inch equals approximately 4 miles) has also been published, in units of 1° of latitude and 2° of longitude, and covers the entire country. Many special maps are published at other scales. Each quadrangle is designated by the name of a city, town, or prominent natural feature within it, and on the margins of the map are printed the names of adjoining quadrangle maps that have been published.

The distinctive characteristic of topographic maps is that the shape of the land is portrayed by contours, which are imaginary lines following the ground surface at a constant elevation. The contour interval is the regular elevation difference separating adjacent contour lines on the map. These intervals depend on the ground slope and the map scale; they vary from 5 to 200 feet. Shaded-relief editions of selected quadrangle maps are also published, on which the relief is shown both by brown contour lines and by shading in brown to further accentuate physical features on the maps.

Published maps of the USGS are colored to distinguish classes of map features. Black is used for manmade or cultural features, such as roads, dams, buildings, names, and boundaries. Blue is used for water or hydrographic features, such as lakes, rivers, canals, and glaciers. Brown is used for relief or hypsographic features—land shape portrayed by contours or shading. Green is used for wooded areas with typical pattern to show scrub, vineyards, and orchards. Red emphasizes important roads and shows built-up urban areas and public land subdivision lines.

In addition to published topographic maps, information of great assistance to engineers is available from the USGS for mapped areas. For example, the locations and true geodetic positions of triangulation stations and elevations of permanent benchmarks are recorded. Also, map manuscripts at the 1:24,000 scale may be available 1½ to 2 years prior to publication of the final map. Figure 44 shows the status of 7½- and 15-minute topographic mapping in the United States. A large index map similar to this is available without charge from the USGS.

River survey maps are often helpful to the designer. These are strip maps which show the course and fall of the stream and nearby topographic and cultural features. River survey maps were prepared largely in connection with the classification of public lands for water resource development, and most of them are of rivers in the western States. If a valley is less than a mile wide, topography is shown to 100 feet or more above the water surface; if the valley is flat and wide, topography is shown for a strip of 1 to 2 miles.

Potential reservoir sites are mapped to the probable flow line of the reservoir. The usual scale is 1:24,000, and the normal contour interval is 20 feet on land and 5 feet on the water surface. Many of these maps include damsites on a large scale and have a profile of the stream. The standard size is 22 by 28 inches.

The availability of river survey maps and other special maps, including national parks and monuments, and a list of agents for topographic maps are indicated on the index to topographic mapping for the various States. These indexes are available without charge from the USGS. Requests for indexes and inquiries concerning published maps and the availability of map manuscripts and related information should be directed to U.S. Geological Survey, Denver Federal Center, Denver, Colo. 80225, or U.S. Geological Survey, Washington, D.C. 20242.

86. *Geologic Maps*.—Considerable useful engineering information is obtainable from geologic maps. These maps identify the rock units and geologic structures directly underlying the reservoir and damsite, and in the surrounding area. The characteristics of rocks are of major importance in the selection of a damsite and in the design of the dam. Conditions beneath the surface can often be correctly deduced from the 3-dimensional information given on geologic maps.

On geologic maps, rocks are identified by their lithologic character and geologic age. The smallest rock unit mapped is generally a formation. This is a body of rock of relatively uniform lithology that extends over a fairly large area and that can be clearly differentiated from overlying or underlying units. The areal extent of these formations is indicated on geologic maps by means of letter symbols, color, and symbolic patterns.

Letter symbols indicate the formation and geologic period. For example, "Jm" might stand for the Morrison Formation of the Jurassic period. In general, standard color and pattern conventions are followed on maps produced by the USGS. Tints of yellow and orange are used for different Cenozoic rocks, tints of green for Mesozoic rocks, tints of blue and purple for Paleozoic rocks, and tints of russet and red for pre-Cambrian rocks. The primary lithologic features of the rock types are depicted, as far as practicable, by conventional patterns. Variations of dot and line patterns are used for sedimentary rocks; wavy lines for metamorphic rocks; and checks, crosses, or crystallike patterns for igneous rocks. Geologic maps indicate the subsurface geologic structure by showing the attitude or orientation of the rock strata or other planar features. The dip-strike symbol, for example, indicates the inclination of a rock bed from the horizontal in degrees, and other symbols refer to flow structure, faults, folds, and other features.

Geologic maps commonly carry one or more geologic structure sections showing the disposition of the various rock units in depth along an arbitrary line usually marked on the map. Geologic sections are somewhat hypothetical and must be used with caution. The vertical scale of some sections is exaggerated, but most are not. Sections prepared solely from surface data are not as accurate as those prepared from subsurface data obtained from drilling or mining records. A columnar section shows only the succession of strata and not the structure of the beds, as does the geologic structure section.

There are several types of geologic maps. A map showing a plan view of the bedrock in the area is a *bedrock geologic map*. Such a map indicates the boundaries of the visible formations and the inferred distribution of those units covered by soil or plant growth, and usually includes one or more geologic sections. Except for indicating thick deposits of alluvium, many bedrock geologic maps do not show soil or unconsolidated mantle. In areas of complex geology where exposures of bedrock are scarce, the location of the contacts between formations is often indicated as approximate or hypothetical. *Surficial geologic maps* differentiate the unconsolidated surface materials of the area according to their origin, such as stream alluvium, glacial gravel, and windblown sand. These maps indicate the areal extent, characteristics, and geologic age of the surface materials. Geologic maps of moderately deformed areas carry enough structure symbols to give an understanding of the geologic structure. In highly complex areas, however, where a great amount of structural data is necessary for an interpretation of the geologic structure, special *structural geologic maps* are prepared.

Most geologic maps are accompanied by explanations giving the relative ages and brief descriptions of the units distinguished on them. Where descriptions of units are lacking, an experienced geologist can determine their characteristics by making analogies with other areas. For more certain identification of the lithology and for details, geologic literature on the whole area must be consulted.

Engineering information can be obtained from geologic maps if the user possesses a knowledge of the fundamentals of geology and an understanding of how engineers use geologic facts in design and construction. By a study of the basic geologic map, together with all the collateral geologic data that pertain to an area, it is possible to prepare a special map that interprets the geology in terms of its engineering characteristics. Suitability of formations for construction materials, foundation

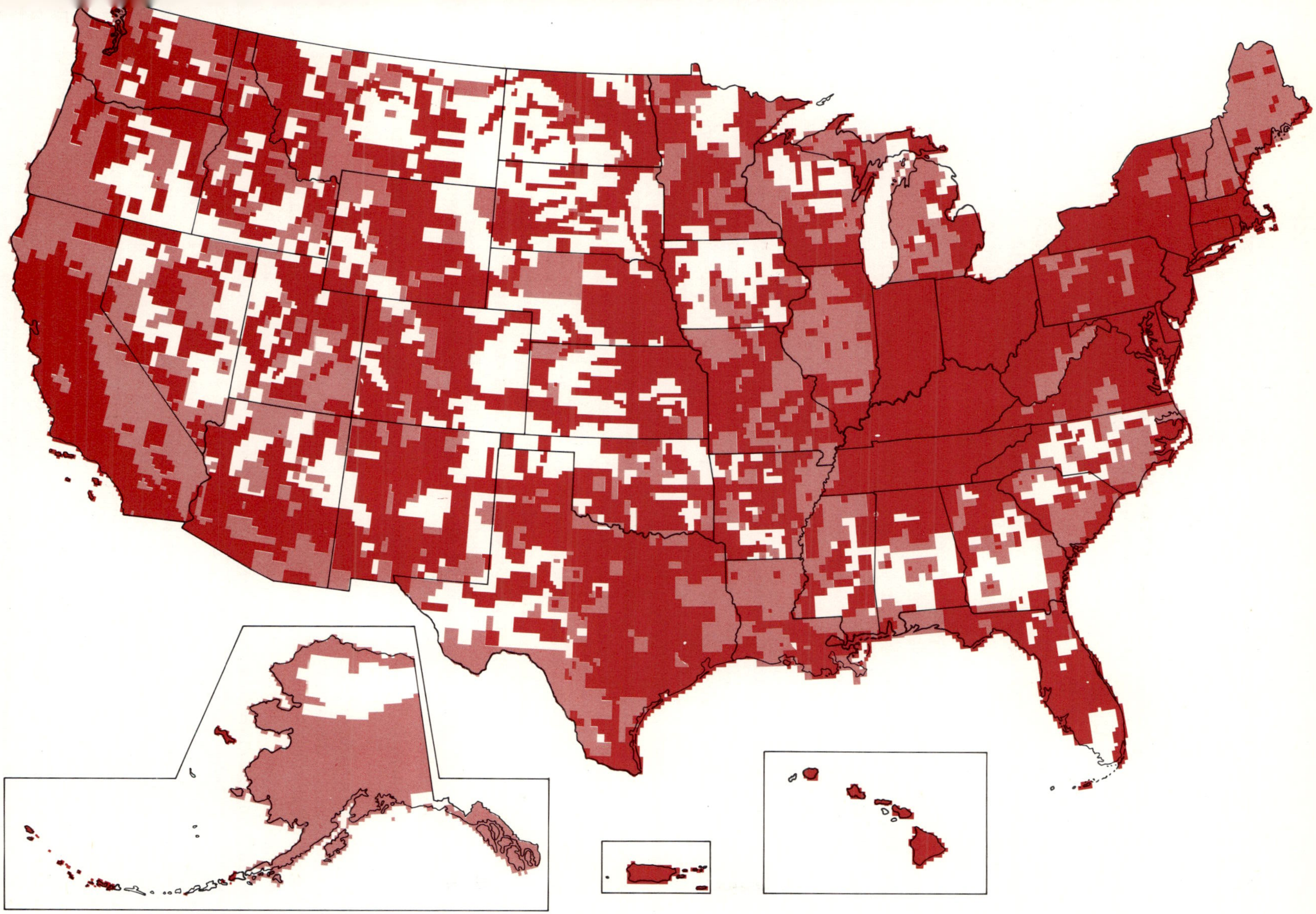

Figure 44. Status of 7-½- and 15-minute topographic quadrangle mapping, 1969. Dark red color indicates 7-½-minute mapping; light red indicates 15-minute mapping. (U.S. Geological Survey.) 288–D–2772, 288–D–2773, and 288–D–2774.

and excavation conditions, as well as surface- and ground-water data, can be interpreted from geologic maps. Such information is valuable in preliminary planning, but is not a substitute for detailed field investigations in the feasibility and specifications stages.

The USGS now publishes a series of Geologic Quadrangle Maps of the United States, which replaces the earlier folios of the Geologic Atlas of the United States, published from 1894 to 1945. Maps in this series are large scale (1:62,500 or larger), and are printed in color. Most have structure sections and other graphic means of presenting geologic data and a brief explanatory text. Full descriptions of the areas shown on these maps and detailed interpretations of geologic history are commonly compiled in other publications, such as the Bulletins and Professional Papers of the USGS.[2] The Geological Survey also publishes geologic maps under series known as "Miscellaneous Geologic Investigations Maps," "Mineral Resources Maps and Charts," "Geophysical Investigations," and "Hydrologic Investigation Atlases." Maps in these series have a wide range of scale and format to meet specific purposes.

There are several geologic maps of special interest to designers of dams. A series of such maps resulted from geologic mapping and general resources investigations conducted by the USGS as part of the Department of the Interior plan for study and development of the Missouri River Basin. These include maps showing construction materials and nonmetallic mineral resources, including sand and gravel deposits, of several of the States in the Missouri River Basin. The USGS has also published a set of six maps entitled "Interpreting Geologic Maps for Engineering Purposes, Hollidaysburg Quadrangle, Pennsylvania," 1953, containing examples of how geologic maps are used to solve engineering problems, including the problem of selection of a damsite. These maps are out of print, but should be available in major public libraries.

Figure 45 shows the status of geologic mapping in the United States. More detailed information about published geologic maps for individual States is given in the series of geologic map indexes available from the USGS. Each published geologic map is outlined on a State base map, with an explanatory key giving the source and date of publication, the author, and the scale. The attention of all those engaged in searching for geological information should be called to the "Directory of Geological Material in North America"[1].[3] This directory includes comprehensive lists of sources of available maps, charts, air photos, logs, cores, etc., for each State and Territory of the United States, as well as for provinces in other countries of North America.

87. Agricultural Soil Maps.—A large portion of the United States has been surveyed by the Department of Agriculture. These investigations are surficial, extending to depths up to 6 feet, and consist of classifying soils according to color, structure, texture, physical constitution, chemical composition, biological characteristics, and morphology. The Department of Agriculture publishes reports of their surveys in which the different soils are described in detail, and their suitability for various crops is given. Included in each report is a map of the area surveyed, usually a county, showing by the pedological classification the various types of soils that occur. In addition to the county soil maps, there are many areas in which individual farms are mapped using the same system of classification.

Figure 46 shows the extent of published soil surveys in the United States. These surveys (if in print) are available for purchase from the Superintendent of Documents, Washington, D.C. Out-of-print maps and other unpublished surveys may be available for examination from the U.S. Department of Agriculture, county extension agents, colleges, universities, and libraries. The Bureau of Reclamation has made surveys for irrigation suitability in most river basins in the 17 western States. These were made for the purpose of identifying and classifying lands for irrigation. Maps and reports on these surveys are available in local project or regional offices of the Bureau

[2] Publications of the U.S.G.S. are available from the Superintendent of Documents, Washington, D.C. 20242.

[3] Numbers in brackets refer to items in the bibliography, section 122.

UNIFIED SOIL CL
INCLUDING IDENTIFICATI

RES s on estimated weights)		GROUP SYMBOLS ⓑ	TYPICAL NAMES
and substantial amounts ticle sizes.		GW	Well graded gravels, gravel-sand mixture little or no fines.
or a range of sizes te sizes missing.		GP	Poorly graded gravels, gravel-sand mixtur little or no fines.
dentification procedures		GM	Silty gravels, poorly graded gravel-sand silt mixtures.
fication procedures		GC	Clayey gravels, poorly graded gravel-san clay mixtures.
s and substantial ediate particle sizes.		SW	Well graded sands, gravelly sands; little o no fines.
or a range of sizes with es missing.		SP	Poorly graded sands, gravelly sands; litt no fines.
entification procedures		SM	Silty sands, poorly graded sand-silt mixture
ication procedures		SC	Clayey sands, poorly graded sand-clay mixt
ER THAN No. 40 SIEVE SIZE			
ANCY TION AKING)	TOUGHNESS (CONSISTENCY NEAR PLASTIC LIMIT)		
to slow	None	ML	Inorganic silts and very fine sands, rock flo or clayey fine sands with slight plasticity
ery slow	Medium	CL	Inorganic clays of low to medium plasticity, clays, sandy clays, silty clays, lean clays
ow	Slight	OL	Organic silts and organic silt-clays of lov plasticity.
none	Slight to medium	MH	Inorganic silts, micaceous or diatomaceous sandy or silty soils, elastic silts.
ne	High	CH	Inorganic clays of high plasticity, fat clays
very slow	Slight to medium	OH	Organic clays of medium to high plasticity
, odor, spongy feel and texture.		Pt	Peat and other highly organic soils.

ristics of two groups are designated by combinations of group symbols. For example GW

NUARY 1952

Figure 56. **Unified Soil Classification Chart. (She**

Major divisions			FIELD IDENTIFICATION PROCEDU (Excluding particles larger than 3 inches and basing fractio	
COARSE GRAINED SOILS More than half of material is larger than No. 200 sieve size [2]. (The No. 200 sieve size is about the smallest particle visible to the naked eye)	GRAVELS More than half of coarse fraction is larger than No. 4 sieve size. (For visual classifications, the $\frac{1}{4}$" size may be used as equivalent to the No. 4 sieve size.)	CLEAN GRAVELS (Little or no fines)	Wide range in grain siz of all intermediate pa	
			Predominantly one siz with some intermedi	
		GRAVELS WITH FINES (Appreciable amount of fines)	Non-plastic fines (for see ML below).	
			Plastic fines (for iden see CL below).	
	SANDS More than half of coarse fraction is smaller than No. 4 sieve size.	CLEAN SANDS (Little or no fines)	Wide range in grain siz amounts of all intern	
			Predominantly one siz some intermediate si	
		SANDS WITH FINES (Appreciable amount of fines)	Non-plastic fines (for i see ML below).	
			Plastic fines (for ident see CL below).	
			IDENTIFICATION PROCEDURES ON FRACTION SMAL	
			DRY STRENGTH (CRUSHING CHARACTERISTICS)	DIL (REA TO S
FINE GRAINED SOILS More than half of material is smaller than No. 200 sieve size.	SILTS AND CLAYS Liquid limit less than 50		None to slight	Quic
			Medium to high	None to
			Slight to medium	S
	SILTS AND CLAYS Liquid limit greater than 50		Slight to medium	Slow
			High to very high	N
			Medium to high	None t
HIGHLY ORGANIC SOILS			Readily identified by co frequently by fibro	

[1] Boundary classifications:- Soils possessing charact

[2] All sieve sizes on this chart are U.S. standard.

ADOPTED BY - CORPS OF ENGINEERS AND BUREAU OF RECLAMATION -

ASSIFICATION
ON AND DESCRIPTION

	INFORMATION REQUIRED FOR DESCRIBING SOILS			LABORATORY CLASSIFICATION CRITERIA	
;	Give typical name; indicate approximate percentages of sand and gravel; max. size; angularity, surface condition, and hardness of the coarse grains; local or geologic name and other pertinent descriptive information; and symbol in parentheses.	Use grain size curve in identifying the fractions as given under field identification	Determine percentages of gravel and sand from grain size curve. Depending on percentage of fines (fraction smaller than No. 200 sieve size) coarse grained soils are classified as follows:- Less than 5% GW, GP, SW, SP, More than 12% GM, GC, SM, SC, 5% to 12% Borderline cases requiring use of dual symbols.	$C_U = \frac{D_{60}}{D_{10}}$ Greater than 4 $C_C = \frac{(D_{30})^2}{D_{10} \times D_{60}}$ Between one and 3	
s,				Not meeting all gradation requirements for GW	
				Atterberg limits below "A" line, or PI less than 4	Above "A" line with PI between 4 and 7 are borderline cases requiring use of dual symbols.
–	For undisturbed soils add information on stratification, degree of compactness, cementation, moisture conditions and drainage characteristics.			Atterberg limits above "A" line with PI greater than 7	
				$C_U = \frac{D_{60}}{D_{10}}$ Greater than 6 $C_C = \frac{(D_{30})^2}{D_{10} \times D_{60}}$ Between one and 3	
or	EXAMPLE:- Silty sand, gravelly; about 20% hard, angular gravel particles $\frac{1}{2}$-in. maximum size; rounded and subangular sand grains coarse to fine; about 15% nonplastic fines with low dry strength; well compacted and moist in place; alluvial sand; (SM)			Not meeting all gradation requirements for SW	
				Atterberg limits below "A" line or PI less than 4	Above "A" line with PI between 4 and 7 are borderline cases requiring use of dual symbols.
res.				Atterberg limits above "A" line with PI greater than 7	
r, silty	Give typical name; indicate degree and character of plasticity, amount and maximum size of coarse grains; color in wet condition, odor if any, local or geologic name, and other pertinent descriptive information; and symbol in parentheses.		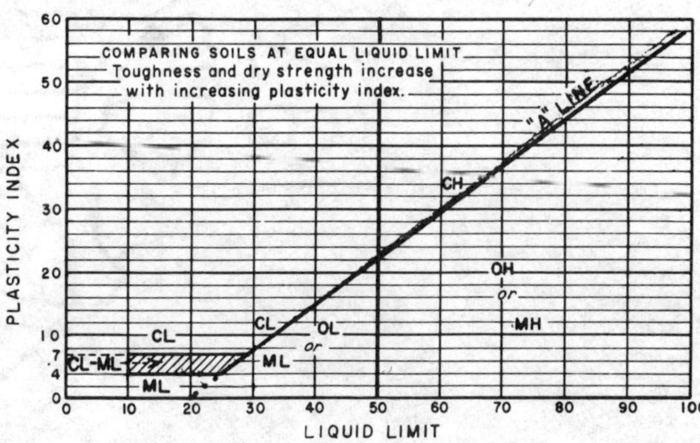		
gravelly					
	For undisturbed soils add information on structure, stratification, consistency in undisturbed and remolded states, moisture and drainage conditions.				
ine					
	EXAMPLE:- Clayey silt, brown; slightly plastic; small percentage of fine sand; numerous vertical root holes; firm and dry in place; loess; (ML)		PLASTICITY CHART FOR LABORATORY CLASSIFICATION OF FINE GRAINED SOILS		

GC, well graded gravel-sand mixture with clay binder.

t 1 of 2.) From drawing no. 103–D–347.

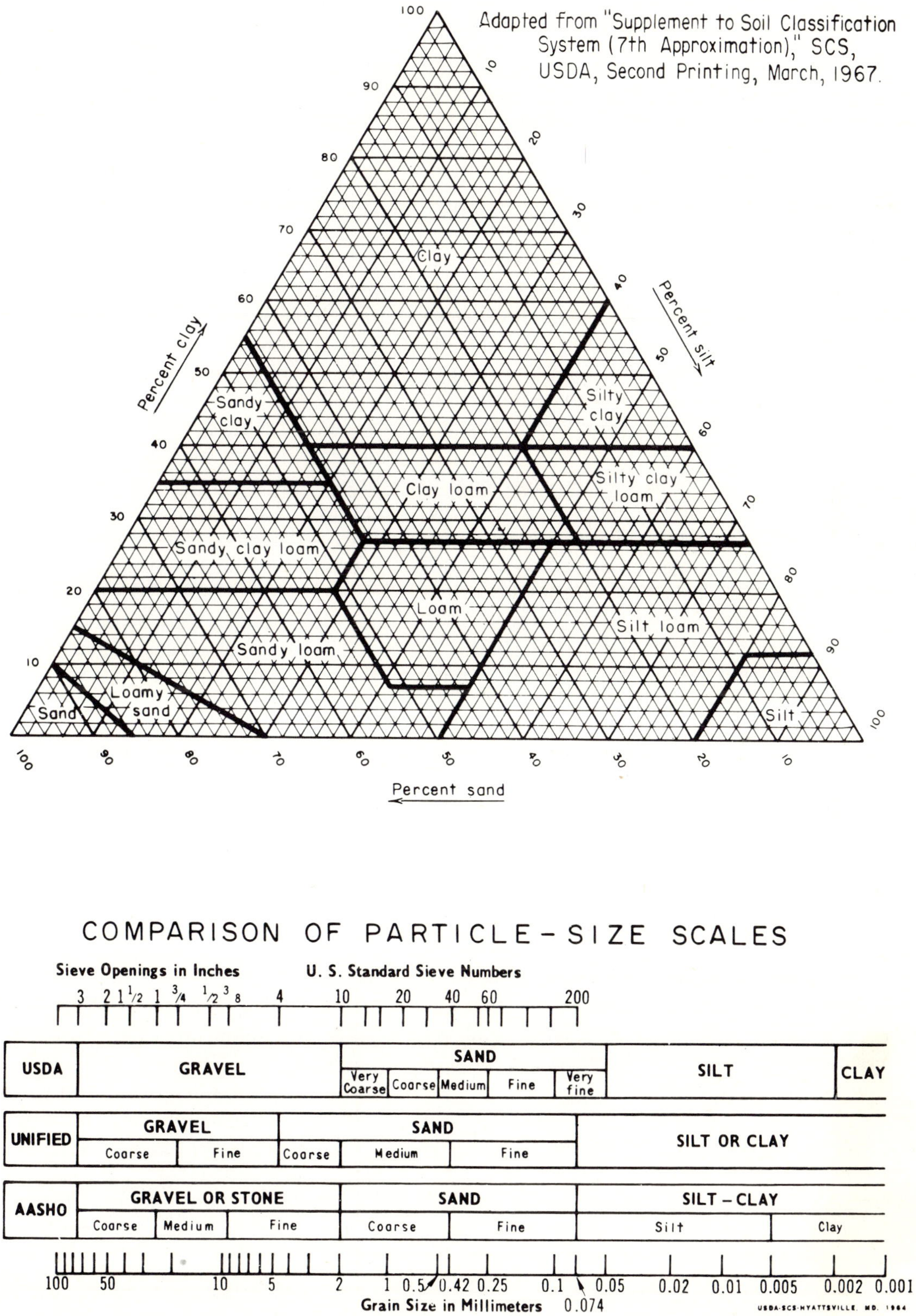

Figure 47. Soil triangle of the basic soil textural classes. (U.S. Soil Conservation Service.) 288—D—2782.

of Reclamation. These data are quite useful for engineering purposes since they provide detailed soil data, along with considerable information relating to the geology and drainage conditions. In addition, the data concerning soils usually include laboratory and field measurements of chemical and physical properties to at least a 10-foot depth and often some information on deeper materials.

In order to apply soil survey maps to explorations of foundations and construction materials, some knowledge of soil classification systems is necessary. For several years, the Department of Agriculture has been developing a new classification system. This system, published in 1960, is titled "Soil Classification, a Comprehensive System, Seventh Approximation," and is referred to as simply the "seventh approximation."

The system represents a new approach to soil classification, and includes a substantial number of new and unusual descriptive words for identifying the various categories included in the system. It does not accept the established Great Order Groups, but creates a new set of composite entities built up from soil characteristics that are defined so as to provide the basis for the groupings.

The category at the highest level of generalization is the "order." There are 10 orders. The second category is the "suborder." This narrows the ranges in soil moisture and temperature regimes, kinds of horizons, and composition, according to which is most important. Additional categories in descending order include "great groups," "subgroups," "families," and "series." The soil "series" is a commonly used term and known to most engineers. It refers to a group of soils having horizons similar in differentiating characteristics and arrangements in the soil profile, except for texture of the surface portion.

The final subdivision of a soil series is the soil "phase." This is a subdivision of soil that has characteristics affecting its use and management, but which does not vary sufficiently to differentiate it as a separate soil series. Phases of soil series are the major components of the soil mapping units currently being shown on detailed soil maps.

Although not a part of the soil classification system presently being used, soil types, a subdivision of soil series which is based on surface texture, are commonly used in most of the published soil survey data.

The textural classification system used in describing soil types or phases in USDA soil survey reports or in Bureau of Reclamation irrigation surveys is shown in figure 47. This is different from the Unified Soil Classification System commonly used by the Bureau of Reclamation for engineering purposes (see part C of this chapter). Figure 47 shows the terminology used for different percentages of clay (defined as particles smaller than 0.002 mm.), silt (0.002 to 0.05 mm.), and sand (0.05 to 2.0 mm.). Note the use of the term "loam", which is defined in the chart as a mixture of sand, silt, and clay within certain percentage limits. Other terms used as adjectives to the names obtained in the triangle classification are: "gravelly" for rounded and subrounded particles from 2 millimeters to 3 inches, "cherty" for gravel sizes of chert, and "stony" for sizes greater than 10 inches.

The textural classification given as part of the soil name in soil types or phases refers to the material in the A horizon only and therefore is not indicative of the entire soil profile. However, the combination of a series name and textural classification provides a considerable amount of significant data. For each soil series, the texture, degree of compaction, presence or absence of hardpan or rock, lithology of the parent material, and chemical composition can be obtained. Similar and frequently more detailed data for specific sites can often be obtained from irrigation suitability surveys made by the Bureau of Reclamation. From the engineering point of view, both the USDA reports and the Bureau of Reclamation surveys have some limitations but can often be used to advantage in reconnaissance studies.

The use and limitation of recent soil survey data for engineering purposes can be judged by the following example taken from "Soil Survey of Hitchcock County, Nebraska"

[4] USDA, Soil Conservation Service, "Soil Survey of Hitchcock County, Nebraska;" published by USDA in cooperation with the University of Nebraska, April 1970.

The "McCook Series" is described in the following manner:

McCook Series (Order—Mollisols; Subgroup—Entic Haplustoll; Family—Mixed Mesic):

The McCook series consists of well-drained, medium-textured, nearly level soils that are underlain by coarse sandy sediments at a depth of 2½ to 12 feet. These soils are on bottom lands along all of the major streams. They are generally stratified. These are the most extensive bottom-land soils in the county.

These soils are very friable and are easily tilled. They have moderate permeability and moderate water-holding capacity. Although they range to strongly alkaline in the lower part, plant growth is not restricted. Erosion is normally not a hazard. Corn and sorghum are the chief crops, but all local crops can be grown. Much of the acreage is irrigated. Water is obtained from shallow wells or from irrigation canals.

A typical profile has a 14-inch surface layer of grayish-brown loam or silt loam. The underlying stratified alluvium ranges from light brownish-gray silt loam to light-gray very fine sandy loam. It has weak, medium and fine, subangular blocky structure. Buried layers are common. Calcium carbonate occurs throughout the profile.

A typical profile, in a cultivated field having a slope of less than 1 percent, 0.3 mile east and 0.2 mile south of the northwest corner of sec. 17, T 2 N, R 34 W:

Ap — 0 to 6 inches, grayish-brown (10YR 5/2)[5] loam, very dark grayish brown (10YR 3/2) when moist; weak, fine, granular structure; soft when dry, very friable when moist; calcareous and slightly effervescent; abrupt, smooth boundary.

Al — 6 to 14 inches, grayish-brown (10YR 5/2) silt loam, very dark grayish brown (10YR 3/2) when moist; weak, medium and fine, subangular blocky structure; soft when dry, very friable when moist; calcacerous and strongly effervescent; gradual, wavy boundary.

C1 — 14 to 26 inches, light brownish-gray (10YR 6/2) silt loam, dark brown (10YR 4/3) when moist; weak, medium and fine, subangular blocky structure; soft when dry, very friable when moist; calcareous and strongly effervescent; gradual, wavy boundary.

C2 — 26 to 35 inches, light brownish-gray (10YR 6/2) loam, dark brown (10YR 4/3) when moist; weak, medium and fine, subangular blocky structure; soft when dry, very friable when moist; calcareous and strongly effervescent; clear, smooth boundary.

Ab — 35 to 44 inches, dark grayish-brown (10YR 4/2) loam, very dark brown (10YR 2/2) when moist; weak, medium and fine, subangular blocky structure; soft when dry, very friable when moist; calcareous and strongly effervescent; clear, smooth boundary.

C3 — 44 to 60 inches +, light-gray (10YR 7/2) very fine sandy loam, grayish brown (10YR 5/2) when moist; weak, medium and fine, subangular blocky structure; soft when dry, very friable when moist; calcareous and violently effervescent.

The A horizon ranges from 8 to 22 inches in thickness. The depth to coarse sandy sediments ranges from 2½ to 12 feet. The depth to the water table ranges from 4 to more than 10 feet.

McCook soils are associated with Haverson, Glenberg, and Las soils. They have a darker colored surface layer than Haverson soils and a finer textured surface layer than Glenberg soils. They are better drained than Las soils and do not contain the salts or alkali that are characteristic of those soils.

Considerable useful information is contained in modern soil profile descriptions, as exemplified in the example shown above, and in modern soil survey reports such as the one cited. Not only is information reported which will be useful to farmers in the agricultural community, but engineering interpretations can be made relating to such properties as suitability of the soil for road subgrades, road fills, building foundations, dikes, levees, and embankments.

[5] The term (10YR 5/2) refers to a standard color code used to describe soils; this code is contained in the "Soil Survey Manual", USDA Handbook No. 18, August 1951, pp. 194-203.

In addition, the modern soil survey reports show not only the USDA textural classification, but also includes the Unified and AASHO classifications for all soils included in the report. Such reports are quite useful for both agricultural and engineering purposes. Figure 47 shows a comparison of particle size scales for the Unified, AASHO, and USDA soil classification systems.

Current soil survey reports also contain an engineering properties section. Engineering test data are given concerning maximum dry density, optimum moisture content, mechanical analysis, liquid limit, plasticity index, permeability, pH value, and shrink-swell potential. These tests are performed by the U.S. Bureau of Public Roads in accordance with standard AASHO procedures. In addition to the USDA textural classification, these reports also give the Unified and AASHO classifications. Figure 47 shows a comparison of particle size scales for the Unified, AASHO, and USDA soil classification systems. Current reports also contain engineering interpretations of the suitability of the various soils for road subgrades, road fills, building foundations, dikes, levees, and embankments. Other sections of the soil survey report concerning "Formation and Classification of the Soils," and "General Nature of the Area" contain very useful engineering information.

88. *Airphotos*.—An airphoto or aerial photograph is a pictorial representation of a portion of the earth's surface taken from the air. It may be a vertical photograph in which the axis of the camera is vertical, or nearly so, or an oblique photograph where the axis of the camera is more or less inclined. High oblique photographs include the horizon; low obliques do not. The vertical photograph is commonly used as the basis for topographic mapping, agricultural soil mapping, and geological interpretations.

Except where dense forest cover obscures large areas from view, the airphoto reveals every natural and manmade detail on the surface of the ground. Relationships are exposed which could not be found on the ground no matter how careful the examination. Identification of features shown on the photo is facilitated by stereoscopic examination. The features are then interpreted for a particular purpose, such as geology, land utilization, or engineering characteristics. The experience and training of the engineer will determine his utilization of aerial photographs. Knowledge of the elements of geology and of soil science will assist him in interpreting airphotos for engineering uses. Airphotos are very often used for locating areas to be examined and sampled in the field and as substitutes for maps.

Virtually the entire area of the United States has been covered by aerial photography. A 28- by 42-inch index map of the United States is available free on application to the U.S. Geological Survey, Washington, D.C. 20242. This map shows which of seven Government agencies can provide prints for particular areas. When ordering photographs, specify contact prints or enlargements, glossy or matte finish, and location. Location should be given by range, township, and section, latitude and longitude, State and county, or shown on an enclosed index map of the area. Stereoscopic coverage should be requested for most uses. Aerial mosaics covering roughly 25 percent of the area of the United States are also available. A mosaic is an assemblage of aerial photographs whose edges have been torn or cut, and matched and mounted to form a continuous representation of the earth's surface. They include halftone photolithographic reproductions from mosaic negatives known as "photo maps." An index map showing the status of aerial mosaics of the United States, including the coverage and the agencies holding mosaic negatives, is available without charge on application to the U.S. Geological Survey.

Airphoto interpretation of earth materials and geologic features is relatively simple and straightforward, but requires experience. The diagnostic features include terrain position, topography, drainage and erosional features, color tones, and vegetative cover. Interpretation is limited mainly to surface and near-surface conditions. There are special cases, however, where features on the photograph permit reliable predictions to be made of deep, underground conditions. Although interpretation can be rendered from any sharp photograph, the scale is a limiting factor, since small-scale photos limit the amount of detailed information that can be obtained. The scale of 1:20,000 has been found satisfactory for engineering and geologic interpretation of surface materials. Large-scale photos often have application to highly detailed work, such as for reservoir clearing estimates, and for geologic reconnaissance mapping of damsites.

Airphotos can be used to identify certain terrain types and land forms. These topographic features are described in part E of this chapter. Stereoscopic photo inspection of an area, taking particular note of regional topography, local terrain features, and drainage conditions, will usually suffice to identify the common terrain types. This permits the possible range in the soil and rock materials to be anticipated, and their characteristics to be defined within broad limits.

Geologic features that may be highly significant to the location or performance of engineering structures sometimes can be identified from airphotos. In many instances these features can be more readily identified on the airphoto than on the ground. It must be recognized, however, that airphoto interpretation is applicable only to those features which develop recognizable surface expressions, such as drainage patterns and alinement of ridges or valleys. Joint systems, landslides, fault zones, folds, and other structural features sometimes

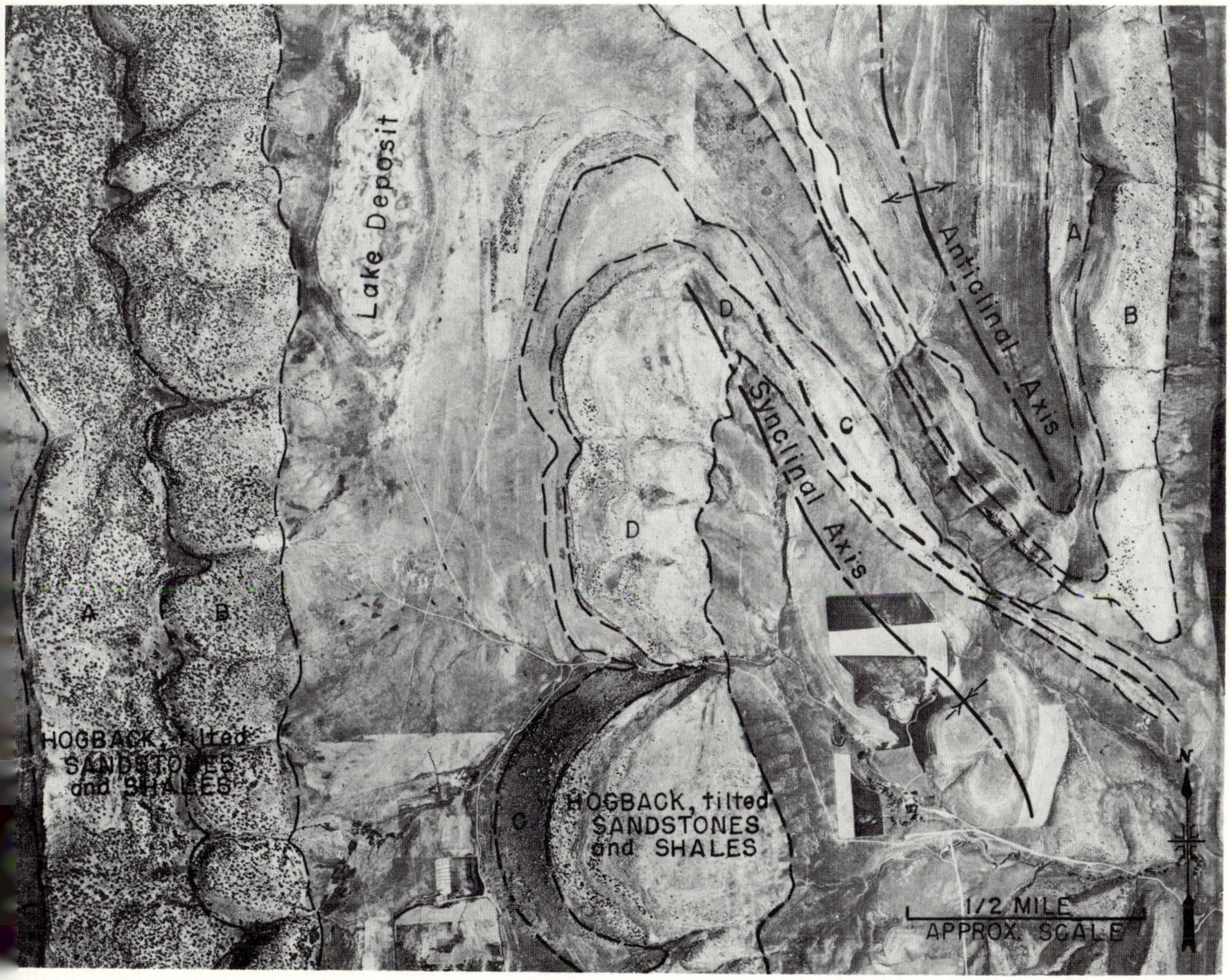

Figure 48. Rock strata illustrating folding in sedimentary rocks. (A) Satanka formation, (B) Lyons formation, (C) Morrison formation, and (D) Lower and Middle Dakota formation. (U.S. Forest Service.) PX–D–16265.

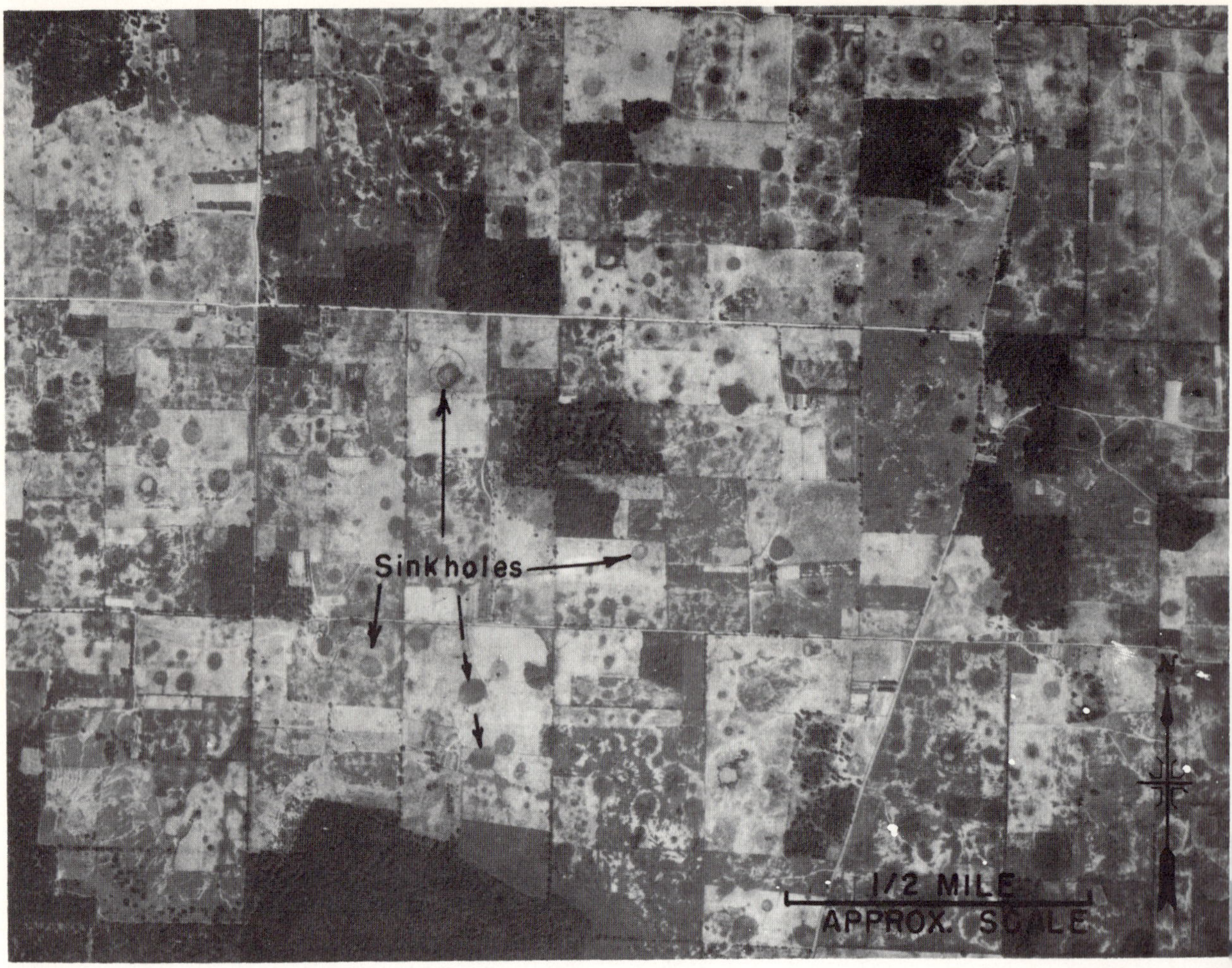

Figure 49. Sinkhole plain indicating deep plastic soils over cavernous limestone, developed in humid climate. (U.S. Agricultural Stabilization and Conservation Service.) PX-D-16264.

are identified quickly in an aerial photo, whereas it may be difficult to find them on the ground. The importance of these items in the location of a dam and its appurtenant works is obvious. The general attitude, bedding, and jointing of exposed rock strata, as well as the presence of dikes and intrusions, often can be interpreted in airphotos. Such information is valuable in appraising the possibilities of landslides into open cuts and of seepage losses in reservoirs.

Drainage patterns, particularly their type and density, provide an indication of the relative permeability of the earth materials. A dense, finely divided drainage pattern indicates an impervious soil area with high runoff and low infiltration. In contrast, the absence of a surface drainage pattern indicates a soil area with low runoff and high infiltration, provided the area is not a desert. The surface drainage pattern in areas of high water table has only limited significance as an indicator of the earth materials present. Definite alinements in the drainage pattern usually indicate control by local geologic structure.

Erosional features have significance in that they often reflect the textural characteristics of the exposed materials. Short, steep, V-shaped gullies with uniform gradients are associated with granular materials; long gullies with uniform gradients of rounded cross-sectional slopes are associated with fine-grained plastic soils. Silts and sand-clay materials usually exhibit gullies having U-shaped cross sections and compound gradients. The significance of gullies as an indicator of soil texture is modified by extreme climatic influence, such as in arid regions where "box" gullies seem to prevail irrespective of soil texture. Regardless of the climatic influence, however, changes in the gradient or cross section of gullies, or changes in the surface slope of eroded surfaces may indicate a change in the exposed soil, rock texture, or geologic structure.

Color tones (relative photographic gray values) have a general significance in that they reflect the soil moisture conditions and often reveal the relative position of the groundwater table. Light color tones are usually associated with well-drained soils, such as gravels, sands, and silts with groundwater levels well below the ground surface. Dark color tones usually indicate poorly drained organic clays and silty clays with groundwater levels near the ground surface. The significance of soil color in airphotos must be appraised from the overall color pattern, since some variation may be expected in the photographic tone quality of individual airphotos. It is also necessary to exclude, visually, the color tones produced by vegetative cover.

Vegetative cover is significant in that the vegetative patterns produced in the airphotos often reflect the nature of soil and moisture conditions. Also, a change in vegetative pattern may indicate a change in the type or texture of the underlying bedrock. The use of vegetative patterns as an indicator of soil conditions will prove most useful in extreme climates, such as in arctic, tropical, and arid regions where the combination of soil and climate is selective of vegetative growth. In arid regions the pattern of vegetation can be used to distinguish between high and low alkali soils and between high and low groundwater levels. The effective use of vegetation as an airphoto materials indicator requires a limited amount of field correlation.

Figures 48 and 49 are examples of readily identifiable geologic features from airphotos. Examples of typical land forms on airphotos are given in part E of this chapter.

C. SOIL CLASSIFICATION

89. *General.*—Most soils are a heterogeneous accumulation of mineral grains that are not cemented together. However, the term "soil" or "earth" as used by engineers includes virtually every type of uncemented or partially cemented inorganic and organic material found in the ground. Only hard rock which remains firm after exposure is wholly excluded. To the engineer engaged in design and construction of foundations and earthworks for dams, the physical properties of soils, such as unit weight, permeability, shearing strength, compressibility, and interaction with water, are of primary importance.

It is advantageous to have a standard method of identifying soils and classifying them into categories or groups which have distinct engineering properties. This enables engineers in the design office and those engaged in field work to speak the same language, thus facilitating exchange of information and experiences. Knowledge of soil classification, including typical engineering properties of soil of the various groups, is especially valuable to the engineer engaged in prospecting for earth materials or investigating foundations for structures. To a limited extent, proper soil classification can be used to estimate numerical values of engineering characteristics of soils for use in low dams where adequate safety factors are provided.

In 1952, the Bureau of Reclamation and the Corps of Engineers, with Prof. Arthur Casagrande of Harvard University as consultant, reached agreement on a modification of Professor Casagrande's airfield classification which they named the "Unified Soil Classification System." This system, which is particularly applicable to the design and construction of dams, takes into account the engineering properties of soils, is descriptive and easy to associate with actual soils, and has the flexibility of being adaptable both to the field and to the laboratory. Probably its greatest advantage is that a soil can be classified readily by visual and manual examination without the necessity for laboratory testing. The Unified Soil Classification System is based on the size of the particles, the amounts of the various sizes, and the characteristics of the very fine grains.

A soil mass consists of solid particles and pore fluids. The solid particles generally are mineral grains of various sizes and shapes, occurring in every conceivable arrangement. These solid particles can be divided into various components, each of which contributes its share to the physical properties of the whole. Soil classification can best be understood by first considering the properties of these soil components. Accordingly, sections 90, 91, and 92 describe the constituents of soil and introduce concepts used in the system. Section 93 gives the essentials of the classification scheme for soils found in nature, as shown in the Unified Soil Classification Chart. Figure 47 gives a comparison of the particle size scales for the Unified, AASHO, and USDA soil classification systems. In addition to proper classification, it is important to include an adequate description of the soil in reports or logs of explorations. The classification chart contains information required for describing soils and includes examples. Additional information on soil descriptions is given in part H. Section 94 contains a comparison of the engineering properties of typical soils of each classification group.

90. Soil Components.—(a) *Size.*—Particles larger than 3 inches are excluded from the Unified Soil Classification System. The amount of each oversized material, however, may be of great importance in the selection of sources for embankment material; hence, logs of exploration always contain information on quantity and size of particles larger than 3 inches. For definitions of terms for materials larger than 3 inches (cobbles, boulders, rock), see appendix D.

Within the size range of the system there are two major divisions; namely, the coarse grains and the fine grains. Coarse grains are those larger than the No. 200 sieve size (0.074 mm.), and they are further divided as follows:

Gravel (G), from 3 inches to No. 4 sieve ($\frac{3}{16}$ inch):
- Coarse gravel—3 inches to ¾ inch.
- Fine gravel—¾ inch to No. 4 sieve.

Sand (S), from No. 4 sieve to No. 200 sieve:
- Coarse sand—No. 4 to No. 10 sieve.
- Medium sand—No. 10 to No. 40 sieve.
- Fine sand—No. 40 to No. 200 sieve.

For visual classification, ¼ inch is considered equivalent to the No. 4 sieve size, and the No. 200 sieve is about the smallest size of particles that can be distinguished individually by the unaided eye.

Fine grains or fines are smaller than the No. 200 sieve size and are of two types: *silt* (M) and *clay* (C). Older classification systems defined clay variously as those particles smaller than 5 microns (0.005 mm.) or 2 microns (0.002 mm.), and they defined silt as fines larger than the clay sizes. (See fig. 47.) It is a mistaken idea, however, that the typical engineering characteristics of silt and clay correspond to particular grain sizes. Natural deposits of rock flour that exhibit all the properties of silt and none of clay may consist entirely of grains smaller than 5 microns. On the other hand, typical clays may consist mainly of particles larger than 5 microns but containing small quantities of extremely fine, colloidal-sized particles. Size distinction is not made between silt and clay in the Unified Soil Classification System; rather, the two materials are differentiated by their behavior.

Organic material (O) is often a component of soil, but it has no specific grain size. It is distinguished by the composition of its particles rather than by their sizes, which range from colloidal-sized particles of molecular dimensions to fibrous pieces of partly decomposed vegetable matter several inches in length.

(b) *Gradation.*—The amounts of the various sizes of grains present in a soil can be determined in the laboratory by means of sieving for the coarse grains, and by sedimentation (wet mechanical analysis) for the fines as described in section 120. The laboratory results are usually presented in the form of a cumulative grain-size curve. For soils consisting mainly of coarse grains, the grain-size distribution reveals something of the physical

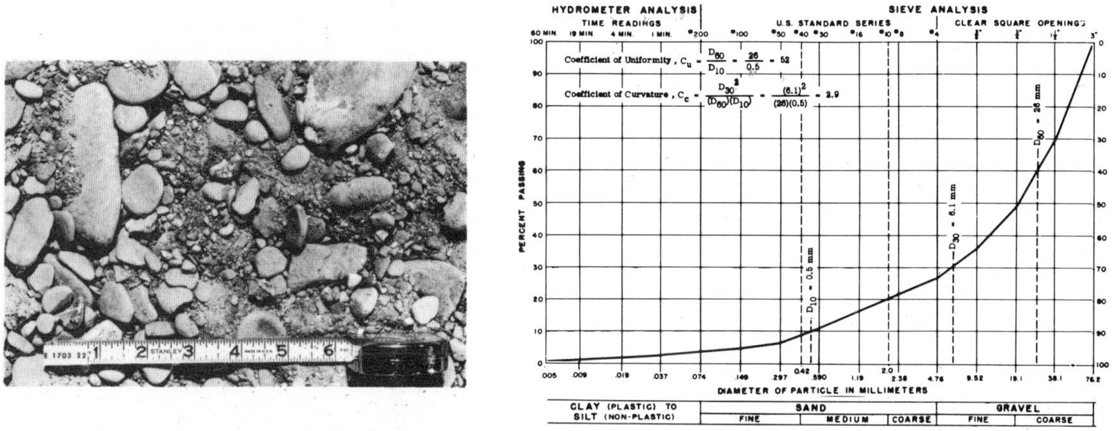

(A) WELL-GRADED GRAVEL (GW), VERMEJO PROJECT, NEW MEXICO

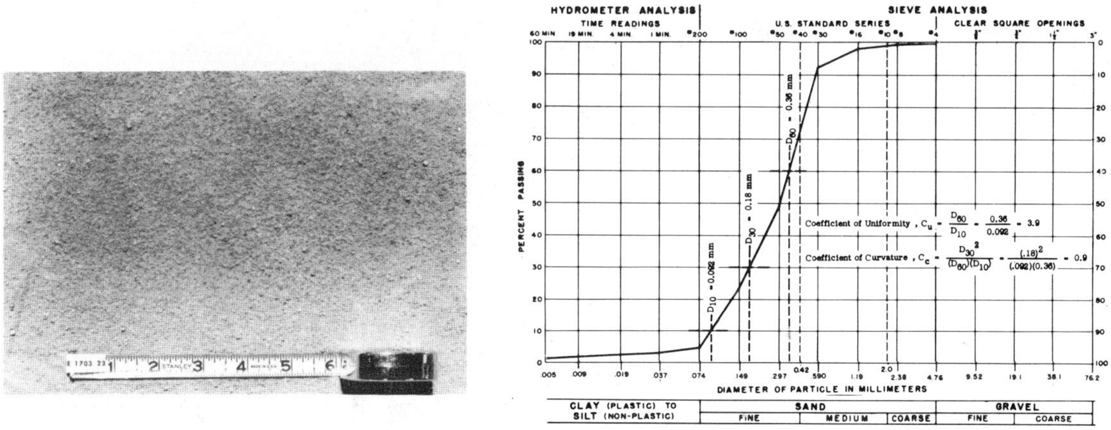

(B) UNIFORM SAND (SP), CHERRY CREEK RESERVOIR, COLORADO

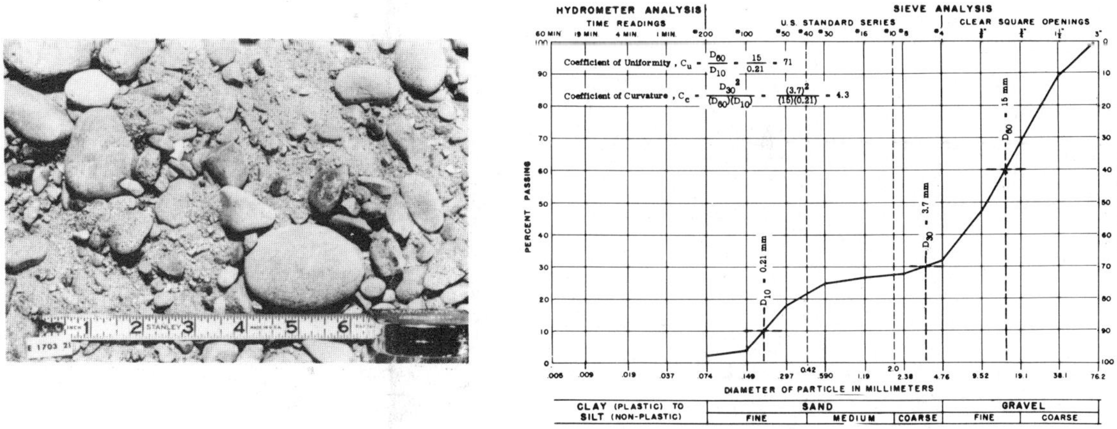

(C) POORLY GRADED GRAVEL (GP), FALCON DAM, TEXAS

Figure 50. Typical soil gradations. POA–4–D–15106.

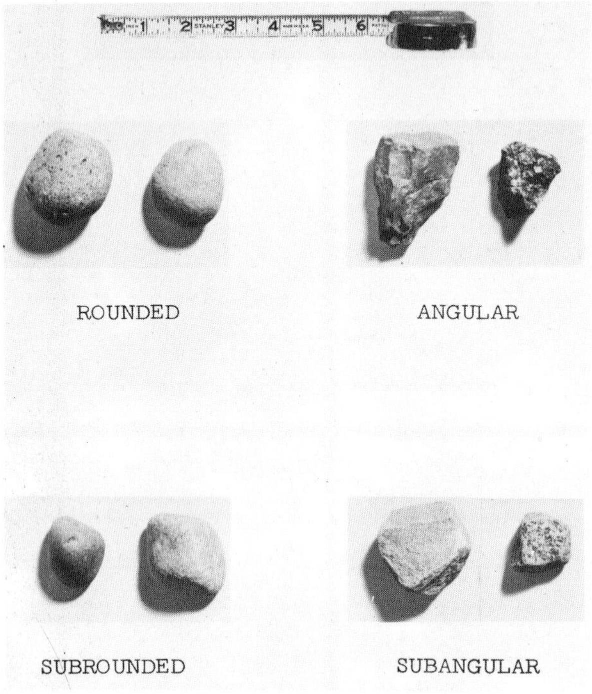

Figure 51. Typical shapes of bulky grains. PX–D–16266.

(A) MICA

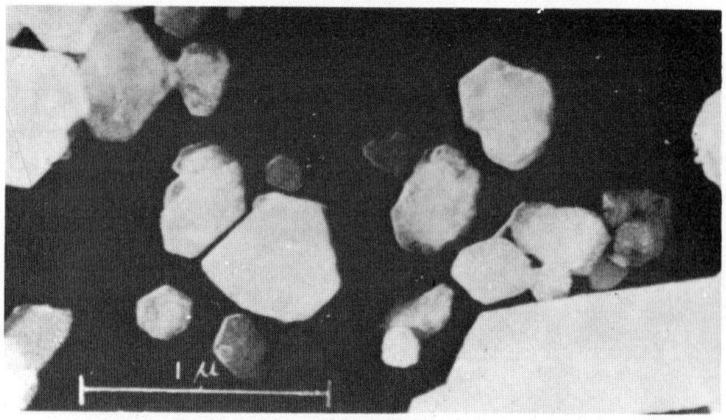

(B) KAOLINITE

(C) HALLOYSITE

Figure 52. Flaky and elongated grains. PX–D–16261.

properties of the material. On the other hand, the grain size is much less significant for soils containing a preponderance of fine grains.

Typical gradations of soils are:

Well graded (W)—Good representation of all particle sizes from largest to smallest.

Poorly graded (P)—Uniform, most particles about the same size; or skip (or gap) gradation—absence of one or more intermediate sizes.

In the field, soil is estimated to be well graded or poorly graded by visual examination. For laboratory purposes the type of gradation can be determined by the use of criteria based on the range of sizes and on the shape of the grain-size curve. The measure of size range is called the coefficient of uniformity, C_u, which is the ratio of the 60-percent-finer-than size (D_{60}) to the 10-percent-finer-than size (D_{10}). The shape of the grain-size curve is given by the coefficient of curvature, C_c, which is the ratio of the square of the 30-percent-finer-than size $(D_{30})^2$ to the product of (D_{60}) by (D_{10}). Photographs of typical gradations and corresponding grain-size curves are shown in figure 50.

(c) *Shape.*—The shape of the particles has an important influence on the physical properties of a soil. The following shapes are most common:

Bulky or equidimensional grains.—These may be further described as rounded, subrounded, subangular, and angular (fig. 51). The coarse-grained components of a soil are usually of the bulky type, consisting chiefly of the minerals quartz and feldspar.

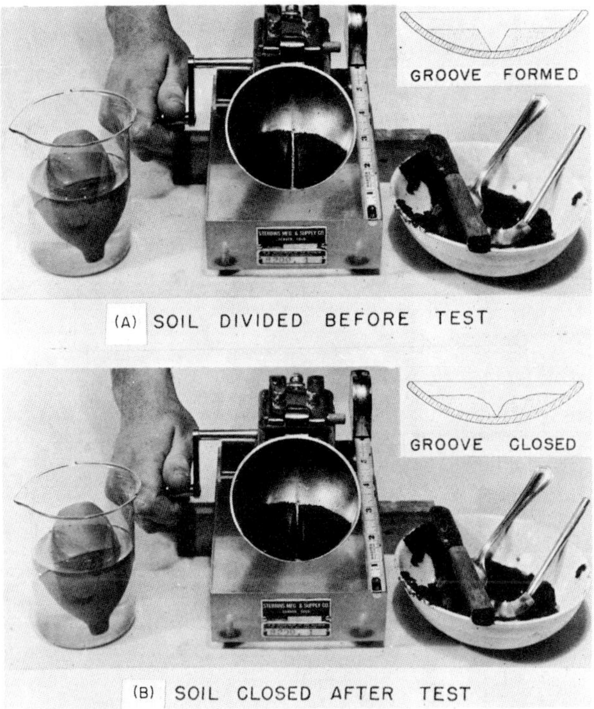

Figure 53. Test for liquid limit. PX–D–17009.

Flaky grains, also called platelike particles.—These are present in appreciable quantities in many fine-grained soils. Mica and some clay minerals have this shape, which is mainly responsible for their high compressibility (fig. 52 (A) and (B)).

Elongated grains and fibers.—The most commonly encountered materials in this class are the clay mineral halloysite (fig. 52(C)), asbestos, some types of volcanic ash, and organic soils such as peat.

91. *Soil Moisture.*—A typical soil mass has three constituents—soil grains, air, and water. In soils consisting largely of fine grains, the amount of water present in the voids has a pronounced effect on the soil properties. Three main states of soil consistency are recognizable:

(1) *Liquid state,* in which the soil is either in suspension or behaves like a viscous fluid;

(2) *Plastic state,* in which the soil can be rapidly deformed or molded without rebounding elastically, changing volume, cracking, or crumbling; and

(3) *Solid state,* in which the soil will crack when deformed or will exhibit elastic rebound.

In describing these soil states it is customary to consider only the fraction of soil smaller than the No. 40 sieve size (the upper limit of the fine sand component). For this soil fraction the water content in percentage of dry weight at which the soil passes from the liquid state into a plastic state is called the liquid limit (LL). A device (fig. 53) which causes the soil to flow under certain conditions is used in the laboratory to determine the liquid limit as described in section 120. Similarly, the water content of the soil at the boundary between the plastic state and the solid state is called the plastic limit (PL). The laboratory test described in section 120 consists of repeatedly rolling threads of the soil to one-eighth-inch diameter until they crumble and then determining the water content (fig. 54). The difference between the liquid limit and the plastic limit corresponds to the range of water contents within which the soil is plastic. This

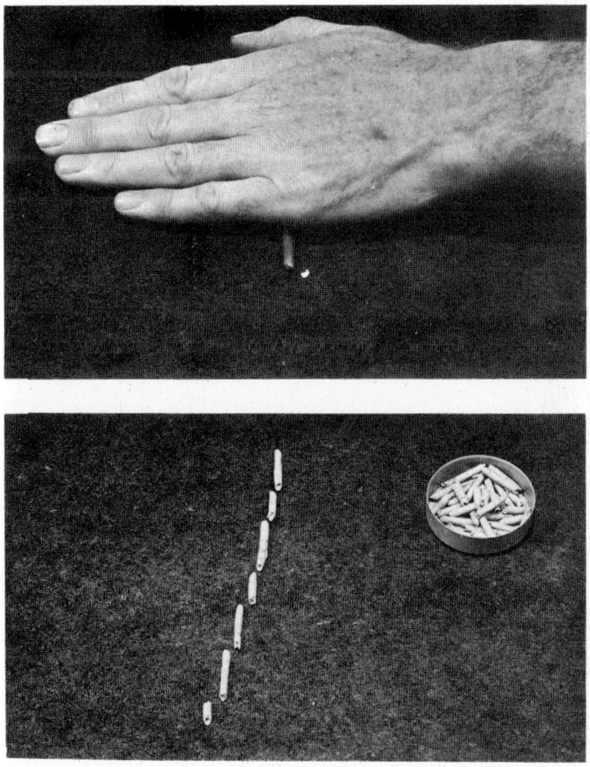

Figure 54. Test for plastic limit. PX–D–16530.

(A) REACTION TO SHAKING

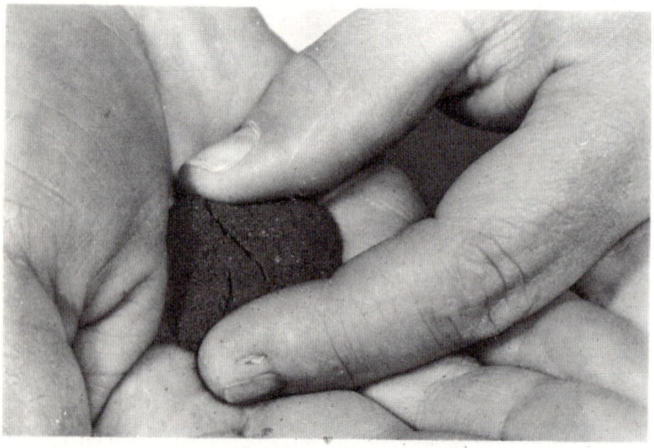

(B) REACTION TO SQUEEZING

Figure 55. Dilatancy test for silt. PX–D–16335.

difference of water content is called the plasticity index (PI). Highly plastic soils have high PI values. In a nonplastic soil the plastic limit and the liquid limit are the same and the PI equals 0.

These limits of consistency, which are called "Atterberg limits" after a Swedish scientist, are used in the Unified Soil Classification System as the basis for laboratory differentiation between materials of appreciable plasticity (clays) and slightly plastic or nonplastic materials (silts). With sufficient experience a soils engineer may acquire the ability to estimate the Atterberg limits of a soil. However, three simple hand tests have been found adequate for field identification and classification of fine soils and for determining whether the fine-grained fraction of a soil is silty or clayey, without requiring estimation of Atterberg limits. These hand tests, which are part of the field procedure in the Unified Soil Classification System, are as follows:

Dilatancy (reaction to shaking).

Dry strength (crushing characteristics).

Toughness (consistency near plastic limit).

They are discussed in the following section.

92. *Properties of Soil Components.*—(a) *Gravel and Sand.*—Both of the coarse-grained components of soil (gravel and sand) have essentially the same engineering properties, differing mainly in degree. The division of gravel and sand sizes by the No. 4 sieve is arbitrary and does not correspond to a sharp change in properties. Well-graded, compacted gravels or sands are stable materials. The coarse-grained soils when devoid of fines are pervious, easy to compact, little affected by moisture, and not subject to frost action. Although grain shape and gradation, as well as size, affect these properties, gravels are generally more pervious, more stable, and less affected by water or frost than are sands, for the same amount of fines.

As a sand becomes finer and more uniform, it approaches the characteristics of silt, with corresponding decrease in permeability and reduction in stability in the presence of water. Very fine, uniform sands are difficult to distinguish visually from silt. Dried sand, however, exhibits no cohesion (does not hold together) and feels gritty in contrast to the very slight cohesion and smooth feel of dried silt.

(b) *Silt and Clay.*—Even small amounts of fines may have important effects on engineering properties of the soils in which they are found. As little as 10 percent of particles smaller than the No. 200 sieve size in sand and gravel may make the soil virtually impervious, especially when the coarse grains are well-graded. Also, serious frost heaving in well-graded sands and gravels may be caused by less than 10 percent of fines. The utility of coarse-grained materials for surfacing roads can be improved by the addition of a small amount of clay to act as a binder for the sand and gravel particles.

Soils containing large quantities of silt and clay are the most troublesome to the engineer. These materials exhibit marked changes in physical properties with change of water con-

FIELD IDENTIFICATION PROCEDURES FOR FINE-GRAINED SOILS OR FRACTIONS

These procedures are to be performed on the minus No. 40 sieve size particles, approximately $\frac{1}{64}$ in. For field classification purposes, screening is not intended; simply remove by hand the coarse particles that interfere with the tests.

DILATANCY (Reaction to shaking)

After removing particles larger than No. 40 sieve size, prepare a pat of moist soil with a volume of about one-half cubic inch. Add enough water if necessary to make the soil soft but not sticky.

Place the pat in the open palm of one hand and shake horizontally, striking vigorously against the other hand several times. A positive reaction consists of the appearance of water on the surface of the pat which changes to a livery consistency and becomes glossy. When the sample is squeezed between the fingers, the water and gloss disappear from the surface, the pat stiffens, and finally it cracks or crumbles. The rapidity of appearance of water during shaking and of its disappearance during squeezing assist in identifying the character of the fines in a soil.

Very fine clean sands give the quickest and most distinct reaction whereas a plastic clay has no reaction. Inorganic silts, such as a typical rock flour, show a moderately quick reaction.

DRY STRENGTH (Crushing characteristics)

After removing particles larger than No. 40 sieve size, mold a pat of soil to the consistency of putty, adding water if necessary. Allow the pat to dry completely by oven, sun, or air drying, and then test its strength by breaking and crumbling between the fingers. This strength is a measure of the character and quantity of the colloidal fraction contained in the soil. The dry strength increases with increasing plasticity.

High dry strength is characteristic for clays of the CH group. A typical inorganic silt possesses only very slight dry strength. Silty fine sands and silts have about the same slight dry strength, but can be distinguished by the feel when powdering the dried specimen. Fine sand feels gritty whereas a typical silt has the smooth feel of flour.

TOUGHNESS (Consistency near plastic limit)

After removing particles larger than the No. 40 sieve size, a specimen of soil about one-half inch cube in size is molded to the consistency of putty. If too dry, water must be added and if sticky, the specimen should be spread out in a thin layer and allowed to lose some moisture by evaporation. Then the specimen is rolled out by hand on a smooth surface or between the palms into a thread about one-eighth inch in diameter. The thread is then folded and rerolled repeatedly. During this manipulation the moisture content is gradually reduced and the specimen stiffens, finally loses its plasticity, and crumbles when the plastic limit is reached.

After the thread crumbles, the pieces should be lumped together and a slight kneading action continued until the lump crumbles.

The tougher the thread near the plastic limit and the stiffer the lump when it finally crumbles, the more potent is the colloidal clay fraction in the soil. Weakness of the thread at the plastic limit and quick loss of coherence of the lump below the plastic limit indicate either inorganic clay of low plasticity, or materials such as kaolin-type clays and organic clays which occur below the A-line.

Highly organic clays have a very weak and spongy feel at the plastic limit.

Figure 56. Unified Soil Classification Chart. (Sheet 2 of 2.) From drawing no. 103—D—347.

tent. A hard, dry clay, for example, may be suitable as a foundation for heavy loads so long as it remains dry, but may turn into a quagmire when wet. Many of the fine soils shrink on drying and expand on wetting, which may adversely affect structures founded upon them or constructed of them. Even when the water content does not change, the properties of fine soils may vary considerably between their natural condition in the ground and their state after being disturbed. Deposits of fine particles which have been subjected to loading in geologic time frequently have a structure which gives the material unique properties in the undisturbed state. When the soil is excavated for use as a construction material or when the natural deposit is disturbed, for example by driving piles, the soil structure is destroyed and the properties of the soil are changed radically.

Silts are different from clays in many important respects, but because of similarity in appearance, they often have been mistaken one for the other, sometimes with unfortunate results. Dry, powdered silt and clay are indistinguishable, but they are easily identified by their behavior in the presence of water. Recognition of fines as being silt or clay is an essential part of the Unified Soil Classification System.

Silts are the nonplastic fines. They are inherently unstable in the presence of water and have a tendency to become "quick" when saturated. Quick silts often are called "bull's liver" by construction men. Silts are fairly impervious, difficult to compact, and are highly susceptible to frost heaving. Silt masses undergo change of volume with change of shape (the property of dilatancy), in contrast to clays which retain their volume with change of shape (the property of plasticity). The dilatancy property, together with the "quick" reaction to vibration, affords a means of identifying typical silt in the loose, wet state. The dilatancy test is illustrated by the photograph of figure 55, and is described on the Classification Chart (fig. 56). When dry, silt can be pulverized easily under finger pressure (indicative of very slight dry strength), and will have a smooth feel between the fingers in contrast to the grittiness of fine sand.

Silts differ among themselves in size and shape of grains, which are reflected mainly in the property of compressibility. For similar conditions of previous geologic loading, the higher the liquid limit of a silt, the more compressible it is. The liquid limit of a typical bulky-grained, inorganic silt is about 30 percent, while highly micaceous or diatomaceous silts (so-called elastic silts), consisting mainly of flaky grains, may have liquid limits as high as 100 percent. The differences in quicking and dilatancy properties afford a means of distinguishing in the field between silts of low liquid limits (L) and those of high liquid limits (H).

Clays are the plastic fines. They have low resistance to deformation when wet, but they dry to hard, cohesive masses. Clays are virtually impervious, difficult to compact when wet, and impossible to drain by ordinary means. Large expansion and contraction with changes in water content are characteristics of clays. The small size, flat shape, and type of mineral composition of clay particles combine to produce a material that is both compressible and plastic. The higher the liquid limit of a clay, the more compressible it will be when compared at equal conditions of previous geologic loading. Hence, in the Unified Soil Classification System, the liquid limit is used to distinguish between clays of high compressibility (H) and those of low compressibility (L). Differences in plasticity of clays are reflected by their plasticity indexes. At the same liquid limit, the higher the plasticity index, the more cohesive is the clay.

Field differentiation among clays is accomplished by the toughness test in which the moist soil is molded and rolled into threads until crumbling occurs, and by the dry strength test which measures the resistance of the clay to breaking and pulverizing. The dry strength and the toughness tests are shown in the Classification Chart (fig. 56). With a little experience in performing these tests, the clays of low compressibility and low plasticity, "lean" clays (L) can be readily distinguished from the highly plastic, highly compressible, "fat" clays (H).

(c) *Organic Matter.*—Organic matter in the form of partly decomposed vegetation is the primary constituent of peaty soils. Varying amounts of finely divided vegetable matter are found in plastic and in nonplastic sediments and often affect their properties sufficiently to influence their classification. Thus, we have organic silts and silt clays of low plasticity and organic clays of medium to high plasticity. Even small amounts of organic material in colloidal form in a clay will result in an appreciable increase in liquid limit of the material without increasing its plasticity index. Organic soils are dark gray or black in color and usually have a characteristic odor of decay. Organic clays feel spongy in the plastic range as compared to inorganic clays. The tendency for soils high in organic content to create voids by decay or to change the physical characteristics of a soil mass through chemical alteration makes them undesirable for engineering use. Soils containing even moderate amounts of organic matter are significantly more compressible and less stable than inorganic soils; hence, they are less desirable for engineering use.

93. *Unified Soil Classification System.*—(a) *General.*—Soils in nature seldom exist separately as gravel, sand, silt, clay, or organic matter but are usually found as mixtures with varying proportions of these components. The Unified Soil Classification System is based on recognition of the type and predominance of the constituents, considering grain size, gradation, plasticity, and compressibility. It divides soils into three major divisions: coarse-grained soils, fine-grained soils, and highly organic (peaty) soils. In the field, identification is accomplished by visual examination for the coarse grains and by a few simple hand tests for the fine-grained soils or fraction. In the laboratory the grain-size curve and the Atterberg limits can be used. The peaty soils (Pt) are readily identified by color, odor, spongy feel, and fibrous texture, and are not further subdivided in the classification system.

(b) *Field Classification.*—A representative sample of soil (excluding particles larger than 3 inches) is first classified as coarse-grained or fine-grained by estimating whether 50 percent, by weight, of the particles can be seen individually by the naked eye. Soils containing more than 50 percent of particles that can be seen are coarse-grained soils; soils containing more than 50 percent of particles smaller than the eye can see are fine-grained soils. If the soil is predominantly coarse-grained, it is then identified as being a gravel or a sand by estimating whether 50 percent or more, by weight, of the coarse grains are larger or smaller than the No. 4 sieve size (about 1/4 inch).

If the soil is a gravel, it is next identified as being "clean" (containing little or no fines), or "dirty" (containing an appreciable amount of fines). For clean gravels final classification is made by estimating the gradation: the well-graded gravels belong to the GW group, and uniform and skip-graded gravels belong to the GP group. Dirty gravels are of two types: those with non-plastic (silty) fines (GM) and those with plastic (clayey) fines (GC). The determination of whether the fines are silty or clayey is made by the three manual tests for fine-grained soils.

If a soil is a sand the same steps and criteria are used as for the gravels in order to determine whether the soil is a well-graded clean sand (SW), poorly graded clean sand (SP), sand with silty fines (SM), or sand with clayey fines (SC).

If a material is predominantly (more than 50 percent by weight) fine-grained, it is classified into one of six groups (ML, CL, OL, MH, CH, OH) by estimating its dilatancy (reaction to shaking), dry strength (crushing characteristics), and toughness (consistency near the plastic limit), and by identifying it as being organic or inorganic. The test procedures and the behavior of the various groups of fine-grained soils for each of the hand tests are shown on the Classification Chart (fig. 56).

Soils that are typical of the various groups are readily classified by the foregoing procedures. Many natural soils, however, will have property characteristics of two groups, because they are close to the borderline between the groups either in percentages of the various sizes or in plasticity characteristics. For this substantial number of soils, boundary classifications are used; that is, the two group symbols most nearly describing the soil are connected by a hyphen, such as GW–GC.

If the percentages of gravel and sand sizes in a coarse-grained soil are nearly equal, the classification procedure is to assume that the soil is a gravel and then continue on the chart until the final soil group, say GC, is reached. Since it could have been assumed that the soil is a sand, the correct field classification is GC–SC, because the criteria for the gravel and sand subgroups are identical. Similarly, within the gravel or sand groupings, boundary classifications such as GW–GP, GM–GC, GW–GM, SW–SP, SM–SC, and SW–SM, can occur.

Proper boundary classification of a soil near the borderline between coarse-grained and fine-grained soils is accomplished by classifying it first as a coarse-grained soil and then as a fine-grained soil. Such classifications as SM–ML and SC–CL are common.

Within the fine-grained division, boundary classifications can occur between low-liquid-limit soils and high-liquid-limit soils as well as between silty and clayey materials in the same range of liquid limits. For example, one may find ML–MH, CL–CH, and OL–OH soils; ML–CL, ML–OL, and CL–OL soils; and MH–CH, MH–OH, and CH–OH soils.

(c) *Laboratory Classification.*—Although most classifications of soil will be done visually and by simple hand tests, the Unified Soil Classification System has provided for precise delineation of the soil groups by mechanical analyses and Atterberg limits tests in the laboratory. Laboratory classifications are often performed on representative samples of soils which are being subjected to extensive testing and to verify field classifications when used in the design of small dams. Laboratory classification can be used to advantage in training the field classifier of soils to improve his ability to estimate percentages and degrees of plasticity.

The grain-size curve is used to classify the soil as being coarse-grained or fine-grained, and if coarse-grained, into gravel or sand by size, using the 50-percent criterion. Within the gravel or sand groupings, soils containing less than 5 percent finer than the No. 200 sieve size are considered "clean" and are then classified as well graded or poorly graded by their coefficients of uniformity and of curvature. In order for a clean gravel to be well graded (GW), it must have *both* a coefficient of uniformity, C_u, greater than 4 and a coefficient of curvature, C_c, between 1 and 3; otherwise, it is classified as a poorly graded gravel (GP). A clean sand having *both* C_u greater than 6 and C_c between 1 and 3 is in the SW group; otherwise, it is a poorly graded sand (SP).

Laboratory classification criteria for coarse-grained soils and for fine-grained soils are given in the Soil Classification Chart, figure 56.

94. *Engineering Characteristics of Soil Groups.* —(a) *General.*—Although there is no satisfactory substitute for actual testing to determine the important engineering properties of a particular soil, approximate values for typical soils of each classification group can be given as a result of statistical analysis of existing information. The attempt to put soils data into quantitative form involves the risk of (1) the data not being representative, and (2) use of the values in design without adequate safety factors. For the design of small dams, however, where investigation has disclosed no complex problems, expensive laboratory tests of permeability, shear, and consolidation of soils appear unwarranted and the use of average values of these properties is permissible. Since the values pertain to the soil groups, proper soil classification becomes of vital importance. Verification of field identification by laboratory gradation and Atterberg limits tests should be made on representative samples of each soil group encountered.

Table 8 is a summary of values obtained on more than 1,500 soil tests performed in the engineering laboratories of the Bureau of Reclamation in Denver, Colo., arranged according to the main soil classification groups and two frequently occurring boundary groups. The data for this table were obtained from reports for which laboratory soil classifications were available. The large majority of soils were from the 17 western States of the United States in which the Bureau operates; however, some foreign soils were included. Although the sampling area of the soils tested is limited, it is believed that the Unified Soil Classification System is relatively insensitive to geographical distribution. The procedure for determining which of many submitted samples should be tested is in itself conducive to obtaining a representative range of values, since samples were selected from the coarsest, finest, and average soil within a potential source.

For each soil property listed, the average and its 90 percent confidence limits are given where sufficient data were available to determine them. Since all laboratory tests, except large-sized permeability tests, were made on the minus No. 4 fraction of the soil, data on average values for the gravels are not available for most properties. However, an indication as to whether these average values will be greater than or less than the average values for the corresponding sand group is given in the table. The averages shown are subject to uncertainties that arise from sampling fluctuations, and they tend to vary from the true averages more widely if the number of observations is small. The plus or minus limits given are determined mathematically from the number of observations and from the standard deviation of the data used to determine the average. These limits imply that the true average, obtained by securing and testing more and more samples under the same essential conditions, lies within the plus or minus values 9 chances out of 10 [4].

The values for Proctor maximum dry density and optimum water content were obtained by tests described in section 120. The other properties are based on tests made on samples compacted to Proctor maximum dry density at optimum water content. The value of void ratio, e_o, is the ratio of the portion of the volume of the soil mass occupied by water and air to the volume of the soil grains. It is derived from the Proctor maximum dry density and the specific gravity of the grains. The MH and CH soil groups have no upper boundary of liquid limits in the classification; hence, it is necessary to give the range of those soils included in the table. The maximum liquid limits for the MH and the CH soils tested were 81 and 88 percent, respectively. Soils with higher liquid limits than these will have inferior engineering properties.

(b) *Permeability.*—The voids in the soil mass provide passages through which water may move. Such passages are variable in size and the paths of flow are tortuous and interconnected. If, however, a sufficiently large number of paths of flow are considered as acting together, an average rate of flow for the soil mass can be determined under controlled conditions that will represent a property of the

TABLE 8.—*Average properties of soils*

Soil classification group	Proctor compaction		Void ratio, e_o	Permeability, k, feet per year	Compressibility		Shearing strength		
	Maximum dry density in pounds per cubic foot	Optimum water content, percent			@ 20 p.s.i., percent	@ 50 p.s.i., percent	C_o p.s.i.	C_{sat} p.s.i.	tan ϕ
GW	>119	<13.3	(*)	27,000±13,000	<1.4	(*)	(*)	(*)	>0.79
GP	>110	<12.4	(*)	64,000±34,000	<0.8	(*)	(*)	(*)	>0.74
GM	>114	<14.5	(*)	>0.3	<1.2	<3.0	(*)	(*)	>0.67
GC	>115	<14.7	(*)	>0.3	<1.2	<2.4	(*)	(*)	>0.60
SW	119±5	13.3±2.5	0.37±*	(*)	1.4±*	(*)	5.7±0.6	(*)	0.79±0.02
SP	110±2	12.4±1.0	0.50±0.03	>15.0	0.8±0.3	(*)	3.3±0.9	(*)	0.74±0.02
SM	114±1	14.5±0.4	0.48±0.02	7.5±4.8	1.2±0.1	3.0±0.4	7.4±0.9	2.9±1.0	0.67±0.02
SM-SC	119±1	12.8±0.5	0.41±0.02	0.8±0.6	1.4±0.3	2.9±1.0	7.3±3.1	2.1±0.8	0.66±0.07
SC	115±1	14.7±0.4	0.48±0.01	0.3±0.2	1.2±0.2	2.4±0.5	10.9±2.2	1.6±0.9	0.60±0.07
ML	103±1	19.2±0.7	0.63±0.02	0.59±0.23	1.5±0.2	2.6±0.3	9.7±1.5	1.3±*	0.62±0.04
ML-CL	109±2	16.8±0.7	0.54±0.03	0.13±0.07	1.0±0.2	2.2±0.0	9.2±2.4	3.2±*	0.62±0.06
CL	108±1	17.3±0.3	0.56±0.01	0.08±0.03	1.4±0.2	2.6±0.4	12.6±1.5	1.9±0.3	0.54±0.04
OL	(*)	(*)	(*)	(*)	(*)	(*)	(*)	(*)	(*)
MH	82±4	36.3±3.2	1.15±0.12	0.16±0.10	2.0±1.2	3.8±0.8	10.5±4.3	2.9±1.3	0.47±0.05
CH	94±2	25.5±1.2	0.80±0.04	0.05±0.05	2.6±1.3	3.9±1.5	14.9±4.9	1.6±0.86	0.35±0.09
OH	(*)	(*)	(*)	(*)	(*)	(*)	(*)	(*)	(*)

The ± entry indicates 90 percent confidence limits of the average value. * Denotes insufficient data, > is greater than, < is less than.

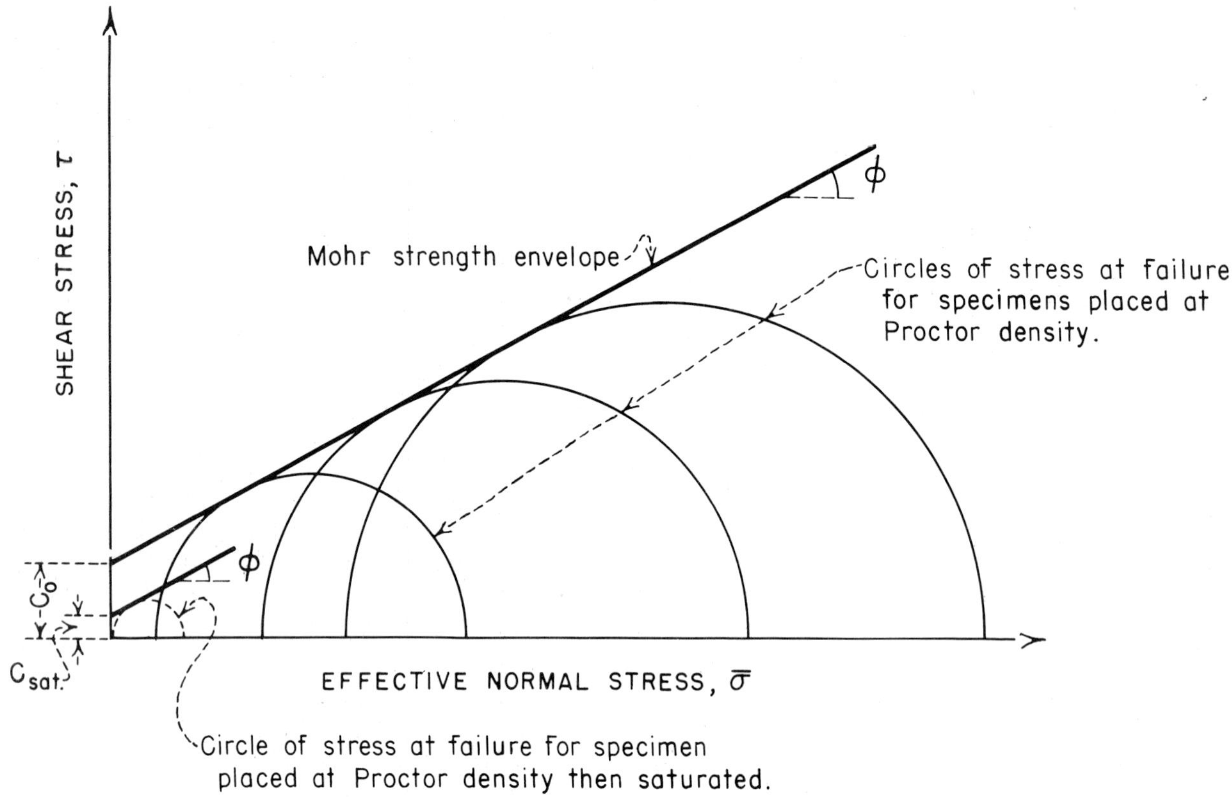

Figure 57. Shear strength of compacted soils. 288–D–2474.

soil. The water movement is called percolation; the measure of it is called permeability; and the factor relating permeability to unit conditions is called the coefficient of permeability, k, which represents the discharge through a unit area at unit hydraulic gradient. The use of k in estimating flow through soils is discussed in section 130(b). There are many units of measurement in common use for expressing the coefficient of permeability. The one used in table 8 is feet per year, or cubic feet per square foot per year at unit gradient. One foot per year is virtually equal to 10^{-6} centimeters per second.

The coefficient of permeability of natural soil deposits ranges from 1 million feet to 0.001 foot per year. In many soil deposits the permeability parallel to the bedding planes may be 100 or even 1,000 times as large as the permeability at right angles to the bedding planes. Permeability in some soils is very sensitive to small changes in density, water content, or gradation. Because of the possible wide variation in permeability, a numerical value of k should be considered only as an order of magnitude. It is customary to describe soils with permeabilities less than 1 foot per year as impervious; those with permeabilities between 1 and 100 feet per year as semipervious; and soils with permeabilities greater than 100 feet per year as pervious. These values, however, are not absolute for the design of dams. Successful structures have been built whose various zones were constructed of soils with permeabilities not within these respective ranges.

(c) *Compressibility.*—Two values are given for compressibility: the value at 20 pounds per square inch effective stress, and the value at 50 pounds per square inch effective stress. These values are for confined compression with drainage permitted. In the confined compression test the soil is prevented from moving laterally by the sides of the container. Porous stones on the top and the bottom permit the water and air in the compacted specimens to drain under the load. The value recorded is the percentage reduction of initial volume at equilibrium under the applied vertical stress. The phenomenon of compressibility is associated with changes in volume in the voids and only to a very limited extent with changes in the solid particles. If the voids are to a large extent filled with air, the addition of a load on the soil mass will result in compression almost immediately. If, on the other hand, the voids are very nearly or completely filled with water, very little or no compression will take place immediately upon application of the load, and only as the water drains from the soil mass will consolidation take place. If the water can drain readily from the soil mass, consolidation may take place in a relatively short period of time, but if the soil is very impervious and the soil mass is large, complete consolidation may require many years.

(d) *Shear Strength.*—Three different values are given for the soil groups under this heading: C_o, C_{sat}, and tan ϕ. The values of C_o and tan ϕ are the vertical intercept and the slope, respectively, of the Mohr strength envelope on an effective stress basis. The Mohr plot is shown in figure 57. The Mohr strength envelope is obtained by testing several sealed specimens of soil, at the Proctor maximum dry density and optimum water content, in a triaxial shear machine in which pore-water pressures developed during the test are measured. The effective stresses are obtained by subtracting the measured pore-water pressures in the specimen from the stresses applied by the machine. No drainage is permitted during the tests; hence, they are sometimes called unconsolidated quick tests. The value C_{sat} was obtained by preparing a specimen at Proctor maximum dry density and optimum water content, saturating it, and shearing it to failure to obtain the small circle shown in figure 57. The value C_{sat} is the intercept on the vertical axis of a line tangent to the circle having an inclination ϕ.

These values for shear strength are applicable for use in Coulomb's equation:

$$s=C+(\sigma-u)\tan\phi \qquad (1)$$

where:

s=shear strength per unit of area,
u=pore-water pressure,
σ=applied normal stress,
tan ϕ is as previously defined, and
C is either C_o or C_{sat}, depending on the water content of the soil.

A discussion of the significance of pore-water pressure in the laboratory tests is beyond the scope of this text. The application of pore pressure measurements to the shear strength of cohesive soils is discussed in reference [17]. The effective-stress principle, which takes the pore-water pressures into account, was used in arriving at recommended slopes given in chapter VI.

D. ROCK CLASSIFICATION

(Adapted from the Army publication, "Geology and its Military Applications" [5])

95. *Rocks and Minerals.*—(a) *Definition and Types.*—In a broad sense, rocks are aggregates of minerals. The principal exceptions to this definition are the products of organic decay such as coal, and volcanic glasses such as obsidian. To the engineer the term "rock" signifies firm and coherent or consolidated substances that cannot normally be excavated by manual methods alone. Based on the principal mode of origin, rocks are grouped into three large classes: igneous, sedimentary, and metamorphic. These are discussed in more detail in sections 96, 97, and 98.

(b) *Mineral Identification.*—The physical properties characteristic of a mineral, controlled by its chemical composition and molecular structure, are valuable aids in its rapid field identification. Those characteristics which can be determined by simple field tests are introduced to aid in the identification of minerals and indirectly in the identification of rocks.

Hardness.—The hardness of a mineral is a measure of its ability to resist abrasion or scratching. A simple scale based on empirical tests for hardness has been universally accepted. The 10 minerals selected to form the standard of comparison, listed in order of increasing hardness from 1 to 10 are:

Talc or mica	1
Gypsum (fingernail about 2)	2
Calcite	3
Fluorite (copper coin between 3 and 4)	4
Apatite (knife blade about 5)	5
Feldspar (window glass about 5.5)	6
Quartz	7
Topaz or beryl	8
Corundum	9
Diamond	10

When testing the hardness of a mineral always use a fresh surface; always rub the mark to make sure it is really a groove made by scratching. If an unknown mineral scratches and in turn is scratched by a member of the scale or a testing medium (copper coin, pocketknife, or window glass), the two are of equal hardness.

Cleavage.—A material is said to have cleavage if smooth, plane surfaces are produced when the mineral is broken. This is a fairly consistent physical property of minerals and when present is of great value in their identification. Cleavage invariably occurs along parallel planes. Some minerals have one cleavage; others have two, three, or even more different cleavage directions which may have varying degrees of perfection. The number of cleavage directions and the angle at which they intersect serve as aids in identification of a mineral (fig. 58).

Fracture.—The broken surface of a mineral, in directions other than those of cleavage planes, is called the fracture. In some cases this property may be very helpful in field identification. The common types of fracture are conchoidal, if the fracture has concentric curved surfaces like the inside of a clamshell; irregular, if the surface is rough; and splintery if it has the appearance of wood.

Figure 58. Mineral cleavage. (U.S. Corps of Engineers.) 288–D–2918.

Luster.—The luster of a mineral is the appearance of its surface due to the quality and intensity of the light reflected. Two major kinds are recognized, metallic and nonmetallic. The main difference between the two is indicated by the name. In addition, metallic minerals are opaque, or nearly so, whereas nonmetallic minerals are transparent on their thin edges. Some of the common nonmetallic lusters are vitreous, having the luster of glass; pearly, having the iridescence of pearl; and adamantine, having brilliant luster like that of a diamond.

Color.—The color of a mineral, as an aid in its identification, must be used with proper precaution, since some show a wide range without perceptible change in composition. Color on the whole, however, is fairly consistent, particularly in the metallic minerals where it is a great help in field identification.

Streak.—The color of the fine powder of a mineral, obtained by rubbing it on some white substance, preferably unglazed porcelain, is known as its streak. The streak of a mineral is quite consistent within a given range, even though its color may vary.

(c) *Common Rock-Forming Minerals.*—Only a dozen or so of the 2,000 known varieties of minerals are found in most common rocks. Descriptions of the most important of these rock-forming minerals or mineral groups follow:

Quartz.—Silicon dioxide. Hardness, 7, scratches glass easily. No cleavage. Fracture, conchoidal. Luster, vitreous. Common varieties, usually white or colorless. Streak, white or colorless. Typical examples are milky quartz and rock crystal quartz.

Feldspar group.—Potassium-aluminum silicate or sodium-calcium-aluminum silicate. Hardness, 6, scratches glass with difficulty. Luster, vitreous. Streak, white. Orthoclase is a common potassium-rich variety which is typically colorless, white, gray, pink, or red, and has two good directions of cleavage that intersect at 90° to each other (① of fig. 58). The sodium-calcium-rich feldspars, commonly referred to as plagioclase feldspar, are typically of various shades of gray, have two cleavage directions that intersect at angles of nearly 90° to each other, and can be distinguished from orthoclase feldspar by the presence of fine, parallel lines that appear on the basal cleavage surface.

Mica group.—Complex potassium-aluminum silicates, often with magnesium, iron and sodium. Hardness, 2 to 3, can be scratched with the thumbnail. Good cleavage in one direction. Luster, vitreous to pearly. Transparent, with varying shades of yellow, brown, green, red, and black in thicker specimens. Streak, white. The true characteristic of this group is that its minerals are capable of being split very easily into extremely thin and flexible sheets. Biotites (black) and muscovites (white) are two representative varieties.

Amphibole group.—Complex calcium-magnesium-iron silicates. Hardness, 5 to 6. Cleavage in two directions at angles 56° and 124°. Color, light to dark green to black. Streak, white to grayish-green. Hornblende is a common variety that is usually distinguishable from other amphiboles by its dark color.

Pyroxene group.—Complex calcium-iron silicates, closely analogous chemically to the amphibole group. Hardness, 5 to 6. Two directions of cleavage, making angles of about 87° and 93°, an important characteristic useful in distinguishing between the minerals of the pyroxene and amphibole groups. Color, light to dark green to black. Streak, white to grayish-green. Augite is a common variety that can be distinguished from hornblende by its cleavage angles.

Olivine.—Magnesium-iron silicate. Hardness, 6.5 to 7. No cleavage. Luster, vitreous. Color, olive to grayish-green to brown. Streak, white to colorless. An important characteristic of this mineral is its friability or tendency to crumble into small grains, which is due to its granular texture.

Calcite and dolomite.—Calcium carbonate and calcium-magnesium carbonate. Hardness, 3 and 3.5 to 4. Perfect cleavage in three directions (② in fig. 58). Luster, vitreous to pearl. Usually white or colorless, but may appear in shades of gray, red, green, blue, or yellow. Streak, white. Calcite may develop in large crystals, whereas dolomite is commonly found in coarse, granular masses.

Clay minerals.—Extremely complex hydrous aluminum silicate. Hardness, 2 to 2.5. Luster, dull to earthy. Color, white, gray, greenish, and yellowish-white. The three most important groups of clay minerals are kaolinite, montmorillonite, and illite. Almost all clays contain one or more of these three groups. Clay minerals can be distinguished only under the microscope and with the aid of X-ray equipment. They occur typically in extremely fine-grained masses of thin, micalike scales.

Limonite and hematite.—Hydrous ferric oxide and ferric oxide. Hardness, 5.5 and 6.5. No cleavage. Color, dark brown to black and reddish-brown to black, depending on the variety. Limonite has a yellowish-brown streak and is characteristically found in dark brown, nodular, earthy masses with no apparent crystal structure. Hematite has a light to dark Indian-red streak, usually occurs in earthy masses, but occasionally is found in botryoidal or reniform shapes known as kidney ore and in foliated masses known as specular iron. Limonite and hematite are important coloring and cementing minerals in many different rocks, especially in the sedimentary group.

96. Igneous Rocks.—(a) *General.*—Igneous rocks are commonly referred to as primary rocks. They are those rocks which have solidified from a molten mass called magma deep within the body of the earth (intrusive rocks), or from lava when extruded on the earth's surface (extrusive rocks). Igneous rocks owe their variation in significant characteristics to

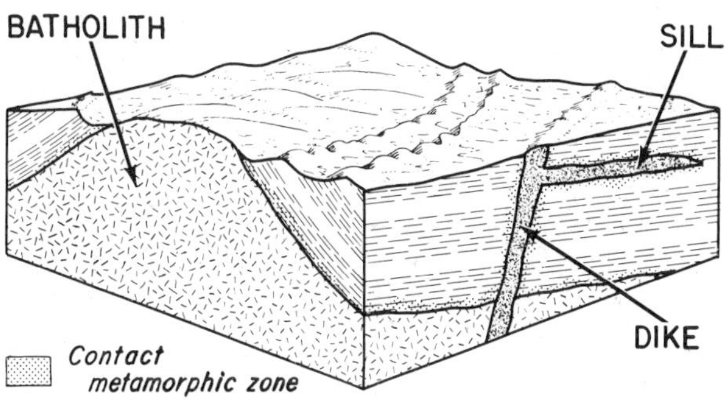

Figure 59. Intrusive igneous masses. (U.S. Corps of Engineers.) 288–D–2919.

Figure 60. Three dikes cutting sedimentary beds. (U.S. Corps of Engineers.) 288–D–2920.

differences in chemical composition of the original molten mass and to differences in physical conditions under which the molten mass solidified.

Intrusive igneous rock masses are shown in figure 59. Dikes are tabular igneous bodies that are commonly intruded at an angle to the bedding of the surrounding formation (fig. 60). Sills are similar bodies which are usually intruded parallel to the bedding planes of the rocks which enclose them. The thickness of a dike or a sill may vary from a few inches to several hundreds of feet, but this dimension is usually quite small in relation to the length and width of the intrusive body. For example, the Palisades sill of New York has a thickness of 1,000 feet and a length of over 100 miles. Very large, irregular masses of intrusive igneous rock covering many square miles of area are called batholiths. Although originally deeply buried beneath the earth's surface, they have become exposed through a process of uplift and erosion. A very striking example of an exposed batholith is the one in central Idaho, which has an estimated area of over 80,000 square miles.

Extrusive igneous rock masses include lava flows and volcanic ejecta. Lava flows are the result of the solidification of lava which has issued from fissures in the earth's crust or poured out of volcanoes. These flows are the most common modes of occurrence of extrusive igneous rocks. Among the most notable of the enormous lava flows in the world is the Columbia River Plateau of Washington, Oregon, and Idaho. The lava sheets cover approximately 200,000 square miles and the succession of flows has a known cumulative thickness of more than 4,000 feet. Explosive volcanoes frequently eject great quantities of broken and pulverized rock material and blobs of molten lava which solidify before striking the ground. These solid volcanic ejecta are termed pyroclastic material, which vary in size from great blocks weighing many tons through small cinders or lapilli to fine dust-sized particles referred to as ash. (See fig. 61.)

(b) *Classification of Igneous Rocks.*—Chemical composition and texture are used to classify igneous rocks. A mass of molten rock material may be regarded as a complex solution containing oxide of silicon which behaves as an acid; and oxides of iron, aluminum, calcium, magnesium, potassium, and sodium which behave as bases. If more acid is available than is necessary to satisfy the bases in the magma, the surplus will show itself as free silicon dioxide (quartz), and the resulting rock is said to be acidic. If the bases are in excess, iron-magnesium minerals will be present and the rock is said to be basic. As a rule acidic rocks are light colored; basic rocks are dark to black. One striking exception to this rule is obsidian, an acid rock which is normally black.

Texture refers to the size and arrangement of the mineral grains in the rock (fig. 62). These factors are influenced primarily by the rate at which the molten mass, magma or lava, cools. A constant rate of cooling produces rocks in which the constituent mineral grains are approximately the same size. In general the slower the molten material cools, the larger the size of the mineral grains. A change in the rate of cooling from an initial slow phase, followed by a more rapid phase usually produces porphyritic texture (④ of fig. 62). These rocks are characterized by mineral grains of two dominant size groups—phenocrysts, or large grains, in a ground mass or

Figure 61. Blocky type of solidified lava flows. Layer of volcanic ejecta (ash) covers area at left and in foreground. (U.S. Corps of Engineers.) 288–D–2921.

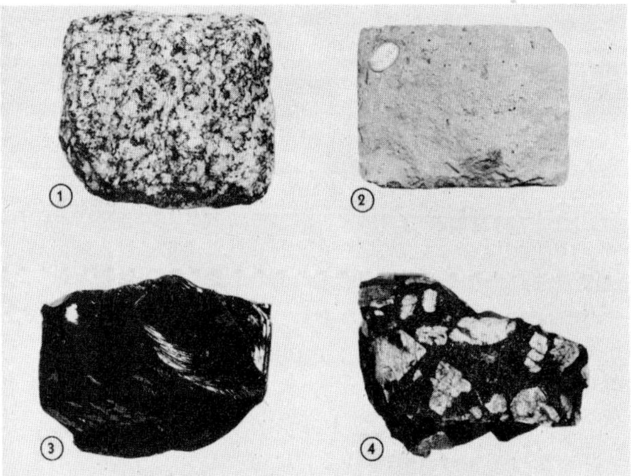

Figure 62. Textures of igneous rocks: ① coarse-grained, ② fine-grained, ③ glassy, and ④ porphyritic. (U.S. Corps of Engineers.) 288–D–2922.

background of smaller grains. Textural terms used in the classification of igneous rocks are:

Coarse-grained in which the individual crystals are easily visible to the naked eye (① of fig. 62).

Fine-grained in which crystals of like or equal size generally form throughout the whole mass of rock, but the individual grains can generally be seen only with a strong hand lens or with a microscope (② of fig. 62).

Glassy (noncrystalline) in which the rock has a noncrystalline or glassy texture (③ of fig. 62).

Table 9 lists the common igneous rocks. Those of similar chemical composition or mineral content are listed in the vertical column; those of similar textures are listed in the horizontal column. The following are the minerals in common igneous rocks:

Granite and *rhyolite* are composed largely of quartz and feldspar (mainly of the orthoclase

TABLE 9.—*Common igneous rocks*

Texture	Composition		
	Acid rocks (more than 50 percent silica)		Basic rocks (less than 50 percent silica)
	Light-colored minerals, chiefly feldspar, predominate		Dark-colored minerals predominate
	Abundant quartz	Little or no quartz	No quartz—abundant amphibole, pyroxene, and plagioclase feldspar
Coarse-grained (mineral crystals easily visible to naked eye).	Granite	Diorite	Gabbro.
Fine-grained (mineral crystals generally invisible to naked eye).	Rhyolite	Andesite (Trap).	Basalt (Trap).
Glassy	Obsidian,[1] pitchstone pumice.		

[1] Contains excess silica, but not as quartz.

variety), and as a rule contain mica (generally the biotite variety).

Diorite and *andesite* are composed of feldspar (mainly plagioclase varieties) and one or more dark minerals (biotite, hornblende, or pyroxene).

Gabbro and *basalt* differ from diorite in that the dark minerals (hornblende, pyroxene, and olivine) predominate. All feldspar is plagioclase; and biotite, although present in some gabbros, is distinctly uncommon.

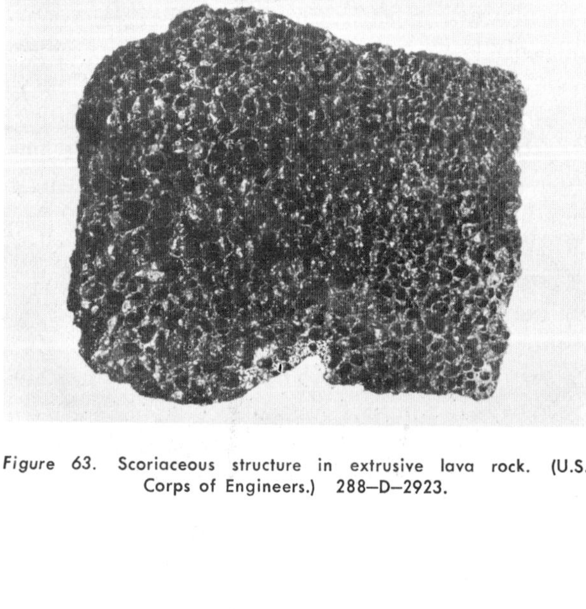

Figure 63. Scoriaceous structure in extrusive lava rock. (U.S. Corps of Engineers.) 288–D–2923.

Obsidian and *pitchstone* correspond in composition to granite and rhyolite. Both are commonly referred to as volcanic glasses. Obsidian is dark-colored to black and with a brilliant luster (③ of fig. 62). Pitchstone is lighter colored and with a dull luster.

Pumice is a porous or cellular glass, usually white or gray in color.

(c) *Primary Structural Features of Igneous Rocks.*—With the exception of those varieties which exhibit a glassy texture, igneous rocks are composed of interlocking grains of different minerals. On this basis they can be distinguished from crystalline sedimentary and massive metamorphic rocks which normally contain crystals of the same mineral. The distinctive structural features common to some, but not all, igneous rocks are as follows:

Flow structure may be exhibited by the glassy-textured igneous rocks, such as obsidian, and by the fine-grained extrusives, such as rhyolite.

Vesicular or *scoriaceous structure* is commonly present in extrusive igneous rocks (fig. 63). Such rock contains tiny spherical to almond-shaped openings called vesicles, formed by gas bubbles in or rising through the lava.

Lamellar or *platy structure,* more or less perfect, may be found in some of the coarser grained igneous rocks. This structure is due to the parallel orientation of such minerals as mica and hornblende, and most commonly occurs near the contacts of intrusive bodies where the friction between the wall rock and the molten material causes the platy minerals to aline themselves in the direction of flow.

Columnar structure is often formed in fine-grained igneous rocks by the development of shrinkage cracks (joints) as the molten mass cools and solidifies. This structural feature is commonly found in basaltic intrusions, such as dikes and sills, that cooled at a moderate rate (fig. 64) and in many lava flows.

97. Sedimentary Rocks.—(a) *General.*— Sedimentary rocks, or stratified rocks, are of secondary origin. They are formed of layerlike masses of sediment that have hardened through cementation, compaction, or incipient recrystallization. The inorganic material entering into the composition of most

Figure 64. Columnar jointing in basaltic sill. Upper photograph is a side view of sill showing regular arrangement of columns and lower photograph is a sectional view showing irregular hexagonal outline of each column. (U.S. Corps of Engineers.) 288–D–2924.

sedimentary rocks is derived from the disintegration and decomposition of preexistent igneous, sedimentary, and metamorphic rocks. This material is then moved from its original position by water, wind, or glaciers in the form of solid particles or dissolved salts. Rock particles dropped from suspension produce deposits of clastic or fragmental sediment. By chemical reaction the dissolved salts become insoluble and form precipitated sediments; or by evaporation of the water medium they form evaporites.

The relative magnitude of pyroclastic deposits is not too well known, but it appears that they constitute only a small part of the sedimentary rocks in the earth's crust. The quantity of material, however, which can be ejected at a single volcanic eruption and transported by the wind is quite large. The organic material entering into the composition of a very small percentage of the total sedimentary rock mass is the result of the activities of plants and animals, either directly or indirectly. Included in this group are certain protective and supporting structures produced by plants and animals which on the death of the organism become sediments and certain precipitated sediments formed by the activities of organisms.

Based on the mode of origin, sediments can be classified as clastic, chemical, and organic. The clastic or fragmental sediments include gravel, sand, silt, and clay, which are differentiated by the dimensions of the particles and their plasticity characteristics. All kinds of rock contribute to clastic material. Each size of clastic particle may be transported by several agencies. The terms "gravel," "sand," "silt," and "clay" have been defined in section 90. The chemically deposited and organic sediments are classified on the basis of chemical composition. The common sediments formed chemically by precipitation and evaporation and through the life processes of organisms are listed in table 10.

TABLE 10.—*Common sedimentary rocks*

Type	Sediment	Rock
Clastic or fragmental	Coarse (gravel)	Conglomerate.
	Medium (sand)	Sandstone.
	Fines (silt and clay)	Siltstone and shale.
Pyroclastic	Coarse (cinder)	Agglomerate.
	Fine (ash)	Tuff.
Chemical precipitates and evaporites.	Calcium carbonate ($CaCO_3$)	Limestone.
	Calcium magnesium carbonate ($Ca(Mg,Fe)(CO_3)$).	Dolomite.
	Silicon dioxide (SiO_2)	Chert.
	Calcium sulfate ($CaSO_4 \cdot 2H_2O$)($CaSO_4$).	Gypsum, anhydrite.
	Sodium chloride (NaCl)	Rock salt.
Organic	Calcium carbonate (animal remains).	Coquina, some coral rock, and chalk.
	Carbon (plant remains)	Coal.

The conversion of sediment into rock, sometimes called lithification, is brought about by a combination of the following processes:

Compaction, in which the rock or mineral particles are brought closer together by pressure of overlying materials such as conversion of clay to shale and conversion of peat to coal.

Cementation, in which porous materials are bound together by minerals precipitated from water solution such as silicon dioxide (quartz), calcium carbonate (calcite), and the iron oxides (limonite and hematite).

Recrystallization, in which a rock with an interlocking crystalline fabric or grain, such as crystalline limestone, is developed by continued growth of the mineral grains in a sediment or the development of new minerals from water.

(b) *Classification of Sedimentary Rocks.*—Sedimentary rocks may be classified as clastic, chemical, or organic, based on the mode of origin of the sediment from which they are derived. The clastic rocks commonly show separate grains. The chemical precipitates and evaporites, on the other hand, either have interlocking crystals or are in earthy masses. The organically formed rocks commonly contain easily recognized animal and plant remains, such as shells, bones, stems, or leaves. Table 10 lists the common sedimentary rocks and the sediment or material from which they have been derived.

To analyze their mineral content, sedimentary rocks are divided into three major types: sand, clay, and lime. The sand and clay types are principally clastic. The lime type includes the precipitates and the calcium carbonates of organic origin.

Sand-type rocks.—The minerals which are commonly found in the predominantly sand-type rocks such as conglomerates (fig. 65) and sandstone, are quartz as grains; feldspar as grains; mica minerals as small plates; clay minerals; and limonite, hematite, calcite, and quartz as cementing material.

Clay-type rocks.—The minerals which are commonly found in predominantly clay-type

Figure 65. Conglomerate. (U.S. Corps of Engineers.) 288–D–2925.

rocks, such as shale and siltstone, are clay minerals; quartz as fine grains; mica minerals as fine plates; and limonite, hematite, calcite, and quartz as cementing materials.

Lime-type rocks.—The minerals commonly found in the predominantly lime-type rocks, such as limestone, chalk, coral rocks, dolomite, and coquina, are calcite as visible grains or crystals; dolomite as visible grains or crystals; quartz as grains; chalcedony or chert as grains; clay minerals; and lime mud, limonite, hematite, and quartz as cementing materials.

(c) *Primary Structural Features of Sedimentary Rocks.*—The primary structural features inherent in the sediment before consolidation are valuable in the field recognition of sedimentary rocks. A universally prevalent structural feature of sedimentary rocks is their *stratification,* as indicated by differences in composition, texture, hardness, or color disposed in approximately parallel bands. These strata may be flat lying, or nearly so, as originally deposited; or they may be tilted or folded as a result of movement within the earth's crust. Each stratum or bed is separated from the one immediately above and below by bedding planes or planes of stratification. The thicknesses of sedimentary strata vary from a few inches (thin bedded) to a few feet (thick bedded). Very thin beds are referred to as laminae.

A primary cleavage structure developed parallel to the stratification of some fine-grained sedimentary rocks is called *bedding* fissility. The ability of these rocks to split along parallel planes is attributed mainly to compositional and grain-size variations between layers. Shale is a sedimentary rock which has bedding fissility. Some sedimentary deposits, usually those composed of granular materials such as sand, commonly exhibit laminae lying at an angle to the true bedding plane. This feature of sedimentary rock is known as *crossbedding* or cross lamination.

Sediment deposited in low, flat places, such as flood plains of rivers and intermittent lakes,

Figure 66. Fossiliferous limestone. (U.S. Corps of Engineers.) 288–D–2926.

may develop *mud cracks* which separate the mass into irregular polygonal blocks and which may become sufficiently hardened to be preserved during lithification of the sediment. Parallel ridges, known as *ripple marks*, developed in sediment moved by wind or water are often preserved when the sediment is consolidated. *Fossils*, the remains or impressions of animals and plants, are not structural features, but they are important to the field identification of sedimentary rocks. (See fig. 66.)

98. *Metamorphic Rocks*.—(a) *General*.—Metamorphic rocks are those formed from preexistent igneous or sedimentary rocks as a result of an enforced adjustment of these rocks to environments different from those in which they were originally formed. This adjustment may include the formation within the rock of new structures, textures, or minerals, or all of these.

Temperature, pressure, and chemically active fluids and gases are the major interrelated factors involved in metamorphism. Each factor is capable of accomplishing metamorphic work individually as follows:

Temperature.—The effect of heat is twofold: it increases the solvent action of fluids, and it helps break up and change chemical compounds. Extremely high temperatures may result from the intrusion of molten masses.

Figure 67. Foliation in metamorphosed sedimentary rocks. (U.S. Corps of Engineers.) 288–D–2927.

TABLE 11.—*Common metamorphic rocks*

FOLIATED

Texture	Rock	Characteristics
Coarse-grained	Gneiss	Streaked or banded; imperfectly foliated.
Medium-grained	Schist	Well foliated; splits easily; generally rich in mica.
Fine-grained	Slate	Splits readily into smooth sheets.

NONFOLIATED OR MASSIVE

Mineral content	Rock	Characteristics
Chiefly quartz	Quartzite	Hard and brittle.
Chiefly calcite (or dolomite).	Marble	Hard and brittle.
Chiefly hydrous magnesium silicate.	Some types of serpentine.	Fairly soft; green.

The zone of altered rock formed adjacent to the molten mass is called the contact metamorphic zone (fig. 59). Heat may also be a normal complement to the depth to which the rocks are buried. In this case the earth's own heat produces metamorphism, and the process is called geothermal metamorphism.

Pressure.—The compressive forces which accompany mountain building and other disturbances in the earth's crust are in the main responsible for the pressures to which many rocks are subjected. By the action of these movements rocks are produced in which the crystals, grains, and rock fragments are flattened and elongated or pulverized as a result of the force.

Fluids and gases.—Water is the most important of the liquids and gases involved in metamorphism. Under heat and pressure, water becomes a powerful chemical agent. It acts as a solvent, promotes recrystallization, and takes part in the composition of minerals for which it is essential. Water may be reinforced locally by carbon dioxide and fluids issuing from igneous magmas.

(b) *Classification of Common Metamorphic Rocks*.—Metamorphic rocks, on the basis of their primary structure, are readily divided into two groups: foliates and nonfoliates.

The foliated metamorphic rocks display a pronounced primary banded or layered structure as a result of the differential pressure to which they have been subjected (fig. 67). The nonfoliated or massive metamorphic rocks do not exhibit primary structural features. Metamorphism has apparently been limited to the process of recrystallization without the action of differential pressure. These structural differences are used as the basis for the simplified classification of the common metamorphic rocks listed in table 11.

Gneiss is characterized by rough, relatively coarse banding or foliation. The bands, often of unlike minerals, commonly appear as alternating light and dark lens-shaped masses in the body of the rock. The common minerals or mineral groups present in gneisses are quartz and the feldspar, mica, amphibole, and pyroxene mineral groups. The specific name assigned is determined by the conspicuous mineral in the rock. For example, gneiss with a predominance of the mineral hornblende would be called hornblende gneiss.

Schist is more homogeneous in appearance and composition than is gneiss. The foliae are much thinner, generally more uniform in thickness, finer textured, and often folded or "crinkled" to a much greater degree than the bands of most gneisses. The minerals are, in general, the same as for gneiss, except that talc, chlorites, serpentine, and graphite may be dominant in some schists. As in gneiss, the specific name of a schist is determined by the predominant mineral present.

Slate is very fine grained and homogeneous. Foliation is developed to a very great degree, thus enabling the slate to split into thin sheets with relatively smooth surfaces. The predominant minerals in slate are quartz, mica, chlorite, and sometimes graphite.

Quartzite is a metamorphic rock derived from sandstone by the recrystallization of or cementation by quartz. Quartzite formed by recrystallization may bear little resemblance to the parent rock. That formed by cementation exhibits much of the same physical appearance of the rock from which it was derived. Differentiation between the quartzite resulting from cementation and the sedimentary rock from which it was derived, therefore, lies in the degree of cementation. The degree of cementation is reflected in the appearance of a fresh fracture. In quartzite, the cementing material is as hard as the sand grains and, therefore, the fracture surfaces are smooth and pass through the sand grains. In sandstone the cementing material is weaker than the sand or silt particles, and therefore the fracture surfaces are rough and pass around the sand grains. The rough surface is produced by the sand or silt grains which stand above the fracture surface of the weaker cementing material.

Marble is a massive metamorphic rock and has essentially the same mineral content as the lime-type sedimentary rocks from which it is derived.

99. *Engineering Properties of Rocks*.—When exposed to the weather long enough, the hardest and most durable rock is ultimately broken down by physical and chemical agents into loose, unconsolidated material or soil. Hence, the physical properties of a rock depend to a large extent on the degree of weathering. For fresh, unweathered rock, the physical properties are affected by the properties of the constituent minerals; the degree to which the mineral grains are bound together; the size and arrangement of the grains which produce such structures as banding and foliation; and the degree of fracturing, jointing, and bedding of the rock mass. The physical properties are the least variable in the igneous rocks, excluding the effects of fracturing. Sedimentary rocks, on the other hand, are so variable that it is difficult to characterize their physical properties, which may range between wide limits. Average or typical ranges in properties can be established for sound, unweathered specimens of the common rock types, but in practice each deposit must be evaluated individually. Some of the important rock properties of engineering significance are weight, porosity, strength and hardness, and durability and toughness.

The heaviest rocks are the dark igneous and metamorphic rocks, such as basalt, gabbro, and some schists, which have an average specific gravity of 2.9 to 3.2. The other dense compact rocks, such as granite, slate, marble, and some limestones, have a specific gravity of about 2.5

to 2.8. Lightest are the sedimentary rocks and volcanic rocks such as chalk, tuff, and pumice, which contain many voids. Pumice is generally so light that it floats in water.

In general, the strongest rocks are most dense, and the weakest rocks are most porous. Porosity of granite and similar igneous rocks and of most metamorphic rocks is low, generally less than 1 percent. Basalt is similarly dense, but in certain areas it may contain many small cavities (vesicles) and in other areas it may be extremely vesicular. Porosity of limestone ranges from 0.5 to 15 percent, and in unusual types, such as coquina, up to 25 percent. The porosity of sandstone is typically high, ranging from 5 to 25 percent.

Among the strongest and hardest rocks are quartzites, igneous rocks like granite and basalt, sound gneiss, and some schist. Compressive strengths of 15,000 to 30,000 pounds per square inch or more can be obtained. Some of the hardest, densest sandstones and siliceous limestones approach these strengths. Most limestones, marbles, dolomites, and sandstones, however, are intermediate in strength and hardness with compressive strengths of about 2,500 to 15,000 or 20,000 pounds per square inch. The weakest and softest rocks include tuff, shale, chalk, soft sandstone, salts and gypsum. The softest rocks are easily cut with hand tools. Most limestones and marbles can be sawed. Sandstones, igneous rocks, and the metamorphic rocks composed of quartz and other hard minerals cannot be sawed or cut readily.

The most durable rocks are igneous rocks and massive quartzite and gneiss, but they do not resist fire which causes cracking and spalling. Of these rocks, the fine-grained varieties, such as basalt, are generally tougher and wear better under abrasion than coarse-grained varieties. Foliated and laminated metamorphic rocks, such as schist and slate, are hard but split readily and fall apart under abrasion. In general, limestones and sandstones are only moderately tough under abrasion. Limestone and sandstone with limy cement are corroded by water or atmosphere containing acids. Chalk and some tuff are soft and easy to handle in construction but harden on exposure. Shale is weak, tends to soften when wet, and disintegrates rapidly when exposed to weather.

E. SURFACE EXPLORATIONS

100. *General*.—A relationship between topographic features or landforms and the characteristics of the subsurface soils has been shown repeatedly. Thus, the ability to recognize terrain features on maps, on airphotos, and during field reconnaissance, combined with an elementary understanding of geological processes, can be of great assistance in locating sources of construction materials and in making a general appraisal of foundation conditions.

The mechanisms which develop soil deposits are water action, ice action, and wind action for transported soils; and the mechanical-chemical action of weathering for residual soils. For the transported soils, each type of action tends to produce a group of typical landforms, modified to some extent by the nature of the parent rock. Soils found in similar locations within similar landforms usually have similar physical properties. The engineer responsible for explorations for foundations and construction materials for dams should become familiar with landforms and with the soils associated with them. Such knowledge is of great assistance during the reconnaissance stage of investigations, and it is useful in controlling the extent of investigation for feasibility and specifications designs.

101. *Fluvial Soils*.—(a) *Definition*.—Soils whose properties are affected predominantly by the action of water to which they have been subjected are designated fluvial soils. Their common characteristic is roundness of individual grains. Frequently, there is considerable sorting action, so that a deposit is likely to be stratified or lensed. Individual strata may be thin or thick, but the material in each stratum will have a small range of grain sizes. The three principal types of fluvial soils, reflecting the water velocity during deposition, are identified as torrential outwash, valley fill, and lake beds.

(b) *Torrential Outwash.*—The typical landforms of this type are alluvial cones and alluvial fans. They vary in size and character from small, steeply sloping deposits of coarse rock fragments to gently sloping plains of fine-grained alluvium several square miles in area. The deposition results from the abrupt flattening of the stream gradient that occurs at the juncture of mountainous terrain and adjacent valleys or plains. Figure 68 is an airphoto and topographic map of an alluvial fan. The coarser material is deposited first; hence, it is found on the steeper slopes at the head of the fan, while the finer material is carried to the outer edges. In arid climates where mechanical rather than chemical weathering predominates, the cones and fans are composed largely of rock fragments, gravel, sand, and silt. In humid climates where the landforms have less steep slopes, the material contains much more sand, silt, and clay.

Sand and gravel from these deposits are generally subrounded to subangular in shape, reflecting movement over relatively short distances, and the deposits have only poorly developed stratification. The torrential outwash deposits are likely sources of sand and gravel for pervious and semipervious embankment materials and for concrete aggregate. The presence of boulders is likely to limit their usefulness for some types of fill materials. The soils are typically skip-graded, resulting in a GP or SP classification. Because this type of deposit is consolidated only by its own weight, settlement should be anticipated when it is used as a foundation for a structure. Normally, torrential outwash deposits are too pervious and do not provide satisfactory abutments for dams. If it is necessary to locate a dam in the vicinity of such a deposit, the dam should be placed along the upstream edge of the fan.

(c) *Valley Fill.*—Valley fill or flood-plain deposits are generally finer, more stratified, and better sorted than are torrential outwash deposits. The degree of variation from the latter depends largely on the volume and on the gradient of the stream. The surface of these stream deposits is nearly flat. The nature of the materials in the deposit can be deduced by the characteristics of the stream. Braided streams usually indicate the presence of silt, sand, and gravel, whereas meandering streams in broad valleys are commonly associated with fine-grained soils (silts and clays).

Flood-plain deposits of sand and gravel are common sources of concrete aggregate and pervious shell materials for dam embankments. The soils in the various strata of river deposits may range from pervious to impervious; hence, the permeability of the resulting material sometimes can be influenced appreciably by the choice of depth of cut. Presence of a high water table is a major difficulty in the use of these deposits, especially as a source of impervious material. Also, removal of materials from the reservoir floor just upstream from a damsite may be undesirable when a positive foundation cutoff is not feasible. When considering borrowing from a river deposit downstream from a dam, it should be remembered that such operations may change the tailwater characteristics of the stream channel and that the spillway and outlet works will have to be designed for the modified channel conditions. If tailwater conditions will be affected, borrow operations must produce a predetermined channel and explorations for the specifications stage must accurately define conditions within the channel.

Stream deposits vary in competency as foundations for dams. Potential difficulties include high water table, variation in soil properties, seepage, consolidation, and possibly low shear strengths. Although valley-fill deposits are usually acceptable as foundations for low dams, their depths and characteristics must be investigated thoroughly during the feasibility stage. An important type of stream deposit is the terrace. It represents an earlier level of valley development into which the river subsequently has become entrenched. Remnants of such deposits are recognized by their flat tops and steep faces, usually persistent over an extended reach of the valley. Examination of the eroded faces facilitates classification and description of the deposits, and the extent of the drainage network on the face is helpful in determining relative permeability. Free-draining material has almost no lateral erosion

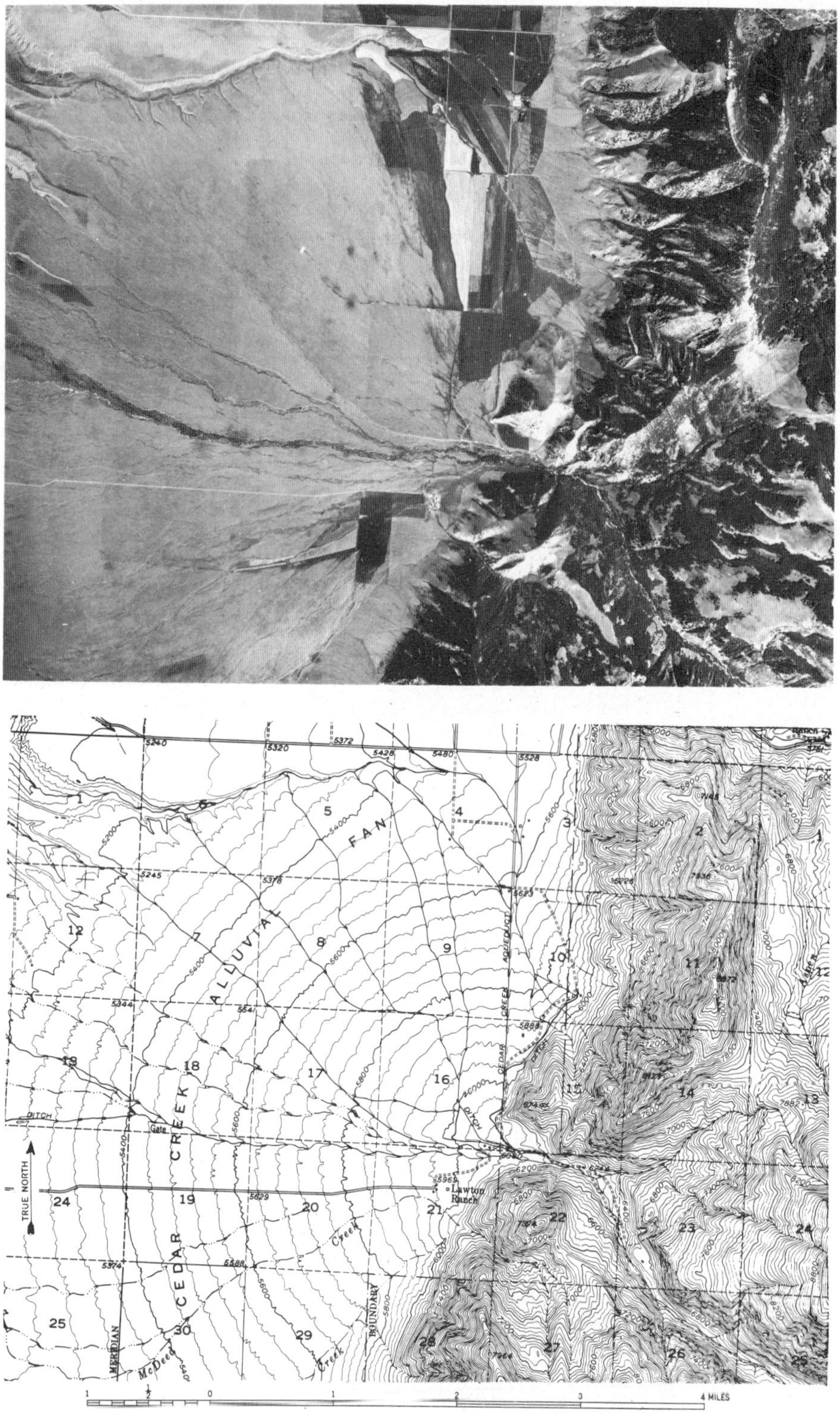

Figure 68. Aerial view and topography of an alluvial fan, a potential source of sand and gravel. 288–D–2928 and 288–D–2929.

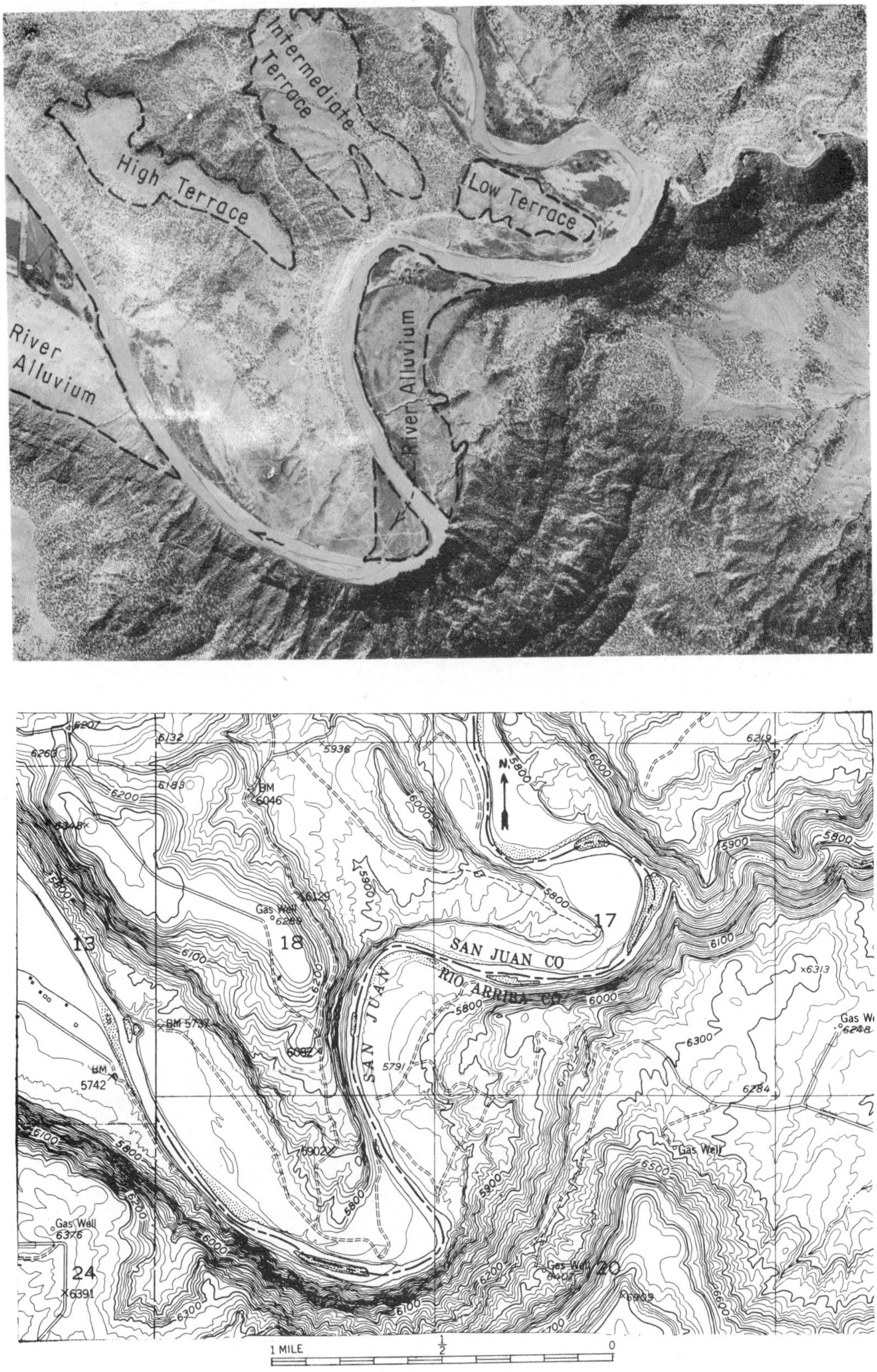

Figure 69. Aerial view and topography of stream deposit showing river alluvium and three levels of gravel terraces. (Photo by Geological Survey.) PX–D–16259.

Figure 70. Aerial view and topography of terminal moraine of continental glaciation. (Photo by U.S. Commodity Stabilization Service.) PX–D–16260.

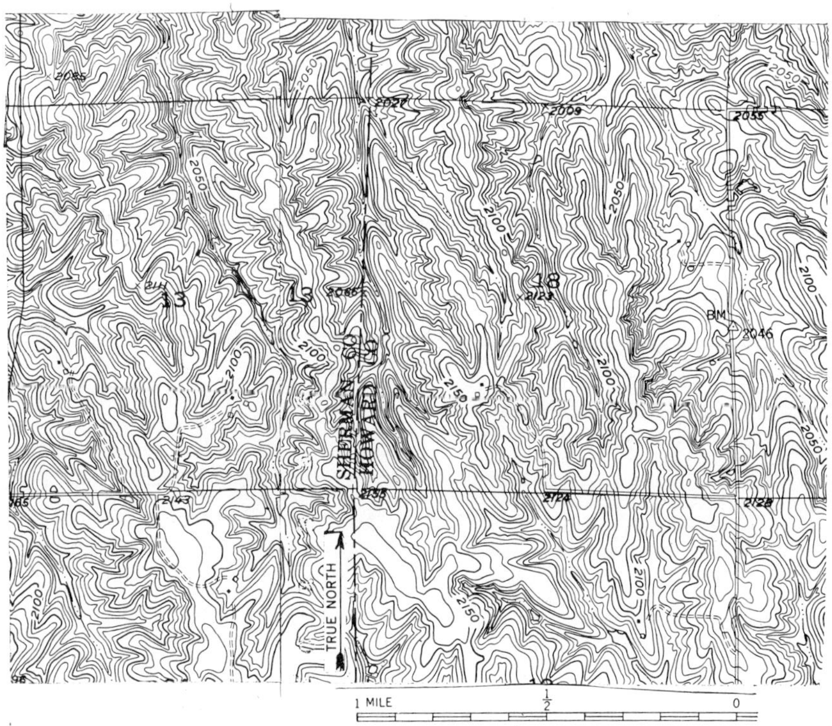

Figure 71. Aerial view and topography of loess identified by smooth silt ridges; usually parallel, right-angle drainage patterns; and steep-sided, flat-bottomed gullies and streams. (Photo by U.S. Commodity Stabilization Service.) PX–D–16263.

channels, whereas impervious clays are finely gullied laterally. These terraces are found along streams throughout the United States and are prevalent in the glaciated regions of the northern sections. Sands and gravels from terrace deposits usually occur in layers and are well graded. They provide excellent sources of construction materials. Figure 69 is an airphoto and topographic map showing river alluvium and terrace deposits.

(d) *Lake Beds.*—Lake sediments, or lacustrine deposits, are the result of sedimentation in still water. Except near the edges of the deposits where alluvial influences are important, the materials are very likely to be fine-grained silt and clay. The stratification is frequently so fine that the materials appear to be massive in structure. Lacustrine deposits are recognizable by their flat surfaces surrounded by high ground. The materials they contain are likely to be impervious, compressible, and low in shear strength. Their principal use is for impervious cores of earthfill dams. Moisture control in these soils is usually a problem, since the water content is very difficult to change. Lake sediments almost certainly will provide very poor foundations for structures. Their use as foundations for dams is beyond the scope of this text, and should not be attempted without special field and laboratory testing and the advice of a specialist.

102. *Glacial Deposits.*—(a) *Definition.*—The results of the advances and retreats of the great North American continental ice sheets during glacial times are represented by recognizable landforms which are important sources of construction materials and which may be encountered in the foundations of dams. Smaller scale evidences of ice action are found in high mountain valleys of the Rocky Mountains and the Sierra Nevadas; in some instances the glaciers still exist. Glacial deposits (glacial drift) are generally heterogeneous and erratic in nature; hence, they are difficult to explore economically. They contain a wide range of particle sizes, from clay or silt up to huge boulders; and the particle shapes of the coarse grains are typically subrounded or subangular, sometimes with flat faces. Deposits of the glacier proper can be distinguished from deposits formed by the glacier melt water.

(b) *Morainal Deposits.*—Glacial till is that part of the glacial drift deposited directly from the ice with little or no transportation by water. It consists of a heterogeneous mixture of boulders, cobbles, gravel, and sand in an impervious matrix of generally nonplastic fines. Gradation, type of rock minerals, and degree of weathering found in till vary considerably, depending on the type of rocks in the path of the ice and the degree of leaching and chemical weathering. Glacial tills usually produce impervious materials with satisfactory shear strength, but removal of boulders will be necessary in order for the soil to be compacted satisfactorily. Fairly high inplace density makes morainal deposits satisfactory for the foundation of low dams. Typical land forms containing till are the *ground moraine* or till plain, which has a flat to slightly undulating poorly drained surface; the *end (or terminal) moraine*, a ridge at right angles to the direction of ice movement, which often curves so that its center is farther downstream than its ends; and *lateral* or *medial-moraines* which occur as ridges parallel to the direction of ice movement. Low, cigar-shaped hills occurring on a ground moraine, with their long axis parallel to the direction of ice movement are called drumlins. They commonly contain unstratified fine-grained soils. Figure 70 shows a typical terminal moraine in an airphoto and topographic map.

(c) *Glacial Outwash.*—Deposits from the glacial melt water are of several types. *Glacial outwash plains* of continental glaciation and their alpine glaciation counterparts, the *valley trains*, commonly contain poorly stratified silt, sand, and gravel similar to the alluvial fans of torrential outwash which they resemble in mode of formation. *Eskers* are prominent winding ridges of sand and gravel which are the remnants of the beds of glacial streams that flowed under the ice. They generally parallel the direction of ice movement, have an irregular crestline, are characterized by steep flanks (about 30°), and are 20 to 100 feet high. Eskers usually contain clean sand and gravel with some boulders and silty strata which are in irregular, poor to fair stratification. They

Figure 72. A 90-foot nearly vertical cut in loess formation—Nebraska. GE–144–11.

are excellent sources of pervious materials and concrete aggregate. *Kames* are low, dome-shaped partially stratified deposits of silt, sand, and gravel formed by hidden glacial streams. They are round to elliptical in plan with the long axis generally at right angles to the direction of ice movement. Their slopes, contents, and uses are similar to eskers. Glacial lake deposits formed in temporary lakes during the ice age are generally similar in character and in engineering uses to fluvial lacustrine deposits. However, they are normally more coarsely stratified (varved) than are the recent lake deposits, and they may contain fine sand.

103. *Aeolian Deposits*.—Soils deposited by the wind are known as Aeolian deposits. The two principal classes that are readily identifiable are dune sands and loess. Dune sand deposits are recognizable as low elongated or crescent-shaped hills, with a flat slope windward and a steep slope leeward of the prevailing winds. Usually, these deposits have very little vegetative cover. The material is very rich in quartz and its characteristics are limited range of grain size, usually in the fine or medium range sand; no cohesive strength; moderately high permeability; and moderate compressibility. They generally fall in the SP group of the Unified Soil Classification System.

Loessial deposits of windblown dust cover extensive areas in the plains regions of the temperate zone. They have a remarkable ability for standing in vertical walls. Figure 71 shows typical loessial topography by map and airphoto. Loess consists mainly of angular particles of silt or fine sand, with a small amount of clay that binds the soil grains together. In its natural state true loess has a characteristic structure formed by remnants of small vertical root holes that makes it moderately pervious in the vertical direction. Although of low density, the naturally dry loessial soils have a fairly high strength because of the clay binder. This, however, may be lost readily upon wetting, and the structure may collapse. When remolded, loessial soils are impervious, moderately compressible, and of low cohesive strength. They usually fall in the ML group or the boundary ML–CL group of the Unified Soil Classification System. Figure 72 shows a 90-foot almost vertical cut in loess.

Aeolian deposits are normally regarded with suspicion, especially as foundations for dams. Such deposits should be avoided if it is practicable to do so; however, they can be used when properly explored and evaluated. Information on the inplace density of Aeolian soils is of vital importance in planning their usefulness for foundations of structures.

104. *Residual Soils.*—As weathering action on rock progresses, the rock fragments are gradually reduced in size until the total material assumes all the characteristics of soil. It is impracticable to define clearly the dividing line between rock and residual soil, but for engineering purposes it may be considered soil if the material can be removed by commonly accepted excavating methods.

A distinguishing feature of many residual soils is that the individual grains are angular but soft. Handling of the material during construction reduces appreciably the grain size, which makes it difficult to predict its performance by laboratory tests. Appreciable settlement and change of material characteristics after utilization also are detrimental factors. As a consequence, such residual soils are best avoided if other types can be readily secured.

It is difficult to recognize and appraise residual soils on the basis of topographic forms. Their occurrence is quite general wherever none of the other types of deposits, with their characteristic shapes, are recognizable and where the material is not clearly rock in place. Talus (fig. 43) and landslides are easily recognizable forms of residual materials moved by gravity. Also, erosional features may be helpful in evaluating a residual deposit. Since the type of parent rock has a very pronounced influence on the character of the residual soil, the rock type should always be determined in assembling data for the appraisal of a residual deposit. The soils usually provide satisfactory foundations for small structures if the parent rock itself is satisfactory. The degree to which alteration has progressed largely governs the strength characteristics.

F. SUBSURFACE EXPLORATORY METHODS

105. *Test Pits, Trenches, and Tunnels.*—(a) *General.*—Open test pits, trenches, and drifts afford the most complete information of the ground penetrated and also may permit examination of the surface of foundation bedrock. When the depth of overburden and groundwater conditions permit their economical use, these methods are recommended for foundation exploration in lieu of relying solely on borings. In prospecting for embankment materials or concrete aggregate containing cobbles and boulders, open pits and trenches may be the only feasible means of obtaining the required information.

(b) *Test Pits.*—Test pitting is an effective means of exploring and sampling earth foundations and construction materials. Their use facilitates inspection, sampling, and making density tests. The depth of a test pit is determined by investigational requirements but is usually limited to a few feet below water table. The minimum recommended cross section for a hand-dug pit is 3 by 5 feet. Dragline, backhoe, clamshell, and bulldozer pits are usually more economical than hand-dug pits for comparatively shallow materials explorations, but are not practicable where a depth of more than about 15 feet is desired. Where the soil is hard, pneumatic paving breakers or spades operated by small trailer-mounted air compressors will facilitate progress in hand-dug pits. The use of explosives to break up large boulders is a common practice.

In hand-dug pits the materials are removed from the hole by buckets operated from a hoist or windlass, which should be equipped with a ratchet device for safety. During excavation, the bottom of the hole should be kept fairly level and of full size so that each lift may represent the corresponding portion of the deposit in quantity and quality.

At the surface the excavated material should be placed in an orderly manner around the pit, and marked stakes should be driven to indicate depth of pit from which the material came in order to facilitate logging and sampling. All hand-dug pits should be cribbed. A convenient method of cribbing, with 3- by 6-inch lumber, is shown in figure 73. In loose materials it is advisable to keep the space between the pit walls and the cribbing at a minimum, also to pack the space with hay or excelsior and to keep the bottom of the cribbing close to the bottom of the pit.

Deep test pits should be ventilated to prevent accumulation of dead air. For this purpose connected lengths of stovepipe, starting slightly above the floor and extending about 3 feet above the mouth of the pit, have been found satisfactory. Test pits left open for inspection should be provided with covers or barricades for safety.

When water is encountered in the pit, a pumping system is required for further progress. Small, portable, gas-powered, self-priming, centrifugal pumps can be used. It is desirable that the suction hose be one-half inch larger in diameter than the discharge opening of the pump and not more than 15 feet long. This requires resetting the pump in the pit (on a frame attached to the cribbing) at intervals of about 12 feet. When the gas engine is in the pit, it is necessary to pipe the exhaust well away from the pit. Unwatering test pits is usually expensive and often is unwarranted. Figures 74 and 75 show different types of test pits.

(c) *Trenches.*—Test trenches are used to provide a continuous exposure of the ground along a given line or section. In general, they

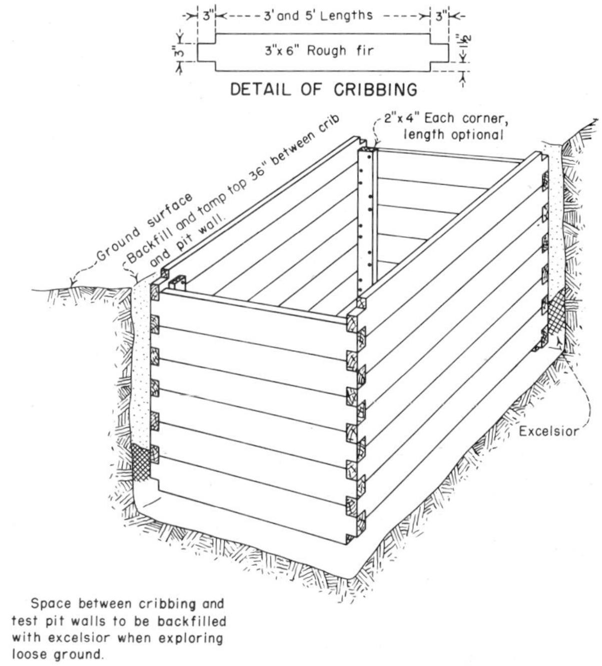

Figure 73. Test pit cribbing. 288–D–2624.

serve the same purpose as the open test pits but have the added advantage of disclosing the continuity or character of particular strata. They are best suited for shallow exploration (10 to 15 feet) on moderately steep slopes, but they have been used advantageously on fairly flat ground.

The fieldwork consists of excavating an open trench from the top to the bottom of the slope to reach representative undisturbed material. Either a single slot trench down the face of the slope or a series of short trenches spaced at appropriate intervals along the slope can be excavated. Depending on the extent of the investigation required, either pick and shovel methods, or bulldozers, ditching machines, or draglines can be used. A trenching layout suitable for materials investigations is shown in figure 76. Figures 77 and 78 are photographs of bulldozer trenches. Safety precautions should be used in deep trenches to prevent accidents caused by caving ground.

The profile exposed by these trenches may represent the entire depth of significant strata in an abutment of a dam; however, their shallow depth may limit exploration to the upper weathered zone of foundations. Trenching

Figure 74. Excavation of hand-dug test pit in borrow area. SB–69–R2.

permits visual inspection of the soil strata which facilitates logging of the profile and selection of samples. It also aids in obtaining large undisturbed samples or large disturbed individual or composite samples. Trenches in sloping ground have the advantage of being self-draining.

(d) *Tunnels.*—Tunnels or drifts have been used to explore areas beneath steep slopes or back of clifflike faces. For such purposes the exploratory tunnel or drift is usually roughly rectangular in shape and approximately 5 feet wide and 7 feet high. The placing of lagging, when required for side and roof supports, should follow excavation as closely as practicable. Excavation of exploratory tunnels may be a slow, expensive process; consequently, this type of investigation should be utilized only when no other method will supply the required information. Logging and sampling of exploratory tunnels should proceed concurrently with excavation operations if possible.

106. *Auger Borings.*—Auger borings often provide the simplest method of soil investigation and sampling. They may be used for any purpose where disturbed samples are satisfactory and are valuable in advancing holes to depths at which undisturbed sampling by thin-walled tubes is required. Depths of auger investigations are, however, limited by the

Figure 75. Equipment-excavated test pit showing elevation of ground, bedrock, and location of samples. LM–166.

groundwater table and by the amount and maximum size of gravel, cobbles, and boulders as compared with the size of equipment used. Hand-operated post-hole augers 4 to 12 inches in diameter can be used for exploration up to about 20 feet deep (figs. 79 and 80). However, with the aid of a tripod, holes up to 80 feet deep have been excavated successfully and economically (fig. 81). Machine-driven augers are of three types: helical augers 3 to 16 inches in diameter; disk augers up to 42 inches in diameter; and bucket augers up to 48 inches in diameter.

Figure 77. Tractor with bulldozer excavating a test trench. SB–61–R2.

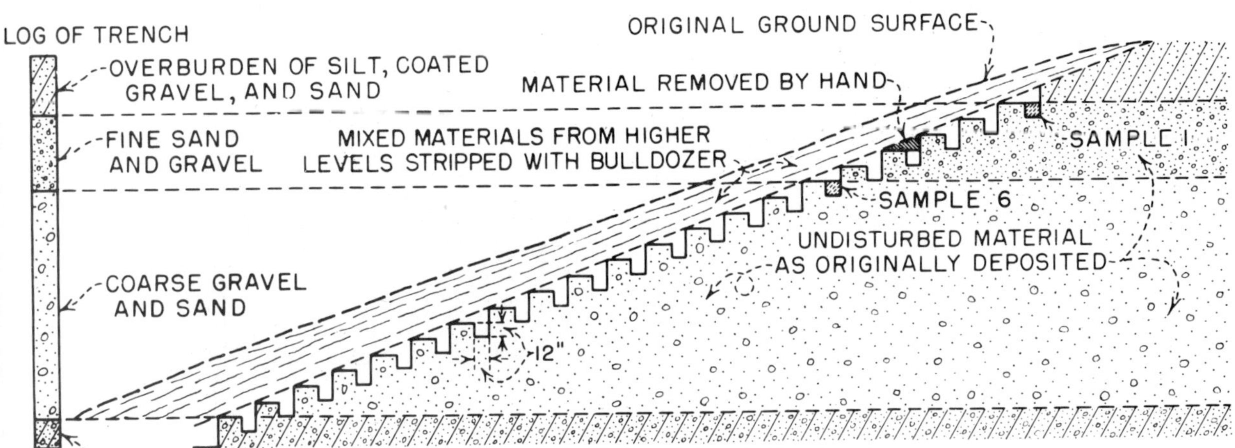

Figure 76. Trenching, a low-cost method of obtaining soil samples. 288–D–2623.

Figure 78. Bulldozer-excavated trench in steep abutment area. 288–D–2931.

Figure 79. Exploring a borrow area with a hand auger. 288–D–2930.

Double-tube auger samplers of the type shown in figure 82 are used to obtain undisturbed samples of fine-grained low density soils from above the groundwater table which have a tendency to be disturbed by the more conventional mechanical sampling methods. The auger sampler does not require drilling fluid or excessive down pressure in the opera-

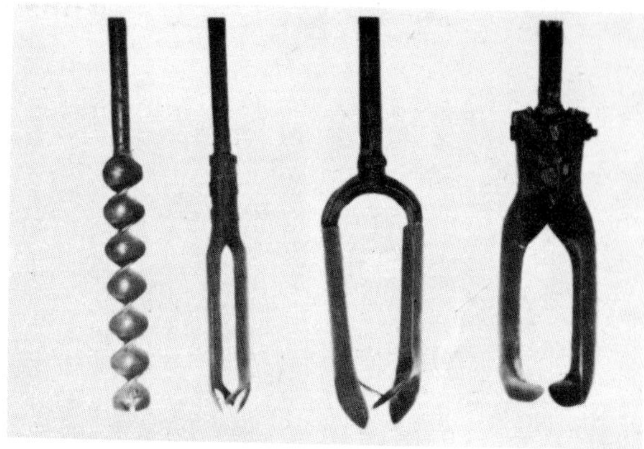

Figure 80. Types of hand augers (2-inch helical; 2- and 6-inch Iwan; and 6-inch Fenn (adjustable)). PX–D–16998.

tion and is suitable for sampling both borrow and foundation materials. The sampler consists of a nonrotating inner barrel, which receives the sample cut by bits attached to the rotating hollow-stem outer auger. Cuttings

Figure 81. Exploration of borrow area with project-designed demountable tripod used to facilitate hand augering of 4-inch holes to a depth of 55 feet. The guide ring is set at 30 feet. 404–1006B.

from the sampling operation are carried up by the auger flights and are deposited in the sludge barrel, which is formed by a continuation of the auger flights. The inner sampler barrels are standard thin-walled drive tubes and serve as shipping and storage containers.

Figure 83 shows a disk auger, and figure 84 shows a bucket auger.

An auger boring is made by turning the auger the desired distance into the soil, withdrawing it, and removing the soil for examination and sampling. The auger is inserted into the hole again, and the process is repeated. Pipe casing is required in unstable soil in which the borehole fails to stay open, and especially where the boring is extended below the groundwater level. The inside diameter of the casing must be slightly larger than the diameter of the auger used.

The soil auger can be used both for boring the hole and for bringing up disturbed samples of the soil encountered. It operates best in somewhat loose, moderately cohesive, moist soil. Holes are usually bored without the addition of water; but in hard, dry soils or in cohesionless sands the introduction of a small amount of water into the hole will aid the drilling and sample extraction. Rock fragments larger than about one-tenth of the diameter of the hole cannot be successfully removed by normal augering methods. Large-sized holes permit examination of the soils in place and, therefore, are preferred for foundation investigations.

107. *Rotary Drilling*.—One of the most important tools for subsurface exploration of dams is the diamond drill. This is a rotary drill with a core barrel, a diamond bit, and a hydraulic or screw feed originally developed for drilling through and sampling either hard or soft rock. Various sizes of rotary drill core bits and accessories are shown in figure 98. The diamond drill may actually be operated with a variety of bits, depending on the hardness of the material to be penetrated. Rotary drill equipment is manufactured in a wide variety of forms, which vary from highly flexible to extremely specialized equipment, from lightweight and highly mobile equipment to heavy stationary plants, and in size of hole and core

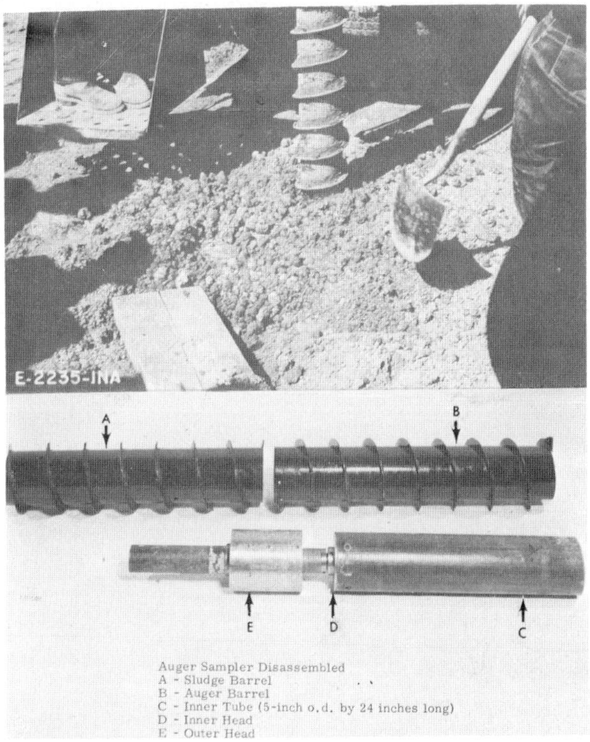

Figure 82. Double-tube auger sampler. E–2235–INA, E–2123–INA.

Figure 83. Disk auger used to explore borrow areas of fine-grained soils. FK–391–CV.

Figure 84. Bucket auger used in exploration of a borrow area containing gravel particles. 288–D–2932.

range from less than 1 inch to 3 feet or more. They are capable of drilling to depths of hundreds of feet.

One of the most advanced rotary drilling developments in recent years has been the introduction of wire-line drill rod and core barrel assemblies. This is especially valuable in deep hole drilling since the method eliminates trips in and out of the hole with the coring equipment. With the wire-line technique, the core barrel becomes an integral part of the drill rod string. The drill rod serves as both a coring device and casing as it is usually not removed except while making bit changes. Core samples are retrieved by removal of the inner barrel assembly from the core barrel through the drill rod. This is accomplished by lowering an overshot or retriever, by wire-line, through the drill rod to release a locking mechanism built into the inner barrel head. The core is removed and the inner barrel returned through the drill rod and coring continued. Present-day wire-line core sizes are limited to a maximum diameter of 3⅜ inches.

Essential accessories for a drill rig are a cathead winch and derrick for driving casing and for hoisting and lowering the drill rods; a pump for circulating water to the bit and for flushing and water testing the hole; a water meter; and the necessary driving weights, bits, drill rods, and core barrels. Cased holes are usually required except when drilling through solid rock or stiff, cohesive soils. A short collar pipe about 5 feet long is commonly used at the top of the hole. The use of drilling mud, including hole stabilizer compounds, sometimes avoids the need of casing in overburden soil, but the foundation cannot be effectively water tested when mud is used.

At least two weights should be available, a 140-pound weight for standard penetration tests and a 250- to 400-pound weight for driving and removing casing pipe. The weights are raised by pulling tight on an attached rope threaded through a sheave at the top of the derrick and wound three or four times on the revolving cathead winch. Sudden loosening of the rope permits the weight to drop on the driving head attached to the casing. Various types of chopping bits are used to facilitate the driving of casing through soils containing cobbles and boulders. Large boulders must be either blasted with explosives or drilled with a diamond bit or a roller rock bit. The casing is raised several feet prior to blasting. Figure 85 shows a diamond drill rig with derrick.

The accuracy and dependability of the records furnished by diamond drilling depend largely on the size of the core in relation to the kind of material drilled, the percentage of core recovery, the behavior during drilling, and the experience of the drill crew. Since a rock that will core well in an NX hole may break up badly in an EX hole, it is important to use the largest practicable size diameter hole and core barrel. Recovery of core is much more important than making rapid progress in drilling the hole. Portions of a core that are lost will probably represent shattered or soft, incompetent rock, whereas the recovered portions represent the best rock, from which an overoptimistic

Figure 85. Diamond-drill rig used in exploration of a dam foundation. E–2255–4NA.

evaluation of the foundation likely will be made. A reasonably large percentage of core recovery, on the other hand, will provide a more continuous section of the materials passed through. The cores provide information on the character and composition of the different formations, with evidence of the spacing and tightness of joints, seams, fissures, and other structural details. When drilling in soft materials, the water circulation must be reduced or stopped entirely, and the core recovered "dry," even though a marked delay in operations may result.

Core barrels are described in section 112. The barrel is fitted with a drill bit and is lowered into the hole by the hollow drill rod. Circulation of the wash water should be started before the core barrel reaches the bottom of the hole to prevent cuttings or sludge from entering the core barrel at the start of coring. The optimum rotation speed of drilling varies with the type of bit used, the diameter of core barrel, and the kind of material to be cored. Excessive rotation speed will result in chattering and rapid wear of the bit and will break the core. Low speed results in less wear and tear on the bit and better cores but lower rates of progress. Approximate ranges of rotation speeds used in medium-hard rock are 300 to 1,500 revolutions per minute for diamond bits and 100 to 500 revolutions per minute for metal bits.

The rate at which the coring bit is advanced depends on the firmness of the material and the amount of pressure applied downward on the bit as well as on its speed of rotation. This pressure must be carefully adjusted by the driller; excessive pressures will cause the bit to plug and may shear the core from its base. The bit pressure is controlled by a hydraulic or a screw feed on the drilling machine. The weight of the column of drill rod will seldom be in excess of the optimum bit pressure for the coring of medium and hard rock, and it is frequently necessary to supply additional downward pressure.

Since the hole left in the rock is clean and the seams and fissures are not sealed off by the action of the drill, opportunity is afforded for making percolation tests to indicate the permeability of the strata and to determine probable leakage through open joints or fissures in the rock. Gravity or pressure tests are made as described in section 118. Large water losses or inflows into the holes during drilling are recorded as indicating, respectively, the pres-

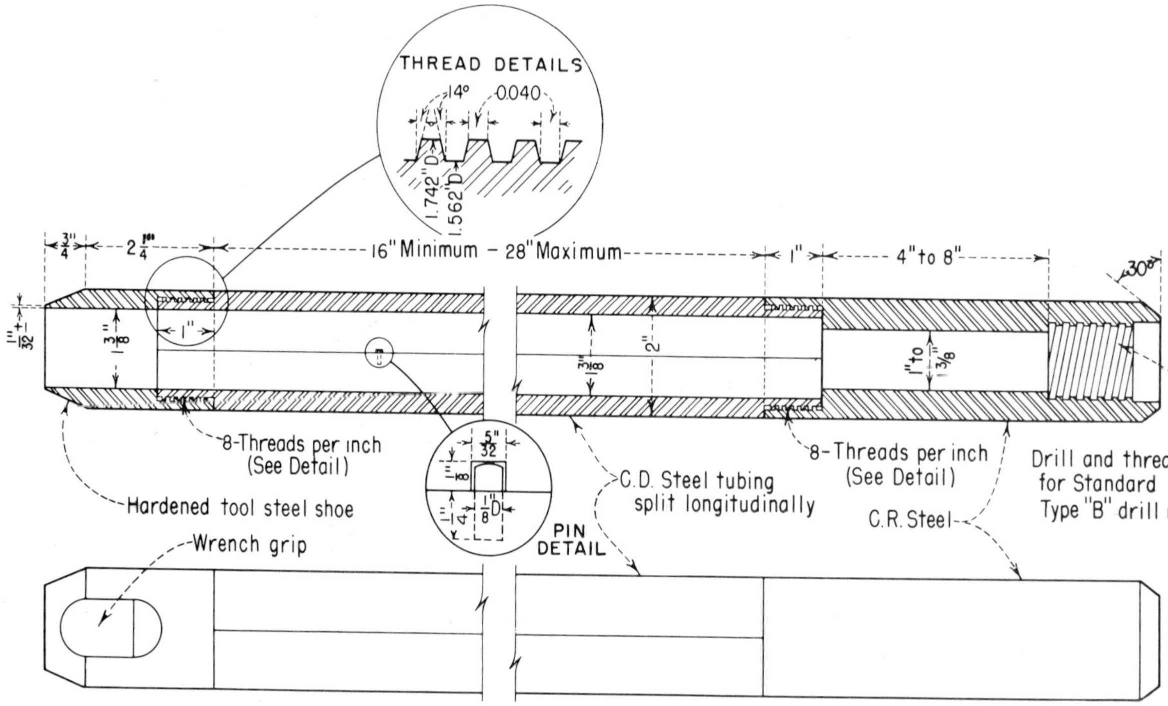

Figure 86. **Standard split-barrel sampler. 101–D–169.**

Figure 87. **Split-barrel** sampler disassembled. 288–D–2933.

ence of large openings in the formation or the existence of an underground flow. Completed holes should be capped to preserve them for use in groundwater level observations or as grout holes, or for reentry if it is later found desirable to deepen the hole. Casing is, of course, required for those sections of the hole in loose material or unconsolidated ground.

Although the rotary drill is designed primarily for penetrating rock rather than soil, sample barrels and cutting bits have been developed which are satisfactory in many kinds of soil deposits. Double-tube soil samplers of the Denison, Denver, and Pitcher types are capable of obtaining undisturbed samples of sands, silts, and clays for laboratory testing [6]. However, for foundations of small dams within the scope of this text, borings for the standard penetration tests described in the next section should preclude the need for Denison-type samples. A detailed explanation of these samplers is given in Designation E-2 of the Bureau's "Earth Manual" [6]. There may be times when simple, undisturbed, thin-wall tube samples would appropriately be taken using drilling equipment. For recommended procedures, reference is made to Designation E-2 of the Bureau of Reclamation's "Earth Manual" [6] and American Society for Testing Materials, Designation: D 1587.

108. *Standard Penetration Borings.*—This is a standardized procedure for taking subsurface soil samples with a split-tube penetration sampler and, at the same time, measuring the in-place strength, firmness, and denseness of the foundation. The split-tube penetration spoon is a thick-wall sampler that collects a core sample $1\frac{3}{8}$ inches in diameter. The core barrel can be separated for examination and removal of the sample, and the resistance to penetration can be measured in terms of the number of blows of a standard weight required to produce 1 foot of penetration. In performing this test it is essential that a standardized procedure be used and that information on soil type, water content, and penetration resistance be recorded. In a range of 5 to 50 blows per foot, fairly reliable correlation with engineering properties has been established for several varieties of material provided the water content is high. The relationship between penetration resistance and relative density for either dry or saturated sands is given in the Bureau's "Earth Manual" [6].

To assure that the material to be tested is undisturbed, use of bottom-discharge fishtail bits or jetting through an open pipe to clean out the hole should not be permitted. Side-discharge bits are permissible below groundwater. Either a rotary drill or an earth auger can be used to advance and clean the hole. Casing is used when drilling in sand, soft clay, or other material that will not allow the hole to stay open. Holes for penetration testing should be at least BX casing size ($2\frac{15}{16}$-inch outside diameter).

The dimensions of the sampler are shown in figure 86, and figure 87 shows the sampler disassembled. The driving shoe is of hardened steel, which must be replaced or repaired when it becomes dented or distorted. Split-tube samplers other than 2 inches in outside diameter may be used; however, all penetration records made with those samplers should be conspicuously marked as to the size of sampler used. The significance of the penetration records in the design of foundations for small dams is discussed in sections 133 and 134.

Penetration tests should be made continuously in exploring foundations for dams, except where the resistance of the soil is too great. Any loss of circulation in drilling fluid during advancing of holes should be noted and recorded on the log. When casing is used, it should not proceed in advance of the sampling operation. Complete groundwater information shall be obtained, including groundwater level and elevations at which water was lost or water under pressure was encountered. Where casing is used, groundwater levels should be measured before and after the casing is pulled.

With the split-tube sampler resting on the bottom of the hole, the sampler is driven 6 inches into the soil. This seats the sampler and assures that the material being sampled has not been disturbed by the drilling procedure. The penetration test is then started, and the number of blows (30-inch drops of the 140-pound drive weight) required to drive the sampler an additional 12 inches is recorded as the penetration resistance. Since 50 or more blows per foot indicates a very dense or very firm material, the penetration test should be carried only to this number of blows. After the sampler has been driven 12 inches or given 50 blows, it is then raised to the surface and opened.

Prior to reinsertion of the sampler, the hole should be cleaned out by an auger or by a rotary drill down to the level of the bottom of the previous test. The test is then repeated. In this manner, an almost continuous record of the material encountered is obtained by sampling from the split-tube, and penetration resistance values are available for alternate 1-foot increments of depth of hole. Figures 88 and 89 show the standard penetration test being performed. Information on sampling and logging of standard penetration holes is given in parts G and H of this chapter, respectively. The penetration resistance values should be recorded and plotted on a log similar to that shown in figure 102. A more detailed description of the standard penetration test is given in Designation E-21 of the Bureau's "Earth Manual" [6].

109. *Geophysical Methods*.—Geophysical methods have been used successfully in foundation investigations as an aid in selecting the best of several damsites and to assist in outlining subsurface conditions. If the various geologic materials which are present are sufficiently different in physical properties, seismic or resistivity methods may be used to secure approximate solutions to such problems as:

(1) Determine depth to bedrock.
(2) Locate buried bedrock channels.
(3) Determine depth of rock weathering.
(4) Determine major layering conditions in soils materials.
(5) Locate boundaries of different materials.

In the seismic method, vibrations artificially created at the ground surface travel downward through the soils until they are refracted back to the surface by a denser material such as bedrock. The vibration travel-time is accurately measured and used in the calculation of the approximate depth of the refracting material. Equipment may range from a small seismograph unit using a hammer blow as the vibration source to larger equipment, having greater depth capability, in which dynamite is used as the energy source.

The resistivity method operates on the principle that different geologic materials may offer different resistances to the passage of an electric current. The current is sent into the ground through two electrodes. The voltage developed is picked up by two other electrodes. By changing the distance between the electrodes, the variation of resistance with depth can be determined.

Both methods require the use of specialized equipment by competent operators; the basic geophysical data must be analyzed and interpreted in terms of subsurface geological conditions by experienced geologists. Both methods may be used without drilling and, by providing approximate data, may serve to expedite reconnaissance investigations and assist in the location of future drilling.

Geophysical survey work conducted in conjunction with limited drilling is the preferred approach. Thus, the geophysical data are calibrated by the facts secured from the drill holes and in turn serve to "fill in" between drill holes. Geophysical results may also help to point out areas which are geologically different from what might have been expected.

G. SAMPLING

110. *General.*—The purposes and uses of samples are numerous. They are required to identify and classify soil or rock properly. They are essential for field density and moisture determinations and for laboratory tests on earth materials, concrete aggregate, and riprap. To a large degree, samples determine the results of explorations for foundations and construction materials for dams. Obviously, erroneous conclusions will be drawn if the samples are not truly representative of the explorations.

Samples are broadly classified as disturbed and relatively undisturbed. Disturbed samples are those in which no effort is made to maintain the soil structure. Such samples may be secured for general observation and inspection, soil classification, moisture determination, or to obtain compaction characteristics. Relatively undisturbed samples vary from moderately disturbed split-barrel samples to almost completely undisturbed hand-cut samples. The importance of obtaining representative samples cannot be overemphasized. This requires considerable care because of the variations in natural deposits of earth materials. Representative samples are comparatively easy to secure in trenches, test pits, and cut banks, because the various undisturbed strata can be inspected visually. Boreholes, however, do not permit a visual inspection of the profile, and it is more difficult to secure representative samples by use of boreholes.

Samples may be either individual or composite. An individual sample is a single sample representing one stratum or type of soil. A group of individual samples representing all the strata from a single test pit may be combined in the same ratio as the thicknesses of the strata from which the samples were taken to form a composite sample. Also, a sample may be composited to any desired ratio of the

Figure 88. Making a standard penetration test by use of a drill rig. PX–D–34356.

soil types involved. It is usually preferable to obtain individual samples from each stratum of a soil deposit rather than to composite the sample in the field, since the most desirable depth of excavation may be determined by trying several composite samples in the laboratory. Representative samples of stockpiles are essentially composite samples in that parts are taken at random over the pile. If the sample is taken from one location, the selected area should be excavated deep enough to avoid segregated materials.

The sizes of samples required depend on the nature of the laboratory tests which may be required. Table 12 gives suggested sizes of samples and the information required on a sample identification tag. Disturbed samples of 75 pounds or more should be placed in bags or other suitable containers which will prevent loss of the fine fraction of the soil and moisture. Samples of silty and clayey soils proposed for borrow which are obtained for laboratory Atterberg limits tests and Proctor compaction tests should be protected against drying and should be shipped in waterproof bags or other suitable containers to retain the natural moisture so far as possible. Samples of sands and gravels may be shipped in closely woven textile

TABLE 12.—*Identification and sizes of samples*

[As identification on sample tags, give project name, feature, area designation, hole number, and depth of sample]

Purpose of material	Sample size	Remarks
Individual and composite samples of disturbed earth materials for classification and Proctor compaction tests.	Sufficient material, all passing the 3-inch sieve, to yield 75 pounds passing the No. 4 sieve.	Include information relative to the percentage by volume 3 inches to 5 inches and plus 5 inches.
Soil-rock permeability tests	300 pounds passing a 3-inch sieve	Air dried.
Relative density test	150 pounds passing a 3-inch sieve	Air dried.
Moisture samples, inspection samples of soil, soil samples for sulfate determination (reaction with concrete).	Sealed pint or quart jar (full)	Individual inspection samples should represent range of moisture and type of materials.
Concrete aggregate	600 pounds of pit-run sand and gravel. If screened: 200 pounds of sand, 200 pounds of No. 4 to ¾-inch size, and 100 pounds of each of the other sizes produced. 400 pounds of quarry rock proposed for crushed aggregate.	For commercial sources, include data on ownership-plant, and service history of concrete made from aggregates.
Riprap	600 pounds, which represents proportionally the quality range from poor to medium to best as found at the source.	Method of excavation used, location of pit and quarry.
Inplace density and water content of fine-grained soils above water table.	8- to 12-inch cubes or cylinders	Sealed in suitable container.

Figure 89. Making a standard penetration test by use of a tripod unit. PX–D–16752.

bags and should be air dried before placing in the bag. When the sack samples are shipped by public carrier, they should be double sacked. The protection and preparation for shipping of undisturbed samples is discussed in section 112. It is recommended that those samples not tested should be stored for possible future examination and testing until the dam is completed and has shown satisfactory operating characteristics.

111. *Disturbed Samples.*—(a) *Sampling Open Excavations.*—An area of the sidewall of the test pit, trench, or open cut should be trimmed to remove all weathered or mixed material. The exposed strata should be examined for changes in gradation, natural water content, plasticity, uniformity, etc., and the representative strata should be selected for sampling.

Either individual or composite samples are obtained by trenching down the vertical face of a pit, trench, or cut bank with a cut of uniform cross section and collecting the soil on a quartering cloth spread below the trench (fig. 90). The minimum cross section of the sampling trench should be at least four times the dimensions of the largest gravel size included in the soil. In taking individual samples it is important to be certain that sufficient representative material is obtained from the stratum and to see that extraneous material is not included. For composite samples, a vertical trench is cut through all strata above any desired elevation.

If the material being sampled is a gravelly soil that contains large percentages (approximately 25 percent or more of the total material) of particles 3 inches or larger in size, it is usually advantageous to take representative proportional parts of the entire excavated material, such as every fifth or tenth bucketful, to be saved as a part of the sample rather than to trim the sample from the sidewall of the excavation. When the samples are larger than required for testing, their size may be reduced by quartering. This is done by piling the total sample in a cone on a canvas or tarpaulin, each shovelful going to the center of the cone and being allowed to run down equally in all directions. The cone is then spread out in a circle by walking around the pile and gradually widening the circle with a shovel until a uniform thickness of material has been obtained.

Individual samples are taken from each layer of soil.
Composite samples are taken from two or more layers of soil.

Figure 90. Sampling trench. PX–D–4784.

The flat pile is then marked into quarters. Two opposite quarters are discarded. The material in the remaining quarters is mixed again by shoveling the material into another conical pile, taking alternate shovelfuls from each of the two quarters. The process of piling, flattening, and rejecting two quarters is continued until the sample is reduced to the desired size.

(b) *Sampling Auger Holes.*—Small auger holes cannot be sampled and logged as accurately as an open trench or a test pit, since the inaccessible auger hole does not permit visual inspection of the total profile and selection of representative strata. Small hand auger (4-

Figure 91. Auger sampling. PX–D–16331.

inch diameter or less) samples are adequate for soil classification, but do not provide sufficient material for testing purposes. An 8-inch-diameter hand auger, however, will provide samples of sufficient size for testing. As the auger hole is advanced, each amount of soil extracted by the auger should be deposited in individual stockpiles nearby to form an orderly depth sequence of removed material. In preparing an individual sample from an auger hole, first, group the various consecutive piles together to form representative samples of the various strata; and second, mix all or equal parts from each of the appropriate stockpiles together to form the desired size of sample from each stratum (fig. 91).

Sampling from large-diameter power augers is accomplished by lifting the auger blade at regular intervals, such as after each 9 to 12 inches of penetration, taking a shovelful of soil from the auger blade, and placing in sacks or stockpiling in an orderly depth sequence as the auger hole is advanced (fig. 83). Consecutive similar stockpiles can be combined to form individual samples.

(c) *Sampling Stockpiles.*—When sampling stockpiles or windrows, care must be taken to see that the samples are not selected from segregated areas. The amount of segregation in materials depends on the gradation of the material and on the methods and equipment used for stockpiling. Even with good control the outer surfaces and fringes of a stockpile are likely to have some segregation, particularly if the slopes are steep and the material contains an appreciable amount of gravel or coarse sand. Representative samples from stockpiles should be obtained by combining and mixing small samples taken from several small test pits or auger holes distributed over the pile. A windrow of soil is best sampled by taking all the material from a narrow cut transverse to the windrow. Samples from either stockpiles or windrows should be fairly large originally and should be thoroughly mixed before they are quartered down to the desired size for testing purposes.

(d) *Riprap Samples.*—The competency and quality of rock for riprap are judged by geologic field conditions, physical properties tests, petrographic examination, and service record of the material. Since the riprap requirements include obtaining proper sizes of rock fragments, quality tests made in the laboratory must be supplemented by data obtained by field examination and the results of blasting tests in proposed quarry sites. The importance of obtaining representative samples of each type of material in a proposed riprap source must be emphasized. If there is more than one type of material in a source, separate samples should be obtained representing each material proposed for use. Intervening layers of soil, shale, or other soft rock obviously unsuitable for riprap need not be sampled, but full descriptions of these materials should appear on the drill logs and in a report of the investigation.

Samples should be obtained by blasting down an open face on the sidewall of a test pit, trench, or exposed ledge to obtain unweathered fragments representing each type of material as it will be quarried and used in riprap. The selection of the exterior weathered rock fragments will not provide a representative sample unless only exterior weathered fragments are to be used as riprap. Figure 92 (A) and (B) show the blasting operation at a rock ledge riprap source for Stampede Dam in California and the resulting riprap material.

Large fields of boulders are sometimes proposed as sources of riprap. The production of riprap from boulder fields is always a costly process, and should be considered only when ledge materials are not available. Field boulders usually do not have the angularity and interlocking properties of quarried riprap. Sampling of boulder sources should include breaking of large boulders with explosives to obtain samples of fragments similar to those likely to result from operations during construction. Talus slopes should be sampled only if the talus itself is proposed for use as riprap. Samples of talus material generally do not represent the material obtainable from the solid rock ledge above the talus slope, because the talus fragments are generally weathered, altered, or case hardened.

(e) *Concrete Aggregate Samples.*—Disturbed samples of concrete aggregate materials are obtained from test pits, trenches, and cased auger holes. Since the gradation of concrete aggregates is of great importance, portable screening apparatus is sometimes used to determine the gradings of the samples in the field, thus permitting an estimate of the processing operations which will be required. Figure 93 shows exploration for concrete aggregate using a portable screening device. Whenever facilities are available, representative samples of the aggregate should be subjected to laboratory tests to determine the physical and chemical properties of the material. In the absence of facilities for laboratory tests, examination of the aggregate by an experienced petrographer will aid considerably in estimating its physical and chemical soundness. The performance record of an aggregate in concrete, when it can be determined, is of great value in appraising the potential source of concrete aggregate and should be employed whenever possible. Laboratory tests on concrete aggregate are discussed in section 121.

112. *Relatively Undisturbed Samples.*—(a) *Penetration Samples.*—The standard penetration boring includes the use of the split-barrel sampler shown in figures 86 and 87. The procedure used in the test under ideal conditions permits the securing of an 18-inch-long sample,

(A)

(B)

Figure 92. Blasting a rock ledge at the riprap source for Stampede Dam in California. (A) Initial blast. (B) Results of blast. The rock is basalt having a specific gravity of 2.6. P949–235–432NA, P949–235–436NA.

Figure 93. Exploration for concrete aggregate including facilities for screening and weighing. 288–D–2934.

the lower 12 inches of which represent the material for which the penetration resistance (in blows) is known. Typical samples of soil from the open split-barrel are placed into jars without ramming. The origin of the sample is noted on the jar, and it is stored in suitable containers for shipment. The jars are sealed with wax or a self-sealing top to prevent evaporation of the soil moisture. The containers should be marked to indicate the sample number, date, project, feature, structure, hole number, location, depth of sample, penetration resistance record, amount of sample recovered, and classification of soil. If more than one type of soil is encountered in the 12-inch drive, separate containers should be used for each type of soil. Samples should be protected from freezing and should not be placed in the sun. The samples are useful for visual classification of the soil and for laboratory tests of water content and Atterberg limits, when required.

(b) *Hand-cut Samples.*—Undisturbed samples in the form of cubes, cylinders, or irregular chunks can be obtained from strata exposed in the sides or bottoms of open excavations, test pits, trenches, and large-diameter auger holes. Such samples are useful for determining inplace density and water content, and for other laboratory tests.

Figures 94, 95, and 96 show procedures commonly used in hand-cut block sampling. Cutting and trimming samples to desired size and shape requires extreme care, particularly when working with easily disturbed soft or brittle materials. The appropriate cutting tool should be used to prevent disturbance and cracking of the sample. Soft, plastic soils require thin, sharp knives, and sometimes a tight, thin, piano wire is advantageous. When climatic conditions are such that rapid drying occurs, wet cloths should be used or other protective means provided while the sample is being cut. After the sample is cut and trimmed to the

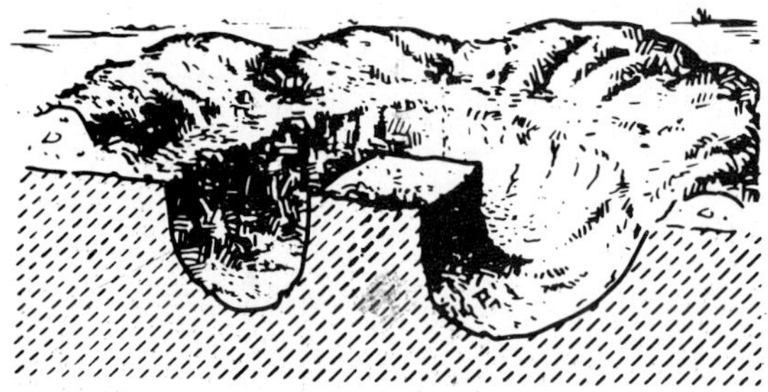

1. Smooth ground surface and mark outline of sample.
2. Carefully excavate trench around sample.

3. Deepen excavation and trim sides of sample to desired size with knife.

4. Cut sample from parent stratum, or encase sample in box before cutting if sample is easily disturbed.

(A)

1. Carefully smooth face surface and mark outline of sample.

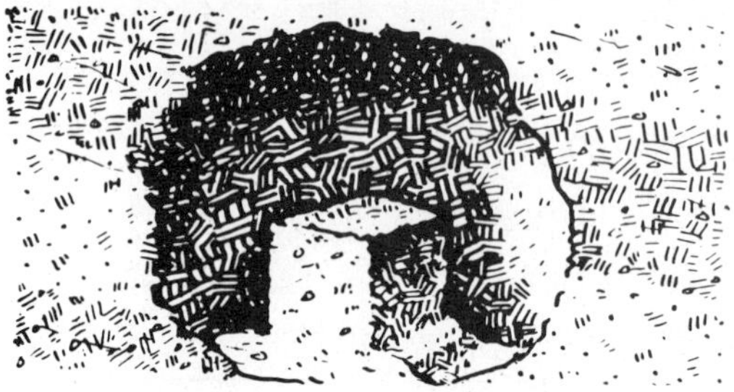

2. Carefully excavate around and in back of sample. Shape sample roughly with knife.

3. Cut sample and carefully remove from hole, or encase sample in box before cutting if sample is easily disturbed.

(B)

Figure 94. Initial steps to obtain a hand-cut undisturbed block sample from (A) bottom of test pit or level surface, (B) cut bank or side of test pit. PX–D–4788.

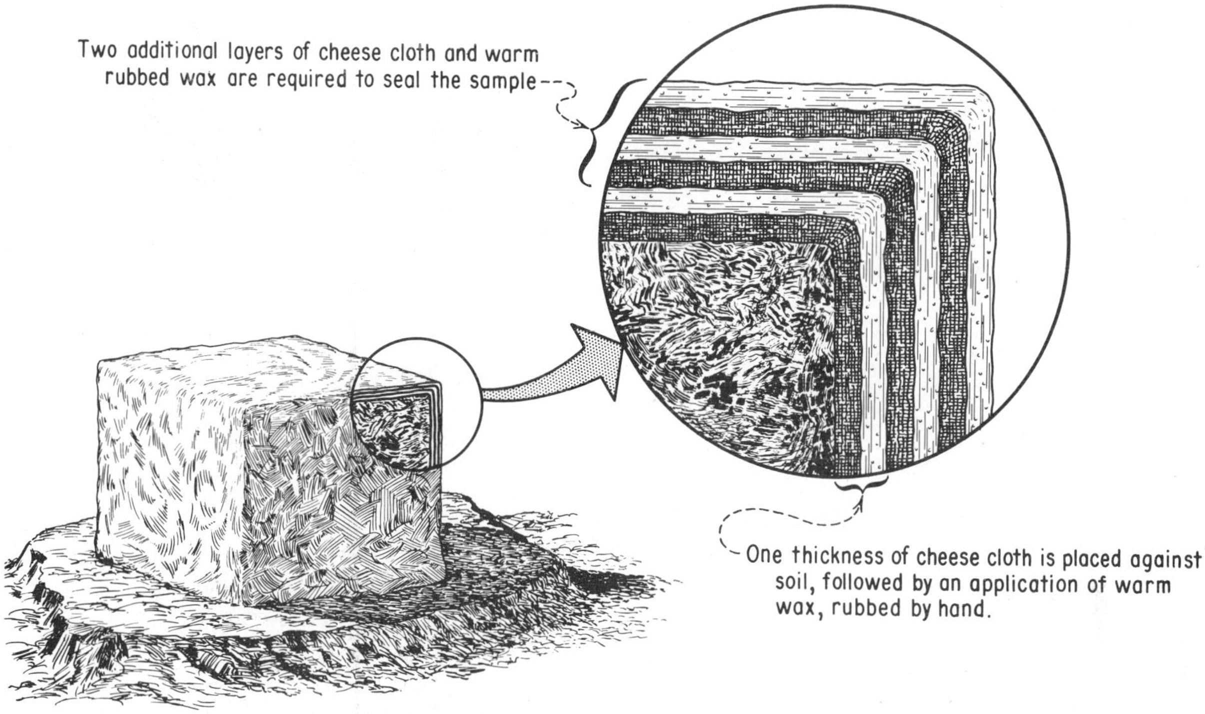

(A.) METHOD FOR SEALING HAND-CUT UNDISTURBED SAMPLES

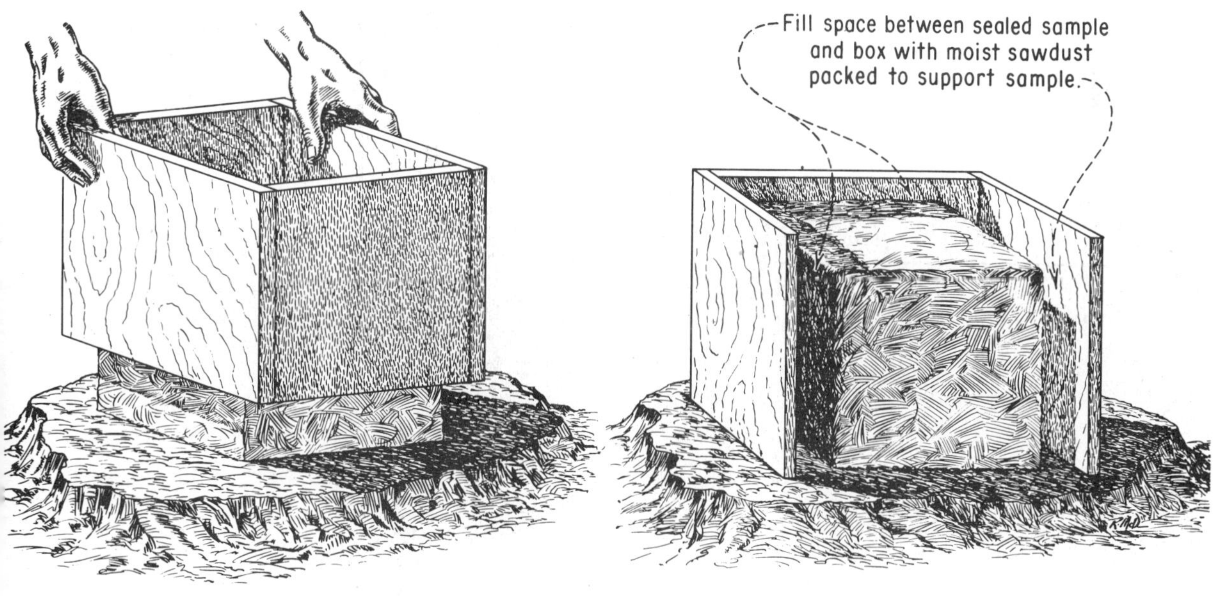

(B.) ENCASE EASILY DISTURBED SAMPLES IN BOX PRIOR TO CUTTING

Figure 95. Final steps to obtain a hand-cut undisturbed block sample. PX–D–4783.

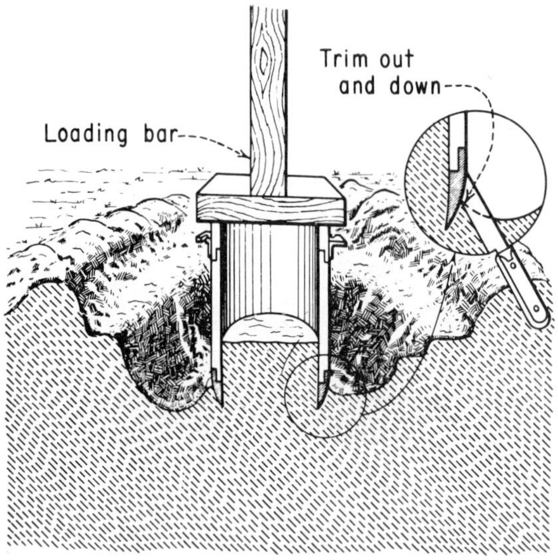

(A)
Level area and drive cylindrical sampler slightly into soil. Carefully excavate trench around cylinder and trim to cutting bit with knife.

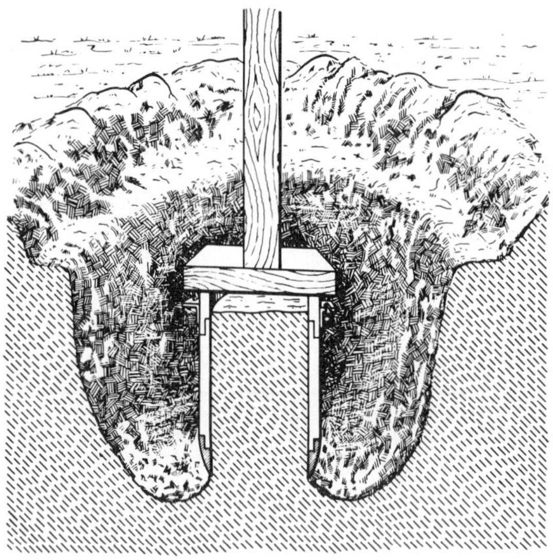

(B)
Continue to drive sampler tube and excavate as shown.

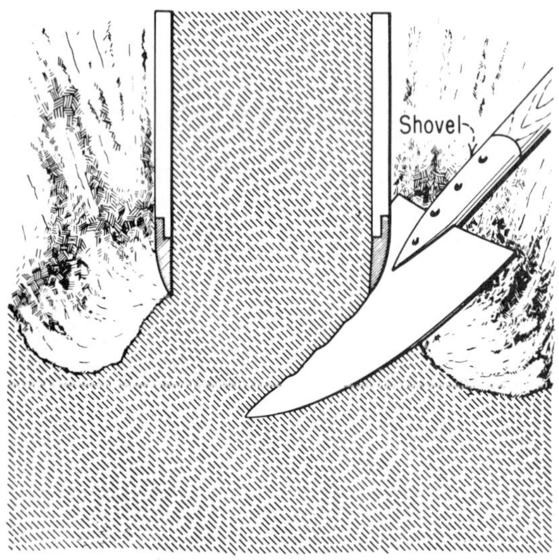

(C)
Carefully cut sample from parent material as shown.

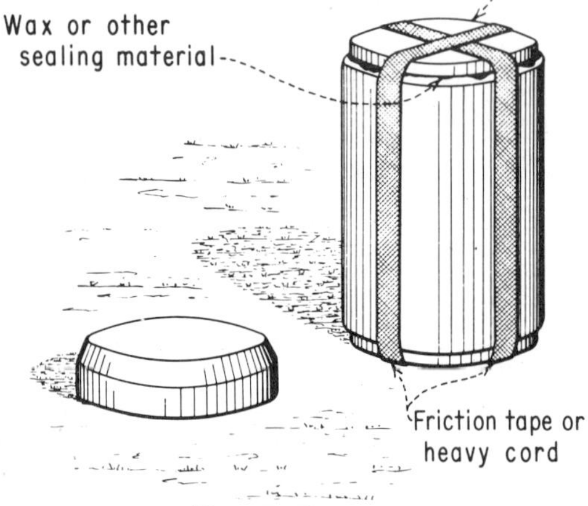

(D)
Seal sample to prevent moisture loss. Pack sample and container in excellsior or moist sawdust for shipment to the laboratory.

Figure 96. Method for obtaining a hand-cut undisturbed cylindrical sample. PX–D–4782.

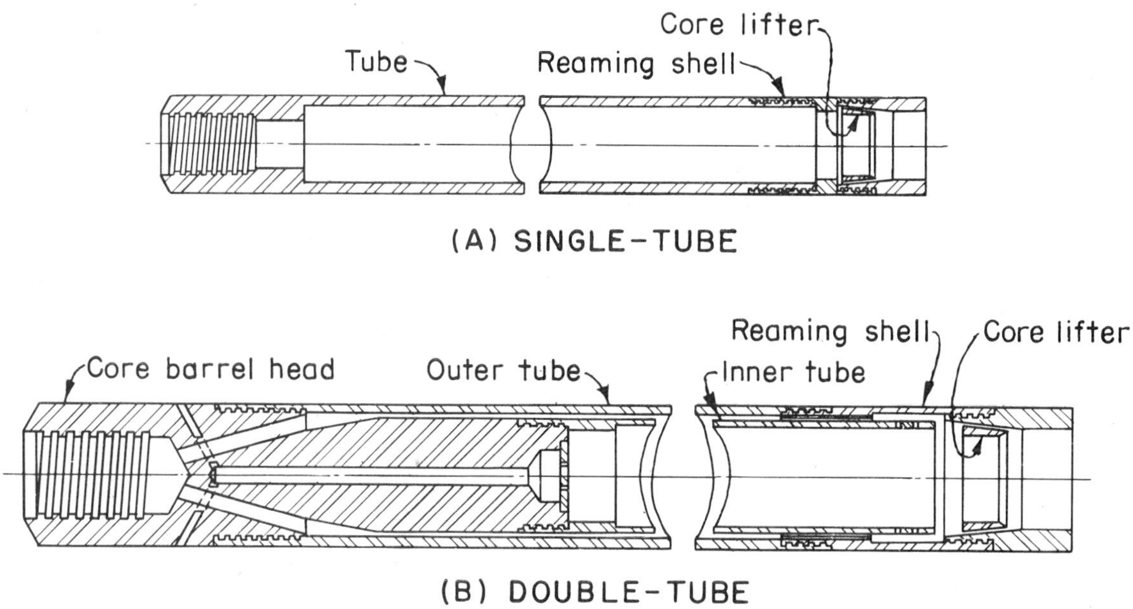

Figure 97. Core barrels used for obtaining samples of rock. 288–D–2514.

desired size and shape, it should be wrapped with a layer of cheesecloth and painted with warm sealing wax. Rubbing the surface with the bare hands will help to seal the pores in the wax. These operations constitute one layer. At least two additional layers of cloth and wax should be applied.

If a soil is easily disturbed, a firmly constructed wooden box with both ends removed should be placed over the sample before it is cut from the parent material and lifted for removal. Space between the sample and the walls should be packed with moist sawdust or similar packing material. The top cover of the box should then be placed over the packing material. After removal, the bottom side of the specimen should be covered with the same number of layers of cloth and wax as the other surfaces, and the bottom of the box placed over the packing material.

Samples may be of various sizes, the most common being 6- or 12-inch cubes. Cylindrical samples 6 to 8 inches in diameter and 6 to 12 inches long are frequently obtained, as shown in figure 96. The metal cylinder shown in the figure may be used to confine the sampler for shipping. Otherwise, the same trimming and sealing procedures described above for boxed samples apply.

(c) *Rock Cores.*—Conventional rotary drilling and sampling were originally developed for drilling through both hard and soft rocks. Core barrels will obtain cores from 3/4 to 6 inches in diameter. There are three principal types of core barrels: single-tube, double-tube, and triple-tube. The single-tube core barrel (see fig. 97) is the simpler in design, consisting of a core barrel head, core barrel, and attached coring bit which cuts an annular groove that permits passage of drilling fluid pumped through the drill rod. This design exposes the core to drilling fluid over its entire length, which results in serious core erosion of unconsolidated or weakly cemented materials. Therefore, it is used primarily to sample hard, solid rock which requires the use of diamond drills and accessories.

Dimensional standards have been established by the Diamond Core Drill Manufacturers Association (DCDMA), composed of members in the United States and Canada, for a series of nesting casings with corresponding sizes for bits and drill rods. The DCDMA standards for core drill bits, casings, and drill rods are given in figure 98. The size combination is such that HX core-barrel bits will pass through flush coupled HX casing (flush coupled casing is denoted by the group letter "X") and

Diamond Core Drill Manufacturers Association Nomenclature

THREE LETTER NAMES		
FIRST LETTER	SECOND LETTER	THIRD LETTER
HOLE SIZE	GROUP	DESIGN
Casing, core barrel, diamond bit, reaming shell and drill rods designed to be used together for drilling an approximate hole size.	Key diameters standardized on an integrated group basis for progressively reducing hole size with nesting casings.	The standardization of other dimensions, including thread characteristics, to permit interchangeability of parts made by different manufacturers.
Letter / Inches / Millimeters R / 1 / 25 E / 1½ / 40 A / 2 / 50 B / 2½ / 65 N / 3 / 75 K / 3½ / 90 H / 4 / 100 P / 5 / 125 S / 6 / 150 U / 7 / 175 Z / 8 / 200	Letters X and W are synonymous when used as the GROUP (second) letter. Any DCDMA standard tool with an X or W as the GROUP letter belongs in that DCDMA integrated group of tools designed using nesting casings and tools of sufficient strength to reach greater depths with minimum reductions in core diameter.	The DESIGN (third) letter designates the specific design of that particular tool. It does not indicate a type of design.

TWO LETTER NAMES	
FIRST LETTER	SECOND LETTER
HOLE SIZE	GROUP AND DESIGN
Approximate hole size, same as in 3-letter names.	GROUP standardization of key diameters for group integration and DESIGN standardization of other dimensions affecting interchangeability.

Figure 98. Nomenclature for diamond core drill equipment. (Diamond Core Drill Manufacturers Association.) (Sheet 1 of 5.) 288–D–2887.

will drill a hole large enough to admit flush coupled NX casing (the next smaller size) and so on to RX size. Flush joint casing (denoted by the group letter "W") is such that 6- by 7¾-inch core barrel bits (fig. 98) will pass through ZW casing and will drill a hole large enough to admit flush jointed UW casing (the next smaller size) and so on to RW size.

The double-tube core barrel sampler (see fig. 97), in addition to the outer rotating barrel, provides an inner stationary barrel which protects the core from the drilling fluid and reduces the torsional forces transmitted to the core. It is used to sample soft or fractured rock and may be used to obtain cores in hard brittle, or partially cemented soils, such as shale and claystone, or cores of soft, partially consolidated, or weakly cemented soils. For

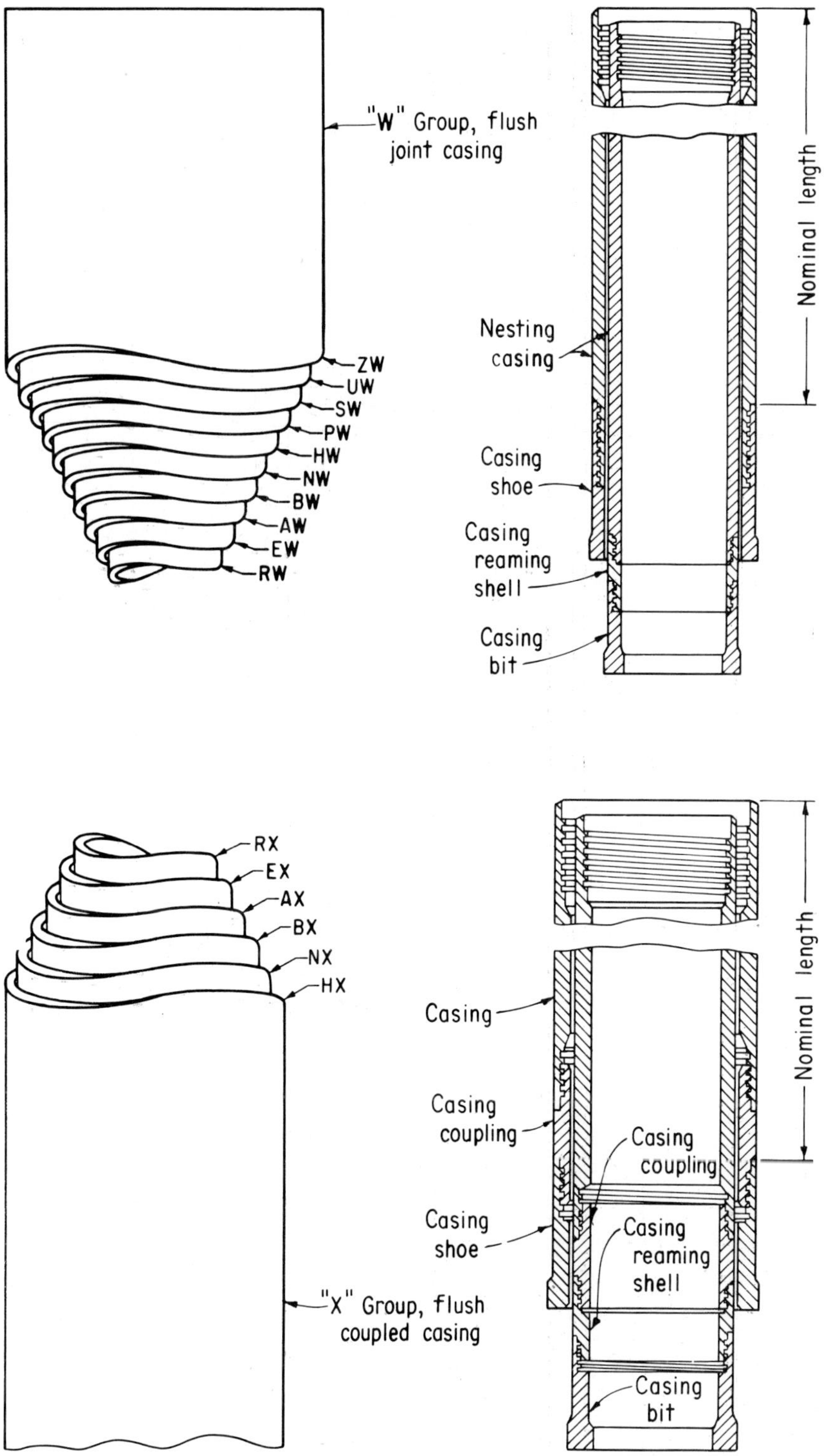

Note: Use of casing shoe allows nesting; use of casing bit does not.

Figure 98. **Size variations for core drill casing. (Diamond Core Drill Manufacturers Association.) (Sheet 2 of 5.) 288–D–2888.**

DIAMOND CORE DRILL MANUFACTURERS ASSOCIATION NOMINAL DIMENSIONS FOR CASINGS AND ACCESSORIES

Size Designations			Casing coupling					Approximate core diameter	
Casing; Casing coupling; Casing bits; Core barrel bits	Rod; Rod couplings	Casing O.D., inches	O.D., inches	I.D., inches	Casing bit, O.D., inches	Core barrel bit O.D., inches*	Drill rod O.D., inches	Normal, inches	Thinwall, inches
RX	RW	1.437	1.437	1.188	1.485	1.160	1.094	—	.735
EX	E	1.812	1.812	1.500	1.875	1.470	1.313	.845	.905
AX	A	2.250	2.250	1.906	2.345	1.875	1.625	1.185	1.281
BX	B	2.875	2.875	2.375	2.965	2.345	1.906	1.655	1.750
NX	N	3.500	3.500	3.000	3.615	2.965	2.375	2.155	2.313
HX	HW	4.500	4.500	3.938	4.625	3.890	3.500	3.000	3.187
RW	RW	1.437	Flush joint	No coupling	1.485	1.160	1.094	—	.735
EW	EW	1.812			1.875	1.470	1.375	.845	.905
AW	AW	2.250			2.345	1.875	1.750	1.185	1.281
BW	BW	2.875			2.965	2.345	2.125	1.655	1.750
NW	NW	3.500			3.615	2.965	2.625	2.155	2.313
HW	HW	4.500			4.625	3.890	3.500	3.000	3.187
PW	—	5.500			5.650	—	—	—	—
SW	—	6.625			6.790	—	—	—	—
UW	—	7.625			7.800	—	—	—	—
ZW	—	8.625			8.810	—	—	—	—
—	AX [1]	—	—	—	—	1.875	1.750	1.000	—
—	BX [1]	—	—	—	—	2.345	2.250	1.437	—
—	NX [1]	—	—	—	—	2.965	2.813	1.937	—

* For hole diameter approximation, assume $\frac{1}{32}$ inch larger than core barrel bit.

[1] Wire line size designation, drill rod only, serves as both casing and drill rod. Wire line core bit, and core diameters vary slightly according to manufacturer.

Figure 98. Nominal dimensions for drill casings and accessories. (Diamond Core Drill Manufacturers Association.) (Sheet 3 of 5.) 288–D–2889.

DIAMOND CORE DRILL MANUFACTURERS ASSOCIATION COMMERCIAL STANDARDS FOR CORING BIT SIZES

Coring bit size	Nominal *		Set size *	
	O. D.	I. D.	O. D.	I. D.
RWT	$1\frac{5}{32}$	$\frac{3}{4}$	1.160	.735
EWT	$1\frac{1}{2}$	$\frac{29}{32}$	1.470	.905
EX, EXL, EWG, EWM	$1\frac{1}{2}$	$\frac{13}{16}$	1.470	.845
AWT	$1\frac{7}{8}$	$1\frac{9}{32}$	1.875	1.281
AX, AXL, AWG, AWM	$1\frac{7}{8}$	$1\frac{3}{16}$	1.875	1.185
BWT	$2\frac{3}{8}$	$1\frac{3}{4}$	2.345	1.750
BX, BXL, BWG, BWM	$2\frac{3}{8}$	$1\frac{5}{8}$	2.345	1.655
NWT	3	$2\frac{5}{16}$	2.965	2.313
NX, NXL, NWG, NWM	3	$2\frac{1}{8}$	2.965	2.155
HWT	$3\frac{29}{32}$	$3\frac{3}{16}$	3.889	3.187
HWG	$3\frac{29}{32}$	3	3.889	3.000
$2\frac{3}{4}$ X $3\frac{7}{8}$	$3\frac{7}{8}$	$2\frac{3}{4}$	3.840	2.690
4 X $5\frac{1}{2}$	$5\frac{1}{2}$	4	5.435	3.970
6 X $7\frac{3}{4}$	$7\frac{3}{4}$	6	7.655	5.970
AX Wire line 1/	$1\frac{7}{8}$	1	1.875	1.000
BX Wire line 1/	$2\frac{3}{8}$	$1\frac{7}{16}$	2.345	1.437
NX Wire line 1/	3	$1\frac{15}{16}$	2.965	1.937

* All dimensions are in inches; to convert to millimeters, multiply by 25.4.
1/ Wire line dimensions and designations may vary according to manufacturer.

Figure 98. Standard coring bit sizes. (Diamond Core Drill Manufacturers Association.) (Sheet 4 of 5.) 288–D–2890.

DIAMOND CORE DRILL MANUFACTURERS ASSOCIATION STANDARD DRILL ROD SIZES

Size designation	Remarks	Rod and coupling O. D.	Rod I. D.	Coupling I. D.	Threads per inch
E	Old standard still in use on many projects	1.313	.844	.438	3
A		1.625	1.266	.563	3
B		1.906	1.406	.625	5
N		2.375	2.000	1.000	4
RW	Present standard	1.094	.719	.406	4
EW		1.375	1.000	.437	3
AW		1.718	1.344	.625	3
BW		2.125	1.750	.750	3
NW		2.625	2.250	1.375	3
HW		3.500	3.062	2.375	3
KWY	Tapered thread (A.P.I.)	2.875	2.312	1.375	4
HWY		3.500	2.875	1.750	4
AQ	Wire line drill rods 1/. Taper thread	1.750	1.375	—	4
BQ		2.188	1.813	—	3
NQ		2.750	2.375	—	3
HQ		3.500	3.063	—	3
PQ		4.500 4.625 2/	4.063	4.063	3

1/ Wire line drill rod dimensions and designations may vary according to manufacturer.

2/ For PQ size designation, rod O.D. = 4.500 inches and coupling O.D. = 4.625 inches.

Figure 98. Standard drill rod sizes. (Diamond Core Drill Manufacturers Association.) (Sheet 5 of 5.) 288–D–2891.

Figure 99. Arrangement of cores in a core box to insure identification of samples. PX–D–16303.

these materials, hard metal drill bits are used.

Many of the double-tube core barrels either can be or have been modified slightly to accept a sample liner inside the inner barrel. This addition serves as a container to ship the core in and also eliminates the possibility of damage to a core upon removal from the inner barrel.

The triple-tube core barrel has been designed with a rotating outer barrel, stationary inner barrel, and a split liner inside the inner barrel to accept the core sample. Removal of the core is accomplished by exerting hydraulic pressure over a piston located at the upper end of the split liner. Plastic or metal sample containers may be used in lieu of the split liner for shipping purposes.

Some of the latest core samplers have been designed with a spring loaded retractable inner barrel. This design enables the same type of core barrel to be used for either soil or hard rock coring. The retractable inner barrel and soil coring bits are replaced with a standard inner barrel and diamond bits for rock coring.

A much more complete explanation of single-tube, double-tube, and triple-tube samplers and their use is given in Designation E-2 of the Bureau's "Earth Manual" [6]. As the rock cores are removed from the core barrels they are placed in core boxes, as shown in figure 99.

H. LOGGING OF EXPLORATIONS

113. ***Identification of Holes.***—To assure completeness of the record and to eliminate confusion, test holes should be numbered in the order they are excavated, and the numbering series should be continuous through the various stages of investigation. If a hole is planned and programed, it is preferable to maintain the hole number in the record as "not drilled" or "abandoned" with an explanatory note rather than to reuse the hole number elsewhere. It is permissible, however, to move holes short distances and retain the program number where such moves are required by local conditions or by changes in engineering plans. When explorations cover several areas, such as alternative damsites and different borrow areas, a new series of numbers for each site or borrow area should be used. A common practice is to start numbering each new area explored at an even 100.

Test holes are prefixed with a 1- or 2-letter designation to describe the type of exploration. The following letter designations are frequently used:

DH Drill hole.
AH Auger hole (hand).
AP Auger hole (power).
TP Open pit.
T Trench.
CH Churn drill hole.
PR Penetration resistance hole.

114. *Log Forms.*—A log is a written record of the data concerning materials and conditions encountered in individual test holes. It provides the fundamental facts on which all subsequent conclusions are based, such as need for additional exploration or testing, feasibility of the site, design treatment required, cost of construction, method of construction, and evaluation of structure performance. A log may represent pertinent and important information that is used over a period of years; it may be needed to delineate accurately a change of conditions with the passage of time; it may form an important part of contract documents; and it may be required as basic evidence in a court of law in case of dispute. Each log, therefore, should be factual, accurate, clear, and complete. It should not be misleading. Log forms are used to record and provide the required information. Examples of logs for three types of exploratory holes are:

Geologic log of drill hole (fig. 100).—this form is suitable for all types of core borings which produce comparatively undisturbed samples.

Log of test pit or auger hole (fig. 101).—This form is suitable for all types of exploratory holes which produce complete but disturbed samples.

Penetration resistance log (fig. 102).—This form can be used for exploratory holes that test the soil in place.

Logs of trenches, tunnels, and shafts are best presented on drawings; these drawings should also contain the pertinent information outlined in figure 100.

The headings on the forms provide spaces for supplying identifying information such as project, feature, hole number, location, elevation, dates started and completed, and the name of the person responsible.

Depth to bedrock and to water table is valuable and important information and should always be reported. When this and any other information called for on the log cannot be obtained, justification in writing stating the reasons why should be required. The body of the log form is divided into a series of columns covering the various kinds of information required according to the type of exploratory hole.

When logging overburden, every stratum of material that is substantially different in composition from either the overlying or the underlying strata should be located by depth interval, separately classified, and described in the body of the log. In holes other than those excavated for exploration of a structure foundation, thin layers or lenses of different material in a relatively uniform stratum of material should be described, but need not be separately classified on the log; for example, "a 1-inch-thick discontinuous lens of fine sand occurs at 7-foot depth." However, logs of holes excavated for exploring of foundations for structures should indicate the location by depth of all lenses and layers of material and include the classification in addition to a detailed description of the material.

Large machine-dug test pits or test trenches may require more than one log to adequately describe the variations in materials found in different portions of the pit or trench. The initial log of such pits or trenches should describe a vertical column of soil at the deepest part of the excavation and is usually taken at the center of one wall of the pit or trench. If this one log will not adequately describe the variations in the different strata exposed by the pit or trench, additional logs for other locations within the test excavation should be prepared to give a true representation of all strata encountered in the test pit or trench. In long

7-1337 (9-69) Bureau of Reclamation GPO 854-401

GEOLOGIC LOG OF DRILL HOLE

SHEET 1 OF 3

FEATURE CASCADE DAM PROJECT JONES RIVER BASIN STATE IDAHO

HOLE NO. 43 LOCATION LEFT SIDE OF VALLEY FLOOR GROUND ELEV. 5,383.2' DIP (ANGLE FROM HORIZ.) VERTICAL

COORDS. N. 21,404 E. 10,253

BEGUN 7-31-72 FINISHED 8-4-72 DEPTH OF OVERBURDEN 17.3' TOTAL DEPTH 100' BEARING N. 30° W.

DEPTH AND ELEV. OF WATER LEVEL AND DATE MEASURED 3.7', El. 5,379.5' LOGGED BY L. DAVIDSON LOG REVIEWED BY G. SANDERS

NOTES ON WATER LOSSES AND LEVELS, CASING, CEMENTING, CAVING, AND OTHER DRILLING CONDITIONS	TYPE AND SIZE OF HOLE	CORE RECOVERY (%)	PERCOLATION TESTS: DEPTH (FEET) FROM (P, Cs, or Cm)	TO	LOSS (G.P.M.)	PRESSURE (P.S.I.)	LENGTH OF TEST (MIN.)	ELEVATION (FEET)	DEPTH (FEET)	SAMPLES FOR TESTING	CLASSIFICATION AND PHYSICAL CONDITION
	NX							5383.2			SAND with excess SILT
		W.S.						5375.2	8'		GRAVEL, well graded
Water level at 3.5'		W.S.	0'	10'	12	G	15	5373.2	10		GRAVEL AND BOULDERS with small amount of SAND AND SILT
		W.S.						5365.9	17.3'		SANDSTONE, white, very fine-grained.
			10'	20'	25	G	15				
		78						5360.6	22.6'		ANDESITE, fine-grained, very hard
		62						5358.7	24.5'	1	SANDSTONE, fine-grained, hard, well cemented, broken
		13						5354.7	28.5'		SILTSTONE, varigated, hard
Water level at 3.7'			P-20'	32'	18	35	15				
		93						5350.9	32.3'	2	Alternating layers of SANDSTONE, CLAYSTONE AND GRIT, hard, well cemented
		100						5346.2	37'		SANDSTONE intruded by ANDESITE
		80						5343.4	39.8'	3	ANDESITE, dense, fine-grained, hard
			P-32'	43'	1.5'	50	15				
		87						5338.3	44.9'		SANDSTONE with streaks of SILTSTONE, fine-grained, hard, well-cemented
			P-43'	46'	2	50	15				
		95									
	BX	78						5321.2	62'	4	ANDESITE, fine-grained, dark gray to black
Water level at 3.8'		100						5317.2	66'		ANDESITE, slightly vesicular, dark gray, CALCITE occurs at 71'-74'.
Lost water return at 71'		100									
			P-46'	72'	1	50	15				
			P-72'	74'	9.8	50	15				
No water return		81									
Water level at 3.7'											
		95	P-74'	100	2.2	50	15	5283.2	100		

EXPLANATION

CORE LOSS / CORE RECOVERY

Type of hole . . . D = Diamond, H = Haystellite, S = Shot, C = Churn
Hole sealed . . . P = Packer, Cm = Cemented, Cs = Bottom of casing
Approx. size of hole (X-series) . . Ex = 1-1/2", Ax = 1-7/8", Bx = 2-3/8", Nx = 3"
Approx. size of core (X-series) . . Ex = 7/8", Ax = 1-1/8", Bx = 1-5/8", Nx = 2-1/8"
Outside dia. of casing (X-series) . Ex = 1-13/16", Ax = 2-1/4", Bx = 2-7/8", Nx = 3-1/2"
Inside dia. of casing (X-series) . . Ex = 1-1/2", Ax = 1-29/32", Bx = 2-3/8", Nx = 3"

FEATURE CASCADE DAM PROJECT JONES RIV. BASIN STATE IDAHO SHEET 1 OF 3 HOLE NO. 43

Figure 100. Geologic log of a drill hole. 288–D–2870.

7-1336 (Combining 7-1336 and 7-1338)
(8-70) Bureau of Reclamation

LOG OF TEST PIT OR AUGER HOLE

FOR BORROW AND FOUNDATION INVESTIGATIONS

Feature: Example — Project: ______ — Area Designation: Borrow Area L-4

Hole No.: AH-455 — Coordinates N. 16,500 E. 6,910 — Ground Elevation: 669.8 — Approx. Dimensions: 24-inch-dia.

Depth to Water Level*: *21.9 feet — Method of Excavation: Auger hole — Date: April 5-12, 19__ — Logged by: ______

CLASSIFICATION SYMBOL: LETTER	CLASSIFICATION SYMBOL: GRAPHIC	DEPTH (FEET)	SIZE AND TYPE OF SAMPLE TAKEN	CLASSIFICATION AND DESCRIPTION OF MATERIAL (SEE CHART - "UNIFIED SOIL CLASSIFICATION"; GIVE GEOLOGIC AND IN-PLACE DESCRIPTION FOR FOUNDATION INVESTIGATIONS)	PERCENTAGE OF COBBLES AND BOULDERS **: VOLUME OF HOLE SAMPLED (CUBIC FEET)	WEIGHT OF 3 TO 5-INCH SAMPLED (LBS.)	PERCENTAGE BY VOLUME OF 3 TO 5-INCH***	WEIGHT OF PLUS 5-INCH SAMPLED (LBS.)	PERCENTAGE BY VOLUME OF PLUS 5-INCH***
ML	Not required for materials explorations		75-lb. sack	0'-2' SILT, slightly organic with some alfalfa and weed roots; nonplastic; small amount of fine sand; dark brown; dry.	40	0	0	0	0
CL			175-lb. sack	2'-8.5' Lean CLAY, medium plasticity, high dry strength; approx. 25% sand and gravel to 3/4-inch size; most of gravel is shale; brown; dry.	130	0	0	0	0
ML-MH				8.5'-16' Micaceous SILT, moderate amount of very fine sand; noticeable mica flakes; very slight plasticity; tan; dry.	150	0	0	0	0
			200-lb. sack	16'-18' SILT, similar to material 8.5 to 16 feet, but contains about 20% shaly gravel to 1-inch size; red; dry.	40	0	0	0	0
ML			90-lb. sack	18'-25' GRAVEL-SAND MIXTURE WITH COBBLES, well-graded; approx. 50% gravel and 50% sand, mostly hard, subrounded; approx. 9% cobbles, by volume to 8-inch maximum size; very small amount of nonplastic fines; black; dry above water table; river terrace gravel. STOPPED BY HARD MATERIAL.	80	2430	19.1	1150	9
GW-SW		G.W.L. 4/12/7_	21.9						

REMARKS: Inplace density test at 8 feet: dry density 89.4 p.c.f., water content 8.9 percent.

Bulk specific gravity of cobbles and boulders: 2.55 by displacement.

NOTES: Record water test and density test data, if applicable, under remarks.
* Record after water has reached its natural level; give date of reading adjacent to graphic symbol or in remarks.
** Applicable only to borrow pits and to foundations which are potential sources of construction materials.
*** $\frac{\text{(Lbs. of rock sampled) 100}}{\text{(Bulk specific gravity of rock) 62.4 (Cubic feet of hole sampled)}}$
Record bulk specific gravity in Remarks, stating how obtained (measured or estimated)

Figure 101. Example of log of test pit in a borrow area. 288–D–2871.

trenches, at least one log for each 50 feet of trench wall should be prepared, regardless of the uniformity of the material or strata. A geologic section of one or both walls of long test trenches may be required to describe variations in strata and material between log locations. When more than one log is needed to describe the material in an exploratory pit or trench, coordinate location and ground surface elevation should be given for each point for which a log is prepared. A plan geologic map and geologic sections should always be prepared for test trenches which encounter bedrock in structure foundations.

A log should always contain information on the size of the hole and on the type of equipment used for boring or excavating the hole. This should include the kind of drilling bit used on drill holes, a description of the penetrating equipment or type of auger used, or method of excavating test pits. The location from which samples are collected should be indicated on the logs, and the amount of material recovered as core should be expressed as a percentage of each length of penetration of the barrel. The logs should also show the extent and the method of support used as the hole is deepened, such as size and depth of casing, location and extent of grouting if used, type of drilling mud, or type of cribbing in test pits. Caving or squeezing material should be noted on the drill hole log as this may represent a low strength stratum.

Information on the presence or absence of water levels and comments on the reliability of these data should be given on all logs. The date measurements are made should be recorded, since water levels fluctuate seasonally. Water levels should be recorded periodically from the time water is first encountered and as the test hole is deepened. Upon completion of drilling, the hole should be bailed and allowed to recover in order to obtain a true water level measurement. Perched water tables and water under artesian pressure are important to note. The extent of water-bearing members should be noted and areas where water is lost as the boring proceeds should be reported, since subsequent work on the hole may preclude duplicating such information. The log should contain information on the water tests made at intervals, as described in section 118. Since it may be desirable to maintain periodic records of water level fluctuations in drilled holes, it should be determined whether this is required before abandoning and plugging the exploratory hole.

Where cobbles and boulders are encountered in explorations for sources of embankment materials, it is important to determine their percentage by volume. The log form for test pit or auger hole (fig. 101) includes a method for obtaining the percentage by volume of 3- to 5-inch rock and rock over 5 inches in size. The method involves weighing the rock, converting weight to solid volume of rock, and measuring the volume of hole containing the rock. This determination can be made either on the total volume of stratum excavated or on a representative portion of the stratum by means of a sampling trench which is described in section 111.

For test holes that penetrate less than 25 feet of potential borrow material, a statement should be made under "Remarks" in the log giving the reason for stopping the hole. For all other types of holes, a statement should be made at the end of the log that the work was completed as required or a statement explaining why the hole was abandoned. Material should not be described as bedrock, slide material, or similar interpretative terminology unless the exploration actually penetrated such conditions and samples were collected to substantiate these conclusions. Terminating statements similar to the following would be considered satisfactory: hole eliminated due to lack of funds; hole caved in, including depth to cave; depth limited by capacity of equipment; encountered water; unable to penetrate hard material in bottom of hole.

115. ***Description of Soils.***—The person logging exploratory holes should be able to identify soils according to the Unified Soil Classification System. The description of a soil in a log should include its typical name, followed by pertinent descriptive data, as listed in table 13. After the soil is described, it should be placed in the appropriate soil classification group by use of letter symbols. These group symbols represent a variety of soils having certain common characteristics; hence, by themselves they are not sufficient to describe a particular soil. Boundary classifications (two sets of symbols separated by a hyphen) should be used when the soil does not fall clearly into one of the groups but has strong characteristics of both groups.

TABLE 13.—*Description of soils*

Items of descriptive data	Borrow		Foundation	
	Coarse-grained soils	Fine-grained soils	Coarse-grained soils	Fine-grained soils
	XX: Req'd information on all logs X: Desired information			
Typical name (examples are shown in Unified Soil Classification chart, fig. 56)	XX	XX	XX	XX
Approximate percentages of gravel and sand	X	X	X	X
Maximum size of particles (including cobbles and boulders)	XX	X	X	X
Shape of the coarse grains—angularity	X	....	X	....
Surface condition of the coarse grains—coatings	X	....	....	....
Hardness of the coarse grains—possible breakdown into smaller sizes	X	....	X	....
Color (in moist condition for fine-grained soils and fraction of fines in coarse-grained soils)	X	X	X	X
Moisture (dry, moist, wet, saturated)	XX	XX	XX	XX
Organic content	X	X	X	X
Plasticity—degree (nonplastic, low, medium, high) and dilatancy, dry strength, and toughness for fine-grained soils and of the fine-grained fraction in coarse-grained soils	X	XX	XX	XX
Structure (stratification, lenses and seams, laminations, giving dip and strike and thickness of layers; honeycomb, flocculent, root holes, etc.)	....	....	XX	XX
Cementation—type	XX	X	XX	XX
Degree of compactness—loose or dense (excepting clays)	X	X	XX	XX
Consistency in undisturbed and remolded states (clays only)	....	....	....	XX
Local or geologic name	X	X	XX	XX
Group symbol	XX	XX	XX	XX

Identification and classification of soils in logs of explorations should be based on visual examination and manual tests. Field logs should not contain refinements that can be determined only by use of laboratory equipment. Laboratory tests may be used to assist untrained personnel in verifying their field classifications.

The purposes for which soils are investigated for small dams can be divided into two categories: (1) Borrow materials for embankments or for backfill, and (2) foundations for the dam and appurtenant structures. The emphasis of various features to be described depends on which of the categories is involved. For many structures, large quantities of soil must be excavated to reach a desired foundation. In the interests of economy, maximum use should always be made of this excavated material in the construction of embankments and for backfill. A foundation area, therefore, often becomes a source of materials, and investigation of the area must take into account this dual purpose. Descriptions of soils encountered in such explorations should contain the essential information required both for borrow materials and for foundation soils.

Soils that are potential sources of borrow material for embankments must be described adequately in the log of the exploratory test pit or auger hole. Since these materials are destined to be disturbed by excavation, transportation, and compaction in the fill, their structure is less important than the amount and characteristics of the soil constituents. However, the recording of their natural water condition is important. Very dry borrow materials require the addition of large amounts of moisture for compaction control, and very wet soils containing appreciable fines may require extensive processing in order to be usable. For simplicity the natural water content of borrow soils should be reported as either *dry*, *moist*, or *wet*. A soil that is reported dry should be one that will definitely require the addition of moisture in order to compact it properly in a fill. A soil should be reported as being moist if it is reasonably close to the Proctor optimum water content. Soils that are reported as wet should be obviously well beyond the optimum water content. Borrow-pit holes are logged so as to indicate divisions between soils of different classification groups. However, within the same soil group significant changes in moisture should be logged.

When soils are being explored as foundations for dams and appurtenant works, their natural structure, compactness, and moisture content are of outstanding importance. Logs of foundation explorations, therefore, must emphasize the inplace condition of a soil in addition to describing its constituents. The natural state of foundation soils is significant because bearing capacity and settlement under load vary tremendously with the consistency or compactness of the soil. Therefore, information that a clay soil is hard and dry, or soft and moist, is important. Changes in consistency of foundation soils due to moisture changes under operating conditions must be considered in the design. Correct classification is needed so that the effect of these moisture changes on foundation properties can be predicted. The use of a general type name such as loess, caliche, adobe, etc., in addition to the soil classification name may be helpful in identifying inplace conditions.

Table 13 lists data that are needed to describe soils for borrow material and for foundations. Under each of these categories, the information desired for coarse-grained soils and for fine-grained soils is indicated by an X. All of these descriptive data are not always needed. Judgment should be used to include pertinent information, to avoid negative information, and to eliminate repetition. The items indicated by XX should always be reported. Examples of soil descriptions are given in the soil classification chart, figure 56, and in the examples on the log forms, figures 100, 101, and 102.

116. *Description of Rock Cores.*—The basic objective of describing rock cores is to provide a concise record of the important geological and physical characteristics of the core materials. The description should be prepared preferably by a geologist; its usefulness will be controlled largely by the individual's experience in logging rock core for engineering purposes and his knowledge of geology. Thus, an experienced individual will include some seemingly minor core features or conditions which he knows have engineering significance and exclude other geologic features having only academic interest.

Description of the rock core should include its typical rock name, followed by data on its lithologic and structural features; physical condition including alteration, and any special geologic, mineralogic, or physical features pertinent to interpretation of the subsurface conditions. Classification of rocks is given in part D of this chapter. Attention should be given to: (1) the attitude and spacing of joints, seams or fractures and whether open or filled, as well as to evidence of shear, crushing, or faulting; (2) planes of bedding, lamination or layering and the ease of splitting along such planes; (3) color, grain size and shape, and (in sedimentary rocks like sandstone) the mineralogy of the grains and cementing material as well as the extent to which the cementing material occupies the intergrain spaces; and (4) the degree of alteration or weathering and hardness of the rock. In the latter case supplementary phrases such as "breaks with sharp hammer blow," "crumbles easily in the fingers," or "hard as common brick" are helpful. Estimates of the average length of core pieces in successive sections of the hole aid in calling attention to changes in formations or rock conditions in the hole not otherwise recognizable but nonetheless useful in evaluating subsurface conditions in the engineering sense.

The purpose of the drilling and logging is to secure evidence of the "inplace" condition of the rock; therefore, care should be taken to note any core condition or damage due to the type of drill bit or core barrel used or to improper conduct of the drilling process. One common cause of core damage or breakage is due to the use of a core catcher in the coring barrel and, therefore, for most rock sampling it is recommended that it not be used unless absolutely necessary. Such factors may have a marked effect on the amount and condition of the core recovered, particularly in soft, friable, or severely fractured rock.

Adequate logs or descriptions of rock core can be prepared solely through visual or "hand specimen" examination of the core with the occasional aid of simple field tests. Detailed microscopic or laboratory testing to define rock type or mineralogy is generally necessary only in special cases. Figure 99 shows how rock cores obtained from a borehole are arranged for logging.

7-1334 (9-71)
Bureau of Reclamation

SUBSURFACE EXPLORATION - PENETRATION RESISTANCE AND LOG

Feature **Example** Project ______ State ______

Hole No. PR- 44 Coordinates N. 1,771,612 E. 411,119 Ground Elevation 19.1 ft.

Depth and Elev. of Water Level* 9.1 ft. 10.1 ft. Location See drawing No. Total Depth 99.5 ft.

*Date Measured 7-16- Date Begun 7-14- Finished 7-15- Logged by ______ Approved by ______

PENETRATION RESISTANCE — Blows per Foot — Weight of Hammer 140 lb. Height of Drop 30 in.

NOTES (Water losses, type and size of hole, drilling method and conditions, Caving and other information)

- Set NX Cs to 29.0'. Used 5" drive tube and auger from 0 to 14'
- Wash bored from 14' to 26.5'
- Used 3" Shelby tube sampler from 26.5' to 47.0'
- Hole squeezed at 42.0'
- Wash bored from 47' to 64'
- Used 3" Shelby tube sampler from 64' to 99.5'
- Water level at 13.0' on 7-15-7__; water level at 10.0' on 7-16-7__. Hole caved to 19.0' on 7-16-7__.
- Pulled Cs on 7-15-__ and bailed hole.
- Filled hole on 7-16-__

CORE RECOVERY-%	BLOWS/FOOT	WATER CONTENT-%	SAMPLES FOR TESTING
	6	17.5	X
	10		X
	12		X
	26		X
	11	34.5	X
	14	35.2	X
	10	27.4	X
	23		X
	26	14.6	X
	44		X
	13		X
	19		X
		22.3	
	22		X
	18	20.6	X
	17		X
	21	23.5	X
	36		X

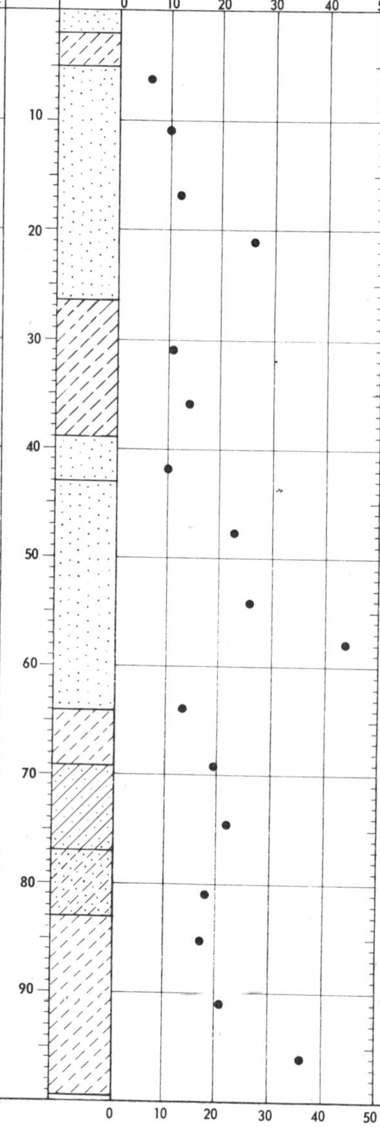

CLASSIFICATION AND DESCRIPTION OF MATERIAL

0-2' Clayey SAND, about 80% fine sand, about 20% fines of medium plasticity, brown, moist, (SC)

2'-5' Lean CLAY, about 80% medium plasticity fines, about 20% fine sand, tan, moist, (CL)

5'-26.5' Poorly-graded SAND, about 95% fine sand, about 5% nonplastic fines, tan, moist to wet, (SP-SM)

26.5'-39.0' Fat CLAY, high plasticity, soft, light gray to light brown, mottled, moist to wet, (CH)

39.0'-43.0' Silty SAND, about 70% fine sand, about 30% nonplastic fines, tan, saturated, (SP-SM)

43.0'-64.0' Poorly-graded SAND, about 95% fine sand, about 5% nonplastic fines, tan, saturated, (SP-SM)

64.0'-69.0' Fat CLAY, high plasticity, about 20% fine sand, light gray to light brown, mottled, small amount of lime nodules, moist to wet, (CH)

69.0'-77.0' Silty SAND, about 50% fine sand, about 50% nonplastic fines, strong reaction to HCl, brown, moist to wet, (SM-ML)

77.0'-83.0' Clayey SAND, about 50% fine sand, about 50% medium plasticity fines, moderate reaction to HCl, brown, moist to wet, (SC-CL)

83.0'-99.5' Fat CLAY, high plasticity, about 20% fine sand, light gray to light brown, mottled, small amount of lime nodules, moist to wet, (CH)

EXPLANATION

BLOWS/FOOT — Record number of blows required for one foot of penetration. If 50 blows result in less than one foot of penetration, record depth penetrated; thus, 50/.4 indicates 0.4 foot penetrated with 50 blows, extrapolated as 125 blows/foot

CLASSIFICATION AND DESCRIPTION — Describe soil type, with emphasis on inplace condition. Include Unified Soil Classification symbol. EXAMPLE: POORLY GRADED SAND. 95% predominantly medium sand, 5% nonplastic fines, max size 1/4 inch, firm, moist, gray, uncemented (SP)

PENETRATION RESISTANCE — Plot, as shown at right. Actual values ●, extrapolated values at 50, open circle with extrapolated value 125○

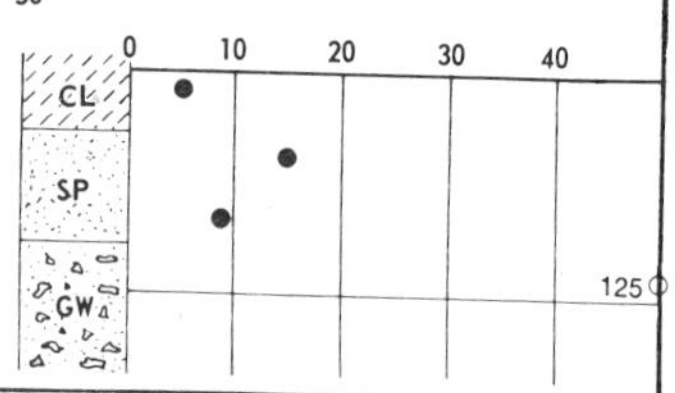

FEATURE **Example** PROJECT ______ STATE ______ HOLE NO. PR- 44

Figure 102. Drill-hole log and penetration resistance data—Gulf Coast Canal, Gulf Basins project, Texas. 288–D–2872.

I. FIELD AND LABORATORY TESTS

117. *General.*—Of the great variety of field and laboratory tests that have been used in the design of dams, only those applicable to the simplified design procedures used in this text are described here. In addition to the quantitative data (the number of blows per foot) obtained during the standard penetration borings described in section 108, two other field tests which obtain values for the natural ground and which are applicable in exploring foundations for small dams are: (1) permeability tests and (2) density-in-place tests. The latter test is used also in borrow areas to determine shrinkage factors between excavation and embankment yardages.

Descriptions of laboratory tests on soils are limited to those required to verify soil classifications and to determine compaction characteristics for comparison with design assumptions made from data of table 8 (sec. 94) and for correlation with construction control tests given in appendix E. The descriptions of the tests are intended to furnish a general knowledge of their scope. For detailed test procedures reference should be made to the Bureau of Reclamation's "Earth Manual" [7] or to the ASTM Designation [9] noted for several of the tests.

The laboratory quality tests of riprap and concrete aggregate, commonly used in specifications for these materials, are described to afford an understanding of the significance of those tests. Details of test procedures can be found in the Bureau of Reclamation's "Concrete Manual" [8] and in the referenced ASTM Designation [18].

118. *Field Permeability Tests.*—(a) *General.*—Approximate values of permeability of individual strata penetrated by borings can be obtained by making water tests in the holes. The reliability of the values obtained depends on the homogeneity of the stratum tested and on certain restrictions of the mathematical formulas used. However, if reasonable care is exercised in adhering to the recommended procedures, useful results can be obtained during ordinary boring operations. Use of the more precise methods of determining permeability by pumping from wells with a series of observation holes to measure drawdown of the water table or by pumping-in tests using large-diameter perforated casing is considered unnecessary for the design of small dams.

The tests described below are of the pumping-in type; that is, they are based on measuring the amount of water accepted by the ground through the open bottom of a pipe or through an uncased section of the hole. These tests are invalid and may be grossly misleading unless clear water is used. The presence of even small amounts of silt or clay in the added water will plug up the test section and give permeability results that are too low. By means of a settling tank or a filter, efforts should be made to assure that only clear water is used. It is desirable for the temperature of the added water to be higher than groundwater temperature so as to preclude the creation of air bubbles in the ground which may greatly reduce the acceptance of water.

(b) *Open-End Tests.*—Figure 103 (A) and (B) shows a test made through the open end of a pipe casing which has been sunk to the desired depth and which has been carefully cleaned out just to the bottom of the casing. When the hole extends below the groundwater table, it is recommended that the hole be kept filled with water during cleaning and especially during withdrawal of tools to avoid squeezing of soil into the bottom of the pipe. After the hole is cleaned to the proper depth, the test is begun by adding clear water through a metering system to maintain gravity flow at a constant head. In tests above the water table (fig. 103 (B)) a stable, constant level is rarely obtained and a surging of the level within a few tenths of a foot at a constant rate of flow for about 5 minutes is considered satisfactory.

If it is desired to apply pressure to the water entering the hole, the pressure, in units of head, is added to the gravity head, as shown in figure 103 (C) and (D). Measurements of constant head, constant rate of flow into the hole, size of casing pipe, and elevations of top and bottom of casing are recorded. The permeability is obtained from the following relation determined by electric analogy experiments:

$$K=\frac{Q}{5.5rH} \quad (2)$$

where:

K=permeability,
Q=constant rate of flow into the hole,
r=internal radius of casing, and
H=differential head of water.

Any consistent set of units may be used. For convenience, equation (2) can be written:

K (in feet per year)

$$=C_1 \frac{Q \text{ (in gallons per minute)}}{H \text{ (in feet)}}$$

Values of C_1 vary with size of casing as follows (see fig. 98):

Size of casing	EX	AX	BX	NX
C_1	204,000	160,000	129,000	102,000

The value of H for gravity tests made below water table is the difference in feet between the level of water in the casing and the groundwater level. For tests above water table, H is the depth of water in the hole. For pressure tests the applied pressure in feet of water (1 p.s.i. = 2.31 feet) is added to the gravity head to obtain H.

For the example shown in figure 103(A):
Given:

NX casing,

Q=10.1 gallons per minute,
H=21.4 feet,

$$K=C_1\frac{Q}{H}=\frac{(102{,}000)(10.1)}{21.4}$$

$$=48{,}100 \text{ feet per year.}$$

For the example shown in figure 103(D):
Given:

NX casing,

Q=7 gallons per minute,
H (gravity)=24.6 feet,
H (pressure)=5 p.s.i.=5×2.31
=11.6 feet of water.

Then H=24.6+11.6=36.2 feet and

$$K=C_1\frac{Q}{H}=\frac{(102{,}000)(7)}{36.2}$$

$$=19{,}700 \text{ feet per year}$$

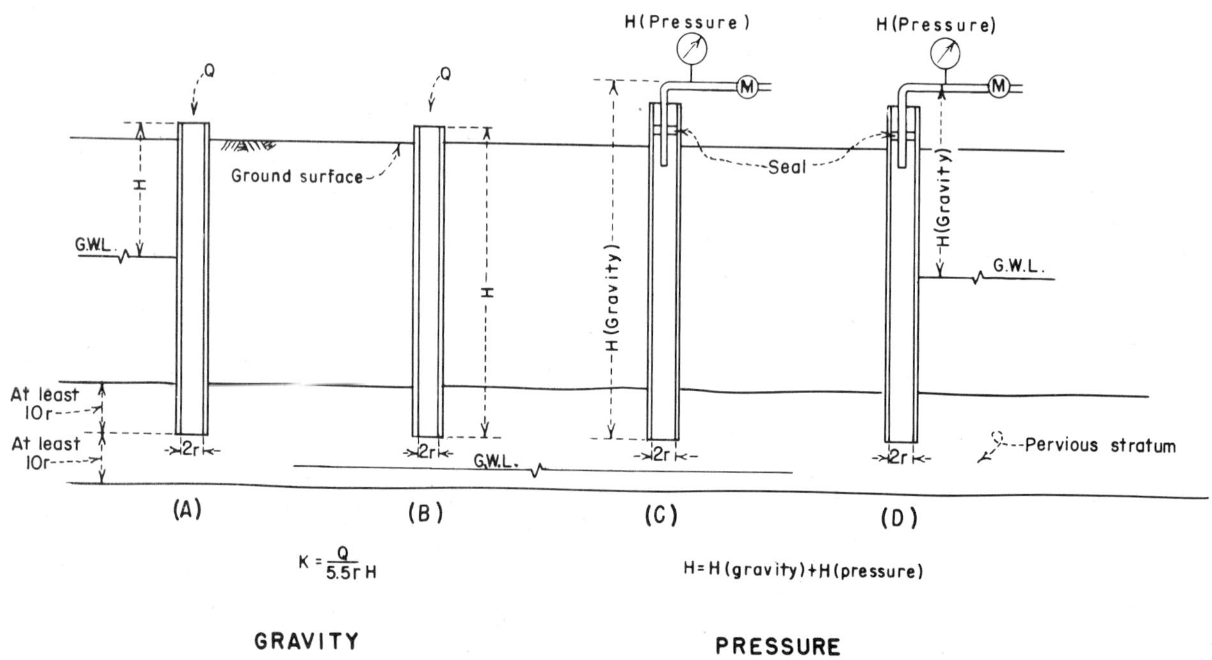

Figure 103. An open-end pipe test for soil permeability which can be made in the field. 288–D–2476.

(c) *Packer Tests.*—Figure 104 shows a permeability test made in a portion of a drill hole below the casing. This test can be made both above and below the water table, provided the hole will remain open. It is commonly used for pressure testing of bedrock using packers, but it can be used in unconsolidated materials where a top packer is placed just inside the casing. When the packer is placed inside the casing, measures must be taken to properly seal the annular space between the casing and drill hole wall to prevent water under pressure from escaping.

The formulas for this test are:

$$K=\frac{Q}{2\pi LH}\log_e\frac{L}{r},\ L\geqq 10r \qquad (3)$$

$$K=\frac{Q}{2\pi LH}\sinh^{-1}\frac{L}{2r},\ 10r>L\geqq r \qquad (4)$$

where:

K=permeability,

Q=constant rate of flow into the hole,

L=length of the portion of the hole tested,

H=differential head of water,

r=radius of hole tested,

$\log_e$=natural logarithm, and

$\sinh^{-1}$=inverse hyperbolic sine.

These formulas have best validity when the thickness of the stratum tested is at least $5L$, and are considered to be more accurate for tests below groundwater table than above it.

For convenience, the formulas can be written:

$$K\text{ (feet per year)}=C_p\frac{Q\text{ (gallons per minute)}}{H\text{ (feet)}}$$

where H is head of water in feet acting on the test length. Where the test length is below

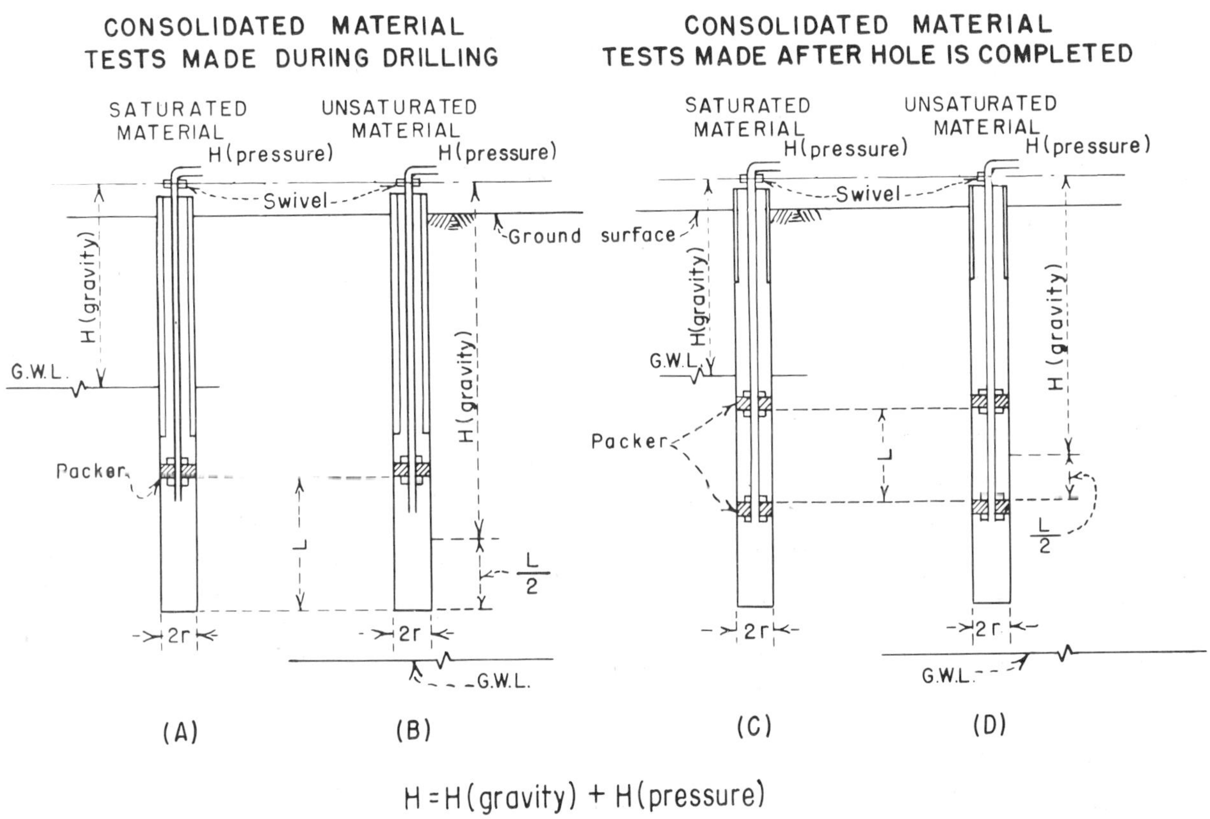

Figure 104. The packer test for soil permeability. 288–D–2477.

the water table, H is the distance in feet from the water table to the swivel plus applied pressure in units of feet of water. Where the test length is above the water table, H is the distance in feet from the center of the length tested to the swivel plus the applied pressure in units of feet of water. For gravity tests (no applied pressure), measurements for H are made to the water level inside the casing (usually the level of the ground).

Values of C_p are given in the following table for various lengths of test section and hole diameters:

Length of test section in feet, L	C_p			
	Diameter of test hole			
	EX	AX	BX	NX
1	31,000	28,500	25,800	23,300
2	19,400	18,100	16,800	15,500
3	14,400	13,600	12,700	11,800
4	11,600	11,000	10,300	9,700
5	9,800	9,300	8,800	8,200
6	8,500	8,100	7,600	7,200
7	7,500	7,200	6,800	6,400
8	6,800	6,500	6,100	5,800
9	6,200	5,900	5,600	5,300
10	5,700	5,400	5,200	4,900
15	4,100	3,900	3,700	3,600
20	3,200	3,100	3,000	2,800

The usual procedure is to drill the hole, remove the core barrel or other tool, seat the packer, make the test, remove the packer, drill the hole deeper, set the packer again to test the newly drilled section, and repeat the test (see fig. 104(A)). If the hole stands without casing, a common procedure is to drill it to final depth, fill with water, surge it, and bail it out. Then set two packers on pipe or drill stem, as shown in figure 104 (C) and (D). The length of packer when expanded should be at least five times the diameter of the hole. The bottom of the pipe holding the packer must be plugged and its perforated portion must be between the packers. In testing between two packers, it is desirable to start from the bottom of the hole and work upward.

Example for figure 104(A):

Given:

NX casing set to depth of 5 feet,

$Q=2.2$ gallons per minute,

$L=1$ foot,

H (gravity) = distance from groundwater level to swivel = 3.5 feet,

H (pressure) = 5 p.s.i. × 2.31 = 11.55 feet of water,

$H = H\text{(gravity)} + H\text{(pressure)} = 15.1$ feet.

From table, $C_p=23{,}300$

$$K=C_p\frac{Q}{H}=\frac{(23{,}300)(2.2)}{15.1}=3{,}400 \text{ feet per year.}$$

119. *Density-in-Place Tests.*—(a) *Sand Density Method.*—This method is used to determine the inplace density in a foundation, a borrow area, or a compacted embankment by excavating a hole from a horizontal surface, weighing the material excavated, and determining the volume of the hole by filling it with calibrated sand. A water content determination on a sample of the excavated soil enables the dry density in the ground to be calculated. Various devices using balloons and water or oil have been used to measure the volume of the hole, but the sand method is most common.

Figure 105. Determining density of rolled embankment material by replacement with a sand of known density. 794–701–401.

About 100 pounds of clean, air-dry, uniform sand passing the No. 16 sieve and retained on the No. 30 sieve has been found to be satisfactory. Clean "blow sand" or dune sand is suitable as well as the sand furnished by commercial suppliers for foundries or fracturing sand used by the oil well drilling industry. When large test holes are used in gravelly

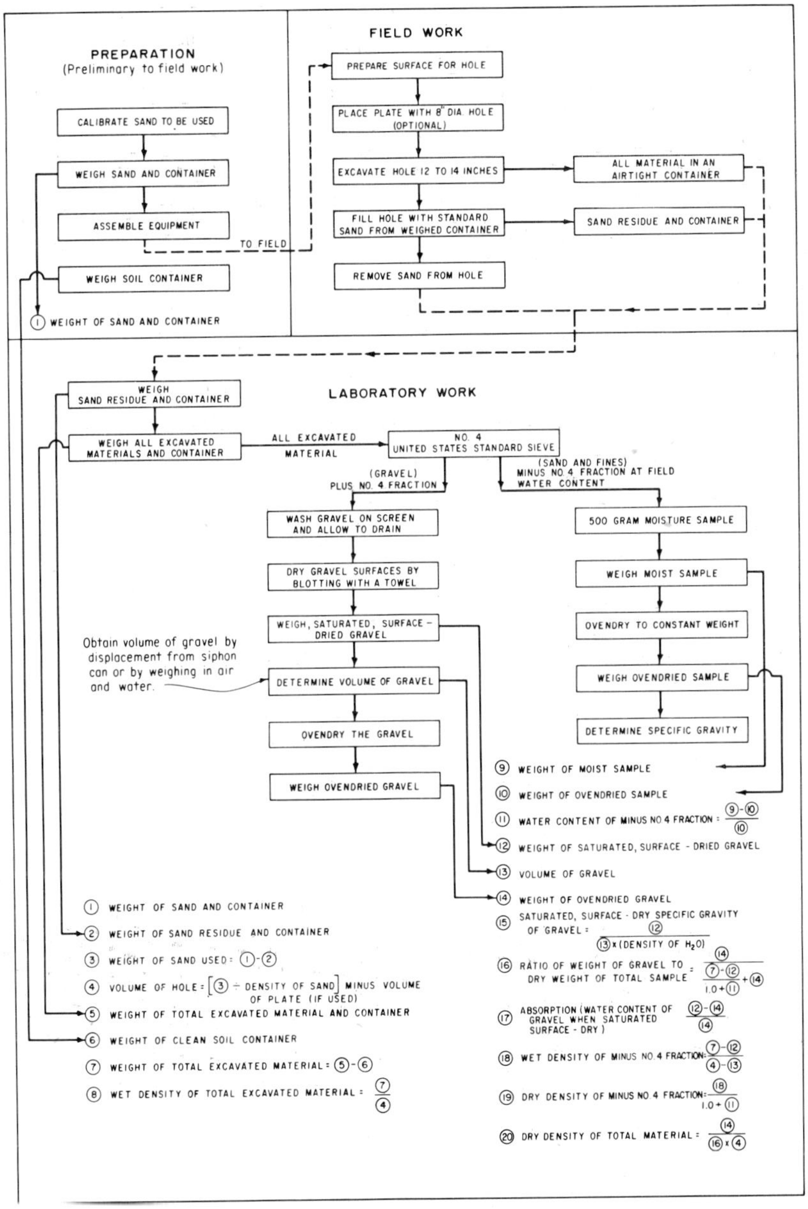

***Figure 106.* Procedure for density-in-place test. 101–D–285.**

soils, coarse sand having rounded particles and passing the No. 4 sieve and retained on the No. 8 sieve is recommended. The sand is calibrated by pouring it into a container of known volume of approximately the size and shape of the type of excavation to be used, weighing it, and calculating its placed unit weight.

At the location to be tested, all loose soil is removed from an area 18 to 24 inches square and the surface is leveled. A working platform supported at least 3 feet from the edge of the test hole should be provided when excavating in soils that may deform and change the dimensions of the hole due to the weight of the operator. An 8-inch-diameter hole 12 to 14 inches deep is satisfactory for cohesive soils which contain little or no gravel. A hole about 12 inches in diameter at the surface, tapering down to about 6 inches at a depth of 12 to 14 inches, is needed for gravelly soils.

A steel or wooden template with the proper sized hole is placed on the ground and the excavation is carefully made with an auger or other handtools. All material taken from the hole is placed in an airtight container for subsequent weighing. To avoid loss of moisture the cover should be kept on the container except when in use; and in hot, dry climates a shade for the test area and a moist cloth over the container should be provided. A plastic bag can be used inside the container to hold the soil removed and sealed to prevent moisture loss.

The volume of the hole is determined by carefully filling it with calibrated sand using the sand-pouring device shown in figure 105 or similar method. The weight of sand used to fill the hole is determined by subtracting the final weight of sand and container (plus the calculated weight of sand occupying the small space in the template), from the initial weight. The volume of the sand (and of the hole) is calculated from the known unit weight of calibrated sand.

The inplace wet density of the soil is the weight of the soil removed from the hole divided by the volume of the hole. For soils containing no gravel, a representative moisture sample is taken and the water content is determined (see section 120 for water content test). The inplace dry density is then calculated.

For soils containing gravel sizes, the wet density of the total material is determined as in the foregoing. In the laboratory the gravel particles are separated from the soil, and their weight and solid volume are determined and subtracted from the total weight of material and volume of the hole, respectively, to obtain the wet density of the minus No. 4 fraction of the soil. This is converted to dry density by a water content determination. The field and laboratory procedures used for den-

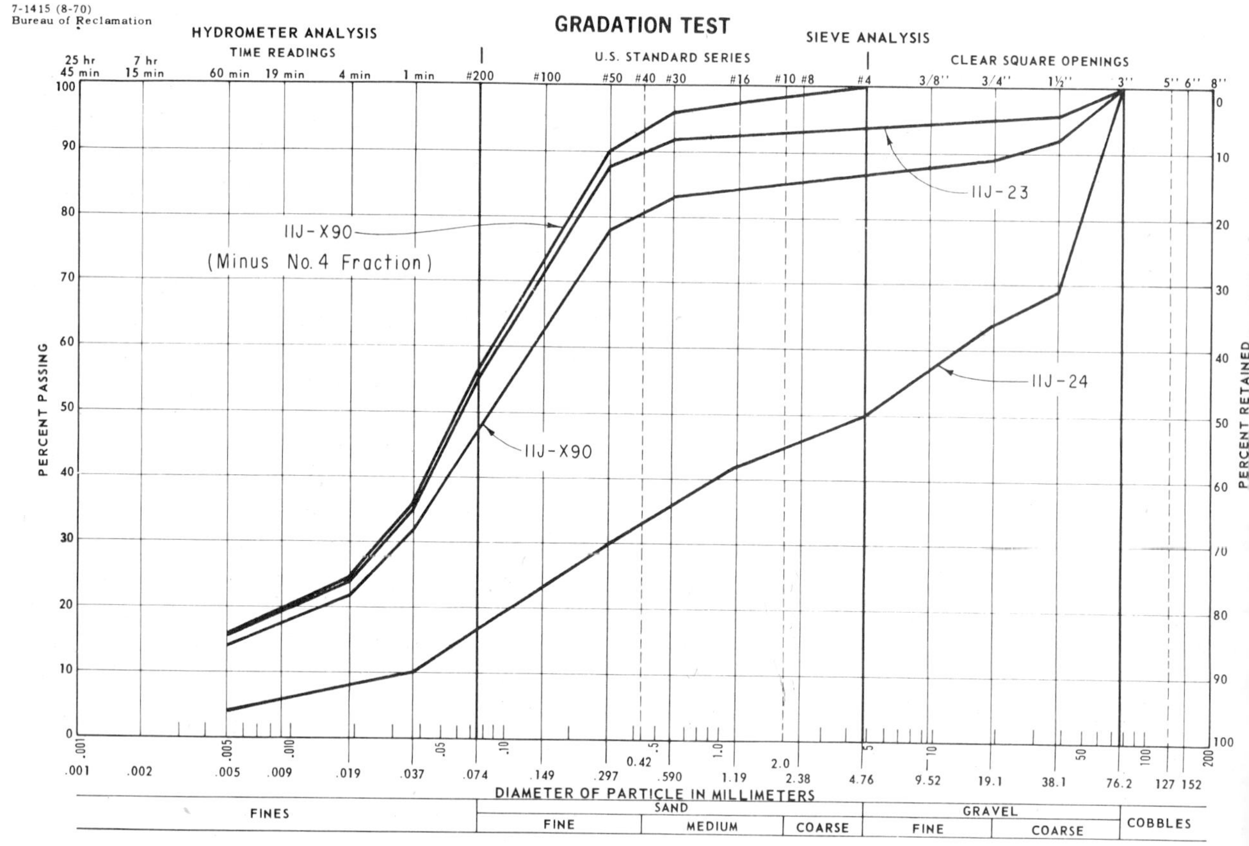

Figure 107. Gradation analysis curves. 101–D–253.

sity inplace tests are shown in figure 106.

Further information concerning the sand density method may be found in Designation E-24 of the Bureau's "Earth Manual" [6].

(b) *Method for Dry, Gravel-Free Soils.*—It is often necessary to determine the inplace dry density and water content of fairly deep foundations of cohesive soils above the water table. The sand density method given in (a) above requires a test pit or large-diameter auger hole to gain access to the foundation. The following simple method has been used successfully to obtain inplace density in stages of depth in foundations and borrow areas of dry gravel-free soils with the use of a hand auger.

A platform should be built on which the investigator can stand without bearing on the soil within 2 feet of the hole he is to auger. A system of sills and joists covered by decking containing a hole about 12 inches in diameter in the center will suffice. Such a device will preclude squeezing of the upper portion of the hole. The procedure is to start a hole with an 8-inch-diameter post-hole auger, penetrating the soil for a distance of between 6 inches and 1 foot, depending on the probable depth of stripping. The soil removed is discarded, and the depth from the surface of the ground to the apex of the cone at the bottom of the hole is measured to within 0.01 foot. The hole is then deepened with the auger to a depth of 3 feet or to any apparent change in soil structure, whichever occurs first; and the soil removed is placed on a clean canvas, sampled for water content, and weighed. The depth from the bottom of the hole to the top of the ground is carefully measured again to the nearest 0.01 foot, and the diameter of the hole is measured at about 1 foot below the surface of the ground.

The volume of the hole sampled is the difference between the two depths measured, multiplied by the area of the hole as computed from the measured diameter. Thus, the wet density and the dry density can be determined for each tested depth below stripping. The tests can be continued to the limit of the practicable hand-auger depth, which is about 20 feet for an 8-inch-diameter auger and which can be extended by use of a tripod to aid in removing the auger from the hole. Since about 1 cubic foot of material is extracted for a 3-foot depth, an accuracy of weighing to about 1 pound is satisfactory. A field scale of 150- or 200-pound capacity can be used.

This method is applicable only to relatively dry, cohesive soils. If the hole tends to cave, this method obviously fails and test pits must be used.

120. *Laboratory Test on Soils.*—(a) *Gradation.*—The gradation or grain size analysis of soils is done by a combination of sieving and wet mechanical analysis. A representative sample of the soil is dried, weighed, and screened on a United States standard No. 4 screen to remove the gravel, which is then passed through a series of screens to determine the amount larger than 3 inches, 1½ inches, ¾ inch, ⅜ inch, and ¼ inch. An oven-dried sample of the minus No. 4 material is used for the remainder of the test. One hundred grams of soil for sands (50 grams for silts and clays) are carefully weighed out and treated with 125 ml. of a 4-percent solution of sodium metaphosphate and distilled water in order to separate the fine grains. After standing for a minimum of 18 hours, the mixture is dispersed by thoroughly mixing in a malted milk type of machine, then transferred to a 1,000-ml. graduate. Distilled water is added to bring it to exactly 1,000 ml. and mixed.

The graduate containing the mixture is placed upon a table, and a stopwatch is started. A soil hydrometer is placed in the mixture and readings are made at 1, 4, 19, and 60 minutes; and also at 7 hours 15 minutes when expansive clays are involved. The hydrometer is of the Bouyoucos type, which is calibrated in grams per liter at 20° C., and its readings are corrected for the meniscus error (the top of the meniscus is read during the test), for difference in temperature from 20° C., and for the amount of deflocculating agent used. On completion of the 1-hour or the 7-hour 15-minute reading the mixture is washed on a No. 200 United States standard sieve and the retained fraction is dried and separated on the Nos. 8, 16, 30, 50, 100, and 200 standard sieves; 15 minutes of shaking in a power sieve shaker is usually done. The residue on each screen is weighed. This procedure is explained in greater detail in

Designation E–6 of the "Earth Manual" [7]. Figure 107 is an example of a resulting gradation analysis curve.

(b) *Water Content.*—The water content of a soil is defined as the weight of water it contains divided by the weight of dry soil. The procedure involves weighing a sample of moist soil and its container and drying it in an oven at 110° C. to constant weight. The time required to attain constant weight varies for different soils, from a few hours for sandy soils to several days for very fat clays. About 16 hours should be the minimum time used. The dried sample and container are placed in a desiccator to cool to room temperature prior to weighing. The water content is calculated as being the difference between the initial and final weights of soil and container, divided by the difference between the weight of the dry soil and container, and the weight of the container. In order to insure accuracy, the following sizes of water content samples are recommended:

Size of sample, grams	*Size of soil particles*
10	Minus No. 40.
200	Minus No. 4.
1,000	Minus ⅜-inch.
2,000	Minus ¾-inch.
3,000	Minus 1½-inch.
3,000 or more [1]	No. 4 to 3-inch gravel.

[1] The sample should be large enough to get a representative sample of the material up to the 3-inch size.

(c) *Atterberg Limits.*—To obtain the liquid limit of a soil, the fraction passing the No. 40 sieve is mixed with water to a puttylike consistency and placed in a brass cup, as shown in figure 53. It is leveled off to a depth of 1 cm. and divided by a grooving tool, as shown in the figure. The crank is turned two rotations per second until the two sides of the sample come in contact at the bottom of the groove for a distance of one-half inch along the groove, and the number of blows is recorded. The water content of soil taken from this portion of the groove is determined. The test is repeated with added water or with less water until a result of 25 blows is bracketed; that is, test results above and below 25 blows are obtained. A "flow curve" is then plotted on a semilogarithmic graph with the number of blows on the logarithmic scale against the water content on the arithmetic scale. The water content corresponding to the 25-blow value is the liquid limit. Detailed procedures for this test and the one-point liquid limit method are given in ASTM Designation D 423–66.

The plastic limit is the lowest moisture content expressed as a percentage of the weight of oven-dried soil at which the soil can be rolled into threads one-eighth inch in diameter without the thread breaking into pieces. In order to determine it, about 15 grams of minus No. 40 fraction of a soil are mixed with enough water to obtain a plastic material and shaped into a ball. The soil is then rolled between the palm of the hand and a ground glass plate or absorbent paper to form the soil into a thread one-eighth inch in diameter. It is then reformed into a ball, kneaded and rolled out again. This procedure is continued until the soil crumbles when the thread becomes one-eighth inch and cannot be reformed. The water content determined for this condition is the plastic limit. Figure 54 shows the test for plastic limit. The plasticity index is the difference between the liquid limit and the plastic limit of a soil. Detailed test procedures are given in ASTM Designation 424–59.

(d) *Specific Gravity.*—Specific gravity is defined as the ratio of the weight in air of a given volume of material to the weight in air of an equal volume of distilled water at a stated temperature. The minus No. 4 fraction of soil is commonly tested for specific gravity by the flask method, as described in designation E–10, Part A, of the "Earth Manual" [7]. In this method a 250-ml., long-necked flask is calibrated for volume at several temperatures. One hundred grams of oven-dried minus No. 4 material is washed into the calibrated flask with distilled water. With the water level well below the neck of the flask, a vacuum is applied to the mixture which causes boiling of entrapped air from the mixture. When the air has been virtually exhausted, distilled water is added to bring the volume to exactly the calibrated volume of the flask, and the vacuum is applied again. When all the air has been removed, the flask and its contents are weighed, and the temperature of the mixture is determined. The volume of the 100 grams of dried soil is determined from the data obtained, and the specific gravity of the soil is then computed.

To determine the specific gravity of gravel and cobbles, the material is immersed in water for a period of 24 hours and then blotted with a towel. This is the saturated surface-dry condition. It is then weighed and carefully placed in a filled siphon can, from which the volume of water it displaces is measured. The bulk specific gravity on a saturated surface-dry basis is the weight of the sample divided by the volume of water displaced. The bulk specific gravity on an oven-dry basis is the oven-dry weight of the material divided by the volume displaced by the saturated surface-dry material. This method is described in detail in designation E-10, Part C, of the "Earth Manual" [7]. See section 121(a) for another method of specific gravity determination.

(e) *Proctor Compaction.*—The Proctor maximum dry density of a soil is the greatest dry unit weight obtainable by the method to be described. The optimum water content of the soil is the water content at this condition. For this test, water is added to about 35 pounds of the minus No. 4 fraction of the soil until its consistency is such that it barely adheres when squeezed firmly in the hand. A sample of the soil is compacted in a 1/20-cubic-foot Proctor mold (with collar attached) in 3 equal layers by 25 uniformly distributed blows per layer of a tamping rod weighing 5.5 pounds dropped freely 18 inches above the layer. The third compacted layer should extend slightly into the collar section. The collar is then removed, and the soil is trimmed to the top of the mold with a straightedge trimmer. The soil and mold are then weighed. The water content of the compacted specimen is determined from a sample taken near the center. This procedure is repeated at least five times using new soil for each specimen and increasing the water added until the resulting compacted wet weight decreases.

The Proctor mold used by the Bureau of Reclamation is one-twentieth cubic foot in volume, and the foregoing procedure with that mold results in a compactive effort of 12,375 foot-pounds per cubic foot of soil. The ASTM Designation D 698–70 and the standard AASHO methods use the same compactive effort, 12,375 foot-pounds per cubic foot, and identical procedures, except that a 1/30-cubic-foot cylinder is used and the free drop is 12 inches instead of 18 inches.

The penetration resistance of the compacted soil for points along the compaction curve, as shown in figure 108, can be obtained by forcing the Proctor needle into each compacted specimen and determining the penetration resistance in pounds per square inch. This method has been extensively used for moisture control of compacted fills. However, the rapid method of compaction control described in appendix E is believed to be a more accurate method which should supplant the Proctor needle for that purpose.

(f) *Relative Density.*—Relative density is defined as the state of compactness of a soil with respect to the loosest and densest states at which it can be placed by specific laboratory procedures. This test is applicable to cohesionless materials which do not have well-defined Proctor curves. The minimum density (zero percent relative density) is obtained by carefully placing dried soil in a container of known size, usually ½ to 1 cubic foot. About 1 inch free fall is permitted for material smaller than ⅜-inch maximum size; material larger than the ⅜-inch size is placed with a scoop. The excess soil is carefully trimmed level to the top and the full container is weighed.

To obtain the maximum density (100 percent relative density), the soil is thoroughly wetted and placed slowly into the container with an attached vibrator operating. After the container is filled, the vibrator is operated for at least 1 minute. The material in the container is then weighed, emptied into a pan, dried, and weighed again. The relative density is defined by the formula:

$$D_d = \frac{e_{max} - e}{e_{max} - e_{min}} = \frac{\gamma_{D_{max}}(\gamma_D - \gamma_{D_{min}})}{\gamma_D(\gamma_{D_{max}} - \gamma_{D_{min}})} \qquad (5)$$

It is usually expressed as a percentage. Figure 109 shows a maximum density test in progress, using a 0.5-cubic-foot container. Designation E–12 of the "Earth Manual" explains this procedure in detail [7].

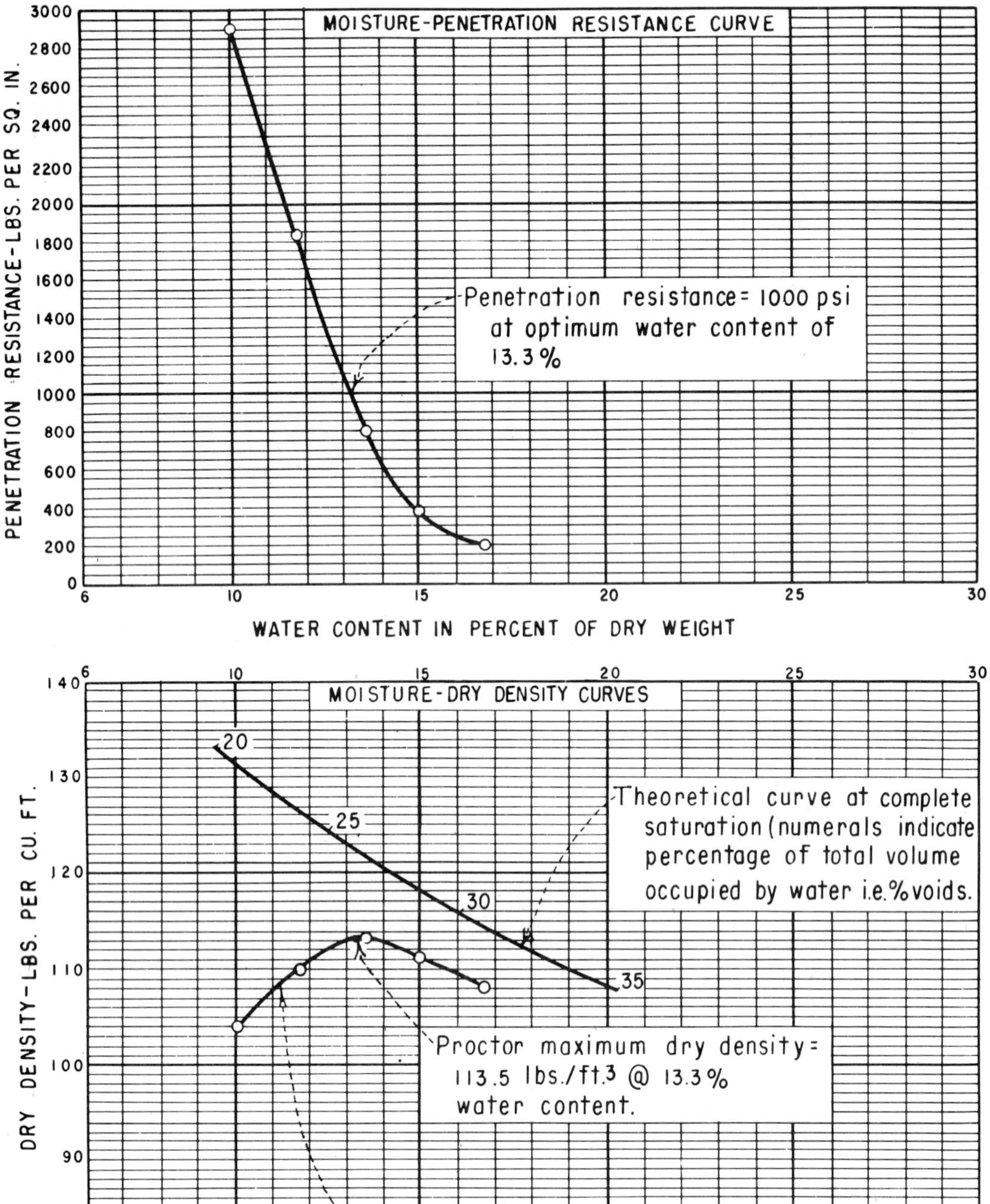

COMPACTION		SOIL PROPERTIES	
25	BLOWS PER LAYER	2.66	SPECIFIC GRAVITY
3	LAYERS	SM	SOIL CLASSIFICATION
5.5	LB. HAMMER	11.2	% LARGER THAN TESTED
18	IN. DROP	113.5	MAX. DRY DENSITY (P.C.F.)
1/20	C.F. CYLINDER	13.3	OPT. WATER CONTENT (%)
		1000	PEN. RES. AT OPT. MOIST. (P.S.I.)

Figure 108. Proctor compaction test curves. **288–D–2478.**

Figure 109. A maximum density test in progress using the 0.5-cubic-foot container. E–2234–INA.

121. *Laboratory Tests on Riprap and Concrete Aggregate.*—(a) *Specific Gravity and Absorption.*—The specific gravity of sand for concrete aggregate can be determined on a saturated surface-dry sample in a manner similar to that given for soil in section 120(d). The specific gravity of coarse aggregates and riprap (crushed to 1½-inch maximum size) is determined by washing the sample to remove dust and other coatings from the surface of the particles, drying to a constant weight and immersing in water at room temperature for 24 hours, blotting with a towel, and weighing. After weighing, the material is placed in a wire basket and is weighed again in water having a temperature of 73.4° F. (23° C.). The sample is then dried to a constant weight in an oven, cooled to room temperature, and weighed again. If A is the weight in grams of the oven-dried sample in air, B the weight in grams of the saturated surface-dried sample in air, and C the weight in grams of the sample in water, then the specific gravity, on a dry basis, equals $\frac{A}{B-C}$, the specific gravity on a saturated surface-dry basis equals $\frac{B}{B-C}$, and the absorption equals $\frac{B-A}{A}$ on a dry basis and $\frac{B-A}{B}$ on a saturated surface-dry basis.
Absorption is usually expressed as a percentage. ASTM Designation C 127–68 describes detailed procedures for these tests.

(b) *Abrasion.*—This test determines the abrasion resistance of crushed rock and natural and crushed gravel. The Los Angeles abrasion machine, which consists of a hollow steel cylinder closed at both ends, having a diameter of 28 inches and a length of 20 inches, is used.

The abrasive charge consists of cast-iron or steel spheres approximately 1⅞ inches in diameter. Twelve spheres are used for an A grading (maximum size of particle 1½ inches), 11 for a B grading (¾ inch maximum), 8 for a C grading (⅜ inch maximum), and 6 for a D grading (No. 4 sieve maximum).

The test sample of 5,000 grams and the proper abrasive charge are placed in the Los Angeles abrasion testing machine, and the machine is rotated for 100 revolutions at a speed of from 30 to 33 revolutions per minute. The material is then removed from the machine, screened on a No. 12 screen, and the material retained on the screen weighed. The entire sample including the dust of abrasion is returned to the testing machine, the machine is rotated an additional 400 revolutions, and the screening and weighing are repeated. The differences between the original weight of the test sample and the weights of the material retained on the screen at 100 revolutions and at 500 revolutions are expressed as percentages of the original weight of the test sample. These values are reported as percentages of wear. ASTM Designation C 131–69 describes detailed procedures for this test.

(c) *Soundness.*—The most commonly used soundness test is the sodium sulfate test. The results of this test are used as an indication of the ability of aggregate and riprap to resist weathering. A carefully prepared saturated solution of sodium sulfate is kept at a temperature of 70° F. (21° C.). After washing and drying in an oven, the material to be tested is screened to provide a specified gradation, usually from 1½ inches to the No. 50 sieve size. Specified weights of the various fractions of the material are placed in separate containers resistant to the action of the solution, and sufficient sodium sulfate solution is poured into the containers to cover the samples. The material is permitted to soak for not less than 16 hours or more than 18 hours, during which the temperature is maintained at 70° F. (21° C.).

After the immersion period, the samples are removed from the solution and dried to constant weight (about 4 hours) at a temperature of 221° to 230° F. (105° to 110° C.) After drying, the sample fractions are cooled to room temperature and the process is repeated. At the end of five cycles, the test sample is inspected and records made of observation. Each fraction is then washed thoroughly to remove the sodium sulfate from the material, dried, and cooled. Each fraction is screened and the quantities of material retained are weighed. The weighted average loss for each fraction is computed and reported. ASTM Designation C 88–69 describes the detailed procedure for this test.

J. BIBLIOGRAPHY

122. *Bibliography.*

[1] Howell, J. V., and Levorsen, A. I., "Directory of Geological Material in North America," American Geological Institute, second edition, 1957, 208 pp.

[2] "Soil Survey Manual," U.S. Department of Agriculture, Handbook No. 18, 1951, p. 209.

[3] ———, p. 473.

[4] "ASTM Manual on Quality Control of Materials," American Society for Testing Materials, Special Publication 15–C, pt. 2, January 1951, pp. 41–51.

[5] "Geology and Its Military Applications," U.S. Department of the Army, Technical Manual TM 5–545, 1952, pp. 3–32.

[6] "Earth Manual," Bureau of Reclamation, first edition—revised, second printing, 1968.

[7] ———, designations E–5, E–6, E–8, E–9, E–10, and E–11.

[8] "Concrete Manual," Bureau of Reclamation, seventh edition, revised 1966, designations 10, 19, and 21.

[9] "Bituminous Materials for Highway Construction and Roofing; Soils; Peats, Mosses, and Humus; Skid Resistance," American Society for Testing Materials, 1970 Annual Book of ASTM Standards, pt. 11, April 1970, 982 pp.

[10] Krynine, D. P., and Judd, W. R., "Principles of Engineering Geology and Geotechniques," McGraw-Hill, 1957.

[11] Gilluly, James, Waters, A. C., and Woodford, A. O., "Principles of Geology," third edition, W. H. Freeman & Co., 1968.

[12] Longwell, C. R., and Flint, R. F., "'Introduction to Physical Geology," second edition, John Wiley & Sons, Inc., 1962.

[13] Longwell, C. R., Flint, R. F., and Sanders, J. E., "Physical Geology," John Wiley & Sons, Inc., 1969.

[14] Parsons, R. B., Simonson, G. H., and Balster, C. A., "Pedogenic and Geomorphic Relationships of Associated Aqualfs, Albolls and Xerolls in Western Oregon," Soil Science America Proc., vol. 32, 1968, pp. 557–560.

[15] "Bureau of Reclamation, Land Classification Handbook," USDI, Bureau of Reclamation, vol. V, pt. 2, 1953.

[16] Maletic, J. T., and Hutchings, T. B., "Selection and Classification of Irrigable Land," Agronomy Monograph 11, Irrigation of Agricultural Land, ASA, Madison, Wisc., 1967, pp. 125–161.

[17] "Application of Pore Pressure Measurements to Shear Strength of Cohesive Soils," Earth Laboratory Report No. EM–761, Bureau of Reclamation, June 1969.

[18] "Concrete and Mineral Aggregates," American Society for Testing Materials, 1970 Annual Book of ASTM Standards, pt. 10, November 1970, 620 pp.

Earthfill Dams

H. G. ARTHUR [1, 2]

A. INTRODUCTION

123. *Origin and Development.*—Earthfill dams for the storage of water for irrigation, as attested both by history and the surviving remnants of ancient structures, have been used since the early days of civilization. Some of the structures built in antiquity were of considerable size. One earthfill dam 11 miles long, 70 feet high, and containing about 17 million cubic yards of embankment was completed in Ceylon in the year 504 B.C. [1].[3] Today, as in the past, the earthfill dam continues to be the most common type of small dam, principally because its construction involves utilization of materials in their natural state with a minimum of processing.

Until modern times all earthfill dams were designed by empirical methods, and engineering literature is replete with accounts of failures [2]. These failures compelled the realization that empirical methods must be replaced by rational engineering procedures in both design and construction. One of the first to suggest that the slopes for earthfill dams be selected on that basis was Bassell in 1907 [3]. However, little progress was made on the development of rational design procedures until the 1930's. The rapid advancement of the science of soil mechanics since that time has resulted in the development of greatly improved procedures for the design of earthfill dams. These procedures include (1) thorough preconstruction investigations of foundation conditions and materials of construction; (2) application of engineering skill and technique to design; and (3) carefully planned and controlled methods of construction.

As a result, earthfill dams are now (1971) being constructed to heights exceeding 1,000 feet above their foundations; and hundreds of large rolled earthfill dams have been constructed in the past 30 years without a single recorded failure. Failures of small earthfill dams, however, continue to be commonplace. Though some of these failures likely are a result of improper design, many of them are caused by lack of care in construction. Proper construction methods include adequate foundation preparation and the placement of materials in the dam embankment with the necessary degree of compaction under an established procedure of testing and control.

The design of an earthfill dam must be realistic. It should reflect the actual foundation conditions at the site and the available materials for embankment construction, and not merely be patterned after a successful design used at a site with dissimilar conditions.

124. *Scope of Discussion.*—This discussion is limited to design procedures for small earthfill dams which are of the rolled-fill type of construction, as defined in section 125. This type of construction is now being used almost entirely for the construction of small dams, to the exclusion of semihydraulic and hydraulic fills.

For the purpose of this discussion, a "small"

[1] Formerly Engineer, Earth Dams Section, now Director of Design and Construction, Bureau of Reclamation.

[2] Supplementary information for the second edition was provided by L. W. Davidson, Civil Engineer, Earth Dams Section, Bureau of Reclamation.

[3] Numbers in brackets refer to items in the bibliography, section 149.

Figure 110. Upstream face of dam and fishscreened inlet structure at Crane Prairie Dam on the Deschutes River in Oregon.. Deschutes 93.

dam is one whose maximum height above the lowest point in the original streambed does not exceed 50 feet, and whose volume is not of such magnitude that significant economies can be obtained by utilizing the more precise methods of designs usually reserved for large dams. A low dam cannot be considered small if its volume exceeds say, 1 million cubic yards. Figures 110 and 37 show typical small dams constructed by the Bureau of Reclamation. Crane Prairie Dam was completed in 1940. It has a height of 31 feet and contains 29,700 cubic yards. Crescent Lake Dam was completed in 1956. It has a height of 22 feet and contains 16,800 cubic yards. Maximum sections of these dams are shown in figures 172 and 173, respectively.

Figures 111 and 112 show dams constructed by the Bureau of Reclamation which are at the upper limit of height for the scope of this text; in fact, Fruitgrowers Dam is slightly above the height limit. It has a maximum height of 55 feet and a volume of 135,500 cubic yards, but is included herein as a matter of interest. Irrigation at this site dates back to 1898; the dam shown here was constructed in 1939 downstream from the original structure, which was breached in June 1937 to forestall failure. A maximum section of Fruitgrowers Dam is shown in figure 176. Shadow Mountain Dam is a 50-foot-high structure containing 168,000 cubic yards of embankment, completed in 1946. Its maximum section is shown in figure 187.

The design procedures given in this text are not sufficiently detailed to permit their sole use for the design of small dams where exceedingly soft or exceedingly pervious foundations are involved, nor where the nature of the only soil

Figure 111. Fruitgrowers Dam, an earthfill storage dam at an offstream location in Colorado. Washington Office 39.

available for construction of the embankment is unusual. In this category are soils of high plasticity, low maximum density, or very high natural water content which cannot be reduced by drainage. These conditions require the services of an engineer specializing in earthfill dam design to direct the investigations, to determine the laboratory testing program, to interpret the laboratory test results, and to supervise the preparation of the design and specifications.

125. Selection of Type of Earthfill Dam.—(a) *General.*—The selection of type of dam (earthfill, rockfill, concrete gravity, or combination of these types) is discussed in chapter IV. When this procedure leads to the selection of an earthfill dam, a further decision must be made as to the type of earthfill dam.

The scope of this text includes only the rolled-fill type of earthfill dam. In this type the major portion of the embankment is constructed in successive, mechanically compacted layers. The material from borrow pits and that which is suitable from excavations for other structures is delivered to the embankment, usually by trucks or scrapers. It is then spread by motor-patrols or bulldozers and sprinkled, if necessary, to form layers of limited thickness having the proper moisture content, which are then thoroughly compacted and bonded with the preceding layer by means of power-operated rollers of proper design and weight. Rolled-fill dams are of three types; namely, diaphragm, homogeneous, and zoned.

(b) *Diaphragm Type.*—In this type of section the bulk of the embankment is constructed of pervious material (sand, gravel, or rock) and a thin diaphragm of impermeable material is provided to form the water barrier. The position of this diaphragm may vary from a blanket on the upstream face to a central vertical core. The diaphragm may consist of earth, portland cement concrete, bituminous concrete, or other material. If the blanket or core is of earth, it is considered to be a "diaphragm" if its horizontal thickness at any elevation is less than 10 feet or less than the height of embankment above any correspond-

Figure 112. Shadow Mountain Dam, an earthfill structure on the Colorado River in Colorado, constructed as part of a large transmountain diversion scheme. SM–175–CBT.

ing elevation in the dam. If the impervious earth zone equals or exceeds these thicknesses, the design is considered to be of the "zoned embankment" type.

If the bulk of material comprising the diaphragm type dam is rock, the dam is classified as a rockfill dam. The design of rockfill dams is discussed in chapter VII.

Although successful dams have been constructed with internal or "buried" diaphragms, this type of construction is not recommended for structures within the scope of this text. The construction of an internal diaphragm of earth with the necessary filters requires a higher degree of precision and closer control than it is feasible to obtain for small dams. Internal diaphragms of rigid material such as concrete also have the disadvantage of not being readily available for inspection or emergency repair if they are ruptured due to settlement of the dam or its foundation.

An earth blanket on the upstream slope of an otherwise pervious dam also is not recommended because of the expense and the difficulty of construction of suitable filters. Furthermore, the earth blanket must be protected from erosion by wave action, so it is buried and not readily available for inspection or repair. If the supply of impervious soil is so limited that the "zoned embankment" type of dam cannot be constructed, a diaphragm of manufactured material placed on the upstream slope of an otherwise pervious embankment is recommended for small dams. The design of suitable impervious pavings is discussed in chapter VII.

(c) *Homogeneous Type.*—A purely homogeneous type of dam is composed of a single kind of material (exclusive of the slope protection). The material comprising the dam must be sufficiently impervious to provide an adequate water barrier and the slopes must be relatively

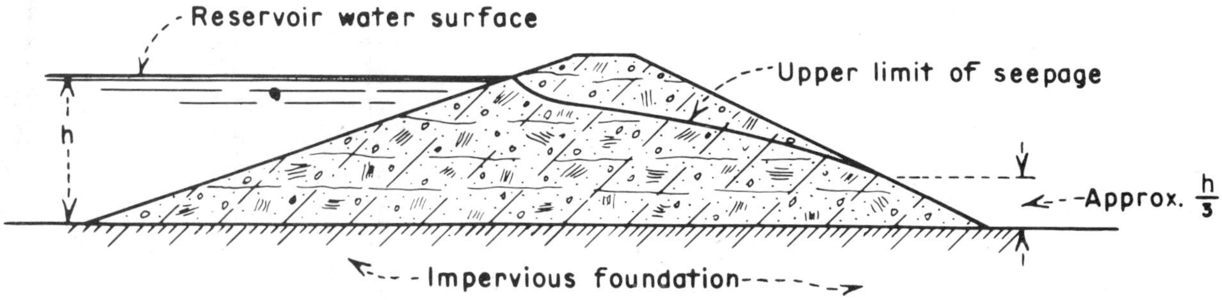

Figure 113. Completely homogeneous dam. 288–D–2479.

flat for stability. To avoid sloughing, the upstream slope must be relatively flat if rapid drawdown of the reservoir after long-term storage is anticipated. The downstream slope must be relatively flat to provide a slope sufficiently stable to resist sloughing when saturated to a high level. For a completely homogeneous section it is inevitable that seepage will emerge on the downstream slope regardless of its flatness and the impermeability of the soil, if the reservoir level is maintained for a sufficiently long time. The downstream slope eventually will be affected by seepage to a height of roughly one-third the depth of the reservoir pool [4], as shown in figure 113.

Although formerly very common in the design of small dams, the purely homogeneous section has been replaced by a modified homogeneous section in which small amounts of carefully placed pervious materials control the action of seepage so as to permit much steeper slopes. The effect of drainage at the downstream toe of the embankment is shown in figure 114.

Rock toes of appreciable size may be provided for drainage, as shown in figure 114(A), or, if suitably graded materials are available, a horizontal drainage blanket shown in figure 114(B) may be used. Another method of providing drainage which has been used is the installation of pipe drains. These are recommended for small dams only when used in conjunction with a horizontal drainage blanket or pervious zones. Reliance should not be placed solely upon pipe drains because of the possibility of the pipe becoming clogged as the result of improper filters, root growth, or deterioration.

Because the modification of the homogeneous type of section by means of drainage furnishes a greatly improved design, the fully homogeneous section should not be used for storage dams; drainage should always be provided when a reservoir pool will be maintained for an appreciable length of time. A homogeneous (or modified-homogeneous) type of dam is applicable in localities where readily available soils show little variation in permeability, and soils of contrasting permeabilities are available only in minor amounts or at considerably greater cost.

(d) *Zoned Embankment Type.*—The most common type of a rolled earthfill dam section is that in which a central impervious core is flanked by zones of materials considerably more pervious, called shells. These pervious zones or shells enclose, support, and protect the impervious core; the upstream pervious zone affords stability against rapid drawdown; and

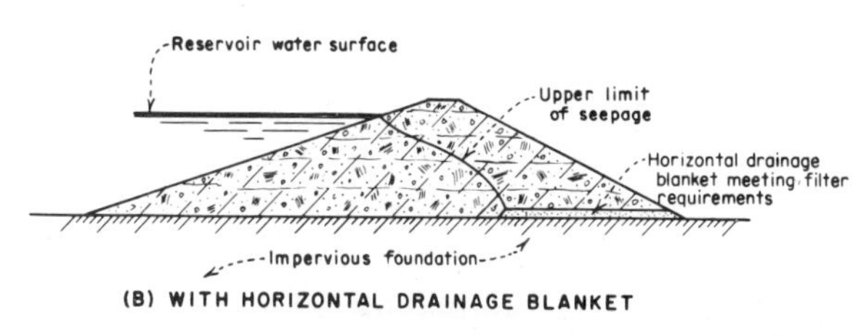

Figure 114. Modified homogeneous dam. 288–D–2480.

the downstream pervious zone acts as a drain to control the line of seepage. For most effective control of through seepage and drawdown seepage, a section through the dam should show, to the extent feasible, a progressive increase in permeability from the center out toward each slope.

The pervious zones may consist of sand, gravel, cobbles, or rock, or mixtures of these materials. For purposes of this text, the dam is considered to be a zoned embankment if the horizontal width of the impervious zone at any elevation equals or exceeds the height of embankment above that elevation in the dam, and is not less than 10 feet. The maximum width of the impervious zone will be controlled by stability and seepage criteria and by the availability of material. A dam with an impervious core of moderate width composed of strong material and with pervious outer shells may have relatively steep outer slopes, limited only by the strength of the foundation, the stability of the embankment itself, and maintenance considerations. Conditions that tend to increase stability may be decisive in the choice of a section even if a longer haul is necessary to obtain required embankment materials.

It may be said that if a variety of soils are readily available, the choice of type of earthfill dam should always be the zoned embankment type because its inherent advantages will lead to economies in cost of construction.

B. DESIGN PRINCIPLES

126. *Design Data.*—The data required for the design of an earthfill dam are discussed in chapter I, and chapter V describes the investigation of foundations and sources of construction materials. The required detail and accuracy of the data will be governed by the nature of the project and the immediate purpose of the design; that is, whether the design is intended as a basis for a cost estimate to determine project feasibility, whether the design is for the purpose of construction, or whether some intermediate purpose is to be served. The extent of investigations of foundations and sources of construction material will also be governed by the complexity of the situation.

127. *Criteria for Design.*—The basic principle of design is to produce a satisfactory functional structure at a minimum total cost. Consideration must be given to maintenance requirements so that economies achieved in the initial cost of construction will not result in excessive maintenance costs. The latter costs will vary with the provisions of upstream and downstream slope protection, drainage features, and the type of appurtenant structures and mechanical equipment. For minimum cost, the dam must be designed for maximum utilization of the most economical materials available, including material which must be excavated for its foundations and for appurtenant structures.

An earthfill dam must be safe and stable during all phases of construction and operation of the reservoir. To accomplish this, the following criteria must be met:

(1) The embankment must be safe against overtopping during occurrence of the inflow design flood by the provision of sufficient spillway and outlet works capacity.

(2) The slopes of the embankment must be stable during construction and under all conditions of reservoir operation, including rapid drawdown of the reservoir in the case of storage dams.

(3) The embankment must be designed so as not to impose excessive stresses upon the foundation.

(4) Seepage flow through the embankment, foundation, and abutments must be controlled so that no internal erosion takes place and so there is no sloughing in the area where the seepage emerges. The amount of water lost through seepage must be controlled so that it does not interfere with planned project functions.

(5) The embankment must be safe against overtopping by wave action.

(6) The upstream slope must be protected against erosion by wave action, and the crest and downstream slope must be protected against erosion due to wind and rain.

(7) If the dam is located in a region subject to earthquakes, the design must be such that the most severe earthquake which can be reasonably anticipated will not impair the function of the structure.

An earthfill dam designed to meet the above criteria will prove permanently safe provided proper construction methods and control are achieved. The design procedure to meet the requirements of criterion (1) above is discussed in chapter IX. Methods for satisfying other criteria for earthfill dams subject to the limitations in scope described in section 124 will be discussed herein. The applicability of the procedures to a specific case will depend upon the purpose of the design, the size and importance of the structure, and the complexity of the problems.

C. FOUNDATION DESIGN

128. *General.*—The term "foundation" as used herein includes both the valley floor and the abutments. The essential requirements of a foundation for an earthfill dam are that it provide stable support for the embankment under all conditions of saturation and loading and that it provide sufficient resistance to seepage to prevent excessive loss of water.

Although the foundation is not actually designed, certain provisions for treatment are made in designs to assure that the essential requirements will be met. No two foundations are exactly alike; each foundation presents its own separate and distinct problems requiring corresponding special treatment and preparation. Various methods of stabilization of weak foundations, reduction of seepage in permeable foundations, and types and locations of devices for the interception of underseepage must depend upon, and be adapted to, local conditions. The importance of adequate foundation treatment is emphasized by the fact that approximately 40 percent of all earthfill dam failures are attributed to failures of the foundation.

Theoretical solutions based on principles of soil mechanics can be made for problems involving pervious or weak foundations. Most of these solutions are relatively complex and they may be relied upon only to the degree that the actual permeabilities in various directions or the strength of the foundation can be determined by expensive, detailed field and laboratory testing. Ordinarily, extensive exploration of this nature and complex theoretical designs are not required for the small dams which are discussed in this text. For these structures it is usually more economical to design foundations on the basis of judgment, deliberately striving for substantial factors of safety. The savings in construction cost that can be achieved by more precise design ordinarily do not warrant the cost of additional exploration, testing, and engineering involved. There are foundations, however, where conditions are unusual to the extent that judgment cannot be relied upon to produce a design with an adequate factor of safety. Such conditions require the services of an engineer specializing in the field of earthfill dam design and are beyond the scope of this text.

Because different types of treatment are appropriate for different conditions, foundations are grouped into three main classes according to their predominant characteristics:

Foundations of rock.

Foundations of coarse-grained material (sand and gravel).

Foundations of fine-grained material (silt and clay).

It is realized that foundations which originate from various sources such as river alluvium, glacial outwash, talus, and other processes of erosion, disintegration, and deposition are characterized by infinite variations in the combinations, structural arrangement, and physical characteristics of their constituent materials. The deposits may be roughly stratified, containing layers of clay, silt, fine sand and gravel; or they may consist of lenticular masses of the same material without any regularity of occurrence and of varying extent and thickness. In spite of this, the character of the foundation, as revealed by exploration, usually can be safely generalized for the design of small dams to fit one of the classes given above, and the nature of the problem requiring treatment will be evident. Ordinarily, coarse-grained, pervious foundations present no difficulties in the matter of settlement or stability for a small dam; conversely, problems of seepage are not associated with fine-grained, weak foundations which are subject to settlement or displacement.

The special treatments required for the different types of foundations listed above are discussed in this chapter. If the foundation material is impervious and comparable to the compacted embankment material in structural characteristics, little foundation treatment will be required. *The minimum treatment for any foundation* is stripping of the foundation area to remove sod, topsoil with high content of organic matter, and other unsuitable material that can be disposed of by open excavation. In many cases where the overburden is comparatively shallow, the entire foundation is stripped to bedrock. *In all soil foundations in which a cutoff trench or partial cutoff trench (see sec. 131) is not used,* a "key" trench should be provided to bond the impervious zone of the embankment to the foundation. The top several feet of the soil foundation, which invariably lack the density of the underlying soil due to frost action, surface runoff, wind, or other cause, should be penetrated by the key trench. A bottom width of 20 feet for the key trench is usually sufficient.

The foundation at any one particular site usually consists of a combination of the three main types of foundations listed above. For example, the stream portion often is a sand-gravel foundation, while the abutments are rock which is exposed on the steep slopes and mantled by deep deposits of clay or silt on the gentle slopes. Therefore, the design of any one dam may involve a variety of foundation design problems.

129. *Rock Foundations.*—Foundations of rock, including hard shale, generally do not present any problems of bearing strength for small dams. Rock foundations should, however, be carefully investigated to determine their permeability. If erosive leakage, excessive uplift pressure, or high water losses will occur through joints, fissures, crevices, permeable strata, or along fault planes, the foundation should be grouted. The determination of whether or not a foundation should be grouted is made by examining the site geology and by analyzing the water loss quantities of foundation exploration holes. A great deal of experience is required to make this decision since every foundation is unique, and the advice of an experienced designer should be sought when questionable conditions exist.

Ordinarily, the design and estimate for a storage dam should provide for foundation grouting. On the other hand, grouting of rock foundations is not generally required for detention dams or for extremely low diversion and storage dams.

Foundation grouting is a process of injecting under pressure a fluid sealing material into the underlying formations through specially drilled holes for the purpose of sealing off or filling joint seams, fissures, or other openings. Unless the geologic conditions dictate otherwise, the foundation should be grouted to a depth below the surface of the rock equal to the reservoir head which lies above the surface of the rock.

The grouting of a dam foundation is usually performed along a single line of grout holes, with center to center distances of from 10 to 20 feet. This creates a deep, impermeable water barrier referred to as a "grout curtain." Multiple lines of grout holes may be used when severely fractured or highly permeable rock is encountered. The grout curtain used on the abutment of Granby Dam in Colorado is shown in figure 115.

In cases where large zones of fractured rock lie at the foundation contact or where the zone of broken rock within a fault has great width, the entire zone of fractured rock or fault gouge

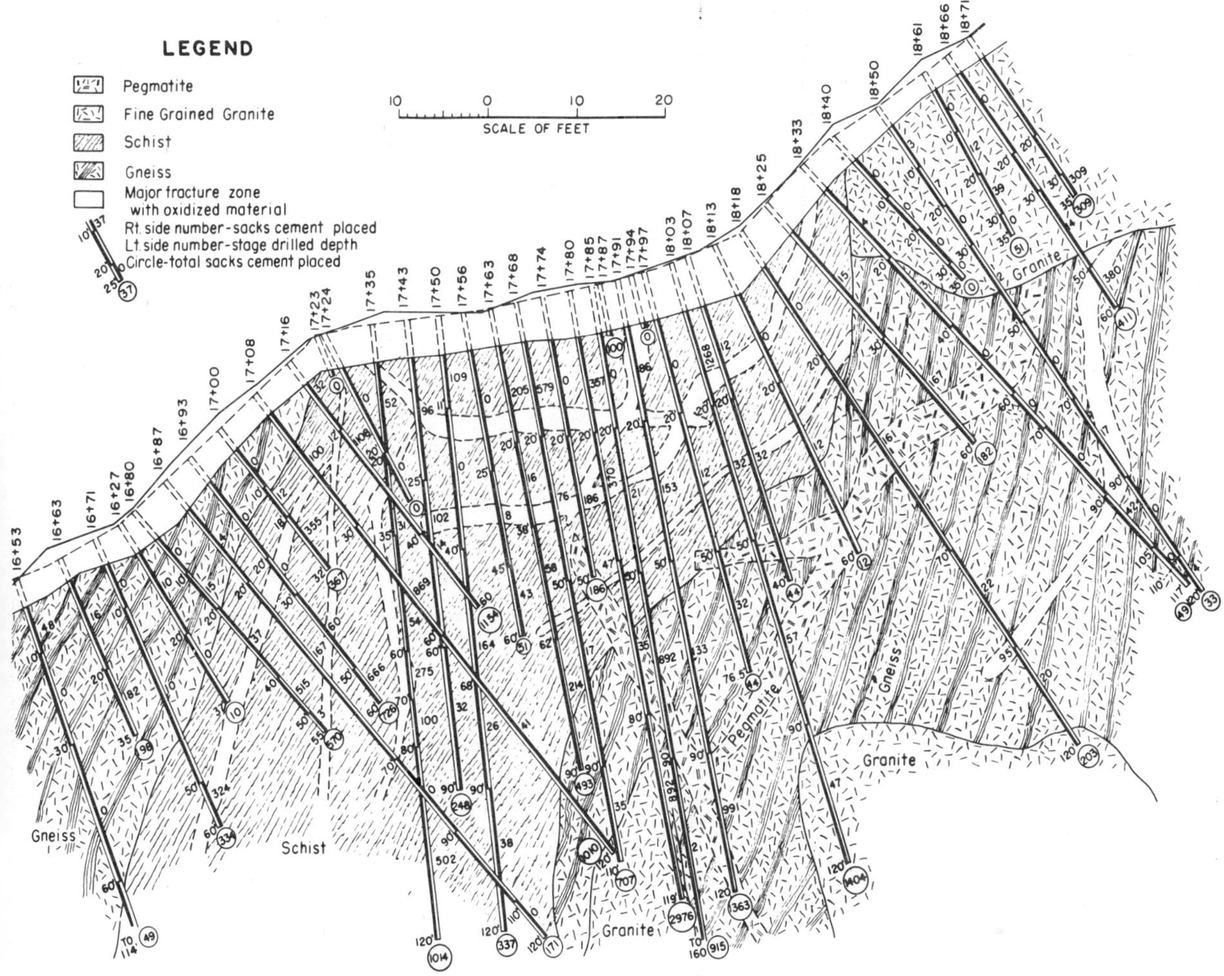

Figure 115. Grout curtain used on the abutment of Granby Dam in Colorado. 101–D–245.

will be grouted to a shallow depth, usually 10 to 30 feet, by using a grid pattern. This type of grouting is referred to as "blanket grouting." It reduces leakage in the fractured zone and provides a more firm foundation for the dam.

Foundation grouting is generally performed with a mixture of cement and water, starting with a ratio of 5:1. If considerable "take" in a hole is experienced, the grout mixture is progressively thickened. Grout mixes usually vary between 10:1 and 0.8:1. If the grout take is excessive, sand is added to give the grout additional bulk. In some cases, bentonite is combined with the sand in small quantities, about 2 percent by weight of the cement, to obtain a more pumpable grout mix.

A suggested gradation of sand, which is used for grouting on Bureau projects, is given in the tabulation below.

Screen No.	Cumulative percent, by weight retained on screen
8	0
16	0 to 5
30	15 to 40
50	50 to 80
100	70 to 90
200	95 to 100

Where the grout hole continues to take a large quantity of grout, it may be advantageous to require intermittent pumping, waiting up to 8 hours between pumping periods to allow grout in the foundation to set.

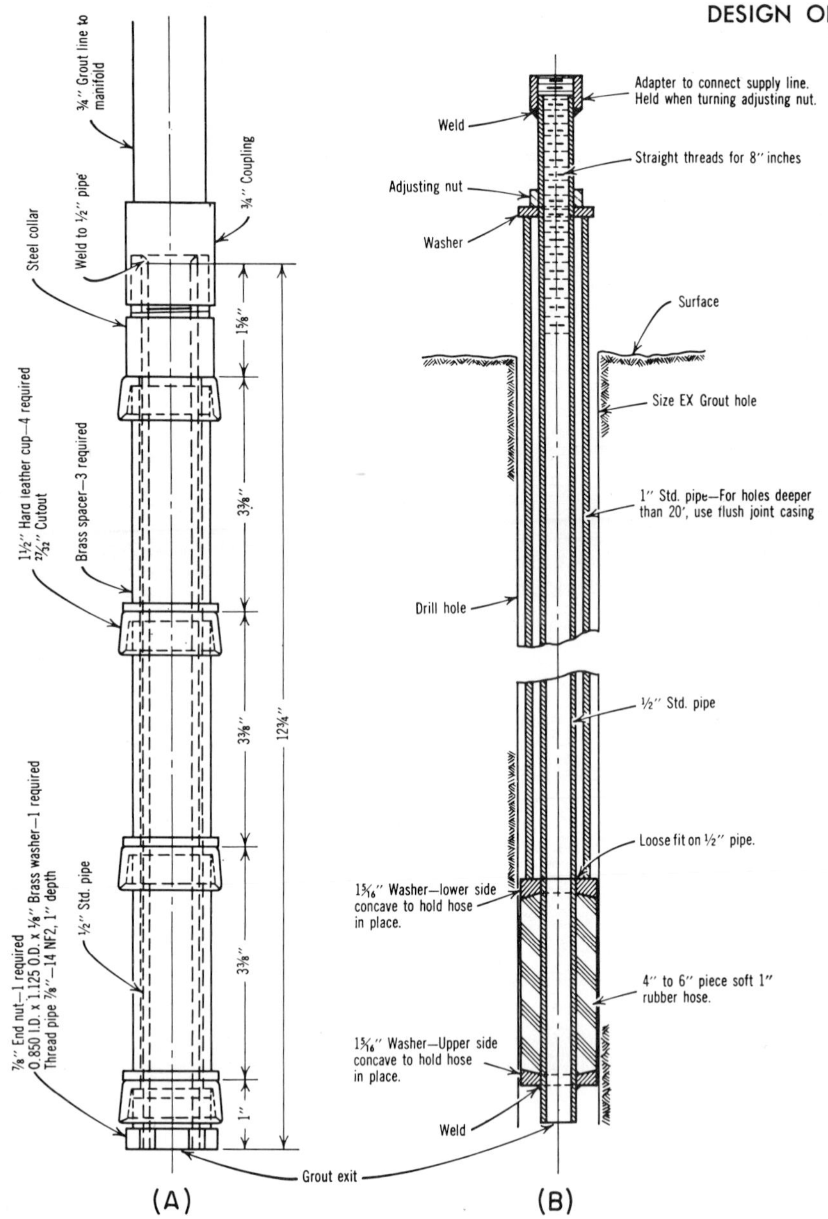

Figure 116. Packers used for grouting by the Bureau of Reclamation. (A) Leather cup. (B) Mechanical. (Sheet 1 of 2.) 288–D–2873.

Grouting is usually performed by one of the following methods:

(1) Staging-down.

(2) Staging-up.

Grouting by the staging-down method consists of drilling the grout hole to predetermined depth, washing the hole, pressure testing with water, and then grouting. After grouting, and before the grout within the hole has hardened, the grout is washed out of the hole and drilling for the second stage is begun. In the second and succeeding stages, the same sequence of operations is used, except that a packer is sealed near the bottom of the previously grouted stage. In this manner, subsequent stages are grouted until the entire length of the hole has been grouted. This method is useful when drill hole caving occurs, when the upper layers of the foundation are extensively cracked, or when the hole suddenly loses drill water.

When grouting by the staging-up method

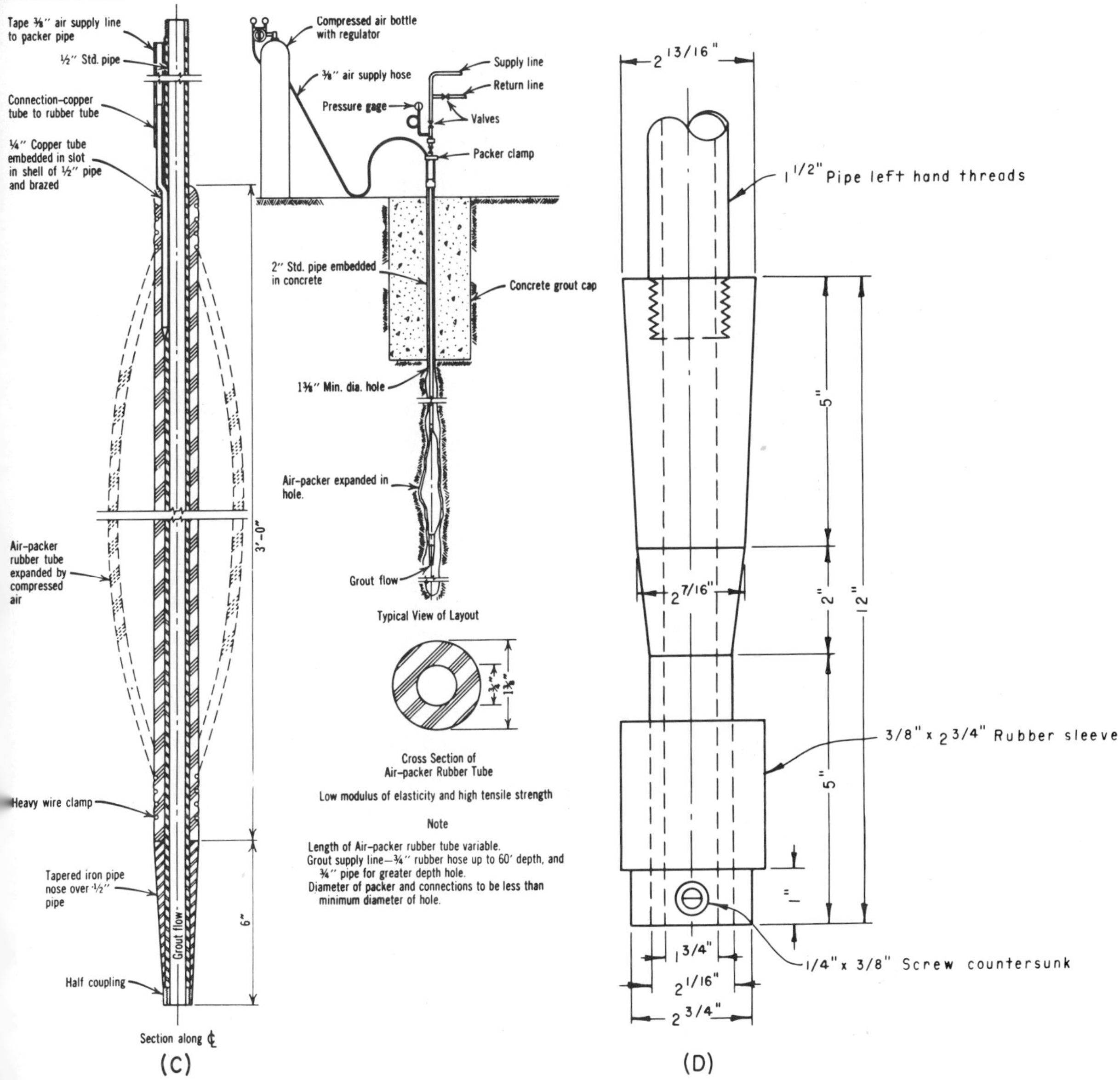

gure 116. Packers used for grouting by the Bureau of Reclamation. (C) Pneumatic. (D) Cone-type. (Sheet 2 of 2.) 288–D–2874.

ıe entire length of the hole is drilled, the hole ashed, and a packer attached to the end of the rout supply pipe, which is then lowered and ated at a predetermined distance above the ottom of the hole, and the grouting performed the required pressure. The grout pipe and ıcker are then withdrawn to the next stage ıd the grouting repeated. This upward stag- g continues until the entire hole is grouted.

Grout holes are usually drilled with the com- ercial standard EX (approximately 1½-inch ameter) drill size and a grout nipple is used introduce the grout into the foundation. A grout nipple is usually a 2-inch-diameter pipe having a length of from 18 inches to 5 feet (depending on rock conditions) which is anchored into the rock by cement grout, oakum, or other suitable caulking material to facilitate drilling and grouting. Rotary-type drills are most commonly used since percussion-type drills tend to seal off joints and fractures intersected by the hole.

Packers are devices which seal off drill holes at any elevation to permit grouting of a selected stage below the packer. The four types of packers most commonly used are shown in fig-

(A)

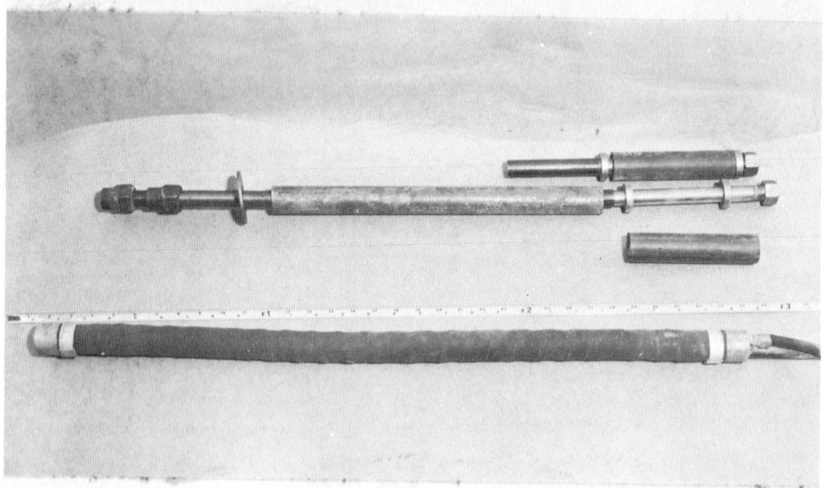

(B)

Figure 117. Types of grout hole packers used by the Bureau of Reclamation. (A) From left to right: leather cup and cone-type. (B) From top to bottom: mechanical and pneumatic. P805–236–1594, P557–420–3459.

ure 116. The leather cup packer seals when the grout forces the cups outward against the drill hole wall and is most commonly used in hard rock. The mechanical packer requires a double pipe arrangement and is seated against the drill hole wall by compressing the annular rubber sleeve at the bottom of the packer pipe by tightening the nut at the top of the pipe; this type of packer is more suitable than the leather cup packer in slightly oversized holes. The pneumatic packer is expanded by compressed air or inert gas and used in poor rock where the drill holes may be considerably oversize. The cone-type packer is seated when grout forces the annular rubber sleeve upward on the cone; it is used in relatively hard rock. Photographs of the four types of packers are shown in figure 117.

A great variety of grouting equipment is available. In general, the equipment consists of a grout mixer, grout agitator, grout pump, and a pipe and/or hose system for circulating the grout. The circulating line and manifold system allows grouting pressures to be controlled at the collar of the hole. Figure 118 illustrates the circulating-type grout system

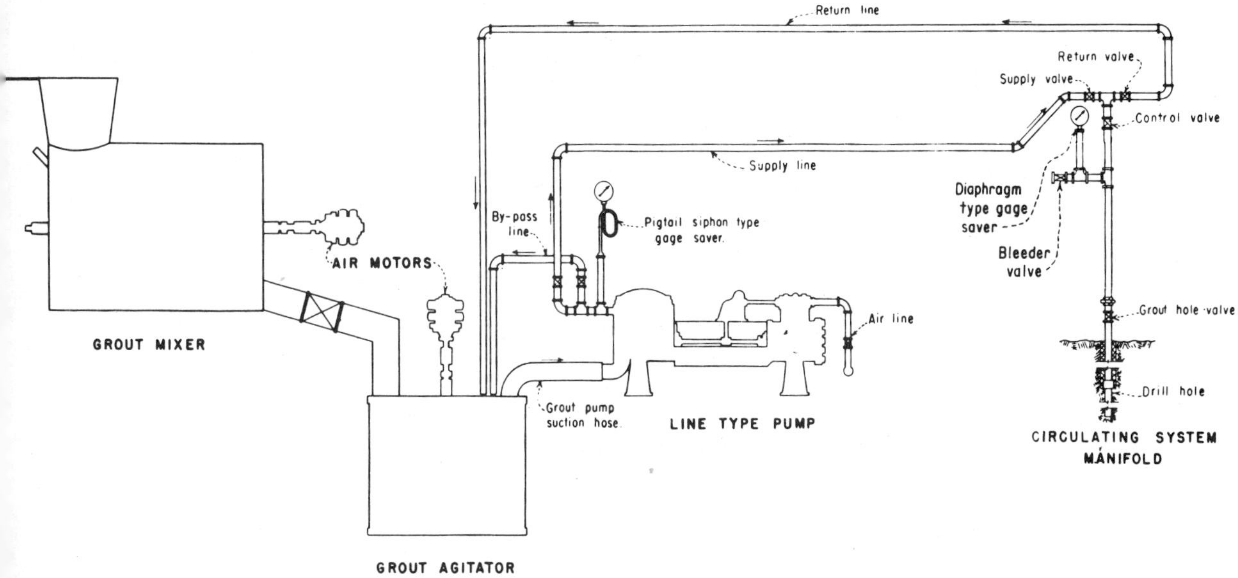

Figure 118. Circulating grouting system. 288–D–2866.

and the equipment generally used for grouting.

Grout is usually pumped with a duplex piston-type pump or a helical-screw rotor-type pump; a standby grout pump should always be required for the grout plant. Figure 119 shows the grout plant used at Ruedi Dam in Colorado.

Grouting pressures are influenced by the following factors:

(1) Type of rock.
(2) Degree to which rock is fractured.
(3) Jointing system within the rock.
(4) Stratification of rock.
(5) Depth of zone being grouted.
(6) Location of hole being grouted.
(7) Weight of overlying material at time of grouting.

The maximum pressure should be such that rock fracture or uplift will not occur. Excessive pressures may weaken the rock strata by fracture or may rupture a portion of the grout curtain already constructed and result in increased permeability. Maximum pressures are difficult to determine since each foundation is unique as to type of rock joint pattern and stratification, and this is usually found by trial at the actual time of foundation grouting or by performing grouting tests prior to foundation treatment.

Unless other criteria are established, one p.s.i. per foot of depth measured from the surface of the foundation to the center of the zone being grouted may be used as the initial grouting pressure and variations determined by observing the grout take. Current Bureau of Reclamation requirements for termination of grouting are given in section G–48(g) of appendix G.

Grout may be introduced to the foundation through grout nipples set directly in the rock; however, bedrock which is found to be badly jointed or broken for some depth below its surface may require a concrete grout cap to facilitate grouting. The grout cap under earth dams is a concrete-filled trench excavated from 3 to 8 feet into the bedrock, depending on the extent of broken rock; the trench usually has a minimum width of 3 feet to facilitate construction. The grout cap performs several functions:

(1) Cuts off seepage in the upper portion of the bedrock which cannot be successfully grouted.
(2) Provides anchorage for the pipe nipple through which the grout is pumped.
(3) Provides stability so that a more effective grout seal can be obtained in the upper portions of the foundation.

Grout pipes (nipples) are normally imbed-

Figure 119. Grouting plant used at Ruedi Dam in Colorado. P382–706–1134NA.

ded at 10-foot centers in the grout cap during the concrete placement. Excavation for the grout cap must be carefully performed so that rock adjacent to the trench is not shattered. Figure 120 shows the construction of a typical grout cap at Navajo Dam in New Mexico.

When grouting foundations in which the surface rock is broken or jointed, grout often rises to the surface through these cracks and prevents complete grouting. Grout caps will sometimes prevent this surface leakage but, if not, the cracks or seams through which grout rises to the surface should be caulked to prevent excessive leakage. Caulking can be done with wooden wedges, cement grout, or burlap. Where extensive surface leakage occurs, concrete can be placed over the area or the surface covered with pneumatically applied mortar. The grout being pumped into the foundation may also be allowed to set within the cracks.

If it is highly probable that the foundation will require extensive grouting, a preliminary test program may be desirable. Such test programs furnish specific data with which the final grouting program may be carefully planned. Test grouting programs can eliminate expensive delays due to large grout overruns and should expedite the completion of the job.

Typical specifications for the performance of foundation grouting and for the excavation of the grout cap are included in sections G–4[?] to G–48 of appendix G. It is recommended that if an extensive grouting program is contemplated, consulting advice should be ob-

Figure 120. Placing concrete in the grout cap at Navajo Dam in New Mexico. The 2-inch-diameter grout nipples are fixed in position at 10-foot intervals. 711-422-250.

tained from an engineer experienced in this type of work. For additional information, the reader is referred to the references at the end of this chapter [5, 35, 68, 69].

Concrete cutoff walls were formerly provided to intercept seepage along the contact of the embankment with the rock foundation. These walls are expensive, liable to cracking, and their usefulness is questionable; they are not recommended for earthfill dams of the type discussed herein. However, in unusual cases where the bedrock is very smooth, a cutoff wall may be warranted. All loose and overhanging rock must be removed from the abutments; rock slopes should not be steeper than 0.5 horizontal to 1 vertical and preferably flatter. Where flattening the rock slopes or overhangs is not practicable, a short section of concrete cutoff wall 5 feet high should be provided.

If the bedrock is a shale which air slakes, it may be necessary to excavate several feet into bedrock to remove the surface disintegration just prior to placement of the embankment; otherwise, excavation into the bedrock (other than for a grout cap) is seldom required. A sample specification for construction on a shale foundation subject to slaking is included in appendix G.

In most instances, bedrock is mantled by overburden of various types and thicknesses. The foundation design then depends on the nature and depth of the overburden, as described in succeeding sections. The above discussion is applicable not only to exposed rock foundations, but also to bedrock which is reached by trenching through the overburden.

130. *Characteristics of Sand and Gravel Foundations.*—(a) *General.*—Often the foundations for dams consist of recent alluvial deposits, composed of relatively pervious sands and gravels overlying impervious geological formations. The pervious materials may range from fine sand to openwork gravels, but more often they consist of stratified heterogeneous mixtures.

Two basic problems are found in pervious foundations; one pertains to the amount of underseepage, and the other is concerned with the forces exerted by the seepage. The type and extent of treatment justified to decrease the amount of seepage will be determined by the purpose of the dam, the streamflow yield in relation to the reservoir conservation capacity, and the necessity for making constant reservoir releases to serve senior water rights or to maintain a live stream for fish, etc. Loss of water through underseepage may be of economic concern for a storage dam but of little consequence for a detention dam. Economic studies of the value of the water and the cost of limiting the amount of underseepage are required in some instances to determine the extent of treatment. On the other hand, adequate measures must be taken to insure the safety of the dam against failure due to piping, regardless of the economic value of the seepage.

A special problem may exist in foundations which are inherently unstable because they consist of clean saturated sand (usually fine and uniform) of very low density. The potential instability is caused by the loose structure of the sand, which is subject to collapse under the action of a dynamic load. Although the loose sand may support small static loads through point-to-point contact of the sand grains, a vibration or shock may cause readjustment of the grains into a more dense structure with a squeezing out of water from the pores. Since drainage cannot take place instantaneously, part of the static load formerly carried by the sand grains is transferred temporarily to the water, and the foundation behaves as a liquid.

Foundations consisting of cohesionless sand of low density are suspect, and special investigations should be made to determine required remedial treatment if the relative density of the foundation is less than 50 percent. The approximate magnitude of the relative density of a cohesionless sand foundation can be determined from the results of standard penetration tests described in section 108. The number of blows per foot is related to the relative density, but is affected by the depth of the test and to some extent by the location of the water table. The following tabulation gives average standard penetration resistance values for 50 percent relative density irrespective of the water table, based on research by the Bureau of Reclamation [6].

Effective overburden pressure in p.s.i. (based on submerged unit weight):	*Number of blows per foot*
0	4
20	12
40	17

Special studies in triaxial shear on undisturbed samples may be required for foundations of cohesionless sand that are indicated to be below 50 percent relative density. Such studies are beyond the scope of this text, and the advice of specialists in dam design should be obtained.

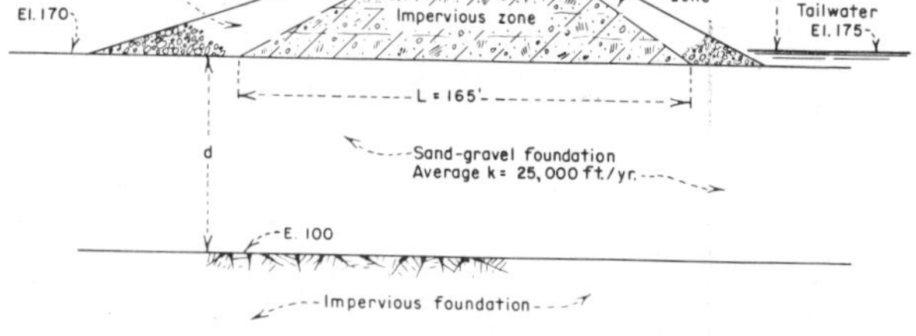

Figure 121. Computation of seepage by Darcy's formula. 288–D–2481.

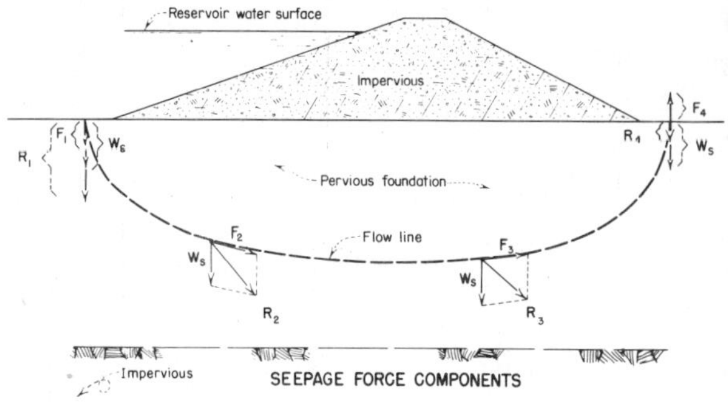

Figure 122. Seepage force components. 288–D–2482.

(b) *Amount of Underseepage.*—To estimate the volume of underseepage which may be expected, it is necessary to determine the coefficient of permeability of the pervious foundation. This coefficient is a function of the size and gradation of the coarse particles, the amount of fines, and the density of the mixture. Three general methods are used in the determination of the coefficient of permeability of foundations: (1) Pumping-out tests in which water is pumped from a well at a constant rate and the drawdown of the water table observed in wells placed on radial lines at various distances from the pumped well; (2) tests conducted by observation of the velocity of flow as measured by the rate of travel of a dye or electrolyte from the point of injection to an observation well; and (3) pumping-in tests in which water is pumped into a drill hole or test pit and the rate of seepage observed under a given head.

The pumping-out tests are relatively high in cost and the results are difficult to interpret. The rate-of-travel method is subject to the same limitations. The pumping-in tests are economical for use in design of small dams because they can be accomplished in conjunction with the usual exploratory drilling; however, the results can be considered as approximations only. Another advantage to the pumping-in tests in drill holes is that the permeability of various layers may be tested rather than only the overall permeability. Pumping-in tests in drill holes are discussed in chapter V.

Upon determination of the coefficient of permeability of the foundation, a rough approximation of the amount of underseepage may be made by use of the Darcy formula:

$$Q=kiA \qquad (1)$$

where:

Q=discharge in given unit of time,

k=coefficient of permeability for the foundation, i.e., discharge through a unit area at unit hydraulic gradient,

i=hydraulic gradient$=\frac{h}{L}$

$=\frac{\text{difference in head}}{\text{length of path}}$, and

A=gross area of foundation through which flow takes place.

The underseepage for the example shown in figure 121 is as follows:

$k=25{,}000$ feet per year

$=\frac{25{,}000}{60\times60\times24\times365}=0.00079$ foot per second

$h=$ elevation 210 $-$ elevation 175 $=35$ feet

$L=165$ feet

$i=\frac{h}{L}=\frac{35}{165}=0.212$

Depth of foundation,

d=elevation 170–elevation 100=70 feet.

For a width of 1 foot, A=70 × 1=70 square feet. Q per foot of width=0.00079 × 0.212 × 70=0.012 second-feet.

For a foundation width of 100 feet, Q=1.2 second-feet; for a foundation width of 1,000 feet, Q=12 second-feet.

The accuracy of the amount of underseepage as determined by the Darcy formula depends on the homogeneity of the foundation and the accuracy with which the coefficient of permeability is determined. The results should be considered as an indication only of the order of magnitude of seepage in evaluation of water loss from a project use viewpoint.

If the foundation is stratified (as is usually the case), the vertical permeability will be much less than the horizontal permeability, and permeable layers at depth will not be fully effective in transmitting underseepage. The quantity of seepage as determined by equation (1) will be liberal if an average coefficient of permeability of the various layers, obtained by weighting each coefficient by the thickness of the layer, is used in the computations.

(c) *Seepage Forces.*—The flow of water through a pervious foundation produces seepage forces as a result of the friction between the percolating water and the walls of the pores of the soil through which it flows, similar to the friction developed by the flow of water through a pipe. Figure 122 shows the flow path of an infinitesimal element of water through the pervious foundation of a dam. The water percolating downward at the upstream toe of the dam increases the submerged weight, W_s, of the soil by the initial seepage force, F_1 with the resultant effective weight of R_1. As the water continues on the seepage path, it continues to exert seepage forces in the direction of flow which are proportional to the friction loss per unit of distance. When the cross-sectional area through which flow takes place is restricted, as under the dam, the velocity of the seepage for a given flow is increased. The increase in velocity is accompanied by an increase in friction loss, and the seepage force is correspondingly increased. This increase is represented in figure 122 by the larger vectors for F_2 and F_3 as compared with F_1 and F_4. As the water percolates upward at the downstream toe of the dam, the seepage force tends to lift the soil, reducing the effective weight to R_4. If F_4 exceeded W_s, the resultant would be acting upward and the soil would be floated out. The erosion would progress backwards along the flow line until a "pipe" would be formed to the reservoir, allowing rapid escape of reservoir storage and subsequent failure of the dam.

Experience has shown that this action can be slow and accumulative and the resulting failure can be a sudden upheaval of the foundation at the downstream toe of the dam. Some engineers [7] refer to this type of piping failure as "failure by heave." Others [8] describe this type of piping failure as "blowouts." It does not follow that piping will always result in a heave or blowout type of failure. If the foundation soil is nonuniform, the fine material may be carried away, leaving the coarse material behind, thus tending to produce a reverse filter which will prevent further piping. Inasmuch as it is difficult to determine whether piping will result in failure or will produce an eventual stabilization in any specific case, it is advisable to design the structure so that piping will not occur.

The magnitude of the seepage forces throughout the foundation and at the downstream toe of the dam where piping must begin depends on the rate of head loss of the percolating water. Relatively impervious foundations or pervious foundations with adequate cutoff trenches are not susceptible to piping because impervious soil offers so much resistance to the flow of water that the reservoir head is largely dissipated in overcoming friction before the downstream toe of the dam is reached; whereas pervious foundations (either homogeneous or stratified) may permit percolation to reach the downstream toe of the dam without a substantial loss of head due to friction. In such instances the designs must be investigated to insure that seepage forces at the downstream toe will not result in blowouts.

Another type of piping failure is due to internal erosion that starts in springs near the downstream toe and proceeds upstream along the base of the dam, the walls of a conduit, a bedding plane in the foundation, an especially pervious stratum, or other weakness which permits a concentration of seepage to reach the area downstream from the dam without high friction losses. This type of failure is termed by some engineers [7] as "failure by subsurface erosion."

The magnitude and distribution of seepage forces in the foundation can be obtained from the flow net, which is a graphical representation of the paths of percolation and lines of equal potential (lines drawn through points of equal total head) in subsurface flow. It consists of flow lines and equipotential lines superimposed upon a cross section of the soil through which flow is taking place. Although the two families of curves may in simple cases be derived mathematically, the graphical solution is more commonly used. The method of drawing and applying the flow net to the solution of problems involving subsurface flow is given in many publications [9, 10, 11, 33].

The analysis of seepage pressures and the safety of the foundation against piping by the flow net method has some serious limitations. It takes considerable experience to draw an accurate flow net, especially where foundations are stratified and where drains or partial cutoffs are installed. For strata and lenses of different permeability, the magnitudes of the permeability coefficient for each layer and in different directions are required. Furthermore, the flow net method of analysis is applicable only to the determination of the safety against blowout piping which, theoretically, is virtually independent of the grain size of the foundation soil and should occur upon first filling of the reservoir. Experience, however, has shown that the grain size and gradation of the foundation material do have an important bearing on piping failures and that piping failures often take place after the dam has been in service for some time. Therefore, it appears that many failures due to piping are of the subsurface erosion type as a result of seepage following minor geological weakness. This type of failure cannot be analyzed by flow nets or other theoretical methods.

For the above reasons and because of the lack of detailed foundation exploration which would make such an analysis meaningful, the construction of flow nets is not required for the design of foundations for small dams. The foundation designs given in the remainder of this chapter are based upon the same theoretical principles which are used in the design of major structures, but the procedures have been simplified so they may be applied to the design of small dams by those who are not specialists in the field of soil mechanics.

131. *Methods of Treatment of Sand and Gravel Foundations.*—(a) *General.*—Various methods of seepage and percolation control can be used, depending on the requirements for preventing uneconomical loss of water and the nature of the foundation as regards stability from seepage forces. Cutoff trenches, sheet piling, mixed-in-place concrete pile curtains, slurry trenches, grouting of alluvium, or combinations of these methods have been used to reduce the flow and to control seepage forces. Blankets of impervious material, extending upstream from the toe of the dam and possibly

Figure 123. Cutoff trench excavation at left abutment of Shadow Mountain Dam, a structure on the Colorado River in Colorado. SM–53–CBT.

covering all or part of the abutments, are frequently used for the same purpose. Horizontal drainage blankets may be incorporated in the downstream toe of a dam or used to blanket the area immediately downstream from the toe of the dam through which percolating water may escape under an appreciable head. The purpose of these blankets is to permit free flow and dissipation of pressure without disruption of the foundation structure and loss of fine soil particles. Pressure relief wells are devices used to relieve pressure in pervious layers which are covered by impervious strata to prevent blowouts downstream from the dam.

The details of these various devices together with an appraisal regarding their effectiveness are contained in the remainder of this section. The application of the various devices to the design of pervious foundations is included in section 132.

(b) *Cutoff Trenches.*—These may be classified into two general types: sloping-side trenches and vertical-side trenches. Sloping-side cutoff trenches are excavated by shovels, draglines, or scrapers and are backfilled with impervious materials which are compacted in the same manner as the impervious zone of the embankment. Vertical-side trenches are also used as cutoffs and may be excavated in open cut by hand or trenching machine, or by stoping where it is necessary to remove and replace breccia or debris in fault zones. Ordinarily, vertical-side trenches are not economical because of the cost of the hand labor involved in placing and compacting the backfill material.

The cutoff trench should be located upstream from the centerline of the crest of the dam, but not beyond a point where the cover of impervious embankment above the trench will fail to provide resistance to percolation at least equal

Figure 124. Placement of compacted fill in the cutoff trench at Sugar Loaf Dam in Colorado. Construction of the outlet works gate chamber can be seen in the middle background. P382–706–2578.

to that offered by the trench itself. The centerline of this trench should be kept parallel to the centerline of the dam across the canyon bottom or valley floor, but it should converge toward the centerline of the dam as it is carried up the abutments in order to maintain the required embankment cover.

Whenever economically possible, seepage through a pervious foundation should be cut off by a trench extending to bedrock or other impervious stratum. This is the most positive means of controlling the amount of seepage and insuring that no difficulty will be encountered by piping through the foundation or by uplift pressures at the downstream toe.

Figure 123 shows the cutoff trench excavation at the left abutment of Shadow Mountain Dam. The designed bottom width and depth of the cutoff trench were 15 and 27 feet, respectively. The completed dam is shown in figure 112, and the maximum section is shown in figure 187. Figure 124 shows placement of compacted fill in the cutoff trench at Sugar Loaf Dam in Colorado; construction of the outlet works gate chamber is shown in the middle background.

In order to provide a sufficient thickness of impermeable material and an adequate contact with the rock or other impervious foundation stratum, the bottom width of the cutoff trench should increase with an increase in reservoir head. However, the cutoff trench bottom width may be decreased as the depth of the trench increases, because the seepage force at the foundation contact will decrease due to loss of head in traveling vertically through the foundation as the depth increases. An adequate width of cutoff trench for a small dam may be determined by the formula:

$$w=h-d \qquad (2)$$

where:

w=bottom width of cutoff trench,

h=reservoir head above ground surface, and

d=depth of cutoff trench excavation below ground surface.

A minimum bottom width of 20 feet should be provided so that excavating and compacting equipment may operate efficiently in trenches which, if below the water table, must be unwatered by well points or sump pumps.

(c) *Partial Cutoff Trenches.*—The Darcy formula for seepage, equation (1), indicates that the amount of seepage is directly proportional to the cross-sectional area of the foundation. It might be concluded from this that the amount of seepage could be reduced 50 percent by extending the impervious zone in figure 121 into the ground so that the depth of the pervious foundation is reduced from 70 to 35 feet; however, this is not the case. The action of a partial cutoff is similar to that of an obstruction in a pipe—the flow is reduced because of the loss of head due to the obstruction, but the reduction in flow is not directly proportional to the reduction in the area of the pipe. Experiments by Turnbull and by Creager on homogeneous isotropic pervious foundations have demonstrated that a cutoff extending 50 percent of the distance to the impervious stratum will reduce the seepage by only 25 percent; an 80 percent cutoff penetration is required to reduce the seepage 50 percent [12].

A partial cutoff trench may be effective in a stratified foundation by intercepting the more pervious layers in the foundation and by substantially increasing the vertical path the seepage must take. Reliance cannot be placed upon a partial cutoff trench in this situation unless extensive subsurface exploration has verified that the more impervious layers are continuous. Pervious foundations also may consist of an impervious foundation stratum of considerable thickness sandwiched between upper and lower pervious layers, and a partial cutoff extending to the impervious layer would cut off only the upper pervious layer. This would be effective if the thicknesses of the impervious and upper pervious layers are sufficient to resist the seepage pressures existing in the lower pervious layers in the vicinity of the downstream toe so that blowouts do not occur.

(d) *Sheet Piling Cutoffs.*—Steel sheet piling is relatively expensive and experience [16, 35, 43, 44] has shown that leakage through the interlocks between the individual piling is considerable. Despite this, it is still used occasionally in combination with a partial cutoff trench as a means of increasing the depth of the cutoff, and under certain conditions it may be used in lieu of a cutoff trench. The sheet piling should be of steel because of its high strength. Sheet piling cutoffs are practically limited to use in foundations of silt, sand, and fine gravel. Where cobbles or boulders are present, or where the material is highly resistant to penetration, driving or jetting not only becomes difficult and costly, but it is highly doubtful that an effective cutoff can be obtained owing to the tendency of the piling to wander and become damaged by breaks in the interlocks or tearing of the steel. A heavy structural section with strong interlocks should be used if the foundation contains gravel.

It is not practicable to drive sheet piling so that it is watertight. Under the best conditions, including the use of compound to seal the interlocks and good contact of the bottom of the piling with an impervious foundation, it can be expected that the piling will be 80 to 90 percent effective in preventing seepage. With poor workmanship, or if the piles cannot be seated in an impervious stratum, they will not be more than 50 percent effective and may be much less.

(e) *Cement-bound Curtain Cutoff.*—The mixed-in-place cement-bound curtain is another means of establishing a cutoff in pervious foundations. It has been used in the construction of Slaterville Diversion Dam near Ogden, Utah; Putah Diversion Dam near Sacramento, Calif.; and recently (1966) at Lower Two Medicine Dam in Montana.

The curtain is constructed by successively overlapping individual piles. Each mixed-in-place pile consists of a column of soil intimately mixed with mortar to form a pilelike structure within the soil. Such a pile is constructed by injecting mortar through a vertical rotating hollow shaft, the lower end of which is equipped with a mixing head for combining the soil with the mortar as the latter is injected. The mortar is introduced into soil that has been loosened by the mixing head as the bit is simultaneously rotated and advanced into, or withdrawn from, the soil. The piles may be reinforced as required. Photographs showing the mixing operation for a single pile and a finished cutoff wall are shown in figure 125 (A) and (B).

Patents on the methods and some of the materials used in construction of the mixed-in-place pile are held by Intrusion-Prepakt, Inc., Cleveland, Ohio. Maximum depths and diameters attainable with present equipment are listed below:

Diameter of pile, inches	*Maximum depth, feet*
12	58
18	50
24	30

While it is not difficult to drill to depths greater than those shown above, minor misalinements associated with normal drilling procedures may occur for greater depths, making the continuity of the piles uncertain.

Current literature [13] states that use of a 15-inch mixing head at 14-inch centers produced piles averaging 18 inches in diameter, with a minimum cross-section thickness of 15 inches.

The mixed-in-place process is designed for use in granular soils, including silt, sand, and gravel, but is not recommended for predominantly clayey materials and is not capable of handling cobbles and boulders. Sample specifications are given in appendix G.

(f) *Slurry Trench Cutoff.*—An important method of constructing positive cutoffs when

Figure 125A. Cement-bound curtain cutoff. (A) Mixing operation. (Intrusion-Prepakt Co.) 288–D–2875.

wet conditions or deep cutoffs in alluvial valleys make conventional construction methods uneconomical is the slurry trench method.

The technique was adapted from well drilling methods used by the oil industry in which bentonite clay mixtures are used to support holes cut in soft soils. The slurry trench method uses a water-bentonite slurry to support the walls of a trench while the excavation is performed. The sequence of construction operations for a slurry trench is shown in figure 126. Vertical-sided trenches, usually from 3 to 10 feet in width, are excavated with either draglines, clam shells, backhoes, or trenching machines. Figure 127 shows excavation by the slurry trench method at the Corps of Engineers' West Point Dam. As material is excavated, the bentonite slurry is pumped into the trench; care is taken to keep the slurry elevation constant and above the water table to avoid caving. The slurry, which weighs slightly more than water, exerts enough pressure on the sides of the trench to prevent sloughing and "cakes" on the walls to prevent seepage. The excavated material is stockpiled next to the trench and the excess slurry allowed to run back into the trench or routed to settling ponds to reduce the sand content. The backfill material (often the same material that was excavated but thoroughly coated with slurry) is then placed in the trench; this displaces the slurry and forms the impermeable barrier.

The slurry is usually made by mixing bentonite, a clay having a high content of the mineral montmorillonite, and water. Wyoming

Figure 125B. Cement-bound curtain cutoff. (B) Finished wall. (Intrusion-Prepakt Co.) 288–D–2876.

bentonite, obtained from deposits in Wyoming and South Dakota, is often specified due to its high content of sodium montmorillonite and its small particle size. Bentonite when added to fresh water swells considerably, resulting in a dispersed mixture with a higher specific gravity and viscosity than water. Bentonite slurries also possess the important property of thixotropy [60], which allows them to become fluid upon agitation and to reacquire shear strength when left standing. The hydrostatic pressure of the slurry is increased due to the suspension of clay and sand particles from the excavation process.

Generally, the ratio of bentonite to water is about 1 to 14 and the bentonite is required to meet standards set forth in section 2, API Standard 13A, "Specification for Oil-Well Drilling Fluid Materials, Fifth Edition," February 1969.

Properties of the slurry which can be controlled are unit weight, viscosity, water loss, shear strength, and sand content. Measurement of the slurry properties listed above can be made by standard tests described in API Code RP 13B, "Standard Procedure for Testing Drilling Fluids, First Edition." Slurry unit weights have varied between 65 and 90 pounds per cubic foot upon initial entry to the trench although the lower range of density is preferred; higher densities sometimes prevent the dragline bucket or clam shell from sinking rapidly, but may be required where large hydrostatic heads or excessive underground currents exist.

Higher slurry densities may be obtained by the addition of finely ground barite (barium sulfate), which has a specific gravity of 4.2. The barite is not soluble in the slurry and does not react with soluble salts.

Determinations of the sand content of the slurry is important. High sand content can cause excessive deposition along the trench bottom, wear on the pumping equipment, and high slurry densities. Sand content is usually controlled by circulating the slurry through centrifugal separators called "desanders", although in some cases settlement ponds have been used effectively. Larger particles may be removed by running excavated materials over shaker screens.

The location and dimensions of the slurry trench should satisfy the specific requirements at each particular site. The cutoff may be placed at any location beneath the dam but the optimum location depends, as with a rolled earthfill cutoff, upon type and location of the core, depth of impervious stratum along the foundation, variations in foundation material, construction sequence, diversion requirements, and reservoir operation, thus requiring separate consideration at each site. An advantage of using a central location for both the core and cutoff is that higher pressures exist along the contact with the impervious core and any settlement of the slurry cutoff tends to be compensated for by the impervious central core. An important disadvantage is that the stability of the structure may be reduced significantly. For this reason, some designers favor placing the slurry trench upstream of the dam and connecting it to the core by an impervious blanket.

Transitions are generally provided between the slurry backfill and the impervious core material to compensate for the effect of differential settlement between the backfill material and the foundation. The transition is accomplished simply by making the trench wider at the top, usually by sloping the walls of the trench. The greater the potential settlement of the slurry backfill compared to the adjacent foundation material, the greater the width of

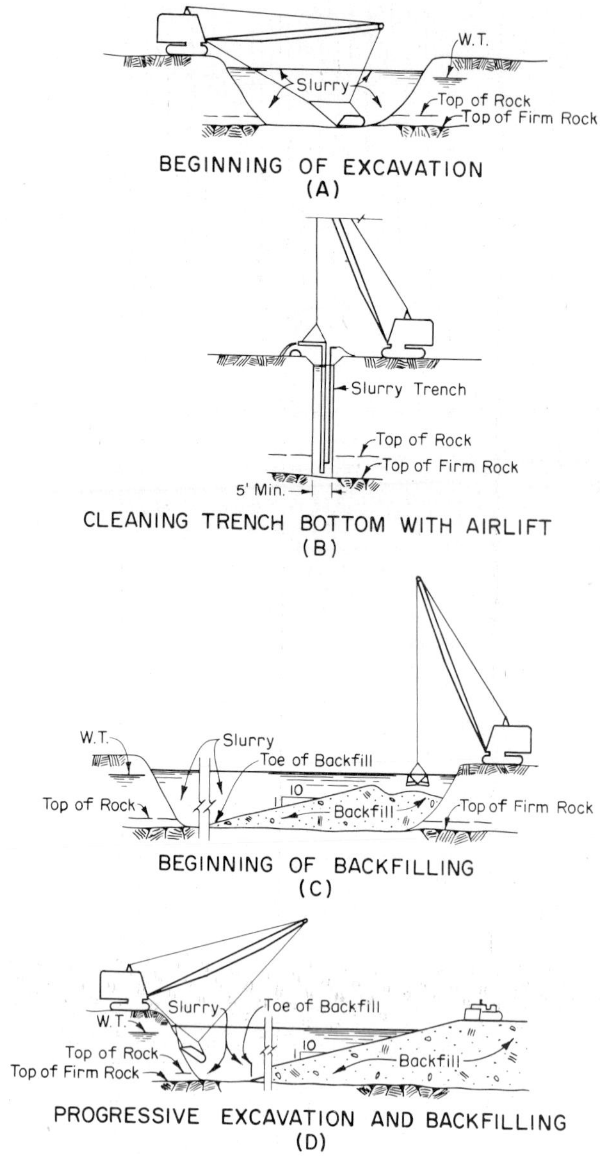

Figure 126. **Sequential operations in the construction of a slurry trench. (U.S. Corps of Engineers.) 288–D–2783.**

the transition should be. Hence, transitions are of great importance when the slurry trench is located directly under the impervious core.

Backfill is usually the same material which has been excavated from the trench. The gradation of this material is often modified by borrow materials to obtain a well-graded mixture, which is then coated lightly with slurry before being replaced in the trench. Initial removal of sand and gravel on the trench bot-

Figure 127. Excavation of a slurry trench. (U.S. Corps of Engineers.) 288–D–2894.

Figure 128. Airlifting sand from bottom of slurry trench. (U.S. Corps of Engineers.) 288–D–2895.

tom may be accomplished by airlift pumps, as shown in figure 128, and any remaining sand can be air-jetted off the bottom ahead of backfill placement and mixed into the slurry.

The use of a slurry trench should be reviewed carefully when cobbles, boulders, or large blocks of rock exist in the lower portions of the trench. In stratified foundations, extreme care must be taken to insure that the bottom of the trench contacts the impervious layer on which the backfill material will rest.

Figure 129. Right abutment blanket construction at Ochoco Dam, on a tributary of the Crooked River in Oregon. Ochoco 51.

Cases have been reported in which one of the harder stratifications is mistaken for the trench bottom and only detected at a later date by coring. This necessitates a great deal of additional construction time and expense, but can be avoided by frequent sounding operations on the trench bottom and prior foundation investigation. For additional information concerning slurry trenches, the reader is referred to the references at the end of the chapter.

Table 14, adapted from Jones [52], gives a general idea of the materials through which slurry trenches have been constructed, the trench widths used, and the hydraulic head for existing or proposed structure.

TABLE 14.—*Comparison of slurry trench cutoffs* [1]

Project	Foundation material	Trench width	Maximum head	Remarks
Kennewick Levee, McNary Dam Project, Columbia River, Washington State, U.S. Owner: Corps of Engineers.	Sandy or silty gravels with zones of open gravel k = 0.4 cm/sec.	6 ft. (1.89 m) Central core.	15 ft. (4.92 m)	Constructed in 1952. Maximum depth 22 ft. (7 m).
Wanapum Dam, Columbia River, Washington State, U.S. Owner: Public Utility District No. 2 of Grant County.	Sandy gravels and gravelly sands underlain by openwork gravels k = (open gravels) = 2.5 cm/sec. Average k = 1 cm/sec.	10 ft. (3.28 m) Central core.	88.5 ft. (27.8 m)	Preconstruction test trench, pump out and laboratory piping tests. Grouting beneath trench. Construction in 1959–62. Maximum depth of cutoff 190 ft. (58 m).
Mangla Closure Dam, Mangla Dam Project, Jhelum River, West Pakistan. Owner: West Pakistan Water and Power Development Authority.	Sandy gravel with cobbles and boulders; gap graded in range of fine gravel and coarse sand. k = 0.4 cm/sec.	10 ft. (3.28 m) Central core.	230 ft. (72.4 m) Construction condition only.	Constructed in 1964. Maximum depth 22 ft. (7 m).
Duncan Lake Dam, Duncan River, British Columbia, Canada. Owner: British Columbia Hydro and Power Authority.	Surface zone of sands and gravels over zone of silt to fine silty sand with some silty clay. k = (surface zone) = 1 cm/sec.	10 ft. (3.28 m) Upstream berm.	102 ft. (32 m) Short term.	Constructed in 1965–66. Maximum depth 60 ft. (18 m).
West Point Dam, Chattahoochee River States of Georgia and Alabama, U.S. Owner: Corps of Engineers.	Upper stratum of alluvial soil, alternating layers of clay, silt, sand, sand and gravel; k varies from 1.8×10^{-2} cm/sec. to 3.5×10^{-5} cm/sec. Lower stratum of residual soil, brown silty sand; k = 0.6×10^{-5} cm/sec.	5 ft. (1.64 m) Upstream blanket.	61 ft. (19.2 m)	Constructed in 1966. Maximum depth 60 ft. (18 m). Grouting in sound rock below the trench. Maximum depth of cutoff 100 ft. (30 m).
Saylorville Dam, Des Moines River, Iowa, U.S. Owner: Corps of Engineers.	Surface zone of impervious alluvial sandy clay. Pervious zone, medium to fine sand and gravelly coarse to fine sand. Average k (gravelly sand) = 0.15 cm/sec.	8 ft. (2.52 m) Upstream berm.	93 ft. (29.2 m) Short term.	Dam under construction. Construction of cutoff scheduled for 1969.[2] Maximum depth approximately 60 ft. (18 m).
Prokopondo Project (Quarry A Cofferdam), Suriname River, Suriname, S.A. Owner: Suriname Aluminum Co.	Uniform fine to coarse sand with some gravel D_{10} = 0.1 mm.	4 ft. (1.26 m)	40 ft. (12.6 m)	Constructed in 1959. Maximum 15 ft. (4.9 m).
Wells Dam, Columbia River, Washington State, U.S. Owner: Public Utility District No. 1 of Douglas County.	Pervious gravels.	8 ft. (2.52 m) Central core.	70 ft. (21.3 m)	Constructed in 1964. Maximum depth greater than 80 ft. (24 m).
Yards Creek Lower Reservoir, New Jersey, U.S. Owner: Public Service Electric & Gas Co., Jersey Central Power & Light, New Jersey Power & Light Co.	Sands, gravels, cobbles, and boulders.	8 ft. (2.52 m) Central core.	55 ft. (16.8 m)	Constructed in 1964. Maximum depth 40 ft. (12 m).
Camanche Dam-Dike 2, Mokelumne River, California, U.S. Owner: East Bay Municipal Utility District.	Alluvial deposit with upper stratum of clayey silts, silts and clayey sands, and a lower stratum of poorly graded medium to fine sand over a thin zone of well-graded gravel. Average k = 7.5×10^{-3} cm/sec.	8 ft. (2.52 m)	135 ft. (41 m)	Constructed in 1966. Maximum depth 95 ft. Maximum head on dam 45 ft. (13.7 m). Head on trench depends on groundwater level downstream.

[1] Adapted from table 1, reference [52].

[2] Construction postponed until 1972.

(g) *Grouting.*—Various materials have been used in attempts to develop grouting procedures which will improve the stability and impermeability of pervious overburden foundations by injecting a substance which will act as a binder and fill the voids. Among these materials have been cement, asphalt, clay, and various chemicals. Cement grouting cannot be successfully used in very fine granular materials because of the comparatively large particle size of the cement, which limits the penetration. However, coarse alluvial material has been successfully grouted in a number of cases using cement or cement-clay grouts. Asphalt grouting is also limited by the particle size. Clay grouting is of doubtful value because the clay is easily carried away by seepage forces. Chemical grouts have about the same viscosity as water and can be injected into pervious soils. However, chemical grouts are too expensive for general use in grouting pervious overburden foundations.

Despite the disadvantages enumerated above, grouting procedures have been developed which will improve the impermeability of pervious overburden materials. Using these procedures, several dams have been built at sites that would have been considered impossible or economically infeasible using other methods. Commonly, a cutoff curtain in overburden consists of several rows of grout holes in which the outer rows of holes are grouted with cement, clay, and cement-clay grouts and one or more inner rows with a chemical grout.

Overburden grouting is a costly process, the injection techniques are complex, and the selection of the grout and appropriate techniques requires considerable field exploration and laboratory and field testing. Furthermore, the results of the injection process are difficult to evaluate. For these reasons, overburden grouting is not considered to be an appropriate treatment for the foundations of small dams within the scope of this text. For information on this subject, the reader is referred to the paper, "Bibliography on Chemical Grouting" [14].

(h) *Upstream Blankets.*—The path of percolation in pervious foundations can be increased by the construction of a blanket of impervious material connecting with the impervious zone of the dam and extending upstream from the toe. Blankets are usually used when cutoffs to bedrock or to an impervious layer are not practicable because of excessive depth; they are also used in conjunction with partial cutoff trenches. The topography just upstream from the dam and the availability of impervious materials are important factors in deciding on the use of blankets. The blanket is generally used for a stream channel or valley floor of sand and gravel but may also be required for portions of the abutments.

Figures 129 and 130 show such an abutment blanket, which was constructed during the rehabilitation of Ochoco Dam by the Bureau of Reclamation in 1949. The purpose of this blanket was to reduce the seepage through the landslide debris which forms the right abutment. It was successful in that it reduced the seepage at full reservoir level from 28 to 12 second-feet.

The blanket is continuous from the impervious zone of the dam and it extends about 400 feet upstream. The abutment was dressed smooth to receive the blanket, which extends from the reservoir floor to an elevation 53 feet above. The blanket was constructed 5 feet thick normal to the approximate 3 to 1 abutment slope, and is protected from erosion by 2 feet of riprap on 12 inches of bedding. Figure 129 shows the earthfill blanket complete and the beginning of riprap placement. Figure 130 shows a general view of the upstream face of the dam and the right abutment blanket completed.

Areas of the foundation which are covered by a natural blanket should be stripped of trees and other vegetation; defective places should be repaired; and the entire surface of the natural blanket should be rolled to seal root holes and other openings. The stripping of a natural blanket upstream from the dam to secure impervious soil for the construction of the dam should be avoided when a positive cutoff is not provided in the design.

Although blankets may be designed by theoretical means [15], a simplified approach may be used for small dams. A suitable thickness for small dams is 10 percent of the depth of the reservoir above the blanket, with a minimum blanket thickness of 3 feet. These thicknesses are for blankets made from materials which are suitable for the construction of the impervious zone of an earthfill dam and similarly compacted.

The length of the blanket will be governed by the desired reduction in the amount of underseepage. From an examination of equation (1) and figure 121 it is apparent that the amount of seepage is inversely proportional to the length of path (for homogeneous isotropic foundations). The blanket should be extended so that the computed seepage loss is reduced to the amount that can be tolerated from a project use standpoint.

An upstream blanket should not be relied upon to reduce the seepage forces in the foundation to the point that piping failures are precluded. Although theoretically an upstream blanket would accomplish this purpose in a homogeneous foundation, the natural stratification that occurs in almost every alluvial foundation makes it possible for high pressures to exist in one or more strata of the foundation at the downstream toe of the dam. Horizontal drainage blankets or pressure relief devices (drains or wells) should always be provided for a dam on a pervious foundation when a complete cutoff trench cannot be secured.

(i) *Downstream Embankment Zones for Pervious Foundations.*—The downstream sections of dams on pervious foundations must be constructed to perform the following functions:

(1) Relieve the uplift pressure from seepage.
(2) Readily permit discharge of seepage water from the foundation.
(3) Prevent piping of the fines from the embankment and/or the foundation.
(4) Provide sufficient weight so that uplift is prevented.
(5) Adequately convey the total amount of seepage to the downstream channel.

These functions are accomplished by providing:

(1) Extended downstream zones to lengthen seepage paths.
(2) Zones with sufficient height above the foundation to prevent uplift.

Figure 130. Upstream slope of Ochoco Dam. Ochoco 72.

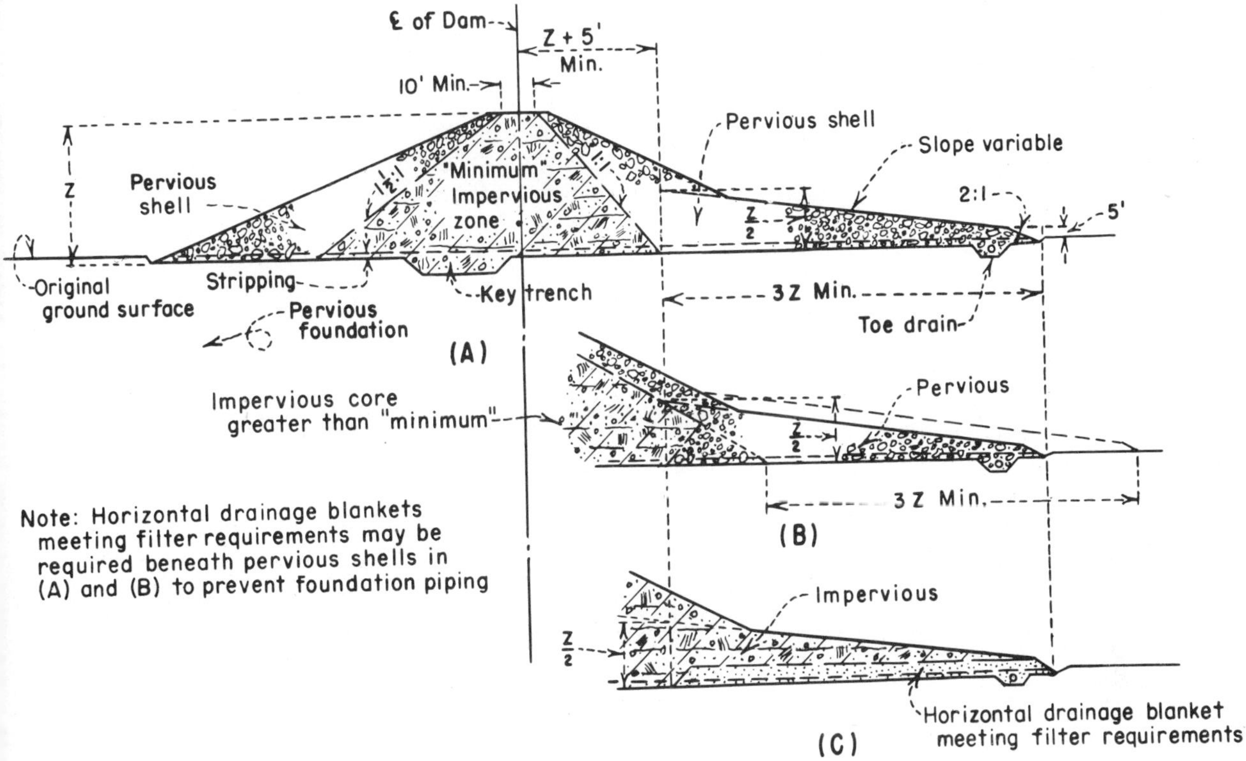

Figure 131. Downstream embankment sections for pervious foundations. 288–D–2483.

(3) Pervious downstream shells or horizontal drainage blankets, which meet filter requirements, to allow water to percolate from the foundation but which prevent piping.

(4) Toe drains with sufficient capacity to conduct the accumulated seepage water back to the stream channel.

Downstream zones incorporating either pervious shells, horizontal drainage blankets, toe drains, or combinations of these should be used in the design of all small dams on relatively homogeneous pervious foundations where positive cutoff trenches are not provided. They may also be used on relatively homogeneous pervious foundations which are overlain by thin impervious layers; the pervious shell will supply weight to stabilize the foundation and will also effectively relieve pressures that may break through the impervious layer. A horizontal drainage blanket meeting filter requirements may be required to prevent piping. Pervious foundations covered by impervious layers may also utilize drainage trenches and pressure relief wells to reduce the uplift pressure and control seepage (see sections 131(j) and (k)).

The required length of the downstream pervious zone may be determined theoretically by means of the flow net (sec. 130 (c)), provided the ratio of the horizontal to the vertical permeability of the foundation is known, by means of the procedure known as transformed sections. This method of dealing with anisotropy in permeable foundations is discussed by Terzaghi and Peck [16] and Cedergren [33]. The method demonstrates that the larger the ratio of horizontal to vertical permeabilities, the farther downstream the seepage emerges from the toe of the impervious zone of the dam and the longer the embankment required downstream.

Because of limitations of the application of the flow net to the design of small dams with relatively meager foundation exploration, the use of a flow net is not required relative to the design criteria presented in this text to determine the lengths of downstream pervious zones.

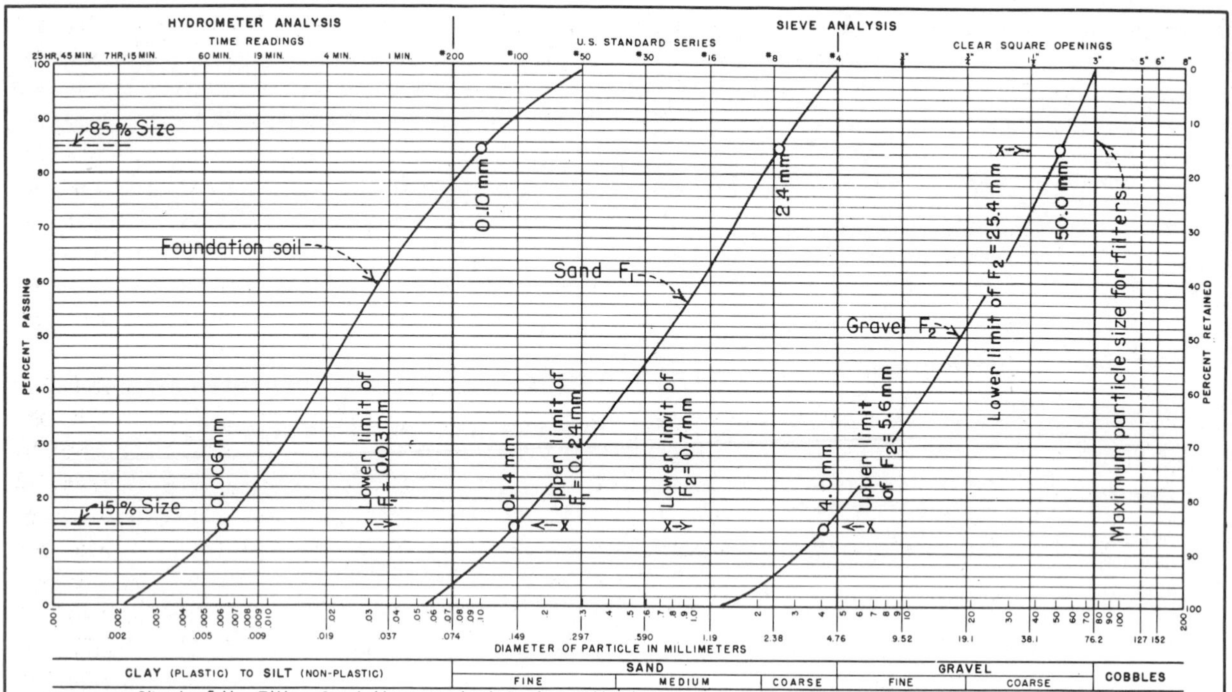

Check of the Filter Gradations Against Design Criteria

For Filter F_1:

$\frac{D_{15} \text{ of } F_1}{D_{15} \text{ of Foundation}} = \frac{0.14}{0.006} = 23$; $5 < 23 < 40$, O.K.

$\frac{D_{15} \text{ of } F_1}{D_{85} \text{ of Foundation}} = \frac{0.14}{0.10} = 1.4$; $1.4 < 5$, O.K.

For Filter F_2:

$\frac{D_{15} \text{ of } F_2}{D_{15} \text{ of } F_1} = \frac{4.0}{0.14} = 29$; $5 < 29 < 40$, O.K.

$\frac{D_{15} \text{ of } F_2}{D_{85} \text{ of } F_1} = \frac{4.0}{2.4} = 1.7$; $1.7 < 5$, O.K.

D_{85} of F_2 = 50 mm; 50 mm > 25.4 mm, O.K.

Figure 132. Typical filter design. 288–D–2877.

As a basis for the design of small dams, it is recommended that the length of the downstream pervious zone be made equal to three times the height of the dam, as shown on figure 131.

Figure 131 shows the typical downstream zoning used with exposed pervious foundations; that is, those which are not overlain by impervious material. Covered pervious foundations are discussed in section 132(e). Each method shown consists of an extended downstream section which lengthens the seepage path. In (A) and (B) of figure 131, the requirement for drainage is provided by the overlying pervious shell. This pervious shell should be designed to prevent piping from the foundation; if the shell material is such that it cannot prevent piping, a horizontal drainage blanket designed to meet filter requirements similar to that shown in figure 131(c) should be used. For homogeneous embankments or where the material permeability of the downstream zone is questionable, horizontal drainage blankets which are designed to prevent piping should be used. These drainage blankets must have sufficient capacity to conduct the total amount of seepage from the embankment and the foundation to the toe drains. Horizontal drainage blankets used with homogeneous embankments convert them to the modified homogeneous type of dam, with the resultant advantages described in section 125.

Figure 131(A) illustrates the recommended minimum length and vertical thickness of the downstream section of a zoned embankment using the recommended "minimum" impervious core (Minimum Core B, fig. 151) for a dam constructed on a pervious foundation without a positive cutoff trench. Figure 131(B) illustrates the recommended design for a zoned dam with an impervious core larger than "Minimum Core B." The reverse slope of the impervious core (fig 131 (B)) is a device used to:

(1) Reduce the length of the downstream pervious shell.

(2) Facilitate construction of the downstream pervious shell if material excavated from the cutoff trench is used.

(3) Reduce the volume of embankment, as shown in figure 131(B).

The dashed outline in figure 131(B) indicates the drainage blanket that would be required if the reverse slope were not used.

The horizontal drainage blanket shown in figure 131(C) must satisfy three requirements:

(1) Gradation must be such that particles of soil from the foundation and the overlying embankment are prevented from entering the filter and clogging it.

(2) Capacity of the filter must be such that it adequately handles the total seepage flow from both the foundation and the embankment.

(3) Permeability must be great enough to provide easy access of seepage water in order to reduce seepage uplift forces.

Requirements for gradation and permeability are closely related and are discussed below.

A minimum drainage blanket thickness of 3 feet is suggested to provide unquestionable capacity for seepage flows.

Multilayer filters for small earthfill dams should in general be avoided; they are more efficient but add to the cost of filter construction. In cases where large seepage quantities must be handled, it has been demonstrated [39] that multilayer filters can provide an economical solution.

If the overlying pervious zones in (A) and (B) of figure 131 are sand-gravel similar in gradation to the sand-gravel of the foundation, there is no danger of flushing of particles of the foundation into the embankment, and no special filters are required. If these zones are constructed of rockfill, a filter must be provided so that the finer foundation material is not carried into the voids of the rockfill.

If sufficient quantities of filter material are available at reasonable cost, it usually will be found economical to provide thicker layers than described above rather than to process material to meet the exact requirements for thin filter design, as subsequently described. The thicker the layer, the greater the permissible deviation from the filter requirements given, especially in the requirement of parallelism of gradation curves between filter and base.

The rational approach to the design of filters is generally credited to Terzaghi [17]. Considerable experimentation has been performed by the Corps of Engineers [18] and the Bureau of Reclamation [19]. Several somewhat different sets of criteria are given by these authorities. The following limits are recommended to satisfy filter stability criteria and to provide ample increase in permeability between base and filter. These criteria are satisfactory for use with filters of either natural sand and gravel or crushed rock and for filter gradations which are either "uniform" or "graded":

(1) $\frac{D_{15}\text{ of the filter}}{D_{15}\text{ of base material}}=5\text{ to }40$, provided that the filter does not contain more than 5 percent of material finer than 0.074 mm. (No. 200 sieve)

(2) $\frac{D_{15}\text{ of the filter}}{D_{85}\text{ of base material}}=5\text{ or less}$

(3) $\frac{D_{85}\text{ of the filter}}{\text{Maximum opening of pipe drain}}=2\text{ or more}$

(4) The grain-size curve of the filter should be roughly parallel to that of the base material.

In the foregoing, D_{15} is the size at which 15 percent of the total soil particles are smaller; the percentage is by weight as determined by mechanical analysis. The D_{85} size is that at which 85 percent of the total soil particles are smaller. If more than one filter layer is required, the same criteria are followed; the finer filter is considered as the "base material" for selection of the gradation of the coarser filter.

In addition to the limiting ratios established

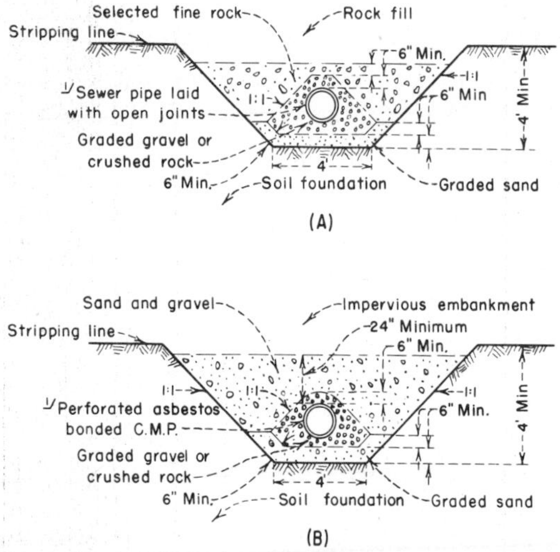

[1] Drain pipes can be either vitrified clay, concrete, or asbestos bonded corrugated metal pipe (C.M.P.). The vitrified clay and concrete pipes can be either plain (laid with open joints) or perforated (laid with closed joints); if asbestos bonded C.M.P. is used it should be perforated and laid with closed joints.

Figure 133. Typical toe drain installations. 288–D–2484.

for adequate filter design, the 3-inch particle size should be the maximum utilized in a filter to minimize segregation and bridging of large particles during placement of filter materials. Also, in designing filters for base materials containing gravel particles, the base material should be analyzed on the basis of the gradation of the fraction smaller than No. 4.

It is important to compact filter material to the same density as that required for construction of sand-gravel zones in embankments, as given in appendix G. Care must be used in placing filter materials to avoid segregation. The construction of thin filter layers requires proper planning and adequate inspection during placement. In many cases, it is possible that concrete sand used in the spillway, outlet works, or appurtenant structures may be used as filter material. This reduces costs by eliminating any special blending requirements.

The following is an example (see fig. 132) of a typical design which would be applicable for thin filters, such as those shown around the toe drains in figure 133.

Given:

Average gradation curve of foundation soil shown on figure 132, with $D_{15}=0.006$ mm. and $D_{85}=0.10$ mm.

Openings in drainpipe, ½ inch.

To find:

Gradation limits of filter materials.

Procedure:

(1) Lower limit of D_{15} of filter$=5\times0.006=0.03$ mm.

(2) Upper limit of D_{15} of filter=the *smaller* of the values: $40\times0.006=0.24$ mm., and $5\times0.10=0.50$ mm.; *use 0.24 mm.*

To meet conditions (1) and (2) and the criterion of parallelism, sand shown as F_1 in figure 132 was selected. For F_1, $D_{15}=0.14$ mm. and $D_{85}=2.4$ mm. This material is too fine to place adjacent to a pipe with ½-inch openings, since the requirement is for D_{85} of the filter to be at least $2\times\frac{1}{2}=1$ inch; hence, a second filter layer of gravel or crushed rock is required.

(3) Lower limit of D_{15} of gravel is $5\times0.14=0.70$ mm.

(4) Upper limit of D_{15} of gravel=the *smaller* of the values: $40\times0.14=5.6$ mm., and $5\times2.4=12.0$ mm.; *use 5.6 mm.*

(5) Least D_{85} of gravel$=2\times\frac{1}{2}=1$ inch$=25.4$ mm.

To meet conditions (3), (4), and (5) and the criterion of parallelism, the gravel shown as F_2 in figure 132 was selected.

The lower and upper limits of the D_{15} for F_1 and for F_2 as well as the lower limit for D_{85} for F_2 are shown on figure 132.

(j) *Toe Drains and Drainage Trenches.*—Toe drains are commonly installed along the downstream toes of dams in conjunction with horizontal drainage blankets in the position shown in figure 131. Beginning with smaller diameter drains laid along the abutment sections, the drains are progressively increased in size, the lines of maximum diameter being placed across the canyon floor. The purpose of these drains is to collect the seepage discharging from the embankment and foundation and lead it to an outfall pipe which discharges into either the spillway or outlet works stilling basin or into the river channel below the dam. Pipes rather than French drains are used to insure adequate capacity to carry seepage flows. Toe drains are also used on impervious foundations to insure that any seepage that may come through the foundation or the embankment is collected and that the groundwater is kept below the surface sufficiently to avoid the creation of unsightly boggy areas below the dam.

The toe drains may be of vitrified clay or concrete tile, or perforated asphalt-bonded corrugated metal pipe. The drainpipes are placed in trenches at a sufficient depth below the ground surface to insure effective interception of the seepage flow. The minimum depth of the trench is normally about 4 feet; the maximum depth is that required to maintain a reasonably uniform gradient though the ground surface undulates. The bottom width of the trench is 3 to 4 feet, depending on the size of the drainpipe. A minimum pipe diameter of 6 inches is recommended for small dams; diameters up to 18 inches may be required for long reaches at flat gradients. The drainpipe should be surrounded by a filter to prevent clogging of the drains by inwash of fine material or piping of foundation material into the drainage system. Two-layer filters are often required; the layer in contact with the pipe must be of sufficient particle size so material will not enter or clog the perforations in metal pipe or the joint openings in clay or concrete tile.

The filter in contact with the toe drain must satisfy criterion number 3 of section 131(i); this criterion requires that the D_{85} size of the filter must be equal to or greater than twice the size of the opening in the pipe. The other design criteria which the filters must satisfy are given in section 131(i). Figure 133 shows typical toe drain installation details used in most Bureau of Reclamation dams. Figure 134 shows construction of the toe drain at Cheney Dam in Kansas.

Drainage trenches are used when pervious foundations are overlain by a thin, impervious layer. A drainage trench is simply an open trench which penetrates the impervious layer and relieves the uplift pressures in the underlying pervious stratum. Toe drains are generally installed in the bottom of the trench, and material satisfying the filter criteria of section 131(i) is used as backfill.

A drainage trench usually is not effective if the underlying pervious foundation is stratified, as it will relieve uplift pressures only in

Figure 134. Toe drain construction at Cheney Dam in Kansas. P835–526–199NA.

the uppermost pervious stratum. More effective drainage of stratified foundations can be accomplished by pressure-relief wells.

(k) *Pressure-Relief Wells.*—In a great number of instances of low dams on pervious foundations overlain by an impervious stratum, the thickness of the top impervious layer is such that there is no danger of piping, either of the blowout or internal-erosion type. Theoretically, piping occurs when the fluid (uplift) pressure at a point at some level in the foundation in the vicinity of the downstream toe reaches the pressure exerted by the combined weight of soil and water above it. For the usual condition of tailwater at the ground surface, the uplift pressure (in feet of water) at the point in question will be equal to the depth, d, of the point below ground plus the reservoir pressure head minus the head lost in seepage through the foundation to the point. The pressure exerted by the weight of soil and water above this point is the saturated unit weight of soil times the depth to the point. If the thickness of the impervious layer is equal to the reservoir head, h, the uplift pressure beneath the layer cannot exceed the weight of the layer. This is so because the saturated weight of soil equals approximately twice the weight of water, and for $h=d$:

$$2\gamma_w \times d \times 1 = (h+d)\gamma_w \qquad (3)$$

(or pressure exerted by saturated weight= uplift pressure).

Actually, there is always appreciable loss of reservoir head because of seepage friction; hence the value h in the right-hand portion of equation (3) is too large and the uplift pressure will be smaller than the pressure exerted by the overlying weight. Therefore, if the thickness of the top impervious stratum is equal to the reservoir head, it may be considered that an appreciable factor of safety against piping is assured. In this situation no further treatment of the foundation is required. However, if the thickness of the top impervious stratum is less than the reservoir head, some preventive treatment is recommended. If the top impervious stratum is less than h but is too thick for treatment by drainage trenches, or if the pervious foundation is stratified, pressure relief wells are required.

The primary requirements for a pressure relief well system are:

(1) The wells should extend into the pervious foundation underlying the impervious top layer to relieve pressures to such a depth that the combined thickness of the impervious layer and drained material is sufficient to provide stability against underlying unrelieved pressures. Depths of wells equal to the height of the dam will usually be satisfactory.

(2) The wells must be spaced sufficiently close together to intercept the seepage and reduce uplift pressures between wells to acceptable limits.

(3) The wells must offer little resistance to infiltration of seepage and discharge thereof.

(4) The wells must be designed so that they will not become ineffective due to clogging or corrosion.

The U.S. Corps of Engineers has carried out extensive research programs on the design and installation of relief wells. Results of these studies have been published in a number of excellent papers [20, 21, 22, 40]. The reader is advised to consult these references for theoretical design methods.

Well spacing usually must be based on judgment because of the lack of detailed information regarding the foundations of small dams. This is an acceptable procedure provided that plans are made to install additional wells after the dam is constructed at the first signs of excessive pressures. When the pervious strata have high rates of permeability, there will be a greater quantity of water at the downstream toe of the dam than if the permeability rate were lower. Suggested well spacing is 25 feet minimum for the most pervious foundations to 100 feet maximum for less pervious foundations.

Experiments have shown that, in general, the well diameter should not be less than 6 inches so that there will be little head loss for infiltrating seepage. It is recommended that a minimum thickness of 6 inches of filter which meets the criteria previously established (sec. 131(i)) be provided between the well screens and the foundation, and that the ratio of the 85 percent size of the filter to the screen opening be greater than 2.0.

Examples of pressure relief wells are shown in figure 135. Figure 135(A) shows the type of pressure relief well used by the Bureau of Reclamation to reduce seepage pressures at Red Willow Dam in Nebraska. The 4-inch stainless steel well screen is surrounded by a minimum of 6 inches of filter material and sealed at the top with an impervious clay-silt mixture to prevent water from rising along the outside of the pipe. The 42-inch precast concrete pipe inspection well allows easy examination of the well's operation and adequate working space for any maintenance required. The relief water is dissipated through the 8-inch outflow pipe to a collector drain along the downstream toe of the dam.

Figure 135(B) shows a pressure relief well developed by the Corps of Engineers [22,40] for use in alluvium adjacent to Mississippi River levees. The well consists of a wooden screen, wooden riser pipe, gravel filter, sand backfill, and a concrete backfill seal near the ground surface. The inside diameter of the riser and screen is 8 inches and the screen slots are 3/16-inch wide and 3¼-inch long; the bottom of the pipe is closed with a wooden plug. Commercially available wooden screens similar to the one shown have a slot area equal to 10 percent of the total circumferential area of the

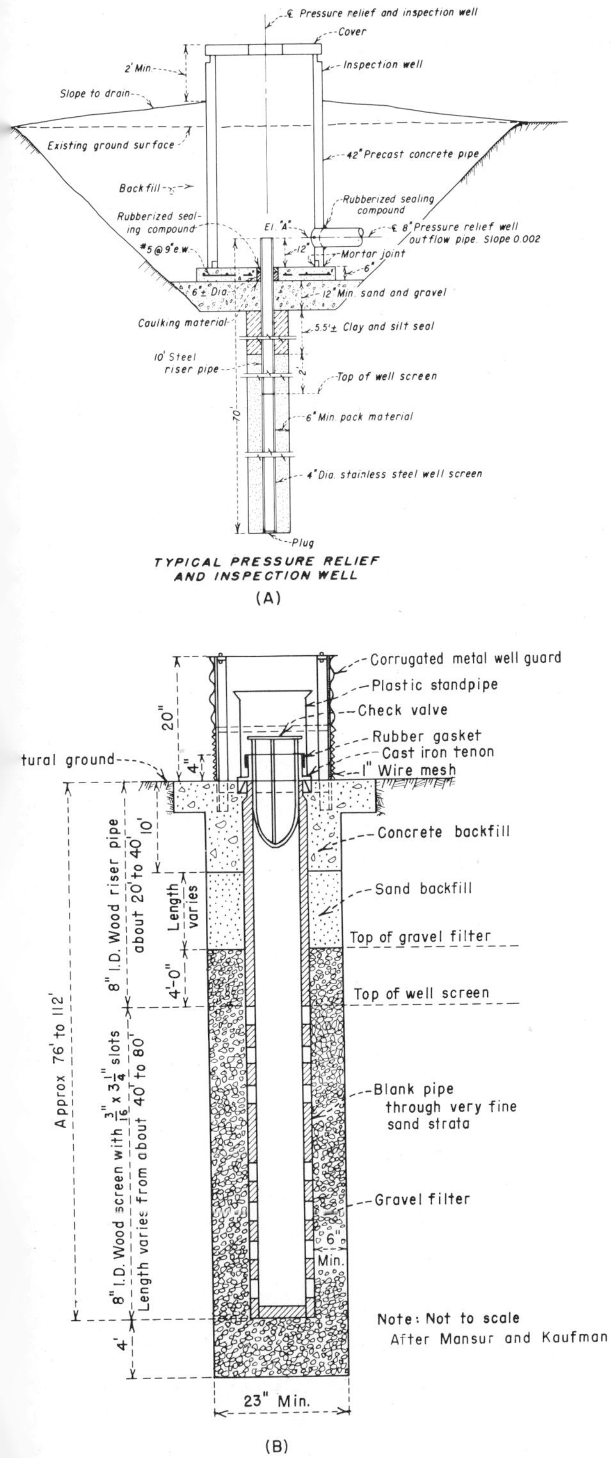

ure 135. Pressure relief wells and appurtenances. 288–D–2485.

reen. It is suggested [40] that the riser pipe ıd screen be constructed of machine-banded ɔod staves of either untreated redwood or treated Douglas fir, white pine, or Southern yellow pine. Heavily galvanized, 6-gage winding wire at 3-inch spacings was used to band the pipe. A "reverse rotary" method of drilling the holes was used. This method is excellent for installing such wells because the material is removed through a suction pipe which helps to prevent sealing of the pervious strata. The walls of the hole are supported by hydrostatic forces acting against a thin film of fine-grained soil on the walls, created by maintaining a head of water in the hole several feet above the groundwater table. Additional information concerning well construction procedure is contained in reference [40].

Another method of installing a pressure relief well requires sinking a casing of appropriate size to the required depth and washing out the soil inside the casing. The assembled well pipe, consisting of the screen and riser, is lowered in the casing and properly alined. The filter is then placed in 6- to 8-inch layers and the casing withdrawn a like amount. This process is repeated until the filter is several feet above the top of the screen section. Above this point, impervious backfill or concrete is placed to prevent leakage along the outside of the pipe. After installation the wells should be cleaned out and pumped or surged to remove any fine soil immediately adjacent. Pressure relief wells should be inspected periodically and cleaned by surging, if necessary.

Relief wells, like other drainage systems, have some limitations. Too great a volume of seepage may require an excessive number of wells. In such cases upstream impervious blanketing of the areas which allows reservoir water to enter the pervious layers may be used to reduce the amount of seepage.

Because the design and installation of pressure relief wells demands specialized knowledge, skill, and the highest quality of construction inspection to insure satisfactory performance, and because such wells also require post-construction supervision and maintenance, they are recommended for use with small dams only when other methods of seepage control are not feasible.

132. *Designs for Sand and Gravel Foundations.* —(a) *General.*—One of the criteria for design

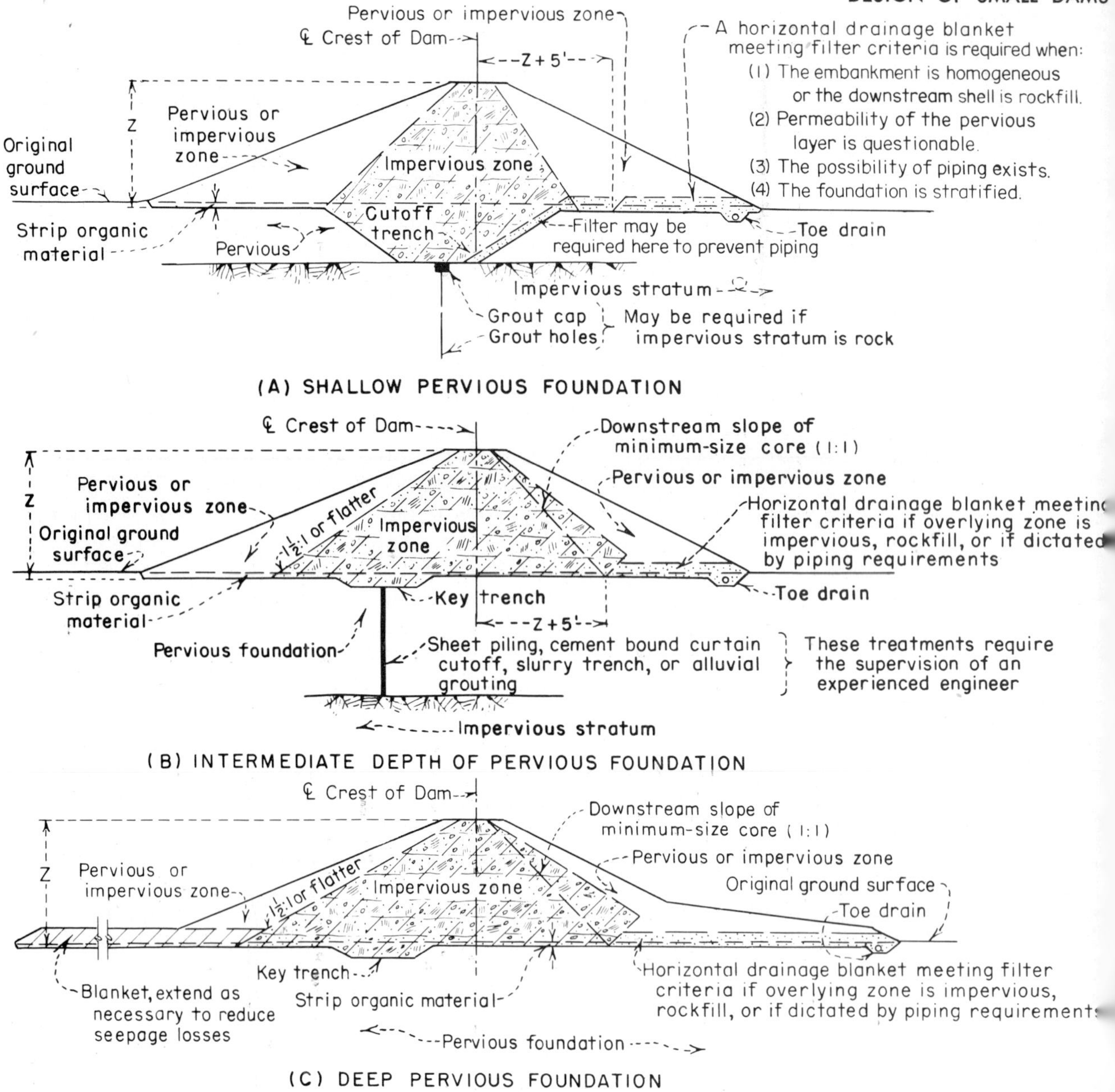

Figure 136. Treatment of Case 1—Exposed pervious foundations. 288–D–2486.

of earthfill dams given in section 127 requires that the flow of seepage through the foundation and abutments be controlled so that no internal erosion takes place and there is no sloughing in the area where the seepage emerges. The criteria also require that the amount of water lost through seepage be controlled so that it does not interfere with planned project functions. Section 128 discusses the basis used for designing foundations for small dams, which requires a generalization of the nature of the foundation in lieu of detailed explorations and the establishment of design procedures that are less theoretical than those used for major structures. Section 128 also cautions against the use of these design procedures for unusual conditions where procedures based largely on judgment and experience are not appropriate.

The purpose of this section is to show the

application of methods of foundation treatment to specific instances. For purposes of discussion, pervious foundations are divided into the following cases:

Case 1. Exposed pervious foundations.

Case 2. Covered pervious foundations. The pervious foundation is overlain by an impervious layer, which may vary in thickness from a few feet to hundreds of feet.

In both of these cases, the pervious foundation may be relatively homogeneous, or it may be strongly stratified with less pervious layers so that the horizontal permeability will be many times greater than the vertical permeability. Stratification will influence selection of the appropriate foundation treatment method.

The treatment of Case 2—Covered Pervious Foundations—is influenced by the thickness of the impervious top layer. The following three conditions, based upon the thickness of the top impervious layer, are considered:

(1) Impervious layer has a thickness of 3 feet or less. It should be assumed that the layer will be largely ineffective as a blanket in preventing seepage because thin surface strata usually lack the density required for impermeability and because they commonly have a large number of openings through them. There also exists the possibility that construction operations near the dam may penetrate the layer or that, while filling the reservoir, unequalized hydrostatic pressure on the surface of the blanket may puncture it. Therefore, a very thin impervious top layer is considered to have little effect on the imperviousness of the foundation.

(2) Impervious layer has a thickness greater than 3 feet but less than the reservoir head. This type of foundation condition is usually treated by using drainage trenches or pressure relief wells to penetrate the impervious layer and relieve the uplift pressures.

(3) Impervious layer thickness is greater than the reservoir head. It can be assumed here that there will be no major problems involved so far as seepage or seepage forces are concerned. This is demonstrated by equation (3), section 131(j).

The treatment of impervious foundations of silt and clay is discussed in section 133.

(b) *Case 1—Exposed Pervious Foundations of Shallow Depth.*—The foundation treatment for an exposed pervious foundation of shallow[4] depth is shown in figure 136(A). A cutoff trench excavated to the impervious stratum, called a positive cutoff, should always be used because it is the most "positive" means of avoiding excessive seepage losses and the possibility of piping. If the stratum is rock, grouting may be required to control the seepage. A horizontal drainage blanket is not necessary if the shallow pervious foundation can act as a filter and provides adequate drainage capacity. For example, if the downstream portion of the embankment is sand and gravel and similar in gradation to the foundation, the horizontal drainage blanket shown in figure 136(A) would not be necessary. Horizontal drainage blankets meeting filter requirements are generally used when the following conditions exist:

(1) The embankment is homogeneous or the downstream shell is rockfill.

(2) The perviousness of the foundation is questionable.

(3) The possibility of piping exists, either from the embankment into the foundation or from the foundation into the downstream zone of the embankment.

(4) The foundation is stratified.

A filter may also be useful against the downstream slope of the cutoff trench to prevent piping.

If the downstream portion of the embankment is rockfill, a horizontal drainage blanket is used to prevent piping from the foundation and should extend from the downstream slope of the dam to the impervious zone, as shown in figure 136(A). If the embankment is homogeneous, the horizontal drainage blanket should be provided because it insures that the seepage line (phreatic line) through the embankment will be drawn down and the stability of the section increased; however, the blanket need not extend upstream closer to the centerline of the

[4] The term shallow is usually meant to infer a depth approximately equal to or less than the height of the dam. This informal definition breaks down when dam heights exceed about 50 feet.

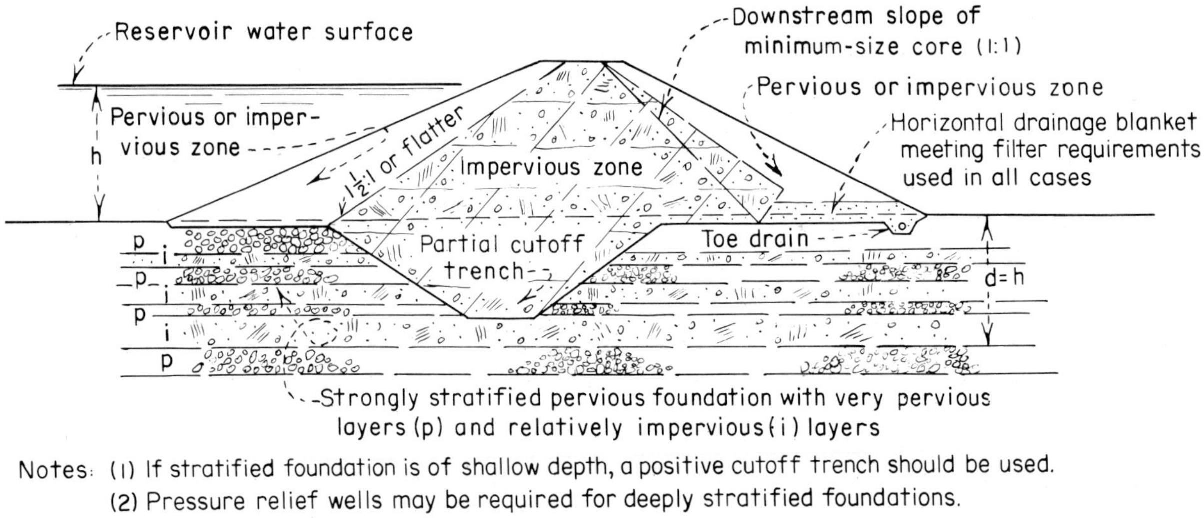

Figure 137. Treatment of stratified foundations. 288–D–2487.

dam than a distance of Z+5 feet. Section 139(d) discusses the extent of the filter required for a homogeneous embankment. Toe drains should be installed to carry away the excess seepage and to keep the downstream toe of the dam dry.

If a shallow foundation is stratified, similar to figure 137, a positive cutoff trench should be constructed. Horizontal drainage blankets are beneficial for stratified foundations, especially if the top layer is impervious or of doubtful permeability. As stated above, grouting may be required if the base stratum is rock of high permeability. Deeply stratified foundations are discussed in section 132(d).

When conditions such as lack of impervious material, short construction seasons, wet climates, high dewatering costs, etc., prevent the use of a trench-type cutoff, it may be decided that other methods of constructing an impermeable barrier such as sheet piling, alluvial grouting, cement-bound curtain cutoffs, or a slurry trench should be used. If one of these methods is used, an experienced dam designer should be consulted.

(c) *Case 1—Exposed Pervious Foundations of Intermediate Depth.*—A foundation is considered to be of intermediate depth when the distance to the impervious layer is too great for a cutoff trench but can be economically reached by another type of positive cutoff. Whether or not a positive cutoff will be economical depends heavily on three items:

(1) The effect of underseepage on the stability of the embankment.

(2) The economic value of the water lost by underseepage.

(3) Whether or not treatment of the foundation as if it were a Case 1—Exposed Pervious Foundation—of great depth (sec. 132(d)) is more economical.

If it is decided that treatment as an intermediate depth foundation is required, the following methods are available to construct a positive cutoff:

(1) Sheet piling.
(2) Cement-bound curtain cutoff.
(3) Slurry trench.
(4) Alluvial grouting.
(5) Concrete cutoff wall.

If one of the methods is used, an experienced engineer should supervise the design and construction of the entire dam. Minimum Core B, for zoned dams on deep pervious foundations (see fig. 151), is the smallest impervious core recommended with this type of foundation treatment. Treatment of the intermediate depth foundation is shown in figure 136(B). As noted in section 132(b), these treatments may also be required when excessively wet climates, short construction seasons, shortages of impervious material for upstream blanket con-

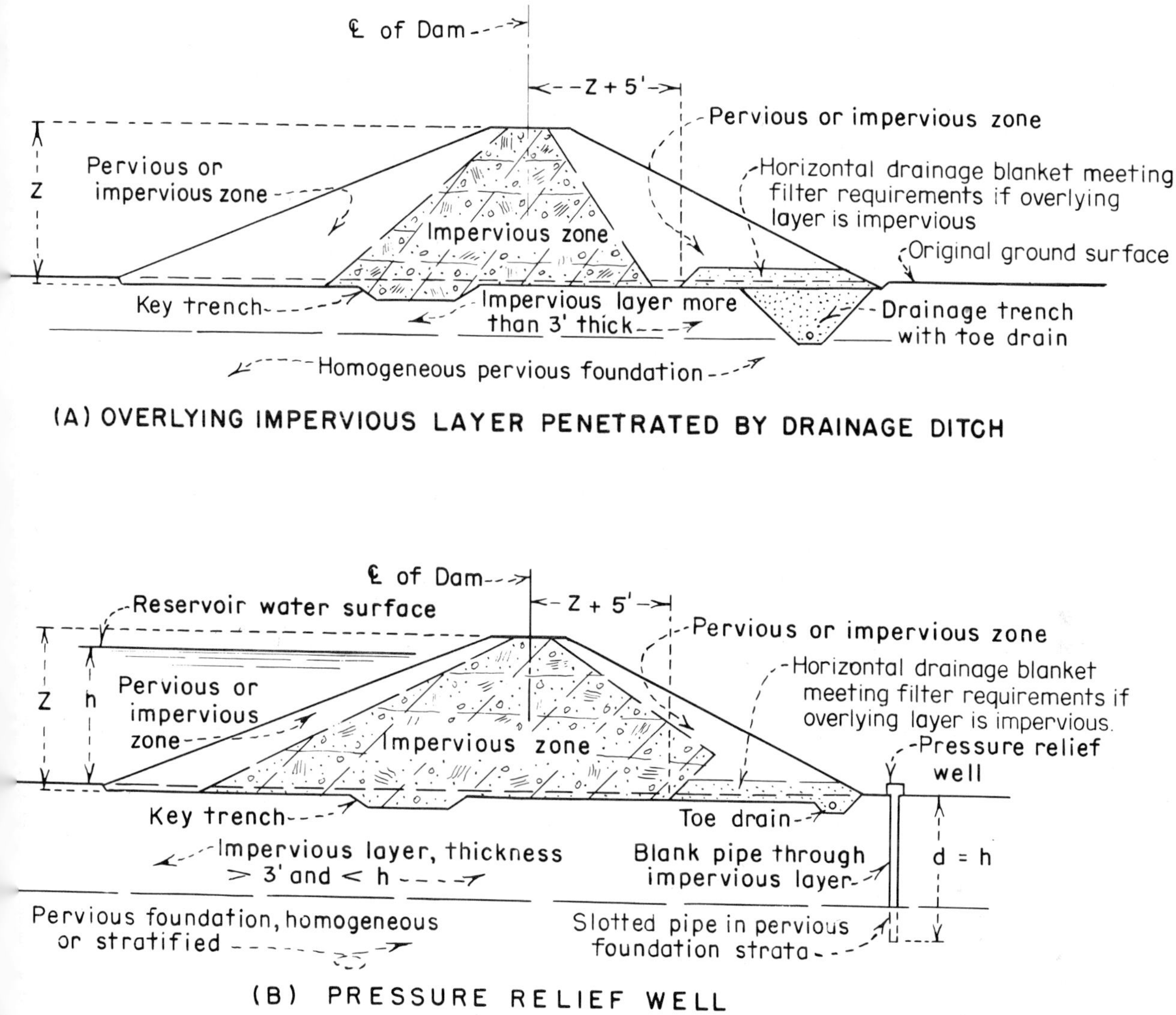

'igure 138. Treatment of Case 2—Covered pervious foundations with overlying impervious layer of thickness more than 3 feet and less than reservoir head. 288–D–2488.

:truction, etc., prevent the use of a positive utoff trench.

(d) *Case 1—Exposed Pervious Foundations f Great Depth.*—The problem in dealing with leep pervious foundations is to determine the atio of horizontal to vertical permeability. f this can be done, an accurate flow net can e constructed, correct seepage quantities deermined, and appropriate drainage measures aken. Because this ratio is difficult to deterine, general assumptions must be made as to 1e homogeneity of the pervious layer and conngency plans made in case the horizontal pereability is much greater than expected or the foundation contains thin layers whose permeability is much greater than the average.

The general treatment for a pervious foundation which is too deep to permit a positive cutoff is shown in figure 136(C). The upstream blanket is provided to lengthen the seepage path and reduce the total quantity of seepage to acceptable limits. If an upstream pervious zone is used in the embankment, the blanket should extend beneath this zone so that it forms a continuation of the impervious core.

The smallest impervious zone recommended for use with zoned embankments on deep pervious foundations is "Minimum Core B," shown in figure 151. The key trench is used to intercept any critical areas of the foundation and should be located upstream from the centerline of the crest of the dam.

To avoid foundation stability problems due to seepage, an adequate thickness of pervious or impervious material must be maintained over the pervious foundation across the valley floor. Suggested dimensions for the length and cover requirements for the downstream section are shown in figure 131. Several sections should be taken along the centerline of the embankment to be sure that adequate material is provided for the most critical section.

If a homogeneous section is used, a horizontal drainage blanket is used to handle seepage from the foundation and to reduce the height of the phreatic line in the embankment. A horizontal drainage blanket should also be used with a downstream pervious zone when the possibility of piping from the foundation exists. The drainage blanket should meet the filter criteria given in section 131(i).

Toe drains are used to collect seepage and convey it to the downstream channel and to prevent wet areas at the downstream toe.

If the permeability is so great as to cause extensive seepage, ponding, or sand boils at the downstream toe, then drainage trenches, pressure relief wells, extension of the downstream toe of the dam, or blanketing the downstream area may be required to control the situation.

In deep stratified foundations, a different treatment is required. This type of foundation consists of alternating layers of relatively pervious and impervious material and is generally treated by constructing a partial cutoff trench. The depth of the partial cutoff must be such that the total combined depth of the cutoff trench and the impervious layer upon which it rests is not less than the reservoir head. Typical embankment and foundation treatment for stratified foundations is shown in figure 137. In this figure, d must be equal to or greater than h. (Note that d is the depth to the uppermost pervious layer which is not cut off and is not the depth of the partial cutoff trench.) If this requirement is met, the foundation should be stable against seepage pressures which may exist in the uppermost pervious layer which is not cut off (refer to equation (3)).

A horizontal drainage blanket meeting filter requirements should be used with stratified foundations in case the downstream embankment zone is impervious or if piping requirements are not met by an overlying pervious layer. Toe drains should also be used. Additional drainage measures, such as the use of pressure relief wells, may be required if high uplift pressures exist below the downstream toe and are not detected until reservoir filling.

(e) *Case 2—Covered Pervious Foundations.*—In the case of a pervious foundation covered by an impervious layer, the type of treatment depends on the thickness and imperviousness of the layer covering the pervious zone and on the permeability of the underlying pervious layer.

If the overlying layer is equal to or less than a few feet in thickness (say 3 feet), its effect is generally ignored due to thickness variations in the vicinity of the damsite and the possibility of a puncture during construction of the dam or a blowout after filling. In this case, the foundation should be designed as a Case 1—Exposed Pervious Foundation, either shallow (sec. 132(b)) or deep (sec. 132(d)).

An overlying impervious layer having a thickness greater than 3 feet and less than the hydraulic head may be assumed to act as an upstream filter blanket if the thickness, continuity, impervious qualities, and upstream distance of the natural deposit have been carefully checked. It will usually be necessary to compact the impervious layer with a heavy roller. If it is not possible to adequately evaluate the suitability of the upstream covering, it should be assumed that the natural blanket is not extensive enough or impermeable enough to prevent seepage or cause substantial head loss to water entering the pervious layer. Provisions should be made to relieve uplift pressures at the downstream toe and to remove the seepage.

If the underlying pervious layer is relatively homogeneous but the top impervious layer is relatively thin, the top layer should be completely penetrated by a drainage trench, a shown in figure 138(A), to relieve the uplift pressures which develop beneath the impervious layer. Additional drainage trenches may be required after construction if excessive uplift pressures still exist or if seepage quantities are large. When the overlying impervious layer is too deep to penetrate economically with a drainage trench, pressure relief wells (sec. 131 (k)) should be used along the downstream toe, as shown in figure 138(B). The correct spacing of relief wells is difficult to determine primarily due to the inaccuracy of estimating the permeability of the strata. Wells are usually spaced from 50 to 100 feet apart for the initial design and then additional wells are added if postconstruction seepage analysis requires them.

When the dam embankment is homogeneous or when the downstream zone is of questionable permeability, a horizontal drainage blanket is constructed, as shown in figure 138, to: (1) lower the height of the seepage line through the embankment, (2) provide adequate embankment drainage, and (3) eliminate piping from the embankment. The blankets should extend from the downstream toe to within a distance of Z+5 feet from the centerline.

Stratified foundations should be treated as discussed in sections 132(b) and (c) and as shown in figure 137.

(f) *Summary of Pervious Foundation Treatments.*—Table 15 is a summary of recommended treatments for various pervious foundation conditions. Foundations are normally considered as either shallow or deep since these are by far the most common conditions encountered; however, if the foundation is determined to be of intermediate depth, special construction methods are required which

TABLE 15.—*Treatment of pervious foundations*

Figure No.	Thickness of overlying impervious layer	Total depth of foundation	Condition of pervious material	Primary device for control of seepage	Additional requirements (other than stripping)
			Case 1—Exposed Pervious Foundations		
136(A)	None	Shallow	Homogeneous	Positive cutoff trench	Toe drain. Horizontal drainage blanket meeting filter requirements may be required. Grouting may be required.
136(C)	None	Deep	Homogeneous	Upstream impervious blanket	Large core required. Horizontal drainage blanket meeting filter requirements may be required. Key trench. Toe drain.
			Case 2—Covered Pervious Foundations		
136(A)	3 feet or less	Shallow	Homogeneous	Treat as Case 1—Exposed pervious foundation (shallow).	
136(C)	3 feet or less	Deep	Homogeneous	Treat as Case 1—Exposed pervious foundation (deep).	
136(A)	More than 3 feet, less than reservoir head	Shallow	Homogeneous	Treat as Case 1—Exposed pervious foundation (shallow).	
138(A) or (B)	More than 3 feet, less than reservoir head	Deep	Homogeneous	Drainage trench or pressure relief wells. Impervious upstream layer.	Key trench. Compaction of the upstream layer.
	More than reservoir head	No treatment required as a pervious foundation.			
			Stratified Foundations		
136(A)	Not important	Shallow	Stratified	Positive cutoff trench	Horizontal drainage blanket meeting filter requirements. Toe drain. Grouting may be required.
137	Not important	Deep	Stratified	Partial cutoff trench	Horizontal drainage blanket meeting filter requirements. Toe drain. Pressure relief wells may be required.

should be supervised by an experienced engineer. Intermediate depth foundations are discussed in section 132(c). The treatment of shallow and deep foundations, either exposed or covered, is discussed in detail in the previous paragraphs.

133. *Methods of Treatment of Silt and Clay Foundations.*—(a) *General.*—Foundations of fine-grained soils are sufficiently impermeable to preclude the necessity of providing design features for underseepage and piping. The main problem with these foundations is stability. In addition to the obvious danger of bearing failure of foundations of saturated silts and clays, the designs must take into account the effect of saturation of the foundations of the dam and appurtenant works by the reservoir.

Methods of foundation treatment are based on the soil type, the location of the water table, and the state of compactness of the soil. For saturated foundations of fine-grained soils (including sands containing sufficient fines to make the material impervious), the standard penetration test described in section 108 will provide an approximate measure of their state of compactness or relative consistency. This test cannot be relied on, however, in fine-grained soils above the water table, especially very dry soils whose resistance to penetration is high although their density is low. In these soils the state of compactness can be determined by density-in-place tests described in section 119.

(b) *Saturated Foundations.*—When the foundation of an earthfill dam consists of saturated fine-grained soils or saturated impervious sands, their ability to resist the shear stresses imposed by the weight of the embankment may be determined by their soil group classification and their relative consistency. Soils that have never been subjected to geologic loads greater than the existing overburden are "normally" consolidated. These soils are relatively weak as compared with strata consolidated by hundreds or thousands of feet of ice or soil which have since been removed. Old lake deposits which have experienced cycles of drying and submergence often exhibit the characteristics of preconsolidated soil due to the capillary forces associated with the shrinkage phenomenon. Soils which have been preconsolidated are recognized by their large resistance to penetration, which is usually more than 20 blows per foot; they provide satisfactory foundations for small dams. On the other hand, the presence of soft, unconsolidated silts and clays represented by a penetration resistance of less than four blows per foot indicates the need for special sampling and testing techniques and requires the advice of specialists. By identifying the soil and determining its resistance to penetration, the standard penetration test can be used to delimit those saturated foundations which can be designed by the approximate methods used in this text and to provide approximate design values.

For cohesionless soils the relative density D_d, which equals $\frac{e_{max}-e}{e_{max}-e_{min}}$ (see sec. 120(f)), is known to be related to the strength of the material. For saturated cohesive soils a similar property, the relative consistency C_r, is also related to strength. C_r is equal to $\frac{LL-w}{LL-PL}=\frac{e_{LL}-e_w}{e_{LL}-e_{PL}}$. At water contents equal to their liquid limits ($C_r=O$), the cohesive strength, C_{LL}, of all remolded saturated soils is about 0.2 pound per square inch and the shear strength can be represented by the Coulomb equation:

$$s_{LL}=0.2 \text{ p.s.i.}+\bar{\sigma} \tan \phi_s \qquad (4)$$

The value of tan ϕ_s can be obtained by "slow" shear tests on saturated soil starting from the liquid limit condition. Drainage is permitted in these tests and the pore-water pressure is zero. Tan ϕ_s is about 0.5 even for fat clays.

At water contents equal to their plastic limits ($C_r=1.0$), the cohesive strengths of saturated soils vary considerably depending on their types, and the shear strength can be represented by the equation:

$$s_{PL}=C_{PL}+\sigma \tan \phi \qquad (5)$$

The value of tan ϕ on an effective stress basis can be obtained from triaxial shear tests on samples compacted at Proctor maximum dry density and optimum water content. This value is usually somewhat smaller than tan ϕ_s. The value of cohesion at the plastic limit, C_{PL}, can be obtained from similar tests made on soil compacted at optimum water content and then saturated. As explained in section 94(d), for these samples the vertical intercept of the tangent to the failure circle making an angle ϕ with the abscissa on the Mohr diagram (fig. 57) is designated C_{sat}. The water content corresponding to C_{sat} is usually close to the plastic limit for clayey soils; that is, C_r is near unity. By assuming a linear variation of cohesion with water content between the liquid and plastic limits,

$$C_{PL}=\frac{C_{sat}-0.2}{C_r}+0.2 \tag{6}$$

where C_r corresponds to C_{sat}.

Using this assumption, Coulomb's equation for shear strength (equation (10)) for a saturated soil at any C_r may be written as follows:

$$s=C_{LL}(1-C_r)+C_rC_{PL}+\bar{\sigma}\tan\phi \tag{7}$$

The last term in the foregoing equation represents the frictional portion of the shear resistance at any point of the potential surface of sliding in the foundation. For the condition of no drainage of the impervious foundation during construction of the embankment, $\bar{\sigma}$ remains constant. The cohesion portion of the equation is a function of C_r. Since C_r cannot increase without drainage, the shear strength of the foundation remains constant while the shear stresses imposed by the embankment increase, thus decreasing the factor of safety against sliding. The methods of treatment applicable to these conditions are (1) to remove the soils of low shear strength, (2) to provide drainage of the foundation to permit increase of strength during construction, and (3) to reduce the magnitude of the average shear stress along the potential surface of sliding by flattening the slopes of the embankment.

Removal of soft foundation soils is sometimes practicable. Relatively thin layers of soft soils overlying firm material may be removed when the cost of excavation and refill is less than the combined cost of special investigations and the flatter embankment slopes required. In the preparation of relatively firm foundations, pockets of material substantially more compressible or lower in strength than the average are usually removed. See appendix E for a discussion of foundation stripping.

There are several instances where vertical drains have been used to facilitate the consolidation so that the strength of a foundation will increase as it is loaded by an embankment. This treatment is applicable primarily to nonhydraulic structures such as highway embankments. Special studies and precautions are required when these drains are used under an earthfill dam, and this device is not recommended for small dams within the scope of this text.

The most practicable solution for foundations of saturated fine-grained soils is to flatten the embankment slopes. This requires the critical surface of sliding to lengthen, thereby decreasing the average shear stress along its path and increasing the factor of safety against sliding. The selection of design slopes is discussed in section 134.

(c) *Relatively Dry Foundations.*—Unsaturated, impermeable-type soils are generally satisfactory for foundations of small dams because the presence of air in the soil voids permits appreciable volume change, increase of normal effective stress, and mobilization of frictional shear resistance without drainage of the pore fluid. That is, for a given void ratio, an impervious soil has greater bearing capacity in the unsaturated condition than in the saturated condition.

In addition, unsaturated soils exhibit the phenomenon of "apparent cohesion," which is the result of less than atmospheric capillary pressures in the water films surrounding the soil particles. The addition of water to these soils first reduces and then destroys the apparent cohesion as saturation is reached. Most soils are sufficiently dense so that reduction of apparent cohesion by saturation will cause no serious difficulties in foundations of small dams.

There is, however, an important group of soils which are of low density and subject to large settlements when saturated by the reservoir, although these soils have high dry strength in the natural state. If proper measures are not taken to control excessive settlement, failure of the dam may occur (1) by differential settlement which causes rupture of the impervious portion of the embankment and thus allows breaching of the dam by the reservoir, or (2) by foundation settlement resulting in a reduction of freeboard and overtopping of the dam, although the impervious portion of the embankment deforms without rupturing.

These low-density soils are typified by but not restricted to loess, a very loose, wind deposited soil which covers vast areas of several continents, including North America. True loess has never been saturated and is generally composed of uniform, silt-sized particles bonded together with a small amount of clay. When its water content is low, loess exhibits sufficient cohesive strength to support 100-foot-high earthfills without large settlement. A substantial increase in water content, however, greatly reduces the cohesion and may result in collapse of the loose structure of the soil under the loading imposed by dams only 20 feet high.

The experiences of the Bureau of Reclamation with the construction of dams on loess in the Missouri River Basin are, in part, described in a publication of the American Society of Civil Engineers [23] and a Bureau monograph [41]. Although the properties of other loessial soils may differ from those found in the Missouri River Basin, a discussion of the Bureau's experience may serve as a guide in other areas.

The typical undisturbed Missouri River Basin loess is a tan to light brown, unstratified, lightweight soil containing many root holes and voids. It consists mostly of silt-sized particles bonded together by a relatively small proportion of clay. The appearance of the loess and the range of gradation are shown in figure 139; 75 percent of the samples investigated were "silty loess," 20 percent were "clayey loess," and the remainder were "sandy loess." The density of the loess ranged from a low of 65 pounds per cubic foot in unusual cases to as high as 100 pounds per cubic foot in areas which had been wetted and consolidated, or where the loess had been eroded and redeposited.

With natural water contents of about 10 percent the supporting capacity of the loess is high, regardless of its density. There is little reduction in bearing capacity for water contents up to about 15 percent. Further increase in moisture is accompanied by an appreciable reduction in supporting capacity for low-density loess, while the increase in moisture has little effect on high-density loess.

Several typical laboratory compression curves for test specimens of loess have been plotted in figure 140 as load versus dry density to demonstrate the effect of inplace density and of wetting on compression characteristics. The low-density loess which is not prewetted (curve A) compresses 5 percent under a load roughly equivalent to a 100-foot-high earthfill dam (100 p.s.i. on figure 140); it compresses an additional 10.5 percent without an increase in load when saturated. The difference between the compression characteristics of low-density loess at the natural moisture and prewetted conditions indicates that dangerous settlement would result even for a 20-foot-high dam. Figure 140 also demonstrates (curve C) that very little postconstruction foundation settlement will take place for a dam constructed on a high-density loess with low natural moisture. Hence, the determination of the inplace density and water content of the loess is of paramount importance in planning its use as a foundation for a dam.

The required treatment of dry, low-density foundations will be dictated by the compression characteristics of the soil. These characteristics are best determined by laboratory tests on undisturbed samples at natural water content to determine whether postconstruction settlement on saturation will be significant (curve A of fig. 140) or of minor amount (curve C of the same figure). For small dams the empirical criteria given in section 134(b) can be used in lieu of laboratory tests.

If the foundation is not subject to appreciable postconstruction settlement upon saturation, little foundation preparation is required. The foundation should be stripped to

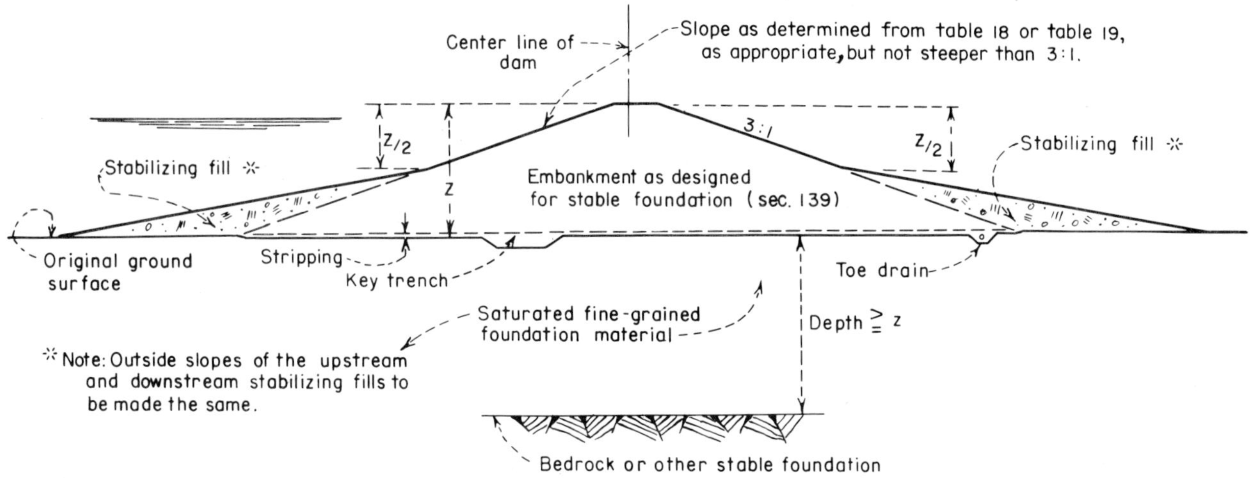

Figure 141. Design of dam on saturated fine-grained foundation. 288–D–2491.

ure 143, will provide an indication of the susceptibility of the soil to collapse on saturation [66]. Figure 143 is primarily applicable to loose, fine-grained soils. The theory is based on soil density. If the density is low enough that void spaces are larger than required to hold the liquid limit moisture content, as shown for Case I in figure 143, the soil can become wetted to above the liquid limit consistency and settle from collapse of the soil skeleton. Conversely, if the density is high enough so that the void space is less than required for the liquid limit moisture, as shown for Case III in figure 143, the soil will not collapse upon sat-

TABLE 16.—*Recommended slopes of stabilizing fills for dams on saturated silt and clay foundations* [1]

Consistency	Average number of blows per foot [2] within foundation depth equal to height of dam	Foundation soil group [3]	Slopes of stabilizing fills for various heights of dam				
			50 feet	40 feet	30 feet	20 feet	10 feet
Soft	Less than 4	Special soil tests and analyses required.					
Medium	4 to 10	SM	4½:1	4:1	3:1	3:1	3:1
		SC	6:1	5:1	4:1	3:1	3:1
		ML	6:1	5:1	4:1	3:1	3:1
		CL	6½:1	5:1	4:1	3:1	3:1
		MH	7:1	5½:1	4½:1	3½:1	3:1
		CH	13:1	10:1	7:1	4:1	3:1
Stiff	11 to 20	SM	4:1	3½:1	3:1	3:1	3:1
		SC	5½:1	4½:1	3½:1	3:1	3:1
		ML	5½:1	4½:1	3½:1	3:1	3:1
		CL	6:1	4½:1	3½:1	3:1	3:1
		MH	6½:1	5:1	4:1	3:1	3:1
		CH	11:1	9:1	6:1	3:1	3:1
Hard	More than 20	SM	3½:1	3:1	3:1	3:1	3:1
		SC	5:1	4:1	3:1	3:1	3:1
		ML	5:1	4:1	3½:1	3:1	3:1
		CL	5:1	4:1	3:1	3:1	3:1
		MH	5½:1	4:1	3:1	3:1	3:1
		CH	10:1	8:1	5½:1	3:1	3:1

[1] Stabilizing fills are not needed when embankment slopes required by tables 18 and 19 are equal to or flatter than the slope listed above.
[2] Standard penetration test (sec. 108).
[3] Unified Soil Classification System (sec. 93).

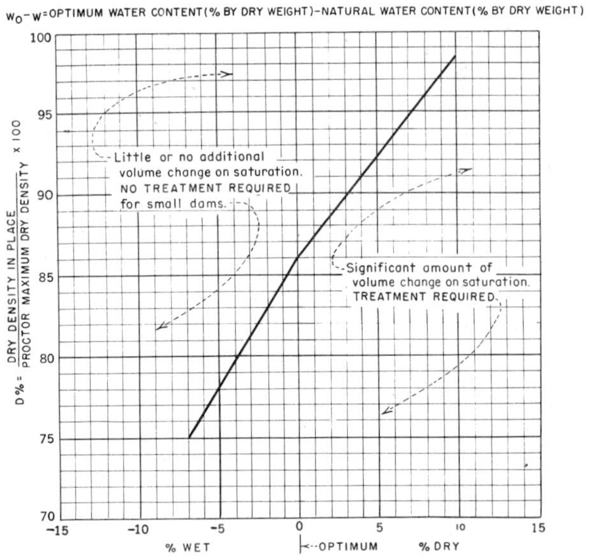

Figure 142. Foundation design criteria for relatively dry fine-grained soils. 288–D–2492.

uration but will reach a plastic state. Therefore, liquid limits and inplace dry densities which plot above the line are indicative of soils susceptible to collapse. Points below the line are indicative of soils that would settle only in a normal manner due to loading.

The foundation treatment at Medicine Creek Dam is typical of results achieved by preirrigation of a loess foundation. This structure is an earthfill dam located in south-central Nebraska in the center of the Missouri River Basin loess area. As shown on figure 144, dry low-density loess occurred on the right abutment of this dam to maximum depths of 60 to 70 feet, with an average depth of about 40 feet. Undisturbed samples were secured by sinking a test pit at a representative location to a depth of 50 feet. Table 17 summarizes partial results of laboratory tests on these samples. These tests indicated a possibility of dangerous postconstruction settlement upon saturation by the reservoir if the dam were constructed on the natural loess. Therefore, the foundation in this area was thoroughly wetted before fill construction by ponding and sprinkling. Figure 145 shows the dikes and ponds full of water; 33 million gallons of water were used over a 2-month period to raise the water content in the critical area to an average value of 28 percent.

TABLE 17.—*Properties of loess in Medicine Creek Dam foundation*

Approximate sample depth, feet	Natural water content, w, percent	Average w_o-w [1]	Inplace dry density, p.c.f.	Average $D(\%)$ [2]	Total load, fill plus overburden, p.s.i.	Compression at total load, percent: With natural moisture	Compression at total load, percent: After being wetted
5	8.8	7.4	79	75	25	8.4	10.9
17	9.7	6.5	77	74	33	1.3	9.1
19	9.6	6.6	81	77	34.5	1.0	3.9
50	6.6	8.7	92	83	55	1.5	6.5

[1] w_o-w=optimum water content for Proctor maximum dry density minus natural water content.

[2] $D\ (\%)=\frac{\text{Inplace dry density}}{\text{Proctor maximum dry density}}\times 100.$

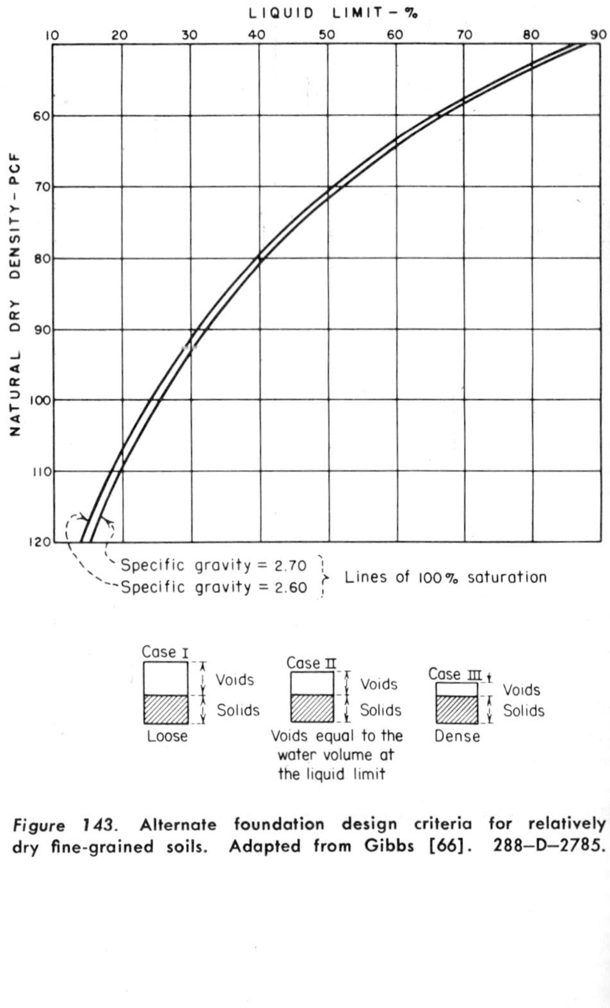

Figure 143. Alternate foundation design criteria for relatively dry fine-grained soils. Adapted from Gibbs [66]. 288–D–2785.

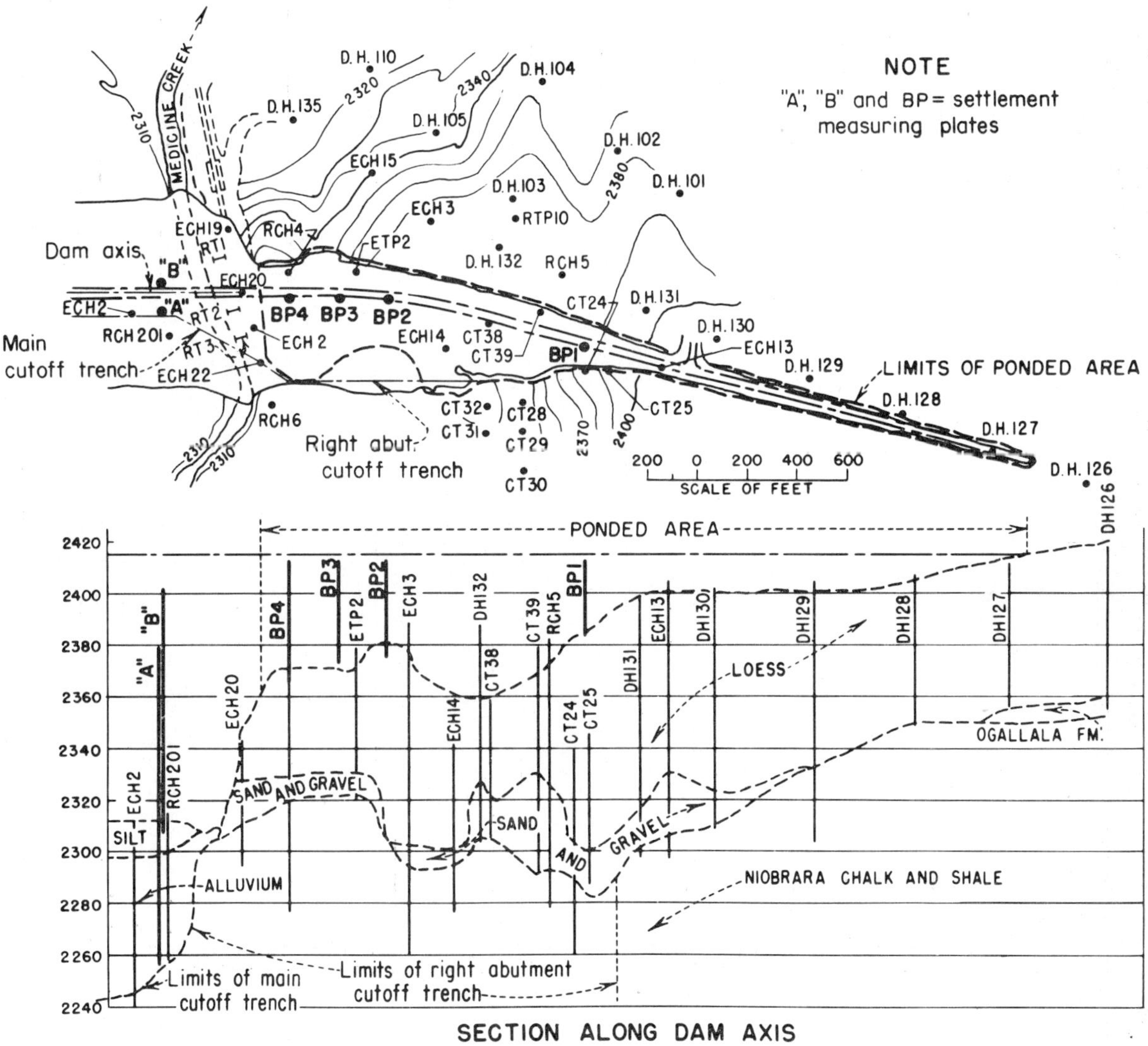

Figure 144. Geology of right abutment of Medicine Creek Dam, an earthfill structure on Medicine Creek in Nebraska. 288–D–2493.

Settlement measuring points throughout the ponded areas revealed that no settlement occurred from saturation alone. Base plate apparatus was installed in the dam to permit measurement of foundation settlement in four locations as the fill was constructed (BP1 to BP4 inclusive, fig. 144). The foundation settlements recorded by the base plate installations are shown in figure 146. Upon completion of the embankment in the fall of 1949, the apparatus indicated a foundation settlement of from 0.41 to 0.66 foot. By mid-1952 the measured foundation settlement ranged from a maximum of 2 feet at BP1 to 0.8 foot at BP4. There was virtually no further increase in the amount of settlement by mid-1954 when measurements were discontinued. The reservoir filled to elevation 2366 (normal water surface) in the spring of 1951 and remained close to that elevation during the period of measurement.

The amount of foundation settlement at Medicine Creek Dam was appreciable, although less than had been anticipated. The foundation consolidation treatment was successful

Figure 145. Ponding on foundation of Medicine Creek Dam. 404—1227B.

in that a large portion of the settlement took place while the embankment was being constructed, and the subsequent settlement was a slow consolidation over a 2-year period, which allowed the dam embankment to undergo the deformation without distress.

D. EMBANKMENTS

135. *Fundamental Considerations.*—Essentially, the design problem for an earthfill dam embankment is to determine that cross section which, when constructed with the available materials, will fulfill its required function with adequate safety at a minimum cost. The designer of an earthfill dam cannot rely on the application of mathematical analyses or formulas to determine the required cross section to the same degree that one can for a concrete dam. Soils occur with infinite combinations of size gradation, composition, and corresponding variation in behavior under different conditions of saturation and loading; further, the stress-strain relationships in an embankment are very complex.

Considerable progress has been made in investigations and studies directed toward the development of methods that will afford a comprehensive analysis of embankment stability.

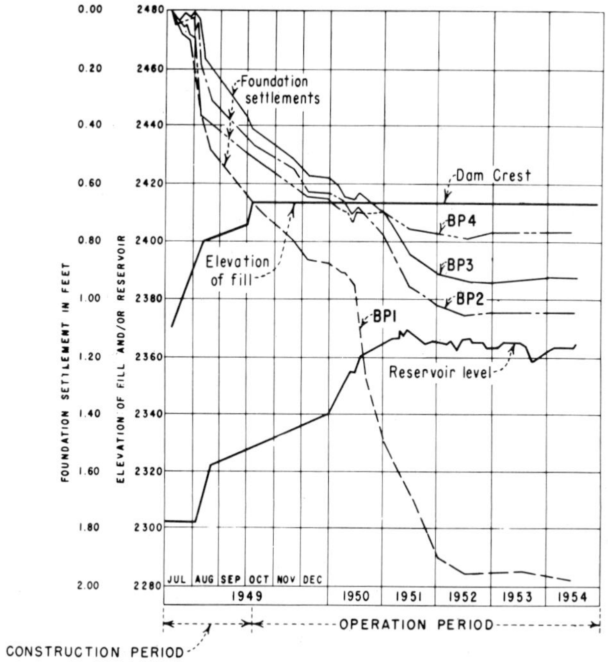

Figure 146. Record of loess foundation settlement at Medicine Creek Dam. 288–D–2494.

These methods provide useful design tools, especially for major structures where the cost of detailed explorations and laboratory testing of available construction materials can be justified on the basis of economies achieved through precise design. Even so, present practice in determining the required cross section of an earthfill dam consists largely of designing to the slopes and characteristics of existing successful dams, making analytical and experimental studies for unusual conditions, and controlling closely the selection and placement of embankment materials. While some modifications are necessarily made in specific designs to adapt them to particular conditions, radical innovations are avoided and fundamental changes in design concepts are developed and adopted gradually through practical experience and trial.

Although the above practice may be criticized as being overly cautious and extravagant, no better method has yet been conclusively demonstrated. Where consideration is given to the possible loss of life, the certainty of costly property damage in many cases, and the waste of money incidental to the failure of a constructed dam, ample justification is provided for conservative procedures. For small dams, where the cost of explorations and laboratory testing of embankment materials for analytical studies together with the cost of the engineering comprise an inordinate proportion of the total cost of the structure, the practice of designing on the basis of successful structures and past experience becomes even more appropriate.

The design criteria for embankments are given in section 127. They require that the slopes of the embankment be stable under all conditions of construction and reservoir operation; that excessive stresses not be induced in the foundation; that seepage through the embankment be controlled; that the embankment be safe against overtopping; that the slopes be protected against erosion; and that the em-

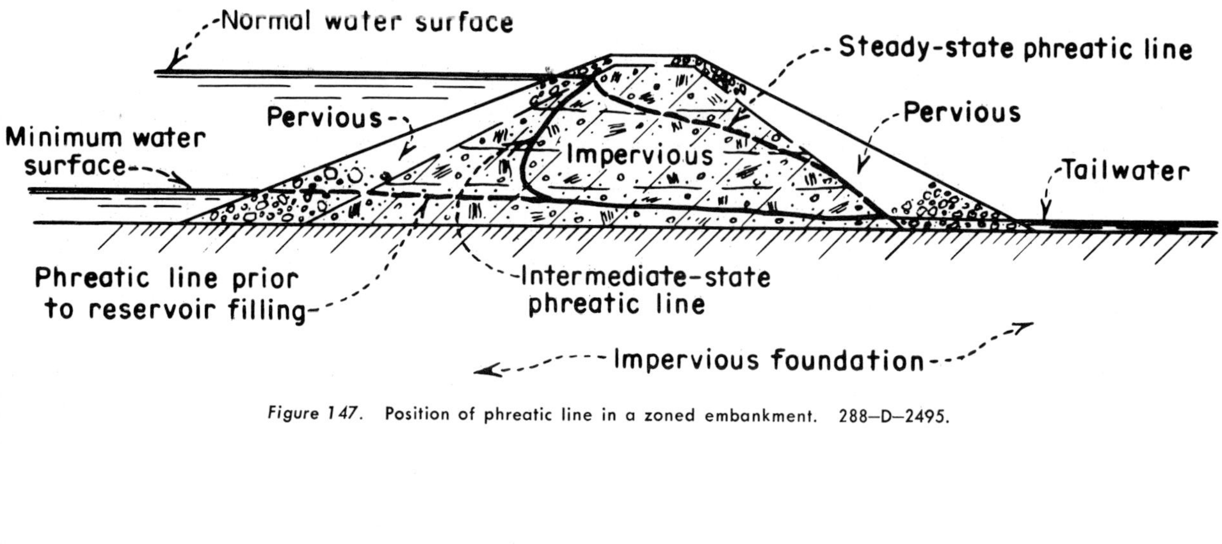

Figure 147. Position of phreatic line in a zoned embankment. 288–D–2495.

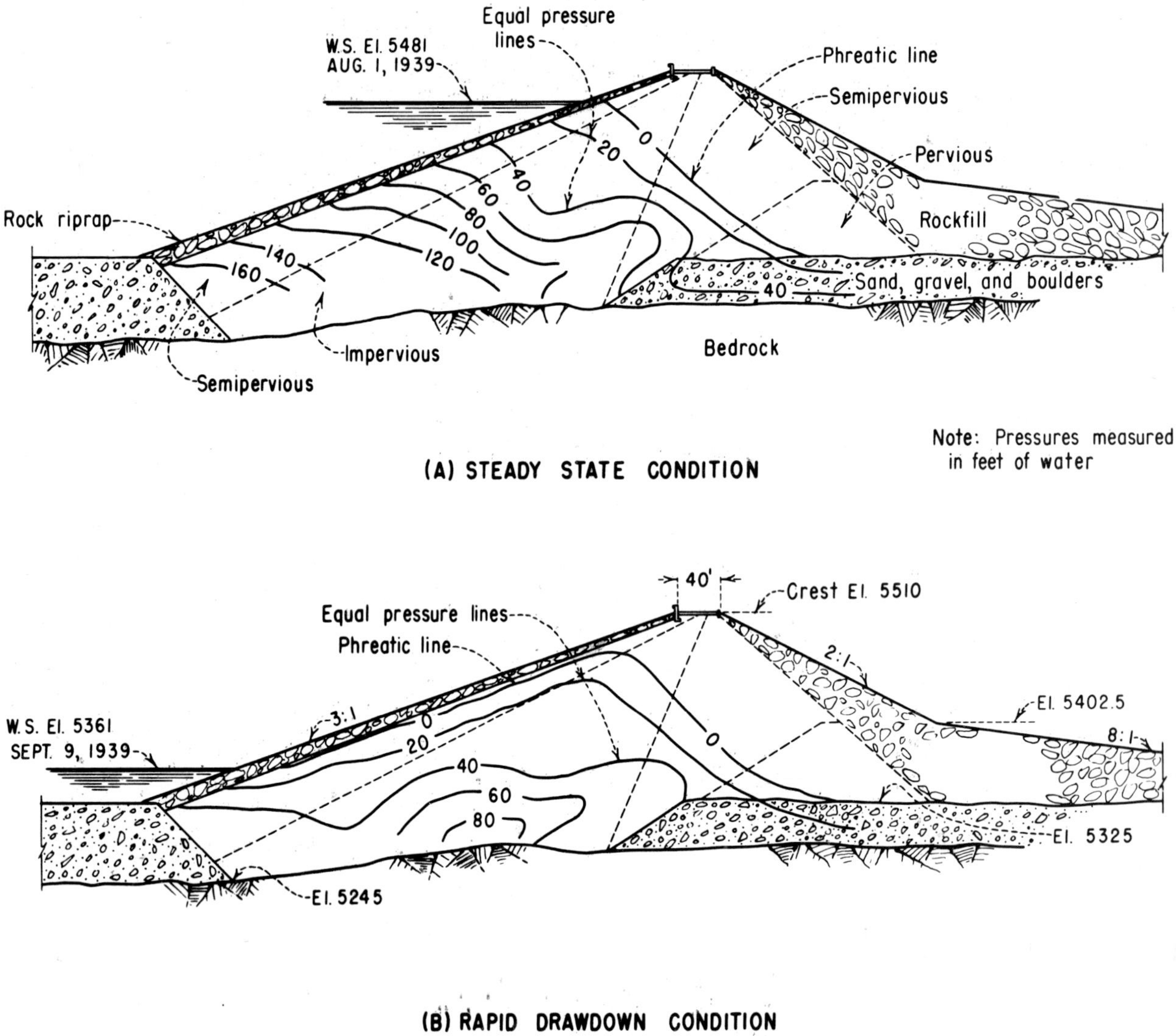

Figure 148. Effect of rapid drawdown on pore pressures at Alcova Dam, an earthfill structure on the North Platte River in Wyoming 288–D–2496.

bankment be stable under appropriate seismic conditions. This part of the chapter is concerned with the slope stability of the embankment due to both static and seismic conditions and with the control of seepage through the embankment. Embankment details concerning the crest, freeboard, slope protection, and surface drainage are discussed in part E.

The stability of an embankment is determined by its ability to resist shear stresses, since failure occurs by sliding along a shear surface. Shear stresses result from externally applied loads, such as reservoir and earthquake, and from internal body forces caused by the weight of the soil and the embankment slopes. The external and internal forces also produce compressive stresses normal to any potential sliding surface. These compressive stresses contribute both to the shear strength of the soil and to the development of destabilizing pore-water pressures.

Embankments of granular or noncohesive materials are more stable than those made of cohesive soils, because granular materials have a higher frictional resistance and because their greater permeability permits rapid dissipation of pore-water pressures resulting from compressive forces. Accordingly, when other conditions permit, somewhat steeper slopes may be adopted for noncohesive soils. Embankments of homogeneous materials of relatively low permeability have slopes generally flatter than those used for zoned embankments, which have free-draining outer zones supporting inner zones of relatively impervious materials.

In brief, it may be stated that the design of an earthfill dam cross section is controlled by the physical properties of the materials available for construction, by the character of the foundation, by the methods of construction that are specified, and by the degree of construction control that is anticipated.

136. *Pore-Water Pressure*.—In 1936, Terzaghi [24] demonstrated that in impervious soils subjected to load, a total stress normal to any plane is made up of an effective stress and a fluid pressure. The concepts of plane surfaces and stresses at a point in soils are not identical with those of an ideal homogeneous isotropic material. The "plane" in soils is a rather wavy surface, touching the soil particles only at their contacts with one another; and the "point" of stress is a small region containing enough of the particles to obtain an average stress. With these qualifications the total normal compressive stress, σ, along a plane in an earth structure can be written:

$$\sigma=\bar{\sigma}+u \tag{8}$$

where u is pore-water pressure. From considerations of equilibrium [25], the shear stress τ along the plane is:

$$\tau=\frac{\sigma_1-\sigma_3}{2}\sin 2\theta \tag{9}$$

where:

σ_1 is the total major principal stress,
σ_3 is the total minor principal stress, and
θ is the angle between the plane considered and the plane on which σ_1 acts.

It is apparent from equation (9) that the shear stress is the same whether σ_1 and σ_3 are used or their effective components $\bar{\sigma}_1$ and $\bar{\sigma}_3$.

The shear strength along a plane can be obtained from Coulomb's equation:

$$s=C+(\sigma-u)\tan\phi \tag{10}$$

which shows that the frictional portion of the resistance along a plane is reduced by pore-water pressure. This equation is discussed and the terms are defined in section 94(d).

Pore-water pressures in compacted cohesive soils caused by compressive stresses occur in the sealed triaxial shear test in the laboratory and in the impervious zone of an embankment during construction. For the laboratory conditions the relation between volume change and fluid pressure in a loaded soil mass consisting of solid particles, water, and air, can be derived by using Boyle's law for compressibility of air and Henry's law for solubility of air and water both at constant temperature. For a soil mass buried in an "impervious" fill where drainage is extremely slow because of the long path of percolation and the very small coefficient of permeability of the material, it is both conservative and reasonable, on the basis of field observations, to use the assumption of no drainage to estimate the magnitude of pore-water pressure for design and control purposes [26]. The concept is that when the moist soil mass is loaded without permitting air or water to escape, part of the load causes the soil grains to deform elastically or to undergo nonelastic rearrangement, but without significant change in their solid volume. This part of the load is carried on the soil skeleton as effective stress. The remaining portion of the load is carried by stress in the air and water contained in the voids and is known as pore-water pressure.

Analysis shows that the magnitude of pore-water pressures from compressive forces depends on the compressibility of the compacted soil and on the amount of air it contains. For given conditions of compressibility and loading, the closer the compacted soil is to saturation, the higher the pore-water pressure will be. This leads to the practice of controlling the water content of materials in order to increase the amount of air in the compacted soil. The water content has been reduced below the optimum water content for compaction at Proctor maximum density in the construction of high earthfill dams, but this procedure is neither necessary nor desirable for the construction of embankments less than 50 feet high. For these heights, compaction of cohesive soils at optimum water content and approximately Proctor maximum dry density will insure sufficient air, even in the most compressible soils, to preclude the development of pore-water pressures of appreciable magnitude.

Placing the material at optimum water content instead of dry of optimum also increases the plasticity of the material and allows it to conform more readily to the shape of the foundation and abutments during post-construction settlement and helps to reduce the probability of tension cracks in the embankment.

For small confining loads, placing material dry of optimum is undesirable because it increases the possibility of:

(1) Low density for the same compactive effort, as shown by the shape of the compaction curve (fig. 108).
(2) Greater permeability in the embankment core.
(3) Excessive softening and settlement after saturation by the reservoir, resulting in possible cracking of the fill.

On the other hand, the water content should not be appreciably greater than the optimum which is obtained at Proctor maximum dry density because difficulties have been experienced with unstable fills when very wet soils are used, even in dams of low height.

The foregoing considerations result in the recommended practice of compacting cohesive soils in the cores of small dams close to the optimum water content at Proctor maximum dry density.

137. *Seepage Through Embankments.*—The core or water-barrier portion of an earthfill dam provides the resistance to seepage which creates the reservoir. Although, as pointed out in section 94(b), soils vary greatly in permeability, even the tightest clays are porous and cannot prevent water from seeping through them.

The progress of percolation of reservoir water through the core depends on the constancy of reservoir level, the magnitudes of permeability of the core material in the horizontal and vertical directions (anisotropy), the amount of remaining pore-water pressures caused by compressive forces during construction, and the element of time. Figure 147 shows the penetration of water into a core shortly after the first filling of the reservoir, and also the penetration when the steady-state seepage condition has finally been reached. The upper surface of seepage is called the phreatic (zero pressure) surface; in a cross section it is referred to as the phreatic line. Although the soil may be saturated by capillarity above this line, giving rise to a "line of saturation," seepage is limited to the portion below the phreatic line.

The position of the phreatic line depends only on the geometry of the section. For soils of vastly different permeabilities but of the same ratio of horizontal to vertical permeability, the phreatic lines eventually will reach identical position. It will take much longer for the steady-state condition to be reached in clay than in sand for the same cross section, and the amount of water emerging at the downstream slope will, of course, be much greater in the more pervious material. The pore-water pressures below the phreatic line reduce the shear strength of the soil mass in accordance with Coulomb's law, equation (10). The steady-state condition, which involves the maximum saturation of the embankment, is the most critical post-construction condition for the stability of the downstream slope.

The most critical operating condition so far as the stability of the upstream slope is concerned is a rapid drawdown following a long period of high reservoir level. Figure 148 shows the effect of rapid drawdown on the pore-water pressures measured in Alcova Dam, Wyoming. It will be noted that the reservoir water surface was lowered 120 feet in 40 days, which is an extremely rapid drawdown for a dam of this height. Figure 148(A) shows the phreatic line and equal-pressure lines under full reservoir conditions; the position of the phreatic line indicates that virtually steady-state conditions were present prior to drawdown. Figure 148(B) shows the pressures under drawdown conditions.

Figure 148 demonstrates that appreciable

pore-water pressures remain in an embankment after drawdown. If a dam is subject to rapid drawdown after long-term storage at high reservoir levels, special provisions for drainage should be made in the design. The upstream slope of an embankment with an appreciable upstream pervious zone usually is not critical for the rapid-drawdown condition. Rapid drawdown may require a flatter slope of a homogeneous embankment than would otherwise be needed for stability.

The use of the flow net in determining the magnitude and distribution of seepage pressures in pervious foundations has been previously described (sec. 130(c)). The flow net is also used as a means of visualizing the flow pattern of percolating water through embankments in order to estimate the magnitude and distribution of pressures due to percolating water, both in the steady state and in the drawdown condition. Analytical methods of stability analyses which are used in the design of major structures require that such pore-water pressures be determined quantitatively. Such a determination is not required for the design procedure given in this text.

138. *Stability Analyses*.—Various methods have been proposed for computing the stability of earthfill dams [35]. In general, these methods are based on the shear strength of the soil and certain assumptions with respect to the character of an embankment failure. The Swedish or "slip-circle" method, which supposes the surface of rupture to be a cylindrical surface, is a comparatively simple method of analyzing embankment stability. Although other and more strictly mathematical solutions have been developed, the slip-circle method of stability analysis is the one most generally accepted. In this method the factor of safety against sliding is defined as the ratio of the average shear strength, as determined from equation (10), to the average shear stress determined by statics on a potential sliding surface.

The force exerted by any segment within the slip-circle is equal to the weight of the segment and acts vertically downward through its center of gravity. The components of this weight acting on a portion of the circle are the force normal to the arc and the force tangent to the arc, as determined by completing the force triangle with lines in the radial and tangential directions. Pore-water pressures acting on the arc result in an uplift force which reduces the normal component of the weight of the segment. Graphical means have been developed by May [27] to facilitate the solution.

The safety factor against sliding for an assumed circle is computed by the equation:

$$\text{Safety factor} = \frac{CL + \tan\phi(N-U)}{T} \qquad (11)$$

where:

N=summation of normal forces along the arc,

U=summation of upift forces due to pore-water pressure along the arc,

T=algebraic summation of tangential forces along the arc;

L=length of arc of slip circle, and

C and $\tan\phi$ are as defined in section 94(d).

Various centers and radii are used and computations repeated until the arc which gives the minimum safety factor is established.

In order to compute the safety factor by means of equation (11), it is necessary to establish the cohesion (C) and angle of internal friction (ϕ) of the soil, and the magnitude of pore-water pressures for construction, steady-state, and drawdown conditions. Furthermore, the strength properties of the foundation must be determined where the overburden above bedrock is silt or clay, as experience has shown that the critical circle will extend into the foundation in such cases. It is therefore apparent that this method of analysis is more suited to the design of major structures, where the cost of foundation exploration and laboratory tests of foundation and embankment materials to determine their average strength properties is justified because of economies which may be achieved by the use of more precise slopes. The recommended designs for small earthfill dams given in this text are based on the Swedish slip-circle method, using average values of soil properties and experience. These designs will result in adequate factors of safety provided proper construction control is obtained. Construction control is discussed in appendix E.

139. Embankment Design.—(a) *Utilization of Materials from Structural Excavation.*—In the discussion of design criteria (sec. 127), it was pointed out that for minimum cost the dam must be designed to make maximum utilization of the most economical materials available, including material which must be excavated for the dam foundation, spillway, outlet works, canals, powerhouse, roadways, and other appurtenant structures. When the yardage from these sources constitutes an appreciable portion of the total embankment quantity, it may strongly influence the design of the dam. Although these materials frequently are less desirable than soil from available borrow areas, economy requires that they be employed to the maximum practicable extent. Available borrow areas and structural excavations must be considered together in arriving at a suitable design.

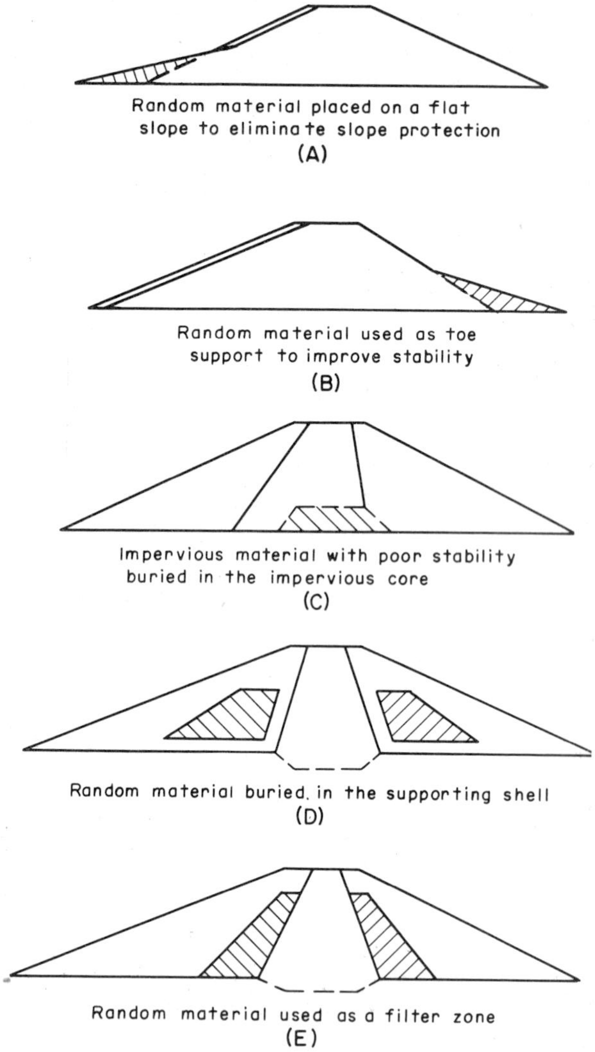

Figure 149. Use of random fill materials within embankment sections. 288–D–2787.

The portion of the cutoff trench excavation above groundwater table may provide limited amounts of material for the impervious core of the dam. Appreciable quantities of sand and gravel are usually obtained in the dewatered portion of the trench from the strata that are being intercepted. When sand and gravel occur in thick, clean beds, this material can be used in the outer zones of the dam. However, pockets or lenses of silt and clay and highly organic material also occur in cutoff trench excavation. These materials contaminate the clean soils and result in wet mixtures of variable permeability and poor workability. Such mixtures should usually be wasted.

Excavation for the spillway provides both overburden soils and formation bedrock. In planning the use of these materials, the designer must recognize that moisture control, processing, and special size requirements will add to the project cost. For these reasons, material from spillway excavations ordinarily is used primarily in the main structural zones of dam embankments.

Tunnel excavations can provide rockfill material for use in the pervious zones of the dam or can provide rock fines which may serve as a transition between the impervious core material and pervious zones.

The feasibility of using materials from structural excavations is influenced by the sequence of construction operations. The construction sequence is in turn influenced by the following items:

(1) Topography of the damsite.
(2) Diversion requirements.
(3) Hydrology of the watershed.
(4) Seasonal climatic changes.
(5) Magnitude of required excavations.

In order to use material from the spillway or cutoff trench in the embankments without having to stockpile and later rehandle large quantities of earth and rock, an adequate placing area must be available. The placing area is usually restricted early in the job; hence, the designer is faced with the choice of specifying that spillway excavation be delayed until space is available for it, of requiring extensive stockpiling, or of permitting large quantities of material to be wasted. The amount of embankment space that can be provided during the early stages of construction depends in part on the diversion requirements and in part on the diversion plan that the contractor will select. Usually, the contractor is allowed considerable flexibility in the method of diversion; this adds to the designer's uncertainty in planning use of materials from structural excavations.

Zoned dams provide an opportunity to specify the use of structural excavation materials and should be used whenever possible. The zoning of the embankment should be based on the most economical utilization of materials which can be devised; however, the zoning must be consistent with the requirements for stability as previously discussed. For example, the use of rockfill sections can allow continual construction throughout the winter, thus effecting an early completion date. An important use of materials from structural excavation has been in portions of the embankment where the permeability and shear strength are not critical and where weight and bulk are the major requirements. The stabilizing fills required for dams on saturated fine-grained foundations, discussed in section 134 (fig. 141), are an illustration of this usage.

Areas within the dam into which such excavated material is placed are called "random zones." Typical locations for these random zones are shown in figure 149.

Since estimates of the percentage of structural excavations usable within the embankment are subject to wide errors, provision should be made to use variable zone boundaries to accommodate any excess or deficiency. In some cases, special laboratory tests or a test embankment may be required before determining the disposition of questionable material or selecting the dimensions of a random zone.

In formulating a design, the designer must anticipate what percentage of the structural excavation will be suitable in the various zones of the embankment and the shrinkage and swell of the material involved. He must then integrate these anticipated quantities with the required borrow area quantities to determine a final design which is both economical and has a reasonable construction sequence. Often, several design schemes are required. The use of a materials distribution chart, such as shown in figure 150, has been found helpful for integrating excavation quantities into the embankment section, for determining the required amounts of borrow material for each zone, and for visualizing the construction sequence. The chart shown is for the Bureau of Reclamation's Wasco Dam, the maximum section of which is shown in figure 192. In addition to showing the sources of all fill materials, the chart contains the assumptions used for shrinkage, swell, and yield on which specifications quantities are based.

(b) *Embankment Slopes, General.*—The design slopes of an embankment may vary widely, depending on the character of the materials available for construction, foundation conditions, and the height of the structure. The embankment slopes as determined in this section are the slopes required for stability of the embankment on a stable foundation. Pervious foundations may require the addition of upstream blankets to reduce the amount of seepage or downstream horizontal drainage blankets for stability against seepage forces. Weak foundations may require the addition of stabilizing fills at either or both toes of the dam. The additional embankments needed because of pervious or weak foundations should be provided beyond the slopes determined herein as required for embankment stability.

The upstream slope may vary from 2:1 to as flat as 4:1 for stability; usually it is 2½:1 or 3 : 1. Flat upstream slopes are sometimes used in order to eliminate expensive slope protection. A berm is often provided at an elevation slightly below the maximum drawdown of the reservoir water surface to form a base for the upstream slope protection, which need not be carried below this point. The upstream slope is often steepened above the elevation where water is stored; that is, in the surcharge range.

A storage dam subject to rapid drawdown of the reservoir should have an upstream zone with permeability sufficient to dissipate pore-water pressures exerted outwardly in the upstream part of the dam. The rate of reservoir drawdown is an important factor which affects the stability of the upstream part of the dam. For a method of designing free draining upstream shells, refer to Cedergren [30]. Where only fine material of low permeability is available, such as that predominating in clays, it is necessary to provide a flat slope if rapid drawdown is a design requirement. Conversely, if free-draining sand and gravel are available to provide a superimposed weight for holding down the fine material of low permeability, a steeper slope may be used. The same result may be secured by utilizing sound and durable rock from required excavations. In the latter case, a layer of sand and gravel or quarry fines must be placed between the superimposed rock and the surface of the impervious embankment to prevent damage and displacement from saturation and wave action.

Flood damage due to failure of the upstream face is very unlikely. Failure can take place only during construction or following a rapid drawdown; in both cases the reservoir should be virtually empty. The weight and seepage forces act as a stabilizing influence on the upstream face when the reservoir is full.

The usual downstream slopes for small earthfill dams are 2 : 1 where a downstream pervious zone is provided in the embankment, and 2½ : 1 where the embankment is impervious. These slopes are stable for soil types commonly used when drainage is provided in the design so that the downstream slope of the embankment does not become saturated by seepage.

The slopes of an earthfill dam depend on the type of dam (that is, diaphragm, modified homogeneous, or zoned embankment), and on the nature of the materials for construction. Of special importance is the nature of the soil which will be used for construction of the modified homogeneous dam or the core of a zoned dam. In the latter case, the relation of the size of the core to the size of the shell is also significant.

In this text, the slopes of the embankment are related to the classification of the soil to be used for construction, especially the impervious soils. The engineering properties of soils in the various classifications are shown in table 8 (sec. 94). The slopes chosen are necessarily conservative and are recommended only for small earthfill dams within the scope of this text, as discussed in section 124.

(c) *Diaphragm Type.*—A diaphragm dam consists of a thin impervious water barrier used in conjunction with a large pervious zone. The diaphragm can be constructed of earth, asphalt, concrete, or metal. If the diaphragm is constructed of impervious earth material, it must have a horizontal thickness at least great enough to accommodate construction equipment. Because it must hold back the full reservoir pressure, its construction must be carefully performed; to prevent piping or erosion it must be protected by graded filters. When an earth diaphragm is centrally located, it is also referred to as a "thin core." A typical earth-diaphragm constructed for Amarillo Regulating Reservoir is shown in figure 166.

Diaphragm-type dams are generally used under the following conditions:

(1) A limited quantity of impervious material is available.
(2) Wet climatic conditions.
(3) Short construction seasons.

A diaphragm should be used only when the design and construction of the dam are performed under the supervision of an experienced earth dam designer. If this type of dam is selected, it is recommended that a diaphragm of manufactured material be placed on the upstream slope of an otherwise pervious embankment in lieu of a soil blanket. If the pervious material is rock, the dam is classified as a rockfill dam, the design of which is discussed in chapter VII.

The pervious material used in the construction of a diaphragm dam must be such that it can be compacted to form a stable embankment which will be subject to only small amounts of post-construction settlement. Poorly graded sands (SP) cannot be satisfactorily compacted; well-graded sand-gravel mixtures (SW-

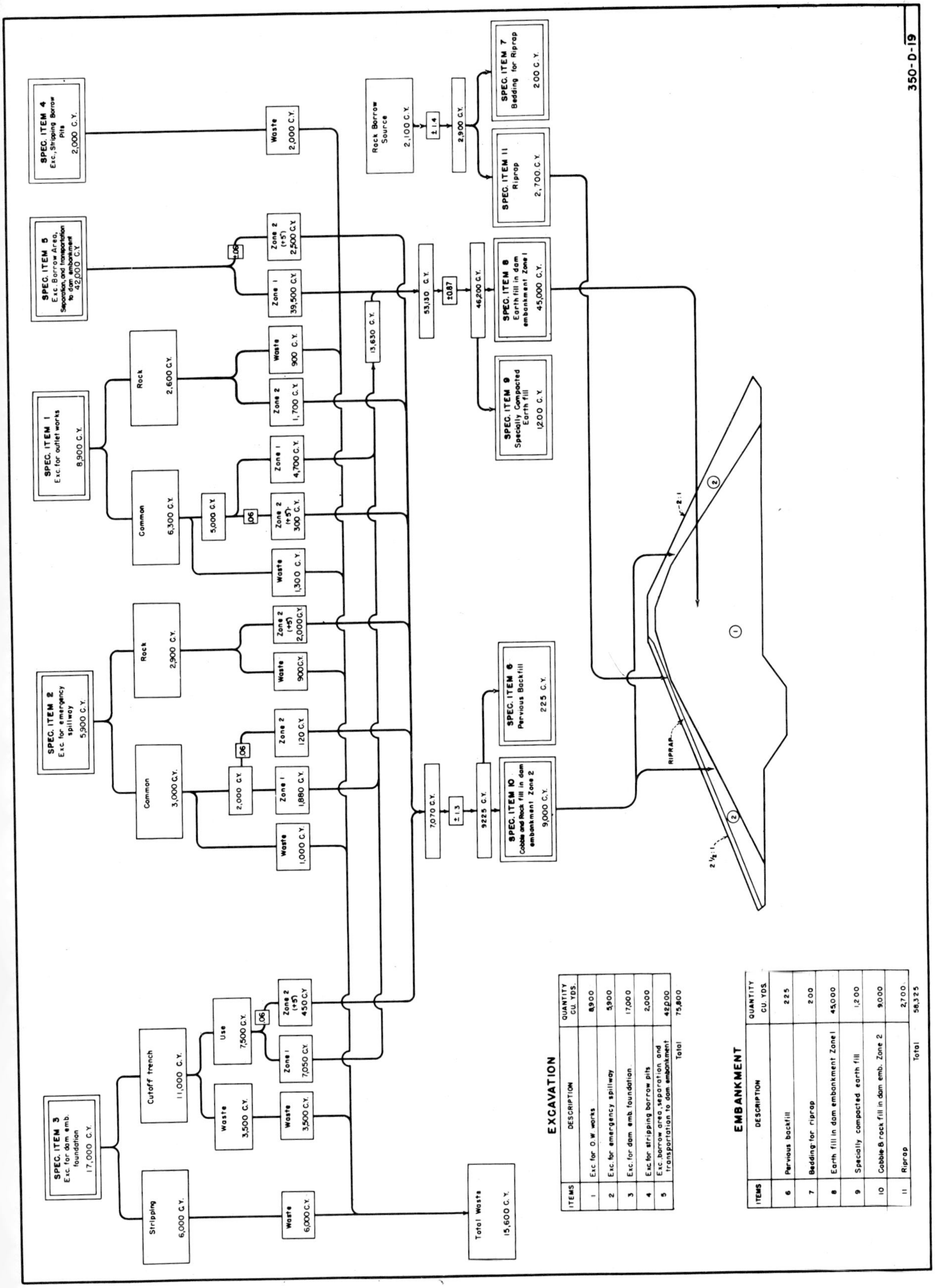

EXCAVATION

ITEMS	DESCRIPTION	QUANTITY CU. YDS.
1	Exc for O.W. works	8,900
2	Exc. for emergency spillway	5,900
3	Exc. for dam emb. foundation	17,000
4	Exc. for stripping borrow pits	2,000
5	Exc. borrow area, separation and transportation to dam embankment	42,000
	Total	75,800

EMBANKMENT

ITEMS	DESCRIPTION	QUANTITY CU. YDS.
6	Pervious backfill	225
7	Bedding for riprap	200
8	Earth fill in dam embankment Zone I	45,000
9	Specially compacted earth fill	1,200
10	Cobble & rock fill in dam emb. Zone 2	9,000
11	Riprap	2,700
	Total	58,325

Figure 150. Materials distribution chart for Wasco Dam, an earthfill structure on Clear Creek in Oregon. 350–D–19.

GW) or well-graded gravels (GW) make satisfactory embankments. Well-graded sand-gravel mixtures which contain more than 5 percent of material finer than the 200-mesh sieve should be tested to determine that they will form free-draining embankments after compaction.

In all respects, except for the use of pervious materials other than rock in construction of the embankment, the diaphragm earthfill dam design as recommended herein for small dams is identical with the design of rockfill dams, which is discussed in chapter VII. That discussion should be referred to for the design of foundations and upstream facings for a diaphragm-type earthfill dam.

(d) *Homogeneous Type.*—The homogeneous-type dam is recommended only where the lack of free-draining materials make the construction of a zone embankment uneconomical, with the further qualification that for storage dams the homogeneous dam must be modified to provide for the inclusion of internal drainage facilities. The recommended drainage facilities for modified homogeneous dams are described in section 125 and are shown on figure 114. If a rockfill toe is provided, a filter must be constructed between the embankment proper and the rockfill toe, as shown in figure 114(A). This filter and the horizontal drainage blanket shown in figure 114(B) should be designed to meet the filter requirements described in section 131(i).

To perform its function of lowering the phreatic line and stabilizing the downstream portion of the dam, the horizontal drainage blanket shown on figure 114(B) should extend from the downstream slope of the dam to well within the body of the embankment. However, it should not extend upstream so far as to reduce the length of the path of percolation through the embankment or the foundation to a dangerous extent. A minimum-length filter blanket is also desirable because filters are expensive to construct. For small dams, it is recommended that the horizontal drainage blanket start at the downstream toe of the embankment and extend upstream to within a distance equal to the height of the dam plus 5 feet from the centerline of the dam. This will afford a blanket of ample extent and yet not reduce the length of the path of percolation beyond desirable limits. The distance of height of dam plus 5 feet is selected on the basis that this will place the upstream limit of the horizontal drainage blanket at the downstream edge of "Minimum Core B," the minimum core required for dams on deep pervious foundations without positive cutoff trenches, as shown on figure 151.

The horizontal drainage blanket should be carried across the valley floor and up the abutments to an elevation corresponding to the highest level at which water will be stored in the reservoir for an appreciable time. It should be a uniformly thick blanket whose upstream position at a given point is downstream from the centerline of the dam a distance equal to the height of the dam at that point plus 5 feet.

Even in the construction of a homogeneous embankment, there is likely to be some variation in the nature of the borrow material. It is important that the coarse and more pervious material be placed at the outer slopes in order to approach, as nearly as possible, the advantages of a zone embankment. It is also important to avoid segregation of the larger particles when the fill is dumped. Segregation leads to the formation of layers of much greater permeability than the other embankment; these layers tend to form drainage channels for the percolating water and to increase the possibility of piping.

The recommended slopes for small homogeneous earthfill dams are shown in table 18 for detention and storage dams on stable foundations with and without rapid drawdown as a design condition. Where more than one soil classification is shown for a set of slopes, it indicates that the dam can be constructed to the slopes shown in table 18, using any one of the soils or any combination thereof.

(e) *Zoned Embankments.*—(1) *General.*—The zoned embankment dam consists of a central impervious core, flanked by zones of material which are considerably more pervious. An excellent example of a zoned dam is Carter Lake Dam No. 3, shown in figure 171. This type of embankment should always be con-

TABLE 18.—*Recommended slopes for small homogeneous earthfill dams on stable foundations*

Case	Type	Purpose	Subject to rapid drawdown [1]	Soil classification [2]	Up-stream slope	Down-stream slope
A	Homogeneous or modified-homogeneous	Detention or storage	No	GW, GP, SW, SP	Pervious, not suitable	
				GC, GM, SC, SM	2½:1	2:1
				CL, ML	3:1	2½:1
				CH, MH	3½:1	2½:1
B	Modified-homogeneous	Storage	Yes	GW, GP, SW, SP	Pervious, not suitable	
				GC, GM, SC, SM	3:1	2:1
				CL, ML	3½:1	2½:1
				CH, MH	4:1	2½:1

[1] Drawdown rates of 6 inches or more per day following prolonged storage at high reservoir levels.
[2] OL and OH soils are not recommended for major portions of homogeneous earthfill dams. Pt soils are unsuitable.

structed where there is a variety of soils readily available because its inherent advantages lead to economies in the cost of construction. Three major advantages in using zoned embankments are listed below:

a. Steeper slopes may be used, with consequent reduction in total volume of embankment material.

b. A wide variety of materials may be utilized.

c. Maximum utilization can be made of material excavated from the foundation, spillway, outlet works, and other appurtenant structures.

(2) *Zoning.*—All zoning schemes are based on the estimated quantities of required excavation and borrow area materials available. The scheme of zoning may divide the dam into two or more zones, depending on the range of variation in the character and gradation of the available materials of construction. In general, the permeability of each zone should increase toward the outer slope. Relatively free-draining materials and, therefore, those with a high degree of inherent stability are used to enclose and support the less stable impervious core. Pervious materials are placed in the downstream sections to avoid building up pressure from percolating water and to permit lowering the phreatic line so as to keep it well within the embankment. Pervious materials are placed in upstream sections to permit dissipation of pressure on rapid drawdown.

Miscellaneous or random zones, as shown in figure 149, are often included in the downstream sections of the embankment to utilize excavated materials of uncertain permeability. Excavated materials not suited for use in any zone and excess excavation may be wasted on the upstream or downstream toes. Section 139(a) discusses more fully the utilization of excavated material.

(3) *Transitions.*—It is important that the gradation of adjacent zones be considered so that materials from one zone are not "piped" into the voids of adjoining zones, either by steady-state or by drawdown seepage forces. Transitions prevent piping and also provide the additional advantage that should the embankment crack, partial sealing of the cracks takes place with subsequent reduction in seepage losses.

A transition of sand-gravel or rock fines must be provided between an impervious zone and a flanking rockfill. Although these transitions need only be a few feet wide, they are usually constructed in widths from 8 to 12 feet to accommodate construction equipment.

If the transition zone is only a few feet thick, it should be designed as a filter in accordance with the procedure given in section 131(i); for thicker transitions, the filter requirements may be made less stringent. Transition zones ordinarily are not required between impervious and sand-gravel zones or between sand-gravel zones and rockfill.

(4) *Impervious Cores for Zoned Embank-*

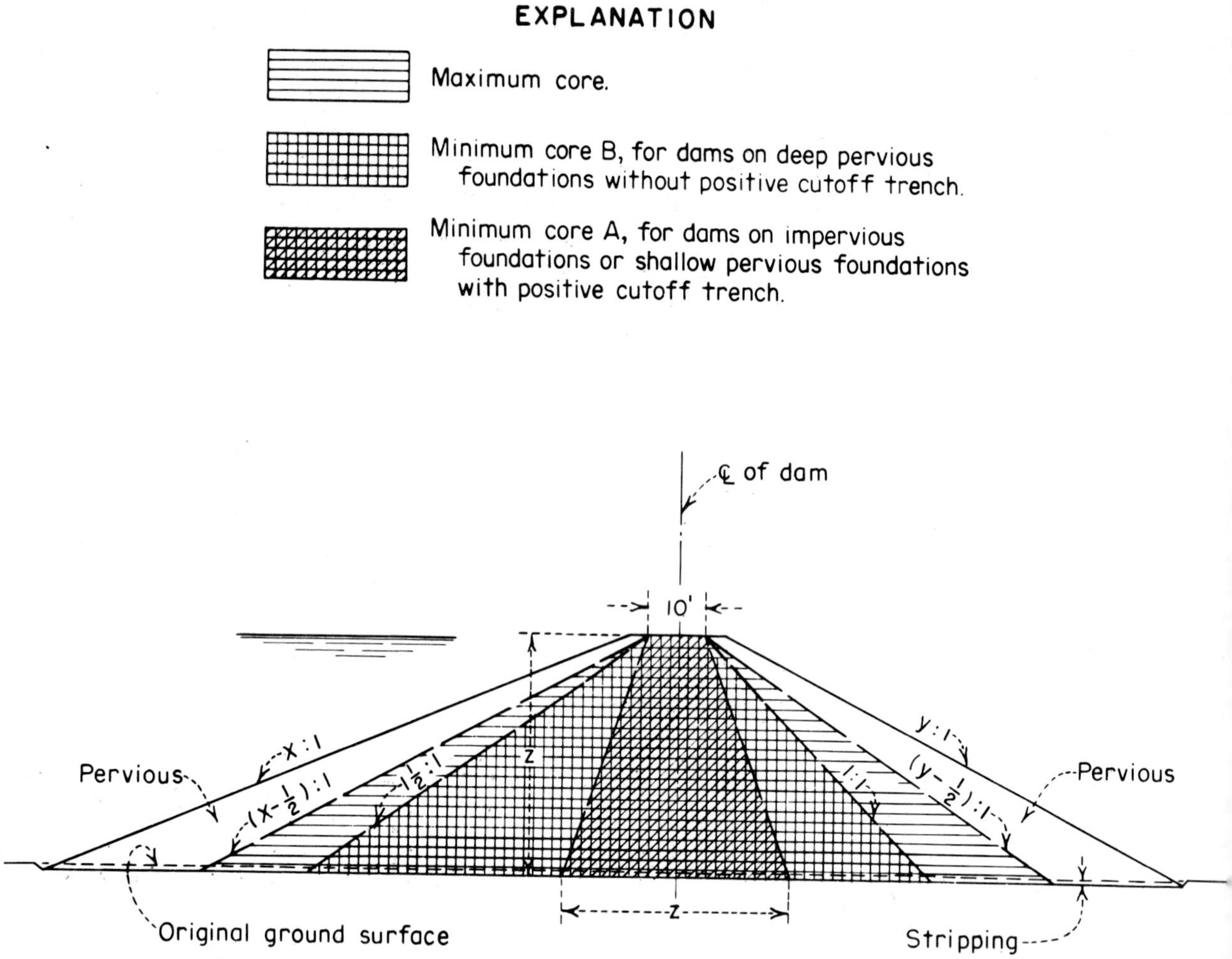

Figure 151. Size range of impervious cores used in zoned embankments. 288–D–2497.

ments.—Figure 151 shows the suggested size of the "minimum core" for the following two conditions:

a. Impervious or pervious foundations of shallow depth penetrated by a positive cutoff trench. This core is hereinafter referred to as "Minimum Core A."

b. Exposed pervious foundations and covered pervious foundations (less than 3 feet of cover) not penetrated by a positive cutoff trench regardless of the depth of pervious material. This core is hereinafter referred to as "Minimum Core B."

The "maximum size" impervious core, which allows the dam to function as a zoned embankment, is also shown.

If the core is smaller than "Minimum Core A," the dam is considered to be of the diaphragm type; if the core is larger than the maximum size shown, the pervious zones are largely ineffectual in stabilizing the core, and the embankment may be considered as the homogeneous type so far as stability is concerned.

The size of "Minimum Core A" shown in figure 151 was selected for both practical and theoretical reasons. The width of 10 feet was taken as the minimum that will permit economical placement and compaction of impervious embankment material by construction equipment such as trucks, dozers, and tamping rollers. The criterion that the thickness of the core at any elevation cannot be less than the height of embankment remaining above that elevation was adopted so that the average hydraulic gradient through the core will be less than unity. Appreciably steeper gradients may result in high seepage forces and the necessity for construction of high-quality filter zones, which for small dams are neither economical nor practicable from a construction control standpoint. Furthermore, if the core were thinner there would be more danger that it could be ruptured by differential settlement of the foundation.

"Minimum Core B" for dams on pervious foundations, as shown on figure 151, is based on consideration of seepage pressures in the foundation. This minimum core applies to dams constructed on "exposed" deep pervious foundations (Case 1), or "covered" deep pervious foundations (Case 2) when the cover is 3 feet or less. It does not apply if the pervious foundations are covered by an impervious layer more than 3 feet thick.

If a positive cutoff trench is not provided in the design, it must be anticipated that regardless of what other type of device is utilized to control seepage the loss of head through the foundation will be relatively gradual and proportional to the length of the seepage path. The minimum length of path suggested for use with deep pervious foundations to reduce the seepage pressure sufficiently is that provided by an impervious zone whose thickness at the contact of the dam with the foundation is at least 2½ times the height of the dam. This impervious zone avoids the possibility of seepage passing under the core of the dam without an appreciable loss of head due to the ineffectiveness of sheet piling, partial cutoff trench, etc., or because no such device is provided. "Minimum Core B," shown on figure 151 for a dam on a pervious foundation, meets these requirements and should be used for all Case 1—Exposed Pervious Foundations and Case 2—Covered Pervious Foundations having 3 feet or less of cover for which positive cutoff trenches are not provided.

If the covered pervious foundation has an impervious layer thickness greater than 3 feet yet less than the reservoir head, the size of core selected will depend on the designer's judgment as to how effective the natural upstream blanket will be. In general, "Minimum Core A," figure 151, should provide a sufficient thickness of core material for use with a natural impervious blanket of this depth.

If the thickness of the impervious cover is greater than the reservoir head, the foundation may be considered as "deep" and the embankment should be designed to utilize stabilizing fills as described in section 134(a) and as shown in figure 141 for saturated fine-grained foundations. "Minimum Core A" should provide sufficient core thickness for this type foundation condition.

With "Minimum Core A" centrally located, as shown in figure 151, the stability of the zoned embankment is not greatly affected by

TABLE 19.—*Recommended slopes for small zoned earthfill dams on stable foundations*

Type	Purpose	Subject to rapid drawdown [2]	Shell material classification	Core material classification [3]	Up-stream slope	Down-stream slope
Zoned with "minimum core A" [1]	Any	Not critical [4]	Rockfill, GW, GP, SW (gravelly), or SP (gravelly)	GC, GM, SC, SM, CL, ML, CH, or MH	2:1	2:1
Zoned with "maximum core" [1]	Detention or storage	No	Rockfill, GW, GP, SW (gravelly), or SP (gravelly)	GC, GM	2:1	2:1
				SC, SM	2¼:1	2¼:1
				CL, ML	2½:1	2½:1
				CH, MH	3:1	3:1
Zoned with "maximum core" [1]	Storage	Yes	Rockfill, GW, GP, SW (gravelly), or SP (gravelly)	GC, GM	2½:1	2:1
				SC, SM	2½:1	2¼:1
				CL, ML	3:1	2½:1
				CH, MH	3½:1	3:1

[1] "Minimum" and "maximum" size cores are as shown on fig. 151.
[2] "Rapid" drawdown is a drawdown rate of 6 inches or more per day following prolonged storage at high reservoir levels.
[3] OL and OH soils are not recommended for major portions of the cores of earthfill dams. Pt soils are unsuitable.
[4] Rapid drawdown will not affect the upstream slope of a zoned embankment which has a large upstream pervious shell.

the nature of the soil comprising the core. The outside slopes are governed largely by the stability of the shell material. Rock, well-graded gravels (GW), and poorly graded gravels (GP) provide suitable material for the shell. Well-graded sand (SW) and poorly graded sand (SP) are suitable if they are gravelly. For any of these shell materials, upstream and downstream slopes of 2 : 1 are stable for dams not exceeding 50 feet in height above the lowest point in the streambed, even if subject to rapid drawdown.

(5) *Embankment Slopes.*—Table 19 shows the recommended upstream and downstream slopes for small, zoned earthfill dams with "Minimum Core A" and with "maximum core." The assumption is made that the foundation is stable; if the foundation is of the saturated fine-grained type, stabilizing fills as described in section 134 should be added. Slopes of small, zoned earthfill dams with cores of intermediate size (including "Minimum Core B" for dams on pervious foundations) will fall between those given in the table for "Minimum Core A" and for the "Maximum Size" core. Where only one slope is shown for more than one soil classification, it indicates that the embankment can be constructed using any one of the soils or any combination of them.

The following example will illustrate the procedure:

Example:

Required: Upstream and downstream slopes for a zoned earthfill storage dam, 50 feet high, on a stable foundation, subject to rapid drawdown.

Given: Foundation = Shallow, exposed, and pervious.
Shell material = SW and SP, both gravelly.
Core material = CL.

Solution: Since the foundation is shallow and a positive cutoff trench can be constructed, "Minimum Core A" will be used. From table 19, select upstream and downstream slopes of 2:1.

140. *Seismic Design.*—The design and construction practices for small earth dams presented herein are considered to be adequate in areas of low seismicity and the safety factors used should preclude major damage for all but the most catastrophic earthquakes.

Although any damsite is subject to earthquake activity, the probability of having an earthquake in some regions is greater than others; this probably is generally determined by the number of previous earthquakes in that region and their intensity. In some cases, maps have been prepared which delineate certain areas having greater earthquake potential. One such "seismic risk map" is shown in figure 152; this map, adapted from Algermissen [36], uses the information collected and abstracted for approximately 28,000 earthquakes in the conterminous United States. If the designer is uncertain about the prospects of an earthquake in any area, a competent geologist or seismologist should be consulted.

After determining that the region is subject to earthquakes, the damsite should be inspected by an experienced engineering geologist to determine if faults or detrimental geologic formations exist which might affect the location of the dam, reservoir, or appurtenant structures. If active faults, unstable alluvial foundations, or the possibility of massive landslides into the reservoir exist, the damsite should be relocated. In general, the designer should assume that a dam within a seismic zone will be shaken by an earthquake.

If foundations consisting of low relative density sands and silts (sec. 130) or uniform fine-grained, cohesionless materials are encountered, serious damage may result to the structure during an earthquake and the assistance of an experienced dam designer is required. If an active fault exists at the proposed damsite, the designs proposed herein are inadequate.

Additional considerations concerning seismic design for earth dams are contained in references at the end of this chapter [31, 35, 47, 49, 50, 51, 67].

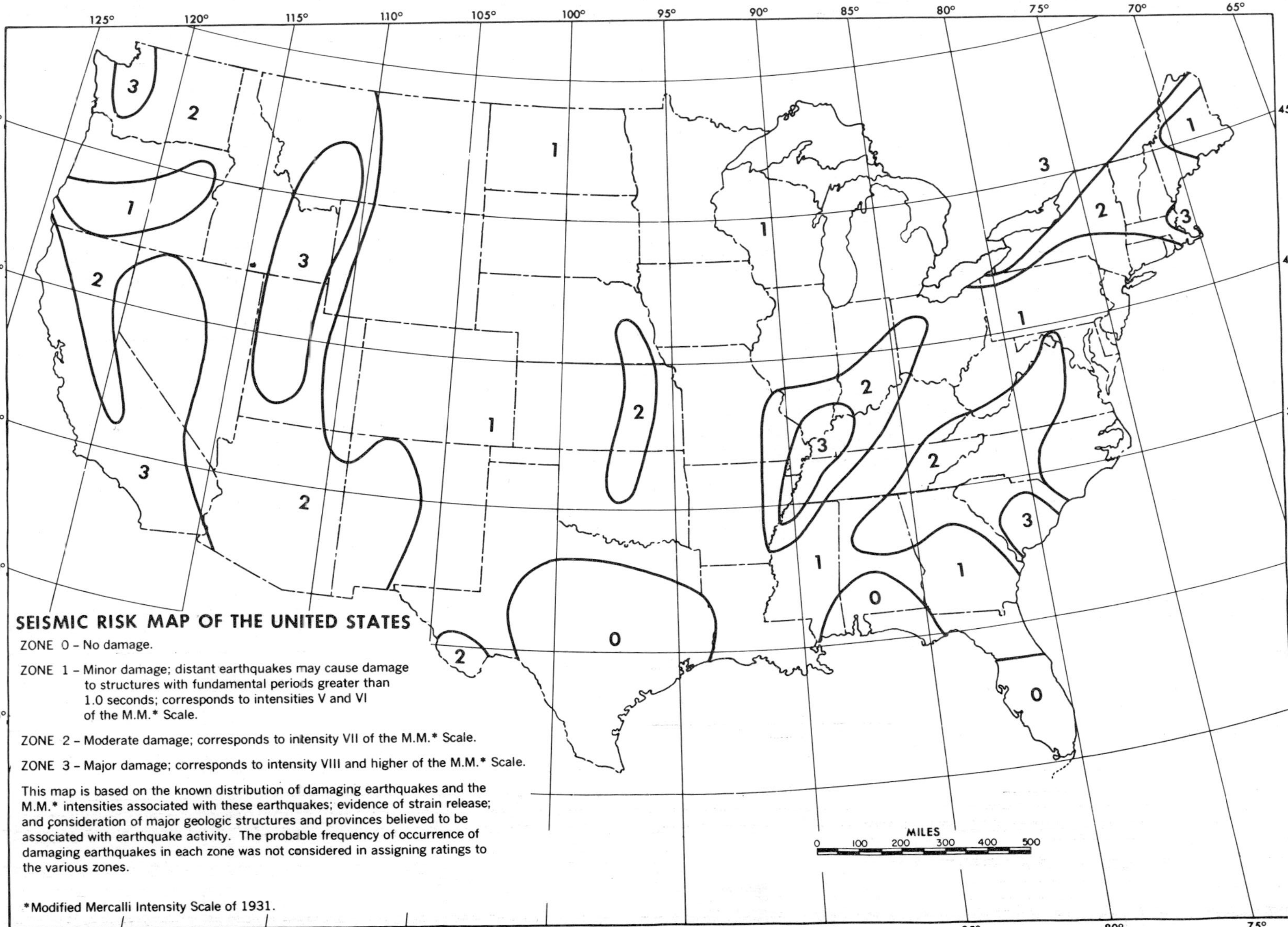

Figure 152. Seismic risk map of the conterminous United States. Adapted from Algermissen [36]. 288–D–2786.

E. EMBANKMENT DETAILS

141. *Crest Design.*—(a) *General.*—In designing the crest of an earth dam, the following items should be considered:

(1) Width.
(2) Drainage.
(3) Camber.
(4) Surfacing.
(5) Safety requirements.
(6) Zoning.

In addition, suitable parking areas should be provided for the convenience of visitors and others at the abutments of the dam, especially for a storage dam whose lake will be used for recreational purposes. A turnaround should be provided where vehicle traffic is permitted on a dam crest which dead ends into the opposite abutment.

(b) *Width.*—The crest width of an earthfill dam depends on considerations such as: (1) nature of embankment materials and minimum allowable percolation distance through the embankment at normal reservoir water level, (2) height and importance of structure, (3) possible roadway requirements, and (4) practicability of construction. A minimum crest width should be that width which will provide a safe percolation gradient through the embankment at the level of the full reservoir. Because of practical difficulties in determining this factor, the crest width is, as a rule, determined empirically and largely by precedent. The following formula is suggested for the determination of crest width for small earthfill dams:

$$w=\frac{z}{5}+10 \qquad (12)$$

where:

w=width of crest in feet, and
z=height of dam in feet above the streambed.

For ease of construction with power equipment, the minimum width should not be less than 12 feet. For many dams, the minimum crest width is determined by the requirements for the roadway which passes across the dam.

(c) *Drainage.*—Surface drainage of the crest should be provided by a crown of at least 3 inches, or by sloping the crest to drain towards the upstream slope. The latter method is preferred unless the downstream slope is protected against erosion by some type of slope protection.

(d) *Camber.*—Camber is ordinarily provided along the crest of earthfill dams to insure that the freeboard will not be diminished by foundation settlement or embankment consolidation. Selection of the amount of camber is necessarily somewhat arbitrary; it is based on the amount of foundation settlement and embankment consolidation expected for a particular dam, with the objective of providing enough extra height so that some residual camber will remain after settlement and consolidation. This residual camber also improves the appearance of the crest.

Impervious embankment materials placed at densities roughly corresponding to the Proctor laboratory maximum will consolidate appreciably when subject to overlying fill loads. It is expected that the major portion of this consolidation will take place during construction before the embankment is completed; therefore, the expected foundation settlement is the more important factor. For dams on relatively noncompressible foundations, cambers of about 1 percent of the height are commonly provided. Several feet of camber may be required for dams constructed on foundations which may be expected to settle. A method of determining foundation settlement is given in reference [46]. Straight-line equations should be used to vary the amount of camber and to make it roughly proportional to the height of the embankment. These equations are easy to use and probably correspond well with the camber lines as constructed in the field.

The additional amount of embankment material required to provide camber in the crest of an embankment is usually nominal, and the increased height of the embankment is provided by pitching the slopes near the crest of the dam, as shown in figure 153. The modifications to the section of the embankment due to the addition of camber are not taken into account in calculating embankment stability.

(e) *Surfacing.*—Some type of surfacing should be placed on top of the crest for protection against damage by wave splash and spray, rainfall runoff, wind, frost action, and traffic wear when the crest is used as a roadway. The usual treatment consists of placing a layer of selected fine rock or gravelly material to a minimum thickness of 4 inches. In the event the crest constitutes a section of a highway, the width of roadway and type of surfacing should conform to those of the highway with which it connects.

(f) *Safety Requirements.*—If the crest of the dam is to be used as a highway, cable- or beam-type guardrails are usually constructed along both shoulders of the crest. If a highway crossing is not anticipated, the crest can be delineated by posts at 25-foot intervals or, on very minor structures, by boulders placed at intervals along the crest, although in many instances no treatment is required.

(g) *Zoning.*—Incorrect zoning of materials at the crest leads to poor construction control, lost time, and possibly local failure of the crest.

In both homogeneous and zoned dams, consideration must be given to the manner in which the slope protection and bedding will intersect the crest. The thickness of the slope protection may have to be reduced by steepening the slopes near the crest to allow construction of the impervious or pervious zones or to facilitate the installation of guard posts. Care must be taken that the remaining slope protection will adequately resist the wave action. In zoned dams, it is common practice to limit the height of the core material to a few feet below the crest because impervious zones extending to the top of the dam are subject to damage by frost action, which causes loosening of the soil, and to the formation of unsightly shrinkage cracks upon drying of the soil. In homogeneous dams where frost action or shrinkage cracks may be problems, crest surfacing should always be provided. The top of the impervious core should also be maintained several feet above the maximum water surface to prevent percolation through the embankment or possible "capillary siphoning" over the top of the core material when the reservoir is full.

The pitching slopes, provided for camber, should avoid being overly steep to facilitate construction.

(h) *Typical Crest Details.*—Figure 192 shows the crest detail for Wasco Dam; 6 inches of camber were provided and a minimum top width of 14 feet was maintained for the impervious zone to insure adequate room for compaction by tamping rollers. The core material was placed 3.5 feet higher than the maximum water surface. Note also that the top of the core material is sloped toward the reservoir to facilitate drainage.

Additional crest details for various Bureau dams are shown in figure 153.

142. *Freeboard.*—Freeboard is the vertical distance between the crest of the embankment (without camber) and the reservoir water surface. The more specific term "normal freeboard" is defined as the difference in elevation between the crest of the dam and the normal reservoir water level as fixed by design requirements. The term "minimum freeboard" is defined as the difference in elevation between the crest of the dam and the maximum reservoir water surface that would result should the inflow design flood occur and should the outlet works and spillway function as planned. The difference between normal and minimum freeboard represents the surcharge head (sec. 188). If the spillway is uncontrolled, there will always be a surcharge head; if the spillway is gated, it is possible for the normal and minimum freeboards to be identical.

A distinction is made between normal and minimum freeboards because of the different requirements for freeboard if surcharge head is involved. The normal freeboard must meet the requirements for longtime storage. It must be sufficient to prevent seepage through a core which has been loosened by frost action or which has cracked due to drying out. This is of particular importance for a dam whose core is a CL or CH material when located in areas where either very cold or very hot dry climates are encountered. It must also be sufficient to prevent overtopping of the embankment by abnormal and severe wave action of rare occurrence that may result from unusual sustained winds of high velocity from a critical direction.

Minimum freeboard is provided to prevent

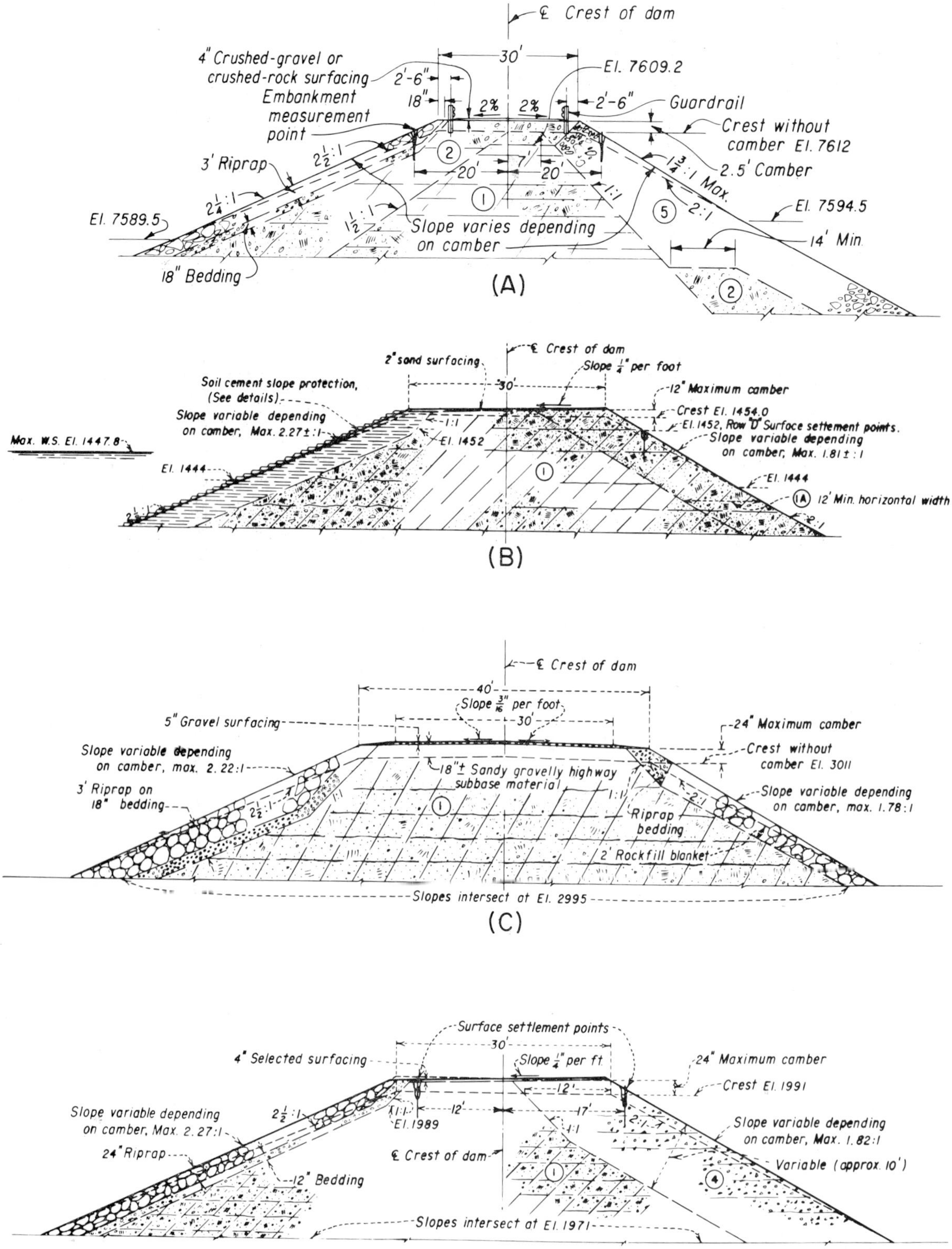

Figure 153. Examples of crest details at maximum camber. 288–D–2788.

overtopping of the embankment by wave action which may coincide with the occurrence of the inflow design flood. Minimum freeboard also provides a safety factor against many contingencies such as settlement of the dam more than the amount anticipated in selecting the camber, occurrence of an inflow flood somewhat larger than the inflow design flood, or malfunction of spillway controls or outlet works with an increase in maximum water surface above that expected. In some instances, especially where the maximum probable inflow is used as a basis for design, the minimum freeboard may be established on the assumption that the dam should not be overtopped as a result of malfunction of the controlled spillway or outlet works which would result from human or mechanical failure to open gates or valves. In such instances, allowances for wave action or other contingencies usually are not made.

The rational determination of freeboard would require a determination of the height and action of waves. The height of waves generated by winds in a reservoir depends on the wind velocity, the duration of the wind, the fetch,[5] depth of water, and the width of the reservoir. The height of the waves as they approach the upstream face of the dam may be altered by the increasing depth of water, or by the decreasing width of reservoir. Upon contact with the face of the dam, the effect of waves is influenced by the angle of the wave train with the dam, the slope of the upstream face, and the texture of the slope surface. The sloping face of an earthfill dam allows the waves to move up the inclined plane and expend part of their energy in raising the water instead of in direct force upon the face itself, as against a vertical wall. The rough surface of dumped riprap reduces wave uprush to approximately 1.5 times the height of the wave, but uprush may be considerably more for smooth surfaces such as concrete. Because there are no specific data on wave height and wave rideup the determination of freeboard requires judgment and consideration of local factors.

A summary of empirical formulas proposed for determination of wave heights is given in an American Society of Civil Engineers report [28], from which the following table was extracted:

Fetch, miles	Wind velocity, miles per hour	Wave height, feet
1	50	2.7
1	75	3.0
2.5	50	3.2
2.5	75	3.6
2.5	100	3.9
5	50	3.7
5	75	4.3
5	100	4.8
10	50	4.5
10	75	5.4
10	100	6.1

All conditions affecting exposure of the dam to the wind must be considered in selecting the maximum wind velocity. It is believed that from a geographical standpoint no locality is safe from an occurrence of winds of up to 100 miles per hour at least once during a period of many years, although a particular site may be topographically sheltered so that the reservoir is protected from sustained winds of high velocity. Under these conditions, velocities of 75 or even 50 miles per hour may be used.

For the design of small dams with riprapped slopes, it is recommended that the freeboard be sufficient to prevent overtopping of the dam due to wave rideup equal to 1.5 times the height of the wave as interpolated from the above tabulation, measured vertically from the still water level. Normal freeboard should be based on a wind velocity of 100 miles per hour, and minimum freeboard on a velocity of 50 miles per hour. Based on these assumptions and on other considerations of the purpose of freeboard, as previously discussed, the following tabulation lists the least amount recommended for both normal and minimum freeboard on

[5] Fetch is the distance over which the wind can act on a body of water. It is generally defined as the normal distance from the windward shore to the structure being designed, although the "effective" fetch may have a slightly curved path, as in the case of the wind sweeping down a winding river valley between land ridges.

riprapped earthfill dams; the design of the dam should satisfy the most critical requirement.

Fetch, miles	Normal freeboard, feet	Minimum freeboard, feet
Less than 1	4	3
1	5	4
2.5	6	5
5	8	6
10	10	7

An increase in the freeboard shown above for dams where the fetch is 2.5 miles and less may be required if the dam is located in very cold or in very hot dry climates, particularly if CL and CH soils are used for construction of the cores. It is also recommended that the amount of freeboard shown in the tabulation be increased by 50 percent if a smooth pavement is to be provided on the upstream slope.

143. *Upstream Slope Protection.*—(a) *General.*—The upstream slopes of earthfill dams must be protected against destructive wave action. In some instances, provision must be made against burrowing animals. Usual types of surface protection for the upstream slope are rock riprap, either dry-dumped or hand-placed, and concrete pavement. Other types of protection that have been used are steel facing, bituminous pavement, precast-concrete blocks, soil-cement pavement, and (on small and relatively unimportant structures) wood and sacked concrete. The upstream slope protection should extend from the crest of the dam to a safe distance below minimum water level (usually several feet). In some cases, it is advantageous to terminate the slope protection on a supporting berm but generally this need not be required.

(b) *Selection of Type of Protection.*—Experience has shown that in the majority of cases dumped riprap furnishes the best type of upstream slope protection at the lowest ultimate cost. Approximately 100 dams, located in various sections of the United States with a wide variety of climatic conditions and wave severity, were examined by the U.S. Corps of Engineers as a basis for establishing the most practical and economical means for slope protection [29]. The dams were from 5 to 50 years old and were constructed by various agencies. This survey found that:

(1) Dumped riprap failed in 5 percent of the cases where used, failures being attributed to improper size of stones.

(2) Hand-placed riprap failed in 30 percent of the cases where used, due to the usual method of single-course construction.

(3) Concrete pavement failed in 36 percent of the cases where used, due generally to the inherent deficiencies of this type of construction.

This survey substantiated the premise that dumped riprap is by far the preferable type of upstream slope protection. The excellent service rendered by dumped riprap is exemplified in the case of Cold Springs Dam, constructed by the Bureau of Reclamation. Figure 154 shows the condition of the riprap on the upstream slope of this dam after 50 years of service. The only maintenance required during that period has been the replacement of some riprap which was dislodged near the center of the dam by a particularly severe storm in 1931. Although some beaching action has taken place subsequently, it has not been severe enough to require further maintenance.

The superiority of dumped rock riprap for upstream slope protection and its low cost of maintenance compared to other types of protection have been demonstrated so convincingly that it has been considered economical to import rock from considerable distances to avoid construction of other types of slope protection for major dams. For example, the Bureau of Reclamation has imported rock from sources which required a rail haul of over 200 miles and a truck haul of 24 miles from the railhead to the dam, and the Corps of Engineers has imported rock from a distance of 170 miles.

When the nearest source of suitable rock is located far from the site, and especially when only small quantities are involved, it may be economical to use hand-placed riprap despite its higher unit cost for labor and material because a lesser thickness may be used. Hand-placed riprap is satisfactory where not exposed to heavy ice conditions, but the rock must be of better quality than the minimum suitable for dumped riprap, and placement must be such that the hand-placed riprap approaches good dry rubble in quality and appearance. It should be recognized that hand-placed riprap is not as flexible as dumped riprap, since it cannot adjust itself as well to foundation or local settlements. Consequently, hand-placed riprap should not be used where considerable settlement is expected.

Concrete paving deserves serious consideration for upstream slope protection where the use of riprap is too expensive because of high transportation costs. The success of concrete pavement as a slope protection medium depends on the evaluation of field conditions and the assumptions made as to the behavior of the embankment and the ability of the paving to resist cracking and deterioration. Concrete pavement has proved to be satisfactory in some cases under moderate wave action. An example is at McKay Dam, constructed by the Bureau of Reclamation near Pendleton, Oreg. This pavement, although exposed to severe weather conditions, is in excellent condition after more than 40 years of service, as shown in figure 155.

Where severe wave action is anticipated, concrete pavement appears practicable only when the settlement within the embankment

Figure 154. Riprap on upstream slope of an earthfill dam in excellent condition after 50 years of service. The structure is Cold Springs Dam, which forms an offstream reservoir on the Umatilla project in Oregon. 10–2194.

Figure 155. Paved upstream slope of an earthfill dam in excellent condition after 40 years of service. The structure is McKay Dam on a tributary of the Umatilla River in Oregon. 288–D–2878.

after construction will be insignificant. In comparing the cost of concrete pavement and riprap, the cost of any additional foundation measures necessary to minimize settlement and the additional freeboard required because of greater wave rideup on the smooth surface should be considered.

Other types of upstream slope protection such as precast concrete blocks, asphaltic-concrete, steel plates, and soil-cement should also be considered. Asphaltic-concrete and soil-cement can often provide economical alternatives for slope protection. Wood and sacked concrete should be used only on very minor structures and then only when the cost of a more permanent type of slope protection is prohibitive.

In this chapter, the following types of slope protection will be discussed:

(1) Dumped rock riprap.
(2) Hand-placed rock riprap.
(3) Concrete pavement.
(4) Soil-cement.

Section 161 discusses asphaltic-concrete slope protection.

(c) *Dumped Rock Riprap.*—Dumped rock riprap consists of stones or rock fragments

Figure 156. Riprap being placed on the upstream slope of Meeks Cabin Dam in Wyoming. P415–432–720.

dumped in place on the upstream slope of an embankment to protect it from wave action. The riprap is placed on a properly graded filter which may be a specially placed blanket or may be the upstream zone of a zoned embankment. Figure 156 shows riprap being placed on Meeks Cabin Dam, constructed by the Bureau of Reclamation. The riprap is feldspathic sandstone having a specific gravity of 2.58.

The efficacy of dumped rock riprap depends on the following characteristics:

(1) Quality of the rock.
(2) Weight or size of the individual pieces.
(3) Thickness of the riprap.
(4) Shape of the stones or rock fragments.
(5) Slopes of the embankment on which the riprap is placed.
(6) Stability and effectiveness of the filter on which the riprap is placed.

Rock for riprap should be hard, dense, and durable, and should be able to resist long exposure to weathering. Most of the igneous and metamorphic rocks, many of the limestones, and some of the sandstones make excellent riprap. Limestones and sandstones that have shale seams are undesirable. The suitability of rock for riprap from a quality standpoint is determined by visual inspection,

Figure 157. Displacement of riprap on a low dike by wave action. 288–D–2936.

by laboratory tests to determine the resistance to weathering and to abrasion, and by petrographic examination to determine the structure of the rock as it affects its durability. The laboratory tests are described in chapter V.

The individual pieces must be of sufficient weight to resist displacement by wave action, which is not necessarily a function of the height of the dam. It is a misconception to consider that large-size rocks are needed only on higher structures, while small-size rocks will afford ample slope protection for low fills, without regard to factors such as wind velocity, wind direction, and fetch distance. This can be demonstrated by comparing figure 157 with figure 154. Cold Springs Dam (fig. 154) is a [illegible]0-foot-high dam whose upstream slope is protected by a 24-inch layer of basalt rock with the average weight of the larger fragments probably not exceeding 100 pounds. The wave action on this reservoir is not severe, and the riprap has given satisfactory service for 60 years with relatively little maintenance required. Figure 157 shows riprap containing relatively large fragments that have been dislodged from the upstream slope of a low dike section of another dam subject to heavier wave action.

The weight or size of the individual pieces required to resist displacement by wave action may be determined theoretically by the methods given in the American Society of Civil Engineers report referred to in the discussion of freeboard requirements [28]. This method is based on the premise that the force a wave exerts on riprap stones on the face of a dam cannot be greater than that of a current flowing at a velocity equal to the velocity of the water particles of the wave. These theoretical methods are in good agreement with the experience and analysis of results obtained on a large number of earthfill dams by the Bureau of Reclamation.

TABLE 20.—*Thickness and gradation limits of riprap on 3:1 slopes*

Reservoir fetch, miles	Nominal thickness, inches	Gradation, percentage of stones of various weights, pounds [1]			
		Maximum size	40 to 50 percent greater than	50 to 60 percent from— to—	0 to 10 percent less than [2]
2.5 and less	30	2,500	1,250	75–1,250	75
More than 2.5	36	4,500	2,250	100–2,250	100

[1] Sand and rock dust shall be less than 5 percent, by weight, of the total riprap material.

[2] The percentage of this size material shall not exceed an amount which will fill the voids in larger rock.

The thickness of the riprap should be sufficient to accommodate the weight and size of stone necessary to resist wave action. The Bureau of Reclamation has found a 3-foot thickness of dumped riprap to be generally most economical and satisfactory for major dams. Lesser thicknesses are used on low dams or on dike sections where wave action will be less severe than on principal structures. Less thickness also has been specified for the upper slopes of dams whose reservoirs are largely allocated to flood control, because of the infrequent and short periods of time that the upper slopes are subject to wave action. Greater thicknesses have been specified in cases where rock having a low specific gravity (less than 2.50) was used. Table 20 shows the recommended thickness and gradation of dumped rock riprap for small dams for fetch lengths equal to or less than 2.5 miles and greater than 2.5 miles, based on theoretical

Figure 158. Completed riprap slope protection on the upstream face of Blue Mesa Dam in Colorado. P662A–427–10496NA.

considerations and the experience and practice of the Bureau of Reclamation.

The shape of the individual stones or rock fragments influences the ability of the riprap to resist displacement by wave action. Angular fragments of quarried rock tend to interlock and resist displacement better than do boulders and rounded cobbles. The values given in table 20 are for angular quarried rock. If boulders or rounded cobbles are to be used, such as shown in figure 37, a thicker layer containing larger sizes may be required, or the slope of the embankment may need to be made flatter than required for stability in order for the boulder and cobble riprap to stay in place, especially if cobbles of relatively uniform diameter are to be used.

Table 20 is for riprap thickness and gradation on 3 : 1 slopes. For 2 : 1 slopes, the 36-inch thickness and the corresponding gradation should be used.

A layer or blanket of graded gravel should be provided underneath the riprap when the compacted material of the underlying earthfill is of such gradation that there is danger that fines may be washed out through the voids in the riprap by wave action, resulting in undermining of the riprap. A blanket is usually not required if the outer zone of a zoned embankment is gravel. Blankets of crushed rock

Figure 159. Hand-placed rock riprap on Indian Creek dike, an off-stream dike for Strawberry Reservoir, part of a transmountain diversion project in Utah. 288–D–2935.

or natural gravel graded from 3/16 to 3½ inches with a thickness equal to one-half the thickness of the riprap (but not less than 12 inches) have proved satisfactory in practice. The blanket gradation may be determined more exactly by the filter criteria given in section 131(i). An example of a finished riprap slope is shown in figure 158. The structure is Blue Mesa Dam in Colorado.

(d) *Hand-Placed Rock Riprap.*—A good example of hand-placed rock riprap is shown in figure 159. The upstream slope protection is in an excellent state of preservation after 36 years of service. Hand-placed riprap consists of stones carefully laid by hand in a more or less definite pattern with a minimum amount of voids and with the top surface relatively smooth. Rounded or irregular stones lay up less satisfactorily and rapidly than stone which is roughly square; stone of a flat, stratified nature should be placed with the principal bedding planes normal to the slope. Joints should be broken as much as possible, and joint openings to the underlying fill should be avoided by carefully arranging the various sizes of stones and closing the openings with spalls or small rock fragments. However, there should be enough openings in the surface of the riprap to vent the subsurface properly.

Stone for hand-placed rock riprap must be of excellent quality. The thickness of hand-placed rock riprap should be one-half of the thickness required for dumped rock riprap but not less than 12 inches, and a filter blanket should be provided underneath the riprap if the underlying zone of the earthfill dam is not gravel.

(e) *Concrete Paving.*—If a complete history could be gathered concerning the numerous instances where concrete paving was used for the protection of the upstream slopes of small dams, it would be found that the number of failures is tremendous. Concrete paving is used on both earthfill and rockfill dams, although its performance on rockfill dams (chapter VII) has been much more satisfactory. However, the fact that some structures protected with concrete facings have withstood the test of time continues to lead engineers to use this type of construction, often without sufficient reference to other past performance records. A properly designed and constructed concrete facing is never cheap. The uncertainty and complexity of the forces which may act on a concrete paving makes it desirable to use conservative treatment whenever this type of slope protection is considered. The recommendations that follow should provide the necessary degree of conservatism, but the number of situations studied is so limited that there is no assurance that adequate consideration has been given to every type of hazard that may be encountered.

Concrete paving used for slope protection should extend from the crest of the dam to several feet below the minimum water surface. It should terminate on a berm and against a concrete curb or header which should extend at least 18 inches below the undersurface of the paving.

For dams approaching 50 feet in height, a paving thickness of 8 inches is recommended; the minimum thickness for lower dams should be 6 inches. Although concrete paving has been constructed in blocks, the generally favored method which has given the best service is to make the paving monolithic to the greatest extent possible, with every measure taken to prevent access of water and consequent development of hydrostatic pressures underneath the concrete. The good service given by the concrete pavement on the upstream slope of McKay Dam (fig. 155) is attributed to monolithic type of construction, durability of concrete, little settlement of the dam or foundation, and pervious nature of underlying fill which prevents development of hydrostatic uplift pressures even though a minor amount of cracking has occurred.

In contrast with the success of concrete paving at McKay Dam is the experience of the Bureau of Reclamation with the concrete paving at Belle Fourche Dam. In this instance monolithic construction was not used. The paving consists of 8-inch-thick blocks, 6 feet 6 inches by 5 feet, placed directly upon the impervious underlying embankment. The condition of the paving after 40 years of service is shown in figure 160. Considerable maintenance of the paving has been required through the years; a number of the blocks have been displaced and broken up by wave action and uplift forces under the slabs. Compared to the general service record of riprap or the concrete pavement in figure 155, this slope protection design cannot be deemed successful.

If monolithic construction is not possible, expansion joints should be kept to a minimum and construction joints should be spaced as widely as possible. The slab should be reinforced with bars in both directions, placed at middepth of the slab, and made continuous through the construction joints. An area of steel in each direction equal to 0.5 percent of the area of the concrete is considered good practice. Joints should be sealed with plastic fillers, and subsequent open cracks in the concrete should be grouted or sealed promptly.

Figure 160. Concrete paving blocks on the upstream slope of Belle Fourche Dam. Note deteriorated condition of this type of paving after 40 years' service in spite of considerable maintenance. Belle Fourche Dam is an earthfill structure on a tributary of the Belle Fourche River in South Dakota. 288–D–2937.

(f) *Soil-cement.*—In recent years, soil-cement as a facing material for earth dams has been found economical where suitable riprap is not available at the site. A reasonably firm foundation is required so that deformation following placement of soil-cement is not significant; however, no unusual design features need be incorporated into the embankment. Normal embankment construction procedures are used, with perhaps special care being taken to insure a minimum of embankment consolidation and foundation settlement after construction. Soil-cement slope protection used on Cheney Dam in Kansas is shown in figure 161.

The soil-cement is generally placed and compacted in stair-step horizontal layers, as shown in figure 162; this promotes maximum construction efficiency and operational effectiveness. With typical embankment slopes of 2 : 1 to 4 : 1, a horizontal layer width of 8 feet will provide minimum protective thicknesses of about 2 to 3½ feet, respectively, measured normal to the slope. Beginning at the lowest layer of soil-cement, each succeeding layer is stepped back a distance equal to the product of the compacted layer thickness in feet times the embankment slope. For example, if the slope is 2 : 1 and the compacted thickness is 6 inches, the step-back is 2 by ½, or 1 foot. The usual compacted layer thickness is 6 inches. Soil-cement layers of these dimensions can be effectively placed and compacted with standard highway equipment.

If the soil-cement facing does not begin at natural ground level, the lower portion of the embankment should be on a flatter slope than the portion protected by the soil-cement; or a berm may be provided at the lowest elevation of the facing. It is essential that the soil-cement extend below the minimum water level and above the maximum water level. The latter elevation should have a freeboard allowance of at least 1.2 times the anticipated maximum wave height, or about 5 feet, whichever is greater. The edges of the completed soil-cement layers should not be trimmed, as the rounded stair-step effect is beneficial in retarding wave runup (fig. 161).

Soil-cement can be built with a wide variety of soils. The principal criterion for determining soil type is gradation. Coarse sandy or gravelly soils containing about 10 to 25 percent material passing the No. 200 sieve are ideal (American Society for Testing and Materials Standard Sieve Series). These soils can be adequately stabilized with from 0.7 to 1.0 barrel of cement per cubic yard of compacted soil-cement. Standard compaction and placement control for soil-cement is used, and is described by Holtz and Walker [42]. If the amount of material smaller than the No. 200 sieve exceeds 35 percent, some effort to find a coarse material may be justified from a processing cost standpoint. Soils containing 50 percent or more material passing the No. 200 sieve are not recommended for use in their natural state.

Any type of portland cement meeting the requirements of the latest American Society for Testing and Materials (ASTM), American Association of State Highway Officials (AASHO), or Federal specifications may be used. Type 1, or normal portland cement, is most commonly used, as the special properties of other types of portland cement are not usually required for soil-cement construction [28]. Standard laboratory tests are necessary to verify the acceptability of the soil and to determine proper cement content, optimum moisture, and maximum density of the soil-cement.

After the soil has been classified by sieve analyses and other tests, the required cement content may be estimated. Moisture-density curves are plotted from test mixtures, using the estimated cement content and at least four moisture contents in order to determine optimum moisture and maximum density of the mixture accurately.

A number of test cylinders are prepared, using the estimated cement content and also cement contents 2 percentage points above and below the estimated content. The results of wet-dry, freeze-thaw, and weight-loss criteria will determine the cement content required. This cement content is then increased by 2 percent points for erosion resistance. If it is necessary to use a soil containing more than 50 percent fines, the cement content increase required for erosion resistance is 4 percentage points. For most soils, a total required cement content of 10 to 12 percent by compacted volume of soil-cement can be considered typical.

Compressive strength tests for soil-cement are considered supplementary to the standard soil-cement tests. Soil-cement mixtures with a compressive strength of about 450 pounds per square inch or more at 7 days will generally pass the wet-dry and freeze-thaw tests. Using cement contents of about 10 percent, 7-day compressive strengths of 500 to 1,000 pounds per square inch are common with a wide range of soils. Figure 163 shows the placement of soil-cement slope protection at Starvation Dam in Utah.

144. *Downstream Slope Protection.*—If the downstream zone of an embankment consists of rock or cobble fill, no special surface treatment of the slope is necessary. Downstream slopes of homogeneous dams or dams with outer sand and gravel zones should be protected against erosion by wind and rainfall runoff by a layer of rock, cobbles, or sod. Because of the uncertainty of obtaining adequate protection by vegetative cover at many damsites, especially in arid regions, protection by cobbles or rock is preferred, and should be used where the cost is not prohibitive. Layers 24 inches thick are easier to place, but a 12-inch-thick layer usually affords sufficient protection.

If grasses are planted, those suitable for a given locality should be selected. Figure 164 shows the native grasses which have protected

Figure 161. Soil-cement paving on the upstream slope of Cheney Dam in Kansas. The soil-cement is in excellent condition after numerous storms. P835–526–1006NA.

the downstream slope of Belle Fourche Dam of the Bureau of Reclamation from erosion for 50 years. Two drainage berms, one of which is shown in the photograph, are located on the downstream slope of this 115-foot-high dam. Usually fertilizer and uniform sprinkling of the seeded areas is necessary to promote the germination and foster the growth of grasses. Appendix G contains sample specifications for placing topsoil, planting seed, and watering the seeded area until completion of construction.

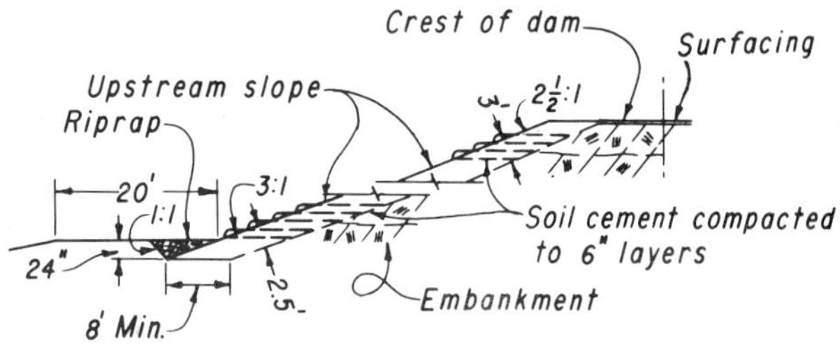

Figure 162. Typical section of soil-cement slope protection used at Cheney Dam in Kansas. 288–D–2795.

145. *Surface Drainage*.—The desirability of providing facilities to take care of surface drainage on the abutments and valley floor is often overlooked in the design of earthfill dams. The result is that, although the upstream and downstream slopes and the crest of the dam are protected against erosion, unsightly gullying takes place at the contact of the embankment with earth abutments from which vegetation has been removed during the construction operations, especially if the abutments are steep.

This condition is most likely to develop along the contact of the downstream slope with the abutments. Gullying can usually be controlled by constructing a gutter along the contact. The gutter may be formed of cobbles or rock used in the downstream surfacing. If the

Figure 163. Placement of soil-cement slope protection at Starvation Dam in Utah. P66–418–3549.

downstream slope is seeded, a concrete, asphalt, or dry-rock paved gutter should be provided. The likelihood of gullying of the abutments and gentle slopes of the valley floor by runoff from the downstream slope of the dam also should be considered; contour ditches or open drains may be needed to control erosion. Figure 165 shows typical sections of a contour ditch and an open drain.

Attention should also be given to the construction of outfall drains or channels to conduct the toe drain discharge away from the downstream toe of the embankment so that an unsightly boggy area will not be created. The need for surface drainage facilities and the most appropriate type for a particular site can usually best be determined by field examination prior to or during construction.

146. *Flared Slopes at Abutments*.—If necessary, the upstream and downstream slopes of the embankment may be flared at the abutments to provide flatter slopes for stability or to control seepage through a longer contact of the impervious zone of the dam with the abutment. If the abutment is pervious and if a positive cutoff cannot be attained economically, it may be possible to obtain the effect of an upstream blanket by flaring the embankment. The design of the transition from normal to flared slopes is governed largely by the topography of the site, the length of contact desired, and the desirability of making a gradual transition without abrupt changes for ease of construction and for appearance.

F. DESIGN EXAMPLES OF SMALL EARTHFILL DAMS

147. *General*.—The designs of 26 Bureau of Reclamation earthfill dams are discussed briefly in the next section. With only a few exceptions these dams are less than 50 feet in height above the original streambed, or are dikes of that size constructed in conjunction with larger dams. The few exceptions were included to illustrate designs for unusual conditions which were not encountered in the construction of any of the smaller dams. These exceptions, however, were chosen so that only dams less than 100 feet in height above the original streambed are involved.

These designs include small earthfill dams constructed by the Bureau of Reclamation since 1930. Two of these dams were constructed prior to 1940, 5 were constructed in the 1940's, 14 in the 1950's, and the remaining 5 in the 1960's. Many minor dikes constructed in conjunction with storage dams or canal systems were omitted because of their similarity to other designs which are shown.

The purpose of these examples is to illustrate the variety of designs that were conceived to meet widely varying conditions in foundations and in availability of construction materials. Without exception, the completed structures have given satisfactory service and no serious or costly maintenance problems have arisen. On the other hand, even the older designs do not appear to be unduly conservative and costly. It is believed that a designer of small earthfill dams can glean valuable ideas from a study of these examples.

148. *Maximum Sections*.—Figures 166 to 192, inclusive, show the maximum sections of small earthfill dams constructed by the Bureau of Reclamation. A brief explanation of each of the designs follows:

(a) *Amarillo Regulating Reservoir (fig. 166)*.—This design illustrates a small dam embankment, together with an earth lining over the entire reservoir floor to prevent excessive seepage through the floor. The earth lining continues up the upstream slope of the embankment and is covered by riprap and gravel to protect it from erosion. A toe drain is provided for drainage. The topsoil protecting the downstream slope came from stripping the structure area.

(b) *Cawker City dike (fig. 167)*.—Several different design concepts are illustrated in this example. Soil-cement was used for upstream slope protection due to the scarcity of good quality rock for riprap. The soil-cement was satisfactorily compacted into lifts thicker than the 6 inches normally considered the standard. The downstream slope was formed by topsoil,

Figure 164. Downstream slope of Belle Fourche Dam protected by grass. 288–D–2938.

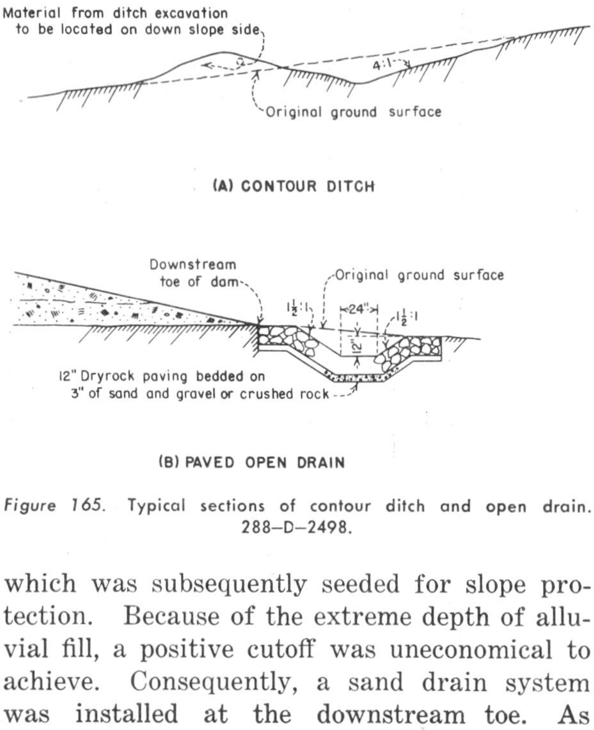

Figure 165. Typical sections of contour ditch and open drain. 288–D–2498.

which was subsequently seeded for slope protection. Because of the extreme depth of alluvial fill, a positive cutoff was uneconomical to achieve. Consequently, a sand drain system was installed at the downstream toe. As shown, a toe drain was also incorporated into the design.

(c) *Big Sandy dike (fig. 168)*.—This dike was constructed in conjunction with Big Sandy Dam (not shown), a 72-foot-high dam of conventional design with a 3 : 1 upstream slope protected by a 3-foot-thick layer of rock riprap. The design of the upstream slope of the dike represents a departure from usual design and was adopted because of the scarcity and expense of rock for riprap. Note that in the surcharge range the upstream slope of the dike is 8 : 1, which is the beaching slope of the embankment material. Freeboard above the maximum water surface is provided by a 3 : 1 slope, which is planted to make it erosion resistant to wave splash and spray to which it will be subjected only rarely. This design is suitable for upstream slope protection of a detention dam, provided the maximum water surface will not be attained more than several times during the expected life of the dam.

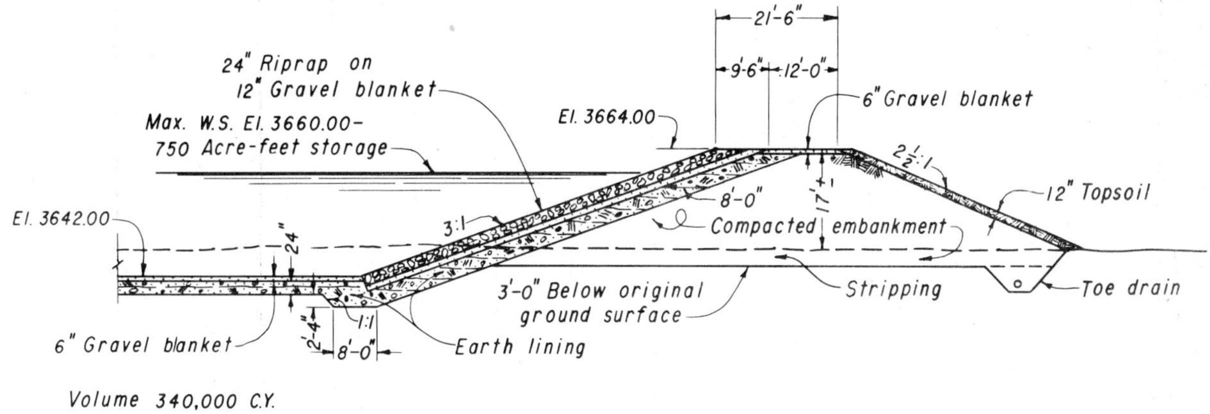

Figure 166. Amarillo Regulating Reservoir, located offstream on the Canadian River Aqueduct, Texas (completed 1965). From drawing No. 662–D–322.

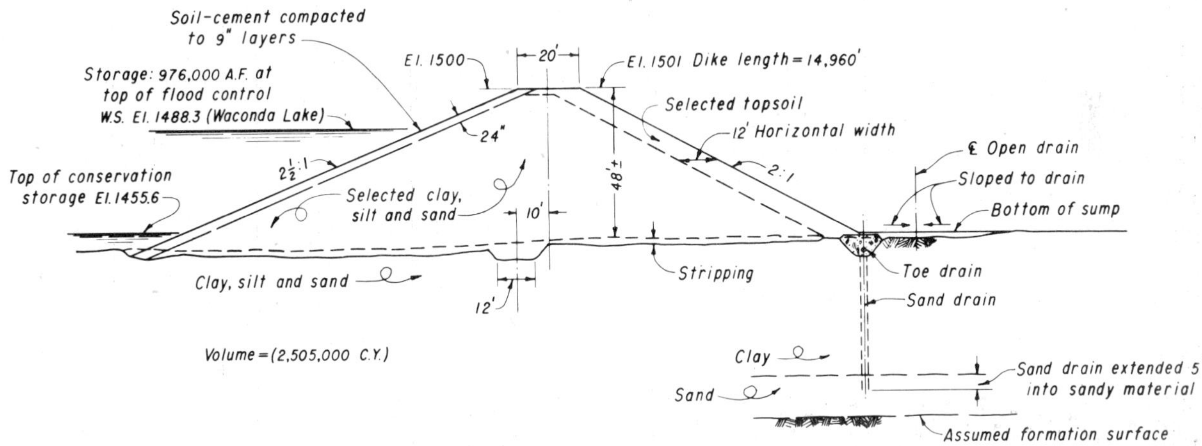

Figure 167. Cawker City dike, protecting Cawker City, Kansas, from the waters of Wacondo Lake (constructed 1967). From drawing No. 495–D–245.

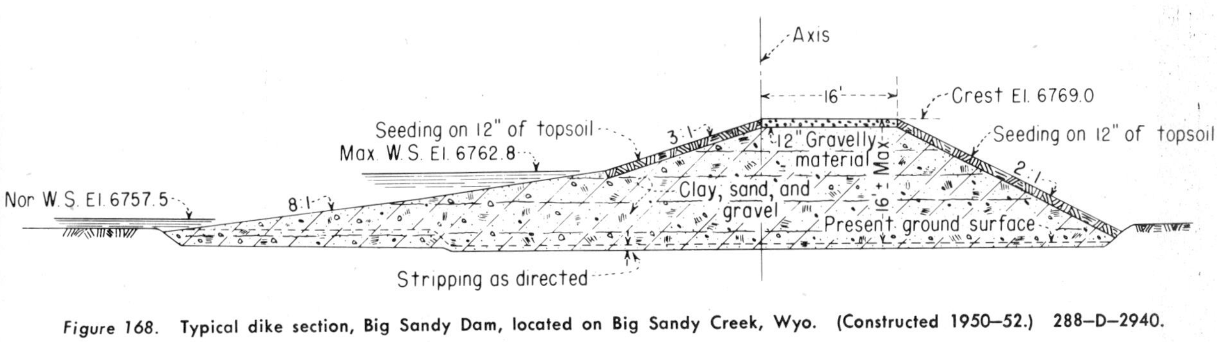

Figure 168. Typical dike section, Big Sandy Dam, located on Big Sandy Creek, Wyo. (Constructed 1950–52.) 288–D–2940.

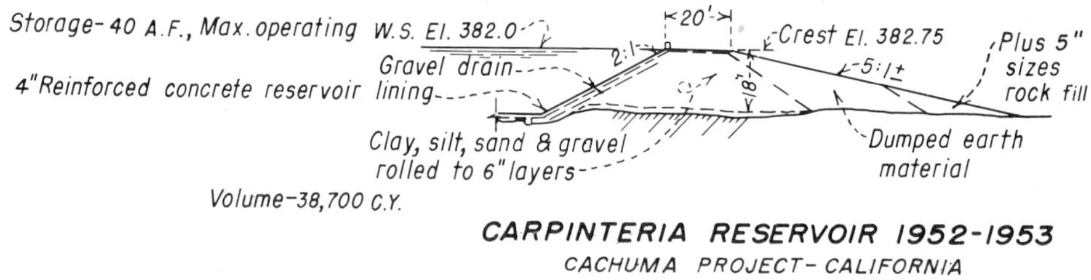

Figure 169. Carpinteria Reservoir dike, terminal reservoir of a distribution system located near Carpinteria, Calif. (Constructed 1952–53.) From drawing No. 103–D–585.

(d) *Carpinteria Reservoir dike (fig. 169).*—Carpinteria Reservoir is a small equalizing reservoir constructed on a gently sloping sidehill by excavating on the uphill side and constructing a dike on the downhill side. A section of this dike is shown in the figure. The concrete lining is provided to prevent seepage which would be serious because of the location of the reservoir with respect to improved property. The concrete lining covers all the side slopes and the bottom of the reservoir. Figure 170 shows the reservoir lining being constructed. Because this reservoir is subject to rapid drawdown, a gravel drain is placed under the side lining to prevent uplift. A pipe drainage system is also provided under the reservoir floor lining.

The embankment was constructed of mate-

Figure 170. Construction of concrete lining at Carpinteria Reservoir. SB–3262–R2.

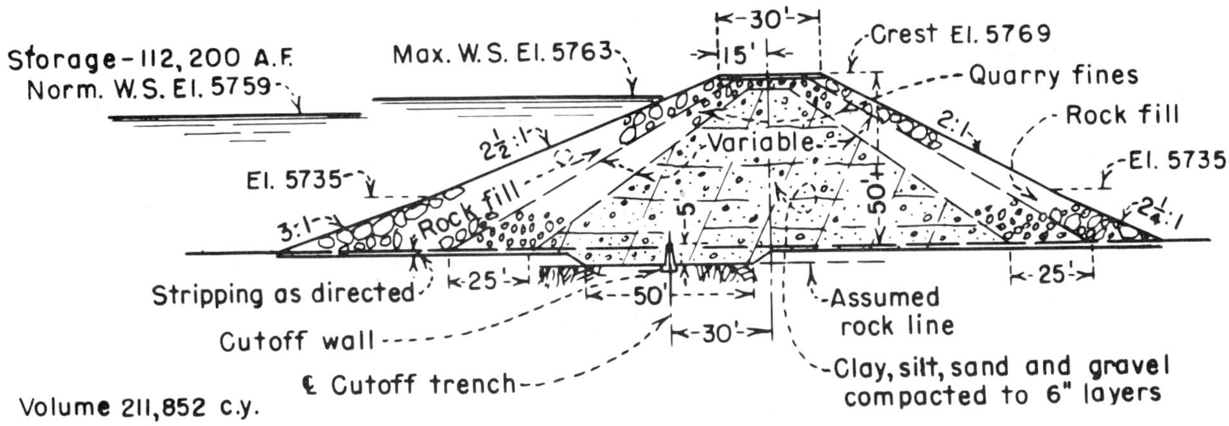

Figure 171. Carter Lake Dam No. 3, located on Dry Creek (a tributary of the Big Thompson River), Colorado. (Constructed 1950–52.) 288–D–2939.

rial from the excavation. This soil contained considerable rock fragments larger than 5 inches in diameter, and separation was required by the specifications in order to obtain an impervious zone which could be compacted satisfactorily. The dumped earth material zone provided for waste disposal of excess excavation. Oversize from the screening operation was used to construct the downstream rockfill toe.

(e) *Carter Lake Dam No. 3 (fig. 171).*—This illustrates the design of a zoned embankment consisting of an earth impervious core and rock shells. At this site there was a limited amount of material for an impervious core, no sand-gravel, but a large amount of rock which could be quarried. Quarrying operations were controlled so as to produce the desired amount and gradation of rock fragments. The rockfill consists of rock with a maximum size of 1 cubic yard and sufficient smaller rocks to fill the voids. The quarry fines zone, which acts as a filter between the rockfill and the impervious core, consists of rock fines not more than 20 percent of which pass a ¼-inch screen with no pieces larger than 8 inches. The cutoff wall shown on this section is no longer used because it is deemed unessential for structures of this size.

(f) *Crane Prairie Dam (fig. 172).*—The design of this small dam is conventional. Except for bottom width of cutoff trench, the design conforms to the recommendations given in this text.

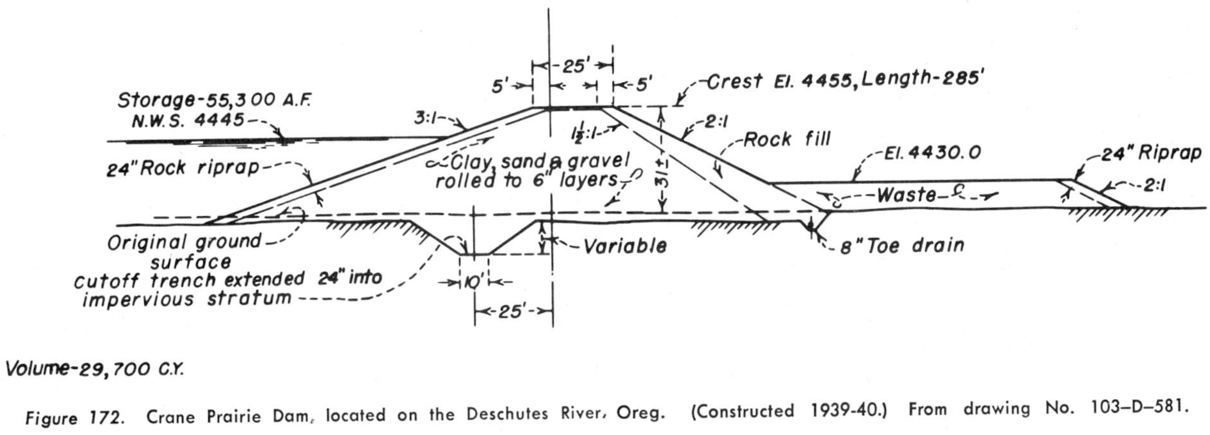

Figure 172. Crane Prairie Dam, located on the Deschutes River, Oreg. (Constructed 1939-40.) From drawing No. 103–D–581.

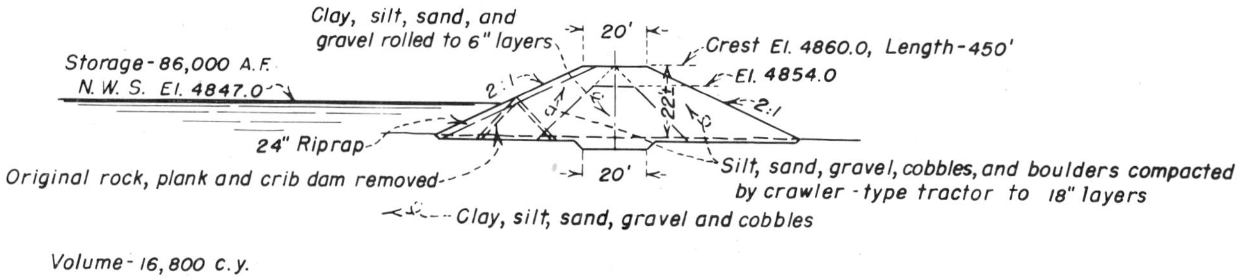

Figure 173. Crescent Lake Dam, located on Crescent Creek, Oreg. (Constructed 1954-56.) From drawing No. 103–D–586.

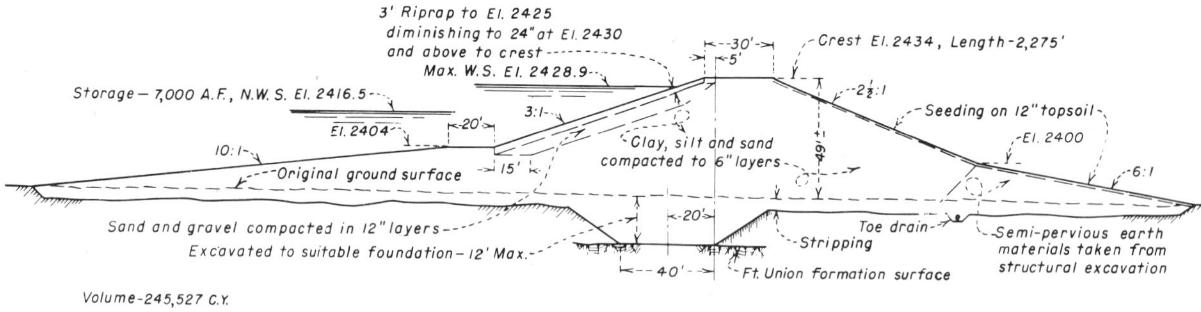

Figure 174. Dickinson Dam, located on the Heart River, N. Dak. (Constructed 1948-50.) From drawing No. 103–D–584.

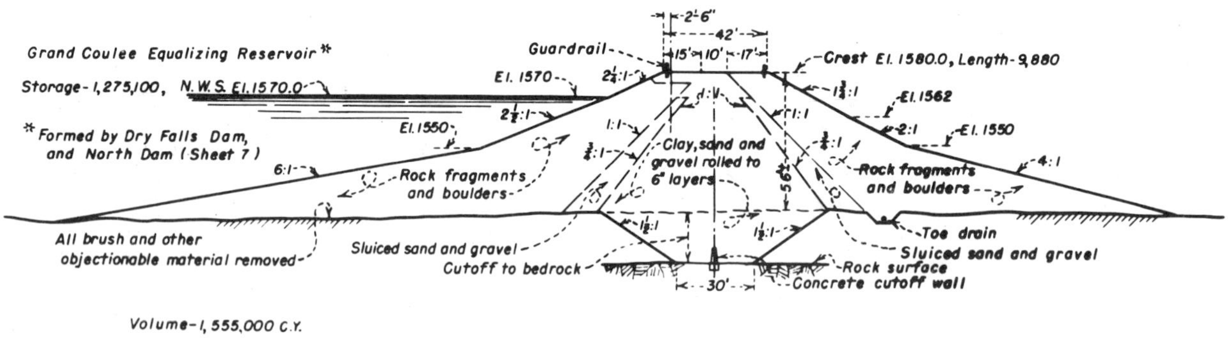

Figure 175. Dry Falls Dam, located near Coulee City, Wash. It forms the south barrier of Grand Coulee Equalizing Reservoir. (Constructed 1946-49.) From drawing No. 103–D–583.

(g) *Crescent Lake Dam (fig. 173)*.—Crescent Lake Dam is a typical modern small zoned earthfill dam. The large pervious shells allow the use of steep slopes on the embankment. Note the key trench and the modification to the zone lines near the crest of the dam to facilitate construction.

(h) *Dickinson Dam (fig. 174)*.—This dam is the modified homogeneous type. The flat slopes at the toes of the dam form stabilizing fills, which were provided because of the unconsolidated and uncemented foundation material. Note the decrease in thickness of riprap near the crest of the dam. This was done to decrease the amount of costly rock, and in view of the infrequent exposure to wave action because of the large surcharge head.

(i) *Dry Falls Dam (fig. 175)*.—This design illustrates a zoned embankment constructed on a soft foundation. This dam is unusual in that rock was used to construct the stabilizing fills formed by flattening the slopes of the dam. Usually, rock is too expensive to be used for this purpose, but in this instance it was excavated for a canal which heads at the dam. The stability afforded to the section by the heavy

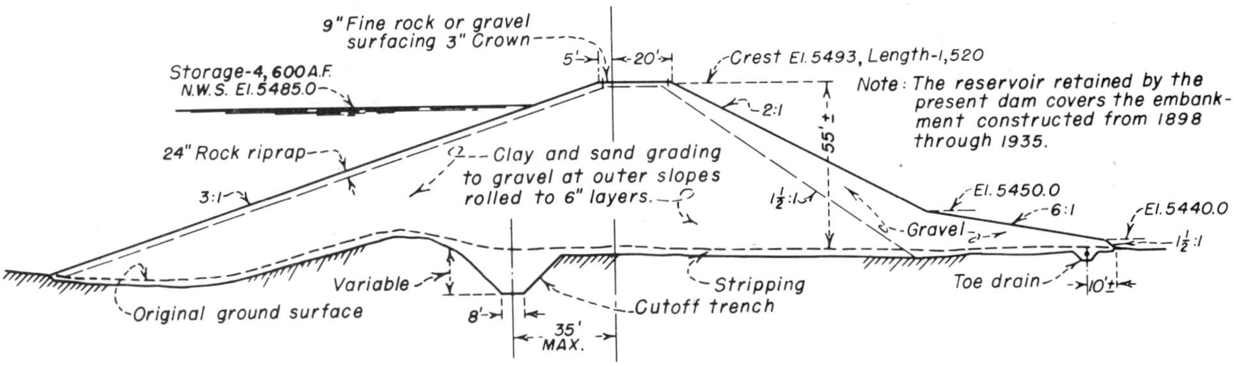

Volume-135,500 C.Y.

Figure 176. Fruitgrowers Dam, located on Alfalfa Run Wash in Colorado. Its main water supply is derived from Currant and Surface Creeks. (Constructed 1938-39.) From drawing No. 103–D–581.

rock zones permits steep slopes for the upper part of the dam. Note the filter zone provided between the core and the rockfills and the modifications made to the zoning lines near the crest of the dam to facilitate construction and preserve a sufficiently long path of percolation through impervious material. The 42-foot-wide crest was required because this dam is used for a major highway crossing.

(j) *Fruitgrowers Dam (fig. 176)*.—This is another example of a small earthfill dam whose design conforms to modern practices except for the narrow bottom width of the cutoff trench. The construction reports note that the bottom width, in general, was made 12 to 14 feet to accommodate equipment. The cutoff trench was extended to shale.

(k) *Howard Prairie Dam (fig. 177)*.—Although this dam is higher than those within the scope of this text, it is included herein as an example of a zoned embankment with a relatively thin impervious core and with heavy rockfill supporting zones. The overburden penetrated by the cutoff trench consisted of topsoil and sand-gravel. Note the transition zones between the impervious core and the rockfill zones, and the modifications made to the zone lines near the crest of the dam to facilitate construction.

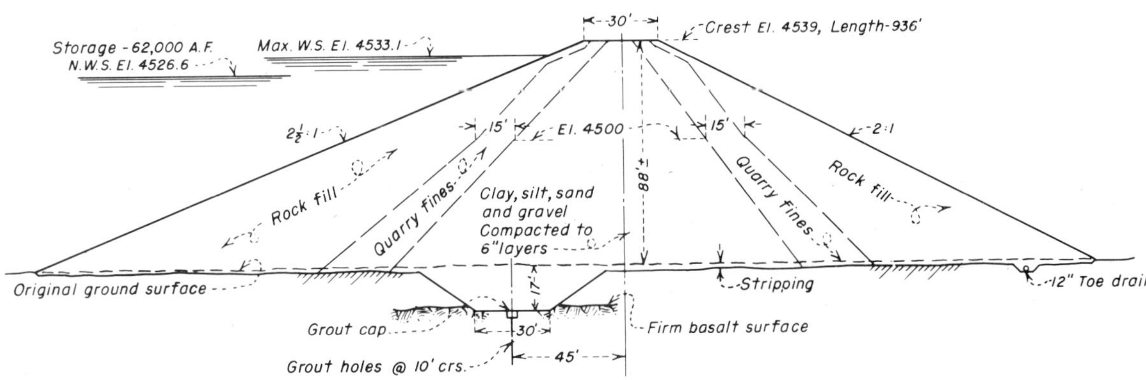

Volume - 406,000 C.Y.

Figure 177. Howard Prairie Dam, located on Beaver Creek (a tributary of Jenny Creek), Oreg. (Constructed 1957–59.) From drawing No. 103–D–587.

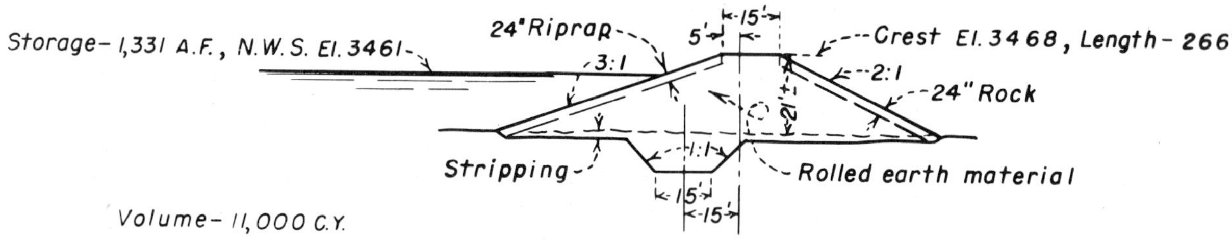

Figure 178. Lion Lake dikes, constructed for a water supply reservoir for Hungry Horse Dam Government camp, Mont. (Constructed 1947.) From drawing No. 103–D–583.

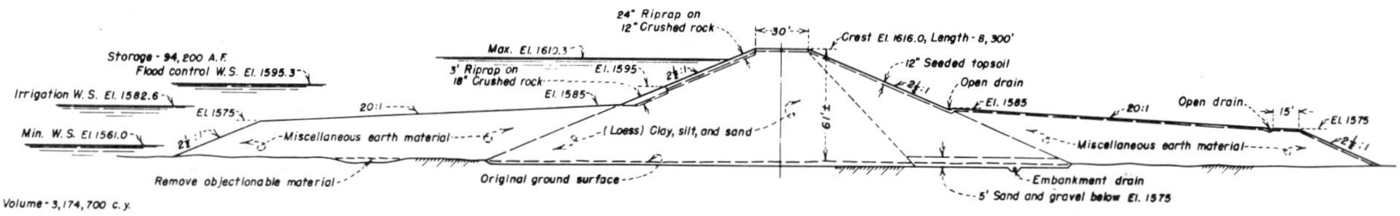

Figure 179. Lovewell Dam, located on the Republican River (offstream from White Rock Creek), Kans. (Constructed 1954–57.) From drawing No. 103–D–586.

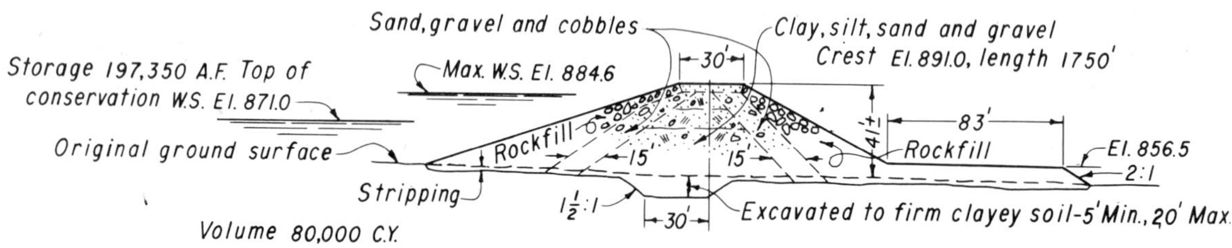

Figure 180. Eklutna Dam, Alaska, built to replace a former dam damaged by the 1964 Alaskan earthquake. (Completed 1965.) From drawing No. 783–D–639.

(l) *Lion Lake dikes (fig. 178)*.—This is illustrative of a very small embankment constructed to impound a water supply reservoir. The trench shown is a relatively deep key extending into glacial deposits of considerable depth.

(m) *Lovewell Dam (fig. 179)*.—Although this dam is somewhat higher than those within the scope of this text, it is included herein to illustrate the use of stabilizing fills on an extremely soft clay foundation. Note also how a minimum amount of riprap is utilized in this design. The 20 : 1 slope of the upstream stabilizing fill does not require protection. Only a minor amount of erosion is expected on the upstream 2½ : 1 slope of the stabilizing fill because the extremely short reservoir fetch below elevation 1575 will produce little wave action. Minor erosion of this extensive stabilizing fill will not be of consequence.

(n) *Eklutna Dam (fig. 180)*.—This dam, located in an area of high seismic risk, embodies a large downstream rockfill zone for stability. In addition, special care was taken to insure that the cutoff trench was excavated to firm clayey soil.

(o) *Lower Two Medicine Dam (fig. 181)*.—This dam replaces a previous dam which was overtopped in 1964. A large concrete overflow section now adequately handles flood flows. The embankment is of standard design except for the riprap and bedding at the downstream toe for protection against erosion by anticipated high tailwater. The downstream drainage was provided in accordance with section 131(i). The key trench shown was excavated through weathered, low-density alluvium down to firm material.

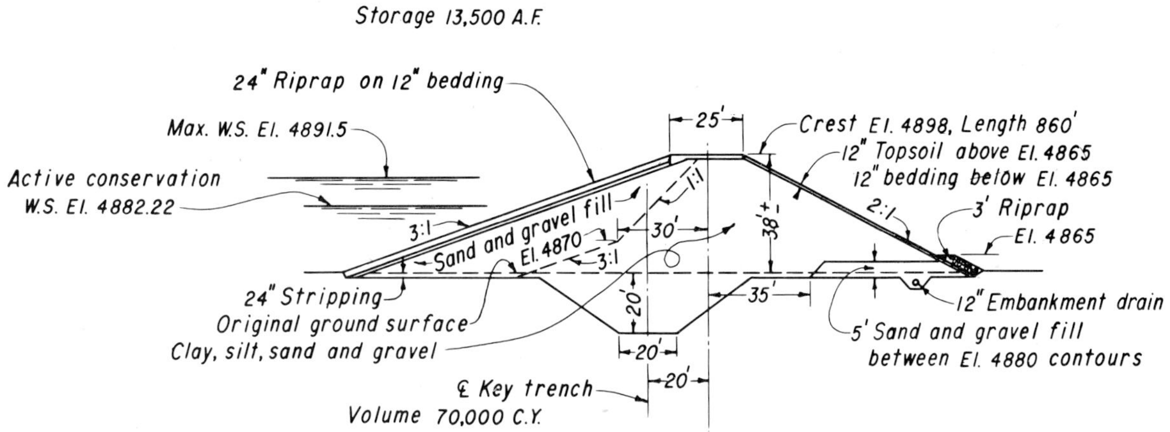

Figure 181. Lower Two Medicine Dam, Montana. This structure replaces a dam which was overtopped in an unprecedented flood. (Completed 1967.) From drawing No. 2–D–24.

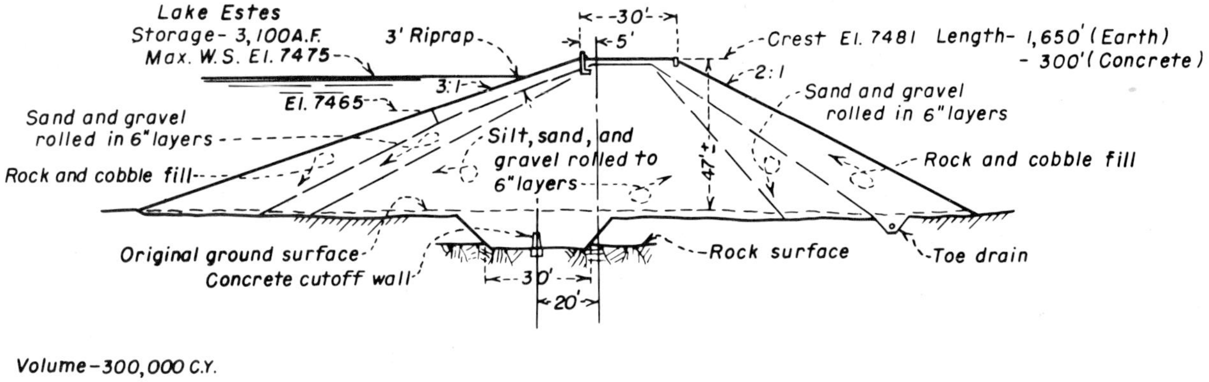

Figure 182. Olympus Dam, located on the Big Thompson River, Colo. (Constructed 1947-49.) From drawing No. 103–D–583.

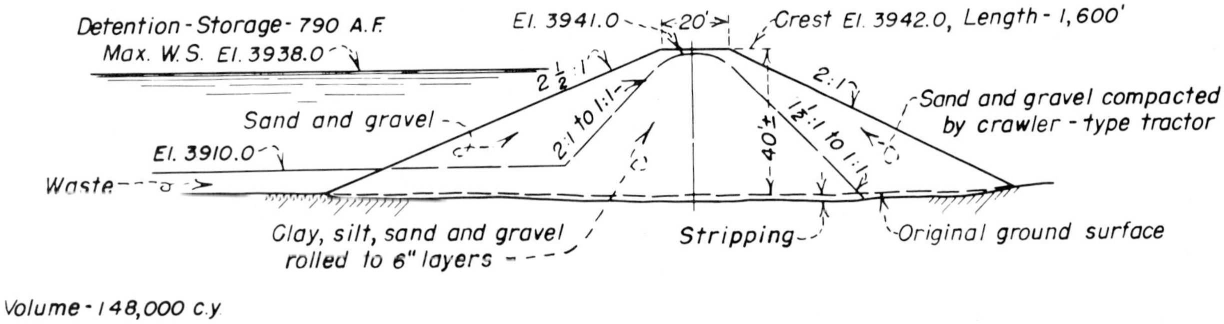

Figure 183. Picacho North Dam, a detention dam located on the North Branch of Picacho Arroyo, N. Mex. (Constructed 1953-54.) From drawing No. 103–D–586.

(p) *Olympus Dam (fig. 182)*.—This design is an example of an earthfill dam with multiple zones. Note the upstream slope protection; the riprap is used only on the upper portion of the slope where heavy wave action is to be expected. The lower portion of the slope is protected by rock and cobble fill because the water level will rarely be lowered into this range.

(q) *Picacho North Dam (fig. 183)*.—This is a detention dam which has no permanent storage pool. It is constructed on a stratified pervious foundation. The impervious zone of the dam was extended to the upstream toe and was made continuous with waste placed upstream from the dam in order to increase the path of percolation through the foundation. To facili-

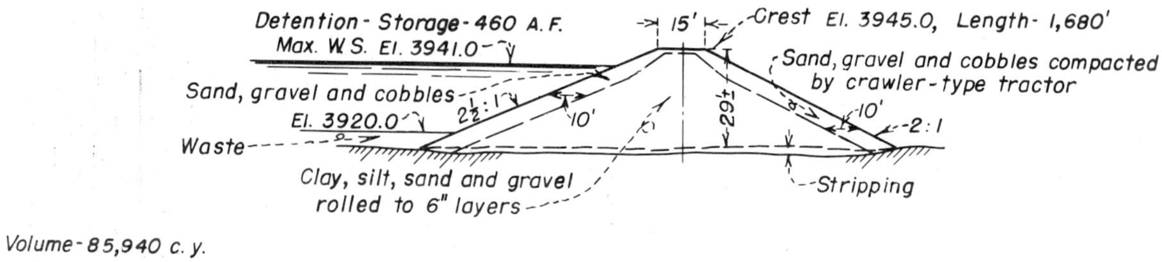

Figure 184. Picacho South Dam, a detention dam located on the South Branch of Picacho Arroyo, N. Mex. (Constructed 1953–54.) From drawing No. 103–D–586.

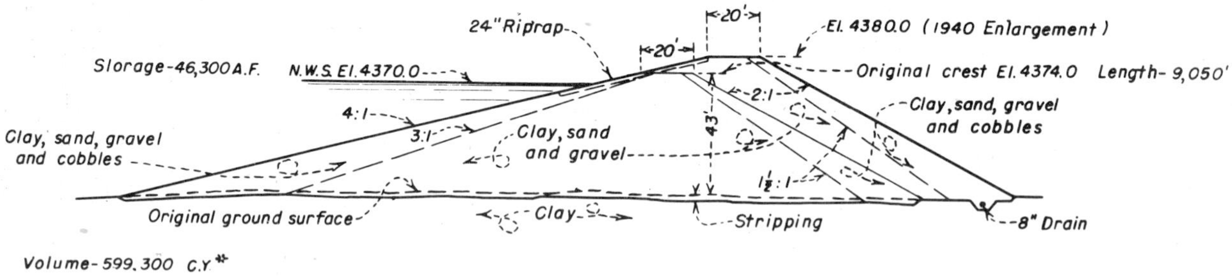

Figure 185. Pishkun dikes, located offstream from the Sun River, Mont. (Constructed 1930–31.) From drawing No. 103–D–580.

tate use of available materials from stratified deposits, the design of the embankment was based on the impervious core varying between the slope limits shown.

(r) *Picacho South Dam (fig. 184).*—This dam is also a detention type dam with no permanent storage pool. At this site the foundation was relatively impervious; only sufficient pervious material for slope protection of the embankment could be found in the vicinity. A 10-foot-wide sand-gravel-cobble zone was specified on both the upstream and downstream slopes to facilitate compaction. Riprap was not considered necessary as the outlet capacity is sufficient to evacuate the reservoir within a period of a few days time.

(s) *Pishkun dikes (fig. 185).*—This is an interesting design in that it illustrates how a 43-foot-high dam was raised 6 feet, ten years after completion of the initial construction. This embankment is essentially homogeneous; a small downstream pervious zone was provided in the original design to contain the more pervious materials found in the borrow pit. Note the sparing use of costly riprap; it is provided only in the operating range of the reservoir where wave action will be most severe.

(t) *Lubbock Regulating Reservoir (fig. 186).*—Soil-cement provides the slope protection for this small regulating reservoir located on the Canadian River Aqueduct near Lubbock, Tex. Soil-cement on an earth lining is also provided to prevent excessive seepage through the reservoir floor. Topsoil to protect the downstream slope came from stripping of the reservoir area.

(u) *Shadow Mountain Dam (fig. 187).*—This dam has a pervious glacial foundation. The design provides a partial cutoff trench and a flat upstream slope which functions as a blanket to reduce seepage, and a drainage blanket in the downstream portion of the dam to control seepage uplift.

(v) *Soda Lake dike (fig. 188).*—In all respects this is a well-designed small dam by the standards given in this text. The filter zones between the impervious core and the outer rockfill zones were made 12 feet wide, and the zone lines were modified near the crest of the dam to facilitate construction.

(w) *Stubblefield Dam (fig. 189).*—The design is typical of a homogeneous dam modified by a horizontal drainage blanket meeting filter requirements and a toe drain to lower the phre-

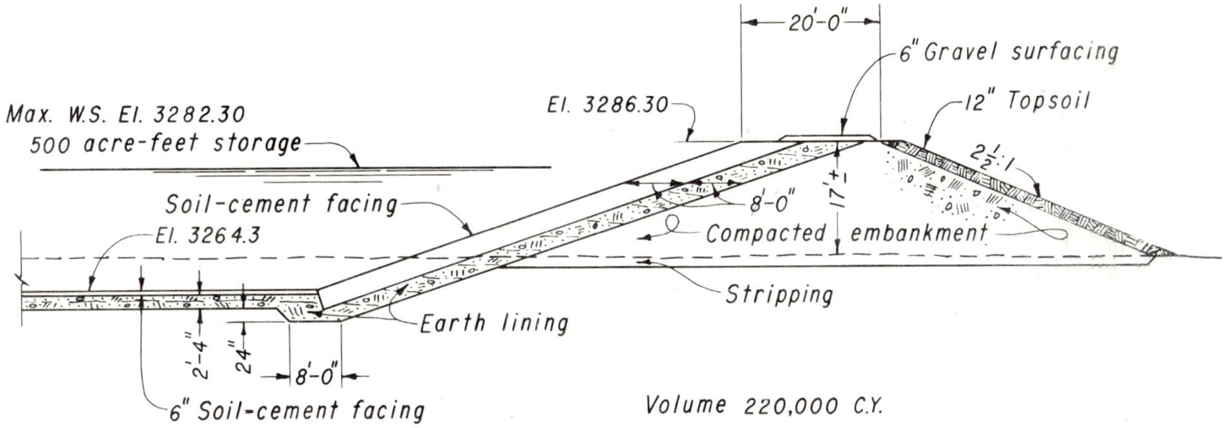

Figure 186. Lubbock Regulating Reservoir, located offstream near Lubbock, Texas. (Completed 1966.) From drawing No. 662–D–409.

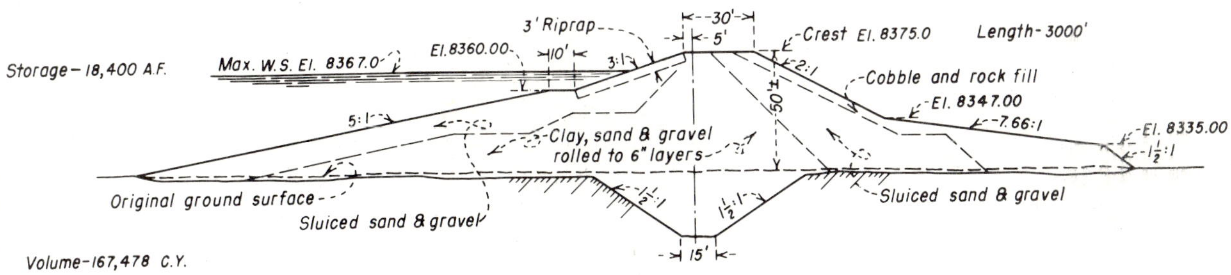

Figure 187. Shadow Mountain Dam, located on the Colorado River, Colo. (Constructed 1943–46.) From drawing No. 103–D–582.

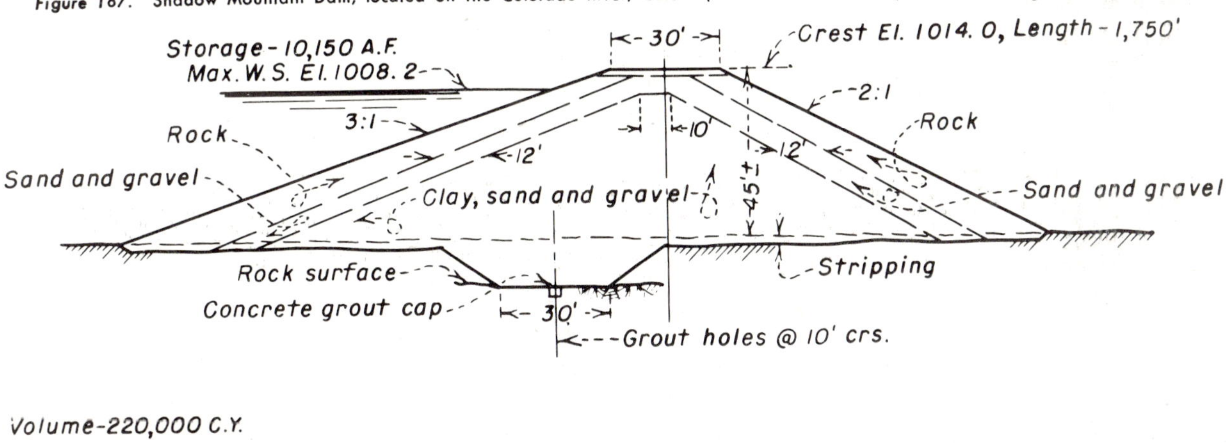

Figure 188. Soda Lake dike, located offstream from the Columbia River, Wash. (Constructed 1950–52.) From drawing No. 103–D–584.

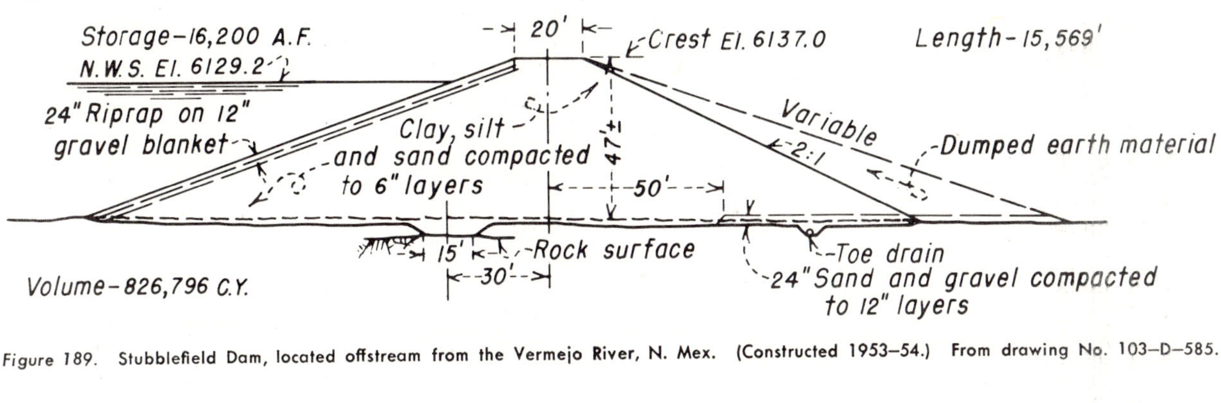

Figure 189. Stubblefield Dam, located offstream from the Vermejo River, N. Mex. (Constructed 1953–54.) From drawing No. 103–D–585.

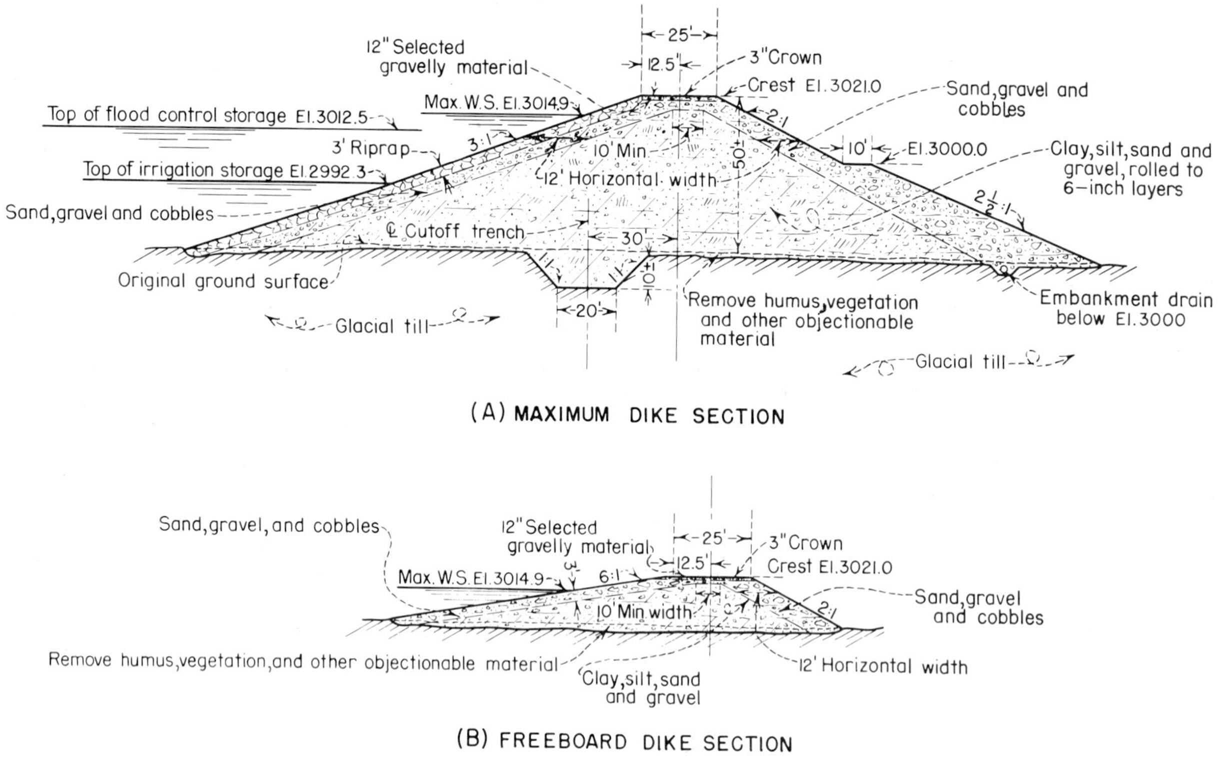

Figure 190. Dike sections of Tiber Dam, located on the Marias River, Mont. (Constructed 1953–56.) 288–D–2501.

atic line in the downstream portion of the embankment. The dumped earth shown outside of the 2 : 1 downstream slope was for disposal of waste material. It was considered that disposal of waste material (including organic material) in this manner not only would flatten the downstream slope but would also develop a vegetative cover which would be adequate for slope protection.

(x) *Tiber dike (fig. 190).*—The upper view of the illustration shows the maximum section of Tiber dike, which is conventional in design.

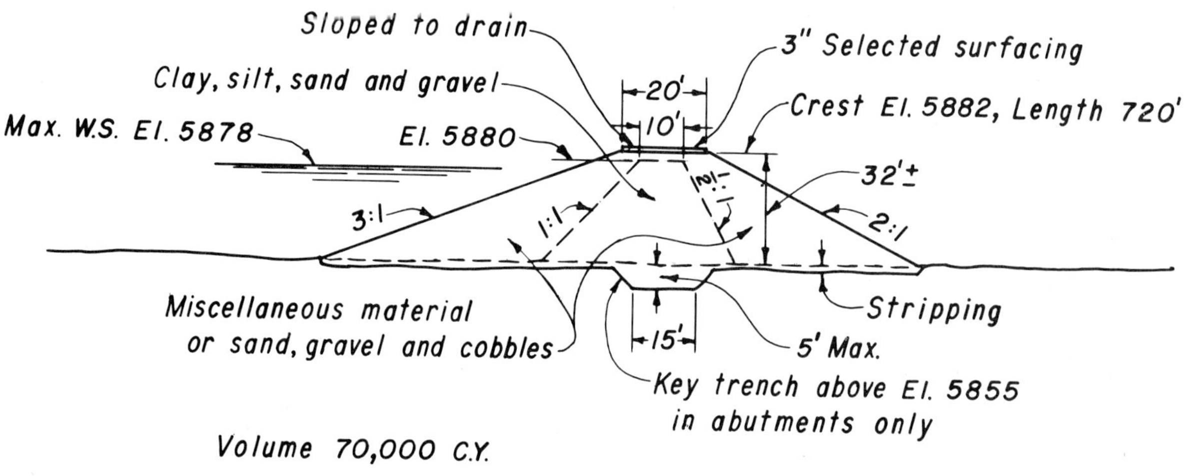

Figure 191. Sheep Creek Barrier Dam, a sediment control dam in southern Utah. (Constructed 1956.) From drawing No. 788–D–12.

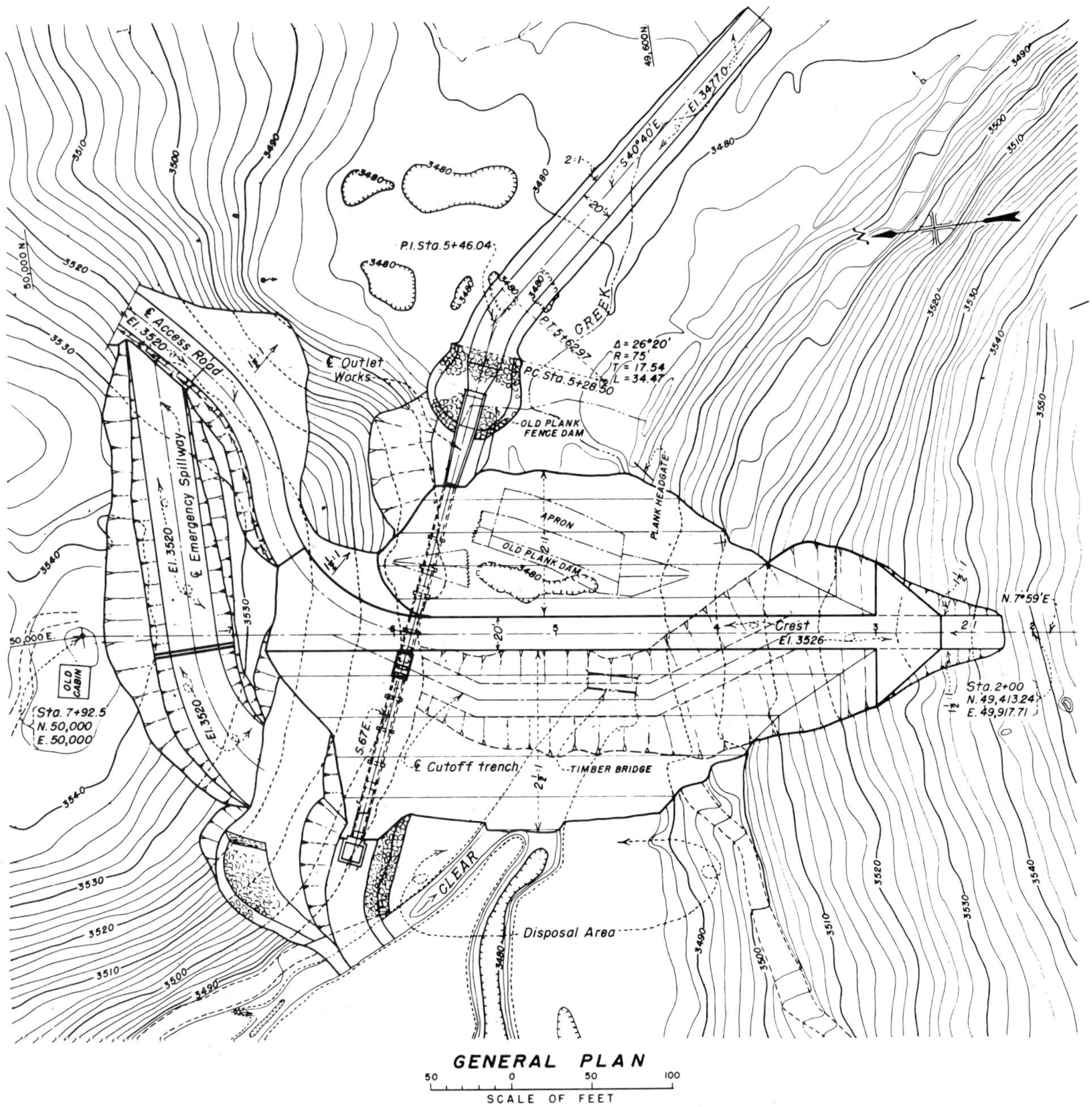

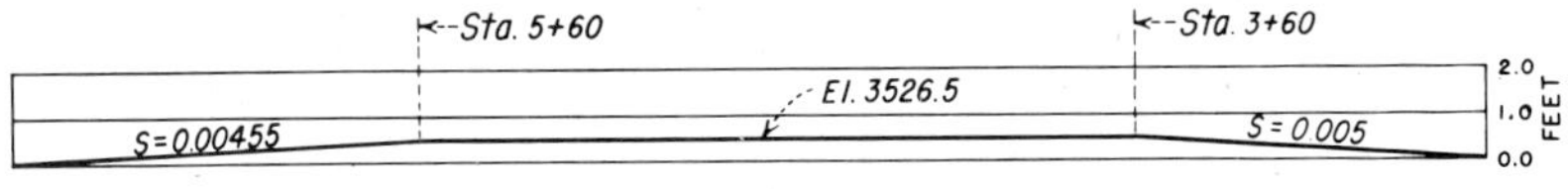

DIAGRAM FOR CAMBER ON CREST OF DAM

Figure 192. General plan and sections, Wasco Dam, located on Clear Creek (a tributary of White River), Oregon. (Completed 1958.) From drawing No. 350–D–4. (Sheet 1 of 2.)

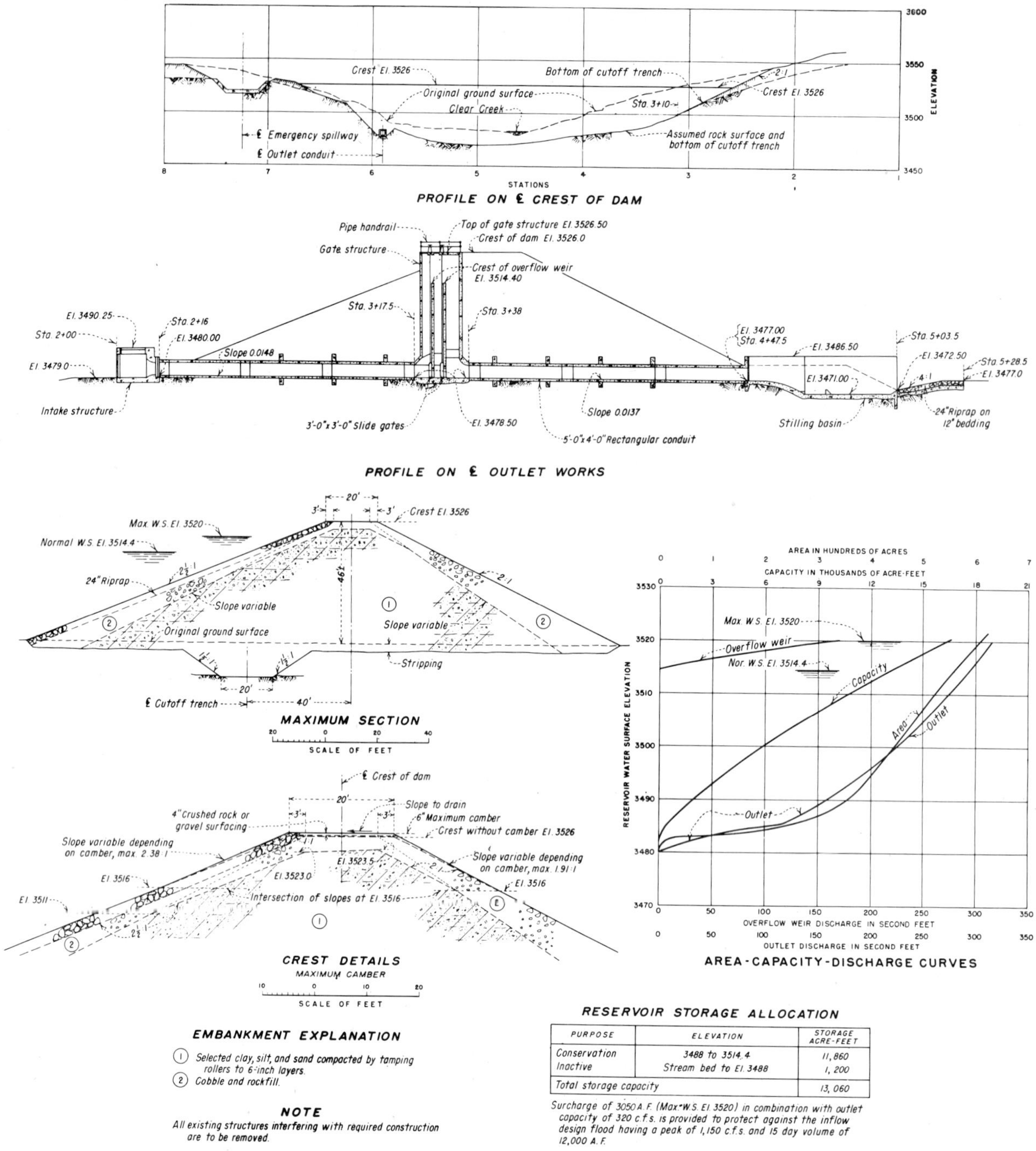

PURPOSE	ELEVATION	STORAGE ACRE-FEET
Conservation	3488 to 3514.4	11,860
Inactive	Stream bed to El. 3488	1,200
Total storage capacity		13,060

Figure 192. General plan and sections, Wasco Dam, located on Clear Creek (a tributary of White River), Oregon. (Completed 1958.) From drawing No. 350–D–4. (Sheet 2 of 2.)

A key trench 10 feet deep was excavated into the glacial till foundation, and the upstream slope is protected by riprap. The lower view shows modifications made to the dike where the original ground is above elevation 2995, which is above the top of the irrigation storage at elevation 2992.3. Because the slope will be subjected to wave action only at time of flood, it was decided that adequate protection would be provided by a 3-foot-thick layer of compacted sand, gravel, and cobbles (maximum size 10 inches) if the slope were flattened to 6 : 1. This modification was economical because of the high cost of riprap. It illustrates design flexibility in achieving the most economical structure.

(y) *Sheep Creek Barrier Dam (fig. 191).*—This dam was constructed to form a barrier to flows of flood water heavily laden with sediment. The sediment will be deposited in the reservoir area, thus clearing up the stream and also helping to reduce the rate of erosion upstream by reducing the hydraulic gradient. Since the reservoir pool is designed to fill with sediment, no upstream slope protection was provided and only a minimum key trench was constructed.

(z) *Wasco Dam (fig. 192).*—Wasco Dam is used as an example throughout this text.[6] It is not unusual in design and it may be considered typical of small zoned earthfill dams constructed on pervious shallow foundations which can economically be cut off by open trench methods. The cobble and rockfill zones utilize oversize rock removed from impervious soil and rock fragments from excavations for appurtenant structures. A toe drain was not used in the design because the foundation overburden is definitely pervious.

[6] The materials distribution chart for Wasco Dam is shown on fig. 150.

G. BIBLIOGRAPHY

149. *Bibliography.*

[1] Schuyler, J. D., "Reservoirs for Irrigation," John Wiley & Sons, Inc., New York, N.Y., 1905.

[2] Justin, J. D., "Earth Dams Projects," John Wiley & Sons, Inc., New York, N.Y., 1932.

[3] Bassell, Burr, "Earth Dams," Engineering News Publishing Co., New York, N.Y., 1907.

[4] Justin, J. D., Hinds, Julian, and Creager, W. P., "Engineering for Dams," John Wiley & Sons, Inc., New York, N.Y., vol. III, p. 675.

[5] Task Committees on Cement and Clay Grouting, "Symposium on Cement and Clay Grouting of Foundations," ASCE Proceedings, vol. 84, February 1958, Journal of the Soil Mechanics and Foundations Division, No. SM1, papers No. 1544 to 1552.

[6] Gibbs, H. J., and Holtz, W. G., "Research on Determining the Density of Sands by Spoon Penetration Testing," Proceedings of the Fourth International Conference on Soil Mechanics and Foundation Engineering, London, 1957, vol. I, p. 38 (adapted from fig. 7).

[7] Terzaghi, Karl, and Peck, R. B., "Soil Mechanics in Engineering Practice," John Wiley & Sons, Inc., New York, N.Y., 1948, p. 230.

[8] Rice, O. L., and Arthur, H. G., "The Most Recent Methods Developed to Avoid Piping or Blow-Outs in Dams," Third Congress on Large Dams, Stockholm, Sweden, 1948, vol. II, Question No. 10, R. 49.

[9] Casagrande, Arthur, "Seepage Through Dams," Journal New England Water Works Association, June 1937, p. 131.

[10] Taylor, D. W., "Fundamentals of Soil Mechanics," John Wiley & Sons, Inc., New York, N.Y., 1948, p. 156.

[11] Creager, W. P., Justin, J. D., and Hinds, Julian, "Engineering for Dams," John Wiley & Sons, Inc., New York, N.Y., vols. I and III, 1945.

[12] Justin, J. D., Hinds, Julian, and Creager, W. P., "Engineering for Dams," John Wiley & Sons, Inc., New York, N.Y., vol. III, p. 695.

[13] "Mixed-in-Place Piles," Special Report No. 109, Intrusion-Prepakt, Inc., Cleveland, Ohio, 1968.

[14] "Bibliography on Chemical Grouting," Third Progress Report—Committee on Grouting, ASCE Proceedings, vol. 92, November 1966, Journal of the Soil Mechanics and Foundations Division, No. SM6, p. 39.

[15] Bennett, P. T., "The Effect of Blankets on Seepage Through Pervious Foundations," Trans. ASCE, vol. III, 1946, p. 215.

[16] Terzaghi, Karl, and Peck, R. B., "Soil Mechanics in Engineering Practice," John Wiley & Sons, Inc., New York, N.Y., 1967, p. 164.

[17] Bertram, G. E., "An Experimental Investigation of Protective Filters," Graduate School of Engineering, Harvard University, Soil Mechanics Series No. 7, January 1940.

[18] "Soil Mechanics Fact Finding Survey; Seepage Studies; Progress Report," Technical Memorandum No. 175–1, U.S. Waterways Experiment Station, Vicksburg, Miss., March 1, 1941.

[19] "The Use of Laboratory Tests to Develop Design Criteria for Protective Filters," Earth Laboratory Report No. EM–425, Bureau of Reclamation, June 20, 1955.

[20] Middlebrooks, T. A., and Jervis, W. H., "Relief Wells for Dams and Levees," Trans. ASCE, vol. 112, 1947, p. 1321.

[21] Turnbull, W. J., and Mansur, C. I., "Relief Well Systems for Dams and Levees," Trans. ASCE, vol. 119, 1954, p. 842.

[22] Mansur, C. I., and Kaufman, R. I., "Control of Underseepage, Mississippi River Levees, St. Louis District," ASCE Proceedings, vol. 82, January 1956, Journal of the Soil Mechanics and Foundations Division, No. SM1, paper No. 864.

[23] Clevenger, W. A., "Experiences With Loess as a Foundation Material," ASCE Proceedings, vol. 82, July 1956, Journal of the Soil Mechanics and Foundations Division, No. SM3, paper No. 1025.

[24] Terzaghi, Karl, "Simple Tests Determine Hydrostatic Uplift," Engineering News-Record, June 18, 1936, p. 872.

[25] Timoshenko, S., and Goodier, J. N., "Theory of Elasticity," McGraw-Hill, New York, N.Y., 1951, p. 14.

[26] Hilf, J. W., "Estimating Construction Pore Pressures in Rolled Earth Dams," Proceedings of the Second Conference on Soil Mechanics and Foundations Engineering, Rotterdam, June 1948, vol. III, p. 234.

[27] May, D. R., "Application of the Planimeter to the Swedish Method of Analyzing the Stability of Earth Slopes," Transactions, Second Congress on Large Dams, Washington, D.C., pt. I of app. A, vol. IV, 1936, p. 540.

[28] "Review of Slope Protection Methods," Subcommittee on Slope Protection, Soil Mechanics and Foundations Division, Proceedings ASCE, vol. 74, June 1948.

[29] "Slope Protection for Earth Dams," Preliminary Report, U.S. Corps of Engineers, Waterways Experiment Station, Vicksburg, Miss., March 1949.

[30] Cedergren, H. R., "Seepage, Drainage, and Flow Nets," John Wiley & Sons, Inc., New York, N.Y., 1967, p. 148.

[31] Woodward, Clyde, Sherard & Associates, "A Study of the Influence of the Earthquake Hazard on the Design of Embankment Dams." A report prepared for the State of California Department of Water Resources, July 1966.

[32] Gibbs, H. J., and Holland, W. Y., "Petrographic and Engineering Properties of Loess," Monograph No. 28, Bureau of Reclamation, November 1960.

[33] Cedergren, H. R., "Seepage, Drainage, and Flow Nets," John Wiley & Sons, Inc., New York, N. Y., 1967, p. 110.

[34] Gardner, W. I., "Dams and Reservoirs in Pleistocene—Eolian Deposit Terrane of Nebraska and Kansas," Symposium Reservoir Leakage and Ground Water Control, Ass. Eng. Geol. Nat. Meeting, Seattle, Wash., 1968.

[35] Sherard, J. L., Woodward, R. J., Gizienski, S. F., and Clevenger, W. A., "Earth and Earth-Rock Dams," John Wiley & Sons, Inc., New York, N.Y., 1963.

[36] Algermissen, S. T., "Seismic Risk Studies in the United States," Proceedings of the Fourth World Conference on Earthquake Engineering, Chilean Association for Seismology and Earthquake Engineering, Santiago, Chile, 1969; also reprinted by U.S. Department of Commerce, ESSA, Coast and Geodetic Survey, 1969, 20 pp.

[37] "Earthquake Activity in Western United States," Dams Branch Report No. DD–8, Bureau of Reclamation, April 1968.

[38] "Soil-Cement Slope Protection for Earth Dams," Portland Cement Assn., Chicago, Ill. (undated).

[39] Cedergren, H. R., "Geological Considerations in the Design of Reservoir Seepage Control Systems," Symposium Reservoir Leakage and Ground Water Control, Ass. Eng. Geol. Nat. Meeting, Seattle, Wash., 1968.

[40] Turnbull, W. J., and Mansur, C. I., "Underseepage and Its Control—A Symposium," Trans., ASCE, vol. 126, pt. I, 1961, pp. 1,428–1,568.

[41] Gibbs, J. J., and Holland, W. Y., "Engineering and Petrographic Properties of Loessial Soils," Engineering Monograph No. 28, Bureau of Reclamation, November 1969.

[42] Holtz, W. G., and Walker, F. C., "Soil-Cement as Slope Protection for Earth Dams," Proceedings of ASCE, Journal of Soil Mechanics and Foundation Division, December 1962, vol. 88, SM 6, pp. 107–134.

[43] Lane, K. S., and Wohlt, P. E., "Performance of Sheet Piling and Blankets for Sealing Missouri River Reservoirs," Proceedings Seventh Congress on Large Dams, 1961, pp. 255–279.

[44] "Report on Steel Sheet Piling Studies," Soil Mechanics and Materials Division, Canadian Department of Agriculture, Saskatoon, Saskatchewan, 1951.

[45] "Symposium on Earth and Rockfill Dams," vols. I and II, Indian National Society of Soil Mechanics and Foundation Engineering, 1968.

[46] Gibbs, H. J., "Estimating Foundation Settlement by One-Dimensional Consolidation Tests," Engineering Monograph No. 13, Bureau of Reclamation, 1953.

[47] "Vibration Effects of Earthquakes on Soils and Foundations," STP 450, American Society for Testing and Materials, 1969.

[48] Gibbs, H. J., and Coffey, C. T., "Application of Pore Pressure Measurements to Shear Strength of Cohesive Soils," Earth Laboratory Report No. EM-761, Bureau of Reclamation, June 1969.

[49] Newmark, N. M., "Effects of Earthquakes on Dams and Embankments," Geotechnique, vol. XV, No. 2, June 1965.

[50] Seed, H. B., "A Method for Earthquake Resistant Design of Earth Dams," Journal, Soil Mechanics and Foundation Division, ASCE, January 1966.

[51] "Efficacy of Partial Cutoffs for Controlling Underseepage Beneath Levees," Technical Memorandum No. 3-267, U.S. Corps of Engineers, Waterways Experiment Station, Vicksburg, Miss., January 1969.

[52] Jones, Jack C., "Deep Cut-Offs in Pervious Alluvium, Combining Slurry Trenches and Grouting," Ninth Congress on Large Dams, vol. 1, 1967, p. 509.

[53] Sherard, James L., "Statistical Survey of the Diaphragm Wall Applications," Specialty Session No. 14, Seventh International Congress on Soil Mechanics and Foundation Engineering, Mexico, 1969.

[54] "Digging a 95-foot Deep Slurry Trench," Western Construction, November 1966, pp. 66–70.

[55] Gerwick, Ben C., "Slurry-Trench Techniques for Diaphragm Walls in Deep Foundation Construction," Civil Engineering, December 1967, pp. 70–72.

[56] Kapp, Martin S., "Slurry-Trench Construction for Basement Wall of World Trade Center," Civil Engineering, April 1969, pp. 36–40.

[57] Gadsby, J. W., and Bares, F. A., "Arrow Project Cofferdam," Canadian Geotechnical Journal, vol. 5, No. 3, August 1968.

[58] Billig, K., "Thixotropic Clay Suspensions and Their Use in Civil Engineering," Civil Engineering and Public Works Review, December 1961, p. 1,573.

[59] Morgenstern, N., and Amir-Tahmasseb, I., "The Stability of a Slurry Trench in Cohesionless Soils," Geotechnique, vol. 15, 1965, pp. 387–395.

[60] "Drilling Mud Reference Manual," Baroid Division, National Lead Co., Houston, Tex., 1965.

[61] "Drilling Mud Data Book," Baroid Division, National Lead Co., Houston, Tex., 1965.

[62] "Mud Technology Handbook," Baroid Division, National Lead Co., Houston, Tex., 1965.

[63] "Construction of Slurry Trench Cutoff," U.S. Army Engineer District, Savannah, Corps of Engineers, Savannah, Ga., 1968.

[64] "Wanapum Hydroelectric Development Final Report," Public Utility District of Grant County, vols. I and II, Ephrata, Wash., 1969.

[65] Lowe, John, "Embankment Dams," Handbook of Applied Hydraulics, edited by C. V. Davis and K. E. Sorensen, third edition, sec. 18, McGraw-Hill, New York, 1969.

[66] "Properties Which Divide Loose and Dense Uncemented Soils," Earth Laboratory Report No. EM-608, Bureau of Reclamation, January 6, 1961.

[67] Wiegel, R. L. (coordinating editor), "Earthquake Engineering," Prentice-Hall, Englewood Cliffs, N. J., 1970.

[68] "Pressure Grouting," Bureau of Reclamation, Technical Memorandum No. 646, June 1957.

[69] "Grouts and Drilling Muds in Engineering Practice," Symposium organized by the British National Society of the International Society of Soil Mechanics and Foundation Engineering, Butterworths, London, 1963.

Chapter VII

Rockfill Dams

R. W. BOCK [1] and L. W. DAVIDSON [2]

A. GENERAL

150. *Origin and Usage.*—Rockfill dams are generally conceded to have had their origin over 100 years ago during the California Gold Rush. From the late 1800's to the middle 1930's, many rockfill dams were constructed; the design and construction of a number of these dams are described by Galloway [1].[3]

Interest in constructing rockfill dams diminished during the following years because of the increased costs of obtaining and placing large amounts of rockfill material, although a number of large dams were constructed in the 1950's [41]. Rockfill dam construction has increased markedly since 1960 and is attributed to the utilization of more remote sites, more economical quarrying and placing operations, the use of excavated material in random zones, better design details, more general knowledge concerning rockfills, and recently the advent of pumped storage projects in mountainous terrain. The excellent performance of an increasing number of rockfill dams is undoubtedly another beneficial factor recommending their use.

Rockfill dams can prove economical when any of the following conditions exist:

(1) Large quantities of rock are readily available or will be excavated in connection with the project such as from a spillway or tunnel.

(2) Earthfill materials are difficult to obtain or require extensive processing to be used.

(3) Short construction seasons prevail.

(4) Excessively wet climatic conditions limit the placement of large quantities of earthfill material.

(5) The dam is to be raised at a later date.

Other factors which favor use of a rockfill dam are the ability to place rockfill throughout the winter and the possibility of grouting the foundation while simultaneously placing the embankment. In addition, uplift pressures and erosion due to seepage through the rockfill material do not generally constitute significant design problems.

Increased interest is also being shown in using "flow through" rockfill sections, used alone or in conjunction with diversion dams, to handle sudden flood flows when the cost of diversion is high [11, 17, 18, 44]. This type of structure requires that a grid work of welded reinforcing rods be placed across the downstream face of the dam, below a given elevation, and anchored to the rockfill so that large flows through the embankment do not displace the rock. The rods are usually ½- to ¾-inch in diameter and are spaced so that rectangles 1 foot vertically by 3 to 4 feet horizontally are blocked out on the downstream face of the embankment. This grid work is then welded to reinforcing rods, which are anchored 10 to 15 feet in the rockfill. The use of "flow through" rockfill dams presents the designer with unique problems concerning the extent of downstream reinforcing to be used and the section's ability to resist overtopping; this type

[1] Assistant Head, Earth Dams Section, Bureau of Reclamation.

[2] Civil Engineer, Earth Dams Section, Bureau of Reclamation.

[3] Numbers in brackets refer to references in the bibliography, section 163.

of design should therefore be used only by an experienced design engineer.

151. *Definition and Types of Rockfill Dams.*—Rockfill dams have been defined as follows [9, 10]: "A dam that relies on rock, either dumped in lifts or compacted in layers, as a major structural element." An impervious membrane is used as the water barrier and can be placed either within the embankment or on the upstream slope. Various materials have been used for this membrane including earth, concrete, steel, asphaltic-concrete, and wood.

Rockfill dams may be classified into three groups, depending on the location of the membrane, as follows: (1) central core, (2) sloping core, and (3) upstream membrane or "decked." Each membrane location has its advantages and disadvantages, which vary according to the type of membrane, materials available at the site, and foundation conditions. Central and sloping cores are referred to as "internal membranes" and these are generally constructed of impervious earth materials. Economic analyses should be made to determine the type of material to use in constructing the membrane, either internal or external. If an internal membrane is to be used, it is recommended that a central vertical core of impervious earth be used; this type of core provides maximum contact pressure with the foundation and requires less strict construction control than a sloping core.

If an external membrane is used, it is recommended that it be constructed of concrete, asphalt, or steel.

Advantages of the internal membrane include (1) less total area exposed to water, (2) shorter grout curtain lengths, and (3) protection from the effects of weathering and external damage. The prime disadvantages of using an internal membrane are the inability to place rockfill material without simultaneous placement of impervious core material and filters, inaccessibility of the membrane for damage inspection, correction of damage should it occur, and the dependence on a smaller section of the dam for stability against sliding. The difference in the ability of internal and up-

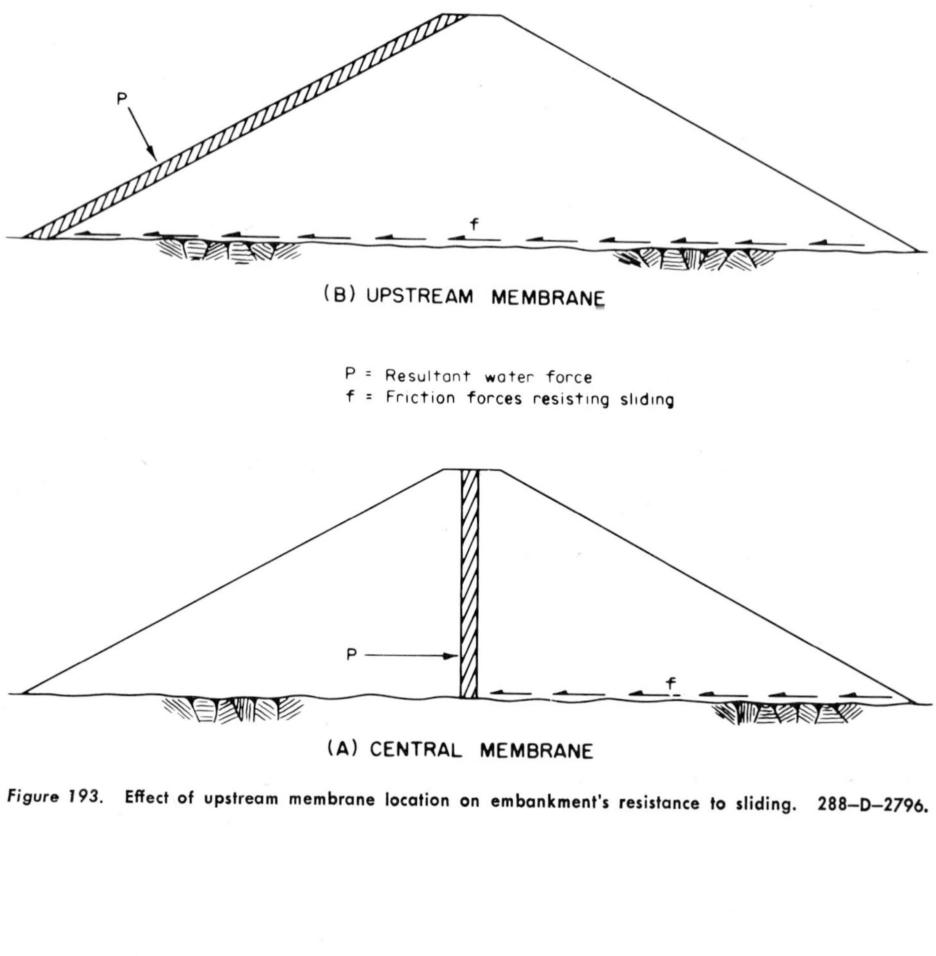

Figure 193. Effect of upstream membrane location on embankment's resistance to sliding. 288–D–2796.

stream membranes in distributing stabilizing reactions against sliding is shown in figure 193.

Upstream membranes have the following advantages:

(1) Readily available for inspection and repair.

(2) Membrane can be constructed after completion of the rockfill section.

(3) Foundation grouting can be performed simultaneously with rockfill placement.

(4) A larger portion of the dam is available for stability against sliding.

(5) Membrane can be used as slope protection.

(6) It is relatively easy to raise the dam at a later date.

If an upstream membrane is used, the reservoir should be capable of being drawn down to an elevation which will permit inspection and repair; TV cameras or audio devices may be used for leak detection and minor repairs may be made by divers.

If an earth-core rockfill dam is used, it requires the use of adequate filters both upstream and downstream; the filters should satisfy the requirements listed in section 131(i). If adequate earth material for either the core or the filter material is not available at the site and separations of impervious material or manufactured filters are required, the earth-core rockfill dam may be uneconomical since filter processing costs can be extreme. Construction control costs of the earth-core rockfill dam will also be increased significantly if several filter layers are required to prevent piping.

B. FOUNDATION DESIGN

152. *Foundation Requirements and Treatment.*—The foundation requirements for rockfill dams are less severe than for a concrete gravity dam, but more severe than for an earthfill dam.

Bedrock foundations which are hard and erosion resistant are the most desirable for rockfill dams. The use of foundations consisting of river gravels or rock fragments is acceptable, but the foundation should be inspected by competent engineers and a positive cutoff to bedrock should be used. The foundation should be selected and treated from the viewpoint of providing minimum settlement to the rockfill embankment. Any materials in cracks, faults, or deep pits which may eventually erode into the rockfill, either from the foundation or the abutment, should be covered with filters (sec. 131(i)) or removed and backfilled with concrete. If an earthcore rockfill dam is to be used, all joints and cracks beneath the core and filters should be cleaned and filled with concrete.

The usual method of treating the foundations to prevent underseepage is cement grouting beneath the cutoff; in addition, potential pervious zones upstream from the impervious membrane can be blanketed with impervious material.

The alinement of the dam should be selected so that either minimum embankment volume or minimum membrane exposure is attained, depending on which criterion is economically more important.

Foundation treatment must be sufficient to satisfy the following criteria:

(1) Minimum leakage.

(2) Prevention of piping.

(3) Limited settlement.

(4) Sufficient friction development between abutments and foundation to insure sliding stability.

153. *Membrane Cutoffs.*—Of critical importance in the functioning of a rockfill dam is the prevention of seepage beneath the dam and the effecting of a watertight seal between the membrane and the foundation. To prevent seepage beneath the dam, foundations are usually grouted; the determination of whether grouting is required and its extent should be based on careful study of the site geology, a visual examination of the drill cores from the rock foundation, and drill-hole water loss values. If no data are available, it should be assumed that grouting will be required; however, where reservoirs are completely drawn down each year,

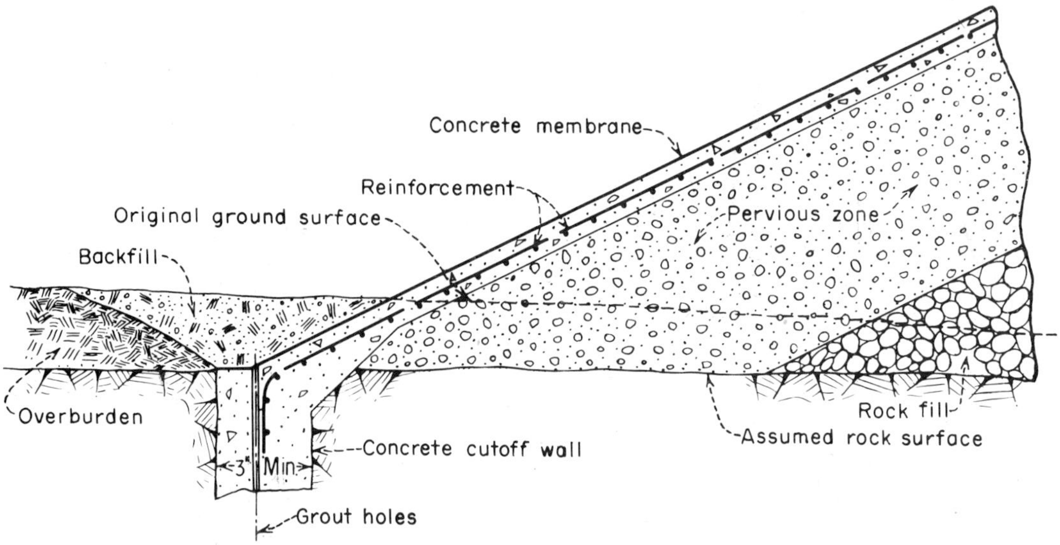

Figure 194. Detail of concrete membrane at cutoff wall. 288–D–2502.

grouting requirements can be based on seepage observations over the first few years' operations.

Cutoff walls excavated to various depths into bedrock are generally used to prevent leakage in the upper few feet of the foundation, to facilitate grouting operations, to provide a watertight seal with the membrane, and to take the downward thrust of the membrane. Figures 194, 195, and 196 illustrate typical cutoff wall details. Drainage galleries are sometimes used in conjunction with cutoffs to facilitate later grouting and to determine seepage locations and quantities, but they are not recommended for use in small dams.

Recently [36], designers have used the doweled slab cutoff shown in figure 197 in conjunction with concrete facings to provide the foundation-membrane seal. Doweled slab cutoffs have the advantage of not requiring extensive excavations in rock, thereby allowing grouting operations to begin earlier, speeding completion time, and reducing design costs. Doweled cutoffs can be used where the bedrock

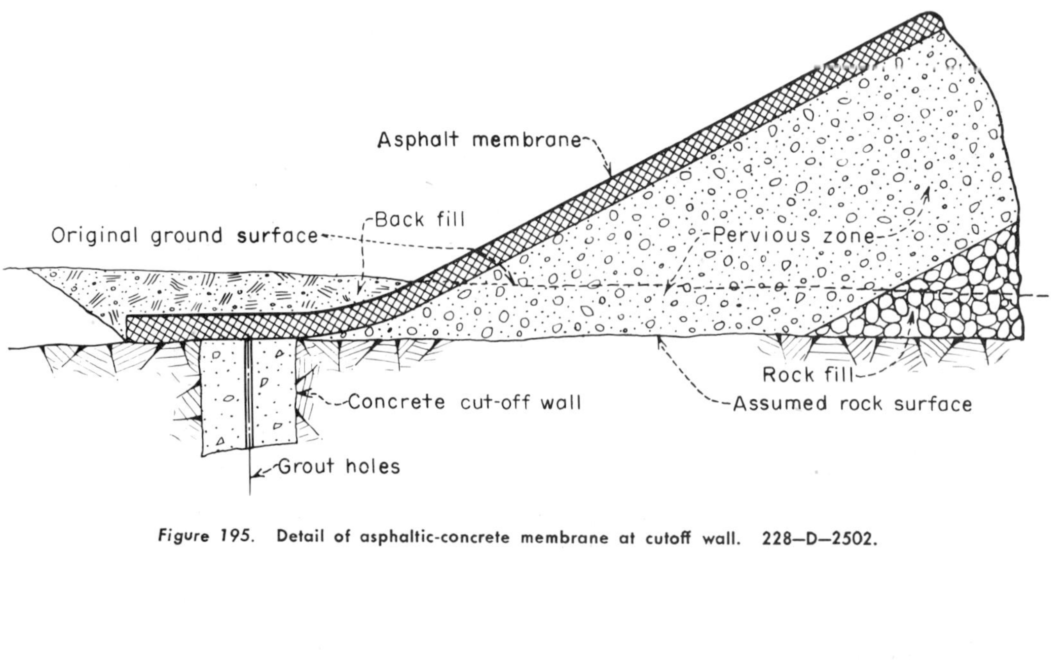

Figure 195. Detail of asphaltic-concrete membrane at cutoff wall. 228–D–2502.

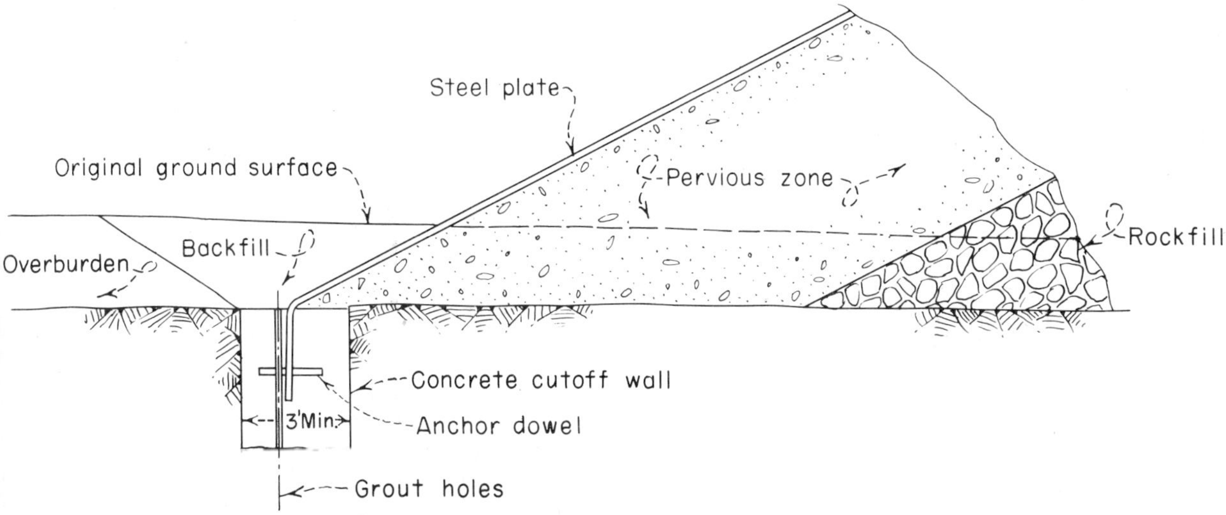

Figure 196. Detail of steel plate membrane at cutoff wall. 288–D–2503.

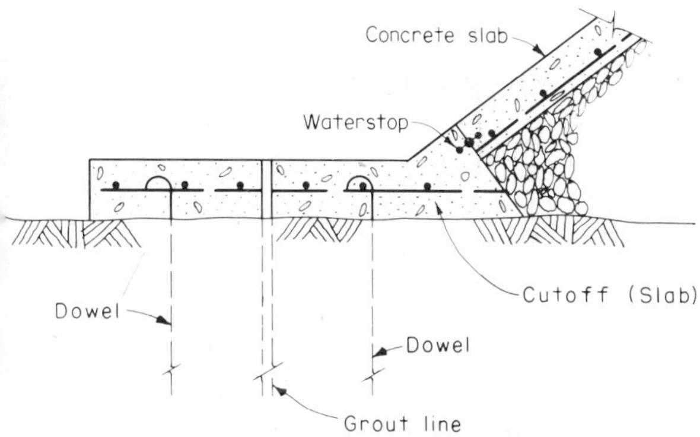

Figure 197. Doweled cutoff slab used with upstream concrete membrane. 288–D–2797.

s sound and few underseepage problems are expected. When uncertainty concerning the permeability of upper portions of the foundation contact exist, such as for soft rock, a cutoff wall into bedrock can provide increased protection and allow an examination of questionable material.

A minimum width and depth of 3 feet is recommended for cutoff walls in sound rock; deeper walls should be used in unsound, broken, or closely jointed rock. The width of the doweled slab will be determined by foundation, construction, or grouting requirements. In addition to its function of preventing underseepage, both the cutoff wall and the doweled cutoff must be designed to provide adequate support for the thrust of the membrane and, in the case of steel membranes, any tension imparted to the cutoff due to embankment settlement.

The cutoff should extend along the entire upstream contact between the membrane and the foundation.

C. EMBANKMENT DESIGN

154. *Selection of Rock Materials.*—A great variety of rock types has been used in the construction of rockfill dams. The types of rock used range from hard, durable, granite and quartzite to weaker materials such as graywacke sandstone and slaty shale. In earlier years, it was thought that only rockfill material of the highest quality should be used; however, with the advent of thinner lifts and more efficient compaction techniques, rock of less desirable characteristics has been made available for use within the embankment section. The use of rock from excavations for spillways, outlet works, tunnels, and other appurtenant structures has reduced the construction cost of rockfill dams without impairing the embankment's usefulness or stability. If small amounts of the less desirable rock types are available, they

Figure 198. Granite rockfill on downstream face of Montgomery Dam in Colorado. CH–520–150.

can be used in random zones within the embankment; the utilization of material in random zones is discussed in section 139(a).

Rock material should preferably be hard, durable, able to withstand disintegration due to weathering, and also to resist excessive breakdown due to quarrying, loading, hauling, and placing operations. Figure 198 shows the granite rockfill on the downstream face of Montgomery Dam. The rock should also be free of unstable minerals that would weather mechanically or chemically, causing the rock to disintegrate. Igneous, metamorphic, and sedimentary rocks have all been used successfully in embankment sections and only general advice can be given concerning rock types since each damsite will present a unique problem in the use of the nearby rock materials. As an aid to the designer, part D of chapter V gives the classification and engineering properties of rocks.

Laboratory tests which measure the abrasion resistance, freeze-thaw characteristics, and percent of water absorption can be used to evaluate the suitability of the rockfill. Petrographic analysis can be used to distinguish minerals known to weather easily. Unconfined or triaxial compression tests will evaluate the strength properties of the rock. One of the best methods of determining a rock's resistance to weathering is simply to examine its *in situ* condition; however, this does not always indicate how the material will perform within the fill. Materials available at the site should be examined by constructing test embankments if economically possible, especially in cases where the material properties are questionable. Test fills can determine the following items:

(1) Whether or not marginal materials can be used.
(2) How selected embankment material will perform during compaction operations.
(3) Correct type of compaction equipment for each material.
(4) Required number of passes for each material.
(5) Correct lift thickness for each material.
(6) The necessity for changing the embankment section to accommodate new materials or different material properties.

As an example, Crisp [13] reports that significant design changes in Carters Dam were proposed from results obtained by placing test embankment sections of quartzite, phyllite, and argillite.

The effect of quarry blasting methods on the gradation of the rock should also be examined, as well as the required extent of quarrying.

Also of great importance to the design engineer when selecting the type of rock is the degree to which small scale triaxial compression tests will provide strength parameters applicable to the actual rockfill. Very limited data are available on this subject; however, Marachi, et al [43] have examined this problem by testing 36-, 12-, and 2.8-inch-diameter specimens in drained triaxial compression tests using parallel grain size curves and identical grain slopes (modeling) to examine the effects of grain size on the strength and deformation characteristics of rockfill material. Also investigated was the effect of particle crushing.

Three types of material were tested as follows:

A. Pyramid Dam: Argillite, a fine-grained sedimentary rock, quarry blasted, angular, with relatively weak particles ($G_S=2.67$).

B. Crushed basalt: Quarry blasted and crushed to the correct size, angular and quite sound ($G_S=2.87$).

C. Oroville Dam: Amphibolite, a meta-volcanic rock, rounded to subrounded particles with some subangular fine sand

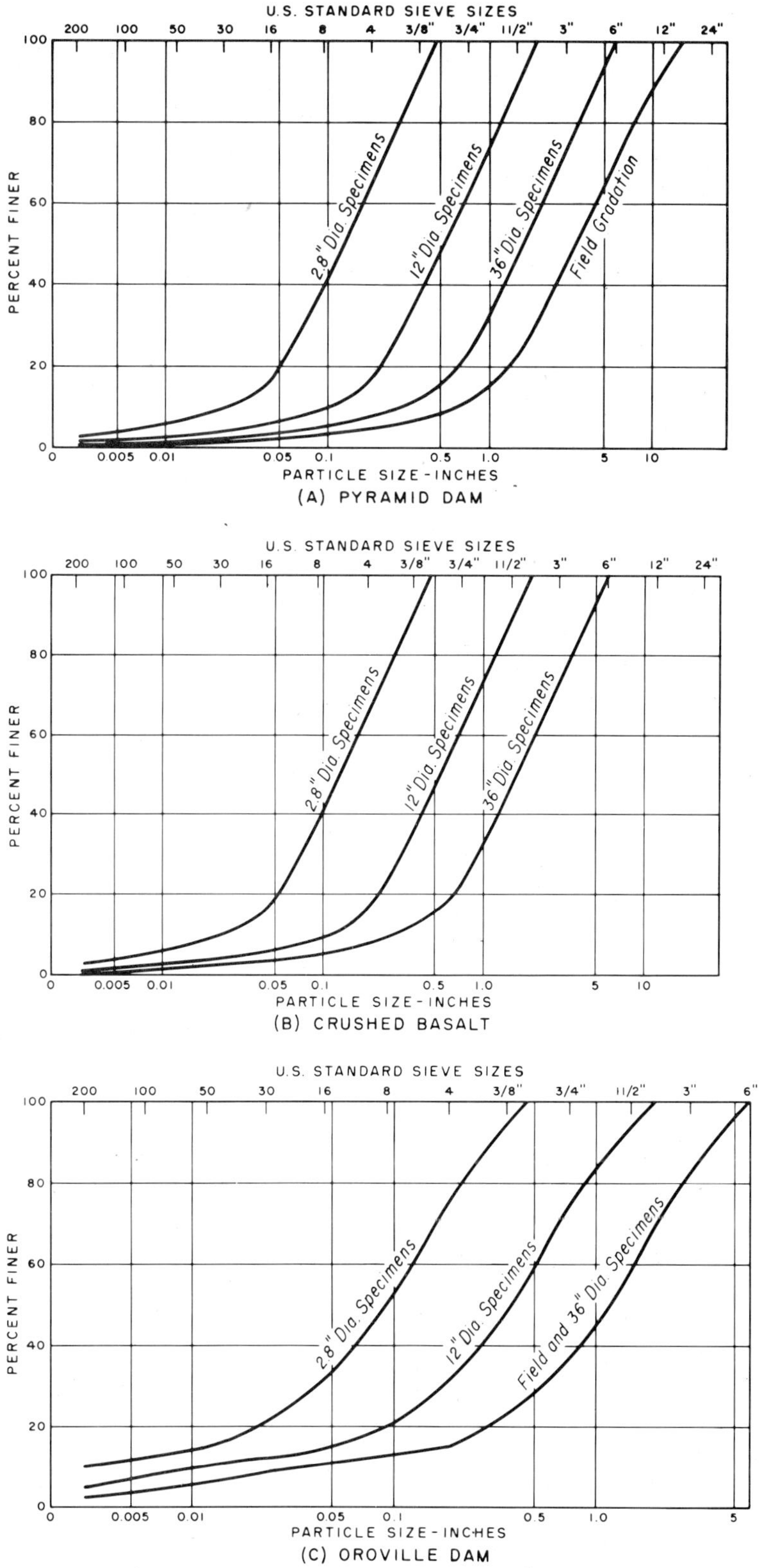

Figure 199. Grain size distribution for modeled rockfill materials. Adapted from Marachi, et al [43]. 288—D—2798.

particles, river dredged material, hard (G_S=2.86 to 2.94).

The gradation curves for the actual rockfill material and the modeled material are shown in figure 199.

Although the report was primarily concerned with the use of rockfill material in high dams, the following general conclusions are applicable to rockfill dams of all sizes:

(1) Rockfill materials can be successfully modeled so that the strength and deformation characteristics of the actual material can be obtained from small scale tests.

(2) At any given confining pressure, as the particle size of the specimen increases, the angle of internal friction decreases a small but significant amount.

(3) Rockfill materials composed of well-graded and well-rounded particles are superior to uniformly-graded angular rockfill materials, especially for high dams.

(4) For any given particle size, as the confining pressure of the sample increases, the angle of internal friction decreases.

Figure 200 illustrates the variation of the angle of internal friction with both particle size and confining pressure. Although most of the confining pressures shown are greater than those attained in small dams, the general reduction in friction angle shown in this figure should be of great interest to designers.

The reader is referred to the report [43] for the details of testing and for further conclusions regarding the strength and deformation properties of rockfill materials and the crushing characteristics of rock subjected to high confining pressures.

155. *Embankment Sections.*—Embankment slopes used for rockfill dams have evolved from very steep slopes, usually 0.5 to 0.75 horizontal to 1 vertical which were used on early rockfill dams, to the flatter slopes of 1.3:1 to 1.7:1 used in current practice. Earlier rockfill dams used upstream membranes exclusively and were constructed with steep upstream and downstream slopes to minimize the volume of rockfill. Since these slopes were considerably steeper than the natural slope of dumped rock, they were stabilized by thick zones of crane-placed, dry rubble masonry which provided the bedding for the upstream slope protection. The rockfill portions of these dams were constructed by dumping the rockfill in thick lifts which ranged from 30 to 165 feet. Later designs eliminated the rubble masonry on the downstream slope by flattening it to the angle of repose of the rock, but the very steep upstream slope was retained. Since most of the upstream zones were constructed by crane placement of large rocks, the cost of the dams continually increased. Gradually, designers found that it was more economical to use slopes approximating the angle of repose of the rock material and eliminate crane placement in favor of compacted rockfills.

The upstream and downstream slopes of the dam will depend upon the type of impervious membrane and its location. Rockfill dams having central or sloping cores will have slopes ranging from 2:1 to 4:1 upstream and downstream, whereas dams in which the membrane is placed on the upstream face usually have upstream slopes of from 1.3 to 1.7 to 1 with downstream slopes approximating the natural slope of the rock.

Most asphaltic-concrete faced dams have been constructed with upstream slopes of 1.6:1 to 1.7:1 to facilitate construction of the membrane, and most steel and concrete faced rockfill dams have used slopes of 1.3:1 to 1.4:1. A review of available literature indicates that very few failures have occurred for these slopes; therefore, small rockfill dams with good foundations could use 1.3:1 to 1.4:1 as an upstream slope for concrete and steel membranes and 1.7:1 as the upstream slope for asphaltic-concrete facings; downstream slopes of 1.3:1 to 1.4:1 may be used in all cases.

The upstream and downstream slopes for central or sloping earth-core rockfill dams will be dependent on the size and soil properties of the earth core, the width of filter zones required, type of foundation material, drawdown requirements, construction sequence, etc., with each site presenting its own unique problems. Generally, upstream and downstream slopes of a typical earth-core rockfill dam will range between 2:1 and 4:1. A typical embankment

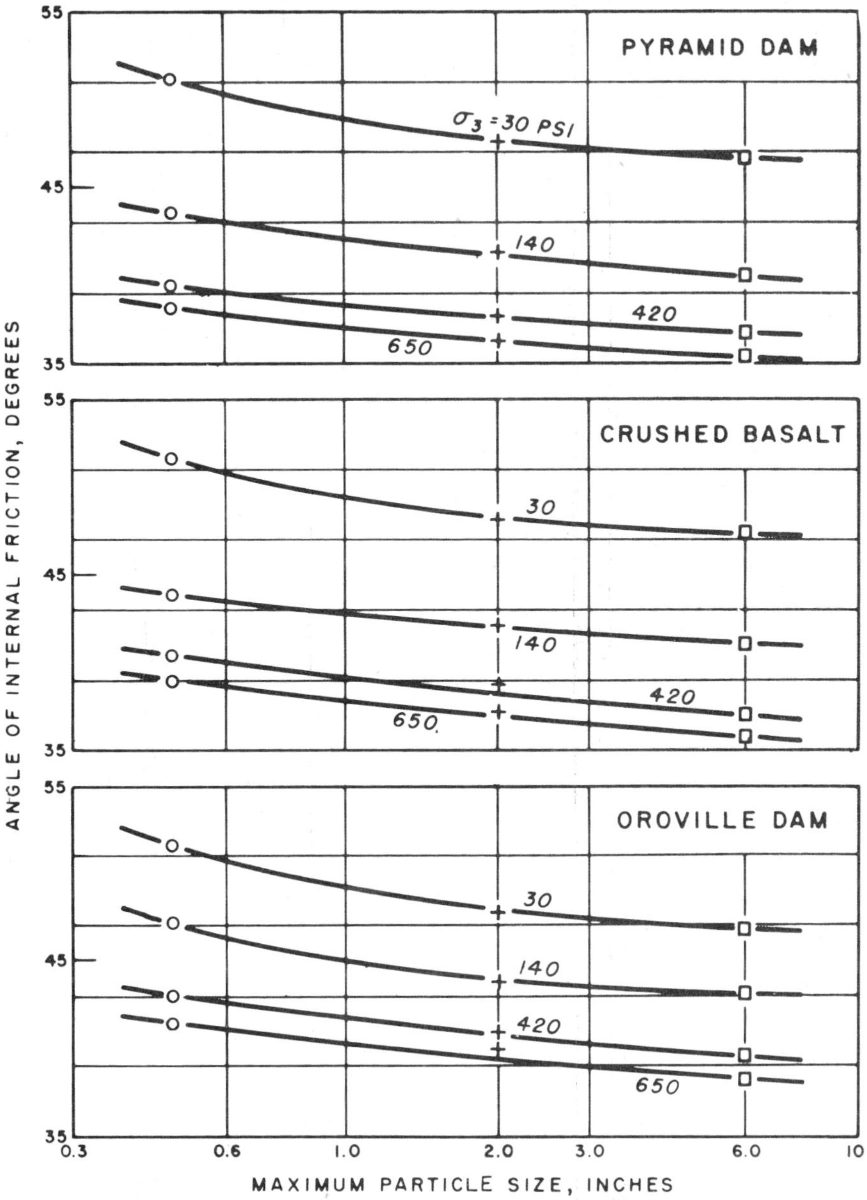

Figure 200. Effect of maximum particle size on the angle of internal friction. Adapted from Marachi, et al [43]. 288–D–2799.

section for a central earth-core rockfill dam is shown in figure 201.

A typical section for a decked rockfill dam is shown in figure 202. The interior section of the decked rockfill dam can be divided into three zones, as shown in figure 202, and can be described as follows:

Zone C: The larger downstream zone of the dam, consisting of the best quality, larger sized, compacted rock; this zone provides high stability to the section.

Zone B: Rock of lesser quality than zone C, such as excavated material from the spillway; used to minimize total dam costs.

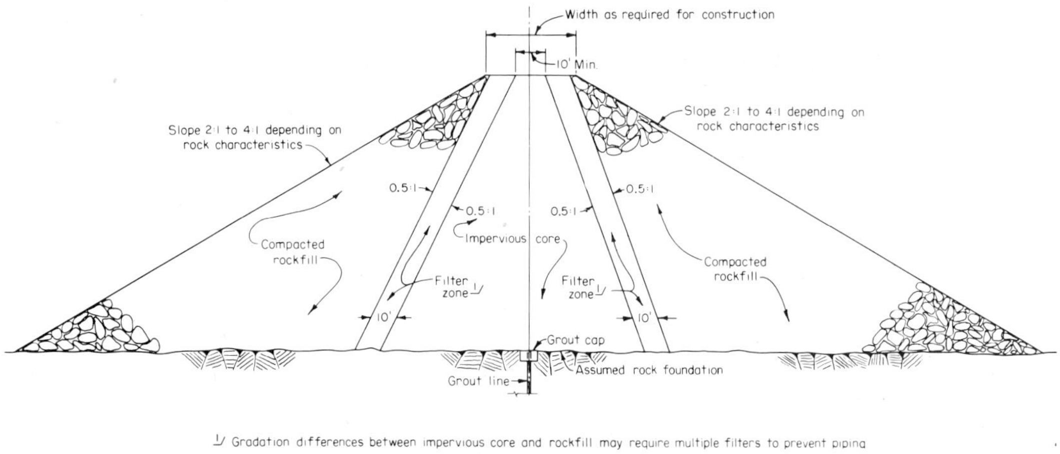

Figure 201. **Typical maximum section of an earth-core rockfill dam using a central core. 288–D–2800.**

Zone A: Well-graded, smaller-sized rock and gravel; used to provide bedding for the upstream membrane and to retard extreme water losses should the membrane crack.

Placement conditions for these three zones are discussed in section 156.

Gradation requirements are difficult to specify since they depend on the type of rock available and the quarrying methods used; as with many aspects of dam design, only general rules apply. For decked rockfill dams, the downstream zone C of the embankment should use the largest and best quality rock available. Large slabby rocks should not be placed in the fill since they tend to bridge, causing large voids which may result in excessive settlement should the rocks break. If possible, rock in zone C should be well graded in size from approximately 1 cubic foot to 1 cubic yard and the finer fraction of the gradation should not be sufficient to fill the voids in the compacted material. Optimally, zone B should be well graded from approximately 3 inches to 1 cubic

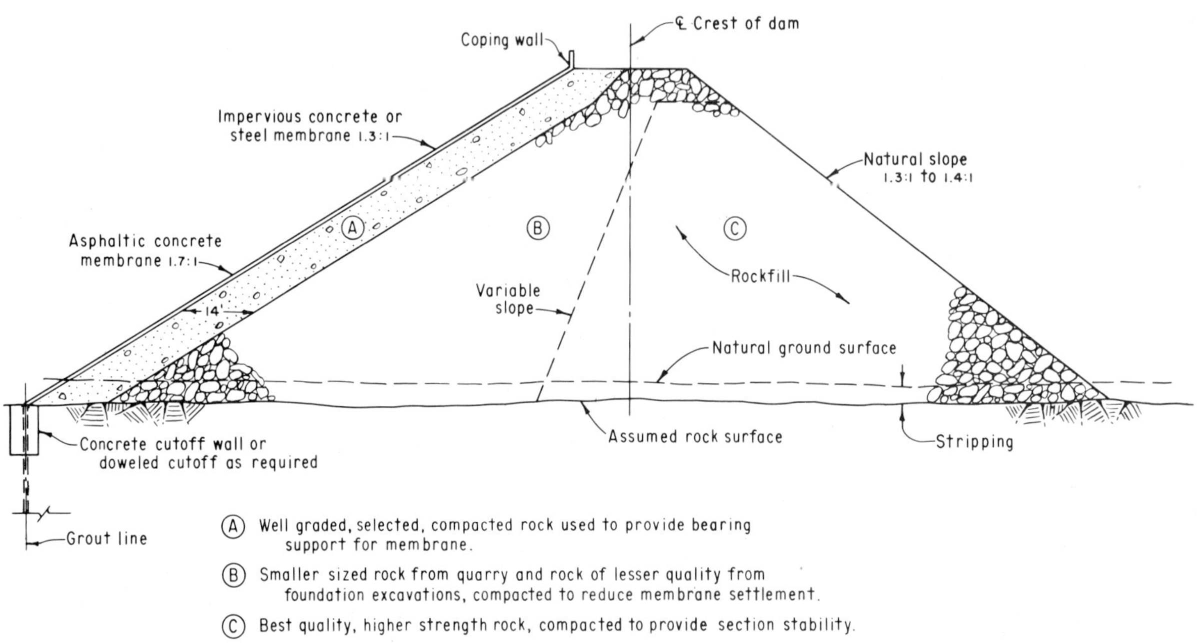

Figure 202. **Typical maximum section of a decked rockfill dam. 288–D–2801.**

foot and zone A well graded from approximately 1/4 to 3 inches. The gradation of zone A will depend on the type of facing used and its method of construction. Zone A material should, however, provide a smooth uniform bearing surface for the facing, yet be graded to retard large water loss should the facing crack.

In general, material in zones B and C should grade from fine rock upstream to coarse rock downstream, with the largest and strongest material placed in the lower downstream portions of zone C. Selection of the rock for each zone should be made at the quarry.

For central, earth-core, rockfill dams the larger and stronger rock should again be placed in the downstream rockfill zone and grade from fine rock next to the filter to coarse rock near the downstream slope. The upstream rockfill zone may be rock of lesser quality than the downstream zone and should grade from fine at the filter to coarse at the upstream face.

The axis of the dam may be either curved (convex upstream) or straight. A curved axis will act to compress the dam as filling occurs, whereas a straight axis has the benefit of easy construction layout and less total dam cost. For small dams, given good foundation and abutment conditions, it is recommended that a straight axis be used. It is also recommended that for upstream-membrane rockfill dams, the layout be such that a minimum area of membrane face be exposed. This expedites face construction, reduces face and cutoff cost, and repair costs should they be necessary.

Random zones constructed of rock with questionable strength or permeability characteristics may also be used within the rockfill embankment if the stability of the section is not affected. The overriding purpose in the layout of any rockfill section is to make maximum economic use of the materials available at the site. Test embankments can be used to determine whether or not materials will be adequate; these are discussed in section 152.

Crest width will be determined by the type of membrane used and by its use after construction. It should, however, be of sufficient width to accommodate construction of the upstream membrane; a minimum width of 15 to 20 feet is recommended. Crest camber should be determined by the amount of foundation and embankment settlement anticipated. Since this is difficult to determine, a value of 1 percent of the embankment height is recommended. A straight line equation may be used to distribute the cambered material on the crest. Additional considerations concerning crest details are given in section 131. Freeboard requirements will depend on maximum wind velocity, fetch, reservoir operating conditions, spillway capacity, and whether or not coping walls are used. If coping walls such as shown in figure 202 are used, freeboard requirements may be the same as those for a riprapped earthfill dam, as discussed in section 142, since wave runup and oversplash need not be considered. If coping walls are not used, the freeboard should be adequate to prevent wave runup from flowing over the crest. Good results have been obtained with coping walls [36] and their use is recommended herein.

156. *Placement of Rockfill Materials.*—Limiting settlement to acceptable limits is critical in the construction of rockfill dams since excessive settlement may rupture the upstream membrane or cause joint separation with subsequent water loss. Early rockfill dams were constructed by dumping the rock in high lifts; it was assumed that the height of drop imparted compaction energy to the fill and decreased the embankment's void space and thus reduced settlement. Many of these high lift embankments have since settled considerably, with concurrent leakage problems, and experience has dictated that rock material dumped in thin layers and compacted by vibratory rollers forms a more stable mass in which settlement is minimal. For decked rockfill dams, the embankment should preferably be completed before construction of the upstream membrane is begun since this reduces the probability of serious membrane cracking by allowing initial settlement to take place.

Settlement of rockfill material has also been correlated with the application of water, and Sowers, et al [16] have shown that rockfill material placed dry and subsequently wetted may settle appreciably. Sluicing has long been advocated as a method which assures that point-

Figure 203. Typical rockfill sluicing operation at Montgomery Dam in Colorado. (Black and Veatch Engineers photo.) 288–D–2879.

to-point bearing occurs between the larger rocks, and also that finer materials are washed into the voids. However, when rockfill material is placed in thin lifts and compacted by vibratory rollers there appears to be no definite proof that sluicing operations provide a significant reduction in the total settlement, especially for the smaller rockfill dams considered here.

The quantity of sluicing water used has varied extensively, but usually ranges from two to four times the volume of rock; dirty rock will require a greater amount of water to wash the fines away. For exceptionally dirty rock the possibility exists that a water-saturated layer of fines may form below the surface of the rock as it is dumped at the edge of the lift, preventing the wetting of all parts of the rock. This situation may necessitate the use of higher lifts, which allows adequate sluicing time for the thinner layers or possibly wetting prior to reaching the fill. Care should also be taken that an accumulation of mud does not form at the bottom of the lift due to sluicing; if mud problems do arise, periodic removal should be mandatory. Sluicing is usually done with monitors having nozzle diameters from 2½ to 4 inches; a typical sluicing operation is shown in figure 203. Sufficient sluicing equipment should be available to handle maximum rock placing rates since its absence may limit the quantity of rockfill placed; the sluicing equipment should be mobile.

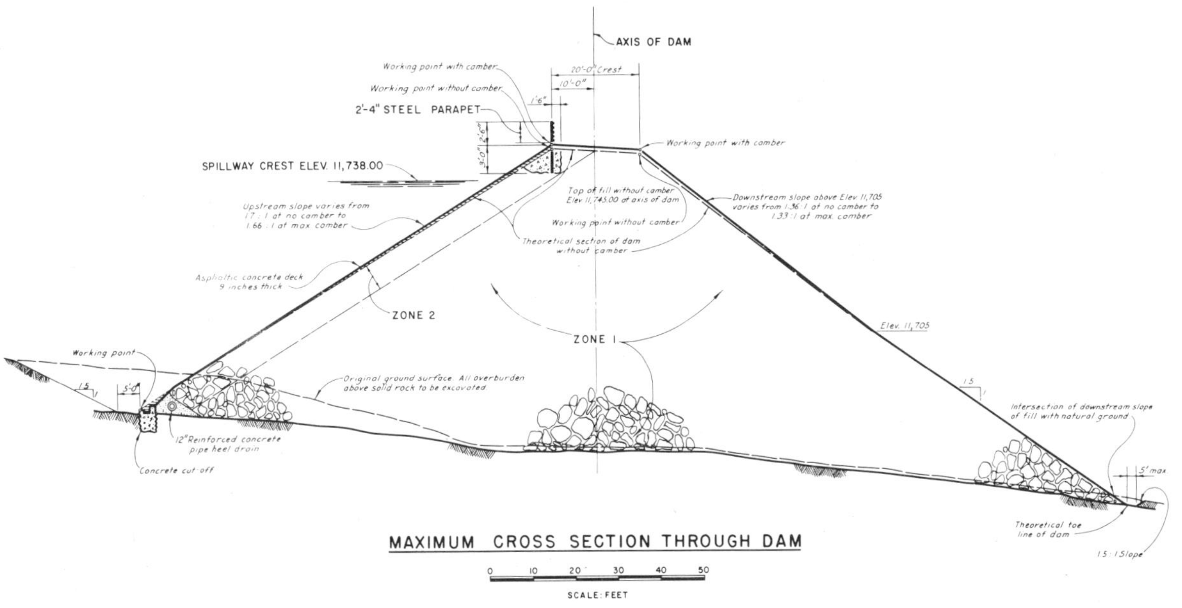

Figure 204. Maximum section of Upper Blue River Dam in Colorado. (Black and Veatch Engineers drawing.) 288–D–2880.

Figure 201 shows a typical decked rockfill dam section consisting of three zones of material. The zone C material should be sound, durable rock of high quality dumped in 3- to 6-foot lifts and compacted by a vibratory roller. Zone B material may be rock of lesser quality than that in zone C, such as spillway excavation or tunnel spoil, and should be dumped in 3-foot lifts and compacted by a vibratory roller. Zone A material provides the bearing surface for the upstream membrane and may be either a processed material or selected material from foundation or borrow pit excavations. Suggested gradations for zones A, B, and C are discussed in section 155. Zone A material should be compacted to 12-inch lifts by either crawler-type tractors or vibratory rollers; the material should be thoroughly wetted prior to compaction. The size of the vibratory roller used for each rockfill zone will depend on the properties of the rock used in that zone and should preferably be established by constructing test embankments. Vibratory rollers from 3 to 10 tons have been the most widely used for rockfill compaction.

In a number of concrete and asphalt-faced rockfill dams [28, 32, 35, 36], the zone A material has been eliminated and only a thin leveling coarse is applied to the face of zone B; in these cases, compaction of the leveling coarse is performed by drawing a smooth-drum vibratory roller up the face. Figure 204 shows the maximum section of Upper Blue River Dam in which zone A has been eliminated and zone B replaced by zone C. The final upstream surface of zone B may also be finished by pulling the vibratory roller up its face. The advice of an experienced dam designer should be obtained before zone A is eliminated.

For central earth-core rockfill dams, such as shown in figure 201, the upstream and downstream rockfills should be compacted in 3-foot lifts by vibratory compactors to provide the most stable section possible. Sluicing operations used with this type dam require that great care be taken to insure that filters are not clogged or impervious material washed away. Filter materials should be compacted to 12-inch lifts either by crawler-type equipment or vibratory rollers. The width of the filter zones should be sufficient to accommodate placing and compacting equipment.

157. *Seismic Design*.—For areas of low seismic activity the designs recommended herein should be adequate. The determination of potential earthquake activity within a given region can be made from a seismic risk map such as figure 152, if available, or by consultation with a seismologist or engineering geologist. If the damsite lies within a zone of high seismic activity the advice of an experienced dam designer should be obtained.

It is the general opinion of many dam designers that the use of large downstream zones of quarried rock compacted in thin layers provides maximum stability against seismic shaking and a maximum resistance to the flow of large quantities of water through the section should cracking occur. Thus, it is recommended that where seismic activity is expected, decked rockfill dams containing large downstream zones of compacted rockfill should be used. The rockfill should preferably be well-graded, angular rock fragments of high strength and durability. To accommodate the larger downstream zones, it is recommended that where questionable earthquake conditions are encountered the downstream slope of the decked rockfill be flattened to 1.7:1 in all cases. The upstream slope of the embankment should also be flattened if additional conservative design measures are warranted.

The foundation of the dam should preferably be firm rock; however, free-draining foundations (cobbles, boulders, rock fragments, etc.) may be used if their state of compaction is similar to the rockfill material and they are approved by a competent dam designer. Trench-type foundation cutoffs are also recommended. In addition, it may be desirable to provide a thicker zone A (fig. 202) beneath the membrane, to require the use of better quality rock in zone B, and to reduce the lift thickness to a maximum of 3 feet within zone C. Still another precaution would be the use of a thicker membrane on the upstream slope and, in the case of a concrete membrane, to place reinforcing in each face.

It should be noted that there are no exact rules for design within earthquake regions and an authority in this field should be consulted when serious seismic conditions exist.

D. MEMBRANE DESIGN

158. *Impervious Central Core*.—A typical earth-core rockfill section using a central impervious earth core is shown in figure 201. Internal membranes of concrete, asphalt, and steel are not recommended due to the inability to inspect or repair them; sloping cores of impervious earth materials, either internal or external, are also not recommended. The rockfill zones of the central core dam are discussed in sections 155 and 156. The upstream rockfill material should be of sufficient size and quality to satisfy riprap requirements, as discussed in section 143(c); riprap bedding requirements need not be met.

Earth-core rockfill dams are economical where site conditions recommend the use of rockfill but prevent the use of a decked rockfill structure. This can be the case when the upstream abutments show highly weathered rock to great depths and thus present questionable cutoff conditions for an upstream membrane, or when the higher elevations of the abutments are covered with deep layers of overburden and prevent the economical installation of a positive trench-type cutoff for the decked rockfill.

The impervious material used in the core should be similar to the material used for earthfill cores, as discussed in chapter VI. The material should be placed at optimum moisture content and compacted to 6-inch lifts by 12 passes of a tamping roller. The plasticity index of the material should be sufficient to allow the core to deform without cracking. Specifications for this type of placement are given in appendix G.

Filter zones should be adequate to prevent any piping of impervious material during steady state or rapid drawdown conditions and it is recommended that the filter criteria in section 131(i) be met. Multiple filters may be required if gradation differences between the core and rockfill materials are large. Figure 205 shows the placement of fine and coarse fil-

Figure 205. Placement of fine and coarse filter material for the 55-foot-high New Exchequer Saddle Dike in California. (Tudor Engineering Co. photo.) 288–D–2881.

ter material for the 55-foot-high New Exchequer Saddle Dike in California.

The foundation and abutments against which the core rests should be carefully treated to prevent piping. Joints, cracks, fissures, and shear zones should be cleaned out to firm material and filled with concrete or grouted. Vertical faces, overhangs, and large rock protrusions should be flattened to slopes not steeper than 0.5 horizontal to 1 vertical by excavation or concrete placement. Trench-type cutoff walls should be used with central impervious earth cores when foundation grouting is required.

Freeboard requirements are the same as for earthfill dams and are discussed in section 142.

159. *Reinforced Concrete.*—A larger number of rockfill dams have been faced with conven-

Figure 206. Placement of concrete by the use of slip forms on the upstream slope of New Exchequer Dam in California. (Tudor Engineering Co. photo.) 288–D–2882.

tionally placed reinforced concrete than with any other type of impervious membrane. In most cases, these facings have performed well, for correctly compacted rockfill embankments, with acceptable limits of leakage and minor repairs. Slab thickness and reinforcing requirements have usually been determined by experience or precedent, with the idea of satisfying the following criteria:

(1) Low permeability.
(2) Sufficient strength to bridge subsided areas of the face.
(3) High resistance to weathering action.
(4) Sufficient flexibility to tolerate small embankment settlement.

Compacted rockfill dams have considerably reduced embankment settlement, and the use of a well-compacted facing layer which acts as

Figure 207. Placement of shotcrete on the upstream face of Taum Sauk Dam in Missouri. (Union Electric Co. photo.) 200—D—2003.

a continuous, firm, bedding surface for the concrete face has reduced the bridging requirements of the face.

It is recommended that for small dams on stable foundations, a reinforced concrete slab with a minimum thickness of 8 inches be used. The concrete should be dense, durable, weather resistant, and of low permeability; concrete mix designs are discussed in appendix E. If foundation settlement may occur or if other factors such as earthquake conditions exist, it would be wise to increase the membrane thickness. The amount of steel reinforcing should meet the generally accepted requirements of 0.5 percent of the concrete area and should be placed both horizontally and vertically. The single layer of reinforcing should be placed in the center of the slab.

Because of the low reservoir head and the small amount of settlement to be expected,

Figure 208. Completed rockfill embankment at Upper Blue River Dam in Colorado prior to membrane placement. (Dept. of Public Utilities, Colorado Springs, Colo. photo.) 288–D–2884.

horizontal or vertical expansion joints normally are not required in the reinforced concrete facings for low dams. Vertical joints may be required to compensate for horizontal expansion on low dams of considerable length and are often used to facilitate construction of the face. Polyvinyl chloride or rubber waterstops should be used to insure impermeability along the joints used.

The type of cutoff between the concrete facing and the foundation will depend on the quality of rock encountered. For sound rock, the doweled cutoff shown in figure 197 has demonstrated its adequacy and economy [36], whereas in closely jointed, weathered rock, or rock of questionable quality, the cutoff wall should be used. Waterstops should be used between the cutoff and facing. Rigid cutoffs, such as shown in figure 194, are not recommended since they restrict the allowable settlement of the face.

Since concrete facings provide little resistance to wave runup, an increased freeboard is required to prevent wave runup and oversplash. Coping or parapet walls similar to that shown in figure 202 can be used to reduce the height of embankment required for freeboard purposes; these walls should be constructed as integral continuations of the concrete face and reinforced accordingly. Coping walls work well and Cooke [36] reports that in one case, walls 10 feet high have stored water to 8 feet without harmful effects, although this procedure is not recommended for the type of structures considered here. When coping walls are used, freeboard requirements should be similar to those for an earthfill dam (sec. 142) and runup and wave splash neglected. Height of the coping wall can be determined either by precedent or the designer's experience. Spillway design should be such that its capacity increases rapidly as the reservoir surface begins to encroach on the coping walls.

Concrete placement has generally been by the same slip-forming process used in road construction, but in some cases shotcrete has been used effectively. Figure 206 shows placement of concrete by the use of slip forms on the upstream slope of New Exchequer Dam in California, and figure 207 shows concrete placement by shotcreting at Taum Sauk Dam near St. Louis, Mo. Placement of the concrete membrane should not begin until the entire embankment has been placed; this allows for maximum construction settlement and reduces the possibility of cracking and excess leakage. If concurrent slab placement is necessary for job completion, the advice of an experienced dam designer should be sought. Concrete overruns will occur due to the voids present in the facing layer and should be accounted for in estimating the quantities.

160. *Asphaltic Concrete*.—The second most widely used facing for rockfill dams is asphaltic concrete. Asphaltic concrete provides more flexibility and can thus tolerate larger settlements than reinforced concrete facings. It offers an economical alternative to concrete and has proved to be dependable when correctly constructed. The upstream slope for asphalt-faced rockfill dams is recommended to be 1.7:1 or flatter, as shown in figure 202. The zone A material should provide a well-graded, free-draining rock layer to eliminate uplift pressures in case of rapid drawdown, but should also provide sufficient resistance to limit water velocities and prevent piping in case a crack forms in the membrane; the gradation of this material should be smaller than zone B material. If zone A material is sufficiently

Figure 209. Placing asphaltic concrete on the face of Montgomery Dam in Colorado. CH–520–81.

coarse to provide a rough bearing surface, or if zone A is not used, a thin levelling layer of smaller, well-graded material, approximately 6 inches in thickness, normal to the slope, should be applied to fill the surface voids and provide a smooth bearing layer for the asphalt membrane and to promote easy travel of the paving machine. Figure 208 shows the completed rockfill section at Upper Blue River Dam prior to membrane placement.

A penetration coat is sometimes applied to the levelling surface prior to membrane placement to bind and stabilize it. The penetration coat does not always stabilize the dam facing since the weight of the paver tends to gouge the levelling surface; the use of a penetration coat on the steeper slopes is therefore of questionable value. Hand placement of asphaltic concrete in the gouged surfaces will be required.

The asphaltic concrete membrane thickness is recommended to be between 9 and 12 inches; it is usually applied by a standard road paver in three approximately equal layers. Figure 209 shows placement of an asphaltic concrete layer at Montgomery Dam in Colorado. A seal coat can be placed on the finished surface of the membrane if desired. The seal coat waterproofs the facing and provides increased durability. Each layer is placed in strips 10 to 12 feet wide, and constructed at right angles

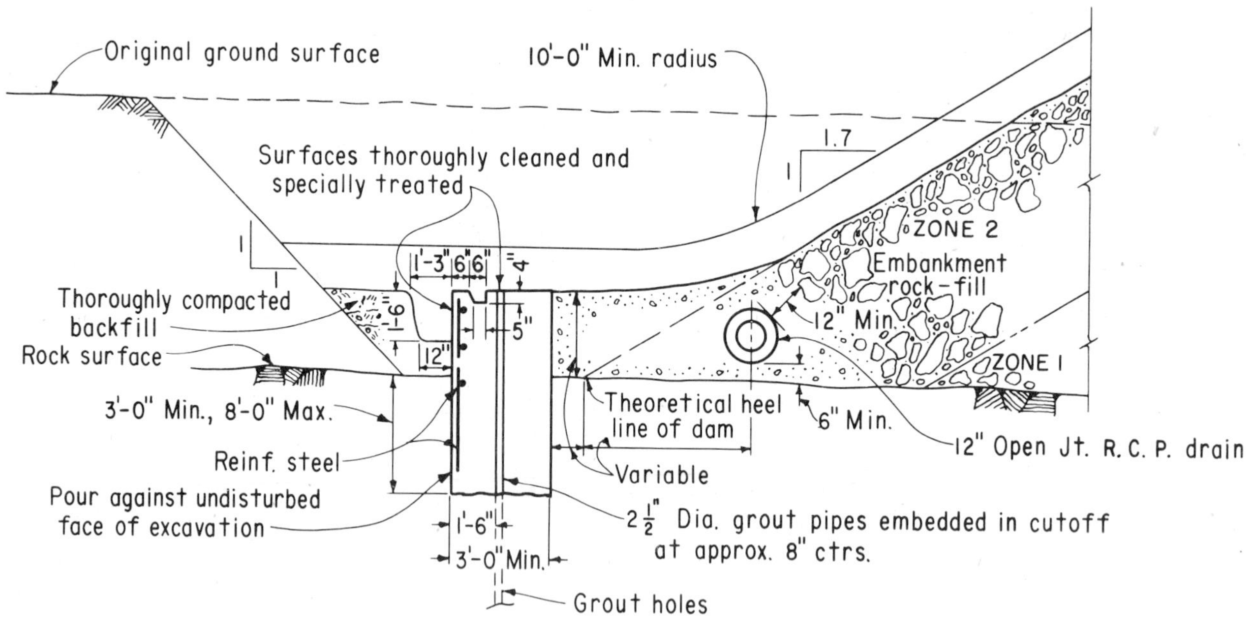

Figure 210. Foundation cutoff used at Montgomery Dam in Colorado. 288–D–2802.

to the axis of the dam. Paving is placed on the upslope pass only, with the machine returned to the bottom and reloaded for each strip. If sufficient asphaltic material is not available for each strip, reloading must be performed on the upstream face; a single paving machine should be capable of placing between 25 and 35 tons of asphaltic concrete per hour. Rolling operations should closely follow the placing operation; smooth wheel rollers, either of the vibratory or standard tandem type, can be used for layer compaction.

Layers should be compacted to a minimum of 97 percent of standard laboratory density.[4] Construction control can be effected by taking cored samples at random locations from the asphalt face and performing asphalt content, density, stability, and permeability tests.

Effecting tight joints between adjacent strips of the facing is important to the imperviousness of the membrane. Transverse joints in the strips should be kept to a minimum and should be hot joints if possible. Cold joints, either between parallel strips or transversely on a single strip, should be treated as follows:

(1) Apply a tack coat of asphaltic cement, the same type used in the mix design.
(2) Place the asphaltic concrete, overlapping the joint 3 to 6 inches.
(3) Reheat the joint with an infrared heater, avoiding open flames.
(4) Compact the joint by rolling, immediately following reheating.

When one layer is placed on top of another, the parallel joints in the strips of the top layer should be offset 3 or 4 feet from the joints of the bottom strip.

The foundation cutoff used with asphalt facings must promote easy placement of the asphalt layer at the contact with the foundation. The trench-type cutoff wall, similar to that shown in figure 195, is recommended. The cutoff used at Montgomery Dam in Colorado is shown in figure 210; the 12-inch-diameter drain was used to reduce uplift pressure during drawdown. The cutoff used at Upper Blue River Dam is visible at the left edge of figure 208.

[4] Standard laboratory compaction as defined by the Bureau of Reclamation.

The upstream asphaltic membrane should be constructed so that it is:

(1) Durable.
(2) Flexible.
(3) Impervious.
(4) Does not creep.
(5) Resists weathering.

Material within an economical distance of the dam should be used in the asphaltic concrete if possible. A number of different materials and gradations ranging from silty sand [31] to graded gravel [32] have been used to construct adequate upstream facings. Clay fines should not be permitted in mixes since the clay tends to ball during the drying process and to crush when compacted, thereby leaving dry material exposed to the reservoir water.

The gradation limits of the material used for the asphaltic concrete facing at Montgomery Dam are shown in the table below:

Sieve designation Square openings (U. S. Standard sieves)	Percent by weighing, passing	
	Specification	Job mix [1]
1½ inch	100	100
¾ inch	80–95	86.4
½ inch	71–89	76.5
No. 4	55–75	58.7
No. 10	40–60	46.5
No. 40	22–36	30.0
No. 80	14–26	19.2
No. 200	7–15	12.8

[1] Average gradation of material used in construction.

Tests should be performed using various gradations, compressive efforts, percent asphaltic cement, and percent lime to determine the mix having maximum density and best satisfying the above criteria. In certain cases, 1 to 3 percent of lime has been shown to reduce the underwater expansion of the asphaltic concrete, thereby increasing its life expectancy.

Tests which may be performed to evaluate the materials and the different mixes are:

(1) Sieve analysis and specific gravity.
(2) Immersion—compression test.
(3) Unconfined—compression test.
(4) Sustained—load test.
(5) Permeability.
(6) Wave-action test.

Tests 1-5 are similar to those described in ASTM Standards, Part 11. Test 6, concerning the effect of wave action, is described in references [29, 30, 31]; this test was developed by the Bureau of Reclamation to simulate wave effect on asphaltic concrete facings and has proved useful in helping to select the correct mix proportions.

Special tests which can be performed are:

(1) Slope-flow.
(2) Coefficient of expansion.
(3) Flexural strength.
(4) Effect of reservoir ice.

All of the above tests, both standard and special, were performed by the Bureau of Reclamation to determine the proper type of asphaltic cement and the correct percentage to be used at Montgomery Dam [29]. These tests led to a mix design incorporating 8.5 percent asphaltic cement; specifications for the asphaltic cement used at Montgomery Dam were as follows:

All asphalt for use in the asphaltic concrete shall be uniform in character, shall not foam when heated to 350° and shall conform to the following specifications and requirements:

a. Penetration (tested per ASTM Specification D 5)
   i. At 77°, 100 grams, 5 seconds 50-60
   ii. At 32°, 200 grams, 60 seconds .... minimum, 12
b. Ductility at 77°, 5 centimeters per minute .... minimum, 140 cm. (Tested per ASTM Specification D 113)
c. Flash point (Cleveland open cup) .... minimum, 450° (Tested per ASTM Specification D 92)
d. Solubility in carbon tetrachloride .... minimum, 99.5 percent (Tested per ASTM Specification D 165)
e. Softening point (ring and ball method) .... minimum, 125° (Tested per ASTM Specification D 36)
f. Spot test (per paragraph 3 of AASHO Specification T 102) .... Negative
g. Results of tests made upon residues after thin-film oven heating, per test method No. Calif. 337-A, January 3, 1956, of the Division of Highways, Department of Public Works, State of California, shall conform

to the following as compared with like tests made on the identical material before such heating. Such method shall be considered to be a part hereof, provided that in lieu of the pertinent AASHO Specifications there shall be used the ASTM Specifications referred to in this item plus ASTM Specifications D-6 and E-11. The Contractor shall cause the producer of the asphalt to supply the Engineer, as required, with data or curves showing the relation between temperature and viscosity representative of the asphalt as furnished for the work.

Test	Percent of result before heating	
	Montgomery Dam	Cawker City Protective Dike [1]
i. Penetration, at 77°, retained, minimum...	65	55
ii. Penetration, at 32°, retained, minimum...	70	45
iii. Ductility, at 77°, retained, minimum...	80	30
iv. Increase in softening point, maximum......	10	15
v. Loss in weight, maximum...........	0.3	0.3

[1] The percentages for Cawker City Dike were added here for comparison with those used for Montgomery Dam.

A very low air-voids content resulting from proper mix design and compaction is required to obtain durable facings; however, a low air-voids ratio cannot be obtained by simply adding additional asphaltic cement. Air-void ratios of 1 percent are commonly obtained and the maximum air-void ratio allowed in the construction of an asphaltic facing should be 5 percent.

Experiences to date have indicated that densely graded aggregates with ample filler (minus No. 200), correctly proportioned with a 50 to 60 penetration, paving grade asphalt cement, will produce a very workable, relatively easily compacted hot mix at about 300° F. The thicker films of asphalt cement obtained with a rich mix of the slightly harder 50 to 60 asphalt cement, as compared with an 85 to 100 or 100 to 150 penetration asphalt cement, can be expected to increase watertightness, stability, and durability. The completed asphaltic concrete facing at Upper Blue River Dam is shown in figure 208.

Parapet walls should be used with asphaltic concrete facings in lieu of increasing the height of the dam to retard wave runup and overlapping. Galvanized corrugated metal has been used for a number of small dams [29, 32] and appears to be performing well; figure 211 shows the parapet wall used at Upper Blue River Dam. When parapet walls are used, freeboard heights should be similar to those required for earthfill dams (section 142) and runup and wave splash neglected. Wall heights can be determined by precedent or design experience.

For further information on asphaltic facings, the reader should consult the references at the end of this chapter.

161. *Steel.*—Steel facings have been used on relatively few dams throughout the world, but their satisfactory performance on these few dams illustrates clearly that they should be given serious economic consideration by dam designers. Few design criteria besides precedent are applicable, and the available literature should be consulted for a complete review of the practices used [4, 33, 34, 35].

Figure 212 shows the upstream face of the Bureau of Reclamation's El Vado Dam; the steel plate is in excellent condition after 35 years of service.

Steel-faced dams can be rapidly constructed and should be capable of tolerating greater embankment movements than either concrete or asphalt. The most prominent disadvantage to steel facings is the probability of corrosion reducing their economic life, although this can be combated effectively by using cathodic protection on both faces of the plate. Experience with the few steel-faced dams now in existence strongly indicates that a corrosion failure of the plate is remote, and that for all practical purposes the facing can be assumed to be permanent if proper maintenance is provided.

Figure 211. Completed asphaltic concrete facing at Upper Blue River Dam in Colorado. (Dept. of Public Utilities, Colorado Springs, Colo. photo.) 288–D–2885.

Steel-faced dams have generally been constructed with upstream slopes from 1.3:1 to 1.7:1. For rockfill dams, the upstream and downstream slopes need not be flatter than the natural slope of the material, which is generally from 1.3:1 to 1.4:1. The steeper slopes lead to reduced costs, but face construction difficulties will be slightly increased.

The portion of the embankment on which the steel plate bears, zone A in figure 202, should in general be constructed of well-graded, pervious gravel to provide a uniform bearing surface for the steel facing. Anchor rods should extend from the facing plates into the embankment to prevent uplift or loosening of the face due to embankment settlement or wave action. Two methods of anchoring the face plate to the em-

Figure 212. Steel facing on the upstream slope of El Vado Dam in New Mexico. The facing is in excellent condition after 35 years of service. 288–D–2886.

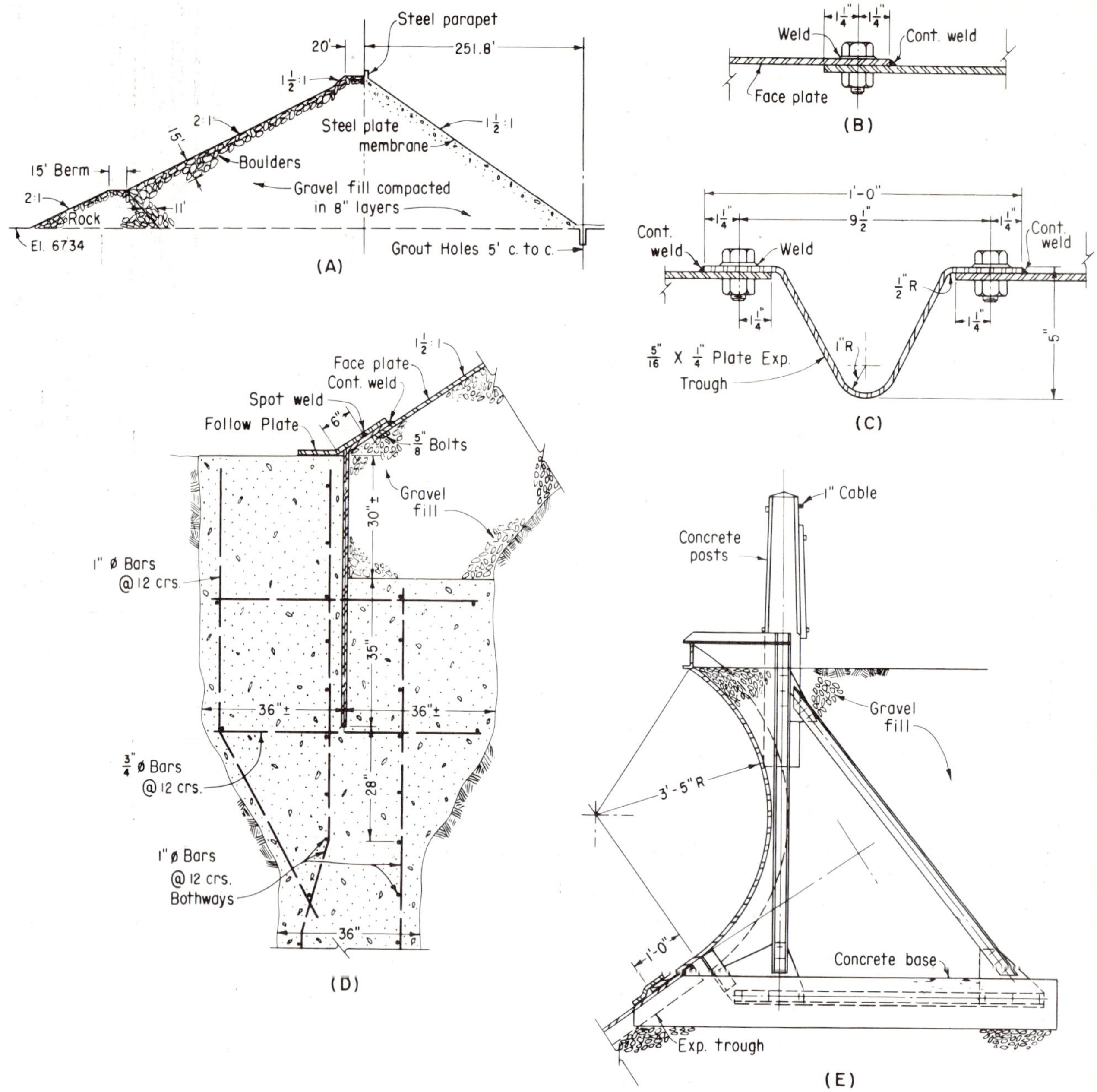

Figure 213. Details of steel membrane used at El Vado Dam in New Mexico. 288–D–2703.

bankment have been used. The first method requires that the steel plate be constructed on a scaffolding grid, which is raised a few feet off the face of the dam; bedding material is then placed between the facing and the embankment either after the plate construction is completed or concurrently with its construction up the face. The second method requires that anchor bar holes be dug in the completed face plate bedding material and the anchor bars grouted in with concrete. The first method provides more adequate plate anchoring but the latter method is less costly.

The steel plates should have a thickness of 1/4 to 3/8 inch, depending on the assumed magnitude of the movements of the dam. All joints and seams should have a continuous fillet weld, and nuts used on alinement or anchor

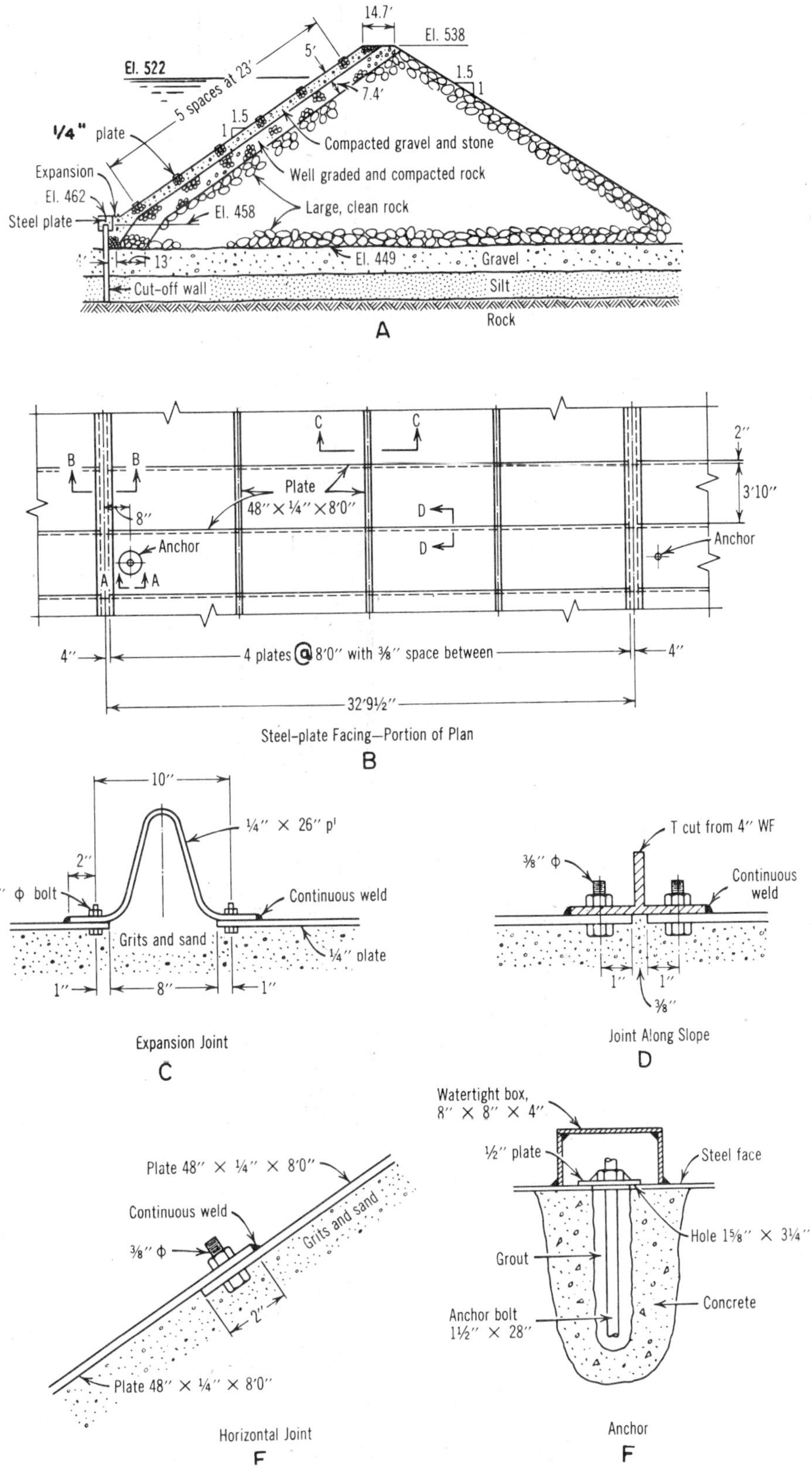

Figure 214. Details of steel membrane used at Rio Lagartijo Dike, Venezuela. Adapted from Sherard [33]. 288–D–2804.

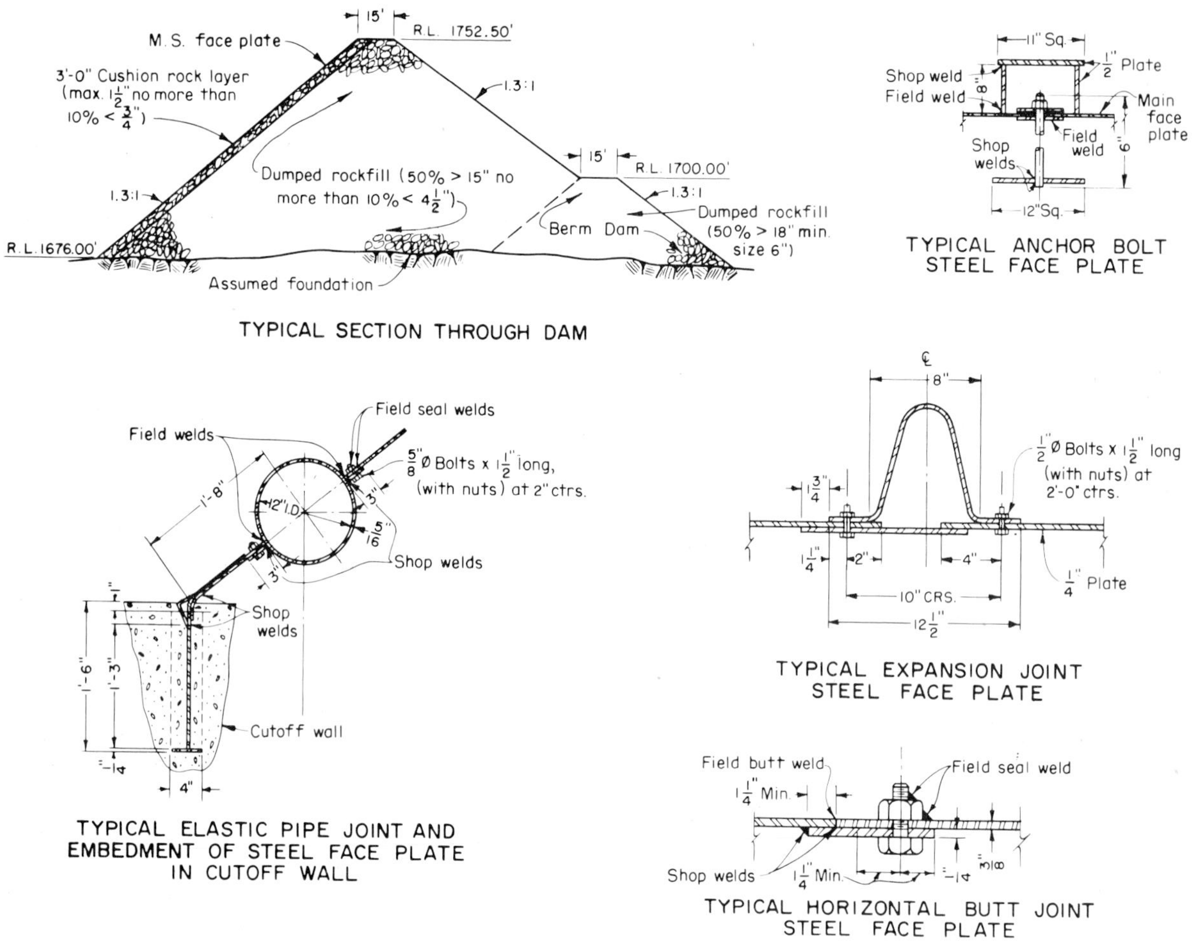

Figure 215. **Details of steel membrane at Sirinumu Dam, New Guinea. Adapted from Fraser [34]. 288–D–2805.**

bolts should be welded to the plate on all sides to prevent leakage. Plate sizes have varied considerably; large plates are more difficult to handle and appear to provide little advantage over smaller plates.

The foundation cutoff used should be the trench-type cutoff shown in figure 196. It should be designed to withstand the tension caused by settlement and stresses imposed by any differential face movement adjacent to the cutoff. Coping walls should be used to retard oversplash. When coping walls are used free-board requirements will be similar to those used for earth dams (sec. 142).

Expansion joints are usually V-shaped metal strips placed perpendicular to the axis of the dam and extending from the crest to the cutoff; the V-strips may have their raised portion placed on either the front or back of the steel face. Design details used for El Vado Dam in New Mexico, Rio Lagartijo Dike in Venezuela, and Sirinumu Dam in New Guinea are shown in figures 213, 214, and 215, respectively.

162. *Timber Planking*.—Timber planking has been used as a temporary type of membrane, but is not recommended for general use even though it is often the most inexpensive type of membrane to construct. The principal objections to this type of construction are the danger of loss by fire at low water and the relatively short life of timber construction when alternately exposed to wetting and drying.

E. BIBLIOGRAPHY

163. *Bibliography.*

[1] Galloway, J. D., "The Design of Rockfill Dams," Trans. ASCE, No. 104, pt. 2, 1939, p. 1.

[2] "Asphalt Deck is Dams Watertight Layer," Engineering News-Record, December 5, 1957, p. 36.

[3] "Abrasive Rock, Asphalt Facing for Water Supply Dam," Engineering News-Record, November 15, 1956, p. 44.

[4] Hovey, O. E., "Steel Dams," American Institute of Steel Construction, New York, 1935, p. 79.

[5] Reid, H. I., "Steel Plates with Welded Joints Seal Rockfill Dam," Engineering News-Record, vol. 108, p. 761.

[6] Seger, C. P., "Steel Used Extensively in Building El Vado Dam," Engineering News-Record, vol. 115, p. 211.

[7] "Asphalt on Earth- and Rock-Fill Dams," Norwegian Geotechnical Institute, Publication No. 48, Oslo, 1962.

[8] Mann, J. N., "Aspects of Rockfill Selection, Placement and Sluicing in the Construction of Moondarra Dam," Journal of the Inst. of Engineers, Australia, April-May, 1963.

[9] Pinkerton, I. L., and Paton, R. J., "Design and Construction of Geehi Dam," Journal of the Inst. of Civil Engineers, Australia, March 1968.

[10] Mitchell, W. R., Fidler, J., and Fitzpatrick, M. D., "Rowallan and Parangana Rockfill Dams," Journal of the Inst. of Civil Engineers, Australia, October-November, 1968.

[11] Fenton, J. D., "Hydraulic and Stability Analysis of Rockfill Dams," Dept. of Civil Engineering, University of Melbourne, July 1968.

[12] Robeson, F. A., and Crisp, R. L., "Rockfill Design — Carters Dam," Proc. ASCE, Journal Construction Div., vol. 92, No. CO3, September 1966.

[13] Crisp, R. L., "Design, Construction, and Instrumentation of Carters Rockfill Dam," Paper presented at USCOLD-CNC/ICOLD meeting, Vancouver, British Columbia, October 1966.

[14] Wilkins, J. K., "Decked Rockfill Dams," Journal Inst. of Engineers, Australia, April 1968.

[15] Rowe, P. W., "The Stress-Dilatancy Relation for Static Equilibrium of an Assembly of Particles in Contact," Proc. Roy. Soc., series A, vol. 269, No. 1339, October 9, 1962.

[16] Sowers, G. F., Williams, R. C., and Wallace, J. S., "Compressibility of Broken Rock and the Settlement of Rockfills," Proc. Sixth Int. Conf. Soil Mechanics and Foundations, vol. 2, Montreal, 1965.

[17] Marsal, R. J., "Large Scale Testing of Rockfill Materials," Journal of the Soil Mechanics and Foundations Division, ASCE, vol. 93, No. SM2, March 1967.

[18] Porkin, A. K., Trollope, D. H., and Lawson, J. D., "Rockfill Structures Subject to Water Flow," Paper 4973, Journal of the Soil Mechanics and Foundations Division, ASCE, November 1966.

[19] Wilkins, J. K., "Rockfill Structures Subject to Water Flow," Discussion of Proceedings, Paper No. 4973, Journal of the Soil Mechanics and Foundations Division, ASCE, vol. 93, No. SM3, May 1967, p. 177.

[20] Givdici, Sergio, "Rockfill Structures Subject to Water Flow," Discussion of Proceedings, Paper No. 4973, Journal of the Soil Mechanics and Foundations Division, ASCE, vol. 93, No. SM3, September 1967, pp. 329–337.

[21] Kjoernsli, B., and Torblaa, I., "Asphalt on Earth and Rockfill Dams," Publication No. 48, Norwegian Geotechnical Institute, Oslo, 1966.

[22] Davis, R. E., "Tests on Models of Sariyar Rockfill Dam," Report to International Engineering Co., Berkeley, Calif., July 1949.

[23] Elsden, O., Keefe, H. G., and Bishop, A. W., "Embankment Dams," in Hydro Electric Engineering Practice, second edition, ch. IX, vol. 1, Glasgow, 1964.

[24] Fumagelli, Emanuele, "Tests on Cohesionless Materials for Rockfill Dams," Journal of the Soil Mechanics and Foundations Division, ASCE, vol. 95, SM1, January 1969, pp. 313–330.

[25] Wesley, L. D., "Settlement of Embankment Materials in Earth and Rockfill Dams," New Zealand Engineering, May 15, 1967, pp. 179–187.

[26] Bertram, G. E., "Rockfill Compaction by Vibratory Rollers," Proc. 2nd Pan-American Conference on Soil Mechanics, Brazil, 1963.

[27] Ellsperman, L. M., and Walker, F. C., "Factors Affecting the Use of Asphaltic Diaphragms in Earth and Rockfill Dams in the Western United States," Seventh International Congress on Large Dams, Rome, Italy, 1961.

[28] Scheidenhelm, F. W., Snethlage, J. B., and Vanderlip, A. N., "Montgomery Dam — Rock Fill with Asphaltic Concrete Deck," Trans. ASCE, vol. 125, pt. 2, 1960.

[29] "Laboratory Investigation of Asphaltic Concrete, Montgomery Dam, Colorado," Association of Asphalt Paving Technologists, Proceedings, 1958.

[30] "Mix Design Investigation of Asphaltic Concrete for Facing Downs Protective Dike," Chemical Engineering Laboratory Report No. ChE–66, Bureau of Reclamation, December 29, 1966.

[31] "Mix Design Investigations of Asphaltic Concrete for Dam Facing—Glen Elder Dam," Chemical Engineering Laboratory Report No. ChE–42, Bureau of Reclamation, February 18, 1965.

[32] Proudfit, D. P., "Asphaltic Concrete Facing for a Rock-Fill Dam," Civil Engineering, March 1968.

[33] Sherard, J. L., "A Steel Faced Rockfill Dam," Civil Engineering, vol. 29, No. 10, October 1959, pp. 698–701.

[34] Fraser, J. B., "A Steel-faced Rockfill Dam for Papua," Civil Engineering Transactions, Institution of Engineers, Australia, vol. CE4, No. 1, March 1962, pp. 35–48.

[35] Sherard, J. L., Woodward, R. J., Gizienski, S. F., and Clevenger, W. A., "Earth and Earth-Rock Dams," John Wiley & Sons, Inc., New York, N.Y., 1963.

[36] Steele, I. C., and Cooke, J. B., "Concrete-Face Rock-Fill Dams," Handbook of Applied Hydraulics, edited by C. V. Davis and K. E. Sorensen, third edition, McGraw-Hill, New York, 1969.

[37] "Question 31—High Rockfill Dams," Eighth International Congress on Large Dams, vol. III, 1964.

[38] Oliver, H., "Through and Overflow Rockfill Dams —New Design Techniques," Proc. Inst. Civil Engrs, Paper 7012, 1967.

[39] "Question 27—Sealing of Earth and Rockfill Dams with Bitumen or Other Materials," Seventh International Congress on Large Dams, vol. IV, Rome, 1961.

[40] Asbeck, Baron W. F. van, "Bitumen in Hydraulic Engineering," vol. II, Elsevier, London, 1964.

[41] "Symposium on Rockfill Dams," Trans. ASCE, vol. 125, pt. 2, 1960.

[42] Terzaghi, K., and Peck, R. B., "Soil Mechanics in Engineering Practice," second edition, art. 62, John Wiley & Sons, Inc., New York, 1967.

[43] Leps, T. M., "Review of Shearing Strength of Rockfill," Journal Soil Mechanics and Foundations Division, ASCE, vol. 96, No. SM4, July 1970, pp. 1,159–1,170.

[44] Shand, N., and Pells, P. J. N., "Experience in the Design and Construction of Reinforced Rockfill Dams," Tenth International Congress on Large Dams, Montreal, June 1970, pp. 270–317.

Chapter VIII

Concrete Gravity Dams

A. T. LEWIS,[1] J. S. CONRAD,[2] L. M. CHRISTIANSEN [2] and E. L. WATSON [3]

A. INTRODUCTION

164. ***Origin and Development.***—A concrete gravity dam is a structure proportioned so that its own weight provides the major resistance to the forces exerted upon it. If the foundation is adequate, and the dam is properly designed and constructed, a solid concrete dam is a permanent structure which requires little maintenance. Gravity dams of uncemented masonry were built several thousand years B.C. Evidence in ruins indicate base widths as much as four times the height. With the passing of centuries, various types of mortar were used to bind the masonry together, thereby increasing the stability and watertightness and permitting steeper slopes. Fi-

[1] Formerly Engineer, Concrete Dams Section, now Chief, Hydraulic Structures Branch, Bureau of Reclamation.

[2] Engineer, Hydraulic Structures Branch, Bureau of Reclamation.

[3] Engineer, Concrete Dams Section, Bureau of Reclamation (retired).

Figure 216. Willwood Diversion Dam, a concrete-gravity structure on the Shoshone River in Wyoming. 2G–1–9.

nally, concrete and cement mortar were used in the construction of cyclopean masonry dams, which were the forerunners of the modern mass concrete gravity dams.

165. *Scope of Discussion.*—This chapter is confined to discussions of "small" concrete gravity dams with limiting dimensions as follows:

(1) Not more than 50 feet high if on rock foundation (firm, hard, solid masses of granite, sandstone, limestone, etc.).

(2) If on a pervious foundation (soil through which water percolates relatively freely, such as gravels, glacial tills, etc.), the permissible height of the structure should be governed by the bearing strength of the soil but should not exceed 30 feet. Maximum net head (headwater to tailwater) is not to exceed 20 feet. For dams on rock where the major part of the dam is overflow section, the concrete gravity wing walls for nonoverflow sections should prove economical. For dams on soil the concrete overflow section is believed to be the safest and most economical type structure; however, earth dikes are usually more economical for the nonoverflow part of the dam. Willwood Diversion Dam on the Shoshone River in Wyoming (fig. 216), although exceeding the 50-foot height, typifies the scope of this discussion for dams on rock foundations. Woodston Diversion Dam on the South Fork of the Solomon River in Kansas (fig. 217) typifies the scope of the discussion for dams on pervious (soil) foundations.

This chapter discusses primarily the forces which act on concrete gravity dams and the requirements for stability. Additional considerations in connection with concrete structures on pervious (soil) foundations are presented and, finally, current practices regarding miscellaneous details of design are briefly described.

Figure 217. Construction view of overflow section of Woodston Diversion Dam, a concrete-gravity structure on the South Fork of the Solomon River in Kansas. 796–701–2944.

B. FORCES ACTING ON THE DAM

166. *General.*—For the design of gravity dams, it is necessary to determine the forces which may be expected to affect the stability of the structure. The forces which must be considered are those due to: (1) external water pressure, (2) internal water pressure (pore pressure or uplift) in the dam and foundation, (3) silt pressure, (4) ice pressure, (5) earthquake, (6) weight of the structure, and (7) forces from gates or other appurtenant structures. In designing the crest of an overflow section, the possibility of subatmospheric pressure developing between the overflowing sheet of water and the concrete should be considered (see chapter IX).

Not all of the above loads will occur or be at maximum intensity at the same time. An expanding ice sheet is not a factor during a maximum flood. The chances of an earthquake and a maximum flood occurring at the same time are extremely remote. The combinations of loadings to be used should be selected by an engineer experienced in the design of dams.

Figure 218(A) shows a nonoverflow section loaded with reservoir water and tailwater. Symbols and definitions for this loading are given below:

ψ=angle between face of element and the vertical.

T=horizontal distance from upstream edge to downstream edge of section.

I=moment of inertia of base of section 1-foot wide about its center of gravity, equal to $\frac{T^3}{12}$.

w_c=unit weight of concrete.

w=unit weight of water.

h or h'=vertical distance from reservoir water or tailwater, respectively, to base of section.

p or p'=reservoir water or tailwater pressure, respectively, at base of section. It is equal to wh or wh'.

W_o=dead load weight above base of section under consideration including the weight of the concrete, W_c, plus such appurtenances as gates and bridges.

W_w or W_w'=vertical component of reservoir water or tailwater load, respectively, on face above base of section.

M_o=moment of W_o about center of gravity of base of section.

M_w or M_w'=moment of W_w or W_w' about center of gravity of base of section.

V or V'=horizontal component of reservoir water or tailwater load, respectively, on face above base of section. This is equal to $\frac{wh^2}{2}$ for V and $\frac{w(h')^2}{2}$ for V' for normal conditions.

M_p or M_p'=moment of V or V' about center of gravity of base of section, equal to $\frac{wh^3}{6}$ for M_p *and* $\frac{w(h')^3}{6}$ for M_p'.

ΣW=resultant vertical force above base of section.

ΣV=resultant horizontal force above base of section.

ΣM=resultant moment of forces above base of section about center of gravity of base of section.

e=distance from center of gravity of base of section to point where resultant of ΣW and ΣV intersects base of section. It is equal to $\Sigma M/\Sigma W$.

U=total uplift force on horizontal section, equal to $T\left(\frac{p+p'}{2}\right)$.

The loadings have been summated into horizontal and vertical components above, which implies that the foundation or potential failure plane is horizontal or nearly horizontal. If the foundation or potential failure plane departs markedly from the horizontal, the loads should be resolved into components normal and parallel to that plane.

167. *Water Pressure.*—(a) *External.*—The external water pressure acting on a nonoverflow

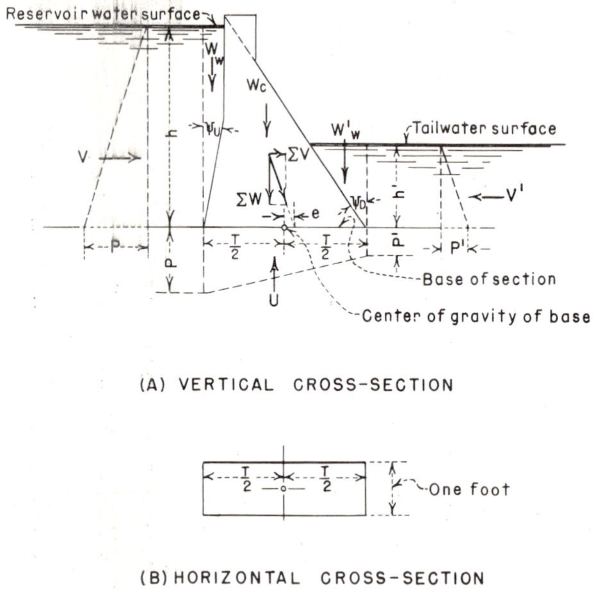

Figure 218. Forces acting on a concrete-gravity dam. 288–D–2505.

dam is illustrated in figure 218(A). On the upstream face, for example, the horizontal force is V and the vertical force is W_w. The weight of water is generally accepted as 62.5 pounds per cubic foot.

On overflow dams without control features, the total horizontal water pressure on the upstream face is closely represented by the trapezoid (*abcd*) in figure 219, in which the unit pressures at the top and at the bottom are $62.5h_1$ and $62.5h$, respectively. The line of action of the force passes through the center of gravity of the trapezoid.

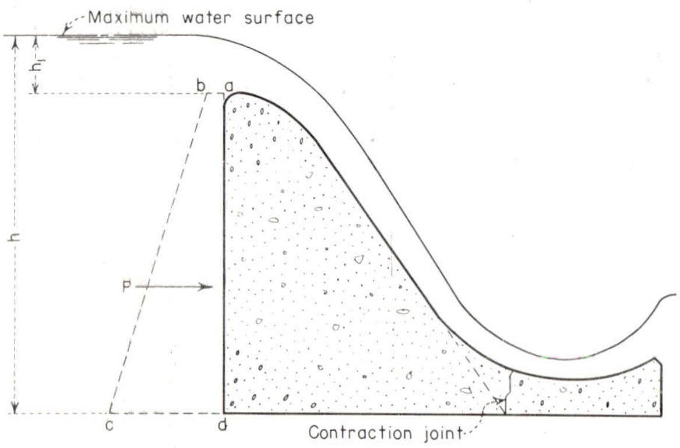

Figure 219. Water pressures acting on an overflow concrete dam. 288–D–2506.

The vertical pressure component of water flowing over the top of the spillway is not used in the analysis because most of the total head has changed to velocity head. Also, the pressure distribution is complex and difficult to determine. The sheet of water flowing down the downstream face generally does not exert enough pressure on the dam to warrant consideration. Where tailwater or backwater stands against the downstream face, it is treated in the same manner as the tailwater in figure 218(A). However, during major overflows of water, the tailwater pressures are involved in the energy dissipating process and may contribute only minor stabilizing forces on the dam.

(b) *Internal or Uplift Pressure.*—Uplift forces occur as internal pressures in pores, cracks, and seams in both the dam and dam foundation. It is evident that these spaces in the dam or foundation will be filled with water, which exerts a pressure in all directions. This pressure may have an important effect on the stability of the dam and must be included in the analysis. It is assumed that uplift pressures are not affected by earthquake forces.

(1) *Dams on rock foundations.*—The intensity of uplift pressure under a concrete dam on a rock foundation is difficult to determine. It is generally assumed that pore pressures in rock or concrete are effective over the entire base of the section. It is evident that under sustained loading the intensity of uplift at the upstream face is equal to the full reservoir pressure and approaches a straight-line variation from this point to tailwater pressure at the downstream face, or zero if there is no tailwater. This is true not only at the contact between the dam and the foundation but within the body of the dam itself.

Uplift pressures can be reduced by forming drains through the concrete of the dam and by drilling drainage holes into the foundation rock. Such drains are usually provided near the upstream face of the dam although care must be exercised to insure that direct piping from the reservoir will not result. Drains of this type are provided in all Bureau of Reclamation dams of considerable height, and actual measurements of uplift under the dams have proved them to be very effective. If the rock of the foundation of a proposed dam were absolutely homogeneous, the effectiveness of the drains could be predetermined. However, owing to the presence of seams and fissures and the uncertainty of intercepting them with the drains, the safest course is to assume the straight-line variation from headwater to tailwater pressures as a measure of uplift. Any other assumptions should be verified by electric analogy or other comparable methods of analysis conducted by engineers experienced in this field.

Other methods used to reduce the uplift at the contact of the dam with the foundation include construction of cutoff walls under the upstream face, construction of drainage channels (usually of sewer pipe) between the dam and the foundation, and pressure grouting the foundation. These methods usually are considered only as additional safety factors in the design of small dams and are not considered as justification for reducing design requirements.

(2) *Dams on pervious (soil) foundations.*—Uplift pressures under a concrete dam on a pervious (soil) foundation are related to seepage through permeable materials. Water percolating through pore spaces in the materials is retarded by frictional resistance, somewhat the same as water flowing through a pipe. The intensity of the uplift can be controlled by construction of properly placed aprons, cutoffs, and other devices. A discussion of these and methods for determining uplift pressures is included in sections 178 through 181.

168. *Silt Pressure.*—Nearly all streams carry an appreciable amount of silt during both normal and flood flows. Where silt is present in a stream on which a dam is built, it will eventually find its way to the reservoir and be deposited in the still water adjacent to the dam. Methods for determining the amount of silt and its deposition in a reservoir are discussed in appendix H. If allowed to accumulate against the upstream face of the dam, the saturated silt will exert loads greater than the hydrostatic pressure of water only. In the absence of reliable test data, a rather common assumption of the magnitude of saturated silt pressure is to consider the horizontal load as equivalent to that of a fluid weighing 85 pounds per cubic foot and the vertical weight as 120 pounds per cubic foot.

Sluiceways are often provided in gravity dams to reduce the accumulation of silt in the proximity of the upstream face of the dam. In diversion dams the main function of the sluiceway is to keep the headworks and canal free from silt, but some benefit may also be derived for the dam by reducing the silt load.

Many gravity dams have been designed without regard to silt load. In general, the silt load against storage dams will be a small factor but against diversion dams it is likely to be more important. In either case there is some basis for neglecting the silt load. Initially the silt load is not present, and by the time it might become a significant factor, the silt has consolidated to some extent and therefore acts less like a fluid. Furthermore, silt deposited in the reservoir will probably be somewhat impervious and will help to minimize the uplift under the dam.

Greater importance must be attributed to silt load when the primary purpose of the dam is silt retention. In this situation the designer may not be satisfied to assume an arbitrary equivalent fluid pressure as mentioned above. More precise computations of silt load can be made by combining the hydrostatic pressure with the horizontal component of the silt load, V_s, as determined by the Rankine formula neglecting cohesion:

$$V_s = \frac{w_s h^2}{2}\left(\frac{1-\sin\phi}{1+\sin\phi}\right) \qquad (1)$$

where:

w_s=submerged weight of material,
h=depth of material, and
ϕ=angle of internal friction.

Here too obstacles will be met in making assumptions regarding submerged weight, angle of internal friction, etc., owing to lack of data. A comprehensive discussion of the Rankine formula and the assumptions involved in its application is beyond the scope of this work, but it may be found in many standard texts.

169. *Ice Pressure.*—Ice pressure is produced by thermal expansion in the ice sheet and by wind drag. The necessary allowance to be made for ice load in the design of a concrete dam is difficult to determine. Data concerning the physical characteristics of ice such as its crushing strength, its modulus of elasticity, and the effects of plastic flow are inadequate and approximate. Furthermore, the thrust exerted by expanding ice depends on the thickness of the sheet, the rate of temperature rise in the ice, fluctuations in the water surface, character of the reservoir shores, wind drag, and other factors. The rate of temperature rise in the ice is a function of rate of rise of the air temperature and the amount of snow cover on the ice. Lateral restraint of the ice sheet depends on the character of the reservoir shores.

In view of all of these variables, the designer is faced with a difficult task in estimating the amount of ice pressure acting against a structure. However, several aids are available to help in the design. In 1947 Rose [1] [4] collected and put into usable form all the known data on ice pressures. His charts, figure 220, show the thrust in kips for thicknesses up to 4 feet for air temperature rises of 5°, 10°, or 15° F. per hour (curves A, B, and C), respectively. The designer may elect to include or disregard the effects of lateral restraint and solar energy.

The Bureau of Reclamation has conducted field and laboratory tests on ice temperatures and pressures. These tests have been summarized by Monfore [2]. With the installed equipment, ice pressures were obtained directly from gages or by computations based on ice temperature readings. Close agreement was obtained by the two methods. In the field tests at Eleven Mile Canyon Reservoir, it was found that most of the expansion in the ice sheet took place in the upper portions and, therefore, the thrust was largely concentrated there. In the computations for thrust exerted by an 18-inch-thick sheet, it was found that no change in temperature occurred at the bottom of the ice; hence, zero pressure was assumed there. It was also demonstrated that a 5-inch snow covering had a remarkable insulating effect in preventing temperature rise and thus minimizing the ice thrust.

Within the United States it is only in the northern and mountainous regions that ice

[4] Numbers in brackets refer to items in the bibliography, sec. 185.

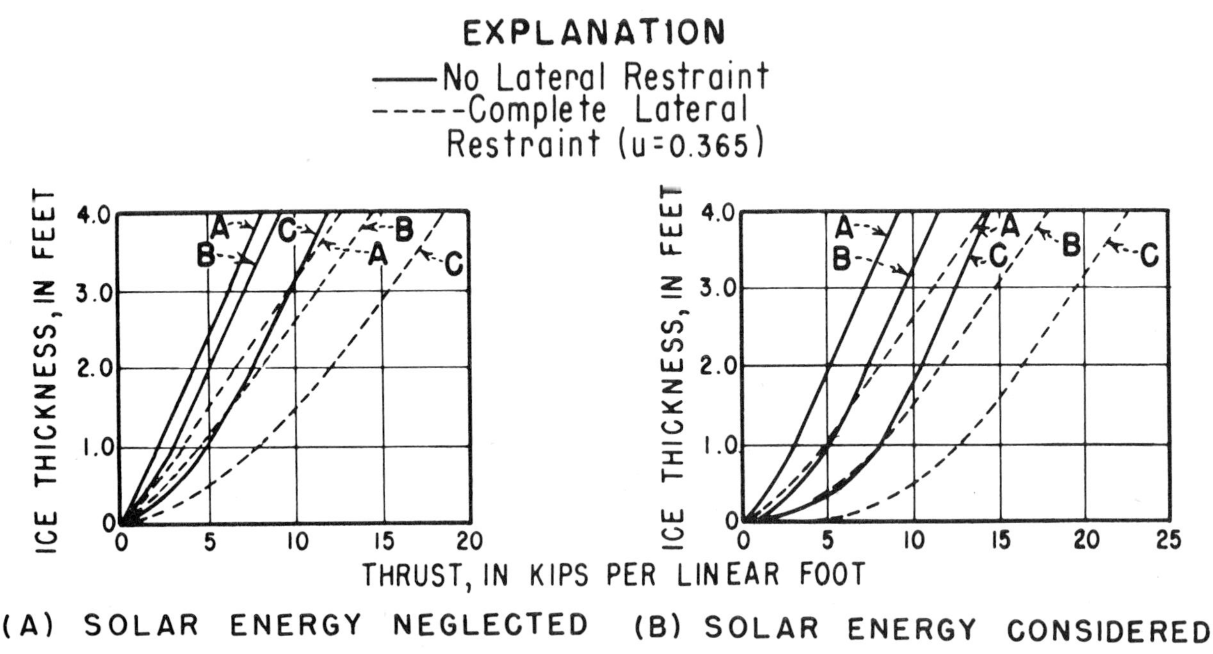

Figure 220. Ice thrust in relation to ice thickness, air-temperature rise, and restraint. 288–D–2507.

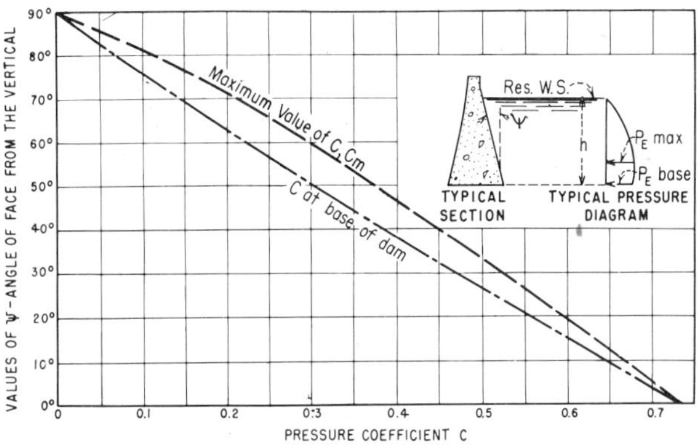

Figure 221. Base and maximum pressure coefficients for constant sloping faces. 288–D–2508.

pressures can be expected to be a significant load on small dams. Where a gravity diversion dam has gates, it is common practice to heat the gates and thus prevent ice forming against the metal, but in the abutment sections ice load can be an important consideration.

170. *Earthquake.*—(a) *General.*—Earthquakes impart accelerations to the dam which usually increase the effective loadings on the dam. Some allowance for earthquake effects must therefore be made in the design of concrete gravity dams.

Both vertical and horizontal earthquake loads should be applied in the direction which produces the least stable structure. For the condition of full reservoir this will be a foundation movement in the upstream direction and a foundation movement downward. The upstream movement increases the downstream force of the water and silt loads and produces a downstream inertial force from the mass concrete in the dam. The downward movement decreases the effective weight of the water above a sloping face and of the concrete in the dam. Both increasing the horizontal loads in a downstream direction and decreasing the effective weights tend to decrease the stability of

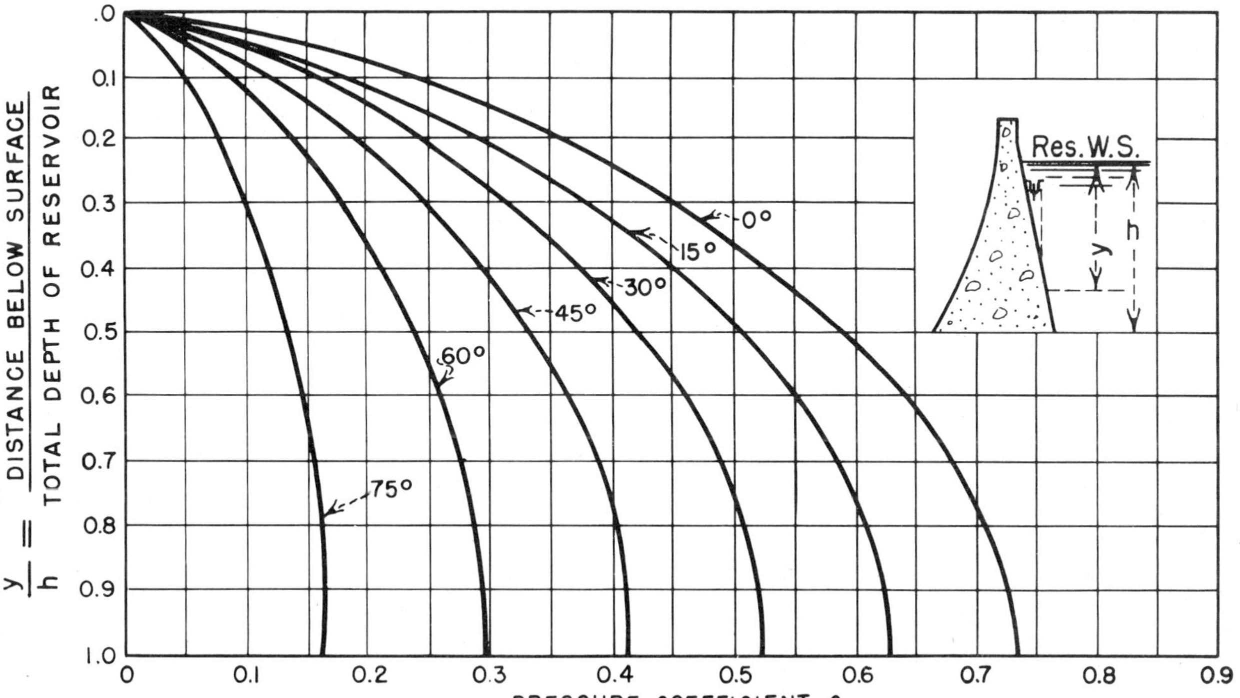

$$C = \frac{1}{2} C_m \left[\frac{y}{h}\left(2 - \frac{y}{h}\right) + \sqrt{\frac{y}{h}\left(2 - \frac{y}{h}\right)} \right]$$

where: C_m is maximum C value from figure 221

Figure 222. Coefficients for pressure distribution for constant sloping faces. 288–D–2509.

the structure. In order to determine the total forces due to an earthquake, it is necessary to establish the earthquake intensity or acceleration. This is usually expressed in relation to the acceleration due to gravity. The accelerations that may be reasonably expected at the damsite are determined from a consideration of the geology of the site, proximity to major faults, previous earthquake history of the region, and such seismic records as are available. In areas not subjected to extreme earthquake conditions a horizontal acceleration of 0.10 of gravity and a vertical acceleration of 0.05 of gravity are generally used.

The internal friction of silt provides considerable damping. However, until more exact data are determined, it is assumed that the dynamic effect of saturated silt is equivalent to that of water only.

Resonance in low dams is not likely to occur during earthquake shock for several reasons. The fundamental period of vibration of a concrete dam 50 feet high and of triangular cross section is between 0.03 and 0.04 of a second. Vibration periods of important earth shocks are estimated by various authorities to lie between 0.15 and 0.6 second; therefore, resonance in the dam is not likely to take place. Furthermore, earthquakes are experimentally and analytically treated as harmonic but ground motions recorded in the destructive zone of the quake do not appear to be harmonic. Also, many forms of damping that are difficult to evaluate act to prevent resonance.

In 1952 Zangar [3] presented formulas for computing the hydrodynamic pressures exerted on vertical and sloping faces by horizontal earthquake. The formulas were derived by electric analogy, based on the assumption that water is incompressible. For low dams the error involved in computing the earthquake force on the water because of this simplifying assumption is probably less than 1 percent.

(b) *Horizontal Earthquake.*—The effect of inertia on the concrete should be applied at the center of gravity of the mass, regardless of the shape of the cross section. For dams with vertical or sloping upstream faces, the increase in water pressure, P_e, in pounds per square foot, at any elevation due to horizontal earthquake is given by the following equation:

$$P_e = C\lambda wh \qquad (2)$$

where:

C (a dimensionless coefficient giving the distribution and magnitude of pressures)

$$= \frac{C_m}{2}\left[\frac{y}{h}\left(2-\frac{y}{h}\right)+\sqrt{\frac{y}{h}\left(2-\frac{y}{h}\right)}\right]$$

λ = the earthquake intensity = $\frac{\text{earthquake acceleration}}{\text{acceleration of gravity}}$,

w = unit weight of water, pounds per cubic foot,

h = total depth of reservoir at section being studied, feet,

y = the vertical distance from the reservoir surface to the elevation in question, feet, and

C_m = maximum value of C for a given constant slope (fig. 221).

Values of C for various degrees of slope and relations of y and h may be obtained from figure 222. It may be shown analytically that the total horizontal force, V_e, above any elevation y distance below the reservoir surface, and the total overturning moment, M_e, above that elevation are:

$$V_e = 0.726\ P_e y \qquad (3)$$

and

$$M_e = 0.299\ P_e y^2 \qquad (4)$$

For dams with a combination vertical and sloping face, the procedure to be used is governed by the relation of the height of the vertical portion to the total height of the dam, as follows:

(1) If the height of the vertical portion of the upstream face of the dam is equal to or greater than one-half the total height of the dam, analyze as if vertical throughout.

(2) If the height of the vertical portion of the upstream face of the dam is less than one-half of the total height of the dam, use the pressures on a sloping line connecting the point of intersection of the upstream face of the dam and reser-

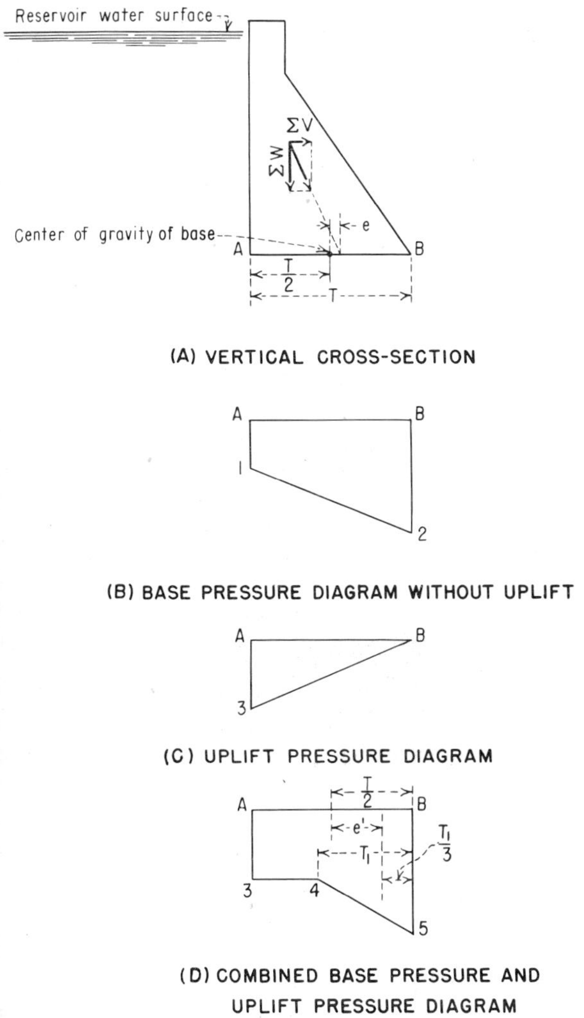

Figure 223. Foundation base pressures for a concrete-gravity dam. 288–D–2510.

voir surface with the point of intersection of the upstream face of the dam and the foundation.

(c) *Vertical Earthquake.*—On sloping faces of dams the weight of the water above the slope should be modified by the appropriate acceleration factor. The weight of the concrete also should be modified by this acceleration factor.

171. *Weight of Structure.*—The weight of the structure includes the weight of the concrete plus appurtenances such as gates and bridges. However, for most low dams only the dead load due to the weight of the concrete is used in the analysis. The unit weight of concrete ordinarily is considered to be 150 pounds per cubic foot. The total weight acts vertically through the center of gravity of the cross section.

172. *Reaction of Foundation.*—Under stable conditions the resultant of the horizontal and vertical loads on the dam will be balanced by an equal and opposite force which constitutes the reaction of the foundation. The vertical reaction of the foundation, computed without uplift, is represented by the trapezoid $A12B$, figure 223 (B).

The stresses $A1$ and $B2$ are determined by the use of eccentric loading formulas and are as follows:

$$A1=\frac{\Sigma W}{T}\left(1-\frac{6e}{T}\right) \tag{5}$$

$$B2=\frac{\Sigma W}{T}\left(1+\frac{6e}{T}\right) \tag{6}$$

When uplift is introduced and the uplift pressure at the upstream face is less than $A1$, the final stresses may be computed by the above formulas. If the uplift pressure at the upstream face is greater than $A1$, the foundation pressure at the base will have to be revised as in figure 223 (D).

To revise the foundation pressure due to excessive uplift, the following procedure should be used:

(1) A horizontal crack is assumed to exist and extend from the upstream face toward the downstream face to a point where the vertical stress of the adjusted diagram is equal to the uplift pressure at the upstream face (fig. 223 (D)).

(2) From figure 223 (A) and (D) and taking moments about the center of gravity of the base, there are obtained:

$$e'=\frac{\Sigma M}{\Sigma W-A3\cdot T} \tag{7}$$

$$T_1=3\left(\frac{T}{2}-e'\right) \tag{8}$$

and

$$B5=\frac{2(\Sigma W-A3\cdot T)}{T_1}+A3 \tag{9}$$

C. REQUIREMENTS FOR STABILITY

173. *General.*—A concrete gravity dam must be designed to resist, with ample factor of safety, these three tendencies to destruction: (1) overturning, (2) sliding, and (3) overstressing.

174. *Overturning.*—There is a tendency for a gravity dam to overturn about the downstream toe at the foundation or about the downstream edge of any horizontal section. If the vertical stress at the upstream edge of any horizontal section computed without uplift exceeds the uplift pressure at that point, the dam is considered safe against overturning. The most critical condition for inducing overturning is when, at the upstream face, the uplift pressure exceeds the vertical stress at any horizontal section and the combined pressure diagram of figure 223(D) is assumed. Under this condition, if *B*5 in figure 223(D) is less than the allowable stress for the concrete in any horizontal section or less than the allowable stress in the concrete and foundation for a horizontal section at the foundation, the dam is considered to be safe against overturning.

175. *Sliding.*—The horizontal force, ΣV, in figure 166 tends to displace the dam in a horizontal direction. This tendency is resisted by the frictional and shear resistance of the concrete or the foundation.

The shear friction factor [4] is the sliding stability criterion for all large concrete dams and should generally be used for small concrete dams on rock foundations. The shear friction factor is:

$$Q=\frac{CA+(\Sigma W-U)\tan\phi}{\Sigma V}$$

where:

C=cohesion value of concrete or rock

A=area of base considered

$\tan\phi$=coefficient of internal friction

The values of cohesion and internal friction of the rock or rock-concrete contact must generally be determined by special laboratory tests. For certain rock types, free from adverse geologic structures, cohesion and internal friction can be estimated from published test data. Rock with infilled jointing or lamination and other adverse geologic structures require investigation and testing of the properties of the rock surfaces and infilling material.

The acceptable factor of safety is dependent on many conditions. For small storage dams where failure would mean loss of life or other catastrophic occurrences, the minimum shear friction factor for normal loading conditions is 4. Under extreme loading conditions, the shear friction factor should be at least 1.5. A typical normal loading condition would include normal headwater, tailwater, uplift, and silt (if applicable). Extreme loading conditions should include the following:

(1) Normal water surface, drains inoperative, and earthquake, or

(2) Maximum water surface and drains inoperative.

For small dams with minimal storage where loss of life, extensive property damage, or any other catastrophic occurrence are not involved in a failure, the acceptable minimum safety factor for rock foundations is 2 for normal loading conditions and 1.25 for extreme loading conditions.

For concrete structures on noncohesive foundation materials, it is usually not feasible to obtain safety factors equivalent to prescribed safety factors for structures on competent rock. However, these structures are usually low and of minimal storage where failure would not involve loss of life or other catastrophic occurrences. A general guide for acceptable factors of safety is 2.0 for normal loading conditions and 1.25 for extreme loading conditions. These safety factors may have to be reduced further in certain cases. The safety factor for these structures is set by selecting a sliding factor, *f*, where:

$$f=\frac{\text{Coefficient of static friction}}{\text{Factor of safety selected}},$$

then, when $(\Sigma W-U)f\geqq\Sigma V$, the factor of safety against a sliding failure is equal to or greater than that selected. Exact values of the coefficient of static friction between surfaces of the assumed sliding plane must generally be determined by tests. Many static friction tests have been made and the results published, but care should be exercised in using them. For certain foundations, such as sand and gravel, published factors [9] can be used as a guide in the selection of the coefficient of static friction; however, for rock foundations consideration must be given to the extent of jointing and the jointing pattern before selecting a published friction coefficient. Determination of shear strength characteristics of shales, silts, and clays usually requires testing.

Concrete cutoff walls are often provided on structures constructed on soil foundations. The cutoff, properly located and designed, engages an additional volume of foundation materials that must be moved before the structure can slide. If a stratum weaker than the overlying strata exists in a foundation of rock or soil, the sliding stability should also be investigated along the top of the weak bed. In this case, however, the weight of the overlying strata and the shear resistance of material downstream from the structure also would be considered in computing the sliding factor.

176. *Overstressing.*—The unit stresses in the concrete and the foundation must be kept within prescribed maximum values. Normally, the stresses in the concrete of gravity dams within the scope of this text will be so low that a concrete mix designed as specified in appendix F to meet other requirements such as durability and workability will attain sufficient strength to insure a factor of safety of at least 4 against overstressing.

The foundation should be investigated and the maximum allowable stress established. Engineering properties of foundation materials and accompanying considerations affecting such properties are discussed in chapter V. Local codes of allowable bearing pressures and engineers qualified in evaluating the adequacy of foundation materials should be consulted as far as possible before final design. Suggested allowable bearing values for footings for structures appurtenant to small dams are given in appendix C. These may be used as a guide in designing small concrete dams. If there is any doubt as to the proper classification and adequacy of the foundation materials, laboratory tests should be made to determine the allowable bearing pressures.

D. DAMS ON PERVIOUS (SOIL) FOUNDATIONS

177. *General.*—Small gravity dams constructed on rock present relatively few difficult foundation problems. The design of dams on pervious (soil) foundations, however, involves problems of erosion of the foundation material, settlement, and seepage under the structure. The complexity of these problems varies greatly and depends on the type, stratification, permeability, homogeneity, and other properties of the foundation materials, as well as the size and physical requirements of the structure itself.

The design of concrete gravity storage dams and diversion dams more than 30 feet high on pervious (soil) foundations usually requires extensive field and laboratory investigations. Such structures are beyond the scope of this text, which for pervious (soil) foundations is limited to gravity dams whose maximum net head (headwater to tailwater) is not appreciably greater than 20 feet.

The control of erosion, seepage, and uplift forces under dams constructed on pervious foundations often requires the use of some, all, or various combinations of the following devices:

(1) Upstream apron, usually with cutoffs at the upstream end.

(2) Downstream apron, with scour cutoffs at the downstream end, and with or without filters and drains under the apron.

(3) Cutoffs at the upstream or downstream end or at both ends of the overflow section, with or without filters or drains under the section.

The locations and extent of these devices to obtain optimum safety and economy of design depend on many conditions. Therefore, only devices or combinations of devices which might be used with relatively simple diversion dams are discussed.

178. *Aprons*.—A concrete apron may be placed upstream of the dam in conjunction with one of the various types of cutoff walls. The function of the apron is to increase the length of the path of percolation in order to reduce uplift under the main portion of the dam. Usually the apron is connected to the dam and to a concrete cap over the cutoff with flexible waterstops, which allow differential movement to take place without accompanying detrimental cracking. The safety of the structure may be further improved by placing an impervious earth blanket over a portion of the concrete apron and the streambed upstream from it.

Downstream concrete aprons have two functions. They lengthen the path of percolation in the foundations and also provide a basin where the energy of the overflowing water may be safely dissipated. Energy dissipation on the concrete helps to prevent dangerous erosion at the toe of the dam. In cases where it is not feasible to construct a concrete apron of sufficient length to avoid erosion completely, additional protection may be gained by placing riprap downstream from the apron.

179. *Cutoff Walls*.—Cutoff walls may be constructed of timber, concrete, cement-bound curtains, steel sheet piling, or impervious earth compacted in a trench. Each type can be effective under appropriate circumstances.

Timber piling may be used as cutoffs under upstream or downstream aprons. Dimension timber piling is not recommended where driving is necessary. Better construction practice consists of erecting overlapping treated timbers in an excavated open trench and then backfilling and compacting impervious material in the trench around the timbers. This method prevents brooming or splitting of the timbers which results when timber is driven through sand and gravel. If the piling is seated on an impervious stratum and is properly connected to the concrete apron, a tight barrier to underseepage is provided.

Concrete cutoffs may be used under aprons or under the overflow section. They may be constructed by trenching with a machine or by hand labor and backfilling against the undisturbed sides of the trench, or they may be constructed by forming the concrete wall in an open excavated trench and then backfilling and compacting impervious material in the trench and around the wall. A concrete cutoff is probably the best type of wall for preventing underseepage and is often used. In addition to acting as a cutoff, such a cutoff can be designed to contribute substantially to the stability (sliding resistance) of the dam when placed under the overflow section.

Cement-bound curtain cutoffs have been used under the upstream apron of a few concrete gravity structures by the Bureau of Reclamation. The construction of this type of cutoff is described in section 131(e).

Sheet piling cutoffs of interlocking steel sections are often used under the aprons of diversion dams. A discussion of sheet piling cutoffs will be found in section 131(d).

180. *Filters and Drains*.—Reduction of uplift pressure under downstream apron or downstream toe of the dam may be accomplished by pipe drains. Drains are often of sewer pipe and are laid in graded material which acts as a filter. They may be perforated pipe or plain pipe laid with open joints. The drains may be located at the downstream toe of the dam, at selected places under the downstream apron, and immediately upstream from the downstream cutoff.

Weep holes are commonly used for reduction of uplift pressure under aprons and excessive pressure behind walls. It is important that the gradation of the filter materials used in conjunction with the weep holes be carefully selected with respect to the gradation of the foundation materials to prevent piping. Both uniform grain-size and graded filters are used. The design of filters is given in section 131(h) and (i).

181. *Uplift and Seepage*.—Cutoff walls, aprons, and drains are installed for two reasons: To control the amount of seepage under the dam, and to limit the intensity of the uplift o that the stability of the structure will not be hreatened. Several factors such as head on he dam, permeability of the foundation, length f upstream and downstream aprons, depths nd tightness of cutoff, and effectiveness of rains enter into consideration of underseep-ge and uplift.

The magnitude and distribution of seepage ressures in the foundation and the amount of nderseepage for a given coefficient of permea-ility can be obtained from a flow net. Flow ets properly constructed are more accurate han Lane's weighted-creep theory (discussed elow), and result in more realistic determina-ions of seepage pressures and piping. An ex-ellent discussion of flow nets is given in ref-rence [33] of chapter VI.

The amount of underseepage can be ap-roximated by use of the Darcy formula, as iven in section 130(b). The weighted-creep heory, as developed by Lane [5] may be used s a means for designing low concrete dams on ervious foundations. Although this is an em-irical method, considerable confidence has een placed in it by many engineers and it has een successfully used for the design of many tructures.

Mr. Lane gives credit for various concepts of reep analysis to the early investigators, Clib-orn, Beresford, Bligh, Griffith, and others. e did, however, test his theory by analysis of ore than 200 dams on pervious foundations, oth failures and nonfailures. His main con-lusions (omitting those regarding the "short-ath" which is not applicable here) were as ollows:

(1) The weighted-creep distance of a cross section of a dam is the sum of the vertical creep distances (steeper than 45°) plus one-third of the horizontal creep distances (less than 45°).

(2) The weighted-creep head ratio is the weighted-creep distance divided by the effective head.

(3) Reverse filter drains, weep holes, and pipe drains are aids to security from underseepage, and recommended safe weighted-creep head ratios may be reduced as much as 10 percent if they are used.

(4) Care must be exercised to insure that cutoffs are properly tied in at the ends so that the water will not outflank them.

(5) The upward pressure to be used in design may be estimated by assuming that the drop in pressure from headwater to tailwater along the contact line of the dam and foundation is proportional to the weighted-creep distance.

Based on his findings, Mr. Lane recommended the weighted-creep ratios shown below:

Material:	*Ratio*
Very fine sand or silt	8.5
Fine sand	7.0
Medium sand	6.0
Coarse sand	5.0
Fine gravel	4.0
Medium gravel	3.5
Coarse gravel including cobbles	3.0
Boulders with some cobbles and gravel	2.5
Soft clay	3.0
Medium clay	2.0
Hard clay	1.8
Very hard clay or hardpan	1.6

Figure 224 is an example of the application of Lane's weighted-creep theory to the design of a concrete dam. In this example, the design is investigated to determine on what types of foundations this dam would be judged safe from piping and the magnitudes of the uplift at various points under the structures are calculated. For the purposes of this example, it is assumed that the maximum head differential occurs with the headwater at the elevation of the crest. All other combinations of head-

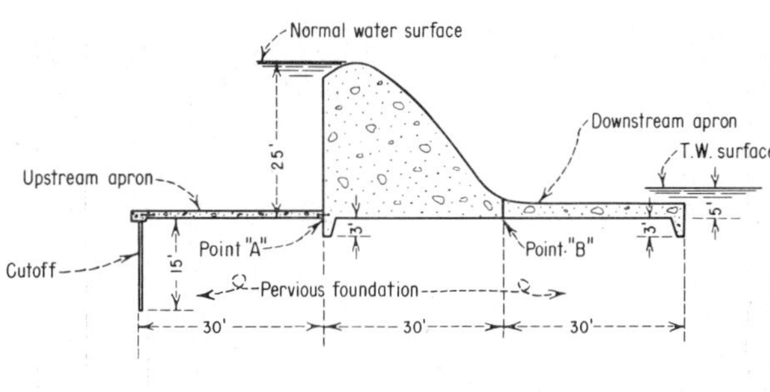

Figure 224. Typical section of a concrete dam on a pervious foundation. 288–D–2511.

water and tailwater should be investigated to establish maximum loading conditions.

For figure 224:

Weighted length of path $=15+15+(4\times3)+\frac{1}{3}(30+30+30)=72$ feet

Head on structure = Headwater − tailwater $=25-5=20$ feet

$$\text{Weighted-creep ratio}=\frac{72}{20}=3.6$$

According to Lane's recommended ratios, this dam would be safe from piping on clay or on medium and coarse gravel, but not on silt, sand, or fine gravel. With properly placed drains and filters, the structure would probably be considered safe on a fine gravel foundation as indicated by conclusion (3).

$$\text{Uplift, point A}=20-\frac{(15+15+10)}{72}\times20 +5 \text{ (depth of tailwater above foundation level)}$$
$$=20-11.1+5$$
$$=13.9 \text{ feet.}$$

$$\text{Uplift, point B}=20-\frac{(15+15+10+3+3+10)}{72}\times20+5$$
$$=20-15.6+5$$
$$=9.4 \text{ feet.}$$

$$\text{Total uplift}=\frac{(13.9+9.4)}{2}\times62.5\times30$$
$$=21{,}840 \text{ pounds per foot of crest length of dam.}$$

The uplift may be computed as indicated by conclusion (5).

The weighted-creep head ratio can be increased by increasing the depth of the upstream cutoff or by increasing the apron length. Either of these alternatives would also decrease the uplift under the structure. The dam shown in figure 224 is a rather common installation and the example was simplified to illustrate the method. Discussion of a more complicated design such as a dam with two or more deep cutoffs, which requires application of the "short-path" theory, has been omitted. Mr. Lane's article [5] should be consulted for details for complex designs.

In addition to the flow net and weighted creep methods of estimating the distribution of uplift pressure are Khosla's method of independent variables [6] and Rao's relaxation method [7] which can be used for making computations of uplift at critical points along the base of the structure. Because these theories are highly mathematical they are not discussed in this text.

Seepage forces and piping in a pervious foundation are discussed in section 130(c). The possibility of piping may be alleviated by several methods. As shown on figure 224, a cutoff may be constructed at the downstream end of the apron. A drain laid in graded material which acts as a filter may be placed immediately upstream of the cutoff, or riprap on a graded blanket of gravel may be placed on the material downstream of the apron to increase the downward forces.

E. DETAILS OF LAYOUT AND DESIGN

182. *General.*—If a concrete dam is appreciably more than 50 feet in length, it is necessary to divide the structure into blocks by providing transverse contraction joints. The spacing of the joints is determined by the capacity of the concreting facilities to be used and considerations of volumetric changes and attendant cracking caused by shrinkage and temperature variations. The possibilities of detrimental cracking can be greatly reduced by the selection of the proper type of cement and by careful control of mixing and placing procedure (see appendix F). In no case, however, is it advisable to exceed 50-foot spacing of contraction joints in constructing small concrete dams. Where foundation conditions are such that undesirable differential settlement or displacement between adjacent blocks will occur, shear keys are formed in the contraction joints. These may be formed vertically, horizontally or a combination of both depending on the direction of the expected displacement. Leakage through the contraction joints is controlled by placing rubber, polyvinyl chloride (PVC), metal, or other type waterstops across the joints near the upstream face of the dam. Typical details of shear keys and waterstops are shown in figures 225 and 226.

183. *Nonoverflow Sections*.—The elevation of the top of a nonoverflow dam is established by assuming a safe freeboard above the maximum high water surface in the reservoir. The freeboard should be sufficient to allow for the maximum wave height as given in the first tabulation of section 142. Although only one-half of the wave height is above the mean water level, the full height is ordinarily used to allow for the run of the waves up the face of the dam. A minimum freeboard of 3 feet is recommended for most small concrete dams.

The top width is determined by such requirements as the need for travel across the dam or for access to gate-operating mechanism, and by climatic conditions. A top width of less than 4 feet is not recommended.

The width of the base and the slope of the downstream face are determined by a stability analysis. The customary method is to assume a section with the slope of the downstream face approximately 0.70 horizontally to 1.0 vertically and intersecting a vertical upstream face at the top of the dam. The assumed section is then analyzed and modified as required by the analysis until it meets the stability requirements. If the dam is stable about its base and also about any section where there is a break in the continuity of slope of either the upstream or downstream face, the portion of the dam between any of these sections is stable and does not require analyzing.

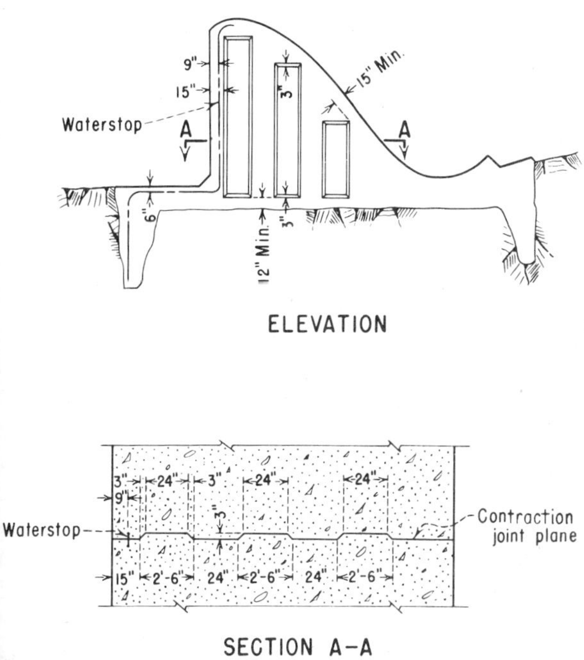

Figure 225. Typical contraction joint shear keys. 288–D–2512.

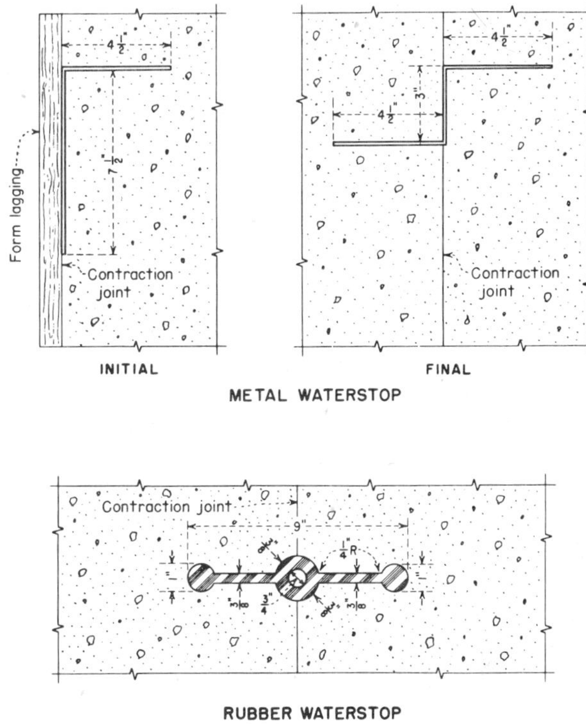

Figure 226. Waterstop installations. 288–D–2513.

184. *Overflow Sections*.—In general, the method for determining the stability of overflow dams is the same as that previously described for nonoverflow dams.

The shape of the crest, the profile of the downstream face, and details of the energy dissipating basin or bucket are discussed in chapter IX. It is customary to provide a longitudinal contraction joint at the downstream toe, as shown in figure 219, and then only that portion of the dam upstream of the joint is used in the stability computations.

In cases where the dissipating device extends only a short distance downstream from the toe and is fairly massive, the contraction joint may be omitted. The structure downstream from the toe is then included in the stability analysis and is so reinforced that it and the gravity portion will act as a unit. Under certain conditions, an upstream apron connected by reinforcement to the upstream face of the dam may be the most economical arrangement to insure stability.

Overflow dams utilizing control features on the crest introduce an additional problem. The forces acting on these features may produce tension in the upper portion of the dam which will require adequate reinforcement.

F. BIBLIOGRAPHY

185. *Bibliography.*

[1] Rose, Edwin, "Thrust Exerted by Expanding Ice Sheet," Trans. ASCE, vol. 112, 1947, p. 871.

[2] Monfore, G. E., "Experimental Investigations by the Bureau of Reclamation," Trans. ASCE, vol. 119, 1954, p. 126.

[3] Zangar, C. N., Engineering Monograph No. 11, "Hydrodynamic Pressures on Dams Due to Horizontal Earthquake Effects," Bureau of Reclamation, May 1952.

[4] Design Standard No. 2, "Concrete Dams," chapter 2, Gravity Dams, Bureau of Reclamation.

[5] Lane, E. W., "Security from Underseepage," Trans. ASCE, vol. 100, 1935, p. 1,235.

[6] Khosla, "Design of Dams on Permeable Foundations," Central Board of Irrigation, India, September 1936.

[7] Rao, "A New Approach to the Design of Weirs on Permeable Foundations," Central Board of Irrigation, India, January 1951.

[8] "Design Criteria for Concrete Gravity and Arch Dams," Engineering Monograph No. 19, Bureau of Reclamation, 1971.

[9] "Earth Manual," Bureau of Reclamation, 1963, p. 51.

Chapter IX

Spillways

C. J. HOFFMAN [1]

A. GENERAL

186. *Function.*—Spillways are provided for storage and detention dams to release surplus or flood water which cannot be contained in the allotted storage space, and at diversion dams to bypass flows exceeding those which are turned into the diversion system. Ordinarily, the excess is drawn from the top of the pool created by the dam and conveyed through an artificial waterway back to the river or to some natural drainage channel. Figure 227 shows a small spillway in operation.

The importance of a safe spillway cannot be overemphasized; many failures of dams have been caused by improperly designed spillways or by spillways of insufficient capacity. Ample capacity is of paramount importance for earthfill and rockfill dams, which are likely to be destroyed if overtopped, whereas concrete dams may be able to withstand moderate overtopping. Usually, increase in cost is not directly proportional to increase in capacity. Very often the cost of a spillway of ample capacity will be only moderately higher than that of one which is obviously too small.

In addition to providing sufficient capacity, the spillway must be hydraulically and structurally adequate and must be located so that spillway discharges will not erode or undermine the downstream toe of the dam. The spillway's bounding surfaces must be erosion resistant to withstand the high scouring velocities created by the drop from the reservoir surface to tailwater, and usually some device will be required for dissipation of energy at the bottom of the drop.

The frequency of spillway use will be determined by the runoff characteristics of the drainage area and by the nature of the development. Ordinary riverflows are usually stored in the reservoir, diverted through headworks, or released through outlets, and the spillway is not required to function. Spillway flows will result during floods or periods of sustained high runoff when the capacities of other facilities are exceeded. Where large reservoir storage is provided, or where large outlet or diversion capacity is available, the spillway will be utilized infrequently. At diversion dams where storage space is limited and diversions are relatively small compared to normal river flows, the spillway will be used almost constantly.

187. *Selection of Inflow Design Flood.*—(a) *General Considerations.*—When floods occur in an unobstructed stream channel, it is considered a natural event for which no individual or group assumes responsibility. However, when obstructions are placed across the channel, it becomes the responsibility of the sponsors either to make certain that hazards to downstream interests are not appreciably increased or to obligate themselves for damages resulting from operation or failure of such structures. Also, the loss of the facility and the loss of project revenue occasioned by a failure should be considered.

If danger to the structures alone were involved, the sponsors of many projects would prefer to rely on the improbability of an ex-

[1] Engineer, Spillways and Outlet Works Section, Bureau of Reclamation (Retired).

Figure 227. A small chute spillway in operation at Shadow Mountain Dam on the Colorado River in Colorado. 288–D–2841.

treme flood occurrence rather than to incur the expense necessary to assure complete safety. However, when the risks involve downstream interests, including widespread damage and loss of life, a conservative attitude is required in the development of the inflow design flood. Consideration of potential damage should not be confined to conditions existing at the time of construction. Probable future development in the downstream flood plain, encroachment by farms and resorts, construction of roads and bridges, etc., should be evaluated in estimating damages and hazards to human life that would result from failure of a dam.

Dams impounding large reservoirs and built on principal rivers with high runoff potential unquestionably can be considered to be in the high-hazard category. For such developments, conservative design criteria are selected on the basis that failure cannot be tolerated because of the possible loss of life and because of the potential damages which could approach disaster proportions. However, small dams built on isolated streams in rural areas where failure would neither jeopardize human life nor create damages beyond the sponsor's financial capabilities can be considered to be in a low-hazard category. For such developments design criteria may be established on a much less conservative basis. There are numerous instances, however, where failure of dams of low heights and small storage capacities has resulted in loss of life and heavy property damage. Most small dams will require a reasonable conservatism in design, primarily because of the criterion that a dam failure must not present a serious hazard to human life.

The selection of the magnitude of an inflow

design flood involves a policy decision. The following spillway capacity requirements, established by the Bureau of Reclamation for consideration of dams to be constructed by means of Federal loans under the Small Reclamation Projects Act of 1956, are offered as a guide in making such a decision:

"(1) In case that failure of the dam would increase the danger to human life, the spillway must have sufficient capacity to accommodate the maximum probable flood when routed through the reservoir.

"(2) In cases where the failure of the dam would not increase the danger to human life but would endanger the continued operation of the responsible organization or would cause heavy property damage, plans involving a reasonable risk will be permitted if the report shows that the risk is clearly understood by the applying organization.

"(3) In case the failure of the dam would not jeopardize human life, the continued operation of the responsible organization, or heavy property damage, the Bureau will not recommend against approval of the loan because of inadequate capacity but should warn the organization of the risk involved and disclaim any responsibility in case of failure."

The foregoing requirements are essentially the same as were agreed to by a number of prominent private consulting engineers in 1946.

(b) *Inflow Design Flood Hydrographs.*—Chapter III discusses the determination of floodflows which may be used as inflow design floods. The procedures presented permit the derivation of inflow flood hydrographs of three magnitudes: the maximum probable flood, the flood for assumption A conditions, and the flood for assumption B conditions (sec. 58). All of these floods are based on the hydrometeorological approach, which requires estimates of storm potential and of the amount and distribution of runoff.

Determination of the *maximum probable flood* is based on rational consideration of the chances of simultaneous occurrence of the maximum of the several elements or conditions which contribute to the flood. Such a flood is the largest that reasonably can be expected and is ordinarily accepted as the inflow design flood for dams where failure of the structure would increase the danger to human life.

The *flood for assumption A* is based on storm values less than probable maximum and on the assumption that the watershed soils are near saturation from a previous storm. In most instances these storm values represent precipitation amounts slightly greater than those which have been observed for respective locations within the United States. The magnitude of this flood is believed to be the minimum appropriate for use where the spillway capacity requirements correspond to (2) in the preceding subsection.

The *flood for assumption B* is based on storm values the same as used for the flood for assumption A, but the watershed soils are considered to be at average moisture content for the time of year when the maximum floods are likely to occur, rather than saturated. The magnitude of this flood is believed to be the minimum appropriate for use where the spillway capacity requirements correspond to (3) in the preceding subsection.

In the case of a minor structure with insignificant storage where it is permissible to anticipate failure within the useful life of the project, a 50- or 100-year frequency flood may be used as the inflow design flood. A discussion of these floods and their determination is given in sections 47 and 59. Estimates of floods of these magnitudes may also be required

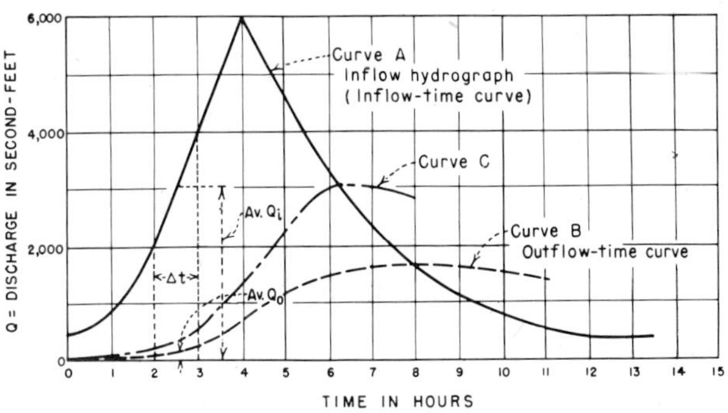

Figure 228. Inflow and outflow hydrographs. 288–D–2399.

to establish the capacity of a service or principal spillway in those cases where an auxiliary spillway will serve to augment the smaller spillway.

188. *Relation of Surcharge Storage to Spillway Capacity*.—Streamflow is normally represented in the form of a hydrograph, which charts the rate of flow in relation to time. A typical hydrograph representing a storm runoff is illustrated in figure 228. The flow into a reservoir at any time and the momentary peak can be read from the curve. The area under the curve is the volume of the inflow, because it represents the product of rate of flow and time.

Where no storage is impounded by a dam, the spillway must be sufficiently large to pass the peak of the flood. The peak rate of inflow is then of primary interest and the total volume in the flood is of lesser importance. However, where a relatively large storage capacity above normal reservoir level can be made available economically by a higher dam, a portion of the flood volume can be retained temporarily in reservoir surcharge space and the spillway capacity can be reduced considerably. If a dam could be made sufficiently high to provide storage space to impound the entire volume of the flood above normal storage level, theoretically no spillway other than an emergency type would be required, provided the outlet capacity could evacuate the surcharge storage in a reasonable period of time in anticipation of a recurring flood. In such a case the maximum reservoir level would depend entirely on the volume of the flood and the rate of inflow would be of no concern. From a practical standpoint, however, there will be relatively few sites that will permit complete storage of an inflow design flood by surcharge storage. Such sites usually will be off-channel reservoirs; that is, reservoirs which are supplied by canal and which have small tributary drainage areas.

In many reservoir projects, economic considerations will necessitate a design utilizing surcharge. The most economical combination of surcharge storage and spillway capacity requires flood routing studies and economic studies of the costs of spillway-dam combinations, subsequently described. However, in making these studies, consideration must be given to the minimum size spillway which must be provided for safety. The inflow design flood hydrographs determined by the methods given in section 58 are for floods resulting from runoff from rain. Normally, such floods will have the highest peak flows but not always the largest volumes. When spillways of small capacities in relation to these inflow design flood peaks are considered, precautions must be taken to insure that the spillway capacity will be sufficient to (1) evacuate surcharge so that the dam will not be overtopped by a recurrent storm and (2) prevent the surcharge from being kept partially full by a prolonged runoff whose peak, although less than the inflow design flood, exceeds the spillway capacity. To meet these requirements, the minimum spillway capacity should be in accord with the following general criteria:

(1) In the case of snow-fed perennial streams, the spillway capacity should never be less than the peak discharge of record that has resulted from snowmelt runoff. (This value may have to be estimated from a study of records on the stream itself or nearby streams. (Secs. 39 and 46.)

(2) The spillway capacity should provide for evacuation of sufficient surcharge storage space so that in routing a succeeding "flood for assumption B," the maximum water surface does not exceed that obtained by routing the inflow design flood. In general, the recurrent storm is assumed to begin 4 days after the time of peak outflow obtained in routing the inflow design flood.

(3) In regions having an annual rainfall of 40 inches or more, the time interval to the beginning of the recurrent storm in criterion (2) should be reduced to [illegible] days.

(4) In regions having an annual rainfall of 20 inches or less, the time interval to the beginning of the recurrent storm in criterion (2) may be increased to [illegible] days.

189. *Flood Routing*.—The accumulation of storage in a reservoir depends on the difference between the rates of inflow and outflow. For an interval of time Δt, this relationship can be expressed by the equation:

$$\Delta S = Q_i \Delta t - Q_o \Delta t \qquad (1)$$

where:

ΔS=storage accumulated during Δt,
Q_i=average rate of inflow during Δt, and
Q_o=average rate of outflow during Δt.

The rate of inflow versus time curve is represented by the inflow design flood hydrograph; the rate of outflow is represented by the spillway discharge versus reservoir-elevation curve; and storage is shown by the reservoir storage versus reservoir-elevation curve. For routing studies, the inflow design flood hydrograph is not variable once selection of the inflow design flood has been made. The reservoir storage capacity also is not variable for a given reservoir site, so far as routing studies are concerned. The spillway discharge curve is variable: It depends not only on the size and type of spillway but also on the manner of operating the spillway (and outlets in some instances) to regulate the outflow.

The quantity of water a spillway can discharge depends on the type of the control device. For a simple overflow crest the flow will vary with the head on the crest, and surcharge will increase with an increase in spillway discharge. For a gated spillway, however, outflow can be varied with respect to reservoir head by operation of the gates. For example, one assumption for an operation of a gate-controlled spillway might be that the gates will be regulated so that inflow and outflow are equal until the gates are wide open; or an assumption can be made to open the gates at a slower rate so that surcharge storage will accumulate before the gates open wide.

Outflows need not necessarily be limited to discharges through the spillway but might be supplemented by releases through the outlets. In all such cases the size, type, and method of operation of the spillway and outlets with reference to the storages or to the inflow must be predetermined in order to establish an outflow-elevation relationship.

If equations could be established for the inflow design flood hydrograph curve, the spillway discharge curve (as may be modified by operational procedures), and the reservoir storage curve, a solution of flood routing could be made by mathematical integration. However, simple equations cannot be written for the flood hydrograph curve and the reservoir storage curve, and such a solution is not practical. Many techniques of flood routing have been devised, each with its advantages and disadvantages. These techniques vary from a strictly arithmetical integration method to an entirely graphical solution. Mechanical and electronic routing machines have been developed, and digital computers have been employed. For simplicity, the arithmetical trial-and-error tabular method is illustrated in this text.

Table 21 is an example of a flood routing for the data given in figures 228, 229, and 230. These data are necessary regardless of the method of flood routing used and they consist of the following:

(1) Inflow hydrograph (rate of inflow versus time), figure 228.
(2) Reservoir capacity (reservoir storage versus reservoir elevation), figure 229.
(3) Discharge curve (rate of outflow versus reservoir elevation), figure 230.

The procedure for computations shown in table 21 is as follows:

(1) Select a time interval, Δt, column (2).

(2) Obtain column (3) from the inflow hydrograph, figure 228.

(3) Column (4) represents average inflow for Δt in second-feet.

(4) Obtain column (5) by converting column (4) values of second-feet for Δt to acre-feet (1 second-foot for 12 hours=1 acre-foot).

(5) Assume trial reservoir water surface in column (6), determine the corresponding rate of outflow from figure 230, and record in column (7).

(6) Average the rate of outflow determined in step (5) and the rate of outflow for the reservoir water surface which existed at the beginning of the period and enter in column (8).

(7) Obtain column (9) by converting column (8) values of second-feet for Δt to acre-feet, similar to step (4).

(8) Column (10)=column (5)−column (9).

(9) The initial value in column (11) rep-

resents the reservoir storage at the beginning of the inflow design flood.

Determine subsequent values by adding ΔS values from column (10) to the previous column (11) value.

(10) Determine reservoir elevation in column (12) corresponding to storage in column (11) from figure 229.

(11) Compare reservoir elevation in column (12) with trial reservoir elevation in column (6). If they do not agree within 0.1 foot, make a second trial elevation and repeat procedure until agreement is reached.

The outflow time curve resulting from the flood routing shown in table 21 has been plotted as curve B on figure 228. As the area under the inflow hydrograph (curve A) indicates the volume of inflow, so will the area under the outflow hydrograph (curve B) indicate the volume of outflow. It follows then that the volume indicated by the area between the two curves will be the surcharge storage. The surcharge storage computed in table 21 can, therefore, be checked by comparing it with the measured area on the graph.

A rough approximation of the relationship of spillway size to surcharge volume can be obtained without making an actual flood routing, by arbitrarily assuming an approximate outflow-time curve and then measuring the area between it and the inflow hydrograph. For example, if the surcharge volume for the problem shown on figure 228 is sought where a 3,000-second-foot spillway would be provided, an assumed outflow curve represented by curve C can be drawn and the area between this curve and curve A can be planimetered. Curve C will reach its apex of 3,000 second-feet where it crosses curve A. The volume represented by the area between the two curves will indicate the approximate surcharge volume necessary for this capacity spillway.

190. *Selection of Spillway Size and Type.*—(a) *General Considerations.*—In determining the

TABLE 21.—*Flood routing computations*

(1) Time t, hours	(2) Δt, hours	(3) Inflow at time, t, second-feet	(4) Average rate of inflow Q_i for Δt, second-feet	(5) Inflow, acre-feet	(6) Trial reservoir storage-elevation at time t	(7) Outflow at time, t, second-feet	(8) Average rate of outflow Q_o for Δt, second-feet	(9) Outflow, acre-feet	(10) Incremental storage Δs, acre-feet	(11) Total storage, acre-feet	(12) Reservoir elevation, end of Δt, feet	(13) Remarks
0		400				0				1,050	300.3	
1	1	800	600	50	~~300.2~~ 300.3	~~5~~ 10	~~3~~ 5	~~0~~ 0	~~50~~ 50	~~1,100~~ 1,100	~~300.3~~ 300.3	~~High~~ OK
2	1	2,000	1,400	117	~~300.3~~ 301.0	~~54~~ 80	~~32~~ 45	~~3~~ 4	~~114~~ 113	~~1,214~~ 1,213	~~301.0~~ 001.0	~~High~~ OK
3	1	4,000	3,000	250	~~302.3~~ 302.1	~~300~~ 260	~~190~~ 170	~~16~~ 14	~~234~~ 236	~~1,447~~ 1,449	~~302.1~~ 302.1	~~Low~~ OK
4	1	6,000	5,000	417	~~303.9~~ 303.8	~~710~~ 690	~~485~~ 475	~~40~~ 40	~~377~~ 377	~~1,826~~ 1,826	~~303.8~~ 303.8	~~Low~~ OK
5	1	4,700	5,350	446	~~305.6~~ 305.3	~~1,060~~ 1,160	~~875~~ 925	~~73~~ 77	~~373~~ 369	~~2,199~~ 2,195	~~305.3~~ 305.3	~~High~~ OK
6	1	3,300	4,000	333	~~306.3~~ 306.2	~~1,500~~ 1,470	~~1,930~~ 1,315	~~111~~ 110	~~222~~ 223	~~2,417~~ 2,418	~~306.2~~ 306.2	~~Low~~ OK
7	1	2,400	2,850	238	306.6	1,610	1,540	128	110	2,528	306.6	OK
8	1	1,600	2,000	167	306.7	1,650	1,630	136	31	2,559	306.7	OK
9	1	1,100	1,350	112	306.6	1,610	1,630	136	—24	2,535	306.6	OK
11	2	500	800	133	306.0 306.1	1,400 1,430	1,505 1,520	251 253	—118 —120	2,417 2,415	306.1 306.1	High OK

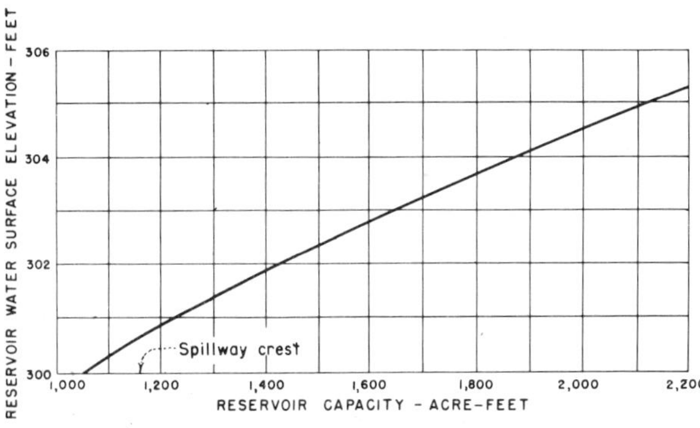

Figure 229. **Reservoir capacity curve. 288–D–2400.**

best combination of storage and spillway capacity to accommodate the selected inflow design flood, all pertinent factors of hydrology, hydraulics, design, cost, and damage should be considered. In this connection and when applicable, consideration should be given to such factors as (1) the characteristics of the flood hydrograph; (2) the damages which would result if such a flood occurred without the dam; (3) the damages which would result if such a flood occurred with the dam in place; (4) the damages which would occur if the dam or spillway were breached; (5) effects of various dam and spillway combinations on the probable increase or decrease of damages above or below the dam (as indicated by reservoir backwater curves and tailwater curves); (6) relative costs of increasing the capacity of spillways; and (7) use of combined outlet facilities to serve more than one function, such as control of releases and control or passage of floods. Service outlet releases may be permitted in passing part of the inflow design flood unless such outlets are considered to be unavailable in time of flood.

The outflow characteristics of a spillway de-

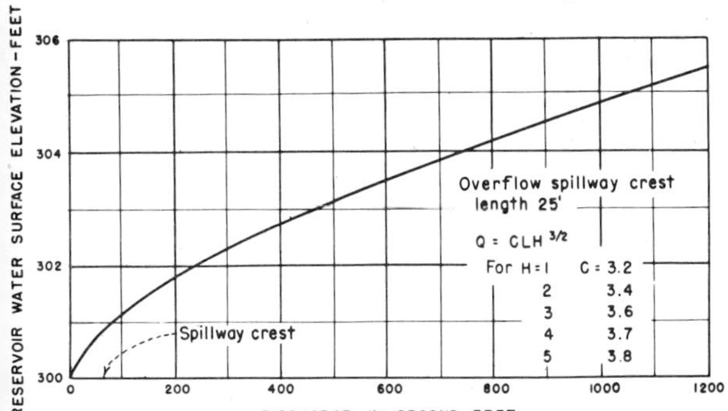

Figure 230. **Spillway discharge-elevation curve. 288–D–2401.**

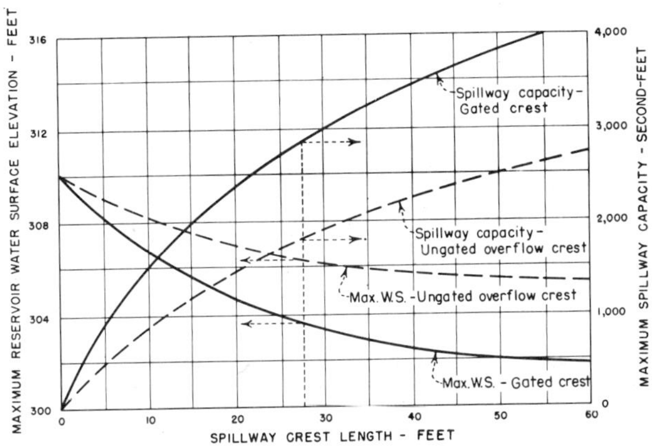

Figure 231. **Spillway capacity-surcharge relationship. 288–D–2402.**

pend on the particular device selected to control the discharge. These control facilities may take the form of an overflow weir, an orifice, a tube, or a pipe. Such devices can be unregulated or they can be equipped with gates or valves to regulate the outflow.

After a spillway control of certain dimensions has been selected, the maximum spillway discharge and the maximum reservoir water level can be determined by flood routing. Other components of the spillway can then be proportioned to conform to the required capacity and to the specific site conditions, and a complete layout of the spillway can be established. Cost estimates of the spillway and

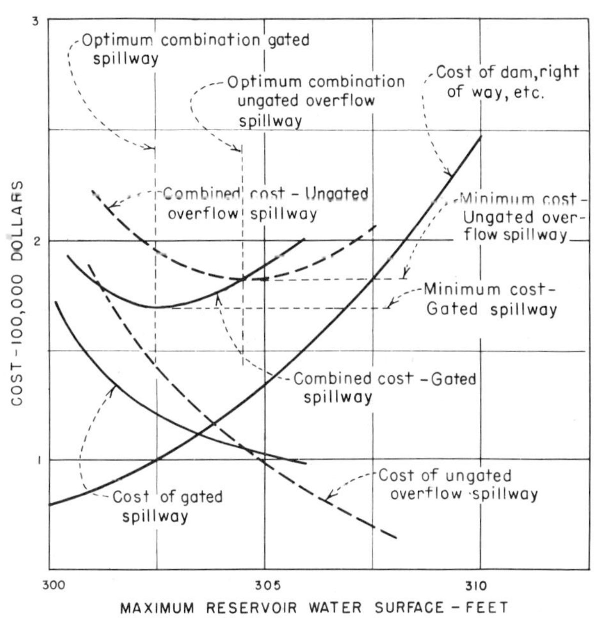

Figure 232. **Comparative costs of spillway-dam combinations. 288–D–2403.**

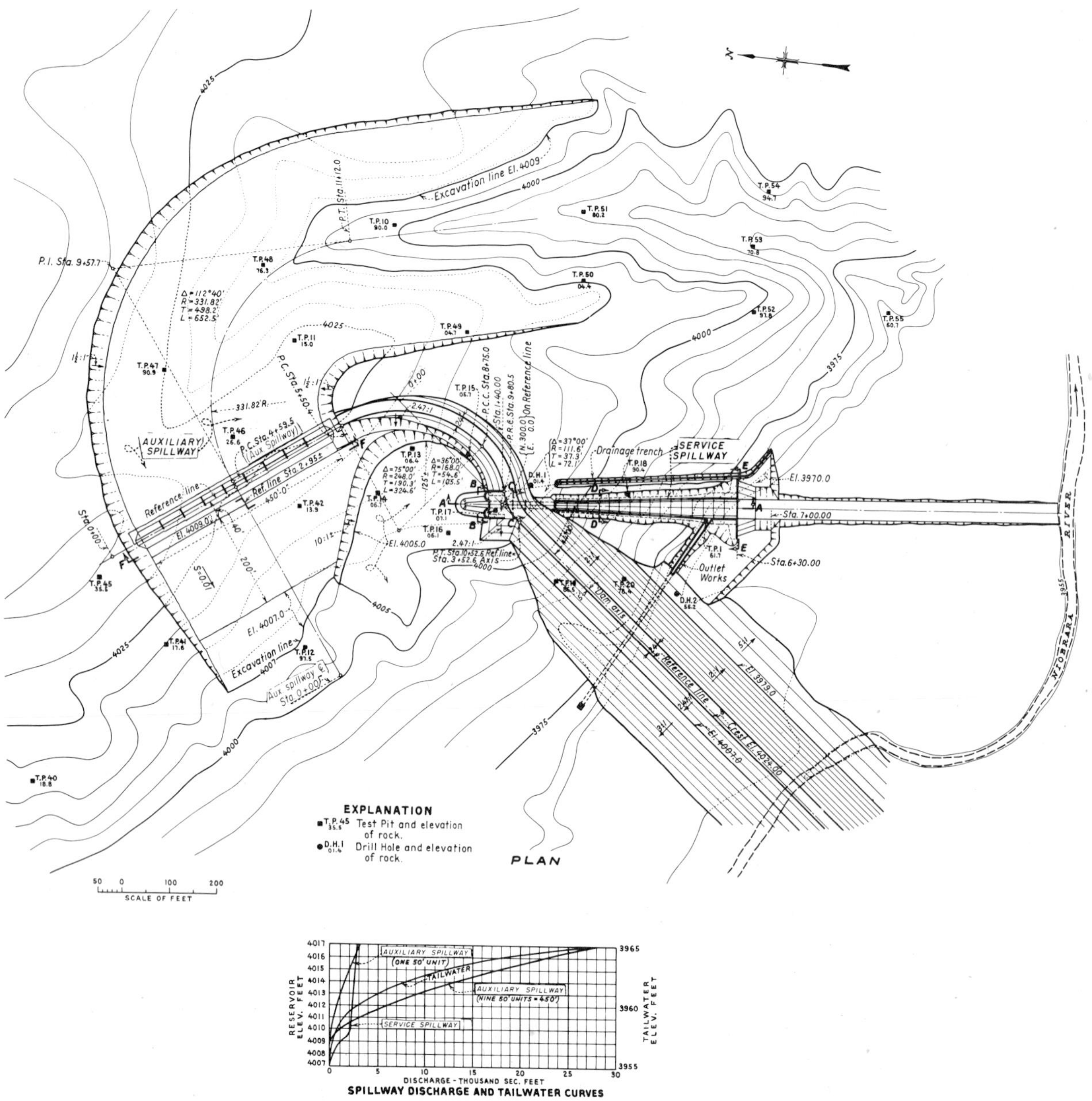

Figure 233. Service and auxiliary spillways for Box Butte Dam—Plan and sections. From drawing 278-D-49. (Sheet 1 of 2.)

dam can then be made. Estimates of various combinations of spillway capacity and dam height for an assumed spillway type, and of alternative types of spillways, will provide a basis for selection of the economical spillway type and the optimum relation of spillway capacity to height of dam. Figures 231 and 232 illustrate the results of such a study. The relationships of spillway capacities to maximum reservoir water surfaces obtained from the flood routings are shown on figure 231 for two spillways. Figure 232 illustrates the comparative costs for different combinations of spillway and dam, and indicates a combination which results in the least total cost.

To make such a study as illustrated requires

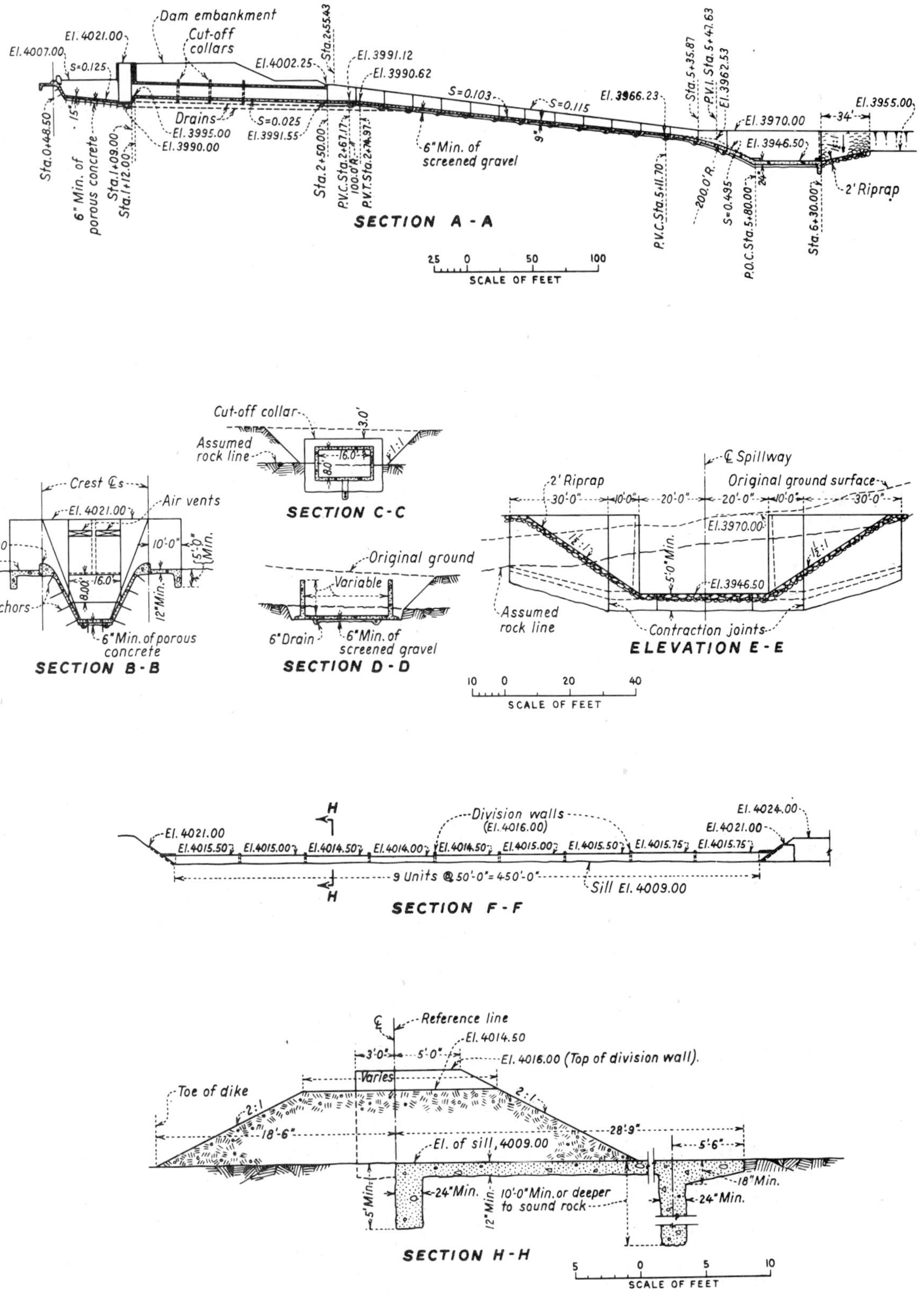

Figure 233. Service and auxiliary spillways for Box Butte Dam—Plan and sections. From drawing 278–D–49. (Sheet 2 of 2.)

many flood routings, spillway layouts, and spillway and dam estimates. Even then, the study is not necessarily complete since many other spillway arrangements could be considered. A comprehensive study to determine alternative optimum combinations and minimum costs may not be warranted for the design of small dams. Judgment on the part of the designer would be required to select for study only the combinations which show definite advantages, either in cost or adaptability. For example, although a gated spillway might be slightly cheaper than an ungated spillway, it may be desirable to adopt the latter because of its less complicated construction, its automatic and trouble-free operation, its ability to function without an attendant, and its less costly maintenance.

(b) *Combined Service and Auxiliary Spillways.*—Where site conditions are favorable, the possibility of gaining overall economy by utilizing an auxiliary spillway in conjunction with a smaller service-type structure should be considered. In such cases the service spillway should be designed to pass floods likely to occur frequently and the auxiliary spillway control set to operate only after such small floods are exceeded. In certain instances the outlet works may be made large enough to serve also as a service spillway. Conditions favorable for the adoption of an auxiliary spillway are the existence of a saddle or depression along the rim of the reservoir which leads into a natural waterway, or a gently sloping abutment where an excavated channel can be carried sufficiently beyond the dam to avoid the possibility of damage to the dam or other structures.

Because of the infrequency of use, it is not necessary to design the entire auxiliary spillway for the same degree of safety as required for other structures; however, at least the control portion must be designed to forestall failure, since its breaching would release large flows from the reservoir. For example, concrete lining may be omitted from an auxiliary spillway channel excavated in competent rock. Where the channel is excavated through less competent material, it might be lined but terminated above the river channel with a cantilevered lip rather than extending to a stilling basin at river level. The design of auxiliary spillways is often based on the premise that some damage to portions of the structure from passage of infrequent flows is permissible. Minor damage by scour to an unlined channel, by erosion and undermining at the downstream end of the channel, and by creation of an erosion pool downstream from the spillway might be tolerated.

Figure 234. Service spillway for Box Butte Dam. 288–D–2896.

An auxiliary spillway can be designed with a fixed crest control, or it can be stoplogged or gated to increase the capacity without additional surcharge head. "Fuse plug" dikes which are designed to breach and wash out when overtopped often are substituted for some or all of the gates. Their advantage over gates is that, if properly designed, breaching becomes automatic whenever overtopping occurs; furthermore, they are cheaper to install and to maintain. Since the chance of their failure from overtopping is contingent on the occurrence of infrequent floods, their cost for replacement is too problematical for evaluation. By dividing the dike into short sections of varying height so that they are not all simultaneously overtopped, smaller floods

Figure 235. Service and auxiliary spillways for Box Butte Dam. 288–D–2897.

might be passed with the failure of one or several of the sections, with total failure occurring only as the probable maximum flood is approached. The breaching of one section at a time will minimize the flood wave brought about by sudden failure of the dike.

Figure 233 shows the general plan and sections of the service and auxiliary spillways at Box Butte Dam. Figure 234 is an aerial photograph showing the service spillway, which consists of a "bathtub-shaped" side channel crest, a culvert conduit under the dam, a diverging concrete-lined chute, and a hydraulic-jump stilling basin. The wide auxiliary spillway channel with fuseplug control structure is shown in the top of figure 235 and the service spillway chute is in the right foreground. Note the outlet works control house and the outlet works channel which empties into the spillway stilling basin.

The Box Butte auxiliary spillway channel was excavated in a soft sandstone with only fair erosion-resistant qualities. To minimize erosion should discharge occur, the channel floor was made level so that velocities would be low. Erosion would start at the downstream end and progress slowly upstream. The control structure consists of a concrete-lined section; the cantilever lip and the downstream cutoff are provided to halt erosion upstream. The division walls and the fuseplug sections of varying crest elevations will insure progressive failure of the dike. The two sections nearest the dam were made the highest so that they will be the last to be overtopped. This was done to keep the flows away from the dam and to make the channel flow distance longer for discharges less than the maximum for which the spillway was designed.

(c) *Emergency Spillways.*—As the name implies, emergency spillways are provided for additional safety should emergencies not contemplated by normal design assumptions arise. Such situations could be the result of an enforced shutdown of the outlet works, a malfunctioning of spillway gates, or the necessity for bypassing the regular spillway because of damage or failure of some part of that structure. An emergency might arise where flood inflows are handled principally by surcharge storage and a recurring flood develops before a previous flood is evacuated by the small service spillway or the outlet works. Emergency spillways would act as auxiliary spillways if a flood greater than the selected inflow design flood occurred.

Under normal reservoir operation, emergency spillways are never required to function. The control crest is, therefore, placed at or above the designed maximum reservoir water surface. The freeboard requirement for the dam is based on a water surface determined by assuming an arbitrary discharge which might result from a possible emergency. Usually, an encroachment on the freeboard provided for the designed maximum water surface is allowed in considering the design of an emergency spillway.

Emergency spillways are provided primarily to avoid an overtopping of the main dam embankment because of an emergency condition. Therefore, to be effective the emergency spillway must offer resistance to erosion greater than does the dam itself. Emergency spillways are often formed by lowering the crest of a dike section below that of the main embankment, by utilizing saddles or depressions along the reservoir rim, or by excavating channels through ridges or abutments. The exit channel of an emergency spillway should be a sufficient distance from the dam to preclude damage to the embankment or appurtenances should the spillway operate.

Figure 192 shows an emergency spillway at Wasco Dam. This spillway was provided to prevent overtopping of the embankment should the combination outlet works-spillway fail to function properly.

B. DESCRIPTION OF SERVICE SPILLWAYS

191. *Selection of Spillway Layout.*—A composite design of a spillway can be prepared by properly considering the various factors influencing the spillway size and type, and correlating alternatively selected components. Many combinations of components can be used in forming a complete spillway layout. After the hydraulic size and outflow characteristics of a spillway are determined by routing of the design flood, the general dimensions of the control can be selected. Then, a specific spillway layout can be developed by considering the topography and foundation conditions, and by fitting the control structure and the various components to the prevailing conditions.

Site conditions greatly influence the selection of location, type, and components of a spillway. The steepness of the terrain traversed by the spillway control and discharge channel, the class and amount of excavation and the possibility for its use as embankment material, the chances of scour of the bounding surfaces and the need for lining, the permeability and bearing capacity of the foundation, and the stability of the excavated slopes must all be considered in the selection.

The adoption of a particular size or arrangement for one of the spillway components may influence the selection of other components. For example, a wide control structure with the crest placed normal to the centerline of the spillway would require a long, converging transition to join it to a narrow discharge channel or to a tunnel; a better alternative might be the selection of a narrower gated control structure or a side channel control arrangement. Similarly, a wide stilling basin may not be feasible for use with a cut-and-cover conduit or tunnel because of the long, diverging transition needed.

A spillway may be an integral part of a dam such as an overflow section of a concrete dam, or it may be a separate structure. In some instances, it may be combined as a common discharge structure with the outlet works or integrated into the river diversion plan for economy. Thus, the location, type, and size of other appurtenances are factors which may influence the selection of a spillway location or its arrangement. The final plan will be governed by overall economy, hydraulic sufficiency, and structural adequacy.

The components of a spillway and common types of spillways are described and discussed herein. Hydraulic design criteria and procedures are discussed in parts C through F of this chapter.

192. *Spillway Components.*—(a) *Control Structure.*—A major component of a spillway is the control device, since it regulates and controls the outflows from the reservoir. This control limits or prevents outflows below fixed reservoir levels, and it also regulates releases when the reservoir rises above that level. The control structure may consist of a sill, weir, orifice, tube, or pipe. The discharge-head relationship may be fixed as in the case of a simple overflow crest or unregulated port, or it may be variable as with a gated crest or a valve-controlled pipe. The control characteristics of a closed conduit might change with the stage relationship. In a culvert spillway, for example, the entrance will act as a weir for low heads when it is not submerged and as an orifice when submerged. As the amount of submergence increases, the flow will be controlled by the conduit acting as a tube, and finally, for greater submergence, the conduit will flow full and the flow will be governed by pressure pipe characteristics.

Figure 236. Semicircular overflow crest for small chute spillway at Fruitgrowers Dam in Colorado. 288–D–2898.

Control structures may take various forms in both positioning and shape. In plan, overflow crests can be straight, curved, semicircular, U-shaped, or round. A semicircular crest for a small spillway is shown in figure 236. Orifice controls can be placed in a horizontal, inclined, or vertical position. Tubes or pipes may be placed vertically, horizontally, or inclined and can be straight, curved, or follow any profile. They can be circular, square, rectangular, horseshoe, or other shape in cross section.

An overflow can be sharp crested, ogee shaped, broad crested, or of varied cross section. Orifices can be sharp edged, round edged, or bellmouth shaped, and can be placed so as to discharge with a fully contracted jet or with a suppressed jet. They may discharge freely or discharge partly or fully submerged. Tubes may have entrance corners which are sharp edged, rounded, or bellmouthed; and they can be of uniform size or be divergent or convergent. Tubes or pipes can operate freely discharging or they can flow full for part or all of their length and can be of uniform or changing size, with the control placed at some point between the inlet and the downstream end. Pipes can be of uniform or changing size, with the control placed either at the downstream end or at some intermediate point along the length. Pipes can flow full under pressure for their entire length or they can flow full and partly full, respectively, above and below their control point.

(b) *Discharge Channel.*—Flow released through the control structure usually is conveyed to the streambed below the dam in a discharge channel or waterway. Exceptions are where the discharge falls free from an arch dam crest or where the flow is released directly along the abutment hillside to cascade down the abutment face. The conveyance structure may be the downstream face of a concrete dam, an open channel excavated along the ground surface, a closed cut-and-cover conduit placed through or under a dam, or a tunnel excavated through an abutment. The profile may be variably flat or steep; the cross section may be variably rectangular, trapezoidal, circular, or of other shape; and the discharge channel may be wide or narrow, long or short.

Discharge channel dimensions are governed primarily by hydraulic requirements, but the selection of profile, cross-sectional shapes, widths, length, etc., is influenced by the geologic and topographic characteristics of the site. Open channels excavated in the abutment usually follow the ground surface profile; steep canyon walls may make a tunnel desirable. In plan, open channels may be straight or curved, with sides parallel, convergent, divergent, or a combination of these. A closed conduit may consist of a vertical or an inclined shaft leading to a nearly horizontal tunnel through the abutment or to a cut-and-cover conduit under or through the dam. Occasionally a combination of a closed conduit and an open channel might be adopted, such as a culvert under an embankment emptying into an open channel leading down the abutment slope. Discharge channels must be cut through or lined with material which is resistant to the scouring action of the accelerating velocities, and which is structurally adequate to withstand the forces from backfill, uplift, waterloads, etc.

(c) *Terminal Structure.*—When spillway flows fall from reservoir pool level to downstream river level, the static head is converted to kinetic energy. This energy manifests itself in the form of high velocities which if impeded result in large pressures. Means of returning the flow to the river without serious scour or erosion of the toe of the dam or damage to adjacent structures must usually be provided.

In some cases the discharge may be delivered at high velocities directly to the stream where the energy is absorbed along the streambed by impact, turbulence, and friction. Such an arrangement is satisfactory where erosion-resistant bedrock exists at shallow depths in the channel and along the abutments or where the spillway outlet is sufficiently removed from the dam or other appurtenances to avoid damage by scour, undermining, or abutment sloughing. The discharge channel may be terminated well above the streambed level or it may be continued to or below streambed.

Upturned deflectors, cantilevered extensions, or flip buckets can be provided to project the jet some distance downstream from the end of the structure. Often, erosion in the streambed at the point of contact of the jet can be minimized by fanning the jet into a thin sheet by the use of a flaring deflector.

Where severe scour at the point of jet impingement is anticipated, a plunge basin can be excavated in the river channel and the sides and bottom lined with riprap or concrete. No definite design criteria except that indicated in section 237 have as yet been established for determining the size or dimensions of a plunge basin which might be necessary to absorb the impact of the flow properly or to avoid scouring velocities. For small installations, it may be expedient to perform a minimum of excavation and to permit the flow to erode a natural pool; protective riprapping or concrete lining may be later provided to halt the scour. In such arrangements an adequate cutoff or other protection must be provided at the end of the spillway structure to prevent it from being undermined.

Where serious erosion to the streambed is to be avoided, the high energy of the flow must be dissipated before the discharge is returned to the stream channel. This can be accomplished by the use of an energy dissipating device, such as a hydraulic jump basin, a roller bucket, a sill block apron, a basin incorporating impact baffles and walls, or some similar energy absorber or dissipator. A description of these devices and a discussion of their hydraulic design is given in part E of this chapter.

(d) *Entrance and Outlet Channels.*—Entrance channels serve to draw water from the reservoir and convey it to the control structure. Where a spillway draws water immediately from the reservoir and delivers it directly back into the river, as in the case with an overflow spillway over a concrete dam, entrance and outlet channels are not required. However, in the case of spillways placed through abutments or through saddles or ridges, channels leading to the spillway control and away from the spillway terminal structure may be required.

Entrance velocities should be limited and channel curvatures and transitions should be made gradual, in order to minimize head loss through the channel (which has the effect of reducing the spillway discharge) and to obtain uniformity of flow over the spillway crest. Effects of an uneven distribution of flow in the entrance channel might persist through the spillway structure to the extent that undesirable erosion could result in the downstream river channel. Nonuniformity of head on the crest may also result in a reduction in the discharge.

The approach velocity and depth below crest level have important influence on the discharge over an overflow crest. As is shown in section 197(a), a greater approach depth with the accompanying reduction in approach velocity will result in a larger discharge coefficient. Thus, for a given head over the crest, a deeper approach will permit a shorter crest length for a given discharge. Within the limits required to secure satisfactory flow conditions and nonscouring velocities, the determination of the relationship of entrance channel depth to channel width is a matter of economics.

Outlet channels convey the spillway flow from the terminal structure to the river channel below the dam. In some instances only a pilot channel is provided, on the assumption that scouring action will enlarge the channel during major spills. Where the channel is in a relatively nonerodible material, it should be excavated to an adequate size to pass the anticipated flow without forming a control which will affect the tailwater stage in the stilling device.

The outlet channel dimensions and its need for protection by lining or riprap will depend on the influences of scour on the tailwater. Although stilling devices are provided, it may be impossible to reduce resultant velocities below the natural velocity in the original stream, and some scouring of the riverbed, therefore, cannot be avoided. Further, under natural conditions the beds of many streams are scoured during the rising stage of a flood and filled during the falling stage by deposition of material carried by the flow. After creation of a reservoir the spillway will normally discharge clear water and the material scoured by the high velocities will not be replaced by deposition. Consequently, there will be a gradual retrogression of the downstream riverbed, which will lower the tailwater stage-discharge relationship. Conversely, scouring where only a pilot channel is provided may build up bars and islands downstream, thereby effecting an aggradation of the downstream river channel which will raise the tailwater elevation with respect to discharges. The dimensions and erosion-protective measures at the outlet channel may be influenced by these considerations.

193. *Spillway Types.*—(a) *General.*—Spillways are ordinarily classified according to their most prominent feature, either as it pertains to the control, to the discharge channel, or to some other component. Spillways often are referred to as controlled or uncontrolled, depending on whether they are gated or ungated. Commonly referred to types are the free overfall (straight drop), ogee (overflow), side channel, open channel (trough or chute), conduit, tunnel, drop inlet (shaft or morning glory) baffled apron drop, culvert, and siphon.

(b) *Free Overfall (Straight Drop) Spillways.*—A free overfall or straight drop spillway is one in which the flow drops freely from the crest. This type is suited to a thin arch or deck overflow dam or to a crest which has a nearly vertical downstream face. Flows may be free discharging, as will be the case with a sharp-crested weir control, or they may be supported along a narrow section of the crest. Occasionally the crest is extended in the form of an overhanging lip to direct small discharges away from the face of the overfall section. In free overfall spillways the underside of the nappe is ventilated sufficiently to prevent a pulsating, fluctuating jet.

Where no artificial protection is provided at the base of the overfall, scour will occur in most streambeds and will form a deep plunge pool. The volume and depth of the hole are related to the range of discharges, the height of the drop, and the depth of tailwater. The erosion-resistant properties of the streambed material including bedrock have little influence on the size of the hole, the only effect being the time necessary to scour the hole to its full depth. Probable depths of scour are discussed in section 210. Where erosion cannot be tolerated, an artificial pool can be created by constructing an auxiliary dam downstream from the main structure, or by excavating a basin which is then provided with a concrete apron or bucket.

If tailwater depths are sufficient, a hydraulic jump will form when a free overfall jet falls upon a flat apron. It has been demonstrated that the momentum equation for the hydraulic jump may be applied to the flow conditions at the base of the fall to determine the elements of the jump.

A free overfall spillway which will be effective over a wide range of tailwater depths can be designed for use with low earthfill dams [1, 2, 3, 4].[2] An artist's conception of such a structure is shown in figure 237. It consists principally of a straight breast wall weir set at the upper end of a rectangular flume section, with its horizontal apron placed at or below streambed level. Floor blocks and an end sill are provided in this case to help in the establishment of the jump and to reduce the downstream scour. This type of structure is not adaptable for high drops on yielding foundations, because of the large impact forces which must be absorbed by the apron at the point of impingement of the jet. Vibrations incident to the impact might crack or displace the struc-

[2] Numbers in brackets refer to the bibliography, sec. 219.

ture, with danger from failure by piping or undermining. Ordinarily, the use of this structure for hydraulic drops from head pool to tailwater in excess of 20 feet should not be considered. The hydraulic design of the free overfall spillway is discussed in section 211.

(c) *Ogee (Overflow) Spillways.*—The ogee spillway has a control weir which is ogee or S-shaped in profile. The upper curve of the ogee ordinarily is made to conform closely to the profile of the lower nappe of a ventilated sheet falling from a sharp-crested weir. Flow over the crest is made to adhere to the face of the profile by preventing access of air to the under side of the sheet. For discharges at designed head, the flow glides over the crest with no interference from the boundary surface and attains near-maximum discharge efficiency. The profile below the upper curve of the ogee is continued tangent along a slope to support the sheet on the face of the overflow. A reverse curve at the bottom of the slope turns the flow onto the apron of a stilling basin or into the spillway discharge channel.

The upper curve at the crest may be made either broader or sharper than the nappe profile. A broader shape will support the sheet and positive hydrostatic pressure will occur along the contact surface. The supported sheet thus creates a backwater effect and reduces the efficiency of discharge. For a sharper shape, the sheet tends to pull away from the crest and to produce subatmospheric pressure along the contact surface. This negative

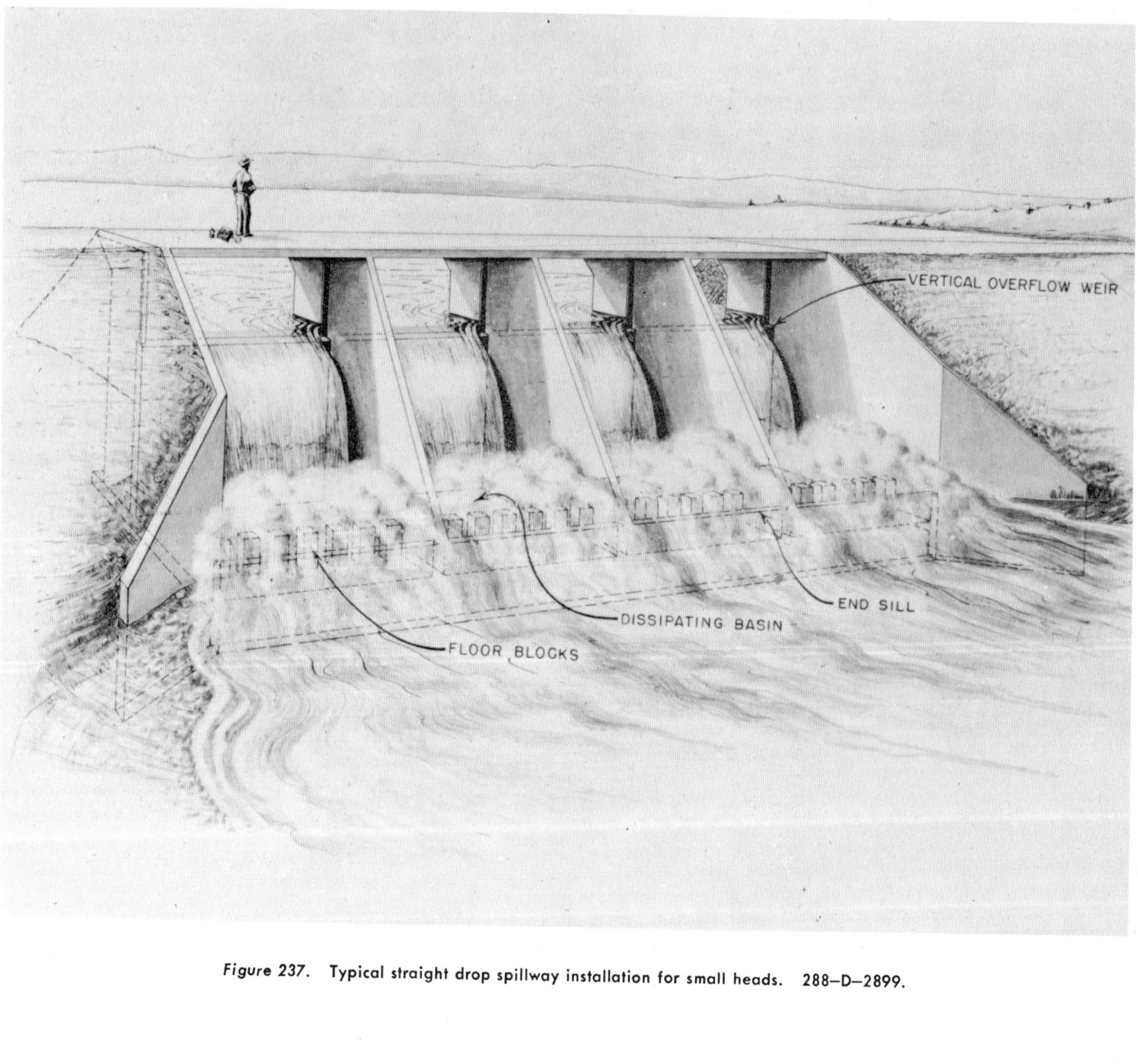

Figure 237. Typical straight drop spillway installation for small heads. 288–D–2899.

pressure effect increases the effective head, and thereby increases the discharge.

An ogee crest and apron may comprise an entire spillway, such as the overflow portion of a concrete gravity dam, or the ogee crest may only be the control structure for some other type of spillway. Because of its high discharge efficiency, the nappe-shaped profile is used for most spillway control crests. Crest shapes and discharge coefficients are discussed in sections 195 through 197.

(d) *Side Channel Spillways.*—The side channel spillway is one in which the control weir is placed along the side of and approximately parallel to the upper portion of the spillway discharge channel. Flow over the crest falls into a narrow trough opposite the weir, turns an approximate right angle, and then continues into the main discharge channel. The side channel design is concerned only with the hydraulic action in the upstream reach of the discharge channel and is more or less independent of the details selected for the other spillway components. Flows from the side channel can be directed into an open discharge channel or into a closed conduit or inclined tunnel. Flow into the side channel might enter on only one side of the trough in the case of a steep hillside location, or on both sides and over the end of the trough if it is located on a knoll or gently sloping abutment. The "bathtub" type of side channel spillway, shown on figures 233 and 234, illustrates the latter type. Figure 238 is an artist's conception of a side channel spillway where flow enters only one side of the trough.

Discharge characteristics of a side channel spillway are similar to those of an ordinary overflow and are dependent on the selected profile of the weir crest. However, for maximum discharges the side channel flow may differ from that of the overflow spillway in that the flow in the trough may be restricted and may partly submerge the flow over the crest. In this case the flow characteristics will be controlled by a constriction in the channel downstream from the trough. The constriction may be a point of critical flow in the channel, an orifice control, or a conduit or tunnel flowing full.

Although the side channel is not hydraulically efficient nor inexpensive, it has advantages which make it adaptable to certain spillway layouts. Where a long overflow crest is desired in order to limit the surcharge head and the abutments are steep and precipitous, or where the control must be connected to a narrow discharge channel or tunnel, the side channel is often the best choice.

The hydraulic design of the side channel spillway is discussed in section 202.

(e) *Chute (Open Channel or Trough) Spillways.*—A spillway whose discharge is conveyed from the reservoir to the downstream river level through an open channel, placed either along a dam abutment or through a saddle, might be called a chute, open channel, or trough type spillway. These designations can apply regardless of the control device used to regulate the flow. Thus, a spillway having a chute-type discharge channel, though controlled by an overflow crest, a gated orifice, a side channel crest, or some other control device, might still be called a chute spillway. However, the name is most often applied when the spillway control is placed normal or nearly normal to the axis of an open channel, and where the streamlines of flow both above and below the control crest follow in the direction of the axis.

The chute spillway has been used with earthfill dams more often than has any other type. Factors influencing the selection of chute spillways are the simplicity of their design and construction, their adaptability to almost any foundation condition, and the overall economy often obtained by the use of large amounts of spillway excavation in the dam embankment. Chute spillways have been constructed successfully on all types of foundation materials, ranging from solid rock to soft clay.

Chute spillways ordinarily consist of an entrance channel, a control structure, a discharge channel, a terminal structure, and an outlet channel. The simplest form of chute spillway has a straight centerline and is of uniform width such as that shown in figure 239. Often, either the axis of the entrance channel or that of the discharge channel must be curved to fit the alinement to the topography. In such cases, the curvature is confined to the entrance channel if possible, because of the low approach velocities. Where the discharge channel must be curved, its floor is sometimes superelevated to guide the high-velocity flow around the bend, thus avoiding a piling up of flow toward the outside of the chute.

Chute spillway profiles are usually influenced by the site topography and by subsurface foundation conditions. The control structure is generally placed in line with or upstream from the centerline of the dam. Usually the upper portion of the discharge channel is carried at minimum grade until it "daylights" along the downstream hillside to minimize excavation. The steep portion of the discharge channel then follows the slope of the abutment.

Flows upstream from the crest are generally at subcritical velocity, with critical velocity occurring when the water passes over the control. Flows in the chute are ordinarily maintained at supercritical stage, either at constant or accelerating rates, until the terminal structure is reached. For good hydraulic performance, abrupt vertical changes or sharp convex or concave vertical curves in the chute profile should be avoided. Similarly, the convergence or divergence in plan should be gradual in order to avoid cross waves, "ride-up" on the walls, excessive turbulence, or uneven distribution of flow at the terminal structure.

The hydraulic design of the chute spillway crest is discussed in part C, determination of hydraulic properties for the discharge channel is given in part D, and stilling basin designs are explained in part E, respectively, of this chapter.

(f) *Conduit and Tunnel Spillways.*—Where a closed channel is used to convey the discharge around or under a dam, the spillway is often called a tunnel or conduit spillway, as appropriate. The closed channel may take the form of a vertical or inclined shaft, a horizontal tunnel through earth or rock, or a conduit constructed in open cut and backfilled with earth materials. Most forms of control structures, including overflow crests, vertical or inclined orifice entrances, drop inlet entrances, and side channel crests, can be used with conduit and tunnel spillways.

With the exception of those with orifice or drop inlet entrances, tunnel and conduit spillways are designed to flow partly full throughout their length. With the drop inlet or orifice control, the tunnel or conduit size is selected so that it flows full for only a short section at the control and thence partly full for its remaining length. Ample aeration must be provided in a tunnel or conduit spillway in order to prevent a make-and-break siphonic action which would result if some part of the tunnel or conduit tends to seal temporarily because of an exhaustion of air caused by surging of the water jet, or by wave action or backwater. To guarantee free flow in the tunnel, the ratio of the flow area to the total tunnel area is often limited to about 75 percent. Air vents may be provided at critical points along the tunnel or conduit to insure an adequate air supply which will avoid unsteady flow through the spillway.

Tunnel spillways may present advantages for damsites in narrow canyons with steep abutments or at sites where there is danger to open channels from snow or rock slides. Conduit spillways may be appropriate at damsites in wide valleys, where the abutments rise gradually and are at a considerable distance from the stream channel. Use of a conduit will permit the spillway to be located under the dam near the streambed.

(g) *Drop Inlet (Shaft or Morning Glory) Spillways.*—A drop inlet or shaft spillway, as the name implies, is one in which the water enters over a horizontally positioned lip, drops through a vertical or sloping shaft, and then flows to the downstream river channel through a horizontal or near horizontal conduit or tunnel. The structure may be considered as being made up of three elements; namely, an overflow control weir, a vertical transition, and a closed discharge channel. Where the inlet is funnel-shaped, this type of structure is often called a "morning glory" or "glory hole" spillway.

Discharge characteristics of the drop inlet spillway may vary with a range of head. The control will shift according to the relative discharge capacities of the weir, the transition, and the conduit or tunnel. For example, as the heads increase on a glory hole spillway, the

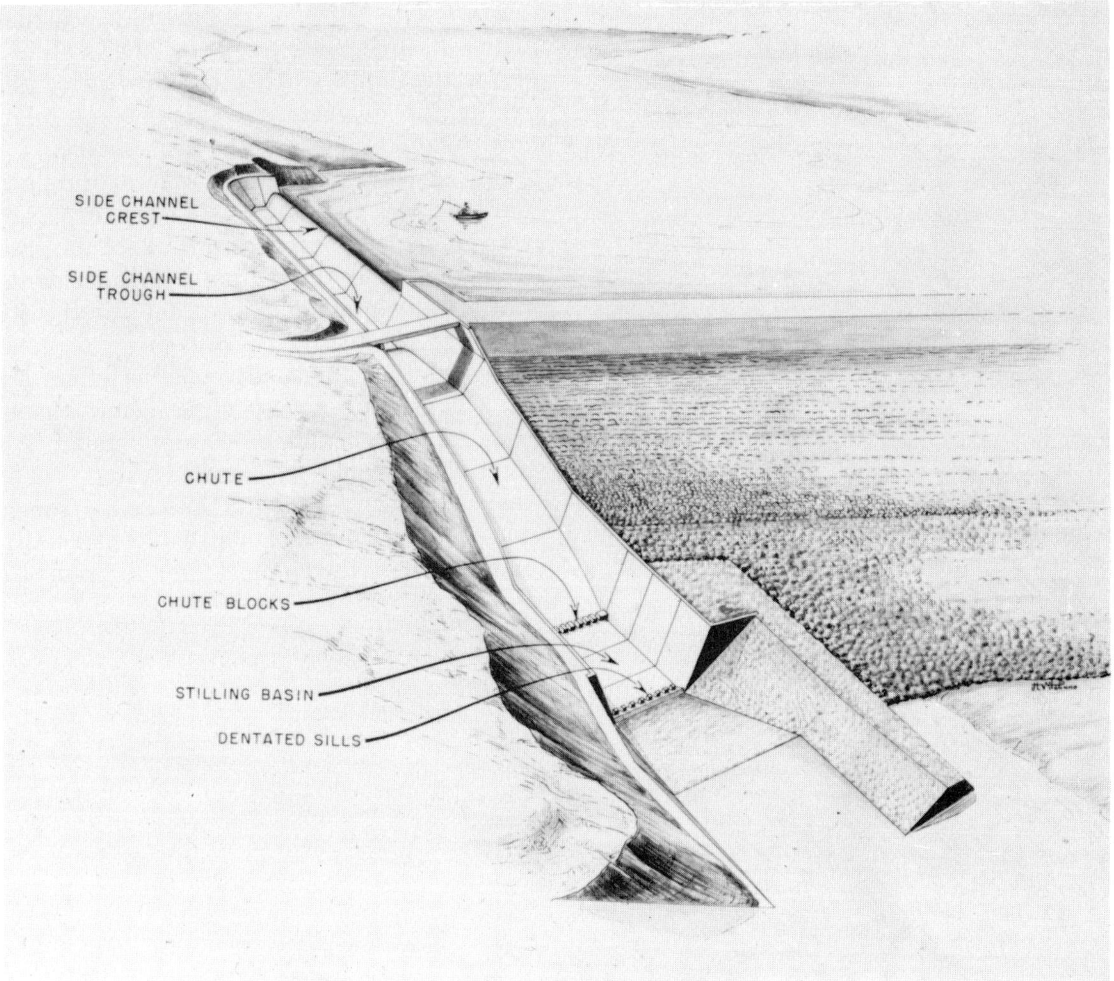

Figure 238. Typical side channel and chute spillway arrangement. 288–D–2900.

control will shift from weir flow over the crest to tube flow in the transition and then to full pipe flow in the downstream portion. Full pipe flow design for spillways except those with extremely low drops is not recommended, as is discussed in section 212(e).

A drop inlet spillway can be used advantageously at dam sites in narrow canyons where the abutments rise steeply or where a diversion tunnel or conduit is available for use as the downstream leg. Another advantage of this type of spillway is that near maximum capacity is attained at relatively low heads; this characteristic makes the spillway ideal for use where the maximum spillway outflow is to be limited. This characteristic also may be considered disadvantageous, in that there is little increase in capacity beyond the designed heads, should a flood larger than the selected inflow design flood occur. This would not be a disadvantage if this type of spillway were used as a service spillway in conjunction with an auxiliary or emergency spillway.

An artist's conception of a drop inlet spillway used with a small earthfill dam is shown in figure 240. Figure 241 shows such a conduit under construction. The hydraulic design is discussed in section 212. Additional information on the design and performance of drop inlet spillways is given in the references listed in the bibliography [5, 6, 23].

(h) *Baffle Apron Drop Spillways.*—Baffled

Figure 239. Chute spillway for Scofield Dam in Utah. 288–D–2901.

aprons or chutes are used in flow ways where water is to be lowered from one level to another and where it is desirable to avoid a stilling basin. The baffle piers partially obstruct the flow, dissipating energy as the water flows down the chute so that the flow velocities entering the downstream channel are relatively low. Advantages of baffled aprons include economy, low terminal velocity of the flows regardless of the height of the drop, downstream degradation does not affect the spillway operation, and there are no requirements for initial tailwater depth in order for the stilling action to be effective.

The chute is normally constructed on a slope of 2:1 or flatter, extending below the outlet channel floor. Chutes having slopes steeper than 2:1 should be model tested [11] and their structural stability checked. The lower end of the chute should be constructed far enough below the channel floor to prevent dam-

Figure 240. Drop inlet spillway for a small dam. 288–D–2902.

Figure 241. Conduit and stilling basin for combined drop inlet spillway and outlet works at Heart Butte Dam during construction. Heart Butte Dam is on the Heart River in North Dakota. 288–D–2903.

age from degradation or scour.

Design capacities of baffled aprons have varied from less than 10 second-feet to over 80 second-feet per foot of width. At Conconully Dam, the spillway baffled apron scale model was designed to represent prototype discharge up to 78 second-feet per foot of width and to operate effectively at 150 second-feet per foot of width. The completed spillway for Conconully Dam is shown in figure 242. The generalized design procedures discussed in this section were obtained from test results on several models of baffled chutes conducted at the Bureau of Reclamation [11].

The typical hydraulic design procedure for a baffle apron drop spillway is given in the numbered steps following and refers to figures 243 and 244:

Figure 242. Baffled apron drop spillway used at Conconully Dam in Washington. The flow shown passing over the spillway is about 50 cfs. P21–141–178NA.

(1) Determine the maximum expected discharge, Q.

(2) Determine unit design discharge $q=Q/W$, where W is the chute width. The chute width may depend on the upstream channel width, the downstream channel width, economy, topography, and frequency of discharge, as well as maximum discharge.

(3) Entrance velocity, V, should be as low as practical. Ideal conditions exist when $V_1=\sqrt[3]{gq}-5$, curve D, figure 244, for discharges up to 69 second-feet per foot of width. Velocities near critical, $V_c=\sqrt[3]{gq}$, curve C, figure 244, or above cause the flow to be thrown into the air after striking the first baffle pier. High velocities may cause the flow to pass completely over the next row or two of baffle piers. It is very important that proper flow conditions be provided at the entrance to the baffle apron because satisfactory performance of the entire structure may hinge on proper entrance flow conditions.

(4) A vertical offset between the approach channel floor and the chute is used to form a desirable uniform entrance velocity, V_1, and will vary in individual installations. A short radius curve provides a crest on the sloping chute. The first row of baffle piers should be placed no more than 12 inches in elevation below the crest. Alternate rows should be staggered to provide a baffle pier below each space and a space below each baffle pier.

(5) The baffle pier height, H, should be about 0.8 D_c or 0.9 D_c, where the critical depth (D_c) for the rectangular chute is given by the formula $D_c=\sqrt[3]{q^2/g}$ and is shown by curve A of figure 244. Baffle pier height is not a critical dimension but should not be less than recommended.

(6) Baffle pier widths and spaces should be equal, preferably about one and one-half H, but not less than H. Other baffle pier dimensions are not critical hydraulically. Suggested cross sectional dimensions are given.

(7) Row spacing of baffle piers along the chute slope should be H divided by the slope, where the slope is given in decimal form. A 2:1 slope, 0.50 in decimal form, makes the row spacing equal to $2H$ parallel to the chute floor.

(8) The baffle piers are usually constructed with the upstream face normal to the chute floor surface; however, piers with vertical faces may be used. Vertical face piers tend to produce more splash and less bed scour, but the differences are minor.

(9) Four rows of baffle piers are needed to establish full control of the flow, although fewer rows have operated successfully. As many additional rows as required beyond the fourth maintain the control established upstream. At least one row of baffles should be buried below the outlet channel grade to protect against scour. Additional rows of

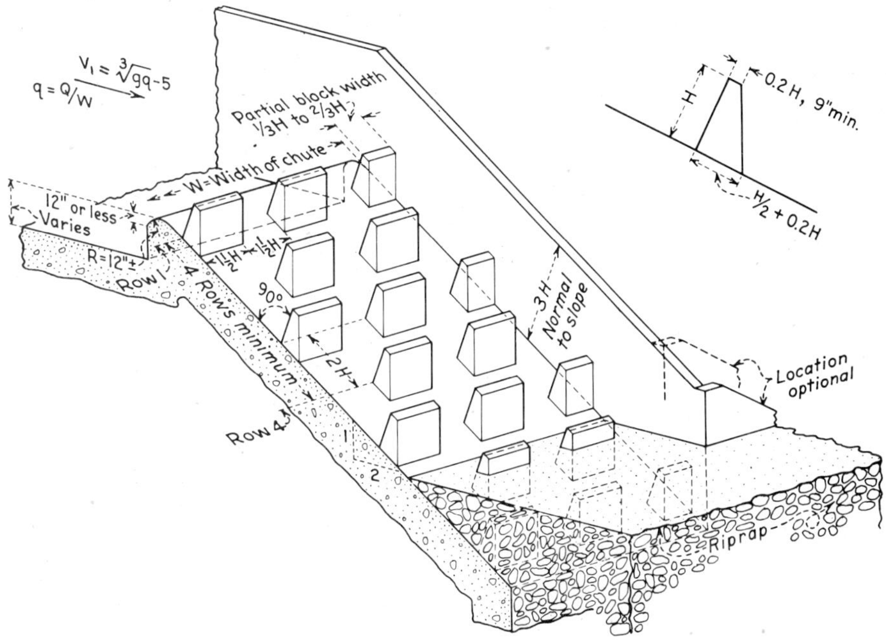

Figure 243. **Basic proportions of a baffled chute spillway. 288–D–2807.**

baffles should be buried as needed to protect against degradation.

(10) The chute training walls should be three times as high as the baffle piers measured normal to the floor. This wall height will contain the main flow and most of the splash. It is not necessary or practical to build the walls high enough to contain all the splash.

(11) Riprap should be placed at the downstream ends of the training walls to prevent erosion of the banks.

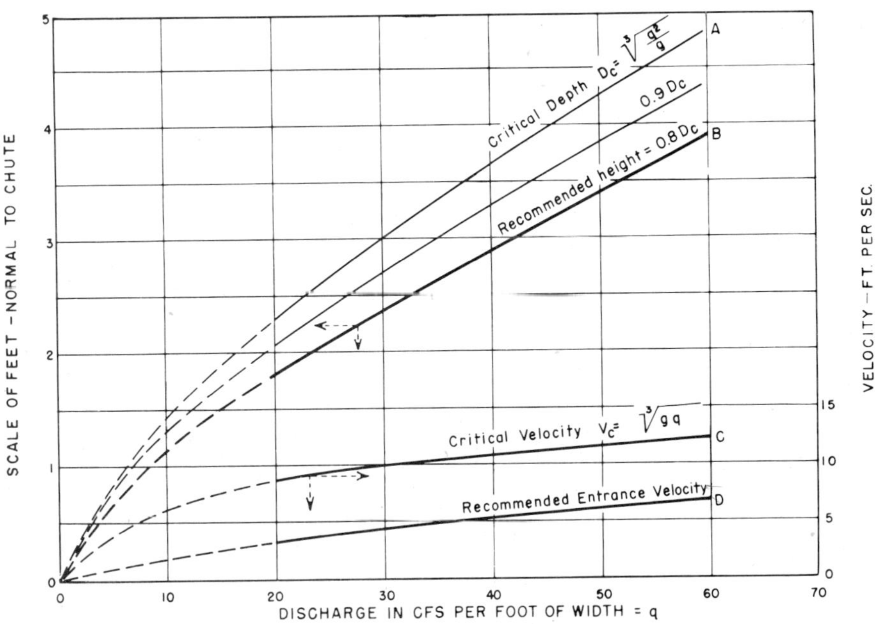

Figure 244. **Recommended baffle pier heights and allowable velocities for baffled chute spillways. 288–D–2806.**

(i) *Culvert Spillways.*—A culvert spillway is a special adaptation of the conduit or tunnel spillway. It is distinguished from the drop inlet and other conduit types in that its inlet opening is placed either vertically or inclined upstream or downstream, and its profile grade is made uniform or near uniform and of any slope. The spillway inlet opening might be sharp edged or rounded, and the approach to the conduit might have flared or tapered sidewalls with a level or sloping floor. If it is desired that the conduit flow partly full for all conditions of discharge, special precautions are taken to prevent the conduit from flowing full; if full flow is desired, bellmouth or streamlined inlet shapes are provided. Special hooded inlets are sometimes added to facilitate the flow passing from part full to full flow conditions as well as to prevent the formation of vortices which would interfere with the full flow action [24].

Culvert spillways operating with the inlet unsubmerged will act similarly to an open channel spillway. Those operating with the inlet submerged, but with the inlet orifice arranged so that full conduit flow is prevented, will act similarly to an orifice-controlled drop inlet spillway, or to an orifice-controlled chute spillway. Where priming action is induced and the conduit flows full, the operation will be similar to that of a siphon spillway. When the culvert spillway is arranged to operate as a siphon, recognition must be taken of the disadvantages of siphon flow, especially those listed as items (4), (5), and (6) in section 193(i).

When culvert spillways placed on steep slopes flow full, reduced or negative pressures prevail along the boundaries of the conduit. Where negative pressures are large, there is danger of cavitation to the surfaces of the conduit or of its collapsing. Where cracks or joints occur along the low-pressure regions, there is the possibility of drawing in soil surrounding the conduit. Culvert spillways, therefore, should not be used for high-head installations where large negative pressures can develop. Further, the transition flow phenomenon, when the flow changes from part-full to full stage, is attended by rather severe pulsations and vibrations which increase in magnitude with increased fall of the culvert. For these reasons, culvert spillways should not be used for hydraulic drops exceeding 25 feet.

For drops not exceeding 25 feet, culvert spillways offer advantages over similar types because of their adaptability for either part-full or full flow operation and because of their simplicity and economy of construction. They might be placed on a bench excavated along the abutment on a relatively steep sidehill location, or they can be placed through the main section of the dam to discharge directly into the downstream river channel. As is the case with a drop inlet or siphon spillway, a principal disadvantage of the culvert spillway is that because its capacity does not substantially increase with increase in head, it does not provide a factor of safety against underestimation of the design flood. This disadvantage would not apply if the culvert type were used as a service spillway in conjunction with an auxiliary or emergency spillway.

The hydraulic design and details for the culvert spillway are discussed in section 213.

(j) *Siphon Spillways.*—A siphon spillway is a closed conduit system formed in the shape of an inverted U, positioned so that the inside of the bend of the upper passageway is at normal reservoir storage level. The initial discharges of the spillway, as the reservoir level rises above normal, are similar to flow over a weir. Siphonic action takes place after the air in the bend over the crest has been exhausted. Continuous flow is maintained by the suction effect due to the gravity pull of the water in the lower leg of the siphon.

Most siphon spillways are composed of five component parts, as shown on figure 245. These include an inlet, an upper leg, a throat or control section, a lower leg, and an outlet. A siphon-breaker air vent is also provided to control the siphonic action of the spillway so that it will cease operation when the reservoir water surface is drawn down to normal level. Otherwise the siphon would continue to operate until air entered the inlet. The inlet is generally placed well below the normal reservoir water surface to prevent entrance of ice and drift and to avoid the formation of vortices and drawdowns which might break the siphon

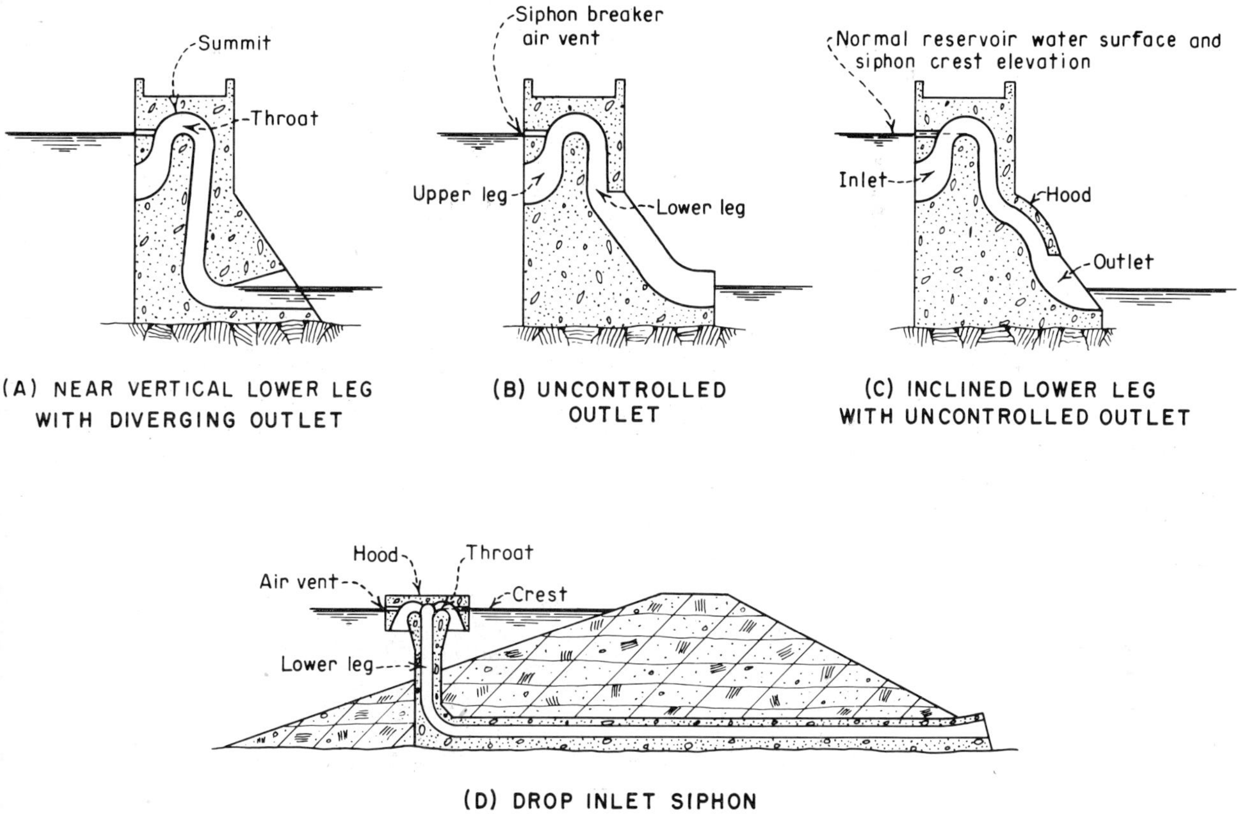

Figure 245. Typical siphon spillways. 288–D–2404.

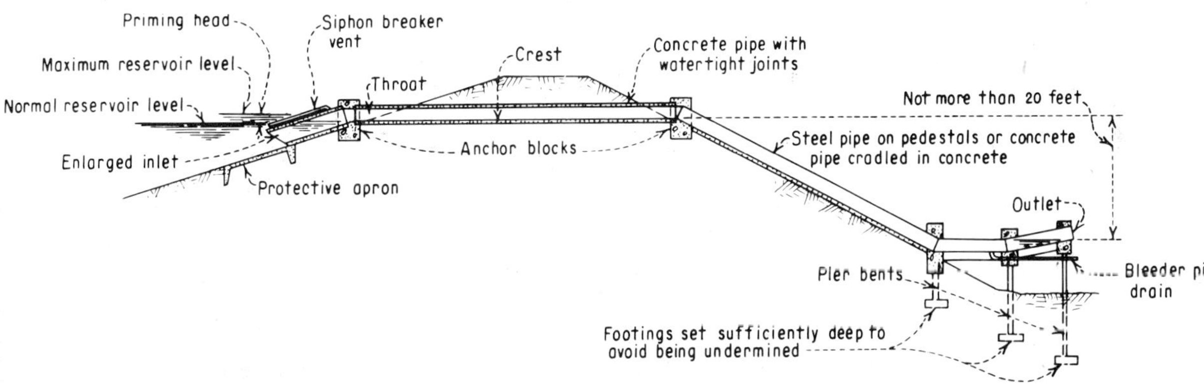

Figure 246. Siphon spillway for a small earthfill dam. 288–D–2405.

action. The upper leg is formed as a bending convergent transition to join the inlet to a vertical throat section. The throat or control section is generally rectangular in cross section and is located at the crest of the upper bend of the siphon. The upper bend then continues to join a vertical or inclined tube which forms the lower leg of the siphon. Often the lower leg is placed on an adverse slope, as shown in figure 245(A), to provide a more positive priming action by forming a flow curtain which seals across the leg. The lower leg can be terminated so as to discharge vertically or along the face of a concrete dam, as shown in (B) and (C), respectively, of figure 245; or it may be provided with a lower bend and diverging outlet tube to release the flow in a horizontal direction, as shown in figure 245(A). The outlet flow can be free discharging or submerged, depending on the arrangement of the lower leg and on tailwater conditions.

A relatively simple siphon spillway which might be used with a small earthfill dam is shown on figure 246. Because of the negative pressures prevalent in the siphon, the pipe should be sufficiently rigid to withstand the collapsing forces. Joints must be made watertight, and measures must be taken to avoid cracking of the pipe from movement or settlement of the embankment. In order to prevent absolute pressures within the conduit from approaching cavitation or collapsing pressures, the total drop of the siphon should be limited to a maximum of 20 feet.

The principal advantage of a siphon spillway is its ability to pass full-capacity discharges with narrow limits of headwater rise. A further advantage is its positive and automatic operation without mechanical devices or moving parts.

In addition to its higher cost, as compared with other types, the siphon spillway has a number of disadvantages, including the following:

(1) The inability of the siphon spillway to pass ice and debris.
(2) The possibility of clogging the siphon passageways and siphon breaker vents with debris or leaves.
(3) The possibility of water freezing in the inlet legs and air vents before the reservoir rises to the crest level of the spillway, thus preventing flow through the siphon.
(4) The occurrence of sudden surges and stoppages of outflow as a result of the erratic make-and-break action of the siphon, thus causing radical fluctuations in the downstream river stage.
(5) The release of outflows in excess of reservoir inflows whenever the siphon operates, if a single siphon is used. Closer regulation which will more nearly balance outflow and inflow can be obtained by providing a series of smaller siphons, with their siphon breaker vents set to prime at gradually increasing reservoir heads.
(6) The more substantial foundation required to resist vibration disturbances, which are more pronounced than in other types of control structures.

As is the case with other types of closed conduit structures, a principal disadvantage of the siphon spillway is its inability to handle flows materially greater than designed capacity although the reservoir head exceeds the design level. Consequently, the siphon spillway is best suited as a service spillway to be used in conjunction with an auxiliary or emergency structure.

194. *Controlled Crests*.—(a) *General*.—The simplest form of control for a spillway is the free or uncontrolled overflow crest which automatically releases water whenever the reservoir water surface rises above crest level. The advantages of the uncontrolled crest are the elimination of the need for constant attendance and regulation of the control devices by an operator, and the freedom from maintenance and repairs of the devices.

A regulating gate or other form of movable crest must be employed if a sufficiently long uncontrolled crest or a large enough surcharge head cannot be obtained for the required spillway capacity. Such devices will also be required if the spillway is to release storages below the normal reservoir water surface. The type and size of the selected control device may be influenced by such conditions as discharge characteristics of a particular device, climate, frequency and nature of floods, winter storage requirements, flood control storage and outflow provisions, the need for handling ice and debris, and special operating requirements. Whether an operator will be in attendance during periods of flood and the availability of electricity, operating mechanisms, operating bridges, etc., are factors which will influence the type of control device employed.

Many types of crest control have been devised. The type selected for a specific installation should be based on a consideration of the factors noted above as well as economy, adaptability, reliability, and efficiency. In the classification of movable crests are such devices as flashboards, stoplogs, bear-trap gates, tilting hinged-leaf gates, and drum gates. Regulating devices include stoplogs, needle beams, bulkheads, vertical or inclined rectangular lift gates, roller gates, and radial gates.

For simplicity of design and operation, only the less complicated control devices are considered appropriate for spillways for small dams. Such devices as flashboards, stoplogs, rectangular gates, and radial gates should be utilized wherever possible, since they can be easily fabricated or obtained commercially.

(b) *Flashboards and Stoplogs.*—Flashboards and stoplogs provide a means of raising the reservoir storage level above a fixed spillway crest level, when the spillway is not needed for releasing floods. Flashboards usually consist of individual boards or panels supported by vertical pins or stanchions anchored to the crest; stoplogs are boards or panels spanning horizontally between grooves recessed into supporting piers. In order to provide adequate spillway capacity, the flashboards or stoplogs must be removed before the floods occur, or they must be designed or arranged so that they can be removed while being overtopped.

Various arrangements of flashboards have been devised. Some must be placed and removed manually, some are designed to fail after being overtopped, and others are arranged to drop out of position either automatically or by being manually triggered after the reservoir exceeds a certain stage. Flashboards provide a simple economical type of movable crest device, and they have the advantage that an unobstructed crest is provided when the flashboards and their supports are removed. They have numerous disadvantages, however, which greatly limit their adaptability. Among these disadvantages are the following: (1) they present a hazard if not removed in time to pass floods, especially where the reservoir area is small and the stream is subject to flash floods; (2) they require the attendance of an operator or crew to remove them, unless they are designed to fail automatically; (3) if they are designed to fail when the water reaches certain stages their operation is uncertain, and when they fail they release sudden and undesirably large outflows; (4) ordinarily they cannot be restored to position while flow is passing over the crest; and (5) if the spillway functions frequently the repeated replacement of flashboards may be costly.

Stoplogs are individual beams or girders set one upon the other to form a bulkhead supported in grooves at each end of the span. The spacing of the supporting piers will depend on the material from which the stoplogs are constructed, the head of water acting against the stoplogs, and the handling facilities provided for installing and removing them. Stoplogs which are removed one by one as the need for increased discharge occurs are the simplest form of a crest gate.

Stoplogs may be an economical substitute for more elaborate gates where relatively close spacing of piers is not objectionable and where removal is required only infrequently. Stoplogs which must be removed or installed in flowing water may require such elaborate hoisting mechanisms that this type of installation may prove to be as costly as gates. A stoplogged spillway requires the attendance of an operating crew for removing and installing the stoplogs. Further, the arrangement may present a hazard to the safety of the dam if the reservoir is small and the stream is subject to flash floods, since the stoplogs must be removed in time to pass the flood.

(c) *Rectangular Lift Gates.*—Rectangular lift gates span horizontally between guide grooves in supporting piers. Although these gates may be made of wood or concrete, they are often made of metal (cast iron or steel). The support guides may be placed either vertically or inclined slightly downstream. The gates are raised or lowered by an overhead hoist. Water is released by undershot orifice flow for all gate openings.

For sliding gates the vertical side members of the gate frame bear directly on the guide members; sealing is effected by the contact pressure. The size of this type of installation is limited by the relatively large hoisting capacity required to operate the gate because of the sliding friction that must be overcome.

Where larger gates are needed, wheels can be mounted along each side of the rectangular lift gates to carry the load to a vertical track on the downstream side of the pier groove. The use of wheels greatly reduces the amount of friction and thereby permits the use of a smaller hoist. Rubber or belting is used along the sides to seal the openings between the upstream leaf plate and the sides of the pier.

(d) *Radial Gates.*—Radial gates are usually constructed of steel or a combination of steel and wood. They consist of a cylindrical segment which is attached to supporting bearings by radial arms. The face segment is made concentric to the supporting pins so that the entire thrust of the waterload passes through the pins; thus, only a small moment need be overcome in raising and lowering the gate. Hoisting loads then consist of the weight of the gate, the friction between the side seals and the piers, and the frictional resistance at the pins. The gate is often counterweighted to partially counterbalance the effect of its weight, which further reduces the required capacity of the hoist.

The small hoisting effort needed to operate radial gates makes hand operation practical on small installations which otherwise might require power. The small hoisting forces involved also make the radial gate more adaptable to operation by relatively simple automatic control apparatus. Where a number of gates are used on a spillway, they might be arranged to open automatically at successively increasing reservoir levels, or only one or two might be equipped with automatic controls, while the remaining gates would be operated by hand or power hoists.

Small radial gates which may be operated either automatically or by hoist operation are available commercially. These gates are fabricated from structural steel members and have either a corrugated-metal or plate-steel faceplate.

C. HYDRAULICS OF CONTROL STRUCTURES

195. *Shape for Uncontrolled Ogee Crest.*—As discussed in section 193(c), crest shapes which approximate the profile of the under nappe of a jet flowing over a sharp-crested weir provide the ideal form for obtaining optimum discharges. The shape of such a profile depends upon the head, the inclination of the upstream face of the overflow section, and the height of the overflow section above the floor of the entrance channel (which influences the velocity of approach to the crest). Crest shapes have been studied extensively in the Bureau of Reclamation hydraulic laboratories, and data from which profiles for overflow crests can be obtained have been published [7]. For most conditions the data can be summarized according to the form shown on figure 247(A), where the profile is defined as it relates to axes at the apex of the crest. That portion upstream from the origin is defined as either a single curve and a tangent or as a compound circular curve. The portion downstream is defined by the equation:

$$\frac{y}{H_o}=-K\left(\frac{x}{H_o}\right)^n \qquad (2)$$

in which K and n are constants whose values depend on the upstream inclination and on the velocity of approach. Figure 247 gives values of these constants for different conditions.

The approximate profile shape for a crest with a vertical upstream face and negligible velocity of approach is shown on figure 248. The profile is constructed in the form of a compounded circular curve with radii expressed in terms of the design head, H_o. This definition is simpler than that shown on figure 247, since it avoids the need for solving an exponential equation; further, it is represented in a form easily used by a layman for constructing forms or templates. For ordinary conditions of design of small spillways and where the approach height, P, is equal to or greater than one-half the maximum head on the crest, this profile is sufficiently accurate to avoid seriously reduced crest pressures and does not materially alter the hydraulic efficiency of the crest. When the approach height is less than one-half the maximum head on the crest, the profile should be determined from figure 247.

196. Discharge Over An Uncontrolled Overflow Ogee Crest.—(a) *General.*—The discharge over an ogee crest is given by the formula:

$$Q=CLH_e^{3/2} \qquad (3)$$

where:

Q=discharge,
C=a variable coefficient of discharge,
L=effective length of crest, and
H_e=total head on the crest, including velocity of approach head, h_a.

The discharge coefficient, C, is influenced by a number of factors, such as (1) the depth of approach, (2) relation of the actual crest shape to the ideal nappe shape, (3) upstream face slope, (4) downstream apron interference, and (5) downstream submergence. The effect of these various factors is discussed in section 197.

The total head on the crest, H_e, does not include allowances for approach channel friction losses or other losses due to curvature of the upstream channel, entrance loss into the inlet section, and inlet or transition losses. Where the design of the approach channel results in appreciable losses, they must be added to H_e to determine reservoir elevations corresponding to the discharges given by the above equation.

(b) *Pier and Abutment Effects.*—Where crest piers and abutments are shaped to cause side contractions of the overflow, the effective length, L, will be less than the net length of the crest. The effect of the end contractions may be taken into account by reducing the net crest length as follows:

$$L=L'-2(NK_p+K_a)H_e \qquad (4)$$

where:

L=effective length of crest,
L'=net length of crest,
N=number of piers,
K_p=pier contraction coefficient,
K_a=abutment contraction coefficient, and
H_e=total head on crest.

The pier contraction coefficient, K_p, is affected by the shape and location of the pier nose, the thickness of the pier, the head in relation to the design head, and the approach velocity. For conditions of design head, H_o, average pier contraction coefficients may be assumed as follows:

	K_p
For square-nosed piers with corners rounded on a radius equal to about 0.1 of the pier thickness	0.02
For round-nosed piers	0.01
For pointed-nose piers	0

The abutment contraction coefficient is affected by the shape of the abutment, the angle between the upstream approach wall and the axis of flow, the head in relation to the design head, and the approach velocity. For conditions of design head, H_o, average coefficients may be assumed as follows:

	K_a
For square abutments with headwall at 90° to direction of flow	0.20
For rounded abutment with headwall at 90° to direction of flow, when $0.5H_o \geqq r \geqq 0.15H_o$	0.10
For rounded abutments where $r>0.5H_o$ and headwall is placed not more than 45° to direction of flow	0.0

where r=radius of abutment rounding.

197. Coefficient of Discharge for Uncontrolled Ogee Crests.—(a) *Effect of Depth of Approach.*—For a high sharp-crested weir placed in a channel, the velocity of approach is small and the under side of the nappe flowing over the weir attains maximum vertical contraction. As the approached depth is decreased, the velocity of approach increases and the vertical contraction diminishes. For sharp-crested weirs whose heights are not less than about one-fifth the heads producing flow over them, the coefficient of discharge remains fairly constant with a value of about 3.3 although the contraction diminishes. For weir heights less than about one-fifth the head, the contraction of the flow becomes increasingly suppressed and the crest coefficient decreases. When the weir height becomes zero, the contraction is entirely suppressed and the overflow weir becomes in effect a channel or a broad-crested weir, for which the theoretical coefficient of discharge is 3.087. If the sharp-crested weir coefficients are related to the head measured from the point of maximum contraction instead of to the head above the sharp crest, coefficients applicable

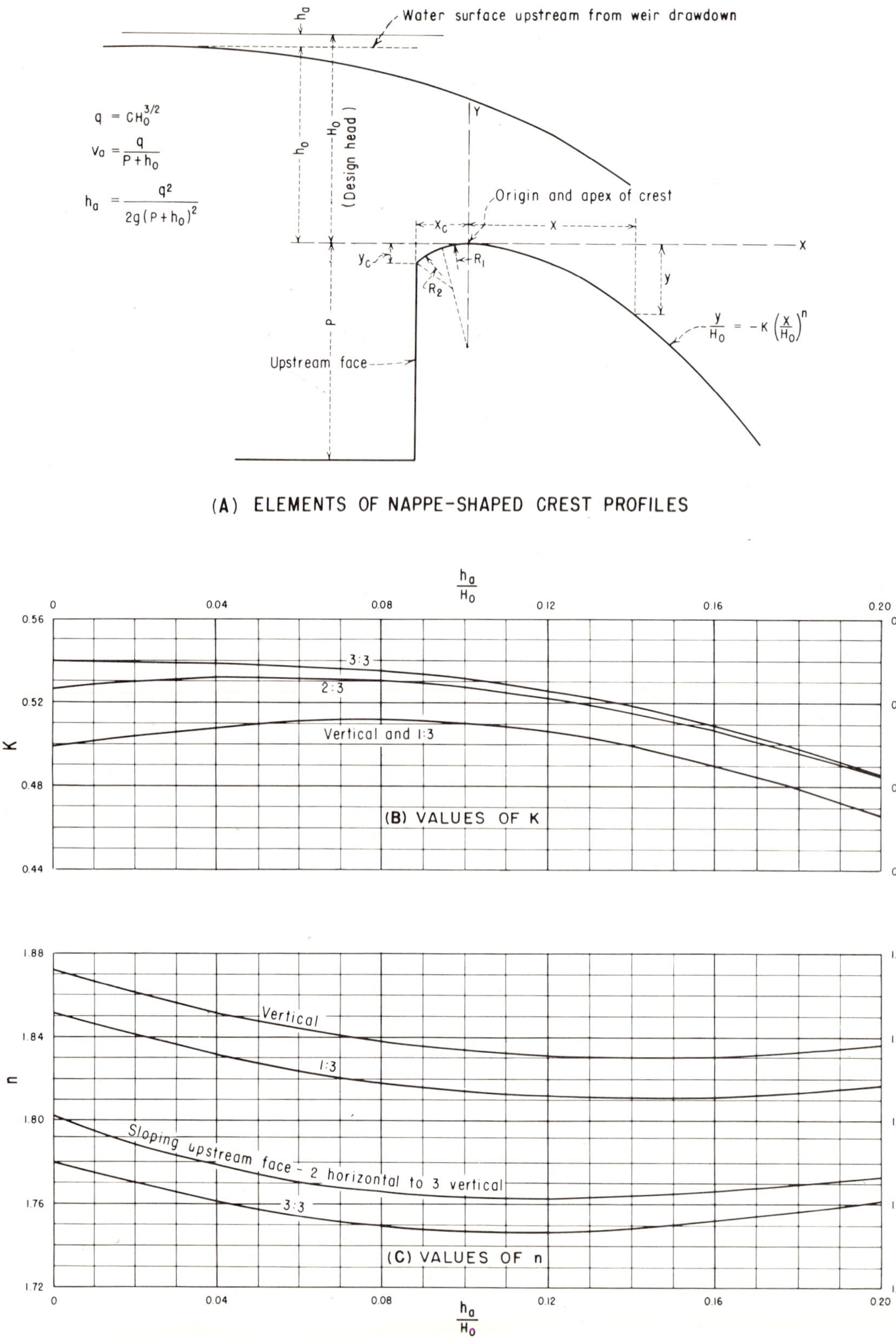

Figure 247. Factors for definition of nappe-shaped crest profiles. 288–D–2406. (Sheet 1 of 2.)

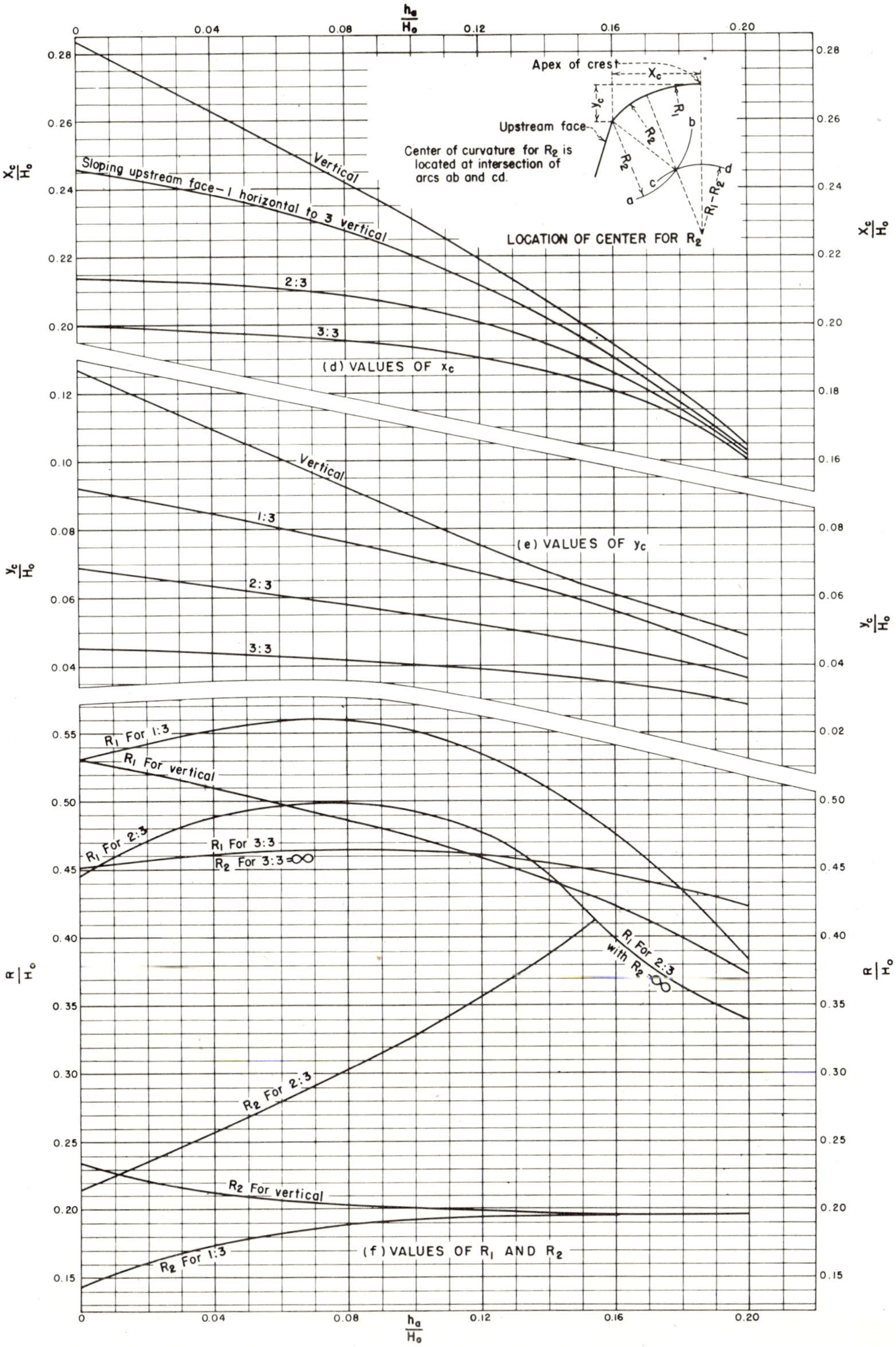

Figure 247. Factors for definition of nappe-shaped crest profiles. 288–D–2407. (Sheet 2 of 2.)

to ogee crests shaped to profiles of under nappes for various approach velocities can be established. The relationship of the ogee crest coefficient, C_o, to various values of $\frac{P}{H_o}$ is shown on figure 249. These coefficients are valid only when the ogee is formed to the ideal nappe shape, that is when $\frac{H_e}{H_o}=1$.

(b) *Effect of Heads Differing from Design Head.*—When the ogee crest is formed to a shape differing from the ideal shape or when the crest has been shaped for a head larger or smaller than the one under consideration, the coefficient of discharge will differ from that shown on figure 249. A widened shape will result in positive pressures along the crest contact surface, thereby reducing the discharge; with a narrower crest shape negative pressures along the contact surface will occur, resulting in an increased discharge. Figure 250 shows the variation of the coefficient as related to values of $\frac{H_e}{H_o}$, where H_e is the actual head being considered.

An approximate coefficient of discharge for an irregularly shaped crest whose profile has not been formed according to the under nappe of the overflow jet can be estimated by finding an ideal shape which most nearly matches it. The design head, H_o, corresponding to the matching shape can then be used as a basis for determining the coefficients.

The coefficients for partial heads on the crest, for preparing a discharge-head relationship, can be determined from figure 250.

(c) *Effect of Upstream Face Slope.*—For small ratios of the approach depth to head on the crest, sloping the upstream face of the overflow results in an increase in the coefficient of discharge. For large ratios the effect is a decrease of the coefficient. Within the range considered in this text, the coefficient of discharge is reduced for large ratios of $\frac{P}{H_o}$ only for relatively flat upstream slopes. Figure 251 shows the ratio of the coefficient for an overflow ogee crest with a sloping face to the coefficient for a crest with a vertical upstream face as obtained from figure 249 (and as adjusted by figure 250 if appropriate), as related to values of $\frac{P}{H_o}$.

(d) *Effect of Downstream Apron Interference and Downstream Submergence.*—When the water level below an overflow weir is high enough to affect the discharge, the weir is said to be submerged. The vertical distance from the crest of the overflow to the downstream apron and the depth of flow in the downstream channel, as it relates to the head pool level, are factors which alter the coefficient of discharge.

Five distinct characteristic flows can occur below an overflow crest, depending on the relative positions of the apron and the downstream water surface: (1) flow will continue at supercritical stage; (2) a partial or incomplete hydraulic jump will occur immediately downstream from the crest; (3) a true hydraulic jump will occur; (4) a drowned jump will occur in which the high-velocity jet will follow the face of the overflow and then continue in an erratic and fluctuating path for a considerable distance under and through the slower water; and (5) no jump will occur—the jet will break away from the face of the overflow and ride along the surface for a short distance and then erratically intermingle with the slow moving water underneath. Figure 252 shows the relationship of the floor positions and downstream submergences which produce these distinctive flows.

Where the downstream flow is at supercritical stage or where the hydraulic jump occurs, the decrease in the coefficient of discharge is due principally to the back-pressure effect of the downstream apron and is independent of any submergence effect due to tailwater. Figure 253 shows the effect of downstream apron conditions on the coefficient of discharge. It will be noted that this curve plots the same data represented by the vertical dashed lines on figure 252 in a slightly different form. As the downstream apron level nears the crest of the overflow $\left(\frac{h_d+d}{H_e}\text{ approaches 1.0}\right)$, the coefficient of discharge is about 77 percent of that for unretarded flow. On the basis of a coefficient of 4.0 for unretarded flow over a high weir, the coefficient when the weir is submerged will be about 3.08, which is virtually the coefficient for

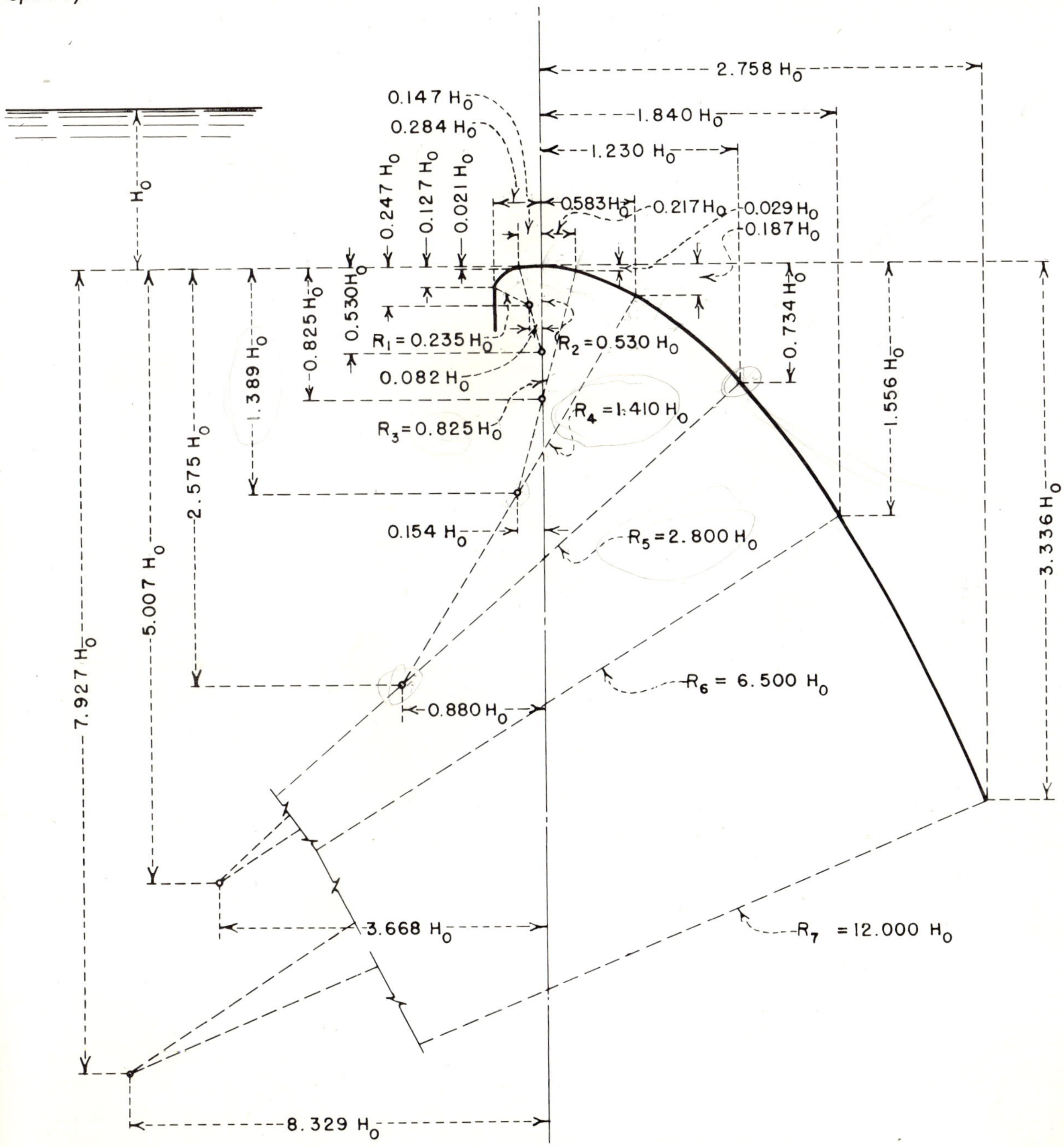

Figure 248. Ogee crest shape defined by compound curves. 288–D–2408.

a broad-crested weir.

From figure 252 it can be seen that when the $\frac{h_a+d}{H_e}$ values exceed about 1.7, the downstream floor position has little effect on the coefficient, but there is a decrease in the coefficient caused by tailwater submergence. Figure 254 shows the ratio of the coefficient of discharge where affected by tailwater conditions, to the coefficient for free flow conditions. This curve plots the data represented by the horizontal dashed lines on figure 252 in a slightly different form. Where the dashed lines on figure 252 are curved, the decrease in the coefficient is the result of a combination of tailwater effects and downstream apron position.

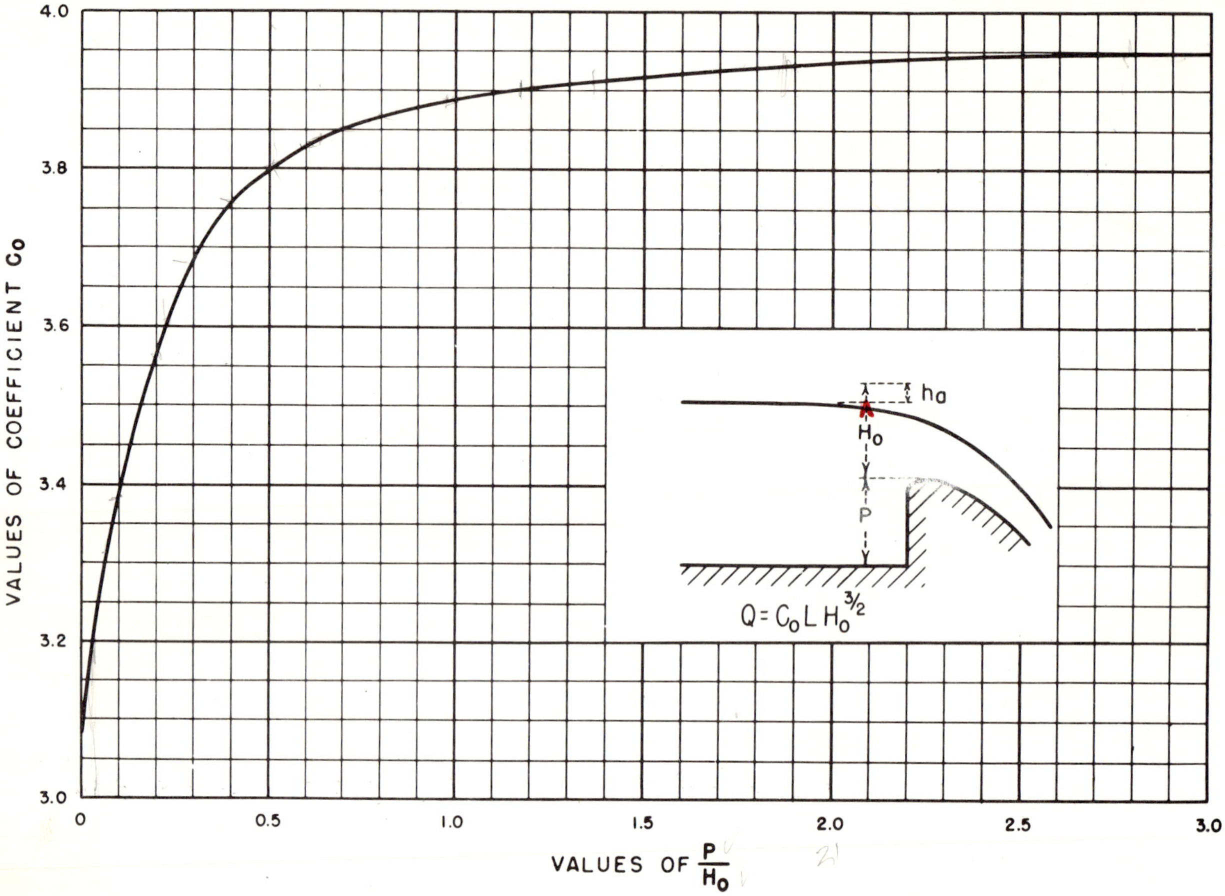

Figure 249. Discharge coefficients for vertical-faced ogee crest. 288–D–2409.

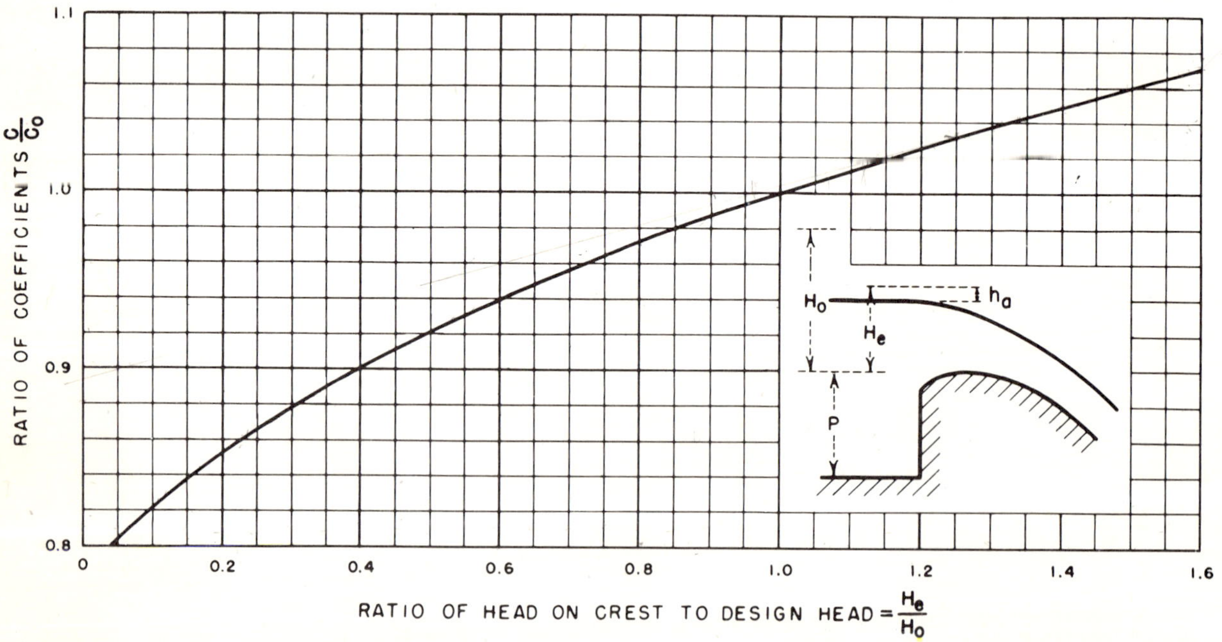

Figure 250. Coefficient of discharge for other than the design head. 288—D—2410.

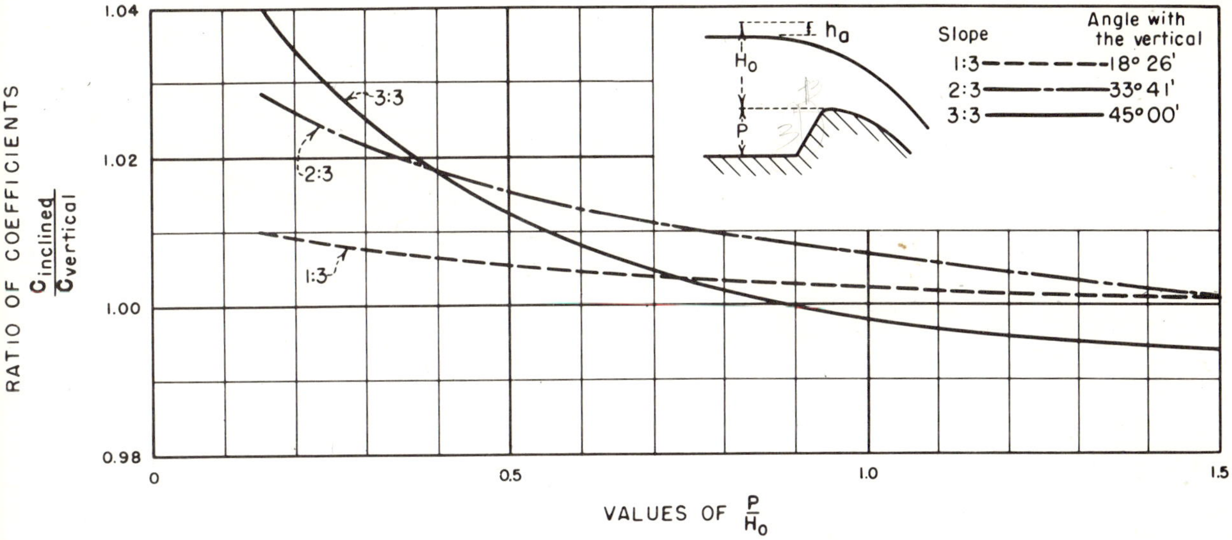

Figure 251. Coefficient of discharge for ogee-shaped crest with sloping upstream face. 288–D–2411.

198. *Examples of Designs of Uncontrolled Ogee Crests.*—The following two examples illustrate the methods of designing uncontrolled ogee crests, including the computation of approach channel losses and velocity head, the determination of the total length of the crest, and the correction of the coefficient of discharge for various effects.

(a) *Example 1.*—Design an uncontrolled overflow ogee crest for a chute spillway, to discharge 2,000 second-feet at a 5-foot head, and prepare a discharge-head curve. The upstream face of the crest is sloped 1:1, and the entrance channel is 100 feet long. A bridge is to span the crest, and 18-inch-wide bridge piers with rounded noses are to be provided. The bridge spans are not to exceed 20 feet. The abutment walls are rounded to a 5-foot radius, and the approach walls are to be placed at 30° with the centerline of the spillway entrance.

To solve the problem, either the approach depth and apron position with respect to the crest must be selected and the appropriate coefficient determined, or an arbitrary coefficient must be selected and the appropriate dimensions determined. The solution will show both procedures.

Procedure 1: First, assume the position of the approach and downstream apron levels with respect to the crest level, say 2 feet below crest level. Then H_e+P is approximately 7 feet.

To evaluate the approach channel losses, assume a value of C to obtain an approximate approach velocity, say $C=3.7$. Then the discharge per unit of crest length, q, is equal to $CH_e^{3/2}=3.7\times5^{3/2}=41$ second-feet. The velocity of approach, v_a, is then equal to $\frac{q}{H_e+P}=\frac{41}{7}=5.9$ feet per second, and the approach velocity head, h_a, is equal to $\frac{v_a^2}{2g}=\frac{5.9^2}{64.4}=0.5$ feet.

Assuming a value of 0.0225 for the friction coefficient, n, in Manning's formula and assuming the hydraulic radius, r, equal to the depth of approach, the friction slope, s, is equal to

$$\left(\frac{v_a n}{1.486 r^{2/3}}\right)^2=\left(\frac{5.9\times0.0225}{1.486\times7^{2/3}}\right)^2=0.0006.$$

Then the total approach channel friction loss, h_f, is equal to $100\times0.0006=0.06$ feet. Assuming an entrance loss into the approach channel equal to 0.1 h_a, the total loss of head in the approach is approximately $0.06+0.1\times0.5=0.11$ feet.

The effective head, H_o, is equal to 5.0–0.11 $=4.89$ feet, and $\frac{P}{H_o}$ is equal to $\frac{2}{4.89}=0.41$. From figure 249, the value of C_o for a $\frac{P}{H_o}$ value of 0.41 is 3.77.

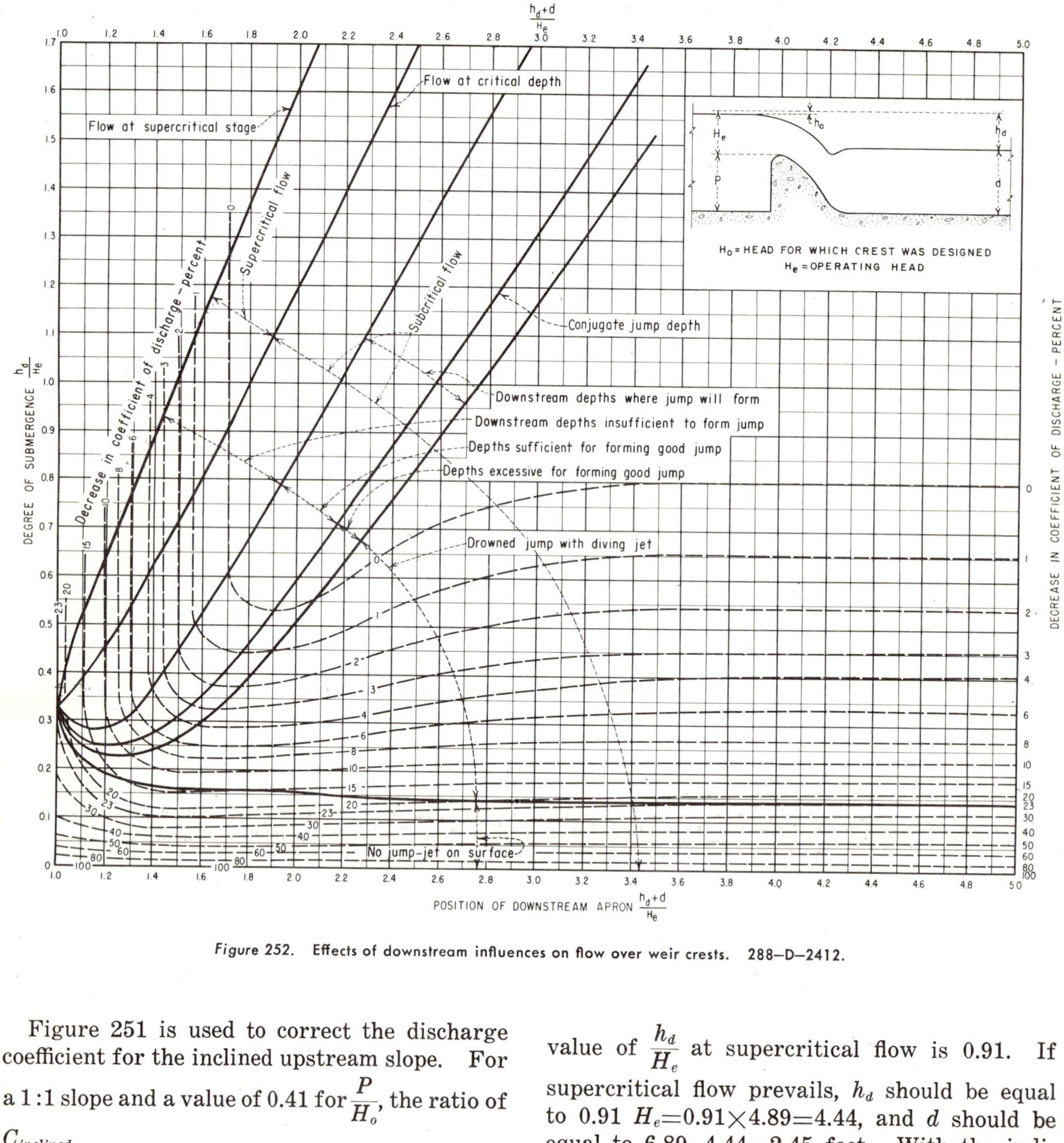

Figure 252. **Effects of downstream influences on flow over weir crests. 288–D–2412.**

Figure 251 is used to correct the discharge coefficient for the inclined upstream slope. For a 1:1 slope and a value of 0.41 for $\frac{P}{H_o}$, the ratio of $\frac{C_{inclined}}{C_{vertical}}$ is 1.018. Then, $C_i=1.018\times3.77=3.84$.

Next, the relationships of $\frac{h_d+d}{H_e}$ and $\frac{h_d}{H_e}$ are evaluated to determine the downstream effects. The value of $\frac{h_d+d}{H_e}$ is approximately $\frac{6.89}{4.89}=1.41$. From figure 252 for an $\frac{h_d+d}{H_o}$ value of 1.41, the value of $\frac{h_d}{H_e}$ at supercritical flow is 0.91. If supercritical flow prevails, h_d should be equal to 0.91 $H_e=0.91\times4.89=4.44$, and d should be equal to 6.89–4.44=2.45 feet. With the indicated unit discharge of approximately 41 second-feet, the downstream velocity will be approximately $\frac{41}{2.45}=16.7$ feet per second, and the velocity head, h_v, will be equal to $\frac{16.7^2}{64.4}=$ 4.4 feet. The closeness of the values of h_d and h_v verifies that the flow is supercritical. From

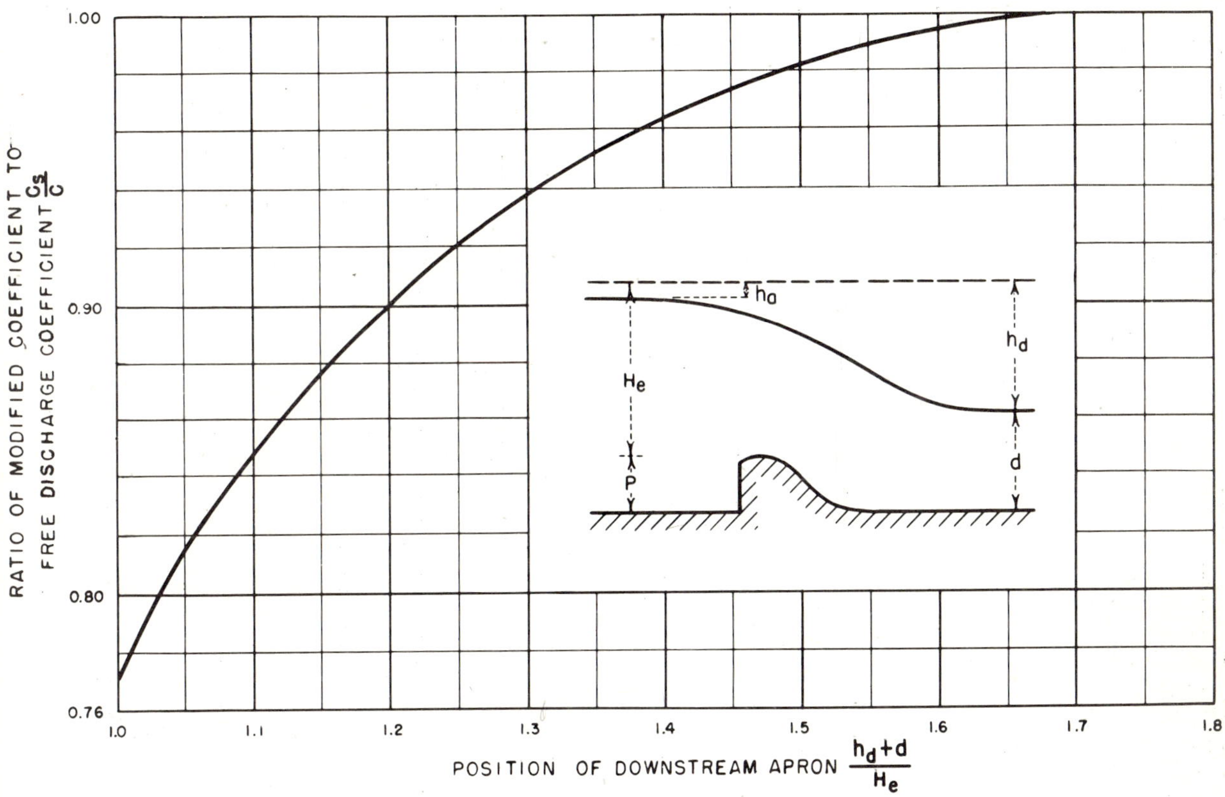

Figure 253. Ratio of discharge coefficients due to apron effect. 288–D–2413.

figure 252 it can be seen that the downstream effect is due to apron influences only, and that the corrections shown on figure 253 will apply. The ratio of the modified C_s to the coefficient C_o for a downstream apron position determined by the $\frac{h_d+d}{H_e}$ ratio of 1.41 is 96.6 percent. The corrected coefficient therefor is $3.84\times0.966=3.71$. This coefficient has now been corrected for all influencing effects.

The next step is to determine the required crest length. For the design head, H_o, of 4.89 feet the required effective crest length, L, is equal to $\frac{Q}{CH_o^{3/2}}=\frac{2{,}000}{3.71\times(4.89)^{3/2}}=49.9$ feet. To correct for pier effects, the net length from equation (4) is:

$$L'=L+2(NK_p+K_a)H_e.$$

If the bridge spans are not to exceed 20 feet, two piers will be required for the approximate 50-foot total span and N will equal 2. Then

$$L'=49.9+2(2\times0.01+0)\,4.89=50.1.$$

The foregoing procedure establishes a coefficient of discharge for the design head. For computing a rating curve, coefficients for lesser heads must be obtained. Since the variations of the different corrections are not consistent, the procedure for correcting the coefficients must be repeated for each lesser head. The variables can be tabulated in a form similar to that used in table 22.

Procedure 2: First, assume an overall coefficient of discharge, say 3.5. The discharge per unit length, q, is then equal to $3.5H_e^{3/2}=39.2$ second-feet for $H_e=5$ feet. Then the required effective length of the crest, L, is equal to $\frac{Q}{q}=\frac{2{,}000}{39.2}=51$ feet.

Next, the approach depth is approximated by use of figure 249. From this figure, for $C=3.5$ the value of $\frac{P}{H_o}$ is approximately 0.2. Thus, the approach depth cannot be less than 1 foot. To

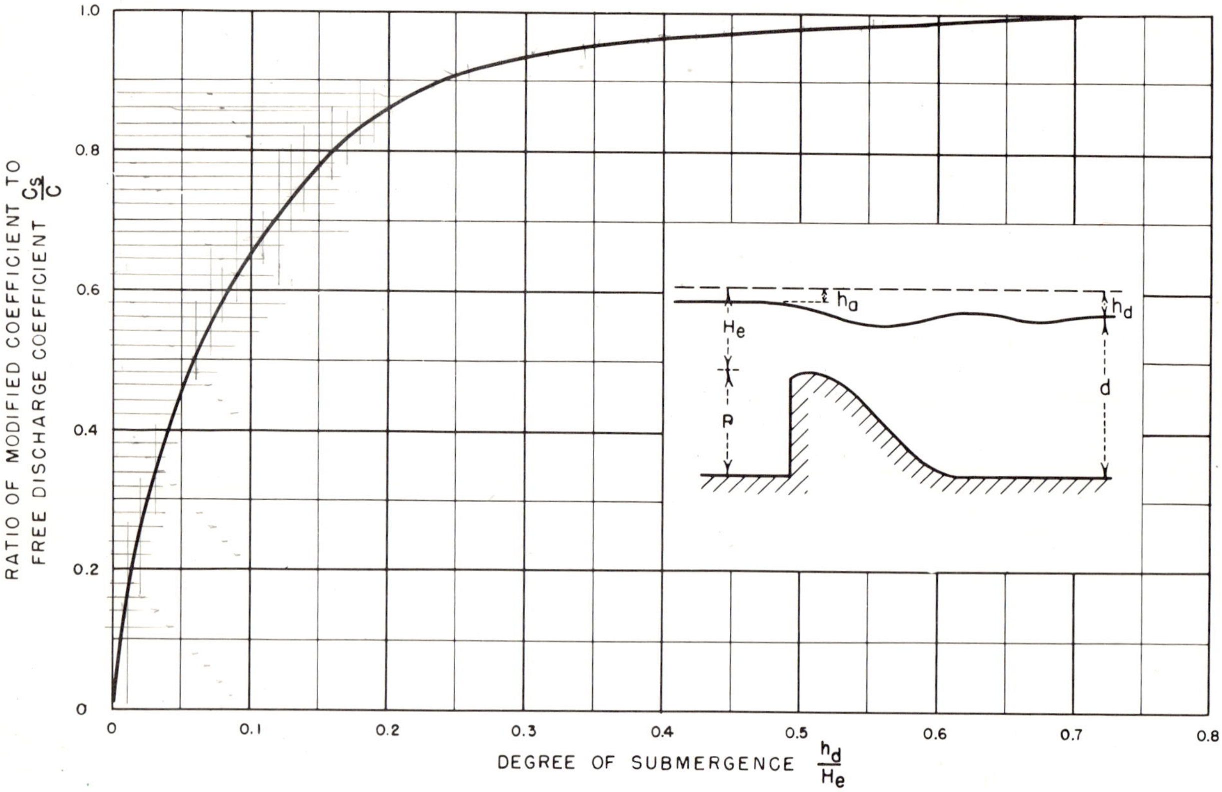

Figure 254. Ratio of discharge coefficients due to tailwater effect. 288–D–2414.

allow for other factors which may reduce the coefficient, an approach depth of about 2 feet might reasonably be assumed.

With a 2-foot approach depth, the computation for approach losses will be the same as in the procedure 1 solution and the effective head, H_o, will become 4.89 feet. Similarly, the value of C_i will be 3.84.

Since the overall coefficient of 3.5 was assumed for the 5-foot gross head, the corre-

TABLE 22.—*Design of an uncontrolled overflow ogee crest—example 1, procedure 1*

[Given l=50 feet.[1] H_o=4.89 feet. P=2 feet]

$\frac{H_e}{H_o}$	H_e feet	(2) $\frac{C}{C_o}$	C_i	h_d+d	$\frac{h_d+d}{H_e}$	(4) $\frac{C_s}{C}$	C_s	$q=C_sH_e^{3/2}$	H_e+P	v_a (approx.)	h_a	s	Entrance loss, 0.1 h_a	Total approach losses, feet	Gross head, feet	Total discharge, second-feet, $Q=C_s LH_e^{3/2}$
0.1	0.49	0.82	3.15	2.49	5.08	1.00	3.15	1.1	2.49	0.44	0.003	0.00001	0	0	0.49	55
.2	.98	.85	3.26	2.98	3.04	1.00	3.26	3.2	2.98	1.07	.02	.00006	0	.01	.99	160
.4	1.96	.90	3.46	3.96	2.02	1.00	3.46	9.5	3.96	2.40	.09	.0002	.01	.03	1.99	475
.6	2.93	.94	3.61	4.93	1.68	1.00	3.61	18.1	4.93	3.67	.21	.0004	.02	.06	2.99	905
.8	3.91	.97	3.73	5.91	1.51	.982	3.66	28.3	5.91	4.79	.36	.0005	.04	.09	4.00	1,415
1.0	4.89	1.0	[3] 3.84	6.89	1.41	.966	3.71	40.0	6.89	5.80	.52	.0006	.05	.11	5.00	2,000
1.2	5.87	1.03	3.96	7.87	1.34	.95	3.76	53.5	7.87	6.80	.72	.0007	.07	.14	6.01	2,675

[1] The effective crest length and the net crest length for H_o are 49.9 feet and 50.1 feet, respectively. Because of the small magnitude of the pier effects, an average length of 50 feet is taken for the effective crest length for all values of H_e. If the pier effects are significant, separate effective crest lengths should be computed for each H_e value.

[2] From fig. 250.

[3] $C_{inclined}$ for H_o.

[4] From fig. 253.

sponding coefficient for the 4.89-foot effective head will be

$$\frac{C_o}{C_{gross\ head}} = \frac{H^{3/2}_{gross\ head}}{H^{3/2}_{effective\ head}}$$

$$\text{or } C_o = C_g\left(\frac{H_g}{H_e}\right)^{3/2} = 3.5\left(\frac{5.0}{4.89}\right)^{3/2} = 1.035 \times 3.5 = 3.62.$$

The submergence ratio, $\frac{C_s}{C_o}$, will then be $\frac{3.62}{3.84} = 0.94$, and, from figure 253, $\frac{h_d+d}{H_e}$ will be 1.3. Thus h_d+d will be $1.3 \times 4.89 = 6.4$ feet. The downstream apron should therefore be placed 1.4 feet below the crest level.

Since it was demonstrated previously that pier and contraction effects are small, they can be neglected in this example, and the net crest length is, therefore, 51 feet. This crest length and downstream apron position can be varied by altering the assumptions of overall coefficient and approach depth.

The discharge rating curve may be developed by a process similar to that used in procedure 1.

(b) *Example 2.*—Design an uncontrolled overflow crest for a diversion dam to pass 2,000 second-feet with a depth of flow upstream from the dam not exceeding 5 feet above the crest. The overflow dam is 8 feet high. The abutment headwall is 90° to the direction of flow and the edge adjacent to the crest is rounded to a 12-inch radius. For 2,000 second-feet flow, the tailwater will rise 3.5 feet above the crest.

For an approximate head, H_e, of 5 feet, a crest height of 8 feet, and a crest submergence of 3.5 feet, $\frac{h_d+d}{H_e} = \frac{13}{5} = 2.6$, and $\frac{h_d}{H_e} = \frac{1.5}{5} = 0.3$. From figure 252 it can be seen that for these relations the downstream flow phenomena will be that of a drowned jump and that the coefficient will be reduced about 6 percent.

Roughly, $\frac{P}{H} = \frac{8}{5} = 1.6$ and the unretarded coefficient from figure 249 is 3.93. Reducing this by 6 percent because of submergence results in an approximate coefficient of 3.7.

The approximate discharge per foot of crest, q, is equal to $CH_o^{3/2} = 3.7 \times 5^{3/2} = 41.5$ second-feet. Then the velocity of approach, v_a, is $\frac{41.5}{13} = 3.2$ feet per second and the approach velocity head, h_a, is 0.16 feet. H_o is $5.0 + 0.16 = 5.16$ feet.

The revised value of $\frac{P}{H_o}$ does not appreciably alter the coefficient obtained from figure 249. The revised value of $\frac{h_d+d}{H_o}$ will be $\frac{13.16}{5.16} = 2.55$ and the revised value of $\frac{h_d}{H_o}$ will be $\frac{1.66}{5.16} = 0.32$. The reduction in coefficient due to submergence effects from figure 252 is 5 percent. The revised coefficient is 95 percent of $3.93 = 3.73$.

The effective crest length, L, is equal to $\frac{Q}{CH_o^{3/2}}$

$$= \frac{2{,}000}{3.73 \times (5.16)^{3/2}} = 45.7 \text{ feet.}$$

The net crest length is determined by use of equation (4). Without piers the net crest length, L', is equal to $L + 2K_aH_e$. For 90° abutment walls rounded to a radius larger than 0.15 H_o, $K_a = 0.10$. Then the net crest length, L', is equal to $45.7 + 2 \times 0.10 \times 5.16 = 46.7$ feet.

199. *Uncontrolled Ogee Crests Designed For Less Than Maximum Head.*—Economy in the design of an ogee crest may sometimes be effected by using a design head which is less than the maximum expected head. As discussed previously, use of a smaller design head results in increased discharges for the full range of heads. The increase in capacity makes it possible to achieve economy by reducing either the crest length or the maximum surcharge head.

Tests have shown that the subatmospheric pressures on a nappe-shaped crest do not exceed about one-half the design head when the design head is not less than about 75 percent of the maximum head. For most conditions in the design of small spillways, these negative pressures will be small, and they can be tolerated because they will not approach absolute pressures which might induce cavitation. Care must be taken, however, in forming the surface of the crest where these negative pressures will occur, since unevenness caused by abrupt offsets, depressions, or projections will amplify the negative pressures to a magnitude where cavitation conditions can develop.

The negative pressure on the crest may be resolved into a system of forces acting both upward and downstream. These forces should be considered in analyzing the structural stability of the crest structure.

An approximate force diagram of the subatmospheric pressures when the design head used to determine the crest shape is 75 percent of the maximum head is shown on figure 255. These data are based on average results of tests made on ideal shaped weirs with negligible velocities of approach. Pressures for intermediate head ratios can be assumed to vary linearly, considering that no subatmospheric pressure prevails when $\frac{H_o}{H_e}$ is equal to 1.

200. *Gate-Controlled Ogee Crests.*—Releases for partial gate openings for gated crests will occur as orifice flow. With full head on the gate and with the gate opened a small amount, a free discharging trajectory will follow the path of a jet issuing from an orifice. For a vertical orifice the path of the jet can be expressed by the parabolic equation:

$$-y=\frac{x^2}{4H} \tag{5}$$

where H is the head on the center of the opening. For an orifice inclined an angle of θ from the vertical, the equation will be:

$$-y=x \tan \theta+\frac{x^2}{4H \cos^2 \theta} \tag{6}$$

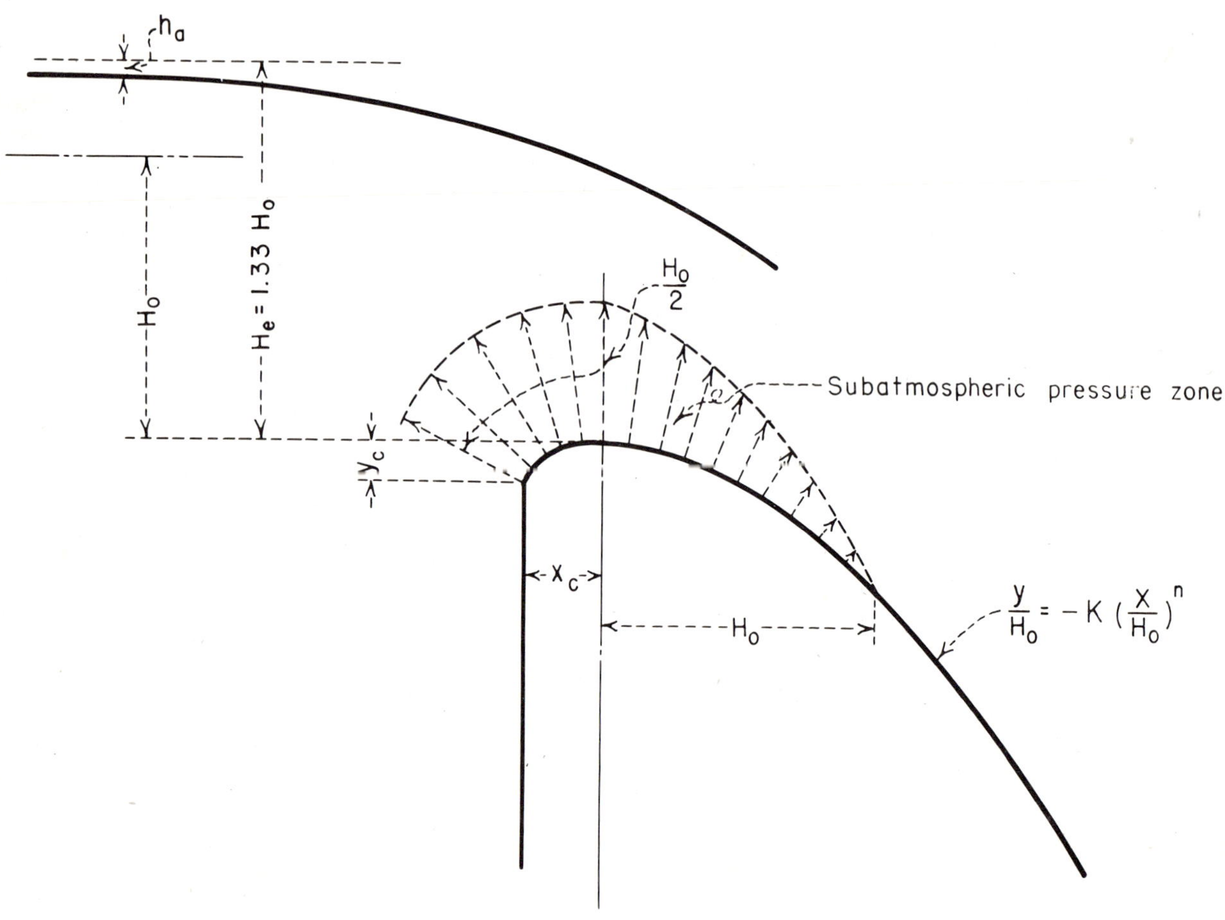

Figure 255. Subatmospheric crest pressures for $\frac{H_o}{H_e} = 0.75$. 288–D–2415.

If subatmospheric pressures are to be avoided along the crest contact, the shape of the ogee downstream from the gate sill must conform to the trajectory profile.

Gates operated with small openings under high heads produce negative pressures along the crest in the region immediately below the gate if the ogee profile drops below the trajectory profile. Tests showed the subatmospheric pressures would be equal to about one-tenth of the design head when the gate is operated at small openings and the ogee is shaped to the ideal nappe profile (equation 2) for maximum head H_o. The force diagram for this condition is shown on figure 256.

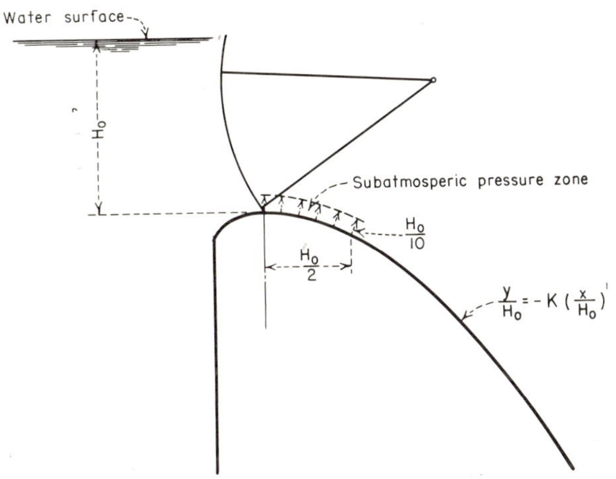

Figure 256. Subatmospheric crest pressures for undershot gate flow. 288–D–2416.

The adoption of a trajectory profile rather than a nappe profile downstream from the gate sill will result in a wider ogee, and reduced discharge efficiency for full gate opening. Where the discharge efficiency is unimportant and where a wider ogee shape is needed for structural stability, the trajectory profile may be adopted to avoid subatmospheric pressure zones along the crest. Where the ogee is shaped to the ideal nappe profile for maximum head, the subatmospheric pressure area can be minimized by placing the gate sill downstream from the crest of the ogee. This will provide an orifice which is inclined downstream for small gate openings and thus will result in a steeper trajectory more nearly conforming to the nappe-shaped profile.

201. *Discharge Over Gate-Controlled Ogee Crests.*—The discharge for a gated ogee crest at partial gate openings will be similar to flow through a low-head orifice and may be computed by the equation:

$$Q=\frac{2}{3}\sqrt{2g}\,CL(H_1^{3/2}-H_2^{3/2}) \qquad (7)$$

where H_1 and H_2 are the total heads (including the velocity head of approach) to the bottom and top of the orifice, respectively. The coefficient, C, will differ with different gate and crest arrangements; it is influenced by the approach and downstream conditions as they affect the jet contractions. Thus, the top contraction for a vertical leaf gate will differ from that for a curved, inclined radial gate; the upstream floor profile will affect the bottom contraction of the issuing jet; and the downstream profile will affect the back pressure and consequently the effective head. Figure 257 shows coefficients of discharge for orifice flows for various ratios of gate opening to total head. The curve represents averages determined for the various approach and downstream conditions described and is sufficiently reliable for determining discharges for small spillway structures.

202. *Side Channel Spillways.*—(a) *General.*—The theory of flow in a side channel spillway [8] is based principally on the law of conservation of linear momentum, assuming that the only forces producing motion in the channel result from the fall in the water surface in the direction of the axis. This premise assumes that the entire energy of the flow over the crest is dissipated through its intermingling with the channel flow and is therefore of no assistance in moving the water along the channel. Axial velocity is produced only after the incoming water particles join the channel stream.

For any short reach of the side channel, the momentum at the beginning of the reach plus any increase in momentum due to external forces must equal the momentum at the end of the reach. If a short reach, Δx in length, is considered and the velocity and discharge at the upstream section are v and Q, respectively, at the downstream section the velocity and discharge will be $v+\Delta v$ and $Q+q(\Delta x)$, where q

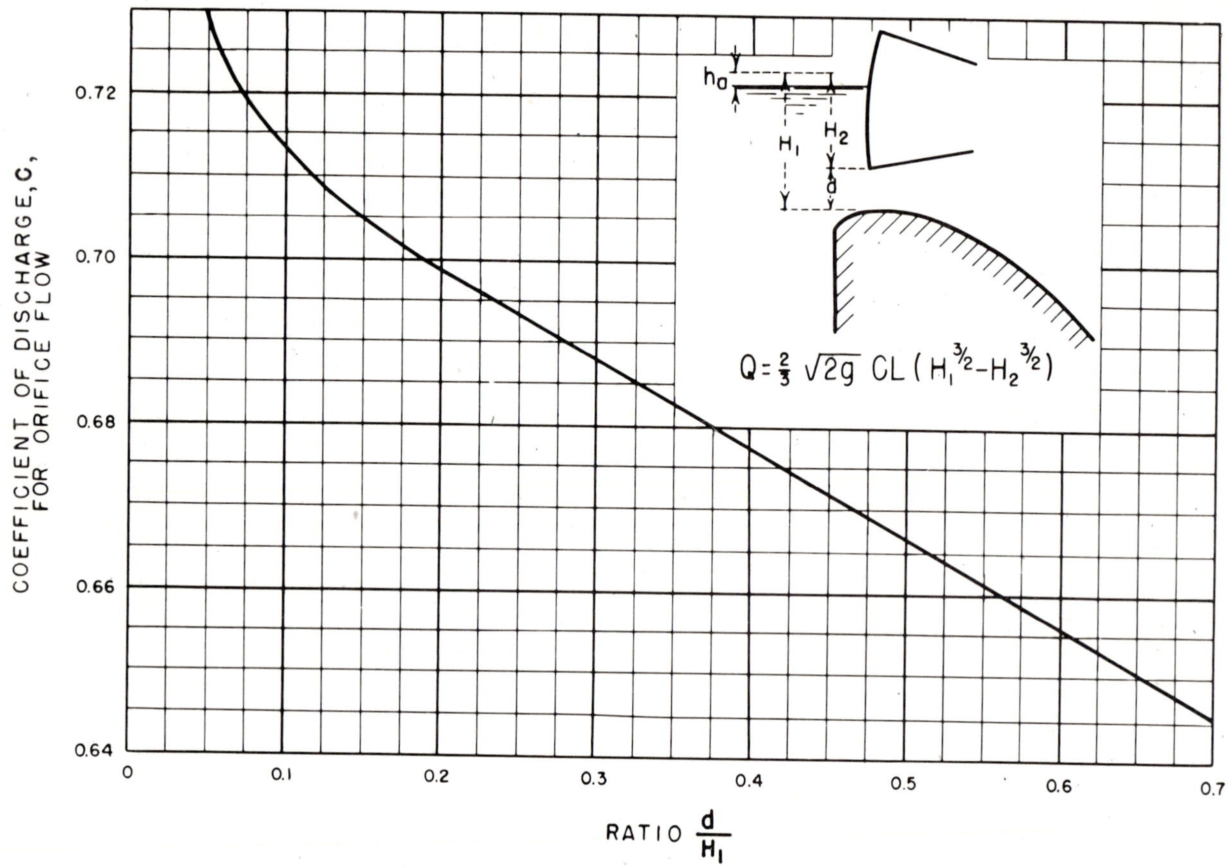

Figure 257. **Coefficient of discharge for flow under gates. 288–D–2417.**

is the inflow per foot of length of weir crest. The momenta[3] at the two sections therefore will be:

$$\text{Upstream, } M_u=\frac{Qv}{g} \tag{8}$$

$$\text{Downstream, } M_d=\frac{[Q+q(\Delta x)]}{g}[v+\Delta v] \tag{9}$$

Subtracting equation (8) from equation (9):

$$\Delta M=\frac{Q(\Delta v)}{g}+\frac{q(\Delta x)}{g}[v+\Delta v] \tag{10}$$

Dividing by Δx:

$$\frac{\Delta M}{\Delta x}=\frac{Q(\Delta v)}{g(\Delta x)}+\frac{q}{g}[v+\Delta v] \tag{11}$$

[3] The weight of 1 cubic foot of water is taken as a unit force to eliminate the necessity of multiplying all forces and momenta by 62.5 to convert them into pounds.

The rate of change of momentum with respect to time being v times the rate of change with respect to x, and considering the average velocity to be $\left[v+\frac{1}{2}(\Delta v)\right]$, equation (11) can be written:

$$\frac{\Delta M}{\Delta t}=\frac{Q(\Delta v)}{g(\Delta x)}\left[v+\frac{1}{2}(\Delta v)\right]+\frac{q}{g}[v+\Delta v]\left[v+\frac{1}{2}(\Delta v)\right] \tag{12}$$

As $\frac{\Delta M}{\Delta t}$ is the accelerating force, which is equal to the slope of the water surface $\frac{\Delta y}{\Delta x}$ times the average discharge, equation (12) becomes:

$$\frac{\Delta y}{\Delta x}\left[Q+\frac{1}{2}(\Delta Q)\right]=\frac{Q(\Delta v)}{g(\Delta x)}\left[v+\frac{1}{2}(\Delta v)\right]+\frac{q}{g}[v+\Delta v]\left[v+\frac{1}{2}(\Delta v)\right] \tag{13}$$

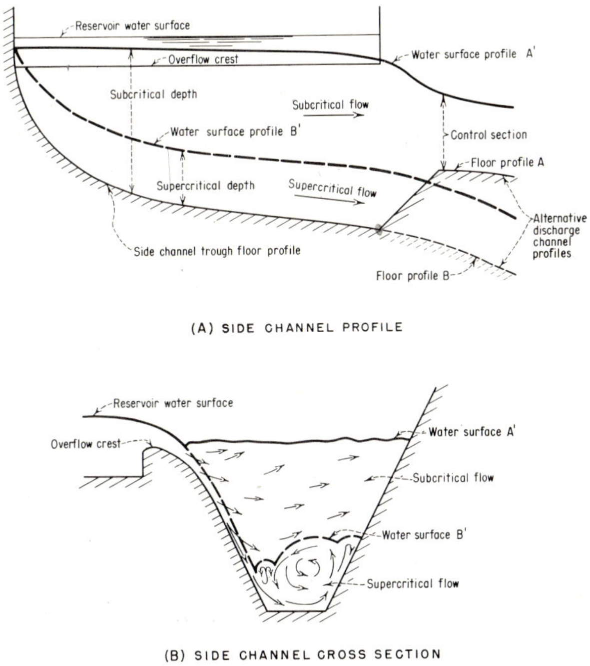

Figure 258. **Side channel flow characteristics. 288–D–2418.**

from which the change in water surface elevation

$$\Delta y = \frac{Q\left[v+\frac{1}{2}(\Delta v)\right]}{g\left[Q+\frac{1}{2}(\Delta Q)\right]}\left\{\Delta v+\frac{q(\Delta x)}{Q}[v+\Delta v]\right\} \qquad (14)$$

If Q_1 and v_1 are values at the beginning of the reach and Q_2 and v_2 are the values at the end of the reach, the equation can be written:

$$\Delta y = \frac{Q_1}{g}\frac{(v_1+v_2)}{(Q_1+Q_2)}\left[(v_2-v_1)+\frac{v_2(Q_2-Q_1)}{Q_1}\right] \qquad (15)$$

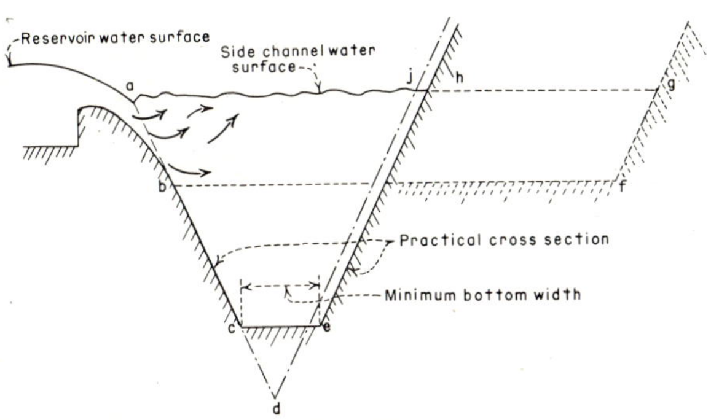

Figure 259. **Comparison of side channel cross sections. 288–D–2419.**

Similarly, the derivation can be developed so that:

$$\Delta y = \frac{Q_2}{g}\frac{(v_1+v_2)}{(Q_1+Q_2)}\left[(v_2-v_1)+\frac{v_1(Q_2-Q_1)}{Q_2}\right] \qquad (16)$$

By use of equation (15) or (16), the water surface profile can be determined for any particular side channel by assuming successive short reaches of channel once a starting point is found. The solution of equation (15) or (16) is obtained by a trial and error procedure. For a reach of length Δx in a specific location, Q_1 and Q_2 will be known. If the depth at one end of the reach has been established, a trial depth at the other end of the reach can be found which will satisfy the indicated and computed values of Δy.

As in other water surface profile determinations, the depth of flow and the hydraulic characteristics of the flow will be affected by backwater influences from some control point, or by critical conditions along the reach of the channel under consideration. The selection of a control for starting the water surface profile computations is treated in the subsequent discussion.

When the bottom of the side channel trough is selected so that its depth below the hydraulic gradient is greater than the minimum specific energy depth, flow will be either at the subcritical or supercritical stage, depending on the relation of the bottom profile to critical slope or on the influences of a downstream control section. If the slope of the bottom is greater than critical and a control section is not established below the side channel trough, supercritical flow will prevail throughout the length of the channel. For this stage, velocities will be high and water depths will be shallow, resulting in a relatively high fall from the reservoir water level to the water surface in the trough. This flow condition is illustrated by profile B′ on figure 258. Conversely, if a control section is established downstream from the side channel trough to increase the upstream depths, the channel can be made to flow at the subcritical stage. Velocities at this stage will be less than critical and the greater depths will result in a smaller drop from the reservoir water surface to the side channel water surface profile. The condition of flow for subcritical depths is illustrated on figure 258 by water surface profile A′.

The effect of the fall distance from the reservoir to the channel water surface for each type of flow is depicted on figure 258(B). It can be seen that for the subcritical stage, the incoming flow will not develop high transverse velocities because of the low drop before it meets the channel flow, thus effecting a good diffusion with the water bulk in the trough. Since both the incoming velocities and the channel velocities will be relatively slow, a fairly complete intermingling of the flows will take place, thereby producing a comparatively smooth flow in the side channel. Where the channel flow is at the supercritical stage, the channel velocities will be high, and the intermixing of the high-energy transverse flow with the channel stream will be rough and turbulent. The transverse flows will tend to sweep the channel flow to the far side of the channel, producing violent wave action with attendant vibrations. It is thus evident that flows should be maintained at subcritical stage for good hydraulic performance. This can be achieved by establishing a control section downstream from the side channel trough.

The cross-sectional shape of the side channel trough will be influenced by the overflow crest on the one side and by the bank conditions on the opposite side. Because of turbulences and vibrations inherent in side channel flow, a side channel design is ordinarily not considered except where a competent foundation such as rock exists. The channel sides will, therefore, usually be a concrete lining placed on a slope and anchored directly to the rock. A trapezoidal cross section is the one most often employed for the side channel trough. The width of such a channel in relation to the depth should be considered. If the width to depth ratio is large, the depth of flow in the channel will be shallow, similar to that depicted by the cross section *abfg* on figure 259. It is evident that for this condition a poor diffusion of the incoming flow with the channel flow will result. A cross section with a minimum width-depth ratio will provide the best hydraulic performance, indicating that a cross section approaching that depicted as *adj* on the figure would be the ideal choice both from the standpoint of hydraulics and economy. Minimum bottom widths are required, however, to avoid construction difficulties due to confined working space. Furthermore, the stability of the structure and the hillside which might be jeopardized by an extremely deep cut in the abutment must also be considered. Therefore, a minimum bottom width must be selected which is commensurate with both the practical and structural aspects of the problem.

A control section downstream from the side channel trough is achieved by constricting the channel sides or elevating the channel bottom to produce a point of critical flow. Flows upstream from the control will be at the subcritical stage and will provide a maximum of depth in the side channel trough. The side channel bottom and control dimensions are then selected so that flow in the trough opposite the crest will be at the greatest depth possible without submerging the flow over the crest. Flow in the discharge channel downstream from the control will be the same as that in an ordinary channel or chute type spillway.

(b) *Design Example.*—A design example is provided to illustrate the procedures for determining the hydraulic design of a side channel spillway control structure. The problem is to design a side channel spillway 100 feet long (station 0+00 to station 1+00) to discharge a maximum of 2,000 second-feet. The spillway crest is at elevation 1000.0. The discharge per foot of length, q, is equal to $\frac{2,000}{100}=20$ second-feet. Assuming the crest coefficient, $C=3.6$, $H_o=\left(\frac{q}{C}\right)^{2/3}=3.1$ feet.

For the side channel trough, assume a trapezoidal section with ½:1 side slopes and a bottom width of 10 feet, whose rise in bottom profile is 1.0 foot in the 100 feet of channel length. (The slope of the channel profile is arbitrary; however, a relatively flat slope will provide greater depths and slower velocities and consequently will insure better intermingling of flows at the upstream end of the channel and avoid the possibility of accelerating or supercritical flows occurring in the channel for smaller discharges.) Also, assume that a control section is placed downstream from the side channel trough with its bottom at the same elevation as the bottom of the side channel floor at the downstream end, and that a transition is made from the ½:1 slopes of the trough section to a rectangular section at the control. Arbitrarily assume a datum for the control section bottom at elevation 100.0.

Then, the critical depth, d_c, for flow at the control is:

$$d_c=\sqrt[3]{\frac{q_1^2}{g}}$$

For this example:

$$q_1=\frac{2,000}{10}=200 \text{ second-feet per foot of width}$$

$$d_c=\sqrt[3]{\frac{200^2}{32.2}}=10.75 \text{ feet}$$

$$v_c=\frac{q_1}{d_c}=\frac{200}{10.75}=18.6 \text{ feet per second}$$

$$h_{v_c}=\frac{v_c^2}{2g}=\frac{18.6^2}{64.4}=5.37 \text{ feet}$$

Assume a transition loss from the end of the side channel trough to the control section (to provide for losses due to contraction, to diffusion of the flows which were not effected in the side channel proper, and to friction losses) equal to 0.2 of the difference in velocity heads between the ends of the transition. The flow characteristics at the downstream end of the side channel are obtained from Bernoulli's theorem (appendix B). For figure 260, Bernoulli's theorem may be written as follows:

$$d_{(1+00)}+h_{v_{(1+00)}}=d_c+h_{v_c}+0.2(h_{v_c}-h_{v_{(1+00)}})$$

This expression must be solved by trial and error. First, assume a value of $d_{(1+00)}$, and solve for $h_{v_{(1+00)}}$. If the use of these values does not result in a balanced equation, a new value must be assumed for $d_{(1+00)}$ and the process repeated. A value of 16.34 feet for $d_{(1+00)}$ was found to satisfy the equation as follows:

For $d_{(1+00)}=16.34$, the area of flow at station 1+00 in the trapezoidal cross section with 10-foot bottom width and ½ : 1 side slopes=297 square feet.

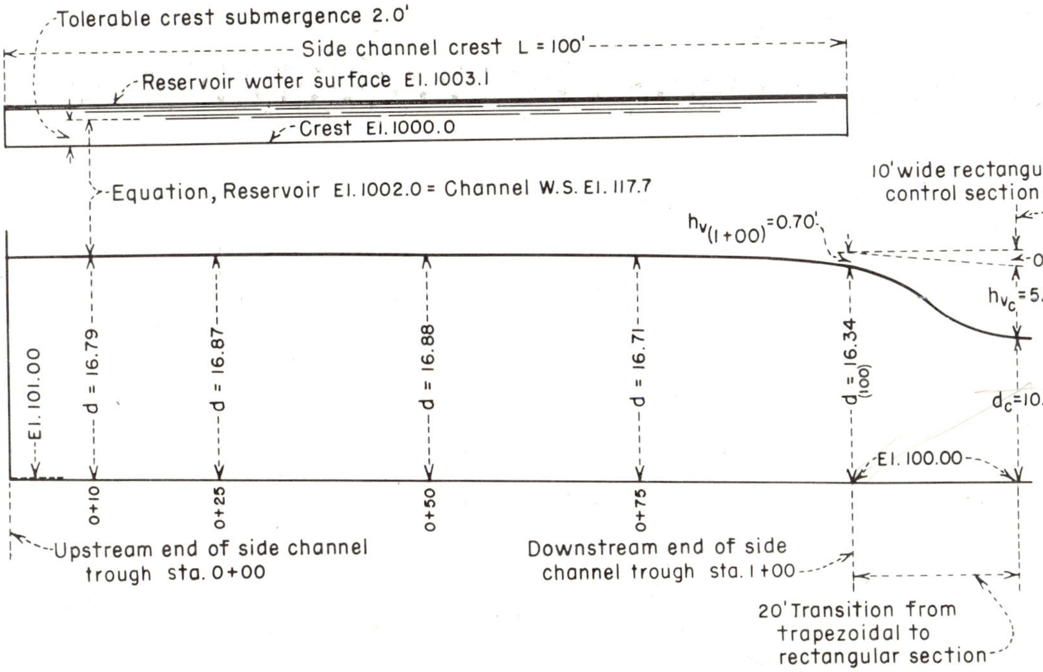

Figure 260. **Example of hydraulic design for side channel spillway. 288—D—2420.**

TABLE 23.—*Side channel spillway computations*
[Q=2,000 second-feet. Bottom width=10 feet. Side slopes=½:1. Bottom slope=1 foot in 100 feet]

[Q=2,000 second-feet. Bottom width=10 feet. Side slopes=½:1. Bottom slope=1 foot in 100 feet]

Station	Δx	Elevation bottom	Trial Δy	Water surface elevation	d	A	Q	v	Q_1+Q_2	$\frac{Q_1}{g(Q_1+Q_2)}$	v_1+v_2	v_2-v_1	Q_2-Q_1	$\frac{Q_2-Q_1}{Q_1}$	$\frac{v_2(Q_2-Q_1)}{Q_1}$	(13)+(16)	Δy=(11)×(12)×(17)	Remarks
(1)	(2)	(3)	(4)	(5)	(6)	(7)	(8)	(9)	(10)	(11)	(12)	(13)	(14)	(15)	(16)	(17)	(18)	(19)
1+00		100.0		116.34	16.34	297	2,000	6.73										
0+75	-25	100.25	1.00	117.34	17.09	317	1,500	4.73	3,500	0.01332	11.46	2.00	-500	-0.333	-2.24	4.24	0.64	Too low.
			.62	116.96	16.71	307		4.89			11.62	1.84				4.08	.63	OK.
0+50	25	100.50	.50	117.46	16.96	313	1,000	3.19	2,500	.01244	8.08	1.70	500	.50	2.44	4.14	.42	Too low.
			.42	117.38	16.88	311		3.22			8.11	1.67				4.11	.41	OK.
0+25	25	100.75	.30	117.68	16.93	313	500	1.60	1,500	.01036	4.82	1.62	500	1.00	3.22	4.84	.24	Too low.
			.24	117.62	16.87	311		1.61			4.83	1.61				4.83	.24	OK.
0+10	15	100.90	.10	117.72	16.82	310	200	.64	700	.00888	2.25	.97	300	1.50	2.41	3.38	.07	Too low.
			.07	117.69	16.79	309		.65			2.26	.96				3.37	.07	OK.

$$v_{(1+00)}=\frac{2{,}000}{297}=6.73 \text{ feet per second}$$

$$h_{v_{(1+00)}}=\frac{6.73^2}{64.4}=0.70 \text{ foot}$$

$$0.2(h_{v_c}-h_{v_{(1+00)}})=0.2(5.37-0.70)=0.93 \text{ foot.}$$

Substituting the values in the equation:

$$16.34+0.70=10.75+5.37+0.93$$

$$17.04=17.05 \text{ (A satisfactory check).}$$

With the hydraulic properties of the side channel at station 1+00 determined, the water surface profile along the side channel trough can be determined from equation (15). The trial-and-error computations are shown in table 23. The resulting water surface profile is shown on figure 260.

The channel profile is next fitted to the crest datum by relating the water surface profile to the reservoir water level. In order to obtain the assumed crest coefficient value of 3.6, excessive submergence of the overflow must be avoided. If it is assumed that a maximum of two-thirds submergence at the upstream end of the channel can be tolerated, the maximum water surface level in the channel will be ⅔ H_0 above the crest, or elevation 1002.0. Then at station 0+10, the channel datum water surface level elevation 117.7 will become elevation 1002.0, placing the channel floor level for station 0+00 at approximately elevation 985.3, and for station 1+00 at approximately elevation 984.3.

The design of the side channel control structure would be completed by designing the uncontrolled ogee crest by the methods shown in section 198, in order to obtain the crest coefficient value of 3.6 which was assumed.

Variations in the design can be made by assuming different bottom widths, different channel slopes, and varying control sections. A proper and economical design can usually be achieved after comparing several alternatives.

D. HYDRAULICS OF FREE-FLOW DISCHARGE CHANNELS

203. *General.*—Discharge generally passes through the critical stage in the spillway control structure and enters the discharge channel as supercritical or shooting flow. To avoid a hydraulic jump below the control, the flow must remain at the supercritical stage throughout the length of the channel. The flow in the channel may be uniform or it may be accelerated or decelerated, depending on the slopes and dimensions of the channel and on the total drop. Where it is desired to minimize the grade to reduce excavation at the upstream end of a channel, the flow might be uniform or decelerating, followed by accelerating flow in the steep drop leading to the downstream river level. Flow at any point along the channel will depend upon the specific energy, $(d+h_v)$, available at that point. This energy will equal the total drop from the reservoir water level to the floor of the channel at the point under consideration, less the head losses accumulated to that point. The velocities and depths of flow along the channel can be fixed by selecting the grade and the cross-sectional dimensions of the channel.

The velocities and depths of free surface flow in a channel, whether an open channel, a conduit, or a tunnel, conform to the principle of the conservation of energy as expressed by the Bernoulli's theorem, which states: "The absolute energy of flow at any cross section is equal to the absolute energy at a downstream section plus intervening losses of energy." As applied to figure 261 this relationship can be expressed as follows:

$$\Delta Z+d_1+h_{v_1}=d_2+h_{v_2}+\Delta h_L \quad (17)$$

When the channel grades are not too steep, for practical purposes the normal depth d_n can be considered equal to the vertical depth d. The term Δh_L includes all losses which occur in the reach of channel, such as friction, turbulence, impact, and transition losses. Since in most channels changes are made gradually, ordinarily all losses except those due to friction can be neglected. The friction loss can then be expressed as:

$$\Delta h_L=s.\Delta_L \quad (18)$$

where s is the average friction slope expressed by either the Chezy or the Manning formula. For the reach ΔL, the head loss can be expressed as $\Delta h_L=\left(\frac{s_1+s_2}{2}\right)\Delta L$. From the Manning formula (equation (30), appendix B),

$$s=\left(\frac{vn}{1.486r^{2/3}}\right)^2.$$

The coefficient of roughness, n, will depend on the nature of the channel surface. For conservative design the frictional loss should be maximized when evaluating depths of flow and minimized when evaluating the energy content of the flow. For determining depths of flow in a concrete-lined channel, a value of n of about 0.018 should be assumed in order to account for

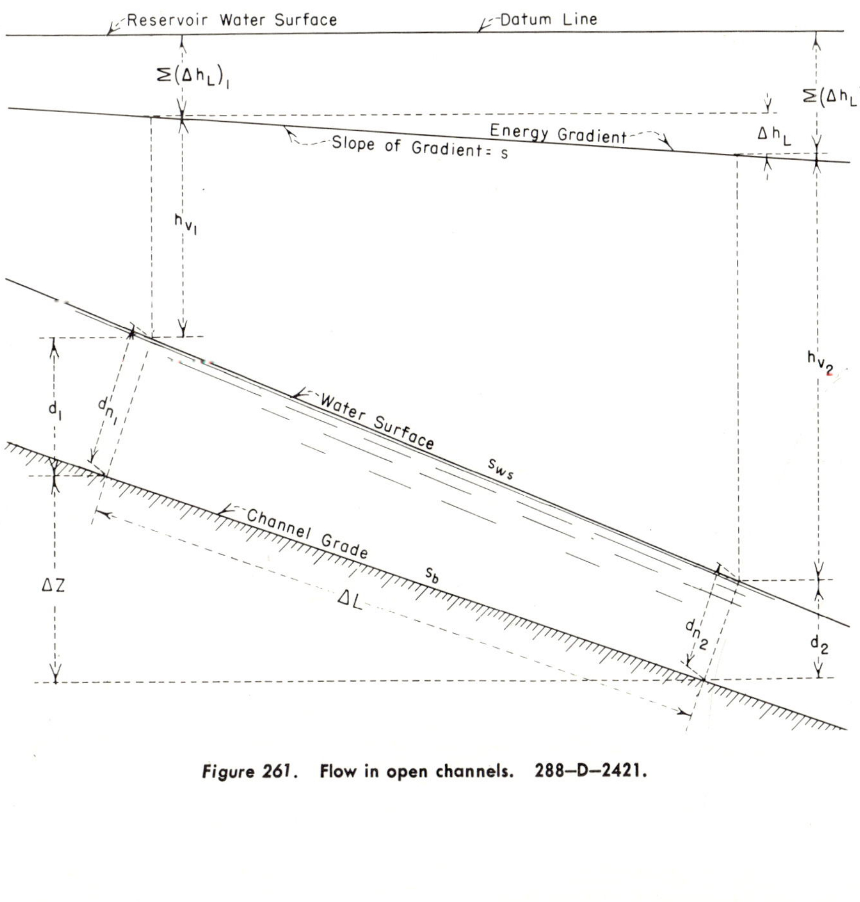

Figure 261. Flow in open channels. 288–D–2421.

air swell, wave action, etc. For determining specific energies of flow needed for designing the dissipating device, a value of n of about 0.008 should be assumed.

Where only rough approximations of depths and velocities of flow in a discharge channel are desired, the total head loss $\Sigma(\Delta h_L)$ to any point along the channel might be expressed in terms of the velocity head. Thus, at any section the relationship can be stated: reservoir water surface elevation minus floor grade elevation $=d+h_v+Kh_v$. For spillways with small drops, K can be assumed as approximately 0.2 for determining depths of flow and 0.1 or less for evaluating the energy of flow. Rough approximations of losses can also be obtained from figure B–5 (app. B).

204. Open Channels.—(a) *Profile.*—The profile of an open channel is usually selected to conform to topographic and geologic site conditions. It is generally defined as straight reaches joined by vertical curves. Sharp convex and concave vertical curves should be avoided to prevent unsatisfactory flows in the channel. Convex curves should be flat enough to maintain positive pressures and thus avoid the tendency for separation of the flow from the floor. Concave curves should have a sufficiently long radius of curvature to minimize the dynamic forces on the floor brought about by the centrifugal force which results from a change in the direction of flow.

To avoid the tendency for the water to spring away from the floor and thereby reduce the surface contact pressure, the floor shape for convex curvature should be made substantially flatter than the trajectory of a free-discharging jet issuing under a head equal to the specific energy of flow as it enters the curve. The curvature should approximate a shape defined by the equation:

$$-y=x\tan\theta+\frac{x^2}{K[4(d+h_v)\cos^2\theta]} \tag{19}$$

where θ is the slope angle of the floor upstream from the curve. Except for the factor K, the equation is that of a free-discharging trajectory issuing from an inclined orifice. To assure positive pressure along the entire contact surface of the curve, K should be equal to or greater than 1.5.

For the concave curvature, the pressure exerted upon the floor surface by the centrifugal force of the flow will vary directly with the energy of the flow and inversely with the radius of curvature. An approximate relationship of these criteria can be expressed in the equations:

$$R=\frac{2qv}{p} \text{ or } R=\frac{2dv^2}{p} \tag{20}$$

where:

R=the minimum radius of curvature measured in feet,

q=the discharge in second-feet per foot of width,

v=the velocity in feet per second,

d=the depth of flow in feet, and

p=the normal dynamic pressure exerted on the floor, in pounds per square foot.

An assumed value of $p=100$ will normally produce an acceptable radius; however, in no event should the radius be less than $10d$. For the reverse curve at the lower end of the ogee crest, radii of not less than $5d$ have been found acceptable.

(b) *Convergence and Divergence.*—The best hydraulic performance in a discharge channel is obtained when the confining sidewalls are parallel and the distribution of flow across the channel is maintained uniform. However, economy may dictate a channel section narrower or wider than either the crest or the terminal structure, thus requiring converging or diverging transitions to fit the various components together. Sidewall convergence must be made gradual to avoid cross waves, "ride ups" on the walls, and uneven distribution of flow across the channel. Similarly, the rate of divergence of the sidewalls must be limited or else the flow will not spread to occupy the entire width of the channel uniformly, which will result in undesirable flow conditions at the terminal structure.

The inertial and gravitational forces of streamlined kinetic flow in a channel can be expressed by the Froude number parameter, $\frac{v}{\sqrt{gd}}$. Variations from streamlined flow due to outside interferences which cause an expansion or a contraction of the flow also can be related to this parameter. Experiments have shown that an angular variation of the flow boundaries not exceeding that produced by the equation,

$$\tan \alpha = \frac{1}{3F} \qquad (21)$$

will provide an acceptable transition for either a contracting or an expanding channel. In this equation, $F = \frac{v}{\sqrt{gd}}$ and α is the angular variation of the sidewall with respect to the channel centerline; v and d are the averages of the velocities and depths at the beginning and at the end of the transition. Figure 262 is a nomograph from which the tangent of the flare angle or the flare angle in degrees may be obtained for known values of depth and velocity of flow.

(c) *Channel Freeboard.*—In a channel conducting flow at supercritical stage, the surface roughness, wave action, air bulking and splash and spray are related to the velocity and energy content of the flow. The energy per foot of width, qh_v, expressed in terms of v and d is $\frac{v^3 d}{2g}$; therefore, the relationship of velocity and depth to the flow energy also can be expressed in terms of v and $\sqrt[3]{d}$. An empirical expression based on this relationship which gives a reasonable indication of desirable freeboard values is as follows:

$$\text{Freeboard (in feet)} = 2.0 + 0.025 v \sqrt[3]{d} \qquad (22)$$

E. HYDRAULICS OF TERMINAL STRUCTURES

205. *Deflector Buckets.*—Where the spillway discharge may be safely delivered directly to the river without providing a dissipating or stilling device, the jet is often projected beyond the structure by a deflector bucket or lip. Flow from these deflectors leaves the structure as a free-discharging upturned jet and falls into the stream channel some distance from the end of the spillway. The path the jet assumes depends on the energy of flow available at the lip and the angle at which the jet leaves the bucket.

With the origin of the coordinates taken at the end of the lip, the path of the trajectory is given by the equation:

$$y = x \tan \theta - \frac{x^2}{K[4(d+h_v)\cos^2 \theta]} \qquad (23)$$

where:

θ=the angle of the edge of the lip with the horizontal, and

K=a factor, equal to 1 for the theoretical jet.

To compensate for loss of energy and velocity reduction due to the effect of air resistance, internal turbulences, and disintegration of the jet, a value for K of about 0.9 should be assumed.

The horizontal range of the jet at the level of the lip is obtained by making y in equation

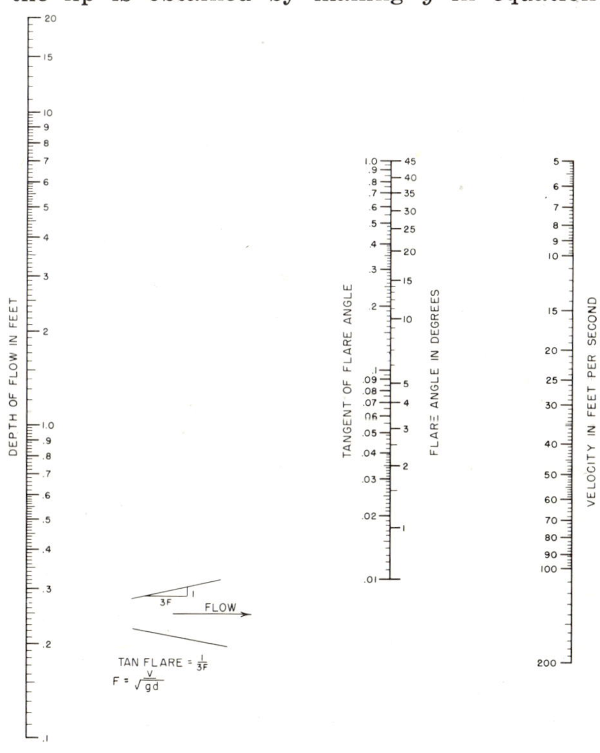

After C. Freeman

Figure 262. Flare angle for divergent or convergent channels. 288–D–2422.

(23) equal to zero. Then:

$$x = 4K(d+h_v) \tan\theta \cos^2\theta$$
$$= 2K(d+h_v) \sin 2\theta$$

The maximum value of x will be equal to $2K(d+h_v)$ where θ is 45°. However, the angle of the lip is influenced by the bucket radius and the height of the lip above the bucket invert; ordinarily the exit angle is limited to not more than 30°.

The bucket radius should be made long enough to maintain concentric flow as the water moves around the curve. The rate of curvature must be limited similar to that of a vertical curve in a discharge channel (sec. 204), so that the floor pressures will not alter the streamline distribution of the flow. The minimum radius of curvature can be determined from equation (20), except that values of p not exceeding 500 pounds per square foot will produce values of the radius which have proved satisfactory in practice. However, the radius should not be less than five times the depth of water. Structurally, the cantilever bucket must be of sufficient strength to withstand this normal dynamic force in addition to the other applied forces.

206. Hydraulic Jump Basins.—(a) *General.*—Where the energy of flow in a spillway must be dissipated before the discharge is returned to the downstream river channel, the hydraulic jump basin is an effective device for reducing the exit velocity to a tranquil state. The jump which will occur in a stilling basin has distinctive characteristics and assumes a definite form, depending on the energy of flow which must be dissipated in relation to the depth of the flow. A comprehensive series of tests have been performed by the Bureau of Reclamation [9] for determining the properties of the hydraulic jump. The jump form and the flow characteristics can be related to the kinetic flow factor, $\frac{v^2}{gd}$, of the discharge entering the basin; to the critical depth of flow, d_c; or to the Froude number parameter, $\frac{v}{\sqrt{gd}}$. Forms of the hydraulic jump phenomena for various ranges of the Froude number are illustrated on figure 263.

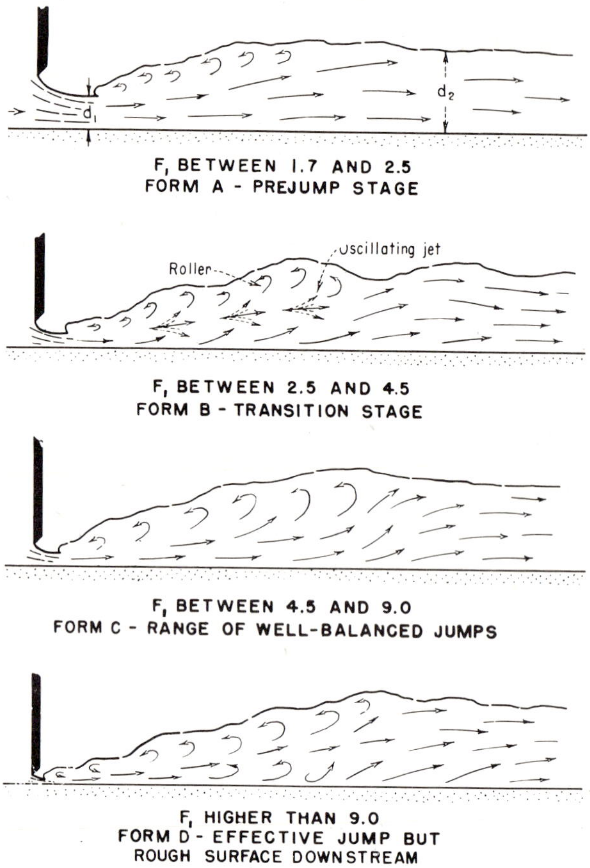

Figure 263. Characteristic forms of hydraulic jump related to the Froude number. 288–D–2423.

When the Froude number of the incoming flow is equal to 1.0, the flow is at critical depth and a hydraulic jump cannot form. For Froude numbers from 1.0 up to about 1.7, the incoming flow is only slightly below critical depth, and the change from this low stage to the high stage flow is gradual and manifests itself only by a slightly ruffled water surface. As the Froude number approaches 1.7, a series of small rollers begin to develop on the surface, which become more intense with increasingly higher values of the number. Other than the surface roller phenomena, relatively smooth flows prevail throughout the Froude number range up to about 2.5. Stilling action for the range of Froude numbers from 1.7 to 2.5 is designated as form A on figure 263.

For Froude numbers between 2.5 and 4.5 an oscillating form of jump occurs, the entering

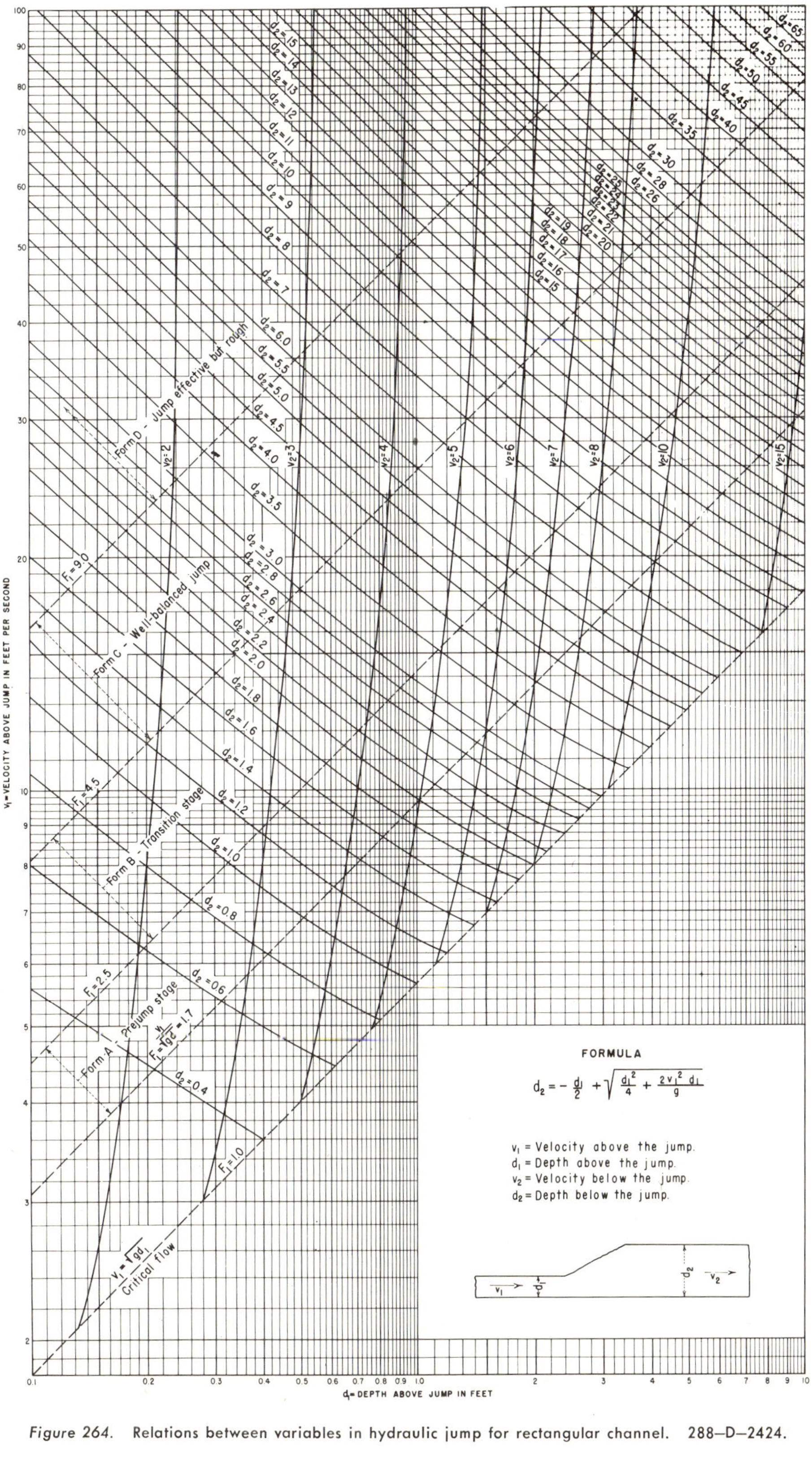

Figure 264. Relations between variables in hydraulic jump for rectangular channel. 288–D–2424.

jet intermittently flowing near the bottom and then along the surface of the downstream channel. This oscillating flow causes objectionable surface waves which carry considerably beyond the end of the basin. The action represented through this range of flows is designated as form B on figure 263.

For the range of Froude numbers for the incoming flow between 4.5 and 9, a stable and well-balanced jump occurs. Turbulence is confined to the main body of the jump, and the water surface downstream is comparatively smooth. As the Froude number increases above 9, the turbulence within the jump and the surface roller becomes increasingly active, resulting in a rough water surface with strong surface waves downstream from the jump. Stilling action for the range of Froude numbers between 4.5 and 9 is designated as form C on figure 263, and that above 9 is designated as form D.

Figure 264 plots relationships of conjugate depths and velocities for the hydraulic jump in a rectangular channel. Also indicated on the figure are the ranges for the various forms of jump described above.

(b) *Basin Design in Relation to Froude Numbers.*—Stilling basin designs suitable to provide stilling action for the various forms of jump are described as follows:

(1) *Basins for Froude numbers less than 1.7.*—For a Froude number of 1.7 the conjugate depth d_2 is about twice the incoming depth, or about 40 percent greater than the critical depth. The exit velocity v_2 is about one-half the incoming velocity, or 30 percent less than the critical velocity. No special stilling basin is needed to still flows where the incoming flow Froude factor is less than 1.7, except that the channel lengths beyond the point where the depth starts to change should be not less than about $4d_2$. No baffles or other dissipating devices are needed. These basins are designated Type I and are not illustrated here (see reference [11]).

(2) *Basins for Froude numbers between 1.7 and 2.5.*—Flow phenomena for basins where the incoming flow-factors are in the Froude number range between 1.7 and 2.5 will be in the form designated as the prejump stage, as illustrated on figure 263. Since such flows are not attended by active turbulence, baffles or sills are not required. The basin should be sufficiently long to contain the flow prism while it is undergoing retardation. Conjugate depths and basin lengths shown in figure B–15 (app. B) will provide acceptable basins. These basins are designated Type I and are not illustrated here (see reference [11]).

(3) *Basins for Froude numbers between 2.5 and 4.5*—Jump phenomena where the incoming flow factors are in the Froude number range between 2.5 and 4.5 are designated as transition flow stage, since a true hydraulic jump does not fully develop. Stilling basins to accommodate these flows are the least effective in providing satisfactory dissipation, since the attendant wave action ordinarily cannot be controlled by the usual basin devices. Waves generated by the flow phenomena will persist beyond the end of the basin and must often be dampened by means apart from the basin.

Where a stilling device must be provided to dissipate flows for this range of Froude number, the basin shown on figure 265, which is designated as type IV basin, has proved to be relatively effective for dissipating the bulk of the energy of flow. However, the wave action propagated by the oscillating flow cannot be entirely dampened. Auxiliary wave dampeners or wave suppressors must sometimes be employed to provide smooth surface flow downstream.

Because of the tendency of the jump to sweep out and as an aid in suppressing wave action, the water depths in the basin should be about 10 percent greater than the computed conjugate depth.

Often the need for utilizing this type of basin in design can be avoided by selecting stilling basin dimensions which will provide flow conditions which fall outside the range of transition flow. For example, with an 800-second-foot capacity spillway where the specific energy at the upstream end of the basin is about 15 feet and the velocity into the basin is about 30 feet per second, the Froude number will be 3.2 for a basin width of 10 feet. The Froude number can be raised to 4.6 by widening the basin to 20 feet. The selection of basin width then becomes a matter of economics as well as hydraulic performance.

(4) *Basins for Froude numbers higher than 4.5.*—For basins where the Froude number value of the incoming flow is higher than 4.5, a true hydraulic jump will form. The elements of the jump vary according to the Froude number as shown on figure B–15 (app. B). The installation of accessory devices such as blocks, baffles, and sills along the floor of the basin produce a stabilizing effect on the jump, which permits shortening the basin and provides a factor of safety against sweep-out due to inadequate tailwater depth.

The basin shown on figure 266, which is designated as a type III basin, can be adopted where incoming velocities do not exceed 50 feet per second. This basin utilizes chute blocks, impact baffle blocks, and an end sill to shorten the jump length and to dissipate the high-velocity flow within the shortened basin length. This basin relies on dissipation of energy by the impact blocks and also on the turbulence of the jump phenomena for its effectiveness. Because of the large impact forces to which the baffles are subjected by the impingement of high incoming velocities and because of the possibility of cavitation along the surfaces of the blocks and floor, the use of this basin must be limited to heads where the velocity does not exceed 50 feet per second.

Cognizance must be taken of the added loads placed upon the structure floor by the dynamic force brought against the upstream face of the baffle blocks. This dynamic force will approximate that of a jet impinging upon a plane normal to the direction of flow. The force, in pounds, may be expressed by the formula:

$$\text{Force}=2wA\,(d_1+h_{v1}) \qquad (24)$$

where:

w=the unit weight of water,

A=the area of the upstream face of the block, and

(d_1+h_{v1})=the specific energy of the flow entering the basin.

Negative pressure on the back face of the blocks will further increase the total load. However, since the baffle blocks are placed a distance equal to $0.8d_2$ beyond the start of the jump, there will be some cushioning effect by the time the incoming jet reaches the blocks and the force will be less than that indicated by the above equation. If the full force computed by equation (24) is used, the negative pressure force may be neglected.

Where incoming velocities exceed 50 feet per second, or where impact baffle blocks are not employed, the basin designated as type II on figure 267 can be adopted. Because the dissipation is accomplished primarily by hydraulic jump action, the basin length will be greater than that indicated for the type III basin. However, the chute blocks and dentated end sill will still be effective in reducing the length from that which would be necessary if they were not used. Because of the reduced margin of safety against sweep-out, the water depth in the basin should be about 5 percent greater than the computed conjugate depth.

(c) *Rectangular Versus Trapezoidal Stilling Basin.*—The utilization of a trapezoidal stilling basin in lieu of a rectangular basin may often be proposed where economy favors sloped side lining over vertical wall construction. Model tests have shown, however, that the hydraulic jump action in a trapezoidal basin is much less complete and less stable than it is in the rectangular basin. In the trapezoidal basin the water in the triangular areas along the sides of the basin adjacent to the jump is not opposed by the incoming high-velocity jet. The jump, which tends to occur vertically, cannot spread sufficiently to occupy the side areas. Consequently, the jump will form only in the central portion of the basin, while areas along the outside will be occupied by upstream-moving flows which ravel off the jump or come from the lower end of the basin. The eddy or horizontal roller action resulting from this phenomenon tends to interfere and interrupt the jump action to the extent that there is incomplete dissipation of the energy and severe scouring can occur beyond the basin. For good hydraulic performance, the sidewalls of a stilling basin should be vertical or as near vertical as is practicable.

(d) *Basin Depths Versus Hydraulic Heads.*—The nomograph shown on figure 268 will aid

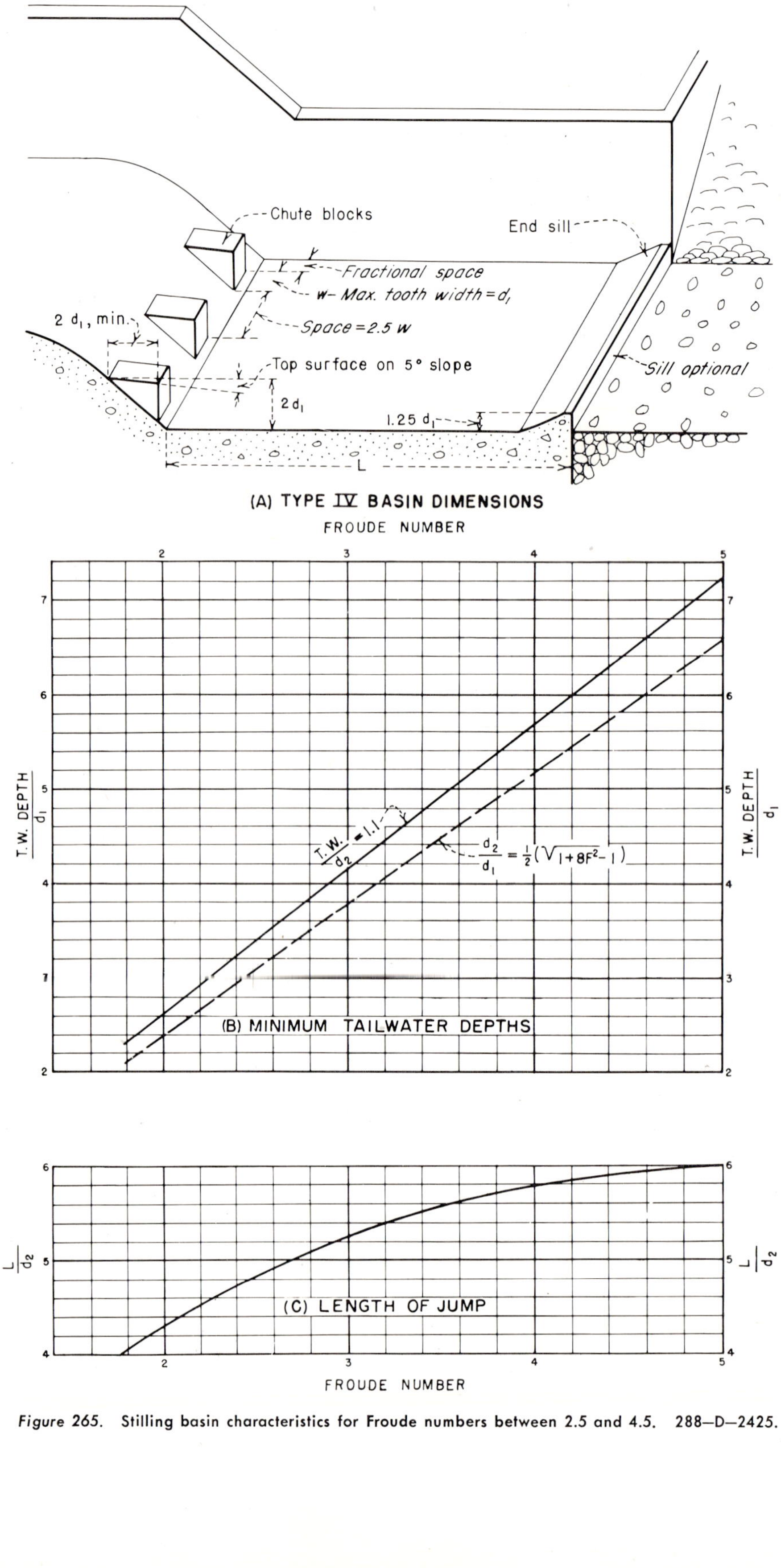

Figure 265. **Stilling basin characteristics for Froude numbers between 2.5 and 4.5. 288–D–2425.**

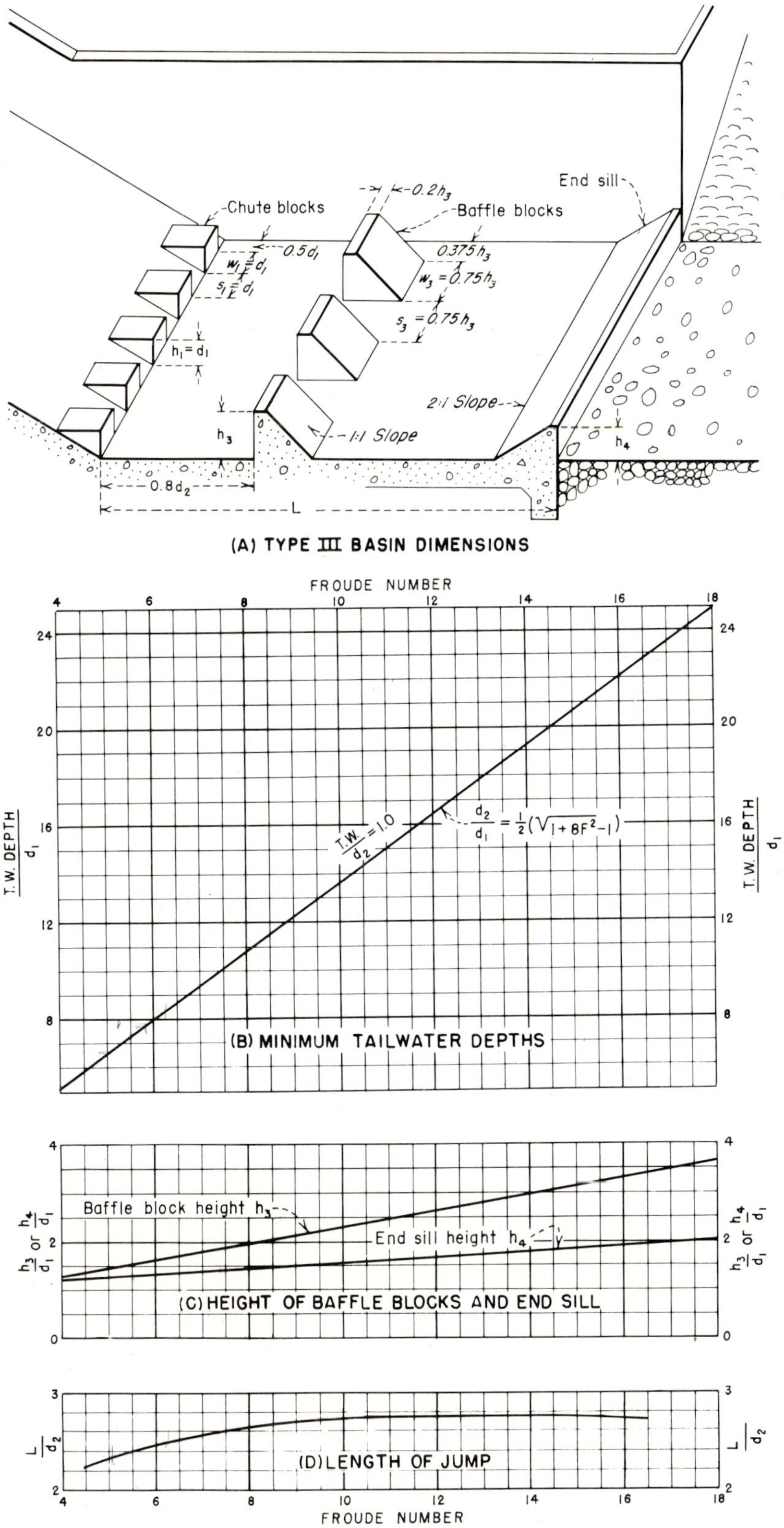

Figure 266. **Stilling basin characteristics for use with Froude numbers above 4.5. 288–D–2426.**

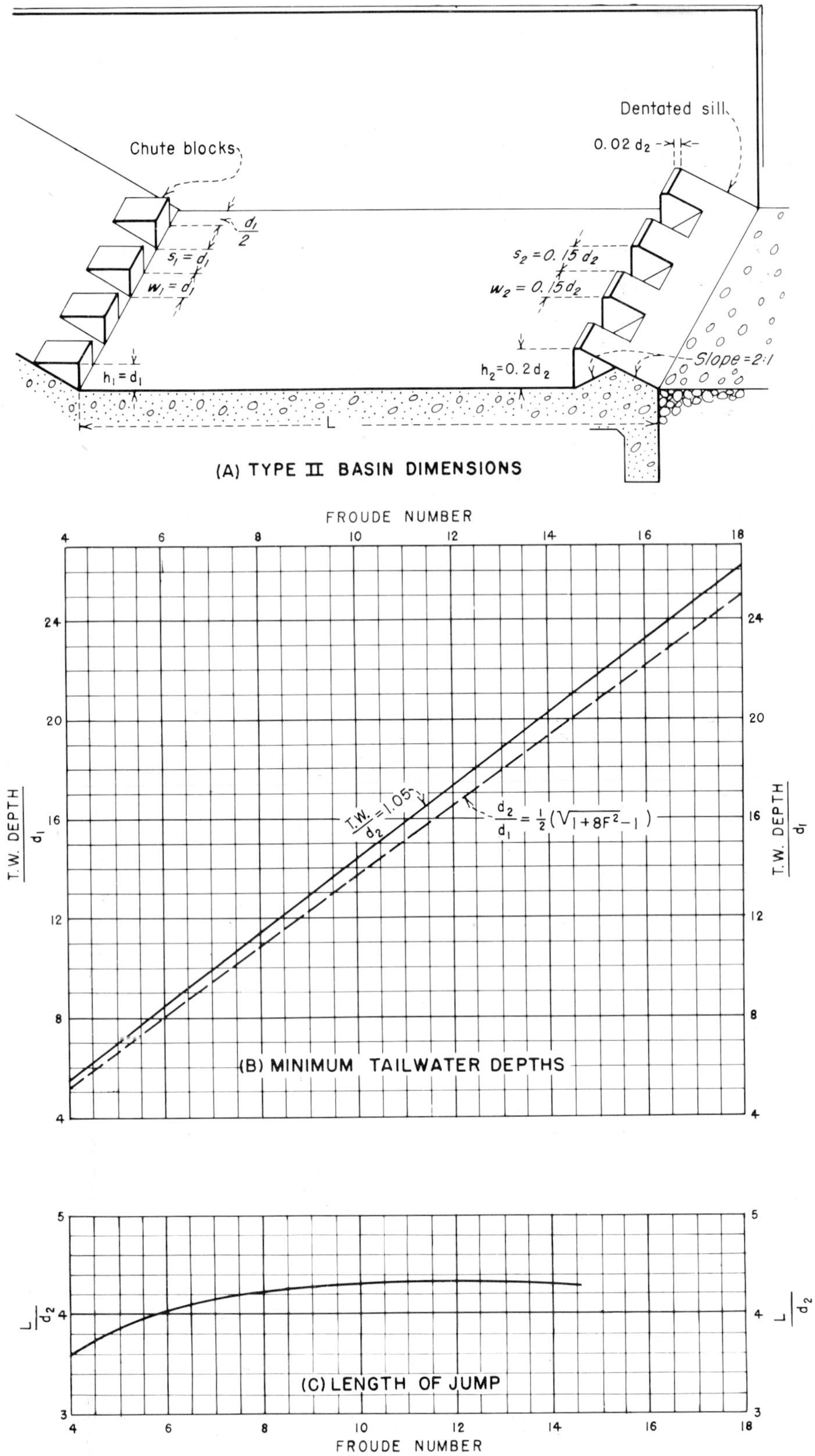

Figure 267. Stilling basin characteristics for use with Froude numbers above 4.5 where incoming velocity (V_1) does not exceed 50-60 feet per second. 288–D–2427.

in determining approximate basin depths for various basin widths and for various differences between reservoir and tailwater levels. Plottings are shown for the condition of no loss of head to the upstream end of the stilling basin, and for 10, 20, and 30 percent loss. (These plottings are shown on the nomographs as scales A, B, C, and D, respectively.) The required conjugate depths, d_2, will depend on the specific energy available at the entrance of the basin, as determined by the procedure discussed in section 203. Where the specific energy is known, the head loss in the channel upstream can be related to the velocity head, the percentage loss can be determined, and the approximate conjugate depth can be read from the nomograph. Where head losses have not been computed, a quick approximation of the head losses can be obtained from figure B-5 (app. B). Where only a rough determination of basin depths is needed, the choice of the loss to be applied for various spillway designs may be generalized as follows:

(1) For a design of an overflow spillway where the basin is directly downstream from the crest, or where the chute is not longer than the hydraulic head, consider no loss of head.

(2) For a design of a channel spillway where the channel length is between one and five times the hydraulic head, consider 10 percent loss of head.

(3) For a design of a spillway where the channel length exceeds five times the hydraulic head, consider 20 percent loss of head.

The nomograph on figure 268 gives values of the conjugate depth of the hydraulic jump. Tailwater depths for the various types of basin described in section 206 should be increased as noted in that section.

(e) *Tailwater Considerations.*—Determination of the tailwater rating curve, which gives the stage-discharge relationship of the natural stream below the dam, is discussed in appendix B, part B. Tailwater rating curves for the regime of river below a dam are fixed by the natural conditions along the stream and ordinarily cannot be altered by the spillway design or by the release characteristics. As discussed in section 192(d), retrogression or aggradation of the river below the dam, which will affect the ultimate stage-discharge conditions, must be recognized in selecting the tailwater rating curve to be used for stilling basin design. Usually river flows which approach the maximum design discharges have never occurred, and an estimate of the tailwater rating curve must either be extrapolated from known conditions or computed on the basis of assumed or empirical criteria. Thus, the tailwater rating curve at best is only approximate, and factors of safety to compensate for variations in tailwater must be included in the design.

For a jump-type stilling basin, downstream water levels for various discharges must conform to the tailwater rating curve, and the basin floor level must therefore be selected to provide jump depths which most nearly agree with the tailwater depths. For a given basin design, the tailwater depth for each discharge seldom corresponds to the conjugate depth needed to form a perfect jump. Thus, the relative shapes and relationships of the tailwater curve to the depth curve will determine the required minimum depth to the basin floor. This is illustrated on figure 269. The tailwater rating curve is shown in (A) as curve 1, and a conjugate depth versus discharge curve for a basin of certain width is represented by curve 3. Since the basin must be made deep enough to provide for full conjugate depth (or some greater depth to include a factor of safety) at the maximum spillway design discharge, the curves will intersect at point D. For lesser discharges the tailwater depth will be greater than the required conjugate depth, thus providing an excess of tailwater which is conducive to the formation of a so-called drowned jump. (With the drowned jump condition, instead of achieving good jump-type dissipation by the intermingling of the upstream and downstream flows, the incoming jet plunges to the bottom and carries along the entire length of the basin floor at high velocity.) If the basin floor is made higher than indicated by the position of curve 3 on the figure, the depth curve and tailwater rating curve will intersect to the left of point D, thus indicating an excess of tailwater for smaller discharges and a deficiency of tail-

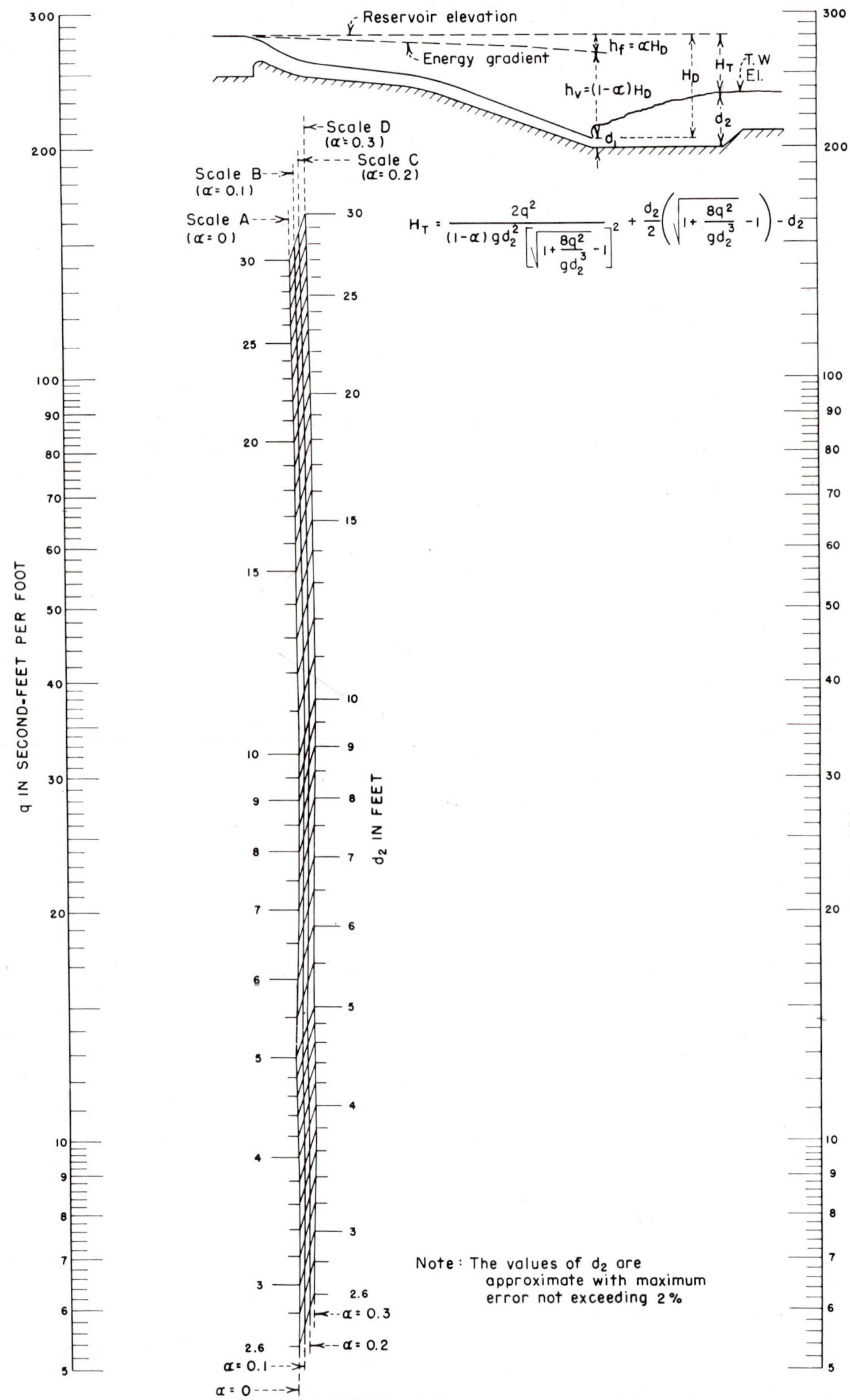

Figure 268. Stilling basin depths versus hydraulic heads for various channel losses. 288–D–2428.

water for higher discharges.

As an alternative to the selected basin which is represented by curve 3, a wider basin might be considered for which the conjugate depth curve 2 will apply. This design will provide a shallower basin, in which the ideal jump depths will more nearly match the tailwater depths for all discharges. The choice of basin widths, of course, involves consideration of economics, as well as hydraulic performance.

Where a tailwater rating curve shaped similar to that represented by curve 4 on figure 269 (B) is encountered, the level of the stilling basin floor must be determined for some discharge other than the maximum design capacity. If the tailwater curve were made to intersect the required water surface elevation at the maximum design capacity, as in figure 269 (A), there would be insufficient tailwater depth for most smaller discharges. In this case the basin floor elevation is selected so that there will be sufficient tailwater depth for all discharges. For the basin of width W, whose required tailwater depth is represented by curve 5, the position of the floor would be selected so that the two curves would coincide at the discharge represented by point E on the figure. For all other discharges the tailwater depth will be in excess of that needed for forming a satisfactory jump. Similarly, if a basin width of $2W$ were considered, the basin floor level would be selected so that curve 6 would intersect the tailwater curve at point F. Here also, the selection of basin widths should be based on economic aspects as well as hydraulic performance.

Where exact conjugate depth conditions for forming the jump cannot be attained, the question of the relative desirability of having insufficient tailwater as compared to having excessive tailwater should be considered. With insufficient tailwater the back pressure will be deficient and sweep-out of the basin will occur. With an excess of tailwater the jump will be formed and energy dissipation within the basin will be quite complete until the drowned jump phenomenon becomes critical. Chute blocks, baffles, and end sills will further assist in energy dissipation, even with a drowned jump.

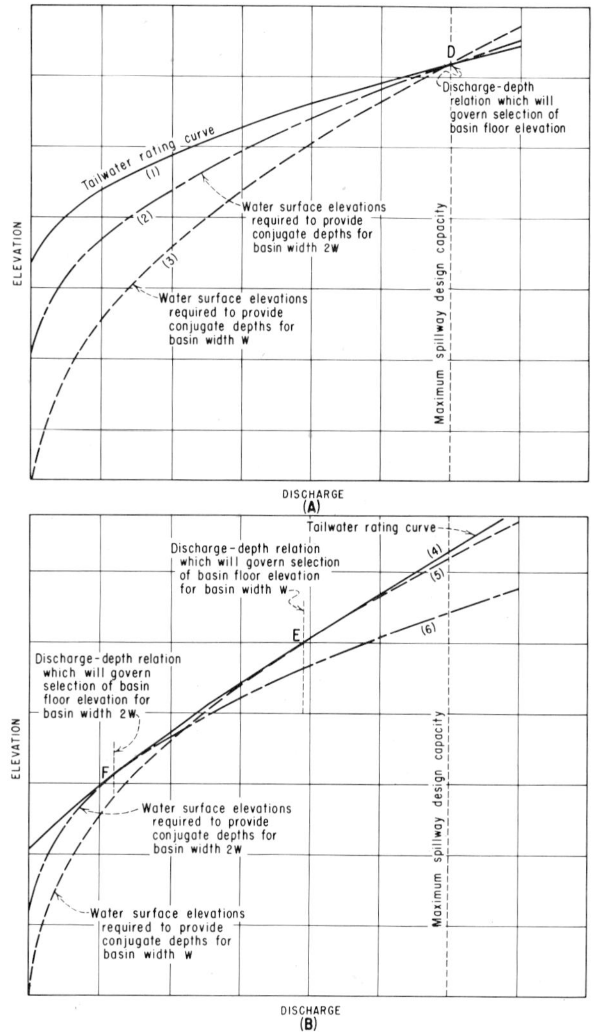

Figure 269. Relationships of conjugate depth curves to tailwater rating curves. 288–D–2429.

(f) *Stilling Basin Freeboard.*—Freeboard is ordinarily provided so that the stilling basin walls will not be overtopped by surges, splash and spray, and wave action set up by the turbulence of the jump. The surface roughness of the flow is related to the energy dissipated in the jump and to the depth of flow in the basin. The following empirical expression provides values which have proved to be satisfactory for most basins:

$$\text{Freeboard in feet} = 0.1\,(v_1 + d_2) \qquad (25)$$

207. *Submerged Bucket Dissipators.*—When the tailwater depth is too great for the formation of a hydraulic jump, dissipation of the high energy of flow can be effected by the use of a submerged bucket deflector. The hydraulic behavior in this type of dissipator is manifested primarily by the formation of two rollers; one is on the surface moving counterclockwise and is contained within the region above the curved bucket, and the other is a ground roller moving in a clockwise direction and is situated downstream from the bucket. The movements of the rollers, along with the intermingling of the incoming flows, effectively dissipate the high energy of the water and prevent excessive scouring downstream from the bucket.

Two types of roller bucket have been developed and model tested [10, 11]. Their shape and dimensional arrangements are shown on figure 270. The general nature of the dissipating action for each type is represented on figure 271. Hydraulic action of the two buckets has the same characteristics, but distinctive features of the flow differ to the extent that each has certain limitations. The high-velocity flow leaving the deflector lip of the solid bucket is directed upward. This creates a high boil on the water surface and a violent ground roller moving clockwise downstream from the bucket. This ground roller continuously pulls loose material back towards the lip of the bucket and keeps some of the intermingling material in a constant state of agitation. A typical scour pattern which results from this action is shown on figure 272. In the slotted bucket the high-velocity jet leaves the lip at a flatter angle, and only a part of the high-velocity flow finds its way to the surface. Thus, a less violent surface boil occurs and there is a better dispersion of flow in the region above the ground roller which results in less concentration of high-energy flow throughout the bucket and a smoother downstream flow.

Use of a solid bucket dissipator may be objectionable because of the abrasion on the concrete surfaces caused by material which is swept back along the lip of the deflector by the ground roller. In addition, the more turbulent surface roughness induced by the severe surface boil carries farther down the river, causing objectionable eddy currents which contribute to riverbank sloughing. Although the slotted bucket provides better energy dissipation with less severe surface and streambed disturbances, it is more sensitive to sweep-out at lower tailwaters and is conducive to a diving and scouring action at excessive tailwaters. This is not the case with the solid bucket. Thus, the tailwater range which will provide good performance with the slotted bucket is much narrower than that of the solid bucket. A solid bucket dissipator should not be used wherever the tailwater limitations of the slotted bucket can be met. Therefore, only the design of the slotted bucket will be discussed.

Flow characteristics of the slotted bucket are illustrated on figure 273. For deficient tailwater depths the incoming jet will sweep the surface roller out of the bucket and will produce a high-velocity flow downstream, both along the water surface and along the riverbed. This action is depicted as stage (A) on figure 273. As the tailwater depth is increased, there will be a depth at which instability of flow will occur, where sweep-out and submergence will alternately prevail. To obtain continuous operation at the submerged

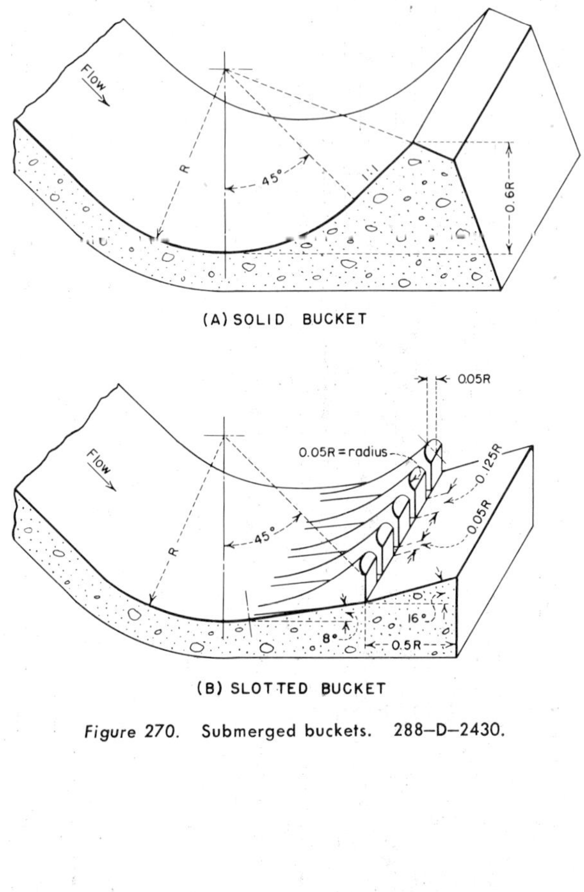

Figure 270. Submerged buckets. 288–D–2430.

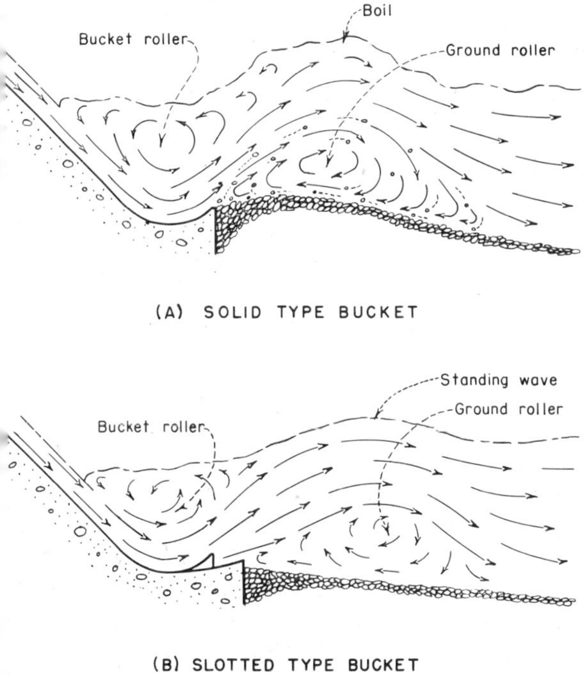

Figure 271. Hydraulic action in solid and slotted buckets. 288–D–2431.

age, the minimum tailwater depth must be bove this instable state. Flow action within he acceptable operating stage is depicted as age (B) on figure 273.

When the tailwater becomes excessively deep, he phenomenon designated as diving flow will ccur. At this stage the jet issuing from the p of the bucket will no longer rise and continue along the surface but intermittently will ecome depressed and dive to the riverbed.

ure 272. Scour pattern downstream from a solid bucket dissipator for an ogee overflow crest—Murdock Diversion Dam in Utah. 288–D–2904.

The position of the downstream roller will change with the change in position of the jet. It will occur at the surface when the jet dives and will form along the river bottom as a ground roller when the jet rides the surface. Scour will occur in the streambed at the point of impingement when the jet dives but will be filled in by the ground roller when the jet rides. The characteristic flow pattern for the diving stage is depicted in (C) and (D) of figure 273. Maximum tailwater depths must be limited to forestall the diving flow phenomenon.

The design of the slotted bucket involves determination of the radius of curvature of the bucket and the allowable range of tailwater depths. These criteria, as determined from experimental results, are plotted on figure 274 in relation to the Froude number parameter. The Froude number values are for flows at the point where the incoming jet enters the bucket. Symbols and criteria are defined on figure 275.

208. ***Examples of Designs of a Stilling Basin and an Alternate Submerged Bucket Dissipator.***—The designs of a stilling basin and of a submerged bucket dissipator are best explained by means of examples. Consider that it is required to make comparative designs of a stilling basin and of a submerged bucket dissipator for an overflow dam whose maximum discharge is 2,000 second-feet and whose controlling dimensions and tailwater conditions are as shown on figure 276.

For a first trial design, assume a crest length of 20 feet. The criteria for different discharges are then as follows:

Total discharge, Q, second-feet	2,000	1,000	500
Discharge per foot, q, second-feet	100	50	25
Assumed coefficient of discharge, C	3.9	3.7	3.5
Head in feet on crest, $H_e=\left(\frac{q}{C}\right)^{2/3}$	8.7	5.7	3.7
Reservoir water level, elevation	1008.7	1005.7	1003.7
Tailwater level, elevation	985.0	981.0	978.0
Reservoir water level minus tailwater level, feet	23.7	24.7	25.7
Velocity head at tailwater level, h_{v_t}, feet (assuming no loss of specific energy)	23.7	24.7	25.7
Velocity of flow in feet per second at tailwater level, $v_t=\sqrt{2gh_{v_t}}$	39.1	39.9	40.7
Depth of flow in feet at tailwater level, $d_t=\frac{q}{v_t}$	2.56	1.25	.61
Froude number at tailwater level, $F_t=\frac{v_t}{\sqrt{gd_t}}$	4.3	6.3	9.2
Specific energy at tailwater level, $d_t+h_{v_t}$	26.3	25.9	26.3

Table 24 shows the computations for a hydraulic jump basin design. Conjugate depths and the required apron elevation for the various discharges are calculated to determine the critical condition. The lowest apron elevation is for the 2,000-second-foot discharge. The Froude number of 6.2 and the incoming velocity not exceeding 50 feet per second, determine that the type III stilling basin shown on figure 266 should be used for this design. The basin length will be 42 feet and the apron elevation will be 968.3.

For the submerged slotted bucket design, the minimum bucket radius for the maximum discharge is determined by use of figure 274. For a Froude number at tailwater level of 4.3, the minimum radius is $0.42(d_t+h_{vt})=0.42\times 26.3=$ 11.0 feet. In this instance the riverbed slopes up, and the use of figure 274 results in the following values for the maximum and minimum tailwater for a Froude number of 4.3 and a $\frac{R}{d_t+h_{v_t}}$ value of 0.42.

$$T_{max}=7.5\ d_t=7.5\times 2.56=19.2 \text{ feet}$$
$$T_{min}=6.5\ d_t=6.5\times 2.56=16.6 \text{ feet}$$

An average tailwater depth of 18 feet will place the bucket invert at elevation 985.0–18.0 =967.0. It is now necessary to check the radius and tailwater conditions for less than maximum flows to determine if the design is satisfactory throughout the range of discharge.

For a unit discharge of 50 second-feet, the minimum radius for F_t of 6.3 is $0.26(d_t+h_{vt})=$ $0.26\times 25.9=6.8$ feet. Therefore, the minimum radius of 11.0 feet determined for the maximum discharge will govern. The maximum and minimum tailwater values for a Froude number

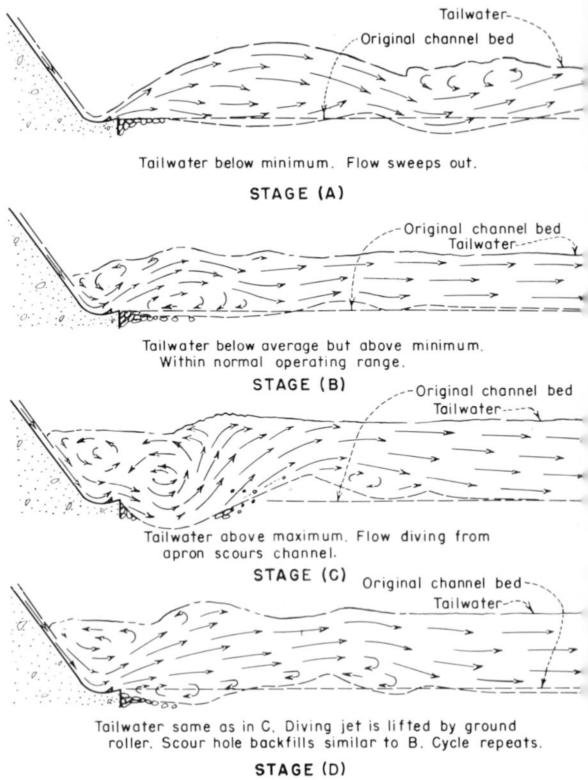

Figure 273. Flow characteristics in a slotted bucket. 288–D–24

of 6.3 and an $\frac{R}{d_t+h_{vt}}$ value of $\frac{11}{25.9}$ or 0.42 ar

$$T_{max}=20.0\ d_t=20.0\times 1.25=25.0 \text{ feet}$$
$$T_{min}=10.1\ d_t=10.1\times 1.25=12.6 \text{ feet}$$

The bucket invert level at elevation 967 as de termined for the maximum discharge will pr vide a tailwater depth of 981.0–967.0= feet, which is within the safe limit for produ ing satisfactory roller action.

TABLE 24.—*Computations for hydraulic jump basin design*

Discharge, Q, second-feet	Discharge per foot, q, second-feet	Reservoir level minus tailwater, feet	Conjugate depth, d_2 feet [1]	Tailwater elevation	Required apron elevation	Specific energy, H_E, at upstream end of basin [2]	Upstream depth of flow at basin floor level [3] d_1	Upstream velocity at basin floor level [3] v_1	Froude number [4] F_1
2,000	100	23.7	16.7	985.0	968.3	40.4	2.01	49.8	6.2
1,000	50	24.7	11.8	981.0	969.2	36.5	1.05	47.6	8.1
500	25	25.7	8.6	978.0	969.4	34.3	0.54	46.3	11.1

[1] From figure 268, assuming no loss in specific energy.
[2] H_E=Reservoir water surface minus apron elevation, assuming no loss in specific energy.
[3] $H_E=d_1+\frac{v_1^2}{2g}$.
[4] $F_1=\frac{v_1}{\sqrt{gd_1}}$.

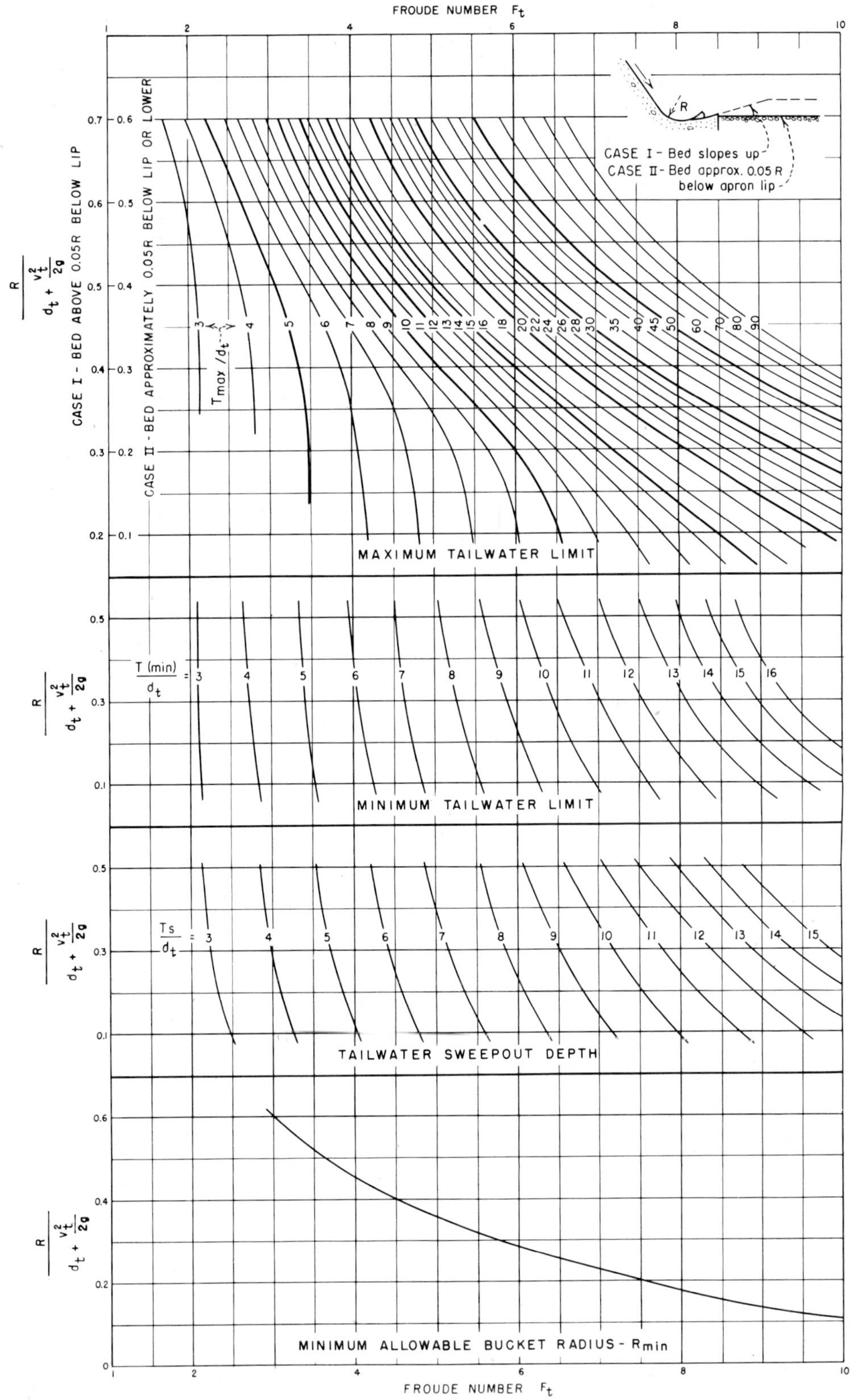

Figure 274. Limiting criteria for slotted bucket design. 288–D–2433.

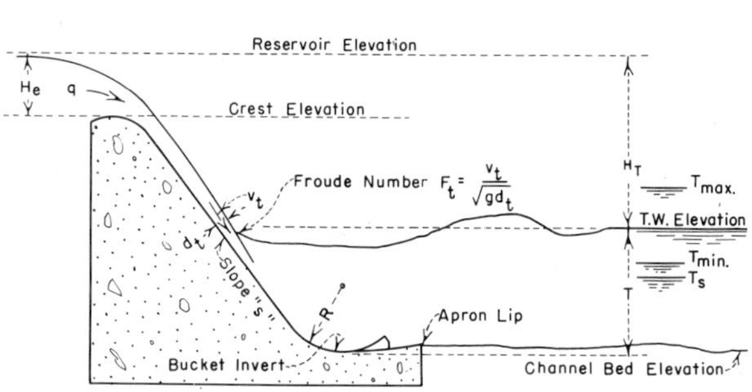

Figure 275. Definition of symbols—submerged buckets. 288–D–2434.

The same procedure should be followed to verify that satisfactory roller action will result for a unit discharge of 25 second-feet. In this case the minimum radius of 11.0 feet determined for the maximum discharge was found to govern. The T_{max} and T_{min} were found to be 50 feet and 10.4 feet, respectively, compared to the 11 feet of tailwater depth provided by the invert elevation placed at 967.0 feet. It may now be considered that the design based on maximum discharges will be satisfactory for all lower discharges.

If a wider range of safe tailwater depths is desired, the radius of curvature of the bucket can be increased. Thus, for a bucket radius of 12 feet, for the maximum discharge, $T_{min}=6.5d_t=16.6$, and $T_{max}=8.5d_t=22.5$ feet. An average tailwater depth of 20 feet, placing the bucket invert at elevation 965, will provide more leeway for tailwater variations.

209. ***Impact Type Stilling Basin.***—An impact type of energy dissipator has been developed [25] which is an effective stilling device that does not depend upon tailwater. The capacity of such a basin is limited to the feasibility of the structural design and to an incoming velocity of about 50 feet per second. This basin can be used with either an open chute or a closed conduit structure. The design shown on figure 277 has been proved for discharges up to about 400 second-feet; for large discharges multiple basins could be placed side by side.

Dissipation is accomplished by the impact of the incoming jet on the vertical hanging baffle, and by eddies which are formed from the changed direction of the jet after it strikes the baffle. Best hydraulic action is obtained when the tailwater height approaches but does not exceed a level halfway up the height of the baffle. For proper performance, the bottom of the baffle should be placed at the same level as the invert of the upstream channel or pipe.

The general arrangement of the basin and the dimensional requirements including riprap are shown on figure 277. Figure 278 shows an impact type stilling basin operating at about 80 percent of its designed capacity. This type of basin is subjected to large dynamic forces and turbulences which must be considered in the structural design. The structure must be made sufficiently stable to resist sliding against the impact load on the baffle wall. The entire structure must resist the severe vibrations inherent in this type of device, and the individual structural members must be sufficiently strong to withstand the large dynamic loads.

Riprapping should be provided along the bottom and sides adjacent to the structure to avoid the tendency for scour of the outlet channel downstream from the end sill when a shallow tailwater exists. Downstream wingwalls placed at 45° may also be effective in reducing scouring tendencies and flow concentrations downstream.

210. ***Plunge Pools.***—When a free-falling overflow nappe drops vertically into a pool in a riverbed, a plunge pool will be scoured to a depth which is related to the height of the fall, the depth of tailwater, and the concentration of the flow [13]. Depths of scour are influenced initially by the erodibility of the stream material or the bedrock and by the size or the gradation of sizes of any armoring material in the pool. However, the armoring or protective surfaces of the pool will be progressively reduced by the abrading action of the churning material to a

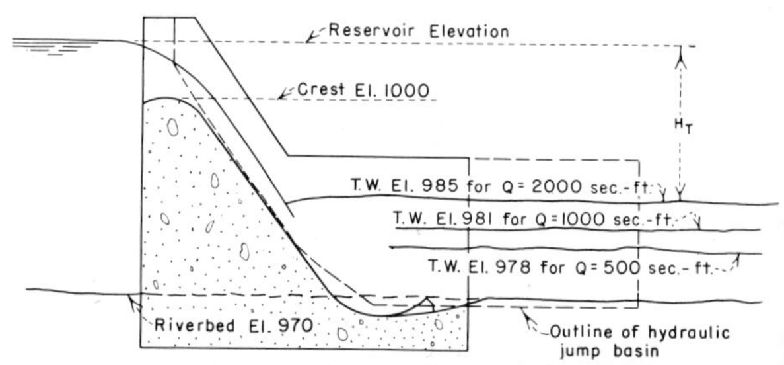

Figure 276. Example of design of stilling device for overflow spillway. 288–D–2435.

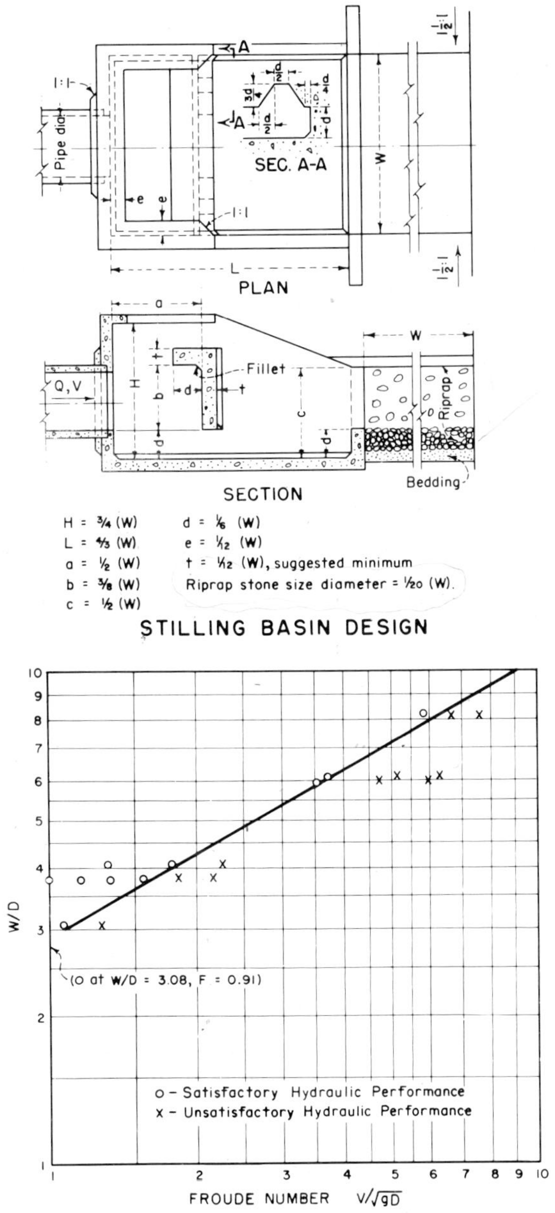

"W" is the inside width of the basin.
"D" represents the depth of flow entering the basin and is the square root of the flow area.
"V" is the velocity of the incoming flow.

Figure 277. **Dimensional criteria for impact type stilling basin. 288–D–2436.**

Figure 278. An impact type stilling basin in operation. 288–D–2905.

size which will be scoured out and the ultimate scour depth will, for all practical considerations, stabilize at a limiting depth irrespective of the material size. An empirical approximation based on experimental data has been developed by Veronese [14] for limiting scour depths, as follows:

$$d_s = 1.32\, H_T^{0.225}\, q^{0.54} \qquad (26)$$

where:

d_s=the maximum depth of scour below tailwater level in feet,

H_T=the head from the reservoir to tailwater levels in feet, and

q=the discharge in second-feet per foot of width.

F. HYDRAULICS OF SPILLWAYS

211. *Free Overfall (Straight Drop) Spillways.*— (a) *General.*—The hydraulic problems of the free overfall spillway are concerned with the characteristics of the control and with the dissipation of flow in the downstream basin. Flow over the control ordinarily is free discharging; air is admitted to the underside of the nappe to avoid the jet being depressed by reduced underneath pressure. Dissipation of the flow in the downstream basin may be obtained by the hydraulic jump, by impact and turbulence induced in a basin with impact blocks, or by a slotted grating dissipator installed immediately downstream from the control.

The control may be either sharp crested to provide a fully contracted vertical jet, broad crested to effect a fully suppressed jet, or shaped to increase the crest efficiency. Coefficients of discharge will approximate those indicated in section 197. The sides of the control usually are arranged to allow for full side contraction, in order to provide side space for the access of air to the underside of the nappe. This contraction is effected by providing square abutment headwalls or by installing square-cornered vertical offsets along the piers or walls opposite the crest. The effective length of the crest is then determined according to equation (4) where K_p and K_a will approximate 0.20.

The dimensions of the stilling basin for the free overfall spillway can be related to two independent variables; namely, the drop distance Y and the unit discharge q. These variables, which are dimensional terms, can be expressed in a dimensionless ratio by expressing q in lineal form by means of the equation for critical depth, $d_c=\sqrt[3]{\frac{q^2}{g}}$, as follows:

$$\frac{d_c}{Y}=\sqrt[3]{\frac{q^2}{gY^3}}$$

From this expression it can be seen that $\frac{q^2}{gY^3}$ is a dimensionless ratio which can be used as an independent variable to which the individual dimensions may be related. This ratio is called the "drop number" and is designated $\overline{D}$. It can be shown that $\overline{D}$ is the product of F_1^2 and $\left(\frac{d_1}{Y}\right)^3$, where F_1 is the Froude number $\frac{v_1}{\sqrt{d_1 g}}$ at the point where the nappe meets the basin floor.

(b) *Hydraulic Jump Basin.*—The jump characteristics of the straight drop basin are basically the same as those for other jump basins, except that the position of the start of the jump cannot be determined as readily as it can for other basins. On figure 279 the point of the start of the jump (point X) will vary with the vertical drop distance and is influenced by the under nappe pool depth, d_f. The basin de-

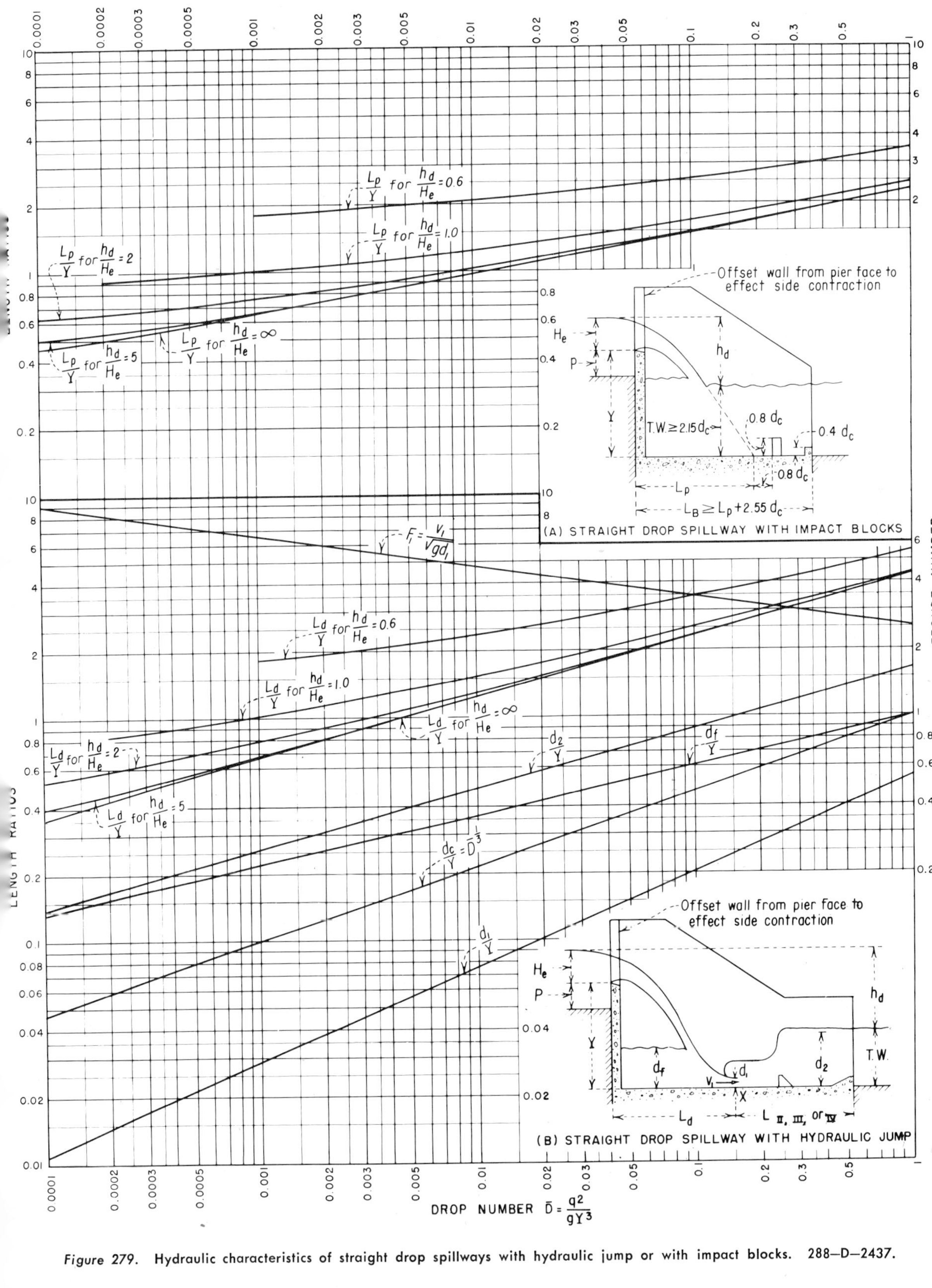

Figure 279. Hydraulic characteristics of straight drop spillways with hydraulic jump or with impact blocks. 288–D–2437.

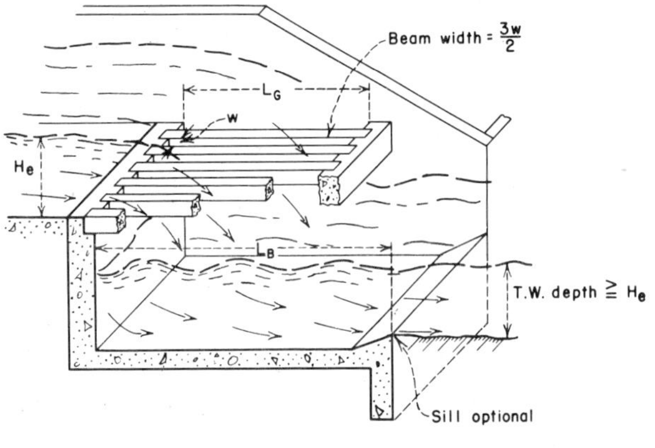

Figure 280. Slotted grating dissipator. 288–D–2438.

sign downstream from point X will be patterned after those discussed in section 206, once distance L_d is determined. Values of the depth d_1, and of the Froude number, F_1, at the start of the jump in relation to the drop number, $\overline{D}$, are shown on figure 279. These relations may be used for determining the basin dimensions.

Where tailwater depths are greater than the conjugate depth d_2, the jump will move back on the free falling nappe raising the depth d_f of the under nappe pool. With greater depths of the under nappe pool, the nappe will not plunge immediately to the floor of the basin but will be deflected upward along the top of the under pool so that it will meet the floor to the right of point X. The distance to the start of the jump, L_d, will become progressively longer as the tailwater depth is increased. Average values of L_d in relation to $\frac{h_d}{H_e}$, as determined from tests, are plotted on figure 279. For a basin with excessive depth the type III basin discussed in section 206 is most adaptable. The impact block type basin, discussed below, also can be adopted for low drop spillways with excessive tailwater depths.

(c) *Impact Block Type Basin.*—An impact block basin has been developed [1] for low heads which gives reasonably good dissipation of energy for a wide range of tailwater depths. The dissipation of the high energy is principally by turbulence induced by the impingement of the incoming flow upon the impact blocks. The required tailwater depths, therefore, become more or less independent of the dro height. The linear proportions are as follows

Minimum basin length, $L_B=L_p+2.55\ d_c$

Minimum length to upstream face of baffl block$=L_p+0.8\ d_c$

Minimum tailwater depth, $d_{tw}=2.15\ d_c$

Optimum baffle block height$=0.8\ d_c$

Width and spacing of baffle blocks$=0.4\ d_c\pm$

Optimum height of end sill$=0.4\ d_c$

(d) *Slotted Grating Dissipator.*—An effec tive dissipator for small drops is illustrated o figure 280. This device has been tested fo values of the Froude number, F_1, as determine at basin apron level, in the range of 2.5 to 4.5 For this arrangement the overfalling sheet i separated into a number of long, thin segment which fall nearly vertically into the basin below where dissipation of energy takes place b turbulence. To be effective the length of th grating, L_G, must be such that the entire incom ing flow will fall through the slots before reach ing the downstream end. The length i therefore a function of the total discharge, th velocity of the incoming flow, and the area c the grating slots. Experimental tests indicat that the following relation gives an effectiv design:

$$L_G=\frac{Q}{0.245wN\sqrt{2gH_e}} \quad (27$$

where:

L_G=the length of the grating in feet,

w=the width of the slot in feet,

N=the number of slots, and

H_e=the depth of flow upstream from th drop.

The length of the basin, L_B, should be approx imately 1.2 L_G. An end sill similar to that fo basin type I, discussed in section 206, can b provided to improve the hydraulic action.

(e) *Example of Design of a Free Overfa Spillway.*—The procedure for designing a fre overfall spillway is best shown by means of a example. Consider that such a spillway is t be designed to discharge 500 second-feet. Th drop from the spillway crest to the tailwate level for a flow of 500 second-feet is 12 feet (Tailwater elevation is 108.0.) The approac channel is 20 feet long and the approach floo is level with the spillway crest which is at ele vation 120.0. Each type of energy dissipator is to be investigated.

The procedure for design of the *hydraulic jump basin* is as follows: First, assume the effective length of the spillway crest to be 15 feet. Assume an approximate value of $C=3.0$. The unit discharge, q, is equal to $\frac{500}{15}=33.3$ second-feet and H_e is equal to $\left(\frac{q}{C}\right)^{2/3}=\left(\frac{33.3}{3.0}\right)^{2/3}=5.0$ feet. The reservoir water surface elevation, therefore, is $120.0+5.0=125.0$. Thus the drop from reservoir level to tailwater level will be approximately 17 feet.

Assume that an offset of 0.5 foot is provided along each side of the weir to effect side contractions for aerating the underside of the sheet, and that the offset is square-cornered. Then the net crest length, which will also be the stilling basin width, is:

$$L'=L+2K_aH_e+2(0.5)=15+2(0.2)(5)+1.0=18.0 \text{ feet.}$$

Figure 268 is used to determine the approximate apron level of the jump basin, assuming the effective width of the basin to be 15 feet and (for the first trial) that there will be no loss of energy between the reservoir and the point where the jet strikes the basin floor. From scale A, the conjugate depth d_2 for $q=33.3$ second-feet and $H_T=17$ feet is 8.8 feet, which places the apron floor at elevation 99.2. Y is equal to elevation 120–elevation 99.2=20.8 feet, and the drop number $\overline{D}$ is equal to $\frac{q^2}{gY^3}=\frac{33.3^2}{32.2\times 20.8^3}=0.0038$. For $\overline{D}=0.0038$, from figure 279, $\frac{d_2}{Y}=0.375$ and $d_2=7.8$ feet. The apron level then must be adjusted to an elevation which is d_2 below the tailwater elevation 108.0, or elevation 100.2.

For the second trial, the adjusted value of Y is 19.8 and $\overline{D}$ is equal to $\frac{33.3^2}{32.2\times 19.8^3}=0.0044$. For $\overline{D}=0.0044$ and $\frac{h_d}{H_e}=\frac{17}{5}=3.4$, from figure 279, $\frac{L_d}{Y}=1.02$ and $L_d=20.2$ feet. Also $d_1=1.1$ feet and $F_1=5.3$.

With the values of $F_1=5.3$, $d_1=1.1$ and $d_2=7.8$, the arrangement of the type III basin shown on figure 266 can be used. From figure 266, $\frac{L}{d^2}=2.37$ and $L=18.5$ feet. The length of the basin measured from the vertical crest is equal to $L_d+L=20.2+18.5=38.7$ feet. The distance of the baffle blocks from the vertical crest for this basin will be 20.2 feet plus 0.8 d_2 or 20.2 plus 0.8 (7.8)=26.4 feet, approximately.

The baffle blocks will be approximately 1.5 d_1 or 1.6 feet high and will be about 14 inches wide and spaced at about 28-inch centers.

For the *impact block basin*, the procedure is as follows: The critical depth, d_c, is equal to $\sqrt[3]{\frac{q^2}{g}}=\sqrt[3]{\frac{33.3^2}{32.2}}=3.3$ feet. Then from figure 279, for $D=0.0044$ and $\frac{h_d}{H_e}=3.4$, $\frac{L_p}{Y}=0.85$ and $L_p=17.0$ feet. The minimum length of the basin, L_B, is equal to $L_p+2.55\ d_c=17.0+2.55\ (3.3)=25.4$ feet, say 26 feet. The minimum tailwater depth of 2.15 d_c will be 7.1 feet which places the basin floor at elevation 100.9. The distance from the vertical crest to the baffle blocks will be $L+0.8\ d_c=17.0+0.8\times 3.3=19.6$ feet, say 20 feet. The baffle blocks will be about 0.8 d_c or 3.0 feet high and about 18 inches wide, spaced at about 3-foot centers. The end sill will be 0.4 d_c or about 1.5 feet high.

It can be seen from the above result that if the impact block basin is used, the basin can be made almost 13 feet shorter than that required for a hydraulic jump basin, and also that the impact block basin will be 0.7 foot shallower. The baffle blocks for the hydraulic jump basin will be smaller and spaced closer together than those for the impact block basin.

This example shows that the impact block basin is considerably smaller than the hydraulic jump basin. However, the impact block basin should be limited to use where the drop distance does not exceed 20 feet. Furthermore, as previously explained, the foundation for an impact block basin must be of better quality because of the concentrated forces involved. The hydraulic jump basin, therefore, has a much wider application of use.

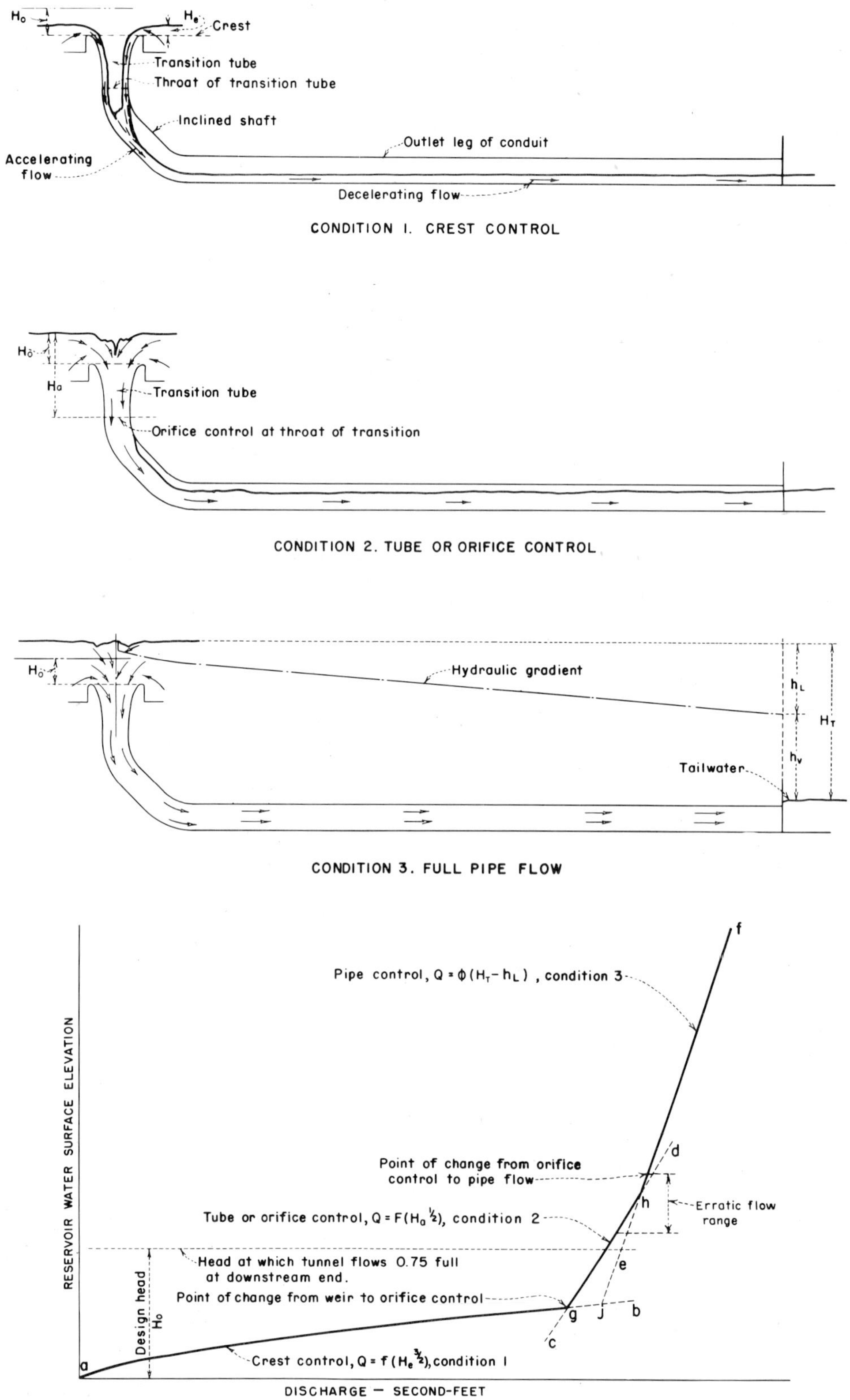

Figure 281. Nature of flow and discharge characteristics of a morning glory spillway. 288–D–2439.

The *slotted grating dissipator* is not suitable in this case because the Froude number of 5.3 is in excess of the 4.5 value, which is the tested limit for a practical slotted grating design.

212. Drop Inlet (Shaft or Morning Glory) Spillways.—(a) *General Characteristics.*—Typical flow conditions and discharge characteristics of a drop inlet spillway are represented on figure 281. As illustrated on the discharge curve, crest control (condition 1) will prevail for heads between the ordinates of *a* and *g*; orifice or tube control (condition 2) will govern for heads between the ordinates of *g* and *h*; and the spillway conduit will flow full for heads above the ordinate of *h* (represented as condition 3).

Flow characteristics of a drop inlet spillway will vary according to the proportional sizes of the different elements. Changing the diameter of the crest will change the curve *ab* on figure 281 so that the ordinate of *g* on curve *cd* will be either higher or lower. For a larger diameter crest, greater outflows can be discharged over the weir at low heads and the transition will fill up and tube control will occur with a lesser head on the crest. Similarly, by altering the size of the throat of the tube, the position of curve *cd* will change, indicating the heads above which tube control will prevail. If the transition is made of such size that curve *cd* is moved to coincide with or lie to the right of point *j*, the control will shift directly from the crest to the downstream end of the conduit. The details of the hydraulic flow characteristics are discussed in following subsections.

(b) *Crest Discharge.*—For small heads, flow over the drop inlet spillway is governed by the characteristics of crest discharge. The vertical transition beyond the crest will flow partly full and the flow will cling to the sides of the shaft. As the discharge over the crest increases, the overflowing annular nappe will become thicker, and eventually the nappe flow will converge into a solid vertical jet. The point where the annular nappe joins the solid jet is called the crotch. After the solid jet forms, a "boil" will occupy the region above the crotch; both the crotch and the top of the boil become progressively higher with larger discharges. For high heads the crotch and boil may almost flood out, showing only a slight depression and eddy at the surface.

Until such time as the nappe converges to form a solid jet, free-discharging weir flow prevails. After the crotch and boil form, submergence begins to affect the weir flow and ultimately the crest will drown out. Flow is then governed either by the nature of the contracted jet which is formed by the overflow entrance, or by the shape and size of the vertical transition if it does not conform to the jet shape. Vortex action must be minimized to maintain converging flow into the drop inlet. Guide piers are often employed along the crest for this purpose [5, 6, 22].

If the crest profile and transition conform to the shape of the lower nappe of a jet flowing over a sharp-crested circular weir, the discharge characteristics for flow over the crest and through the transition can be expressed as:

$$Q=CLH^{3/2} \qquad (3)$$

where H is the head measured either to the apex of the under nappe of the overflow, to the spring point of the circular sharp-crested weir, or to some other established point on the overflow. Similarly, the choice of the length L is related to some specific point of measurement such as the length of the circle at the apex, along the periphery at the upstream face of the crest, or along some other chosen reference line. The value of C will change with different definitions of L and H. If L is taken at the outside periphery of the overflow crest (the origin of the coordinates in figure 282) and if the head is measured to the apex of the overflow shape, equation (3) can be written:

$$Q=C_o(2\pi R_s)H_o^{3/2} \qquad (28)$$

It will be apparent that the coefficient of discharge for a circular crest differs from that for a straight crest because of the effects of submergence and back pressure incident to the joining of the converging flows. Thus C_o must be related to both H_o and R_s, and can be expressed in terms of $\frac{H_o}{R_s}$. The relationship of C_o, as determined from model tests [15], to values of $\frac{H_o}{R_s}$ for three conditions of approach depth is plotted on figure 283. These coefficients are valid only if the crest profile and transition shape conform to that of the jet flowing over a sharp-crested circular weir at H_o head and if aeration is provided so that subatmospheric pressures do not exist along the lower nappe surface contact.

When the crest outline and transition shape conform to the profile of the nappe shape for an H_o head over the crest, free flow prevails for $\frac{H_o}{R_s}$ ratios up to approximately 0.45, and weir control governs. As the $\frac{H_o}{R_s}$ ratio increases above 0.45, the weir partly submerges and flow showing characteristics of a submerged weir is the controlling condition. When the $\frac{H_o}{R_s}$ ratio approaches 1.0, the water surface above the weir is completely submerged. For this and higher stages of $\frac{H_o}{R_s}$, the flow phenomenon is that of orifice flow. The weir formula, $Q=CLH^{3/2}$, is used as the measure of flow through the drop inlet entrance regardless of the submergence, by using a coefficient which reflects the flow conditions through the various $\frac{H}{R_s}$ ranges. Thus, from figure 283 it will be seen that the weir coefficient is only slightly changed from that normally indicated for values of $\frac{H_o}{R_s}$ less than 0.45, but reduces rapidly for the higher $\frac{H_o}{R_s}$ ratios.

It will be noted that for most conditions of flow over a circular weir the coefficient of discharge increases with a reduction of the approach depth, whereas the opposite is true for a straight weir. For both weirs a shallower approach lessens the upward vertical velocity component and consequently suppresses the contraction of the nappe. However, for the circular weir the submergence effect is reduced because of a depressed upper nappe surface, giving the jet a quicker downward impetus, which lowers the position of the crotch and increases the discharge.

Coefficients for partial heads of H_e on the crest can be determined from figure 284 to prepare a discharge-head relationship. The designer must be cautious in applying the above criteria, since subatmospheric pressure or submergence effects may alter the flow conditions differently for variously shaped profiles. This criteria, therefore, should not be applied for flow conditions where $\frac{H_e}{R_s}$ exceeds 0.4.

(c) *Crest Profiles.*—Values of coordinates to define the shape of the lower surface of a nappe flowing over an aerated sharp-crested circular weir for various conditions of $\frac{P}{R_s}$ and $\frac{H_s}{R_s}$ are shown in tables 25, 26, and 27. These data are based on experimental tests [15] conducted by the Bureau of Reclamation. The relationships of H_s to H_o are shown on figure 285. Typical upper and lower nappe profiles for various values of $\frac{H_s}{R_s}$ are plotted on figure 286 in terms of $\frac{x}{H_s}$ and $\frac{y}{H_s}$ for the condition of $\frac{P}{R_s}=2.0$.

Illustrated on figure 287 are typical lower nappe profiles, plotted for various values of H_s for a given value of R_s. In contrast to the straight weir where the nappe springs farther from the crest as the head increases, it will be seen from figure 287 that the lower nappe profile for the circular crest springs farther only in the region of the high point of the trace, and then only for $\frac{H_s}{R_s}$ values up to about 0.5.

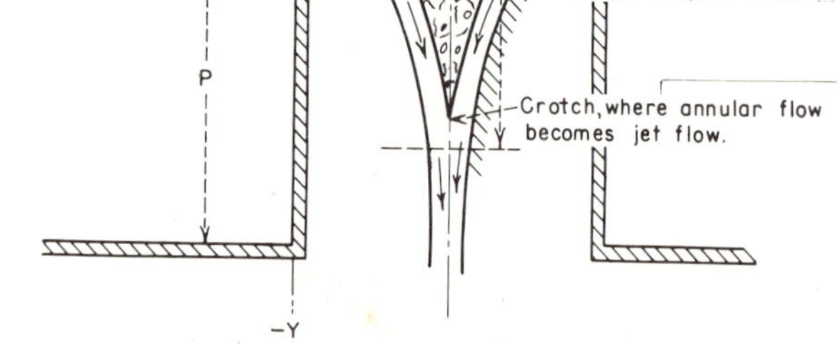

Figure 282. Elements of nappe-shaped profile for circular weir. 288–D–2440.

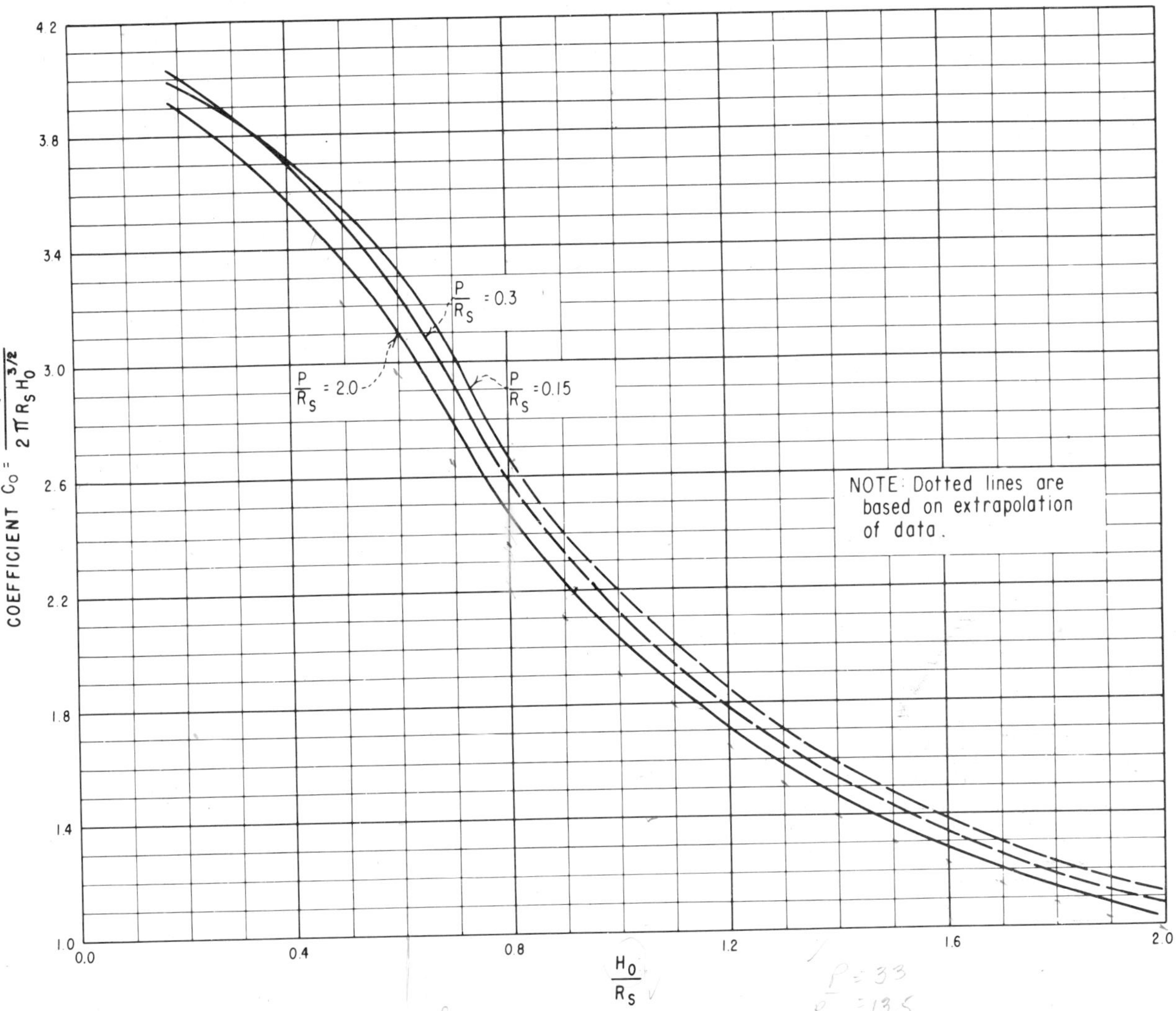

Figure 283. Relationship of circular crest coefficient C_o to $\frac{H_o}{R_s}$ for different approach depths (aerated nappe). 288–D–2441.

The profiles become increasingly suppressed for larger $\frac{H_s}{R_s}$ values. Below the high point of the profile the traces cross and the shapes for the higher heads fall inside those for the lower heads. Thus, if the crest profile is designed for heads where $\frac{H_s}{R_s}$ exceeds about 0.25 to 0.3, it appears that subatmospheric pressure will occur along some portion of the profile when heads are less than the designed maximum. If subatmospheric pressures are to be avoided along the crest profile, the crest shape should be selected so that it will give support to the overflow nappe for the smaller $\frac{H_e}{R_s}$ ratios. Figure 288 shows the approximate increase in radius required to minimize subatmospheric pressures on the crest. The crest shape for the enlarged crest radius is then based on a $\frac{H'_s}{R'_s}$ ratio of 0.3.

(d) *Transition Design.*—The diameter of a jet issuing from a horizontal orifice can be determined for any point below the water surface if it is assumed that the continuity equation, $Q=av$, is valid and if friction and other losses are neglected.

TABLE 25.—*Coordinates of lower nappe surface for different values of* $\frac{H_s}{R}$ *when* $\frac{P}{R}=2$

[Negligible approach velocity and aerated nappe]

$\frac{H_s}{R}$	0.00	0.10*	0.20	0.25	0.30	0.35	0.40	0.45	0.50	0.60	0.80	1.00	1.20	1.50	2.00
$\frac{X}{H_s}$	$\frac{Y}{H_s}$ For portion of the profile above the weir crest														
0.000	0.0000	0.0000	0.0000	0.0000	0.0000	0.0000	0.0000	0.0000	0.0000	0.0000	0.0000	0.0000	0.0000	0.0000	0.0000
.010	.0150	.0145	.0133	.0130	.0128	.0125	.0122	.0119	.0116	.0112	.0104	.0095	.0086	.0077	.0070
.020	.0280	.0265	.0250	.0243	.0236	.0231	.0225	.0220	.0213	.0202	.0180	.0159	.0140	.0115	.0090
.030	.0395	.0365	.0350	.0337	.0327	.0317	.0308	.0299	.0289	.0270	.0231	.0198	.0168	.0126	.0085
.040	.0490	.0460	.0435	.0417	.0403	.0389	.0377	.0363	.0351	.0324	.0268	.0220	.0176	.0117	.0050
.050	.0575	.0535	.0506	.0487	.0471	.0454	.0436	.0420	.0402	.0368	.0292	.0226	.0168	.0092	
.060	.0650	.0605	.0570	.0550	.0531	.0510	.0489	.0470	.0448	.0404	.0305	.0220	.0147	.0053	
.070	.0710	.0665	.0627	.0605	.0584	.0560	.0537	.0514	.0487	.0432	.0308	.0201	.0114	.0001	
.080	.0765	.0710	.0677	.0655	.0630	.0603	.0578	.0550	.0521	.0455	.0301	.0172	.0070		
.090	.0820	.0765	.0722	.0696	.0670	.0640	.0613	.0581	.0549	.0471	.0287	.0135	.0018		
.100	.0860	.0810	.0762	.0734	.0705	.0672	.0642	.0606	.0570	.0482	.0264	.0089			
.120	.0940	.0880	.0826	.0790	.0758	.0720	.0683	.0640	.0596	.0483	.0195				
.140	.1000	.0935	.0872	.0829	.0792	.0750	.0 05	.0654	.0599	.0460	.0101				
.160	.1045	.0980	.0905	.0855	.0812	.0765	.0710	.0651	.0585	.0418					
.180	.1080	.1010	.0927	.0872	.0820	.0766	.0705	.0637	.0559	.0361					
.200	.1105	.1025	.0938	.0877	.0819	.0756	.0688	.0611	.0521	.0292					
.250	.1120	.1035	.0926	.0850	.0773	.0683	.0596	.0495	.0380	.0068					
.300	.1105	.1000	.0850	.0764	.0668	.0559	.0446	.0327	.0174						
.350	.1060	.0930	.0750	.0650	.0540	.0410	.0280	.0125							
.400	.0970	.0830	.0620	.0500	.0365	.0220	.0060								
.450	.0845	.0700	.0450	.0310	.0170	.000									
.500	.0700	.0520	.0250	.0100											
.550	.0520	.0320	.0020												
.600	.0320	.0080													
.650	.0090														
$\frac{Y}{H_s}$	$\frac{X}{H_s}$ For portion of the profile below the weir crest														
0.000	0.668	0.615	0.554	0.520	0.487	0.450	0.413	0.376	0.334	0.262	0.158	0.116	0.093	0.070	0.048
−.020	.705	.652	.592	.560	.526	.488	.452	.414	.369	.293	.185	.145	.120	.096	.074
−.040	.742	.688	.627	.596	.563	.524	.487	.448	.400	.320	.212	.165	.140	.115	.088
−.060	.777	.720	.660	.630	.596	.557	.519	.478	.428	.342	.232	.182	.155	.129	.100
−.080	.808	.752	.692	.662	.628	.589	.549	.506	.454	.363	.250	.197	.169	.140	.110
−.100	.838	.784	.722	.692	.657	.618	.577	.532	.478	.381	.266	.210	.180	.150	.118
−.150	.913	.857	.793	.762	.725	.686	.641	.589	.531	.423	.299	.238	.204	.170	.132
−.200	.978	.925	.860	.826	.790	.745	.698	.640	.575	.459	.326	.260	.224	.184	.144
−.250	1.040	.985	.919	.883	.847	.801	.750	.683	.613	.490	.348	.280	.239	.196	.153
−.300	1.100	1.043	.976	.941	.900	.852	.797	.722	.648	.518	.368	.296	.251	.206	.160
−.400	1.207	1.150	1.079	1.041	1.000	.944	.880	.701	.700	.502	.400	.322	.271	.220	.168
−.500	1.308	1.246	1.172	1.131	1.087	1.027	.951	.849	.753	.598	.427	.342	.287	.232	.173
−.600	1.397	1.335	1.260	1.215	1.167	1.102	1.012	.898	.793	.627	.449	.359	.300	.240	.179
−.800	1.563	1.500	1.422	1.369	1.312	1.231	1.112	.974	.854	.673	.482	.384	.320	.253	.184
−1.000	1.713	1.646	1.564	1.508	1.440	1.337	1.189	1.030	.899	.710	.508	.402	.332	.260	.188
−1.200	1.846	1.780	1.691	1.635	1.553	1.422	1.248	1.074	.933	.739	.528	.417	.340	.266	
−1.400	1.970	1.903	1.808	1.748	1.653	1.492	1.293	1.108	.963	.760	.542	.423	.344		
−1.600	2.085	2.020	1.918	1.855	1.742	1.548	1.330	1.133	.988	.780	.553	.430			
−1.800	2.196	2.130	2.024	1.957	1.821	1.591	1.358	1.158	1.008	.797	.563	.433			
−2.000	2.302	2.234	2.126	2.053	1.891	1.630	1.381	1.180	1.025	.810	.572				
−2.500	2.557	2.475	2.354	2.266	2.027	1.701	1.430	1.221	1.059	.838	.588				
−3.000	2.778	2.700	2.559	2.428	2.119	1.748	1.468	1.252	1.086	.853					
−3.500	--------	2.916	2.749	2.541	2.171	1.777	1.489	1.267	1.102						
−4.000	--------	3.114	2.914	2.620	2.201	1.796	1.500	1.280							
−4.500	--------	3.306	3.053	2.682	2.220	1.806	1.509								
−5.000	--------	3.488	3.178	2.734	2.227	1.811									
−5.500	--------	3.653	3.294	2.779	2.229										
−6.000	--------	3.820	3.405	2.812	2.232										
$\frac{H_s}{R}$	0.00	0.10	0.20	0.25	0.30	0.35	0.40	0.45	0.50	0.60	0.80	1.00	1.20	1.50	2.00

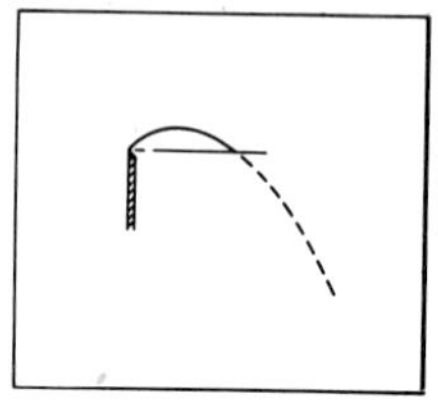

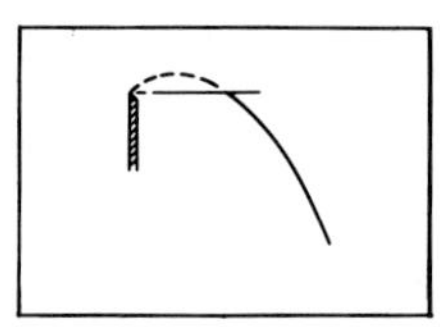

*The tabulation for $\frac{H_s}{R}=0.10$ was obtained by interpolation between $\frac{H_s}{R}=0$ and 0.20.

After Wagner [15]

TABLE 26.—*Coordinates of lower nappe surface for different values of* $\frac{H_s}{R}$ *when* $\frac{P}{R} = 0.30$

$\frac{H_s}{R}$	0.20	0.25	0.30	0.35	0.40	0.45	0.50	0.60	0.80
$\frac{X}{H_s}$	$\frac{Y}{H_s}$ For portion of the profile above the weir crest								
0.000	0.0000	0.0000	0.0000	0.0000	0.0000	0.0000	0.0000	0.0000	0.0000
.010	.0130	.0130	.0130	.0125	.0120	.0120	.0115	.0110	.0100
.020	.0245	.0242	.0240	.0235	.0225	.0210	.0195	.0180	.0170
.030	.0340	.0335	.0330	.0320	.0300	.0290	.0270	.0240	.0210
.040	.0415	.0411	.0390	.0380	.0365	.0350	.0320	.0285	.0240
.050	.0495	.0470	.0455	.0440	.0420	.0395	.0370	.0325	.0245
.060	.0560	.0530	.0505	.0490	.0460	.0440	.0405	.0350	.0250
.070	.0610	.0575	.0550	.0530	.0500	.0470	.0440	.0370	.0245
.080	.0660	.0620	.0590	.0565	.0530	.0500	.0460	.0385	.0235
.090	.0705	.0660	.0625	.0595	.0550	.0520	.0480	.0390	.0215
.100	.0740	.0690	.0660	.0620	.0575	.0540	.0500	.0395	.0190
.120	.0800	.0750	.0705	.0650	.0600	.0560	.0510	.0380	.0120
.140	.0840	.0790	.0735	.0670	.0615	.0560	.0515	.0355	.0020
.160	.0870	.0810	.0750	.0675	.0610	.0550	.0500	.0310	
.180	.0885	.0820	.0755	.0675	.0600	.0535	.0475	.0250	
.200	.0885	.0820	.0745	.0660	.0575	.0505	.0435	.0180	
.250	.0855	.0765	.0685	.0590	.0480	.0390	.0270		
.300	.0780	.0670	.0580	.0460	.0340	.0220	.0050		
.350	.0660	.0540	.0425	.0295	.0150				
.400	.0495	.0370	.0240	.0100					
.450	.0300	.0170	.0025						
.500	.0090	−.0060							
.550									
$\frac{Y}{H_s}$	$\frac{X}{H_s}$ For portion of the profile below the weir crest								
−0.000	0.519	0.488	0.455	0.422	0.384	0.349	0.310	0.238	0.144
−.020	.560	.528	.495	.462	.423	.387	.345	.272	.174
−.040	.598	.566	.532	.498	.458	.420	.376	.300	.198
−.060	.632	.601	.567	.532	.491	.451	.406	.324	.220
−.080	.664	.634	.600	.564	.522	.480	.432	.348	.238
−.100	.693	.664	.631	.594	.552	.508	.456	.368	.254
−.150	.760	.734	.701	.661	.618	.569	.510	.412	.290
−.200	.831	.799	.763	.723	.677	.622	.558	.451	.317
−.250	.893	.860	.826	.781	.729	.667	.599	.483	.341
−.300	.953	.918	.880	.832	.779	.708	.634	.510	.362
−.400	1.060	1.024	.981	.932	.867	.780	.692	.556	.396
−.500	1.156	1.119	1.072	1.020	.938	.841	.745	.595	.424
−.600	1.242	1.203	1.153	1.098	1.000	.891	.780	.627	.446
−.800	1.403	1.359	1.301	1.227	1.101	.970	.845	.672	.478
−1.000	1.549	1.498	1.430	1.333	1.180	1.028	.892	.707	.504
−1.200	1.680	1.622	1.543	1.419	1.240	1.070	.930	.733	.524
−1.400	1.800	1.739	1.647	1.489	1.287	1.106	.959	.757	.540
−1.600	1.912	1.849	1.740	1.546	1.323	1.131	.983	.778	.551
−1.800	2.018	1.951	1.821	1.590	1.353	1.155	1.005	.797	.560
−2.000	2.120	2.049	1.892	1.627	1.380	1.175	1.022	.810	.569
−2.500	2.351	2.261	2.027	1.697	1.428	1.218	1.059	.837	
−3.000	2.557	2.423	2.113	1.747	1.464	1.247	1.081	.852	
−3.500	2.748	2.536	2.167	1.778	1.489	1.263	1.099		
−4.000	2.911	2.617	2.200	1.796	1.499	1.274			
−4.500	3.052	2.677	2.217	1.805	1.507				
−5.000	3.173	2.731	2.223	1.810					
−5.500	3.290	2.773	2.228						
−6.000	3.400	2.808							
$\frac{H_s}{R}$	0.20	0.25	0.30	0.35	0.40	0.45	0.50	0.60	0.80

After Wagner [15]

TABLE 27.—*Coordinates of lower nappe surface for different values of* $\frac{H_s}{R}$ *when* $\frac{P}{R} = 0.15$

$\frac{H_s}{R}$	0.20	0.25	0.30	0.35	0.40	0.45	0.50	0.60	0.80
$\frac{X}{H_s}$	$\frac{Y}{H_s}$ For portion of the profile above the weir crest								
0.000	0.0000	0.0000	0.0000	0.0000	0.0000	0.0000	0.0000	0.0000	0.0000
.010	.0120	.0120	.0115	.0115	.0110	.0110	.0105	.0100	.0090
.020	.0210	.0200	.0195	.0190	.0185	.0180	.0170	.0160	.0140
.030	.0285	.0270	.0265	.0260	.0250	.0235	.0225	.0200	.0165
.040	.0345	.0335	.0325	.0310	.0300	.0285	.0265	.0230	.0170
.050	.0405	.0385	.0375	.0360	.0345	.0320	.0300	.0250	.0170
.060	.0450	.0430	.0420	.0400	.0380	.0355	.0330	.0265	.0165
.070	.0495	.0470	.0455	.0430	.0410	.0380	.0350	.0270	.0150
.080	.0525	.0500	.0485	.0460	.0435	.0400	.0365	.0270	.0130
.090	.0560	.0530	.0510	.0480	.0455	.0420	.0370	.0265	.0100
.100	.0590	.0560	.0535	.0500	.0465	.0425	.0375	.0255	.0065
.120	.0630	.0600	.0570	.0520	.0480	.0435	.0365	.0220	
.140	.0660	.0620	.0585	.0525	.0475	.0425	.0345	.0175	
.160	.0670	.0635	.0590	.0520	.0460	.0400	.0305	.0110	
.180	.0675	.0635	.0580	.0500	.0435	.0365	.0260	.0040	
.200	.0670	.0625	.0560	.0465	.0395	.0320	.0200		
.250	.0615	.0560	.0470	.0360	.0265	.0160	.0015		
.300	.0520	.0440	.0330	.0210	.0100				
.350	.0380	.0285	.0165	.0030					
.400	.0210	.0090							
.450	.0015								
.500									
.550									
$\frac{Y}{H_s}$	$\frac{X}{H_s}$ For portion of the profile below the weir crest								
−0.000	0.454	0.422	0.392	0.358	0.325	0.288	0.253	0.189	0.116
−.020	.499	.467	.437	.404	.369	.330	.292	.228	.149
−.040	.540	.509	.478	.444	.407	.368	.328	.259	.174
−.060	.579	.547	.516	.482	.443	.402	.358	.286	.195
−.080	.615	.583	.550	.516	.476	.434	.386	.310	.213
−.100	.650	.616	.584	.547	.506	.462	.412	.331	.228
−.150	.726	.691	.660	.620	.577	.526	.468	.376	.263
−.200	.795	.760	.729	.685	.639	.580	.516	.413	.293
−.250	.862	.827	.790	.743	.692	.627	.557	.445	.319
−.300	.922	.883	.843	.797	.741	.671	.594	.474	.342
−.400	1.029	.988	.947	.893	.828	.749	.656	.523	.381
−.500	1.128	1.086	1.040	.980	.902	.816	.710	.567	.413
−.600	1.220	1.177	1.129	1.061	.967	.869	.753	.601	.439
−.800	1.380	1.337	1.285	1.202	1.080	.953	.827	.655	.473
−1.000	1.525	1.481	1.420	1.317	1.164	1.014	.878	.696	.498
−1.200	1.659	1.610	1.537	1.411	1.228	1.059	.917	.725	.517
−1.400	1.780	1.731	1.639	1.480	1.276	1.096	.949	.750	.531
−1.600	1.897	1.843	1.729	1.533	1.316	1.123	.973	.770	.544
−1.800	2.003	1.947	1.809	1.580	1.347	1.147	.997	.787	.553
−2.000	2.104	2.042	1.879	1.619	1.372	1.167	1.013	.801	.560
−2.500	2.340	2.251	2.017	1.690	1.423	1.210	1.049	.827	
−3.000	2.550	2.414	2.105	1.738	1.457	1.240	1.073	.840	
−3.500	2.740	2.530	2.153	1.768	1.475	1.252	1.088		
−4.000	2.904	2.609	2.180	1.780	1.487	1.263			
−4.500	3.048	2.671	2.198	1.790	1.491				
−5.000	3.169	2.727	2.207	1.793					
−5.500	3.286	2.769	2.210						
−6.000	3.396	2.800							
$\frac{H_s}{R}$	0.20	0.25	0.30	0.35	0.40	0.45	0.50	0.60	0.80

After Wagner [15]

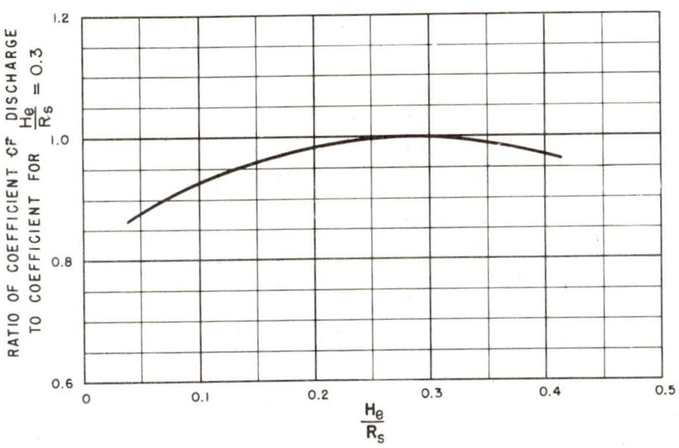

Figure 284. Circular crest coefficient of discharge for other than design head. 288–D–2442.

For a circular jet the area is equal to πR^2. The discharge is equal to $av=\pi R^2\sqrt{2gh_r}$. Solving for R, $R=\frac{Q_a^{1/2}}{5H_a^{1/4}}$ where H_a is equal to the difference between the water surface and the elevation under consideration. The diameter of the jet thus decreases with the distance of the free vertical fall for normal design applications.

If an assumed total loss (to allow for jet contraction losses, friction losses, velocity losses due to direction change, etc.) is taken as $0.1H_a$, the equation for determining the approximate required shaft radius can be written:

$$R=0.204\frac{Q^{1/2}}{H_a^{1/4}} \tag{29}$$

Since this equation is for the shape of the jet, its use for determining the shape of the shaft will result in the minimum size which will accommodate the flow without restrictions and without developing pressures along the side of the shaft.

A typical shaft profile obtained by equation (29) is shown by the dot-dash lines, *abc*, on figure 289. If the shaft profile, *abc*, is enlarged above selected points *b*, as shown by the dotted lines *db*, the section at A-A will flow under pressure; below section A-A the free jet profile should follow lines *bc*.

Aeration is required at the control either through introduction of air to a sudden enlargement of the shaft or the installation of a

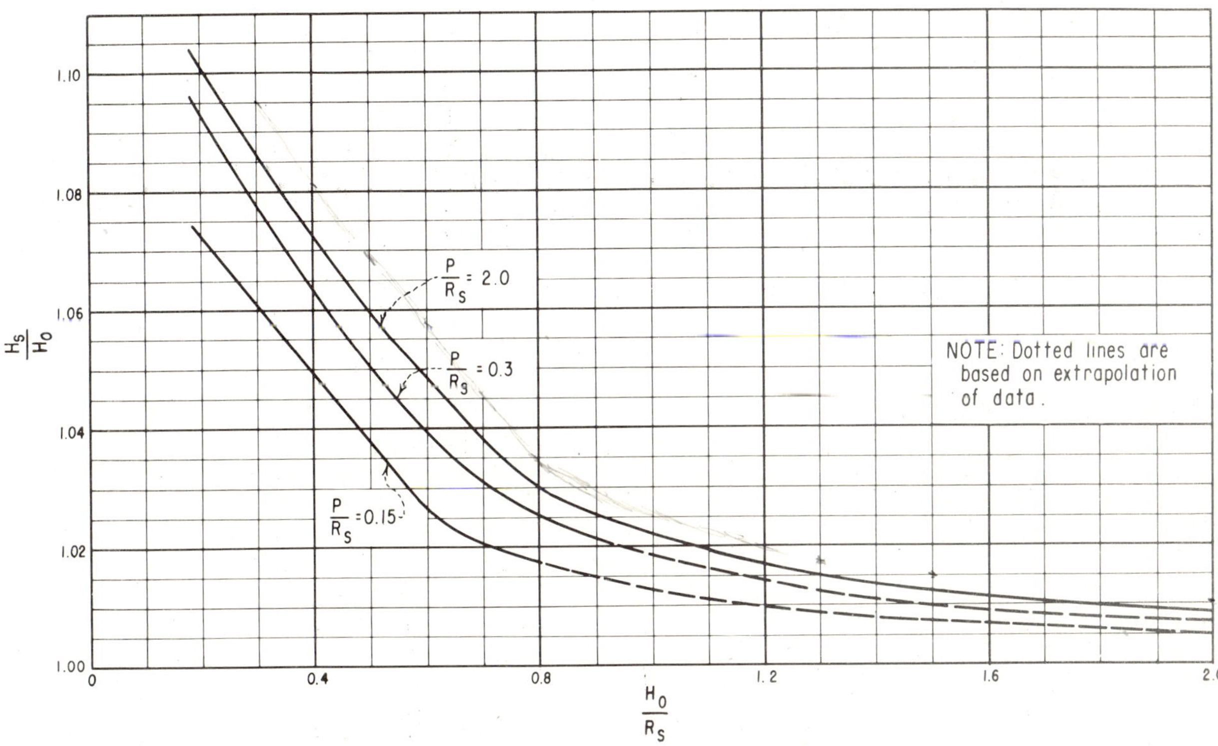

Figure 285. Relationship of $\frac{H_s}{H_o}$ to $\frac{H_o}{R_s}$ for circular sharp-crested weirs. 288–D–2443.

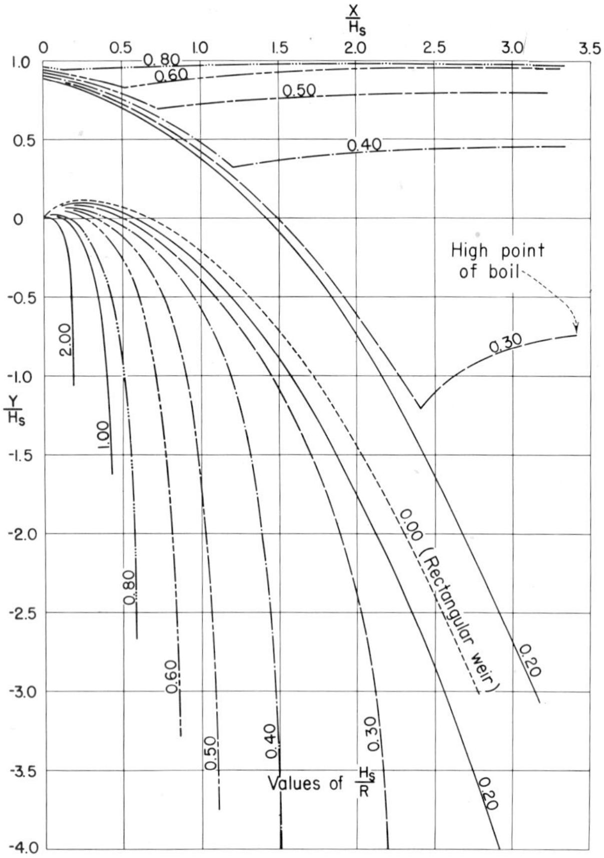

Figure 286. Upper and lower nappe profiles for circular weir (aerated nappe and negligible approach velocity). 288–D–2444.

deflector to insure free flow below the control section A-A. Elbows and passageway sizes and slopes must be such that free flow is maintained below the point of control. Failure to provide adequate aeration at the point of control could introduce cavitational tendencies and make and break siphonic action, which could cause severe vibrations. For a profile, *abe*, established for a specific head, the control must remain at section A-A for any higher head so that above the section pressure flow will prevail. The flow below section A-A must be kept free flow. If the profile *dbe* is adopted, once a head is reached to make the shaft flow full at point *b*, section A-A will be the point of control and pressure flow above the control will prevail for that and all greater heads.

For submerged crest flow, the corresponding nappe shape as determined from subsection

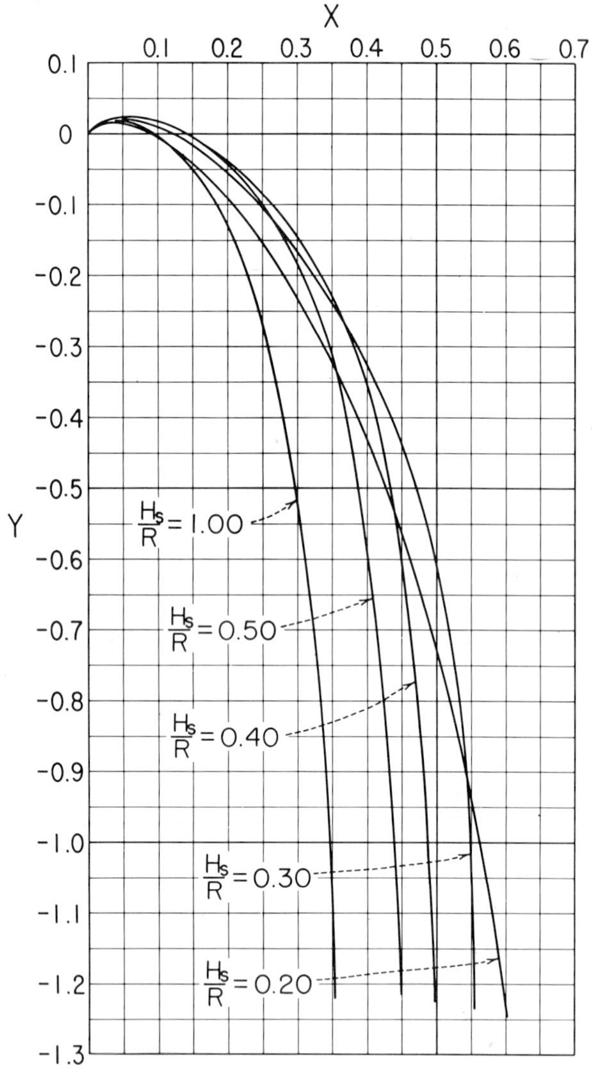

Figure 287. Comparison of lower nappe shapes for a circular weir for different heads. 288–D–2445.

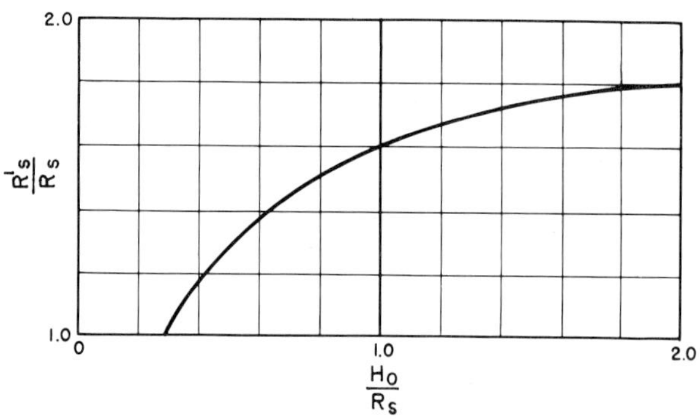

Figure 288. Increased circular crest radius needed to minimize subatmospheric pressure along crest. 288–D–2446.

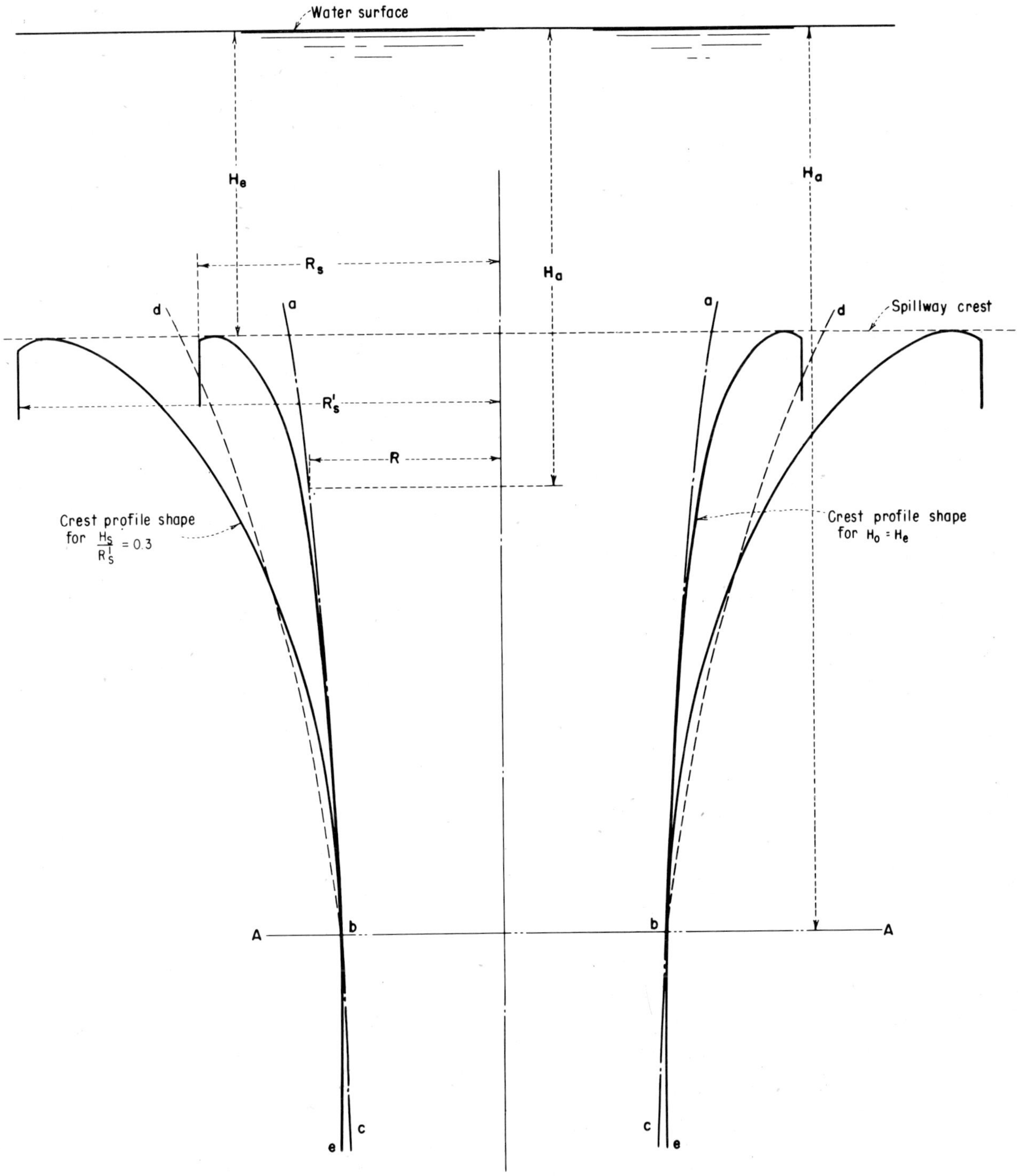

Figure 289. Comparison of drop inlet profiles for various flow conditions. 288–D–2447.

212(c) for a design head H_o will be such that along its lower levels it will closely follow the profile determined from equation (29) if H_e approximates H_o. It must be remembered that on the basis of the losses assumed in equation (29), profile *abc* will be the minimum shaft size which will accommodate the required flow and that no part of the crest shape should be permitted to project inside this profile. As has been noted in section 199, small subatmospheric crest pressures can be tolerated if proper precautions are taken to obtain a smooth surface and if the negative pressure forces are recognized in the structural design. The choice of the minimum crest and transition shapes in preference to some wider shape then becomes a matter of economics, structural arrangement, and layout adaptability.

Where the transition profile corresponds to the continuation of the crest shape as determined by tables 25, 26, and 27, the discharge can be computed from equation (28) using a coefficient from figure 283. Where the transition profile differs from the crest shape profile so that a constricted control section is established, the discharge must be determined from equation (29). On figure 281 the discharge-head relationship curve *ag* will then be computed from the coefficients determined from figure 284 while the discharge-head relationship curve *gh* will be based on equation (29).

(e) *Conduit Design.*—If, for a designated discharge, the conduit of a drop inlet spillway were to flow full below the transition without being under pressure, the required size of the shaft and outlet leg would vary according to the available net head along its length. So long as the slope of the hydraulic gradient which is dictated by the hydraulic losses is flatter than the slope of the conduit, the flow will accelerate and the conduit will decrease in size. When the conduit slope becomes flatter than the slope of the hydraulic gradient, flow will decelerate and the conduit size will increase. All points along the conduit will act simultaneously to control the rate of flow. For heads in excess of that used to proportion the conduit, it will flow under pressure with the control at the downstream end; for heads less than that used to determine the size, the conduit will flow partly full for its entire length and the control will remain in the transition upstream. On figure 281, the head at which the conduit just flows full is represented by point *h*. At heads above point *h* the conduit flows full under pressure; at heads less than *h* the conduit flows partly full with controlling conditions dictated by the transition design.

Because it is impractical to build a conduit with a varying diameter, it is ordinarily made of constant size beyond the inlet transition. Thus, the conduit from the control point in the transition to the downstream end will have an excess of area. If atmospheric pressure can be maintained along the portion of the conduit flowing partly full, it will continue to flow at that stage even though the downstream end fills. Progressively greater discharges will not alter the part full flow in the upper part of the conduit, but full flow conditions under pressure will occupy increasing lengths of the downstream end of the conduit. At the discharge represented by point *h* on figure 281, the full flow condition has moved back to the transition control section and the conduit will flow full for its entire length.

If the conduit flows at such a stage that the downstream end flows full, both the inlet and outlet will be sealed. To forestall siphon action by the withdrawal of air from the conduit would require an adequate venting system. Unless venting is effected over the entire length of conduit, it may prove inadequate to prevent subatmospheric pressures along some portion of the length, because of the possibility of sealing at any point by surging, wave action, or eddy turbulences. Thus, if no venting is provided or if the venting is inadequate, a make-and-break siphon action will attend the flow in the range of discharges approaching full flow conditions. This action is accompanied by erratic discharges, by thumping and vibrations, and by surges at the entrance and outlet of the spillway.

To avoid the possibility of siphonic flow conditions, the downstream conduit size for ordinary designs (and especially for those for higher heads) is chosen so that it will never flow full beyond the inlet transition. To allow for air bulking, surging, etc., the conduit size is ordinarily selected so that it will not flow more than 75 percent full (in area) at the downstream end at maximum discharge. Under this limitation, air will be able to pass up the conduit from the downstream portal and thus prevent the formation of subatmospheric pressure along the conduit length. Precautions must be taken, however, in selecting vertical or horizontal curvature of the conduit profile and alinement to prevent sealing along some portion by surging or wave action.

(f) *Design Example.*—The following example illustrates the procedure for designing a morning-glory type of drop inlet spillway. Consider that an ungated drop inlet spillway which is to operate under a maximum surcharge head of 10 feet, but which must limit the outflow to 2,000 second-feet, is to be designed. Determine alternate crest shapes and discharge head relationships, considering (1) that the crest radius must be minimized because the intake is formed as a tower away from the abutment and that subatmospheric pressures along the overflow crest can be tolerated, and (2) that the crest can be of any size because it is located on a knoll at the abutment and that it is desired to keep subatmospheric pressures along the crest to a minimum. In either case the conduit must not flow more than 75 percent full at the downstream end. The controlling dimensions are as shown on figure 290.

Case 1.—Radius of crest to be minimized and subatmospheric pressures may be tolerated.

Assume $\frac{P}{R_s} \geq 2$ (see fig. 282). The value of R_s is determined by means of a trial and error procedure of assuming values of R_s and computing the discharge.

Assume $R_s = 7.0$ feet. $\frac{H_o}{R_s} = \frac{10}{7} = 1.43$. For $\frac{H_o}{R_s} = 1.43$ and $\frac{P}{R_s} \geq 2$, from figure 283, C_o—1.44. Then $Q = C_o(2\pi R_s)H_o^{3/2} = 1.44\ (2\pi)\ (7.0)\ (10)^{3/2} = 2{,}010$ second-feet, which is approximately the required discharge. From figure 285 for $\frac{H_o}{R_s} = 1.43$ and $\frac{P}{R_s} \geq 2, \frac{H_s}{H_o} = 1.014$; H_s is equal to $1.014\ H_o = 1.014(10) = 10.14$ feet. Then $\frac{H_s}{R_s}$ is equal to $\frac{10.14}{7.0} = 1.45$.

Using table 25, points on the profile of the crest shape which conforms to the lower nappe surface for an $\frac{H_s}{R_s} = 1.45$ are computed by interpolation and plotted as shown on figure 291.

The next step is to determine the transition shape which will be required to pass 2,000 second-feet with an H_o of 10 feet above the crest (water surface elevation 110). This shape is determined by the use of equation (29):

$$R = 0.204\frac{Q^{1/2}}{H_a^{1/4}} = 0.204\frac{(2{,}000)^{1/2}}{H_a^{1/4}} = \frac{9.12}{H_a^{1/4}}$$

Points on the transition are computed as shown in the following table and plotted on the same sheet on which points for the crest shape have already been plotted (fig. 291).

Elevation of section	H_a	$H_a^{1/4}$	$R = \frac{9.12}{H_a^{1/4}}$
100	10	1.78	5.13
98	12	1.86	4.90
96	14	1.93	4.72
94	16	2.00	4.56
92	18	2.06	4.43
88	22	2.17	4.20

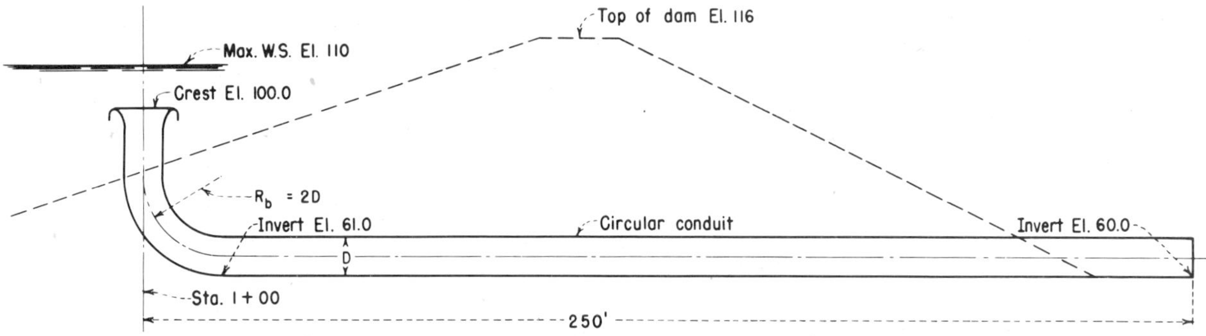

Figure 290. Drop inlet spillway—profile for illustrative example. 288–D–2448.

A smooth curve is drawn through the controlling points on the crest and transition shapes, determining the final shape of the crest and transition.

The last step is to determine the minimum uniform conduit diameter which will pass the flow from the transition section to the conduit portal without the conduit flowing more than 75 percent full. The procedure is as follows: (1) Select a trial conduit and throat diameter and find the corresponding throat location, (2) compute the length from transition throat to outlet portal, (3) approximate the friction losses in the conduit by assuming the conduit flows three-fourths full for its entire length, and (4) check the elevation of the invert at the outlet portal required to pass the design discharge through the selected size conduit. After an approximate conduit size has been determined in this manner, it should be checked by computing the water surface profile through the conduit by means of open channel flow computations.

For this problem assume a conduit diameter of 9.0 feet. From figure 291, a radius of 4.5 feet is found to be at 6.9 feet below the crest. Thus the elevation of the 9.0-foot-diameter throat is 93.1. The tunnel length may be scaled or calculated by approximate methods. In this example the approximate tunnel length is 270 feet. Assuming that the conduit flows 75 percent full,

$$\text{Area}=0.75(\pi)(4.5^2)=47.7 \text{ square feet}$$

$$\text{Velocity}=\frac{2{,}000}{47.7}=41.9 \text{ feet per second}$$

$$h_v=\frac{41.9^2}{64.4}=27.3 \text{ feet}$$

From table *B–3* (appendix B), for 75 percent full flow, $\frac{d}{D}=0.702$ and $r=0.2964(9.0)=2.67$. Using a value of n of 0.018 to maximize the losses, by Manning's equation (equation (*30*), appendix B),

$$s=\left(\frac{vn}{1.486r^{2/3}}\right)^2=\left[\frac{(41.9)(0.018)}{(1.486)(2.67)^{2/3}}\right]^2=0.07,$$

and $h_f=0.07\times270=18.9$ feet.

The invert elevation at the downstream portal of the conduit will then be equal to (1) the elevation of the throat, plus (2) the velocity head at the throat, minus (3) the velocity head in the conduit flowing 75 percent full, minus (4) the friction losses in the conduit, minus (5) the depth of flow at the downstream portal. The required portal invert elevation for this trial conduit diameter is approximately:

$$\text{Invert elevation}=93.1+\frac{1}{1.1}(110-93.1)-27.3-18.9-0.702(9.0)$$
$$=93.1+15.4-27.3-18.9-6.3=56.0$$

Since the outlet portal invert elevation required for the 9.0-foot-diameter conduit is lower than the established elevation 60.0, this conduit is too small and a larger size must be selected.

Assume a new trial conduit diameter=9.50 feet at throat elevation 96.4.

The approximate tunnel length equals 273 feet. Assuming that the conduit flows 75 percent full,

$$\text{Area}=0.75(\pi)(4.75)^2=53.2 \text{ square feet}$$

$$\text{Velocity}=\frac{2{,}000}{53.2}=37.6 \text{ feet per second}$$

$$h_v=\frac{(37.6)^2}{64.4}=21.9 \text{ feet}$$

$$\frac{d}{D}=0.702 \text{ and } r=0.2964(9.5)=2.82.$$

Using Manning's equation,

$$h_f=273\left[\frac{(37.6)(0.018)}{1.486(2.82)^{2/3}}\right]^2=14.2 \text{ feet}$$

The required portal invert elevation for a 9.50-foot conduit diameter is approximately:

$$\text{Elevation}=96.4+\frac{1}{1.1}(110-96.4)-21.9-14.2-0.702(9.5)$$
$$=96.4+12.4-21.9-14.2-6.7=66.0$$

Although this elevation is somewhat higher than the established portal invert elevation 60, actual losses through the conduit will be larger than those estimated on the basis of the conduit flowing 75 percent full throughout its length.

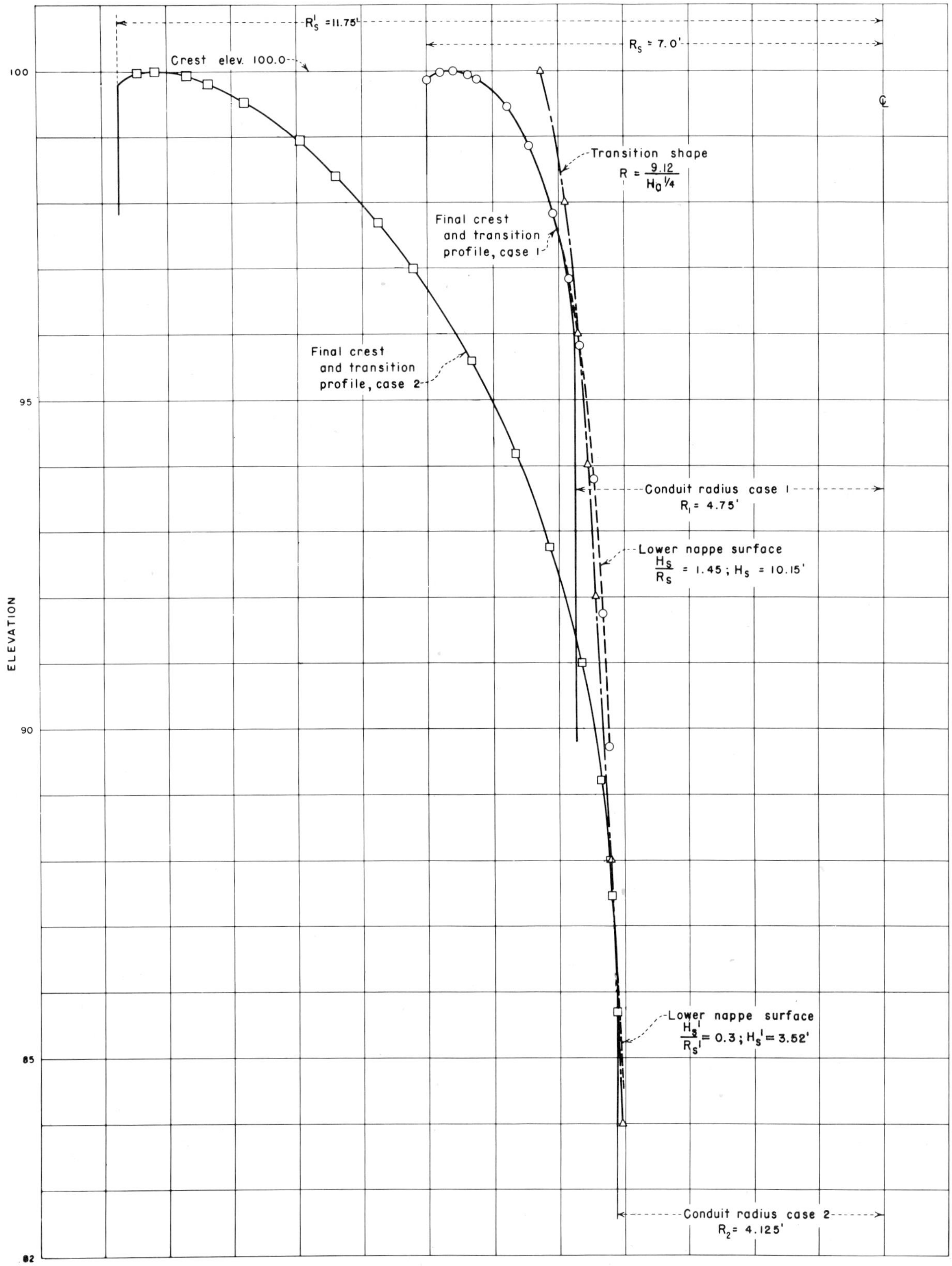

Figure 291. Drop inlet crest, transition, and conduit plottings for illustrative example. 288–D–2515.

TABLE 28.—*Water surface profile computations for case 1 (9.5-foot-diameter conduit)*

[Q=2,000 second-feet. n=0.018]

Station	ΔL	Trial d/D	d	a	v	h_v	r	$r^{2/3}$	s	$\frac{s_1+s_2}{2}$	Δh_L	$\Sigma\Delta h_L$	$d_2+h_{v_2}+\Sigma\Delta h_L$	Invert elevation	Datum gradient	Remarks
1+00		1.00		70.9	28.2	12.4	2.37	1.78	0.037					96.4	108.8	
1+19	42	0.56	5.3	40.8	49.0	37.3	2.54	1.86	.102	0.070	2.9	2.9	45.5	61.0	106.5	Too low.
		.55	5.2	40.0	50.0	38.8	2.52	1.85	.107	.072	3.0	3.0	47.0		108.0	OK.
2+30	111	.60	5.7	44.4	45.1	31.6	2.64	1.91	.082	.095	10.5	13.5	50.8	60.5	111.3	Too high.
		.61	5.8	45.3	44.2	30.3	2.66	1.92	.078	.092	10.2	13.2	49.3		109.8	OK.
3+50	120	.70	6.6	53.0	37.7	22.1	2.81	1.99	.053	.066	7.9	21.1	49.8	60.0	109.8	Too high.
		.71	6.7	53.8	37.2	21.5	2.83	2.00	.051	.065	7.8	21.0	49.2		109.2	OK.

TABLE 29.—*Computations for discharge curve for case 1* R_s=*7.0 feet*

Head on crest, feet	Crest control			Throat control	
	$\frac{H_e}{R_s}$	C*	$Q=C(2\pi R_s)H_e^{3/2}$	H_a	$Q=\left(\frac{R}{0.204}\right)^2 H_a^{1/2}$
1	0.14	3.56	157		
2	.29	3.75	368		
3	.43	3.58	820		
4				7.6	1,500
6				9.6	1,680
8				11.6	1,850
10				13.6	2,000

*Coefficient of 3.75 assumed for $\frac{H_e}{R_s}$=0.3 (from fig. 283) Coefficients for $\frac{H_e}{R_s}$ values other than 0.3 based on ratios shown on fig. 284.

Therefore, the 9.50-foot-diameter conduit appears to be, for all practical purposes, the minimum uniform diameter conduit meeting the requirements of the problem. Computations of the water surface profile through the 9.50-foot-diameter conduit shown in table 28 are then performed in order to verify the approximate solution given above. These computations are based on Bernoulli's theorem (equation (32), app. B).

Discharge-head computations for this design are shown in table 29. For the lower range of heads the coefficient relationships of various $\frac{H_e}{R_s}$ values are obtained from figure 284, assuming a coefficient of 3.75 for an $\frac{H_e}{R_s}$ value of 0.3. For the higher ranges of head the discharges can be obtained from equation (29) using a throat radius of 4.75 at elevation 96.4. Smooth curves are then plotted for both head range computations. The intersection of the curves is replaced by an approximate transition curve to more nearly represent actual conditions. The discharge curve is plotted on figure 292. The computations show that the conduit will be only 76 percent full at the downstream end; therefore the design is satisfactory.

Case 2.—Radius of crest to be determined to minimize subatmospheric pressures along crest.

First determine the minimum crest radius for the given H_o=10 feet and Q=2,000 second-feet similar to case 1. Assume $\frac{P}{R_s}$=0.15 and, as in case 1, by trial and error determine value of R_s.

Assume R_s=7.0 feet, $\frac{H_o}{R_s}=\frac{10}{7}$=1.43. For $\frac{H_o}{R_s}$=1.43 and $\frac{P}{R}$=0.15 from figure 283, C_o=1.55. Then $Q=C_o(2\pi R_s)H_o^{3/2}$=1.55 (2π) (7.0) $(10)^{3/2}$=2,155 second-feet. Since a 2,000-second-foot discharge is required, the assumed value of R_s is too large.

Assume R_s=6.7 feet. $\frac{H_o}{R_s}=\frac{10}{6.7}$=1.49. From figure 283 C_o=1.49 and Q=1,985 second-feet, which is approximately the required discharge.

Using the value of $\frac{H_o}{R_s}$ thus computed, enter figure 288 and find the approximate increased crest radius required to minimize subatmospheric pressures. For $\frac{H_o}{R_s}$=1.49, $\frac{R'_s}{R_s}$=1.74, and R'_s=1.74(6.7)=11.7 feet; use 11.75 feet. Points on the profile of the crest shape which conform to the lower nappe surface for an $\frac{H'_s}{R'_s}$=0.30 and R'_s=11.75 are computed using values from table 27 and are plotted as shown on figure 291.

Computations for the required transition shape to pass 2,000 second-feet with a head of 10 feet on the crest are identical to those given in case 1. Figure 291 shows the plotted points and the crest and transition curves.

From an inspection of the plotting of the transition and crest shapes for case 2, it can be seen that the conduit diameter for case 1 is too large for case 2. If the 9.5-foot-diameter conduit used in case 1 were used in case 2, a smooth transition connecting the crest and conduit would be considerably outside the transition shape determined by equation (29). This means that for a head of 10 feet on the crest, the discharge would no longer be limited to 2,000 second-feet by the transition but would increase because of the larger size transition. The resulting discharge would be approximately 2,400 second-feet. This discharge not only exceeds the maximum discharge assumed in the problem, but also would require a larger uniform-diameter conduit in order to pass the discharge and not flow more than 75 percent full. A still larger uniform-diameter conduit with a still larger maximum discharge would finally have to be adopted to obtain a satisfactory hydraulic design. If, however, a smaller uniform-diameter conduit is chosen, it is apparent that the conduit will flow more than 75 percent full at the downstream end.

The simplest solution to this problem is to vary the diameter of the conduit. An upstream diameter is chosen based on the crest profile and transition where they converge. This procedure establishes a throat of the size necessary to limit the maximum discharge to 2,000 second-feet. At some suitable location downstream from the throat, the conduit is enlarged to prevent it from flowing more than 75 percent full. Location of the enlargement is determined by economic or construction considerations to meet hydraulic requirements.

For this problem, a conduit diameter of 8.25 feet is selected with the throat at elevation 86.0. It will be assumed that the most economical design is obtained by extending the 8.25-foot-diameter conduit to the point where it flows 75 percent full, at which point the conduit is enlarged to the diameter needed to make it flow 75 percent full at the downstream portal. In order to determine the point at which the tunnel must be enlarged, water surface profiles are run downstream by the step method as shown in table 30. A bend radius of 16.5 feet ($2D$) is used. The table shows that the conduit must increase in size starting at the P.T. of the vertical bend, station 1+16.5. The downstream size of the conduit may be approximated by assuming a given size conduit flowing 75 percent full at the downstream portal and using the distance from enlargement to portal as one reach in the water surface profile computations. Although this method results in losses slightly larger than would be obtained

TABLE 30.—*Water surface profile computations for case 2 (varying diameter conduit)*

[Q=2,000 second-feet. n=0.018]

Station	ΔL	Trial d/D	d	a	v	h_v	r	$r^{2/3}$	s	$\frac{s_1+s_2}{2}$	Δh_L	$\Sigma\Delta h_L$	$d_2+h_{v_2}+\Sigma\Delta h_L$	Invert elevation	Datum gradient	Remarks
1+00		1.00		53.5	37.4	21.7	2.06	1.62	0.078					86.0	107.7	
1+16.5	30	0.6	5.0	33.5	59.7	55.4	2.29	1.74	.173	0.125	3.8	3.8	64.2	61.0	125.2	
		.7	5.8	40.0	50.0	38.9	2.44	1.81	.112	.095	2.9	2.9	47.6		108.6	OK.
Try 9.50-foot-diameter conduit flowing 75 percent full at the portal																
3+50	234	.702	6.7	53.1	37.7	22.1	2.81	1.99	.052	.082	19.2	22.1	50.9	60.0	110.9	Too high.
Try 9.75-foot-diameter conduit flowing 75 percent full at the portal																
3+50	234	.702	6.8	56.0	35.7	19.8	2.89	2.03	.045	.078	18.3	21.2	47.8	60.0	107.8	OK.

TABLE 31.—*Computations for discharge curve for case 2. R_s'=11.75 feet*

Head on crest, feet	Crest control: $\frac{H_e}{R'_s}$	Crest control: C*	Crest control: $Q=C(2\pi R'_s)H_e^{3/2}$	Throat control: H_a	Throat control: $Q=\left(\frac{R}{0.204}\right)^2 H_a^{1/2}$
1	0.09	3.55	260		
2	.17	3.74	780		
3	.26	3.85	1,480	17	1,680
4	.34	3.82	2,260	18	1,730
6				20	1,830
8				22	1,920
10				24	2,000

*Coefficient of 3.86 assumed for $\frac{H_e}{R_s}=0.3$ (from fig. 283). Coefficients for $\frac{H_e}{R_s}$ values other than 0.3 based on ratios shown on fig. 284.

by using shorter reaches, it is sufficiently accurate to determine conduit size if the length of the conduit downstream from the expansion is not excessively long. Use of shorter steps and an assumed minimum value of n would be required to determine the depth and velocity at the downstream portal for use in designing an energy dissipator. The transition from the smaller to the larger diameter conduit should be proportioned as explained in section 204(b).

Discharge-head relationships for this case are computed similarly to those for case 1. The throat radius in this instance is 4.13 feet at elevation 86.0. Computations are shown in

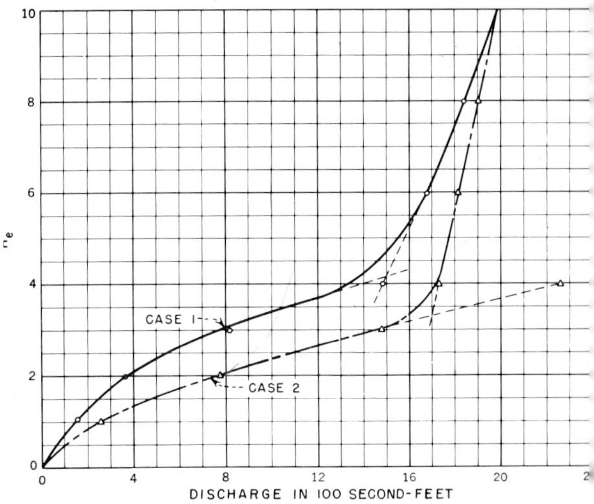

Figure 292. Drop inlet spillway—discharge curves for illustrative example. 288–D–2516.

table 31. This discharge curve is plotted on figure 292.

213. *Culvert Spillways.*—(a) *General.*—As described in section 193(h), the culvert spillway ordinarily consists of a simple culvert conduit placed through the dam or along the abutment, generally on a uniform grade with the entrance placed vertically or inclined. The culvert cross section might be round if it is constructed of fabricated or precast pipe, or i

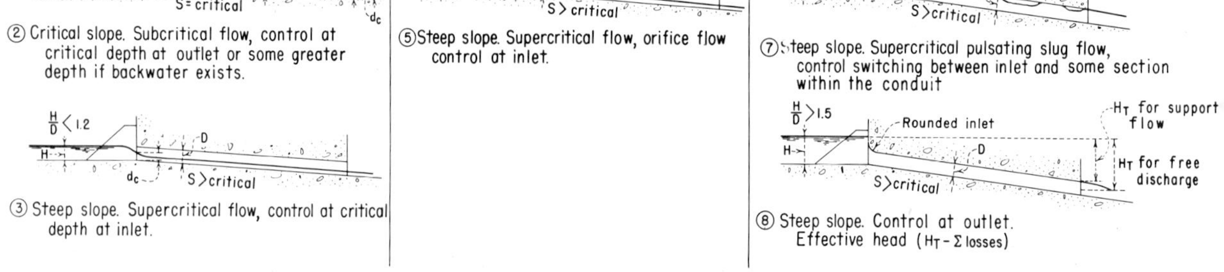

Figure 293. Typical flow conditions—culvert spillways on mild and steep slopes. 288–D–2517.

might be square, rectangular, or of some other shape if cast in place. The culvert can freely discharge, or it can empty into an open channel so that the outflowing jet is supported along the channel floor.

The factors which combine to determine the nature of flow in a culvert spillway include such variables as the slope, size, shape, length, and roughness of the conduit barrel, and the inlet and outlet geometry. The combined effect of these factors determines the location of the control which in turn determines the discharge characteristics of the conduit. The location of the control dictates whether the conduit flows partly full or full and thereby establishes the head-discharge relationship.

The grade of the conduit might be mild or steep; that is, its slope may be flatter or steeper than one which for a given discharge will just support flow at the critical stage. For either the mild or steep slope the control may be either at the inlet or the outlet, depending on the entrance geometry and head relationship and on the flow conditions at the outlet. The various conditions which may govern a particular flow are illustrated on figure 293.

If the inlet is not submerged the control for a conduit on a mild slope flowing partly full will usually be at the outlet. If the outlet discharges freely the flow at that point will pass through critical depth. This condition is shown as ① in figure 293. If the tailwater is

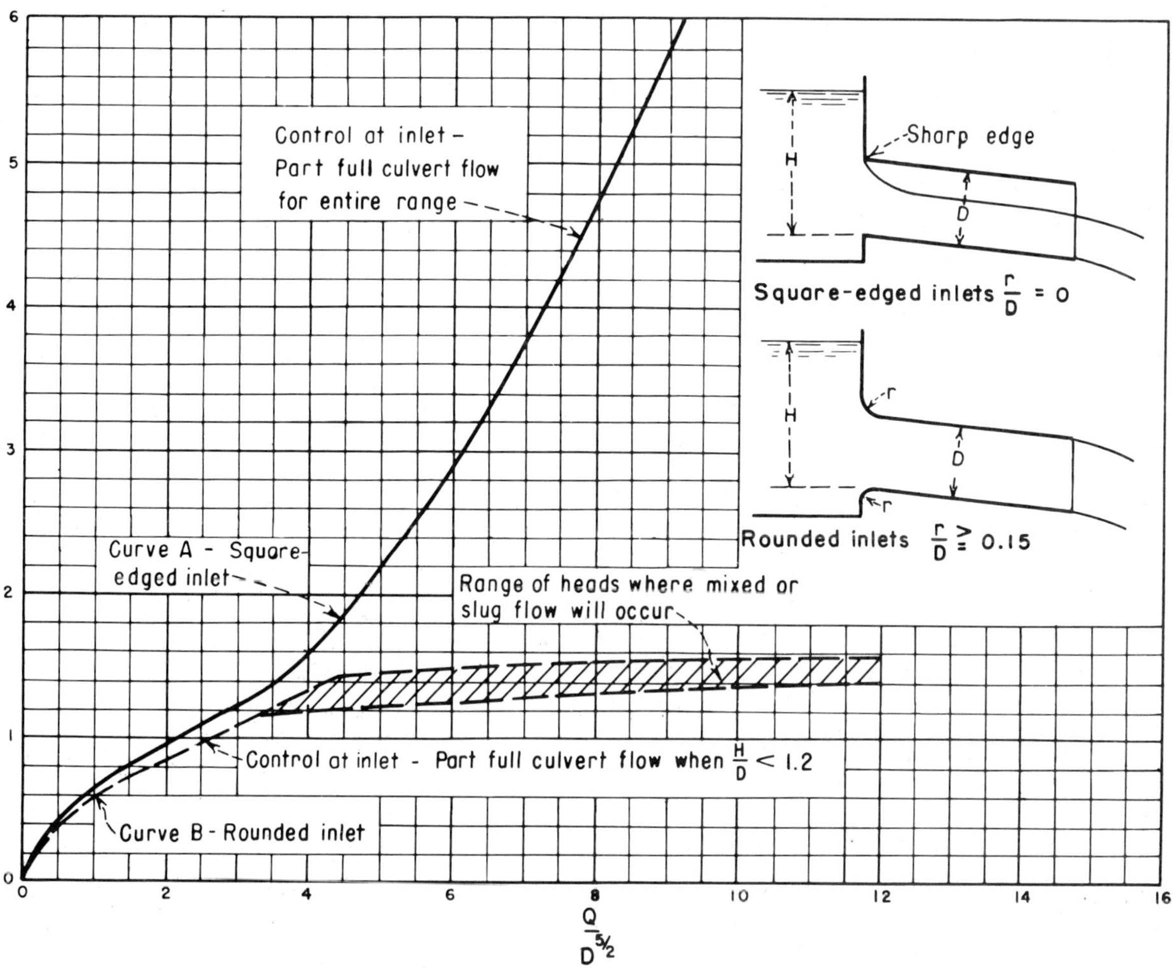

Figure 294. Head-discharge curves for square-edged and rounded inlets for circular culverts on steep slopes. 288–D–2518.

high enough to maintain a depth greater than critical, the tailwater level will control the flow in the upstream barrel. If the tailwater submerges the outlet, the conduit might flow full for its entire length and thus submerge the inlet. This flow condition is depicted as ⑥ on figure 293. Until the conduit flows full, the flow ordinarily will be at subcritical stage and the discharge relationships will be determined according to the Bernoulli equation with computations starting at the outlet where the reservoir level submerges the inlet and where $\frac{H}{D}$ exceeds 1.2, the control at critical depth may occur at the inlet, if the culvert is relatively short so that a jump does not form within the barrel. This condition is illustrated as ④.

When the conduit is on a steep slope and the entrance is not submerged, the flow will be controlled by critical depth at the inlet, as indicated by condition ③ on figure 293. The water surface will drop rapidly to critical depth at the entrance, and open channel flow at supercritical velocities will exist throughout the conduit barrel. Discharge for a given reservoir level will be governed by channel or weir flow, assuming critical depth to occur at the culvert entrance.

After the inlet has been submerged or where H exceeds about $1.2D$, it is still possible to have open channel flow at supercritical stage in the conduit barrel, as depicted for condition ⑤, if the control remains at the entrance. In this case flow at the inlet is analogous to orifice or sluice flow. This flow condition is contingent on the formation of a contraction at the top of the entrance so that an airspace is maintained along the top of the barrel to permit part full flow throughout the conduit length.

As the head at the entrance and the resulting discharge increase, channel friction or local disturbances may force the barrel to flow full near the outlet, thus sealing the conduit at the downstream end. The high-velocity flow in the culvert will carry away some of the air trapped at the top of the barrel, thus reducing the pressure in the conduit to less than atmospheric. Furthermore, if the entrance is shaped to eliminate the inlet contraction, the barrel will start to flow full near the inlet, after which the full flow zone will extend rapidly down the conduit toward the outlet. The effect of the full flow condition wil be a draft tube action (similar to siphonic action) which will increase the discharge. The increased discharge will cause a deeper drawdown just upstream from the inlet. A vortex will form and air which will thus be introduced into the culvert will break the draft tube action. The reduction in discharge will result in the return to orifice control at the inlet. Immediately, the full flow action will begin again and the cycle will be repeated. This alternate priming and breaking action will cause a pulsating flow stage with the slug flow phenomenon indicated by condition ⑦ on figure 293. When the reservoir stage condition is such that the $\frac{H}{D}$ ratio exceeds 1.5, the entrance drawdown may be insufficient to interfere with the full flow action, and a steady state of full pipe flow indicated by condition ⑧ will prevail.

If it is intended that the spillway conduit not flow full, the geometry of the inlet becomes an important consideration. The inlet must be shaped to obtain a maximum discharge efficiency and yet maintain a top contraction which will provide a freely aerated surface in the conduit barrel for all reservoir stages. The sharp-edged square inlet produces the desired contraction without materially reducing the discharge capacity. The inlet contraction can also be formed (but at reduced hydraulic capacity) by a projecting inlet, a mitered inlet with a downstream sloping face, an inlet orifice ring which is made smaller than the remainder of the conduit, or a curtain wall closing off the top of the conduit entrance.

If the conduit is permitted to flow full at the higher reservoir stages, the control will be at the outlet and the geometry of the inlet has much less significance. For this case the inlet must be shaped to minimize the jet contraction to avoid separation of the incoming flow from the conduit barrel, since it is desired to obtain full pipe flow for all conditions except when the inlet is not submerged. The more streamlined shape will reduce entrance losses for the full pipe flow condition. The suppression of the contraction is achieved by rounding the inlet or by providing a gradually tapering transition to the conduit barrel.

Culvert inlets may have many variations in approach conditions, in cross-sectional shapes, and in entrance arrangements. For example, an entrance may be rounded, beveled, square or bellmouthed; it may be installed either flush with or protruding through a vertical or sloping headwall. The approach to the inlet may or may not be a well-defined channel. Wing walls or warped transition approaches may be utilized. In cross section, a culvert entrance may be round, square, rectangular, or arch shaped. All such variations have a marked bearing on the culvert performance, as they affect orifice discharge, inlet contractions, and the entrance losses for full pipe flow.

A common arrangement for a circular pipe culvert installation provides for a vertical headwall with the pipe end placed flush with the wall. Similarly, box culvert arrangements usually involve a trapezoidal approach channel with vertical or warped approach walls leading to the culvert entrance. The hydraulic design of these two types of installation is discussed in detail.

(b) *Circular Conduit with Vertical Headwall.*—Figure 294 shows head-discharge relationships for a circular conduit placed flush with a vertical headwall, for both square-edged and rounded inlets. This plotting is based on an average of numerous experimental tests [16, 17, 18, 19] of pipe culvert entrances with the conduit placed on steep slopes. The head-discharge relationships for the square-edged inlet are based on the control remaining at the inlet for all reservoir heads. Where the $\frac{H}{D}$ values are less than about 1.2, the flow characteristics are those of critical depth flow in a circular pipe, modified only by the effects of the jet contraction. For $\frac{H}{D}$ ranges above 1.2 the flow characteristics are those of orifice or sluice flow. Since the conduit is considered to be flowing partly full at supercritical stage for all $\frac{H}{D}$ ranges indicated, the downstream conditions will have no influence on the discharge.

On figure 294 the head-discharge relationship for the rounded inlet for values of $\frac{H}{D}$ less than about 1.2 lie to the right of those for the square-edged inlet, indicating slightly greater discharges for equal size conduits. The increased discharge capacity through the critical-depth flow range is the result of improved streamlined flow brought about by the suppression of the inlet contractions. For $\frac{H}{D}$ values exceeding 1.2 the pulsating flow characteristics will begin and the discharge-head relationship in this range of flow will be uncertain; it cannot be determined until the flow stabilizes as full flow stage. Since full pipe flow is governed by control at the outlet, the head-discharge relationship can be determined by the application of Bernoulli's theorem. Referring to figure 295,

$$H_T=H+L\sin\theta-\frac{D}{2} \tag{30}$$

Similarly,

$$H_T=h_v+h_e+h_f, \text{ or}$$

$$H_T=\left(1+K_e+f\frac{L}{D}\right)\frac{v^2}{2g} \tag{31}$$

where K_e is the entrance loss coefficient and f is the friction loss coefficient in the Darcy-Weisbach formula, $h_f=f\frac{L}{D}\frac{v^2}{2g}$ (see secs. 234 and 235).

Combining equations (30) and (31), dividing by D, and stating the equation in terms of Q instead of v,

$$\frac{H}{D}+\frac{L}{D}\sin\theta-\frac{1}{2}=0.0252\left(1+K_e+f\frac{L}{D}\right)\left(\frac{Q}{D^{5/2}}\right)^2 \tag{32}$$

In the above equation, it is assumed that the culvert discharges freely at the outlet and that the pressure line at the outlet is approximately at the center of the pipe. If the outlet discharges into a channel so that the outflowing jet is supported, equation (32) becomes:

$$\frac{H}{D}+\frac{L}{D}\sin\theta-1.0=0.0252\left(1+K_e+f\frac{L}{D}\right)\left(\frac{Q}{D^{5/2}}\right)^2 \tag{33}$$

Equations (32) and (33) are for full flow conditions. They are expressed in terms of $\frac{H}{D}$ and $\frac{Q}{D^{5/2}}$ so that by referring to figures 293 and 294 it can be determined whether or not the full flow condition exists.

Where friction losses are to be based on values of Manning's friction factor, n, the value of f in the Darcy-Weisbach formula can be determined from figure B–7 (app. B). Appropriate values of n are given in section 235(b). Values of K_e for various entrance conditions have been determined by different experimenters, with results ranging in the values shown below:

	Range	*Average*
For square-edged inlets installed flush with vertical headwalls.	0.43 to 0.70	0.50
For rounded inlets installed flush with vertical headwalls, $\frac{r}{D} \geq 0.15$	0.08 to 0.27	.10
For grooved or socket-ended concrete pipe installed flush with vertical headwall.	0.10 to 0.33	.15
For projecting concrete pipe with grooved or socket ends.		.20
For projecting steel or corrugated metal pipes.	0.5 to 0.9	.85

Nomographs for determining flow in circular pipes with entrance control and flowing full, which have been developed by the U.S. Bureau of Public Roads, are included in appendix B. These can be used as design aids in determining flow in circular culvert spillways. Figure B–8 is for flow in concrete pipe culverts having entrance control and the following types of entrances: (1) headwall with square-edge entrance, (2) headwall with groove-end pipe, and (3) headwall with groove end of pipe projecting. Figure B–9 is for flow in corrugated metal pipe culverts having entrance control and the following types of entrances: (1) flush headwall, (2) end mitered to conform to slope, and (3) projecting pipe. Figure B–10 is for concrete pipe culverts flowing full, based on a value of $n=0.012$ and entrance loss coefficients of 0.1, 0.2, and 0.5. Figure B–11 is for corrugated metal pipe flowing full, based on a value of $n=0.024$ and entrance loss coefficients of 0.5 and 0.9.

(c) *Box Culvert with Vertical or Warping Inlet Walls.*—If the inlet is such that the bottom and side contractions will be suppressed, flow through a box culvert on a steep slope can alternately go through the three distinct phases of flow described previously, depending on submergence conditions and on the factors which dictate flow conditions within the conduit.

For conditions when the inlet is not submerged, critical flow will occur in the region of the inlet, in which case for a rectangular section,

$$d_c=\sqrt[3]{\frac{q^2}{g}} \text{ or } H=1.5\sqrt[3]{\frac{q^2}{g}}.$$

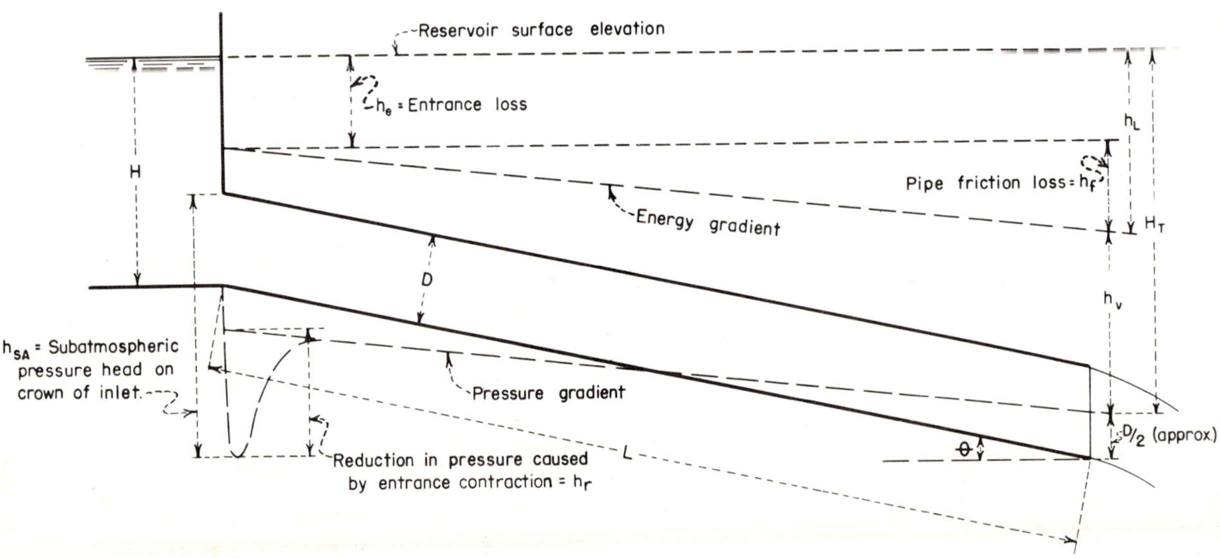

Figure 295. Hydraulic characteristics of full pipe flow for culvert spillways. 288–D–2519.

Relating this equation of critical flow to the discharge Q,

$$Q=w\sqrt{g}\left(\frac{H}{1.5}\right)^{3/2} \qquad (34)$$

where w is the width at the culvert entrance.

When the conduit entrance becomes submerged, the flow may be considered analogous to that of a sluice if the entrance has a square edge at the top. For this condition, top contraction of the jet will occur, and flow can be computed according to orifice flow, or $Q=CA\sqrt{2gh}$. The coefficient C depends on whether the area A is defined as the area of the opening, the area of the contracted jet, or some similar referenced area. Similarly, the coefficient will depend on the definition of the head h, whether it is measured to the top, center, or bottom of the opening. Ordinarily, for a square-edged orifice in a vertical headwall the area is measured at the plane of the headwall face (designated as a). If the head is measured from the water surface to the bottom of the opening (designated as H), the discharge can be computed by the equation:

$$Q=C_d a\sqrt{2gH}, \text{ or } Q=C_d wD\sqrt{2gH} \qquad (35)$$

where D is the height of the opening. Values of C_d as determined from experiments [20] are plotted on figure 296.

As with circular culverts, full flow in box culverts depends on suppression of the top contraction. Full culvert flow will be governed by control at the outlet, and discharge-head relationships can be computed according to the equation:

$$Q=a\sqrt{2g(H_T-h_L)} \qquad (36)$$

where a is the area of the culvert barrel and H_t and h_L are heads as indicated on figure 295.

Reducing the equation and expressing it in terms of the entrance loss coefficient, K_e, and of the friction loss coefficient, Manning's n,

$$Q=a\sqrt{2g}\left[\frac{H+L\sin\theta-\frac{D}{2}}{1+K_e+\frac{29.1n^2L}{r^{4/3}}}\right]^{1/2} \qquad (37)$$

where r is the hydraulic radius of the culvert flowing full. The above equation is based on free discharge at the outlet. If the outflowing jet is supported, equation (37) will become:

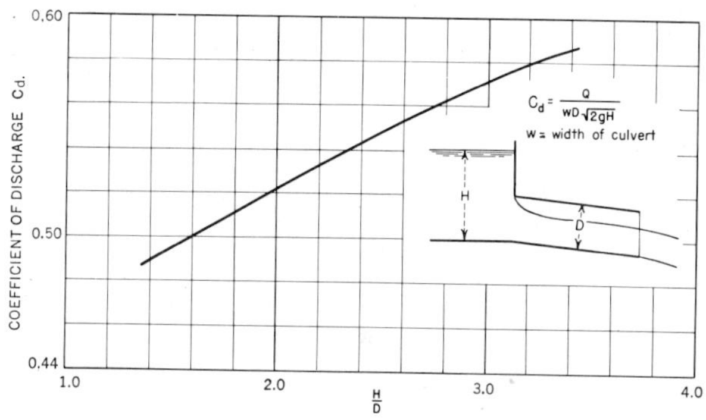

Adapted from reference [20]

Figure 296. Coefficient of discharge for submerged box culvert spillway with square-edged top opening. 288–D–2520.

$$Q=a\sqrt{2g}\left[\frac{H+L\sin\theta-D}{1+K_e+\frac{29.1n^2L}{r^{4/3}}}\right]^{1/2} \qquad (38)$$

Bureau of Public Roads nomographs for solution of flow in box culverts are also included in appendix B. Figure B–12 is for box culverts with entrance control for various positions of the wingwalls. The discharges are based on discharge coefficients which approximate those shown on figure 296. Figure B–13 is for concrete box culverts flowing full, based on a value of $n=0.013$, and entrance loss coefficients of 0.1, 0.2, 0.5, and 0.7.

(d) *Conduit Pressures.*—When the grade of a culvert spillway is greater than the friction slope, for full pipe flow the pressure gradient will lie below the center of the pipe, as indicated on figure 295. The difference in head between this hydraulic gradient and any point on the pipe vertically above it will be the measure of the subatmospheric pressure that will exist at the point. Cavitation will occur when the subatmospheric pressure approaches one atmosphere, so that the residual absolute pressure is near vapor pressure. Collapsing can occur whenever the residual pressure in the pipe is less than atmospheric. To avoid cavitation along the pipe surfaces, the minimum absolute pressure must be limited to some value greater than vapor pressure. The pressure reduction in the pipe will be greatest at the crown immediately downstream from the entrance. It can be reduced further by any pressure drop caused by an inlet contraction, such as a sharp-edged or constricted opening.

From figure 295 it can be seen that

$$h_v+h_e+h_r=h_{SA}+(H-D) \qquad (39)$$

where:

h_r=the reduction of pressure head due to contraction, and

h_{SA}=the resulting subatmospheric pressure head.

Vapor pressure of water varies with temperature. It is equivalent to about 0.2 foot of head at 32° F. and about 1.4 feet of head at 85° F. To make sure that cavitation is avoided and to allow for other uncertainties, the residual pressure ordinarily should not be significantly less than 10 feet absolute. Based on probable minimum atmospheric pressures at different elevations above sea level, the limiting subatmospheric pressures indicated in table 32 are recommended.

TABLE 32.—*Allowable subatmospheric pressures for conduits flowing full*

Elevation above sea level	Allowable subatmospheric pressure, h_{SA}, in feet of water
0	22
2000	20
4000	18
6000	16
8000	14

The reduction in pressure head due to jet contraction will depend on the geometry of the inlet. For streamlined entrances very little reduction will be effected, but for sharp-edged projecting inlets the reduction can be almost equal to the velocity head. For sharp-edged square inlets the reduction in pressure may approach $0.7h_v$. Equation (39) written in terms of loss coefficients (sec. 234) will be:

$$\frac{v^2}{2g}(K_v+K_e+K_r)=h_{SA}+(H-D), \text{ or} \qquad (40)$$

$$\frac{v^2}{2g}=h_v=\frac{h_{SA}+(H-D)}{K_v+K_e+K_r} \qquad (41)$$

where K_r is the pressure reduction coefficient.

For a square-edged entrance where K_e=0.5, K_r=0.7, K_v=1.0, H=1.5D, and h_{SA} (for an installation at 6,000 feet above sea level) is 16 feet, equation (41) will be written:

$$h_v=\frac{16+0.5D}{1.0+0.5+0.7}=\frac{16+0.5D}{2.2}$$

For a 4-foot-diameter conduit, the velocity head is 8.2 feet and the velocity in the conduit would have to be limited to about 23 feet per second. From equation (31) the total drop from the reservoir water surface to the centerline of the downstream end of a 200-foot-long conduit with an f value of 0.023 for D=4 (figure B–7, appendix B, for a value of n=0.014) is:

$$H_T=8.2\left(1.5+f\frac{L}{D}\right), \text{ or}$$

$$H_T=8.2\left(1.5+\frac{0.023\times 200}{4.0}\right)=21.7 \text{ feet.}$$

(e) *Antivortex Devices.*—Although experimental tests have shown that for a properly rounded entrance the culvert begins to flow full after the $\frac{H}{D}$ ratio exceeds 1.2, the full pipe flow condition could not be stabilized until an $\frac{H}{D}$ value of 1.5 or higher was reached. This condition was due to the "slug flow action" which resulted from the introduction of air into the conduit by entrance drawdown and by vortices immediately upstream from the inlet. To reduce the range where slug flow action prevails, antivortex devices have been used above conduit entrances. These devices not only stabilize the flow condition at a lower $\frac{H}{D}$ ratio, but they also help to start the priming action sooner. The devices have consisted of grillages, rafts, or fixed solid hoods placed so as to break up the vortices or to prevent their formation where they could feed air into the conduit [24]. In order to be effective, the hood or grillage must be placed immediately above the entrance and must extend at least two diameters in front of and to each side of the inlet.

(f) *Energy Dissipators.*—A culvert spillway may discharge freely or it may empty into an open channel chute which then conveys the ow to a downstream terminal structure. The ow from a freely discharging conduit might mpty directly into the natural stream channel, nto a trapezoidal plunge basin described in ection 237(b), or into an impact basin de- cribed in section 209. Where the discharge rom the full flowing culvert empties into an pen chute, the hydraulics beyond the portal ill be according to open channel flow, as dis- ussed in section 204. Stilling devices such as hose described in sections 205 through 210 an be utilized to dissipate the energy of flow efore returning the discharge to the river hannel.

(g) *Design Examples.*—To illustrate the rocedures for a culvert spillway design, sev- al typical examples are presented.

Example 1.—The size of a culvert spillway quired to discharge 100 second-feet at reser- oir elevation 110 is to be determined. The ormal sill level of the spillway entrance is at evation 100. The culvert is to flow partly ll for all heads. If a circular conduit is se- cted, the design procedure is as follows:

The head-discharge-diameter relationship for circular conduit with entrance placed flush ith a vertical headwall can be obtained from gure 294. Curve A is used because the con- it is to flow partly full. By assuming vari- s sizes of conduit, a size can be found which eets the requirements, as follows:

Assume a conduit 3.5 feet in diameter, then $^{/2}=23$ and $\frac{Q}{D^{5/2}}=4.35$. For a $\frac{Q}{D^{5/2}}$ value of 5, $\frac{H}{D}=1.75$ and $H=6.1$. Since an H of 10 et is available, the culvert can be made aller.

As a second trial, assume a 3-foot conduit. en $D^{5/2}=15.6$ and $\frac{Q}{D^{5/2}}=6.41$. From curve $\frac{H}{D}=3.2$ and $H=9.6$, which approximates the feet available.

f a box culvert is selected, the procedure is follows:

$$Q=C_d wD\sqrt{2gH} \text{ (equation (35))}$$

$$wD=\frac{Q}{C_d\sqrt{2gH}}$$

Assuming a 2.5-foot-high culvert, $\frac{H}{D}=\frac{10}{2.5}=4.0$, and C_d from figure 296 is approximately 0.6. Then

$$2.5w=\frac{100}{0.6\times 8.02\times\sqrt{10}}=6.6, \text{ and } w=2.6 \text{ feet.}$$

Example 2.—Find the discharge through the conduits in the previous example if the en- trances are shaped to provide full conduit flow. The conduit length is 200 feet and the invert grade at the outlet is elevation 80.0. The con- duit discharges freely at the outlet end. The procedure is as follows:

Equation (32) may be written:

$$\left(\frac{Q}{D^{5/2}}\right)^2=\frac{1}{0.0252}\left[\frac{\frac{H}{D}+\frac{L}{D}\sin\theta-\frac{1}{2}}{1+K_e+f\frac{L}{D}}\right]$$

For a 3-foot circular conduit with K_e of 0.10 and $f=0.025$ (from figure B-7, appendix B, for value of $n=0.014$),

$$\left(\frac{Q}{15.6}\right)^2=\frac{1}{0.0252}\left[\frac{\frac{10}{3}+\frac{200}{3}\times\frac{20}{200}-\frac{1}{2}}{1+0.10+\frac{0.025\times 200}{3.0}}\right]=136$$

and $\frac{Q}{15.6}=11.67$. Then $Q=182$ second-feet.

This flow will provide a velocity of 25.7 feet per second in the conduit.

Equation (40) may be written as:

$$h_{SA}=\frac{v^2}{2g}(K_r+K_e+K_r)-(H-D).$$

The subatmospheric pressure in the conduit, based on a pressure reduction coefficient, K_r, of 0.1 and a value of K_e of 0.1 for a rounded en- trance is equal to:

$$\frac{(25.7)^2}{64.4}(1.0+0.1+0.1)-(10.0-3.0)=5.3 \text{ feet.}$$

This subatmospheric pressure is less than that allowed in table 32; therefore the design is satisfactory.

For the box culvert spillway, from equation (37), assuming $K_e=0.1$ and $n=0.014$,

$$Q=2.5\times 2.6\times 8.02\left[\frac{10+200\times\frac{20}{200}-1.25}{1+0.1+\frac{29.1\times 200(0.014)^2}{(0.64)^{4/3}}}\right]^{1/2}$$

$$=157 \text{ second-feet.}$$

The velocity in the culvert will then be about 24 feet per second.

214. *Siphon Spillways*.—After it primes, flow in a siphon spillway is comparable to that in a culvert spillway and the hydraulic design criteria for the latter will apply with only slight variations. As with the culvert spillway, the design of the siphon spillway requires (1) the establishment of the minimum allowable absolute pressures, (2) the determination of the maximum corresponding throat velocity, and (3) the evaluation of various losses throughout the closed pipe system.

To determine the pressure conditions at the crest of the throat of the siphon, equation (39) can be rewritten as follows:

$$h_{vs}=h_{SA}+h_s-\Sigma h_{L_u} \qquad (42)$$

where:

h_{vs}=the velocity head at the crest of the throat,

h_{SA}=the subatmospheric pressure head at the crest of the throat (the maximum allowable subatmospheric pressure head is equal to the probable minimum atmospheric pressure head at the location under consideration, minus the minimum tolerable residual pressure head),

h_s=the head measured from the reservoir water level to the crest of the throat, for the condition when the siphon primes, and

Σh_{L_u}=the summation of all head losses upstream from the throat.

Subatmospheric pressure heads in excess of those indicated in table 32 should be avoided at the throat because lower pressures increase the tendency of the flow to part from the crest, forming vapor pressure pockets which induce cavitation along the siphon barrel walls. Such action is accompanied by undesirable vibration.

Flow around the upper bend of a siphon is similar to the action of free vortex flow i which the water stream rotates concentricall around a central axis. For vortex flow, th velocities of the flow elements vary inversel with increased distances from the center (rotation. Expressed mathematically,

$$vR=\text{A constant} \qquad (43$$

Thus it can be seen that the velocity in th upper bend of the siphon will be greatest at th inside of the bend which is at the crest of th throat.

Equation (43) can be written:

$$v_xR_x=v_sR_c \text{ or } v_x=\frac{v_sR_c}{R_x} \qquad (4$$

where v_x is the velocity at R_x distance from th center of curvature, and v_s is the velocity at th crest of the throat where the radius is R_c. Th discharge, dq, of an element of flow of dep dR_x in the bend is equal to v_xdR_x. Substituti the value of v_x from equation (44),

$$dq=\frac{v_sR_c}{R_x}dR_x.$$

Integrating between the limits R_c and R_s, whe R_s is the radius of curvature at the summit the throat, results in the following equation f unit discharge:

$$q=v_sR_c\int_{R_c}^{R_s}\frac{1}{R_x}dR_x=v_sR_c\log_e\frac{R_s}{R_c} \qquad (4$$

Substituting $\sqrt{2gh_{v_s}}$ for v_s, equation (45) c be written:

$$Q=8.02\,B\sqrt{h_{v_s}}R_c\log_e\frac{R_s}{R_c} \qquad (4$$

where B is the width of the throat section.

Thus it can be seen that the maximum allo able discharge for a given value of h_{v_s} will (pend on the curvature of the upper be Once the curvature is established, the rema ing siphon components must be proportion so that the limiting residual pressure head the crest is maintained. Table 32 gives allo able subatmospheric pressures. For simplifi tion, the allowable subatmospheric press can be taken as 0.7 of the probable minim atmospheric pressure, h_{AT}. Equation (46) then may be written as:

$$Q=8.02\, B\sqrt{0.7h_{AT}}R_c \log_e \frac{R_s}{R_c} \qquad (47)$$

The discharge through a siphon depends on the available head from reservoir level to outlet level, less the accumulated head losses, including entrance loss, friction loss, losses due to transitions and bends, and the head lost at the exit. The above can be stated and solved with expressions similar to equations (5) and (7) in chapter X, and by relating the velocity heads for the various component losses to that at the throat. The throat area is used for area a_1 in the equations.

Equation (36) expresses the discharge-head relationship of a closed conduit flowing full, where the area of the passageway is constant. For a rectangular siphon passageway where the areas of the upper and lower legs are the same as that of the throat, equation (36) can be written:

$$q=D\sqrt{2g\,(H_T-h_L)} \qquad (48)$$

where D is the height of the throat and siphon barrel.

Loss coefficients ordinarily assumed for a typical rectangular siphon similar to that shown on figure 245, where the main siphon waterway is of constant section, are as follows:

Entrance and converging transition loss,

$$h_e=0.2\frac{v_1^2}{2g}$$

Friction loss, $h_f=0.25\,\dfrac{v_1^2}{2g}$

Bend losses for upper and lower bends where the centerline radius=2.5D, $h_b=0.42\,\dfrac{v_1^2}{2g}$

Transition losses—diverging outlets,

$$h_{ex}=0.2\left(\frac{v_1^2}{2g}-\frac{v_o^2}{2g}\right)$$

Transition losses—diverging outlets,

$$h_{ex}=0.1\left(\frac{v_o^2}{2g}-\frac{v_1^2}{2g}\right)$$

Exit losses, $h_v=\left(\dfrac{a_1}{a_o}\right)^2\dfrac{v_1^2}{2g}$

where a_1 and v_1 are the area and velocity, respectively, at the throat, and a_o and v_o are the area and velocity, respectively, at the outlet.

Equation (48), $q=D\sqrt{2g\,(H_T-h_L)}$, can be written in the form:

$$q=CD\sqrt{2gH_T} \qquad (49)$$

where C is a coefficient which will account for the various losses through the siphon passageway. Equating these two expressions for q and solving for C,

$$C=\sqrt{\frac{H_T-h_L}{H_T}} \qquad (50)$$

The arrangement and design of a typical low-head siphon spillway is shown on figure 297. The figure includes an explanation of the design arrangement, a recommended design procedure, and approximate maximum values for the coefficients of discharge, C, for given assumptions of losses. The use of these discharge coefficients will provide satisfactory residual pressures at the crest of the throat.

G. STRUCTURAL DESIGN DETAILS

215. *General.*—The structural design of a spillway and the selection of specific structural details follow the determination of the spillway type and arrangement of components and the completion of the hydraulic design.

Usually the foundation material of a spillway is not competent to resist the destructive action of high-velocity flows; therefore, a nonerodible lining ordinarily must be provided along the spillway waterway. This lining may be of wood, steel, handlaid grouted riprap, rubble masonry, or concrete. Such lining serves to prevent erosion, reduces friction losses by providing smooth bounding surfaces for the channel (which also permits smaller hydraulic sections), and provides a relatively watertight conveyance channel for directing flow past the dam. Economy and durability most often favor concrete as the appropriate lining material for water conveyance structures.

A spillway can be constructed on almost any foundation capable of sustaining applied loads without undue deformation. Although it is not usually advisable, a spillway can be placed on the face of an earthfill dam or through the dam, provided precautions are taken in the selection of design details to accommodate settlement and to prevent leakage from the structure. The type of walls, linings, and associated structures of a spillway and the details of the design will depend on the nature of the foundation. For instance, the details of the design for a spillway founded entirely in rock will differ from one constructed on clay. Structural details will differ according to foundation bearing capabilities, settlement or heave characteristics, and permeability and seepage features. Although concrete walls, linings, and associated structures may be adequate to withstand normal hydrostatic and earth loadings, they must also be arranged to allow for movements due to temperature change, and for unequal or large foundation settlements and heavings because of frost action. Provisions must also be made for handling leakage from the channel or underseepage from the foundation which might cause saturation of the underlying materials and large uplifts against the structure undersurfaces.

Subsequent sections discuss the structural designs of open channel spillways, including crest structures, walls, channel linings, and miscellaneous details. The structural designs of spillway conduits and tunnels are similar to those for outlet conduits and tunnels which are discussed in chapter X.

216. *Crest Structures and Walls.*—Spillway control structures and overflow crests against which reservoir heads act are essentially overflow dams, and spillway abutment structures or flanking dikes are similar to concrete nonoverflow dams or earthfill embankments. The design of earthfill dams is described in chapter VI; the design of overflow and nonoverflow concrete dams is discussed in chapter VIII.

The nature or type of confining side walls for open channel spillways will depend on the material upon which they are founded and on the loading to which they will be subjected. For spillway channels excavated in rock or firm material, and where sloping of the wall faces is permissible, lining placed directly against the excavated slopes may offer sufficient stability for forming the channel sidewalls. Otherwise, self-supporting retaining walls of the gravity, cantilever, or counterforted type will be required. A monolithic flume-type section with the walls continuous with the floor and heels may also be considered.

The design of a gravity or reinforced concrete retaining wall is similar to that for a gravity dam, in that the stability against sliding and overturning and the magnitude and distribution of the foundation reaction resulting from the weight and applied loads must be determined. Methods of analyzing gravity structures for stability, including allowable sliding factors, and for determining foundation reactions are discussed in chapter VIII. Suggested allowable bearing values are presented in appendix C.

Earth loadings are assumed on the basis of equivalent fluid pressures, based on cohesionless soil, as given by Rankine. Figure C–1 (app. C) gives criteria for determining earth loadings on vertical and inclined walls. Wall footings must be safeguarded against frost heaving, and the wall panels must be articulated to provide for adjustments in the event of foundation yielding or unequal settlement. To avoid differential settlement in soft or yielding foundations, wall footing dimensions should be selected to minimize foundation load concentrations and to provide nearly uniform bearing reactions across the base areas.

Inlet channel and chute walls may be subject to various combinations of loading. When flow is occurring through the spillway, hydrostatic loads on the channel side of the wall tend to offset the loads caused by backfill. If, however, the fill has shrunk away from the walls the wall members may be subject to full channel-side waterload before the deflection is sufficient to gain support from the backfill. This condition is more likely to exist where the

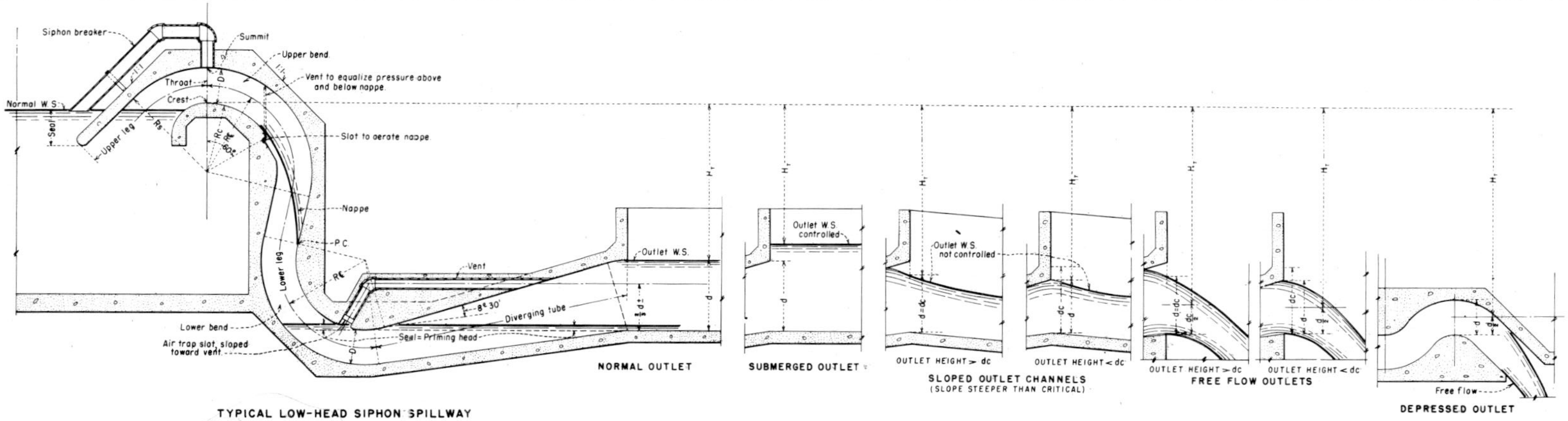

TYPICAL LOW-HEAD SIPHON SPILLWAY

(Maximum head = atmospheric pressure equivalent at site)

DESIGN DATA

$$q = CD\sqrt{2gH} \leqq Rc\sqrt{0.7h_{AT}(2g)} \times \log_e \frac{Rs}{Rc} = \text{(Vortex flow} \approx \text{Max. q)}$$

q = Cubic ft. per sec. per ft. of crest width.
Rc = Radius of crest at throat in ft.
Rs = Radius of summit at throat in ft.
D = Throat size in ft.
d = Depth of water at outlet in ft.
h_{AT} = Atmospheric pressure at site in ft. of water.
H_T = Available head = (See outlet details.)
C = Discharge coefficient based on $\frac{d}{D}$ and $\frac{R℄}{D}$
g = 32.16

SECTION	DESIGN	REASONS	REMARKS
Upper leg	A well proportioned transition providing a gradually contracting area. Inlet area should be 2 or 3 times the area of the throat.	Provides accelerating velocity. Reduces entrance losses.	Entrance loss for form shown suggested as 0.2 Δ hv.
Throat	Size depends upon required capacity. Recommended ratio of R℄ to D is 2.5. For forming with timber D is 2'-0" min.; metal forms left in place may be used for less than 2'-0". Recommended width of throat = 2 D.	Keeps bend losses low.	Wide crests and shallow depths reduce priming time.
Upper bend	Recommended ratio of R℄ to D is 2.5. Aerating slot is set at 60° from crest. Pipe vent, set in sidewalls, connects slot to summit area.	Keeps bend losses low. Slot prevents nappe from clinging to crest. Vent insures nappe trajectory.	Area in upper and lower bend is same as area at throat.
Lower bend	P.C. is set to contact nappe. P.C. is a point on trajectory of nappe as determined from desired priming head at crest. Recommended ratio of R℄ to D is 2.5. Top of siphon at vertical section is set at elevation of outlet bottom. Air trap elevation is elevation of outlet bottom plus desired priming head. Vent area = throat area ÷ 48 (approx.)	Provides seal and exhausts air from upper bend. Provides seal of lower bend and insures creation of partial vacuum. Vent exhausts entrained air with negligible loss of discharge.	Vent decreases priming time, increases priming head slightly.
Diverging tube	Outlet end set at desired elevation. Length of tube is set by 8°30' angle of divergence. Vent outlet set at $\frac{2}{3}$ depth (±) of outlet at max. capacity.	Provides transition. Increases discharge.	Discharge loss for form shown suggested as 0.2 Δ hv.
Siphon breaker	A pipe vent set at summit with inlet end set at or slightly below normal water surface. Area = throat area ÷ 24 (Min.)	To break siphonic action slowly; prevents severe vibration.	If practicable, adjustable inlet on siphon breaker is desirable.
Seal at intake	Lip of upper leg is submerged an arbitrary distance, but should be set deep enough to prevent excessive draw down.	Prevents "gulping" of air, and provides more normal action of siphon breaker.	

Note: Design recommendations given above are presumed to give minimum losses according to experiments. Deviation therefrom should include consideration of additional losses.

SIPHON SPILLWAY COEFFICIENTS OF DISCHARGE

VALUES OF C FOR USE IN EQUATION $q = CD\sqrt{2gH_T}$

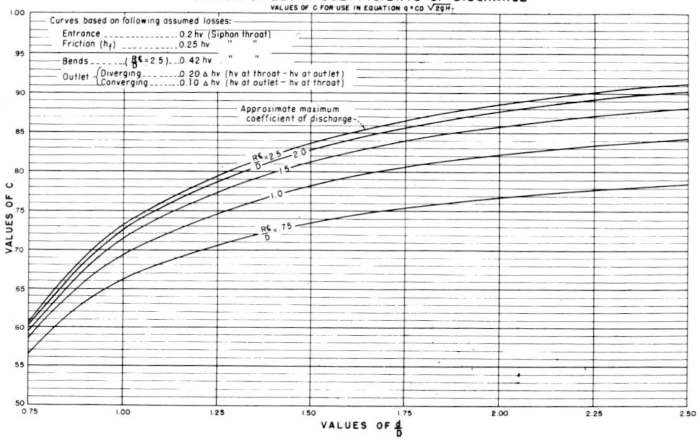

REFERENCES

"Design of A High-Head Siphon Spillway" by Elmer Rock A.S.C.E. Transactions, Vol. 105, Page 1050.
"A Study of Low-Head Siphons" by T.G. Owen - 1939.
U.S.B.R. Data on Model Tests, Wasteway No. 2 (Yakima Project) 1942.
A.S.C.E. Transactions, - 1934, Page 986.
A.S.C.E. Transactions, 1939, Page 1789.
"Siphon Spillways" by A.H. Naylor - 1935.

RECOMMENDED DESIGN PROCEDURE

1. Determine crest Elevation.
2. Determine head available (crest EL. − W.S. EL. at outlet.)
3. Select preliminary values for D and R℄ and solve for max. q using Vortex flow formula: $q = Rc\sqrt{0.7h_{AT}(2g)}\,\mathrm{Log}_e \frac{Rs}{Rc}$ (Recommend D = 2'-0" Min. and $\frac{R℄}{D}$ = 2.5)
4. Select values of d and H_T and obtain value of C from graph. Solve equation $q = CD\sqrt{2gH_T}$ (q must not be greater than value obtained in (3)).
5. Solve for required crest width = $\frac{Q}{q}$. Select crest width and number of siphons to be used for total Q and determine q for this crest width.
6. Repeat (3) and (4) until $q = CD\sqrt{2gH_T} \leqq Rc\sqrt{0.7h_{AT}(2g)}\,\mathrm{Log}_e \frac{Rs}{Rc}$
7. Select priming head by determining the reservoir W.S. EL. at which operation is desired. If priming head is less than $\frac{D}{3}$ locate P.C. of lower bend on trajectory of nappe.
8. Determine size and inlet EL. of siphon breaker (Area of siphon breaker = Area of siphon throat ÷ 24).
9. Select and design priming aids to be used (Slot to aerate nappe, Vent to equalize pressure above and below nappe, and vent at lower bend).

Figure 297. Typical low-head siphon spillway. From drawing 103–D–46.

wall leans into the backfill. On the other hand, when the reservoir is drawn down below spillway level, there is no spill through the structure and the walls will be subject to full backfill loads without any support from waterloads. The structural design of wall members must recognize all these possibilities of loading. In the case of the assumption that the backfill may not be tight against the wall to help support the wall against water pressures, an increase in allowable stresses may be considered.

When permeable backfill is placed behind stilling basin walls or when the back of the wall is partly exposed to tailwater, the water pressure resulting from tailwater will need to be added to the backfill loading. For higher spillway discharges, the water level inside the basin will be depressed by the profile of the jump and an unbalanced hydrostatic load acting to overturn the walls will occur. The design loading assumptions must recognize this condition of unbalanced pressures as well as the increased uplift forces when sliding and overturning analyses are considered.

217. *Open Channel Linings.*—Floor pavings are provided primarily to form a reasonably watertight protective surfacing over the channel to prevent erosion or damage to the foundation. During spillway flows, the floor is subjected to hydrostatic forces due to the weight of the water in the channel, to boundary drag forces due to frictional resistance along the surface, to dynamic forces due to flow impingement, to uplift forces due to reduction of pressure along the boundary surface, or to uplift pressure caused by leakage through joints or cracks. When there are no spills, the floor is subject to the action of the elements including expansion and contraction due to temperature variations, alternate freezing and thawing, and weathering and chemical deterioration; to the effects of settlement and buckling; and to uplift pressures brought about by underseepage or high ground-water conditions. Since it is not always possible to evaluate the various forces which might occur nor to make the lining heavy enough to resist them, the thickness of the lining is most often established on a more or less arbitrary basis; and underdrains, anchors, cutoffs, etc., are utilized to stabilize the floor.

To provide a relatively watertight lining which will withstand reasonable weathering and abrasion, and which will hold up against ordinarily experienced forces caused by expansion, contraction, frost heave, and settlement of the foundation, a nominal minimum thickness of 8 inches is recommended for small spillways when the lining is placed directly on rock. When the lining is placed on earth or on an intervening gravel layer, a somewhat thicker slab should be provided to forestall cracking or buckling if expansion and contraction can move or displace the slab.

When a spillway channel is excavated in rock, the paving slab is cast directly on the excavated surface. Anchor bars grouted into holes drilled into the rock can be provided to tie the paving to the foundation. A slab which is bonded to the foundation may not move as the result of expansion and contraction. Instead, numerous cracks which in effect divide the slab into a series of small individual blocks will occur. Reinforcement therefore must be provided to tie these individual blocks together and to distribute the cracking and minimize the crack openings. Typical details for paving slabs on rock are shown on figure 298(A). The anchorage provided increases the effective weight of the slab against displacement by the amount of foundation rock to which the anchors can be tied. Depth and spacing of anchors will depend on the nature of the bedrock, its stratification, jointing, weathering, etc. The anchor should be of sufficient size to hold the weight of the foundation to which it is anchored without exceeding the yield stress of the steel. A gridwork of underdrains laid with open joints in gravel-filled trenches is provided to prevent a buildup of uplift under the paving. When leakage through the joints is to be minimized, metal or rubber waterstops are provided.

When a spillway channel is excavated through earth, the paving slab may be cast directly on the excavated surface, or an intervening pervious blanket may be required, depending on the nature of the foundation as related to its permeability, susceptibility to heaving from frost action, and heterogeneity as it may affect differential settlement. Because the

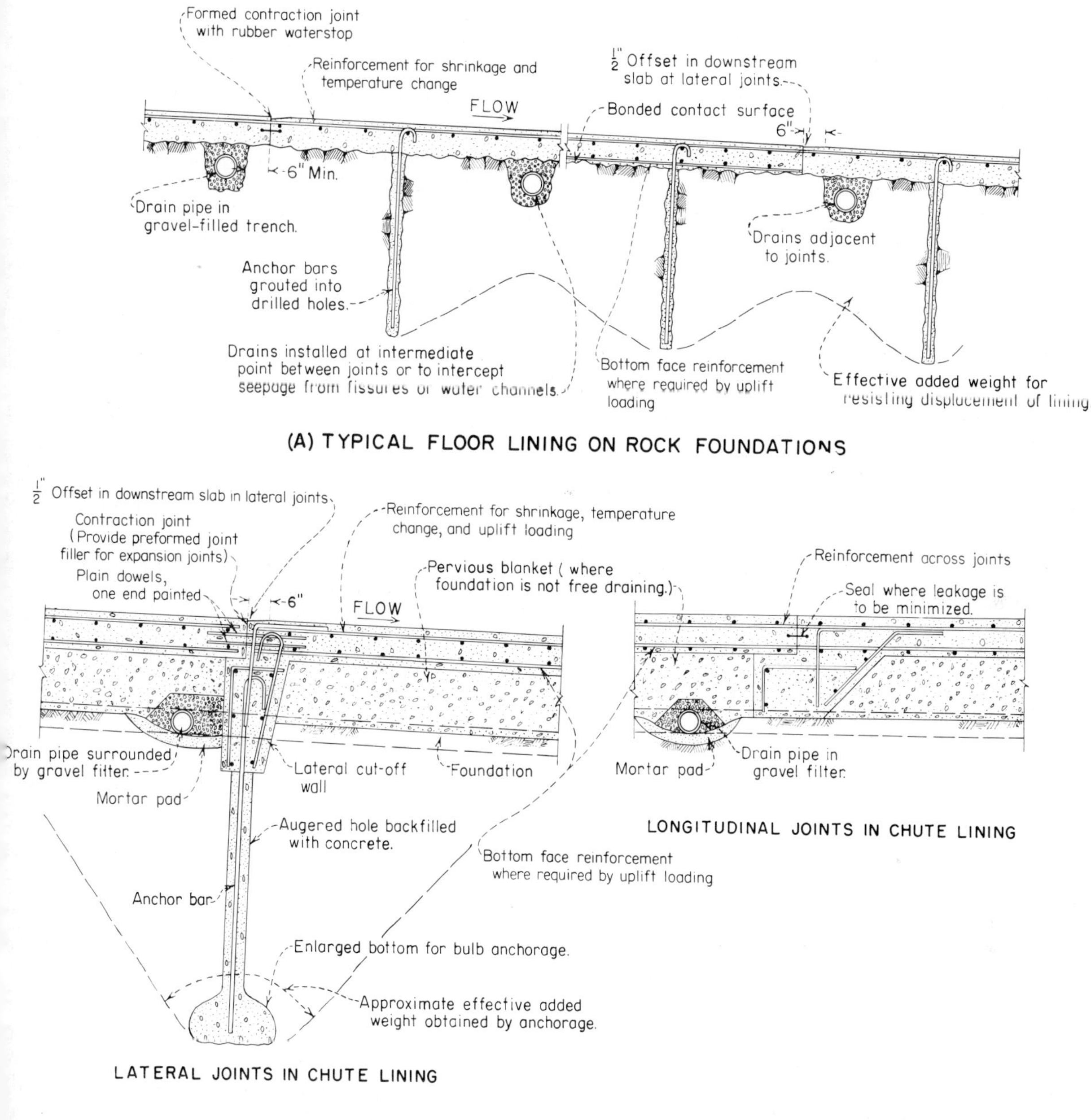

Figure 298. **Floor lining details for spillway channels. 288–D–2521.**

ab is not bonded to the foundation, it is subject to movement from expansion or contraction, and it must be restrained from creeping hen it is constructed on a slope. This restraint is best achieved by cutoffs which can held in a more or less fixed position with spect to the slab and to the foundation, or tieing the slab to walls, piles, or similar gid members of the spillway structure.

Since the slab is relatively free to move upon the foundation, the movement will take place from the fixed edges and the paving should be reinforced sufficiently to permit its sliding without cracking of the concrete or yielding of the reinforcement. To assist further in holding the slab to the foundation, bulb anchors are sometimes employed as shown on figure 298(B). The anchor in this instance in effect ties the slab to a cone of earth, the volume of which will depend on the anchor depth and spacing and on the angle of internal friction of the soil.

A pervious gravel blanket is often provided between the slab and the foundation when the foundation is sufficiently impervious to prevent leakage from draining away, or where it is subject to capillarity which will draw moisture to the underside of the lining. The blanket serves not only as a free-draining medium but also aids in insulating the foundation from frost penetration. The thickness of the blanket thus depends on the climate at the site and on the susceptibility of the foundation to frost heaving. A gridwork of underdrains laid with open joints in gravel and bedded on a mortar pad to prevent the foundation material from being leached into the pipe is provided as a collecting system for the seepage. The network of drainage pipe empties into one or more trunk drains which carry the seepage flows to outlets through the channel floor or walls.

In stratified foundations, ground water or seepage can cause uplift on layers below the floor lining, and drainage holes are sometimes augered into the underlying material and backfilled with gravels to relieve the underpressure.

When watertightness of the paving against exterior water heads is required, metal or rubber waterstops are installed to seal the joints. Such seals are provided in floor slabs upstream from the control structure if watertightness is desired to increase the percolation path under the structure. They are commonly provided at transverse joints along concave curved portions of the downstream channel where the dynamic pressures on the floor cause a high head for introducing water into the joint. Seals may be desirable along longitudinal joints in a stilling basin on a permeable base. Differential heads resulting from the sloping water surface of the jump can cause a circulating flow under the slab if leakage is allowed to enter the joint at the downstream end of the basin and to flow out of the joint at the upstream end.

Lateral joints over which flow velocities are high are arranged so that the upstream edge of the lower slab cannot heave without moving the lower edge of the upper slab a like amount further, the lower slab edge is constructed about one-half inch lower than the upper slab edge. These provisions are made to forestall a high buildup of dynamic head at the joint which would result if the surface of the downstream slab were to project above the surface of the upstream slab. The dynamic head could introduce water at high pressure under the slab which would result in uplift or dislodgement.

Contraction joints are generally placed from 25 to 50 feet apart in both the floor and walls. Joints are also provided where angular changes of the floor surface occur and where they are required to avoid reentrant angles in the slab which often cause cracking of the paving. The use of joint fillers in contraction joints should be minimized because deterioration of the filler will result in an open joint which is difficult to maintain. If joints are provided at the indicated spacings, the contraction or expansion movements may not be severe and filler material in the joint may not be necessary. Floor slabs can be constructed in alternate panels; the initial placement shrinkage of the concrete may then afford sufficient joint opening for subsequent expansion. Keyed joints in thin floors and walls which may be subjected to differential movement are unsatisfactory, since inequalities in deflection across the joint will place high stress on the keys or keyways and cause them to spall. An unkeyed joint with slip dowels is a better detail.

Water normally will stand in a stilling basin whose floor is at a lower level than the river channel. With this condition the foundation under the basin will be permanently saturated. When the water is lowered in the basin, the floor can be subjected to an uplift equal to the tailwater head or higher if the pressure is augmented by head from a higher source. During times of spillway discharge, the water weight in the basin will be reduced because of the slope of the jump profile, and at the upstream end of the basin the uplift will far exceed the downward weight. The basin floor must be heavy enough to withstand this unbalanced waterload, unless an adequate drainage system is installed to relieve the uplift pressure when jump sweep-out occurs. Since drainage system cannot be considered entirely effective because of the possibility of clogging or silting of the drain outlets, the floor slab is usually made sufficiently heavy to resist the flotation effect on the floor.

For design, the stilling basin floor is considered to be a free body in static equilibrium with foundation reactions balancing active loads. Uplift forces caused by hydrostatic head on the bottom of the slab are counterbalanced by the weight of the concrete and the effective weight of the water in the basin. Differential horizontal hydrostatic forces are opposed by the sliding resistance of the horizontal leg of the slab on the foundation. Equilibrium against rotation is achieved by equating any unbalanced forces with a foundation reaction force positioned so that the mo-

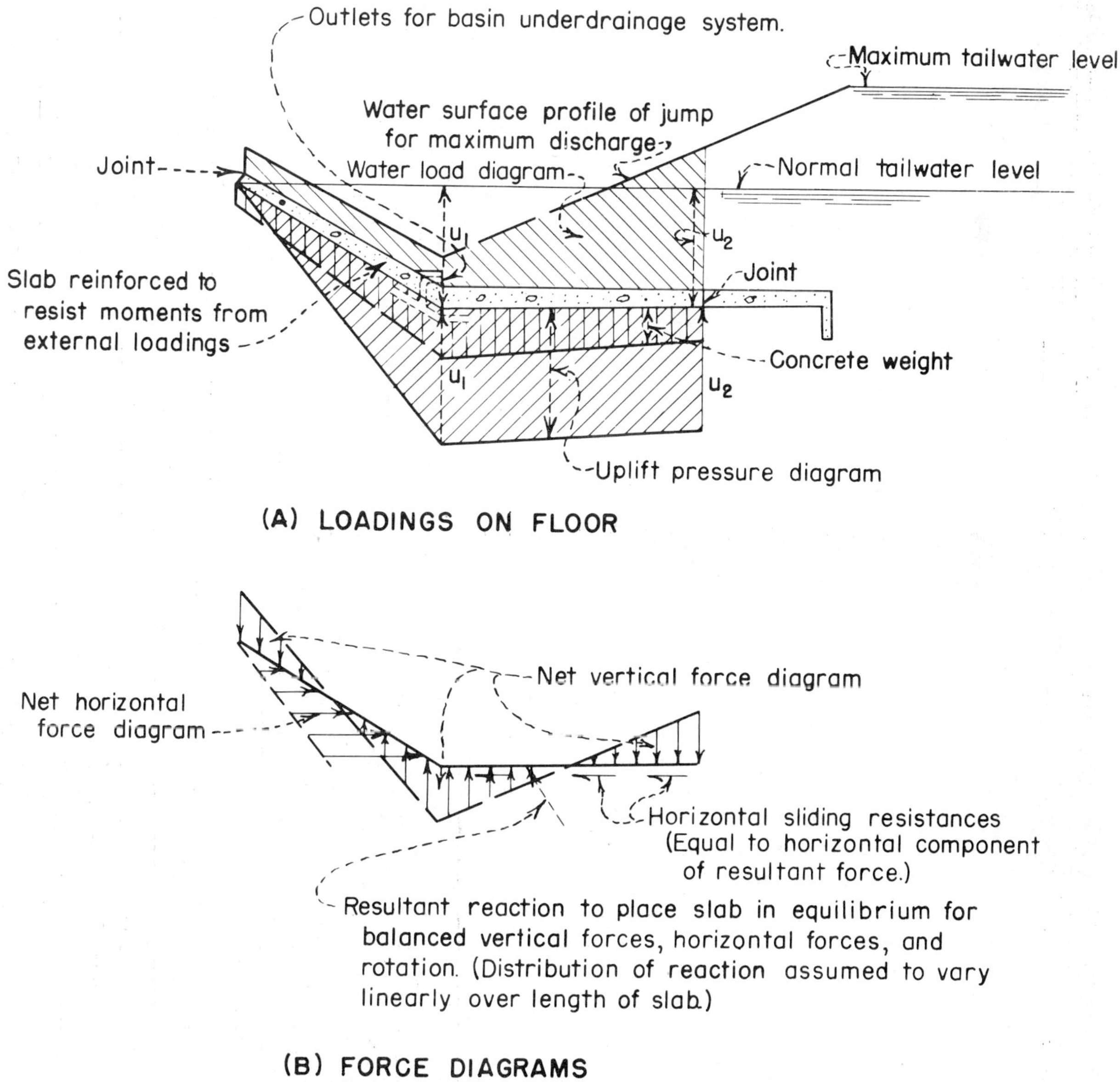

Figure 299. Illustration of uplift forces acting on a stilling basin floor. 288–D–2522.

ments will be zero about any point.

An illustrative force diagram for a typical basin analysis is shown on figure 299. Note that the dynamic force which will occur because of impingement of the incoming high-velocity flow on the horizontal apron is not included. Where chute blocks or dentates are used at the upstream end of the apron, reduced pressure zones opposite the blocks will occur. Because the reduced pressures and the impingement forces tend to offset each other, they are both neglected in the analysis.

It will be seen that the weight required in the floor will vary in direct relation to the amount of assumed uplift. If uplift equal to the head from maximum tailwater is considered, the required floor thickness will be almost prohibitive. Therefore, an uplift based on a lesser tailwater level, such as that from the normal tailwater condition, is generally assumed and reliance is placed on the drainage system for relief of greater uplift pressures. Ordinarily the outlets for the basin underdrainage system are located in the sills near the upstream end of the basin floor. The head on these outlets will be reduced when the jump occurs, which facilitates relief of hydrostatic pressures under the slab. The stilling basin floor slab must be designed to withstand the internal moments resulting from the external loadings indicated by the force diagram. The slab will be thickest at the junction of the sloping leg and the horizontal apron. Reinforcement will be required in the slab to withstand the computed internal stresses.

218. *Miscellaneous Details.*—(a) *Cutoffs.*—One or more cutoffs are generally provided at the upstream end of a spillway for various purposes. They may form a watertight curtain against seepage under the structure, or they may be used to increase the path of percolation under the structure and thus reduce uplift forces. Cutoffs also can be used to intercept permeable strata in the foundation so as to minimize seepage and prevent a buildup of uplift pressure under the spillway or adjacent areas. When the cutoff trench for the dam extends to the spillway, it is generally joined to the upstream spillway cutoff to provide a continuous barrier across the abutment area. In jointed rock the cutoff acts as a grout cap for a grout curtain which is often extended across the spillway foundation.

A cutoff is usually provided at the downstream end of a spillway structure as a safeguard against erosion and undermining of the end of the structure. Cutoffs at intermediate points along the length of a spillway are sometimes provided as barriers against water flowing along the contact between the structure and the foundation and to lengthen the path of percolation under the structure. When the spillway is a conduit under the dam, the cutoff takes the shape of collars placed at intervals around the conduit barrel. Wherever possible, cutoffs in rock foundations are placed in vertical trenches. In earth foundations where the cutoffs must be formed in a trench with sloping sides, care must be taken to compact the trench backfill properly with impervious material to obtain a reasonably watertight barrier.

(b) *Backfill.*—When a spillway is placed adjacent to a dam so that the impervious zone of the embankment abuts the spillway walls, the wall backfill is actually the impervious zone of the dam and is similarly compacted. Backfill elsewhere along the spillway walls ordinarily should be free-draining material to minimize hydrostatic pressures against the walls. Backfill other than that adjacent to the dam may be either compacted or uncompacted. The choice of backfill material and the compaction methods used in placing such material will affect the design loadings on the walls.

(c) *Riprap.*—When the spillway approach channel is excavated in material that will be eroded by the approach velocities, a zone of riprap is often provided immediately upstream from the inlet lining to prevent scour of the channel floor and side slopes adjacent to the spillway concrete. The riprap is generally a continuation of that along the upstream face of the dam, is of similar size and gradation, and has similar bedding. Riprap is normally used in the outlet channel adjacent to the downstream cutoff to prevent excessive erosion and undermining of the downstream end of the structure. To resist scour from high exit velocities, the riprap should be the largest size possible and should be bedded on a graded material. The voids should be filled with spalls to prevent the underlying material from washing out, which would cause the riprap to settle or to be displaced.

H. BIBLIOGRAPHY

219. *Bibliography.*

[1] Donnelly, C. A., and Blaisdell, F. W., "Straight Drop Spillway Stilling Basin," University of Minnesota, Saint Anthony Falls Hydraulic Laboratory, Technical Paper No. 15, series B, November 1954.

[2] Morris, B. T., and Johnson, D. C., "Hydraulic Design of Drop Structures for Gully Control" Trans. ASCE, vol. 108, 1943, p. 887.

[3] Moore, W. L., "Energy Loss at the Base of a Free Overfall," Trans. ASCE, vol. 108, 1943, p. 1343.

[4] Rand, Walter, "Flow Geometry at Straight Drop Spillways," ASCE Proceedings, Paper No. 791, September 1955.

[5] Peterka, A. J., "Morning-Glory Shaft Spillways," Trans. ASCE, vol. 121, 1956, p. 385.

[6] Bradley, J. N., "Morning-Glory Shaft Spillways: Prototype Behavior," Trans. ASCE, vol. 121, 1956, p. 312.

[7] Bureau of Reclamation, "Studies of Crests of Overfall Dams," Bulletin 3, pt. VI, Hydraulic Investigations, Boulder Canyon Project Final Reports, 1948.

[8] Hinds, Julian, "Side Channel Spillways," Trans. ASCE, vol. 89, 1926, p. 881.

[9] Bradley, J. N., and Peterka, A. J., "The Hydraulic Design of Stilling Basins," ASCE Proceedings, vol. 83, October 1957, Journal of the Hydraulics Division, No. HY5, Papers No. 1401 to 1406, inclusive.

[10] Warnock, J. E., "Experiments Aid in Design at Grand Coulee," Civil Engineering, vol. 6, 1936, p. 737.

[11] Bureau of Reclamation, "Hydraulic Design of Stilling Basin and Bucket Energy Dissipators," Engineering Monograph No. 25, 1964.

[12] Bradley, J. N., and Peterka, A. J., "Hydraulic Design of Stilling Basins: Small Basins for Pipe or Open Channel Outlets—No Tailwater Required (Basin VI)," ASCE Proceedings, vol. 83, October 1957, Journal of the Hydraulics Division, No. HY5, Paper No. 1406.

[13] Doddiah, D., Albertson, M. L., and Thomas, R. A., "Scour From Jets," Proceedings, Minnesota International Hydraulics Convention (Joint Meeting of International Association for Hydraulic Research and Hydraulics Division, ASCE), Minneapolis, Minn., August 1953, p. 161.

[14] Veronese, Alessandro, "Erosioni Di Fondo A Valle Di Uno Scarico," (Downstream Bed Erosion Due to a Discharge), Annali dei Lavori Pubblici, vol. 75, No. 9, September 1937, p. 717.

[15] Wagner, W. E., "Morning Glory Shaft Spillways: Determination of Pressure-Controlled Profiles," Trans. ASCE, vol. 121, 1956, p. 345.

[16] Mavis, F. T., "The Hydraulics of Culverts," Pennsylvania State College Engineering Experiment Station, Bulletin 56, vol. XXXVII, No. 7, February 12, 1943.

[17] Straub, L. G., Anderson, A. G., and Bowers, C. E., "Importance of Inlet Design on Culvert Capacity, Culvert Hydraulics," Highway Research Board of the National Academy of Sciences, National Research Council, Publication 287, Research Report No. 15–B, 1953, p. 53.

[18] Karr, M. H., and Clayton, L. A., "Model Studies of Inlet Designs for Pipe Culverts on Steep Grades," Engineering Experiment Station, Oregon State College, Bulletin No. 35, June 1954.

[19] Schiller, R. E. Jr., "Tests on Circular Pipe Culvert Inlets," Culvert Flow Characteristics, Highway Research Board of the National Academy of Sciences, National Research Council, Publication 413, Bulletin No. 126, 1956, p. 11.

[20] Shoemaker, R. H. Jr., and Clayton, L. A., "Model Studies of Tapered Inlets for Box Culverts," Culvert Hydraulics, Highway Research Board of the National Academy of Sciences, National Research Council, Publication 287, Research Report No. 15–B, p. 1.

[21] Blaisdell, F. W., "Hydraulics of Closed Conduit Spillways—Part I—Theory and Its Application," University of Minnesota, Saint Anthony Falls Hydraulic Laboratory, Technical Paper No. 12, series B, January 1952, revised February 1958.

[22] Blaisdell, F. W., "Hydraulics of Closed Conduit Spillways—Parts II through VII—Results of Tests on Several Forms of the Spillway," University of Minnesota, Saint Anthony Falls Hydraulic Laboratory, Technical Paper No. 18, series B, March 1958.

[23] Blaisdell, F. W., "Hydraulics of Closed Conduit Spillways—Part VIII—Miscellaneous Laboratory Tests and Part IX—Field Tests," University of Minnesota, Saint Anthony Falls Hydraulic Laboratory, Technical Paper No. 19, series B, March 1958.

[24] Blaisdell, F. W., and Donnelly, C. A., "Hydraulics of Closed Conduit Spillways—Part X—The Hood Inlet," University of Minnesota, Saint Anthony Falls Hydraulic Laboratory, Technical Paper No. 20, series B, April 1958.

[25] Beichley, G. L., "Hydraulic Design of Stilling Basin for Pipe or Channel Outlets," Bureau of Reclamation, Research Report No. 24, 1971.

Chapter X

Outlet Works

C. J. HOFFMAN[1]

A. GENERAL

220. *Functions of Outlet Works.*—An outlet works serves to regulate or release water impounded by a dam. It may release incoming flows at a retarded rate, as in the case of a detention dam; divert incoming flows into canals or pipelines, as in the case of a diversion dam; or release stored waters at such rates as may be dictated by downstream needs, evacuation considerations, or a combination of multiple-purpose requirements.

Outlet works structures can be classified according to their purpose, their physical and structural arrangement, or their hydraulic operation. An outlet works which empties directly into the river could be designated a river outlet; one which discharges into a canal could be classed as a canal outlet; and one which delivers water into a closed pipe system could be termed a pressure pipe outlet. An outlet works may be described according to whether it consists of an open channel or closed conduit waterway, or whether the closed waterway is a conduit in cut-and-cover or in tunnel. The outlet works may also be classified according to its hydraulic operation, whether it is gated or ungated or, for a closed conduit, whether it flows under pressure for part or all of its length or only as a free-flow waterway. Typical outlet works installations are shown on figures 300 through 306.

Occasionally the outlet may be placed at a higher level to deliver water to a canal, and a bypass extended to the river to furnish necessary flows below the dam. Such flows may be required to satisfy prior right uses downstream from the site; or they may be required for the maintenance of a live stream for abatement of stream pollution, preservation of aquatic life, or stock watering purposes. For dams constructed to provide reservoirs principally for recreation or fish and wildlife, a fairly constant lake level is desired and an outlet works may be needed only to release the minimum flows which will provide a live stream below the dam.

In certain instances the outlet works of a dam may be used in lieu of a service spillway in conjunction with an auxiliary or secondary spillway. In this event the usual outlet works installation might be modified to include a bypass overflow, so that the structure can serve as both an outlet works and a spillway. Such structures are typified by figures 192 and 306, for Wasco Dam and Lion Lake dikes, respectively. In these installations the overflow weirs in the control shaft automatically bypass surplus inflows whenever the reservoir rises above normal storage level.

An outlet works may also act as a flood control regulator, to release waters temporarily stored in flood control storage space or to evacuate storage in anticipation of flood inflows. Further, the outlets may serve to empty the reservoir to permit inspection, to make needed repairs, or to maintain the upstream face of the dam or other structures normally inundated. The outlets may also aid in lowering the reser-

[1] Engineer, Spillways and Outlet Works Section, Bureau of Reclamation (Retired).

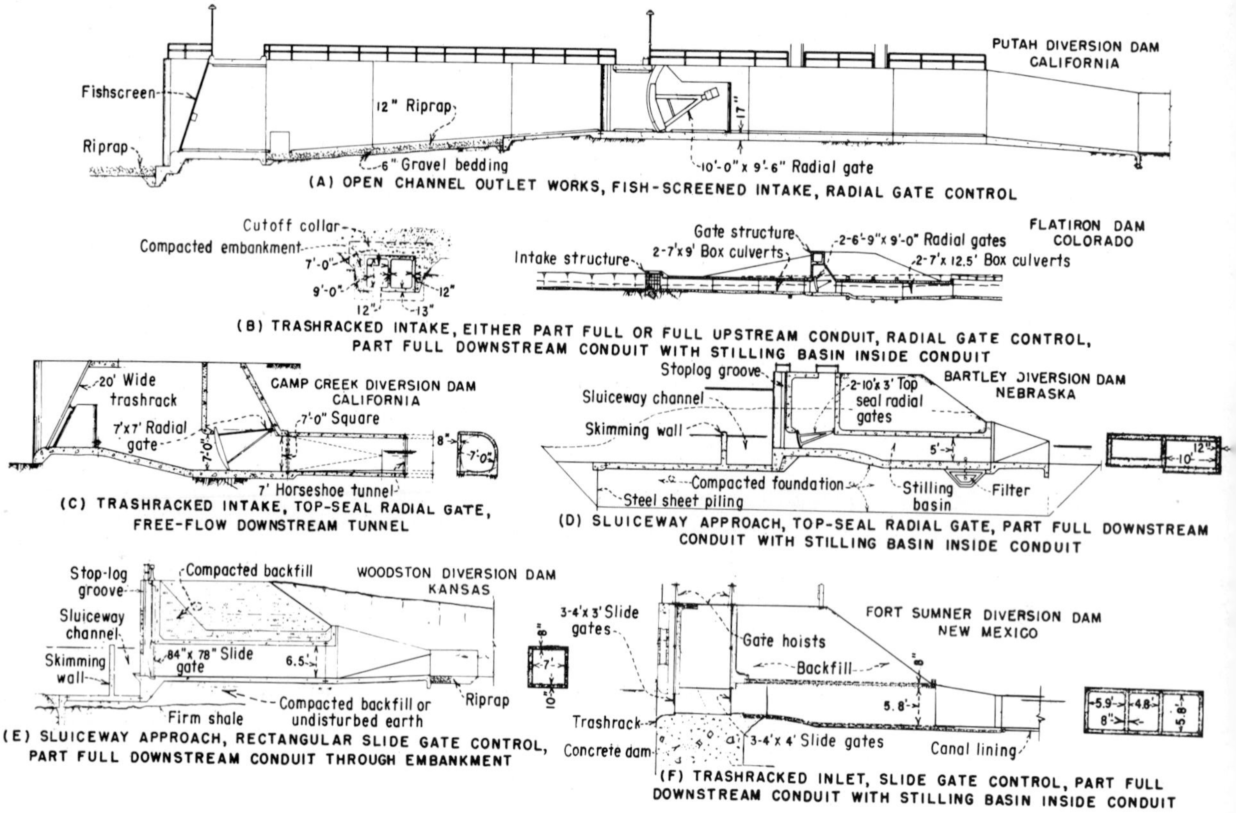

Figure 300. Typical low-head outlet works installations. 288–D–2523.

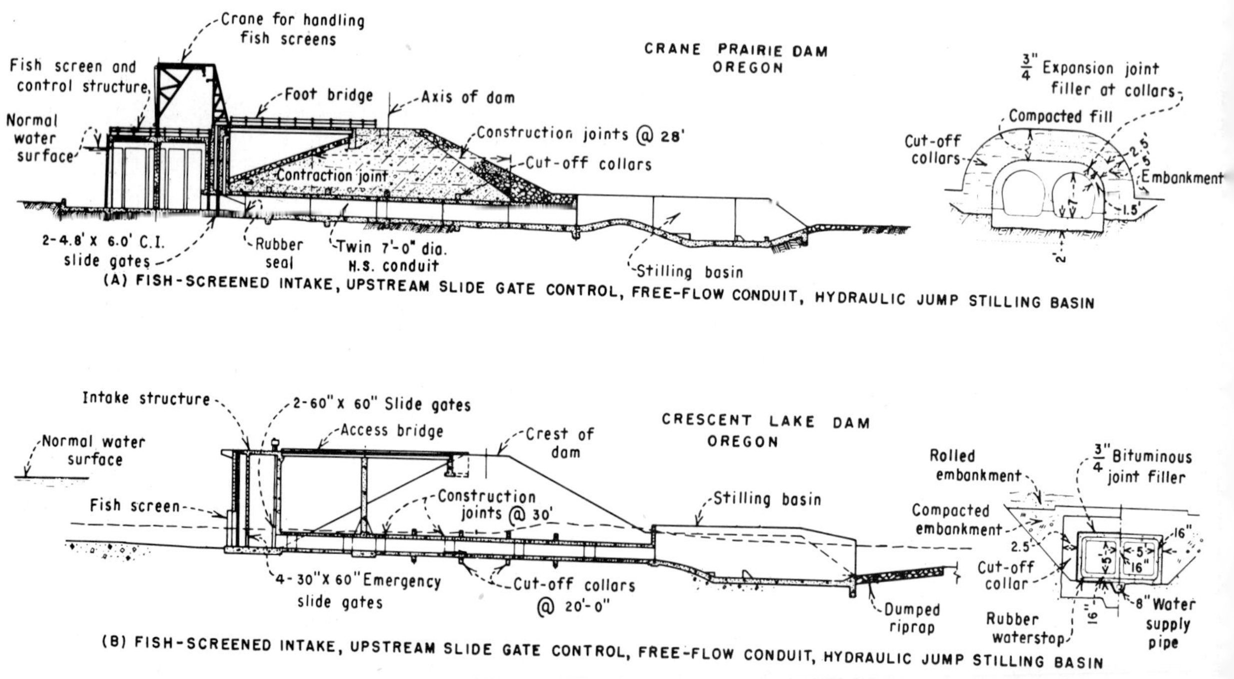

Figure 301. Typical free-flow conduit outlet works installations. 288–D–2524.

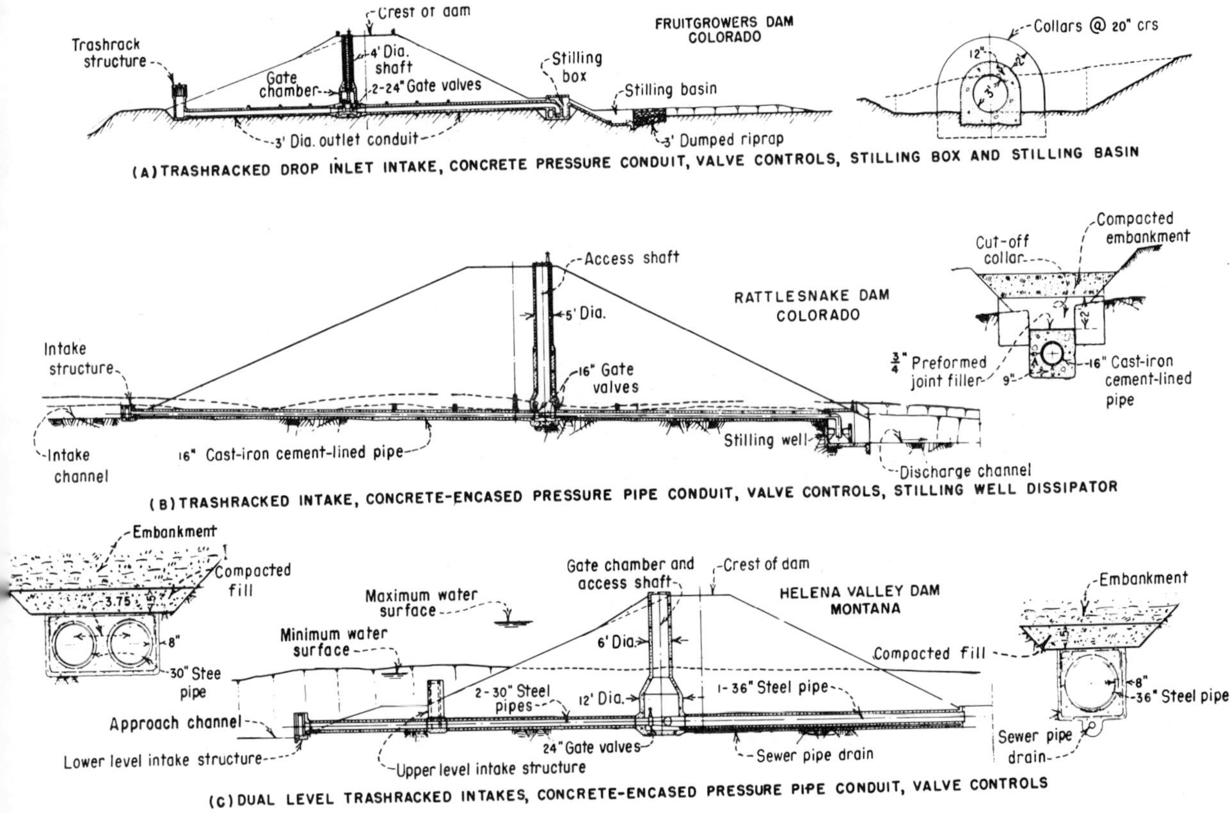

Figure 302. Typical pressure conduit outlet works installations. 288–D–2525.

voir storage when it is desired to control or to poison scrap fish or other objectionable aquatic life in the reservoir.

221. *Determination of Required Capacities.*— Outlet works controls are designed to release water at specific rates, as dictated by downstream needs, flood control regulation, storage considerations, or legal requirements. Delivery of irrigation water is usually determined from project or farm needs and is related to the consumptive use and to any special water requirements of the irrigation system. Delivery for domestic use can be similarly established. Releases of flows to satisfy prior rights must generally be included with other needed releases. Minimum downstream flows for pollution abatement, fish preservation, and other companion needs may often be accommodated through other required releases.

Irrigation outlet capacities are determined from reservoir operation studies and must be based on a consideration of a critical period of low runoff when reservoir storages are low and daily irrigation demands are at their peak. The most critical draft from the reservoir, considering such demands (commensurate with remaining reservoir storage) together with prior rights or other needed releases, generally determines the minimum irrigation outlet capacity. These requirements are stated in terms of discharge at either a given reservoir content or water surface elevation. Occasionally outlet capacity requirements are established for several reservoir contents or alternate water surfaces. For example, outlet requirements may be set forth as: 20 second-feet capacity at reservoir content 500 acre-feet, and 100 second-feet capacity at reservoir content 3,000 acre-feet.

Evacuation of waters stored in an allocated flood control storage space of a reservoir can be accomplished through a gated spillway at

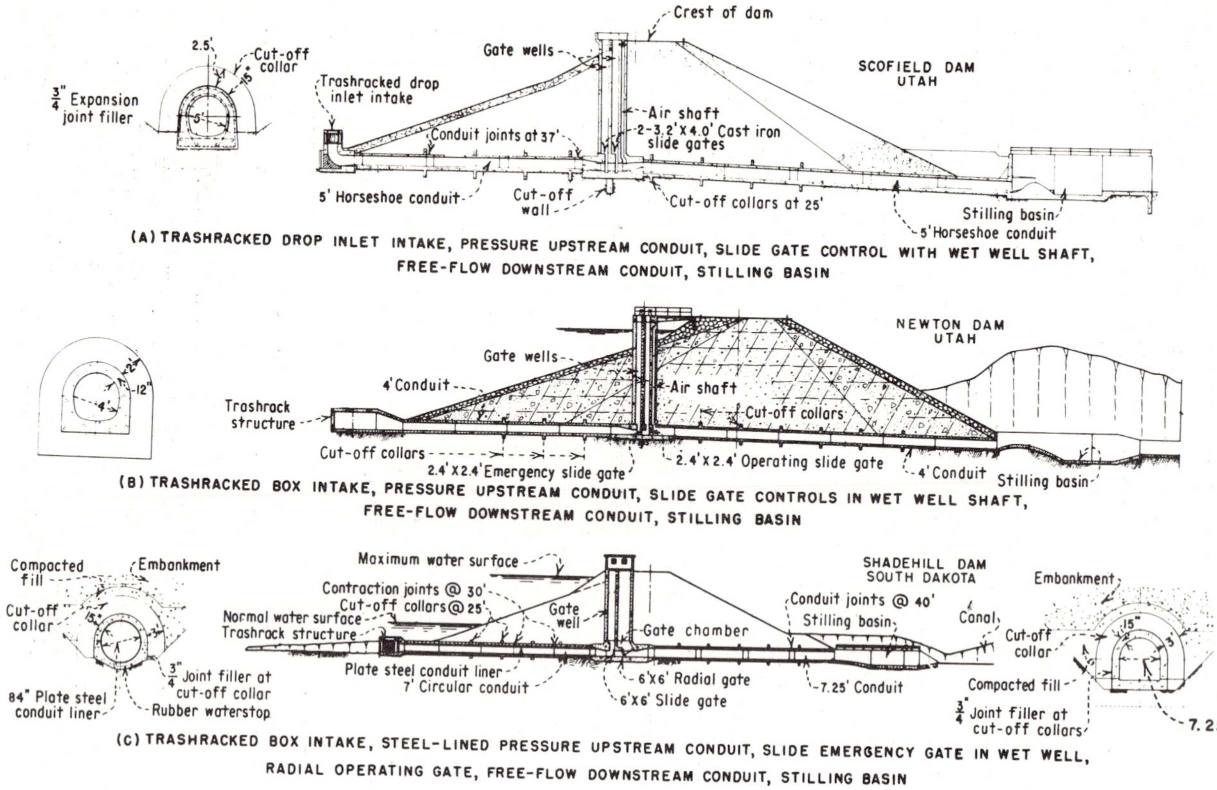

Figure 303. Typical combined pressure and free-flow conduit outlet works installations. 288–D–2526.

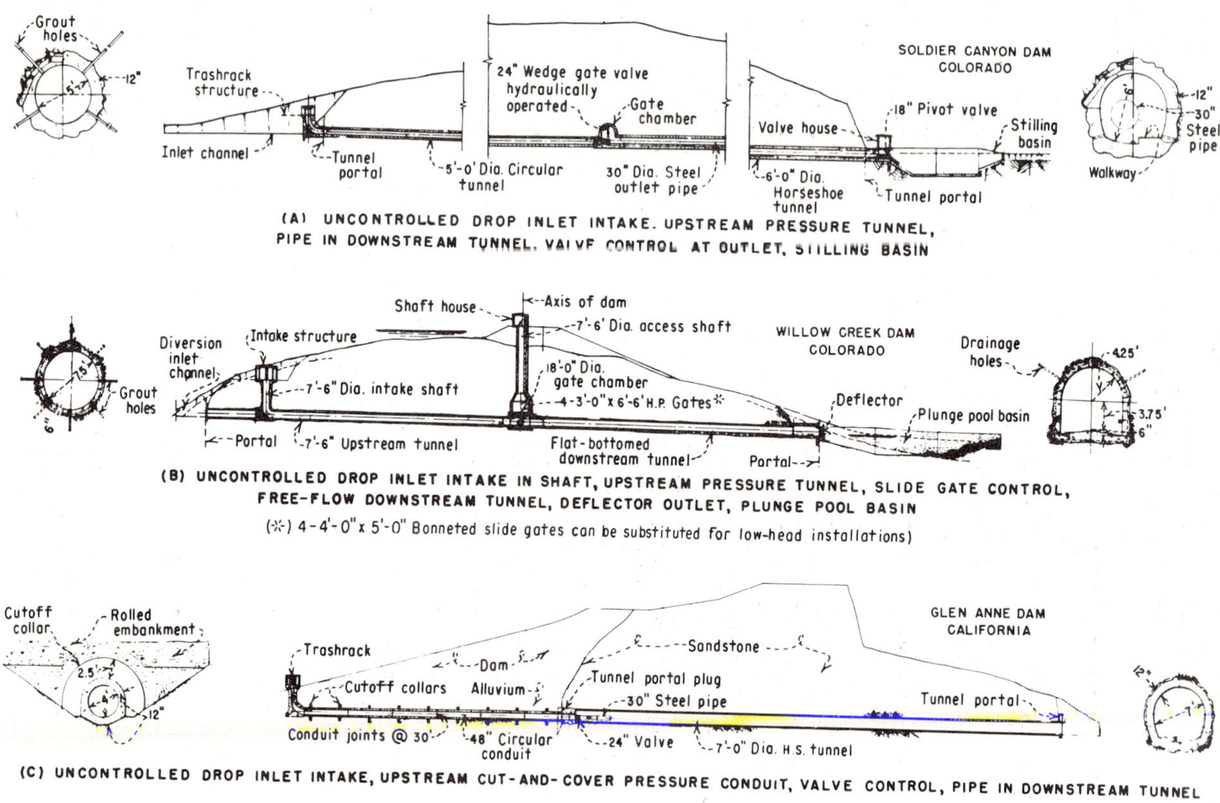

Figure 304. Typical tunnel outlet works installations. 288–D–2527.

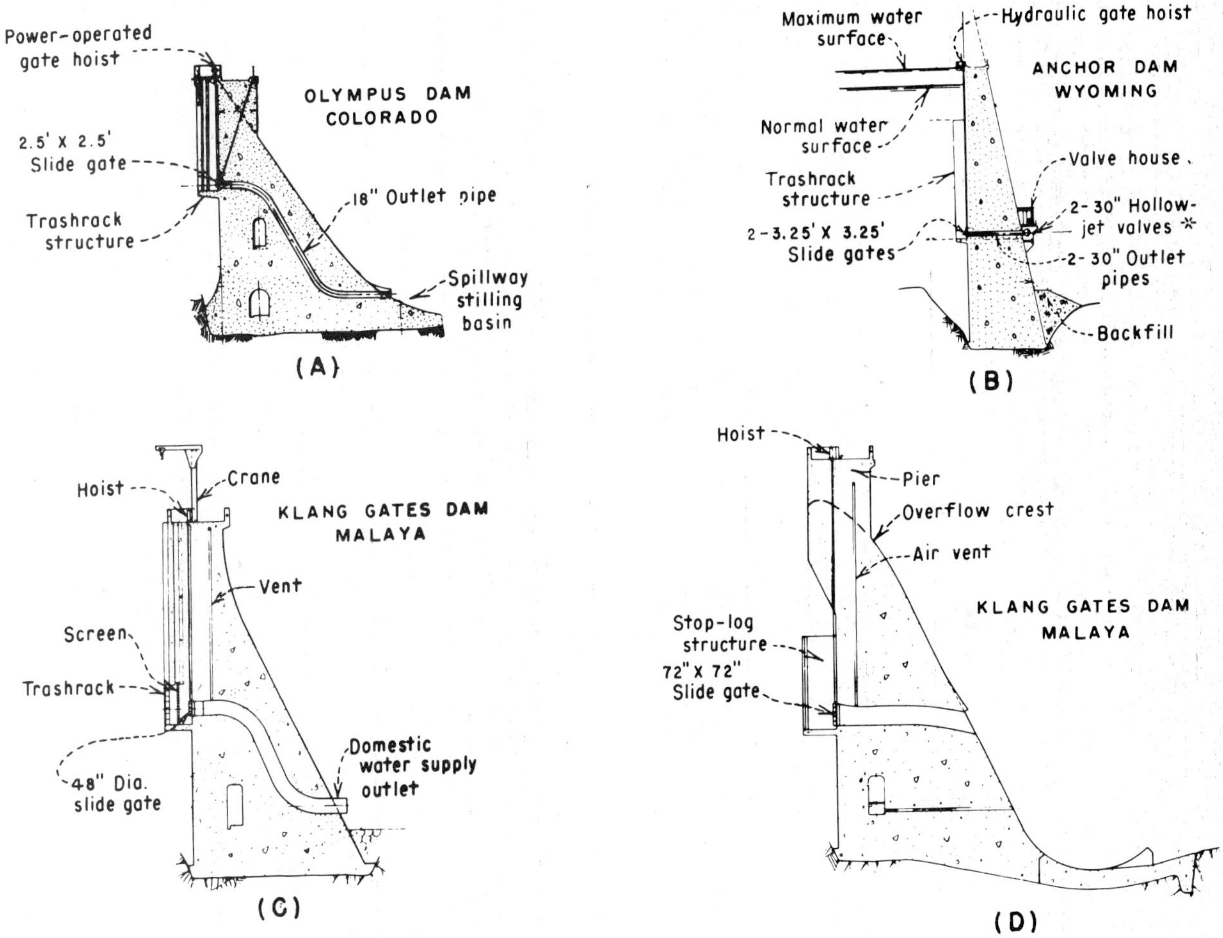

(A) Sluice through non-overflow section emptying into spillway stilling basin. Trashracked intake, slide gate control at upstream face of dam. Free-flow pipe.

(B) Outlet pipe through non-overflow section. Trashracked intake, upstream emergency slide gate. Downstream valve control, freely discharging.

(C) Outlet pipe through non-overflow section. Trashracked intake, upstream slide gate control. Pressure pipe.

(D) Sluice through spillway section. Upstream slide gate control, downstream free-flow conduit

(E) Sluice through spillway section of dam controlled by slide gate. Gate operated from gallery in dam. Upstream pressure conduit, downstream free-flow conduit.

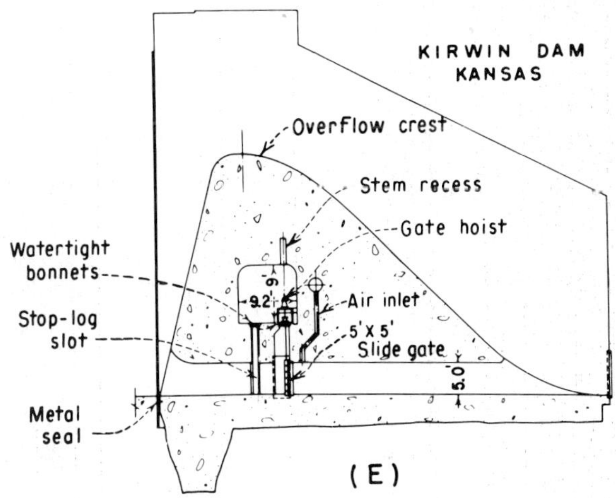

(* Butterfly valves can be substituted for low-head installations)

Figure 305. Typical outlet works installations for concrete dams. 288–D–2941.

the higher reservoir levels or through an outlet at the lower levels. Flood control releases generally can be combined with the irrigation outlet releases if the outlet empties into a river instead of into a canal. The capacity of the flood control outlet is determined by the required time of evacuation of a given storage space, considering the inflow into the reservoir during this emptying period. The combined flood control and irrigation releases ordinarily must not exceed the safe channel capacity of the river downstream from the dam and must allow for any anticipated inflows immediately below the dam. These inflows may be the natural runoffs or may result from releases from other storage developments along the river or from adjacent developments on tributaries emptying into the river.

If an outlet is to serve as a service spillway in releasing surplus inflows from the reservoir, the required discharge for this purpose may fix the outlet capacity. Similarly, for emptying the reservoir for inspection or repair, the volume of water to be evacuated and the allotted emptying period may be the determining conditions for establishing the minimum outlet capacity. Here again, the inflow into the reservoir during the emptying period must be considered. The capacity at low reservoir level should be at least equal to the average inflow expected during the maintenance or repair period. It can, of course, be assumed that any required repair work might be delayed until service demands are light and that it will be done at times of low inflow and at seasons favorable to such construction.

An outlet works cut-and-cover conduit or tunnel often may be utilized for diverting the riverflow during the construction period, thus avoiding the necessity for supplementary installations for that purpose. The outlet structure size dictated by this use rather than the size

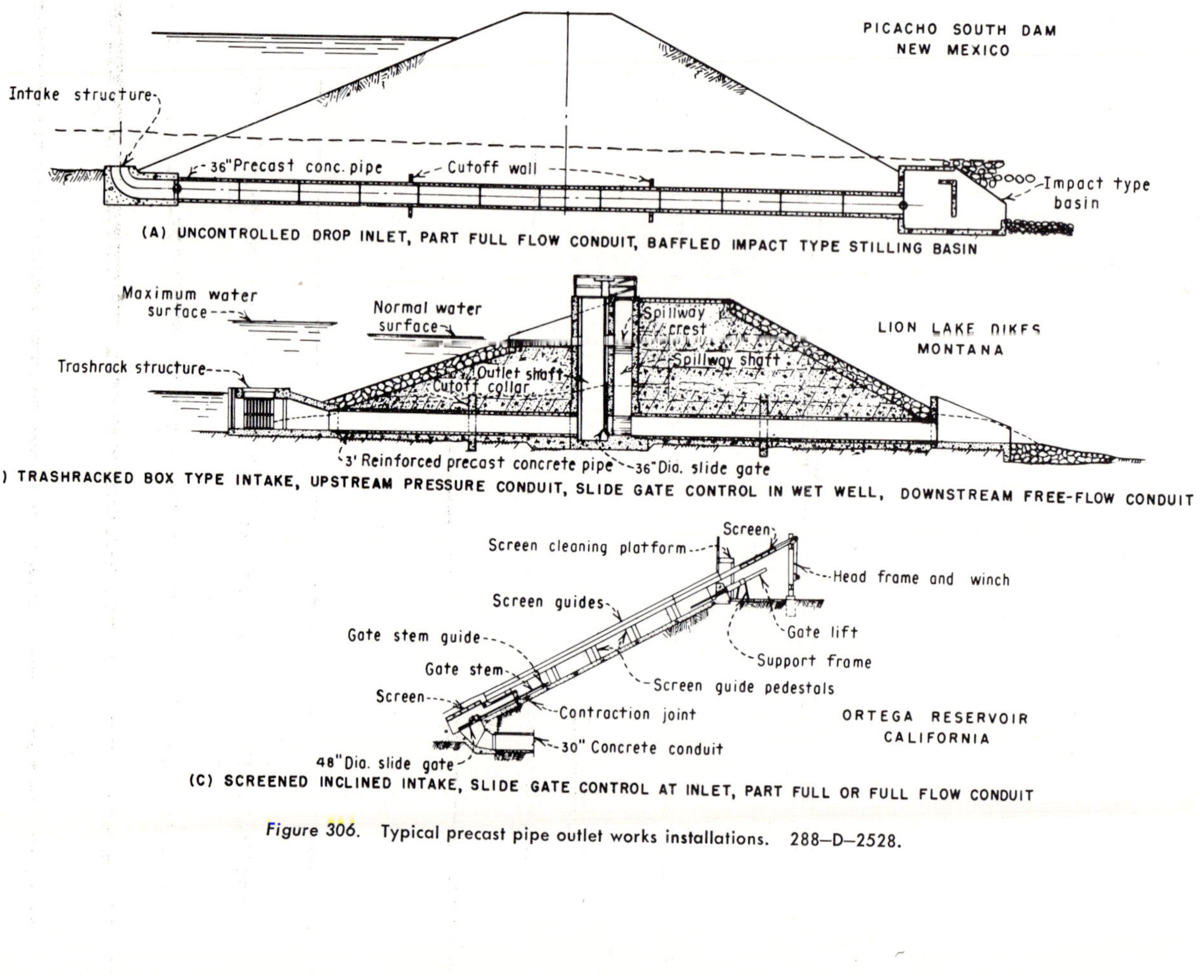

Figure 306. Typical precast pipe outlet works installations. 288–D–2528.

indicated for ordinary outlet works requirements may determine the final outlet works capacity.

222. *Outlet Works Position in Relation to Reservoir Storage Levels.*—The establishment of the intake level and the elevations of the outlet controls and the conveyance passageway, as they relate to the reservoir storage levels, are influenced by many considerations. Primarily, in order to attain the required discharge capacity, the outlet must be placed sufficiently below minimum reservoir operating level to provide head for effecting outlet works flows.

Outlet works for small detention dams are generally constructed near riverbed level since permanent storage space, except for silt retention, is ordinarily not provided. (These outlet works may be ungated in order to retard the outflow while the reservoir temporarily stores the bulk of the flood runoff, or they may be gated in order to regulate the releases of the temporarily stored waters.) If the purpose of the dam is only to raise and divert incoming flows, the main outlet works generally is a headworks or regulating structure at a high level, and a sluiceway or small bypass outlet is provided to furnish water to the river downstream or to drain the water from behind the dam during off-season periods. For dams which impound waters for irrigation, domestic use, or other conservation purposes, the outlet works must be placed low enough to draw the reservoir down to the bottom of the allocated storage space; however, it might be placed at some level above the riverbed, depending on the elevation of the established minimum reservoir storage level.

It is usual practice to make an allowance in a storage reservoir for inactive storage for sediment deposition, fish and wildlife conservation, and recreation. The positioning of the intake sill then becomes an important consideration, since it must be high enough to prevent interference from the sediment deposits, but at the same time low enough to permit either a partial or a complete drawdown below the top of the inactive storage.

As is discussed in section 234, the size of an outlet conduit for a required discharge varies according to an inverse relationship with the available head for producing the discharge. This relationship may be expressed by the following equation:

$$H_T = K_1 h_v, \text{ or } H_T = K_2 \frac{Q^2}{a^2},$$

where:

H_T=the total available head for producing flow,

Q=the required outlet works discharge, and

a=the required area of the conduit.

The above relationship for a particular design is illustrated on figure 307(A). In this example, if the head available for the required outlet works discharge is increased from 1.6 to 4.6 feet, the corresponding conduit diameter can be decreased from 6 to 4.75 feet. This shows that the conduit size can be significantly reduced if the inactive storage level can be increased. The reduction in active storage capacity resulting from increasing the inactive storage level 3 feet would have to be compensated by adding an equivalent amount of capacity to the top of the pool. By referring to the reservoir capacity curve, figure 307(B), it will be apparent that for equivalent storages (represented by *de* and *gh*) the 3 feet of head represented by ordinate *cd* added to obtain a reduced outlet works size would require a much smaller increase (represented by the ordinate *fg*) in the height of the dam. Thus, economic studies can be utilized to determine the proper outlet size in relation to the minimum reservoir storage level.

Where an outlet is placed at riverbed level to accommodate the construction diversion plan or to drain the reservoir, the operating sill can be placed at a higher level to provide a sediment and debris basin and other desired inactive storage space, or the intake can be designed to permit raising the sill as sediment accumulates. During the construction period, a temporary diversion opening can be formed in the base of the intake for handling diversion flows and later closed with a plug. For emptying the reservoir, a bypass can be installed around the intake at riverbed level, either emptying into the lower portion of the conduit or passing under it. Delivery of water to a canal at a higher level can be made by a pressure riser pipe connecting the conduit to the canal.

223. *Conditions Which Determine Outlet Works Layout.*—The layout of a particular outlet works will be influenced by many conditions relating to the hydraulic requirements, to the site adaptability and the interrelation of the outlet works to the construction procedures, and to other appurtenances of the development. Thus, an outlet works leading to a high-level canal or into a closed pipeline might differ from one emptying into the river. Similarly, a scheme in which the outlet works is used for diversion might vary from one where diversion is effected by other means. In certain instances, the proximity of the spillway may permit combining some of the outlet works and spillway components into a single structure. As an example, the spillway and outlet works layout might be arranged so that discharges from both structures will empty into a common stilling basin. An interesting arrangement in which a spillway and outlet works are combined into a single structure is illustrated on figure 308. In this installation the outlet works intake encircles the drop inlet tower of the spillway, and the outlet conduit extends along the top of the spillway conduit and empties into the latter some distance downstream from the spillway inlet.

The topography and geology of a site may have a great influence on the layout selection. Some sites may be suited only for a cut-and-cover conduit type of outlet works, while at other sites either a cut-and-cover conduit or a tunnel can be selected. Unfavorable foundation geology, such as deep overburdens or inferior foundation rock, will obviate the selection of a tunnel scheme. On the other hand, sites in narrow canyons with steep abutments may make a tunnel outlet the only choice. Because of confined working space and excessive costs where hand construction methods must be employed, it is not practicable to make a tunnel smaller than about 6 feet in diameter. If constructed of precast material or if cast-in-place with the inside bore formed by a prefabricated liner, a cut-and-cover conduit can be constructed to almost any size. Thus the minimum size dictated by construction conditions, as compared to the size established by hydraulic requirements, will have considerable influence on the choice of alternative cut-and-cover conduit or tunnel schemes.

Some sites favorable for a tunnel outlet may have unfavorable portal conditions which make it difficult to fit the inlet and exit structures to the remainder of the outlet works. In this situation, a central tunnel with cut-and-cover conduits leading to and away from the tunneled portion of the outlet may prove to be feasible. Such an arrangement is shown on figure 304 for Glen Anne Dam.

If water is to be taken from a reservoir for domestic use, special consideration must be given to the positioning of the intake. To assure the proper quality of the water, it may be necessary to draw from different levels of the reservoir at different seasons of the year or to restrict the draft to specific levels, depending on the reservoir stage. To prevent silt from being carried into the outlet system, intake locations at low points or pockets in the reservoir must be avoided. Similarly, intakes must not be placed at points in the reservoir where stagnant water or algae can accumulate or where prevailing winds will drift debris or undesirable trash to the intake entrance.

224. *Arrangement of Outlet Works.*—An outlet works for a low dam, whether it is to divert water into a canal or release it to the river, often may consist of an open channel or a cut-and-cover structure placed at the dam abutment. The structure might consist of a conventional open flume or rectangular channel with a gate similar to that used for ordinary spillway installations, or it might be regulated by a submerged-type gate placed to close off openings in a curtain or head wall. Where the outlet is to be placed through a low earthfill embankment, a closed-type structure might be used which may consist of single or multiple units of buried pipe or box culverts placed through or under the embankment. Flow for such an installation could be controlled by gates placed at the inlet or placed at an intermediate point along the conduit, such as at the crest of the embankment, where a shaft would be provided for gate operation. Downstream from the control structure, the channel would

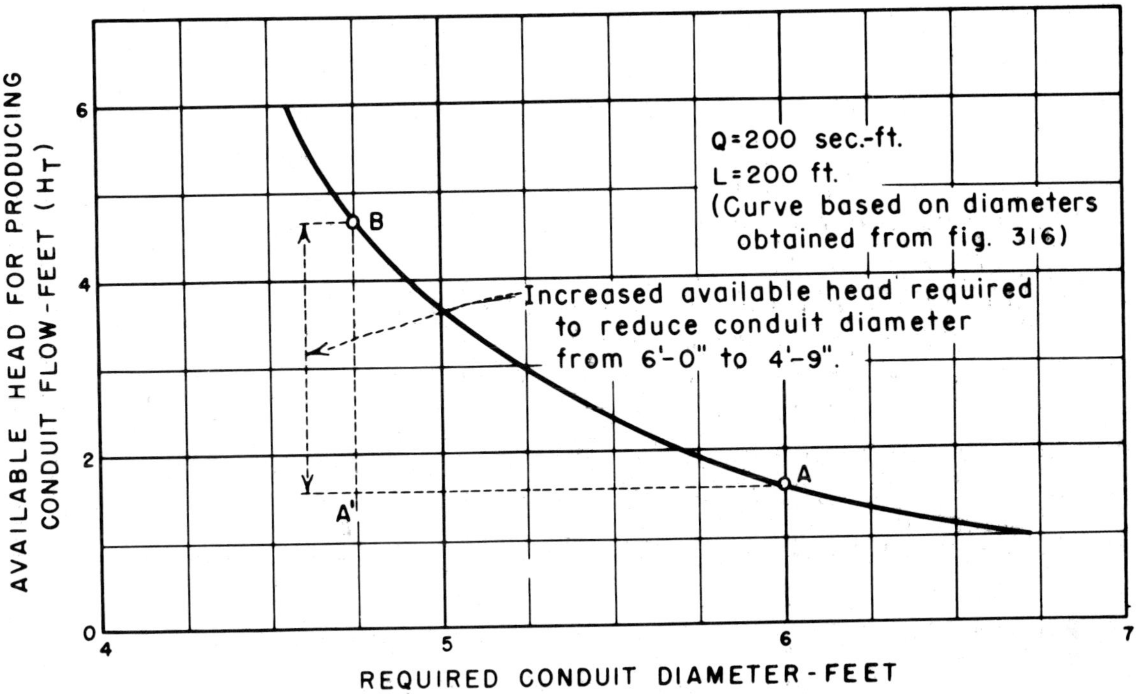

(A) RELATION OF CONDUIT SIZE TO AVAILABLE HEAD

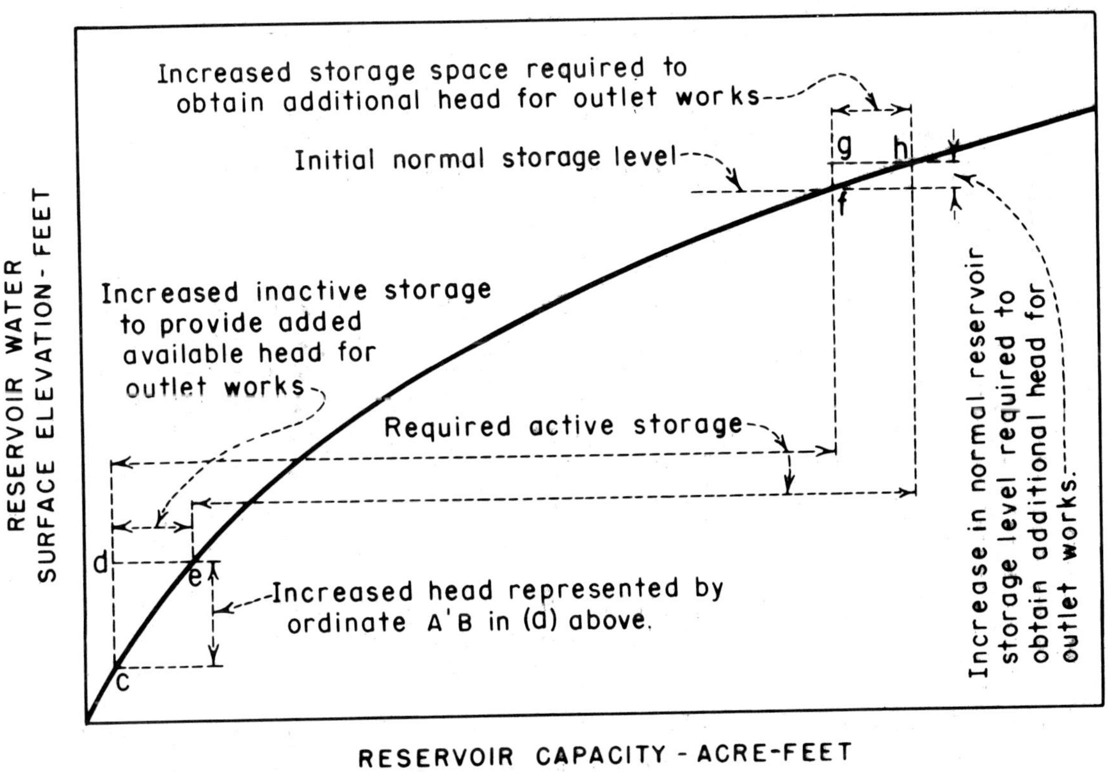

(B) RELATION OF CONDUIT SIZE TO NORMAL STORAGE LEVEL

Figure 307. Relation of minimum design head to conduit size. 288–D–2529.

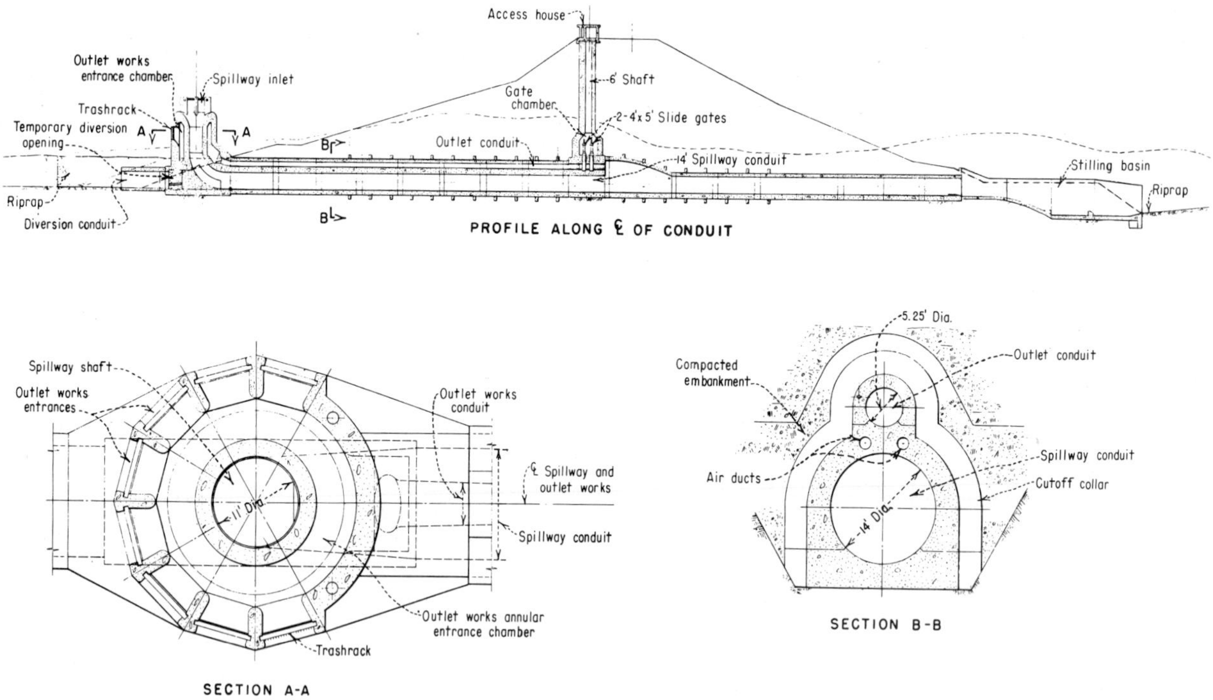

Figure 308. **Combined spillway and outlet works for Heart Butte Dam in North Dakota. 288–D–2530.**

continue to the canal or to the river where, depending on the exit velocities which might prevail for the particular installation, a stilling device similar to one described in chapter IX might be employed. Figure 300 shows typical installations of arrangements as described above.

For higher earthfill dams where an open channel outlet structure would not prove feasible, the outlet might be carried through, under, or around the dam as a cut-and-cover conduit or through the abutment as a tunnel. Depending on the position of the control device, the conduit or tunnel could be free flowing, flowing under pressure for a portion of its length, or flowing under pressure for its entire length. Intakes might be arranged to draw water from the bottom of the reservoir, or the inlet sills might be placed at some higher reservoir level. Dissipating devices similar to those described in chapter IX could be utilized at the downstream end of the conduit. The outlet works also may discharge into the spillway stilling basin. Depending on the method of control and the flow conditions in the structure, access to the operating gates might be by bridge to an upstream intake tower, by shaft from the crest level of the dam, by walkway within the conduit or tunnel with entrance from the downstream end, or by a separate conduit or tunnel access adit. Typical arrangements as described above are illustrated on figures 301 through 304.

For a concrete dam the outlet works installation is usually carried through the dam as a formed conduit or a sluice, or as a pipe embedded in the concrete mass. Intakes and terminal devices can be attached to the upstream and downstream faces of the dam. Often the outlet is formed through the spillway overflow section, using a common stilling basin to dissipate both spillway and outlet works flows. Where an outlet works conduit is installed in the nonoverflow section of the dam or where an outlet must empty into a canal, a separate dissipating device will, of course, be necessary. Instead of a large single conduit, multiple smaller conduits might be utilized in a concrete dam to provide a less expensive as well as a more feasible arrangement for handling the outlet works releases. The conduits might be placed at a single level, or for added flexibility they may be positioned at several levels. Such an arrangement would reduce the cost of the control gates, because of the lower heads on the upper level gates. Typical outlet works installations for concrete dams are shown on figure 305.

Where a diversion tunnel is utilized during the construction of a concrete dam, it is often feasible to convert the tunnel into a permanent outlet works by providing outlet sluices or conduits through the tunnel plug. Ordinarily, the diversion tunnel for a concrete dam will be in good quality rock and will therefore require a minimum of lining protection. Further, the outlet portal of the tunnel will generally be located far enough downstream from the dam so that no dissipating structure will be needed, or at most only a deflector will be required to direct the flow to the downstream river channel.

225. *Location of Outlet Works Controls.*—(a) *General.*—Where an outlet works is ungated, as will be the case with many detention dams, flow in the conduit will be similar to that in a culvert spillway, as described in section 213. Where water must be stored and the release regulated at specific rates, control gates or valves will need to be installed at some point along the conduit.

Gates and valves for outlet works are categorized according to their functional use in the structure. Operating gates and regulating valves are used to control and regulate the outlet works flow and are designed to operate in any position from closed to fully open. Guard or emergency gates are designed to be utilized only to effect closure in the event of failure of the operating gates, or where unwatering is desired either to inspect the conduit below the guard gates or to inspect and repair the operating gates. Occasionally slots are provided at the conduit entrance to accommodate stoplogs or bulkheads so that the conduit can be closed off during an emergency period. For such installations, guard gates may or may not be provided, depending on whether or not the stoplogs can be placed readily if an emergency arises during normal reservoir operating periods.

The control gate for an outlet works can be placed at the upstream end of the conduit, at an intermediate point along its length, or in some instances at the lower end of the structure. Where flow from a control gate is released directly into the open as free discharge, only that portion of the conduit upstream from the gate will be under pressure. Where a control gate or valve discharges into a closed pressure pipe, the control will serve only to regulate the releases; full pipe flow will occur in the conduit both upstream and downstream from the control gate. For the pressure-pipe type, the location of the gate or valve will have little influence on the design insofar as internal pressures are concerned. However, where a control discharges into a free-flowing conduit, the location of the control gate becomes an important consideration in the design of the outlet. The effects of locating the control at various positions in a conduit are discussed in the following subsections.

(b) *Control at Upstream End of Conduit.*—For an outlet works with an upstream control discharging into a free-flow conduit, part full flow will occur throughout the length of the structure. Ordinarily, the operating head and the conduit slope will result in flow at supercritical stage. The structural design of the conduit and the safety and practical aspects of the layout will then be concerned only with the effects of external loadings and outside water pressures on the structure. Along the upstream portion of the conduit and extending until sufficient rock cover is available over a tunnel or until an adequate thickness of impervious embankment is obtained over a cut-and-cover conduit, practically full reservoir head will be exerted against the outside of the conduit barrel. The conduit walls must therefore be designed to withstand such pressures, and details of design must be selected to preserve the watertightness of the conduit. For a cut-and-cover conduit where settlement of the structure (due to foundation consolidation with increasing embankment load) must be anticipated, special care must be taken in design details to prevent the cracking of the conduit barrel and to seal any formed joints, since cracks and open joints will invite excessive leakage or piping of surrounding embankment material into the conduit.

With the controls placed at the upstream end of a conduit, fishscreens, stoplog slots, trashracks, guard gates, and regulating gates or valves can all be combined in a single intake structure. This arrangement will simplify outlet works operation by centralizing all control features at one point. Further, the entire conduit may be readily unwatered for inspection or repair. The intake will consist of a tower rising from the base of the outlet conduit to an operating deck placed above maximum reservoir water level, with the tower located in the reservoir area near the upstream toe of the dam. Access to the structure operating deck will then be possible only by boat, unless an access bridge is provided from the reservoir shore or from the crest of the dam. The intakes at Crane Prairie and Crescent Lake Dams (fig. 301) illustrate typical tower arrangements. Figure 37 is a photograph of the intake tower and access bridge at Crescent Lake Dam. Figure 110 shows the intake tower at Crane Prairie Dam.

(c) *Control at Intermediate Point Along Conduit.*—Where a control gate is placed at an intermediate point along a conduit and discharges freely into the downstream section or where the flow is conveyed in a separate downstream pipe, the internal pressure upstream from the control will be approximately equal to full reservoir head. The structural design and safety aspects of the upstream portion will then be concerned with the effects of both the external loadings and the internal hydrostatic pressure acting on the conduit shell. The watertightness of the conduit in the extreme upstream section will be of less importance because the external and internal hydrostatic pressures will closely balance, and leakage into or out of the conduit will be minimized. However, the external pressure around the conduit will normally diminish with increasing distances from the reservoir. At downstream portions of the pressure conduit, there may be an excess of internal pressure which could cause leakage through joints or cracks into the material surrounding the conduit barrel. The flow from such leaks might follow along the outside of the conduit to the section not under pressure where piping through joints could occur. Where a pressure conduit is carried through an embankment, the development of piping with eventual failure of the dam is a possibility. Where such a conduit comprises a tunnel, leakage through seams in the rock might saturate the hillside overburden above the tunnel and cause a sloughing or landslide on the abutment.

To minimize the possibilities of failures such as those described above, it is normal practice to limit the length of the pressure portion of a cut-and-cover conduit to that part of the outlet upstream from the crest of the dam, or, in some instances, to approximately the upstream one-third of the dam only. Where there is concern regarding the watertightness of a pressure conduit in the upstream portion of a dam, but there are compelling reasons why the control cannot be located near the upstream end of the conduit, that portion upstream from the control may be provided with a steel liner. Such a detail was employed at Shadehill Dam (fig. 303).

For a tunnel installation, except for the possibilities of leakage discussed previously, the location of the control gate is not as critical as it is for cut-and-cover outlets. However, the pressure portion of the tunnel ordinarily should not extend downstream beyond a point where the weight of the column of rock above the tunnel or the side resistance to a blowout is less than the internal pressure forces, unless the tunnel lining is properly reinforced to withstand the internal pressure and a waterproof liner is provided to prevent a buildup of hydrostatic pressures outside the lining.

There may be cases where neither pressure nor free flow is desirable, either for a portion of a conduit or for its entire length. Such instances may occur where it is expected that excessive settlement or movement of the conduit will occur and that cracking and opening of joints cannot be avoided. In this situation, to forestall serious leakage that would occur if a free-flow or pressure conduit were used, a separate small steel pipe can be installed inside the larger conduit to convey the flow. The control gate or valve could be installed at the upper end of the pipe, at some intermediate location, or at the downstream end. If the control gates are not placed at the upstream end, guard gates might be provided at the upstream end of the pipe to effect closure in the event of a leak or failure along any part of the pipe.

Where a control gate discharges into a free-flow conduit, an access and operating shaft extending from the conduit to a level above high water surface in the reservoir will be required. For a cut-and-cover conduit under an earthfill dam, the location of the control gates is usually selected so that the operating shaft is positioned immediately upstream from the crest of the dam (fig. 303). Where flows in the downstream portion of the conduit are carried in a separate pipe, a control chamber is usually provided at the upstream end of the pipe (fig. 302).

The control gates or valves for a conduit or sluice through a concrete dam can be positioned at any point, either upstream to afford free flow in the sluice or at the downstream end to provide pressure pipe flow. Where the sluices are placed in the overflow section of the dam, upstream gates controlling the entrance or valves operated from an interior gallery in the dam are ordinarily employed. Where the outlets are placed in the nonoverflow section, either upstream gates or downstream valves are utilized (fig. 305).

B. OUTLET WORKS COMPONENTS

226. *General.*—For an open channel outlet works or for a conduit-type outlet where part full flow prevails, the control gates or valves are the determining factors which establish the outlet works capacity. Where an outlet works operates as a pressure pipe, the size of the waterway as well as that of the control device determines the capacity.

The overall size of an outlet works is determined by its hydraulic head and the required discharge capacity. The selection of the size of some of the component parts of the structure, such as the tunnel, is dictated by practical considerations or by collateral requirements such as diversion. Since, as discussed in part C of this chapter, the capacity of a closed system outlet is influenced by the hydraulic losses through the components, the sizes of various features can be changed in relation to one another for a given capacity. For example, a streamlined inlet may permit the installation of a smaller gate for a given size conduit, or a larger gate may allow the use of a smaller conduit. Or, for a given discharge, enlargement of the upstream pressure conduit of a closed pipe system may permit reduction in the size of the downstream pressure pipe and consequently in the size of the downstream conduit. The determination of the best overall layout to achieve economy in the design may, therefore, require alternative studies involving various trial sizes of the different components of the outlet works.

When the type of waterway is chosen and the method of control is established, the associated structures to complete the layout can be selected. The type of intake structure will depend on its location and function and on the various appurtenances such as fishscreens, trashracks, stoplog arrangements, or operating platforms which must be furnished. A means for dissipating the energy of flow before returning the discharge to the river may have to be provided. This might be accomplished by a deflector lip, a stilling basin, or a similar dissipator device. Gate chambers, control platforms, or enclosures may be required to provide operating space and protective housing for the control devices. An outlet works also may require an entrance channel to lead diversion flows or flows when the reservoir is low to the intake structure, and an outlet channel to return releases to the river.

227. *Waterways.*—(a) *Open Channels.*—Open channel waterways for outlet works are similar to those for a canal headworks structure, a sluiceway through a dam, or an ordinary spillway. The waterway will usually consist of a channel or flume placed through the embankment to carry the flow from the reservoir to a canal or to the downstream river level. Details of the design are comparable to those for a gated orifice-controlled spillway.

(b) *Tunnels.*—Because of its inherent advantages, a tunnel outlet works is preferred where abutment and foundation conditions will permit its utilization and if it is economical compared with other types. A tunnel is not in direct contact with the dam embankment, and therefore it provides a much safer and more durable layout than can be achieved with either a cut-and-cover conduit or an open channel structure. A minimum of foundation settlement, differential movement, and structural displacement will be experienced with a tunnel which has been bored through competent abutment material, and seepage along the outer surfaces of the tunnel lining or leakage into the material surrounding the tunnel will be less serious. Furthermore, there is less likelihood that failure of some portion of the tunnel would cause failure of the dam than if a cut-and-cover conduit passed under or through the dam.

Ordinarily, pressure tunnels in competent rock do not require lining reinforced to withstand full internal hydrostatic pressures, since the surrounding rock normally can assume such stresses. If the rock cover has sufficient weight and enough side resistance to prevent blowouts, only an unreinforced lining is necessary to provide watertightness in seamy rock and smoother surfaces for better hydraulic flow.

Where pressure tunnels are placed through less competent foundations, such as jointed or yielding rock, the tunnel lining must be designed to withstand external hydrostatic and rock loadings in addition to internal hydrostatic pressures. At the extreme upstream end of an outlet works tunnel, where external hydrostatic pressures may nearly balance the internal pressures, the lining will need to be reinforced to withstand rock loads only. However, if provision is made for unwatering the tunnel by use of intake gates, bulkheads or stoplogs, an unbalanced hydrostatic condition will exist. At the downstream portions of the tunnel, where outside water pressures diminish, the design of the tunnel lining will need to consider both external loads from the rock and internal water pressures.

For free-flow tunnels in competent rock, lining might be provided only along the sides and bottom to form a smooth waterway. In less competent material, lining of the complete cross section may be necessary to prevent caving. For that portion of a free-flow tunnel immediately adjacent to the reservoir or just downstream from a pressure tunnel, cognizance must be taken of the possibility of hydrostatic pressure buildup behind the lining due to leakage through the walls of the pressure tunnel or to seepage from the reservoir. Ordinarily, such external water pressure can be reduced by grouting and by providing drain holes through the lining of the free-flow tunnel.

The need for lining a tunnel in which an independent pipe is installed depends entirely on the competency of the rock to stand unsupported. Since such a tunnel is used to house the pressure pipe and provide access to an upstream gate, lining sufficient to avoid rock falls might be provided for protection of the pipe and operating personnel.

For a pressure tunnel a circular cross-sectional shape is the most efficient, both hydraulically and structurally. For a free-flow tunnel a horseshoe shape or a flat bottom tunnel will provide better hydraulic flow, but it is not as efficient as the circular shape for carrying external loads. For small tunnels under only moderate heads the horseshoe-shaped pressure tunnel and either the horseshoe or the flat-bottomed free-flow tunnel may be permissible, depending on foundation conditions. As discussed in section 223, it is not practicable to provide tunnels much smaller than about 6 feet in diameter. The structural design of tunnels, including reinforcement of linings, is discussed in section 241.

(c) *Cut-and-Cover Conduits.*—If a closed conduit is to be provided and if foundation conditions are not suitable for a tunnel, or if the required size of the waterway is too small to justify the minimum-size tunnel, a cut-and-cover conduit must be used. Since such a conduit passes through or under the dam, conservative and safe designs must be used. Numerous failures of earthfill dams caused by improperly designed or constructed cut-andover outlet conduits have demonstrated the need for conservative procedures.

A conduit should be placed on the most competent portion of the dam foundation. Details of the design must allow for expected settlement, shrinkage, and lateral or longitudinal displacement without interfering with the continuity of the structure which must provide a safe and leakproof waterway.

Where bedrock occurs at the site, every attempt should be made to place the entire conduit on such a foundation. If this is not physically or economically feasible, the structure should be located where overburden is shallow so there will be a minimum of foundation settlement. If a uniform foundation exists and it is determined that foundation settlement will not be excessive, the excavation for the conduit should be to exact grade and the conduit supported on undisturbed material. Where the conduit foundation in its natural state is not suitable, the unsuitable material should be excavated to a depth where a material competent to support the load is reached, and the excavation should be refilled with compacted material of desired stability and impermeability. Unsuitable foundation materials include those which are so permeable as to permit excessive seepage, those subject to excessive settlement on loading, and those subject to settlement on saturation of the foundation by the reservoir. These materials are described in chapter VI. In all cases, regardless of the nature of the foundation, the contact of the conduit with the foundation must provide a watertight bond, free of void spaces or unconsolidated areas.

Cut-and-cover conduits must be designed with sufficient strength to withstand the load of the fill overlying the structure. Pressure conduits must also be designed to resist an internal hydrostatic pressure loading equal to full reservoir head. Design loadings for conduits are further discussed in section 242.

The adaptability of a cut-and-cover conduit and the desirability of utilizing such a conduit as a pressure pipe or as a free-flow waterway are discussed in section 225. Since a cut-and-cover conduit in most instances must be constructed before the embankment, conduit settlement will follow the foundation settlement resulting from the embankment loading. The conduit settlement therefore will be maximum at the point of highest fill and will diminish toward each end. Structure details must be selected to allow for such settlement, and conduit profiles must be adjusted to take account of the drop in grade near the center of the dam. Joint treatment and reinforcement requirements are discussed in section 242.

228. Controls.—(a) *Control Devices.*—Selection of the outlet works arrangement for small dams should be based on the use of commercially available gates and valves or relatively simple gate designs, rather than on the use of special devices which will involve expensive design and fabrication costs. Cast-iron slide gates, which may be used for control and guard gates, are available for both rectangular and circular openings and for design heads in excess of 50 feet. Simple radial gates are available for ordinary surface installations, and top-seal radial gates can be secured from manufacturers on the basis of simple designs and specifications. For low heads, commercial gate valves and butterfly valves are suitable for control at the downstream end of pressure pipes if they are designed to operate under free discharging conditions. They are also suitable as inline guard valves for wide-open operation, and they can be adopted for inline control valves if air venting of the pipe is provided immediately downstream from the valve.

(b) *Arrangement of Controls.* — Flows through low-head outlet works can be controlled by various devices, as shown on figure 300. A surface radial gate may be installed in an open channel, as shown for Putah Diversion Dam. Top-seal radial gates installed at the entrance or within a culvert outlet works are shown for Flatiron Dam and for Camp Creek and Bartley Diversion Dams. Slide gates, similar to those shown for Woodston and Fort Sumner Diversion Dams, may be used to control flows through either open channel or culvert outlet works which are provided with headwall structures.

Upstream gate controls for conduits are generally placed in a tower structure, with the gate hoists mounted on the operating deck (fig. 301). With this arrangement the tower must extend above the maximum water surface.

If controls are located at some intermediate point along the conduit, slide gates or top-seal radial gates can be used, operating in a wet wall shaft which extends vertically from the conduit level to the level of the crest of the dam. These arrangements are typified by the installations shown on figure 303.

A variation of the slide gate control which will eliminate the need for a wet well shaft is possible. In this instance watertight bonnets are provided over the gate slots and the gates are operated either from a dry shaft or from an operating chamber located above the conduit level. Watertight bushings are provided where the gate stems extend through the bonnets.

Valves also can be used as controls at intermediate points along conduits. A dry well is provided and the valve is placed in a length of pipe whose upstream end is encased in a concrete plug. This type of installation is illustrated in figure 302. If the flow is carried by separate pipe in a conduit sufficiently large to afford access along the pipe from the downstream end, a domed chamber can be used rather than a dry well shaft. Such a chamber is provided at Soldier Canyon Dam as shown on figure 304.

If a concrete dam utilizes a slide gate control on its upstream face, the gate frame and stem guides can be mounted directly on the concrete face and the hoist can be placed on a platform cantilevered from the crest of the dam. If the gate is placed at an intermediate point along a conduit formed through the concrete dam, the gate can be operated either in a wet well with the hoist placed at the crest of the dam, or from a gallery if the watertight bonnet cover arrangement is provided over the gate well. Inline valves can also be operated from the gallery or from a chamber formed inside the dam. A control valve placed on the end of the conduit at the downstream face of the dam can be operated from a platform extending from the face of the dam. Typical installations are illustrated on figure 305.

(c) *Control and Access Shafts.*—Where a free-flow conduit is provided downstream from the control devices, access for operating is usually from a shaft located directly over the controls. If the wet well arrangement is utilized, a shaft of sufficient width and breadth to accommodate the several wells must be provided. When the type of controls permits dry well installations, only sufficient space to provide operating room at the bottom of the shaft is needed. A smaller access shaft, either directly above or offset from the chamber, and just large enough to permit passage of removable and replaceable gate parts, will then be needed.

The operating or access shaft for a tunnel outlet works can be sunk into the undisturbed hillside and lined with concrete as necessary to maintain the shaft walls intact. Where such a shaft is used for access and ventilation only, a minimum of wall lining will be needed. Where an access shaft is to be used for a wet well arrangement, adequate lining to make the shaft reasonably watertight will be required. If a cut-and-cover conduit scheme is used, the shaft must be constructed through the dam embankment. The structural design must consider the possibility of settlement and lateral displacement as a result of movement of the embankment. Where a wet well shaft is employed, care must be taken in the design to prevent cracks and the opening of joints which would permit leakage from the interior of the shaft into the surrounding embankment. The walls of the wet well shaft must be designed to resist internal hydrostatic pressure from full reservoir head in addition to the external embankment loading. If a shaft extends through the embankment and projects into the reservoir, external hydrostatic loads must also be considered. The protruding portion of the shaft constitutes a tower which is subject to the ice loads discussed in section 229.

(d) *Control Houses.*—A housing around the outlet controls is sometimes provided where operating equipment would otherwise be exposed or where adverse weather conditions will prevail during operating periods. A house is sometimes provided to enclose the top of an access shaft, although the controls may be located elsewhere. Such houses are usually made sufficiently large to accommodate auxiliary equipment such as ventilating fans, heaters, flow measuring and recording meters, air pumps, small power-generator sets, and equipment needed for maintenance.

229. *Intake Structures.*—In addition to forming the entrance into the outlet works, an intake structure may accommodate control devices. It also supports necessary auxiliary appurtenances (such as trashracks, fish screens, and bypass devices), and it may include temporary diversion openings and provisions for installation of bulkhead or stoplog closure devices.

An intake structure may take on many forms, depending on the functions it must serve as noted above, on the range in reservoir head under which it must operate, on the discharge it must handle, on the frequency of reservoir drawdown, on the trash conditions in the reservoir which will determine the need for or the frequency of cleaning of the trashracks, on reservoir ice conditions or wave action which could affect the stability, and on other such considerations. An intake structure may either be submerged or extended as a tower to some height above the maximum reservoir water surface, depending on its function. A tower must be provided if the controls are placed at the intake, or if an operating platform is needed for trash raking, maintaining and cleaning of fish screens, or installing stoplogs. Where the structure serves only as an entrance to the outlet conduit and where trash cleaning ordinarily will not be required, a submerged structure can be adopted.

The conduit entrance can be placed vertically, inclined, or horizontally, depending on intake requirements. Where a sill level higher than the conduit level is desired, the intake can be a drop inlet similar to the entrance of a drop inlet spillway. A vertical entrance is usually provided for inlets at the conduit level. In certain instances at small installations where the gate is placed and operated on the upstream slope of a low dam, an inclined entrance can be adopted. Such an arrangement is typified by the Ortega Reservoir outlet shown on figure 306. In most cases conduit entrances should be rounded or bellmouthed to reduce hydraulic entrance losses.

The necessity for trashracks on an outlet works depends on the size of the sluice or conduit, the type of control device used, the nature of the trash burden in the reservoir, the utilization of the water, the need for excluding small trash from the outflow, and other factors. These factors will determine the type of trashracks and the size of the openings. Where an outlet consists of a small conduit with valve controls, closely spaced trashbars will be needed to exclude small trash. Where an outlet involves a large conduit with large slide gate controls, the racks can be more widely spaced. If there is no danger of clogging or damage from small trash, a trashrack may consist simply of struts and beams placed to exclude only the larger trees and such floating debris. The rack arrangement will also depend on accessibility for removing accumulated trash. Thus, a submerged rack which seldom will be unwatered must be more substantial than one which is at or near the surface. Similarly, an outlet with controls at the entrance where the gates can be jammed by trash protruding through the rack bars must have a more substantial rack arrangement than if the controls are not at the entrance.

Trash bars usually consist of thin, flat steel bars which are placed on edge from 3 to 6 inches apart and assembled in rack sections. The required area of the trashrack is fixed by a limiting velocity through the rack, which in turn depends on the nature of the trash which must be excluded. Where the trashracks are inaccessible for cleaning, the velocity through the racks ordinarily should not exceed 2 feet per second. A velocity of up to approximately 5 feet per second may be tolerated for racks which are accessible for cleaning.

Trashrack structures also may take on varied shapes, depending on how they are mounted or arranged on the intake structure. Trashracks for a drop inlet intake are generally formed as a cage surmounting the entrance. They may be arranged as an open box placed in front of a vertical entrance or they may be positioned along the front side of a tower structure. Figures 300 through 306 illustrate various arrangements of trashracks at entrances to outlet works.

At some reservoir sites it may be desirable or required to screen the inlet entrance to prevent fish from being carried through the outlet works. Several such installations are illustrated on figure 301. Because small openings must be used to exclude fish, the screens can easily become clogged with debris. Provisions must therefore be made for periodically removing the screens and cleaning them by brooming or water jetting.

Where the control is placed at an intermediate point along a conduit, some means of unwatering the upstream pressure section of the conduit and the intake is desirable to make inspections and needed repairs. Stoplog or bulkhead slots are generally provided for this purpose in the intake or immediately downstream from the intake. In intake towers containing control devices, the stoplog slots are placed upstream from the controls. A circular, flat bulkhead which can drop down over the entrance is generally provided for a drop inlet structure. The bulkhead can be stored on supports near the top of the structure. Closure can then be effected under water by lowering the bulkhead with a cable winch operated from a barge, or from the top of the structure if the reservoir is low enough to expose the upper portion.

For an intake structure with inlet sill above the invert of the conduit, it may be desirable for various reasons to draw the reservoir down below that level. In such an instance a bypass can be provided near the base of the structure to connect the reservoir to the conduit downstream. In other instances, where flow must be maintained while installing or maintaining the control gates and outlet pipes or while repairing or maintaining the free-flow conduit concrete, it may be desirable to carry a separate pipe under or alongside the conduit to bypass it entirely. In either case, the bypass inlet can be placed in the intake structure and usually can be controlled by a slide gate mounted on one of the faces of the structure and operated from some higher level.

Where winter reservoir storage is maintained and the surface ices over, the effect of such conditions on the intake structure must be considered. Where reservoir surface ice can freeze around an intake structure, there is danger to the structure not only from the ice pressures acting laterally but also from the uplift forces if a filling reservoir lifts the ice mass vertically. These effects must be considered when the advantages or disadvantages of a tower are compared with those of a submerged intake.

If a tower is constructed where icing conditions present a hazard, ice may be prevented from forming around the structure by the subsurface release of air. The air causes the slightly warmer water at lower depths to rise and mix with the colder surface water, thus preventing freezing. If an insufficient supply of warmer water is available, as when the approach channel to the tower is shallow or the reservoir storage is small, the compressed air may actually cause freezing around the structure.

230. *Terminal Structures and Dissipating Devices.*—The discharge from an outlet, whether through gates, valves, or free-flow conduits, will emerge at a high velocity, usually in a nearly horizontal direction. For a free-flow conduit, deflector devices might be employed to direct the high-velocity flow away from the outlet structure and past the downstream toe of the dam if erosion-resistant bedrock exists at shallow depths in the downstream channel. Where softer foundations exist, a dissipating device might be provided to absorb the energy of flow before it is returned to the river or canal. The flow from valves at the end of an outlet will generally be in the form of a jet, which can be discharged directly into the river, into a plunge basin downstream from the outlets, or into a hydraulic jump-type basin.

Where an outlet is terminated as a submerged pipe, a stilling well dissipator is sometimes employed to dissipate the flow energy. This device consists of a vertical water-filled well in which dissipation is achieved by turbulence and by diffusion of the incoming flow. The incoming flow can be directed horizontally into the well near the bottom, or it may be directed vertically downward into the well through a vertical pipe and released near the bottom. In both cases the flow rises upward and emerges out of the top of the well.

Terminal structures for free-flow conduit outlet works are essentially the same as those for spillways. The hydraulic designs of such basins are discussed in part E of chapter IX. The design of basins to dissipate jet flow and the design of stilling wells to dissipate submerged pipe flow are discussed in section 237.

231. *Entrance and Outlet Channels.*—An entrance channel leading to the outlet works intake and an outlet channel to deliver flow to the river downstream, are often required with a tunnel or cut-and-cover conduit layout. An entrance channel may be required to convey diversion flows to a conduit placed in an abutment, or to deliver water to the outlet works intake during low reservoir stage. Outlet channels may be required to convey discharges from the end of the outlet works to the river downstream or to a canal. All such channels should be excavated to stable slopes and to dimensions which will provide nonscouring velocities. Entrance channel velocities are usually made less than those through the trashracks, and the entrance channel is often widened near the intake structure to permit a smooth, uniform flow into all trashrack openings.

The outlet channel dimensions and the need for lining or riprap protection will depend on the nature of the material through which the channel is excavated. Occasionally a control or a measuring station is placed in the outlet channel, in which event the selection of the grade and cross section of the channel becomes an important consideration. The effects of aggradation or degradation of the main river channel must be considered in selecting the outlet works outlet channel dimensions.

C. HYDRAULIC DESIGN

232. *Nature of Flow in Outlet Works.*—The hydraulics of outlet works usually involve either one or both of two conditions of flow—open channel (or free) flow and full conduit (or pressure) flow. Analysis of open channel flow in outlet works, either in an open waterway or in a part full conduit, is based on the principle of steady nonuniform flow conforming to the law of conservation of energy. Full pipe flow in closed conduits is based on pressure flow, which involves a study of hydraulic losses to determine the total heads needed to produce the required discharges.

Hydraulic jump basins, baffle or impact block dissipators, or other stilling devices normally are employed to dissipate the energy of flow at the downstream end of the outlet works. Many of these devices are designed on the basis of the law of conservation of momentum.

233. *Open Channel Flow in Outlet Works.*—Flow in an open channel outlet works will be similar to that in open channel spillways, which is discussed in chapter IX. Where unsubmerged radial or slide gates are used, discharges through the control with the gates completely raised will be open crest flow as computed by equation (3) of chapter IX:

$$Q=CLH^{3/2}$$

Discharge coefficients applicable to various crest arrangements are discussed in section 197.

When open channel outlet flow is controlled by partly open surface gates, or where top-seal radial gates or submerged slide gates control the flow, sluice flow will result. Discharges for such flow are given by equation (7) of chapter IX:

$$Q=\frac{2}{3}\sqrt{2g}\,CL\,(H_1^{3/2}-H_2^{3/2})$$

Discharge coefficients for sluice control can be determined from figure 257 or table 33 (sec. 235).

In instances where there is high tailwater due to canal water surfaces or to downstream influences in the streambed, the control openings may be partly or entirely submerged. For such conditions the discharge through the control will be in accordance with submerged

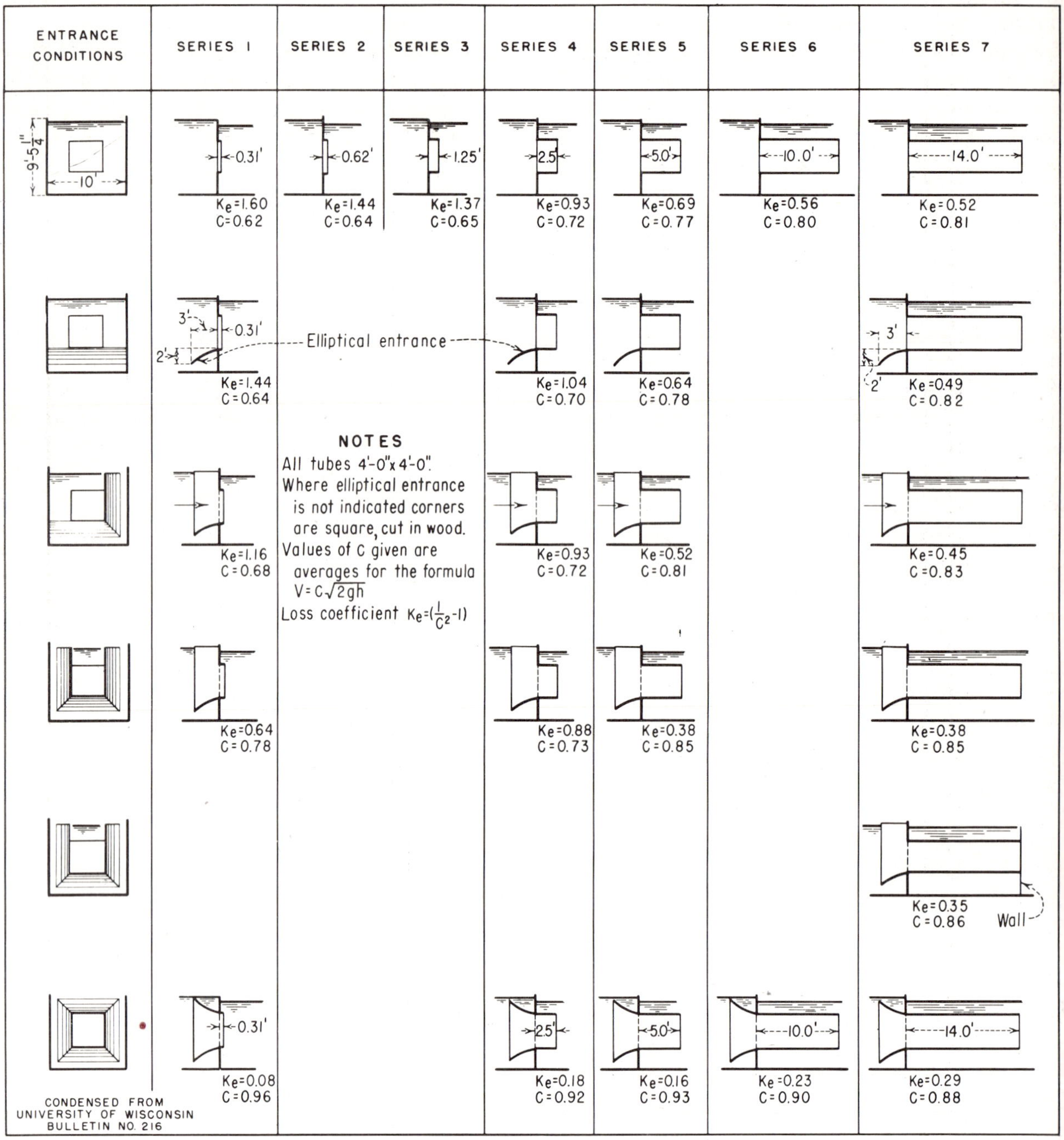

Figure 309. Flow through submerged tubes. 288–D–2531.

orifice or tube flow as computed by the equation.

$$Q=CA\sqrt{2gH} \quad (1)$$

where:

A=the area of the opening,

H=the difference between the upstream and downstream water levels, and

C=the coefficient of discharge for submerged orifice or tube flow.

Coefficients for various conditions of orifice

suppression and tube geometry can be evaluated from figure 309, or from published data in various hydraulic handbooks [1, 2][2] and textbooks.

Flow in an open channel downstream from the headworks will be at either subcritical or supercritical stage, depending on the flow conditions through the control structure. In either case, flow depths and velocities throughout the channel can be determined from Bernoulli's equation, as discussed in section 203.

Flow in ungated outlet conduits will be similar to that in a culvert spillway, as discussed in section 213. Where the inlet geometry and the conduit slope are such that the control will remain at the inlet, part full flow will prevail and flow depths and velocities will be in accordance with the Bernoulli theorem for open channel flow. When flow from a pressure conduit discharges into a free-flow conduit, the flow in the latter most often will be at supercritical stage, with flow depths and velocities comparable to those which would prevail in an open channel. Computation procedures to determine the flow conditions according to the Bernoulli equation are presented in section 203.

Outlet conduits flowing partly full should be analyzed using maximum and minimum assumed values of the coefficient of roughness, n, when evaluating the required conduit size and the energy content of the flow as is done for spillway design (see sec. 203). To be assured of a sufficient conduit size to allow for air swell and surges, values of n of about 0.018 should be assumed in computing the depth or area of flow in a concrete-lined conduit. For computing the energy of flow at the end of the conduit to determine dissipator design, a value of n of about 0.008 should be assumed. To assure a free surface in the conduit for all stages of flow, and to guarantee against sealing of some portion from splashing or surging, the conduit should be designed to flow not more than 75 percent full at maximum capacity.

Terminal deflectors or energy dissipating devices placed at the downstream end of free-flow outlet conduits will be similar to those discussed in part E of chapter IX for spillways. Transitions to diverge the flow from the conduit portal to the stilling device and the allowable convex curvature of the floor entering the stilling device are determined as discussed in section 204.

[2] Numbers in brackets refer to items in the bibliography, sec. 244.

234. *Pressure Flow in Outlet Conduits.*—If a control gate is placed at some point downstream from the conduit entrance, that portion above the control gate will flow under pressure. An ungated conduit may also flow full depending on the inlet geometry. The phenomena and the hydraulic equations for flow through an ungated conduit under pressure are discussed in section 213. The hydraulic design of a gated pressure conduit is similar to that for an ungated pressure conduit, discussed in section 213.

For flow in a closed pipe system, as shown on figure 310, Bernoulli's equation can be written as follows:

$$H_T = h_L + h_{v_2} \tag{2}$$

where:

H_T=the total head needed to overcome the various head losses to produce discharge, and

h_L=the cumulative losses of the system.

Equation (2) can be expanded to list each loss, as follows:

$$H_T = h_t + h_e + h_{b_5} + h_{f_5} + h_{ex_{(5-4)}} + h_{f_4} + h_{c_{(4-3)}} + h_{g_3} + h_{ex_{(3-1)}} + h_{f_1} + h_{b_1} + h_{c_{(1-2)}} + h_{g_2} + h_{v_2} \tag{3}$$

where:

h_t=trashrack losses,
h_e=entrance losses,
h_b=bend losses,
h_c=contraction losses,
h_{ex}=expansion losses,
h_g=gate or valve losses,
h_f=friction losses, and
h_v=velocity head exit loss at the outlet.

In the above equation the number subscripts refer to the various components, transitions, and reaches to which head losses apply.

For a free-discharging outlet, H_T is measured from the reservoir water surface to the center of the outlet gate or the outlet opening. If the

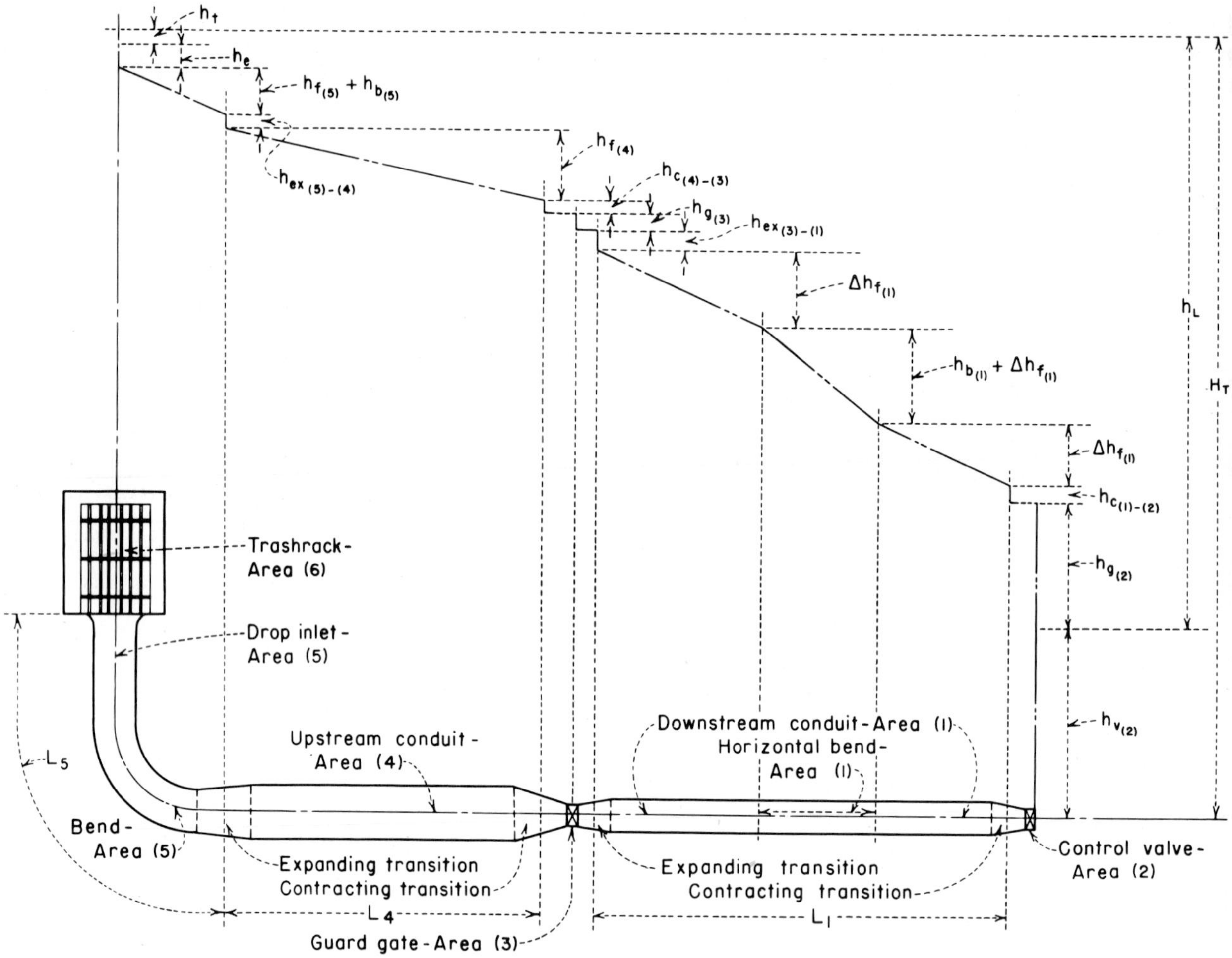

Figure 310. Pictorial representation of head losses in conduit flowing under pressure. 288–D–2532.

outflowing jet is supported on a downstream floor the head is measured to the top of the emerging jet at the point of greatest contraction; if the outlet portal is submerged the head is measured to the tailwater level.

Where the various losses are related to the individual components, equation (3) can be written:

$$H_T=K_t\left(\frac{v_6^2}{2g}\right)+K_e\left(\frac{v_5^2}{2g}\right)+K_{b_5}\left(\frac{v_5^2}{2g}\right)+\frac{fL_5}{D_5}\left(\frac{v_5^2}{2g}\right)$$
$$+K_{ex}\left(\frac{v_5^2}{2g}-\frac{v_4^2}{2g}\right)+\frac{fL_4}{D_4}\left(\frac{v_4^2}{2g}\right)+K_c\left(\frac{v_3^2}{2g}-\frac{v_4^2}{2g}\right)$$
$$+K_g\left(\frac{v_3^2}{2g}\right)+K_{ex}\left(\frac{v_3^2}{2g}-\frac{v_1^2}{2g}\right)+\frac{fL_1}{D_1}\left(\frac{v_1^2}{2g}\right)+K_{b_1}\left(\frac{v_1^2}{2g}\right)$$
$$+K_c\left(\frac{v_2^2}{2g}-\frac{v_1^2}{2g}\right)+K_g\left(\frac{v_2^2}{2g}\right)+K_v\left(\frac{v_2^2}{2g}\right) \quad (4)$$

where:

K_t=trashrack loss coefficient,
K_e=entrance loss coefficient,
K_b=bend loss coefficient,
f=friction factor in the Darcy-Weisbach equation for pipe flow (discussed in sec. 235(b)),
K_{ex}=expansion loss coefficient,
K_c=contraction loss coefficient,
K_g=gate loss coefficient, and
K_v=exit velocity head coefficient at the outlet.

Equation (4) can be simplified by expressing the individual losses in terms of an arbitrarily chosen velocity head. This velocity head is usually selected as that in a significant section of the system. If the various velocity heads for the system shown on figure 310 are related to that in the downstream conduit, area (1), the conversion for "x" area is found as follows:

Since:

$$Q=a_1v_1=a_xv_x,\ a_1^2v_1^2=a_x^2v_x^2,\ \text{and}\ \frac{a_1^2v_1^2}{2g}=\frac{a_x^2v_x^2}{2g}$$

then:

$$\frac{v_x^2}{2g}=\left(\frac{a_1}{a_x}\right)^2\frac{v_1^2}{2g}$$

Equation (4) then can be written:

$$\begin{aligned}H_T=\frac{v_1^2}{2g}\Bigg[&\left(\frac{a_1}{a_6}\right)^2K_t+\left(\frac{a_1}{a_5}\right)^2\left(K_e+K_{b_5}+\frac{fL_5}{D_5}+K_{ex}\right)\\&+\left(\frac{a_1}{a_4}\right)^2\left(\frac{fL_4}{D_4}-K_{ex}-K_c\right)+\left(\frac{a_1}{a_3}\right)^2(K_c+K_g+K_{ex})\\&+\left(\frac{fL_1}{D_1}-K_{ex}+K_{b_1}-K_c\right)\\&+\left(\frac{a_1}{a_2}\right)^2(K_c+K_g+K_v)\Bigg]\end{aligned}\qquad(5)$$

If the bracketed part of the expression is represented by K_L, the equation can be written:

$$H_T=K_L\frac{v_1^2}{2g}\qquad(6)$$

Then:

$$Q=a_1\sqrt{\frac{2gH_T}{K_L}}\qquad(7)$$

235. *Pressure Flow Losses in Conduits.*—(a) *General.*—Head losses in outlet works conduits are caused primarily by the frictional resistance to flow along the conduit side walls. Additional losses result from trashrack interferences, entrance contractions, contractions and expansions at gate installations, bends, gate and valve constrictions, and other interferences in the conduit. As with free-flow conduits, greater than average loss coefficients should be assumed for computing required conduit and component sizes, and smaller loss coefficients should be used for computing energies of flow at the outlet. The major contributing losses of a conduit or pipe system are discussed in the remainder of this section.

(b) *Friction Losses.*—For flow in large pipes, the Darcy-Weisbach formula is most often employed to determine the energy losses due to frictional resistances of the conduit. The loss of head is stated by the equation:

$$h_f=\frac{fL}{D}\frac{v^2}{2g}\qquad(8)$$

where f is the friction loss coefficient. This coefficient varies with the conduit surface roughness and with the Reynolds number. The latter is a function of the diameter of the pipe, and the velocity, viscosity, and density of the fluid flowing through it. Data and procedures for evaluating the loss coefficient are presented in a Bureau of Reclamation engineering monograph [3]. Since f is not a fixed value, many engineers are unfamiliar with its variations and would rather use Manning's coefficient of roughness, n, which has been more widely defined. If the effect of the Reynolds number influence is neglected, and if the roughness factor in relation to the pipe size is assumed constant, the relation of f in the Darcy-Weisbach equation to n in the Manning equation will be:

$$f=\frac{116.5n^2}{r^{1/3}}=\frac{185n^2}{D^{1/3}}\qquad(9)$$

Relationships between the Darcy-Weisbach and Manning's coefficients can be determined from figure B–7 (appendix B).

Where the conduit cross section is horseshoe or rectangular in shape, the Darcy-Weisbach formula does not apply because it is for circular pipes, and the Manning equation may be used to compute the friction losses. Manning's equation as applied to closed conduit flow is:

$$h_f=29.1n^2\frac{L}{r^{4/3}}\frac{v^2}{2g}\qquad(10)$$

Maximum and minimum values of n which may be used to determine the conduit size and the energy of flow are as follows:

	Maximum value	*Minimum value*
Concrete pipe or cast in place conduit	0.014	0.008
Steel pipe with welded joints	.012	.008
Unlined rock tunnel	.035	.020

(c) *Trashrack Losses.*—Trashrack structures which consist of widely spaced structural members without rack bars will cause very little head loss, and trashrack losses in such a case might be neglected in computing conduit losses. When the trash structure consists of racks of bars, the loss will depend on the bar thickness, depth, and spacing. An average approximation can be obtained [2] from the equation:

$$\text{Loss} = K_t \frac{v_n^2}{2g}$$

where:

$$K_t = 1.45 - 0.45\frac{a_n}{a_g} - \left(\frac{a_n}{a_g}\right)^2 \tag{11}$$

In the above:

K_t=the trashrack loss coefficient (empirical),
a_n=the net area through the rack bars,
a_g=the gross area of the racks and supports, and,
v_n=the velocity through the net trashrack area.

Where maximum loss values are desired, assume that 50 percent of the rack area is clogged. This will result in twice the velocity through the trashrack. For minimum trashrack losses, assume no clogging of the openings when computing the loss coefficient, or neglect the loss entirely.

(d) *Entrance Losses.*—The loss of head at the entrance of a conduit is comparable to the loss in a short tube or in a sluice. If H is the head producing the discharge, C is the coefficient of discharge, and a is the area, the discharge

Q is equal to $Ca\sqrt{2gH}$

and the velocity

v is equal to $C\sqrt{2gH}$.

Or,

$$H = \frac{1}{C^2}\frac{v^2}{2g} \tag{12}$$

Since H is the sum of the velocity head h_v and the head lost at the entrance h_e, equation (12) may be written:

$$\frac{v^2}{2g} + h_e = \frac{1}{C^2}\frac{v^2}{2g} \text{ or } h_e = \left(\frac{1}{C^2} - 1\right)\frac{v^2}{2g}$$

Then:

$$K_e = \left(\frac{1}{C^2} - 1\right) \tag{13}$$

Coefficients of discharge for square sluice entrances are shown on figure 309. Coefficients of discharge and loss coefficients for typical entrances for conduits, as given in various texts and technical papers, are listed in table 33.

TABLE 33.—*Coefficients of discharge and loss coefficients for conduit entrances*

	Coefficient C			Loss coefficient K_e		
	Maximum	Minimum	Average	Maximum	Minimum	Average
(a) Gate in thin wall—unsuppressed contraction	0.70	0.60	0.63	1.80	1.00	1.50
(b) Gate in thin wall—bottom and sides suppressed.	.81	.68	.70	1.20	0.50	1.00
(c) Gate in thin wall—corners rounded	.95	.71	.82	1.00	.10	0.50
(d) Square-cornered entrances	.85	.77	.82	.70	.40	.50
(e) Slightly rounded entrances	.92	.79	.90	.60	.18	.23
(f) Fully rounded entrances $\frac{r}{D} \geq 0.15$	.96	.88	.95	.27	.08	.10
(g) Circular bellmouth entrances	.98	.95	.98	.10	.04	.05
(h) Square bellmouth entrances	.97	.91	.93	.20	.07	.16
(i) Inward projecting entrances	.80	.72	.75	.93	.56	.80

(e) *Bend Losses.*—Bend losses in closed conduits in excess of those due to friction loss through the length of the bend are a function of the bend radius, pipe diameter, and the angle through which the bend turns. Although experimental data on bend losses in large pipes are meager, the loss can be related to those determined for smaller pipe. Figure 311(A) shows the coefficients found by various investigators for 90° bends for various ratios of radius of bend to diameter of pipe, and an adjusted curve assumed to be suitable for large pipes.

Figure 311(B) indicates the correction factor to be applied to the values indicated in figure 311(A) for other than 90° bends. The value of the loss coefficient, K_b for various values of $\frac{R_b}{D}$ can be applied directly for circular conduits; for rectangular conduits D is taken as the height of the section in the plane of the bend.

(f) *Transition Losses.*—Head losses in gradual contractions or expansions in a conduit can be considered in relation to the increase or decrease in velocity head, and will vary according to the rate of change of the area and the length of the transition. For contractions the loss of head, h_c, will be approximately equal to $K_c\left(\frac{v_2^2}{2g}-\frac{v_1^2}{2g}\right)$, where K_c varies from 0.1 for gradual contractions to 0.5 for abrupt contractions. Where the flare angle does not exceed that indicated in section 236(b), the loss coefficient can be assumed as 0.1. For greater flare angles, the loss coefficient can be assumed to vary in a straight-line relationship to a maximum of 0.5 for a right angle contraction.

For expansions, the loss of head, h_{ex}, will be approximately equal to $K_{ex}\left(\frac{v_1^2}{2g}-\frac{v_2^2}{2g}\right)$, where K_{ex} is as follows:

Flare angle α	2°	5°	10°	12°	15°	20°	25°	30°	40°	50°	60°
K_{ex} [1]	0.03	0.04	0.08	0.10	0.16	0.31	0.40	0.49	0.60	0.67	0.72
K_{ex} [4]	.02	.12	.16	-----	.27	.40	.55	.66	.90	1.00	----

(g) *Gate and Valve Losses.*—No gate loss need be assumed where a gate is mounted at the entrance to a conduit so that when wide open it does not interfere with the entrance flow conditions. Where a gate is mounted at either the upstream or downstream side of a thin headwall such that the sides and bottom of the jet are suppressed but the top is contracted, loss coefficients shown as item b in table 33 will apply. Where a gate is mounted in a conduit so that the floor, sides, and roof both upstream and downstream are continuous with the gate opening, only the losses due to the slot will need to be considered, for which a value of K_g not exceeding 0.1 might be assumed. For partly open gates, the coefficient of loss will depend on the top contraction; for smaller openings it will approach the value of 1.0 as shown for item b in table 33.

For wide open gate valves K_g will approximate 0.19. Similar to partly open gates, values of the loss coefficient will increase for smaller valve openings. Indicated loss coefficients for partly open gate valves are 1.15 for three-fourths open, 5.6 for one-half open, and 24.0 for one-fourth open. Average values of K_g for butterfly valves in the wide open position are about 0.15; values vary between 0.1 and 0.5, depending on the thickness of the gate leaf in relation to the gross area. Losses in spherical valves are negligible.

(h) *Exit Losses.*—No recovery of velocity head will occur where the release from a pressure conduit freely discharges, or is submerged or supported on a downstream floor. The velocity head loss coefficient, K_v, in these instances is equal to 1.0. When a diverging tube is provided at the end of a conduit, recovery of a portion of the velocity head will be obtained if the tube expands gradually and if the end of the tube is submerged. The velocity head loss coefficient will then be reduced from the value of 1.0 by the degree of velocity head recovery. If a_1 is the area at the beginning of the diverging tube and a_2 is the area at the end of the tube, K_v is equal to $\left(\frac{a_1}{a_2}\right)^2$.

236. *Transition Shapes.*—(a) *Entrances.*—To minimize head losses and to avoid zones where cavitation pressures can develop, the entrance to a pressure conduit should be streamlined to provide smooth, gradual changes in the flow. To obtain the best inlet efficiency, the shape of the entrance should simulate that of a jet discharging into air. As with the nappe-shaped weir, the entrance shape should guide and support the jet with minimum interference until it is contracted to the dimensions of the conduit. If the entrance curve is too sharp or too short, subatmospheric pressure areas which may induce cavitation will develop. A bellmouth entrance which conforms to or slightly encroaches upon the free-jet profile will provide the best entrance shape. For a circular entrance, this shape can be approximated by an elliptical entrance curve represented by the equation:

$$\frac{x^2}{(0.5D)^2}+\frac{y^2}{(0.15D)^2}=1 \qquad (14)$$

where x and y are coordinates whose x-x axis is parallel to and $0.65D$ from the conduit centerline and whose y-y axis is normal to the conduit-centerline and $0.5D$ downstream from the entrance face. The factor D is the diameter of the conduit at the end of the entrance transition.

The jet issuing from a square or rectangular opening is not as easily defined as one issuing from a circular opening; the top and bottom curves may differ from the side curves both in length and curvature. Consequently, it is more difficult to determine a transition which will eliminate subatmospheric pressures. An elliptical curved entrance which will tend to minimize the negative pressure effects is defined by the equation:

$$\frac{x^2}{D^2}+\frac{y^2}{(0.33D)^2}=1 \qquad (15)$$

where D is the vertical height of the conduit for defining the top and bottom curves, and is the horizontal width of the conduit for defining the side curves. The major and minor axes are positioned similarly to those indicated for the circular bellmouth.

For a rectangular entrance with the bottom placed even with the upstream floor and with curved guide piers at each side of the entrance opening, both the bottom and side contractions will be suppressed and a sharper contraction will take place at the top of the opening. For this condition the top contraction curve is defined by the equation:

$$\frac{x^2}{D^2}+\frac{y^2}{(0.67D)^2}=1 \qquad (16)$$

where D is the vertical height of the conduit downstream from the entrance shape.

(b) *Contractions and Expansions.*—To minimize head losses and to avoid cavitation tendencies along the conduit surfaces, contraction and expansion transitions to and from gate control sections in a pressure conduit should be gradual. For contractions, the maximum convergent angle should not exceed that indicated by the relationship:

$$\tan \alpha=\frac{1}{U} \qquad (17)$$

where:

α=an angle of the conduit wall surfaces with respect to its centerline, and

U=an arbitrary parameter $\frac{v}{\sqrt{gD}}$.

The values of v and D are the average of the velocities and diameters at the beginning and end of the transition.

Expansions should be more gradual than contractions because of the danger of cavitation where sharp changes in the side walls occur. Furthermore, as has been indicated in section 235(f), loss coefficients for expansions increase rapidly after the flare angle exceeds about 10°. Expansions should be based on the relationship:

$$\tan \alpha=\frac{1}{2U} \qquad (18)$$

The notations are the same as for equation (17). For usual installations, the flare angle should not exceed about 10°.

The criteria for establishing maximum contraction and expansion angles for conduits flowing partly full are the same as those for open channel flow, as given in section 204(b).

(c) *Exit Transitions.*—When a circular conduit flowing partly full empties into a chute, the transition from the circular section to one with a flat bottom can be made in the open channel downstream from the conduit portal, or it can be made within the conduit so that the bottom will be flat at the portal section.

Two types of transition are commonly used. The first type is made by constructing a straight line of intersection between the 45° points on each side of the invert of the upstream circular section to the corners of the flat bottom at the downstream end of the transition. The radius of curvature between the

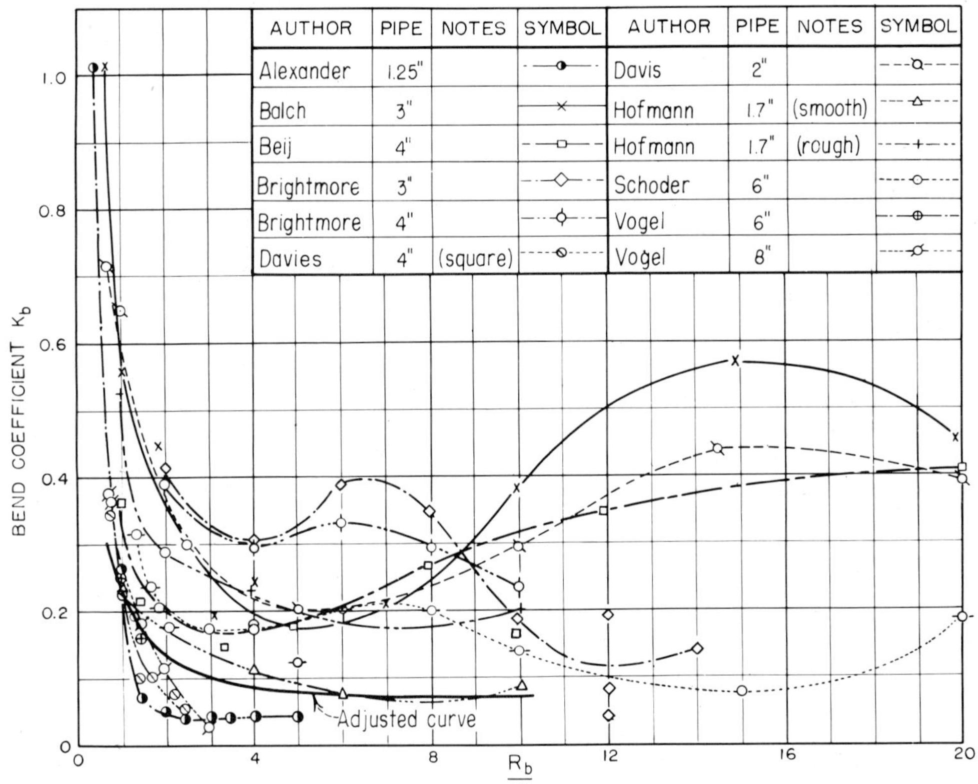

(A) VARIATION OF BEND COEFFICIENT WITH RELATIVE RADIUS FOR 90° BENDS OF CIRCULAR CROSS SECTION, AS MEASURED BY VARIOUS INVESTIGATORS

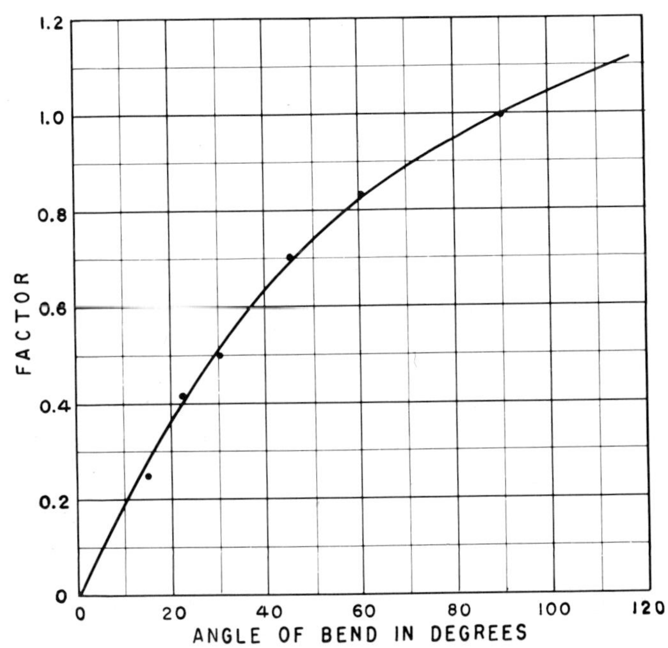

(B) FACTORS FOR OTHER THAN 90° BENDS

Figure 311. Bend loss coefficients. 288–D–2533.

springline and the intersection line increases from the radius of the conduit at the upstream end to a radius of infinity along the vertical wall at the downstream end. The radius of curvature of the bottom of the transition also increases from the radius of the conduit at the upstream end to infinity along the flat bottom of the downstream end. Figure 312 shows a typical example of this type of transition, including the conduit transition table. The distance "X" in figure 312 denotes the flair from the upstream end of the transition to the wider section downstream as in the case of a stilling basin.

The second type of transition is made by gradually decreasing the radius of the circular quadrants in the lower half of the pipe from their initial radius at the upstream end to a radius of zero at the corners of the flat bottom at the downstream end. In both transitions the roof of the conduit remains curved. These transitions must satisfy the expansion criteria established in section 236 (b), above.

For usual installations the length of the transition can be related to the exit velocity. An empirical rule which will give a satisfactory transition is:

$$L \text{ (in feet)} = \frac{vD}{5} \tag{19}$$

where:
D=the exit velocity in feet per second, and
v=the conduit diameter in feet.

Downstream from a free-flow conduit the chute sections, including the transition into a stilling basin, will be governed by open channel flow criteria. Floor curvatures and maximum flare angles should be determined by equations (19) and (21), respectively, of chapter IX. To reduce the length of the open channel portion from the conduit portal to the stilling basin, the beginning of the flare and of the convex curve may be located inside the conduit. This transition may be combined with the transition of the bottom shape.

In certain instances, as illustrated on figure 301, Crane Prairie Dam, and figure 303, Scofield and Newton Dams, an adverse slope and a hump have been employed immediately downstream from the portal to permit more rapid widening of the channel before it enters the basin. No firm criteria have been established for the design of these devices; the details were determined by model tests. Certain inherent disadvantages to this type of design are: (1) care must be taken to avoid a hump of such height that back pressure will cause a hydraulic jump to occur inside the conduit, (2) the floor section at the hump must be made structurally sufficient to withstand the large dynamic forces resulting from impingement of the flow on the rising floor, (3) during periods of no flow a pond which can freeze during the winter is formed in the conduit unless provision is made to drain the sump, and (4) access into the downstream conduit is difficult unless drainage is provided. Depending on tailwater conditions, pumping may be required to provide drainage.

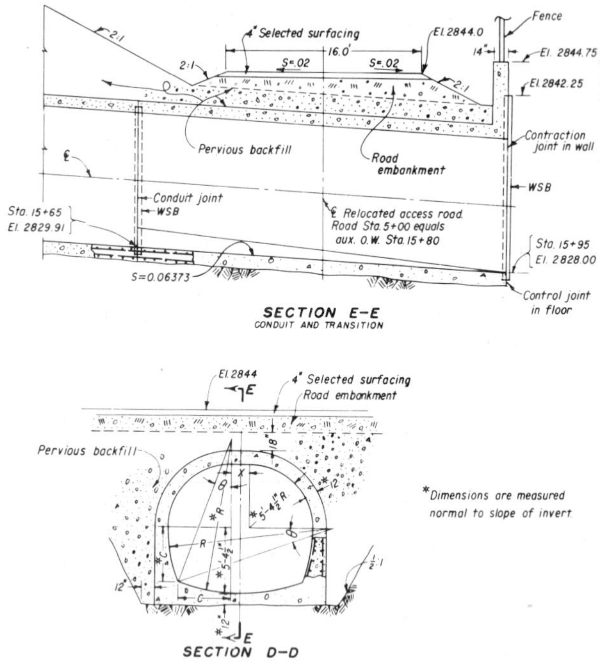

STA.	θ	R	C	X
15+65	45°	5′-4 1/2″	3′-9 5/8″	0
+67	41°31′45″	5′-10 9/16″	3′-10 3/4″	0-2 11/16″
+69	37°59′10″	6′-5 15/16″	4′-0″	0-5 3/8″
+71	34°33′21″	7′-2 3/4″	4′-1 3/16″	0-8 1/16″
+73	31°13′39″	8′-1 1/4″	4′-2 7/16″	0-10 3/4″
+75	27°59′30″	9′-2″	4′-3 5/8″	1′-1 7/16″
+77	24°51′12″	10′-5 3/4″	4′-4 7/8″	1′-4 1/8″
+79	21°47′56″	12′-1 5/8″	4′-6 1/16″	1′-6 3/4″

STA.	θ	R	C	X
15+81	18°49′53″	14′-3 [illegible]″	4′-7 5/16″	1′-9 7/16″
+83	15°56′38″	17′-1 15/16″	4′-8 9/16″	2′-0 1/8″
+85	13°7′42″	21′-2 11/16″	4′-9 7/8″	2′-2 13/16″
+87	10°23′18″	27′-3 7/8″	4′-11 1/8″	2′-5 1/2″
+89	7°42′48″	37′-6 1/4″	5′-0 7/16″	2′-8 3/16″
+91	5°6′22″	57′-9 3/4″	5′-1 3/4″	2′-10 7/8″
+93	2°33′32″	117′-9 3/16″	5′-3 1/16″	3′-1 9/16″
+95	0°	—	5′-4 1/2″	3′-4 1/4″

CONDUIT TRANSITION TABLE
STATIONS ARE AT INVERT

Figure 312. A typical conduit transition. 288–D–2809.

237. *Terminal Structures.*—(a) *General.*—Deflector buckets, hydraulic jump basins, and impact type stilling basins are suitable terminal structures for free-flow conduits, when appropriately used. These structures are commonly used in conjunction with spillways, and their hydraulic designs are discussed in part E of chapter IX. Other types of stilling devices employed more often with outlet works than with spillways are plunge basins and stilling wells. The hydraulic designs of these structures are discussed in this section.

The hydraulic jump stilling basin is most often used for energy dissipation of outlet works discharges. However, where flow emerges from the outlet in the form of a free jet, as will be the case with valve-controlled outlets of pressure conduits, it must be directed onto the transition floor approaching the basin so it will become uniformly distributed before entering the basin. Otherwise proper dissipation of energy will not be obtained.

To evaluate the energy which must be dissipated by the stilling device, the losses through the outlet system should be minimized, as discussed in sections 233 and 235(b). The specific energy immediately downstream from a gate or valve control will equal the exit velocity head based on minimum losses through the pressure system, as measured above the outflowing water surface. If specific energies have not been computed, approximate basin depths can be obtained from figure 268, as discussed in section 206(d).

(b) *Plunge Basins.*—Where the outlet conduit ends with a flip bucket or where flows issue from a downstream control valve or freely discharging pipe, a riprap- or concrete-lined trapezoidal plunge basin might be utilized. Such a basin should be employed only where the jet discharges into the air and then plunges downward into the basin. Tests have shown that if the angle of impingement is too flat the jet will ride and skip across the surface at high velocity. This will cause waves and eddies in the basin sufficient to erode the side slopes, and there will be high exit velocities.

As indicated in section 210, no fixed criteria have yet been established for plunge basins which will provide satisfactory dissipation for all heads, discharges, and incoming jet conditions. However, criteria that were established for several small outlet works plunge basins which have operated reasonably satisfactorily are herewith presented for use only as a preliminary guide to determine approximate basin geometry. The general arrangement of this basin is represented on figure 313. The basin depths were made about one-fifth of the difference in elevation between maximum reservoir water surfaces and maximum tailwater levels. The minimum bottom widths were made the width of the incoming jet, or the width required to limit the average velocity at the end of the basin to about 3 feet per second, whichever was greater.

(c) *Stilling Wells.*—Stilling well designs as described in section 230 are illustrated on figures 314 and 315. The well dimensions and performance criteria for these designs were established from model tests, and general criteria for such designs applicable to various conditions were not determined. The hydraulic stilling action in these devices results from turbulence and diffusion of the incoming high-energy flow into the water bulk in the well, and successful stilling is aided materially by special fillets and diffuser blocks incorporated along the sides and in the corners of the well. The net area of the well is generally selected by limiting the average rising velocity to between 1 and 3 feet per second. The total depth of the well will be dictated by the energy of the incoming flow which must be dissipated, and by the effectiveness of the diffuser blocks and fillets in diffusing the rising flow. Basins with similar criteria can be patterned after those illustrated in the figures. Basins for considerably different conditions should be model tested.

238. *Chart for Estimating Pressure Conduit Sizes.*—Figure 316 is a nomograph for the solution of the equation for pipe flow. By use of this figure, the required conduit diameter can be determined for a given length and gross head from reservoir water surface to the end of the pipe. As will be noted from the equation, an average total head loss coefficient of 0.5 has been assumed for all contributing losses except

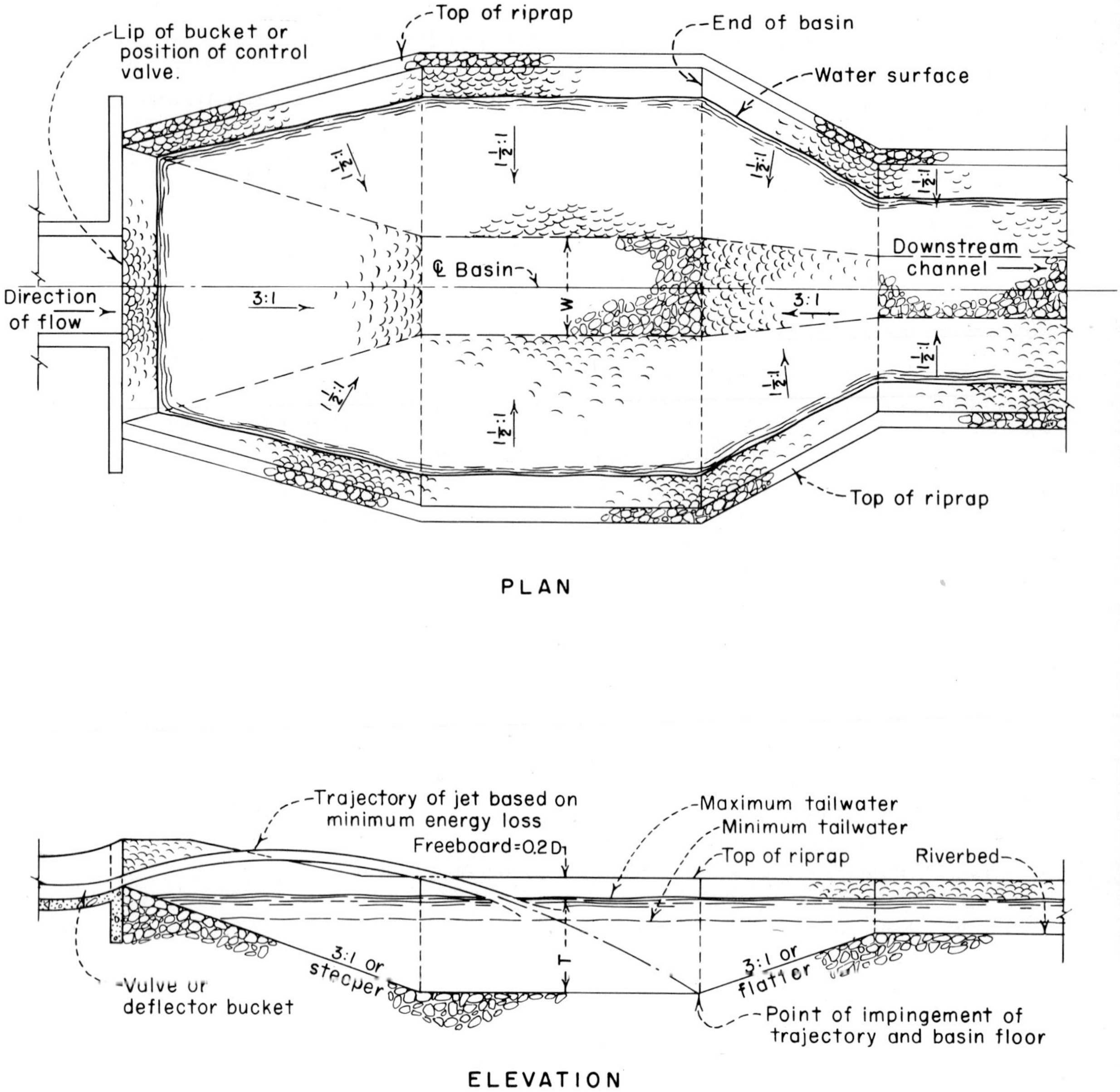

Figure 313. Plunge basin energy dissipator. 288–D–2534.

friction. It will also be noted that a value of n of 0.013 was used in determining friction losses. Adjustments for other values of n can be made by using a compensated value for L determined as shown on the figure.

To illustrate the use of the nomograph, assume the value of n to be 0.013, an available gross head of 5 feet, a conduit length of 300 feet, and a required discharge of 300 second-feet. First, find the intersection of the discharge and length on the right-hand portion of the figure. Lay a straightedge extending from this intersection to the 5-foot mark on the gross head scale. Where the straightedge intersects the 300-foot length on the left portion of the diagram, read the diameter of 5.75 feet. Similarly, for a 40-foot gross head, a length of 300 feet, and a discharge of 300 second-feet,

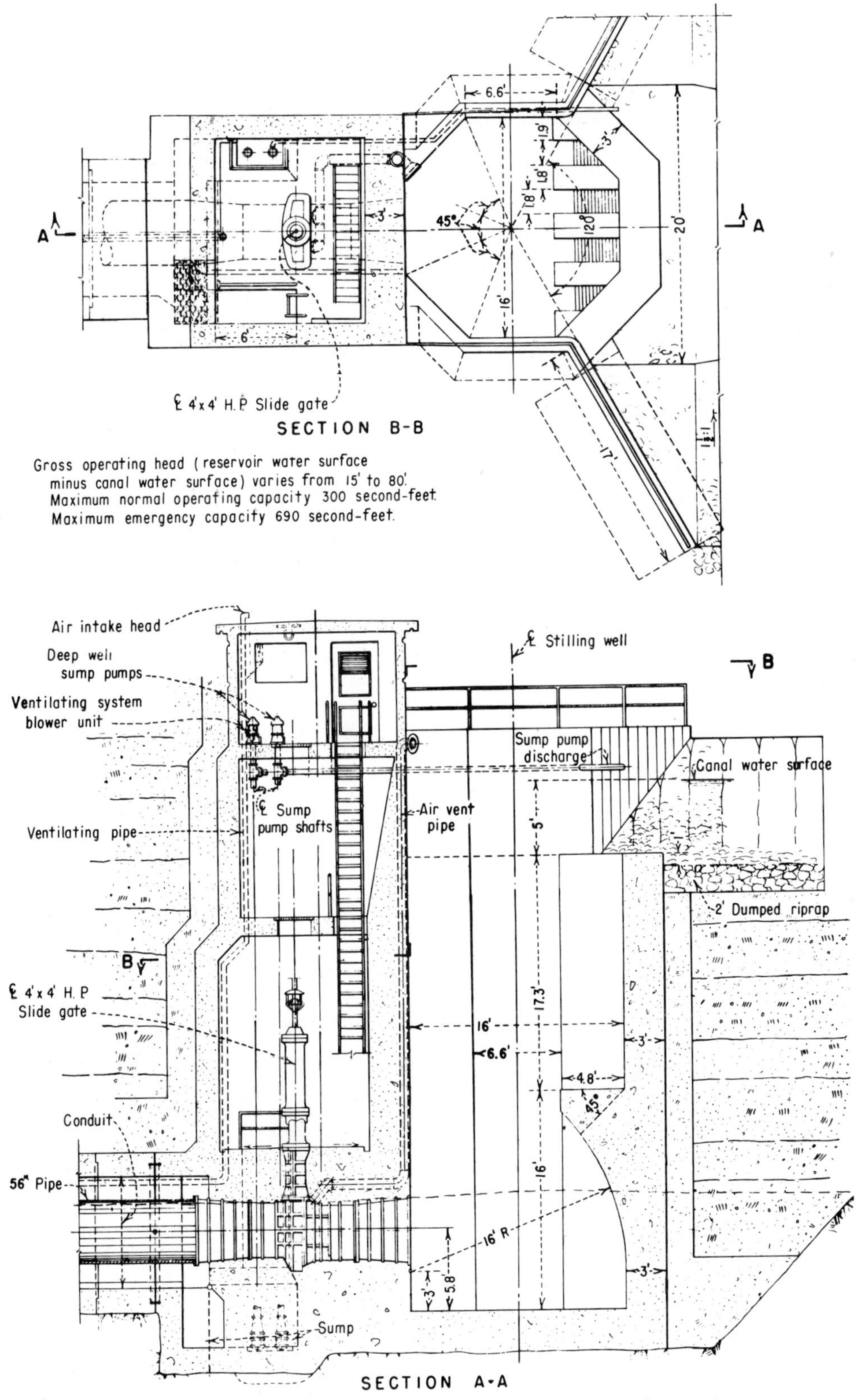

Figure 314. Riser well energy dissipator installation at Trenton Dam, Neb. 288—D—2535.

a 3.7-foot-diameter pipe is found to be needed.

If the size of an unlined tunnel having an estimated value of n of 0.0225 is desired for the same length and discharge as before, the adjusted length of conduit to be used will be equal to $\left(\frac{n}{0.013}\right)^2 L=\left(\frac{0.0225}{0.013}\right)^2 300=900$ feet. The required diameter for a 5-foot gross head is found to be approximately 6.5 feet, from the chart. For a 40-foot head, the required diameter is found to be about 4.3 feet.

The chart can also be used to estimate the sizes of pipe in a compound system by considering each size separately and then adjusting for the recovery of velocity head. For example, assume an upstream conduit 200 feet long and 7 feet in diameter discharging 300 second-feet into a downstream pipe 300 feet long and 4 feet in diameter. On the basis of an n of 0.013, the gross head for a separate pipe of the dimensions of the downstream pipe will be about 29 feet. Similarly, the required head for a separate pipe of the dimensions of the upstream conduit will be about 2 feet. The velocity head, h_v, in the upstream pipe will be equal to:

$$\frac{1}{2g}\left(\frac{Q}{a}\right)^2=\frac{1}{64.4}\left[\frac{(4)(300)}{\pi(7^2)}\right]^2=0.95 \text{ feet.}$$

Then the gross head required will be 29 feet +2 feet −0.95 foot, or approximately 30 feet.

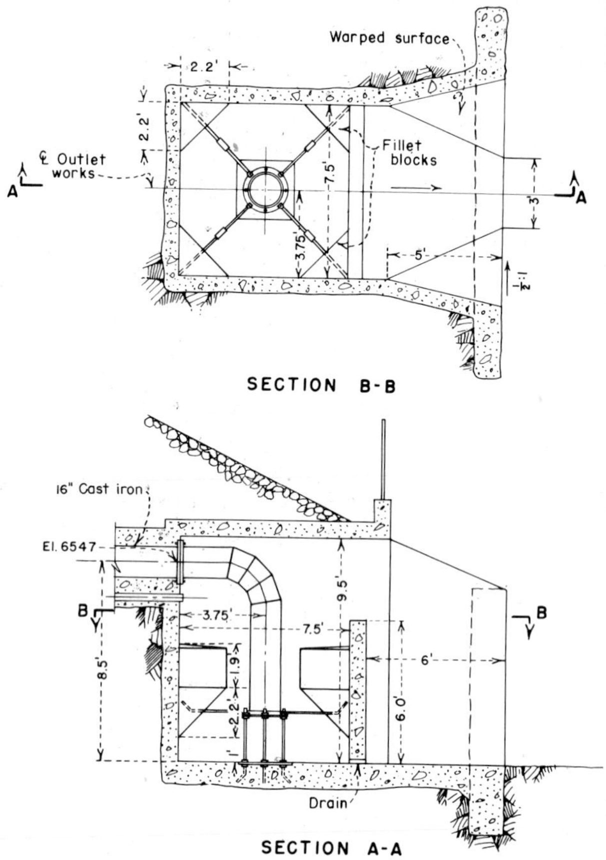

Figure 315. Stilling well energy dissipator installation at Rattlesnake Dam, Colo. 288–D–2536.

TABLE 34.—*Computation of total loss coefficients—Example 1*

Element	Area of element, square feet	$\left(\frac{a_1}{a_x}\right)^2$	Loss	Coefficient symbol	Maximum loss considerations		Minimum loss considerations	
					Coefficient	$\left(\frac{a_1}{a_x}\right)^2\times$coefficient	Coefficient	$\left(\frac{a_1}{a_x}\right)^2\times$coefficient
Trashrack	Gross, 326; net, 272	0.005	Trashrack	[1] K_t	1.09	0.01	0.36	0.00
Entrance	20.0	1.00	Entrance	[2] K_e	.20	.20	.07	.07
Conduit	20.0	1.00	Friction	[3] K_f	.51	.51	.17	.17
			Transition	[4] K_c	.20	−.20	.10	−.10
Gate	9.0	4.9	do	K_c	.20	.98	.10	.49
			Gate	[5] K_g	1.20	5.88	.50	2.45
			Exit	K_v	1.00	4.90	1.00	4.90
Total loss coefficient, K_L					12.28		7.98	

a_1=area of conduit.

a_x=area of element.

[1] From equation (11):

For maximum loss, $K_t=1.45-0.45\left(\frac{136}{326}\right)-\left(\frac{136}{326}\right)^2=1.09$

For minimum loss, $K_t=1.45-0.45\left(\frac{272}{326}\right)-\left(\frac{272}{326}\right)^2=0.36$

[2] From table 33, item h.

[3] $K_f=29.1\ n^2\left(\frac{L}{r^{4/3}}\right)$, L=101 feet, $r=\frac{20}{18}$=1.11, n_{max}=0.014, n_{min}=0.008

[4] Because of the wet well immediately above the transition, assume a maximum K_c of 0.2 and a minimum of 0.1.

[5] From table 33, item b. Note that when both gates are wide open, the downstream gate will not be submerged because of the top contraction of the issuing stream through the upstream gate and therefore will not affect the flow.

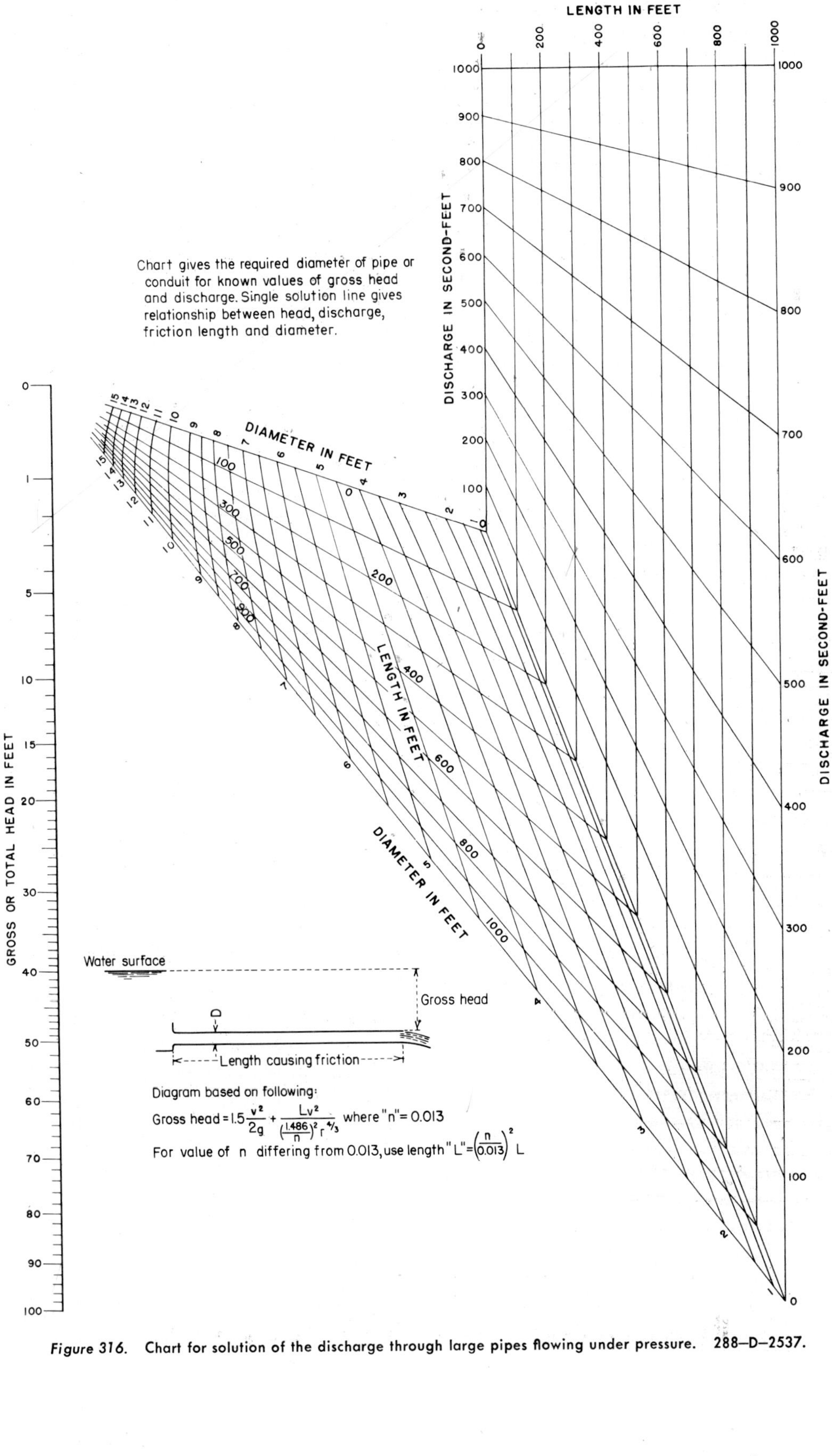

Figure 316. **Chart for solution of the discharge through large pipes flowing under pressure. 288–D–2537.**

239. *Design Examples.*—To illustrate the procedures for hydraulic design of outlet works two examples are presented.

(a) *Example 1.*—The problem is to compute a discharge curve for the conduit for Wasco Dam shown on figure 192, and to check the stilling basin for the condition of design discharge. The outlet works is to discharge 300 second-feet with the reservoir at normal water surface. The solution is as follows:

First determine the total head needed to produce flow in terms of the component losses from equation (4), listing both maximum and minimum assumed losses and relating the loss coefficients to the area of the conduit barrel. These assumptions and computations are tabulated in table 34.

From equation (7), $Q=a_1\sqrt{\frac{2gH_T}{K_L}}$, for maximum loss conditions Q is equal to $20\sqrt{\frac{64.4H_T}{12.28}}=45.8\sqrt{H_T}$. For minimum loss conditions Q is equal to $20\sqrt{\frac{64.4H_T}{7.98}}=56.8\sqrt{H_T}$. Discharges based on the average for these two extremes will be represented approximately by the equation $Q=51\sqrt{H_T}$. A discharge curve for this relationship as shown on figure 192 can be computed if the value of H_T is determined. Since the jet issuing from the gate opening is supported, H_T will be measured from the normal water surface to the top of the jet.

The depth of water downstream from the gate opening may be estimated by use of the coefficient of discharge for the gate, which in this case is an approximate measure of the top contraction. From item b of table 33, the maximum, minimum, and average values of the coefficient of discharge are 0.81, 0.68, and 0.70, respectively. The approximate depth of water downstream from the gate will be these values multiplied by the height of the gate (3.0 feet), or 2.4, 2.0, and 2.1 feet, respectively. Values of H_T for maximum, minimum, and averages losses are found by subtracting the elevation of the corresponding downstream water surfaces from the normal water surface elevation of 3514.4, as follows:

H_T (maximum losses) $=3514.4-(3478.5+2.0)=33.9$ feet.

H_T (minimum losses) $=3514.4-(3478.5+2.4)=33.5$ feet.

H_T (average losses) $=3514.4-(3478.5+2.1)=33.8$ feet.

The corresponding discharges are:

Q (maximum losses) $=45.8\ (33.9)^{1/2}=267$ second-feet.

Q (minimum losses) $=56.8\ (33.5)^{1/2}=329$ second-feet.

Q (average losses) $=51\ (33.8)^{1/2}=297$ second-feet.

The computed discharge for average losses corresponds closely with the 300-second-foot discharge for which the outlet works was to be designed; therefore that portion of the system which flows under pressure can be considered to meet the hydraulic design requirements.

In order to analyze the downstream free-flow portion of the outlet works, the hydraulic gradient immediately below the gate must be determined. The hydraulic gradients for both maximum and minimum losses are computed, and the average hydraulic gradient thereby obtained is used with the average discharge which has already been computed.

With minimum losses, the discharge is equal to 329 second-feet at normal reservoir water surface elevation 3514.4. If the gate and exit loss coefficients in the last column of table 31 are subtracted from K_L, the losses upstream from the gate are found to be $0.63h_{v_1}$. For a discharge of 329 second-feet through the conduit barrel, the velocity will be $\frac{329}{20}=16.5$ feet per second and the velocity head will be 4.2 feet. The loss up to the gate is equal to (0.63) (4.2) $=2.7$ feet, indicating the hydraulic gradient just upstream from the gates to be at elevation $3514.4-2.7=3511.7$. This provides a net head to the center of the gate of $3511.7-3480.0=31.7$ feet.

For the usual assumption that the coefficient of velocity through a standard orifice is 0.98, the velocity at the contracted section downstream from the gate is equal to $0.98\sqrt{(64.4)\ (31.7)}=44.3$ feet per second ($h_v=30.9$ feet). The area of the contracted section is equal to

$\frac{329}{44.3}=7.4$ square feet. For the 4-foot-wide free-flow downstream conduit, the depth will be $\frac{7.4}{4}=1.9$ feet, thus providing a gradient which is $d+h_v$ or $1.9+30.9=32.8$ feet above the invert of the conduit and establishing the hydraulic gradient at elevation 3511.3.

The discharge at normal water surface assuming maximum loss conditions is 267 second-feet. Following the same procedure as above, the gradient is found to be at elevation 3509.3. The average elevation of the hydraulic gradient is $\frac{3511.3+3509.3}{2}=3510.$

The next step is to compute the flow through the downstream free-flow conduit for the design discharge of 300 second-feet, which previous computations have shown to be approximately equal to the discharge through the pressure portion of the outlet works with average loss conditions. Here again, the losses should be maximized and minimized to determine extreme conditions at the downstream portal. Computations can be tabulated as shown in tables 35 and 36. (For procedure see sec. 203.)

For the stilling basin design, from figure 267 for a value of F_1 of 9.4 and a d_1 of 0.68, as indicated from table 35, the required tailwater depth will be 8.7 feet, which closely matches an actual tailwater depth of 9 feet. From table 36 the depth indicated at the downstream portal will be about 3 feet, indicating that the conduit will be about 0.6 full. This condition of flow should provide ample air space to forestall sealing from splash or wave action. The design of the free-flow portion of the conduit is therefore satisfactory for the design discharge of 300 second-feet.

(b) *Example 2.*—The problem is to design an outlet works system similar in layout to that shown for Soldier Canyon Dam on figure 304, capable of discharging 100 second-feet at reservoir elevation 100.0. The centerline of the downstream control valve is at elevation 75.0, and the sill of the drop inlet intake is at elevation 85.0. The length of the pressure conduit upstream from the emergency closure gate valve is 300 feet, including the vertical length and the length around the bend at the inlet. The length of the downstream pressure pipe is 200 feet. The solution is as follows:

First, an evaluation of the approximate size of the conduit can be obtained by the use of the chart shown on figure 316 and the procedure illustrated in section 238. A pipe approximately 2.75 feet in diameter will be required for the entire length of 500 feet for the available head of 25 feet, if the losses other than for friction can be approximated at 0.5 h_v. Because of the installation of gates or valves in the system, these losses will likely exceed that amount and a somewhat larger pipe may be required.

For the indicated pipe size, a 30-inch butterfly control valve might be considered. Assuming that K_g is 0.5, the discharge coefficient of the valve is equal to $\sqrt{\frac{1}{1.5}}=0.818$, say 0.8. The head needed to discharge 100 second-feet will then be $h_g=\frac{1}{2g}\left(\frac{Q}{aC}\right)^2=\frac{1}{64.4}\left[\frac{100}{(4.91)\,(0.8)}\right]^2=$ 10.1 feet. This will leave about 15 feet for all other losses.

Next, consider the pipe size downstream from the gate chamber. A size equal to the 2.75-foot uniform diameter determined previously for the entire length might be used. The loss, h_f, through the 200 feet of length will then equal $\frac{fL}{D}\frac{v^2}{2g}$; or for an n of 0.012, from figure B–7 (appendix B), $f=0.0192$, and $h_f=\left[\frac{(0.0192)\,(200)}{2.75}\right]\left[\frac{100}{5.94}\right]^2\left[\frac{1}{64.4}\right]=6.2$ feet. This loss plus the required head for the valve discharge will leave about 8.7 feet for upstream and other losses.

Next, select the conduit size upstream from the pipe. Assuming a 3.5-foot-diameter conduit with an n of 0.014, $f=0.024$; and the loss through the 300-foot length is $h_f=\left[\frac{(0.024)\,(300)}{3.5}\right]\left[\frac{100}{9.62}\right]^2\left[\frac{1}{64.4}\right]=3.4$ feet. This loss plus the pipe and regulating valve losses of 6.2 and 10.1, respectively, total approximately 20 feet, leaving about 5 feet for other losses. This seems reasonable enough to warrant evaluation.

Assuming, then, a 3.5-foot-diameter upstream conduit, a 24-inch emergency gate valve,

TABLE 35.—*Hydraulic computation for free-flow portion of conduit—Example 1*
Minimum Losses

[n=0.008; Q=300 second-feet]

Station	ΔL	Trial d	Width	a	v	h_v	r	$r^{2/3}$	s	$\frac{s_1+s_2}{2}$	Δh_L	$\Sigma\Delta h_L$	$d_2+h_{v2}+\Sigma\Delta h_L$	Invert elevation	Datum gradient	Remarks
3+38		1.71	4.0	6.84	43.9	29.9	0.92	0.95	0.062				31.5	3478.5	3510.0	
3+93	55	1.9	4.0	7.6	39.5	24.2	.97	.98	.047	.054	3.0	3.0	29.1	3477.8	3506.8	Too low.
		1.8		7.2	41.7	27.0	.95	.97	.054	.058	3.2	3.2	31.9		3509.7	OK.
4+48	55	1.9	4.0	7.6	39.5	24.2	.97	.98	.047	.050	2.8	6.0	32.1	3477.0	3509.1	OK.
4+68	20	.7	10.0	7.0	42.9	28.5	.61	.72	.103	.075	1.5	7.5	36.7	3471.0	3507.7	Too low.
		.68		6.8	44.1	30.2	.60	.71	.112	.080	1.6	7.6	38.5		3509.5	OK.

Then at the upstream end of the hydraulic jump basin, d_1=0.68, v_1=44.1 and $F_1=\frac{v_1}{\sqrt{gd_1}}=9.4$.

TABLE 36.—*Hydraulic computation for free-flow portion of conduit—Example 1*
Maximum Losses

[n=0.018; Q=300 second-feet]

Station	ΔL	Trial d	Width	a	v	h_v	r	$r^{2/3}$	s	$\frac{s_1+s_2}{2}$	Δh_L	$\Sigma\Delta h_L$	$d_2+h_{v2}+\Sigma\Delta h_L$	Invert elevation	Datum gradient	Remarks
3+38		1.71	4.0	6.84	43.9	29.9	0.92	0.95	0.31				31.6	3478.5	3510.0	
3+93	55	2.0	4.0	8.0	37.5	21.8	1.0	1.0	.21	0.26	14.3	14.3	38.1	3477.8	3515.9	Too high.
		2.2		8.8	34.1	18.0	1.05	1.03	.16	.23	12.7	12.7	32.9		3510.7	OK.
4+48	55	2.5	4.0	10.0	30.0	14.0	1.11	1.07	.12	.14	7.7	20.3	36.8	3477.0	3513.8	Too high.
		2.9		11.6	25.9	10.4	1.18	1.12	.08	.12	6.6	19.3	32.6		3509.6	OK.

TABLE 37.—*Computation of total loss coefficient—Example 2*

Element	Designated area subscript	Area, square feet	$\left(\frac{a_1}{a_x}\right)^2$	Item	Loss	Loss symbol	Loss coefficient	$\left(\frac{a_1}{a_x}\right)^2$ times loss coefficient
Trashrack	2	Gross, 60; [1] net, 25	0.06	a	Trashrack	[2] K_t	1.09	0.06
Entrance	3	9.62	.38	b	Entrance	[3] K_e	.10	.04
Upstream conduit	3	9.62	.38	c	Bend	[4] K_b	.10	.04
		9.62	.38	d	Friction	[5] K_f	2.06	.78
		9.62	.38	e	Contraction	K_c	.10	—.04
Gate valve	4	3.14	3.58	f	do	K_c	.10	.36
		3.14	3.58	g	Gate valve	K_g	.19	.68
		3.14	3.58	h	Expansion	K_{ex}	.10	.36
Downstream pipe	1	5.94	1.0	i	do	K_{ex}	.10	—.10
		5.94	1.0	j	Friction	[6] K_f	1.40	1.40
		5.94	1.0	k	Contraction	K_c	.10	—.10
Regulating valve	2	4.91	1.46	l	do	K_c	.10	.15
		4.91	1.46	m	Valve	K_g	.50	.73
		4.91	1.46	n	Exit	K_v	1.00	1.46
Total loss coefficient, K_L							5.82	

a_1=area of conduit.

a_x=area of element.

[1] Assuming trashracks designed for 2 feet per second velocity and 50 percent clogged, gross area=net area × 1.2=60.

[2] $K_t=1.45-0.45\left(\frac{25}{60}\right)-\left(\frac{25}{60}\right)^2=1.09.$

[3] From table 33, item g.

[4] From fig. 311.

[5] $K_f=\frac{fL}{D}$, D=3.5, n=0.014, f from fig. B-7 (appendix B)=0.024

$$K_f=\frac{(0.024)\,(300)}{3.5}=2.06$$

[6] $K_f=\frac{fL}{D}$, D=2.75, n=0.012, f from fig. B-7 (appendix B)=0.0192

$$K_f=\frac{(0.0192)\,(200)}{2.75}=1.40$$

a 33-inch downstream pipe, and a 30-inch regulating valve, a detailed analysis of the losses can be made. The losses will be based on the maximum loss coefficients as discussed previously. Table 37 shows the results in tabular form.

Then from equation (7), $Q=a_1\sqrt{\frac{2gH_T}{K_L}}$, for a value of K_L of 5.82, Q is equal to $5.94\sqrt{\frac{64.4H_T}{5.82}}=19.8\sqrt{H_T}$; or for a 25-foot head, $Q=99$ second-feet. This value is slightly less than the design requirement, and one or more of the elements must be enlarged to reduce the total loss. If the emergency gate size is increased to 30 inches, the area designated as 4 in the table will change from 3.14 to 4.91. Then the items f, g, and h, which total 1.40, will change to $\left(\frac{3.14}{4.91}\right)^2$ (1.40), or 0.57, and the total value of K_L will be reduced to 4.99. Then $Q=5.94\sqrt{\frac{64.4H_T}{4.99}}=21.3\sqrt{H_T}$; or for a head of 25 feet, $Q=106$ second-feet.

D. STRUCTURAL DESIGN DETAILS

240. *General.*—The same types of structures may be used for either spillways or outlet works. Spillways utilize open channels more often than do outlet works, and the structural design details for open channels are therefore discussed in part G of chapter IX. The details of the design of walls, open channel linings, and floors, discussed as spillway structures, are applicable to these structures when used for outlet works. Also, the headworks of open channel outlet works are similar to gated crest structures for spillways as regards structural design details.

On the other hand, closed conduit waterways are more commonly used for outlet works than they are for spillways, and the design details of closed conduits are therefore discussed in this chapter. The design details are the same in either case.

A closed conduit waterway might be a cast-in-place cut-and-cover culvert or conduit, a precast or prefabricated pipe, or a tunnel bored through the abutment. Waterways for a spillway will most often be free flowing, while those for outlet works may either flow full under pressure or partly full. The security of earthfill and rockfill dams depends to a large degree on the safety of the spillway and outlet structures, especially when conduits pass through the embankment. In those cases where all or part of a conduit is under internal pressure due to reservoir head, any leakage or failure of the conduit may cause openings through the dam which may gradually be enlarged until partial or complete failure results. There is also the danger of seepage along the contact surfaces between the conduit and the earthfill, which may result in serious damage. A third danger is the possibility of structural collapse of the conduit which would be almost certain to result in failure of an earthfill dam. These facts emphasize the importance of using durable materials, conservative design procedures, proper design details, and construction methods that will insure safe structures.

Replacement of a conduit through either an earthfill or rockfill dam is usually a difficult and expensive operation which can be avoided by the use of permanent material, such as cast iron for small size pipes and reinforced concrete cast-in-place conduit or precast concrete pipe for larger sizes. For small reservoirs with comparatively low heads, exceptions may be made only where the possible damage from failure is of little or no consequence. In such cases, where it is economically advantageous, the use of iron or steel pipe protected by galvanizing or bituminous coating or by some other rust-resisting treatment may be justifiable. It should be recognized that buried steel pipe is vulnerable to deterioration from electrolytic action, even where rust-resisting treatments or protective coatings are provided. Limited service with a possibility of eventual failure must therefore be expected. The use of iron or steel pipe as a watertight liner in a concrete conduit is permissible. However, unencased metal pipe should not be used if loss of life or serious damage to property will result from failure of the dam which might follow deterioration of the pipe.

Conduit joints must be made watertight to prevent leakage into the surrounding embankment. Joints of concrete cast-in-place conduits must be sealed with waterstops, and rubber-gasketed joints must be used for precast concrete pipe. For metal pipe, couplings are required which will remain watertight after movement or settlement of the pipe. Corrugated metal pipe with riveted seams should not be used for conduits unless the seams are welded or adequately calked and sealed with durable materials. If used for pressure conduits, corrugated metal pipe should be pretested for watertightness with pressures equal to twice the operating head.

When the outlet conduit consists of prefabricated pipe, whether of reinforced concrete, cast iron, or steel, the methods of bedding the pipe and backfilling around it should be such as to insure, insofar as possible, against unequal settlement and to secure the most uniform possible distribution of load on the foundation. When filling around these structures, extreme care should be taken to secure tight contact between the fill and the conduit surface and to obtain proper densities of the earthfill material. (See sec. E-4, app. E.) This is important not only for the prevention of seepage along the conduit, but also to insure that the fill develops a lateral restraint on the structure which will prevent excessive stresses in the conduit shell.

When the outlet consists of precast reinforced concrete or metal pipe, it should be set carefully on a good foundation and well bedded in concrete, as shown on figure C-2 (app. C). The concrete bedding not only aids in distributing the conduit load on the foundation, but also guarantees against uncompacted zones and void spaces under the pipe which could induce leakage along the undersurface of the structure. Void spaces or inadequate compaction of impervious materials at the invert of pipes have been the cause of numerous failures of small earthfill dams. The practice of supporting pipes on piers or collars without a concrete bedding should be avoided, because the greater foundation reaction at the concentrated support points will cause unequal stress distribution in the pipe. Furthermore, if the foundation settles from under the conduit between piers, the unsupported conduit will sag and crack. If the conduit is sufficiently strong to sustain the fill load, the earth shrinking away from the underside will leave voids which will permit the free passage of water.

Details of designs for cut-and-cover conduits are discussed in section 242.

241. *Tunnel Details.*—Linings are provided in tunnel waterways for both hydraulic and structural reasons. The smooth boundary surfaces reduce frictional resistance and permit a smaller diameter tunnel for a required capacity. Lining also is used to prevent saturation of the surrounding ground by seepage. Structural lining is used to support the tunnel walls against raveling or yielding ground.

Where the purpose of the lining is to provide a smooth hydraulic flow surface or to reduce seepage, the thickness may be determined by requirements for shrinkage, temperature change, and concrete placement. For ordinary linings where reasonably stable ground is encountered and where a minimum of tunnel support is required, an average lining thickness between $3/4$ and 1 inch per foot of tunnel diameter is ordinarily used. The minimum thickness usually provided is 6 inches. Yielding ground or areas through water-bearing strata may require thicker linings to resist external rock loads and hydrostatic pressures. A full circular lining is the most efficient shape to withstand such external loads.

Where the tunnel lining is to be reinforced, it must be made sufficiently thick both to accommodate the reinforcement mat and to provide sufficient room for placing the concrete in the confined space behind the forms. A minimum thickness of 6 inches is suggested for tunnel linings with a single layer of reinforcement. Where two layers of reinforcement are required, a minimum thickness of 9 inches may be desirable. In either case, the contractor may need to provide additional space outside the reinforcement to accommodate the concrete discharge pipe.

The portions of a tunnel which must be reinforced and the amount of reinforcement required depend on the tunnel shape, external and internal loadings, requirements for watertightness, and many geological factors. For a nonpressure tunnel, reinforcement may be required to resist external loads due to unstable ground or to grout or water pressures. Pressure tunnels with high internal hydrostatic loads must have lining reinforced sufficiently to withstand bursting where inadequate cover or unstable supporting rock prevails.

A suggested general guide for determining reinforcement requirements in tunnels is as follows:

(1) A pressure tunnel should ordinarily be reinforced whenever the depth of cover is adequate to withstand the unbalanced internal pressure head or whenever leakage control is important. The reinforcement should be sufficient to provide the required structural strength and leakage control for the maximum internal hydrostatic pressure reduced by a conservative estimate of the external hydrostatic pressure that is expected along the length of the tunnel. Restraint from the surrounding rock should only be considered when properties of the rock are known and a complete analysis is made. Where there are provisions for unwatering, the external pressure head should be the maximum that can be expected along the length of the tunnel.

(2) The transition from a pressure tunnel to a nonpressure tunnel should be specially reinforced to prevent excessive cracking which would permit leakage from the pressure portion of the tunnel to enter behind the lining of the nonpressure portion. Reinforcement for the pressure portion for a distance upstream from the junction equal to five times the diameter of the tunnel should be based on full internal hydrostatic head with no allowance for restraint from the surrounding rock. The nonpressure portion of the tunnel should be reinforced for a distance downstream from the control equal to from two to five times the tunnel diameter, for an external static head equal to the internal head just upstream from the control.

(3) A nominal amount of both longitudinal and circumferential reinforcement should be provided near the portals of both pressure and nonpressure tunnels to resist loads resulting from loosened rock above the tunnel or from sloughing of the portal cuts. This reinforcement should extend back from the portals for a distance equal to at least twice the tunnel diameter.

(4) If in competent rock, the tunnel, other than at the portals and at the transition from pressure to nonpressure, may be unreinforced where the rock cover is adequate to withstand the unbalanced internal pressure head. If in unstable ground, the lining should be reinforced to support probable rock loadings. Methods of estimating loadings for tunnel supports given in the publication, "Rock Tunneling With Steel Supports" [5], can be used to estimate requirements for the reinforced lining. Where the properties of the rock are known, an analytical determination can be made of the stresses that will theoretically exist in the rock surrounding the excavated tunnel and permanent supports and concrete lining, either acting separately or in combination, can be designed. Figure 317 shows typical tunnel supports used for the outlet works tunnel at Ruedi Dam in Colorado.

Permanent supports consist of steel ribs, steel lagging, steel-liner plates, shotcrete, reinforcement sheets, rock bolts with or without chain link fabric, or a combination of these. The choice of one or a combination of these types of supports should depend upon geologic conditions, groundwater levels, excavation methods to be used, length of time between excavating and placing permanent lining, and economic factors.

Supporting ribs, rock bolts, or other support types must be capable of holding up large blocks of material whose natural support was removed in excavating the tunnel. The lagging must be closely spaced where the rock is highly fractured or slacks off in small pieces; elsewhere it may be more widely spaced or even omitted. Methods of assuming and computing the size of supports are given by Proctor and White [5]. Loadings will be based on the

Figure 317. Typical tunnel supports used for the outlet works tunnel at Ruedi Dam in Colorado. P382–706–699NA.

nature of the ground encountered and, unless the exact underground conditions are known beforehand, the design of the ground-support system can only be approximated. Because of the uncertainties that may exist, the required size and spacing of supports are often determined by trial as the work progresses. In permanently supported sections of the tunnel, all spaces outside of the lagging or liner plates should be filled as completely and compactly as possible with clean gravel or rock spalls, and the spaces thoroughly filled with grout after the lining has been placed.

For tunnels through jointed rock or where seepage is to be minimized, the areas surrounding the tunnel are usually grouted both to consolidate the material and to fill open fissures in the rock and the voids between the lining and the rock. This grouting is accomplished by drilling holes through the lining, or through pipes placed in the lining for this purpose, into the surrounding rock and then injecting grout under pressure, as described in part E of appendix G. Permissible grouting pressures will depend on the nature of the surrounding ground and on the lining thickness. For small tunnels, rings of grout holes are spaced at about 20-foot centers, depending on the nature of the rock. Each ring consists of grout holes distributed at about 90° around the periphery, with alternate rings placed on vertical and 45° axes.

Drainage holes are often provided in other than pressure tunnels to relieve external pressures caused by seepage along the outside of the tunnel lining. The drainage holes also are spaced at about 20-foot centers, at intermediate locations between the grout hole rings. At successive sections, one vertical hole is drilled near the crown alternating with two drilled horizontal holes, one in each side wall. In free-flow tunnels, drainage holes are provided only above the water surface; if flow through the tunnel is conveyed in a separate pipe, the horizontal holes are drilled near the invert.

Typical tunnel supports for circular, horseshoe, modified horseshoe, and miscellaneous tunnels are shown in figures 318, 319, and 320. The "A" lines shown on the typical sections are lines within which no unexcavated material of any kind, no timbering, and no metallic or other supports shall be permitted to remain. The "B" lines are the outside limits to which measurement, for payment, of excavation will be made.

Suggested "A" line to "B" line dimensions are shown in figure 321(D). Specifications regarding tunnels are given in part D of appendix G.

242. *Cut-and-Cover Conduit Details.*—(a) *General.*—The design of a cut-and-cover conduit to be constructed through or under an earthfill embankment must include details which will provide for movement and settlement without excessive cracking or leakage. To obtain a safe structure, the following factors must be considered:

(1) Provide devices to minimize seepage along the contact of the conduit and the impervious embankment.

(2) Provide details to forestall cracking which might result in leakage of water into the fill surrounding a pressure conduit and to prevent piping of embankment material into a free-flow conduit.

(3) Select and treat foundation to minimize differential settlement which is a cause of cracking.

(4) Provide a structure to safely carry the loads to which the conduit will be subjected.

Selection of designs and details to accomplish the above purposes is discussed in this section.

(b) *Cutoff Collars.*—Foundation preparation and compaction around conduits must be equivalent to foundation preparation for the dam and to compaction of the impervious earthfill. Projecting fins or cutoff collars are provided to minimize seepage along the contact between the outside surface of the conduit and the embankment. These collars should be made of reinforced concrete, generally from 2 to 3 feet high, 12 to 18 inches wide, and spaced from 7 to 10 times their height along that portion of the conduit which lies within the impervious zone of the dam. The length of the percolation path along the contact is thereby increased by 20 to 30 percent. Figure 322 shows typical cutoff collars used on the outlet works at Silver Jack Dam in Colorado.

For a conduit on an earth foundation, the collar should completely encircle the conduit barrel. Where the foundation is sound rock, good contact along the base may be expected and the collars need extend only to be keyed into the rock foundation. The collars should be separated from the conduit to avoid introducing concentrated stresses into the conduit walls, which would alter the normal stress in the barrel. This is accomplished by constructing the collars with watertight fillers between the collars and the barrel. The structural separation permits lateral slipping of the collar on the barrel, eliminates secondary stresses in the conduit which would otherwise be caused by the stiffening effect of the collars, and avoids the introduction of torsional stresses in the conduit if horizontal movement or displacement of the embankment should occur. The joint filler material can be several layers of graphite-coated paper if only slight movement is expected, or premolded bituminous fillers where greater movement is expected.

Although cutoff collars usually are located between joints in the conduit, there are cases where collars have been constructed to span the joints. When so located they also serve as watertight covers for the joints. Where the collar is not placed at a conduit joint or where it is placed over a joint which is restrained from movement by keyways or by reinforcement extending across it, the collar ordinarily

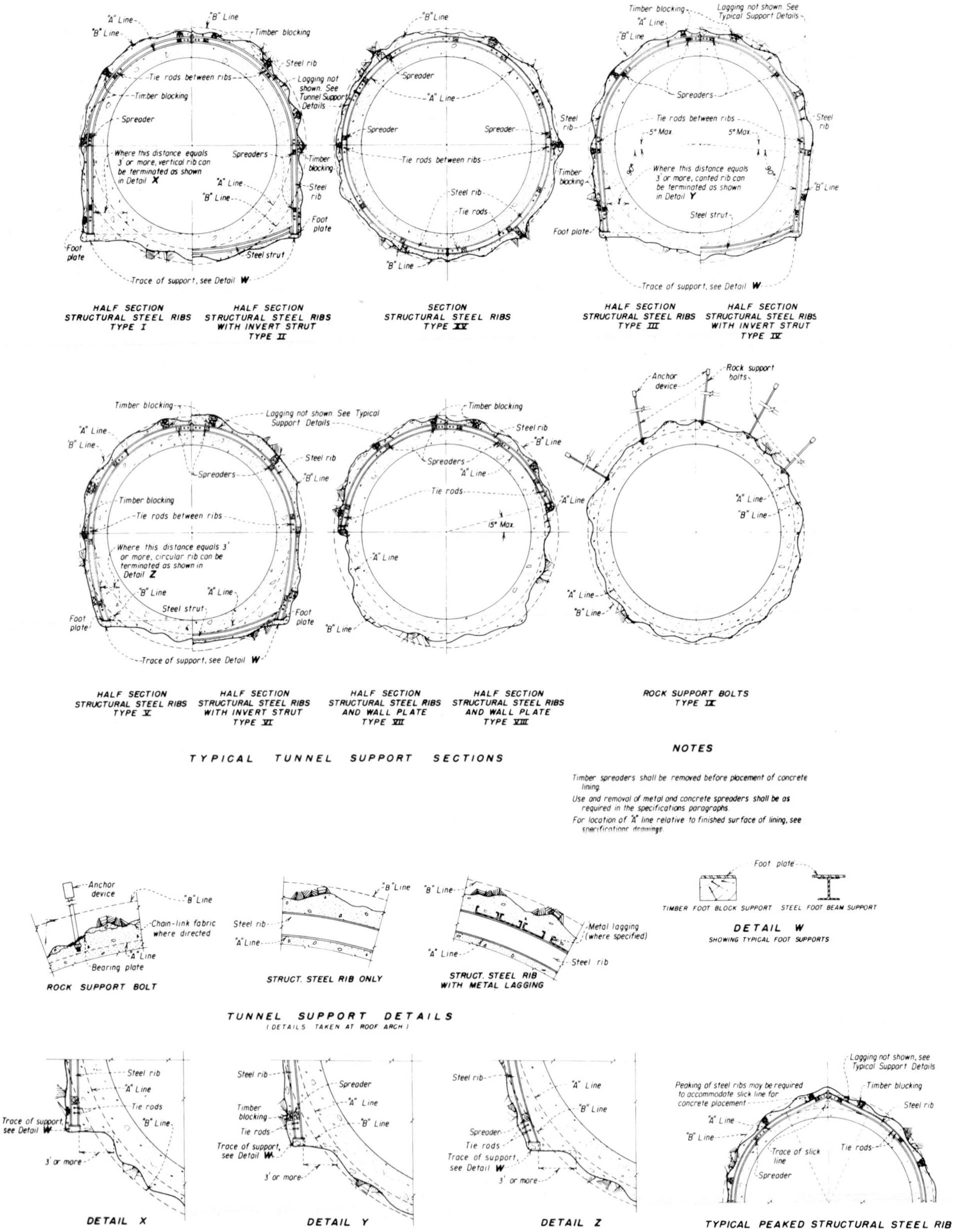

Figure 318. Sections and details for circular tunnels. From drawings 885–D–9, 288–D–2810.

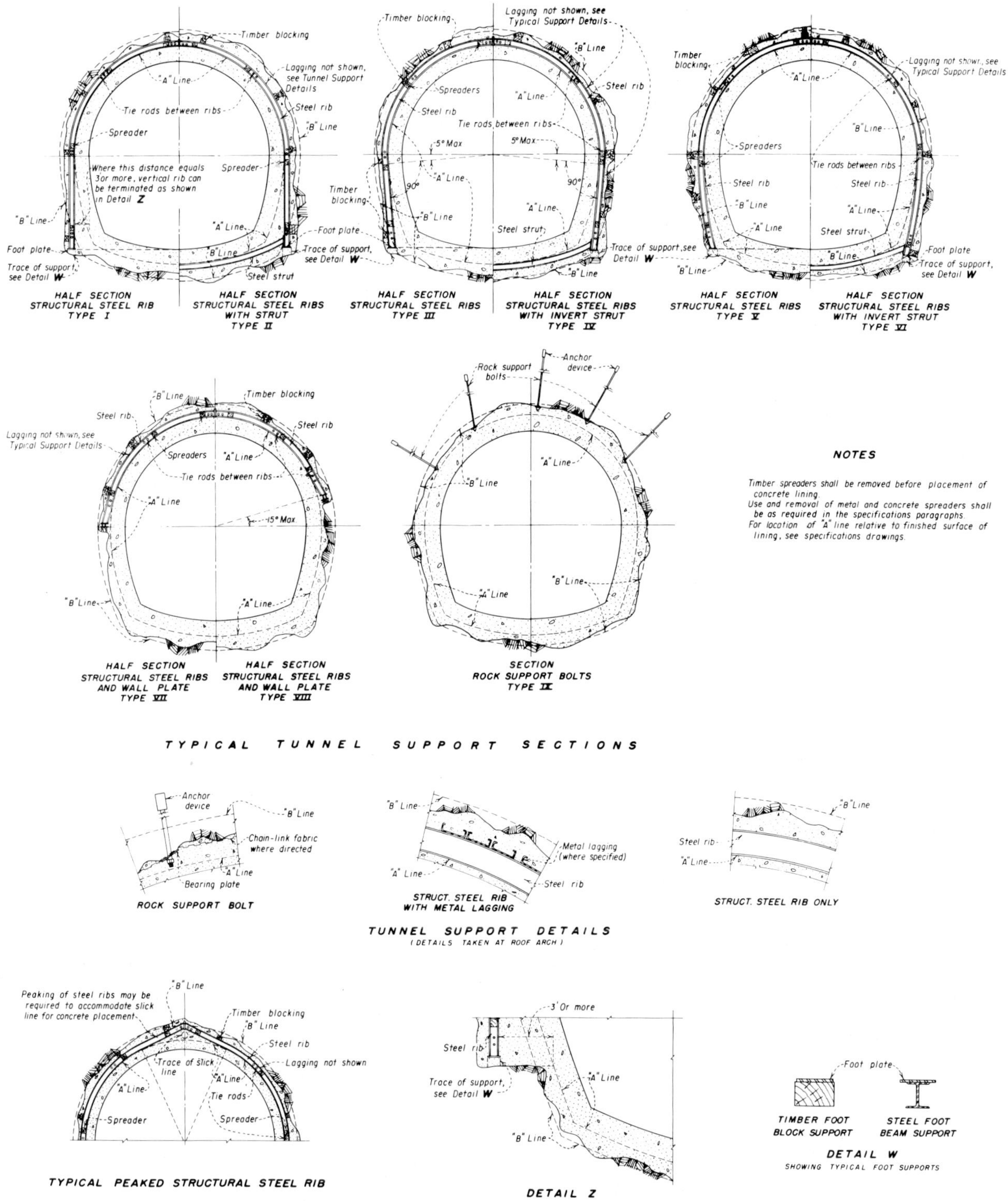

Figure 319. Sections and details for horseshoe tunnels. From drawings 885–D–10, 288–D–2811.

will not be subjected to large lateral loadings. In such cases it will need to be only strong enough to resist the superimposed fill load. When a collar covers a joint designed to permit differential movement, either the collar must be designed sufficiently strong to restrain such

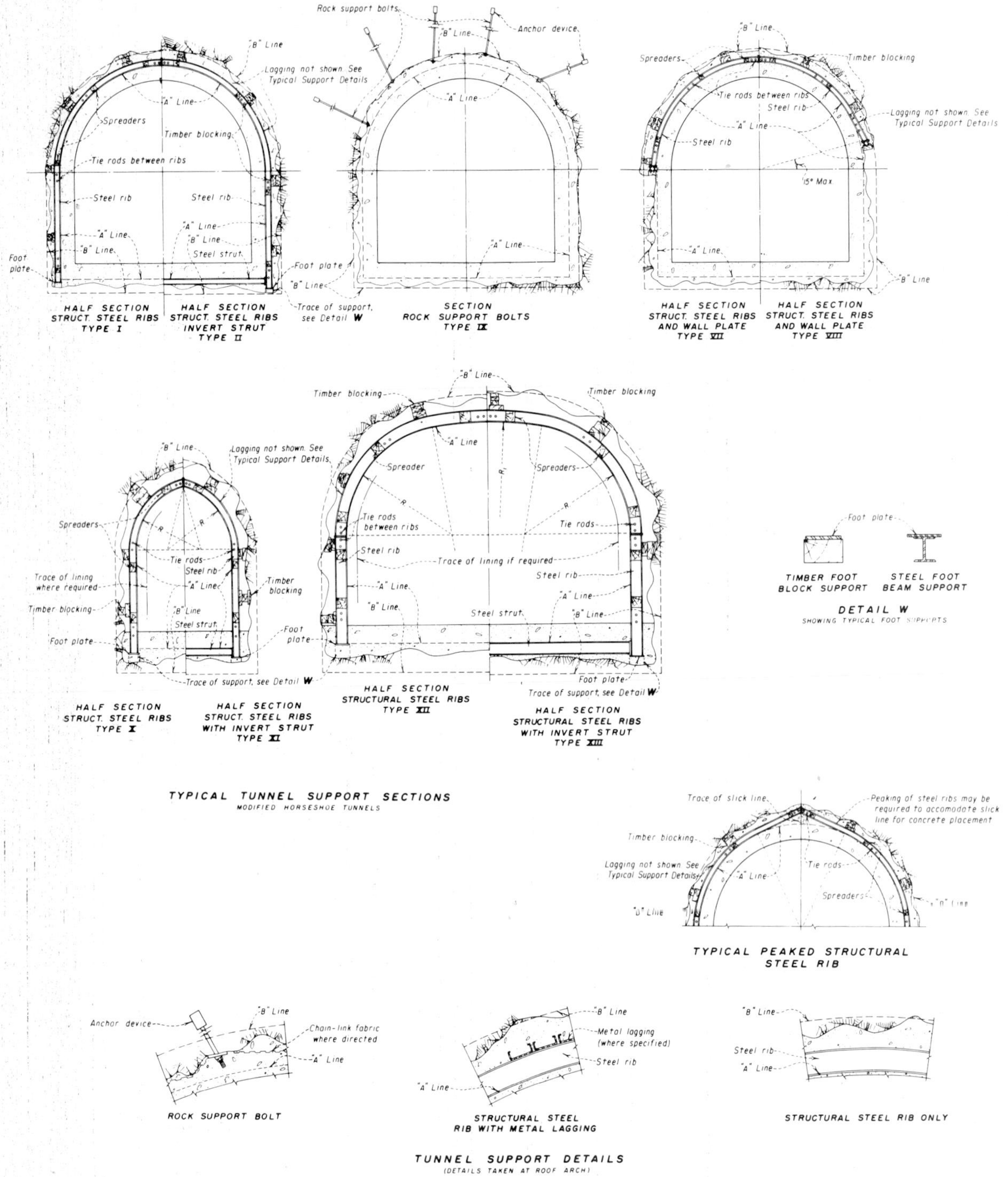

Figure 320. Sections and details for modified horseshoe and miscellaneous tunnels. From drawings 885—D—11, 288—D—2812.

movement, or the collar must adjust to the movement without losing the watertight contact.

(c) *Conduit Joints.*—Major cracking of cast-in-place conduit can be avoided by placing the conduit in short sections, usually 12 to 16

"A" LINE TO "B" LINE DIMENSIONS

For tunnels excavated by conventional blasting method

"A" LINE DIAMETERS IN FEET AND INCHES	RIB SET SUPPORTED			UNSUPPORTED AND ROCK BOLTED (ALL SHAPES) (INCHES)
	CIRCULAR (INCHES)	HORSESHOE (INCHES)	MODIFIED HORSESHOE (INCHES)	
7'-0" TO 9'-6"	14	13	12	10
9'-7" TO 13'-6"	15	14	13	10
13'-7" TO 18'-6"	15	15	13	10
18'-7" TO 22'-0"	16	16	14	10
22'-1" TO 30'-0"	16	17	14	10
30'-1" TO 36'-0"	17	18	15	10
36'-1" TO 50'-0"	17	18	16	10

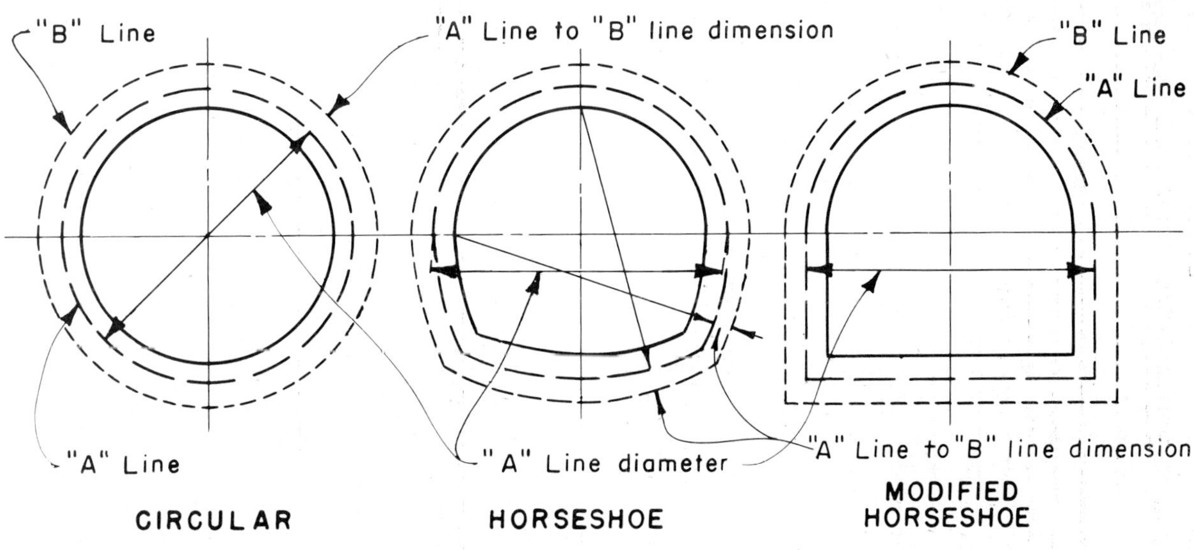

NOTE: Enter table with "A" line shape.
For strutted ribs increase table dimensions 1-inch.
For full circle ribs increase table dimensions :
1-inch for "A" line diameters from 22'-1" to 36'-0"
2-inches for "A" line diameters from 36'-1" to 50'-0"

Figure 321. Definitions of "A" and "B" line dimensions for different shapes of tunnels. 288—D—2813.

Figure 322. Typical cutoff collars used on the outlet works conduit at Silver Jack Dam in Colorado. P860–427–964NA.

feet, by continuing the longitudinal reinforcement through the joints, and by the placement of adjoining sections of the conduit barrel after the major volume change from shrinkage of first placed section or sections has taken place. Transverse joints in cast-in-place conduits are called conduit joints. Waterstops placed across the joint provide a watertight seal. The ends of the first placed sections are painted with sealing compound to break the bond between sections. The longitudinal reinforcement is continuous through the conduit joints. The waterstops and painted joints are used to confine any cracking from settlement to the joints to prevent leakage. The longitudinal reinforcement prevents longitudinal forces from opening the joints between sections. Details of this type of joint construction are shown on figure 323.

Precast pipe used for outlet works or spillway conduits under or through earthfill dams should be bedded in a concrete base (see appendix C). The concrete base is used to prevent percolation along the underside of the pipe, as well as giving structural support to the precast pipe both laterally and longitudinally. The concrete base should be reinforced with the longitudinal reinforcement continuous through the transverse joints. Differential lateral displacement of precast pipe conduit sections at the joints is ordinarily restrained by a bell-and-spigot joint or by a reinforced collar en-

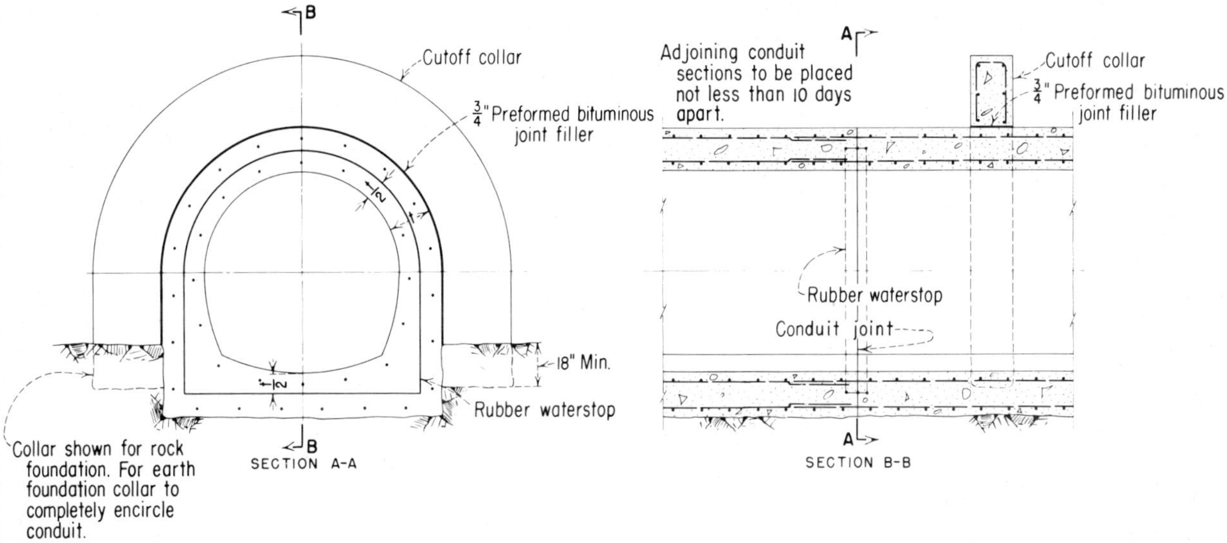

Figure 323. **Typical conduit joint and cutoff collar details. 288–D–2539.**

circling a plain joint. Rubber-gasketed joints similar to those shown on figure 324 can be adopted for joining individual lengths of concrete pipes.

Specifications for pipe and pipe joints indicated on figure 324 can be found in the Bureau of Reclamation publication, "Standard Specifications for Reinforced Concrete Pressure Pipe," February 1, 1969.

(d) *Design Loads.*—Embankment loads on conduits may vary over a wide range depending on many factors relating to the foundation, method of bedding, flexibility or rigidity of the conduit; and to the soil characteristics of the embankment such as angle of internal friction, unit weight, homogeneity, consolidation properties, cohesiveness, and moisture content. All possible combinations of these various factors must be considered to evaluate their overall effect. The loads must be considered not only as they may occur during construction but also as they may be altered after embankment completion, reservoir loading, and embankment saturation.

The "Marston Theory" of embankment pressures is usually adopted for precast conduits under relatively low fills. This theory is discussed in many bulletins published by the Iowa State College Experiment Station and is abstracted in various handbooks [6, 7] which contain bibliographies of publications dealing with this subject. On the basis of the Marston theory, the vertical load on a conduit is considered to be a combination of the weight of the fill directly above the conduit and the frictional forces acting either upward or downward due to the adjacent fill. A settlement of adjacent fill greater than the overlying fill induces frictional forces acting downward which increase the resultant load on the conduit; a greater settlement immediately above the conduit will result in an arching condition which reduces the load on the conduit. Thus a conduit laid in trench excavated in a compact natural soil will practically never receive the full weight of the backfill above it, because of the development of arching action when the backfill starts to settle. On the other hand, if the conduit is placed so that it projects in whole or in part above the natural ground surface, the embankment load which may come upon it can in some cases be as much as 50 percent greater than the weight of the fill directly above it. The designs of precast concrete pressure pipe discussed in appendix C are based on loads as derived by the Marston theory.

For cast-in-place conduits under relatively high fills, where the conduit is placed in cut so that neither a full trench nor a complete projecting condition exists, a loading assumption

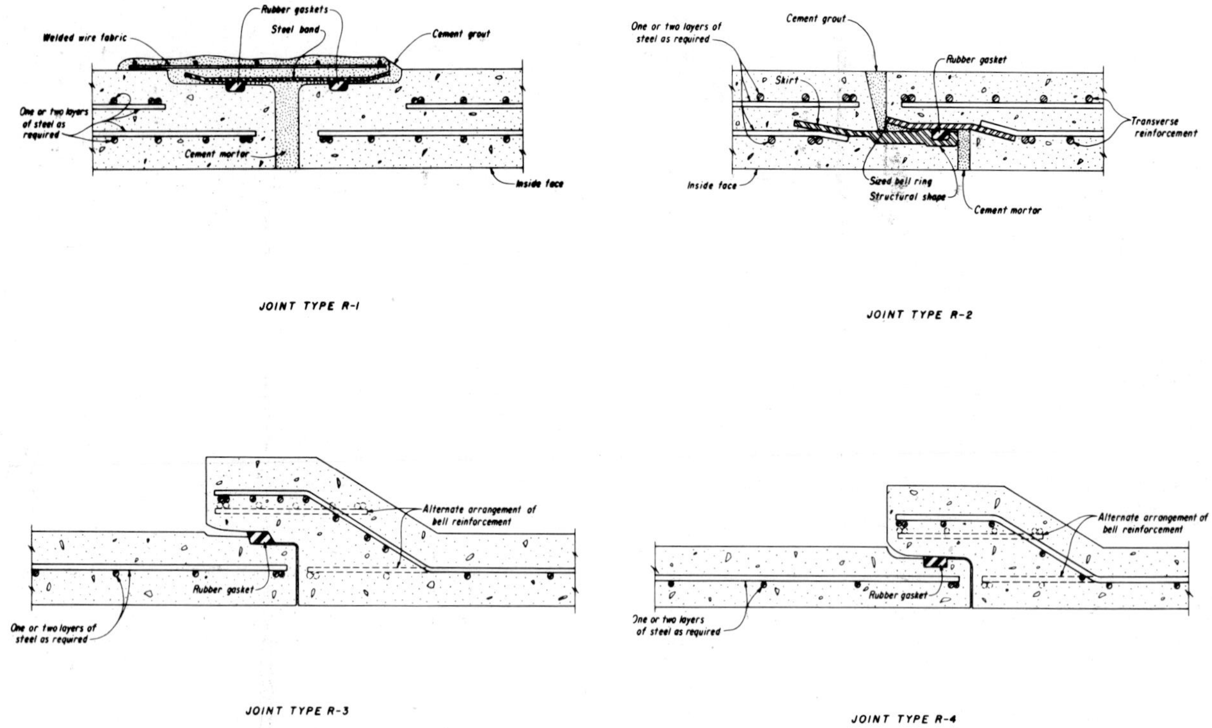

Figure 324. Typical precast concrete pipe joint details. 288–D–2814, 288–D–2815, 288–D–2816, and 288–D–2817.

which averages the extremes noted above is assumed. For this case the load on the conduit is assumed to be the weight of the column of fill directly above it. The load over that portion of a conduit under the upstream part of the dam includes both the weight of the saturated fill and the weight of the reservoir water above the fill. The conduit barrel is designed on the basis of a given factor of safety, considering that the unit horizontal lateral load on the conduit is one-third of the unit vertical load. The design is then checked on the basis of a reduced factor of safety considering no horizontal lateral load exists. The vertical reaction of the base of the conduit is taken equal to the vertical load plus the weight of the conduit. On an earth foundation, the base reaction is assumed to be distributed uniformly across the width of the conduit; on a rock foundation it is assumed to be distributed triangularly, varying from twice the average unit reaction at the outside edges to zero at the center of the base. External hydrostatic pressures are assumed to act equally in all directions, vertically downward as an increased load, upward as uplift, and laterally on the sides of the conduit.

Procedures for designing concrete box culverts and circular conduits are comprehensively discussed in "Concrete Culverts and Conduits" [8]. Data for selecting a cast-in-place conduit design based on design procedures using Beggs Deformeter coefficients [9] are included in appendix C. These data list required conduit thickness and reinforcement requirements for various fill heights and hydrostatic loading conditions. Also included in appendix C are data for selecting precast concrete pipe for use as conduits under limited fill loads.

243. *Details of Typical Structures.*—Figures 325 through 332 show arrangements for outlet works intakes, shafts, and stilling basins constructed at various small Bureau of Reclamation dams. These are presented as examples which can be used as guides in the design of similar structures.

E. BIBLIOGRAPHY

244. *Bibliography.*

[1] King, W. H., "Handbook of Hydraulics," fourth edition, McGraw-Hill, New York, N.Y., 1954.

[2] Creager, W. P., and Justin, J. D., "Hydroelectric Handbook," second edition, John Wiley & Sons, Inc., New York, N.Y., 1950.

[3] "Friction Factors for Large Conduits Flowing Full," Engineering Monograph No. 7, Bureau of Reclamation, 1965.

[4] Rouse, Hunter, "Engineering Hydraulics," John Wiley & Sons, Inc., New York, N.Y., 1950.

[5] Procter, R. V., and White, T. L., "Rock Tunneling with Steel Supports," Commercial Shearing & Stamping Co., Youngstown, Ohio, 1946.

[6] "Concrete Pipe Handbook," American Concrete Pipe Association, 228 North La Salle Street, Chicago, Ill., 1951.

[7] "Handbook of Drainage and Construction Products," Armco Drainage & Metal Products, Inc., Middletown, Ohio, 1958.

[8] "Concrete Culverts and Conduits," Portland Cement Association, 33 West Grand Avenue, Chicago, Ill.

[9] Phillips, H. B., "Beggs Deformeter Stress Analysis of Single-Barrel Conduits," Engineering Monograph No. 14, Bureau of Reclamation, 1968.

[10] "Building Code Requirements for Reinforced Concrete," (ACI 318–63), American Concrete Institute, Detroit, Mich.

[11] "Supplement for ACI 318–63 Code," Bureau of Reclamation, Denver, Colo., 1968.

[12] "Ultimate-Strength Design Handbook," Special Publication No. 17, American Concrete Institute, Detroit, Mich.

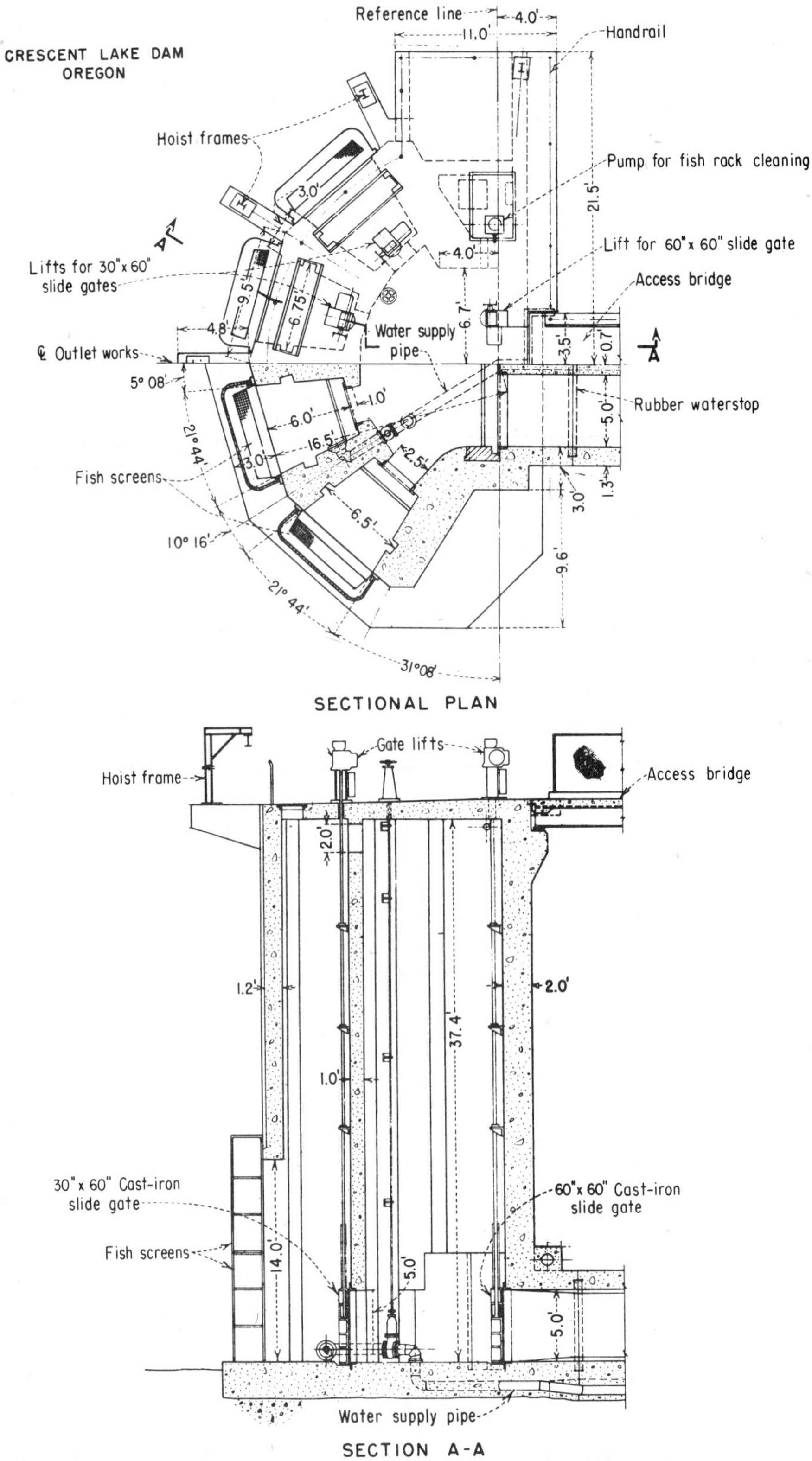

Figure 325. Typical intake tower arrangement. 288–D–2541.

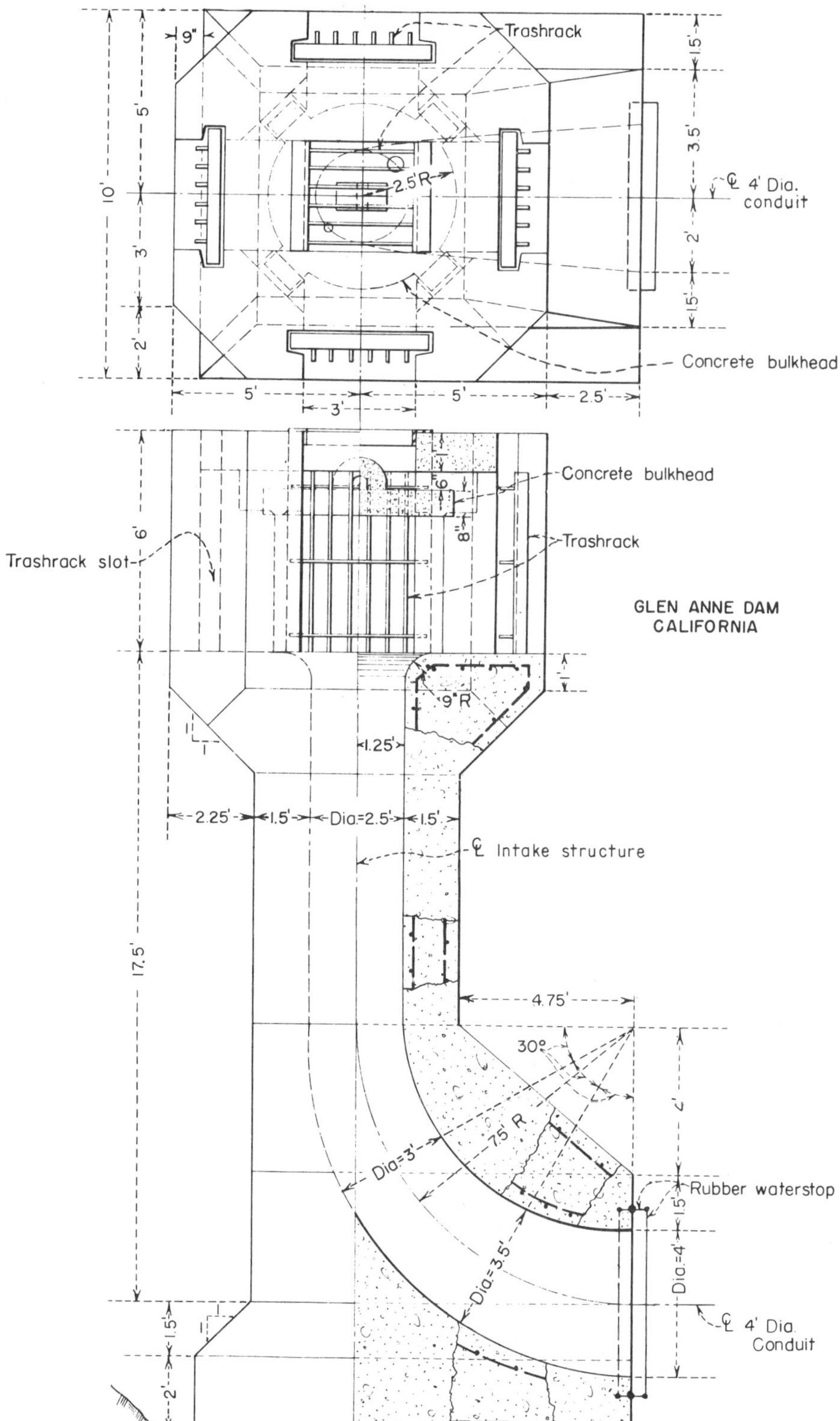

Figure 326. Typical drop inlet intake. 288—D—2542.

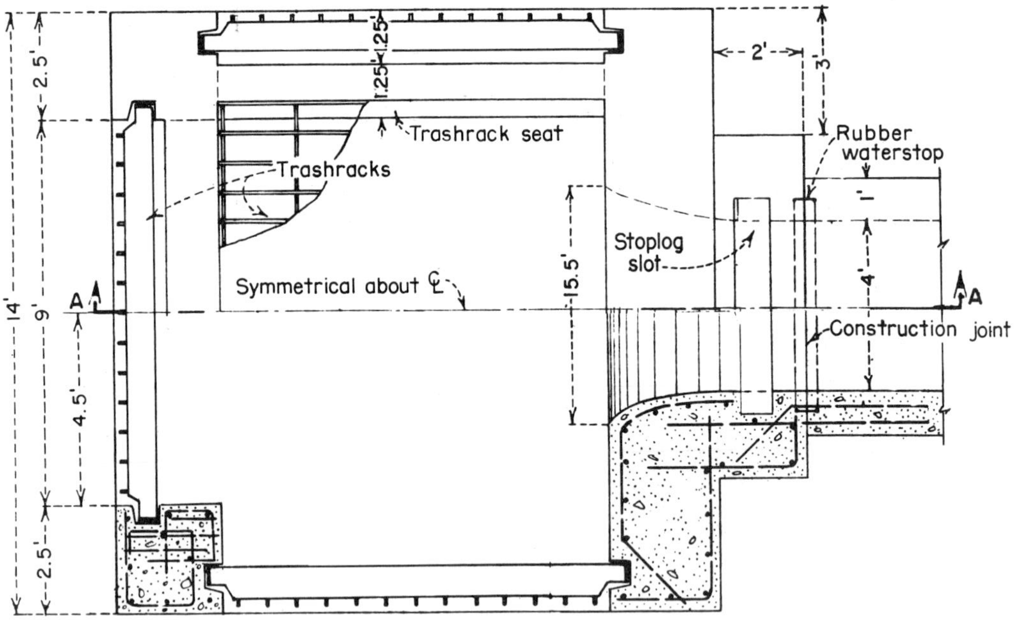

SECTIONAL PLAN

WASCO DAM
OREGON

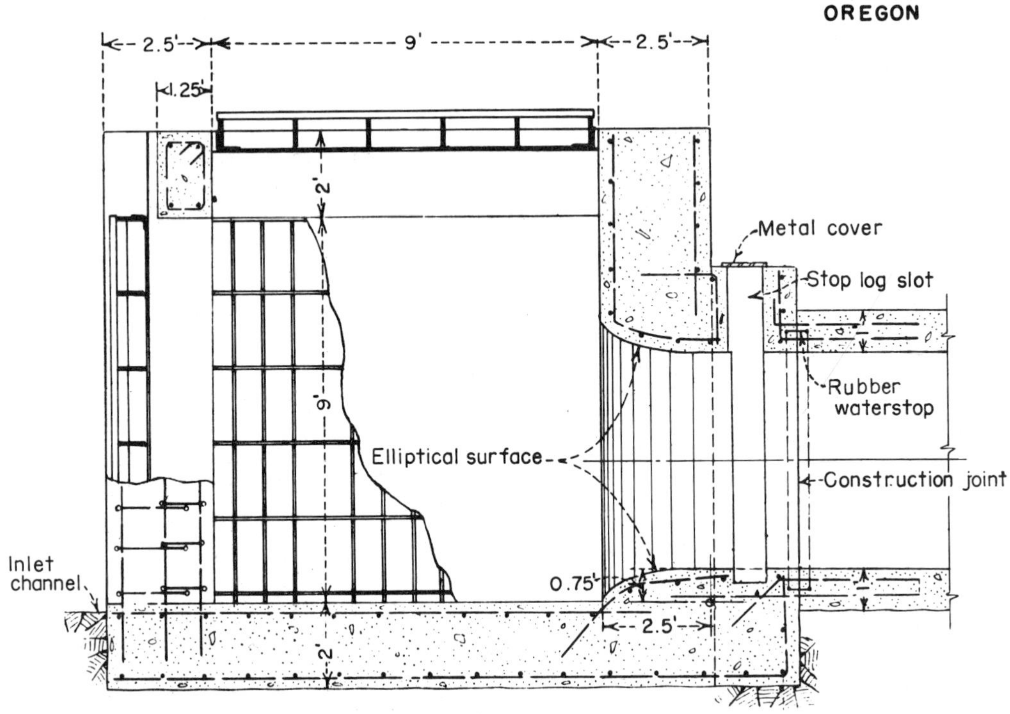

SECTION A–A

Figure 327. Typical trashracked box intake. 288–D–2543.

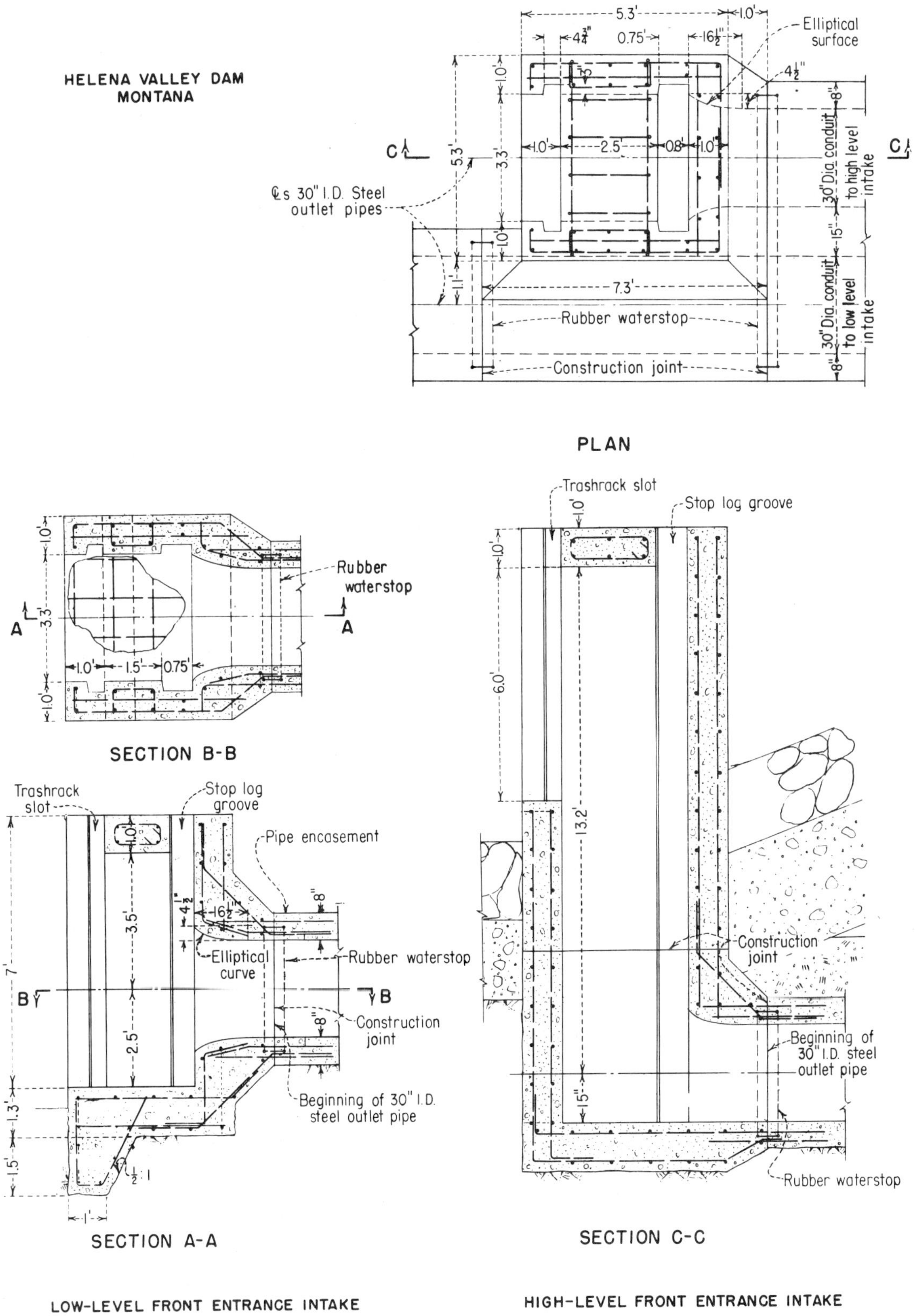

Figure 328. Typical front-entrance intake structures. 288—D—2544.

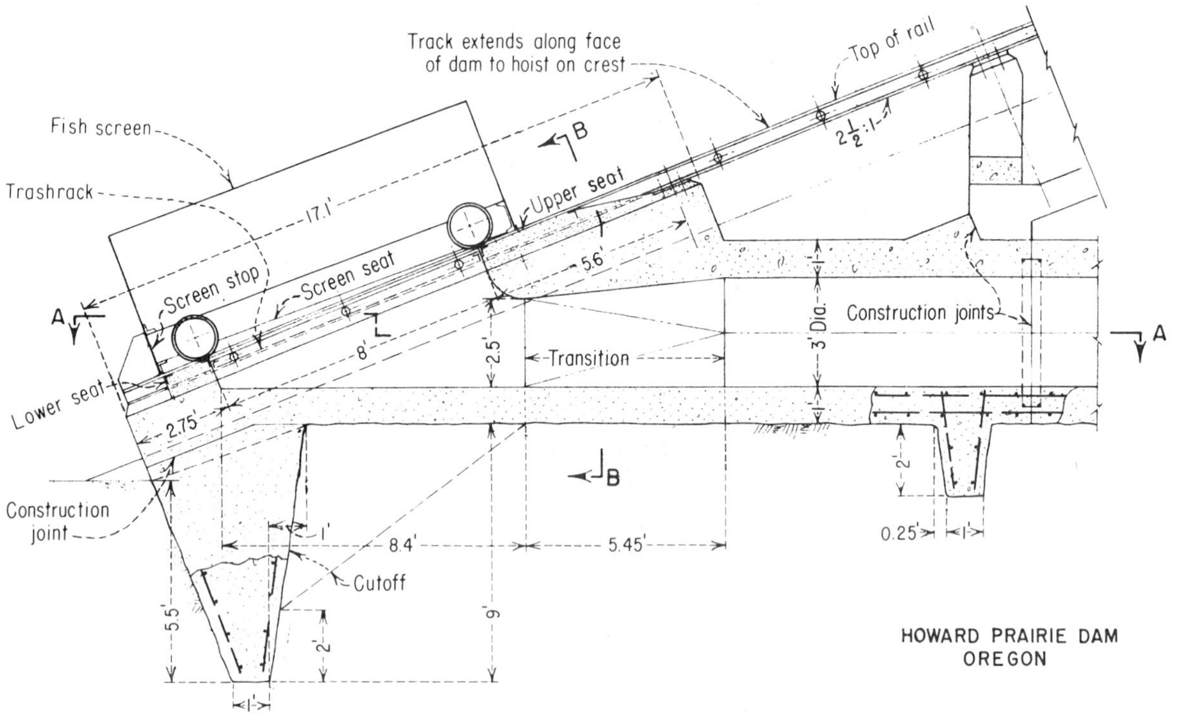

LONGITUDINAL SECTION OF INTAKE

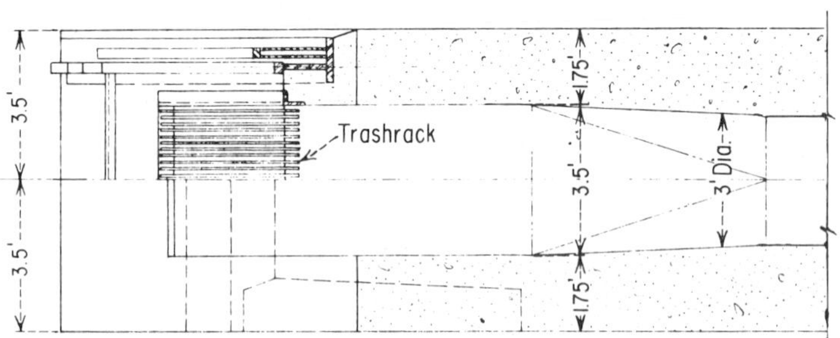

SECTIONAL PLAN A–A

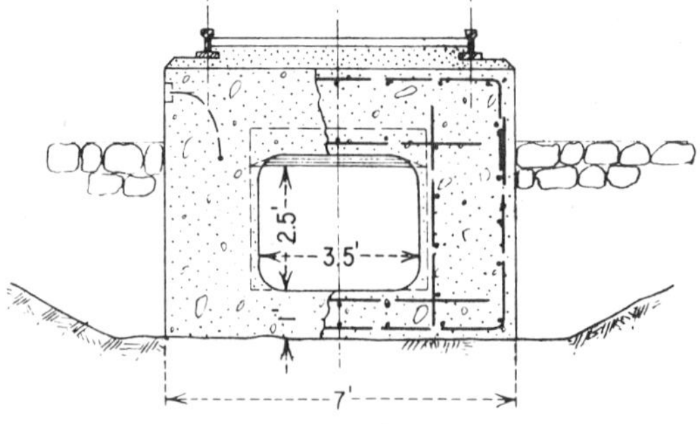

SECTION B–B

Figure 329. Typical intake with sloping entrance. 288–D–2545.

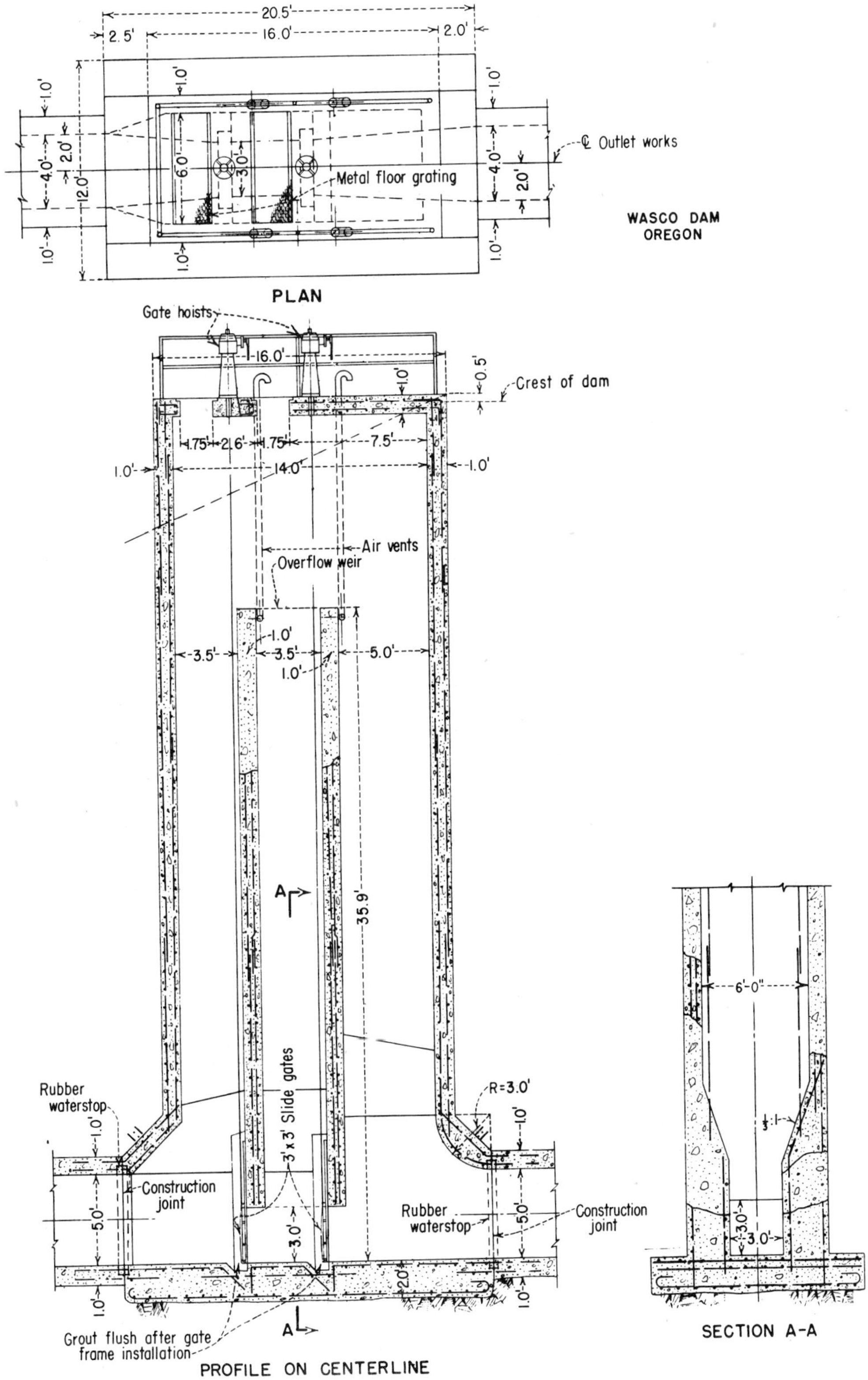

Figure 330. Typical shaft for slide gate control. 288—D—2546.

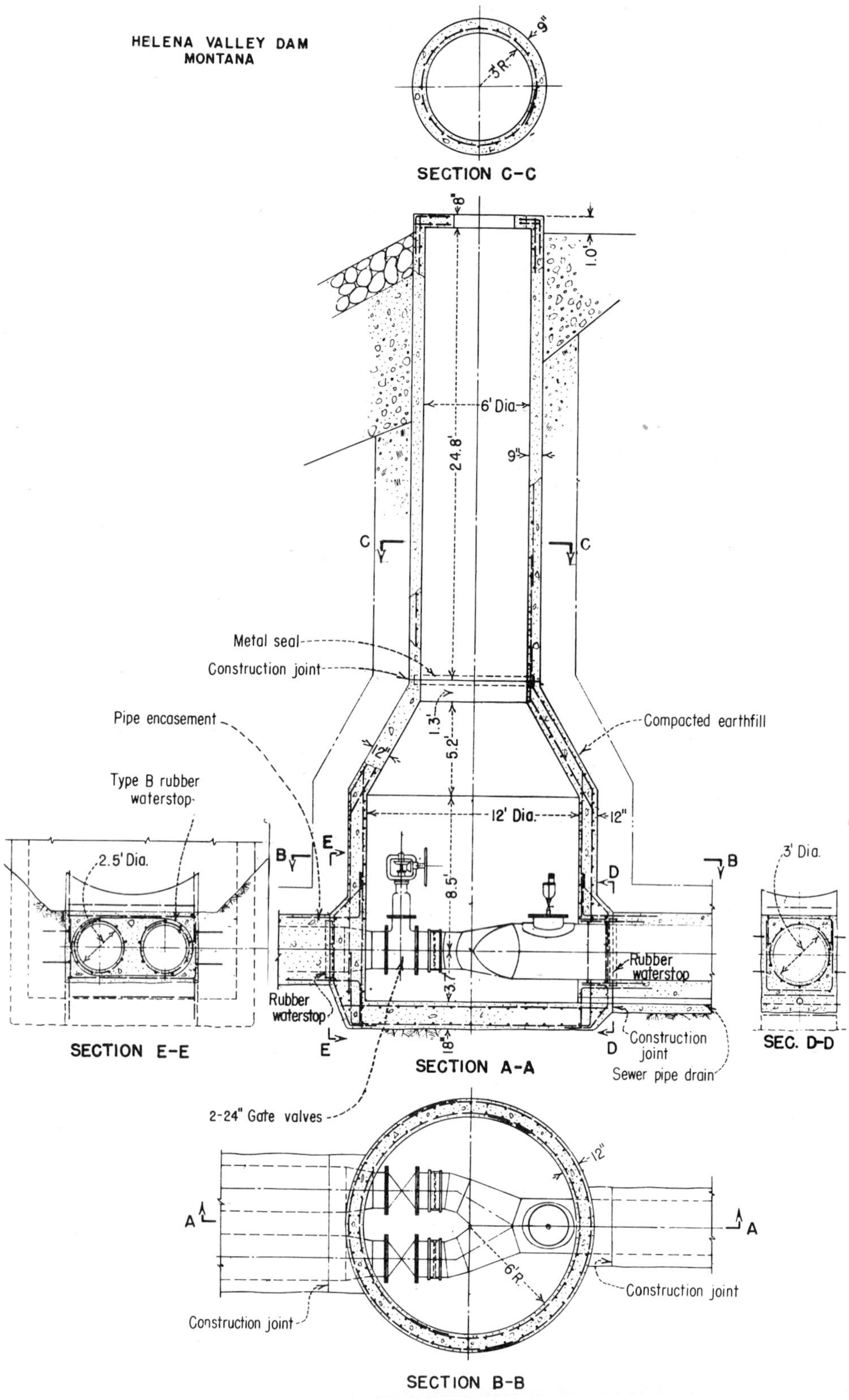

Figure 331. Dry well shaft for gate-valve control installation. 288–D–2547.

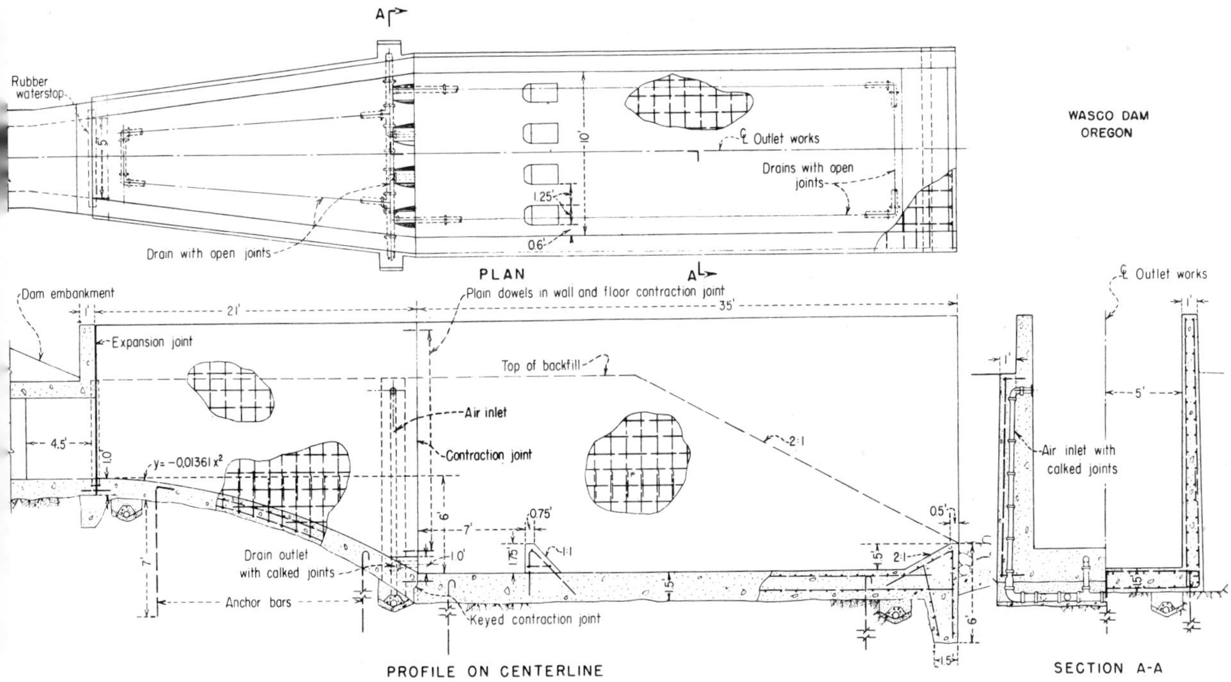

Figure 332. **Typical hydraulic jump stilling basin. 288–D–2548.**

Chapter XI

Diversion During Construction

E. R. LEWANDOWSKI [1]

A. DIVERSION REQUIREMENTS

245. *General.*—The design for a dam which is to be constructed across a stream channel must consider diversion of the streamflow around or through the dam site during the construction period. The extent of the diversion problem will vary with the size and flood potential of the stream; at some dam sites diversion may be costly and time-consuming and may affect the scheduling of construction activities, while at other sites it may not offer any great difficulties. However, a diversion problem exists to some extent at all sites except those located off-stream, and the selection of the most appropriate scheme for handling the flow of the stream during construction is important to secure economy in the cost of the dam. The scheme selected ordinarily will represent a compromise between the cost of the diversion facilities and the amount of risk involved. The proper diversion plan will minimize serious potential flood damage to the work in progress at a minimum of expense. The following factors should be considered in a study to determine the best diversion scheme:

(1) Characteristics of streamflow.
(2) Size and frequency of diversion flood.
(3) Methods of diversion.
(4) Specifications requirements.

246. *Characteristics of Streamflow.*—Streamflow records provide the most reliable information regarding streamflow characteristics, and should be consulted whenever available.

Depending upon the size of the drainage area and its geographical location, floods on a stream may be the result of snowmelt, seasonal rains, or cloudbursts. Because each of these types of runoff have their peak flows and their periods of low flow at different times of the year, the nature of runoff will influence the selection of the diversion scheme. A site subject only to snowmelt floods will not have to be provided with elaborate measures for use later in the construction season. A site where seasonal rains may occur will require only the minimum of diversion provisions for the rest of the year. A stream subject to cloudbursts which may occur at any time is the most unpredictable and probably will require the most elaborate diversion scheme, since the contractor must be prepared to handle both the low flows and floodflows at all times during the construction period.

247. *Selection of Diversion Flood.*—Usually, it is not economically feasible to plan on diverting the largest flood that has ever occurred or may be expected to occur at the site, and consequently some lesser requirement must be decided upon. This, therefore, brings up the question as to how much risk is involved in the diversion scheme under consideration. In the case of an earthfill dam, where considerable areas of foundation and structure excavation are exposed, or where overtopping of the embankment while under construction may result in serious damage or loss of the partially completed work, the importance of eliminating the risk of flooding is relatively great. This consideration is not as important in the case of a concrete dam since the floodwaters may, if the location of appurtenant structures permits,

[1] Formerly Engineer, Earth Dams Section, now Regional Engineer, Region 2, Bureau of Reclamation.

overtop the dam with little or no adverse effect.

In selecting the flood to be used in the diversion designs, consideration should be given to the following:

(1) How long the work will be under construction, to determine the number of flood seasons which will be encountered.

(2) The cost of possible damage to work completed or still under construction if it is flooded.

(3) The cost of delay to completion of the work, including the cost of forcing the contractor's equipment to remain idle while the flood damage is being repaired.

(4) The safety of workmen and possibly the safety of downstream inhabitants in case the failure of diversion works results in unnatural flooding.

After an analysis of these factors is made, the cost of increasing the protective works to handle progressively larger floods can be compared to the cost of damages resulting if such floods occurred without the increased protective work. Judgment can then be used in determining the amount of risk that is warranted.

For small dams which will be constructed in a single season, only the floods which may occur during that season need be considered. For most small dams, involving at the most two construction seasons, it should be sufficiently conservative to provide for the largest flood likely to occur in a 5-year period. However, to minimize the risk, the diversion capacity might be increased to handle the 10-year or larger flood if it can be done at little additional cost. The methods for determining floods of 5-year, 10-year or less frequency are discussed in section 47.

It should be considered that floods may be recurrent; therefore, if the diversion scheme involves temporary storage of cloudburst-type runoff, facilities must be provided to evacuate such storage within a reasonable period of time, usually a few days.

B. METHODS OF DIVERSION

248. *General*.—The method or scheme of diverting floods during construction depends on the magnitude of the flood to be diverted; the physical characteristics of the site; the type of dam to be constructed; the nature of the appurtenant works, such as the spillway, penstocks, or outlet works; and the probable sequence of construction operations. The objective is to select the optimum scheme considering practicability, cost, and the risks involved. The diversion works should be such that they may be incorporated into the overall construction program with a minimum of loss, damage, or delay.

Common practice for diverting streams during construction utilizes one or a combination of the following provisions: tunnels driven through the abutments, conduits through or under the dam, temporary channels through the dam, or multiple-stage diversion over the tops of alternate construction blocks of a concrete dam. Outlet works conduits or tunnels frequently are made sufficiently large to carry the diversion flow. On a small stream the flow may be bypassed by the installation of a temporary wood or metal flume or pipeline, or the flow may be impounded behind the dam during its construction, pumps being used if necessary to control the water surface. Figures 333 and 334 show flumes used to divert the streamflow during the construction of an earthfill dam and a concrete dam, respectively. In any case, barriers are constructed across or along the stream channel in order that the site, or portions thereof, may be unwatered and construction can proceed without interruption.

A common problem is the meeting of downstream requirements when the entire flow of the stream is stopped during closure of the diversion works. Downstream requirements may demand that a small flow be maintained at all times, in which case the contractor must provide the required flow by pumping or by other means (bypasses or siphons) until water is stored in the reservoir to a sufficient level so that it may be released by gravity flow through the outlet works.

Figure 335 shows how diversion of the river

Figure 333. Temporary diversion flume through an earthfill dam site. (Willow Creek Dam, CBT 245–704–3287.)

was accomplished during the construction of Folsom Dam and Powerplant on the American River in California. Although this structure is considerably larger than the dams discussed in this text, this photograph is included because it illustrates many of the diversion principles discussed in this chapter. The river, flowing from top to bottom in the picture, is being diverted through a tunnel; "a" and "b" mark the inlet and outlet portals, respectively. Construction is proceeding in the original river channel between earthfill cofferdams "c" and "d." Discharge from pipe "e" at the lower left in the photograph is from unwatering of the foundation. Since it was impracticable to provide sufficient diversion tunnel capacity to handle the large anticipated spring floods, the contractor made provisions to minimize damage that would result from overtopping of the cofferdam. These provisions included the following:

(1) Placing concrete in alternate low blocks in the dam "f" to permit overflowing with a minimum of damage;

(2) Construction of an auxiliary rockfill and cellular steel sheet-piling cofferdam "g" to protect the powerplant excavation "h" from being flooded by overtopping of the cofferdam; and

(3) Early construction of the permanent training wall "i" to take advantage of the protection it affords.

249. *Tunnels.*—It is usually not feasible to do a significant amount of foundation work in a narrow canyon until the stream is diverted. In this situation a tunnel may prove the most feasible means for diversion, either for a concrete dam or for an earthfill dam. The streamflow is bypassed around the construction area through tunnels in one or both abutments. If tunnel spillways or tunnel outlet works are provided in the design, it usually proves economical to utilize them in the diversion plan. If the upstream portion of the permanent tunnel is above the streambed elevation, a temporary upstream diversion adit can be provided to effect a stream-level bypass. Figure 336 shows such an adit which was constructed at Seminoe Dam to permit diversion through the spillway tunnel. The diversion adit leads from the streambed to the intersection of the horizontal portion of the spillway tunnel and the inclined shaft leading from the spillway gate structure. First stages of construction of the spillway gate structure can be seen in the upper right-hand portion of the photograph.

The advisability of lining the diversion tunnel will be influenced by the cost of a lined tunnel compared with that of a larger unlined tunnel of equal carrying capacity; the nature of the rock in the tunnel, as to whether it can stand unsupported and unprotected during the passage of the diversion flows; and the permeability of the material through which the tunnel is carried, as it will affect the amount of leakage through or around the abutment.

Some means of shutting off diversion flows must be provided; in addition, some means of regulating the flow through the diversion tunnel may be necessary. Closure devices may consist of a timber, concrete, or steel bulkhead gate; a slide gate; or stoplogs. Regulation of flow to satisfy downstream needs after storage of water in the reservoir has been started can be accomplished by the use of a slide gate, wheel-type regulating gate, or temporary bypass until the water surface in the reservoir

Figure 334. Temporary diversion flume used during construction of a concrete dam. (Horsetooth Feeder Canal Tunnel No. 1, CBT 245–704–330.)

Figure 335. Diversion of the river during construction of Folsom Dam and Powerplant (Corps of Engineers and Bureau of Reclamation.) (AR–1627–CV.)

reaches the level of the outlet works intake. Figure 337 shows temporary closure of the upstream portal of a diversion tunnel by means of a wooden bulkhead. The pipes in the bulkhead were provided to pass flows to meet downstream requirements until the reservoir level rose to the lip of the outlet works intake, when the pipes were closed by the flap gates shown. The pipes extended through the diversion tunnel plug and ultimately they were grouted to complete the closure.

Permanent closure of the diversion tunnel is made by placing a concrete plug in the tunnel. Where a temporary diversion adit joins a permanent tunnel, the plug is usually placed immediately upstream from the intersection as shown on figure 338. Keyways may be excavated into the rock or formed into the lining to insure adequate shear resistance between the plug and the rock or lining. After the plug has been placed and the concrete cooled, grout is forced through previously installed grout connections into the contact between the plug and the surrounding rock or concrete lining to insure a watertight joint. Sample specifications for this grouting are contained in appendix G.

250. *Conduits.*—The outlet works for an earthfill dam often entails the construction of a conduit which may be utilized for diversion during construction of the dam. This method for handling the diversion flows is an economical one, especially if the size of the conduit required for the outlet works is adequate to carry the diverson flows. Where diversion requirements exceed the capacity of the completed outlet works, an increase in capacity can be obtained by delaying the installation of gates, valves, pipe, and trashracks until the need for diversion is over. Increased capacity also can be obtained by increasing the height of the cofferdam, thereby increasing the head. In some instances the storage capacity of the reservoir at lower elevations may be such that much of the diversion design flood may be

Figure 336. Diversion adit and upstream cofferdam at Seminoe Dam. (891, Kendrick.)

temporarily retained and then evacuated through a diversion conduit of smaller capacity than would be required to discharge the peak of the flood.

In cases where the intake to the outlet works conduit is above the level of the streambed, an auxiliary stream-level conduit may be provided to join the lower portion of the permanent conduit. Such an auxiliary conduit is shown in figure 339. Permanent closure of this auxiliary conduit after diversion is completed can be accomplished in the same manner as that outlined in section 249. A concrete plug in an auxiliary diversion conduit is shown in figure 340.

Diversion conduits at stream level are sometimes provided through a concrete dam. These openings are provided with keyways, metal seals, and grouting systems and must be permanently closed throughout their entire length in the manner prescribed for placing tunnel plugs.

251. *Temporary Diversion Channel—Earthfill Dams.*—At sites where it may not be economical to provide a tunnel or conduit large enough to pass the diversion design flood, a temporary channel involving a gap through the earthfill dam may be utilized to divert streamflows while the remainder of the embankment is being constructed, as shown in figure 341. Though the dam shown is larger than the size of structures considered in this text, the methods and requirements for diversion through a temporary gap in the dam are the same. This method is adaptable to wide sites; obviously it cannot be used in narrow canyons. However, it is in the wider valleys where the diversion flows are likely to be too large to be economically carried in tunnels or conduits.

Before the stream is diverted, the foundation preparation required for the dam should be completed in the area where the temporary opening will be left through the embankment.

This preparation would include excavation and refilling of a cutoff trench, if one is to be constructed. The stream is then channeled through this area, after which the foundation work in the remainder of the streambed is completed. The portion of the embankment to either side of the diversion opening may then be completed. The side slopes of the opening should not be steeper than 4 to 1 to facilitate filling of the gap at the end of the construction period and to decrease the danger of cracking of the embankment due to differential settlement. The flat slope also provides a good bonding surface between the previously constructed embankment and the material to be placed.

The bottom grade of the temporary channel through the embankment should be the same as the original streambed so that erosion in the channel will be minimized. The width of opening will depend on the magnitude of the diversion design flood and on considerations of the equipment capabilities for filling the gap in the time which will be available.

The diversion is carried through the opening in the dam until sufficient progress is made in the completion of the embankment and appurtenant works so that floods can be carried safely through the completed or partially completed spillway. Closure of the gap in the embankment can then be made. To reduce the risk of the rising water surface in the reservoir overtopping the embankment being placed in the closure section, a period should be selected during the construction season when large floods are least likely to occur. Construction equipment should be mobilized so that the gap can be filled as quickly as possible to an eleva-

Figure 337. Bulkhead closure of diversion tunnel with pipes to pass flows for downstream requirements. (Granby Dam, 400–GL–216.)

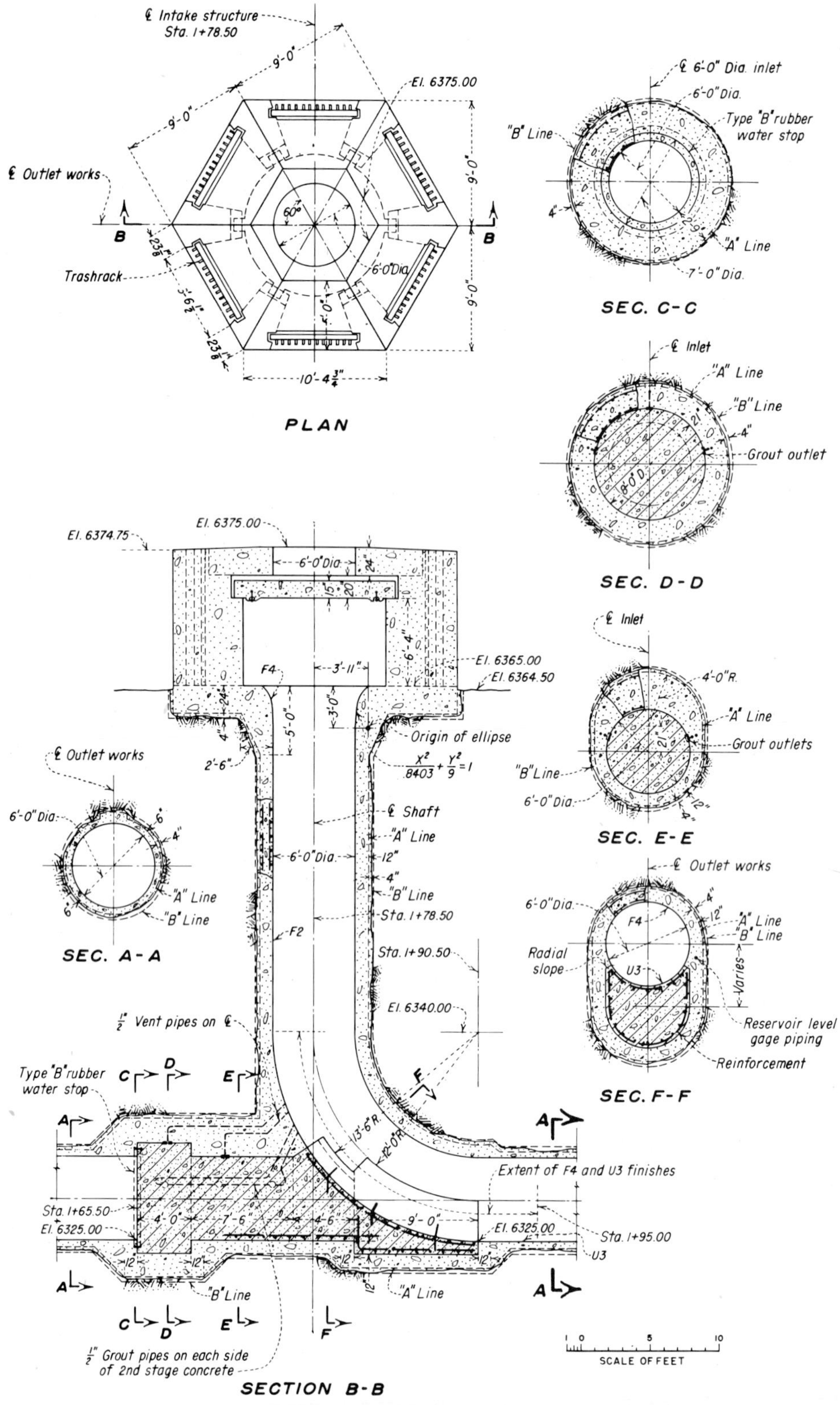

Figure 338. Concrete plug in diversion adit. From drawing No. 339–D–199.

Figure 339. Diversion through auxiliary stream-level conduit. Construction is in progress on outlet works intake. (Heart Butte Dam, DR–P–319.)

tion which will permit discharge of a flood, should one occur, through the spillway. The average rate of embankment placement must be such that the gap can be filled faster than the water rises in the reservoir. The capability of the contractor's plant to meet this requirement may be gaged by considering the average rate of embankment placing he will have to attain in order to complete the entire dam within the contract period, taking into account the number of days during the contract period that the weather will likely be suitable for embankment construction.

Care must be exercised during the filling of the gap so that the quality of work is not sacrificed due to the exigency of the situation. This is of great importance because frequently the diversion gap is in the area where the dam will be of maximum height. Extreme care must be used in order to obtain required densities and thus avoid excessive settlement of the completed embankment. Special attention must also be given to bonding of the newly placed material with the earthfill previously placed.

252. *Multiple-Stage Diversion for Concrete Dams.*—The multiple-stage method of diversion over the tops of alternate low construction blocks or through diversion conduits in a concrete dam requires shifting of the cofferdam from one side of the river to the other during construction. During the first stage, the flow is restricted to one portion of the stream chan-

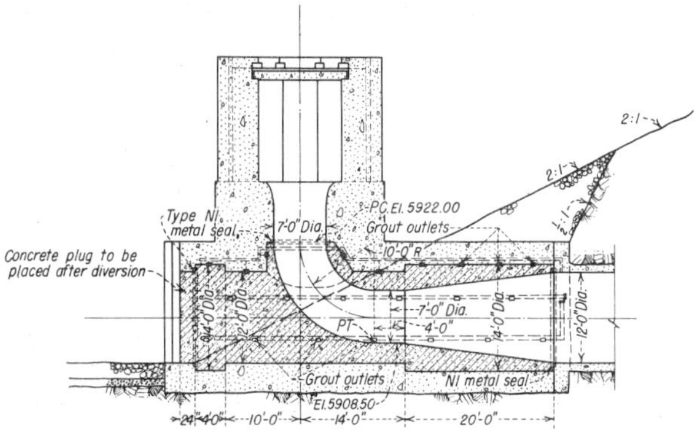

Figure 340. Detail of a concrete plug in an auxiliary diversion conduit. From drawing 526–D–146.

nel while the dam is constructed to a safe elevation in the remainder of the channel. In the second stage, the cofferdam is shifted and the stream is carried over low blocks or through diversion conduits in the constructed section of the dam while work proceeds on the unconstructed portion of the dam. The dam is then carried to its ultimate height, with diversion ultimately being made through the spillway, penstock, or permanent outlets. Figure 342 shows diversion through a conduit in a concrete dam, with excess flow over the low blocks.

253. *Cofferdams.*—A cofferdam is a temporary dam or barrier used to divert the stream or to enclose an area during construction. The design of an adequate cofferdam involves the problem of construction economics. Where the construction is timed so that the foundation work can be executed during the low water season, use of cofferdams can be held to a minimum. Where the streamflow characteristics are such that this is not practicable, the cofferdam must be so designed that it is not only

Figure 341. Temporary diversion channel through an earthfill dam. (Bonny Dam, 414–289C.)

Figure 342. Flows through diversion opening and over low blocks of a concrete dam. (Olympus Dam, EPA–PS–330–CBT.)

safe, but also of the optimum height. The height to which a cofferdam should be constructed may involve an economic study of cofferdam height versus diversion works capacity, including routing studies of the diversion design flood, especially when the outlet works requirements are small. If outlet works requirements dictate a relatively large outlet conduit or tunnel, diversion flows ordinarily may be accommodated without a high cofferdam. It should be remembered that floodwater accumulated behind the cofferdam must be evacuated in time to accommodate a recurrent storm. The maximum height to which it is feasible to construct the cofferdam without encroaching upon the area to be occupied by the dam must also be considered. Furthermore, the design of the cofferdam must take into consideration the effect that excavation and unwatering of the foundation of the dam will have on its stability, and must anticipate removal, salvage, and other factors.

Generally, cofferdams are constructed of materials available at the site. The two types normally used in the construction of dams are earthfill cofferdams and rockfill cofferdams, the design considerations of which closely follow those for permanent small dams of the same type. Figure 336 shows the construction of an earth and rockfill cofferdam. Other types not as common include timber or concrete cribs filled with earth or rock, and cellular steel cofferdams filled with earth or rock. Figure 343 shows a combination of several types. In this case, the major portion of the cofferdam consists of an earth and rock embankment, and steel sheet piling was used to effect final closure in swift water. Figure 335 shows the use of both earthfill cofferdams and cofferdams formed of steel-piling cells. Cellular steel cofferdams and steel sheet piling are adaptable to confined areas where currents are swift.

If the cofferdam can be designed so that it is permanent and adds to the structural stability of the dam, it will have a decided economic advantage. In some earthfill dams the cofferdam can be incorporated into the main embankment. In such instances, the saving is twofold—the amount saved by reducing the amount of embankment material required, and the amount saved by not having to remove the cofferdam when no longer needed.

C. SPECIFICATIONS REQUIREMENTS

254. *Contractor's Responsibilities.*—It is general practice to require the contractor to assume responsibility for the diversion of the stream during the construction of the dam and appurtenant structures. The requirement should be defined by appropriate paragraphs in the specifications which describe the contractor's responsibilities and inform him as to what provisions, if any, have been incorporated in the design to facilitate construction. Usually the specifications should not prescribe the capacity of the diversion works, nor the details of the diversion method to be used, but hydrographs prepared from streamflow records, if available, should be included. Also, the specifications usually require that the contractor's diversion plan be subject to the owner's approval.

In some cases, such as in constructing a concrete gravity dam in a wide canyon, the entire diversion scheme might be left in the contractor's hands, with the expectation that the flexibility afforded to the contractor's operations by allowing him to choose the scheme of diversion will be reflected in low bids. Since various contractors will usually present different schemes, the schedule of bids in such instances should require diversion of the river to be included as a lump-sum bid. Sometimes a few pertinent paragraphs are appropriate in the specifications giving stipulations which affect the contractor's construction procedures. For example, for an earthfill dam where diversion by means of a temporary channel is feasible or contemplated, the specifications may permit the contractor to divert the stream over the embankment placed in the completed cutoff trench, but usually would prohibit him from making final closure of the diversion works until the dam has been constructed to an elevation well above the spillway crest. Also, the contractor may be required to have the concrete in the spillway and outlet works essentially completed prior to closure of the temporary channel.

These, or similar restrictions, tend to guide the contractor toward a safe diversion plan. However, to further define the contractor's responsibility, other statements should be made to the effect that the contractor shall be responsible for and shall repair at his expense any damage to the foundation, structures, or any other part of the work caused by flood, water, or failure of any part of the diversion or protective works. The contractor should also be cautioned concerning the use of the hydrographs by a statement to the effect that the contracting authority does not guarantee the reliability or accuracy of any of the hydrographs and assumes no responsibility for any deductions, conclusions, or interpretations that may be made from them.

Sample specifications regarding diversion during construction are included in appendix G.

255. *Designer's Responsibilities.*—For difficult diversion situations, it may prove economical for the owner to assume the responsibility for the diversion plan. One reason for this is that contractors tend to increase bid prices for diversion of the stream if the specifications contain many restrictions and there is a large amount of risk involved. Where a dam is to be constructed in a narrow gorge, a definite scheme of cofferdams and tunnels might be specified, since here the loss of life and property damage might be heavy if a cofferdam built at the contractor's risk were to fail.

Another point to consider is that many times the orderly sequence of constructing various stages of the entire project depends on a particular diversion scheme being used. If the responsibility for diversion rests on the contractor, he may pursue a different diversion scheme, with possible delay to completion of the entire project. This could result in a delay in delivery of irrigation water or in generation

Figure 343. Upstream cofferdam of steel sheet piling and earthfill diverting streamflow into tunnel. (Green Mountain Dam, GM–283–CBT.)

of power, or both, with a subsequent loss in revenue.

If the owner assumes responsibility for the diversion scheme, it is important that the diversion scheme be realistic in all respects, and compatible with the probable ability and capacity of the contractor's construction plant.

D. BIBLIOGRAPHY

256. *Bibliography.*

[1] Ackerman, A. J., and Locher, C. H., "Construction Planning and Plant," McGraw-Hill, New York, 1940.

[2] Stafford, J. P., "Diversion Works for Construction," ASCE National Meeting on Water Resources Engineering, New Orleans, February 1969.

Chapter XII

Maintenance and Operation

H. G. ARTHUR [1]

A. GENERAL

257. *Inspections and Schedule of Observations.*—Arrangements should be made, immediately following the completion of a dam, for periodic inspection of the structure and all of the operating equipment. Adequate measures to accomplish this are usually taken on the more important structures but frequently are neglected on small dams. General responsibility for such structures may lie with a State, county, municipality, or a special board or commission endowed with administrative powers. The owner should advise such authority of the definite arrangements being made for periodic inspection and report by a responsible person who is informed of the hazards. For more important structures, the inspection should be conducted by an engineer. In remote locations, arrangements may be made with a forester, a minor county official, or a nearby rancher for inspection of small structures.

258. *Maintenance and Operating Instructions.*—Written instructions for maintenance and operation of the structures and equipment should be prepared as part of the design function and furnished to the owner or operating agency. These instructions should establish the frequency of and describe the extent and nature of inspections.

The instructions should also provide for the routine servicing of mechanical equipment where gates and valves are provided, and should include such instructions furnished by the manufacturer. The inspections and servicing should be accomplished in accordance with the principles discussed in part B of this chapter.

The instructions should also include detailed discussions of the proper operation of gates and valves from both mechanical and functional viewpoints. If a spillway is controlled by manually operated gates, specific instructions should be given regarding the operation of the gates during flood inflows into the reservoir.

[1] Formerly Engineer, Earth Dams Section, now Director of Design and Construction, Bureau of Reclamation.

B. INSPECTION AND MAINTENANCE OF DAMS

259. *Earthfill Dams.*—(a) *General Information.*—Routine maintenance of embankment slopes and the crests of earth embankments can be expected. However, any unusual conditions which may adversely affect the safety of an earthfill dam should be reported promptly. Any abnormal requirements for maintenance should also be reported.

(b) *Inspection of Embankments and Foundations.*—The embankment, abutments, and visible portions of the foundation adjacent to an earth embankment should be inspected at regular intervals for evidence of development of unfavorable conditions.

During rapid filling of the reservoir, the downstream slope of the embankment and the foundation downstream from the embankment should be carefully inspected at frequent intervals for indications of cracks, slides, sloughs, subsidences, impairment of slope protection, springs, seeps, or boggy areas caused by seepage from the reservoir. The upstream slope of the embankment should also be carefully inspected after periods of sustained high-velocity winds, and as the reservoir water surface is being drawn down, for evidence of cracks, slides, sloughs, subsidences, or damages to the slope protection, such as displacement of riprap or other signs of serious erosion.

During periods of low reservoir level, the exposed portions of the abutments and the reservoir floor should be carefully examined for sinks or seepage holes, unusual beaching conditions, or cracking. During periods of sustained high reservoir level, monthly inspections should be made of the embankment, with particular attention being given to the crest of the dam, to the visible portions of the upstream slope protection, to the downstream slopes, and to areas downstream from the dam for evidence of abnormal development. The frequency of inspections may be decreased after several seasons of operation have disclosed no abnormal conditions.

260. Concrete Dams and Concrete Structures.—(a) *General Information.*—All concrete structures, including dams, should be periodically inspected, with the following objectives in mind:

(1) To verify the safety of structures and facilities.
(2) To disclose conditions which might cause disruption or failure of operation.
(3) To determine the adequacy of structures and facilities to serve the purpose for which they were designed and are being used.
(4) To note the extent of deterioration as a basis for planning maintenance, repair, and rehabilitation.

(b) *Inspection of Concrete Structures.*—Annual inspections should be made by operating personnel and, at intervals of not greater than 6 years, major concrete structures should be inspected by a board of qualified engineers. Annual inspections should be made of all portions of the structures readily accessible and all other portions where there is reason to believe that damage may have occurred. Inspections by the board of engineers should be more thorough and detailed than those made by operating forces and should include portions of the structures not ordinarily accessible, such as penstocks, conduits, etc. These inspections should be scheduled during periods of low water to check the condition of structures normally submerged and during periods of maximum water level to check structural behavior under full load, or during maximum spillway discharges. The inspections should cover such items as:

(1) Abnormal settlements, heaving, deflections, or lateral movement of concrete structures.
(2) Cracking or spalling of concrete and opening of contraction joints.
(3) Deterioration, erosion, or cavitation of concrete.
(4) Abnormal leakage through foundation or formed drains or through concrete surfaces, construction joints, or contraction joints.
(5) Possible undermining of the downstream toe or other foundation damage.
(6) Unusual or inadequate operational behavior.

A written report of this inspection should be made to the owner or operating agency discussing the structures and conditions observed and recommendations for corrective measures where required.

(c) *Inspection of Channels and Surrounding Areas.*—Channels and surrounding areas, including backfill adjacent to concrete structures and embankment not included in the limits of an earthfill dam, should be inspected in conjunction with the concrete structures. These inspections should cover such items as:

(1) Channel bank or bed erosion and silting.
(2) Condition of riprap areas.
(3) Presence and condition of undergrowth in bottoms and on sides of channels and estimated effect on tailwater levels.
(4) River aggradation or degradation and possible effect on hydraulic operation of structures involved.
(5) Abnormal subsidence of backfill or embankment areas.

(6) Unusual or inadequate operational behavior.

261. *Mechanical Equipment.*—Periodic inspection of spillway gates and tests of operating equipment should be made by an engineer or a mechanic familiar with the purposes of the equipment. Inlet and outlet gates and valves should be tested regularly to see that they work freely. All gates and valves should be exercised at least annually, to determine that they are in good operating condition. Mechanical equipment should be lubricated and serviced in accordance with the manufacturer's instructions.

Trashracks should be cleaned of debris and accumulated sediment; metalwork should be painted to prevent rusting.

262. *Reporting of Abnormal Developments.*—The occurrence of unusual conditions should be reported immediately to the operating agency or owner of the dam by letter, telegram, or telephone, depending upon the nature of the development and the apparent urgency for repair. The description of slides, sloughs, or sudden subsidences should include the location, extent, rate of subsidence, effects of adjoining structures, reservoir and tailwater elevations, prevailing weather conditions, and other facts believed to be pertinent.

Information regarding the development of springs, seeps, and boggy areas should include such data as the location and size of the affected areas, the estimated discharge, the nature of discharge (whether clear or cloudy water), and the reservoir and tailwater elevations. To facilitate analysis of conditions, a map should be prepared showing the extent of all seeped areas, springs, and data such as dates and reservoir levels at the time of observation. This map should be revised periodically.

C. OPERATION

263. *Storage Dams.*—The dams discussed in this text normally will be used to store water for supplementary irrigation, domestic water supply, recreational purposes, stock ponds, or auxiliary flood control in tributaries of main streams. Their operation will rarely require continuous attention, except at seasonal intervals. If warranted, there should be an operator's house with telephone service, at or near the control works of dams. Also, the operator should have available a supply of small tools, sandbags, and other maintenance and emergency equipment.

Besides control for purposes of distribution, the water level in the reservoir may require a change at regular intervals to prevent propagation of malarial or pest mosquitoes, and to evade algae or other aquatic growth. The pond level on storage dams may also have to be drawn down upon receipt of storm warnings to provide storage for floodwaters.

The stimulation and protection of growth of vegetative cover to retard erosion on the slopes of the reservoir, on the borrow pits used in construction, and on the slopes of earthfill dams not otherwise protected is an important item of maintenance to which careful attention should be given. This cover is an essential item of protection against erosion and sloughing of banks, as well as of beautification of the structure, and may have an important influence on the cost of repairs.

Expert advice on suppression of algae growth in reservoirs should be obtained and followed, and no chemicals should be introduced into a reservoir without competent advice.

Instructions for operation of mechanical equipment should be followed closely so as to prevent damage to any of the installations through improper operation. Instructions given for the manual control of spillway gates during the occurrence of floodflows into the reservoir should be followed in detail, and operating personnel should not presume to interfere with the operation of the spillway in the manner which was contemplated in the design as explained in the maintenance and operating instructions.

264. *Diversion Dams.*—Diversion dams are

dams built for the purpose of raising the level of the stream and not for purposes of storage or equalization of flow. Such dams may divert flows into canals for irrigation of lowlands in the stream valleys or to spreading grounds for replenishment of ground-water storage.

Diversion dams are almost invariably overflow dams or have long overflow sections. Control gates are usually supplied so that the required diversion level may be maintained in spite of fluctuations in streamflow or in order to pass portions of the flow as needed to satisfy downstream water rights. This mechanical equipment should be operated and maintained in accordance with the instructions furnished as a part of the design function.

Diversion dams are often founded on sandy or gravelly streambed materials. In such cases, their stability may be insured by a broad base with cutoff walls. Such dams must be safeguarded by frequent inspections for evidence of piping or boils below the dams or for an increase in volume of seepage appearing at the downstream toe. The downstream apron is usually protected at the toe by heavy riprap. After floods, the streambed should be examined and the riprap renewed and repaired if necessary.

265. *Flood Detention Reservoirs.*—Flood detention reservoirs serve to reduce flood peaks by the temporary storage of that part of the flow which is in excess of the capacity of the spillway or outlet works of the dam. All reservoirs or pools produce some detention effect.

Structures built for the specific purpose of flood control by detention may be built with outlets which will automatically control the rate of release within safe limits. Overflow spillways are also provided in order to protect the dams even at the expense of possible flood damage below the structure, should a flood occur larger than that for which the dam was designed to control.

In addition to general inspections, the outlet works of such structures should be kept clear of soil deposits and debris which might affect their proper functioning.

266. *Changes in Operating Plan.*—A dam built for purposes of flood control may be diverted from its intended use by reason of the demand of a community for full storage for irrigation or water supply. Such demands, if acceded to, may result in a dangerous situation and possibly in the complete loss of the dam by overflow in the event of excessive runoff. A change in the operation of a structure or dam should not be made without a complete appraisal of the effects which such a change will have upon the structural and functional design of the structure.

Raising the height of a dam is sometimes undertaken without due consideration of the relation of the increased pressures to the limitations of the original design. No structural changes should be made without reference to the original plans or without the advice of an experienced engineer, preferably the designer.

The capacity of a storage reservoir should not be increased by placing stoplogs or other obstructions in an open-crest spillway without reference to the original plans and contemplated method of operation or without the advice of an experienced engineer. Such devices may operate in such a manner as to effectively reduce the ability of the reservoir and dam to safely store and pass the predicted inflow design flood.

The ability of a dam and spillway to safely store and pass the inflow design flood should be reviewed where storm and runoff occurrence and hydrologic technique developments since design of the structure indicate that a newly determined inflow design flood would be significantly larger than that on which the design was based. Operational or structural modifications, such as enlargement of the spillway, should be made where necessary to provide for safe handling of the newly determined inflow design flood.

267. *Coordination of Multiple Uses.*—Storage dams may be operated for more than one purpose. Multiple use may be made of the same storage space, or various head ranges in the same reservoir may be utilized for different purposes, such as flood control, power, irrigation, recreation, water supply, or navigation. Such combined operation requires very careful planning and control, as some of the uses are not compatible with others. For example, a power user, in order to be sure of the maximum amount of firm power, may wish to have the storage full when flood hazard is imminent and the need for available reservoir capacity greatest. Such combined operation is possible only with loss of some measure of benefit to one or all of the participants. In spite of these difficulties, conditions often make it desirable to permit such multiple use. Where such allocations are known in advance, they may have an influence on the design of the control devices. Careful management is important in the operation of multiple-use reservoirs in order to maintain a balanced perspective in the matter of relative values.

The method of operation of a multiple-use reservoir should not be arbitrarily changed without consideration of the effect of such a change upon the operation of the dam and upon the maximum water surface which will result, should the design inflow flood occur.

Estimating Rainfall Runoff from Soil and Cover Data

A-1. *Introduction*.—Although an engineer attempting to estimate the amount of runoff that will result from a given amount of precipitation on a specific area must exercise considerable judgment, the more information he has available the more accurate will be his estimate. The information presented in this appendix provides a means whereby an engineer may obtain an estimate of direct runoff from a given amount of rainfall. It is taken from the "Hydrology Guide for Use in Watershed Planning,"[1] published in 1956 by the Soil Conservation Service, with some editorial changes to meet the purpose of this text. The SCS National Engineering Handbook, Section 4, "Hydrology, Part I—Watershed Planning," issued August 1964 (see sec. 44), supersedes the previous publication. However, only a few changes have been made in the material pertaining to estimating runoff from soil and cover data. These changes are included in this text, the main change being the addition of about 2,000 soil names classified into respective hydrologic groups.

A-2. *Hydrologic Soil Groups*.[2]—(a) *Purpose*.—Watershed soil determinations are used in the preparation of *hydrologic soil-cover complexes* (sec. A-4), which in turn are used in estimating direct runoff. Four major soil groups are used. The soils are classified on the basis of intake of water at the end of long-duration storms occurring after prior wetting and opportunity for swelling, and without the protective effects of vegetation.

In the definitions that follow, the *infiltration rate* is the rate at which water enters the soil at the surface and which is controlled by surface conditions, and the *transmission rate* is the rate at which the water moves in the soil and which is controlled by the soil horizons. The hydrologic soil groups, as defined by SCS soil scientists, are as follows:

Group A. (Low runoff potential.) Soils having high infiltration rates even when thoroughly wetted and consisting chiefly of deep, well- to excessively-drained sands or gravels. These soils have a high rate of water transmission.

Group B. Soils having moderate infiltration rates when thoroughly wetted and consisting chiefly of moderately deep to deep, moderately well to well-drained soils with moderately fine to moderately coarse textures. These soils have a moderate rate of water transmission.

Group C. Soils having slow infiltration rates when thoroughly wetted and consisting chiefly of soils with a layer that impedes downward movement of water, or soils with moderately fine to fine texture. These soils have a slow rate of water transmission.

Group D. (High runoff potential.) Soils having very slow infiltration rates when thoroughly wetted and consisting chiefly of clay soils with a high swelling potential, soils with a permanent high water table, soils with a claypan or clay layer at or near the surface, and shallow soils over nearly impervious

[1] "Hydrology Guide for Use in Watershed Planning," National Engineering Handbook, Section 4, Hydrology, Supplement A, U.S. Department of Agriculture, Soil Conservation Service.

[2] The text of section A-2 is based on material that was largely prepared by G. W. Musgrave, Staff Specialist (Infiltration), Soil Conservation Service, for the "Hydrology Guide."

material. These soils have a very slow rate of water transmission.

The classification of soils into the major hydrologic soil groups may be done on the basis of the soil array, as discussed in the next section, provided agricultural soil maps are available. If such maps are not available, the soils will have to be classified on the basis of judgment.

(b) *Determination of Hydrologic Soil Groups.*—Each watershed is classified as being in one of the four major hydrologic groups given above. The soil array, table A–1, is used to find the average classification of soils. For example, a given watershed has 80 percent of its area in the B soil group and 20 percent in the C. The C soils are interspersed with the B soils. The watershed is classed as a B soil group area. The C soil group cannot be handled separately unless it is a compact area, in which case the watershed is divided into two parts which are handled as individual watersheds during the estimates of direct runoff volumes. In general, however, the watersheds should be used as units.

(c) *Soil Arrays.*—The relative hydrologic response of different soils or soil groups is an essential item in many engineering determinations. The list of major soils of the United States shown in table A–1 (i.e., either of major importance locally or of major extent) is a guide to the relative base rating of any soil. A local soil not included in the list can be compared with those that are included and its relative position thus determined. The capital letter following a name designates the hydrologic soil group classification.

The soil names will be found on agricultural soil maps which are available for a large portion of the United States. Section 87 discusses the availability of agricultural soil maps and surveys, and the type of information that is shown in them. Figure 46 shows the extent of published agricultural soil mapping in the United States.

The array is based on the premise that soils of similar profile characteristics (particularly depth, texture, organic matter content, structure, and degree of swelling when saturated) will respond in essentially similar manner under a long storm of appreciable intensity. In making comparisons it is assumed that the soils have minimum cover (bare); maximum swelling has taken place; and the applied rainfall exceeds potential infiltration. Since different types of infiltrometers are known to give different magnitudes of f_c [3] for the same soils, it is necessary to avoid using data from diverse techniques of measurement in developing the array. One soil cannot be placed on the basis of one technique of measurement and another soil placed according to a different method. All types of information are justifiably used in placing a soil in its proper relative position among other soils.

The measurement of infiltration rates through the artificial application of water, as by "rainfall simulators" or by one of several procedures for flooding the soil surface without rain impact, has been a common procedure for comparing soils or the effects of vegetation. These procedures are valuable for the *relative* comparisons they provide, but they are not satisfactory as a means of expressing infiltration rates quantitatively.

Extensive comparisons under well replicated and controlled conditions show:

(1) Rainfall simulators providing rainfall impact and turbidity of surface water give lower rates for the same soils and vegetation than do flooding types. (Thus Types F and FA infiltrometers give lower rates than do tubes or rings.)

(2) Infiltrometers of larger ground area give lower rates than do those covering a smaller area where the proportion of border effects is greater.

(3) Generally, infiltrometers also tend to give larger rates than those derived from watersheds.

Conversion factors for adjusting the results from different techniques to an equivalent base have been found to be impractical.

To overcome these difficulties, use is made of the extensive knowledge of soil profile characteristics possessed by the soil scientists. On this basis the major soils of the country are placed in an array, certain members of which

[3] Minimum infiltration rate.

already have been rated by watershed studies.

When the major soils of the United States are arranged in proper relative order, the range of f_c will begin with tight clays—essentially zero rates under these conditions—and extend to the maximum rates of deep, well-aggregated silts or to those of the deep sands such as are found in sandhill areas.

The curve of the array is essentially that of figure A–1. A possible normal range of variation due to variation within a given soil by depth, structure, or texture is expressed by the dash line—25 percent above and 25 percent below the mean. The mean is indicated by the heavy line. On this curve a number of points have already been established and additional ones will be forthcoming from evaluation of watersheds, analyses of existing research data, and new experimental work. Thus the magnitude of the curve can be fixed on the basis of minimum watershed performance. The problem of converting different kinds of infiltrometer data to a watershed basis is thus solved.

It is clear that the curve presenting this array of United States soils can be regarded as a base or "floor" upon which may be superimposed the effects, for example, of lesser soil moisture, varying kinds of vegetation, and the accumulation with time or organic matter derived from better vegetation. Supplementary information on the divergent effects of soil moisture on sands, clays, or laterites, of the aggregating effects of organic matter on these different kinds of soil, and the resultant effect on infiltration, will aid in the practical application of the basic soils rating.

The major soils in each of four hydrologic groups are listed in table A–1.

Acknowledgments are due the numerous soil scientists and soil correlators who have so freely provided basic information on the physical properties of these soil profiles. It is their detailed knowledge that has made possible this array of about 4,000 major soils of the continental United States.

A-3. *Land Use and Treatment Classes.*— (a) *Purpose.*—These classes are used in the preparation of hydrologic soil-cover complexes (sec. A–4), which in turn are used in estimating direct runoff. Types of land use and treatment are classified on a flood runoff-producing basis. The greater the ability of a given land use or treatment to increase total retention, the lower it is on a flood runoff-production scale. Land use or treatment types not described here may be classified by interpolation, as discussed in section A–4.

(b) *Crop Rotations.*—The sequence of crops on a watershed must be evaluated on the basis of its hydrologic effects. Rotations range from *poor* (or weak) to *good* (or strong) largely in proportion to the amount of dense vegetation in the rotation. *Poor rotations* are those in which a row crop or small gain is planted in the same field year after year. A poor rotation may combine row crops, small grains, or fallow, in various ways. *Good rotations* will contain alfalfa or other close-seeded legumes or grasses, to improve tilth and increase infiltration. For example, a 2-year rotation of wheat and fallow may be a good rotation for crop production where low annual rainfall is a limiting factor, but hydrologically it is a poor rotation.

(c) *Native Pasture and Range.*—Three conditions are used, based on hydrologic considerations, not on forage production. *Poor pasture or range* is heavily grazed, has no mulch, or has plant cover on less than about 50 percent of the area. *Fair pasture or range* has between about 50 and 75 percent of the area with plant cover and is not heavily grazed. *Good pasture or range* has more than about 75 percent of the area with plant cover, and is lightly grazed.

(d) *Farm Woodlots.*—The classes are based on hydrologic factors, not on timber production. *Poor woodlots* are heavily grazed and regularly burned in a manner that destroys litter, small trees, and brush. *Fair woodlots* are grazed but not burned. These woodlots may have some litter, but usually these woods are not protected. *Good woodlots* are protected from grazing so that litter and shrubs cover the soil.

(e) *Commercial Forest.*—The hydrologic condition classes are determined on the basis of depth and quality of litter, humus, and compactness of humus. The U.S. Forest Service procedure for determining the classes is given in section A–4.

(f) *Miscellaneous.*—Usually only very small parts of a watershed are in farmsteads, roads, and urban areas. When this is so, the areas

TABLE A–1.—*Soil names and hydrologic soil group classifications***. 288–D–2818 (1 of 5).*

Aastad B
Aberdeen D
Abilene C
Abington B
Acadia D
Acme B
Acton B
Adair D
Adams A
Adamsville C
Ade A
Adel B
Adelanto B
Adelphia B
Adler C
Adolph D
Afton D
Agar B
Agate C
Agawam B
Agency C
Agnew B
Aguadilla A
Aguilita C
Aguirre D
Ahmeek B
Ahnberg C
Ahtanum C
Aiken B
Airmont B
Akaka B
Akan C
Akaska B
Alachua B
Alaeloa B
Alama C
Alamance C
Alamosa C
Albaton D
Albemarle B
Albertville C
Albia C
Albion B
Alcester B
Alcoa B
Alden D
Alderwood B
Aldino D
Alexandria B
Alexis B
Alford B
Algarrobo D
Algiers D
Alicel B
Allard B
Allegheny B
Allen B
Allendale C
Allenwood B
Alligator D
Allis C
Allison C
Allouez B
Alma C
Almena C
Almirante D
Almo D
Almy B
Alonso C
Alps C
Alsen C
Altamaha D
Altamont C
Altavista B
Alto C
Alton B
Altoona C
Altura B
Altvan B
Alvin B
Alvira C
Amalu D
Amarillo B
Amboy B
Amelia C
Amenia B
Americus A
Ames C
Amite B
Amity C
Amsterdam B
Andover D
Andres B
Angie C
Angola C
Ankeny A
Annandale B
Anoka A
Anselmo B
Anthony B
Antigo B
Apache B
Apakuie A
Apishapa C
Apison C

Applegate C
Appling B
Arch B
Archer C
Arenzville B
Argyle B
Ark C
Arkport B
Arland B
Armagh D
Armer C
Armour B
Armuchee C
Arnold A
Arnot C
Arredondo A
Artesia C
Arvada D
Arveson D
Arzell D
Asa B
Asbury B
Ascalon B
Ashby C
Ashe B
Ashkum C
Ashley A
Ash Springs C
Ashton B
Ashuelot C
Ashwood D
Asotin B
Assumption B
Astoria B
Athelwold B
Athena B
Atherton D
*Athol B
Atkins D
Atterberry B
Atwood B
Auburn C
Auburndale D
Au Gres C
Augusta C
Aurora C
Austin B
Ava C
Avalanche B
Avery B
Avon B
Avonburg D
Axtell D
Ayr B

Babylon A
Baca B
Bagnell D
Bainville B
Baker C
Balch A
Baldock B
Baldwin C
Balfour B
Balm A
Balmorhea D
Bangor B
Banks A
Bannerville C
Baraboo B
Barbour B
Barbourville B
Barclay C
Barnard C
Barnes B
Barneston A
Barnett D
Barney A
Barnhardt A
Barnstead B
Barrancas D
Barron C
Borronett C
Barth B
Bartle D
Basher B
Bass B
Bastrop B
Batavia B
Bates B
Bath C
Baudette B
Baugh C
Baxter B
Bayamon B
Bayard A
Bayboro D
Bayside C
Beadle C
Bearden C
Beardstown C
Bear Lake D
Bear Prairie B
Beatty B
Beaucoup C

Beauford D
Beaumont D
Beauregard C
Beaver B
Beaverhead B
Beaverton B
Becket C
Beckton D
Beckwith C
Bedford C
Bedington B
Beecher C
Beechy D
Belfast D
Belfore C
Belgrade B
Belknap C
Bell D
Belle B
Bellingham D
Belmont B
Beltrami B
Beltsville C
Belvoir C
Benevola B
Benfield C
Benld D
Bennington C
Benoit D
**Benson C
Bentonville C
Beotia B
Berg C
Bergland D
Berkeley C
Berks C
Berks[illegible]ire B
Berm[illegible] B
Bernar[illegible]
Bernardston C
Berrien B
Berthoud B
Bertie C
Bertolotti A
Bertrand B
Berwyn C
Bethany C
Bethel D
Beulah B
Beverly A
Bewleyville B
Bibb D
Bickleton B
Biddeford D
Bienville B
Biggs A
Biggsville B
Big Horn C
Billett A
Billings D
Binnsville C
Bippus B
Birds C
Birdsall D
Birdsboro B
Birkbeck B
Biscay D
Bitterroot C
Blacklock D
Blackwater D
Bladen D
Blago D
Blaine C
Blair C
Blairton C
Blakeland A
Blakely B
Blanchard A
Blanco B
Bland C
Blandford C
Blanding B
Blanket D
Blanton A
Blencoe C
Blichton C
Blockton C
Blodgett B
Blomford B
Bloomfield A
Bloomington B
Blount C
Blue Earth D
Bluffton D
Bluford D
Bobtail C
Bodine B
Bogota C
Bohemian B
Bold B
Bolivia B
Bolton B
Bombay C
Bonaccord D
Bonaparte A

Bonham B
Bonilla B
Bonita D
Bonner B
Bonneville A
Bonnie D
Bono D
Bonpas B
Boomer C
Boone A
Bordeaux B
Bosket B
Boswell D
Bow D
Bowdoin D
Bowdre C
Bowie B
Bowmansville D
Boyd D
Boyer B
Boynton C
Bozarth C
Bozeman B
Braceville B
Bracken D
Brackett C
Braddock B
Braden C
Bradenton C
Bradley B
Brady B
Braham B
Brallier A
Brandon B
Brandywine C
Branford B
Brashear C
Bratton B
Braxton B
Brayton C
Brazito A
Brazos B
Brecknock C
Breece B
Breese D
Bremer B
Bremo C
Brennan B
Brenner C
Brenton B
Bresser B
Brewer D
Brewster C
Brickton C
Bridgehampton B
Bridgeport B
Bridger B
Bridgeville B
Briggs A
Briggsdale C
Brill B
**Brimfield C
Brimley B
Brinkerton D
Briscoe B
Brittain C
Broadbook C
Broadview C
Brockport C
Brooke C
Brookfield B
Brookings B
Brooklyn D
Brookston B
Brooksville D
Broughton D
Broward C
Brownfield A
Browning B
Brownlee C
Bruno A
Bryce D
Bub C
Buchanan C
Buckingham C
Buckland C
Buckley D
Buckner A
Bucks B
Bucoda D
Bude C
Buncombe A
Bunkerville D
Burchard B
Burdett C
Burgess B
Burgin D
Burke B
Burkhardt B
Burleson D
Burnham D
Burnside D
Burnsville B
Burnt Fork B
Burrell C

Burton B
Buse B
Butler D
Butlertown C
Butte B
Buxin D
Buxton C
Byars D
Byrds D

Cabinet C
Cabo Rojo C
Cabot C
Cacapon B
Caddo C
Cagey D
Caguas D
Cahaba B
Cajon A
Calais B
Caldwell B
Calhoun D
Califon C
Calloway C
Calvert D
Calverton C
Calvin C
Camaguey D
Camas A
Cambridge C
Camden B
Cameron C
Camillus C
Camp B
Campo C
Camroden C
**Canaan C
Canadian B
Canadice C
Canandaigua C
Canaseraga B
Cane B
Caneadea C
Caneyville C
Canfield B
Canoncito D
Canyon C
Cape Fear D
Capron B
Capshaw C
Captina C
Capulin B
Carbo C
Cardiff B
Cardington C
Carey B
Caribou B
Carlisle D
Carlsborg A
Carlton B
Carmi B
Carnegie B
Carnero C
Caroline C
Carrington B
Carrizo A
Carroll D
Carson D
Carstairs C
Carver A
Carytown D
Casa B
Casa Grande C
Cascade C
Casco A
Casey C
Cashmere A
Cashton D
Cass A
Cassville B
Castana B
Castle D
Castner C
Catalina C
Catalpa C
Catano A
Cataula C
Cathcart C
Catlett C
Catlin B
Catoctin C
Catron D
Catskill C
Cattaraugus B
Cave C
Cavode C
Cavot B
Cavour D
Cayagua C
Caylor B
Cayucos D
Cayuga C
Cazenovia C
Cecil B
Celina B

Center C
Centerton B
Central A
Chagrin B
Chalfont D
Chama B
Chamber D
Chamberino C
Chamokane B
Chandler C
Channahon B
Chariton C
Charleston C
Charlos B
Charlotte D
Charlton B
Chaseburg B
Chastain D
**Chatfield C
Chatsworth D
Chattahoochee B
Chauncey C
Chehalis B
Chelsea A
Chemawa B
Chenango B
Cheney B
Chenoweth B
Cherette A
Cherokee D
Cherry B
Cherryhill B
Cheshire B
Chester B
Chesterfield B
Chetek A
Chewacla C
Chewelah B
Cheyenne B
Chickasha B
Chiefland A
Chigley C
Chilgren C
Chilhowie D
Chili B
Chillisquaque C
Chillum C
Chilmark C
Chilo D
Chipeta D
Chippewa D
Chiricahua D
Choctaw B
Choptank A
Choteau C
Christian B
Christiana C
Christianburg D
Churchill D
Ciales B
Cialitos C
Cicero C
Cincinnati B
Cinebar A
Cintrona D
Cisne D
Cispus A
Clackamas C
Claiborne B
Clallam C
Clarence D
Clareville C
Clarinda D
Clarion B
Clark Fork B
Clarksburg C
Clarksdale C
Clarkson C
Clarksville B
Clary B
Clatsop D
Claverack B
Clawson C
Clayton C
Cleaver D
Cleburne B
Cle Elum B
Cleman B
Clement C
Cleora B
Clermont D
Clifton B
Climax D
Clinton B
Clio C
Clipper C
Clodine B
Cloquallum C
Cloquet B
Cloud D
Clover Creek C
Clovis B
Clyde C
Clymer B
Coamo C

* The hydrological grouping for the rocky phases of these soils should be reduced one group (i.e., C reduces to B).
** When these soils are classified as having class 3 or 4 rockiness, the hydrological grouping should be reduced one group.
*** Enlargement and revision of list by G. W. Musgrave in "How Much of the Rain Enters the Soil." Water, The Yearbook of Agriculture, 1955, U.S. Dept. of Agri., Washington, D.C.

TABLE A–1.—*Soil names and hydrologic soil group classifications***.*
288–D–2818 (2 of 5).

Cobb B
Cochise B
Cocoa B
Cody A
Coeburn C
Cogswell C
Cokdale C
Coker D
Cokesbury D
Colbert D
Colby B
Colden D
Coldwater D
Colebrook Loamy Fine Sand A
Colebrook Fine Sandy Loam B
Coleman B
Colemantown D
Colfax C
Colinas C
Collamar B
Collington B
Collins C
Collinsville C
Colo C
Coloma A
Colonie A
Coloso D
Colosse A
Colp D
Colrain B
Colton A
Colts Neck B
Columbia B
Colville B
Colwood D
Colyer C
Comfrey C
Comly C
Commerce C
Comoro A
Compton B
Conant B
Conasauga C
Concord D
Condit D
Condon B
Conestoga B
Conesus C
Congaree B
Conley C
Conotton B
Conover B
Conowingo C
Constable A
Continental C
Conway C
Cook D
Cookeville B
Cookport C
Coolidge B
Coolville C
Cooney C
Coope
Copake B
Copalis C
Copas C
Copeland C
Coplay D
Coral B
Corcega D
Corduroy B
Corkindale C
Corley C
Cornutt D
Corvallis B
Corwin B
Corydon C
Cossayuna C
Cotaco C
Coto C
Cottonwood C
Cougar C
Couparle D
Coupeville B
Courtland B
Courtney D
Couse C
Cove D
Coveland C
Coveytown C
Covington D
Cowden D
Cowiche B
Cowling B
Coxville D
Crago B
Craig C
Crandon B
Crane B
Craven C
Crawford D
Creal D
Creedmoor C
Cresbard C
Crescent B
Crestmore C
Crete D
Crevasse A
Crider B

Crockett D
Crofton B
Croghan B
Croom C
Crosby C
Crossville B
Croton D
Crow C
Crowder D
Crowley D
Crown B
Crystal B
Culleoka B
Cullo C
Culpeper B
Culvers C
Cumberland B
Curran C
Curtis B
Cushman B
Custer D
Cut Bank B
Cuthbert C
Cypremort C

Dade A
Daggett B
Dakota B
Dalbo B
Dale C
Dalhart B
Dalton C
Dana B
Dandridge C
Daniels B
Danley C
Dannemora C
Danvers C
Darien C
Darling C
Darnell C
Darret B
Darwin D
Davidson B
Davie D
Dawes C
Dayton D
Deary C
Decatur B
Deckerville D
Decorra B
Defiance D
*Dekalb B
Delanco C
Delfina C
Dell C
Dellrose B
Delphi B
Delpine D
Delray D
Del Rey C
Demers C
Denham D
Dennis C
Denny D
Denrock D
Denson C
Denton C
Depew C
Derby A
Descalabrado B
Deschutes A
De Sota C
Detour C
Detroit B
Dewart B
Dewey B
Dexter B
Diablo D
Dick A
Dickey A
Dickinson Fine Sandy Loam A
Dickinson Loam B
Dickson C
Dill B
Dilldown B
Dillinger B
Dillon D
Dilman B
Dimmick D
Disco B
Dixmont B
Dodgeville Deep Phase B
Dodgeville Shallow Phase C
Doland B
Dominguito D
Dominic B
Donerail C
Donlonton C
Dorchester B
Dorsey D
Dos Cabezas B
Doty B
Dougherty B
Douglas B
Dover C
Dowellton D
Dowling D

Downs B
Doylestown D
Dragston C
Drake B
Dresden B
Dripping Springs D
Drummer B
Drummond D
Drury B
Dryad C
Duane B
Dubbs B
Dubois C
Dubuque Deep Phase B
Dubuque B
Dubuque Shallow Phase C
Duffield B
Duffy C
Dukes A
Dulac C
Dunbar C
Duncan C
Duncannon B
Duncom D
Dundas C
Dundee C
Dune Sand A
Dunellen B
Dungeness B
Dunham B
Dunkirk B
Dunlap C
Dunmore C
Dunning D
Du Page B
Duplin B
Dupo C
Dupont D
Durant D
Durham B
Durkee C
*Dutchess B
Dutson C
Duval B
Dwight D
Dwyer A
Dyke B

Eakin B
Easton C
Eastonville A
Ebbert D
Ebbs B
Ebeys B
Eckman B
Ector C
Edalgo D
Eddy C
Eden C
Edenton C
Edge D
Edgeley C
Edgemont B
Edgewick A
Edgington C
Edina D
Edinburg C
Edisto C
Edith A
Edmonds D
Edna D
Edneyville B
Edom C
Edwards D
Eel C
Efland C
Egam C
Egeland B
Eifort C
Elbert D
Elburn B
Elco B
Eld B
Eldon C
Eldorado C
Elfrida C
Elioak B
Elk B
Elkins D
Elkinsville B
Elkton D
Ellery D
Elliber A
Ellington C
Elliott C
Ellis D
Ellison B
Ellsberry C
Ellsworth D
Elmo C
Elmore C
Elmwood C
Elsinboro B
Elsmere A
Elwha C
Emmert A
Emmet B
Emory B
Empey C

Empeyville C
Enders C
Enfield B
Englund D
Ennis B
Enon C
Ensenada C
Ensley D
Enstrom B
Enterprise A
Enumclaw B
Ephrata A
Epping D
Era B
Eram D
Erie C
Ernest C
Escondido C
Espinosa B
Esquatzel B
Essex C
Estacion B
Estelline B
Estevan C
Estherville B
Esto C
Etowah B
Ettrick D
Eubanks B
Eufaula A
Eulonia C
Eustis A
Eutaw D
Evans B
Evendale C
Everett A
Everson C
Ewing A
Exline D
Eylar D

Faceville B
Fahey B
Fairfax B
Fairhaven B
Fairhope C
Fairmount D
Fajardo C
Falaya C
Falcon B
Falkner D
Fall B
Fallbrook B
Fallsburg C
Fallsington D
Falun B
Fannin B
Fargo D
Farland B
**Farmington C
Farnum C
Farragut C
Faunce A
Fauquier B
Fawcett C
Faxon D
Fayette B
Fe D
Felda D
Felida B
Fellowship C
Fergus C
Fiander C
Fidalgo C
Fillmore D
Fincastle C
Fitch A
Fitchville C
Fitzhugh B
Flamingo D
Flanagan B
Flandreau B
Flasher A
Flathead B
Fleetwood B
Fletcher C
Flint C
Flora D
Florence C
Florsheim D
Floyd B
Fluvanna C
Foard D
Foley C
Folsom C
Fonda D
Fordney A
Fordville B
Fore D
Forestdale D
Forrest D
Fort Collins B
Fort Lyon B
Fort Meade A
Fort Pierce B
Fortuna D
Fox B
Foxhome B
Frankfort D
Franklinton C
Frankstown B

Fraternidad D
Frederick B
Fredon C
Freehold B
Freeland C
Freeon B
Freer C
Fremont C
Frenchtown D
Freneau D
Frio B
Frost D
Fruita B
Frye C
Fullerton B
Fulton C

Gage D
Gainesville A
Gale B
Galen B
Galestown A
Gallatin D
Gallion B
Galveston A
Galvin B
Gann B
Gannett D
Gara C
Gardnerville C
Garfield D
Garner D
Garrison A
Garwin D
Gasconade D
Gaviota A
Gayville D
Gearhart A
Geary B
Geer C
Geiger D
Gem C
Genessee B
Genoa D
Genola B
Georgetown C
Georgeville B
Gerald D
Germania B
Geronimo B
Gila B
Gilcrest B
Gilead C
Giles A
Gilford D
Gilligan B
Gilman B
Gilpin C
Gilson B
Gilt Edge D
Ginat D
Gird B
Givin C
Glasgow C
Glenbar C
Glencoe D
Glendale C
Glendive A
Glenelg B
Glenfield C
Glenford C
Glenoma B
Glenville C
Gloucester B
Godwin B
Goessel D
Gogebic G
Goldridge A
Goldsboro B
Goldston C
Goldvein C
Goliad C
Gooch C
Gore D
Gorus B
Goshen B
Gosport C
Gothard D
Gowen C
Grady D
Graham C
Grail C
Granby D
Grande Ronde C
Grant B
Grantsburg C
Grantsdale B
Granville B
Grayling A
Great Bend B
Greeley B
Green Bluff B
Greenbush B
Greendale B
Greenfield B
Greenport C
Green River B
Greensboro B
Greenville B
Greenwater A
Greer C

Grenada C
Grenville B
Gresham C
Greybull C
Greys B
Griffin C
Grimstad C
Grosclose C
Groton A
Grove A
Groveland B
Grover B
Groveton B
Grundy C
Guadalupe B
Guanica D
Guayabo C
Guayama C
Guckeen C
Gudrid B
Guelph B
Guernsey C
Guin A
Guthrie D

Habersham B
Haccke D
Hackers B
Hackettstown B
Hadley B
Hagener A
Hagerstown B
Haig C
Haiku C
Haines B
Halawa C
Halesakala A
Halewood B
Half Moon C
Halfway C
Halii C
Haliimaile B
Hall B
Halsey D
Hamburg B
Hammerly C
Hamilton B
Hamlin B
Hammond D
Hampshire C
Hampton B
Hanalei C
Hanceville B
Hand B
Hanford B
Hanipoe B
Hannahatchee B
Hanover B
Hanska C
Hanson B
Harbin B
Harbourton C
Harlem B
Harley C
Harlingen D
Harmon D
Harmony C
Harpster B
Harriet D
Harris D
Harrisburg C
Harrison C
Harstine A
Hartford B
Hartland B
Hartleton B
Hartsburg B
Hartsells B
Harwood B
Haskill A
Hassel D
Hastings C
Hatchie C
Haven B
Havre B
Haxtun A
Hayden B
Hayesville B
Haymond B
Haynie B
Hayter B
Hazel C
Hazen B
Heath B
Hebo D
Hecla B
Hector B
Hedville C
Heisler B
Heitt C
Helena C
Hemmi C
Hempstead B
Henderson D
Hennepin B
Henry D
Henshaw C
Herbert B
Herkimer B
Hermiston B
Hermitage B

* The hydrological grouping for the rocky phases of these soils should be reduced one group (i.e., C reduces to B).
** When these soils are classified as having class 3 or 4 rockiness, the hydrological grouping should be reduced one group.

TABLE A–1.—*Soil names and hydrologic soil group classifications***. 288–D–2818 (3 of 5).*

Hermon B
Hermosa C
Hernando B
Herndon C
Hero B
Herrick C
Hershal B
Hesch B
Hesseltine A
Hesson C
Hialeah D
Hiawatha A
Hibbard C
Hibbing B
Hickory C
Hicks B
Hidalgo B
Hidewood C
Highfield B
Higley B
Hiko Springs D
Hilger B
Hilliard B
Hillsboro B
Hillsdale B
Hilo B
Hilton C
Hinckley A
Hinman D
Hiwassee B
Hiwood A
Hixton B
Hobble B
Hobbs B
Hockley C
Hoffman C
Hogansburg B
Hoko C
Holbrook D
Holcomb D
Holdrege B
Holland B
Hollinger C
**Hollis C
Hollister C
Holloway B
Holly C
Hollywood D
Holmdel B
Holston B
Holt B
**Holyoke C
Homer C
Hondo C
Honeoye B
Honokaa B
Honolua B
Honomanu C
Honouliuli C
Hood B
Hoodsport A
Hooker B
Hoosic B
Hopewell C
Hopper B
Hoquiam B
Hord B
Hornell C
Hortman C
Horton B
Hosmer C
Houdek B
Houghton D
Houlka D
Houlton C
Housatonic C
Houseville C
Houston D
Houston Black D
Hovde B
Hoven D
Howard B
Howell B
Hoye C
Hoyleton C
Hoypus A
Hoytville D
Hubbard A
Huckabee A
Huckleberry C
Hudson C
Huey D
Huff B
Huffine B
Huggins C
Hugo B
Huikau A
Humacao C
Humbarger B
Humeston C
Humphreys B
Hunt D
Hunters B
Huntington B
Huntsville B
Hurley D
Hurst D
Hutchinson C
Hyattsville B
Hyde D
Hymon C

Iao B
Iberia D
Ida B
Idana C
Ihlen D
Ilion C
Illiopolis B
Ima B
Immokalee C
Imperial D
Ina C
Independence A
Inglefield C
Ingomar D
Inman C
Inola D
Io B
Iola A
Iona B
Ipava B
Iredell D
Irion D
Irish D
Iron River B
Irurena D
Irving D
Irvington C
Irwin D
Isanti D
Islote B
Isom B
Issaquah B
Istokpoga D
Iuka C
Ivanhoe D
Ives B
Izagora C

Jacana C
Jackson B
Jacob D
Jaffrey A
Jaucas A
Jayuya C
Jeanerette C
Jefferson B
Jerauld D
Jerome D
Jessup C
Joe Creek B
Johnston D
Joliet C
Jonesville A
Joplin B
Josefa D
Josephine C
Joy B
Juana Diaz B
Judith B
Judson B
Jules B
Juliaetta A
Juncos C
Juniata B
Junius C
Juno A

Kaena D
Kahana B
Kalamazoo B
Kalihi D
Kalispell B
Kalkaska A
Kalmia B
Kaloko D
Kamananui C
Kanab C
Kanapaha B
Kaneohe B
Kapapala B
Karnak D
Karnes B
Karro B
Kars B
Kasota C
Kasson C
Katemcy D
Kato B
Katy D
Kaufman D
Kawaihae C
Kawaihapai A
Keaau D
Kealakekua B
Keansburg D
Keating C
Kedron C
Keene C
Keith B
Kelly D
Kelso C
Kelton A
Kempsville B
Kempton B
Kenansville B
Kendaia C
Kendall B
Kennebec B
Kennedy B
Kenney A
Kenspur B
Kent D
Keomah C
Kerby B
Keri C
Kerrtown B
Kershaw A
Kettle B
Kettleman C
Keyesport C
Keyport C
Keystone A
Kibbie C
Kickerville B
Kilauea A
Kilbourne A
Kimbrough D
Kinghurst B
Kings C
Kinross D
Kipling D
Kipp B
Kipson D
Kirkland D
Kirvin C
Kistler B
Kitsap C
Kittitas B
Kittson B
Kiwanis B
Klaberg C
Klamath C
Klaus A
Klej B
Kline A
Klinesville C
Knappa B
Knight C
Knox B
Koch C
Koehler B
Kohala B
Kokokahi D
Kokomo D
Kolekole C
Konokti A
Koolau D
Kopiah D
Kosmos D
Koster C
Kranzburg B
Krause B
Kreamer C
Kresson C
Krum C
Kukaiau B
Kunia B
Kutztown C

La Belle C
Labette C
Labounty C
La Brier C
Lacamas D
La Casa C
Lackawanna B
Ladd C
La Delle B
Ladoga C
Ladysmith D
Lafe D
Lagonda D
La Grande C
La Hogue B
Lahontan D
Laidig C
Laidlaw A
Laie D
Lairdsville B
Lajas C
Lake Charles D
Lake Creek B
Lakehurst A
Lakeland A
Lakemont C
Lakeville
 Sandy Loam A
Lakeville Loam B
Lakewood A
Lakin A
La Lande B
Lamington D
Lamont Fine
 Sandy Loam A
Lamont Loam B
Lamonta C
Lamoure C
Lamson C
Lanark B
Lancaster B
Land C
Landes B
Landisburg C
Lane C
Langford C
Langley B
Langrell B
Lanham D
Lansdale B
Lansdowne C
Lansing B
Lantz D
La Palma B
Lapine A
Lapon D
Laporte D
La Prairie B
Laredo B
Lares C
Largent C
Largo C
Larimer B
Larkin B
La Rose B
Larry D
Las Animas A
Lashley B
Las Lucas D
Las Piedras D
Lassen C
Las Vegas D
Latah C
Lauderdale C
Laurel B
Lauren A
Laveen B
La Verkin B
Lawhorn D
Lawrence C
Lawrenceville C
Lawson B
Lawton B
Lax C
Lea B
Leadvale C
Leaf D
Leal B
Leavenworth A
Leavitt B
Leavittville B
Lebanon D
LeBar B
Leck Kill B
Lee D
Leeds B
Leeper D
Leetonia B
Legore C
Lehigh C
Leicester C
Lela D
Lempster D
Lena D
Lenoir Fine
 Sandy Loam B
Lenoir
 Silt Loam C
Lenox C
Leon C
Leona D
Leonardtown D
Leota C
Leshara B
Lester B
LeSueur B
Letcher D
Letort B
Levan A
Lewisberry B
Lewiston C
Lewisville B
Lexington B
Liberty C
Lick B
Lick Creek B
Lickdale D
Lightning D
Lignum C
Lihen A
Likes A
Lima C
Limerick C
Lincoln A
Lincroft A
Lindley C
Lindsborg D
Lindside C
Lindstrom B
Linganore C
Link B
Linker B
Linneus B
Lino C
Lintonia B
Lisbon B
Lismas D
Lismore B
Littlefield D
Little Horn B
Littleton B
Litz C
Livingston D
Llave C
Lloyd B
Lobdell B
Lobelville C
Lockhard C
Lockhart B
Lockport C
Locust C
Lodi B
Lofton C
Logan C
Logandale C
Lolo B
Lomax B
Lonepine B
Lone Rock B
Longford C
Longrie C
Lonoke B
Lookout C
Loon A
Loradale B
Lorain C
Lordstown C
Lorella C
Lorenzo A
Loring B
Los Guineos D
Los Osos C
Loudon C
Loudonville C
Louisa B
Louisburg B
Loup D
Lowell C
Loy D
Loysville C
Lualualei D
Lubbock C
Lucas C
Lucien C
Ludlow C
Lufkin D
Lukin C
Lummi C
Lun C
Lunt C
Lupton D
Lura D
Luray C
Luton D
Luverne B
Luzena D
Lycoming C
*Lyman C
Lynchburg C
Lynden A
Lynndyl C
Lyons D
Lystair A

Mabi D
Machete C
Mack B
*Macomber C
Macon D
Madalin C
Maddock A
Maddox B
Madison B
Madras C
Madrid B
Maginiis C
Magnolia B
Mahaska B
Mahnomen B
Mahoning D
Maiden C
Maile B
Makalapa D
Makawao B
Makena B
Malaga B
Malaya C
Maleza B
Mamala C
Manalapan D
Manana C
Manassa B
Manassas B
Manastash C
Manatee D
Manchester A
Mangus C
Manhattan B
Manheim C
Manlius C
Manor B
Mansfield D
Mansic C
Mansker B
Mantachie C
Manteo D
Manvel B
Maple D
**Mapleton C
Marble A
Marcus D
Marcy D
Mardin C
Marengo C
Mariana C
Marias D
Marietta B
Marina A
Marion D
Marissa C
Markland C
Marksboro B
Marlboro B
Marlow C
Marlton C
Marna D
Marquette B
Marshall B
Martha C
Martin Pena D
Martinsdale B
Martinton C
Masada B
Mason B
Massena C
Massillon B
Matanzas C
Matapeake B
Matawan C
Matlock D
Matmon C
Matney B
Mattapex C
Maumee D
Maunabo D
Maury B
Maverick D
May B
Mayhew D
Maynard Lake B
Mayo B
Mayodan B
Maytown C
Mazeppa B
McAfee B
McAllister C
McBride B
McDonald C
McDowell D
McEwen C
McGary C
McKamie C
McKay D
McKenna B
McKenzie D
McLain C
McMurray A
McNeal C
McPaul B
McPherson D
Meadin A
Meadowville B
Meadville B
Mecklenburg C
Meda B
Medary C
Medford C
Medina B
Medio C
Meeteetse C
Mehlhorn C
Meigs C
Melbourne C
Mellenthin B
Melrose C
Melvern C
Melvin D
Memphis B
Menahga A
Mench B
Menfro B
Menlo D
Mentor B
Mercedita D
Mercer C
Mereta D
Meridian B
Meros A
Merrimac B
Mertz B
Mesa B
Meskill D
Metea B
Methow B
Metolius A
Mexico D
Mhoon C
Miami B
Middlebury B
Midland D
Midway D
Mifflinburg B
Miguel D
Milaca B
Milam B
Miles B
Milford C
Millbrook B
Mill Creek B
Miller D
Millington B
Millsdale B
Milo D
Milroy D
Mimosa C
Minatare D
Minco B
Mineola B
Miner D
Minnequa B
Minora C
Minvale B
Mires A
Mission C
Mitchell B
Moca D

* The hydrological grouping for the rocky phases of these soils should be reduced one group (i.e., C reduces to B).
** When these soils are classified as having class 3 or 4 rockiness, the hydrological grouping should be reduced one group.

TABLE A–1.—*Soil names and hydrologic soil group classifications***. 288–D–2818 (4 of 5).*

Modale C
Moenkopie C
Moffat B
Mohave C
Mohawk B
Moiese B
Moira C
Mokena C
Molena A
Moline D
Monarda C
Monee D
Monmouth C
Monona B
Monongahela C
Monroeville C
Montalto B
Montara D
Montell D
Monteola D
Montesano B
Montevallo C
Montgomery D
Monticello B
Montoya D
Moody B
Moreau D
Morley C
Mormon Mesa D
Moro Bay D
Morocco C
Moro Cojo A
Morrill B
Morris C
Morrison B
Morrow C
Morse D
Morton B
Moscow B
Moshannon B
Mossyrock A
Mottsville A
Mount Carroll B
Mount Lucas C
Mountview B
Mucara B
Muir B
Muirkirk B
Mukilteo A
Mullins D
Munising B
Munuscong D
Murrill B
Muscatine B
Muse B
Muskingum C
Muskogee C
Musselshell B
Myatt D
Myersville B

Naalehu B
Naches B
Nacimiento C
Nacogdoches C
Naiwa B
Nakelele B
Nantucket C
Nanum B
Napa D
Napier B
Nappanee D
Naranjito C
Narcisse B
Narragansett C
Nasel C
Nason C
**Nassau C
Natalie C
Natchez B
Natchitoches D
National A
Navajo D
Navasota D
Navesink B
Naylor C
Nebish B
Neble A
Needmore C
Negley B
Nehalem B
Nellis B
Neosho D
Neptune A
Nereson B
Neshaminy B
Nesika B
Nester D
Neubert B
Nevada D
Neville B
Newark C
Newart B
Newberg B
Newberry C
New Cambria C
Newfane B
Newkirk B
Newport B
Newton D
Newtonia B

Nicholson B
Nicholville B
Nickel D
Nicollet B
Niles C
Nimrod C
Ninigret B
Niota D
Nipe B
Nisqually A
Nixa C
Nixon B
Nixonton B
Noble B
Nobscot A
Nodaway B
Nogales C
Nohili D
Nolan C
Nolichucky B
Nolo C
Nonopahu C
Nookachamps D
Nooksack C
Nora B
Norden B
Norfolk B
Norge C
Norma C
Northport C
North Powder C
Northumberland C
Northville D
Norton B
Norwich D
Norwood B
Nosbig D
Nuby C
Nuckolls C
Nueces A
Nunda C
Nunn C
Nutley D
Nymore A

Oahe B
Oakford B
Oakland B
Oasis C
Ochlockonee B
Ochopee D
Ockley B
Oconee C
Odessa C
Odin C
O'Fallon D
Ogemaw B
Okaw D
Okeechobee D
Okeelanta D
Okemah C
Okenee D
Okoboji B
Oktibbeha D
Olaa B
Olequa C
Olinda B
Olivier C
Olmitz B
Olmsted C
Olympic B
Omaha B
Omega A
Ona C
Onalaska B
Onamia B
Onarga B
Onaway B
Ondawa B
Oneida C
O'Neill B
Onslow B
Ontario B
Ontonagon C
Onyx B
Ookala B
**Oquaga C
Oquawka A
Ora C
Oracle B
Orange D
Orangeburg B
Orcas A
Orchard B
Ordway D
Orelia D
Orella D
Orient B
Orienta B
Orio C
Orion C
Orlando
High Phase A
Orlando
Low Phase B
Orman D
Orrville C
Ortello A
Orting C
Osage D
Osceola D

Oshawa D
Oshtemo B
Osmund B
Oso A
Ostrander B
Otero B
Othello D
Otisville A
Otsego C
Ottawa A
Otter D
Otterholt B
Ottokee A
Otway D
Ovid C
Owaneco D
Owen Creek C
Owens D
Ozona C

Paaloa B
Paauhau B
Pace B
Paden C
Page D
Pahranagat C
Pahroc D
Paia B
Painesville B
Paiso C
Paiute B
Palatine C
Palestine B
Palmas Altas D
Palm Beach A
Palmdale C
Palmyra B
Palouse B
Pana B
Pandura B
Panton D
Papago B
Papakating D
Papineau C
Parishville C
Parkdale A
Parke B
Parker B
Parkwood C
Parnell D
Parr B
Parsons D
Pasco B
Paso Seco C
Pasquotank D
Patent C
Patit Creek B
Patoutville C
Patrick B
Patton C
Paulding D
Pauwela C
Pawlet B
Pawnee D
Paxton C
Paymaster B
Payne D
Peace River D
Peacham D
Pearman C
Pearson C
Pecatonica B
Pedernales C
Pekin C
Pelan B
Pella B
Pembroke B
Pence B
Penn C
Pennington B
Penoyer C
Penrose C
Penwood A
Peoh C
Peone C
Peotone C
Pequea C
Perkinsville B
Perks A
Perrine D
Perry D
Persayo D
Pershing C
Peru C
Peshastin A
Petoskey A
Petrolia D
Pettis B
Pheba C
Phelps B
Phillips D
Philo C
Picacho D
Pickaway C
Pickford D
Pickwick B
Pierce B
Pierre D
Piihonua B
Pilchuck A
Pilot B

Pilot Rock C
Pima C
Pinal D
Pinckney C
Pinones D
Pinson B
Pintura A
Pisgah B
**Pittsfield C
Pittstown C
Pittwood B
Placentia D
Plainfield A
Plano B
Plata C
Platea C
Platner C
Plattsmouth B
Plattville B
Pledger D
Plummer D
Plymouth B
Pocomoke D
Podunk B
Poinsett B
Poland B
Polson C
Pomello A
Pompano D
Pomroy B
Poncena D
Pond Creek C
Pontotoc B
Pope B
Poppleton A
Poquonock C
Port B
Portales B
Port Byron B
Porters B
Portland D
Portsmouth D
Portugues D
Poskin C
Post D
Potamo D
Potter C
Pottsville D
Poultney B
Poverty C
Powder B
Powell C
Poygan D
Pozo Blanco C
Prather C
Pratt A
Pretiss C
Prescott D
Presque Isle B
Preston A
Prewitt C
Prieta B
Princeton B
Pring B
Proctor B
Progresso B
Promise D
Prospect B
Prosser B
Providence C
Provo B
Prowers B
Ptarmigan B
Puget B
Puhi B
Pulaski B
Pulehu A
Pullman D
Purdy D
Purgatory D
Puu Oo B
Puu Pa B
Puyallup B

Quamba D
Quandahl B
Quay C
Quicksell C
Quincy A
Quinlan B
Quonset A

Raber C
Rabun B
Racine B
Racoon D
Radford B
Radnor D
Ragnar A
Rago C
Rainbow C
Rains D
Ralston B
Ramona C
Ramsey B
Randall D
Ranger C
Rankin C
Rantoul D
Rapidan C
Rarden C

Raritan B
Raub B
Rauville D
Ravalli D
Ravenna C
Ravola B
Ray B
Rayne B
Redington C
Reagan C
Reaton B
Reaville C
Rebuck C
Red Bay B
Redfield B
Red Hook C
Redlands B
Redmond B
Reed D
Reeser C
Reeves C
Regent C
Regnier D
Reinach B
Reliance C
Renfrow D
Reno D
Renohill C
Renova B
Renshaw B
Renslow B
Rentide C
Reparada D
Retsof C
Rex C
Reynolds B
Rhinebeck C
Rhoades D
Richfield B
Richland C
Richview C
Richwood B
Ridgebury C
Ridgely B
Ridgeville B
Riesel D
Riffe A
Riga C
Riggs D
Riley B
Rillito B
Rimer C
Rinard D
Ringling C
Ringold B
Rio Arriba D
Rio Canas C
Rio Lajas A
Rio Piedras D
Ritchey B
Rittman C
Ritzville B
Riverside A
Riverton C
Roane C
Roanoke D
Robbs D
Robertsville D
Robinsonville B
Roby C
Rockaway B
Rockbridge B
Rockdale D
Rockmart C
Rockport B
Rockton B
Rockwood B
Rocky Ford B
Rodman A
Roe B
Roebuck D
Rogers D
Rohrersville D
Rokeby D
Rolfe C
Romeo C
Romulus D
Rusachi D
Rosario C
Roscoe C
Roscommon D
Rosebud B
Rosedell D
Roselms D
Rosemount B
Roseville B
Rositas A
Roslyn C
Ross B
Rossmoyne C
Round Butte D
Routon D
Rowe D
Rowland C
Rowley C
Rox B
Roy B
Royalton C
Roza D
Rozetta B
Ruark C

Rubicon A
Rubio C
Rucker B
Rudyard C
Rumford B
Rumney C
Rupert A
Rushtown A
Rushville D
Ruskin C
Russell B
Russellville C
Ruston B
Rutlege D
Ryder C

Sabana C
Sabana Seca D
Saco D
Saffell B
Sage D
Sagemoor C
St. Albans B
St. Charles B
St. Clair D
St. Helens A
St. Joe B
St. Johns D
St. Lucie A
St. Marys C
St. Paul C
Salal B
Salem B
Salemsburg B
Salisbury D
Salix B
Salkum D
Salmon B
Salol D
Saltillo C
Saluvia C
Salvisa D
Samish D
Sammamish B
Sams B
San Anton C
San Antonio D
San German C
Sango C
San Joaquin D
San Jose B
San Juan B
San Saba D
Santa C
Santa Clara D
Santa Isabel D
Santa Lucia C
Santiago B
Sargeant D
Sarpy A
Sassafras B
Sauble A
Saugatuck C
Sauk B
Savage C
Savannah C
Sawmill C
Sawtooth C
Sawyer C
Saybrook B
Scandia B
Scantic C
Scarboro D
Schapville C
Schoharie C
Schooley D
Schumacher C
Schuylkill B
Scio B
Sciotoville C
Scipio D
Scituate C
Scobey B
Scott D
Scott Lake B
Scowlale B
Scranton C
Searing B
Seaton B
Sebeka D
Sebewa D
Sebring D
Sedan C
Segal D
Segno B
Sediu C
Selah C
Selkirk D
Selle A
Selma B
Semiahmoo B
Senecaville C
Sequatchie B
Sequoia C
Serrano D
Sexton D
Seymour C
Shannon B
**Shapeleigh C
Sharkey D
Sharon B

* The hydrological grouping for the rocky phases of these soils should be reduced one group (i.e., C reduces to B).
** When these soils are classified as having class 3 or 4 rockiness, the hydrological grouping should be reduced one group.

TABLE A–1.—*Soil names and hydrologic soil group classifications***. 288–D–2818 (5 of 5).*

- Sharpsburg B
- Shavano B
- Shelburne C
- Shelby C
- Shelbyville C
- Shelmadine C
- Shelocta B
- Shelton B
- Sheppard A
- Sheridan B
- Sherman B
- Shiloh C
- Shoals C
- Shook B
- Shoshone B
- Shouns B
- Shrewsbury D
- Shubuta C
- Shuwah C
- Sicily B
- Sidell B
- Sierra C
- Sifton B
- Signal C
- Siler A
- Silerton B
- Silver Creek D
- Simcoe B
- Simla B
- Sims D
- Sinai C
- Sinclair B
- Singsaas B
- Sioux B
- Sipple B
- Siskiyou B
- Sites C
- Skaggs C
- Skagit B
- Skalkaho B
- Skamania B
- Skames C
- Skerry C
- Skiyou B
- Skokomish A
- Skyberg C
- Skykomish A
- Sleeth C
- Sloan D
- Slocum D
- Smoky Butte C
- Smolan C
- Snow B
- Soda Lake B
- Sodus C
- Sogn C
- Soller D
- Solomon D
- Somers B
- Somerset B
- Sonoita B
- Sontag D
- Souva D
- Sparta A
- Spearfish B
- Spencer B
- Sperry C
- Spilo D
- Spooner C
- Spottswood B
- Spring D
- Spring Creek D
- Springer A
- Springfield D
- Springtown B
- Spur B
- Staatsburg C
- Stambaugh B
- Stamford D
- Stanfield C
- Stanton D
- Starks C
- Starr B
- Staser B
- State B
- Stecum D
- Steekee C
- Steihauer B
- Steinsburg B
- Stendal C
- Stephensburg C
- Stephenville B
- Stetson B
- Stevenson B
- Stewart D
- Stidham B
- Stimson B
- Stissing C
- Stockbridge C
- Stockland B
- Stockton B
- Stoneham B
- Stonington B
- Stono C
- Stookey B
- Storden B
- Story C
- Stough C
- Stoy D
- Strasburg C
- Strauss C
- Strawn B
- Stronghurst B
- Stryker D
- Stukel C
- Stumpp D
- Sudbury B
- Suffield C
- Sula B
- Sulphura C
- Sultan B
- Sumas C
- Summerville C
- Summit C
- Sumner A
- Sumter D
- Sun D
- Sunbury B
- Suncook A
- **Sunderland C
- Sunniland C
- Sunnyside B
- Sunrise B
- Sunsweet C
- Superstition A
- Surry B
- Susquehanna D
- Sutherlin C
- Sutphen D
- Sutton B
- Swaim C
- Swanton C
- Swantown A
- Swartswood C
- Sweden B
- Sweeney C
- Sweetwater D
- Swims B
- Switzer D
- Swygert C
- Sylvan B
- Symerton B

- Tabernash B
- Tabler D
- Tabor D
- Taft C
- Tahoe C
- Talante D
- Talbott D
- Talcot D
- Talihina D
- Talladega C
- Tallula B
- Tally B
- Taloka D
- Tama B
- Tamms C
- Tanama C
- Tanberg D
- Taneum C
- Tanwax A
- Taos B
- Tarrant D
- Tate B
- Tatum B
- Taylor C
- Teague D
- Teas C
- Tebo B
- Tedrow A
- Teja C
- Tell B
- Teller B
- Tellico B
- Tenino A
- Tepee D
- Teresa D
- Terril B
- Terry B
- Trescott B
- Teton B
- Tetonka C
- Thackery B
- Thatuna C
- Thayer B
- Thomasville C
- Thompson A
- **Thorndike C
- Thornton D
- Thornwood A
- Thoroughfare A
- Thorp C
- Thurman A
- Thurmont B
- Thurston B
- Tiburones D
- Tice C
- Tickfaw D
- Tieton B
- Tifton B
- Tijeras B
- Tilden C
- Tillman B
- Tilsit C
- Timmer B
- Timmerman A
- Timpahute D
- Timula B
- Tinton B
- Tiocano D
- Tioga B
- Tippah D
- Tippecanoe B
- Tipperary A
- Tipton B
- Tirzah B
- Tisbury B
- Tisch C
- Tishomingo C
- Titusville C
- Tivoli A
- Toa C
- Tobin B
- Tobosa D
- Todd B
- Toddville B
- Tokul A
- Toledo D
- Tolley B
- Tolo B
- Toltec B
- Tombigbee A
- Tonawanda C
- Tongue River C
- Tonopah B
- Toppenish C
- Topton B
- Torres C
- Tortugas D
- Tours C
- Toutle A
- Tovey B
- Tower D
- Townsbury B
- Townsend B
- Toxaway D
- Toyah B
- Traer C
- Transylvania B
- Trapper B
- Travessilla D
- Travis C
- Treadway D
- Trego C
- Trempealeau B
- Trenary B
- Trent B
- Trexler C
- Trinity D
- Tripp B
- Tromp B
- Trout River A
- Trowbridge B
- Troxel B
- Troy C
- Truman B
- Trumbull D
- Tubac D
- Tucker D
- Tucumcari B
- Tuffit B
- Tughill D
- Tujunga A
- Tuller D
- Tully C
- Tumacacori B
- Tumbez D
- Tumwater A
- Tunica D
- Tunkhannock B
- Tupelo D
- Turbotville C
- Turin C
- Turkey Creek B
- Turnbow D
- Turner B
- Turnerville C
- Tuscan B
- Tuscarora C
- Tuscola B
- Tusculum B
- Tusquitee B
- Tuxedo D
- Twin Creek B
- Twin Lakes B
- Two Dot B
- Tyler C

- Udolpho C
- Uinta B
- Ulen B
- Ulm B
- Ulupalakua B
- Ulysses B
- Umapine B
- Una D
- Unadilla B
- Uncompahgre B
- Ungers B
- Union C
- Unison B
- Unity A
- Upshur C
- Urbana B
- Urbo D
- Usine D
- Ursula D
- Utica B
- Utuado B
- Uvalde B
- Vader B
- Vaiden D
- Vale B
- Valentine A
- Valera C
- Vallecitos C
- Valois C
- Vance C
- Vandalia C
- Vanderville C
- Vanoss B
- Varna C
- Vaucluse C
- Vayas D
- Vebar B
- Vega Alta C
- Vega Baja D
- Vekol C
- Velma B
- Venango D
- Venedy D
- Verdel D
- Verdigris B
- Verdun D
- Vergennes D
- Verhalen D
- Vernon D
- Verona B
- Veyo D
- Via C
- Vicksburg B
- Victor B
- Victoria D
- Vienna B
- Vilas A
- Vinton A
- Viola D
- Vira B
- Virden C
- Virgil B
- Virgin River D
- Virtue C
- Vista B
- Vives C
- Vivi B
- Vlasaty C
- Volga D
- Volin B
- Volke C
- Volney B
- Volperie C
- Volusia C
- Vona A
- Vrooman B

- Wacousta C
- Wade D
- Wadell D
- Wadena B
- Wadesboro B
- Wadsworth D
- Wagner D
- Waha C
- Wahee C
- Wahiawa B
- Wahtum C
- Waialua B
- Waikaloa B
- Waikapu B
- Wailea B
- Waimanalo D
- Waimea B
- Waipahu C
- Waiska B
- Waits B
- Wakeland B
- Wakonda C
- Wallace B
- Walla Walla B
- Waller C
- Wallington C
- Wallkill C
- Wallpack B
- Walpole C
- Walsh B
- Walters D
- Walton C
- Wampsville B
- Wann A
- Wapato C
- Wapping B
- Ward D
- Warden B
- Warman D
- Warne D
- Warners C
- Warrenton A
- Warrior C
- Warsaw B
- Wartrace C
- Warwick B
- Washburn D
- Washington B
- Washoe C
- Washougal C
- Washtenaw C
- Wassaic C
- Wassuk D
- Watauga B
- Watchaug B
- Waterboro D
- Waterloo C
- Waterville C
- Watseka A
- Watson C
- Watsonville D
- Watt D
- Watton C
- Waubay B
- Waugh C
- Waukegan B
- Waukesha B
- Waukon B
- Waumbek B
- Wauseon D
- Waverly D
- Wayland C
- Wayne B
- Waynesboro B
- Wea B
- Weaver C
- Webb C
- Webster C
- Weeksville B
- Wehadkee D
- Weikert C
- Weinbach C
- Weir D
- Weld C
- Weller D
- Wellington D
- Wellman C
- Wellsboro C
- Wellston B
- Wemple B
- Wenas C
- Wenatchee B
- Wesley C
- Wessington B
- Westbury C
- Westfall C
- Westland D
- **Westminster C
- Westmoreland C
- Weston C
- Westphalia B
- West Point D
- Westport A
- Westville C
- Wethersfield C
- Weymouth C
- Whalan B
- Wharton C
- Whatcom D
- Whately D
- Wheeling B
- Whidbey A
- Whippany C
- Whitefish B
- Whiteford C
- White House C
- Whitelaw C
- Whitesburg C
- Whiteson D
- White Store D
- White Swan C
- Whitetail B
- Whitlock A
- Whitman D
- Whitson D
- Whitwell C
- Wibaux C
- Wichita C
- Wickersham C
- Wickham B
- Wickiup A
- *Wilbraham C
- Wilcox D
- Wildwood D
- Wilkes C
- Wilkeson B
- Will D
- Willamette B
- Willard C
- Williams B
- Williamsburg C
- Williamson B
- Willoughby C
- Willow Creek B
- Wilson D
- Winchester A
- Windom B
- Wind River A
- Windsor A
- Windthorst C
- Winema B
- Winfield C
- Wingville B
- Winifred D
- Winlock D
- Winnett D
- Winooski B
- Winslow B
- Winston A
- Winterset C
- Witt B
- Wolcottsburg C
- Woldale D
- Wolf B
- Wolftever C
- Woodbridge C
- Woodglen D
- Woodinville B
- Woodlyn C
- Wood River D
- Woodson D
- Woodstown C
- Woodward B
- Wooster B
- Woostern B
- Worland B
- Worsham D
- Worth C
- Worthen B
- Worthington B
- Wortman C
- Wrightsville D
- Wurtsboro C
- Wykoff B
- Wynoose D

- Xenia B

- Yabucoa D
- Yadkin B
- Yahola B
- Yakima A
- Yale C
- Yauco D
- Yeoman B
- Yoder B
- Yonaba C
- Yordy B
- York C
- Yunes D

- Zaca D
- Zahl B
- Zaleski B
- Zaneis C
- Zapata D
- Zell B
- Zimmerman A
- Zion C
- Zipp C
- Zita B
- Zook D
- Zuber B
- Zwingle D

* The hydrological grouping for the rocky phases of these soils should be reduced one group (i.e., C reduces to B).
** When these soils are classified as having class 3 or 4 rockiness, the hydrological grouping should be reduced one group.

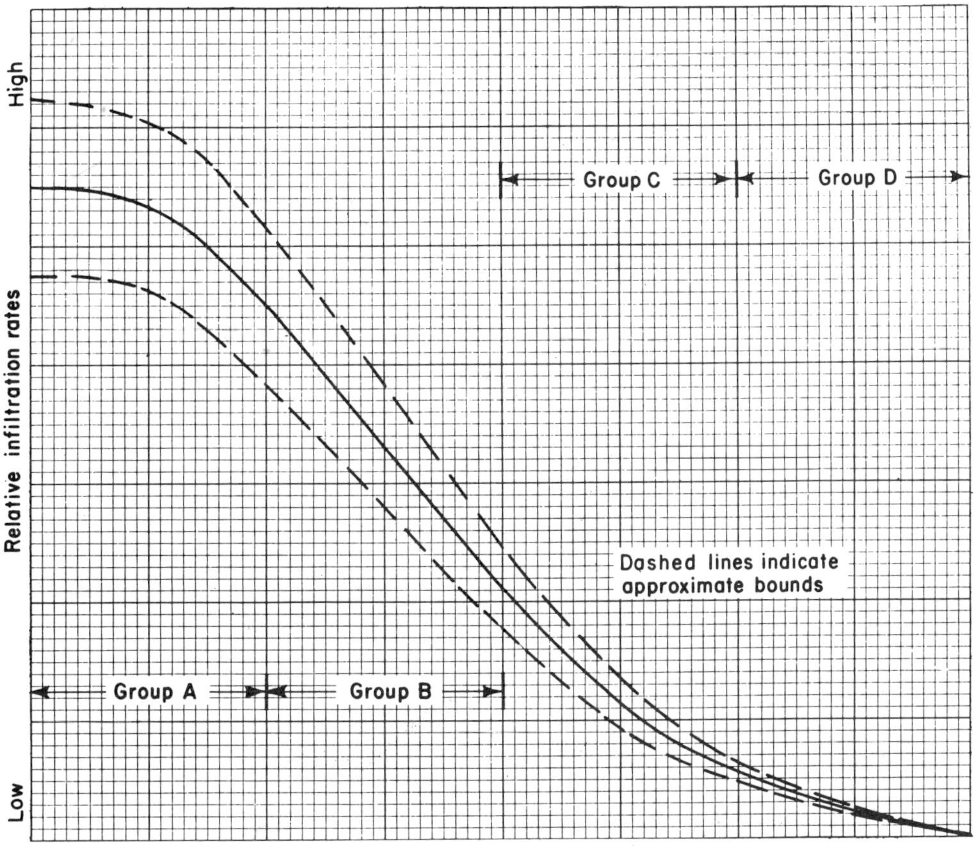

Figure A–1. Relative infiltration rates of hydrologic soil groups. (U.S. Soil Conservation Service.) 288–D–2906.

may be included with one of the other land use cover types (such as fallow or small grain) in the computation of runoff (see sec. A–4).

Provision is made in table A–2 (sec. A–4) for farmsteads and roads. These land uses are generalized, since they vary so much. Where it is necessary to work with more detail (as sometimes in a very small watershed, or with superhighway, airport, or urban areas) the impervious areas are considered an individual class with 100 percent runoff, and the remaining land uses are handled as usual (see sec. A–4).

(g) *Straight-Row Farming.*—This class includes up-and-down and cross-slope farming in straight rows. In areas of 1 or 2 percent slope, cross-slope farming in straight rows is almost the same as contour farming. Where the proportion of cross-slope farming is believed to be significant, it may be classed halfway between straight-row and contour farming in the table A–2.

(h) *Contouring.*—Contour furrows used with small grains and legumes are made while planting, are generally small, and tend to disappear due to climatic action. Contour furrows, and beds on the contour, as used with row crops are generally large. They may be made in planting and later reduced in size by cultivation, or they may be insignificant after planting and become large from cultivation. Average conditions are used in table A–2.

Surface runoff reductions due to contour farming are greater as land slopes decrease. The curve numbers for contouring shown in table A–2 were obtained using data from experimental watersheds having slopes of 3 to 8 percent.

Contour furrows in pasture or range land are usually of the permanent type. Their dimensions and spacing generally vary with climate and topography. Table A–2 considers average conditions in the Great Plains.

(i) *Terracing.*—Terraces may be graded, open-end level, or closed-end level. The effects of graded and open-end level terraces are considered in table A–2, and the effects of both contouring and the grass waterway outlets are included.

Closed-end level terraces should be handled like contour furrows.

A-4. *Hydrologic Soil-Cover Complexes.*—(a) *Purpose.*—Table A–2 combines soil groups and land use and treatment classes into *hydrologic soil-cover complexes.* The numbers show the relative value of the complexes as direct runoff-producers (see sec. A–5). The higher the number, the greater the amount of direct runoff to be expected from a storm.

(b) *Table A–2.*—Table A–2 was prepared using data from gaged watershed with known soils and cover. Storm rainfall appropriate for antecedent moisture condition[4] AMC-II was plotted versus direct runoff for annual floods for respective watersheds of one soil group and one cover type. The curve from figure A–4 best fitting the plotted points was determined and the curve numbers for table A–2 obtained. Related curve numbers for above average (AMC-III) and below average (AMC-I) points were similarly developed and are shown in table A–7 next to the CN values for AMC-II.

For several of the soil-cover complexes shown in table A–2, curve numbers (CN) were estimated or computed from relations developed by the Soil Conservation Service since hydrologic data were not available for all given soil-cover complexes.

(c) *Forest Service Procedure.*—Chapter 4 of "Forest and Range Hydrology Handbook," U.S. Forest Service, Washington, D.C., 1959, describes how CN are determined for national and commercial forests in the eastern United States. Section 1 of "Handbook on Methods of Hydrologic Analysis," U.S. Forest Service, Washington, D.C., 1959, describes how CN are determined for forest-range regions in the western United States. Selections from these handbooks, which are included in the Soil Conservation Service National Engineering Handbook, issued in 1964, are given here.

[4] Antecedent moisture conditions are defined in section A-5(a).

TABLE A–2.—*Runoff curve numbers (CN) for hydrologic soil-cover complexes*
(FOR WATERSHED CONDITION AMC-II AND $I_a=0.2S$)[3]

Land use or cover	Treatment or practice	Hydrologic condition for infiltrating	Hydrologic soil group			
			A	B	C	D
Fallow	SR		77	86	91	94
Row crops	SR	Poor	72	81	88	91
	SR	Good	67	78	85	89
	C	Poor	70	79	84	88
	C	Good	65	75	82	86
	C&T	Poor	66	74	80	82
	C&T	Good	62	71	78	81
Small grain	SR	Poor	65	76	84	88
	SR	Good	63	75	83	87
	C	Poor	63	74	82	85
	C	Good	61	73	81	84
	C&T	Poor	61	72	79	82
	C&T	Good	59	70	78	81
Close-seeded legumes[1] or rotation meadow.	SR	Poor	66	77	85	89
	SR	Good	58	72	81	85
	C	Poor	64	75	83	85
	C	Good	55	69	78	83
	C&T	Poor	63	73	80	83
	C&T	Good	51	67	76	80
Pasture or range		Poor	68	79	86	89
		Fair	49	69	79	84
		Good	39	61	74	80
	C	Poor	47	67	81	88
	C	Fair	25	59	75	83
	C	Good	6	35	70	79
Meadow (permanent).		do	30	58	71	78
Woods (farm woodlots).		Poor	45	66	77	83
		Fair	36	60	73	79
		Good	25	55	70	77
Farmsteads			59	74	82	86
Roads (dirt)[2]			72	82	87	89
(hard surface).[2]			74	84	90	92

[1] Close-drilled or broadcast. (U.S. Soil Conservation Service.)
[2] Including right-of-way.
[3] See sec. A–5.
SR=Straight row.
C=Contoured.
T=Terraced.
C&T=Contoured and terraced.

(1) *Forest in Eastern United States.*—In the humid forest regions of the eastern United States, soil group, humus type, and humus depth are the principal factors used in the Forest Service method of determining CN. The undecomposed leaves or needles, twigs, bark, and other vegetative debris on the forest floor form the litter from which humus is derived. Natural litter protects humus from oxidation and therefore indirectly enters into

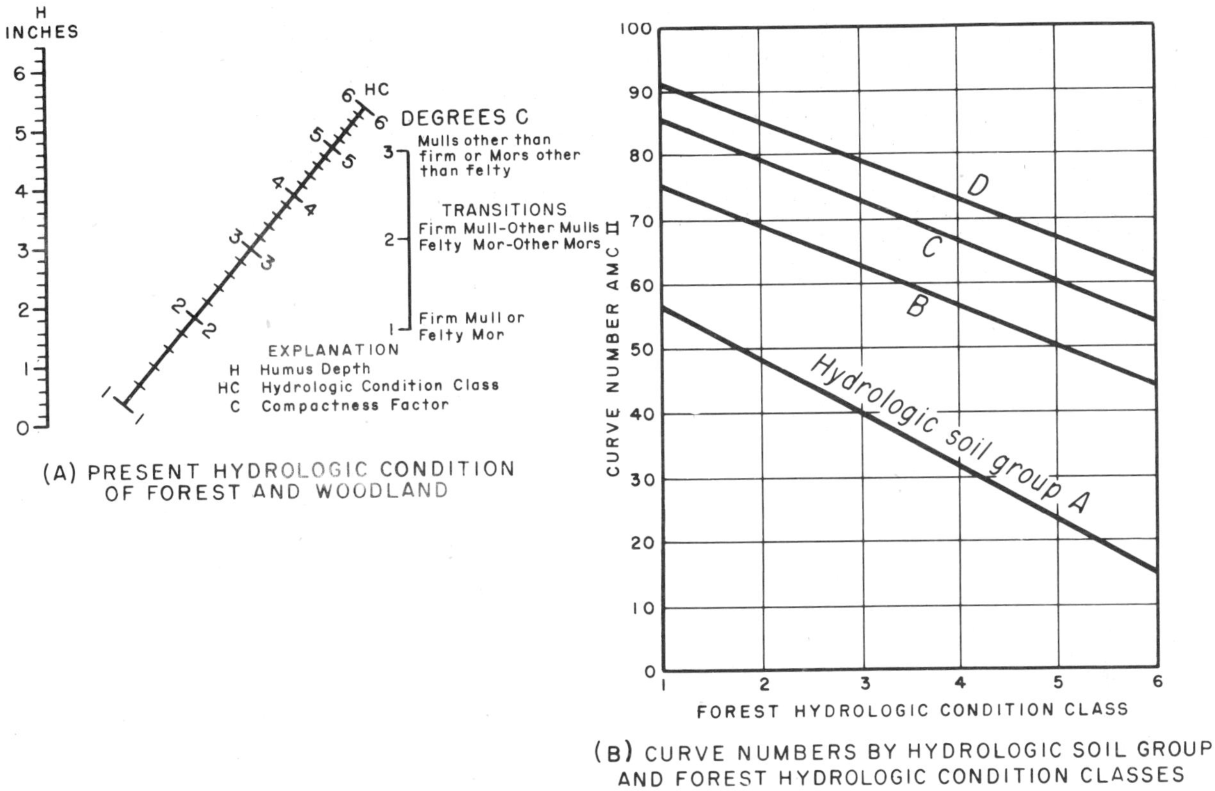

Figure A–2. Chart for determining CN for humid forest regions. 288–D–2823.

the determination; if the depth of litter is less than ½ inch, the humus is considered unprotected and the hydrologic condition class (fig. A–2(a)) is reduced by 0.5.

Humus is the organic layer immediately below the litter layer from which it is derived. It may consist of *mull,* which is an intimate mixture of organic matter and mineral soil, or of *mor,* which is practically pure organic matter unrecognizable as to origin from material lying on the forest floor. Humus depth increases with age of forest stand until an equilibrium is reached between the processes that build up humus and those that break it down. As much as 12 inches of humus may be produced under favorable conditions, but a depth of 5 or 6 inches is considered the maximum attainable under average conditions. Under good management practices (proper use, protection, and improvement), humus is porous and has high infiltration and storage capacities. Under poor management practices (burning, overcutting, or overgrazing), humus is compact enough to impede the absorption of water.

Humus is evaluated by means of degrees of compaction, which are:

(1) *Compact.* Mulls are firm; mors are felty.

(2) *Moderately compact.* A transition stage.

(3) *Loose or friable.* Mulls are not firm; mors are not felty.

Frost in compact humus is of the concrete form, which inhibits infiltration, and in loose humus it is the granular or stalactite form, which does not. Because of the correlation between humus type and frost, a separate determination of the effects of frost is unnecessary.

The *hydrologic condition* of a forest area is the runoff-producing potential. The condition class is indicated by a number ranging from 1 to 6, the lower the number the higher the runoff-producing potential. The relation between classes and humus type and depth is shown in figure A–2(a). If field data as to forest watershed conditions are meager, table

A–3 may be used to estimate a runoff curve number.

The *CN* for the present hydrologic condition of a forest area is determined as follows: sample plots are located in the area; soil group, litter depth, humus type, and humus depth are determined by means of shallow soil wells dug in the plots; the nomograph, figure A–2(a), gives the hydrologic condition class of the plot; and the network chart, figure A–2(b) gives the *CN*.

TABLE A–3.—*Runoff curve numbers (CN) for hydrologic soil-cover complexes*

I. COMMERCIAL OR NATIONAL FOREST. FOR WATERSHED CONDITION AMC-II AND I_a=0.2*S*

Hydrologic condition class	Hydrologic soil group			
	A	B	C	D
I. Poorest	56	75	86	91
II. Poor	46	68	78	84
III. Medium	36	60	70	76
IV. Good	26	52	62	69
V. Best	15	44	54	61

(2) *Forest-Range in Western United States.*—In the forest-range regions of the western United States, soil group, cover type, and cover density are the principal factors used in estimating *CN*. Figure A–3 shows the relationship between these factors and *CN* for soil-cover complexes used to date. The figures are based on information in table 2.1, part 2, of the Forest Service "Handbook on Methods of Hydrologic Analysis." The covers are defined as follows:

Herbaceous.—Grass-weed-brush mixtures, with brush the minor element.

Oak-Aspen.—Mountain brush mixtures of oak, aspen, mountain mahogany, bitter brush, maple, and other brush.

Juniper-Grass.—Juniper or pinon with an understory of grass.

Sage-Grass.—Sage with an understory of grass.

The amount of litter is taken into account when estimating the density of cover.

If data pertaining to ground cover density are unavailable, a runoff curve number may be obtained from table A–4.

TABLE A–4.—*Runoff curve numbers (CN) for forest-range areas in western United States (AMC–II)*

Cover	Condition	Soil groups			
		A	B	C	D
Herbaceous	Poor	....	78	85	92
	Fair	....	68	81	88
	Good	....	59	71	84
Sagebrush	Poor	....	64	78	....
	Fair	....	46	67	....
	Good	....	35	46	....
Oak-Aspen	Poor	....	63	71	....
	Fair	....	40	54	....
	Good	....	30	40	....
Juniper	Poor	....	73	84	....
	Fair	....	54	70	....
	Good	....	40	59	....

(d) *Supplementary Information.*—Table A–5 gives *CN* for complexes in a typical watershed in Contra Costa County, California. The *CN* were obtained by the Contra Costa County Flood Control District and the SCS, using streamflow data from the watershed and a trial-and-error process. The range in *CN* for a particular cover and soil group indicates the variation for soil subgroups.

TABLE A–5.—*Runoff curve numbers (CN) for hydrologic soil-cover complexes of a typical watershed in Contra Costa County, California (AMC–II and I_a=0.2S)*

Cover	Condition	Hydrologic soil group			
		A	B	C	D
Scrub (native brush)	....	25–30	41–46	57–63	66
Grass-oak (native oaks with understory of forbs [1] and annual grasses)	Good	29–33	43–48	59–65	67
Irrigated pasture	Good	32–37	46–51	62–68	70
Orchard (winter period with understory of cover crop)	Good	37–41	50–55	64–69	71
Range (annual grass)	Fair	46–49	57–60	68–72	74
Small grain (contoured)	Good	61–64	69–71	76–80	81
Truck crops (straight-row)	Good	67–69	74–76	80–83	84
Urban areas:					
Low density (15 to 18 percent impervious surfaces)	....	69–71	75–78	82–84	86
Medium density (21 to 27 percent surfaces)	....	71–73	77–80	84–86	88
High density (50 to 75 percent impervious surfaces)	....	73–75	79–82	86–88	90

[1] Forbs are defined as any herb other than grass.

(e) *Determination of Curve Numbers (CN) for Mixed Areas.*—Table A–6 shows the process by which a weighted soil-cover complex number

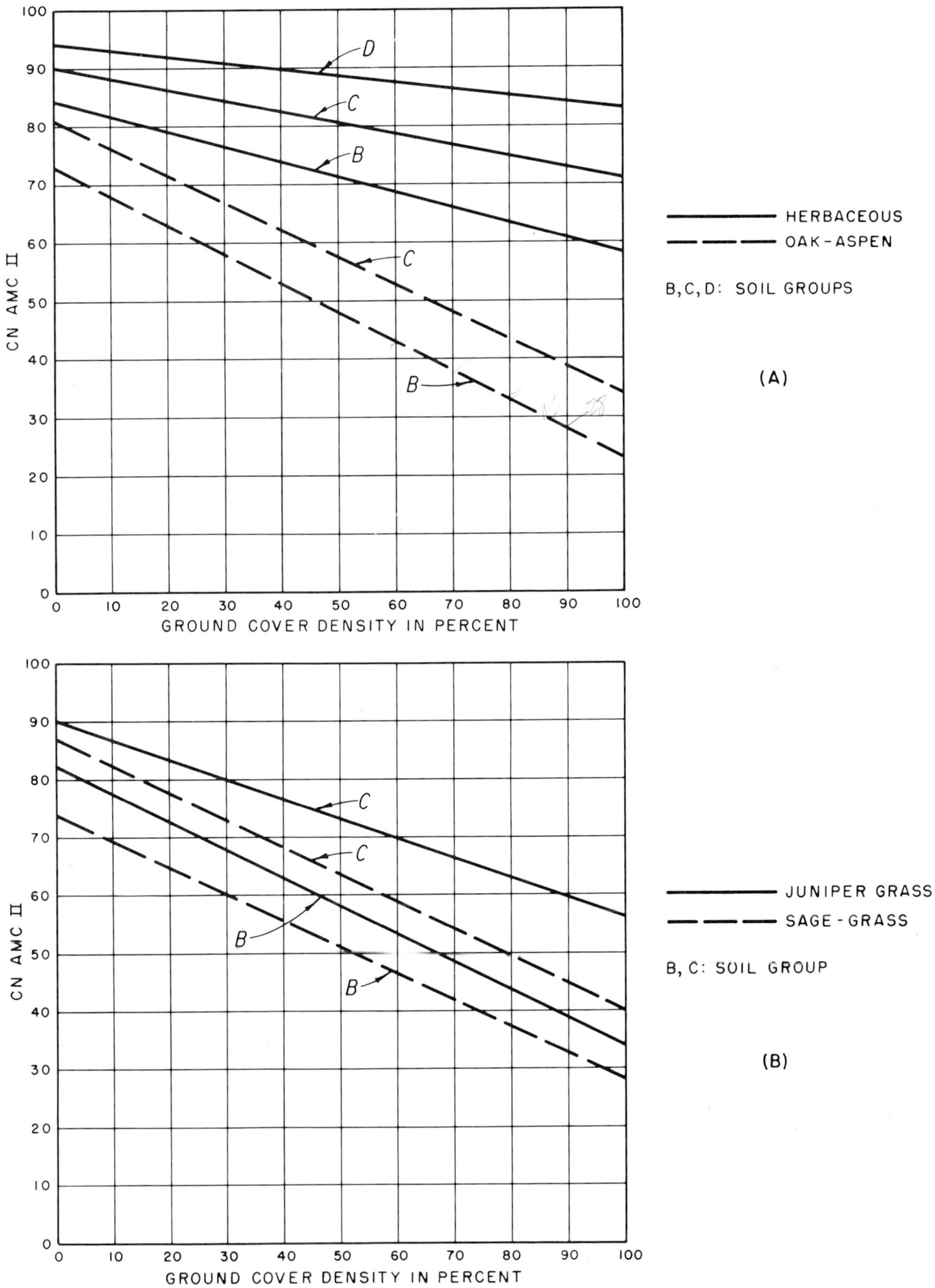

Figure A—3. **Determining CN for forest-range in the western United States. 288—D—2824.**

is obtained for areas having several soil-cover complexes. The example area contains a Group B soil. The weighted number can be more easily obtained by accumulative multiplication on a calculator.

TABLE A–6.—*Sample computation, weighting of hydrologic soil-cover complex numbers*

Complex	Curve number	Percent of area	Number times percent
Row crop, straight row, good rotation	78	56.2	4,384
Legumes, contoured, good rotation	69	37.5	2,588
Meadow, permanent	58	6.3	365
Total		100	7,337

Weighted number $=\frac{7,337}{100}=73.37$ (U.S. Soil Conservation Service.)
Round-off to 73.

A-5. Estimation of Direct Runoff From Rainfall.—(a) *General Storm Rainfall.*—The method uses three variables in estimating runoff: rainfall, antecedent moisture condition, and the hydrologic soil-cover complex.

(1) *Runoff equation.*—The curves of figure A–4 are obtained using the equation:

$$Q=\frac{(P-0.2S)^2}{P+0.8S} \tag{1}$$

where:

Q=direct runoff, in inches

P=storm rainfall, in inches, and

S=maximum potential difference between P and Q, in inches, at time of storm's beginning.

Equation (1) is derived by starting with the proportion:

$$\frac{P-Q}{S}=\frac{Q}{P} \tag{2}$$

where $\frac{P-Q}{S}$ is visualized as the ratio of actual to potential difference between P and Q, and $\frac{Q}{P}$ is visualized as the ratio of actual to potential runoff.

Solving for Q gives:

$$Q=\frac{P^2}{P+S} \tag{3}$$

Equation (3) is useful under conditions where there is a possibility of runoff whenever there is rainfall. For the condition that $Q=0$ at a value of P greater than zero, use of an initial abstraction, I_a, is required (see the diagram on fig. A–4). With the condition that I_a cannot be greater than P, equation (2) then becomes:

$$\frac{(P-I_a)-Q}{S}=\frac{Q}{(P-I_a)} \tag{4}$$

And solving for Q gives:

$$Q=\frac{(P-I_a)^2}{(P-I_a+S)} \tag{5}$$

Since S includes I_a, an empirical relation between the variables can be developed to simplify equation (5). Data from watershed in various parts of the country give:

$$I_a=0.2S \tag{6}$$

Substituting ($0.2S$) for I_a in equation (5) gives equation (1).

Equation (5) can be further expanded to recognize other given conditions or factors. However, such expansions in themselves do not mean greater accuracy in runoff estimating, since this would require that the terms expressing additional factors be of a high order of accuracy.

(2) *Significance of S.*—Plottings of direct runoff, Q, versus storm rainfall, P, on natural watersheds show that Q approaches P as P continues to increase in the storm. The same data show that (P–Q) approaches a constant as P continues to increase. The quantities can be shown together as in equation (2), above, and it is apparent that the constant S is the maximum difference (P–Q) that could occur for the given storm and watershed conditions. The proportion can be made more complex (as discussed above), but not all the additional factors can be related to S.

The variable S is therefore a maximum potential (P–Q). During a storm, the actual (P–Q) that occurs is limited by either soil-water storage or an infiltration rate as P increases. The maximum potential (P–Q) or S, therefore, is dependent on soil-water storage and the infiltration rates of a watershed.

(3) *Significance of I_a.*—The insert on figure A–4 shows that I_a is equal to the rainfall that occurs before runoff starts. Physically, I_a consists principally of interception, infiltration, and surface storage. Equation (6), which relates I_a to S, is based on data from large and small watersheds in various parts of the country. Further refinement of equation (6) is not recommended, since the data needed to break I_a into components of interception, infiltration, and surface storage are seldom available on a watershed basis. For the same reason, adjustment of the coefficient 0.2 in equation (6) is not recommended.

(4) *System of curve numbering.*—For convenience in interpolation, the curves of figure A–4 are numbered from 100 to zero. The numbers are related to S as follows:

$$\text{Curve number} = \frac{1{,}000}{10+S} \qquad (7)$$

A curve for the case $I_a=0$, equation (3), is displaced to the right for the case $I_a=0.2S$, equation (1), by the amount of $0.2S$. Therefore, the curve numbers given in table A–2 should be used only with figure A–4 or with equation (1).

(5) *Antecedent moisture conditions.*—The amount of rainfall in a period of 5 to 30 days preceding a particular storm is referred to as antecedent rainfall, and the resulting condition of the watershed in regard to potential runoff is referred to as an antecedent condition. In general, the heavier the antecedent rainfall, the greater the direct runoff that occurs from a given storm. The effects of infiltration and evapo-transpiration during the antecedent period are also important, as they may increase or lessen the effect of antecedent rainfall.

Because of the difficulties of determining antecedent storm conditions from data normally

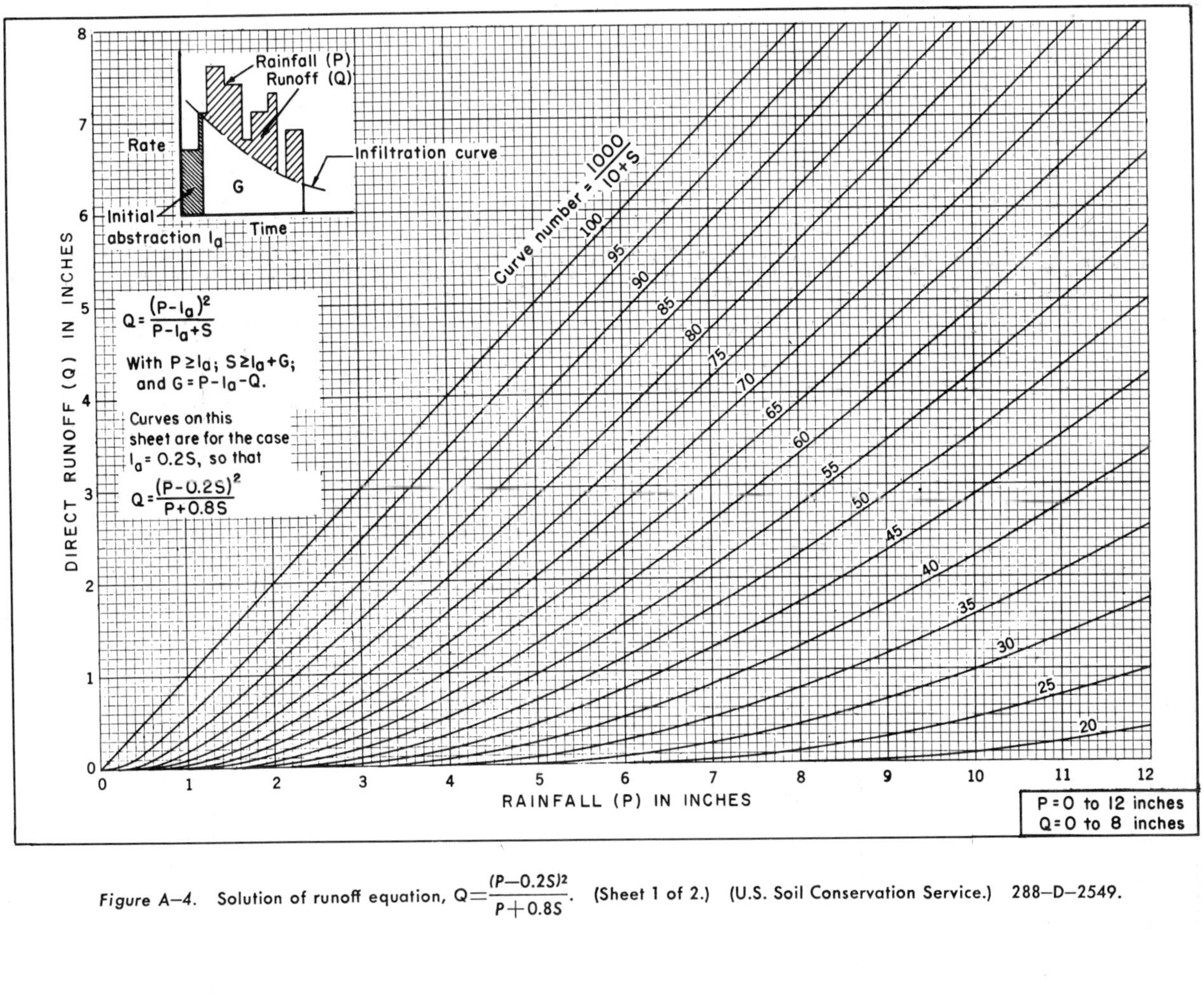

Figure A–4. Solution of runoff equation, $Q=\frac{(P-0.2S)^2}{P+0.8S}$. (Sheet 1 of 2.) (U.S. Soil Conservation Service.) 288–D–2549.

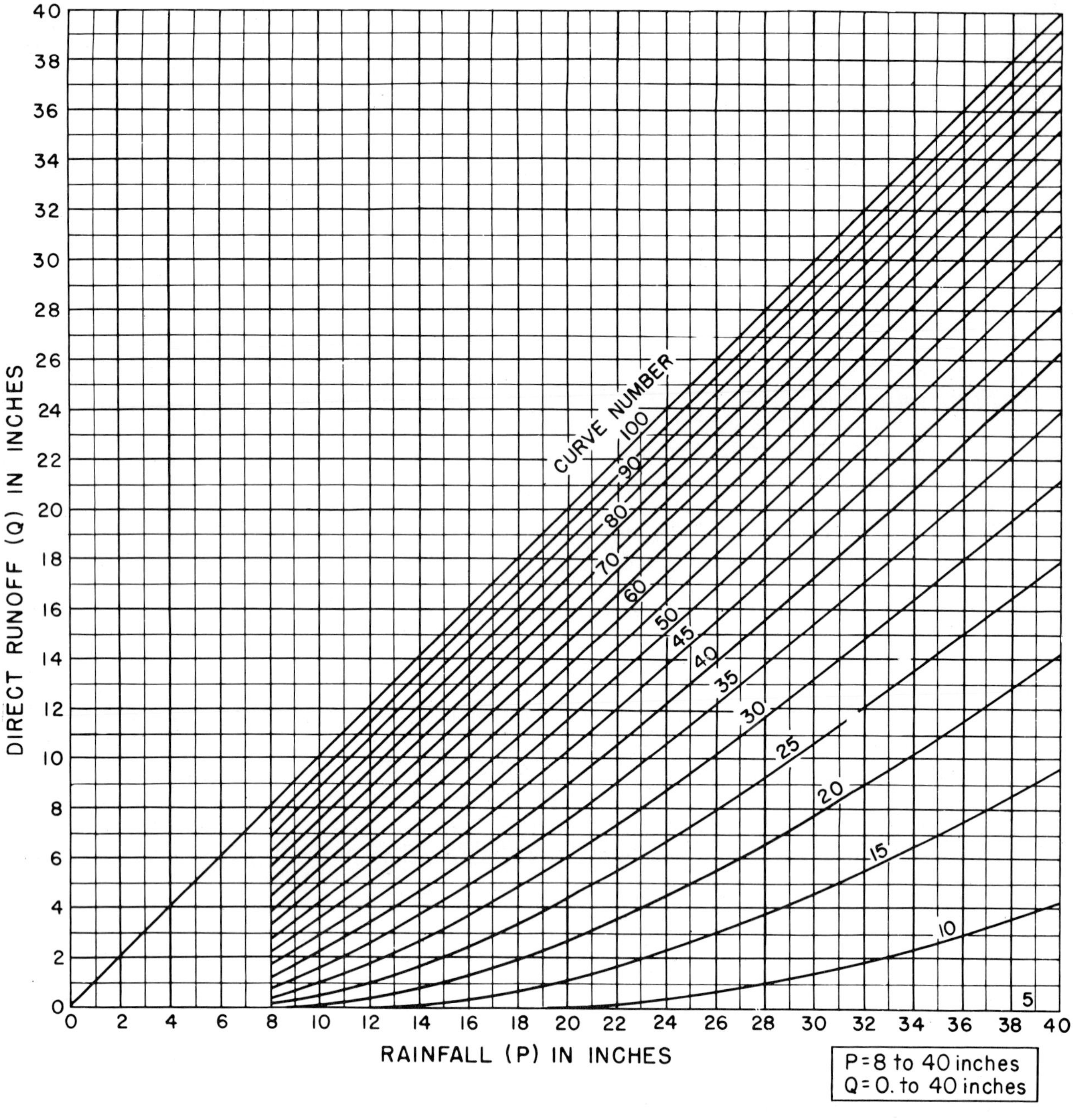

Figure A–4. Solution of runoff equation, $Q=\frac{(P-0.2S)^2}{P+0.8S}$. (Sheet 2 of 2.) (U.S. Soil Conservation Service.) 288–D–2549.

available, the conditions are reduced to the following three cases:

AMC–I.—A condition of watershed soils where the soils are dry but not to the wilting point, and when satisfactory plowing or cultivation takes place. (This condition is *not* considered applicable to the design flood computation methods presented in this text.)

AMC-II.—The average case for *annual floods*, that is, an average of the conditions which have preceded the occurrence of the maximum annual flood on numerous watersheds.

AMC-III.—When heavy rainfall or light rainfall and low temperatures have occurred during the 5 days previous to the given storm, and the soil is nearly saturated.

The numbers on tables A–2, A–3, A–4 and A–5 and figures A–2 and A–3 are for the average watershed condition, AMC-II. Curve numbers for one antecedent moisture condition may be converted to a different antecedent moisture condition by the use of table A–7. For example, the computations in table A–6 give an AMC-II number of 73. The corresponding curve number for AMC-I and AMC-III can be obtained from columns 2 and 3 of table A–7. The curve numbers for AMC-I and AMC-III are 54 and 87, respectively.

TABLE A–7.—*Curve numbers (CN) and constants*
[*For the case $I_a=0.2S$*]

(1)	(2)	(3)	(4)	(5)	(1)	(2)	(3)	(4)	(5)
CN for condition II	*CN* for conditions I	III	*S* values[1] inches	Curve[1] starts where *P* = , inches	*CN* for condition II	*CN* for conditions I	III	*S* values[1] inches	Curve[1] starts where *P* = , inches
100	100	100	0.	0.	60	40	78	6.67	1.33
99	97	100	.101	.02	59	39	77	6.95	1.39
98	94	99	.204	.04	58	38	76	7.24	1.45
97	91	99	.309	.06	57	37	75	7.54	1.51
96	89	99	.417	.08	56	36	75	7.86	1.57
95	87	98	.526	.11	55	35	74	8.18	1.64
94	85	98	.638	.13	54	34	73	8.52	1.70
93	83	98	.753	.15	53	33	72	8.87	1.77
92	81	97	.870	.17	52	32	71	9.23	1.85
91	80	97	.989	.20	51	31	70	9.61	1.92
90	78	96	1.11	.22	50	31	70	10.0	2.00
89	76	96	1.24	.25	49	30	69	10.4	2.08
88	75	95	1.36	.27	48	29	68	10.8	2.16
87	73	95	1.49	.30	47	28	67	11.3	2.26
86	72	94	1.63	.33	46	27	66	11.7	2.34
85	70	94	1.76	.35	45	26	65	12.2	2.44
84	68	93	1.90	.38	44	25	64	12.7	2.54
83	67	93	2.05	.41	43	25	63	13.2	2.64
82	66	92	2.20	.44	42	24	62	13.8	2.76
81	64	92	2.34	.47	41	23	61	14.4	2.88
80	63	91	2.50	.50	40	22	60	15.0	3.00
79	62	91	2.66	.53	39	21	59	15.6	3.12
78	60	90	2.82	.56	38	21	58	16.3	3.26
77	59	89	2.99	.60	37	20	57	17.0	3.40
76	58	89	3.16	.63	36	19	56	17.8	3.56
75	57	88	3.33	.67	35	18	55	18.6	3.72
74	55	88	3.51	.70	34	18	54	19.4	3.88
73	54	87	3.70	.74	33	17	53	20.3	4.06
72	53	86	3.89	.78	32	16	52	21.2	4.24
71	52	86	4.08	.82	31	16	51	22.2	4.44
70	51	85	4.28	.86	30	15	50	23.3	4.66
69	50	84	4.49	.90					
68	48	84	4.70	.94	25	12	43	30.0	6.00
67	47	83	4.92	.98	20	9	37	40.0	8.00
66	46	82	5.15	1.03	15	6	30	56.7	11.34
65	45	82	5.38	1.08	10	4	22	90.0	18.00
64	44	81	5.62	1.12	5	2	13	190.0	38.00
63	43	80	5.87	1.17	0	0	0	infinity	infinity
62	42	79	6.13	1.23					
61	41	78	6.39	1.28					

[1] For CN in column 1.

(b) *Thunderstorm Rainfall.*—Estimates of direct runoff from thunderstorm rainfall are more difficult than for general storm rainfall because data for thunderstorm-caused floods are meager. Type and density of vegetative cover on a watershed, as well as the time-sequence of rainfall, appear to have more influence on thunderstorm runoff than do antecedent moisture condition and soil group.

A definite time-sequence of 15-minute incremental rainfall has been assigned for each designated geographical zone for thunderstorm precipitation values given in this text. A simplified "Thunderstorm Cover Index" relating to vegetative cover type and ground cover density for watersheds west of the 105° meridian has been developed. After design thunderstorm precipitation values have been arranged in respective time-sequence magnitude (see table A–8), direct runoff increments are computed by the procedure in the examples, section 58(c), using appropriate runoff curve number, table A–9, and minimum 15-minute retention loss rate for respective hydrologic soil group, table A–10.

TABLE A–8.—*Design thunderstorm precipitation time-sequence in 15-minute intervals*

Time, hour	Order of magnitude: Zones I and III	Zone II
0.0	0	0
0.25	4	3
.50	3	2
.75	2	1
1.00	1	[1]4
1.25	5	
1.50	6	
1.75	7	
2.00	8	
2.25	9	
2.50	10	
2.75	11	
3.00	[1]12	

[1] Duration limit of zone rainfall.

TABLE A-9.—*Design thunderstorm precipitation direct runoff curve number*

Thunderstorm cover index		CN for zone and type flood								
Cover type	Cover density at ground surface, percent	Zone I			Zone II			Zone III		
		Type of flood			Type of flood			Type of flood		
		IDF [1]	A [2]	B [3]	IDF [1]	A [2]	B [3]	IDF [1]	A [2]	B [3]
Tree, including understory [4]	More than 50	70	78	85	70	78	85	70	78	85
	50 or less	75	89	91	75	82	88	75	83	91
Brush, sage, grass or combinations	More than 50	75	82	88	80	86	91	70	79	88
	50 or less	80	86	91	85	90	94	75	85	94

[1] IDF—inflow design flood, probable maximum thunderstorm rainfall.
[2] A thunderstorm—Flood for assumption A.
[3] B thunderstorm—Flood for assumption B.
[4] Understory can be defined as a layer of foliage, in a forest, below the level of the main canopy; or the trees forming such a layer.

TABLE A-10.—*Minimum 15-minute retention loss rates*

Hydrologic	Minimum loss rate, inch per 15 minutes
A	0.10
B	0.06
C	0.03
D	0.02

Hydraulic Computations

R. I. STRAND,[1] J. M. LARA,[1] and C. J. HOFFMAN [2]

A. HYDRAULIC FORMULAS

B-1. *Lists of Symbols and Conversion Factors.*— The following list includes symbols used in hydraulic formulas given in chapters IX and X and in this appendix. Standard mathematical notations and symbols having only very limited applications have been omitted.

Symbol	*Description*
A, a	An area; area of a surface; cross-sectional area of flow in an open channel; cross-sectional area of a closed conduit
a_g	Gross area of a trashrack
a_n	Net area of a trashrack
B	Width of a siphon throat
b	Bottom width of a channel
C	A coefficient; coefficient of discharge
C_d	Coefficient of discharge through an orifice
C_i	Coefficient of discharge for an ogee crest with inclined upstream face
C_o	Coefficient of discharge for a nappe-shaped ogee crest designed for an H_o head
C_s	Coefficient of discharge for a partly submerged crest
D	Diameter; conduit diameter; height of a rectangular conduit or passageway; height of a square orifice
$\overline{D}$	"Drop number" parameter for defining the dimensions of a straight drop spillway, $\overline{D}=\frac{q^2}{gY^3}$
d	Depth of flow in an open channel; height of an orifice or gate opening
d_c	Critical depth
d_f	Depth of the pool under a free overfall nappe
d_H	Depth for high (subcritical) flow stage (alternate to d_L)
d_i	Height of a hydraulic jump (difference in the conjugate depths)
d_L	Depth for low (supercritical) flow stage (alternate to d_H)
d_m	Mean depth of flow
d_{m_c}	Critical mean depth
d_n	Depth of flow measured normal to channel bottom
d_s	Depth of scour below tailwater in a plunge pool
d_t	Depth of flow in a chute at tailwater level
E	Energy
E_m	Energy of a particle of mass
F	Froude number parameter for defining flow conditions in a channel, $F=\frac{v}{\sqrt{gd}}$
F_t	Froude number parameter for flow in a chute at the tailwater level
f	Friction loss coefficient in the Darcy-Weisbach formula, $h_f=\frac{fL}{D}\frac{v^2}{2g}$
G	Drop in water surface level in a reach of a natural channel
g	Acceleration due to the force of gravity
H	Head over a crest; head on center of an orifice opening; head on the bottom of a culvert entrance; head difference at a gate (between the upstream and downstream water surface levels)
H_A	Absolute head above a datum plane, in channel flow
H_{AT}	Probable minimum atmospheric pressure at the site under consideration
H_a	Head above a section in the transition of a drop inlet spillway
H_1	Head measured to bottom of an orifice opening

[1] Hydraulic Engineer, Sedimentation Section, Bureau of Reclamation.

[2] Engineer, Spillways and Outlet Works Section, Bureau of Reclamation (Retired).

Symbol	Description
H_2	Head measured to top of an orifice opening
h	Head; height of baffle block; height of end sill
h_a	Approach velocity head
h_b	Head loss due to bend
h_c	Head loss due to contraction
H_D	Head from reservoir water surface to water surface at a given point in the downstream channel
h_d	Difference in water surface level, measured from reservoir water surface to the downstream channel water surface
H_E	Specific energy head
H_{E_C}	Specific energy head at critical flow
H_e	Total head on a crest, including velocity of approach
h_e	Head loss due to entrance
h_{ex}	Head loss due to expansion
h_f	Head loss due to friction
Δh_f	Incremental head loss due to friction
h_g	Head loss due to gates or valves
h_L	Head losses from all causes
Σh_{L_u}	Sum of head losses upstream from a section
Δh_L	Incremental head loss from all causes
$\Sigma(\Delta h_L)$	Sum of incremental head losses from all causes
H_o	Design head over ogee crest
h_o	Head measured from the crest of an ogee to the reservoir surface immediately upstream, not including the velocity of approach (crest shaped for design head H_o)
h_r	Reduction of pressure head due to inlet contraction
H_s	Total head over a sharp-crested weir
h_s	Head over a sharp-crested weir, not including velocity of approach
h_S	Priming head on a siphon spillway
h_{SA}	Subatmospheric pressure head
H_T	Total head from reservoir water surface to tailwater, or to center of outlet of a free-discharging pipe
h_t	Head loss due to trashrack
h_v	Velocity head; head loss due to exit
h_{v_c}	Critical velocity head
h_{v_S}	Velocity head at throat of siphon spillway
h_{v_t}	Velocity head at tailwater level
K	A constant factor for various equations; a coefficient
k	A constant
K_a	Abutment contraction coefficient
K_b	Bend loss coefficient
K_c	Contraction loss coefficient
K_d	Conveyance capacity factor in the Manning formula, $K_d = \frac{1.486}{n} ar^{2/3}$
K_e	Entrance loss coefficient
K_{ex}	Expansion loss coefficient
K_g	Gate or valve loss coefficient
K_L	A summary loss coefficient for losses due to all causes
K_p	Pier contraction coefficient
K_r	Coefficient of pressure reduction due to inlet contraction
K_t	Trashrack loss coefficient
K_v	Velocity head loss coefficient
L	Length; length of a channel or a pipe; effective length of a crest; length of a hydraulic jump; length of a stilling basin; length of a transition
ΔL	Incremental length; incremental channel length
L_I, L_{II}, L_{III}	Stilling basin lengths for different hydraulic jump stilling basins
L'	Net length of a crest
L_B	Length of a basin for a straight drop spillway; length of a slotted grating dissipator basin
L_d	Distance from the upstream face of an overflow weir to the start of a hydraulic jump in a straight drop spillway stilling basin
L_G	Length of a slotted grating dissipator
L_m	Length of a meandering reach in a natural channel
L_p	Distance from the upstream face of an overflow weir to the point of impingement on the basin floor of a straight drop spillway
L_s	Length of a straight reach in a natural channel
M	Momentum
M_d	Momentum in a downstream section
M_u	Momentum in an upstream section
ΔM	Difference in momentum between successive sections
m	Mass
N	Number of piers on an overflow crest; number of slots in a slotted grating dissipator
n	Exponential constant used in equation for defining crest shapes; coefficient of roughness in the Manning equation
P	Approach height of an ogee weir; hydrostatic pressure of a water prism cross section
p	Unit pressure intensity; unit dynamic pressure on a spillway floor; wetted perimeter of a channel or conduit cross section
Q	Discharge; volume rate of flow
ΔQ	Incremental change in rate of discharge
q	Unit discharge
Q_c	Critical discharge
q_c	Critical discharge per unit of width
Q_i	Average rate of inflow

Symbol	*Description*
Q_o	Average rate of outflow
R	Radius; radius of a cross section; crest profile radius; vertical radius of curvature of the channel floor profile; radius of a terminal bucket profile
r	Hydraulic radius; radius of abutment rounding; radius of rounding of a culvert inlet opening
R_b	Radius of a bend in a channel or pipe
R_c	Radius of curvature at the crest of a siphon throat
R_s	Radius of curvature at the summit of a siphon throat; radius of a circular sharp-crested weir
S	Storage
ΔS	Increment of storage
s	Friction slope in the Manning equation; spacing
s_b	Slope of the channel floor, in profile
s_{ws}	Slope of the water surface
T	Tailwater depth; width at the water surface in a cross section of an open channel
T_{max}	Limiting maximum tailwater depth
T_{min}	Limiting minimum tailwater depth
t	Time
Δt	Increment of time
T_s	Tailwater sweep-out depth
$T.W.$	Tailwater; tailwater depth
U	A parameter for defining flow conditions in a closed waterway, $U=\frac{v}{\sqrt{gD}}$
v	Velocity
Δv	Incremental change in velocity
v_a	Velocity of approach
v_c	Critical velocity
v_S	Velocity at the crest of a siphon throat
v_t	Velocity of flow in a channel or chute, at tailwater depth
W	Weight of a mass; width of a stilling basin
w	Unit weight of water; width of a culvert entrance; width of a slot for a slotted grating dissipator; width of chute and baffle blocks in a stilling basin
x	A coordinate for defining a crest profile; a coordinate for defining a channel profile; a coordinate for defining a conduit entrance
Δx	Increment of length
x_c	Horizontal distance from the break point, on the upstream face of an ogee crest, to the apex of the crest
x_s	Horizontal distance from the vertical upstream face of a circular sharp-crested weir to the apex of the undernappe of the overflow sheet
Y	Drop distance measured from the crest of the overflow to the basin floor, for a free overfall spillway
y	A coordinate for defining a crest profile; a coordinate for defining a channel profile; a coordinate for defining a conduit entrance
$\overline{y}$	Depth from water surface to the center of gravity of a water prism cross section
Δy	Difference in elevation of the water surface profile between successive sections in a side channel trough
y_c	Vertical distance from the break point, on the upstream face of an ogee crest, to the apex of the crest
y_s	Vertical distance from the crest of a circular sharp-crested weir to the apex of the undernappe of the overflow sheet
Z	Elevation above a datum plane
ΔZ	Elevation difference of the bottom profile between successive sections in an open channel
z	Ratio, horizontal to vertical, of the slope of the sides of a channel cross section
α	A coefficient; angular variation of the side wall with respect to the structure centerline
θ	Angle from the horizontal; angle from vertical of the position of an orifice; angle from the horizontal of the edge of the lip of a deflector bucket

Table B–1 presents conversion factors most frequently used by the designer of small dams to convert from one set of units to another—for example, to convert from cubic feet per second to acre-feet. Also included are some basic conversion formulas such as the ones for converting flow for a given time to volume.

B-2. Flow in Open Channels.—(a) *Energy and Head.*—If it is assumed that streamlines of flow in an open channel are parallel and that

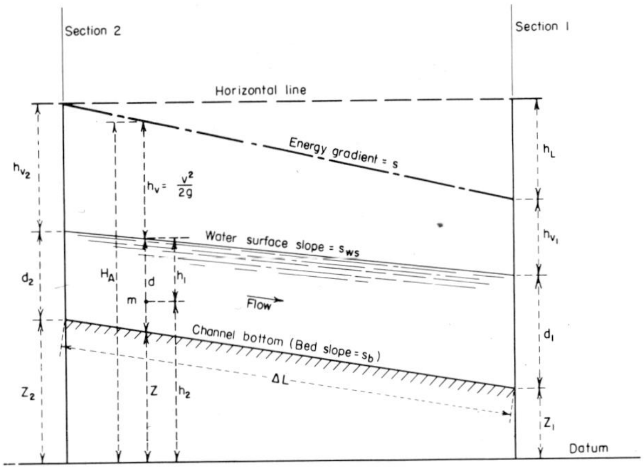

Figure B–1. Characteristics of open-channel flow. 288–D–2550.

TABLE B–1.—*Conversion factors and formulas*

[To reduce units in column 1 to units in column 4, multiply column 1 by column 2
To reduce units in column 4 to units in column 1, multiply column 4 by column 3]

CONVERSION FACTORS

Column 1	Column 2	Column 3	Column 4
LENGTH			
In.	2.54	0.3937	Cm.
	0.0254	39.37	M.
Ft.	0.3048	3.2808	M.
Miles	1.609	0.621	Km.
AREA			
Sq. in.	0.4510	0.1550	Sq. cm.
Sq. m.	10.764	.0929	Sq. ft.
Sq. miles	27.8784×10^{6}	0.3587×10^{-7}	Sq. ft.
	640.0	$.15625\times10^{-2}$	Acres (1 section).
	30.976×10^{5}	$.3228\times10^{-6}$	Sq. yd.
	2.59	.386	Sq. km.
Acre	43,560.0	0.22957×10^{-4}	Sq. ft.
	4,046.9	$.2471\times10^{-3}$	Sq. m.
	4,840.0	$.2066\times10^{-3}$	Sq. yd.
VOLUME			
Cu. ft.	1,728.0	0.5787×10^{-3}	Cu. in.
	7.4805	.13368	Gal.
	6.2321	.16046	Imperial gal.
Cu. m.	35.3145	0.028317	Cu. ft.
	1.3079	.76456	Cu. yd.
Gal.	231.0	0.4329×10^{-2}	Cu. in.
	3.7854	.26417	Liters.
Million gal.	133,681.0	0.74805×10^{-5}	Cu. ft.
	3.0689	.32585	Acre-ft.
Imperial gal.	1.2003	0.83311	Gal.
Acre-in.	3,630.0	$.27548\times10^{-3}$	Cu. ft.
Acre-ft.	1,233.5	0.81071×10^{-3}	Cu. m.
	43,560.0	$.22957\times10^{-4}$	Cu. ft.
In. on 1 sq. mile	232.32×10^{4}	0.43044×10^{-6}	Cu. ft.
	53.33	.01875	Acre-ft.
Ft. on 1 sq. mile	278.784×10^{5}	0.3587×10^{-7}	Cu. ft.
	640.0	$.15625\times10^{-2}$	Acre-ft.
VELOCITY AND GRADE			
Miles/hr.	1.4667	0.68182	Ft./sec.
M./sec.	3.2808	.3048	Ft./sec.
	2.2369	.44704	Miles/hr.
Fall in ft./mile	189.39×10^{-6}	5.28×10^{3}	Fall/ft.

CONVERSION FACTORS

Column 1	Column 2	Column 3	Column 4
FLOW			
Cu. ft./sec. (second-feet) (sec.-ft.).	60.0	0.016667	Cu. ft./min.
	86,400.0	$.11574\times10^{-4}$	Cu. ft./day.
	31.536×10^{6}	$.31709\times10^{-7}$	Cu. ft./yr.
	448.83	$.2228\times10^{-2}$	Gal./min.
	646,317.0	$.15472\times10^{-5}$	Gal./day.
	1.98347	.50417	Acre-ft./day.
	723.98	$.13813\times10^{-2}$	Acre-ft./365 days.
	725.78	$.13778\times10^{-2}$	Acre-ft./366 days.
	55.54	.018005	Acre-ft./28 days.
	57.52	.017385	Acre-ft./29 days.
	59.50	.016806	Acre-ft./30 days.
	61.49	.016262	Acre-ft./31 days.
	50.0	.020	Miner's inch in Idaho, Kans., Nebr., N. Mex., N. Dak., S. Dak., and Utah.
	40.0	.025	Miner's inch in Ariz., Calif., Mont., Nev., and Oreg.
	38.4	.026042	Miner's inch in Colo.
	35.7	.028011	Miner's inch in British Columbia.
	0.028317	35.31	Cu. m./sec.
	1.699	.5886	Cu. m./min.
	0.99173	1.0083	Acre-in./hr.
Cu. ft./min.	7.4805	0.13368	Gal./min.
	10,772.0	$.92834\times10^{-4}$	Gal./day.
10^{6} gal./day	1.5472	0.64632	Cu. ft./sec.
	694.44	$.1440\times10^{-2}$	Gal./min.
	3.0689	.32585	Acre-ft./day.
In. depth/hr.	645.33	0.15496×10^{-2}	Sec.-ft./sq. mile.
In. depth/day	26.889	0.03719	Sec.-ft./sq. mile.
	53.33	.01878	Acre-ft./sq. mile.
Sec.-ft./sq. mile	1.0413	0.96032	In. depth/28 days.
	1.0785	.92720	In. depth/29 days.
	1.1157	.89630	In. depth/30 days.
	1.1529	.86738	In. depth/31 days.
	13.574	.073668	In. depth/365 days.
	13.612	.073467	In. depth/366 days.
Acre-ft./day	226.24	0.442×10^{-2}	Gal./min.
	20.17	.0496	Miner's inch in Calif.
	19.36	.0517	Miner's inch in Colo.
Gal./sec.	5.347	0.187	Miner's inch in Calif.
	5.128	.195	Miner's inch in Colo.
PERMEABILITY			
Meinzer (gal./day through 1 sq. ft. under unit gradient).	48.8	0.02049	Bureau of Reclamation (cu. ft./yr. through 1 sq. ft. under unit gradient).

TABLE B–1.—*Conversion factors and formulas*—Continued

CONVERSION FACTORS

Column 1	Column 2	Column 3	Column 4
POWER AND ENERGY			
Hp	555.0 0.746 6,535. 42.4 1.0	0.18182×10^{-2} 1.3405 0.15303×10^{-3} .0236 1.0	Ft.-lb./sec. Kw. Kw.-hr./yr. B.t.u./min. Sec.-ft. falling 8.8 ft.
Hp.-hr	0.746 198.0×10^{4} 2,545.0	1.3405 0.505×10^{-6} $.393\times10^{-3}$	Kw.-hr. Ft.-lb. B.t.u.
Kw	8,760.0 737.56 11.8 3,412.0	0.11416×10^{-3} $.1354\times10^{-2}$ .0846 $.29308\times10^{-3}$	Kw.-hr./yr. Ft.-lb./sec. Sec.-ft. falling 1 ft. B.t.u./hr.
Kw.-hr	0.975	1.025	Acre-ft. falling 1 ft.
B.t.u	778.0 0.1×10^{-3} to $.834\times10^{-4}$	0.1285×10^{-2} 10,000 to 12,000	Ft.-lb. Lb. of coal.
PRESSURE			
Ft. water at max. density	62.425 0.4335 .0295 .8826 773.3	0.01602 2.3087 33.93 1.133 0.1293×10^{-2}	Lb./sq./ft. Lb./sq. in. Atm. In. Hg at 30° F. Ft. air at 32° F. and atm. pressure.
Ft. avg. sea water	1.026	0.9746	Ft. pure water.
Atm., sea level, 32° F	14.697	.06804	Lb./sq. in.
Millibars	295.299×10^{-4} 75.008×10^{-2}	33.863 1.3331	In. Hg. Mm. Hg.
Atm	29.92	33.48×10^{-3}	In. Hg.
WEIGHT			
P.p.m	0.00136 .0584 8.345	735.29 17.123 0.1198	Tons/acre-ft. Gr./gal. Lb./10^{6} gal.
Lb	7.0×10^{3}	0.14286×10^{-3}	Gr.
Gm	15.432	.064799	Gr.
Kg	2.2046	.45359	Lb.
Lb. water at 39.1° F	27.6812 0.11983 .09983 .453617 .01602 .01560	0.03612 8.345 10.016 2.204 62.425 64.048	Cu. in. Gal. Imperial gal. Liters. Cu. ft. pure water. Cu. ft. sea water.
Lb. water at 62° F	0.01604 .01563	62.355 63.976	Cu. ft. pure water. Cu. ft. sea water.

FORMULAS

VOLUME

Average depth in inches, or acre-inch per acre

$$=\frac{(\text{cu. ft./sec.})\ (\text{hr.})}{\text{acres}}$$

$$=\frac{(\text{gal./min.})\ (\text{hr.})}{450\ (\text{acres})}$$

$$=\frac{(\text{miner's in.})\ (\text{hr.})}{(40^{*})\ (\text{acres})}$$

*Where 1 miner's in. = 1/40 sec.-ft.

Use 50 where 1 miner's in. = 1/50 sec.-ft.

Conversion of inches depth on area to sec.-ft.

$$\text{sec.-ft.}=\frac{(645)\ (\text{sq. miles})\ (\text{in. on area})}{(\text{time in hr.})}$$

POWER AND ENERGY

$$\text{hp.}=\frac{(\text{sec.-ft.})\ (\text{head in ft.})}{8.8}$$

$$=\frac{(\text{sec.-ft.})\ (\text{pressure in lb./sq. in.})}{3.8}$$

$$=\frac{(\text{gal./min.})\ (\text{head in ft.})}{3{,}960}$$

$$=\frac{(\text{gal./min.})\ (\text{pressure in lb./sq. in.})}{1{,}714}$$

$$\text{b. hp.}=\frac{\text{water hp.}}{\text{pump efficiency}}$$

kw.-hr./1,000 gal. pumped/hr.

$$=\frac{(\text{head in ft.})\ (0.00315)}{(\text{pump efficiency})\ (\text{motor efficiency})}$$

Kw.-hr. = (plant efficiency) (1.025) (head in ft.) (water in acre-ft.)

$$\text{Load factor}=\frac{(\text{kw.-hr. in time } t)}{(\text{kw. peak load})\ (\text{time } t \text{ in hr.})}$$

SEDIMENTATION

Tons/acre-ft. = (unit weight/cu. ft.) (21.78)

Tons/day = (sec.-ft.) (p.p.m.) (0.0027)

TEMPERATURE

$$°\,\text{C.}=\frac{5}{9}\,(°\,\text{F.}-32°) \qquad °\,\text{F.}=\frac{9}{5}\,°\,\text{C.}+32°$$

velocities at all points in a cross section are equal to the mean velocity v, the energy possessed by the water is made up of two parts: kinetic (or motive) energy and potential (or latent) energy. Referring to figure B–1, if W is the weight of a mass m, the mass possesses Wh_2 foot-pounds of energy with reference to the datum. Also, it possesses Wh_1 foot-pounds of energy because of the pressure exerted by the water above it. Thus, the potential energy of the mass m is $W(h_1+h_2)$. This value is the same for each particle of mass in the cross section. Assuming uniform velocity, the kinetic energy of m is $W\left(\frac{v^2}{2g}\right)$.

Thus, the total energy of each mass particle is:

$$E_m=W\left(h^1+h^2+\frac{v^2}{2g}\right) \tag{1}$$

Applying the above relationship to the whole discharge Q of the cross section in terms of the unit weight of water w,

$$E=Qw\left(d+Z+\frac{v^2}{2g}\right) \tag{2}$$

where E is total energy per second at the cross section.

The portion of equation (2) in the parentheses is termed the absolute head, and is written:

$$H_A=d+Z+\frac{v^2}{2g} \tag{3}$$

Equation (3) is called the Bernoulli equation.

The energy in the cross section, referred to the bottom of the channel, is termed the specific energy. The corresponding head is referred to as the specific energy head and is expressed as:

$$H_E=d+\frac{v^2}{2g} \tag{4}$$

Where $Q=av$, equation (4) can be stated:

$$H_E=d+\frac{Q^2}{2ga^2} \tag{5}$$

For a trapezoidal channel where b is the bottom width and z defines the side slope, if q is expressed as $\frac{Q}{b}$ and a is expressed as $d(b+zd)$, equation (5) becomes:

$$H_E=d+\frac{q^2}{2gd^2\left(1+\frac{zd}{b}\right)^2} \tag{6}$$

Equation (5) is represented in diagrammatic form on figure B–2 to show the relationships between discharge, energy, and depth of flow in an open channel. The diagram is drawn for several values of unit discharge in a rectangular channel.

It can be seen that there are two values of d, d_H and d_L, for each value of H_E, except at the point where H_E is minimum, where only a single value exists. The depth at energy $H_{E_{min}}$ is called the critical depth, and the depths for other values of H_E are called alternate depths. Those depths lying above the trace through the locus of minimum depths are in the subcritical flow range and are termed subcritical depths, while those lying below the trace are in the supercritical flow range and are termed supercritical depths.

Figure B–3 plots the relationships of d to H_E as stated in equation (6), for various values of unit discharge q and side slope z. The curves can be used to quickly determine alternate depths of flow in open channel spillways.

(b) *Critical Flow.*—Critical flow is the term used to describe open channel flow when certain relationships exist between specific energy and discharge and between specific energy and depth. As indicated in section B–2(a) and as demonstrated on figure B–2, critical flow terms can be defined as follows:

(1) *Critical discharge.*—The maximum discharge for a given specific energy, or the discharge which will occur with minimum specific energy.

(2) *Critical depth.*—The depth of flow at which the discharge is maximum for a given specific energy, or the depth at which a given discharge occurs with minimum specific energy.

(3) *Critical velocity.*—The mean velocity when the discharge is critical.

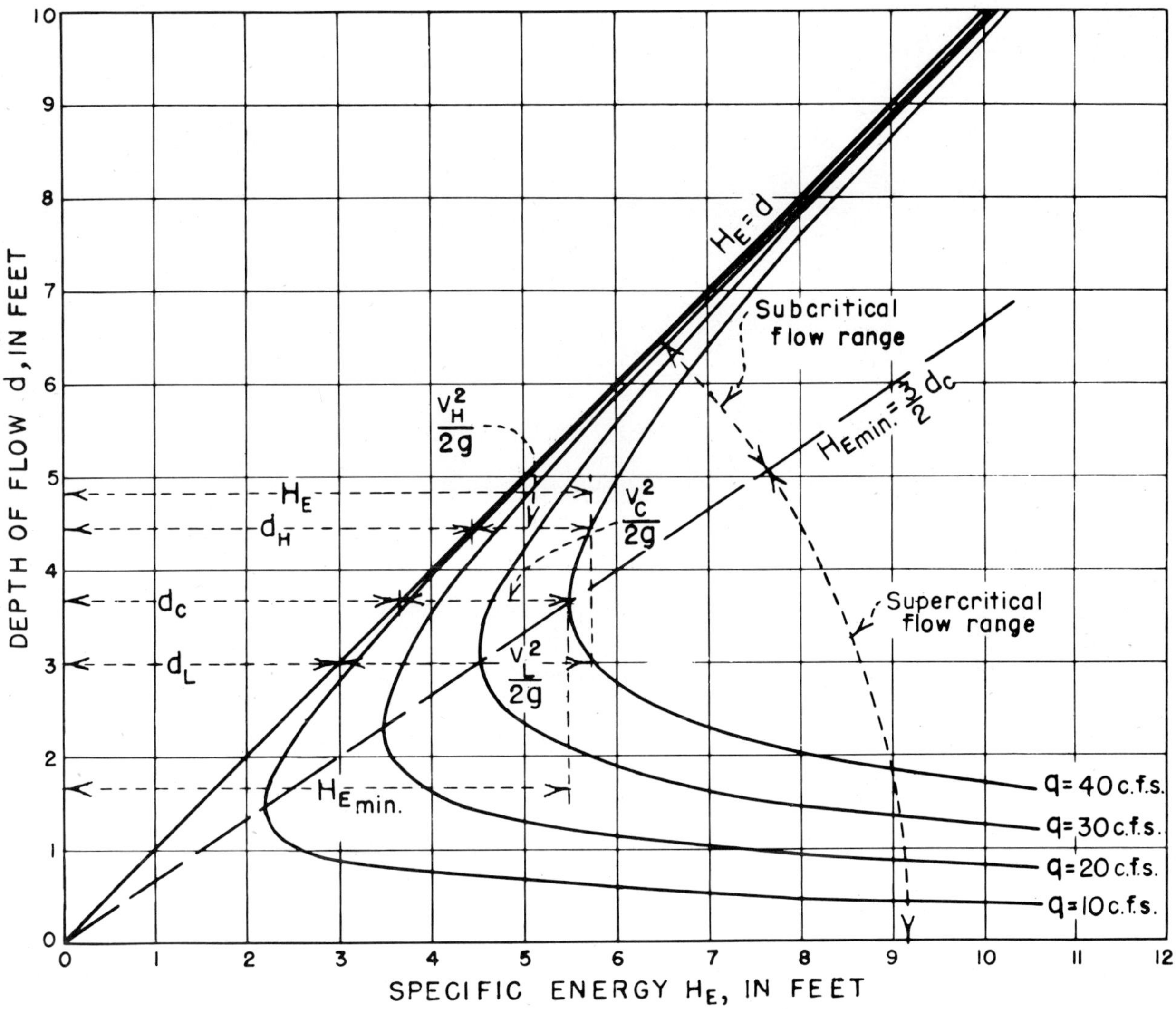

$$H_E = d + \frac{v^2}{2g} = d + \frac{q^2}{2gd^2} \quad \text{where } q = \text{discharge per unit width.}$$

$$d_c = \left(\frac{q_c}{\sqrt{g}}\right)^{\frac{2}{3}} = \frac{2}{3} H_{E_{min.}} \quad \text{where } d_c = \text{critical depth}$$

q_c = critical discharge per unit width

$H_{E_{min.}}$ = minimum energy content.

Figure B–2. Depth of flow and specific energy for rectangular section in open channel. 288–D–2551.

(4) *Critical slope.*—That slope which will sustain a given discharge at uniform critical depth in a given channel.

(5) *Subcritical flow.*—Those conditions of flow for which the depths are greater than critical and the velocities are less than critical.

(6) *Supercritical flow.*—Those conditions of flow for which the depths are less than critical and the velocities are greater than critical.

More complete discussions of the critical

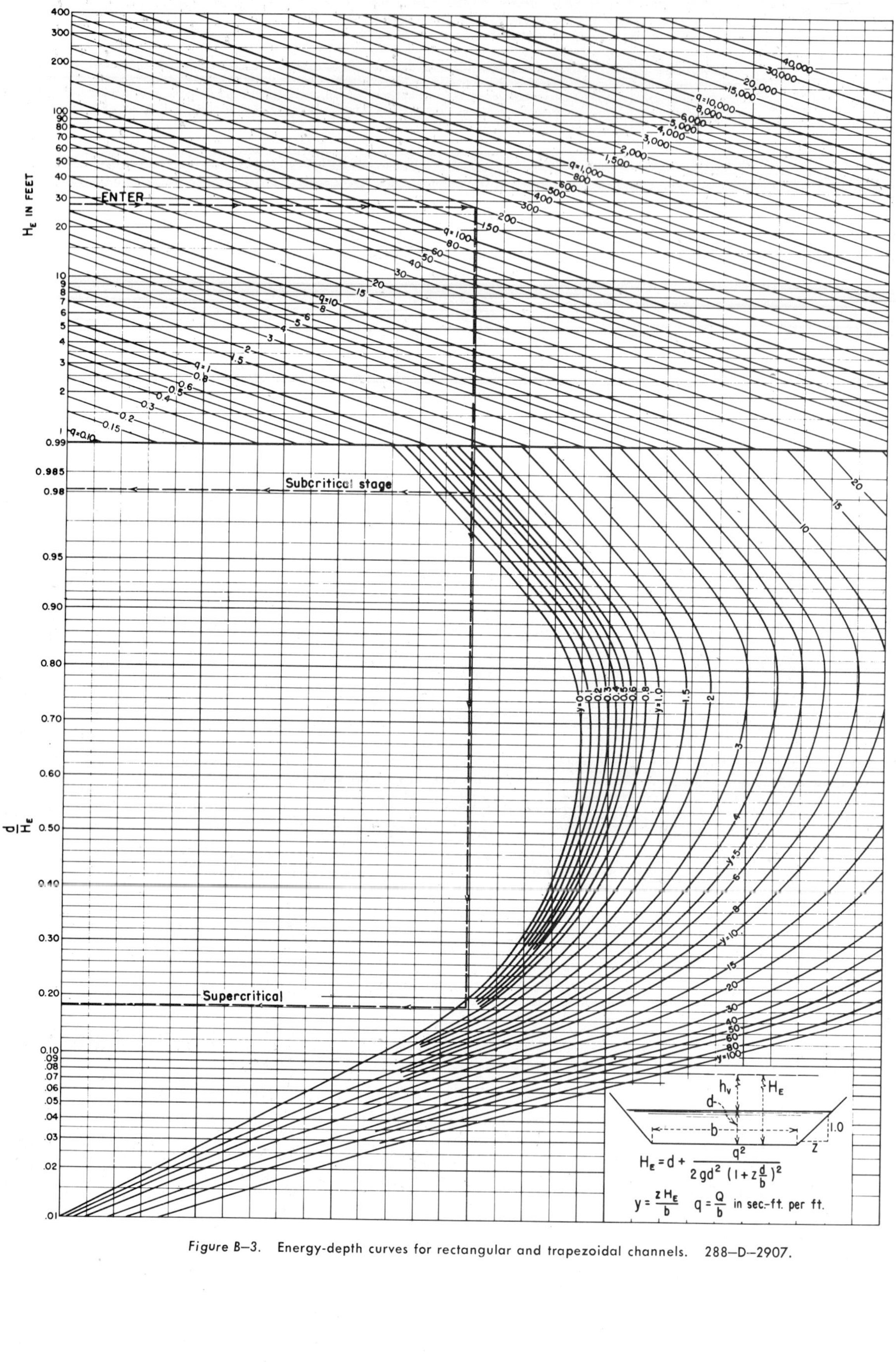

Figure B–3. Energy-depth curves for rectangular and trapezoidal channels. 288–D–2907.

flow theory in relationship to specific energy are given in most hydraulic textbooks [1, 2, 3, 4, 5].[3] The relationship between cross section and discharge which must exist in order that flow may occur at the critical stage is:

$$\frac{Q^2}{g}=\frac{a^3}{T} \quad (7)$$

where:
a=cross-sectional area in square feet, and
T=water surface width in feet.

Since $Q^2=a^2v^2$, equation (7) can be written:

$$\frac{v_c^2}{2g}=\frac{a}{2T} \quad (8)$$

Also, since $a=d_mT$, where d_m is the mean depth of flow at the section, and $\frac{v_c{}^2}{2g}=h_{v_c}$, equation (8) can be rewritten:

$$h_{v_c}=\frac{d_{m_c}}{2} \quad (9)$$

Then equation (4) can be stated

$$H_E=d_c+\frac{d_{m_c}}{2} \quad (10)$$

From the foregoing, the following additional relations can be stated:

$$d_{m_c}=\frac{v_c^2}{g} \quad (11)$$

$$d_{m_c}=\frac{Q_c^2}{a^2g} \quad (12)$$

$$v_c=\sqrt{gd_{m_c}} \quad (13)$$

$$v_c=\sqrt{\frac{ag}{T}}=5.67\sqrt{\frac{a}{T}} \quad (14)$$

$$Q_c=a\sqrt{gd_{m_c}} \quad (15)$$

For rectangular sections, if q is the discharge per foot width of channel, the various critical flow formulas are:

$$H_{E_c}=\frac{3}{2}d_c \quad (16)$$

$$d_c=\frac{2}{3}H_{E_c} \quad (17)$$

$$d_c=\frac{v_c^2}{g} \quad (18)$$

$$d_c=\sqrt[3]{\frac{q_c^2}{g}} \quad (19)$$

$$d_c=\sqrt[3]{\frac{Q_c^2}{b^2g}} \quad (20)$$

$$v_c=\sqrt{gd_c} \quad (21)$$

$$v_c=\sqrt[3]{gq_c} \quad (22)$$

$$v_c=\sqrt[3]{\frac{gQ_c}{b}} \quad (23)$$

$$q_c=d_c{}^{3/2}\sqrt{g} \quad (24)$$

$$Q_c=5.67bd_c{}^{3/2} \quad (25)$$

$$Q_c=3.087bH_{E_c}{}^{3/2} \quad (26)$$

The critical depth for trapezoidal sections is given by the equation:

$$d_c=\frac{v_c^2}{g}-\frac{b}{2z}+\sqrt{\frac{v_c^4}{g^2}+\frac{b^2}{4z^2}} \quad (27)$$

where z=the ratio, horizontal to vertical, of the slope of the sides of the channel.

Similarly, for the trapezoidal section,

$$v_c=\sqrt{\left(\frac{b+zd_c}{b+2zd_c}\right)d_cg} \quad (28)$$

and

$$Q_c=d_c{}^{3/2}\sqrt{\frac{g(b+zd_c)^3}{b+2zd_c}} \quad (29)$$

The solutions of equations (25) and (29) are simplified by use of figure B–4.

A general equation for critical depth cannot be expressed for natural channels. However, a check for the existence of critical flow in these channels is discussed in part B of this appendix.

(c) *Manning Formula.*—The formula developed by Manning for flow in open channels

[3] Numbers in brackets refer to items in the bibliography, section B-11.

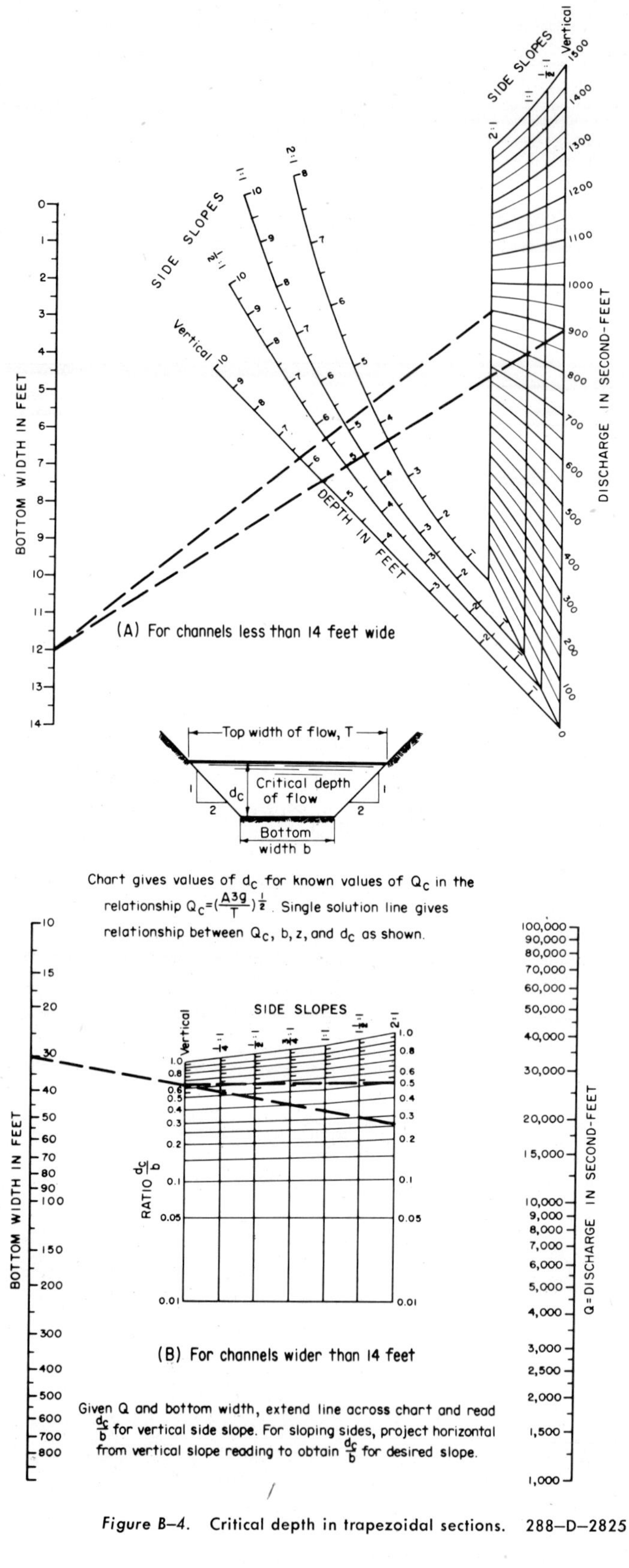

Example No. 1

Q_c = 900 cfs

Bottom width "b" = 12'

Side slope	Critical depth "d_c" (feet)
2:1	4.4
Vertical	5.6

Example No. 2

Q = 15,000 cfs

Bottom width "b" = 30'

Side slope = 2:1

Vertical "d_c" = .68b = 20'+

"d_c" for 2:1 = (0.5)(30) = 15'

Figure B–4. Critical depth in trapezoidal sections. 288–D–2825 After C. Freeman is used in most of the hydraulic analyses discussed in this text. It is a special form of Chezy's formula; the complete development is contained in most textbooks on elementary fluid mechanics. The formula is written as follows:

$$v=\frac{1.486}{n}r^{2/3}s^{1/2} \tag{30}$$

or

$$Q=\frac{1.486}{n}ar^{2/3}s^{1/2} \tag{31}$$

where:

Q=discharge in cubic feet per second,
a=the cross section of flow area in square feet,
v=the velocity in feet per second,
n=a roughness coefficient,
r=the hydraulic radius= $\frac{\text{area } (a)}{\text{wetted perimeter } (p)}$, and
s=the slope of the energy gradient.

The value of the roughness coefficient, n, varies according to the physical roughness of the sides and bottom of the channel and is influenced by such factors as channel curvature, size and shape of cross section, alinement, and type and condition of the material forming the wetted perimeter.

Values of n commonly used in design of artificial channels are as follows:

Description of channel	Values of n		
	Minimum	Maximum	Average
Earth channels, straight and uniform	0.017	0.025	0.0225
Dredged earth channels	.025	0.033	.0275
Rock channels, straight and uniform	.025	.035	.033
Rock channels, jagged and irregular	.035	.045	.045
Concrete lined	.012	.018	.014
Neat cement lined	.010	.013	....
Grouted rubble paving	.017	.030	....
Corrugated metal	.023	.025	.024

The determination of n values for natural channels is discussed in part B of this appendix.

(d) *Bernoulli Theorem.*—The Bernoulli theorem, which is the principle of conservation of energy applied to open channel flow, may be stated: The absolute head at any section is equal to the absolute head at a section downstream plus intervening losses of head. Expressed in terms of equation (3), from figure B–1:

$$Z_2+d_2+h_{v_2}=Z_1+d_1+h_{v_1}+h_L \tag{32}$$

where h_L represents all losses in head between section 2 (subscript 2) and section 1 (subscript 1). Such head losses will consist largely of friction loss, but may include minor other losses such as those due to eddy, transition, obstruction, impact, etc.

When the discharge at a given cross section of a channel is constant with respect to time, the flow is steady. If steady flow occurs at all sections in a reach, the flow is continuous and

$$Q=a_1v_1=a_2v_2 \tag{33}$$

Equation (33) is termed the equation of continuity. Equations (32) and (33), solved simultaneously, are the basic formulas used in solving problems of flow in open channels.

(e) *Hydraulic and Energy Gradients.*—The hydraulic gradient in open channel flow is the water surface. The energy gradient is above the hydraulic gradient a distance equal to the velocity head. The fall of the energy gradient for a given length of channel represents the loss of energy, either from friction or from friction and other influences. The relationship of the energy gradient to the hydraulic gradient reflects not only the loss of energy, but also the conversion between potential and kinetic energy. For uniform flow the gradients are parallel and the slope of the water surface represents the friction loss gradient. In accelerated flow the hydraulic gradient is steeper than the energy gradient, indicating a progressive conversion from potential to kinetic energy. In retarded flow the energy gradient is steeper than the hydraulic gradient, indicating a conversion from kinetic to potential energy. The Bernoulli theorem defines the progressive relationships of these energy gradients.

For a given reach of channel ΔL, the average slope of the energy gradient is $\frac{\Delta h_L}{\Delta L}$, where Δh_L is the cumulative losses through the reach. If these losses are solely from friction, Δh_L will become Δh_f and

$$\Delta h_f=\left(\frac{s_2+s_1}{2}\right)\Delta L \qquad (34)$$

Expressed in terms of the hydraulic properties at each end of the reach and of the roughness coefficient,

$$\Delta h_f=\frac{n^2}{4.41}\left[\left(\frac{v_2}{r_2^{2/3}}\right)^2+\left(\frac{v_1}{r_1^{2/3}}\right)^2\right]\Delta L \qquad (35)$$

If the average friction slope, s_f, is equal to $\frac{s_2+s_1}{2}=\frac{\Delta h_f}{\Delta L}$, and s_b is the slope of the channel floor, by substituting $s_b\Delta L$ for Z_2-Z_1, and H_E for $(d+h_v)$, equation (32) may be written:

$$\Delta L=\frac{H_{E_1}-H_{E_2}}{s_b-s_f} \qquad (36)$$

(f) *Chart for Approximating Friction Losses in Chutes.*—Figure B–5 is a nomograph from which approximate friction losses in a channel can be evaluated. To generalize the chart so that it can be applied for differing channel conditions, several approximations are made. First, the depth of flow in the channel is assumed equal to the hydraulic radius; the results will therefore be most applicable to wide, shallow channels. Furthermore, the increase in velocity head is assumed to vary proportionally along the length of the channel. Thus, the data given in the chart are not exact and are intended to serve only as a guide in estimating channel losses.

The chart plots the solution of the equation $s=\frac{dh_f}{dx}$, integrated between the limits from zero to L, or

$$h_f=\int_0^L s\,dx,$$

where, from the Manning equation,

$$s=\frac{v^2}{\left(\frac{1.486}{n}\right)^2 r^{4/3}}$$

B-3. *Flow in Closed Conduits.*—(a) *Part Full Flow in Conduits.*—The hydraulics of part full flow in closed conduits is similar to that in open channels, and open channel flow formulas are applicable. Hydraulic properties for different flow depths in circular and horseshoe conduits are tabulated in tables B–2 through B–5 to facilitate hydraulic computations for these sections.

Tables B–2 and B–4 give data for determining critical depths, critical velocities, and hydrostatic pressures of the water prism cross section for various discharges and conduit diameters. If the area at critical flow, a_c, is represented as k_1D^2 and the top width of the water prism, T, for critical flow is equal to k_2D, equation (7) can be written:

$$\frac{Q_c^2}{g}=\frac{(k_1D^2)^3}{k_2D}, \text{ or } Q_c=k_3D^{5/2} \qquad (37)$$

Values of k_3, for various flow depths, are tabulated in column 3. The hydrostatic pressure, P, of the water prism cross section is $wa\bar{y}$, where $\bar{y}$ is the depth from the water surface to the center of gravity of the cross section. If $a_c=k_1D^2$ and $\bar{y}=k_4D$, then

$$P=k_5D^3 \qquad (38)$$

Values of k_5, for various flow depths, are tabulated in column 4. Column 2 gives the values of h_{v_c} in relation to the conduit diameter, for various flow depths.

The use of tables B–2 and B–4 can be illustrated by an example. Suppose that it is desired to find the critical depth for 650 second-feet flowing free in a 9-foot-diameter circular conduit. For this case, $\frac{Q}{D^{5/2}}=\frac{243}{650}=2.675$. From column 3, by interpolation, the corresponding value in column 1 is $\frac{d}{D}=0.701$. The critical depth is then $9\times0.701=6.31$ feet. The critical velocity from column 2 is $h_{v_c}=0.3212\times 9=2.89$ feet, which gives a critical velocity of 13.6 feet per second.

Column 4 gives hydrostatic pressure upon the cross section of the water prism. The tabular value multiplied by D^3 gives pressure in cubic units of water. The pressure in pounds is 62.5 times the tabular value. For this example, $P=62.5\times9^3\times0.1822=8{,}300$ pounds.

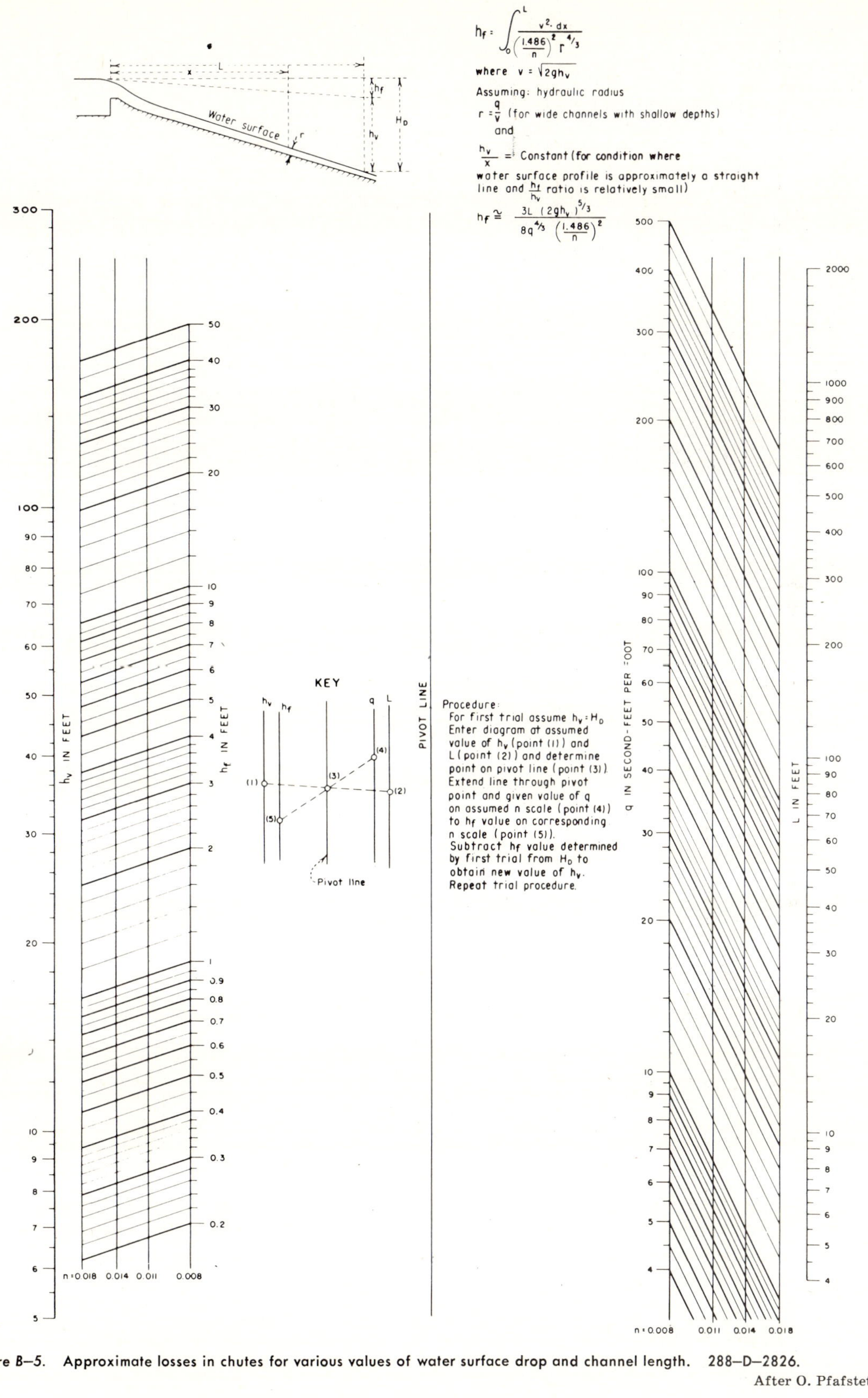

Figure B–5. **Approximate losses in chutes for various values of water surface drop and channel length. 288–D–2826.**

After O. Pfafstetter

TABLE B-2.—*Velocity head and discharge at critical depths and static pressures in circular conduits partly full*

D=Diameter of pipe.
d=Depth of flow.
h_{v_c}=Velocity head for a critical depth of d.
Q_c=Discharge when the critical depth is d.
P=Pressure on cross section of water prism in cubic units of water. To get P in pounds, when d and D are in feet, multiply by 62.5.

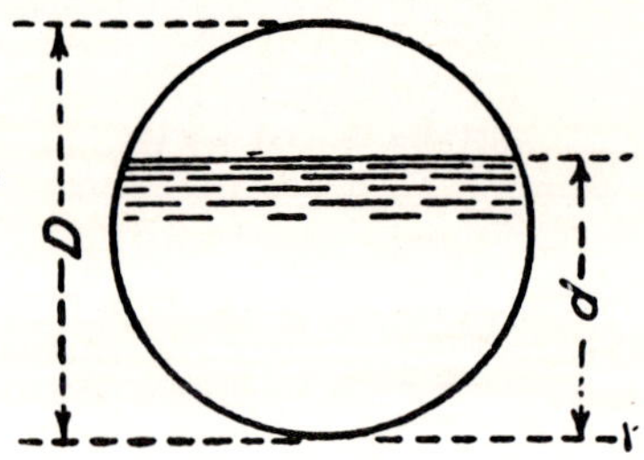

$\frac{d}{D}$	$\frac{h_{v_c}}{D}$	$\frac{Q_c}{D^{5/2}}$	$\frac{P}{D^3}$	$\frac{d}{D}$	$\frac{h_{v_c}}{D}$	$\frac{Q_c}{D^{5/2}}$	$\frac{P}{D^3}$	$\frac{d}{D}$	$\frac{h_{v_c}}{D}$	$\frac{Q_c}{D^{5/2}}$	$\frac{P}{D^3}$
1	2	3	4	1	2	3	4	1	2	3	4
0.01	0.0033	0.0006	0.0000	0.34	0.1243	0.6657	0.0332	0.67	0.2974	2.4464	0.1644
.02	.0067	.0025	.0000	.35	.1284	.7040	.0356	.68	.3048	2.5182	.1700
.03	.0101	.0055	.0001	.36	.1326	.7433	.0381	.69	.3125	2.5912	.1758
.04	.0134	.0098	.0002	.37	.1368	.7836	.0407	.70	.3204	2.6656	.1816
.05	.0168	.0153	.0003	.38	.1411	.8249	.0434	.71	.3286	2.7414	.1875
.06	.0203	.0220	.0005	.39	.1454	.8671	.0462	.72	.3371	2.8188	.1935
.07	.0237	.0298	.0007	.40	.1497	.9103	.0491	.73	.3459	2.8977	.1996
.08	.0271	.0389	.0010	.41	.1541	.9545	.0520	.74	.3552	2.9783	.2058
.09	.0306	.0491	.0013	.42	.1586	.9996	.0551	.75	.3648	3.0607	.2121
.10	.0341	.0605	.0017	.43	.1631	1.0458	.0583	.76	.3749	3.1450	.2185
.11	.0376	.0731	.0021	.44	.1676	1.0929	.0616	.77	.3855	3.2314	.2249
.12	.0411	.0868	.0026	.45	.1723	1.1410	.0650	.78	.3967	3.3200	.2314
.13	.0446	.1016	.0032	.46	.1769	1.1899	.0684	.79	.4085	3.4112	.2380
.14	.0482	.1176	.0038	.47	.1817	1.2399	.0720	.80	.4210	3.5050	.2447
.15	.0517	.1347	.0045	.48	.1865	1.2908	.0757	.81	.4343	3.6019	.2515
.16	.0553	.1530	.0053	.49	.1914	1.3427	.0795	.82	.4485	3.7021	.2584
.17	.0589	.1724	.0061	.50	.1964	1.3955	.0833	.83	.4638	3.8061	.2653
.18	.0626	.1928	.0070	.51	.2014	1.4493	.0873	.84	.4803	3.9144	.2723
.19	.0662	.2144	.0080	.52	.2065	1.5041	.0914	.85	.4982	4.0276	.2794
.20	.0699	.2371	.0091	.53	.2117	1.5598	.0956	.86	.5177	4.1465	.2865
.21	.0736	.2609	.0103	.54	.2170	1.6164	.0998	.87	.5392	4.2721	.2938
.22	.0773	.2857	.0115	.55	.2224	1.6735	.1042	.88	.5632	4.4056	.3011
.23	.0811	.3116	.0128	.56	.2279	1.7327	.1087	.89	.5900	4.5486	.3084
.24	.0848	.3386	.0143	.57	.2335	1.7923	.1133	.90	.6204	4.7033	.3158
.25	.0887	.3667	.0157	.58	.2393	1.8530	.1179	.91	.6555	4.8725	.3233
.26	.0925	.3957	.0173	.59	.2451	1.9146	.1227	.92	.6966	5.0603	.3308
.27	.0963	.4259	.0190	.60	.2511	1.9773	.1276	.93	.7459	5.2726	.3384
.28	.1002	.4571	.0207	.61	.2572	2.0409	.1326	.94	.8065	5.5183	.3460
.29	.1042	.4893	.0226	.62	.2635	2.1057	.1376	.95	.8841	5.8118	.3537
.30	.1081	.5225	.0255	.63	.2699	2.1716	.1428	.96	.9885	6.1787	.3615
.31	.1121	.5568	.0266	.64	.2765	2.2386	.1481	.97	1.1410	6.6692	.3692
.32	.1161	.5921	.0287	.65	.2833	2.3067	.1534	.98	1.3958	7.4063	.3770
.33	.1202	.6284	.0309	.66	.2902	2.3760	.1589	.99	1.9700	8.8263	.3848
								1.00	------	------	.3927

Tables B–3 and B–5 give areas and hydraulic radii for partially full conduits and coefficients which can be applied in the solution of the Manning equation. If $A=k_6\frac{\pi D^2}{4}$ and $r=k_7D$, Manning's equation can be written:

$$Q=\frac{1.486}{n}\left(k_6\frac{\pi D^2}{4}\right)(k_7D)^{2/3}\,s^{1/2},$$

or

$$\frac{Qn}{D^{8/3}s^{1/2}}=k_6\frac{1.486\pi}{4}(k_7)^{2/3}=k_8 \qquad (39)$$

TABLE B-3.—*Uniform flow in circular sections flowing partly full*

d=Depth of flow.
D=Diameter of pipe.
A=Area of flow.
r=Hydraulic radius.

Q=Discharge in second-feet by Manning's formula.
n=Manning's coefficient.
s=Slope of the channel bottom and of the water surface.

$\frac{d}{D}$	$\frac{A}{D^2}$	$\frac{r}{D}$	$\frac{Qn}{D^{8/3}s^{1/2}}$	$\frac{Qn}{d^{8/3}s^{1/2}}$	$\frac{d}{D}$	$\frac{A}{D^2}$	$\frac{r}{D}$	$\frac{Qn}{D^{8/3}s^{1/2}}$	$\frac{Qn}{d^{8/3}s^{1/2}}$
1	2	3	4	5	1	2	3	4	5
0.01	0.0013	0.0066	0.00007	15.04	0.51	0.4027	0.2531	0.239	1.442
.02	.0037	.0132	.00031	10.57	.52	.4127	.2562	.247	1.415
.03	.0069	.0197	.00074	8.56	.53	.4227	.2592	.255	1.388
.04	.0105	.0262	.00138	7.38	.54	.4327	.2621	.263	1.362
.05	.0147	0325	.00222	6.55	.55	.4426	.2649	.271	1.336
.06	.0192	.0389	.00328	5.95	.56	.4526	.2676	.279	1.311
.07	.0242	.0451	.00455	5.47	.57	.4625	.2703	.287	1.286
.08	.0294	.0513	.00604	5.09	.58	.4724	.2728	.295	1.262
.09	.0350	.0575	.00775	4.76	.59	.4822	.2753	.303	1.238
.10	.0409	.0635	.00967	4.49	.60	.4920	.2776	.311	1.215
.11	.0470	.0695	.01181	4.25	.61	.5018	.2799	.319	1.192
.12	.0534	.0755	.01417	4.04	.62	.5115	.2821	.327	1.170
.13	.0600	.0813	.01674	3.86	.63	.5212	.2842	.335	1.148
.14	.0668	.0871	.01952	3.69	.64	.5308	.2862	.343	1.126
.15	.0739	.0929	.0225	3.54	.65	.5404	.2882	.350	1.105
.16	.0811	.0985	.0257	3.41	.66	.5499	.2900	.358	1.084
.17	.0885	.1042	.0291	3.28	.67	.5594	.2917	.366	1.064
.18	.0961	.1097	.0327	3.17	.68	.5687	.2933	.373	1.044
.19	.1039	.1152	.0365	3.06	.69	.5780	.2948	.380	1.024
.20	.1118	.1206	.0406	2.96	.70	.5872	.2962	.388	1.004
.21	.1199	.1259	.0448	2.87	.71	.5964	.2975	.395	0.985
.22	.1281	.1312	.0492	2.79	.72	.6054	.2987	.402	.965
.23	.1365	.1364	.0537	2.71	.73	.6143	.2998	.409	.947
.24	.1449	.1416	.0585	2.63	.74	.6231	.3008	.416	.928
.25	.1535	.1466	.0634	2.56	.75	.6319	.3017	.422	.910
.26	.1623	.1516	.0686	2.49	.76	.6405	.3024	.429	.891
.27	.1711	.1566	.0739	2.42	.77	.6489	.3031	435	.873
.28	.1800	.1614	.0793	2.36	.78	.6573	.3036	.441	.856
.29	.1890	.1662	.0849	2.30	.79	.6655	.3039	.447	.838
.30	.1982	.1709	.0907	2.25	.80	.6736	.3042	.453	.821
.31	.2074	.1756	.0966	2.20	.81	.6815	.3043	.458	.804
.32	.2167	.1802	.1027	2.14	.82	.6893	.3043	.463	.787
.33	.2260	.1847	.1089	2.09	.83	.6969	.3041	.468	.770
.34	.2355	.1891	.1153	2.05	.84	.7043	.3038	.473	.753
.35	.2450	.1935	.1218	2.00	.85	.7115	.3033	.477	.736
.36	.2546	.1978	.1284	1.958	.86	.7186	.3026	.481	.720
.37	.2642	.2020	.1351	1.915	.87	.7254	.3018	.485	.703
.38	.2739	.2062	.1420	1.875	.88	.7320	.3007	.488	.687
.39	.2836	.2102	.1490	1.835	.89	.7384	.2995	.491	.670
.40	.2934	.2142	.1561	1.797	.90	.7445	.2980	.494	.654
.41	.3032	.2182	.1633	1.760	.91	.7504	.2963	.496	.637
.42	.3130	.2220	.1705	1.724	.92	.7560	.2944	.497	.621
.43	.3229	.2258	.1779	1.689	.93	.7612	.2921	.498	.604
.44	.3328	.2295	.1854	1.655	.94	.7662	.2895	.498	.588
.45	.3428	.2331	.1929	1.622	.95	.7707	.2865	.498	.571
.46	.3527	.2366	.201	1.590	.96	.7749	.2829	.496	.553
.47	.3627	.2401	.208	1.559	.97	.7785	.2787	.494	.535
.48	.3727	.2435	.216	1.530	.98	.7817	.2735	.489	.517
.49	.3827	.2468	.224	1.500	.909	.7841	.2666	.483	.496
.50	.3927	.2500	.232	1.471	1.00	.7854	.2500	.463	.463

TABLE B-4.—*Velocity head and discharge at critical depths and static pressures in horseshoe conduits partly full*

D=Diameter of horseshoe.
d=Depth of flow.
h_{v_c}=Velocity head for a critical depth of d.
Q_c=Discharge when the critical depth is d.
P=Pressure on cross section of water prism in cubic units of water. To get P in pounds, when d and D are in feet, multiply by 62.5.

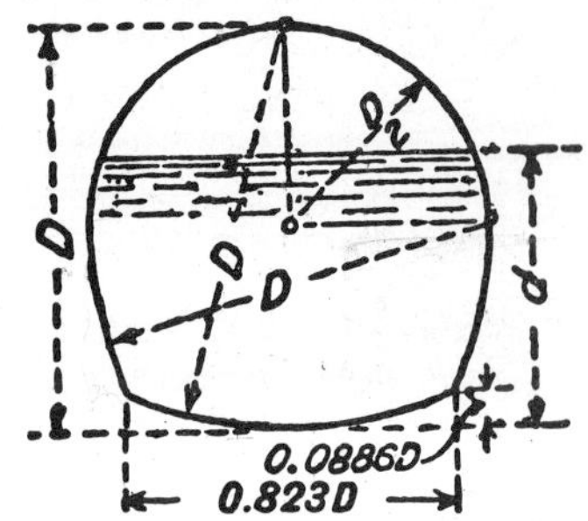

$\frac{d}{D}$	$\frac{h_{v_c}}{D}$	$\frac{Q_c}{D^{5/2}}$	$\frac{P}{D^3}$	$\frac{d}{D}$	$\frac{h_{v_c}}{D}$	$\frac{Q_c}{D^{5/2}}$	$\frac{P}{D^3}$	$\frac{d}{D}$	$\frac{h_{v_c}}{D}$	$\frac{Q_c}{D^{5/2}}$	$\frac{P}{D^3}$
1	2	3	4	1	2	3	4	1	2	3	4
0.01	0.0033	0.0009	0.0000	0.35	0.1472	0.8854	0.0449	0.69	0.3362	2.8922	0.1999
.02	.0067	.0035	.0000	.36	.1518	.9296	.0478	.70	.3443	2.9702	.2062
.03	.0100	.0079	.0001	.37	.1563	.9746	.0508	.71	.3528	3.0499	.2125
.04	.0134	.0139	.0002	.38	.1609	1.0205	.0540	.72	.3615	3.1311	.2190
.05	.0168	.0217	.0004	.39	.1655	1.0673	.0572	.73	.3707	3.2140	.2255
.06	.0201	.0312	.0007	.40	.1702	1.1148	.0605	.74	.3802	3.2987	.2321
.07	.0235	.0425	.0010	.41	.1749	1.1633	.0639	.75	.3902	3.3853	.2385
.08	.0269	.0554	.0014	.42	.1795	1.2125	.0675	.76	.4006	3.4740	.2457
.09	.0305	.0703	.0018	.43	.1843	1.2626	.0711	.77	.4116	3.5650	.2525
.10	.0351	.0879	.0024	.44	.1890	1.3135	.0748	.78	.4232	3.6584	.2595
.11	.0397	.1069	.0030	.45	.1938	1.3652	.0786	.79	.4354	3.7544	.2666
.12	.0443	.1272	.0037	.46	.1986	1.4178	.0825	.80	.4484	3.8534	.2737
.13	.0489	.1487	.0045	.47	.2035	1.4712	.0865	.81	.4623	3.9557	.2809
.14	.0534	.1714	.0054	.48	.2084	1.5253	.0907	.82	.4771	4.0616	.2882
.15	.0579	.1953	.0063	.49	.2133	1.5803	.0949	.83	.4930	4.1716	.2956
.16	.0624	.2203	.0074	.50	.2183	1.6361	.0992	.84	.5102	4.2863	.3030
.17	.0669	.2465	.0085	.51	.2234	1.6928	.1036	.85	.5289	4.4063	.3105
.18	.0714	.2736	.0098	.52	.2285	1.7505	.1081	.86	.5494	4.5325	.3181
.19	.0758	.3019	.0111	.53	.2337	1.8092	.1127	.87	.5719	4.6660	.3258
.20	.0803	.3312	.0125	.54	.2391	1.8688	.1174	.88	.5969	4.8080	.3335
.21	.0847	.3615	.0140	.55	.2445	1.9294	.1223	.89	.6251	4.9605	.3413
.22	.0891	.3928	.0156	.56	.2500	1.9911	.1272	.90	.6570	5.1256	.3492
.23	.0936	.4251	.0173	.57	.2557	2.0537	.1322	.91	.6939	5.3065	.3572
.24	.0980	.4583	.0191	.58	.2615	2.1174	.1373	.92	.7371	5.5077	.3653
.25	.1024	.4926	.0210	.59	.2674	2.1821	.1425	.93	.7889	5.7354	.3733
.26	.1069	.5277	.0229	.60	.2735	2.2479	.1478	.94	.8528	5.9996	.3813
.27	.1113	.5638	.0250	.61	.2797	2.3148	.1532	.95	.9345	6.3157	.3894
.28	.1158	.6009	0271	.62	.2861	2.3828	.1587	.96	1.0446	6.7114	.3976
.29	.1202	.6389	.0294	.63	.2926	2.4519	.1643	.97	1.2053	7.2417	.4058
.30	.1247	.6777	.0317	.64	.2994	2.5221	.1700	.98	1.4742	8.0892	4140
.31	.1292	.7175	.0342	.65	.3063	2.5936	.1758	.99	2.0804	9.5780	.4223
.32	.1337	.7582	.0367	.66	.3134	2.6663	.1817	1.00	----------	----------	.4306
.33	.1382	.7997	.0393	.67	.3208	2.7402	.1877				
.34	.1427	.8421	.0421	.68	.3283	2.8155	.1937				

Values of k_8, for various flow depths, are tabulated in column 4. If $D=k_9d$, equation (39) can be written:

$$\frac{Qn}{d^{8/3}s^{1/2}}=\frac{1.486\pi}{4}k_6(k_7)^{2/3}(k_9)^{8/3}=k_{10} \qquad (40)$$

Values of k_{10}, for various flow depths, are tabulated in column 5.

(b) *Pressure Flow in Conduits.*—Since factors affecting head losses in conduits are independent of pressure, the same laws apply to flow in both closed conduits and open channels,

TABLE B–5.—*Uniform flow in horseshoe sections flowing partly full*

d=Depth of flow.
D=Diameter.
A=Area of flow.
r=Hydraulic radius.
Q=Discharge in second-feet by Manning's formula.
n=Manning's coefficient.
s=Slope of the channel bottom and of the water surface.

$\frac{d}{D}$	$\frac{A}{D^2}$	$\frac{r}{D}$	$\frac{Qn}{D^{8/3}s^{1/2}}$	$\frac{Qn}{d^{8/3}s^{1/2}}$	$\frac{d}{D}$	$\frac{A}{D^2}$	$\frac{r}{D}$	$\frac{Qn}{D^{8/3}s^{1/2}}$	$\frac{Qn}{d^{8/3}s^{1/2}}$
1	2	3	4	5	1	2	3	4	5
0.01	0.0019	0.0066	0.00010	21.40	0.51	0.4466	0.2602	0.2705	1.629
.02	.0053	.0132	.00044	14.93	.52	.4566	.2630	.2785	1.593
.03	.0097	.0198	.00105	12.14	.53	.4666	.2657	.2866	1.558
.04	.0150	.0264	.00198	10.56	.54	.4766	.2683	.2946	1.524
.05	.0209	.0329	.00319	9.40	.55	.4865	.2707	.303	1.490
.06	.0275	.0394	.00473	8.58	.56	.4965	.2733	.311	1.458
.07	.0346	.0459	.00659	7.92	.57	.5064	.2757	.319	1.427
.08	.0421	.0524	.00876	7.37	.58	.5163	.2781	.327	1.397
.09	.0502	.0590	.01131	6.95	.59	.5261	.2804	.335	1.368
.10	.0585	.0670	.01434	6.66	.60	.5359	.2824	.343	1.339
.11	.0670	.0748	.01768	6.36	.61	.5457	.2844	.351	1.310
.12	.0753	.0823	.02117	6.04	.62	.5555	.2864	.359	1.283
.13	.0839	.0895	.02495	5.75	.63	.5651	.2884	.367	1.257
.14	.0925	.0964	.02890	5.47	.64	.5748	.2902	.374	1.231
.15	.1012	.1031	.0331	5.21	.65	.5843	.2920	.382	1.206
.16	.1100	.1097	.0375	4.96	.66	.5938	.2937	.390	1.181
.17	.1188	.1161	.0420	4.74	.67	.6033	.2953	.398	1.157
.18	.1277	.1222	.0467	4.52	.68	.6126	.2967	.405	1.133
.19	.1367	.1282	.0516	4.33	.69	.6219	.2981	.412	1.109
.20	.1457	.1341	.0567	4.15	.70	.6312	.2994	.420	1.087
.21	.1549	.1398	.0620	3.98	.71	.6403	.3006	.427	1.064
.22	.1640	.1454	.0674	3.82	.72	.6493	.3018	.434	1.042
.23	.1733	.1508	.0730	3.68	.73	.6582	.3028	.441	1.021
.24	.1825	.1560	.0786	3.53	.74	.6671	.3036	.448	1.000
.25	.1919	.1611	.0844	3.40	.75	.6758	.3044	.454	0.979
.26	.2013	.1662	.0904	3.28	.76	.6844	.3050	.461	.958
.27	.2107	.1710	.0965	3.17	.77	.6929	.3055	.467	.938
.28	.2202	.1758	.1027	3.06	.78	.7012	.3060	.473	.918
.29	.2297	.1804	.1090	2.96	.79	.7094	.3064	.479	.898
.30	.2393	.1850	.1155	2.86	.80	.7175	.3067	.485	.879
.31	.2489	.1895	.1220	2.77	.81	.7254	.3067	.490	.860
.32	.2586	.1938	.1287	2.69	.82	.7332	.3066	.495	.841
.33	.2683	.1981	.1355	2.61	.83	.7408	.3064	.500	.822
.34	.2780	.2023	.1424	2.53	.84	.7482	.3061	.505	.804
.35	.2878	.2063	.1493	2.45	.85	.7554	.3056	.509	.786
.36	.2975	.2103	.1563	2.38	.86	.7625	.3050	.513	.768
.37	.3074	.2142	.1635	2.32	.87	.7693	.3042	.517	.750
.38	.3172	.2181	.1708	2.25	.88	.7759	.3032	.520	.732
.39	.3271	.2217	.1781	2.19	.89	.7823	.3020	.523	.714
.40	.3370	.2252	.1854	2.13	.90	.7884	.3005	.526	.696
.41	.3469	.2287	.1928	2.08	.91	.7943	.2988	.528	.678
.42	.3568	.2322	.2003	2.02	.92	.7999	.2969	.529	.661
.43	.3667	.2356	.2079	1.973	.93	.8052	.2947	.530	.643
.44	.3767	.2390	.2156	1.925	.94	.8101	.2922	.530	.625
.45	.3867	.2422	.2233	1.878	.95	.8146	.2893	.529	.607
.46	.3966	.2454	.2310	1.832	.96	.8188	.2858	.528	.589
.47	.4066	.2484	.2388	1.788	.97	.8224	.2816	.525	.569
.48	.4166	.2514	.2466	1.746	.98	.8256	.2766	.521	.550
.49	.4266	.2544	.2545	1.705	.99	.8280	.2696	.513	.527
.50	.4366	.2574	.2625	1.667	1.00	.8293	.2538	.494	.494

and the formulas for each take the same general form. Thus, the equation of continuity, equation (33), $Q=a_1v_1=a_2v_2$, also applies to pressure flow in conduits.

A mass of water as such does not have pressure energy. Pressure energy is acquired by contact with other masses and is, therefore, transmitted to or through the mass under consideration. The pressure head $\frac{p}{w}$ (where p is the pressure intensity in pounds per square foot and w is unit weight in pounds per cubic foot), like velocity and elevation heads, also expresses energy. Thus for pressure pipe flow, the Bernoulli equation for flow in open channels, equation (3), can be written:

$$H_A=\frac{p}{w}+Z+\frac{v^2}{2g} \qquad (41)$$

The Bernoulli theorem for flow in a reach of pressure pipe (as shown on fig. B–6) is:

$$\frac{p_1}{w}+Z_1+h_{v_1}=\frac{p_2}{w}+Z_2+h_{v_2}+\Delta h_L \qquad (42)$$

where Δh_L represents the head losses within the reach from all causes. If H_T is the total head and v is the velocity at the outlet, Bernoulli's equation for the entire pipe is:

$$H_T=\Sigma(\Delta h_L)+h_v \qquad (43)$$

As in open channel flow, the Bernoulli theorem and the continuity equation are the basic formulas used in solving problems of pressure conduit flow.

(c) *Energy and Pressure Gradients.*—If piezometer stand pipes were to be inserted at various points along the length of a conduit flowing under pressure, as illustrated on figure B–6, water would rise in each pipe to a level equal to the pressure head in the conduit at those points. The level may be equal to the pipe grade or be above or below it; that is, the pressure at that point may be equal to, greater than, or less than the local atmospheric pressure. The height to which the water would rise in a piezometer is termed the pressure gradient. The energy gradient is above the pressure gradient a distance equal to the velocity head. The fall of the energy gradient for a given length of pipe represents the loss of energy, either from friction or from friction and other influences. The relationship of the energy gradient to the pressure gradient reflects the variations between kinetic energy and pressure head.

(d) *Friction Losses.*—Many empirical formulas have been developed for evaluating the flow of fluids in pipes. Those in most common use are the Manning equation, which, written in terms of the pipe length and diameter (in feet), is:

$$h_f=185n^2\frac{L}{D^{4/3}}\frac{v^2}{2g} \qquad (44)$$

and the Darcy-Weisbach equation, which, written in similar terms, is:

$$h_f=\frac{fL}{D}\frac{v^2}{2g} \qquad (45)$$

The Manning equation assumes that the energy loss depends only on the velocity, the dimensions of the conduit, and the magnitude of wall roughness as defined by the friction coefficient n. The n value is related to the physical roughness of the conduit wall and is independent of the size of the pipe or of the density and viscosity of the water.

The Darcy-Weisbach equation assumes the loss to be related to the velocity, the dimensions of the conduit, and the friction factor f. The factor f is a dimensionless variable based on the viscosity and density of the fluid and on the roughness of the conduit walls as it relates to the size of the conduit.

Data and criteria for determining f values for large pipe are given in a Bureau of Reclamation engineering monograph [6]. A value of the Manning coefficient, n, corresponding to the Darcy coefficient, f, can be obtained by combining equations (44) and (45). Thus:

$$n=\left(\frac{fD^{1/3}}{185}\right)^{1/2} \text{ or } f=\frac{185\,n^2}{D^{1/3}} \qquad (46)$$

The relationship between these coefficients is shown in nomograph form on figure B–7.

(e) *Design Charts for Flow in Culverts.*—Figures B–8 through B–13 are nomographs prepared by the Bureau of Public Roads, presenting data which can be used for determining flow in circular and box culvert spillways. Figures B–8 and B–9 give discharges for pipes for conditions where the control is at the inlet; these charts are based on experimental data. Figures B–10 and B–11 give discharges for pipes flowing full, and represent the solution of equation (33) of chapter IX. Similarly, figure B–12 gives discharges through box culverts with entrance control representing the solution of equation (35) of chapter IX, with the value of the coefficient C_d determined experimentally. Figure B–13 gives discharges for box culverts flowing full, representing the solution of equation (38) of chapter IX.

B-4. *Hydraulic Jump*.—The hydraulic jump is an abrupt rise in water surface which may occur in an open channel when water flowing at high velocity is retarded. The formula for the hydraulic jump is obtained by equating the unbalanced forces acting to retard the mass of flow to the rate of change of the momentum of flow. The general formula for this relationship is:

$$v_1^2 = g\frac{a_2\bar{y}_2 - a_1\bar{y}_1}{a_1\left(1-\frac{a_1}{a_2}\right)} \quad (47)$$

where:

v_1=the velocity before the jump,

a_1 and a_2=the areas before and after the jump, respectively, and

$\bar{y}_1$ and $\bar{y}_2$=the corresponding depths from the water surface to the center of gravity of the cross section.

The general formula expressed in terms of discharge is:

$$Q^2 = g\frac{a_2\bar{y}_2 - a_1\bar{y}_1}{\frac{1}{a_1}-\frac{1}{a_2}} \quad (48)$$

or:

$$\frac{Q^2}{ga_1} + a_1\bar{y}_1 = \frac{Q^2}{ga_2} + a_2\bar{y}_2 \quad (49)$$

For a rectangular channel, equation (47) can be reduced to $v_1^2 = \frac{gd_2}{2d_1}(d_2+d_1)$, where d_1 and d_2 are the flow depths before and after the jump, respectively. Solving for d_2:

$$d_2 = -\frac{d_1}{2} + \sqrt{\frac{2v_1^2 d_1}{g} + \frac{d_1^2}{4}} \quad (50)$$

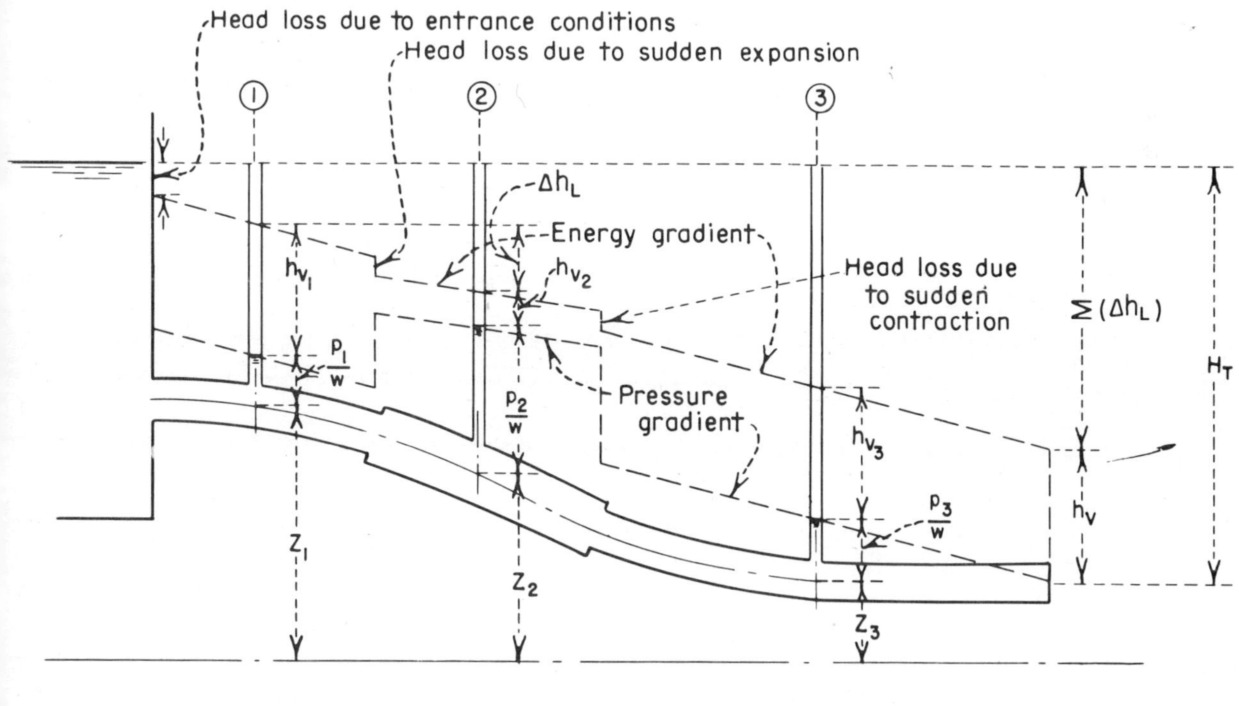

Figure B–6. Characteristics of pressure flow in conduits. 288–D–2555.

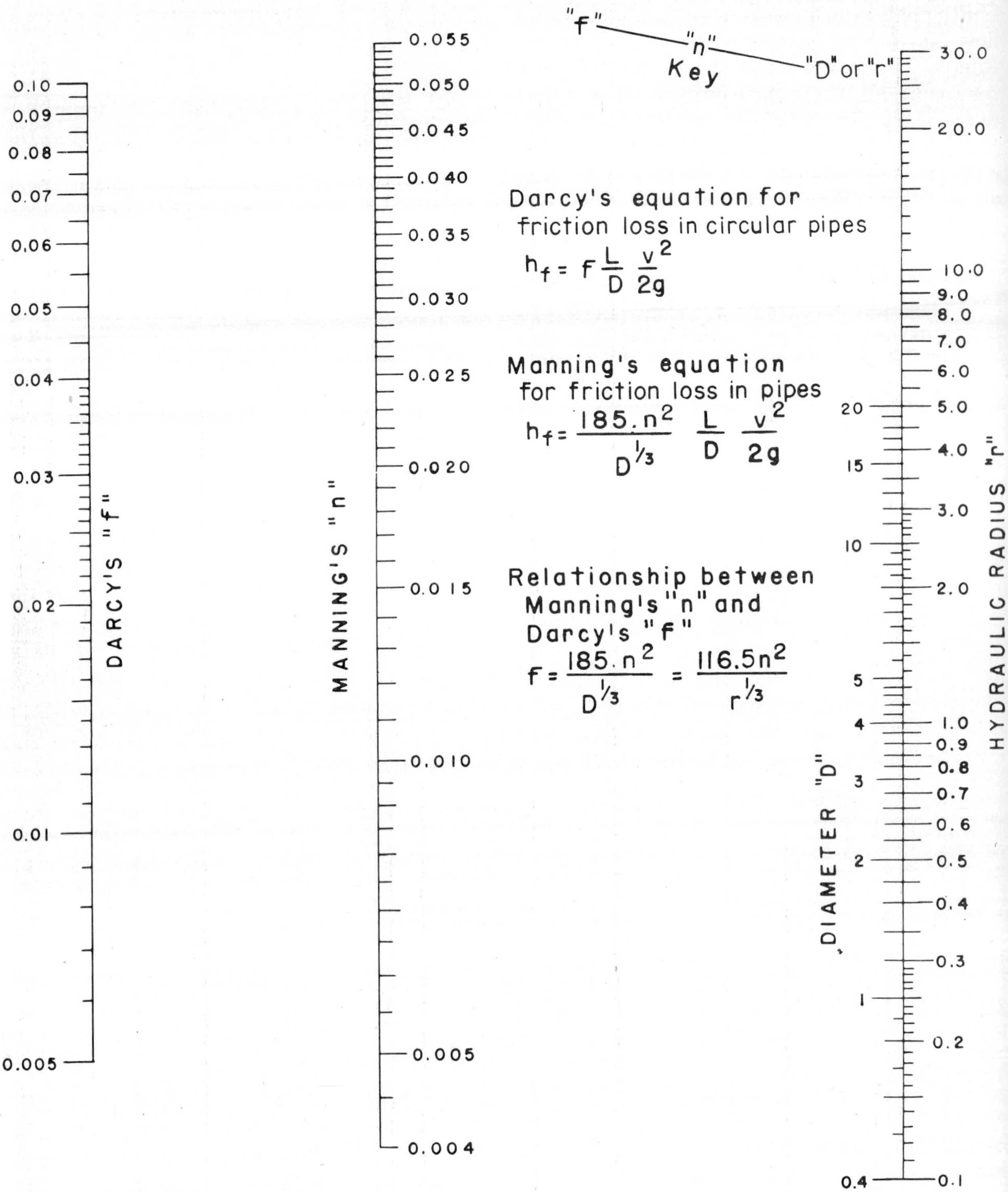

Figure B–7. Relationship between Darcy's *f* and Manning's *n* for flow in pipes. 288–D–2556.

To use scale (2) or (3), project horizontally to scale (1), then use straight inclined line through D and Q scales, or reverse as illustrated.

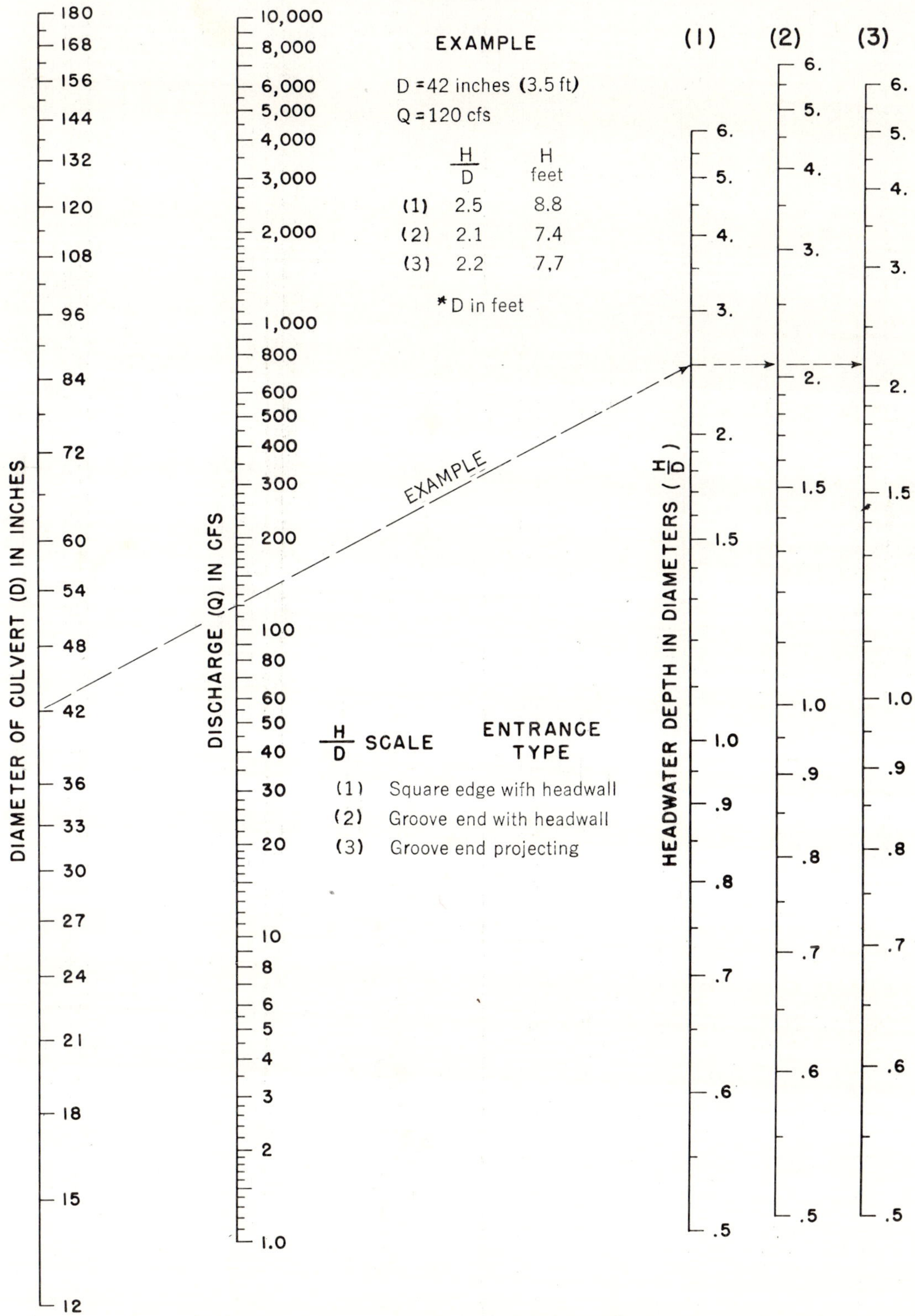

Figure B–8. Headwater depth for concrete pipe culverts with entrance control. (U.S. Bureau of Public Roads.) 288–D–2908.

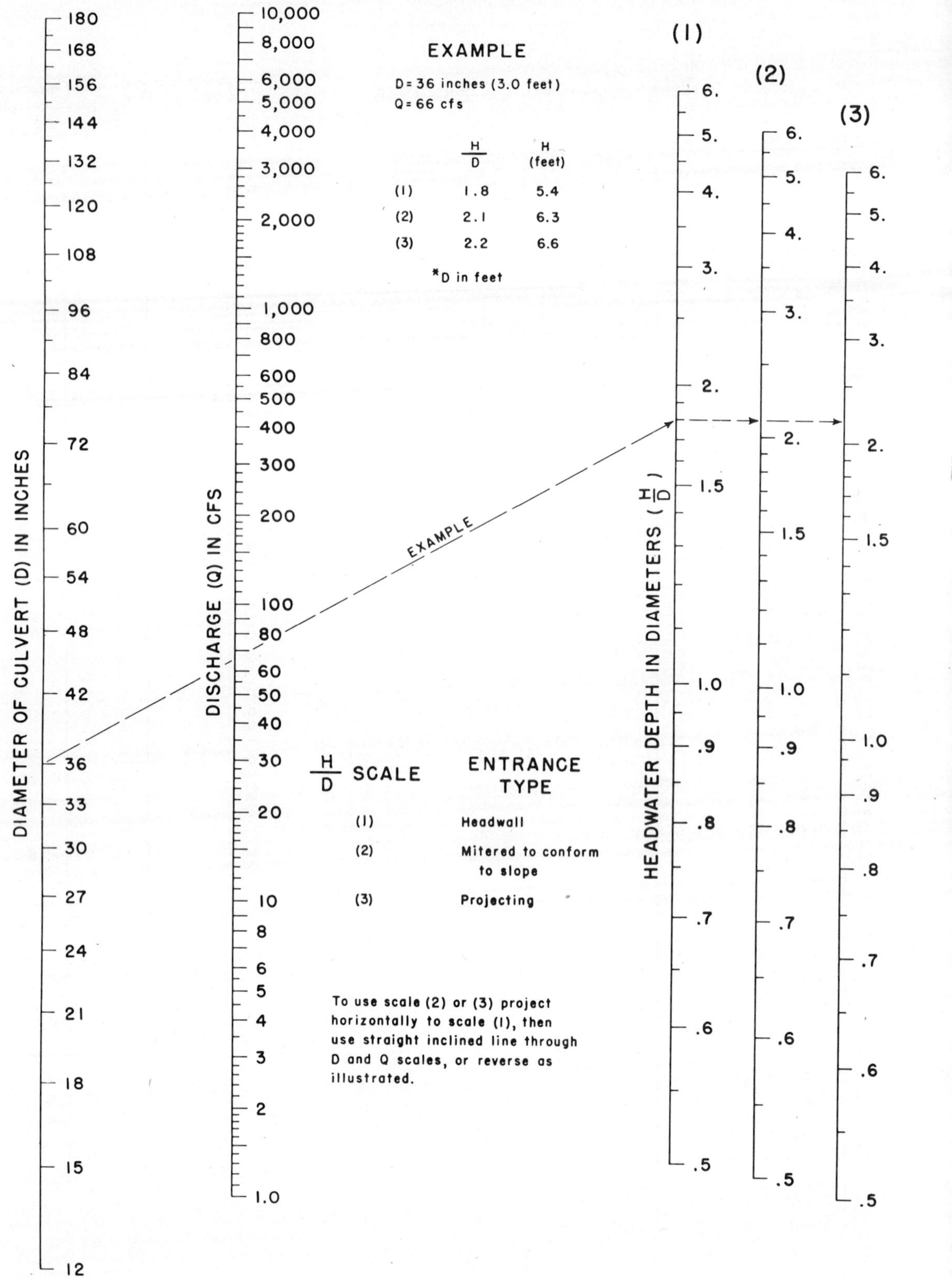

Figure B–9. Headwater depth for corrugated-metal pipe culverts with entrance control. (U.S. Bureau of Public Roads.) 288–D–2909.

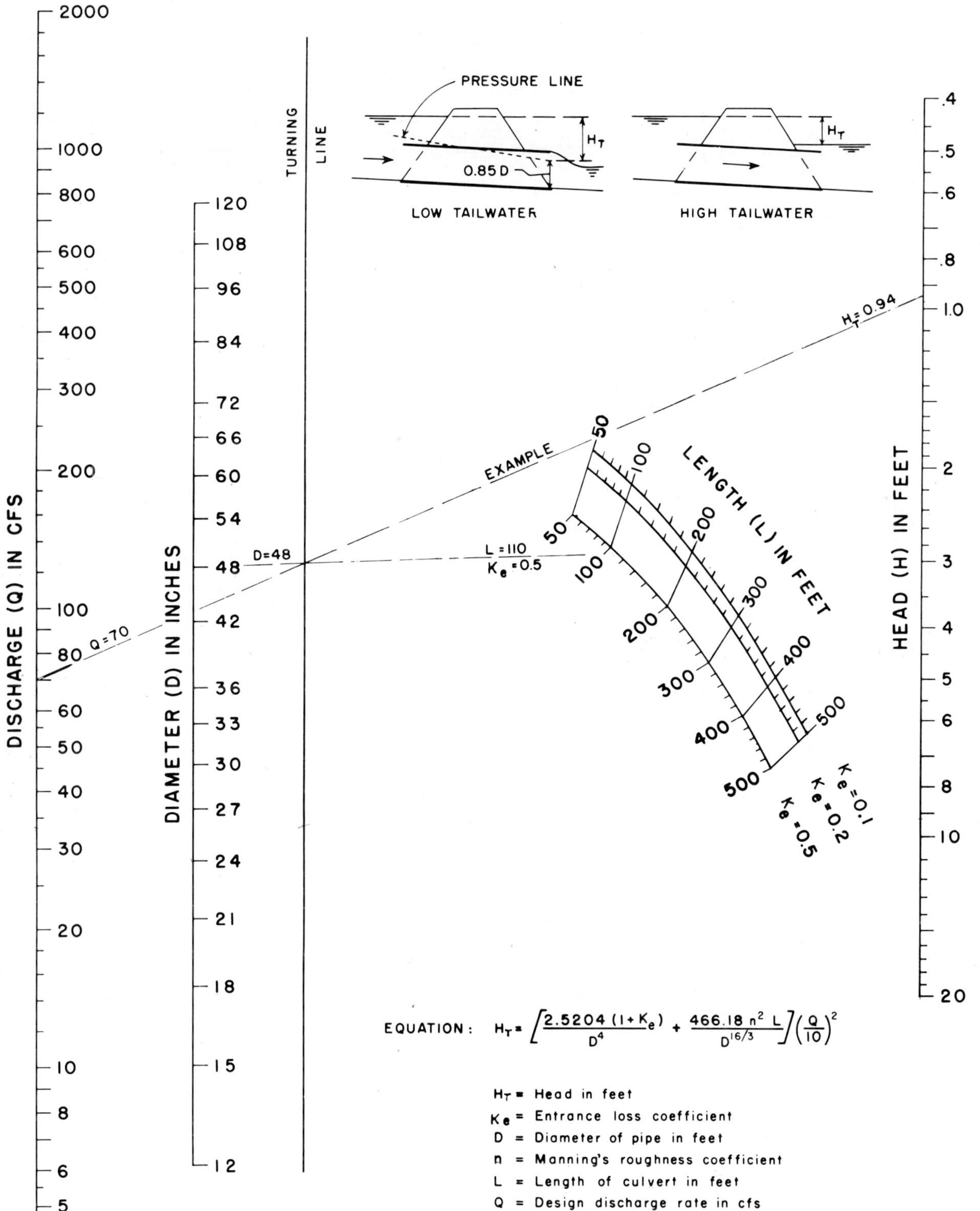

EQUATION: $$H_T = \left[\frac{2.5204\,(1+K_e)}{D^4} + \frac{466.18\,n^2\,L}{D^{16/3}}\right]\left(\frac{Q}{10}\right)^2$$

H_T = Head in feet
K_e = Entrance loss coefficient
D = Diameter of pipe in feet
n = Manning's roughness coefficient
L = Length of culvert in feet
Q = Design discharge rate in cfs

Figure B–10. Head for concrete pipe culverts flowing full, $n=0.012$. (U.S. Bureau of Public Roads.) 288–D–2910.

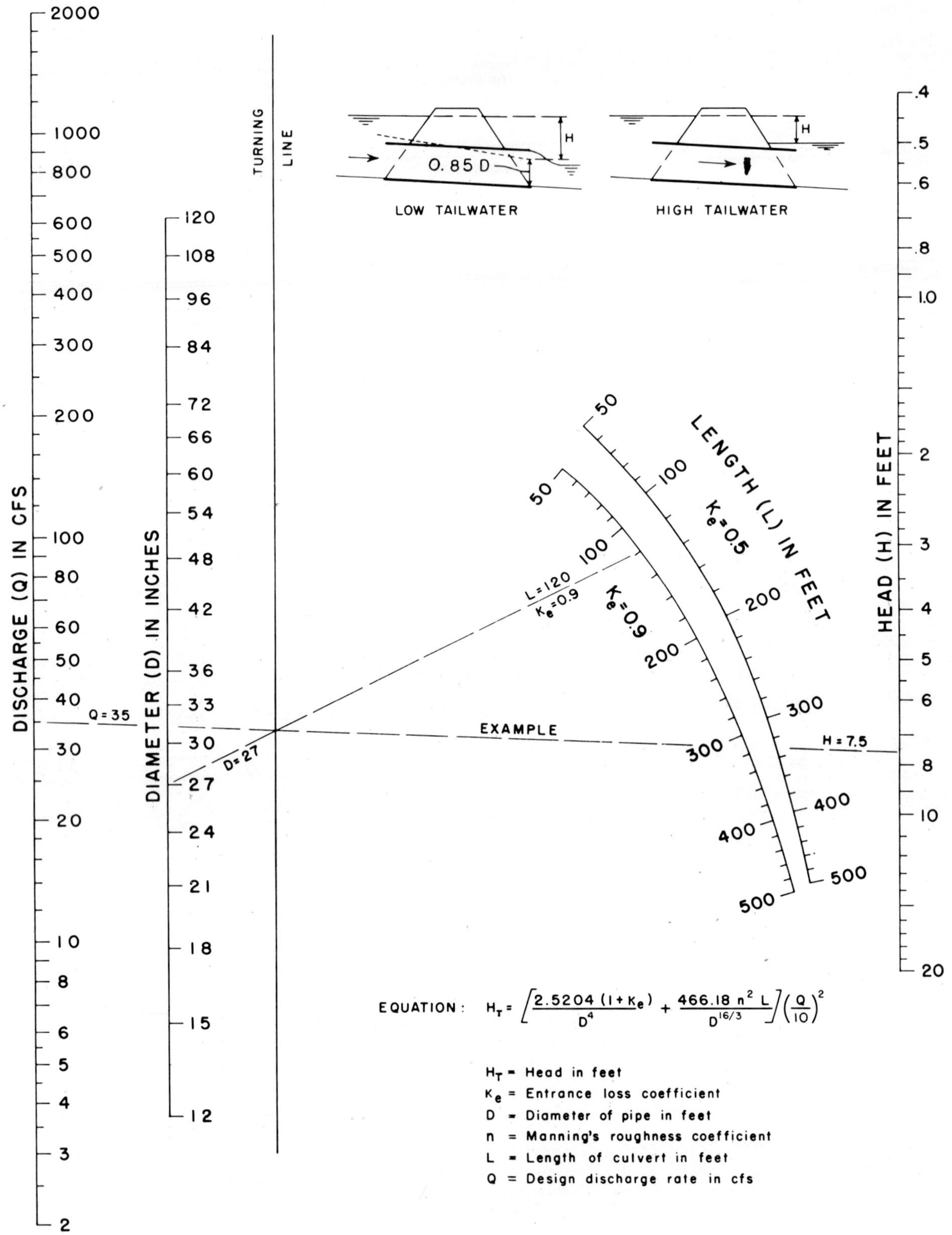

Figure B–11. Head for corrugated-metal pipe culverts flowing full, $n=0.024$. (U.S. Bureau of Public Roads.) 288–D–2911.

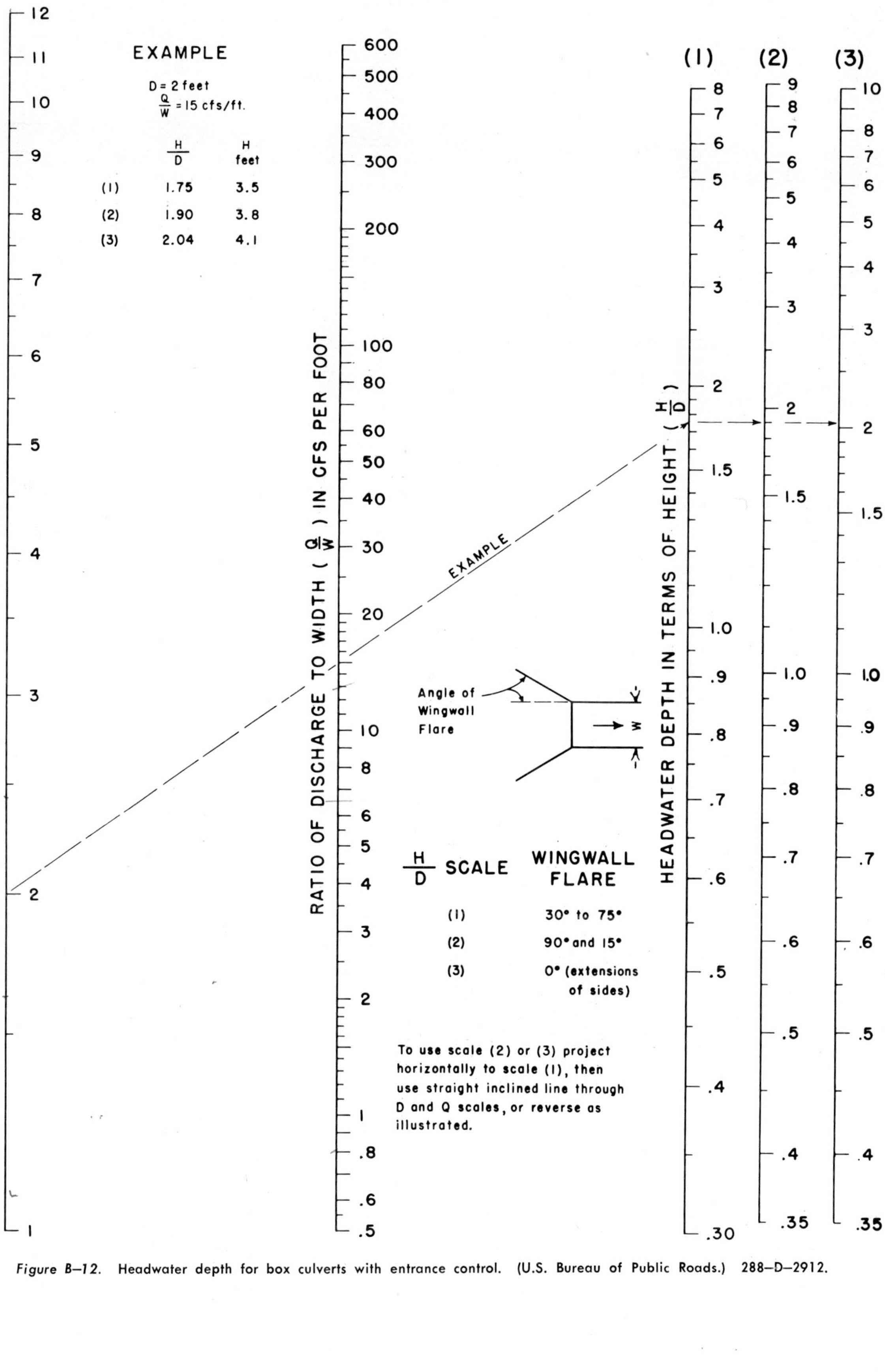

Figure B–12. Headwater depth for box culverts with entrance control. (U.S. Bureau of Public Roads.) 288–D–2912.

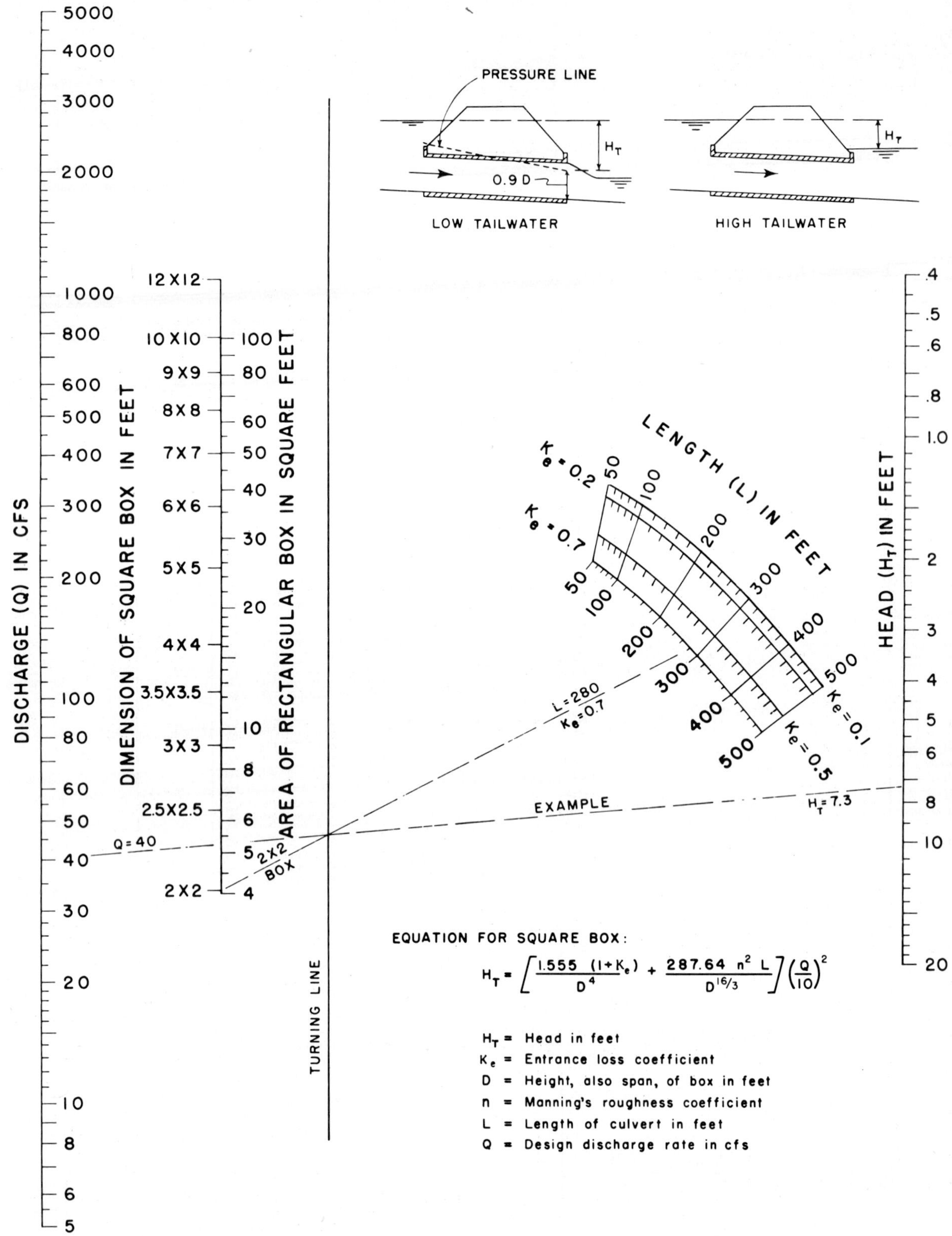

Figure B–13. Head for concrete box culverts flowing full, n=0.013. (U.S. Bureau of Public Roads.) 288–D–2913.

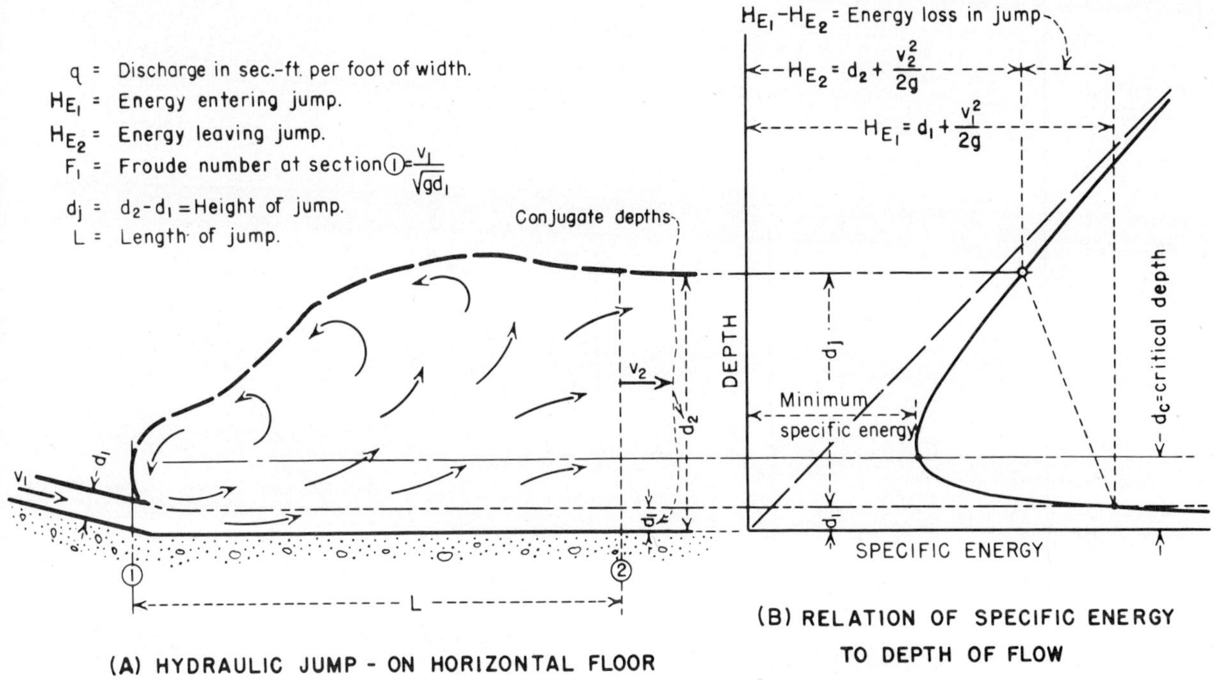

Figure B–14. Hydraulic jump symbols and characteristics. 288–D–2557.

Similarly, expressing d_1 in terms of d_2 and v_2:

$$d_1 = -\frac{d_2}{2} + \sqrt{\frac{2v_2^2 d_2}{g} + \frac{d_2^2}{4}} \tag{51}$$

A graphic solution of equation (50) is shown on figure B–16.

If the Froude number $F_1 = \frac{v_1}{\sqrt{gd_1}}$ is substituted in the equation (50):

$$\frac{d_2}{d_1} = \frac{1}{2}(\sqrt{8F_1^2 + 1} - 1) \tag{52}$$

Figure B–14 shows a graphical representation of the characteristics of the hydraulic jump. Figure B–15 shows the hydraulic properties of the jump in relation to the Froude number, as determined from experimental data [7]. Data are for jumps on a flat floor with no chute blocks, baffle piers or end sills. Ordinarily, the jump length can be shortened by incorporation of such devices in the designs of a specific stilling basin.

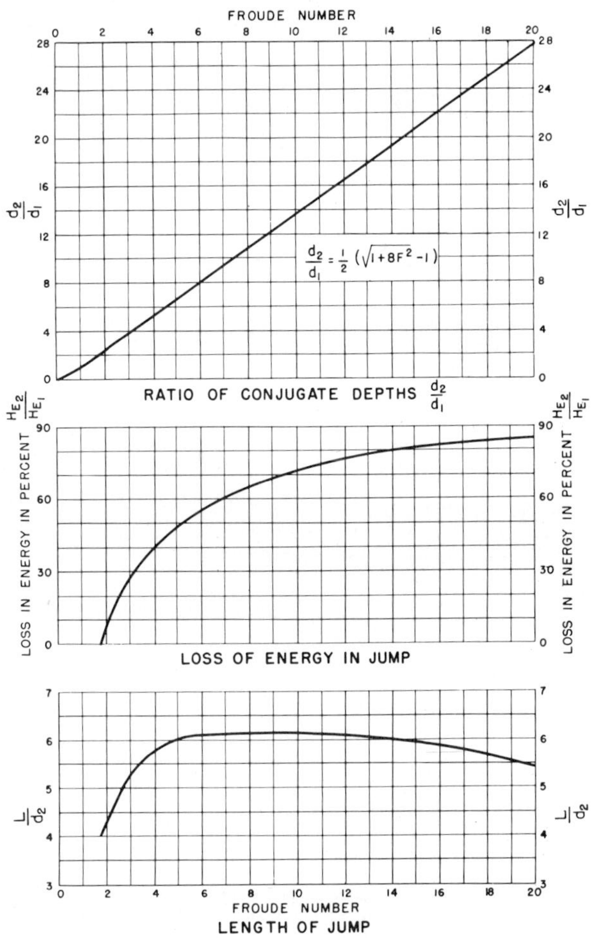

Figure B–15. Hydraulic jump properties in relation to Froude number. 288–D–2558.

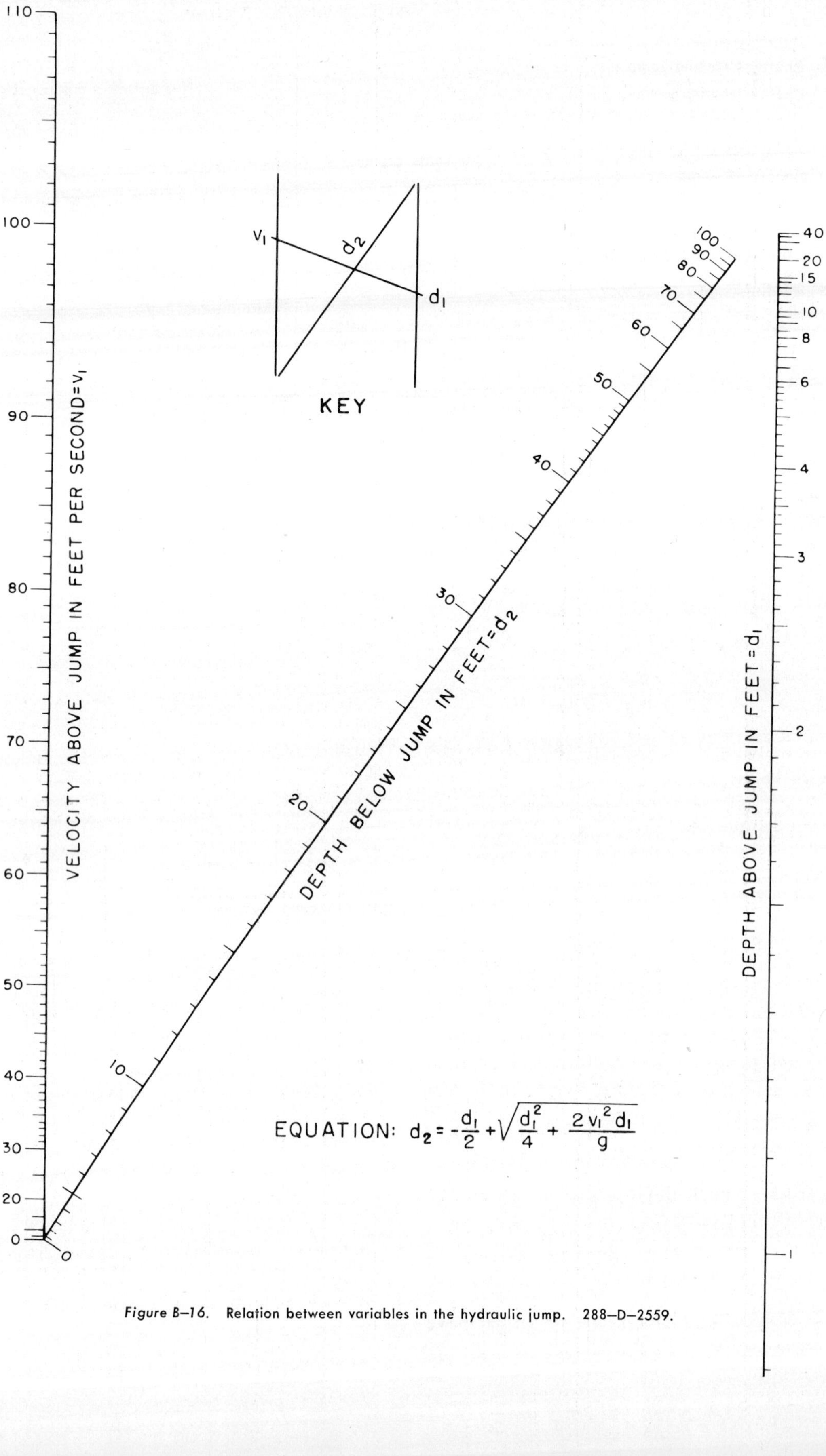

Figure B–16. Relation between variables in the hydraulic jump. 288–D–2559.

B. FLOW IN NATURAL CHANNELS

B-5. *General.*—This portion of the appendix presents briefly the hydraulic theory and analyses of natural stream channel flow as related to the design of small dams. These analyses are all directed toward the goal of establishing rating curves (stage-discharge relation) at some desired location within the stream channel. One of the more important uses of the rating curve is in connection with establishing tailwater conditions in the design of stilling basins.

The hydraulic conditions discussed herein involve only the conditions of steady and nonuniform flow. A steady flow condition is said to prevail when the discharge is the same at all sections along the channel and remains constant with respect to time. The flow is defined as nonuniform when the grade of the water surface is different from that of the channel bottom, which implies accelerating or decelerating flow. Both conditions generally prevail in natural channels. The flows may be confined to the main channel portion of the stream, but generally overbank or flood plain discharge will occur at design discharges.

Suggested procedures are given for the collection of an adequate set of field data to define each of the components in Manning's formula or Bernoulli's theorem which are used in the hydraulic computations. The accuracy of the results is very much dependent on gathering the field data representative of the prevailing hydraulic conditions and analyzing them through reasonable assumptions and interpretations. An accuracy of 0.5 foot is ordinarily a reasonable limit for the computation of water surfaces for low flows. The degree of accuracy expected for higher discharges involving overbank flow is about 1.0 foot. The loss in accuracy for overbank flow conditions is due to more difficult definition of overbank n values and flow distances between sections.

B-6. *Collection of Data.*—(a) *Streamflow Records.*—Records of streamflow taken from established U.S. Geological Survey (USGS) gaging stations furnish valuable information for the required hydraulic analysis. A check of the USGS water supply papers and the water resources data published annually for each State will show the available streamflow data for each station. Detailed descriptions of gaging stations can be studied in Water Supply Paper 888 [10].

Damsites are occasionally located near gaging stations. The rating curve of a station can be transposed to a cross section at or near the damsite if the hydraulic conditions of the reach between the damsite and the station are relatively uniform. A reasonable distance between the damsite and the gaging station for transposing the curve probably should not exceed 1,000 feet. Extreme changes in grade of the streambed, cross-sectional dimensions, and n values destroy the uniformity of the channel reach and consequently reduce the accuracy of the transposed rating curve.

(b) *Topographic Maps and Aerial Photographs.*—A topographic map and aerial photographs, if available, also provide useful information for rating curve development or water surface profile computation. For rough studies, the hydraulic properties of the stream at the damsite can be determined for a cross section as plotted from the topographic map. An example is included in section B-8 showing how this is done. The topographic map is also useful in studies involving water surface profile computations, where it can be used to locate a series of cross sections below the damsite. Flow distances between such cross sections can be measured from the map.

Aerial photographs can be used to assist in selecting the location of the cross sections. Further, the n coefficients can be evaluated by observing the areal coverage of vegetation and the meandering pattern of the stream.

(c) *Field Surveys.*—The field work required to define the hydraulic dimensions of a single cross section is relatively simple and generally inexpensive. Essentially, the procedure involves setting up a level or transit in direct line with the cross section and taking intermittent soundings and distance measurements across the section. Concurrently, levels can be run along the stream thalweg (lowest points in the streambed) or water surface (intermittent shots may be taken at water's edge along both banks) to define a slope for use in Manning's formula; a distance of 200 to 300 feet downstream and upstream from the cross section will usually be sufficient. Elevations to the nearest tenth of a foot are satisfactory for cross-sectional data; however, it may be necessary to establish water surface elevations to the nearest hundredth of a foot on streams having relatively flat gradients. Horizontal distances measured to the nearest foot are usually adequate for the other cross-sectional coordinate.

If water surface profiles are to be computed for a more precise determination of a rating curve, a series of cross sections downstream from the dam will be needed. The sections should be spaced so that there will be a drop of no more than 2 feet in water surface between cross sections and a total drop of 8 to 10 feet in the study reach.

High water marks may be used to define the water surface slope. Field observations of these marks should be made immediately following the occurrence of any flows of sufficient magnitude to leave discernible high water marks. An example of their use in a slope-area computation is shown in the next section.

(d) *Determination of* n *Values.*—The selection of n values for use in the Manning formula requires considerable judgment. Table B-6 gives values of n for average channels of various conditions. Table B-7 presents a procedure for computing a mean n value by systematically considering the factors which are involved.

The following publications will serve as guides in the proper selection of n values:

1. "Hydraulic and Excavation Tables," Bureau of Reclamation, eleventh edition, 1957, U.S. Government Printing Office, Washington, D.C.

2. "Handbook of Hydraulics," King, H. W. and Brater, E. F., fifth edition, 1963, McGraw-Hill, New York, N.Y.

3. "Hydrologic and Hydraulic Analyses, Computation of Backwater Curves in River Channels," Part CXIV, ch. 9, Engineering Manual, Civil Works Construction, May 1952, Department of the Army, Corps of Engineers, Office of the Chief of Engineers.

4. "Flow of Water in Drainage Channels," Ramser, C. E., Technical Bulletin No. 129, November 1929, U.S. Department of Agriculture, Washington, D.C.

5. "Flow of Water in Irrigation and Similar Canals," Scobey, F. C., Bureau of Agricultural Engineering, U.S. Department of Agriculture, Washington, D.C., February 1939.

6. "Design Criteria for Interrelated Highway and Agricultural Drainage and Erosion Control," Tentative ASAE Recommendation, Agricultural Engineers Yearbook, 1958, American Society of Agricultural Engineers.

7. "Roughness Characteristics of Natural Channels," Water Supply Paper 1849, U.S. Department of the Interior, Geological Survey, 1967.

B-7. *Slope-Area Method of Computing Streamflow.*—The slope-area method is utilized primarily to determine the discharge of a stream from specific field data. However, if the discharge is known, the method can be used to compute the value of n. Field procedures required to obtain needed data for the slope-area method include:

1. Selecting a representative reach of river channel: (a) The length of the reach should be at least 75 times the mean depth in the channel.
   (b) The fall in the reach should be equal to or greater than the velocity head or at least 0.50 foot.
2. Defining channel cross sections:
   (a) A minimum of three cross sections is recommended.
3. Measuring the water surface slopes from observed high water marks:
   (a) Average the elevations on both banks at each cross section.
4. Selecting a suitable roughness factor, n.

With these data, the discharge can be determined by Manning's formula,

$$Q = \frac{1.486}{n} a r^{2/3} s^{1/2} \quad (53)$$

where:

Q=total discharge in cubic feet per second,
n=a roughness coefficient,
a=the cross-sectional area of the channel in square feet,
r=hydraulic radius in feet, and
s=energy gradient.

This procedure involves combining such factors as the area, hydraulic radius, and n to compute the conveyance capacity, K_d, for each section, defined by the equation:

$$K_d=\frac{1.486}{n}ar^{2/3} \tag{54}$$

From equations (53) and (54), it can be seen that

$$Q=K_d\, s^{1/2} \tag{55}$$

The Manning equation was developed for conditions of uniform flow, but it has been assumed that the equation is also valid for nonuniform reaches that are characteristic of natural channels, if the energy gradient is modified to reflect only the losses due to boundary friction. Figure B-17 is a definition sketch of a two-section slope-area reach, from which the energy equation can be written:

$$Z_2+d_2+h_{v_2}=Z_1+d_1+h_{v_1}+h_f+k(\Delta h_v) \tag{56}$$

where:

Z=channel bottom elevation,
d=depth of water,
h_v=velocity head $=\frac{\alpha V^2}{2g}$,
h_f=energy loss due to boundary friction,
Δh_v=upstream velocity head minus the downstream velocity head, and
$k(\Delta h_v)$=energy loss due to acceleration or deceleration in a contracting or expanding reach.

Adding $Z+d$ gives the water surface elevation, h, and equation (56) may be written:

$$h_2+h_{v_2}=h_1+h_{v_1}+h_f+k(\Delta h_v) \tag{57}$$

The friction slope may now be solved:

$$s=\frac{h_f}{L}=\frac{\Delta h+\Delta h_v-k(\Delta h_v)}{L} \tag{58}$$

where Δh is the difference in water surface elevation at the two sections and L is the length of the reach.

The reach discharge can now be computed

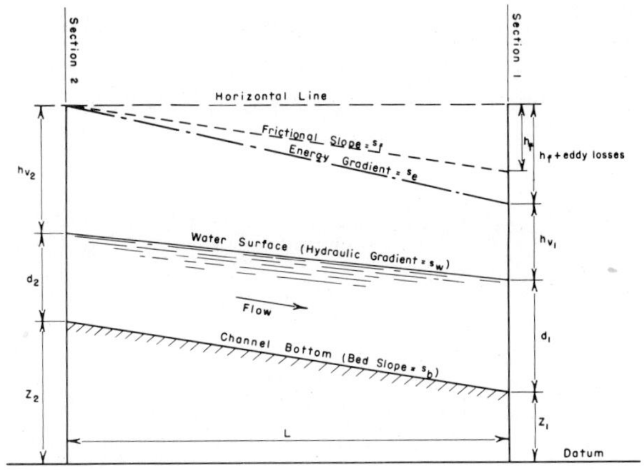

Figure B–17. Energy of open-channel flow. 288–D–2827.

using the friction slope and the geometric mean of the cross-section conveyances, or

$$Q=\sqrt{K_{d_1}\cdot K_{d_2}\cdot s} \tag{59}$$

The velocity head (h_v) at each section is computed as

$$h_v=\frac{\alpha V^2}{2g} \tag{60}$$

where V is the mean velocity in the section and α is the velocity-head coefficient. The value of α is assumed to be 1.0 if the section is not subdivided. The value of α in subdivided channels is computed as

$$\alpha=\frac{\Sigma(Kd_i/a_i^2)}{K_{d_T}{}^3/a_T{}^2} \tag{61}$$

where the subscript i refers to the conveyance or area of the individual subsections and T to the area or conveyance of the entire cross section.

The energy loss due to contraction or expansion of the channel in the reach is assumed to be equal to the difference in velocity heads at the two sections (Δh_v) times a coefficient, k. The value of k has been set at zero for contracting reaches and 0.5 for expanding reaches by the U.S. Geological Survey [12]. The value of Δh_v is computed as the upstream velocity head minus the downstream velocity head; therefore, the friction slope to be used in the

Manning equation is computed algebraically as

$$s=\frac{\Delta h+(\Delta h_v/2)}{L} \quad \text{(when } \Delta h_v \text{ is positive)} \qquad (62)$$

and

$$s=\frac{\Delta h+\Delta h_v}{L} \quad \text{(when } \Delta h_v \text{ is negative)} \qquad (63)$$

The U.S. Geological Survey publication [12] contains discharge equations for slope-area computations developed from the above basic equations. The general equation and the equation for a two-section reach are presented here.

Two sections:

$$Q=K_{d_1}\sqrt{\frac{\Delta h}{\frac{K_{d_1}}{K_{d_2}}L+\frac{K_{d_1}{}^2}{2gA_1{}^2}\left[-\alpha_2\left(\frac{A_1}{A_2}\right)^2(1-k)+\alpha_1(1-k)\right]}} \qquad (64)$$

where:

Q=Total discharge in cubic feet per second, and

A=Total cross-sectional area in square feet.

Multiple sections (n=number of sections):

$$Q=K_{d_1}\sqrt{\frac{\Delta h}{X+Y}} \qquad (65)$$

where:

$$X=K_{d_1}{}^2\frac{L_{[n-(n-1)]}}{K_{d_n}K_{d_{(n-1)}}}+K_1{}^2\frac{L_{[(n-1)-(n-2)]}}{K_{d_{(n-1)}}K_{d_{(n-2)}}}+\cdots$$

$$\frac{K_{d_1}{}^2 L_{3-2}}{K_{d_3}K_{d_2}}+\frac{K_{d_1}{}^2 L_{2-1}}{K_{d_2}K_{d_1}}, \text{ and}$$

$$Y=\frac{K_{d_1}{}^2}{A_1{}^2 2g}\left[-\alpha_n\left(\frac{A_1}{A_n}\right)^2(1-k_{n-(n-1)})+\right.$$

$$\alpha_{n-1}\left(\frac{A_1}{A_{n-1}}\right)^2(k_{(n-1)-(n-2)}-k_{n-(n-1)})+$$

$$\alpha_{(n-2)}\left(\frac{A_1}{A_{n-2}}\right)^2(k_{(n-2)-(n-3)}-k_{(n-1)-(n-2)})+\cdots+$$

$$\left.\alpha_2\left(\frac{A_1}{A_2}\right)^2(k_{2-1}-k_{3-2})+\alpha_1(1-k_{2-1})\right].$$

Using the data from the USGS publication [12] (see figures B-18, B-19, and B-20), the following procedure is used to compute the discharge. The conveyance (K_d), the velocity head coefficient for each cross section (α), and the weighted conveyance (K_w) of each subreach are computed and tabulated on the form shown in figure B-21. It should be noted that the subscripting is in reverse order of that used in the definition sketch and discharge formulas. Next, use the two-section formula given above to compute directly the discharge for each two-section subreach. The computed values will most likely differ for each subreach. Then, using the appropriate discharge as the "assumed" value on the form, complete for each subreach the computation of the various heads, slope, and "computed" discharge. The "computed" discharge must agree exactly with the "assumed" if all computations have been made correctly.

Using the multiple section equation, the final value of discharge can then be computed for the total reach. After this value is determined, it is used to compute the subsection discharges for subdivided sections, the corresponding velocities, and the mean velocities for all sections. Enter the computations in the two columns at the right of the computation form (fig. B-21). Gross errors can be often recognized if velocities are greatly different from those anticipated based on inspection of the channel after the flood.

A check should be made for critical or supercritical flow conditions at each section. This matter will be discussed in further detail in section B-9.

If the discharge is known, an analysis similar to the above can be used to arrive at the appropriate n value to assign to the channel sections. A trial and error solution will be necessary.

B-8. Development of Rating Curves.—(a) *Approximate Method.*—The field data required for development of a rating curve are similar to those used in a slope-area analysis. A determination must be made of the hydraulic properties for the section under consideration. Data required include channel cross-section geometry, channel bottom or water surface

TABLE B-6.—*Coefficient of roughness, average channels*

Value of n	*Channel condition*
0.016–0.017	Smoothest natural earth channels, free from growth, with straight alinement.
0.020	Smooth natural earth channels, free from growth, little curvature.
0.0225	Average, well-constructed, moderate-sized earth channels in good condition.
0.025	Small earth channels in good condition, or large earth channels with some growth on banks or scattered cobbles in bed.
0.030	Earth channels with considerable growth. Natural streams with good alinement, fairly constant section. Large floodway channels, well maintained.
0.035	Earth channels considerably covered with small growth. Cleared but not continuously maintained floodways.
0.040–0.050	Mountain streams in clean loose cobbles. Rivers with variable section and some vegetation growing in banks. Earth channels with thick aquatic growths.
0.060–0.075	Rivers with fairly straight alinement and cross section, badly obstructed by small trees, very little underbrush or aquatic growth.
0.100	Rivers with irregular alinement and cross section, moderately obstructed by small trees and underbrush. Rivers with fairly regular alinement and cross section, heavily obstructed by small trees and underbrush.
0.125	Rivers with irregular alinement and cross section, covered with growth of virgin timber and occasional dense patches of bushes and small trees, some logs and dead fallen trees.
0.150–0.200	Rivers with very irregular alinement and cross section, many roots, trees, bushes, large logs, and other drift on bottom, trees continually falling into channel due to bank caving.

TABLE B-7.—*A method of computing mean n value for a channel*

(Used by U.S. Soil Conservation Service)

Steps

1. Assume basic n
2. Select modifying n for roughness or degree of irregularity
3. Select modifying n for variation in size and shape of cross section
4. Select modifying n for obstructions such as debris deposits, stumps, exposed roots, and fallen logs
5. Select modifying n for vegetation
6. Select modifying n for meandering
7. Add items 1 through 6

Aids in Selecting Various n Values

1. Recommended basic n values

Channels in earth	0.010	Channels in fine gravel	0.014
Channels in rock	0.015	Channels in coarse gravel	0.028

2. Recommended modifying n value for degree of irregularity

Smooth	0.000	Moderate	0.010
Minor	0.005	Severe	0.020

3. Recommended modifying n value for changes in size and shape of cross section

Gradual	0.000	Frequent	0.010 to 0.015
Occasional	0.005		

4. Recommended modifying n value for obstructions such as debris, roots, etc.

Negligible effect	0.000	Appreciable effect	0.030
Minor effect	0.010	Severe effect	0.060

5. Recommended modifying n values for vegetation

Low effect	0.005 to 0.010	High effect	0.025 to 0.050
Medium effect	0.010 to 0.025	Very high effect	0.050 to 0.100

6. Recommended modifying n value for channel meander

L_s=Straight length of reach L_m= Meander length of reach

L_m/L_s	n
1.0–1.2	0.000
1.2–1.5	0.15 times n_s
>1.5	0.30 times n_s

where n_s=items 1+2+3+4+5

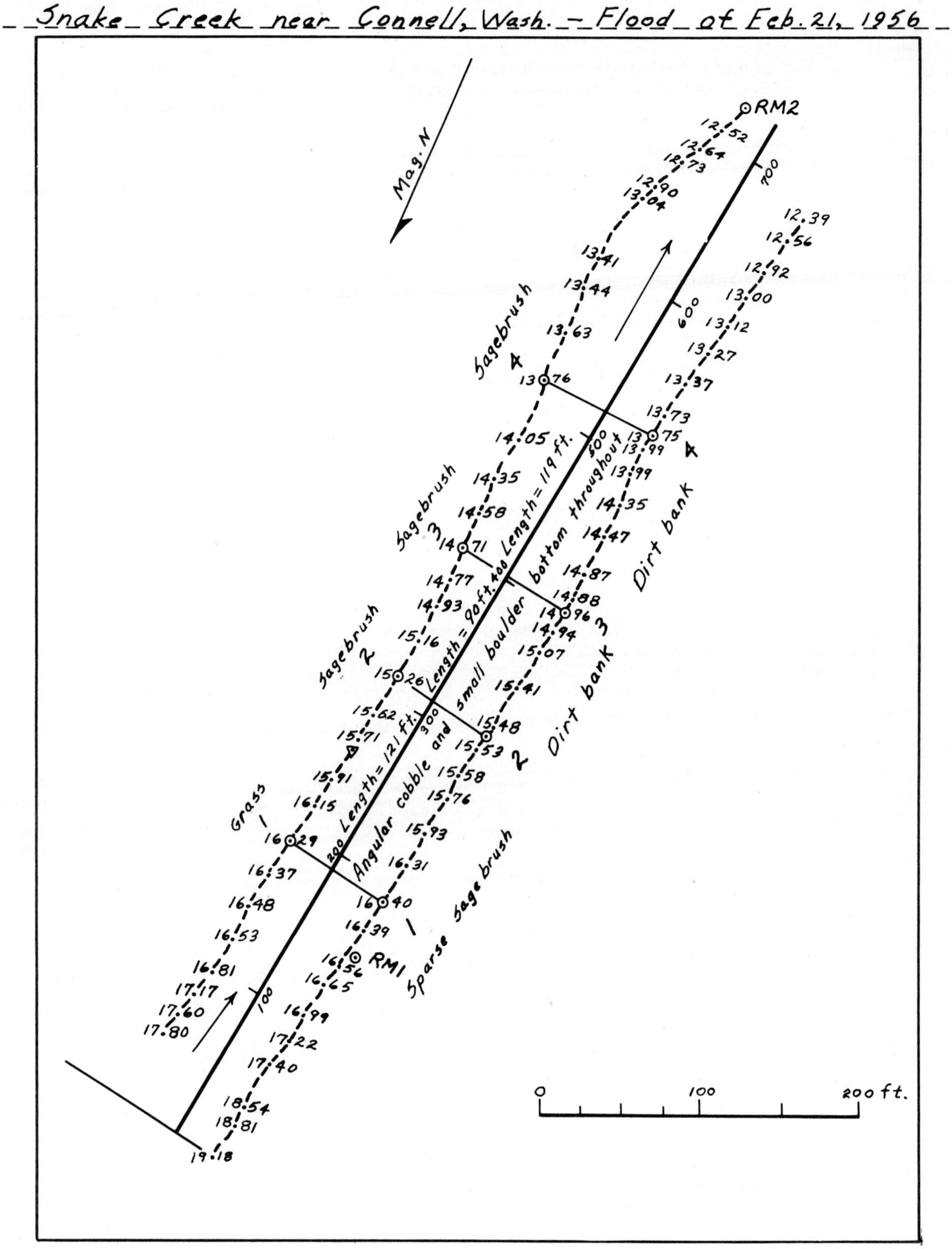

Figure B–18. Sample slope-area computation, plan view of reach. 288–D–2828.

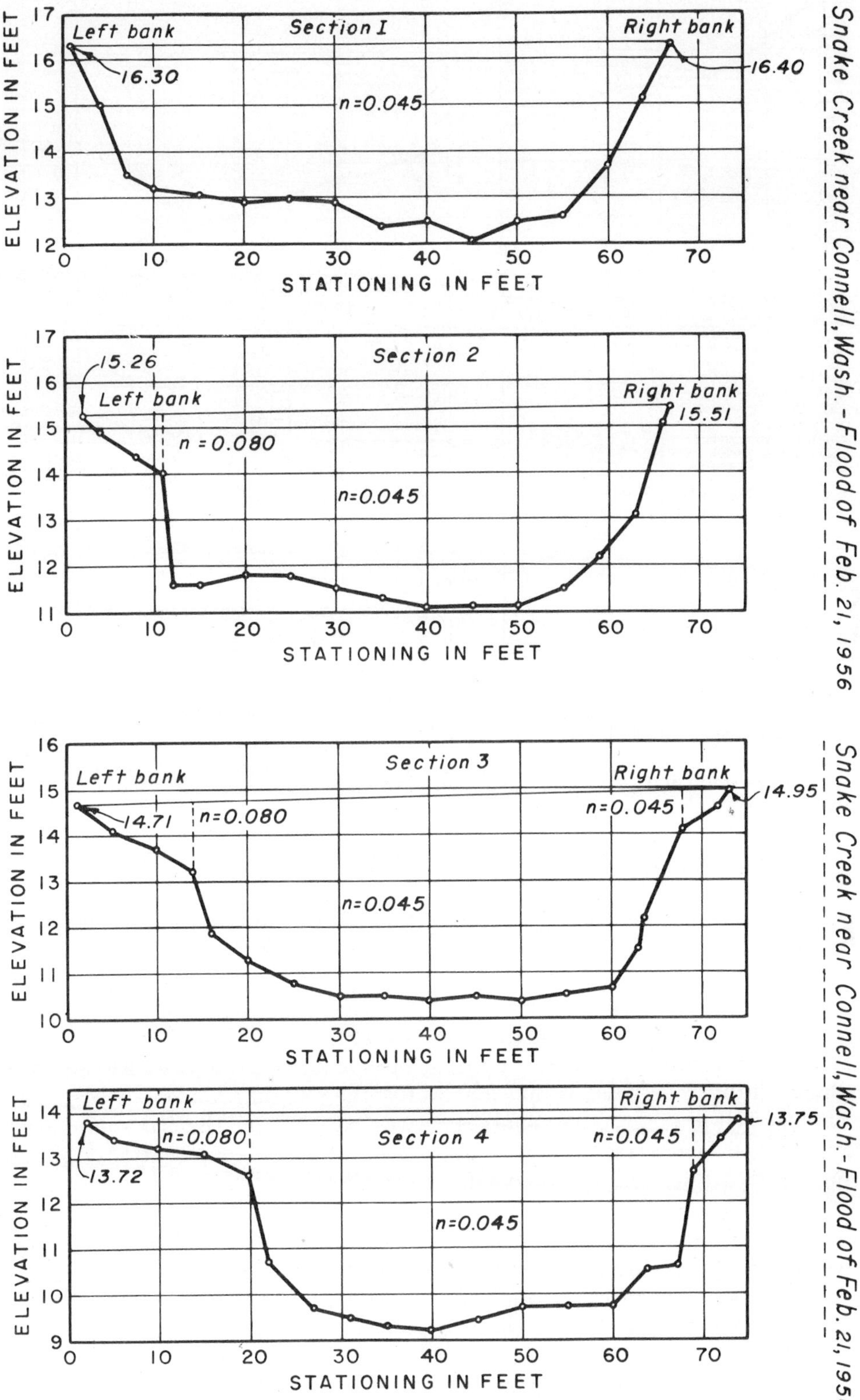

Figure B–19. Sample slope-area computation, cross sections. 288–D–2829.

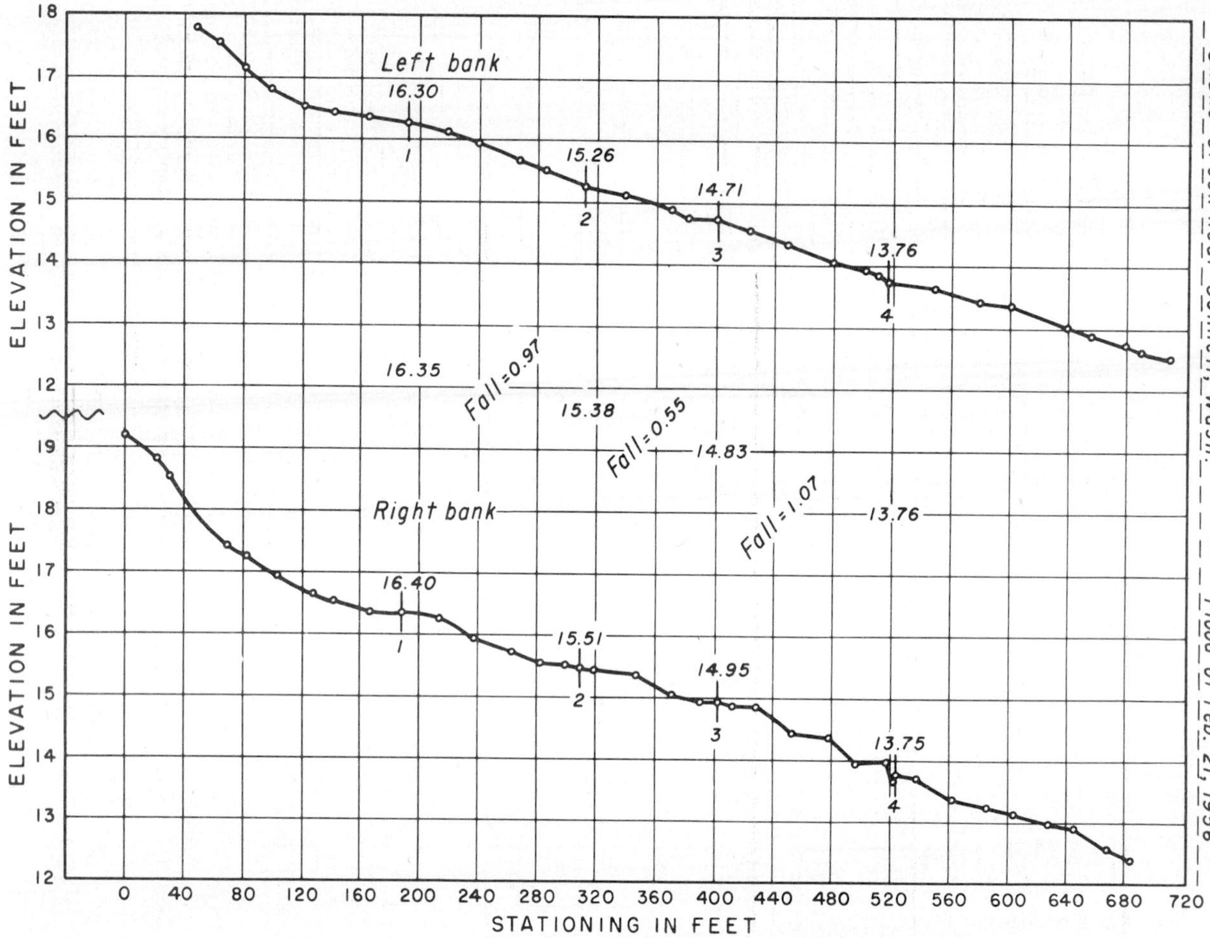

Figure B–20. Sample slope-area computation, high-water profile. 288–D–2830.

slope, and assignment of n value(s) to the affected area.

When economics or time will not permit detailed field measurements to be taken, the curve must be developed from whatever pertinent data are available. A topographic map might be used for determining the cross section and possibly the channel bed profile. Other sources of information, such as photographs, may provide data from which a selection of an n value can be made.

Figures B-22 and B-23 and table B-8 illustrate the preparation of a rating curve from a topographic map. The procedures shown are also applicable if the cross section is established from a field survey. First, the centerline of the dam was located on the topographic map available for the area, as shown in (A) of figure B–22. The cross section in (B) was then developed by scaling the distances between contours on the map. The low point in the streambed was determined by interpolation between contours. The mean bed slope of the stream was obtained from map-scaled measurements of distances between contours crossing the stream channel. A value of n of 0.030 was selected on the basis of various descriptions and field observations that considerable growth of vegetation was present in a stream of relatively straight alinement. The computations were then performed as shown in table B–8 on page 582.

The conveyance capacity method illustrated under the slope-area discussion has been utilized in the computations. From the geometric properties of the section and an n of 0.030, the conveyance capacity, K_d, is computed for various elevations. These values are multiplied

9–193
Slope-area measurement
December 1960

UNITED STATES DEPARTMENT OF THE INTERIOR
GEOLOGICAL SURVEY
WATER RESOURCES DIVISION

File ________

Meas. No. ________

Slope-area measurement of Snake Creek near Connell, Wash.
(Miscellaneous site for flood of Feb. 21, 1956

Reach between sections	1–2	2–3	3–4	Gage height, —— ft
Length of reach (L), ft	121	90	119	Discharge[1] 1,380 cfs
Fall in reach (Δh), ft	0.97	0.55	1.07	Drainage area, 115 sq mi

SECTION PROPERTIES

Section	n	$\frac{1.486}{n}$	a	r	$r^{2/3}$	$K=\frac{1.486}{n}ar^{2/3}$	K^3/a^2		[2] α	[3] q	v
1 Sta. 1-67	.045	33.0	208	3.09	2.12	14,550			1.00	1,380	6.64
2 2-11	.080	18.6	6.1	.68	.77	90	0.	196		8	1.31
11-67	.045	33.0	203	3.47	2.29	15,340	876			1,370	6.75
			209.1			15,430	876	.196	1.04		
							842				
3 1-14	.080	18.6	10.0	.77	.84	160		.410		13	1.30
14-68	.045		212	3.83	2.45	17,140	1 122			1,360	6.42
68-73	.045		2.6	.51	.64	50		.185		4	1.54
			224.6			17,350	1 122	.595	1.08		
							1 035				
4 2-20	.080	18.6	11.1	.62	.73	150		.274		13	1.17
20-69	.045		193	3.79	2.43	15,480	996			1,360	7.05
69-74	.045		2.8	.55	.67	60		.275	1.10	4	1.43
			206.9			15,690	996	549			
							901				

Weighted conveyance[4] (K_w), 1–2 14,930 2–3 16,360 3–4 16,500

COMPUTATION OF DISCHARGE

Reach	Assumed Q	[5] h_v	[6] Δh_v	[7] h_f	$s=h_f/L$	$s^{1/2}$	Computed Q $=K_w s^{1/2}$
1–2	1,330	Upstr. .636 Dwnstr. .655	-0.019	0.951	0.00786	0.0887	1,330
2–3	1,320	Upstr. .645 Dwnstr. .580	.065	.582	.00647	.0804	1,320
3–4	1,460	Upstr. .710 Dwnstr. .852	-.142	.928	.00780	.0883	1,460

[1]DISCHARGE (by formula) 1,380, or ________

FORMULAS

2 $\alpha = \Sigma(K^3/a^2) \div K^3_{total}/A^2_{total}$

3 $q = Q\,(K/K_{total})$

4 $K_w = \sqrt{K_{Upstr.} \times K_{Dwnstr.}}$

5 $h_v = \alpha\, V^2/2g$

6 $\Delta h_v = \text{Upstr. } h_v - \text{Dwnstr. } h_v$

7 When Δh_v is positive, $h_f = \Delta h + \tfrac{1}{2}\Delta h_v$
When Δh_v is negative, $h_f = \Delta h + \Delta h_v$

Summary of factors influencing measuring conditions (floodmarks, surge, scour, fill, channel configuration, angle of flow, selection of n, etc.): ________

Sheet No. 11 of 13 Sheets. Prepared by E.L.S. Date 3-19-56 Checked by E.G.P. Date 5-14-56

Figure B–21. Sample slope-area computation, discharge. 288–D–2831.

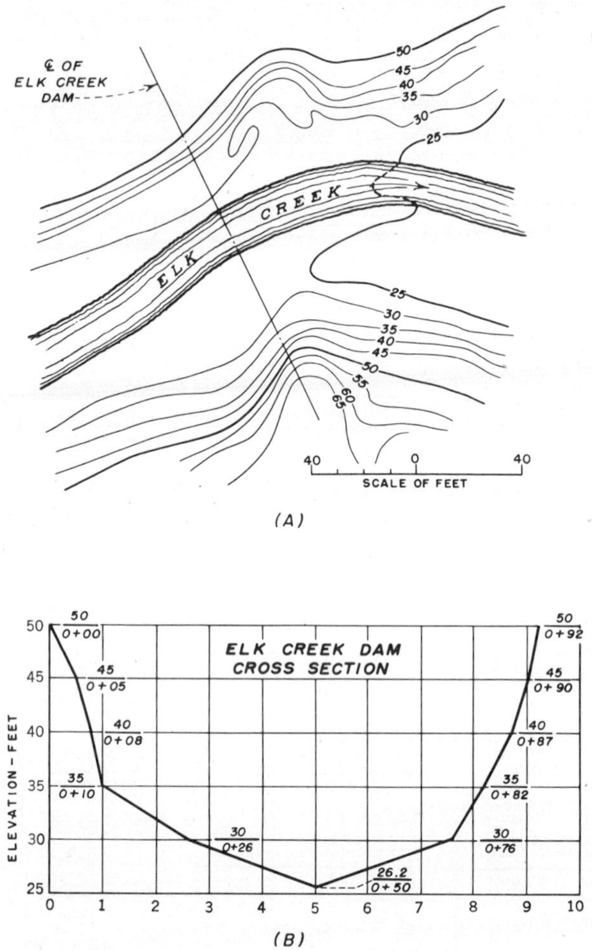

Figure B–22. Plan and cross section of Elk Creek Dam site. 288–D–2832.

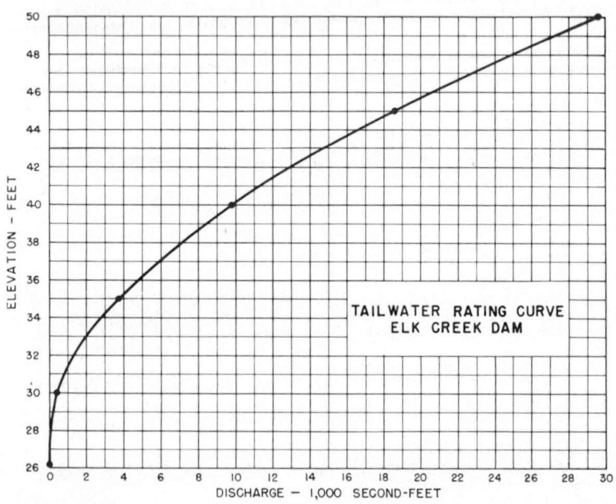

Figure B–23. Tailwater rating curve for Elk Creek Dam. 288–D–2560.

TABLE B-8.—*Computations for tailwater rating curve*

ELK CREEK DAM

Data: $n=0.030$ s_b=mean bed slope=0.00395

$$C=\frac{1.486}{n}=49.53 \quad s_b^{1/2}=0.0628$$

Elevation	a	p	r	$r^{2/3}$	C	K_d	$s_b^{1/2}$	Q
26.2......	0	0	0	0	......	....	....	......
30........	95	50.6	1.88	1.523	49.53	7,170	0.0628	450
35........	400	75.1	5.32	3.047	49.53	60,400	.0628	3,790
40........	753	87.6	8.60	4.198	49.53	156,000	.0628	9,800
45........	1160	99.2	11.70	5.154	49.53	296,000	.0628	18,600
50........	1610	111.7	14.40	5.919	49.53	472,000	.0628	29,600

by the one-half power of the mean bed slope to compute the discharge at each elevation. The tailwater rating curve derived from the computations is shown in figure B-23.

(b) *Water Surface Profile Method.*—In studies where more exact tailwater curves are required, water surface profiles may be developed for a range of discharges. The computations in such studies are more involved and require a series of cross sections downstream from the damsite.

Several methods [1, 2, 8, 9] have been developed for computing water surface profiles; however, this discussion will be limited to Bureau of Reclamation Method A. This method is adaptable to irregular channels having various roughness segments and large variations in cross-section geometry. It is limited to the assumption that flow paths between cross sections are equal in length for all roughness segments. Other methods [8, 9] must be used for variable travel distances.

In Method A, a number of cross sections are selected at intervals below the damsite. The selection of the cross-section locations is probably the most important factor in preparing a good study. The sections should be located so the average area, hydraulic radius, and n values of any two sections will be representative of the reach length between them. Channels that are alternately wide and narrow should have sections located in the narrow and wide points so the average condition can be defined. Special attention should also be given to locating a section at each point that indicates a definite change in grade or a control such as a natural constriction or bridge. The cross sections should be surveyed and segmented according to roughness (n) characteristics. A minimum of five cross sections should be selected, with no more than 2 feet of fall between the sections.

Reference is again made to figure B-17 and the basic hydraulic equations presented in section B-7. Equation (55) can be rearranged to give

$$s_f=\left(\frac{Q}{K_d}\right)^2 \tag{66}$$

in which K_d is the total conveyance capacity of the main channel and overbank areas. Now the discharge in each subdivision of the cross section can be determined as follows:

$$Q=K_d\, s_f^{1/2} \tag{67}$$

$$\text{and } Q_s=K_{d_s}\, s_f^{1/2} \tag{68}$$

where:

Q=total discharge
s_f=slope of the friction gradient
K_d=total conveyance
and Q_s and K_{d_s}=corresponding elements for partial discharge and conveyance capacity of the subdivision under consideration.

Dividing one equation by the other

$$\frac{Q}{Q_s}=\frac{K_d}{K_{d_s}}$$

and solving for Q_s:

$$Q_s=\frac{Q\cdot K_{d_s}}{K_d} \tag{69}$$

The friction head, h_f, is determined by averaging the computed friction slopes at sections 1 and 2 and multiplying by the length as below:

$$h_f=L\left(\frac{s_{f_1}+s_{f_2}}{2}\right) \tag{70}$$

The velocity head, h_v, is derived by a weighting process using the partial discharges occurring in each subdivision of the cross section. Velocities in each segment are computed by the equation $V_s=Q_s/a_s$ where a_s is the area of the segment. The velocity head results from the following equation:

$$h_v=\frac{\Sigma(V_s^2\, Q_s)}{2gQ} \tag{71}$$

It should be noted that this equation gives the same result as equation (60).

Method A deviates slightly at this point from the slope-area computation in that a coefficient of 0.1 is used to determine the eddy losses in contracting reaches. A coefficient of 0.5 is also used here to compute losses in expanding reaches.

The following example shows the procedure required to establish a rating curve by Method A:

(1) Seven cross sections designated by numbers 1 through 7 were selected from figure B-24, and their geometric properties were determined by field survey.
(2) Area and hydraulic radius curves were developed for each cross section, similar to the one shown in figure B-25. It should be noted that a separate curve is drawn for each subsection.
(3) A set of conveyance curves was developed for each section similar to the one shown in figure B-26.
(4) A rating curve for the lowermost cross section, section 1 (fig. B-27), was determined by the approximate method described in section B-8(a) of this appendix.
(5) Water surface profiles were computed for several discharges to define the required tailwater rating curve. The computations shown in table B-9 are for Q=11,100 cubic feet per second.
(6) Section 1 is the starting point of the profile computation. The starting elevation 5714.0 in column 2 was taken from the rating curve in figure B-27. The area values in column 3 were read from the area curve for section 1 (not shown). The distance between sections 1 and 2 was entered in column 4 and the K_d values were read from

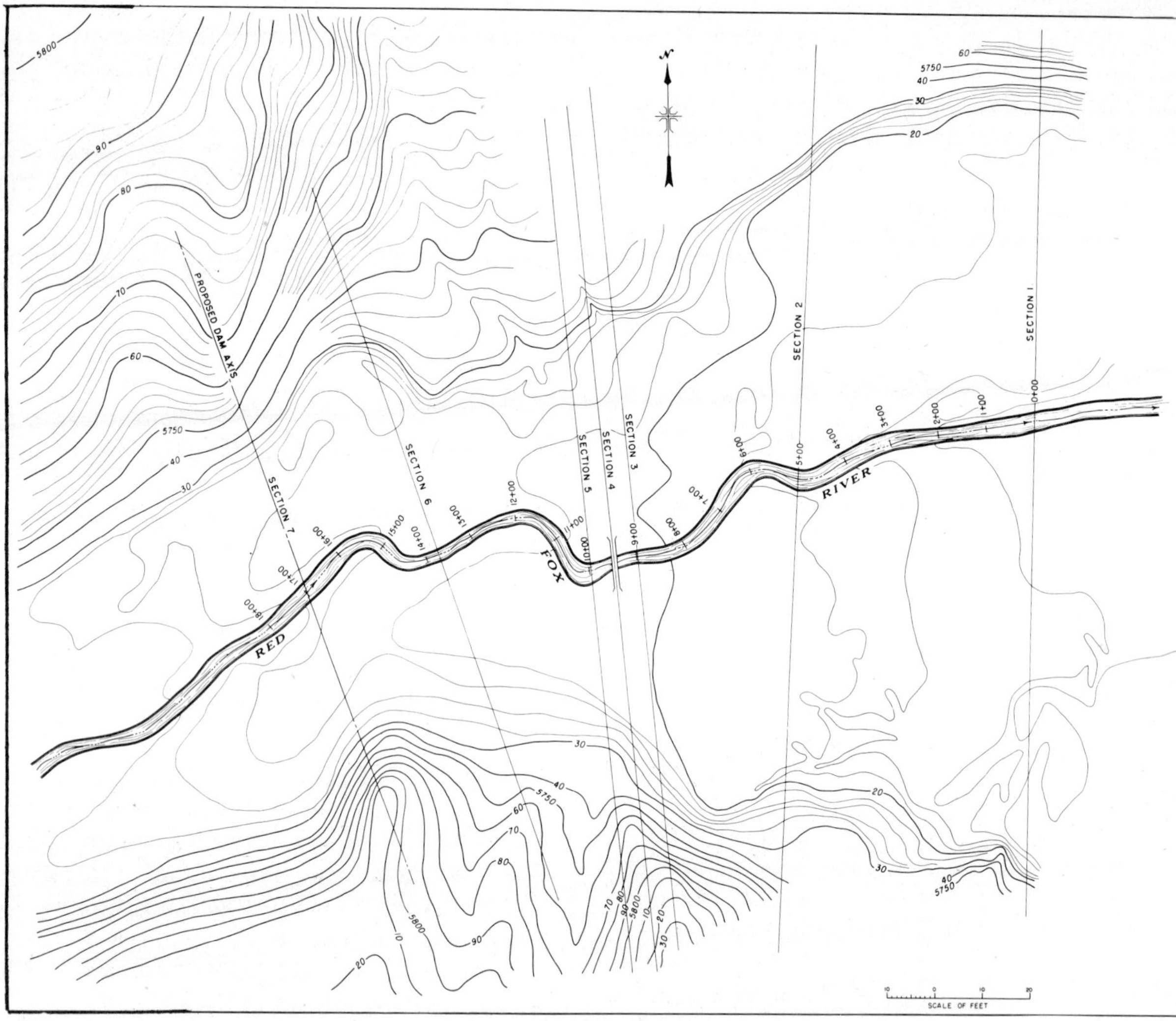

Figure B–24. Cross sections used to establish a rating curve by Method A. 288–D–2833.

the conveyance capacity curve for section 1 (not shown) at elevation 5714.0.

(7) Next, the s_f value was computed for section 1 using equation (66).

(8) Q_s values were computed for each subsection of section 1 using equation (69).

(9) The velocity in each subsection was computed using $V=Q/A$, and the values V^2Q were computed and placed in column 11 and totaled.

(10) The h_v value was next derived for section 1 using equation (71).

(11) The elevation 5714.0 was repeated in column 17.

(12) A trial water surface elevation was then placed in column 2 for section 2 and steps 6 and 7 were repeated based on the data for section 2.

(13) A mean friction slope was computed for the reach by averaging the friction slopes computed for sections 1 and 2.

(14) The friction head loss, h_f, between sections 1 and 2 was then arrived at by multiplying the mean friction slope by the distance L.

(15) Steps 8, 9, and 10 were then repeated for section 2 data.

(16) The algebraic difference in velocity head was then entered in column 13.

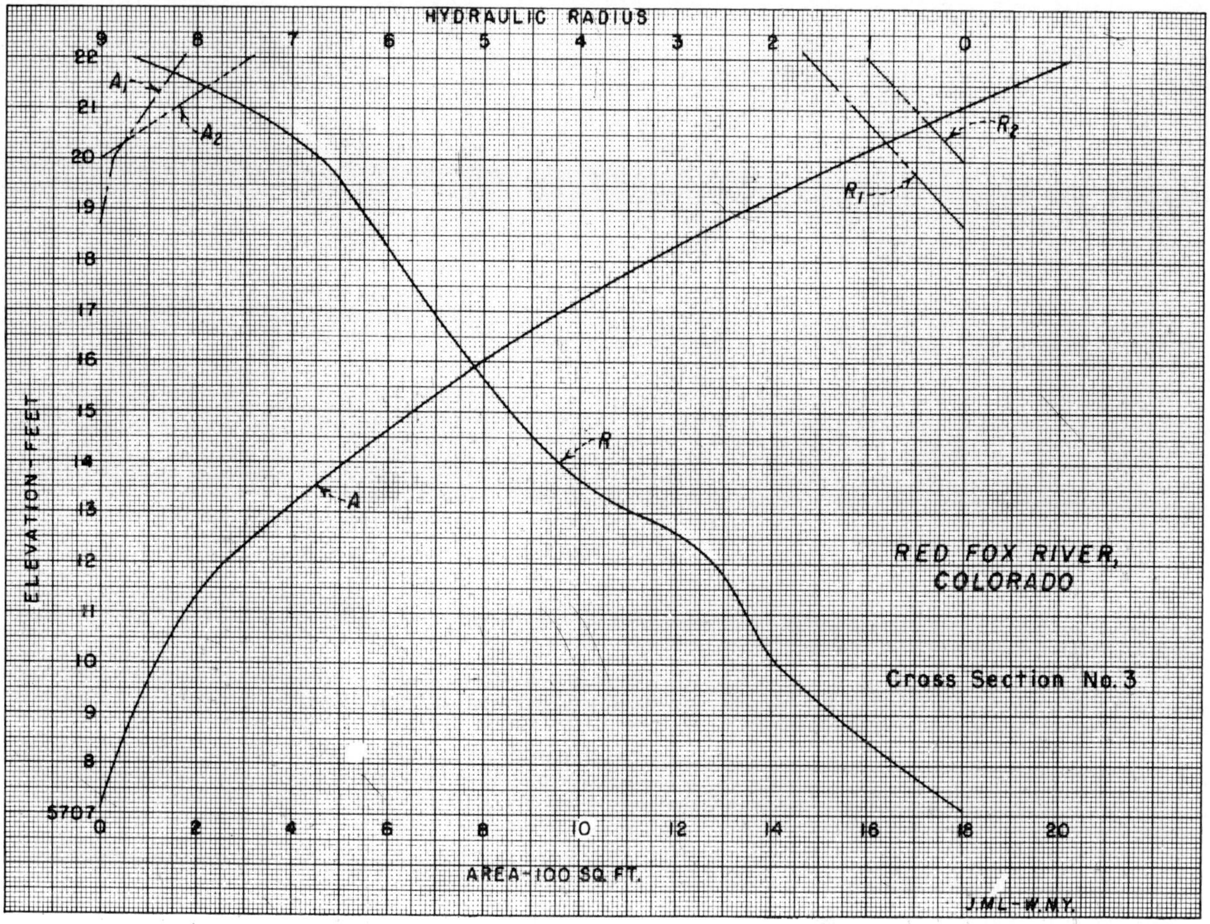

Figure B–25. Area and hydraulic radius curves—Section 3—Red Fox River. 288–D–2834.

The upstream velocity head is subtracted from the downstream value.

(17) The eddy loss is assumed to be 0.1 and 0.5 of the absolute value of the difference in velocity heads for contracting and expanding reaches, respectively. Thus, if the Δh_v is -1.0 foot, the eddy loss is 0.5 foot and if the Δh_v is $+1.0$ foot, the eddy loss is 0.1 foot. The eddy loss value was placed in column 14.

(18) The total loss is equal to the friction head loss plus the eddy loss. The ΔH value is the algebraic sum of the total head loss and the change in velocity head.

(19) The water surface elevation in column 17 is the sum of the water surface elevation at section 1 and the ΔH. If the computed water surface elevation in column 17 agrees with the assumed elevation within a tenth of a foot, the computation is completed and computations are begun for the next upstream section.

When a sufficient number of profiles have been computed, a rating curve can be developed from the computed water surface elevations at any given section. This method is more reliable than the approximate method previously described, because if the first section does not typify the average stream channel hydraulics the variations are recognized as the profiles are continued upstream. This may be proven by changing the starting elevation at the first section and computing a new profile. It will be found that a considerable change in the starting elevation at section 1 will make much

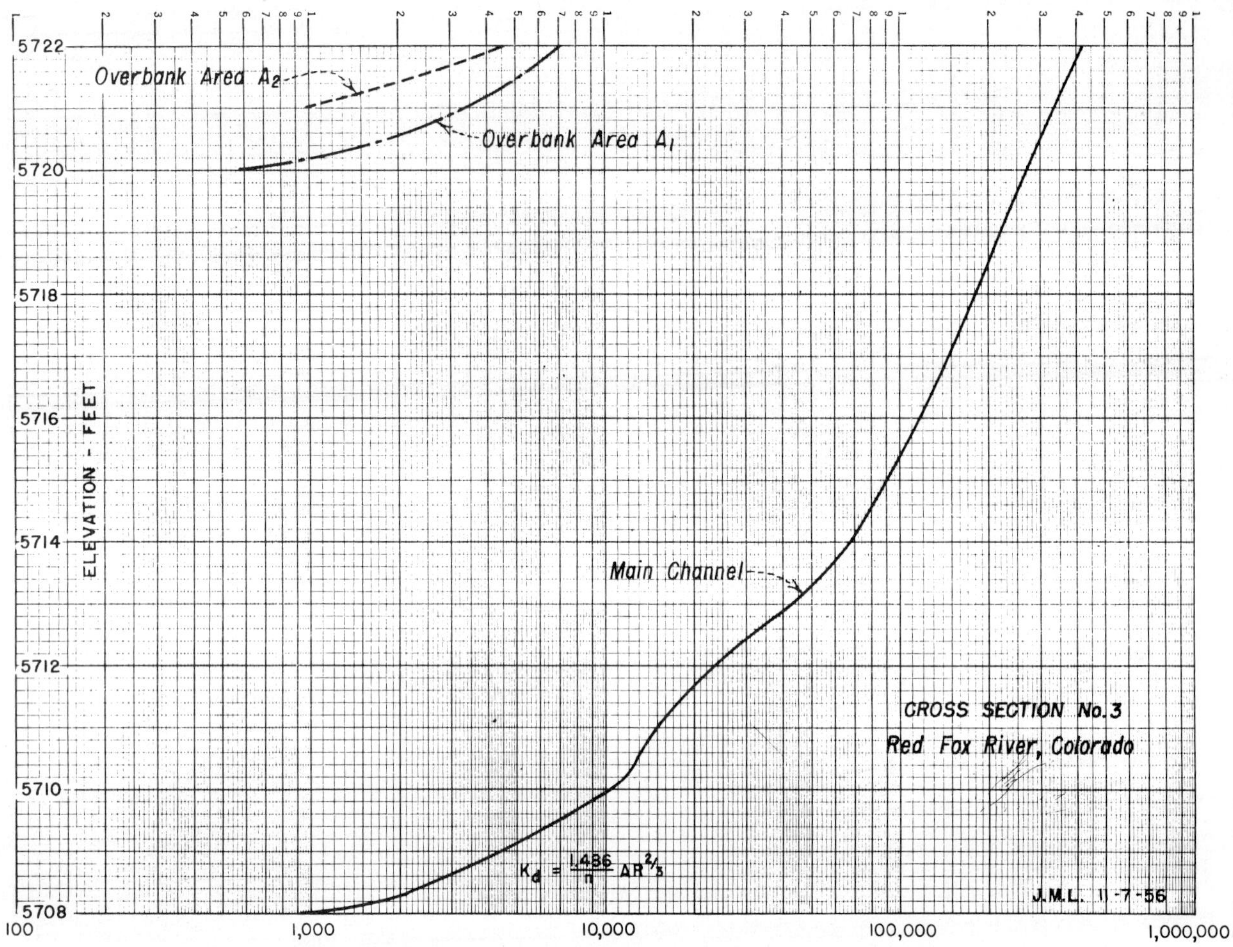

Figure B–26. Conveyance (K_d) curves—Section 3—Red Fox River. 288–D–2835.

less difference in the computed elevation for section 7.

B-9. *Critical Flow.*—The hydraulic analysis of flow in open channels becomes more complex when critical conditions can occur at some point along the river reach under consideration. The conditions of critical flow can be commonly observed at a "control" section in the channel. Such controls occur at locations where there is a material change in the cross section causing a constriction of the flow. These constrictions may be natural, or artificial such as bridges. Another cause may be a significant change in bottom grade. Vortices, eddies, cross currents, and large standing waves are some of the characteristics indicating critical flow conditions. A field reconnaissance of the hydraulic reach under investigation should include the location of any critical sections.

Whenever a computed velocity appears very high, a check should be made to see if the velocity exceeds critical velocity. This is true if the elevation has been arrived at by the approximate method or by an energy balance. Equation (14) of this appendix,

$$V_c = 5.67\sqrt{\frac{a}{T}}, \tag{14}$$

can be used to determine the critical velocity and, when the velocity at the computed elevation exceeds critical, it may be assumed that a control exists. Other methods may be used for checking critical flow but they are not discussed because all are based on different ways of analyzing equation (7)

$$\frac{Q^2}{g} = \frac{a^3}{T} \tag{7}$$

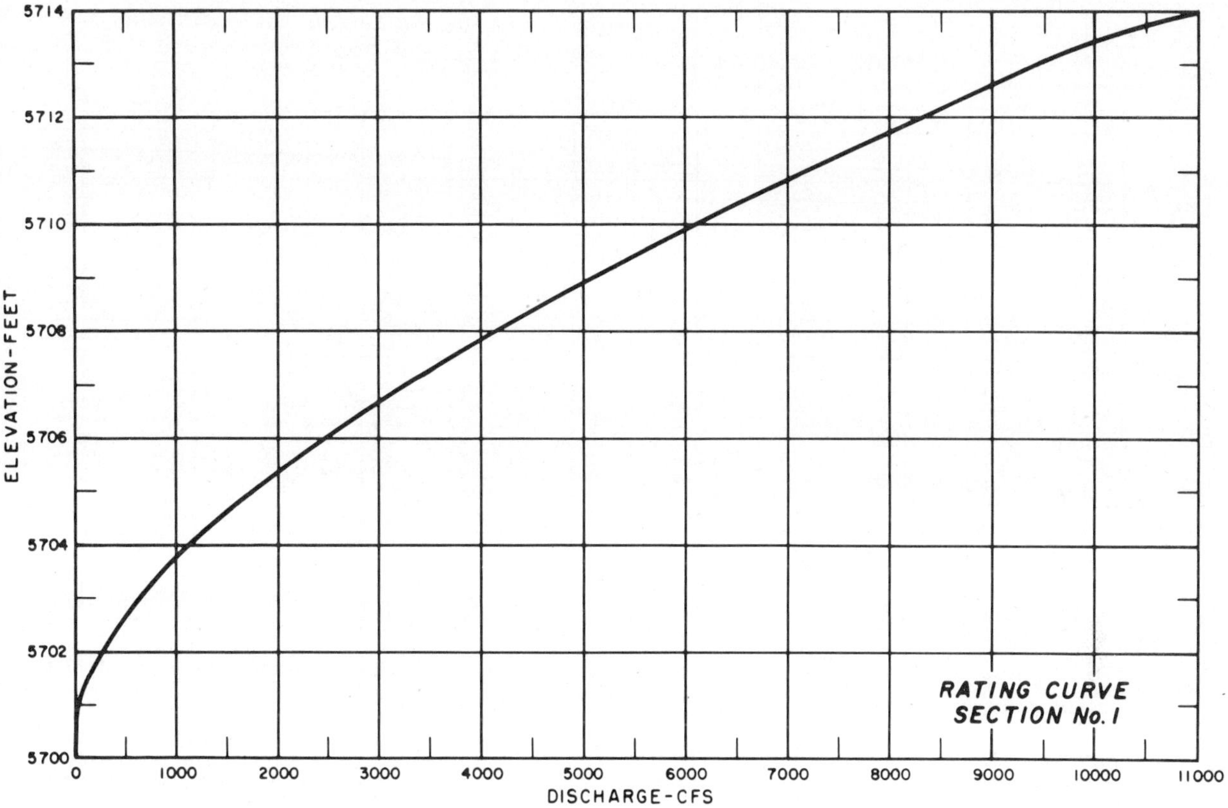

Figure B-27. Rating curve—Section No. 1. 288-D-2836.

When the depth of flow is greater than critical depth throughout the reach under study and a control point is not evident at some relatively close distance downstream from the reach, the computed water surface profiles are acceptable. If a control point is located downstream from the reach, the reach should be extended to the control, and profile computations started at that point. When the control point is encountered within the study reach, the critical depth elevation is determined and the profile computation is started again at the critical discharge elevation. Generally, in most streams, supercritical flow conditions will occur only in the immediate vicinity of a control section.

The above discussion involves the analysis of critical flow as applied to water surface profile computations. It may be required to develop a critical rating curve for a control section which is located at or near a damsite. In this event the critical velocities are computed by equation (14) and multiplied by the area to determine the discharge. This is done for several elevations and a curve of critical flow stage versus discharge plotted.

It is advisable to consider only the main channel portion of subdivided cross sections because the overbank areas will tend to lower the mean cross-section velocity and not depict the velocities attained in the channel itself.

B-10. *Computer Applications.*—The necessary trial-and-error solution of the energy balance equation involving iterative processes is ideally suited to digital computer applications. The Bureau of Reclamation has made use of several computer programs for computing water surface profiles; however, all are based upon an energy balance procedure and differ only in degree of sophistication.

TABLE B-9.—*Water surface profile computations—Method A.* 288-D-2859.

WATER SURFACE PROFILE COMPUTATIONS - METHOD A

SUBJECT: Red Fox River, Colorado Discharge = 11,100 cfs

COMPUTED BY: ______ DATE: ______ CHECKED BY: ______ DATE: ______ SHEET ___ OF ___

1	2	3	4	5	6	7	8	9	10	11	12	13	14	15	16	17
STATION	Assumed Water Surface Elev.	Area a	Dist. L	$K_d = \frac{1.486}{n} ar^{2/3}$	$s_f = \left(\frac{Q}{K_d}\right)^2$	Mean s_f	Mean s_f x L = h_f	Q	V	V^2Q	h_v	$h_{v_1} - h_{v_2}$	Eddy Loss	Total Loss	ΔH	W.S. El.
Section 1	5714.0	594		125,000				10,700	18.0	3,460,000						
		192		4,150				400	2.08	1,730						
				129,150	0.00735			11,100		3,461,730	4.85					5714.0
			500													
Section 2	5720.9	863		230,000				5,960	6.80	276,000						
		1,170		130,000				3,370	2.88	28,000						
		1,000		25,000				650	0.65	274						
		1,500		43,000				1,120	0.75	630						
				428,000	0.000673	0.00401	2.00	11,100		304,904	0.43	+4.42	0.44	2.44	6.86	5720.86
			400													
Section 3	5721.2	1,832		360,000				11,000	6.00	396,000						
		115		2,700				70	0.61	26						
		185		1,250				30	0.16	8						
				363,950	0.00108	0.000877	0.35	11,100		396,034	0.56	-0.13	0.07	0.42	0.29	5721.15
			100													
Section 4	5721.2	1,300		158,000	0.00493	0.00301	0.30	11,100	8.54		1.13	-0.57	0.27	0.57	0.00	5721.20
			100													
Section 5	5722.1	1,681		220,000	0.00254	0.00374	0.37	11,100	6.60		0.68	+0.45	0.05	0.42	0.87	5722.07
			400													
Section 6	5722.8	1,018		174,000				10,900	10.7	1,250,000						
		130		3,380				190	1.46	405						
		20		161				10	0.50	3						
				177,541	0.00391	0.00328	1.31	11,100		1,250,408	1.75	-1.07	0.54	1.85	0.78	5722.85
			500													
Section 7	5725.3	844		124,000				10,500	12.5	1,640,000						
		98		1,400				110	1.12	138						
		220		5,800				490	2.23	2,440						
				131,200	0.00714	0.00553	2.77	11,100		1,642,578	2.30	-0.55	0.28	3.05	2.50	5725.35

C. BIBLIOGRAPHY

B-11. *Bibliography.*—

[1] King, H. W., and Brater, E. F., "Handbook of Hydraulics," fifth edition, McGraw-Hill Book Co., Inc., New York, N.Y., 1963.

[2] Woodward, S. B., and Posey, C. J., "Steady Flow in Open Channels," John Wiley & Sons, Inc., fourth printing, September 1949.

[3] Bakhmeteff, B. A., "Hydraulics of Open Channels," McGraw-Hill Book Co., Inc., New York, N.Y., 1932.

[4] Binder, R. C., "Fluid Mechanics," third edition, Prentice-Hall, Inc., Englewood Cliffs, N.J., 1955.

[5] Rouse, Hunter, "Engineering Hydraulics," John Wiley & Sons, Inc., New York, N.Y., 1950.

[6] "Friction Factors for Large Conduits Flowing Full," Engineering Monograph No. 7, Bureau of Reclamation, 1965.

[7] Bradley, J. N., and Peterka, A. J., "The Hydraulic Design of Stilling Basins," ASCE Proceedings, vol. 83, October 1957, Journal of Hydraulics Division, No. HY 5, Papers No. 1401 to 1404, inclusive.

[8] "Guide for Computing Water Surface Profiles," Bureau of Reclamation, November 1957.

[9] "Hydrologic and Hydraulic Analyses—Computation of Backwater Curves in River Channels," Engineering Manual, Department of the Army, Corps of Engineers, pt. CXIV, ch. 9, May 1952.

[10] "Stream Gaging Procedure," Water Supply Paper 888, U.S. Department of the Interior, Geological Survey, 1945.

[11] Ramser, C. E., "Flow of Water in Drainage Channels," Technical Bulletin No. 129, U.S. Department of Agriculture, 1929.

[12] Dalrymple, Tate, and Benson, M. A., "Measurement of Peak Discharge by the Slope-Area Method," Techniques of Water Resources Investigations of the U.S. Geological Survey—ch. A2, U.S. Department of the Interior, Geological Survey, 1967.

[13] "Water Measurement Manual," Bureau of Reclamation, second edition, 1967.

Structural Design Data

C. J. HOFFMAN [1] and P. K. BOCK [2]

C-1. ***Introduction.***—This appendix presents structural design data peculiar to hydraulic structures for the design of concrete appurtenances to small dams. It is presumed that the user of this text is familiar with reinforced concrete design or that he will consult other texts for information on this subject. A major portion of this appendix is concerned with the design of reinforced concrete conduits for use as spillways or outlet works under or through small earthfill dams.

C-2. ***Earth Pressures on Retaining Walls.***—Figure C–1 presents a method for obtaining active earth loads on retaining walls when the properties of the fill material behind the wall are known. The curves are based on Coulomb's theory of active earth pressure against retaining walls [1].[3] In applying Coulomb's theory, the angle of friction between the earth and the back of the wall has been assumed equal to zero. Detailed discussions of methods available for determining earth pressure and designing retaining walls can be found in many texts [1, 2, 3, 4].

C-3. ***Allowable Bearing Values for Structure Footings.***—Table C-1 gives suggested allowable bearing values for footings of structures appurtenant to small dams. These values are based on an evaluation of several sources of published data [1, 5, 6], in the light of the problems peculiar to hydraulic structures. The allowable bearing values listed for foundations of soils are smaller than are generally given in building codes and, with the exception of the gravels, vary according to the relative density and relative consistency of cohesionless and cohesive soils, respectively, rather than with the soil classifications group.

C-4. ***Precast Concrete Pipe Conduits.***—(a) *General.*—When precast concrete pipes are used for outlet works conduits (or for spillways) under or through earthfill dams, they should be bedded in a concrete base as shown on figure C–2. The primary purpose of the concrete base is to prevent percolation along the underside of the pipe where tightly compacted earth bedding is difficult to obtain. For this reason the concrete base should be placed concurrently with or after the pipe is in position.

Cutoff collars should be provided to prevent percolation along the pipe. These collars should be identical to those used with cast-in-place concrete conduits, as discussed in section 242(b).

Precast concrete pipe joints should have rubber gaskets, as shown in figure 266. The joints should not be filled with cement mortar but should be filled with a preformed plastic sealing compound conforming to Interim Federal Specification SS-S-00210, so that greater flexibility will be provided should settlements due to superimposed embankment loads occur.

(b) *Design of Precast Concrete Pipe.*—When precast concrete pipe is used for outlet works, it should be designed for internal pressure, super-imposed embankment loads, and a combination of the two as conditions require. The embankment loads should be computed in accordance with Marston's theory, as described in section 242(d). The precast concrete pipe should be assumed as a rigid conduit. In determining embankment loads for design purposes, the projection condition is most likely to govern. The various conditions of densities of material and settlements in foundation and em-

[1] Engineer, Spillways and Outlet Works Section, Bureau of Reclamation (Retired).

[2] Engineer, Operations Branch, Bureau of Reclamation.

[3] Numbers in brackets refer to items in the bibliography, section C-6.

CASE (a):

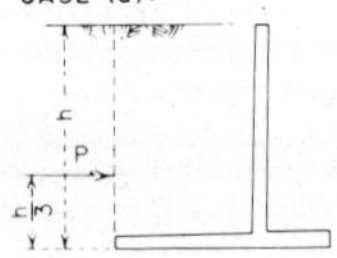

$$P = K\,\frac{wh^2}{2}$$

where K is obtained from chart below, for β is 0 and w is the weight of backfill in lbs per cubic foot.

CASE (d):

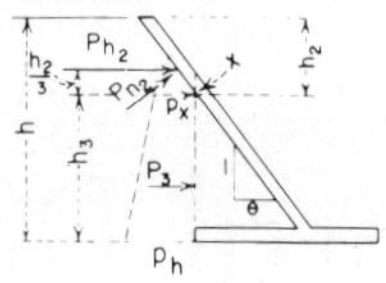

P_{h_2} and P_{n_2} computed above point x as in case (c)

$$P_3 = \left(\frac{p_x + p_h}{2}\right) h_3 \text{ where}$$

$p_x = K\ wh_2$ and $p_h = K\ wh$

P_3 is applied at $\frac{h_3}{3}\left(\frac{2p_x + p_h}{p_x + p_h}\right)$ above base

CASE (b):

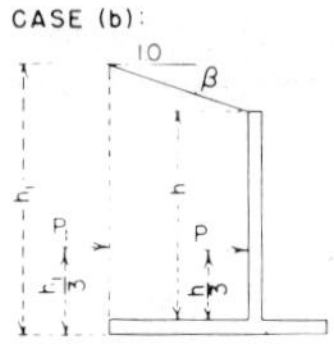

$$P = K\,\frac{wh^2}{2}$$

$$P_1 = K\,\frac{wh_1^2}{2}$$

where K is obtained from chart below, for $\beta > 0$

CASE (e):

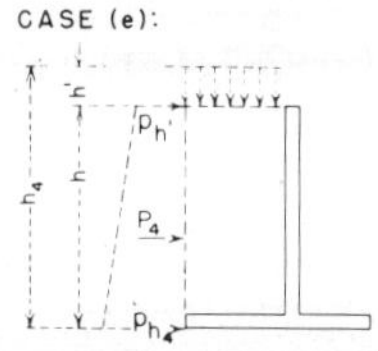

For uniform surcharge, convert to equivalent height of fill h'

Then $P_4 = \left(\frac{p_{h_4} + p_{h'}}{2}\right) h$ where

$p_{h_4} = K\ wh_4$ and $p_{h'} = K\ wh'$

P_4 is applied at $\frac{h}{3}\left(\frac{2p_{h'} + p_{h_4}}{p_{h'} + p_{h_4}}\right)$ above base.

CASE (c):

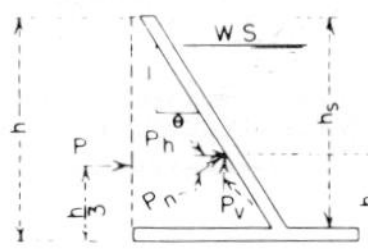

P same as for case (a)

$$P_n = K\,\frac{wh^2}{2}$$

where K is obtained from chart below, for appropriate value of θ

$$P_h = \frac{P_n}{\sqrt{1+\theta^2}}$$

$$P_v = \frac{\theta P_n}{\sqrt{1+\theta^2}}$$

If stem is subjected to hydrostatic load on the channel side, assume fill has shrunk away from wall and check stem design for hydrostatic load only, using allowable steel stress of $1.4\ f_s$

NOTE: For fully saturated pervious backfill or backfill retaining water:

Vertical unit weight = unit weight of dry fill plus 62.5 lbs. per cu. ft. x percent of voids in the fill.

Horizontal unit pressure p = full hydrostatic pressure plus 0.4 of dry equivalent fluid pressure.

EARTH PRESSURES ON RETAINING WALLS

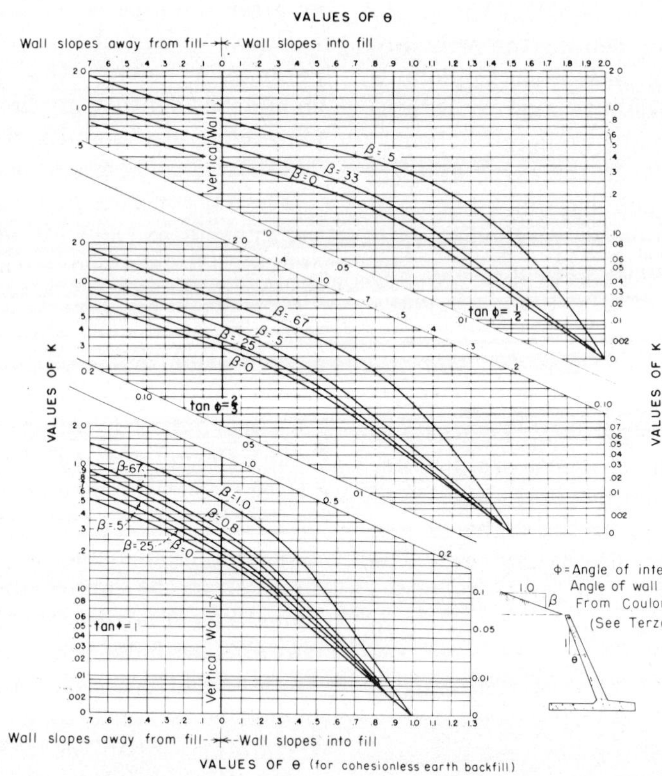

φ = Angle of internal friction
Angle of wall friction = 0
From Coulomb's equations
(See Terzaghi - "Theoretical Soil Mechanics" p. 80)

Active pressure coefficients for earthfill loads on walls

Figure C–1. Earth pressures on retaining walls. 288–D–2892.

TABLE C-1.—*Suggested allowable bearing values for footings of structures appurtenant to small dams*

Material	Condition, relative density, or relative consistency	Average standard penetration values in fine-grained soils: Effective overburden pressure, pounds per square inch	Average standard penetration values in fine-grained soils: Number of blows per foot	Allowable bearing pressure, tons per square foot
Massive igneous metamorphic or sedimentary hard rock	Sound (minor cracks allowed)			100
Hard laminated rock such as slate	Sound (minor cracks allowed)			35
Residual deposits of bedrock of any kind except shale. (Unsound shale is treated as clay.)	Shattered or broken			10
Gravel (GW, GP, GM, GC)				4
Cohesionless sands (SW, SP)	Loose	0 20 40	4 12 17	(2)
	Medium	0 20 40	4 to 8 12 to 24 17 to 40	1
	Dense	0 20 40	8 24 40	2
Saturated [1] sands, silts, and clays (SM, SC, ML, CL, MH, CH)	Soft		4	0.25
	Medium		4 to 10	0.5
	Stiff		11 to 20	1.0
	Hard		20	1.5

[1] Values are for foundations that are almost or completely saturated during the construction period. Bearing values can be increased by one-third if the foundation is relatively dry, provided that the criteria in fig. 142 for "no treatment required" are met.

[2] Requires compaction.

bankment must be given due consideration. It is well to remember that even where natural ground and embankment have equal densities, the settlements in constructed embankment will be greater than in the natural foundation.

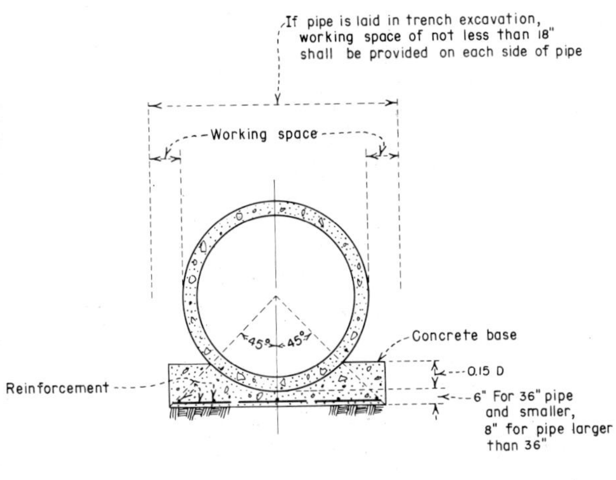

Figure C-2. Precast concrete pipe on concrete base for conduit under or through earthfill dams. 288-D-2914.

Precast concrete pipe shells and reinforcement should be designed in accordance with the formulas given below and using the ultimate strength design methods [11], with a load factor of 1.8 and without consideration of compressive reinforcement. In order to insure watertightness, a hypothetical case for bursting due to hydrostatic head only should be calculated and the unit stress in the reinforcement should not exceed 16,000 pounds per square inch for hydrostatic heads up to 50 feet. The formulas for the moments, shears, and thrust are based on bulb-like distribution of earth pressures [7]. Figure C-3 shows the location of the critical design sections and the notations. Dimensions r, r_i, t, and H in the

TABLE C–2.—*Reinforcement and wall thicknesses for reinforced concrete pressure pipe for 12-inch through 108-inch diameter*

	Minimum circumferential reinforcement in square inches per linear foot of pipe											
Internal diameter of pipe in inches	12		15		18				21			
Type of reinforcement	Circular		Circular		Circular		Elliptical		Circular		Elliptical	
Nominal wall thickness, inches	2	3	2	3	2¼	3	2¼	3	2⅜	3	2⅜	3
Layers of reinforcement	Single	Single	Single	Single	Single	Single	Single	Single	Single	Single	Single	Single
CLASS												
A-25	0.07	0.06	0.10	0.08	0.12	0.11	0.12	0.12	0.15	0.13	0.14	0.14
B-25	.10	.08	.14	.11	.18	.15	.16	.12	.23	.19	.19	.14
C-25	.13	.09	.19	.14	.25	.19	.21	.14	.32	.26	.27	.19
D-25	.16	.11	.25	.17	.32	.24	.28	.17	.42	.33	.35	.23

	Minimum circumferential reinforcement in square inches per linear foot of pipe											
Internal diameter of pipe in inches	24				27							
Type of reinforcement	Circular		Elliptical		Circular						Elliptical	
Nominal wall thickness, inches	2½	3	2½	3	2⅝	3⅛	3¼		4¼		2⅝	3¼
Layers of reinforcement	Single	Single	Single	Single	Single	Single	Inner	Outer	Inner	Outer	Single	Single
CLASS												
A-25	0.18	0.16	0.16	0.16	0.21	0.19	0.13	0.09	0.11	0.07	0.18	0.18
B-25	.29	.25	.23	.18	.35	.30	.20	.11	.15	.08	.27	.20
C-25	.40	.33	.32	.24	.49	.41	.27	.14	.20	.09	.37	.27
D-25	.54	.43	.41	.30	....	.54	.34	.16	.24	.10	.49	.34

	Minimum circumferential reinforcement in square inches per linear foot of pipe									
Internal diameter of pipe in inches	30									
Type of reinforcement	Circular								Elliptical	
Nominal wall thickness, inches	2¾	3⅛	3¼		3½		4¾		2¾	3½
Layers of reinforcement	Single	Single	Inner	Outer	Inner	Outer	Inner	Outer	Single	Single
CLASS										
A-25	0.24	0.23	0.16	0.10	0.15	0.10	0.12	0.08	0.20	0.20
B-25	.41	.37	.25	.14	.22	.13	.17	.09	.31	.22
C-25	.60	.51	.33	.17	.30	.15	.21	.10	.43	.30
D-25	....	.69	.43	.21	.38	.19	.26	.11	.59	.38

	Minimum circumferential reinforcement in square inches per linear foot of pipe									
Internal diameter of pipe in inches	33									
Type of reinforcement	Circular								Elliptical	
Nominal wall thickness, inches	2⅞	3⅛	3¼		3¾		4¾		2⅞	3¾
Layers of reinforcement	Single	Single	Inner	Outer	Inner	Outer	Inner	Outer	Single	Single
CLASS										
A-25	0.28	0.26	0.18	0.12	0.16	0.11	0.13	0.09	0.22	0.22
B-25	.48	.44	.29	.17	.25	.14	.19	.10	.35	.25
C-25	....	.64	.40	.22	.33	.17	.25	.12	.50	.33
D-25	....	....	.53	.27	.43	.21	.32	.14	.67	.43

TABLE C–2.—*Reinforcement and wall thicknesses for reinforced concrete pressure pipe for 12-inch through 108-inch diameter*—Continued

	Minimum circumferential reinforcement in square inches per linear foot of pipe								
Internal diameter of pipe in inches	36								
Type of reinforcement	Circular							Elliptical	
Nominal wall thickness, inches	3⅛	3¼		4		5		3⅛	4
Layers of reinforcement	Single	Inner	Outer	Inner	Outer	Inner	Outer	Single	Single
CLASS									
A-25	0.31	0.21	0.14	0.17	0.12	0.15	0.10	0.24	0.24
B-25	.53	.34	.20	.27	.15	.22	.12	.36	.27
C-25	....	.49	.26	.37	.19	.28	.14	.52	.37
D-25	....	.64	.33	.46	.23	.35	.16	.69	.46

	Minimum circumferential reinforcement in square inches per linear foot of pipe																
Internal diameter of pipe in inches	39								42								
Type of reinforcement	Circular						Elliptical		Circular						Elliptical		
Nominal wall thickness, inches	3½		4¼		5¼		3½	4¼	3¾		4½		5½		3¾	4½
Layers of reinforcement	Inner	Outer	Inner	Outer	Inner	Outer	Single	Single	Inner	Outer	Inner	Outer	Inner	Outer	Single	Single
CLASS																
A-25	0.22	0.15	0.19	0.13	0.16	0.11	0.26	0.26	0.24	0.16	0.20	0.14	0.17	0.12	0.28	0.28
B-25	.37	.22	.29	.17	.24	.13	.37	.29	.40	.23	.32	.18	.26	.14	.40	.32
C-25	.52	.28	.40	.21	.31	.15	.52	.40	.54	.30	.43	.22	.34	.17	.54	.43
D-25	.67	.35	.50	.25	.39	.18	.67	.50	.73	.38	.55	.27	.43	.20	.73	.55

	Minimum circumferential reinforcement in square inches per linear foot of pipe															
Internal diameter of pipe in inches	45								48							
Type of reinforcement	Circular						Elliptical		Circular						Elliptical	
Nominal wall thickness, inches	3⅞		4¾		5¾		3⅞	4¾	4⅛		5		5¾		4⅛	5
Layers of reinforcement	Inner	Outer	Inner	Outer	Inner	Outer	Single	Single	Inner	Outer	Inner	Outer	Inner	Outer	Single	Single
CLASS																
A-25	0.26	0.18	0.22	0.15	0.19	.13	0.30	0.30	0.28	0.19	0.24	0.16	0.21	0.14	0.31	0.31
B-25	.43	.25	.34	.19	.28	.15	.43	.34	.45	.27	.36	.21	.32	.18	.45	.36
C-25	.60	.33	.46	.24	.37	.19	.60	.46	.66	.36	.51	.27	.44	.22	.66	.51
D-25	.80	.42	.59	.30	.47	.22	.80	.59	.86	.45	.65	.33	.54	.26	.80	.65

	Minimum circumferential reinforcement in square inches per linear foot of pipe															
Internal diameter of pipe in inches	51								54							
Type of reinforcement	Circular						Elliptical		Circular						Elliptical	
Nominal wall thickness, inches	4¼		5¼		6		4¼	5¼	4½		5½		6¼		4½	5½
Layers of reinforcement	Inner	Outer	Inner	Outer	Inner	Outer	Single	Single	Inner	Outer	Inner	Outer	Inner	Outer	Single	Single
CLASS																
A-25	0.30	0.20	0.25	0.17	0.23	0.15	0.33	0.33	0.32	0.21	0.27	0.18	0.25	0.16	0.35	0.35
B-25	.49	.29	.38	.22	.34	.19	.49	.38	.50	.30	.40	.23	.36	.20	.50	.40
C-25	.72	.40	.54	.29	.47	.24	.72	.54	.75	.41	.58	.31	.50	.26	.75	.58
D-25	.97	.50	.69	.35	.58	.28	......	.69	1.00	.52	.74	.37	.63	.30	......	.74

TABLE C-2.—*Reinforcement and wall thicknesses for reinforced concrete pressure pipe for 12-inch through 108-inch diameter*—Continued

	Minimum circumferential reinforcement in square inches per linear foot of pipe															
Internal diameter of pipe in inches.	57								60							
Type of reinforcement	Circular						Elliptical		Circular						Elliptical	
Nominal wall thickness, inches	4¾		5¾		6½		4¾	5¾	5		6		6¾		5	6
Layers of reinforcement	Inner	Outer	Inner	Outer	Inner	Outer	Single	Single	Inner	Outer	Inner	Outer	Inner	Outer	Single	Single
CLASS																
A-25	0.34	0.22	0.29	0.19	0.26	0.17	0.37	0.37	0.35	0.23	0.30	0.20	0.28	0.18	0.39	0.39
B-25	.53	.32	.43	.25	.38	.22	.53	.43	.55	.33	.45	.26	.41	.23	.55	.45
C-25	.78	.43	.61	.33	.53	.28	.78	.61	.82	.45	.65	.35	.56	.29	.82	.65
D-25	1.03	.55	.78	.40	.67	.33	......	.78	1.09	.57	.82	.43	.71	.35	......	.82

	Minimum circumferential reinforcement in square inches per linear foot of pipe													
Internal diameter of pipe in inches.	63								66					
Type of reinforcement	Circular						Elliptical		Circular				Elliptical	
Nominal wall thickness, inches	5¼		6¼		7		5¼	6¼	5½		6½		5½	6½
Layers of reinforcement	Inner	Outer	Inner	Outer	Inner	Outer	Single	Single	Inner	Outer	Inner	Outer	Single	Single
CLASS														
A-25	0.37	0.25	0.32	0.22	0.30	0.20	0.41	0.41	0.39	0.26	0.34	0.23	0.43	0.43
B-25	.58	.35	.48	.28	.43	.25	.58	.48	.60	.36	.50	.30	.60	.50
C-25	.87	.48	.69	.37	.60	.31	.87	.69	.89	.49	.72	.38	.89	.72
D-25	1.15	.60	.88	.46	.76	.38	....	.88	1.19	.63	.94	.48	....	.94

	Minimum circumferential reinforcement in square inches per linear foot of pipe											
Internal diameter of pipe in inches	69						72					
Type of reinforcement	Circular				Elliptical		Circular				Elliptical	
Nominal wall thickness, inches	5¾		6¾		5¾	6¾	6		7		6	7
Layers of reinforcement	Inner	Outer	Inner	Outer	Single	Single	Inner	Outer	Inner	Outer	Single	Single
CLASS												
A-25	0.41	0.28	0.36	0.24	0.45	0.45	0.46	0.30	0.40	0.26	0.48	0.47
B-25	.62	.37	.53	.31	.63	.53	.69	.41	.58	.34	.69	.58
C-25	.91	.51	.74	.40	.91	.74	1.01	.56	.81	.44	1.01	.81
D-25	1.23	.66	.98	.50	....	.98	1.39	.73	1.08	.56	....	1.08

	Minimum circumferential reinforcement in square inches per linear foot of pipe											
Internal diameter of pipe in inches	78				84				90			
Type of reinforcement	Circular				Circular				Circular			
Nominal wall thickness, inches	6½		7½		7		8		7½		8	
Layers of reinforcement	Inner	Outer	Inner	Outer	Inner	Outer	Inner	Outer	Inner	Outer	Inner	Outer
CLASS												
A-25	0.49	0.33	0.44	0.29	0.53	0.35	0.48	0.32	0.57	0.38	0.54	0.36
B-25	.73	.44	.63	.37	.78	.47	.68	.40	.83	.50	.77	.46
C-25	1.06	.59	.87	.48	1.10	.62	.94	.51	1.16	.65	1.07	.59
D-25	1.47	.78	1.17	.61	1.54	.82	1.26	.66	1.61	.85	1.44	.77

TABLE C–2.—*Reinforcement and wall thicknesses for reinforced concrete pressure pipe for 12-inch through 108-inch diameter*—Continued

	Minimum circumferential reinforcement in square inches per linear foot of pipe											
Internal diameter of pipe in inches	96				102				108			
Type of reinforcement	Circular				Circular				Circular			
Nominal wall thickness, inches	8		8½		8½		9		9		9½	
Layers of reinforcement	Inner	Outer	Inner	Outer	Inner	Outer	Inner	Outer	Inner	Outer	Inner	Outer
CLASS												
A-25	0.62	0.41	0.59	0.39	0.67	0.44	0.64	0.42	0.71	0.47	0.68	0.45
B-25	.88	.53	.83	.50	.94	.56	.88	.53	.99	.60	.94	.56
C-25	1.23	.69	1.13	.64	1.29	.73	1.19	.67	1.35	.77	1.27	.71
D-25	1.67	.90	1.51	.81	1.73	.94	1.60	.85	1.79	.98	1.67	.90

NOTE: Designations A, B, C, and D for class of pipe denote 5, 10, 15 and 20 feet of cover, respectively. The number "25" for class of pipe denotes design hydrostatic pressure head in feet measured to centerline of pipe.

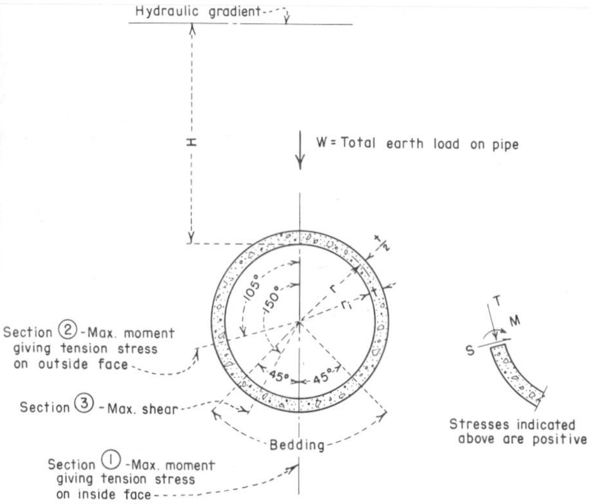

Figure C–3. Location of critical sections in design of precast concrete pressure pipe. 288–D–2915.

formulas are in feet; W=total earth load on the pipe in pounds per linear foot, M=moment in foot-pounds, S=shear in pounds, and T=thrust in pounds.

Section 1: $M=-115r^2t-24rr_i^2-0.126r\ W$

$T=+195rt-53r_i^2+0.324\ W$

Section 2: $M=+83r^2t+17rr_i^2+0.089r\ W$

$T=+280rt-12r_i^2+0.539\ W$

Section 3: $S=-244rt-51r_i^2-0.273\ W$

T due to hydrostatic pressure= $-62.4r_i\,H$

Reinforcing steel: Reinforcement should consist of steel wire conforming to ASTM Designation: A82; of steel bars or rods conforming to Federal Specification QQ-S-632b, Type I or II, Class B40; ASTM designation: A615, Grade 40; or welded wire fabric conforming to Federal Specification RR-W-375 or to ASTM Designation: A185. The ultimate tensile strength of the reinforcement is 40,000 pounds per square inch.

Concrete pipe should be manufactured in accordance with the Bureau of Reclamation publication, "Standard Specifications for Reinforced Concrete Pressure Pipe, February 1, 1969" [8]. The pipe shell thickness selected should be in accordance with these specifications since most manufacturers have pipe forms for the thickness specified therein. The concrete compressive strength used in design is 4,500 pounds per square inch. Should design requirements be such as to indicate the need for shell thicknesses greater than available for precast concrete pipe, it would be desirable to use cast-in-place type of construction.

Table C–2 gives the minimum circumferential reinforcement and nominal wall thicknesses for reinforced concrete pressure pipe from 12 inches to 108 inches in diameter and for overfills of up to 20 feet above the top of the pipe. In each case, the internal pressure is measured to the centerline of the pipe. Although this table was developed for the design of precast concrete pressure pipelines, it may be used for the design of precast concrete pipe conduits under earthfill dams.

TABLE C–3.—*Standardized design for cast-in-place conduits with external loads only*

Conduit diameter, feet	Height of fill above top of conduit, feet	Required conduit wall thickness, inches	Transverse reinforcement: Hoop reinforcement per foot of conduit length	Transverse reinforcement: Additional reinforcement required per foot of length throughout — Sec. 1	Sec. 2	Sec. 3	Longitudinal reinforcement spaced at 18 inches ±
10	50	15	1 No. 6 e.f.	1 No. 9 i.f.	1 No. 6 o.f.	1 No. 11 i.f.	No. 7 e.f.
	40	14	1 No. 6 e.f.	1 No. 8 i.f.	1 No. 6 o.f.	1 No. 10 i.f.	No. 6 e.f.
	30	12	1 No. 6 e.f.	1 No. 8 i.f.	1 No. 6 o.f.	1 No. 10 i.f.	No. 5 e.f.
	20	10	1 No. 6 e.f.	1 No. 7 i.f.	1 No. 6 o.f.	1 No. 9 i.f.	No. 5 e.f.
8	50	12	1 No. 6 e.f.	1 No. 8 i.f.	1 No. 5 o.f.	1 No. 10 i.f.	No. 6 e.f.
	40	11	1 No. 6 e.f.	1 No. 7 i.f.	1 No. 5 o.f.	1 No. 9 i.f.	No. 6 e.f.
	30	10	1 No. 6 e.f.	1 No. 7 i.f.	1 No. 5 o.f.	1 No. 8 i.f.	No. 5 e.f.
	20	9	1 No. 5 e.f.	1 No. 6 i.f.	1 No. 5 o.f.	1 No. 8 i.f.	No. 5 e.f.
6	50	10	1 No. 5 e.f.	1 No. 7 i.f.	1 No. 5 o.f.	1 No. 8 i.f.	No. 6 e.f.
	40	9	1 No. 5 e.f.	1 No. 7 i.f.	1 No. 5 o.f.	1 No. 8 i.f.	No. 5 e.f.
	30	8	1 No. 5 c.s.	1 No. 8 c.s.	1 No. 6 c.s.	1 No. 9 c.s.	No. 5 c.s.
	20	7	1 No. 5 c.s.	1 No. 6 c.s.	1 No. 5 c.s.	1 No. 8 c.s.	No. 5 c.s.
4	50	8	1 No. 5 c.s.	1 No. 7 c.s.	1 No. 5 c.s.	1 No. 7 c.s.	No. 5 c.s.
	40	7	1 No. 5 c.s.	1 No. 6 c.s.	1 No. 5 c.s.	1 No. 7 c.s.	No. 5 c.s.
	30	6	1 No. 5 c.s.	1 No. 6 c.s.	...	1 No. 6 c.s.	No. 5 c.s.
	20	6	1 No. 5 c.s.	1 No. 5 c.s.	...	1 No. 5 c.s.	No. 5 c.s.

NOTE: i.f. denotes inside face, o.f. denotes outside face, and e.f. denotes each face for conduits having 2 layers of reinforcement. c.s. denotes center of section for conduits having 1 layer of reinforcement.

C-5. *Cast-in-Place Concrete Conduits*.—This section presents wall thickness and required reinforcement for cast-in-place circular conduits with flat bases. The details are for conduits with inside diameters ranging from 4 to 10 feet. The conduits have been designed for loads resulting from fills whose surfaces are from 20 to 50 feet above the top of the conduits, and for the interior hydrostatic pressures which could exist in conduits in combination with fill loads. Details for intermediate conduit diameters and fill heights not shown in the tables can be obtained by interpolation. The thickness of the conduit wall should be interpolated between values shown in the tables by assuming a straight line variation for both fill height and diameter. The reinforcement shown for the next larger conduit diameter in combination with the next higher fill loading should be used.

The conduits have been designed and tables prepared for two conditions of loading. Details given in table C–3 are based on the assumption of no internal pressures inside the conduits, and the conduits have been designed to resist fill loads only. Details given in table C–4 are based on the assumption that hydrostatic pressure will exist inside the conduits, and the conduits have been designed to resist both internal pressure and external fill loads. The internal hydrostatic pressure was assumed to be uniform and to be produced by a head of water equal to the height of fill above the centerline of the conduit.

Each condition of loading was investigated for two assumed exterior-fill pressure distributions. The "normal condition" of fill pressure distribution considers a uniform vertical load on the top of the conduit and a uniform foundation reaction on the bottom, in combination with a uniform horizontal load equal to one-third of the vertical load. The vertical pressure was assumed to be equal to the height of fill above the centerline of the conduit, multiplied by the unit weight of fill. (See sec. 242(d).) For the "unusual condition" of fill pressure distribution, the vertical fill pressure was left unchanged and the horizontal fill pressure was assumed equal to zero. The factors of safety used in design for the unusual fill load distribution were reduced. The saturated weight of the fill was taken as 130 pounds per cubic foot in all cases.

Moments, thrusts, and shears for use in the design of the conduit were determined using coefficients obtained from Beggs deformeter

TABLE C–4.—*Standardized design for cast-in-place conduits with external loads and internal pressures*

Conduit diameter, feet	Height of fill above top of conduit, feet	Required conduit wall thickness, inches	Transverse reinforcement				Longitudinal reinforcement spaced at 18 inches ±
			Hoop reinforcement per foot of conduit length	Additional reinforcement required per foot of length throughout			
				Sec. 1	Sec. 2	Sec. 3	
10	50	18	1 No. 6 e.f.	1 No. 9 i.f.	1 No. 6 o.f.	1 No. 11 i.f.	No. 7 e.f.
	40	16	1 No. 6 e.f.	1 No. 8 i.f.	1 No. 6 o.f.	1 No. 10 i.f.	No. 6 e.f.
	30	13	1 No. 6 e.f.	1 No. 8 i.f.	1 No. 6 o.f.	1 No. 9 i.f.	No. 5 e.f.
	20	11	1 No. 6 e.f.	1 No. 7 i.f.	1 No. 6 o.f.	1 No. 9 i.f.	No. 5 e.f.
8	50	15	1 No. 6 e.f.	1 No. 8 i.f.	1 No. 5 o.f.	1 No. 9 i.f.	No. 6 e.f.
	40	13	1 No. 6 e.f.	1 No. 7 i.f.	1 No. 5 o.f.	1 No. 8 i.f.	No. 6 e.f.
	30	11	1 No. 6 e.f.	1 No. 7 i.f.	1 No. 5 o.f.	1 No. 8 i.f.	No. 5 e.f.
	20	9	1 No. 6 e.f.	1 No. 6 i.f.	1 No. 5 o.f.	1 No. 7 i.f.	No. 5 e.f.
6	50	12	1 No. 6 e.f.	1 No. 6 i.f.		1 No. 7 i.f.	No. 6 e.f.
	40	11	1 No. 5 e.f.	1 No. 6 i.f.	1 No. 5 o.f.	1 No. 7 i.f.	No. 5 e.f.
	30	9	1 No. 5 e.f.	1 No. 6 i.f.	1 No. 5 o.f.	1 No. 7 i.f.	No. 5 e.f.
	20	8	1 No. 5 c.s.	1 No. 6 c.s.	1 No. 5 c.s.	1 No. 7 c.s.	No. 5 c.s.
4	50	10	1 No. 5 e.f.	1 No. 5 i.f.		1 No. 5 i.f.	No. 5 e.f.
	40	9	1 No. 5 e.f.	1 No. 5 i.f.		1 No. 5 i.f.	No. 5 e.f.
	30	8	1 No. 5 c.s.	1 No. 5 c.s.		1 No. 5 c.s.	No. 5 c.s.
	20	6	1 No. 5 c.s.	1 No. 5 c.s.		1 No. 5 c.s.	No. 5 c.s.

NOTE: i.f. denotes inside face, o.f. denotes outside face, and e.f. denotes each face for conduits having 2 layers of reinforcement. c.s. denotes center of section for conduits having 1 layer of reinforcement.

stress analysis [9]. The conduit was designed on the basis of ACI Standard 318-63 [10], with certain exceptions. The concrete was assumed to have a 28-day strength of 3,750 pounds per square inch, and the reinforcing steel was assumed to have a yield strength of 60,000 pounds per square inch.

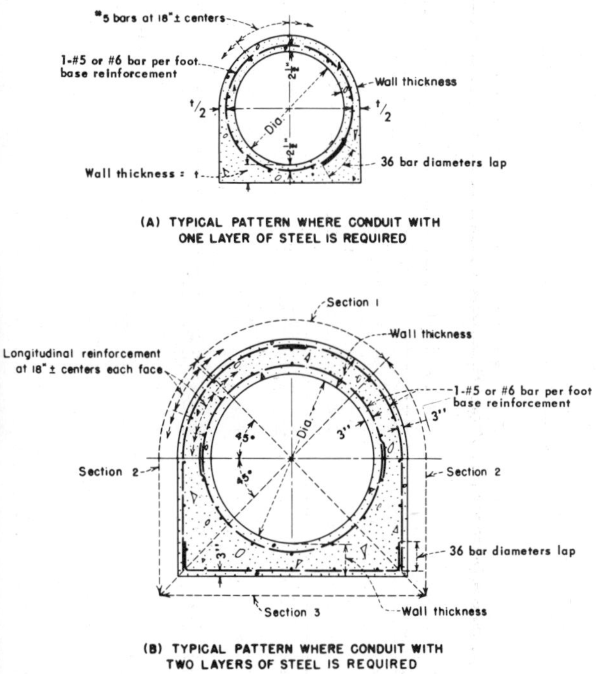

Figure C–4. Typical reinforcement patterns for circular cast-in-place concrete conduits. 288–D–2837.

A computer program was used to determine the thickness of the conduit wall so that the concrete would be sufficient to resist the shear and diagonal tension stresses without the use of stirrups. The diagonal tension stresses were not allowed to exceed 1.75 $\sqrt{f^1_c}$ for normal loads. For unusual loads the stresses were increased one-third. Bond stresses were computed by standard formula [10]. The ultimate strength method of design [11, 12] was used to determine the required areas of reinforcement to resist moments and thrusts. A load factor of 1.8 was used in designing for the normal fill load distribution, and a load factor of 1.25 was used in designing for the unusual fill load distribution. Generally, the conduit design was governed by the unusual fill load distribution. Typical reinforcement patterns for the circular conduit are shown in figure C-4. The additional reinforcement of tables C-3 and C-4 must be extended at least 12 bar diameters or a distance, "d" (effective depth), equal to the depth from the compressing face to the tension steel, whichever is greater, beyond the specified section boundaries.

C-6. *Bibliography.*

[1] Terzaghi, Karl, "Theoretical Soil Mechanics," John Wiley & Sons, Inc., New York, N.Y., 1943.

[2] Terzaghi, Karl, and Peck, R. B., "Soil Mechanics in Engineering Practices," John Wiley & Sons, Inc., New York, N.Y., 1967.

[3] Tschebotarioff, G. P., "Soil Mechanics, Foundations, and Earth Structures," McGraw-Hill, New York, N.Y., 1951.

[4] Huntington, W. C., "Earth Pressures and Retaining Walls," John Wiley & Sons, Inc., New York, N.Y., 1957.

[5] "Final Revised and Adopted Form of Part 29 Excavation and Foundations of the Boston Building Code," Journal of the Boston Society of Civil Engineers, vol. 50, No. 3, July 1963, pp. 149–171.

[6] Gibbs, H. J., and Holtz, W. G., "Research on Determining the Density of Sands by Spoon Penetration Testing," Proceedings, Fourth International Conference on Soil Mechanics and Foundation Engineering, vol. I, London, 1957.

[7] Olander, H. C., "Stress Analysis of Concrete Pipe," Engineering Monograph No. 6, Bureau of Reclamation, Denver, Colo., October 1950.

[8] "Standard Specifications for Reinforced Concrete Pressure Pipe," Bureau of Reclamation, Denver, Colo., February 1, 1969.

[9] Phillips, H. B., and Allen, I. E., "Beggs Deformeter Stress Analysis of Single-Barrel Conduits," Engineering Monograph No. 14, Bureau of Reclamation, Denver, Colo., June 1965.

[10] "Building Code Requirements for Reinforced Concrete," (ACI 318–63), American Concrete Institute, Detroit, Mich.

[11] "Bureau of Reclamation Supplement for ACI 318–63 Code," Bureau of Reclamation, Denver, Colo., 1968.

[12] "Ultimate-Strength Design Handbook," Special Publication No. 17, American Concrete Institute, Detroit, Mich., 1967.

Appendix D

Soil Mechanics Nomenclature

DR. J. W. HILF [1]

D-1. ***Introduction.***—The following definitions of terms and symbols were selected from ASTM Designation D 653-67, "Standard Definition of Terms and Symbols Relating to Soil and Rock Mechanics," prepared by Subcommittee G-3 on Nomenclature and Definitions of ASTM Committee D-18 on Soil and Rock for Engineering Purposes, in cooperation with the Committee on Glossary of Terms and Definitions in Soil Mechanics of the Soil Mechanics and Foundations Division of the American Society of Civil Engineers. The list that follows is an abbreviated version of the ASTM designation, in which most of the cross-references and terms which have little or no relation to the subject matter of this text were omitted.

Units, where applicable, are indicated in capital letters on the right-hand side, under the item, and immediately above the definition. The letters denote:

F, force, such as pound, ton, gram, kilogram.

L, length, such as inch, foot, centimeter.

T, time, such as second, minute.

D, dimensionless.

Positive exponents designate multiples in the numerator.

Negative exponents designate multiples in the denominator.

Degrees of angles are indicated as "degrees" (°).

Expressing the unit either in the metric (C.G.S.) or English system has been purposely omitted in order to leave the choice of the system and specific unit to the engineer and the particular application. For example, FL^{-2} may be expressed in pounds per square inch, kilograms per square centimeter, tons per square foot, etc.; LT^{-1} may be expressed in feet per minute, centimeters per second, etc. No significance should be placed on the order in which symbols are presented where two or more are given.

The following letters of the Greek alphabet are used in this nomenclature:

Greek letter	*Greek name*	*Greek letter*	*Greek name*
α	Alpha	μ	Mu
β	Beta	σ	Sigma
γ	Gamma	τ	Tau
Δ, δ	Delta	ϕ	Phi
ε	Epsilon	ψ	Psi
θ	Theta		

D-2. ***Definitions, Symbols, and Units.***

ABSORBED WATER:

Water held mechanically in a soil mass and having physical properties not substantially different from ordinary water at the same temperature and pressure.

ADHESION:

Unit: c_a — FL^{-2}

Total: C_a — F or FL^{-1}

Shear resistance between soil and another material under zero externally applied pressure.

ADSORBED WATER:

Water in a soil mass, held by physicochemical forces, having physical properties substantially

[1] Formerly Engineer, Earth Dams Section, now Chief, Division of Design, Bureau of Reclamation.

different from absorbed water or chemically combined water, at the same temperature and pressure.

AEOLIAN DEPOSITS:

Wind-deposited material such as dune sands and loess deposits.

AIRSPACE RATIO:

G_a D

Ratio of (1) volume of water that can be drained from a saturated soil under the action of force of gravity to (2) total volume of voids.

AIR-VOID RATIO:

G_v D

The ratio of (1) the volume of airspace to (2) the total volume of voids in a soil mass.

ALLOWABLE BEARING VALUE (ALLOWABLE SOIL PRESSURE):

q_a, p_a FL^{-2}

The maximum pressure that can be permitted on foundation soil, giving consideration to all pertinent factors, with adequate safety against rupture of the soil mass or movement of the foundation of such magnitude that the structure is impaired.

ALLOWABLE PILE BEARING LOAD:

Q_a, P_a F

The maximum load that can be permitted on a pile with adequate safety against movement of such magnitude that the structure is endangered.

ALLUVIUM:

Soil the constituents of which have been transported in suspension by flowing water and subsequently deposited by sedimentation.

ANGLE OF EXTERNAL FRICTION (ANGLE OF WALL FRICTION):

δ Degrees (°)

Angle between the abscissa and the tangent of the curve representing the relationship of shear resistance to normal stress acting between soil and surface of another material.

ANGLE OF INTERNAL FRICTION:

ϕ Degrees (°)

Angle between the abscissa and the tangent of the curve representing the relationship of shear resistance to normal stress acting within a soil.

ANGLE OF OBLIQUITY:

α, B, θ, ψ Degrees (°)

The angle between the direction of the resultant stress or force acting on a given plane and the normal to that plane.

ANGLE OF REPOSE:

α Degrees (°)

Angle between the horizontal and the maximum slope that a soil assumes through natural processes. For dry granular soils the effect of the height of slope is negligible; for cohesive soils the effect of height of slope is so great that the angle of repose is meaningless.

ANISOTROPIC MASS:

A mass having different properties in different directions at any given point.

AQUIFER:

A water-bearing formation that provides a ground-water reservoir.

ARCHING:

The transfer of stress from a yielding part of a soil mass to adjoining less-yielding or restrained parts of the mass.

AREA OF INFLUENCE OF A WELL:

a L^2

Area surrounding a well within which the piezometric surface has been lowered when pumping has produced a maximum steady rate of flow.

AREA RATIO OF A SAMPLING SPOON, SAMPLER, OR SAMPLING TUBE:

A_r D

$A_r = \frac{D_e^2 - D_i^2}{D_i^2} \times 100$ where D_e represents the maximum external diameter of the sampling spoon and D_i represents the minimum internal diameter of the sampling spoon at the cutting edge. The area ratio is an indication of the volume of soil displaced by the sampling spoon (tube).

BASE COURSE (BASE):

A layer of specified or selected material of planned thickness constructed on the subgrade or subbase for the purpose of serving one or more functions such as distributing load, providing drainage, minimizing frost action, etc.

BASE EXCHANGE:
The physicochemical process whereby one species of ions adsorbed on soil particles is replaced by another species.

BEARING CAPACITY (OF A PILE):
Q_p, P_p F
The load per pile required to produce a condition of failure.

BEDROCK (LEDGE):
Rock of relatively great thickness and extent in its native location.

BENTONITIC CLAY:
A clay with a high content of the mineral montmorillonite, usually characterized by high swelling on wetting.

BERM:
A shelf that breaks the continuity of a slope.

BINDER (SOIL BINDER):
Portion of soil passing No. 40 United States standard sieve.

BOULDER:
A rock fragment, usually rounded by weathering or abrasion, with an average dimension of 12 inches or more.

BOULDER CLAY:
A geological term used to designate glacial drift that has not been subjected to the sorting action of water and therefore contains particles from boulders to clay sizes.

BULKING:
The increase in volume of a material due to manipulation. Rock bulks upon being excavated; damp sand bulks if loosely deposited, as by dumping, because the "apparent cohesion" prevents movement of the soil particles to form a reduced volume.

CALIFORNIA BEARING RATIO:
CBR D
The ratio of (1) the force per unit area required to penetrate a soil mass with a 3-square-inch circular piston (approximately 2-inch-diameter) at the rate of 0.05 inch per minute to (2) that required for corresponding penetration of a standard material. The ratio is usually determined at 0.1-inch penetration, although other penetrations are sometimes used. Original California procedures required determination of the ratio at 0.1-inch intervals to 0.5 inch. Corps of Engineers' procedures require determination of the ratio at 0.1 inch and 0.2 inch. Where the ratio at 0.2 inch is consistently higher than at 0.1 inch, the ratio at 0.2 inch is used.

CAPILLARY ACTION (CAPILLARITY):
The rise or movement of water in the interstices of a soil due to capillary forces.

CAPILLARY FRINGE ZONE:
The zone above the free water elevation in which water is held by capillary action.

CAPILLARY HEAD:
h L
The potential, expressed in head of water, that causes the water to flow by capillary action.

CAPILLARY MIGRATION (CAPILLARY FLOW):
The movement of water by capillary action.

CAPILLARY RISE (HEIGHT OF CAPILLARY RISE):
h_c L
The height above a free water elevation to which water will rise by capillary action.

CAPILLARY WATER:
Water subject to the influence of capillary action.

CLAY (CLAY SOIL):
Fine-grained soil or the fine-grained portion of soil that can be made to exhibit plasticity (putty-like properties) within a range of water contents, and which exhibits considerable strength when air-dry. The term has been used to designate the percentage finer than 0.002 mm. (0.005 in some cases), but it is

strongly recommended that this usage be discontinued, since there is ample evidence that from an engineering standpoint the properties described in the above definition are many times more important.

CLAY SIZE:

That portion of the soil finer than 0.002 mm. (0.005 mm. in some cases). (See discussion under Clay.)

COBBLE (COBBLESTONE):

A rock fragment usually rounded or semi-rounded with an average dimension between 3 and 12 inches.

COEFFICIENT OF COMPRESSIBILITY (COEFFICIENT OF COMPRESSION):

a_v L^2F^{-1}

The secant slope, for a given pressure increment, of the pressure-void ratio curve. Where a stress-strain curve is used, the slope of this curve is equal to $\frac{(a_v)}{1+e}$.

COEFFICIENT OF CONSOLIDATION:

c_v L^2T^{-1}

A coefficient utilized in the theory of consolidation, containing the physical constants of a soil affecting its rate of volume change.

$c_v = \frac{k(1+e)}{a_v \cdot \gamma_w}$, wherein

k=coefficient of permeability, LT^{-1}

e=void ratio, D

a_v=coefficient of compressibility, L^2F^{-1}

γ_w=unit weight of water, FL^{-3}

NOTE.—In the literature published prior to 1935, the coefficient of consolidation, usually designated c, was defined by the equation $c = \frac{k}{a_v \cdot \gamma_w \ (1+e)}$. This original definition of the coefficient of consolidation may be found in some more recent papers and care should be taken to avoid confusion.

COEFFICIENT OF EARTH PRESSURE:

K D

The principal stress ratio at a point in a soil mass.

ACTIVE:

K_A D

The minimum ratio of (1) the minor principal stress to (2) the major principal stress. This is applicable where the soil has yielded sufficiently to develop a lower limiting value of the minor principal stress.

AT REST:

K_o D

The ratio of (1) the minor principal stress to (2) the major principal stress. This is applicable where the soil mass is in its natural state without having been permitted to yield or without having been compressed.

PASSIVE:

K_p D

The maximum ratio of (1) the major principal stress to (2) the minor principal stress. This is applicable where the soil has been compressed sufficiently to develop an upper limiting value of the major principal stress.

COEFFICIENT OF INTERNAL FRICTION:

The tangent of the angle of internal friction. (See Internal Friction.)

COEFFICIENT OF PERMEABILITY (PERMEABILITY):

k LT^{-1}

The rate of discharge of water under laminar flow conditions through a unit cross-sectional area of a porous medium under a unit hydraulic gradient and standard temperature conditions (usually 20° C.).

COEFFICIENT OF SUBGRADE REACTION (MODULUS OF SUBGRADE REACTION):

k FL^{-3}

Ratio of (1) load per unit area of horizontal surface of a mass of soil to (2) corresponding settlement of the surface. It is determined as the slope of the secant, drawn between the point corresponding to zero settlement and the point of 0.05-inch settlement, of a load settlement curve obtained from a plate load test on a soil using a 30-inch or greater diameter loading plate. It is used in the design of concrete pavements by the Westergaard method.

COEFFICIENT OF UNIFORMITY:

C_u D

The ratio D_{60}/D_{10}, where D_{60} is the particle diameter corresponding to 60 percent finer on the grain-size curve, and D_{10} is the particle diameter corresponding to 10 percent finer on the grain-size curve.

COEFFICIENT OF VISCOSITY (COEFFICIENT OF ABSOLUTE VISCOSITY):

μ FTL^{-2}

The shearing force per unit area required to maintain a unit difference in velocity between two parallel layers of a fluid a unit distance apart.

COEFFICIENT OF VOLUME COMPRESSIBILITY (MODULUS OF VOLUME CHANGE):

m_v L^2F^{-1}

The compression of a soil layer per unit of original thickness due to a given unit increase in pressure. It is numerically equal to the coefficient of compressibility divided by one plus the original void ratio: $\frac{a_v}{1+e}$

COHESION:

The portion of the shear strength of a soil indicated by the term c in Coulomb's equation, $s=c+\bar{\sigma}\tan\phi$.

APPARENT COHESION:

Cohesion in granular soils due to capillary forces.

COHESIONLESS SOIL:

A soil that when unconfined has little or no strength when air-dried, and that has little or no cohesion when submerged.

COHESIVE SOIL:

A soil that when unconfined has considerable strength when air-dried, and that has significant cohesion when submerged.

COLLOIDAL PARTICLES:

Soil particles that are so small that the surface activity has an appreciable influence on the properties of the aggregate.

COMPACTION:

The densification of a soil by means of mechanical manipulation.

COMPACTION CURVE (PROCTOR CURVE) (MOISTURE-DENSITY CURVE):

The curve showing the relationship between the dry unit weight (density) and the water content of a soil for a given compactive effort.

COMPACTION TEST (MOISTURE-DENSITY TEST):

A laboratory compacting procedure whereby a soil at a known water content is placed in a specified manner into a mold of given dimensions, subjected to a compactive effort of controlled magnitude, and the resulting unit weight determined. The procedure is repeated for various water contents sufficient to establish a relation between water content and unit weight.

COMPRESSIBILITY:

Property of a soil pertaining to its susceptibility to decrease in volume when subjected to load.

COMPRESSION INDEX:

C_c D

The slope of the linear portion of the pressure-void ratio curve on a semilog plot.

COMPRESSIVE STRENGTH (UNCONFINED COMPRESSIVE STRENGTH):

p_c, q_u FL^{-2}

The load per unit area at which an unconfined prismatic or cylindrical specimen of soil will fail in a simple compression test.

CONCENTRATION FACTOR:

n D

A parameter used in modifying the Boussinesq equations to describe various distributions of vertical stress.

CONSISTENCY:

The relative ease with which a soil can be deformed.

CONSISTENCY INDEX:

See Relative Consistency.

CONSOLIDATED DRAINED TEST (SLOW TEST):

A soil test in which essentially complete consolidation under the confining pressure is followed by additional axial (or shear) stress applied in such a manner that even a fully saturated soil of low permeability can adapt itself completely (fully consolidate) to the changes in stress due to the additional axial (or shear) stress.

CONSOLIDATED UNDRAINED TEST (CONSOLIDATED QUICK TEST):

A test in which complete consolidation under the vertical load (in a direct shear test) or under the confining pressure (in a triaxial test) is followed by a shear at constant water content.

CONSOLIDATION:

The gradual reduction in volume of a soil mass resulting from an increase in compressive stress.

INITIAL CONSOLIDATION (INITIAL COMPRESSION):

A comparatively sudden reduction in volume of a soil mass under an applied load due principally to expulsion and compression of gas in the soil voids preceding primary consolidation.

PRIMARY CONSOLIDATION (PRIMARY COMPRESSION) (PRIMARY TIME EFFECT):

The reduction in volume of a soil mass caused by the application of a sustained load to the mass and due principally to a squeezing out of water from the void spaces of the mass and accompanied by a transfer of the load from the soil water to the soil solids.

SECONDARY CONSOLIDATION (SECONDARY COMPRESSION) (SECONDARY TIME EFFECT):

The reduction in volume of a soil mass caused by the application of a sustained load to the mass and due principally to the adjustment of the internal structure of the soil mass after most of the load has been transferred from the soil water to the soil solids.

CONSOLIDATION RATIO:

U_z D

The ratio of (1) the amount of consolidation at a given distance from a drainage surface and at a given time to (2) the total amount of consolidation obtainable at that point under a given stress increment.

CONSOLIDATION TEST:

A test in which the specimen is laterally confined in a ring and is compressed between porous plates.

CONSOLIDATION-TIME CURVE (TIME CURVE) (CONSOLIDATION CURVE) (THEORETICAL TIME CURVE):

A curve that shows the relation between (1) the degree of consolidation and (2) the elapsed time after the application of a given increment of load.

CREEP:

Slow movement of rock debris or soil usually imperceptible except to observations of long duration.

CRITICAL CIRCLE (CRITICAL SURFACE):

The sliding surface assumed in a theoretical analysis of a soil mass for which the factor of safety is a minimum.

CRITICAL DENSITY:

The unit weight of a saturated granular material below which it will lose strength and above which it will gain strength when subjected to rapid deformation. The critical density of a given material is dependent on many factors.

CRITICAL HEIGHT:

H_c L

The maximum height at which a vertical or sloped bank of soil will stand unsupported under a given set of conditions.

CRITICAL SLOPE:

The maximum angle with the horizontal at which a sloped bank of soil of given height will stand unsupported.

DEFLOCCULATING AGENT (DEFLOCCULANT) (DISPERSING AGENT):

An agent that prevents fine soil particles in suspension from coalescing to form flocs.

DEFORMATION:

Change in shape.

DEGREE OF CONSOLIDATION (PERCENT CONSOLIDATION):

U D

The ratio, expressed as percentage, of (1) the amount of consolidation at a given time within a soil mass, to (2) the total amount of consolidation obtainable under a given stress condition.

DENSITY:

See Unit Weight.

NOTE.—Although it is recognized that density is defined as mass per unit volume, in the field of soil mechanics the term is frequently used in place of unit weight.

DEVIATOR STRESS:

Δ, σ FL^{-2}

The difference between the major and minor principal stresses in a triaxial test.

DILATANCY:

The expansion of cohesionless soils when subject to shear deformation.

DIRECT SHEAR TEST:

A shear test in which soil under an applied normal load is stressed to failure by moving one section of the soil container (shear box) relative to the other section.

DISCHARGE VELOCITY:

v LT^{-1}

Rate of discharge of water through a porous medium per unit of total area perpendicular to the direction of flow.

DRAWDOWN:

L

Vertical distance the free water elevation is lowered or the reduction of the pressure head due to the removal of free water.

EARTH PRESSURE:

Unit: p FL^{-2}

Total: P F or FL^{-1}

The pressure or force exerted by soil on any boundary.

ACTIVE EARTH PRESSURE:

P_A, p_A

The minimum value of earth pressure. This condition exists when a soil mass is permitted to yield sufficiently to cause its internal shear resistance along a potential failure surface to be completely mobilized.

AT REST:

P_o, p_o

The value of the earth pressure when the soil mass is in its natural state without having been permitted to yield or without having been compressed.

PASSIVE EARTH PRESSURE:

P_P, p_P

The maximum value of earth pressure. This condition exists when a soil mass is compressed sufficiently to cause its internal shear resistance along a potential failure surface to be completely mobilized.

EFFECTIVE DIAMETER (EFFECTIVE SIZE):

D_{10}, D_e L

Particle diameter corresponding to 10 percent finer on the grain-size curve.

EFFECTIVE FORCE:

$\overline{F}$ F

The force transmitted through a soil mass by intergranular pressures.

EFFECTIVE POROSITY (EFFECTIVE DRAINAGE POROSITY):

n_e D

The ratio of (1) the volume of the voids of a soil mass that can be drained by gravity to (2) the total volume of the mass.

ELASTIC STATE OF EQUILIBRIUM:

State of stress within a soil mass when the internal resistance of the mass is not fully mobilized.

EQUIPOTENTIAL LINE:

Line along which water will rise to the same elevation in piezometric tubes.

EQUIVALENT DIAMETER (EQUIVALENT SIZE):

D L

The diameter of a hypothetical sphere composed of material having the same specific gravity as that of the actual soil particle and of such size that it will settle in a given liquid at the same terminal velocity as the actual soil particle.

EQUIVALENT FLUID:

A hypothetical fluid having a unit weight such that it will produce a pressure against a lateral support presumed to be equivalent to that produced by the actual soil. This simplified approach is valid only when deformation conditions are such that the pressure increases linearly with depth and the wall friction is neglected.

EXCHANGE CAPACITY:

The capacity to exchange ions as measured by the quantity of exchangeable ions in a soil.

FILL:

Manmade deposits of natural soils and waste materials.

FILTER (PROTECTIVE FILTER):

A layer or combination of layers of pervious materials designed and installed in such a manner as to provide drainage, yet prevent the movement of soil particles due to flowing water.

FINES:

Portion of soil finer than a No. 200 United States standard sieve.

FLOC:

Loose, open-structured mass formed in a suspension by the aggregation of minute particles.

FLOCCULATION:

The process of forming flocs.

FLOW CHANNEL:

The portion of a flow net bounded by two adjacent flow lines.

FLOW CURVE:

The locus of points obtained from a standard liquid limit test and plotted on a graph representing water content as ordinate on an arithmetic scale and the number of blows as abscissa on a logarithmic scale.

FLOW FAILURE:

Failure in which a soil mass moves over relatively long distances in a fluidlike manner.

FLOW INDEX:

F_w, I_f D

The slope of the flow curve obtained from a liquid limit test, expressed as the difference in water contents at 10 blows and at 100 blows.

FLOW LINE:

The path that a particle of water follows in its course of seepage under laminar flow conditions.

FLOW NET:

A graphical representation of flow lines and equipotential lines used in the study of seepage phenomena.

FLOW SLIDE:

The failure of a sloped bank of soil in which the movement of the soil mass does not take place along a well-defined surface of sliding.

FLOW VALUE:

N_ϕ D

A quantity equal to $\tan^2\left(45°+\frac{\phi}{2}\right)$.

FOOTING:

Portion of the foundation of a structure that transmits loads directly to the soil.

FOUNDATION:

Lower part of a structure that transmits the load to the earth.

FOUNDATION SOIL:

Upper part of the earth mass carrying the load of the structure.

FREE WATER (GRAVITATIONAL WATER) (GROUND WATER) (PHREATIC WATER):

Water that is free to move through a soil mass under the influence of gravity.

FREE WATER ELEVATION (WATER TABLE) (GROUND-WATER SURFACE) (FREE WATER SURFACE) (GROUND-WATER ELEVATION):

Elevations at which the pressure in the water is zero with respect to the atmospheric pressure.

FROST ACTION:

Freezing and thawing of moisture in materials and the resultant effects on these materials and on structures of which they are a part or with which they are in contact.

FROST BOIL:

(1) Softening of soil occurring during a thawing period due to the liberation of water from ice lenses or layers.

(2) Breaking of a highway or airfield pavement under traffic and the ejection of subgrade soil in a soft and soupy condition caused by the melting of ice lenses formed by frost action.

FROST HEAVE:

The raising of a surface due to the accumulation of ice in the underlying soil.

GLACIAL TILL (TILL):

Material deposited by glaciation, usually composed of a wide range of particle sizes, which has not been subjected to the sorting action of water.

GRADATION (GRAIN SIZE DISTRIBUTION) (SOIL TEXTURE):

Proportion of material of each grain size present in a given soil.

GRAIN SIZE ANALYSIS (MECHANICAL ANALYSIS):

The process of determining gradation.

GRAVEL:

Rounded or semirounded particles of rock that will pass a 3-inch and be retained on a No. 4 United States standard sieve.

HARDPAN:

Layer of extremely dense soil.

HEAVE:

Upward movement of soil caused by expansion or displacement resulting from phenomena such as: moisture absorption, removal of overburden, driving of piles, and frost action.

HOMOGENEOUS MASS:

A mass that exhibits essentially the same physical properties at every point throughout the mass.

HORIZON (SOIL HORIZON):

One of the layers of the soil profile, distinguished principally by its texture, color, structure, and chemical content.

A HORIZON:

The uppermost layer of a soil profile from which inorganic colloids and other soluble materials have been leached. Usually contains remnants of organic life.

B HORIZON:

The layer of a soil profile in which material leached from the overlying A horizon is accumulated.

C HORIZON:

Undisturbed parent material from which the overlying soil profile has been developed.

HUMUS:

A brown or black material formed by the partial decomposition of vegetable or animal matter; the organic portion of soil.

HYDRAULIC GRADIENT:

i, s D

The loss of hydraulic head per unit distance of flow, $\frac{dh}{dL}$.

CRITICAL HYDRAULIC GRADIENT:

i_c D

Hydraulic gradient at which the intergranular pressure in a mass of cohesionless soil is reduced to zero by the upward flow of water.

HYDROSTATIC PRESSURE:

u_o FL^{-2}

The pressure in a liquid under static conditions; the product of the unit weight of the liquid and the difference in elevation between the given point and the free water elevation.

EXCESS HYDROSTATIC PRESSURE (HYDROSTATIC EXCESS PRESSURE):

$\bar{u}, u$ FL^{-2}

The pressure that exists in pore water in excess of the hydrostatic pressure.

HYGROSCOPIC CAPACITY (HYGROSCOPIC COEFFICIENT):

w_c D

Ratio of (1) the weight of water absorbed by a dry soil in a saturated atmosphere at a given temperature to (2) the weight of the oven-dried soil.

HYGROSCOPIC WATER CONTENT:

w_H D

The water content of an air-dried soil.

INTERNAL FRICTION:

FL^{-2}

The portion of the shear strength of a soil indicated by the terms $p \tan \phi$ in Coulomb's equation $s=c+p \tan \phi$. It is usually considered to be due to the interlocking of the soil grains and the resistance to sliding between the grains.

ISOCHRONE:

A curve showing the distribution of the excess hydrostatic pressure at a given time during a process of consolidation.

ISOTROPIC MASS:

A mass having the same property (or properties) in all directions.

KAOLIN:

A variety of clay containing a high percentage of kaolinite.

LAMINAR FLOW (STREAMLINE FLOW) (VISCOUS FLOW):

Flow in which each water particle moves in a direction parallel to every other particle, and in which the head loss is proportional to the first power of the velocity.

LANDSLIDE (SLIDE):

The failure of a sloped bank of soil in which the movement of the soil mass takes place along a surface of sliding.

LEACHING:

The removal of soluble soil material and colloids by percolating water.

LINE OF CREEP (PATH OF PERCOLATION):

The path that water follows along the surface of contact between the foundation soil and the base of a dam or other structure.

LINE OF SEEPAGE (SEEPAGE LINE) (PHREATIC LINE):

The upper free water surface of the zone of seepage.

LINEAR EXPANSION:

L_E D

The increase in one dimension of a soil mass, expressed as a percentage of that dimension at the shrinkage limit, when the water content is increased from the shrinkage limit to any given water content.

LINEAR SHRINKAGE:

L_S D

Decrease in one dimension of a soil mass, expressed as a percentage of the original dimension, when the water content is reduced from a given value to the shrinkage limit.

LIQUID LIMIT:

LL, L_w, w_L D

(1) The water content corresponding to the arbitrary limit between the liquid and plastic states of consistency of a soil.

(2) The water content at which a pat of soil, cut by a groove of standard dimensions, will flow together for a distance of one-half inch under the impact of 25 blows in a standard liquid limit apparatus.

LIQUEFACTION (SPONTANEOUS LIQUEFACTION):

The sudden large decrease of the shear resistance of a cohesionless soil. It is caused by a collapse of the structure by shock or other type of strain and is associated with a sudden but temporary increase of the pore-fluid pressure. It involves a temporary transformation of the material into a fluid mass.

LIQUIDITY INDEX (WATER PLASTICITY RATIO) (RELATIVE WATER CONTENT):

B, R_w, I_L D

The ratio, expressed as a percentage, of (1) the natural water content of a soil minus its plastic limit to (2) its plasticity index.

LOAM:

A mixture of sand, silt, or clay, or a combination of any of these, with organic matter (humus). It is sometimes called topsoil in contrast to the subsoils that contain little or no organic matter.

LOESS:

A uniform aeolian deposit of silty material having an open structure and relatively high cohesion due to cementation of clay or calcareous material at grain contacts. A characteristic of loess deposits is that they can stand with nearly vertical slopes.

MODULUS OF ELASTICITY (MODULUS OF DEFORMATION):

E, M FL^{-2}

The ratio of stress to strain for a material under given loading conditions; numerically equal to the slope of the tangent or the secant of a stress-strain curve. The use of the term Modulus of Elasticity is recommended for materials that deform in accordance with Hooke's law; the term Modulus of Deformation for materials that deform otherwise.

MOHR CIRCLE:

A graphical representation of the stresses acting on the various planes at a given point.

MOHR ENVELOPE (RUPTURE ENVELOPE) (RUPTURE LINE):

The envelope of a series of Mohr Circles representing stress conditions at failure for a given material. According to Mohr's rupture hypothesis, a rupture envelope is the locus of points the coordinates of which represent the combinations of normal and shear stresses that will cause a given material to fail.

MOISTURE CONTENT (WATER CONTENT):

w D

The ratio, expressed as a percentage, of (1) the weight of water in a given soil mass to (2) the weight of solid particles.

MOISTURE EQUIVALENT:

CENTRIFUGE MOISTURE EQUIVALENT:

W_c, CME D

The water content of a soil after it has been saturated with water and then subjected for one hour to a force equal to 1,000 times that of gravity.

FIELD MOISTURE EQUIVALENT:

FME

The minimum water content, expressed as a percentage of the weight of the oven-dried soil, at which a drop of water placed on a smoothed surface of the soil will not immediately be absorbed by the soil but will spread out over the surface and give it a shiny appearance.

MUCK:

An organic soil of very soft consistency.

MUD:

A mixture of soil and water in a fluid or weakly solid state.

MUSKEG:

Level, practically treeless areas supporting dense growth consisting primarily of grasses. The surface of the soil is covered with a layer of partially decayed grass and grass roots which is usually wet and soft when not frozen.

NORMALLY CONSOLIDATED SOIL DEPOSIT:

A soil deposit that has never been subjected to a pressure greater than the existing overburden pressure.

OPTIMUM MOISTURE CONTENT (OPTIMUM WATER CONTENT):

OMC, W_o D

The water content at which a soil can be compacted to the maximum dry unit weight by a given compactive effort.

ORGANIC CLAY:

A clay with a high organic content.

ORGANIC SILT:

A silt with a high organic content.

ORGANIC SOIL:

Soil with a high organic content. In general, organic soils are very compressible and have poor load-sustaining properties.

OVERCONSOLIDATED SOIL DEPOSIT:

A soil deposit that has been subjected to pressure greater than the present overburden pressure.

PARENT MATERIAL:

Material from which a soil has been derived.

PEAT:

A fibrous mass of organic matter in various stages of decomposition, generally dark brown to black in color and of spongy consistency.

PENETRATION RESISTANCE (STANDARD PENETRATION RESISTANCE) (PROCTOR PENETRATION RESISTANCE):

p_R, N FL^{-2} or blows L^{-1}

(1) Number of blows of a hammer of specified weight falling a given distance required to produce a given penetration into soil of a pile, casing, or sampling tube.

(2) Unit load required to maintain constant rate of penetration into soil of a probe or instrument.

(3) Unit load required to produce a specified penetration into soil at a specified rate of a probe or instrument. For a Proctor needle, the specified penetration is 2½ inches and the rate is ½ inch per second.

PENETRATION RESISTANCE CURVE (PROCTOR PENETRATION CURVE):

The curve showing the relationship between (1) the penetration resistance and (2) the water content.

PERCENT COMPACTION:

The ratio, expressed as a percentage, of (1) dry unit weight of a soil to (2) maximum unit weight obtained in a laboratory compaction test.

PERCENT SATURATION (DEGREE OF SATURATION):

S D

The ratio, expressed as a percentage, of (1) the volume of water in a given soil mass to (2) the total volume of intergranular space (voids).

PERCHED WATER TABLE:

A water table usually of limited area maintained above the normal free water elevation by the presence of an intervening relatively impervious confining stratum.

PERCOLATION:

The movement of gravitational water through soil. (See Seepage.)

PERMAFROST:

Perennially frozen soil.

pH

pH D

An index of the acidity or alkalinity of a soil in terms of the logarithm of the reciprocal of the hydrogen ion concentration.

PIEZOMETER:

An instrument for measuring pressure head.

PIEZOMETRIC SURFACE:

The surface at which water will stand in a series of piezometers.

PILE:

Relatively slender structural element which is driven, or otherwise introduced, into the soil, usually for the purpose of providing vertical or lateral support.

PIPING:

The movement of soil particles by percolating water leading to the development of channels.

PLASTIC EQUILIBRIUM:

State of stress within a soil mass, or a portion thereof, which has been deformed to such an extent that its ultimate shear resistance is mobilized.

ACTIVE STATE OF PLASTIC EQUILIBRIUM:

Plastic equilibrium obtained by an expansion of a mass.

PASSIVE STATE OF PLASTIC EQUILIBRIUM:

Plastic equilibrium obtained by a compression of a mass.

PLASTICITY:

The property of a soil which allows it to be deformed beyond the point of recovery without cracking or appreciable volume change.

PLASTIC FLOW (PLASTIC DEFORMATION):

The deformation of a plastic material beyond the point of recovery, accompanied by continuing deformation with no further increase in stress.

PLASTIC LIMIT:

w_p, PL, P_w D

(1) The water content corresponding to an arbitrary limit between the plastic and the semisolid states of consistency of a soil.

(2) Water content at which a soil will just begin to crumble when rolled into a thread approximately one-eighth inch in diameter.

PLASTIC SOIL:

A soil that exhibits plasticity.

PLASTIC STATE (PLASTIC RANGE):

The range of consistency within which a soil exhibits plastic properties.

PLASTICITY INDEX:

I_p, PI, I_w D

Numerical difference between the liquid limit and the plastic limit.

PORE PRESSURE (PORE WATER PRESSURE):

See Neutral Stress under Stress.

POROSITY:

n D

The ratio, usually expressed as a percentage, of (1) the volume of voids of a given soil mass to (2) the total volume of the soil mass.

POTENTIAL DROP:

Δh L

The difference in pressure head between two equipotential lines.

PRECONSOLIDATION PRESSURE (PRESTRESS):

p_c FL^{-2}

The greatest pressure to which a soil has been subjected.

PRESSURE:

p FL^{-2}

The load divided by the area over which it acts.

PRESSURE BULB:

The zone in a loaded soil mass bounded by an arbitrarily selected isobar of stress.

PRESSURE—VOID RATIO CURVE (COMPRESSION CURVE):

A curve representing the relationship between pressure and void ratio of a soil as obtained from a consolidation test. The curve has a characteristic shape when plotted on a semilog paper with pressure on the log scale. The various parts of the curve and extensions to the parts have been designated as recompression, compression, virgin compression, expansion, rebound, and other descriptive names by various authorities.

PRINCIPAL PLANE:

Each of three mutually perpendicular planes through a point in a soil mass on which the shearing stress is zero.

INTERMEDIATE PRINCIPAL PLANE:

The plane normal to the direction of the intermediate principal stress.

MAJOR PRINCIPAL PLANE:

The plane normal to the direction of the major principal stress.

MINOR PRINCIPAL PLANE:

The plane normal to the direction of the minor principal stress.

PROCTOR COMPACTION CURVE:

See Compaction Curve.

PROCTOR PENETRATION CURVE:

See Penetration Resistance Curve.

PROCTOR PENETRATION RESISTANCE:

See Penetration Resistance.

PROGRESSIVE FAILURE:

Failure in which the ultimate shear resistance is progressively mobilized along the failure surface.

QUICK CONDITION (QUICKSAND):

Condition in which water is flowing upward with sufficient velocity to reduce significantly the bearing capacity of the soil through a decrease in intergranular pressure.

QUICK TEST:

See Unconsolidated Undrained Test.

RADIUS OF INFLUENCE OF A WELL:

Distance from the center of the well to the closest point at which the piezometric surface is not lowered when pumping has produced the maximum steady rate of flow.

RELATIVE CONSISTENCY:

I_c, C_r D

Ratio of (1) the liquid limit minus the natural water content to (2) the plasticity index.

RELATIVE DENSITY:

D_d D

The ratio of (1) the difference between the void ratio of a cohesionless soil in the loosest state and any given void ratio to (2) the difference between its void ratios in the loosest and in the densest states.

REMOLDED SOIL:

Soil that has had its natural structure modified by manipulation.

REMOLDING INDEX:

I_R D

The ratio of (1) the modulus of deformation of a soil in the undisturbed state to (2) the modulus of deformation of the soil in the remolded state.

REMOLDING SENSITIVITY (SENSITIVITY RATIO):

S_t D

The ratio of (1) the unconfined compressive strength of an undisturbed specimen of soil to (2) the unconfined compressive strength of a specimen of the same soil after remolding at unaltered water content.

RESIDUAL SOIL:

Soil derived in place by weathering of the underlying material.

ROCK:

Natural solid mineral matter occurring in large masses or fragments.

ROCK FLOUR:

See Silt.

SAND:

Particles of rock that will pass the No. 4 United States standard sieve and be retained on the No. 200 sieve.

SAND BOIL:

The ejection of sand and water resulting from piping.

SEEPAGE (PERCOLATION):

The slow movement of gravitational water through the soil.

SEEPAGE FORCE:

J F

The force transmitted to the soil grains by seepage.

SEEPAGE VELOCITY:

v_s, v_1 LT^{-1}

The rate of discharge of seepage water through a porous medium per unit area of void space perpendicular to the direction of flow.

SENSITIVITY:

The effect of remolding on the consistency of a cohesive soil.

SHAKING TEST:

A test used to indicate the presence of significant amounts of rock flour, silt, or very fine sand in a fine-grained soil. It consists of shaking a pat of wet soil, having a consistency of thick paste, in the palm of the hand; observing the surface for a glossy or livery appearance; then squeezing the pat; and observing if a rapid apparent drying and subsequent cracking of the soil occurs.

SHEAR FAILURE (FAILURE BY RUPTURE):

Failure in which movement caused by shear

stresses in a soil mass is of sufficient magnitude to destroy or seriously endanger a structure.

GENERAL SHEAR FAILURE:

Failure in which the ultimate strength of the soil is mobilized along the entire potential surface of sliding before the structure supported by the soil is impaired by excessive movement.

LOCAL SHEAR FAILURE:

Failure in which the ultimate shearing strength of the soil is mobilized only locally along the potential surface of sliding at the time the structure supported by the soil is impaired by excessive movement.

SHEAR STRENGTH:

s FL^{-2}

The maximum resistance of a soil to shearing stresses.

SHEAR STRESS (SHEARING STRESS) (TANGENTIAL STRESS):

See Stress.

SHRINKAGE INDEX:

SI D

The numerical difference between the plastic and shrinkage limits.

SHRINKAGE LIMIT:

SL D

The maximum water content at which a reduction in water content will not cause a decrease in volume of the soil mass.

SHRINKAGE RATIO:

R D

The ratio of (1) a given volume change, expressed as a percentage of the dry volume, to (2) the corresponding change in water content above the shrinkage limit, expressed as a percentage of the weight of the oven-dried soil.

SILT (INORGANIC SILT) (ROCK FLOUR):

Material passing the No. 200 United States standard sieve that is nonplastic or very slightly plastic and that exhibits little or no strength when air-dried.

SILT SIZE:

That portion of the soil finer than 0.02 mm. and coarser than 0.002 mm. (0.05 mm. and 0.005 mm. in some cases).

SKIN FRICTION:

f FL^{-2}

The frictional resistance developed between soil and a structure.

SLAKING:

The process of breaking up or sloughing when an indurated soil is immersed in water.

SLOW TEST:

See Consolidated-Drained Test.

SOIL (EARTH):

Sediments or other unconsolidated accumulations of solid particles produced by the physical and chemical disintegration of rocks, and which may or may not contain organic matter.

SOIL CEMENT:

A tightly compacted mixture of pulverized soil, portland cement and water that, as the cement hydrates, forms a hard, durable, low-cost paving material.

SOIL MECHANICS:

The application of the laws and principles of mechanics and hydraulics to engineering problems dealing with soil as an engineering material.

SOIL PHYSICS:

The organized body of knowledge concerned with the physical characteristics of soil and with the methods employed in their determinations.

SOIL PROFILE (PROFILE):

Vertical section of a soil, showing the nature and sequence of the various layers, as developed by deposition or weathering, or both.

SOIL STABILIZATION:

Chemical or mechanical treatment designed to increase or maintain the stability of a mass of soil or otherwise to improve its engineering properties.

SOIL STRUCTURE:

The arrangement and state of aggregation of soil particles in a soil mass.

FLOCCULENT STRUCTURE:

An arrangement composed of flocs of soil particles instead of individual soil particles.

HONEYCOMB STRUCTURE:

An arrangement of soil particles having a comparatively loose, stable structure resembling a honeycomb.

SINGLE-GRAINED STRUCTURE:

An arrangement composed of individual soil particles; characteristic structure of coarse-grained soils.

SOIL SUSPENSION:

Highly diffused mixture of soil and water.

SOIL TEXTURE:

See Gradation.

SPECIFIC GRAVITY:

SPECIFIC GRAVITY OF SOLIDS:

G_s, S_s D

Ratio of (1) the weight in air of a given volume of soil solids at a stated temperature to (2) the weight in air of an equal volume of distilled water at a stated temperature.

APPARENT SPECIFIC GRAVITY:

G_a, S_a D

Ratio of (1) the weight in air of a given volume of the impermeable portion of a permeable material (that is the solid matter including its impermeable pores or voids) at a stated temperature to (2) the weight in air of an equal volume of distilled water at a stated temperature.

BULK SPECIFIC GRAVITY (SPECIFIC MASS GRAVITY):

G_m, S_m D

Ratio of (1) the weight in air of a given volume of a permeable material (including both permeable and impermeable voids normal to the material) at a stated temperature to (2) the weight in air of an equal volume of distilled water at a stated temperature.

SPECIFIC SURFACE:

L^{-1}

The surface area per unit of volume of soil particles.

STABILITY FACTOR (STABILITY NUMBER):

N_s D

A pure number used in the analysis of the stability of a soil embankment. Defined by the following equation:

$$N_s=\frac{H_c \cdot \gamma_e}{c}$$

where: H_c=critical height of the sloped bank,
γ_e=the effective unit weight of the soil, and
c=the cohesion of the soil.

NOTE.—Taylor's stability number is the reciprocal of Terzaghi's stability factor.

STICKY LIMIT:

T_w D

The lowest water content at which a soil will stick to a metal blade drawn across the surface of the soil mass.

STONE:

Crushed or naturally angular particles of rock that will pass a 3-inch sieve and be retained on a No. 4 United States standard sieve.

STRAIN:

ε D

The change in length per unit of length in a given direction.

STRESS:

σ, p, f FL^{-2}

The force per unit area acting within the soil mass.

EFFECTIVE STRESS (EFFECTIVE PRESSURE) (INTERGRANULAR PRESSURE):

$\bar{\sigma}$, $\bar{f}$ FL^{-2}

The average normal force per unit area transmitted from grain to grain of a soil mass. It is the stress that is effective in mobilizing internal friction.

NEUTRAL STRESS (PORE PRESSURE) (PORE WATER PRESSURE):

u, u_w FL^{-2}

Stress transmitted through the pore water (water filling the voids of the soil).

NORMAL STRESS:

σ, p FL^{-2}

The stress component normal to a given plane.

PRINCIPAL STRESS:

σ_1, σ_2, σ_3 FL^{-2}

Stress acting normal to three mutually perpendicular planes intersecting at a point in a body on which the shear stress is zero.

MAJOR PRINCIPAL STRESS:

σ_1 FL^{-2}

The largest (with regard to sign) principal stress.

MINOR PRINCIPAL STRESS:

σ_3 FL^{-2}

The smallest (with regard to sign) principal stress.

INTERMEDIATE PRINCIPAL STRESS:

σ_2 FL^{-2}

The principal stress whose value is neither the largest nor the smallest (with regard to sign) of the three.

SHEAR STRESS (SHEARING STRESS) (TANGENTIAL STRESS):

τ, s FL^{-2}

The stress component tangential to a given plane.

TOTAL STRESS:

σ, f FL^{-2}

The total force per unit area acting within a mass of soil. It is the sum of the neutral and effective stresses.

SUBBASE:

A layer used in a pavement system between the subgrade and base course, or between the subgrade and portland-concrete pavement.

SUBGRADE:

The soil prepared and compacted to support a structure or a pavement system.

SUBGRADE SURFACE:

The surface of the earth or rock prepared to support a structure or a pavement system.

SUBSOIL:

(1) Soil below a subgrade or fill.

(2) That part of a soil profile occurring below the A horizon.

TALUS:

Rock fragments mixed with soil at the foot of a natural slope from which they have been separated.

THERMO-OSMOSIS:

The process by which water is caused to flow in small openings of a soil mass due to differences in temperature within the mass.

THIXOTROPY:

The property of a material that enables it to stiffen in a relatively short time on standing, but upon agitation or manipulation to change to a very soft consistency or to a fluid of high viscosity, the process being completely reversible.

TILL:

See Glacial Till.

TIME FACTOR:

T_v, T D

Dimensionless factor, utilized in the theory of consolidation, containing the physical constants of a soil stratum influencing its time-rate of consolidation, expressed as follows:

$$T = \frac{k(1+e)t}{a_v \gamma_w \cdot H^2} = \frac{c_v t}{H^2}$$

where: k=coefficient of permeability (LT^{-1}),
e=void ratio (dimensionless),
t=elapsed time that the stratum has been consolidated (T),
a_v=coefficient of compressibility (L^{2}F^{-1}),
γ_w=unit weight of water (FL^{-3}),
H=thickness of stratum drained on one side only (if stratum is drained on both sides, its thickness equals $2H$ (L)), and
c_v=coefficient of consolidation (L^{2}T^{-1}).

TOPSOIL:

Surface soil usually containing organic matter.

TORSIONAL SHEAR TEST:

A shear test in which a relatively thin test specimen of solid circular or annular cross section, usually confined between rings, is subjected to an axial load and to shear in torsion. In-place torsion shear tests may be performed by pressing a dentated solid circular or annular plate against the soil and measuring its resistance to rotation under a given axial load.

TOUGHNESS INDEX:

I_T, T_w

The ratio of (1) the plasticity index to (2) the flow index.

TRANSFORMED FLOW NET:

A flow net whose boundaries have been properly modified (transformed) so that a net consisting of curvilinear squares can be constructed to represent flow conditions in an anisotropic porous medium.

TRANSPORTED SOIL:

Soil transported from the place of its origin by wind, water, or ice.

TRIAXIAL SHEAR TEST (TRIAXIAL COMPRESSION TEST):

A test in which a cylindrical specimen of soil encased in an impervious membrane is subjected to a confining pressure and then loaded axially to failure.

TURBULENT FLOW:

That type of flow in which any water particle may move in any direction with respect to any other particle, and in which the head loss is approximately proportional to the second power of the velocity.

ULTIMATE BEARING CAPACITY:

q_o, q_{ult} FL^{-2}

The average load per unit of area required to produce failure by rupture of a supporting soil mass.

UNCONFINED COMPRESSIVE STRENGTH:

See Compressive Strength.

UNCONSOLIDATED-UNDRAINED TEST (QUICK TEST):

A soil test in which the water content of the test specimen remains practically unchanged during the application of the confining pressure and the additional axial (or shear) force.

UNDERCONSOLIDATED SOIL DEPOSIT:

A deposit that is not fully consolidated under the existing overburden pressure.

UNDISTURBED SAMPLE:

A soil sample that has been obtained by methods in which every precaution has been taken to minimize disturbance to the sample.

UNIT WEIGHT:

γ FL^{-3}

Weight per unit volume.

DRY UNIT WEIGHT (UNIT DRY WEIGHT):

γ_d, γ_o FL^{-3}

The weight of soil solids per unit of total volume of soil mass.

EFFECTIVE UNIT WEIGHT:

γ_e FL^{-3}

That unit weight of a soil which, when multiplied by the height of the overlying column of soil, yields the effective pressure due to the weight of the overburden.

MAXIMUM UNIT WEIGHT:

γ_{max} FL^{-3}

The dry unit weight defined by the peak of a compaction curve.

SATURATED UNIT WEIGHT:

γ_G, γ_{sat} FL^{-3}

The wet unit weight of a soil mass when saturated.

SUBMERGED UNIT WEIGHT (BUOYANT UNIT WEIGHT):

γ_m, γ', γ_{sub} FL^{-3}

The weight of the solids in air minus the weight of water displaced by the solids per unit of volume of soil mass; the saturated unit weight minus the unit weight of water.

UNIT WEIGHT OF WATER:

γ_w FL^{-3}

The weight per unit volume of water; nominally equal to 62.4 pounds per cubic foot or 1 gram per cubic centimeter.

WET UNIT WEIGHT (MASS UNIT WEIGHT):

γ_m, γ_{wet} FL^{-3}

The weight (solids plus water) per unit of total volume of soil mass, irrespective of the degree of saturation.

ZERO AIR VOIDS UNIT WEIGHT:

γ_z FL^{-3}

The weight of solids per unit volume of a saturated soil mass.

UPLIFT:

Unit: u — FL^{-2}

Total: U — F or FL^{-1}

The upward water pressure on a structure.

VANE SHEAR TEST:

An inplace shear test in which a rod with thin radial vanes at the end is forced into the soil and the resistance to rotation of the rod is determined.

VARVED CLAY:

Alternating thin layers of silt (or fine sand) and clay formed by variations in sedimentation during the various seasons of the year, often exhibiting contrasting colors when partially dried.

VOID:

Space in a soil mass not occupied by solid mineral matter. This space may be occupied by air, water, or other gaseous or liquid material.

VOID RATIO:

e — D

The ratio of (1) the volume of void space to (2) the volume of solid particles in a given soil mass.

CRITICAL VOID RATIO:

e_c — D

The void ratio corresponding to the critical density.

VOLUMETRIC SHRINKAGE (VOLUMETRIC CHANGE):

V_s — D

The decrease in volume, expressed as a percentage of the soil mass when dried, of a soil mass when the water content is reduced from a given percentage to the shrinkage limit.

WALL FRICTION:

f' — FL^{-2}

Frictional resistance mobilized between a wall and the soil in contact with the wall.

WATER CONTENT:

See Moisture Content.

WATER-HOLDING CAPACITY:

D

The smallest value to which the water content of a soil can be reduced by gravity drainage.

ZERO AIR VOIDS CURVE (SATURATION CURVE):

The curve showing the zero air voids unit weight as a function of water content.

ZERO AIR VOIDS DENSITY (ZERO AIR VOIDS UNIT WEIGHT):

See Unit Weight.

Appendix E

Construction of Embankments

DR. J. W. HILF [1]

E-1. ***General.***—The necessity for control of construction of embankments to impound water has been recognized for many years. In 1932, Justin [1][2] wrote:

"An entirely safe and substantial design may be entirely ruined by careless and shoddy execution, and the failure of the structure may very possibly be the result. Careful attention to the details of construction is, therefore, fully as important as the preliminary investigation and design."

The consequences of ignoring control are exemplified by the large number of earthfill dams built in the United States during the first quarter of this century which did not survive the first filling of the reservoir. Records show that most of these dams were constructed without moistening the soil and without applying special compactive effort.

The rapid increase in knowledge of soil mechanics since the year 1925 has resulted in substantial progress toward understanding the factors involved in transforming loose earth into a structural material. During this same period, however, the development of huge economical earth-moving machines has increased the placing rate of earthfill many times, thereby intensifying the problem of quality control. Future progress in design economy in the field of earthwork depends not only on advances in soil mechanics and foundation engineering, but also to a large extent on the insistence of inspection personnel on good construction practices in accordance with proper specifications and on their ability to understand and conscientiously apply sound control techniques.

Construction control is obtained by inspection, testing, and reports. The inspector of foundations and earthwork is charged with the responsibility of assuring that work assigned to him is completed in compliance with the specifications. To discharge this responsibility efficiently, he should be fully informed of the design and specifications provisions relating to the work. The attributes of fairness, courtesy, and firmness, coupled with initiative and good judgment, are highly desirable in an inspector. The inspector's diary, containing data on conditions and progress of the work and records of conversations and instructions given to the contractor, is a valuable document that should be carefully preserved.

Proper control of earthwork requires the use of laboratory facilities. For small dams these can be of the portable variety or a small field laboratory can be set up in the vicinity of the site. In most cases commercial laboratory facilities can be used. The control procedures recommended in this text will minimize the cost of control testing consistent with assuring a satisfactory job.

Discoveries of remnants of earthfill dams indicate that man's first engineering structures probably were made of earth. The ancient earthfill dams were constructed by armies of workmen carrying baskets loaded with soil. Excavation was done manually, and some incidental compaction was obtained on the fill by the tramping feet of the porters. Available records do not indicate that there was any intentional moistening or compacting of soil prior to the 19th century. The importance of compaction of earthfill was first evident in England

[1] Formerly Engineer, Earth Dams Section, now Chief, Division of Design, Bureau of Reclamation.

[2] Numbers in brackets refer to items in the bibliography, section E-11.

where, by the year 1820, cattle and sheep were used for this purpose. By the middle of the 19th century, heavy smooth rollers made of concrete or metal were in use in Europe and had also been introduced in the United States.

The first sheepsfoot roller, the "Petrolithic" roller, was patented in the United States in 1906 for use in compacting oil-treated road surfacing. The most notable early use of the sheepsfoot roller for compaction of fills started in 1912 in the construction of storage reservoirs by the oil companies of southern California. This type of roller was found to be the only one which compacted the fill in layers and gave uniform compaction without producing laminations. Largely because of the development of the automobile and the airplane, which require roadbeds and airport subgrades of great strength, larger and heavier rollers were developed by the construction industry in the first half of the 20th century.

Published material on moisture control for rolled fills dates back to 1907 when Bassell [2], wrote:

"Too much or too little (water) is equally bad and is to be avoided. It is believed that only by experience is it possible to determine just the proper quantity of water to use with different classes of materials and their varying conditions. In rolling and consolidating of the bank, all portions that have a tendency to quake must be removed at once * * *,"

It was not until 1933 that a definite procedure for moisture and compaction control was established. In a series of articles published in 1933, Proctor [3] gave the principles of soil compaction and their application. Figure 108 shows the Proctor compaction curve, which indicates that for a given compactive effort there is one water content, called the optimum water content, which produces the maximum density or smallest amount of total voids for a given

Figure E–1. Embankment placing operations in impervious, sand-gravel, and cobble-rockfill zones. The structure is Shadow Mountain Dam on the Colorado River in Colorado. SM–156–CBT.

cohesive soil. For greater compactive efforts on the same soil, different moisture-density curves are obtained, whose optimum points occur at smaller water contents and at greater densities than for lesser compactive efforts.

Figure E–1 shows embankment placing operations at Shadow Mountain Dam. The maximum section of this dam is shown in figure 187, and photographs of the completed structure are shown in figure 112 and in the frontispiece.

E-2. *Soil Mechanics of Control.*—Compaction of cohesive soils has been definitely proved to follow the principles stated by Proctor. Although there are many kinds of compactive effort used as compaction standards and for compacting cohesive soils, the effect of variation of water content on the resulting unit weight is similar for all methods. Each compactive effort has its own optimum water content. The laboratory standard of compaction used by the Bureau of Reclamation, which has the same intensity of effort as ASTM Designation D–698–70 (see sec. 120(e)), has been found to approximate the actual compaction achieved in the field by 12 passes of the 20-ton dual-drum tamping roller

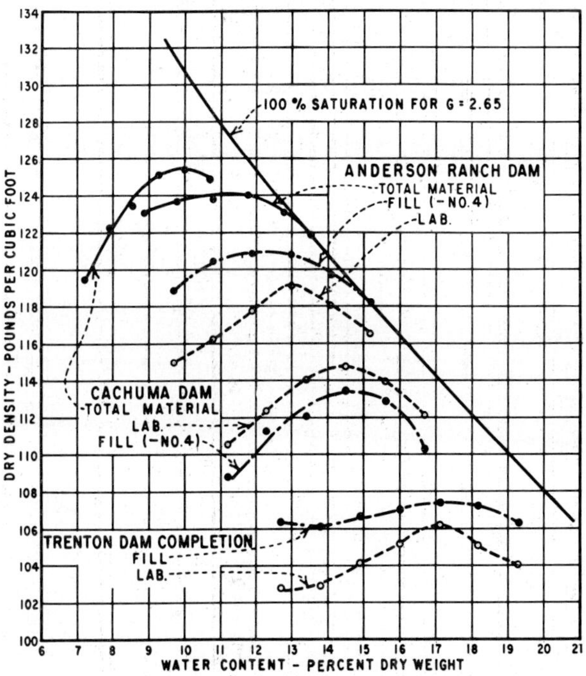

Figure E–2. Average field and laboratory compaction curves for three dam embankment soils. 101–D–248.

specified in section G–21 (appendix G) on 8- to 9-inch loose lifts (6-inch compacted lifts). The relation between the moisture-density curve for this roller effort on the fill and the standard laboratory compaction curve varies for different soils, but it is close enough so that the laboratory standard can be used for control purposes. Figure E–2 [4] shows the average roller curves for three very different soils used in Bureau dams with their respective laboratory curves.

In compacted cohesive soils, permeability, shear strength, and compressibility are of major concern. It has been shown, both theoretically and by experiment, that an increase in dry unit weight reduces the permeability of a given soil because of the corresponding reduction in amount of voids in the soil mass. From the standpoint of imperviousness, it is desirable to obtain maximum practicable compaction. Extreme impermeability, however, is not always required in the design and, especially in the case of clays, only moderate compactive effort is needed to assure impermeability. On the other hand, well-graded sands and gravels, or even formation rock, can be made quite impervious by the crushing and compacting effort of heavy tamping rollers.

The embankment designs given in chapter VI are based on values of angles of internal friction and cohesion obtained by laboratory tests on typical soils. Compaction control has the objective of securing a dry density of soil in the fill sufficient to obtain strength values comparable to those used in these designs. Although the effect of density on cohesion is small in comparison with the effect of variation of water content, test data indicate that the true angle of internal friction of a given soil varies with the compactness of the soil. The angle of internal friction varies among soils because of differences in mineral composition and in the size, shape, and gradation of the soil grains.

For cohesive soils the maximum strength obtainable with a given compactive effort does not occur at that water content which results in the maximum density, but rather at a slightly reduced water content with a density somewhat less than maximum. This condition results because pore pressures produced by compressive forces act to reduce the apparent strength of the soil. These pore pressures increase rapidly with increases in water content in the vicinity of the peak of the compaction curve. Compaction of the soil at water contents slightly less than optimum results in a net increase in strength, because the slight reduction in shear strength (which accompanies the reduction in density) is more than compensated by the comparatively large reduction in the destabilizing pore pressures which is thereby obtained.

The compressibility of a soil is the relation between effective stress on the soil skeleton and the volume change. Impermeable-type soils vary in compressibility, depending on the amount and character of the fines and according to the amount and gradation of coarse particles they contain. For a particular soil at a given water content, the greater the density the less compressible it will be. The relation between compressibility and development of construction pore-water pressure is such that, for a particular air and water content, the pore pressure increases rapidly with increase in compressibility In general, a very compressible cohesive soil will develop high pore pressures when loaded unless there is an appreciable amount of air in the compacted soil. The most efficient means of obtaining air in the soil and still have a fairly high density is to compact the soil at a water content slightly less than Proctor optimum.

Coarse-grained, permeable-type soils, also known as cohesionless or free-draining soils, are commonly used as major zones in earthfill dams and as backfill around conduits or behind retaining walls. This type of soil is also used as filter material for drainage in wells and around hydraulic structures. These soils are inherently permeable and have fairly high shear strengths when compacted. However, in the uncompacted state they are compressible and may be subject to liquefaction upon satura-

Figure E–3. Special compaction of earthfill at contact with irregular surface of rock abutment. This structure is Green Mountain Dam on the Blue River in Colorado. GM–375–CBT.

tion. The desirable properties of high strength and low compressibility can be greatly improved by compaction of permeable-type soils. Although permeability is thereby decreased, the reduction is usually allowable from a design standpoint.

The most efficient method of compacting cohesionless soils is by vibration when the material is either perfectly dry or when it is nearly saturated with water. The latter method is usually the only practicable one in the field, since perfectly dry material seldom occurs naturally. The shear strength of permeable materials, such as fairly clean sands and gravels or rockfills, depends almost entirely on the angle of internal friction. Cohesion is negligible, and pore-water pressures are never greater than hydrostatic pressure because of free drainage of the soil. The angle of internal friction is a function of the size, shape, and gradation of the grains, but for a given cohesionless soil its magnitude varies significantly with the void ratio. The state of compactness of the soils is given by their relative density, which is defined in section 120(f).

E-3. *Preparation of Foundations.*—Foundation design features have been discussed in chapter VI, part C. The weak points in earthfill dam construction are generally within the foundation and at the contact of the natural ground surface with placed embankment. Construction of the foundation seepage control and stability features included in the design must be carefully supervised by the inspection force in accordance with the specifications. Dewatering methods used in connection with excavating cutoff trenches or stabilizing the foundations should be carefully checked to see that fine material is not being washed out of the overburden because of improper screening of wells. Whenever possible, well points and sumps should be located outside the area to be excavated to avoid

Figure E–4. Special compaction of earthfill at contact with rock abutment using a small vibrating roller. The structure is Blue Mesa Dam on the Gunnison River in Colorado. P622A–427–1886.

creation of a "live" bottom due to upward flow of water. Concrete footings for cutoff walls or concrete ground caps should be founded on unfractured rock. Blasting for the excavation of these structures should be prohibited or strictly controlled in accordance with the specifications to avoid shattering the foundation.

When overburden is stripped to rock foundations, the rock surface including all pockets or depressions should be carefully cleaned of soil or rock fragments before placing the embankment upon it. This may require handwork and compressed-air cleaning. Rock surfaces which disintegrate rapidly on exposure should be covered immediately with embankment material.

When the foundation is earth, all organic or other unsuitable materials, such as stumps, brush, sod, and large roots, should be stripped and wasted. Stripping operations should be carefully performed to assure removal of all material that may be rendered unstable by saturation, of all material that may interfere with the creation of a proper bond between the foundation and the embankment, and of all pockets of soils significantly more compressible than the average foundation material. Stripping of pervious materials under the pervious or semipervious zones of an embankment should be limited to the removal of surface debris and grass roots. Test pits for further exploration should be excavated if the stripping operations indicate the presence of unstable or otherwise unsuitable material, and an inspection should be made by an experienced engineer.

Prior to placing the first layer of embankment on an earth foundation, moistening and compacting the surface by rolling with a tamping roller is necessary to obtain proper bond. Rock foundation surfaces should be moistened, but no standing water should be permitted when the first lift is placed. Sometimes the foundation surface requires scarification by harrows to assure proper bonding; however, no additional scarification is usually necessary if it is penetrated by tamping rollers. Where the rock foundation would be injured by penetration of the tamping roller feet, it is permissible to use a thicker-than-specified layer of earthfill for the first compacted layer. However, the first layer should never exceed 15 inches loose lift for 9-inch-long tamper feet, and additional roller passes are required to insure that proper compaction is obtained. Special compaction methods such as hand tamping should be used in pockets that cannot be compacted by the specified roller, instead of permitting an unusually thick initial lift to obtain a uniform surface for compaction.

Figures E-3 and E-4 show special compaction along the contact between the earthfill portion of a dam and the rock abutments. The irregular surface of the rock prevents proper compaction by means of rollers.

The use of very wet soil for the first lift against the foundation should generally be avoided; rather, the foundation should be properly moistened. On steep, irregular rock abutments material wetter than optimum may be necessary or desirable in order to obtain bond. However, the use of such material should be permitted only in special cases with the approval of the contracting authority. Care must be exercised when special compaction is used to insure that bond is created between successive layers of material. This may require light scarification between lifts of tamped material. Appendix G contains sample specifications pertinent to items of work required for preparation of foundations.

E-4. *Earthfill.*—Specifications provisions for control of placement, water content, and compaction of earthfill are given in appendix G. These must be implemented by establishing procedures to insure their attainment. For the construction of small dams within the scope of this text, the plan of control for embankments of cohesive soil is to place the material at the Proctor optimum water content and maximum laboratory density. Optimum water content rather than a water content slightly less than optimum is selected for the reasons given in section 136. The most important variables affecting construction of earthfill embankments are distribution of soils, placement, water content and its uniformity throughout the spread material, water content of the borrow material, methods for correcting borrow material water content if too wet or too dry, roller characteristics, number of roller passes, thickness of layers, maximum size and quantity of gravel sizes in the material, condition of the surface of layers after rolling, and effectiveness of power tamping in places inaccessible or undesirable for roller operation.

Figure E–5 shows placing, leveling, and compacting of the semipervious zone of the embankment at Olympus Dam. Compacting was done by tamping rollers, as the material was not pervious enough to permit compaction as a pervious fill in the manner described in section E–6. The maximum section of this dam is shown on figure 182, and a photograph of the completed structure is shown as figure 40.

Adequate inspection and laboratory testing are essential to maintain the control of earthfill construction. Attention is called to the fact that it is impossible even for an experienced soils engineer to determine visually the degree of compaction of cohesive soil, especially when it is dry of optimum. The apparent cohesion of these soils makes them firm and gives them the appearance of denseness which disappears when the soil becomes saturated. There is no satisfactory substitute for control testing to determine the degree of compactness for these soils. The testing must include all critical areas where seepage or loss of strength may induce failure.

Borrow pit inspection includes controlling and recording all earthwork operations prior to dumping the material on the embankment. Areas to be excavated are selected, depths of cut are determined, and the zone of the dam in which a particular material should be placed is predetermined. The borrow pit inspector checks the adequacy of any mixing or separation methods used by the contractor. As required, he cooperates with the contractor in determining the amount of water to be added to the borrow pit by irrigation or to be removed by drainage to attain proper water content of the materials

Figure E–5. Placing, leveling and compacting the fill on Olympus Dam. A combination earthfill and concrete gravity dam on the Big Thompson River in Colorado. 375–EPA–PS.

prior to placing. Either the rapid method of compaction control given in this appendix, or the Proctor needle value (fig. 95) obtained in a cylinder of compacted minus No. 4 fraction of soil, can be used to determine the status of natural moisture conditions in the borrow pit with respect to the optimum water content. Every effort should be made to have the excavated material as close as possible to the optimum water content prior to delivery on the embankment.

The embankment inspector should be provided with a means for determining the location and elevation of tests made on the embankment and for reporting the location of the contractor's operations. Horizontal control by means of coordinates or stations and offsets should be established. It has been found desirable to establish vertical control by benchmarks and by the use of stadia rods from which the inspector can establish an elevation anywhere on the fill by using a hand level. When materials are brought on the embankment, the inspector should see that they are placed in the proper zones. If a zoned embankment is being constructed, lines of demarcation can be painted on rock abutments or marked by flags. Within a particular zone or on homogeneous embankments, the objective is to direct the placing of materials so that the most impervious soils are located in the center of the impervious zone and the coarser, more pervious soils are placed toward the slopes of the embankment, so that the permeability and stability of materials will increase toward the outer slopes. In general, when materials differ in dry density but have about the same percolation rate, the material having the greater dry density should be placed in the outer sections of the zone or of the dam as the case may be.

After the materials are placed in the proper location, the embankment inspector determines whether they contain the proper amount of moisture prior to compaction. *This is of utmost importance.* The rapid compaction control method or the Proctor needle value should be used for this determination. Should the materials arrive on the embankment too dry, it will be necessary to condition them by sprinkling prior to, during, or after spreading. The contractor's operations in sprinkling and mixing the moisture with the soil will vary, but it is of paramount importance that the proper water content be uniformly distributed throughout the spread layer prior to rolling.

Another important inspection task is the determination of the thickness of the compacted layer. A layer that is spread too thick will not give the desired density for given compaction conditions. Initial placing operations will determine the proper spread thickness of a layer that will compact to the specified thickness. This is usually 8 to 9 inches for a 6-inch compacted lift of earthfill. A method of determining average thickness of placed layers is to take and plot daily a cross section of the fill at a reference station. The inspector's report for that day should contain the number of layers placed at that section, from which the average thickness can be determined.

The removal of oversized rock from earthfill embankment material when the oversized rock content is greater than about 1 percent is most efficiently done prior to delivery of the soil on the embankment. This procedure was used at Crescent Lake Dam, shown in figure E–6. The shovel at the left in the photograph is excavating and mixing the borrow material. The scoopmobile transports the material to electrically operated screens, which separate the oversize rock from the soil; trucks are loaded by means of a conveyor belt.

Smaller amounts of oversized rock can be removed by hand picking or, under favorable conditions, by various kinds of rock rakes. Oversized rock that has been overlooked prior to rolling can generally be detected by the inspector during rolling, by his observing the bounce which occurs when the roller passes over the hidden rock. The inspector should assure that the rock is removed from the fill.

The inspector is charged with the responsibility of assuring that the specified number of roller passes is made on each lift. An oversight in maintaining the proper number of passes may lead to considerable drop in the desired degree of compaction. The insistence of orderliness on the fill and the establishment of routine construction operations will minimize trouble from too few roller passes.

Figure E–6. Operations in impervious borrow area for Crescent Lake, a small earthfill storage dam on Crescent Creek in Oregon. Oversize rock is being removed by screening pit-run material. P806–126–55.

The final check on the degree of compaction attained is done by the rapid method of compaction control given in section E–5. If the field dry density of the material passing the No. 4 screen is above the minimum allowable density as given in section E–10, and if the water content is within the allowable limits, the embankment will be ready for the next layer after such scarifying and moistening as may be necessary to secure a good bond between the layers.

Mechanical tamping, when used around structures, along abutments and in areas inaccessible to the rolling equipment, should be carefully watched and checked by frequent density tests. The procedure to be followed for mechanical tamping will depend greatly on the type of tamper used. Some of the factors affecting density are thickness of the layer being placed, time of tamping, air pressure if air tampers are used, water content of the material, and weight of the tamping unit.

An important function of inspection is to determine when and where to make field density tests. These tests should be made (1) in areas where the degree of compaction is doubtful, (2) in areas where embankment operations are concentrated, and (3) for every 2,000 cubic yards of embankment when no doubtful or concentrated areas occur. Areas susceptible to insufficient compaction include the junction between mechanical tamping and rolled embankment, along abutments or cutoff walls; areas where rollers turn during rolling operations; areas where too thick a layer is being compacted; areas where improper water content exists in a material; and areas where less than the specified number of roller passes are made.

When embankment operations are concentrated in a small area (that is, if many layers of material are being placed one over the other in a single day), tests should be made in this area in every third or fourth layer to assure

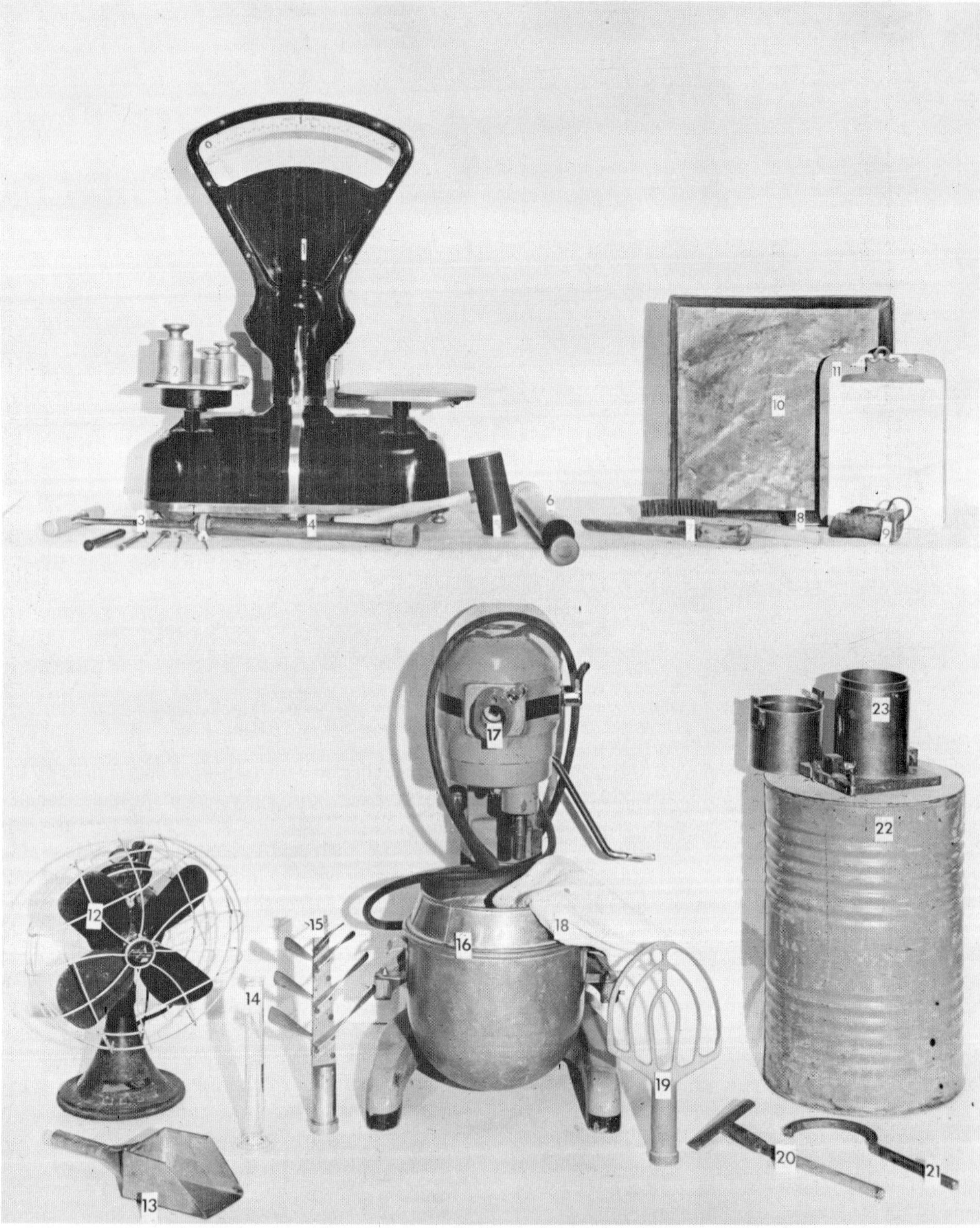

Figure E–7. Equipment used for rapid compaction control method. 288–D–2893.

that the desired density is being attained. If areas of doubtful compaction do not exist and no tests are required because of concentrated areas, at least one field density test should be made for each 2,000 cubic yards of compacted embankment and it should be representative of the degree of compaction being obtained.

E-5. ***Rapid Compaction Control Method.***—The density-in-place test described in section 119 determines the dry density and the water content of the compacted fill. For control purposes these values must be compared with the laboratory maximum dry density and the optimum water content. The rapid control procedure described herein yields the exact percentage of laboratory maximum dry density and a very close approximation of the difference between fill water content and optimum water content of a field density sample, without requiring knowledge of water contents. The theoretical basis for this method was described by Hilf [5]. With this method, compaction control can be effected within 1 hour from the time the field test is made. For record purposes one water content (the fill water content) is measured, and after it is known (usually the following day), the values of fill dry density, cylinder dry density at fill water content, laboratory maximum dry density, and optimum water content are computed.

The equipment required for the rapid compaction control method, using mechanical mixing and drying, is listed below and is shown on figure E–6. Each item is listed according to its number on the figure. The asterisks indicate mechanical equipment not required when hand mixing methods are used. Items 3 and 4 are needed only when it is desired to control water content with the Proctor needle rather than the rapid method procedure.

1. Scale, 30-pound capacity, graduated in 0.01 pound.
2. Scale weights.
3. Penetration resistance needles.
4. Penetration resistance gage (penetrometer).
5. Wood mallet.
6. 5.5-pound compaction hammer with 18-inch drop guide.
7. Large butcherknife.
8. Wire brush.
9. Small hand scoop.
10. Mixing pan.
11. Clipboard for data sheet.

*12. Electric fan.

13. Hand scoop.
14. Glass graduate, 100-ml. capacity.

*15. Mixing paddle, type III.
*16. Mixer bowl with collar, 3-gallon capacity.
*17. Mixer (with ⅓-horsepower, 60-cycle, 115-volt motor).

18. Towel.

*19. Mixing paddle type II.

20. Straightedge trimmer.
21. Spanner wrench for compaction cylinder.
22. Base for compaction cylinder, concrete-filled carbide can (or equivalent).
23. 1/20-cubic-foot compaction cylinder with base plate and collar.

Forms suggested for use in obtaining rapid control values are shown in figures E–8 and E–9. Figure E–9 is used for soils of unusually low density. A 1-minute mixing period with mechanical equipment using a paddle speed of 90 revolutions per minute has provided uniform distribution of water through the 7.5-pound sample. A type III paddle shown in figure E–7 may be used for all soils; a type II paddle provides satisfactory mixing of low-plasticity silts. Water loss during a 1-minute mixing period will be negligible in most cases. For temperatures greater than 80° F. and low relative humidities, a check of water loss should be made by weighing the soil, bowl, and paddle after mixing. Water is most efficiently added to the sample by pouring it on the surface before starting the mixer. Hand mixing in the pans should be done as rapidly and thoroughly as practicable. Drying a soil sample when necessary to remove water may be accomplished rapidly by blowing air across the bowl while slowly mixing the sample. The standard compaction method given in section 120(e) is used.

It should be noted that the forms shown in figures E–8 and E–9 are designed for the Bureau of Reclamation's laboratory compaction standard using the 1/20-cubic-foot cylinder. By changing the weights of soil sample and of water added in proportion to the size of cylin-

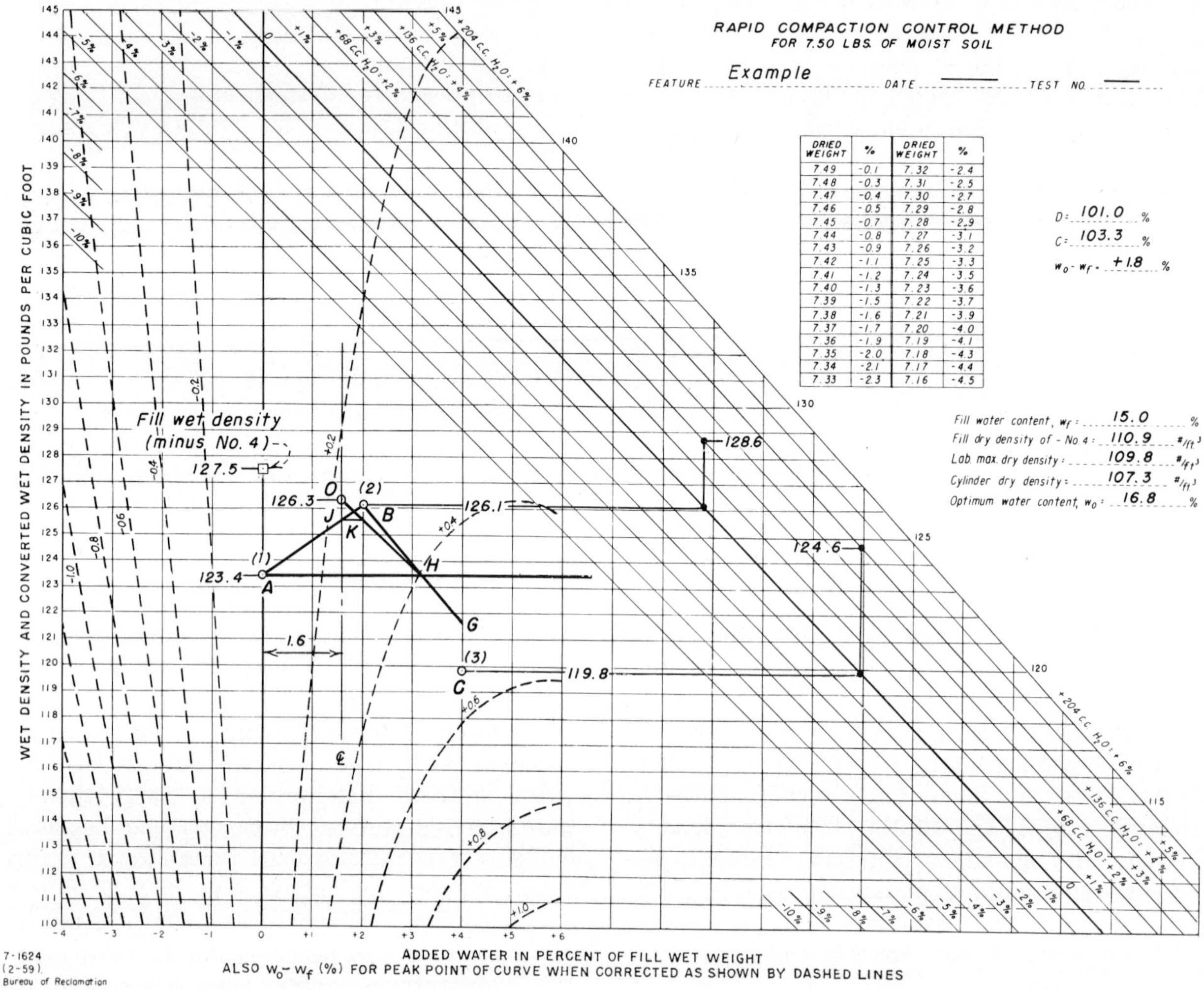

Figure E–8. Form used for rapid compaction control method. 101–D–324.

der used, this form is suitable for density control for any compaction standard. Also, for most soils the correction curves for moisture control can be used for other compaction standards of approximately the same compactive effort, such as the ASTM Designation D 698–70 and the standard AASHO T99–70 method. For moisture control of soils with very low specific gravities (lower than 2.50) the curves may have to be modified or the optimum water content may have to be estimated using the equations of the rapid compaction control method [5].

The rapid compaction control method involves the determination of the following:

$$D = \frac{\text{Fill dry density}}{\text{Laboratory maximum dry density}}$$

$$C = \frac{\text{Fill dry density}}{\text{Cylinder dry density at fill water content}}$$

$w_o - w_f$ (optimum water content minus field water content) in percent

Ratios D and C are actually obtained on the basis of corresponding wet densities. As a first step the fill wet density of the minus No. 4 fraction is obtained (see fig. 106). Then three or four points are computed and plotted on the moisture versus "converted wet density" curve as follows:

(1) To obtain point 1: Compact soil at

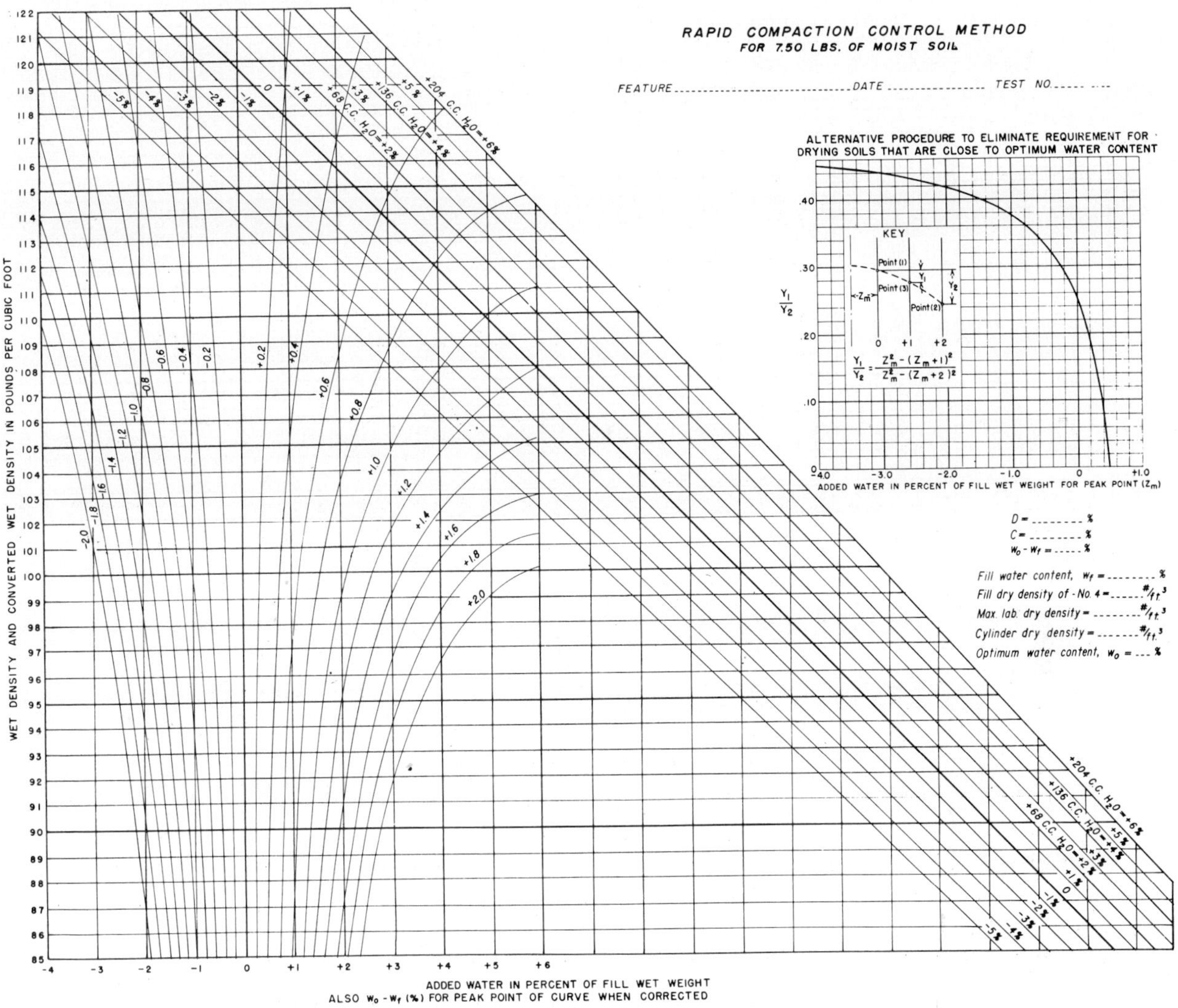

Figure E–9. Form used for rapid compaction control method for low density soils. 288–D–2838.

fill water content into a standard cylinder. Plot the resulting wet density on the 0 percent vertical line on the form shown in figure E-8 or on figure E-9 as appropriate.

(2) To obtain point 2: to 7.50 pounds of soil at fill water content add 68 cc. (2 percent) water, and mix and compact into a cylinder to determine corresponding wet density. Convert the wet density to wet density at fill water content (called the converted wet density) by dividing the wet density by 1.02. Plot the converted wet density on the +2 percent vertical line. The converted wet density may also be found by graphical division as follows: find the point on the +2 percent diagonal line corresponding to this wet density; project vertically to the 0 percent diagonal line, thence horizontally to plot point 2 on the +2 percent vertical line.

(3) To obtain point 3 if the converted wet density of point 2 is greater than point 1: to 7.50 pounds of soil at fill water content add 136 cc. (4 percent) water, mix, and compact into a cylinder. Convert the wet density to wet density at fill water content by dividing the wet density by 1.04 and plot the converted wet density on the +4 percent vertical line. The converted wet density may be found graphically as fol-

7-1671 (2-70)

Values of Y_2		-6.0	-5.9	-5.8	-5.7	-5.6	-5.5	-5.4	-5.3	-5.2	-5.1	-5.0	-4.9	-4.8	-4.7	-4.6	-4.5	-4.4	-4.3	-4.2	-4.1	-4.0	-3.9	-3.8	-3.7	-3.6	-3.5	-3.4	-3.3	-3.2	-3.1	-3.0	-2.9	-2.8	-2.7	-2.6	-2.5	-2.4	-2.3	-2.2	-2.1	-2.0	-1.9	-1.8	-1.7	-1.6	-1.5	-1.4	-1.3	-1.2	-1.1	-1.0	-0.9	-0.8	-0.7	-0.6	-0.5	-0.4	-0.3
0	x_m	1.0	1.0	1.0	1.0	1.0	1.0	1.0	1.0	1.0	1.0	1.0	1.0	1.0	1.0	1.0	1.0	1.0	1.0	1.0	1.0	1.0	1.0	1.0	1.0	1.0	1.0	1.0	1.0	1.0	1.0	1.0	1.0	1.0	1.0	1.0	1.0	1.0	1.0	1.0	1.0	1.0	1.0	1.0	1.0	1.0	1.0	1.0	1.0	1.0	1.0	1.0	1.0	1.0	1.0	1.0	1.0	1.0	1.0
	Y_m	0.8	0.7	0.7	0.7	0.7	0.7	0.7	0.7	0.6	0.6	0.6	0.6	0.6	0.6	0.6	0.6	0.6	0.5	0.5	0.5	0.5	0.5	0.5	0.5	0.5	0.5	0.4	0.4	0.4	0.4	0.4	0.4	0.4	0.3	0.3	0.3	0.3	0.3	0.3	0.3	0.2	0.2	0.2	0.2	0.2	0.2	0.2	0.2	0.1	0.1	0.1	0.1	0.1	0.1	0.1	0.1	0	0
0.1	x_m	1.0	1.0	1.0	1.0	1.0	1.0	1.0	1.0	1.0	1.0	1.0	1.0	1.0	1.0	1.0	1.0	1.0	1.0	1.0	1.0	1.0	1.0	1.1	1.1	1.1	1.1	1.1	1.1	1.1	1.1	1.1	1.1	1.1	1.1	1.1	1.1	1.1	1.1	1.1	1.1	1.1	1.1	1.1	1.1	1.1	1.1	1.1	1.1	1.1	1.2	1.2	1.2	1.2	1.2	1.3	1.3	1.3	1.4
	Y_m	0.8	0.8	0.8	0.8	0.8	0.8	0.8	0.7	0.7	0.7	0.7	0.7	0.7	0.7	0.7	0.6	0.6	0.6	0.6	0.6	0.6	0.6	0.6	0.5	0.5	0.5	0.5	0.5	0.5	0.5	0.5	0.4	0.4	0.4	0.4	0.4	0.4	0.4	0.4	0.3	0.3	0.3	0.3	0.3	0.3	0.3	0.3	0.2	0.2	0.2	0.2	0.2	0.2	0.2	0.2	0.1	0.1	0.1
0.2	x_m	1.1	1.1	1.1	1.1	1.1	1.1	1.1	1.1	1.1	1.1	1.1	1.1	1.1	1.1	1.1	1.1	1.1	1.1	1.1	1.1	1.1	1.1	1.1	1.1	1.1	1.1	1.1	1.1	1.1	1.1	1.1	1.1	1.1	1.1	1.1	1.1	1.1	1.1	1.2	1.2	1.2	1.2	1.2	1.2	1.2	1.2	1.2	1.2	1.3	1.3	1.3	1.3	1.3	1.4	1.4	1.4	1.5	1.6
	Y_m	0.9	0.9	0.9	0.9	0.9	0.8	0.8	0.8	0.8	0.8	0.8	0.8	0.8	0.7	0.7	0.7	0.7	0.7	0.7	0.7	0.7	0.6	0.6	0.6	0.6	0.6	0.6	0.6	0.6	0.5	0.5	0.5	0.5	0.5	0.5	0.5	0.5	0.4	0.4	0.4	0.4	0.4	0.4	0.4	0.4	0.3	0.3	0.3	0.3	0.3	0.3	0.3	0.3	0.3	0.2	0.2	0.2	0.2
0.3	x_m	1.1	1.1	1.1	1.1	1.1	1.1	1.1	1.1	1.1	1.1	1.1	1.1	1.1	1.1	1.1	1.1	1.1	1.1	1.1	1.1	1.1	1.1	1.1	1.1	1.1	1.1	1.2	1.2	1.2	1.2	1.2	1.2	1.2	1.2	1.2	1.2	1.2	1.2	1.2	1.2	1.2	1.2	1.3	1.3	1.3	1.3	1.3	1.3	1.3	1.4	1.4	1.4	1.4	1.5	1.5	1.5	1.6	1.7
	Y_m	1.0	1.0	1.0	0.9	0.9	0.9	0.9	0.9	0.9	0.9	0.9	0.8	0.8	0.8	0.8	0.8	0.8	0.8	0.8	0.7	0.7	0.7	0.7	0.7	0.7	0.7	0.7	0.6	0.6	0.6	0.6	0.6	0.6	0.6	0.6	0.5	0.5	0.5	0.5	0.5	0.5	0.5	0.5	0.4	0.4	0.4	0.4	0.4	0.4	0.4	0.4	0.4	0.4	0.3	0.3	0.3	0.3	0.3
0.4	x_m	1.1	1.1	1.1	1.1	1.1	1.1	1.1	1.1	1.1	1.1	1.1	1.1	1.1	1.1	1.1	1.2	1.2	1.2	1.2	1.2	1.2	1.2	1.2	1.2	1.2	1.2	1.2	1.2	1.2	1.2	1.2	1.2	1.2	1.2	1.2	1.2	1.3	1.3	1.3	1.3	1.3	1.3	1.3	1.3	1.3	1.3	1.4	1.4	1.4	1.4	1.4	1.5	1.5	1.5	1.6	1.6	1.7	1.7
	Y_m	1.1	1.1	1.0	1.0	1.0	1.0	1.0	1.0	1.0	1.0	0.9	0.9	0.9	0.9	0.9	0.9	0.9	0.9	0.8	0.8	0.8	0.8	0.8	0.8	0.8	0.8	0.7	0.7	0.7	0.7	0.7	0.7	0.7	0.7	0.6	0.6	0.6	0.6	0.6	0.6	0.6	0.6	0.6	0.5	0.5	0.5	0.5	0.5	0.5	0.5	0.5	0.5	0.5	0.4	0.4	0.4	0.4	0.4
0.5	x_m	1.1	1.1	1.1	1.1	1.2	1.2	1.2	1.2	1.2	1.2	1.2	1.2	1.2	1.2	1.2	1.2	1.2	1.2	1.2	1.2	1.2	1.2	1.2	1.2	1.2	1.2	1.2	1.2	1.2	1.2	1.3	1.3	1.3	1.3	1.3	1.3	1.3	1.3	1.3	1.3	1.3	1.3	1.4	1.4	1.4	1.4	1.4	1.4	1.5	1.5	1.5	1.5	1.6	1.6	1.6	1.7	1.7	1.8
	Y_m	1.1	1.1	1.1	1.1	1.1	1.1	1.1	1.1	1.0	1.0	1.0	1.0	1.0	1.0	1.0	1.0	0.9	0.9	0.9	0.9	0.9	0.9	0.9	0.9	0.8	0.8	0.8	0.8	0.8	0.8	0.8	0.8	0.8	0.7	0.7	0.7	0.7	0.7	0.7	0.7	0.7	0.7	0.6	0.6	0.6	0.6	0.6	0.6	0.6	0.6	0.6	0.6	0.5	0.5	0.5	0.5	0.5	0.5
0.6	x_m	1.2	1.2	1.2	1.2	1.2	1.2	1.2	1.2	1.2	1.2	1.2	1.2	1.2	1.2	1.2	1.2	1.2	1.2	1.2	1.2	1.2	1.2	1.2	1.2	1.3	1.3	1.3	1.3	1.3	1.3	1.3	1.3	1.3	1.3	1.3	1.3	1.3	1.3	1.4	1.4	1.4	1.4	1.4	1.4	1.4	1.4	1.5	1.5	1.5	1.5	1.5	1.6	1.6	1.6	1.7	1.7	1.8	1.8
	Y_m	1.2	1.2	1.2	1.2	1.2	1.2	1.2	1.1	1.1	1.1	1.1	1.1	1.1	1.1	1.1	1.0	1.0	1.0	1.0	1.0	1.0	1.0	1.0	0.9	0.9	0.9	0.9	0.9	0.9	0.9	0.9	0.9	0.8	0.8	0.8	0.8	0.8	0.8	0.8	0.8	0.8	0.7	0.7	0.7	0.7	0.7	0.7	0.7	0.7	0.7	0.7	0.6	0.6	0.6	0.6	0.6	0.6	0.6
0.7	x_m	1.2	1.2	1.2	1.2	1.2	1.2	1.2	1.2	1.2	1.2	1.2	1.2	1.2	1.2	1.2	1.2	1.2	1.2	1.3	1.3	1.3	1.3	1.3	1.3	1.3	1.3	1.3	1.3	1.3	1.3	1.3	1.3	1.3	1.3	1.4	1.4	1.4	1.4	1.4	1.4	1.4	1.4	1.4	1.5	1.5	1.5	1.5	1.5	1.5	1.6	1.6	1.6	1.6	1.7	1.7	1.7	1.8	1.8
	Y_m	1.3	1.3	1.3	1.3	1.2	1.2	1.2	1.2	1.2	1.2	1.2	1.2	1.2	1.2	1.1	1.1	1.1	1.1	1.1	1.1	1.1	1.1	1.0	1.0	1.0	1.0	1.0	1.0	1.0	1.0	1.0	0.9	0.9	0.9	0.9	0.9	0.9	0.9	0.9	0.9	0.8	0.8	0.8	0.8	0.8	0.8	0.8	0.8	0.8	0.8	0.7	0.7	0.7	0.7	0.7	0.7	0.7	0.7
0.8	x_m	1.2	1.2	1.2	1.2	1.2	1.2	1.2	1.2	1.2	1.2	1.2	1.2	1.3	1.3	1.3	1.3	1.3	1.3	1.3	1.3	1.3	1.3	1.3	1.3	1.3	1.3	1.3	1.3	1.3	1.3	1.3	1.4	1.4	1.4	1.4	1.4	1.4	1.4	1.4	1.4	1.4	1.5	1.5	1.5	1.5	1.5	1.5	1.6	1.6	1.6	1.6	1.6	1.7	1.7	1.7	1.8	1.8	1.8
	Y_m	1.4	1.4	1.4	1.4	1.3	1.3	1.3	1.3	1.3	1.3	1.3	1.3	1.2	1.2	1.2	1.2	1.2	1.2	1.2	1.2	1.2	1.1	1.1	1.1	1.1	1.1	1.1	1.1	1.1	1.1	1.0	1.0	1.0	1.0	1.0	1.0	1.0	1.0	1.0	0.9	0.9	0.9	0.9	0.9	0.9	0.9	0.9	0.9	0.9	0.8	0.8	0.8	0.8	0.8	0.8	0.8	0.8	0.8
0.9	x_m	1.2	1.2	1.2	1.2	1.2	1.2	1.3	1.3	1.3	1.3	1.3	1.3	1.3	1.3	1.3	1.3	1.3	1.3	1.3	1.3	1.3	1.3	1.3	1.3	1.3	1.3	1.3	1.4	1.4	1.4	1.4	1.4	1.4	1.4	1.4	1.4	1.4	1.4	1.5	1.5	1.5	1.5	1.5	1.5	1.5	1.5	1.6	1.6	1.6	1.6	1.6	1.7	1.7	1.7	1.8	1.8	1.8	1.9
	Y_m	1.5	1.5	1.5	1.4	1.4	1.4	1.4	1.4	1.4	1.4	1.4	1.3	1.3	1.3	1.3	1.3	1.3	1.3	1.3	1.3	1.2	1.2	1.2	1.2	1.2	1.2	1.2	1.2	1.2	1.1	1.1	1.1	1.1	1.1	1.1	1.1	1.1	1.1	1.1	1.0	1.0	1.0	1.0	1.0	1.0	1.0	1.0	1.0	1.0	1.0	0.9	0.9	0.9	0.9	0.9	0.9	0.9	0.9
1.0	x_m	1.3	1.3	1.3	1.3	1.3	1.3	1.3	1.3	1.3	1.3	1.3	1.3	1.3	1.3	1.3	1.3	1.3	1.3	1.3	1.3	1.3	1.3	1.3	1.4	1.4	1.4	1.4	1.4	1.4	1.4	1.4	1.4	1.4	1.4	1.4	1.4	1.5	1.5	1.5	1.5	1.5	1.5	1.5	1.5	1.6	1.6	1.6	1.6	1.6	1.6	1.7	1.7	1.7	1.7	1.8	1.8	1.8	1.9
	Y_m	1.6	1.6	1.5	1.5	1.5	1.5	1.5	1.5	1.5	1.5	1.4	1.4	1.4	1.4	1.4	1.4	1.4	1.4	1.4	1.3	1.3	1.3	1.3	1.3	1.3	1.3	1.3	1.3	1.2	1.2	1.2	1.2	1.2	1.2	1.2	1.2	1.2	1.2	1.1	1.1	1.1	1.1	1.1	1.1	1.1	1.1	1.1	1.1	1.1	1.0	1.0	1.0	1.0	1.0	1.0	1.0	1.0	1.0
1.1	x_m	1.3	1.3	1.3	1.3	1.3	1.3	1.3	1.3	1.3	1.3	1.3	1.3	1.3	1.3	1.3	1.3	1.3	1.3	1.3	1.3	1.4	1.4	1.4	1.4	1.4	1.4	1.4	1.4	1.4	1.4	1.4	1.4	1.4	1.4	1.5	1.5	1.5	1.5	1.5	1.5	1.5	1.5	1.6	1.6	1.6	1.6	1.6	1.6	1.6	1.7	1.7	1.7	1.7	1.8	1.8	1.8	1.8	1.9
	Y_m	1.6	1.6	1.6	1.6	1.6	1.6	1.6	1.6	1.6	1.5	1.5	1.5	1.5	1.5	1.5	1.5	1.5	1.5	1.4	1.4	1.4	1.4	1.4	1.4	1.4	1.4	1.4	1.3	1.3	1.3	1.3	1.3	1.3	1.3	1.3	1.3	1.3	1.2	1.2	1.2	1.2	1.2	1.2	1.2	1.2	1.2	1.2	1.2	1.2	1.1	1.1	1.1	1.1	1.1	1.1	1.1	1.1	1.1
1.2	x_m	1.3	1.3	1.3	1.3	1.3	1.3	1.3	1.3	1.3	1.3	1.3	1.3	1.3	1.3	1.3	1.3	1.4	1.4	1.4	1.4	1.4	1.4	1.4	1.4	1.4	1.4	1.4	1.4	1.4	1.4	1.4	1.5	1.5	1.5	1.5	1.5	1.5	1.5	1.5	1.5	1.5	1.6	1.6	1.6	1.6	1.6	1.6	1.6	1.7	1.7	1.7	1.7	1.8	1.8	1.8	1.8	1.9	1.9
	Y_m	1.7	1.7	1.7	1.7	1.7	1.7	1.7	1.7	1.6	1.6	1.6	1.6	1.6	1.6	1.6	1.6	1.6	1.5	1.5	1.5	1.5	1.5	1.5	1.5	1.5	1.5	1.4	1.4	1.4	1.4	1.4	1.4	1.4	1.4	1.4	1.4	1.4	1.3	1.3	1.3	1.3	1.3	1.3	1.3	1.3	1.3	1.3	1.3	1.2	1.2	1.2	1.2	1.2	1.2	1.2	1.2	1.2	1.2
1.3	x_m	1.3	1.3	1.3	1.3	1.3	1.3	1.3	1.3	1.3	1.3	1.3	1.3	1.4	1.4	1.4	1.4	1.4	1.4	1.4	1.4	1.4	1.4	1.4	1.4	1.4	1.4	1.4	1.4	1.4	1.5	1.5	1.5	1.5	1.5	1.5	1.5	1.5	1.5	1.5	1.6	1.6	1.6	1.6	1.6	1.6	1.6	1.7	1.7	1.7	1.7	1.7	1.7	1.8	1.8	1.8	1.8	1.9	1.9
	Y_m	1.8	1.8	1.8	1.8	1.8	1.8	1.8	1.7	1.7	1.7	1.7	1.7	1.7	1.7	1.7	1.7	1.6	1.6	1.6	1.6	1.6	1.6	1.6	1.6	1.6	1.6	1.5	1.5	1.5	1.5	1.5	1.5	1.5	1.5	1.5	1.5	1.4	1.4	1.4	1.4	1.4	1.4	1.4	1.4	1.4	1.4	1.4	1.3	1.3	1.3	1.3	1.3	1.3	1.3	1.3	1.3	1.3	1.3
1.4	x_m	1.3	1.3	1.3	1.3	1.3	1.3	1.3	1.3	1.4	1.4	1.4	1.4	1.4	1.4	1.4	1.4	1.4	1.4	1.4	1.4	1.4	1.4	1.4	1.4	1.4	1.4	1.5	1.5	1.5	1.5	1.5	1.5	1.5	1.5	1.5	1.5	1.5	1.5	1.6	1.6	1.6	1.6	1.6	1.6	1.6	1.7	1.7	1.7	1.7	1.7	1.7	1.8	1.8	1.8	1.8	1.8	1.9	1.9
	Y_m	1.9	1.9	1.9	1.9	1.9	1.9	1.8	1.8	1.8	1.8	1.8	1.8	1.8	1.8	1.8	1.7	1.7	1.7	1.7	1.7	1.7	1.7	1.7	1.7	1.7	1.6	1.6	1.6	1.6	1.6	1.6	1.6	1.6	1.6	1.6	1.5	1.5	1.5	1.5	1.5	1.5	1.5	1.5	1.5	1.5	1.5	1.5	1.5	1.4	1.4	1.4	1.4	1.4	1.4	1.4	1.4	1.4	1.4
1.5	x_m	1.3	1.3	1.3	1.3	1.3	1.4	1.4	1.4	1.4	1.4	1.4	1.4	1.4	1.4	1.4	1.4	1.4	1.4	1.4	1.4	1.4	1.4	1.4	1.4	1.5	1.5	1.5	1.5	1.5	1.5	1.5	1.5	1.5	1.5	1.5	1.5	1.6	1.6	1.6	1.6	1.6	1.6	1.6	1.6	1.7	1.7	1.7	1.7	1.7	1.7	1.8	1.8	1.8	1.8	1.8	1.9	1.9	1.9
	Y_m	2.0	2.0	2.0	2.0	2.0	1.9	1.9	1.9	1.9	1.9	1.9	1.9	1.9	1.9	1.8	1.8	1.8	1.8	1.8	1.8	1.8	1.8	1.8	1.8	1.7	1.7	1.7	1.7	1.7	1.7	1.7	1.7	1.7	1.7	1.7	1.6	1.6	1.6	1.6	1.6	1.6	1.6	1.6	1.6	1.6	1.6	1.6	1.5	1.5	1.5	1.5	1.5	1.5	1.5	1.5	1.5	1.5	1.5
1.6	x_m	1.3	1.4	1.4	1.4	1.4	1.4	1.4	1.4	1.4	1.4	1.4	1.4	1.4	1.4	1.4	1.4	1.4	1.4	1.4	1.4	1.4	1.5	1.5	1.5	1.5	1.5	1.5	1.5	1.5	1.5	1.5	1.5	1.5	1.5	1.6	1.6	1.6	1.6	1.6	1.6	1.6	1.6	1.6	1.7	1.7	1.7	1.7	1.7	1.7	1.7	1.8	1.8	1.8	1.8	1.8	1.9	1.9	1.9
	Y_m	2.1	2.1	2.1	2.1	2.0	2.0	2.0	2.0	2.0	2.0	2.0	2.0	2.0	1.9	1.9	1.9	1.9	1.9	1.9	1.9	1.9	1.9	1.8	1.8	1.8	1.8	1.8	1.8	1.8	1.8	1.8	1.8	1.8	1.8	1.7	1.7	1.7	1.7	1.7	1.7	1.7	1.7	1.7	1.7	1.7	1.7	1.7	1.6	1.6	1.6	1.6	1.6	1.6	1.6	1.6	1.6	1.6	1.6
1.7	x_m	1.4	1.4	1.4	1.4	1.4	1.4	1.4	1.4	1.4	1.4	1.4	1.4	1.4	1.4	1.4	1.4	1.4	1.4	1.4	1.5	1.5	1.5	1.5	1.5	1.5	1.5	1.5	1.5	1.5	1.5	1.5	1.5	1.5	1.6	1.6	1.6	1.6	1.6	1.6	1.6	1.6	1.7	1.7	1.7	1.7	1.7	1.7	1.7	1.7	1.8	1.8	1.8	1.8	1.8	1.9	1.9	1.9	1.9
	Y_m	2.2	2.2	2.2	2.1	2.1	2.1	2.1	2.1	2.1	2.1	2.1	2.1	2.1	2.0	2.0	2.0	2.0	2.0	2.0	2.0	2.0	2.0	2.0	1.9	1.9	1.9	1.9	1.9	1.9	1.9	1.9	1.9	1.9	1.8	1.8	1.8	1.8	1.8	1.8	1.8	1.8	1.8	1.8	1.8	1.8	1.8	1.7	1.7	1.7	1.7	1.7	1.7	1.7	1.7	1.7	1.7	1.7	1.7
1.8	x_m	1.4	1.4	1.4	1.4	1.4	1.4	1.4	1.4	1.4	1.4	1.4	1.4	1.4	1.4	1.4	1.4	1.5	1.5	1.5	1.5	1.5	1.5	1.5	1.5	1.5	1.5	1.5	1.5	1.5	1.5	1.5	1.6	1.6	1.6	1.6	1.6	1.6	1.6	1.6	1.6	1.6	1.7	1.7	1.7	1.7	1.7	1.7	1.7	1.8	1.8	1.8	1.8	1.8	1.8	1.9	1.9	1.9	1.9
	Y_m	2.3	2.3	2.2	2.2	2.2	2.2	2.2	2.2	2.2	2.2	2.2	2.2	2.1	2.1	2.1	2.1	2.1	2.1	2.1	2.1	2.1	2.1	2.0	2.0	2.0	2.0	2.0	2.0	2.0	2.0	2.0	2.0	2.0	1.9	1.9	1.9	1.9	1.9	1.9	1.9	1.9	1.9	1.9	1.9	1.9	1.9	1.8	1.8	1.8	1.8	1.8	1.8	1.8	1.8	1.8	1.8	1.8	1.8
1.9	x_m	1.4	1.4	1.4	1.4	1.4	1.4	1.4	1.4	1.4	1.4	1.4	1.4	1.4	1.5	1.5	1.5	1.5	1.5	1.5	1.5	1.5	1.5	1.5	1.5	1.5	1.5	1.5	1.5	1.5	1.6	1.6	1.6	1.6	1.6	1.6	1.6	1.6	1.6	1.6	1.6	1.7	1.7	1.7	1.7	1.7	1.7	1.7	1.7	1.8	1.8	1.8	1.8	1.8	1.8	1.9	1.9	1.9	1.9
	Y_m	2.4	2.3	2.3	2.3	2.3	2.3	2.3	2.3	2.3	2.3	2.3	2.2	2.2	2.2	2.2	2.2	2.2	2.2	2.2	2.2	2.2	2.1	2.1	2.1	2.1	2.1	2.1	2.1	2.1	2.1	2.1	2.1	2.0	2.0	2.0	2.0	2.0	2.0	2.0	2.0	2.0	2.0	2.0	2.0	2.0	2.0	1.9	1.9	1.9	1.9	1.9	1.9	1.9	1.9	1.9	1.9	1.9	1.9
2.0	x_m	1.4	1.4	1.4	1.4	1.4	1.4	1.4	1.4	1.4	1.4	1.4	1.4	1.5	1.5	1.5	1.5	1.5	1.5	1.5	1.5	1.5	1.5	1.5	1.5	1.5	1.5	1.5	1.5	1.6	1.6	1.6	1.6	1.6	1.6	1.6	1.6	1.6	1.6	1.6	1.7	1.7	1.7	1.7	1.7	1.7	1.7	1.7	1.8	1.8	1.8	1.8	1.8	1.8	1.9	1.9	1.9	1.9	1.9
	Y_m	2.5	2.4	2.4	2.4	2.4	2.4	2.4	2.4	2.4	2.4	2.3	2.3	2.3	2.3	2.3	2.3	2.3	2.3	2.3	2.3	2.3	2.2	2.2	2.2	2.2	2.2	2.2	2.2	2.2	2.2	2.2	2.2	2.1	2.1	2.1	2.1	2.1	2.1	2.1	2.1	2.1	2.1	2.1	2.1	2.1	2.1	2.0	2.0	2.0	2.0	2.0	2.0	2.0	2.0	2.0	2.0	2.0	2.0
2.1	x_m	1.4	1.4	1.4	1.4	1.4	1.4	1.4	1.4	1.4	1.5	1.5	1.5	1.5	1.5	1.5	1.5	1.5	1.5	1.5	1.5	1.5	1.5	1.5	1.5	1.5	1.5	1.6	1.6	1.6	1.6	1.6	1.6	1.6	1.6	1.6	1.6	1.6	1.6	1.7	1.7	1.7	1.7	1.7	1.7	1.7	1.7	1.8	1.8	1.8	1.8	1.8	1.8	1.8	1.9	1.9	1.9	1.9	1.9
	Y_m	2.5	2.5	2.5	2.5	2.5	2.5	2.5	2.5	2.5	2.5	2.4	2.4	2.4	2.4	2.4	2.4	2.4	2.4	2.4	2.4	2.3	2.3	2.3	2.3	2.3	2.3	2.3	2.3	2.3	2.3	2.3	2.2	2.2	2.2	2.2	2.2	2.2	2.2	2.2	2.2	2.2	2.2	2.2	2.2	2.2	2.1	2.1	2.1	2.1	2.1	2.1	2.1	2.1	2.1	2.1	2.1	2.1	2.1
2.2	x_m	1.4	1.4	1.4	1.4	1.4	1.4	1.4	1.5	1.5	1.5	1.5	1.5	1.5	1.5	1.5	1.5	1.5	1.5	1.5	1.5	1.5	1.5	1.5	1.5	1.6	1.6	1.6	1.6	1.6	1.6	1.6	1.6	1.6	1.6	1.6	1.6	1.6	1.7	1.7	1.7	1.7	1.7	1.7	1.7	1.7	1.7	1.8	1.8	1.8	1.8	1.8	1.8	1.8	1.9	1.9	1.9	1.9	1.9
	Y_m	2.6	2.6	2.6	2.6	2.6	2.6	2.6	2.6	2.6	2.5	2.5	2.5	2.5	2.5	2.5	2.5	2.5	2.5	2.5	2.4	2.4	2.4	2.4	2.4	2.4	2.4	2.4	2.4	2.4	2.4	2.4	2.3	2.3	2.3	2.3	2.3	2.3	2.3	2.3	2.3	2.3	2.3	2.3	2.3	2.3	2.2	2.2	2.2	2.2	2.2	2.2	2.2	2.2	2.2	2.2	2.2	2.2	2.2
2.3	x_m	1.4	1.4	1.4	1.4	1.5	1.5	1.5	1.5	1.5	1.5	1.5	1.5	1.5	1.5	1.5	1.5	1.5	1.5	1.5	1.5	1.5	1.5	1.5	1.6	1.6	1.6	1.6	1.6	1.6	1.6	1.6	1.6	1.6	1.6	1.6	1.6	1.7	1.7	1.7	1.7	1.7	1.7	1.7	1.7	1.7	1.8	1.8	1.8	1.8	1.8	1.8	1.8	1.9	1.9	1.9	1.9	1.9	1.9
	Y_m	2.7	2.7	2.7	2.7	2.7	2.7	2.7	2.7	2.6	2.6	2.6	2.6	2.6	2.6	2.6	2.6	2.6	2.6	2.6	2.5	2.5	2.5	2.5	2.5	2.5	2.5	2.5	2.5	2.5	2.5	2.4	2.4	2.4	2.4	2.4	2.4	2.4	2.4	2.4	2.4	2.4	2.4	2.4	2.4	2.4	2.3	2.3	2.3	2.3	2.3	2.3	2.3	2.3	2.3	2.3	2.3	2.3	2.3
2.4	x_m	1.4	1.4	1.5	1.5	1.5	1.5	1.5	1.5	1.5	1.5	1.5	1.5	1.5	1.5	1.5	1.5	1.5	1.5	1.5	1.5	1.5	1.6	1.6	1.6	1.6	1.6	1.6	1.6	1.6	1.6	1.6	1.6	1.6	1.6	1.6	1.7	1.7	1.7	1.7	1.7	1.7	1.7	1.7	1.7	1.8	1.8	1.8	1.8	1.8	1.8	1.8	1.8	1.9	1.9	1.9	1.9	1.9	1.9
	Y_m	2.8	2.8	2.8	2.8	2.8	2.8	2.8	2.7	2.7	2.7	2.7	2.7	2.7	2.7	2.7	2.7	2.7	2.7	2.6	2.6	2.6	2.6	2.6	2.6	2.6	2.6	2.6	2.6	2.6	2.6	2.5	2.5	2.5	2.5	2.5	2.5	2.5	2.5	2.5	2.5	2.5	2.5	2.5	2.5	2.4	2.4	2.4	2.4	2.4	2.4	2.4	2.4	2.4	2.4	2.4	2.4	2.4	2.4
2.5	x_m	1.5	1.5	1.5	1.5	1.5	1.5	1.5	1.5	1.5	1.5	1.5	1.5	1.5	1.5	1.5	1.5	1.5	1.5	1.5	1.6	1.6	1.6	1.6	1.6	1.6	1.6	1.6	1.6	1.6	1.6	1.6	1.6	1.6	1.6	1.7	1.7	1.7	1.7	1.7	1.7	1.7	1.7	1.7	1.8	1.8	1.8	1.8	1.8	1.8	1.8	1.8	1.9	1.9	1.9	1.9	1.9	1.9	1.9
	Y_m	2.9	2.9	2.9	2.9	2.9	2.9	2.8	2.8	2.8	2.8	2.8	2.8	2.8	2.8	2.8	2.8	2.8	2.7	2.7	2.7	2.7	2.7	2.7	2.7	2.7	2.7	2.7	2.7	2.7	2.6	2.6	2.6	2.6	2.6	2.6	2.6	2.6	2.6	2.6	2.6	2.6	2.6	2.6	2.6	2.5	2.5	2.5	2.5	2.5	2.5	2.5	2.5	2.5	2.5	2.5	2.5	2.5	2.5
2.6	x_m	1.5	1.5	1.5	1.5	1.5	1.5	1.5	1.5	1.5	1.5	1.5	1.5	1.5	1.5	1.5	1.5	1.5	1.6	1.6	1.6	1.6	1.6	1.6	1.6	1.6	1.6	1.6	1.6	1.6	1.6	1.6	1.6	1.7	1.7	1.7	1.7	1.7	1.7	1.7	1.7	1.7	1.7	1.7	1.8	1.8	1.8	1.8	1.8	1.8	1.8	1.8	1.9	1.9	1.9	1.9	1.9	1.9	1.9
	Y_m	3.0	3.0	3.0	3.0	3.0	3.0	2.9	2.9	2.9	2.9	2.9	2.9	2.9	2.9	2.9	2.9	2.9	2.8	2.8	2.8	2.8	2.8	2.8	2.8	2.8	2.8	2.8	2.8	2.8	2.7	2.7	2.7	2.7	2.7	2.7	2.7	2.7	2.7	2.7	2.7	2.7	2.7	2.7	2.7	2.6	2.6	2.6	2.6	2.6	2.6	2.6	2.6	2.6	2.6	2.6	2.6	2.6	2.6
2.7	x_m	1.5	1.5	1.5	1.5	1.5	1.5	1.5	1.5	1.5	1.5	1.5	1.5	1.5	1.5	1.5	1.5	1.6	1.6	1.6	1.6	1.6	1.6	1.6	1.6	1.6	1.6	1.6	1.6	1.6	1.6	1.6	1.7	1.7	1.7	1.7	1.7	1.7	1.7	1.7	1.7	1.7	1.7	1.8	1.8	1.8	1.8	1.8	1.8	1.8	1.8	1.8	1.9	1.9	1.9	1.9	1.9	1.9	1.9
	Y_m	3.1	3.1	3.1	3.1	3.1	3.0	3.0	3.0	3.0	3.0	3.0	3.0	3.0	3.0	3.0	3.0	2.9	2.9	2.9	2.9	2.9	2.9	2.9	2.9	2.9	2.9	2.9	2.9	2.8	2.8	2.8	2.8	2.8	2.8	2.8	2.8	2.8	2.8	2.8	2.8	2.8	2.8	2.8	2.8	2.7	2.7	2.7	2.7	2.7	2.7	2.7	2.7	2.7	2.7	2.7	2.7	2.7	2.7
2.8	x_m	1.5	1.5	1.5	1.5	1.5	1.5	1.5	1.5	1.5	1.5	1.5	1.5	1.5	1.5	1.5	1.6	1.6	1.6	1.6	1.6	1.6	1.6	1.6	1.6	1.6	1.6	1.6	1.6	1.6	1.6	1.7	1.7	1.7	1.7	1.7	1.7	1.7	1.7	1.7	1.7	1.7	1.7	1.8	1.8	1.8	1.8	1.8	1.8	1.8	1.8	1.8	1.9	1.9	1.9	1.9	1.9	1.9	1.9
	Y_m	3.2	3.2	3.2	3.2	3.2	3.1	3.1	3.1	3.1	3.1	3.1	3.1	3.1	3.1	3.1	3.1	3.0	3.0	3.0	3.0	3.0	3.0	3.0	3.0	3.0	3.0	3.0	3.0	2.9	2.9	2.9	2.9	2.9	2.9	2.9	2.9	2.9	2.9	2.9	2.9	2.9	2.9	2.9	2.8	2.8	2.8	2.8	2.8	2.8	2.8	2.8	2.8	2.8	2.8	2.8	2.8	2.8	2.8
2.9	x_m	1.5	1.5	1.5	1.5	1.5	1.5	1.5	1.5	1.5	1.5	1.5	1.5	1.5	1.6	1.6	1.6	1.6	1.6	1.6	1.6	1.6	1.6	1.6	1.6	1.6	1.6	1.6	1.6	1.6	1.7	1.7	1.7	1.7	1.7	1.7	1.7	1.7	1.7	1.7	1.7	1.7	1.8	1.8	1.8	1.8	1.8	1.8	1.8	1.8	1.8	1.9	1.9	1.9	1.9	1.9	1.9	1.9	2.0
	Y_m	3.3	3.3	3.3	3.3	3.2	3.2	3.2	3.2	3.2	3.2	3.2	3.2	3.2	3.2	3.2	3.1	3.1	3.1	3.1	3.1	3.1	3.1	3.1	3.1	3.1	3.1	3.1	3.0	3.0	3.0	3.0	3.0	3.0	3.0	3.0	3.0	3.0	3.0	3.0	3.0	3.0	3.0	3.0	2.9	2.9	2.9	2.9	2.9	2.9	2.9	2.9	2.9	2.9	2.9	2.9	2.9	2.9	2.9
3.0	x_m	1.5	1.5	1.5	1.5	1.5	1.5	1.5	1.5	1.5	1.5	1.5	1.6	1.6	1.6	1.6	1.6	1.6	1.6	1.6	1.6	1.6	1.6	1.6	1.6	1.6	1.6	1.6	1.6	1.7	1.7	1.7	1.7	1.7	1.7	1.7	1.7	1.7	1.7	1.7	1.7	1.8	1.8	1.8	1.8	1.8	1.8	1.8	1.8	1.8	1.8	1.9	1.9	1.9	1.9	1.9	1.9	1.9	2.0
	Y_m	3.4	3.4	3.4	3.3	3.3	3.3	3.3	3.3	3.3	3.3	3.3	3.3	3.3	3.3	3.2	3.2	3.2	3.2	3.2	3.2	3.2	3.2	3.2	3.2	3.2	3.2	3.2	3.1	3.1	3.1	3.1	3.1	3.1	3.1	3.1	3.1	3.1	3.1	3.1	3.1	3.1	3.1	3.1	3.0	3.0	3.0	3.0	3.0	3.0	3.0	3.0	3.0	3.0	3.0	3.0	3.0	3.0	3.0
3.1	x_m	1.5	1.5	1.5	1.5	1.5	1.5	1.5	1.5	1.5	1.5	1.6	1.6	1.6	1.6	1.6	1.6	1.6	1.6	1.6	1.6	1.6	1.6	1.6	1.6	1.6	1.6	1.6	1.7	1.7	1.7	1.7	1.7	1.7	1.7	1.7	1.7	1.7	1.7	1.7	1.7	1.8	1.8	1.8	1.8	1.8	1.8	1.8	1.8	1.8	1.8	1.9	1.9	1.9	1.9	1.9	1.9	1.9	2.0
	Y_m	3.5	3.5	3.5	3.4	3.4	3.4	3.4	3.4	3.4	3.4	3.4	3.4	3.4	3.4	3.3	3.3	3.3	3.3	3.3	3.3	3.3	3.3	3.3	3.3	3.3	3.3	3.3	3.2	3.2	3.2	3.2	3.2	3.2	3.2	3.2	3.2	3.2	3.2	3.2	3.2	3.2	3.2	3.1	3.1	3.1	3.1	3.1	3.1	3.1	3.1	3.1	3.1	3.1	3.1	3.1	3.1	3.1	3.1
3.2	x_m	1.5	1.5	1.5	1.5	1.5	1.5	1.5	1.5	1.6	1.6	1.6	1.6	1.6	1.6	1.6	1.6	1.6	1.6	1.6	1.6	1.6	1.6	1.6	1.6	1.6	1.6	1.7	1.7	1.7	1.7	1.7	1.7	1.7	1.7	1.7	1.7	1.7	1.7	1.7	1.8	1.8	1.8	1.8	1.8	1.8	1.8	1.8	1.8	1.8	1.9	1.9	1.9	1.9	1.9	1.9	1.9	1.9	2.0
	Y_m	3.6	3.6	3.5	3.5	3.5	3.5	3.5	3.5	3.5	3.5	3.5	3.5	3.5	3.4	3.4	3.4	3.4	3.4	3.4	3.4	3.4	3.4	3.4	3.4	3.4	3.4	3.3	3.3	3.3	3.3	3.3	3.3	3.3	3.3	3.3	3.3	3.3	3.3	3.3	3.3	3.3	3.3	3.3	3.2	3.2	3.2	3.2	3.2	3.2	3.2	3.2	3.2	3.2	3.2	3.2	3.2	3.2	3.2
3.3	x_m	1.5	1.5	1.5	1.5	1.5	1.5	1.6	1.6	1.6	1.6	1.6	1.6	1.6	1.6	1.6	1.6	1.6	1.6	1.6	1.6	1.6	1.6	1.6	1.6	1.6	1.7	1.7	1.7	1.7	1.7	1.7	1.7	1.7	1.7	1.7	1.7	1.7	1.7	1.8	1.8	1.8	1.8	1.8	1.8	1.8	1.8	1.8	1.8	1.8	1.9	1.9	1.9	1.9	1.9	1.9	1.9	1.9	2.0
	Y_m	3.7	3.6	3.6	3.6	3.6	3.6	3.6	3.6	3.6	3.6	3.6	3.6	3.6	3.5	3.5	3.5	3.5	3.5	3.5	3.5	3.5	3.5	3.5	3.5	3.5	3.5	3.4	3.4	3.4	3.4	3.4	3.4	3.4	3.4	3.4	3.4	3.4	3.4	3.4	3.4	3.4	3.4	3.3	3.3	3.3	3.3	3.3	3.3	3.3	3.3	3.3	3.3	3.3	3.3	3.3	3.3	3.3	3.3
3.4	x_m	1.5	1.5	1.5	1.5	1.5	1.6	1.6	1.6	1.6	1.6	1.6	1.6	1.6	1.6	1.6	1.6	1.6	1.6	1.6	1.6	1.6	1.6	1.6	1.6	1.7	1.7	1.7	1.7	1.7	1.7	1.7	1.7	1.7	1.7	1.7	1.7	1.7	1.7	1.8	1.8	1.8	1.8	1.8	1.8	1.8	1.8	1.8	1.8	1.9	1.9	1.9	1.9	1.9	1.9	1.9	1.9	1.9	2.0
	Y_m	3.8	3.7	3.7	3.7	3.7	3.7	3.7	3.7	3.7	3.7	3.7	3.7	3.6	3.6	3.6	3.6	3.6	3.6	3.6	3.6	3.6	3.6	3.6	3.6	3.6	3.5	3.5	3.5	3.5	3.5	3.5	3.5	3.5	3.5	3.5	3.5	3.5	3.5	3.5	3.5	3.5	3.5	3.4	3.4	3.4	3.4	3.4	3.4	3.4	3.4	3.4	3.4	3.4	3.4	3.4	3.4	3.4	3.4
3.5	x_m	1.5	1.5	1.5	1.6	1.6	1.6	1.6	1.6	1.6	1.6	1.6	1.6	1.6	1.6	1.6	1.6	1.6	1.6	1.6	1.6	1.6	1.6	1.6	1.7	1.7	1.7	1.7	1.7	1.7	1.7	1.7	1.7	1.7	1.7	1.7	1.7	1.7	1.8	1.8	1.8	1.8	1.8	1.8	1.8	1.8	1.8	1.8	1.9	1.9	1.9	1.9	1.9	1.9	1.9	1.9	1.9	1.9	2.0
	Y_m	3.8	3.8	3.8	3.8	3.8	3.8	3.8	3.8	3.8	3.8	3.8	3.8	3.7	3.7	3.7	3.7	3.7	3.7	3.7	3.7	3.7	3.7	3.7	3.7	3.7	3.6	3.6	3.6	3.6	3.6	3.6	3.6	3.6	3.6	3.6	3.6	3.6	3.6	3.6	3.6	3.6	3.6	3.5	3.5	3.5	3.5	3.5	3.5	3.5	3.5	3.5	3.5	3.5	3.5	3.5	3.5	3.5	3.5
3.6	x_m	1.5	1.5	1.6	1.6	1.6	1.6	1.6	1.6	1.6	1.6	1.6	1.6	1.6	1.6	1.6	1.6	1.6	1.6	1.6	1.6	1.6	1.6	1.7	1.7	1.7	1.7	1.7	1.7	1.7	1.7	1.7	1.7	1.7	1.7	1.7	1.7	1.8	1.8	1.8	1.8	1.8	1.8	1.8	1.8	1.8	1.8	1.8	1.9	1.9	1.9	1.9	1.9	1.9	1.9	1.9	1.9	1.9	2.0
	Y_m	3.9	3.9	3.9	3.9	3.9	3.9	3.9	3.9	3.9	3.9	3.9	3.8	3.8	3.8	3.8	3.8	3.8	3.8	3.8	3.8	3.8	3.8	3.8	3.8	3.7	3.7	3.7	3.7	3.7	3.7	3.7	3.7	3.7	3.7	3.7	3.7	3.7	3.7	3.7	3.7	3.7	3.7	3.6	3.6	3.6	3.6	3.6	3.6	3.6	3.6	3.6	3.6	3.6	3.6	3.6	3.6	3.6	3.6
3.7	x_m	1.6	1.6	1.6	1.6	1.6	1.6	1.6	1.6	1.6	1.6	1.6	1.6	1.6	1.6	1.6	1.6	1.6	1.6	1.6	1.6	1.6	1.7	1.7	1.7	1.7	1.7	1.7	1.7	1.7	1.7	1.7	1.7	1.7	1.7	1.7	1.7	1.8	1.8	1.8	1.8	1.8	1.8	1.8	1.8	1.8	1.8	1.8	1.9	1.9	1.9	1.9	1.9	1.9	1.9	1.9	1.9	1.9	2.0
	Y_m	4.0	4.0	4.0	4.0	4.0	4.0	4.0	4.0	4.0	4.0	4.0	3.9	3.9	3.9	3.9	3.9	3.9	3.9	3.9	3.9	3.9	3.9	3.9	3.9	3.8	3.8	3.8	3.8	3.8	3.8	3.8	3.8	3.8	3.8	3.8	3.8	3.8	3.8	3.8	3.8	3.8	3.7	3.7	3.7	3.7	3.7	3.7	3.7	3.7	3.7	3.7	3.7	3.7	3.7	3.7	3.7	3.7	3.7
3.8	x_m	1.6	1.6	1.6	1.6	1.6	1.6	1.6	1.6	1.6	1.6	1.6	1.6	1.6	1.6	1.6	1.6	1.6	1.6	1.6	1.6	1.7	1.7	1.7	1.7	1.7	1.7	1.7	1.7	1.7	1.7	1.7	1.7	1.7	1.7	1.7	1.8	1.8	1.8	1.8	1.8	1.8	1.8	1.8	1.8	1.8	1.8	1.9	1.9	1.9	1.9	1.9	1.9	1.9	1.9	1.9	1.9	2.0	2.0
	Y_m	4.1	4.1	4.1	4.1	4.1	4.1	4.1	4.1	4.1	4.1	4.0	4.0	4.0	4.0	4.0	4.0	4.0	4.0	4.0	4.0	4.0	4.0	4.0	3.9	3.9	3.9	3.9	3.9	3.9	3.9	3.9	3.9	3.9	3.9	3.9	3.9	3.9	3.9	3.9	3.9	3.9	3.8	3.8	3.8	3.8	3.8	3.8	3.8	3.8	3.8	3.8	3.8	3.8	3.8	3.8	3.8	3.8	3.8
3.9	x_m	1.6	1.6	1.6	1.6	1.6	1.6	1.6	1.6	1.6	1.6	1.6	1.6	1.6	1.6	1.6	1.6	1.6	1.6	1.7	1.7	1.7	1.7	1.7	1.7	1.7	1.7	1.7	1.7	1.7	1.7	1.7	1.7	1.7	1.7	1.8	1.8	1.8	1.8	1.8	1.8	1.8	1.8	1.8	1.8	1.8	1.8	1.8	1.9	1.9	1.9	1.9	1.9	1.9	1.9	1.9	1.9	2.0	2.0
	Y_m	4.2	4.2	4.2	4.2	4.2	4.2	4.2	4.2	4.2	4.2	4.1	4.1	4.1	4.1	4.1	4.1	4.1	4.1	4.1	4.1	4.1	4.1	4.1	4.0	4.0	4.0	4.0	4.0	4.0	4.0	4.0	4.0	4.0	4.0	4.0	4.0	4.0	4.0	4.0	4.0	4.0	3.9	3.9	3.9	3.9	3.9	3.9	3.9	3.9	3.9	3.9	3.9	3.9	3.9	3.9	3.9	3.9	3.9
4.0	x_m	1.6	1.6	1.6	1.6	1.6	1.6	1.6	1.6	1.6	1.6	1.6	1.6	1.6	1.6	1.6	1.6	1.6	1.7	1.7	1.7	1.7	1.7	1.7	1.7	1.7	1.7	1.7	1.7	1.7	1.7	1.7	1.7	1.7	1.8	1.8	1.8	1.8	1.8	1.8	1.8	1.8	1.8	1.8	1.8	1.8	1.8	1.9	1.9	1.9	1.9	1.9	1.9	1.9	1.9	1.9	1.9	2.0	2.0
	Y_m	4.3	4.3	4.3	4.3	4.3	4.3	4.3	4.3	4.3	4.2	4.2	4.2	4.2	4.2	4.2	4.2	4.2	4.2	4.2	4.2	4.2	4.2	4.2	4.1	4.1	4.1	4.1	4.1	4.1	4.1	4.1	4.1	4.1	4.1	4.1	4.1	4.1	4.1	4.1	4.1	4.1	4.0	4.0	4.0	4.0	4.0	4.0	4.0	4.0	4.0	4.0	4.0	4.0	4.0	4.0	4.0	4.0	4.0
4.1	x_m	1.6	1.6	1.6	1.6	1.6	1.6	1.6	1.6	1.6	1.6	1.6	1.6	1.6	1.6	1.6	1.6	1.7	1.7	1.7	1.7	1.7	1.7	1.7	1.7	1.7	1.7	1.7	1.7	1.7	1.7	1.7	1.7	1.8	1.8	1.8	1.8	1.8	1.8	1.8	1.8	1.8	1.8	1.8	1.8	1.8	1.8	1.9	1.9	1.9	1.9	1.9	1.9	1.9	1.9	1.9	1.9	2.0	2.0
	Y_m	4.4	4.4	4.4	4.4	4.4	4.4	4.4	4.4	4.4	4.3	4.3	4.3	4.3	4.3	4.3	4.3	4.3	4.3	4.3	4.3	4.3	4.3	4.3	4.2	4.2	4.2	4.2	4.2	4.2	4.2	4.2	4.2	4.2	4.2	4.2	4.2	4.2	4.2	4.2	4.2	4.1	4.1	4.1	4.1	4.1	4.1	4.1	4.1	4.1	4.1	4.1	4.1	4.1	4.1	4.1	4.1	4.1	4.1
4.2	x_m	1.6	1.6	1.6	1.6	1.6	1.6	1.6	1.6	1.6	1.6	1.6	1.6	1.6	1.6	1.6	1.7	1.7	1.7	1.7	1.7	1.7	1.7	1.7	1.7	1.7	1.7	1.7	1.7	1.7	1.7	1.7	1.7	1.8	1.8	1.8	1.8	1.8	1.8	1.8	1.8	1.8	1.8	1.8	1.8	1.8	1.8	1.9	1.9	1.9	1.9	1.9	1.9	1.9	1.9	1.9	1.9	2.0	2.0
	Y_m	4.5	4.5	4.5	4.5	4.5	4.5	4.5	4.5	4.4	4.4	4.4	4.4	4.4	4.4	4.4	4.4	4.4	4.4	4.4	4.4	4.4	4.4	4.3	4.3	4.3	4.3	4.3	4.3	4.3	4.3	4.3	4.3	4.3	4.3	4.3	4.3	4.3	4.3	4.3	4.3	4.2	4.2	4.2	4.2	4.2	4.2	4.2	4.2	4.2	4.2	4.2	4.2	4.2	4.2	4.2	4.2	4.2	4.2
4.3	x_m	1.6	1.6	1.6	1.6	1.6	1.6	1.6	1.6	1.6	1.6	1.6	1.6	1.6	1.6	1.7	1.7	1.7	1.7	1.7	1.7	1.7	1.7	1.7	1.7	1.7	1.7	1.7	1.7	1.7	1.7	1.7	1.7	1.8	1.8	1.8	1.8	1.8	1.8	1.8	1.8	1.8	1.8	1.8	1.8	1.8	1.9	1.9	1.9	1.9	1.9	1.9	1.9	1.9	1.9	1.9	1.9	2.0	2.0
	Y_m	4.6	4.6	4.6	4.6	4.6	4.6	4.6	4.6	4.5	4.5	4.5	4.5	4.5	4.5	4.5	4.5	4.5	4.5	4.5	4.5	4.5	4.4	4.4	4.4	4.4	4.4	4.4	4.4	4.4	4.4	4.4	4.4	4.4	4.4	4.4	4.4	4.4	4.4	4.4	4.4	4.3	4.3	4.3	4.3	4.3	4.3	4.3	4.3	4.3	4.3	4.3	4.3	4.3	4.3	4.3	4.3	4.3	4.3
4.4	x_m	1.6	1.6	1.6	1.6	1.6	1.6	1.6	1.6	1.6	1.6	1.6	1.6	1.6	1.7	1.7	1.7	1.7	1.7	1.7	1.7	1.7	1.7	1.7	1.7	1.7	1.7	1.7	1.7	1.7	1.7	1.7	1.8	1.8	1.8	1.8	1.8	1.8	1.8	1.8	1.8	1.8	1.8	1.8	1.8	1.8	1.9	1.9	1.9	1.9	1.9	1.9	1.9	1.9	1.9	1.9	1.9	2.0	2.0
	Y_m	4.7	4.7	4.7	4.7	4.7	4.7	4.7	4.6	4.6	4.6	4.6	4.6	4.6	4.6	4.6	4.6	4.6	4.6	4.6	4.6	4.6	4.5	4.5	4.5	4.5	4.5	4.5	4.5	4.5	4.5	4.5	4.5	4.5	4.5	4.5	4.5	4.5	4.5	4.5	4.5	4.4	4.4	4.4	4.4	4.4	4.4	4.4	4.4	4.4	4.4	4.4	4.4	4.4	4.4	4.4	4.4	4.4	4.4
4.5	x_m	1.6	1.6	1.6	1.6	1.6	1.6	1.6	1.6	1.6	1.6	1.6	1.6	1.7	1.7	1.7	1.7	1.7	1.7	1.7	1.7	1.7	1.7	1.7	1.7	1.7	1.7	1.7	1.7	1.7	1.7	1.8	1.8	1.8	1.8	1.8	1.8	1.8	1.8	1.8	1.8	1.8	1.8	1.8	1.8	1.8	1.9	1.9	1.9	1.9	1.9	1.9	1.9	1.9	1.9	1.9	1.9	2.0	2.0
	Y_m	4.8	4.8	4.8	4.8	4.8	4.8	4.8	4.7	4.7	4.7	4.7	4.7	4.7	4.7	4.7	4.7	4.7	4.7	4.7	4.7	4.7	4.6	4.6	4.6	4.6	4.6	4.6	4.6	4.6	4.6	4.6	4.6	4.6	4.6	4.6	4.6	4.6	4.6	4.6	4.5	4.5	4.5	4.5	4.5	4.5	4.5	4.5	4.5	4.5	4.5	4.5	4.5	4.5	4.5	4.5	4.5	4.5	4.5
4.6	x_m	1.6	1.6	1.6	1.6	1.6	1.6	1.6	1.6	1.6	1.6	1.6	1.7	1.7	1.7	1.7	1.7	1.7	1.7	1.7	1.7	1.7	1.7	1.7	1.7	1.7	1.7	1.7	1.7	1.7	1.7	1.8	1.8	1.8	1.8	1.8	1.8	1.8	1.8	1.8	1.8	1.8	1.8	1.8	1.8	1.9	1.9	1.9	1.9	1.9	1.9	1.9	1.9	1.9	1.9	1.9	1.9	2.0	2.0
	Y_m	4.9	4.9	4.9	4.9	4.9	4.9	4.9	4.8	4.8	4.8	4.8	4.8	4.8	4.8	4.8	4.8	4.8	4.8	4.8	4.8	4.8	4.7	4.7	4.7	4.7	4.7	4.7	4.7	4.7	4.7	4.7	4.7	4.7	4.7	4.7	4.7	4.7	4.7	4.7	4.6	4.6	4.6	4.6	4.6	4.6	4.6	4.6	4.6	4.6	4.6	4.6	4.6	4.6	4.6	4.6	4.6	4.6	4.6
4.7	x_m	1.6	1.6	1.6	1.6	1.6	1.6	1.6	1.6	1.6	1.6	1.7	1.7	1.7	1.7	1.7	1.7	1.7	1.7	1.7	1.7	1.7	1.7	1.7	1.7	1.7	1.7	1.7	1.7	1.7	1.8	1.8	1.8	1.8	1.8	1.8	1.8	1.8	1.8	1.8	1.8	1.8	1.8	1.8	1.8	1.9	1.9	1.9	1.9	1.9	1.9	1.9	1.9	1.9	1.9	1.9	1.9	2.0	2.0
	Y_m	5.0	5.0	5.0	5.0	5.0	5.0	4.9	4.9	4.9	4.9	4.9	4.9	4.9	4.9	4.9	4.9	4.9	4.9	4.9	4.9	4.9	4.8	4.8	4.8	4.8	4.8	4.8	4.8	4.8	4.8	4.8	4.8	4.8	4.8	4.8	4.8	4.8	4.8	4.8	4.7	4.7	4.7	4.7	4.7	4.7	4.7	4.7	4.7	4.7	4.7	4.7	4.7	4.7	4.7	4.7	4.7	4.7	4.7
4.8	x_m	1.6	1.6	1.6	1.6	1.6	1.6	1.6	1.6	1.6	1.7	1.7	1.7	1.7	1.7	1.7	1.7	1.7	1.7	1.7	1.7	1.7	1.7	1.7	1.7	1.7	1.7	1.7	1.7	1.8	1.8	1.8	1.8	1.8	1.8	1.8	1.8	1.8	1.8	1.8	1.8	1.8	1.8	1.8	1.8	1.9	1.9	1.9	1.9	1.9	1.9	1.9	1.9	1.9	1.9	1.9	2.0	2.0	2.0
	Y_m	5.1	5.1	5.1	5.1	5.1	5.0	5.0	5.0	5.0	5.0	5.0	5.0	5.0	5.0	5.0	5.0	5.0	5.0	5.0	5.0	4.9	4.9	4.9	4.9	4.9	4.9	4.9	4.9	4.9	4.9	4.9	4.9	4.9	4.9	4.9	4.9	4.9	4.9	4.9	4.8	4.8	4.8	4.8	4.8	4.8	4.8	4.8	4.8	4.8	4.8	4.8	4.8	4.8	4.8	4.8	4.8	4.8	4.8
4.9	x_m	1.6	1.6	1.6	1.6	1.6	1.6	1.6	1.6	1.7	1.7	1.7	1.7	1.7	1.7	1.7	1.7	1.7	1.7	1.7	1.7	1.7	1.7	1.7	1.7	1.7	1.7	1.7	1.7	1.8	1.8	1.8	1.8	1.8	1.8	1.8	1.8	1.8	1.8	1.8	1.8	1.8	1.8	1.8	1.9	1.9	1.9	1.9	1.9	1.9	1.9	1.9	1.9	1.9	1.9	1.9	2.0	2.0	2.0
	Y_m	5.2	5.2	5.2	5.2	5.2	5.1	5.1	5.1	5.1	5.1	5.1	5.1	5.1	5.1	5.1	5.1	5.1	5.1	5.1	5.1	5.0	5.0	5.0	5.0	5.0	5.0	5.0	5.0	5.0	5.0	5.0	5.0	5.0	5.0	5.0	5.0	5.0	5.0	5.0	4.9	4.9	4.9	4.9	4.9	4.9	4.9	4.9	4.9	4.9	4.9	4.9	4.9	4.9	4.9	4.9	4.9	4.9	4.9
5.0	x_m	1.6	1.6	1.6	1.6	1.6	1.6	1.6	1.7	1.7	1.7	1.7	1.7	1.7	1.7	1.7	1.7	1.7	1.7	1.7	1.7	1.7	1.7	1.7	1.7	1.7	1.7	1.7	1.8	1.8	1.8	1.8	1.8	1.8	1.8	1.8	1.8	1.8	1.8	1.8	1.8	1.8	1.8	1.9	1.9	1.9	1.9	1.9	1.9	1.9	1.9	1.9	1.9	1.9	1.9	1.9	2.0	2.0	2.0
	Y_m	5.3	5.3	5.3	5.3	5.2	5.2	5.2	5.2	5.2	5.2	5.2	5.2	5.2	5.2	5.2	5.2	5.2	5.2	5.2	5.1	5.1	5.1	5.1	5.1	5.1	5.1	5.1	5.1	5.1	5.1	5.1	5.1	5.1	5.1	5.1	5.1	5.1	5.1	5.1	5.0	5.0	5.0	5.0	5.0	5.0	5.0	5.0	5.0	5.0	5.0	5.0	5.0	5.0	5.0	5.0	5.0	5.0	5.0

VALUES OF Y_4

Figure E–10. Tabulation of coordinates of maximum density point. (Sheet 1 of 2.) 101–D–518, 288–D–2839.

RAPID METHOD OF COMPACTION CONTROL

TABULATION OF COORDINATES OF MAXIMUM DENSITY POINT

Values in the table are coordinates of the maximum density point of the compaction curve from Point A, the extreme left point. The upper value is X_m or the value in the direction of water added, the lower value is Y_m or the value in the direction of density. Table can be used only if the three points A, B and C are spaced 2 % apart horizontally.

Y_2 is the density of point B minus the density of point A.
Y_4 is the density of point C minus the density of point A, Y_4 may be negative.

EXAMPLE

Y_2 = 119.2 − 117.6 = +1.6
Y_4 = 114.2 − 117.6 = −3.4

From table, X_m = 1.5
Y_m = 1.8

Value of maximum density point O =
Density of point A + Y_m =
117.6 + 1.8 = 119.4

Added water at maximum density point O, Z_m = Added water at Point A + X_m =
0 + 1.5 = 1.5 %

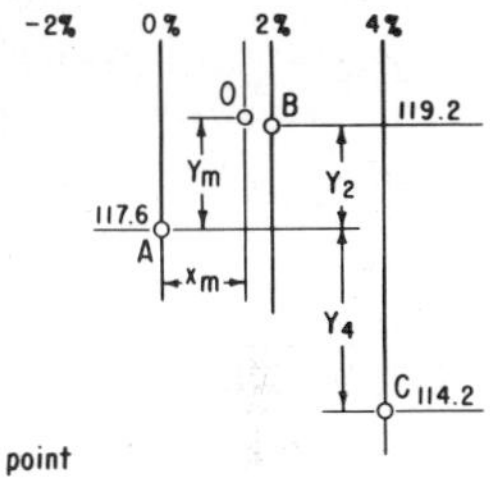

NOTE

If point A were at − 2 %, Z_m would be −2.0 + 1.5 = −0.5 %
then B would be at 0%
and C would be at +2%

VALUES OF Y_2 (rows) / VALUES OF Y_4 (columns)

Y_2		-0.2	-0.1	0	+0.1	+0.2	+0.3	+0.4	+0.5	+0.6	+0.7	+0.8	+0.9	+1.0	+1.1	+1.2	+1.3	+1.4	+1.5	+1.6	+1.7	+1.8	+1.9	+2.0	+2.1	+2.2	+2.3	+2.4	+2.5	+2.6	+2.7	+2.8	+2.9	+3.0	+3.1	+3.2	+3.3	+3.4	+3.5	+3.6	+3.7	+3.8	+3.9	+4.0	+4.1	+4.2	+4.3	+4.4	+4.5	+4.6	+4.7	+4.8	+4.9	+5.0
0	X_m	0
	Y_m	0
0.1	X_m	1.5	1.7	2.0	3.0
	Y_m	0.1	0.1	0.1	0.1
0.2	X_m	1.7	1.8	2.0	2.3	3.0
	Y_m	0.2	0.2	0.2	0.2	0.2
0.3	X_m	1.8	1.9	2.0	2.2	2.5	3.0
	Y_m	0.3	0.3	0.3	0.3	0.3	0.3
0.4	X_m	1.8	1.9	2.0	2.1	2.3	2.6	3.0
	Y_m	0.4	0.4	0.4	0.4	0.4	0.4	0.5
0.5	X_m	1.8	1.9	2.0	2.1	2.3	2.4	2.7	3.0
	Y_m	0.5	0.5	0.5	0.5	0.5	0.5	0.5	0.6
0.6	X_m	1.9	1.9	2.0	2.1	2.2	2.3	2.5	2.7	3.0
	Y_m	0.6	0.6	0.6	0.6	0.6	0.6	0.6	0.6	0.7
0.7	X_m	1.9	1.9	2.0	2.1	2.2	2.3	2.4	2.6	2.7	3.0
	Y_m	0.7	0.7	0.7	0.7	0.7	0.7	0.7	0.7	0.8	0.8
0.8	X_m	1.9	1.9	2.0	2.1	2.1	2.2	2.3	2.5	2.6	2.8	3.0
	Y_m	0.8	0.8	0.8	0.8	0.8	0.8	0.8	0.8	0.8	0.9	0.9
0.9	X_m	1.9	1.9	2.0	2.1	2.1	2.2	2.3	2.4	2.5	2.6	2.8	3.0
	Y_m	0.9	0.9	0.9	0.9	0.9	0.9	0.9	0.9	0.9	1.0	1.0	1.0
1.0	X_m	1.9	2.0	2.0	2.1	2.1	2.2	2.3	2.3	2.4	2.5	2.7	2.8	3.0
	Y_m	1.0	1.0	1.0	1.0	1.0	1.0	1.0	1.0	1.0	1.0	1.1	1.1	1.1
1.1	X_m	1.9	2.0	2.0	2.0	2.1	2.2	2.2	2.3	2.4	2.5	2.6	2.7	2.8	3.0
	Y_m	1.1	1.1	1.1	1.1	1.1	1.1	1.1	1.1	1.1	1.1	1.2	1.2	1.2	1.2
1.2	X_m	1.9	2.0	2.0	2.0	2.1	2.1	2.2	2.3	2.3	2.4	2.5	2.6	2.7	2.8	3.0
	Y_m	1.2	1.2	1.2	1.2	1.2	1.2	1.2	1.2	1.2	1.2	1.3	1.3	1.3	1.3	1.4
1.3	X_m	1.9	2.0	2.0	2.0	2.1	2.1	2.2	2.2	2.3	2.4	2.4	2.5	2.6	2.7	2.9	3.0
	Y_m	1.3	1.3	1.3	1.3	1.3	1.3	1.3	1.3	1.3	1.3	1.3	1.4	1.4	1.4	1.4	1.5
1.4	X_m	1.9	2.0	2.0	2.0	2.1	2.1	2.2	2.2	2.3	2.3	2.4	2.5	2.6	2.6	2.7	2.9	3.0
	Y_m	1.4	1.4	1.4	1.4	1.4	1.4	1.4	1.4	1.4	1.4	1.4	1.5	1.5	1.5	1.5	1.5	1.6
1.5	X_m	1.9	2.0	2.0	2.0	2.1	2.1	2.2	2.2	2.3	2.3	2.4	2.4	2.5	2.6	2.7	2.8	2.9	3.0
	Y_m	1.5	1.5	1.5	1.5	1.5	1.5	1.5	1.5	1.5	1.5	1.5	1.5	1.6	1.6	1.6	1.6	1.7	1.7
1.6	X_m	1.9	2.0	2.0	2.0	2.1	2.1	2.1	2.2	2.2	2.3	2.3	2.4	2.5	2.5	2.6	2.7	2.8	2.9	3.0
	Y_m	1.6	1.6	1.6	1.6	1.6	1.6	1.6	1.6	1.6	1.6	1.6	1.6	1.7	1.7	1.7	1.7	1.7	1.8	1.8
1.7	X_m	1.9	2.0	2.0	2.0	2.1	2.1	2.1	2.2	2.2	2.3	2.3	2.4	2.4	2.5	2.5	2.6	2.7	2.8	2.9	3.0
	Y_m	1.7	1.7	1.7	1.7	1.7	1.7	1.7	1.7	1.7	1.7	1.7	1.7	1.8	1.8	1.8	1.8	1.8	1.8	1.9	1.9
1.8	X_m	1.9	2.0	2.0	2.0	2.1	2.1	2.1	2.2	2.2	2.2	2.3	2.3	2.4	2.4	2.5	2.6	2.6	2.7	2.8	2.9	3.0
	Y_m	1.8	1.8	1.8	1.8	1.8	1.8	1.8	1.8	1.8	1.8	1.8	1.8	1.8	1.9	1.9	1.9	1.9	1.9	2.0	2.0	2.0
1.9	X_m	2.0	2.0	2.0	2.0	2.1	2.1	2.1	2.2	2.2	2.2	2.3	2.3	2.4	2.4	2.5	2.5	2.6	2.7	2.7	2.8	2.9	3.0
	Y_m	1.9	1.9	1.9	1.9	1.9	1.9	1.9	1.9	1.9	1.9	1.9	1.9	1.9	2.0	2.0	2.0	2.0	2.0	2.0	2.1	2.1	2.1
2.0	X_m	2.0	2.0	2.0	2.0	2.1	2.1	2.1	2.1	2.2	2.2	2.3	2.3	2.3	2.4	2.4	2.5	2.5	2.6	2.6	2.7	2.8	2.9	3.0
	Y_m	2.0	2.0	2.0	2.0	2.0	2.0	2.0	2.0	2.0	2.0	2.0	2.0	2.0	2.1	2.1	2.1	2.1	2.1	2.1	2.2	2.2	2.2	2.3
2.1	X_m	2.0	2.0	2.0	2.0	2.0	2.1	2.1	2.1	2.2	2.2	2.2	2.3	2.3	2.4	2.4	2.4	2.5	2.6	2.6	2.7	2.7	2.8	2.9	3.0
	Y_m	2.1	2.1	2.1	2.1	2.1	2.1	2.1	2.1	2.1	2.1	2.1	2.1	2.1	2.1	2.2	2.2	2.2	2.2	2.2	2.2	2.3	2.3	2.3	2.4
2.2	X_m	2.0	2.0	2.0	2.0	2.0	2.1	2.1	2.1	2.2	2.2	2.2	2.3	2.3	2.3	2.4	2.4	2.5	2.5	2.6	2.6	2.7	2.8	2.8	2.9	3.0
	Y_m	2.2	2.2	2.2	2.2	2.2	2.2	2.2	2.2	2.2	2.2	2.2	2.2	2.2	2.2	2.3	2.3	2.3	2.3	2.3	2.3	2.4	2.4	2.4	2.4	2.5
2.3	X_m	2.0	2.0	2.0	2.0	2.0	2.1	2.1	2.1	2.1	2.2	2.2	2.2	2.3	2.3	2.4	2.4	2.4	2.5	2.5	2.6	2.6	2.7	2.8	2.8	2.9	3.0
	Y_m	2.3	2.3	2.3	2.3	2.3	2.3	2.3	2.3	2.3	2.3	2.3	2.3	2.3	2.3	2.4	2.4	2.4	2.4	2.4	2.4	2.4	2.5	2.5	2.5	2.6	2.6
2.4	X_m	2.0	2.0	2.0	2.0	2.0	2.1	2.1	2.1	2.1	2.2	2.2	2.2	2.3	2.3	2.3	2.4	2.4	2.5	2.5	2.5	2.6	2.7	2.7	2.8	2.8	2.9	3.0
	Y_m	2.4	2.4	2.4	2.4	2.4	2.4	2.4	2.4	2.4	2.4	2.4	2.4	2.4	2.4	2.4	2.5	2.5	2.5	2.5	2.5	2.5	2.6	2.6	2.6	2.6	2.7	2.7
2.5	X_m	2.0	2.0	2.0	2.0	2.0	2.1	2.1	2.1	2.1	2.2	2.2	2.2	2.3	2.3	2.3	2.4	2.4	2.4	2.5	2.5	2.6	2.6	2.7	2.7	2.8	2.9	2.9	3.0
	Y_m	2.5	2.5	2.5	2.5	2.5	2.5	2.5	2.5	2.5	2.5	2.5	2.5	2.5	2.5	2.5	2.6	2.6	2.6	2.6	2.6	2.6	2.6	2.7	2.7	2.7	2.7	2.8	2.8
2.6	X_m	2.0	2.0	2.0	2.0	2.0	2.1	2.1	2.1	2.1	2.2	2.2	2.2	2.2	2.3	2.3	2.3	2.4	2.4	2.4	2.5	2.5	2.6	2.6	2.7	2.7	2.8	2.9	2.9	3.0
	Y_m	2.6	2.6	2.6	2.6	2.6	2.6	2.6	2.6	2.6	2.6	2.6	2.6	2.6	2.6	2.6	2.7	2.7	2.7	2.7	2.7	2.7	2.7	2.8	2.8	2.8	2.8	2.9	2.9	2.9
2.7	X_m	2.0	2.0	2.0	2.0	2.0	2.1	2.1	2.1	2.1	2.1	2.2	2.2	2.2	2.3	2.3	2.3	2.3	2.4	2.4	2.5	2.5	2.5	2.6	2.6	2.7	2.7	2.8	2.9	2.9	3.0
	Y_m	2.7	2.7	2.7	2.7	2.7	2.7	2.7	2.7	2.7	2.7	2.7	2.7	2.7	2.7	2.7	2.8	2.8	2.8	2.8	2.8	2.8	2.8	2.8	2.9	2.9	2.9	2.9	3.0	3.0	3.0
2.8	X_m	2.0	2.0	2.0	2.0	2.0	2.1	2.1	2.1	2.1	2.1	2.2	2.2	2.2	2.2	2.3	2.3	2.3	2.4	2.4	2.4	2.5	2.5	2.6	2.6	2.6	2.7	2.7	2.8	2.9	2.9	3.0
	Y_m	2.8	2.8	2.8	2.8	2.8	2.8	2.8	2.8	2.8	2.8	2.8	2.8	2.8	2.8	2.8	2.8	2.9	2.9	2.9	2.9	2.9	2.9	2.9	3.0	3.0	3.0	3.0	3.1	3.1	3.1	3.2
2.9	X_m	2.0	2.0	2.0	2.0	2.0	2.1	2.1	2.1	2.1	2.1	2.2	2.2	2.2	2.2	2.3	2.3	2.3	2.3	2.4	2.4	2.4	2.5	2.5	2.6	2.6	2.7	2.7	2.8	2.8	2.9	2.9	3.0
	Y_m	2.9	2.9	2.9	2.9	2.9	2.9	2.9	2.9	2.9	2.9	2.9	2.9	2.9	2.9	2.9	2.9	3.0	3.0	3.0	3.0	3.0	3.0	3.0	3.0	3.1	3.1	3.1	3.1	3.2	3.2	3.2	3.3
3.0	X_m	2.0	2.0	2.0	2.0	2.0	2.1	2.1	2.1	2.1	2.1	2.2	2.2	2.2	2.2	2.3	2.3	2.3	2.3	2.4	2.4	2.4	2.5	2.5	2.5	2.6	2.6	2.7	2.7	2.8	2.8	2.9	2.9	3.0
	Y_m	3.0	3.0	3.0	3.0	3.0	3.0	3.0	3.0	3.0	3.0	3.0	3.0	3.0	3.0	3.0	3.0	3.1	3.1	3.1	3.1	3.1	3.1	3.1	3.1	3.2	3.2	3.2	3.2	3.2	3.3	3.3	3.3	3.4
3.1	X_m	2.0	2.0	2.0	2.0	2.0	2.1	2.1	2.1	2.1	2.1	2.1	2.2	2.2	2.2	2.2	2.3	2.3	2.3	2.3	2.4	2.4	2.4	2.5	2.5	2.5	2.6	2.6	2.7	2.7	2.8	2.8	2.9	2.9	3.0
	Y_m	3.1	3.1	3.1	3.1	3.1	3.1	3.1	3.1	3.1	3.1	3.1	3.1	3.1	3.1	3.1	3.1	3.2	3.2	3.2	3.2	3.2	3.2	3.2	3.2	3.3	3.3	3.3	3.3	3.3	3.4	3.4	3.4	3.5	3.5
3.2	X_m	2.0	2.0	2.0	2.0	2.0	2.0	2.1	2.1	2.1	2.1	2.1	2.2	2.2	2.2	2.2	2.3	2.3	2.3	2.3	2.4	2.4	2.4	2.5	2.5	2.5	2.6	2.6	2.6	2.7	2.7	2.8	2.8	2.9	2.9	3.0
	Y_m	3.2	3.2	3.2	3.2	3.2	3.2	3.2	3.2	3.2	3.2	3.2	3.2	3.2	3.2	3.2	3.2	3.2	3.3	3.3	3.3	3.3	3.3	3.3	3.3	3.3	3.4	3.4	3.4	3.4	3.4	3.5	3.5	3.5	3.6	3.6
3.3	X_m	2.0	2.0	2.0	2.0	2.0	2.0	2.1	2.1	2.1	2.1	2.1	2.2	2.2	2.2	2.2	2.2	2.3	2.3	2.3	2.3	2.4	2.4	2.4	2.5	2.5	2.5	2.6	2.6	2.6	2.7	2.7	2.8	2.8	2.9	2.9	3.0
	Y_m	3.3	3.3	3.3	3.3	3.3	3.3	3.3	3.3	3.3	3.3	3.3	3.3	3.3	3.3	3.3	3.3	3.3	3.4	3.4	3.4	3.4	3.4	3.4	3.4	3.4	3.5	3.5	3.5	3.5	3.5	3.6	3.6	3.6	3.6	3.7	3.7
3.4	X_m	2.0	2.0	2.0	2.0	2.0	2.0	2.1	2.1	2.1	2.1	2.1	2.2	2.2	2.2	2.2	2.2	2.3	2.3	2.3	2.3	2.4	2.4	2.4	2.4	2.5	2.5	2.5	2.6	2.6	2.7	2.7	2.7	2.8	2.8	2.9	2.9	3.0
	Y_m	3.4	3.4	3.4	3.4	3.4	3.4	3.4	3.4	3.4	3.4	3.4	3.4	3.4	3.4	3.4	3.4	3.4	3.5	3.5	3.5	3.5	3.5	3.5	3.5	3.5	3.5	3.6	3.6	3.6	3.6	3.6	3.7	3.7	3.7	3.8	3.8	3.8
3.5	X_m	2.0	2.0	2.0	2.0	2.0	2.0	2.1	2.1	2.1	2.1	2.1	2.1	2.2	2.2	2.2	2.2	2.3	2.3	2.3	2.3	2.3	2.4	2.4	2.4	2.5	2.5	2.5	2.6	2.6	2.6	2.7	2.7	2.7	2.8	2.8	2.9	2.9	3.0
	Y_m	3.5	3.5	3.5	3.5	3.5	3.5	3.5	3.5	3.5	3.5	3.5	3.5	3.5	3.5	3.5	3.5	3.5	3.6	3.6	3.6	3.6	3.6	3.6	3.6	3.6	3.6	3.7	3.7	3.7	3.7	3.7	3.8	3.8	3.8	3.8	3.9	3.9	3.9
3.6	X_m	2.0	2.0	2.0	2.0	2.0	2.0	2.1	2.1	2.1	2.1	2.1	2.1	2.2	2.2	2.2	2.2	2.2	2.3	2.3	2.3	2.3	2.4	2.4	2.4	2.4	2.5	2.5	2.5	2.6	2.6	2.6	2.7	2.7	2.8	2.8	2.8	2.9	2.9	3.0
	Y_m	3.6	3.6	3.6	3.6	3.6	3.6	3.6	3.6	3.6	3.6	3.6	3.6	3.6	3.6	3.6	3.6	3.6	3.6	3.7	3.7	3.7	3.7	3.7	3.7	3.7	3.7	3.8	3.8	3.8	3.8	3.8	3.8	3.9	3.9	3.9	3.9	4.0	4.0	4.1
3.7	X_m	2.0	2.0	2.0	2.0	2.0	2.0	2.1	2.1	2.1	2.1	2.1	2.1	2.2	2.2	2.2	2.2	2.2	2.3	2.3	2.3	2.3	2.3	2.4	2.4	2.4	2.5	2.5	2.5	2.5	2.6	2.6	2.6	2.7	2.7	2.8	2.8	2.8	2.9	2.9	3.0
	Y_m	3.7	3.7	3.7	3.7	3.7	3.7	3.7	3.7	3.7	3.7	3.7	3.7	3.7	3.7	3.7	3.7	3.7	3.7	3.8	3.8	3.8	3.8	3.8	3.8	3.8	3.8	3.8	3.9	3.9	3.9	3.9	3.9	4.0	4.0	4.0	4.0	4.1	4.1	4.1	4.2
3.8	X_m	2.0	2.0	2.0	2.0	2.0	2.0	2.1	2.1	2.1	2.1	2.1	2.1	2.2	2.2	2.2	2.2	2.2	2.2	2.3	2.3	2.3	2.3	2.4	2.4	2.4	2.4	2.5	2.5	2.5	2.6	2.6	2.6	2.7	2.7	2.7	2.8	2.8	2.9	2.9	2.9	3.0
	Y_m	3.8	3.8	3.8	3.8	3.8	3.8	3.8	3.8	3.8	3.8	3.8	3.8	3.8	3.8	3.8	3.8	3.8	3.8	3.9	3.9	3.9	3.9	3.9	3.9	3.9	3.9	3.9	4.0	4.0	4.0	4.0	4.0	4.0	4.1	4.1	4.1	4.1	4.2	4.2	4.2	4.3
3.9	X_m	2.0	2.0	2.0	2.0	2.0	2.0	2.1	2.1	2.1	2.1	2.1	2.1	2.1	2.2	2.2	2.2	2.2	2.2	2.3	2.3	2.3	2.3	2.3	2.4	2.4	2.4	2.4	2.5	2.5	2.5	2.6	2.6	2.6	2.7	2.7	2.7	2.8	2.8	2.9	2.9	2.9	3.0
	Y_m	3.9	3.9	3.9	3.9	3.9	3.9	3.9	3.9	3.9	3.9	3.9	3.9	3.9	3.9	3.9	3.9	3.9	3.9	4.0	4.0	4.0	4.0	4.0	4.0	4.0	4.0	4.0	4.0	4.1	4.1	4.1	4.1	4.1	4.2	4.2	4.2	4.2	4.3	4.3	4.3	4.4	4.4
4.0	X_m	2.0	2.0	2.0	2.0	2.0	2.0	2.1	2.1	2.1	2.1	2.1	2.1	2.1	2.2	2.2	2.2	2.2	2.2	2.3	2.3	2.3	2.3	2.3	2.4	2.4	2.4	2.4	2.5	2.5	2.5	2.5	2.6	2.6	2.6	2.7	2.7	2.7	2.8	2.8	2.9	2.9	3.0	3.0
	Y_m	4.0	4.0	4.0	4.0	4.0	4.0	4.0	4.0	4.0	4.0	4.0	4.0	4.0	4.0	4.0	4.0	4.0	4.0	4.0	4.1	4.1	4.1	4.1	4.1	4.1	4.1	4.1	4.1	4.2	4.2	4.2	4.2	4.2	4.2	4.3	4.3	4.3	4.3	4.4	4.4	4.4	4.5	4.5
4.1	X_m	2.0	2.0	2.0	2.0	2.0	2.0	2.1	2.1	2.1	2.1	2.1	2.1	2.1	2.2	2.2	2.2	2.2	2.2	2.2	2.3	2.3	2.3	2.3	2.3	2.4	2.4	2.4	2.4	2.5	2.5	2.5	2.5	2.6	2.6	2.6	2.7	2.7	2.7	2.8	2.8	2.9	2.9	3.0	3.0
	Y_m	4.1	4.1	4.1	4.1	4.1	4.1	4.1	4.1	4.1	4.1	4.1	4.1	4.1	4.1	4.1	4.1	4.1	4.1	4.1	4.2	4.2	4.2	4.2	4.2	4.2	4.2	4.2	4.2	4.3	4.3	4.3	4.3	4.3	4.3	4.4	4.4	4.4	4.4	4.5	4.5	4.5	4.5	4.6	4.6
4.2	X_m	2.0	2.0	2.0	2.0	2.0	2.0	2.0	2.1	2.1	2.1	2.1	2.1	2.1	2.2	2.2	2.2	2.2	2.2	2.2	2.3	2.3	2.3	2.3	2.3	2.4	2.4	2.4	2.4	2.4	2.5	2.5	2.5	2.6	2.6	2.6	2.6	2.7	2.7	2.7	2.8	2.8	2.9	2.9	3.0	3.0
	Y_m	4.2	4.2	4.2	4.2	4.2	4.2	4.2	4.2	4.2	4.2	4.2	4.2	4.2	4.2	4.2	4.2	4.2	4.2	4.2	4.3	4.3	4.3	4.3	4.3	4.3	4.3	4.3	4.3	4.3	4.4	4.4	4.4	4.4	4.4	4.4	4.5	4.5	4.5	4.5	4.6	4.6	4.6	4.7	4.7	4.7
4.3	X_m	2.0	2.0	2.0	2.0	2.0	2.0	2.0	2.1	2.1	2.1	2.1	2.1	2.1	2.1	2.2	2.2	2.2	2.2	2.2	2.2	2.3	2.3	2.3	2.3	2.3	2.4	2.4	2.4	2.4	2.5	2.5	2.5	2.5	2.6	2.6	2.6	2.7	2.7	2.7	2.8	2.8	2.8	2.9	2.9	3.0	3.0
	Y_m	4.3	4.3	4.3	4.3	4.3	4.3	4.3	4.3	4.3	4.3	4.3	4.3	4.3	4.3	4.3	4.3	4.3	4.3	4.3	4.4	4.4	4.4	4.4	4.4	4.4	4.4	4.4	4.4	4.4	4.5	4.5	4.5	4.5	4.5	4.5	4.6	4.6	4.6	4.6	4.6	4.7	4.7	4.7	4.8	4.8	4.8
4.4	X_m	2.0	2.0	2.0	2.0	2.0	2.0	2.0	2.1	2.1	2.1	2.1	2.1	2.1	2.1	2.2	2.2	2.2	2.2	2.2	2.2	2.3	2.3	2.3	2.3	2.3	2.4	2.4	2.4	2.4	2.4	2.5	2.5	2.5	2.5	2.6	2.6	2.6	2.7	2.7	2.7	2.8	2.8	2.8	2.9	2.9	3.0	3.0
	Y_m	4.4	4.4	4.4	4.4	4.4	4.4	4.4	4.4	4.4	4.4	4.4	4.4	4.4	4.4	4.4	4.4	4.4	4.4	4.4	4.5	4.5	4.5	4.5	4.5	4.5	4.5	4.5	4.5	4.5	4.5	4.6	4.6	4.6	4.6	4.6	4.6	4.7	4.7	4.7	4.7	4.8	4.8	4.8	4.8	4.9	4.9	5.0
4.5	X_m	2.0	2.0	2.0	2.0	2.0	2.0	2.0	2.1	2.1	2.1	2.1	2.1	2.1	2.1	2.2	2.2	2.2	2.2	2.2	2.2	2.3	2.3	2.3	2.3	2.3	2.3	2.4	2.4	2.4	2.4	2.5	2.5	2.5	2.5	2.6	2.6	2.6	2.6	2.7	2.7	2.7	2.8	2.8	2.8	2.9	2.9	3.0	3.0
	Y_m	4.5	4.5	4.5	4.5	4.5	4.5	4.5	4.5	4.5	4.5	4.5	4.5	4.5	4.5	4.5	4.5	4.5	4.5	4.5	4.5	4.6	4.6	4.6	4.6	4.6	4.6	4.6	4.6	4.6	4.6	4.7	4.7	4.7	4.7	4.7	4.7	4.8	4.8	4.8	4.8	4.8	4.9	4.9	4.9	5.0	5.0	5.0	5.1
4.6	X_m	2.0	2.0	2.0	2.0	2.0	2.0	2.0	2.1	2.1	2.1	2.1	2.1	2.1	2.1	2.1	2.2	2.2	2.2	2.2	2.2	2.2	2.3	2.3	2.3	2.3	2.3	2.4	2.4	2.4	2.4	2.4	2.5	2.5	2.5	2.5	2.6	2.6	2.6	2.6	2.7	2.7	2.7	2.8	2.8	2.8	2.9	2.9	3.0	3.0
	Y_m	4.6	4.6	4.6	4.6	4.6	4.6	4.6	4.6	4.6	4.6	4.6	4.6	4.6	4.6	4.6	4.6	4.6	4.6	4.6	4.6	4.7	4.7	4.7	4.7	4.7	4.7	4.7	4.7	4.7	4.7	4.8	4.8	4.8	4.8	4.8	4.8	4.8	4.9	4.9	4.9	4.9	5.0	5.0	5.0	5.0	5.1	5.1	5.1	5.2
4.7	X_m	2.0	2.0	2.0	2.0	2.0	2.0	2.0	2.1	2.1	2.1	2.1	2.1	2.1	2.1	2.1	2.2	2.2	2.2	2.2	2.2	2.2	2.3	2.3	2.3	2.3	2.3	2.3	2.4	2.4	2.4	2.4	2.4	2.5	2.5	2.5	2.5	2.6	2.6	2.6	2.6	2.7	2.7	2.7	2.8	2.8	2.8	2.9	2.9	3.0	3.0
	Y_m	4.7	4.7	4.7	4.7	4.7	4.7	4.7	4.7	4.7	4.7	4.7	4.7	4.7	4.7	4.7	4.7	4.7	4.7	4.7	4.7	4.8	4.8	4.8	4.8	4.8	4.8	4.8	4.8	4.8	4.8	4.8	4.9	4.9	4.9	4.9	4.9	4.9	5.0	5.0	5.0	5.0	5.0	5.1	5.1	5.1	5.2	5.2	5.2	5.3	5.3
4.8	X_m	2.0	2.0	2.0	2.0	2.0	2.0	2.0	2.1	2.1	2.1	2.1	2.1	2.1	2.1	2.1	2.2	2.2	2.2	2.2	2.2	2.2	2.2	2.3	2.3	2.3	2.3	2.3	2.4	2.4	2.4	2.4	2.4	2.5	2.5	2.5	2.5	2.6	2.6	2.6	2.6	2.7	2.7	2.7	2.7	2.8	2.8	2.8	2.9	2.9	3.0	3.0
	Y_m	4.8	4.8	4.8	4.8	4.8	4.8	4.8	4.8	4.8	4.8	4.8	4.8	4.8	4.8	4.8	4.8	4.8	4.8	4.8	4.8	4.9	4.9	4.9	4.9	4.9	4.9	4.9	4.9	4.9	4.9	4.9	5.0	5.0	5.0	5.0	5.0	5.0	5.1	5.1	5.1	5.1	5.1	5.2	5.2	5.2	5.2	5.3	5.3	5.3	5.4	5.4
4.9	X_m	2.0	2.0	2.0	2.0	2.0	2.0	2.0	2.1	2.1	2.1	2.1	2.1	2.1	2.1	2.1	2.2	2.2	2.2	2.2	2.2	2.2	2.2	2.3	2.3	2.3	2.3	2.3	2.3	2.4	2.4	2.4	2.4	2.4	2.5	2.5	2.5	2.5	2.6	2.6	2.6	2.6	2.7	2.7	2.7	2.7	2.8	2.8	2.8	2.9	2.9	3.0	3.0
	Y_m	4.9	4.9	4.9	4.9	4.9	4.9	4.9	4.9	4.9	4.9	4.9	4.9	4.9	4.9	4.9	4.9	4.9	4.9	4.9	4.9	5.0	5.0	5.0	5.0	5.0	5.0	5.0	5.0	5.0	5.0	5.0	5.1	5.1	5.1	5.1	5.1	5.1	5.1	5.2	5.2	5.2	5.2	5.2	5.3	5.3	5.3	5.3	5.4	5.4	5.4	5.5	5.5
5.0	X_m	2.0	2.0	2.0	2.0	2.0	2.0	2.0	2.1	2.1	2.1	2.1	2.1	2.1	2.1	2.1	2.1	2.2	2.2	2.2	2.2	2.2	2.2	2.3	2.3	2.3	2.3	2.3	2.3	2.4	2.4	2.4	2.4	2.4	2.4	2.5	2.5	2.5	2.5	2.6	2.6	2.6	2.6	2.7	2.7	2.7	2.8	2.8	2.8	2.9	2.9	2.9	3.0	3.0
	Y_m	5.0	5.0	5.0	5.0	5.0	5.0	5.0	5.0	5.0	5.0	5.0	5.0	5.0	5.0	5.0	5.0	5.0	5.0	5.0	5.0	5.0	5.1	5.1	5.1	5.1	5.1	5.1	5.1	5.1	5.1	5.1	5.1	5.2	5.2	5.2	5.2	5.2	5.2	5.3	5.3	5.3	5.3	5.3	5.4	5.4	5.4	5.4	5.5	5.5	5.5	5.6	5.6	5.6

Figure E–10. Tabulation of coordinates of maximum density point. (Sheet 2 of 2.) 101–D–518, 288–D–2840.

lows: find the point on the +4 percent diagonal line corresponding to the wet density; project vertically to the 0 percent diagonal line, thence horizontally to plot point 3 on the +4 percent vertical line.

(4) To obtain point 3 if the converted wet density of point 2 is less than point 1: permit 7.50 pounds of soil at fill water content to dry without loss of soil; then weigh.[3] The table on the right-hand portion of figure E–8 gives the percentage of water loss corresponding to the dried weight. Compact the dried soil into a cylinder. Convert the wet density to wet density at fill water content by dividing by 1 plus the negative percentage (e.g., 1+(−0.02)=0.98). Plot the converted wet density on the vertical line corresponding to the negative percentage. This division may also be done graphically as follows: find the point on the diagonal line (interpolate if necessary) corresponding to the wet density; project vertically to the 0 percent diagonal line, thence horizontally to plot point 3 on the vertical line corresponding to the correct percentage.

(5) Three plotted points are sufficient if both the left and right points are lower in density than the center point; if not, a fourth point is necessary. The maximum density point of the converted wet density curve can now be easily found from the coordinates obtained from figure E–10. An example of the use of figure E–10 is given at the end of this explanation. Note that figure E–10 can be used only if points (1), (2), and (3) are equally spaced horizontally in 2-percent increments, with the left and right points being less in density than the center point.

The maximum density point may also be found by the parabola method shown in figure E–11, or by sketching the curve if the number and locations of the points permit accuracy without use of the parabola method.

(6) Plot the fill wet density of minus No. 4 fraction on the 0 percent vertical line.

(7) *To obtain D:* Divide the fill wet density of the minus No. 4 fraction by the maximum density point of the converted wet density curve. This may be done graphically by projecting the maximum density point (obtained in step 5) horizontally to the 0 percent diagonal line, thence vertically to the value of the fill wet density, D=100 percent plus the interpolated percentage given by the diagonal lines, taking minus signs into account.

(8) *To obtain C:* Divide the fill wet density by the wet density of point 1 or determine graphically as follows: project point 1 horizontally to the 0 percent diagonal line, thence vertically to the value of the fill wet density. C equals 100 percent plus the interpolated percentage given by the diagonal lines, taking minus signs into account.

(9) To obtain $w_o - w_f$: This value is the horizontal distance from the zero vertical line to the maximum density point (abscissa), corrected by adding the value shown in red (the dashed vertical curves in figure E–8 or the solid vertical curves in figure E–9) on the chart nearest to the maximum density point, interpolating where necessary and taking minus signs into account. Added water and correction values should be taken to the nearest 0.1 percent.

(10) Rapid control values obtained in steps (7), (8), and (9) yield sufficient information to accept or reject the work.

For record purposes, dry a sample of minus No. 4 fraction in an oven at 110° C. to obtain actual fill water content, w_f. Then:

Fill dry density of minus No. 4= (Fill wet density of minus No. 4) ÷ $(1+w_f)$

Laboratory maximum dry density= (Maximum density point) ÷ $(1+w_f)$

Cylinder dry density= (Wet density of point 1) ÷ $(1+w_f)$

Optimum water content= $w_f + (1+w_f)$ x (added water at maximum density point).

[3] If the converted wet density of point 2 is within 3 pounds per cubic foot of point 1, the requirement for drying may be eliminated by using the alternative procedure given at the end of this section.

GRAPHICAL PARABOLA METHOD

Graphical solution for the maximum density point, O, of the converted wet density curve, given three points A, B, and C. If more than three points are available, use the three closest to optimum.

NOTE: Solution based on the assumption that converted wet density curve is a parabola whose axis is vertical.

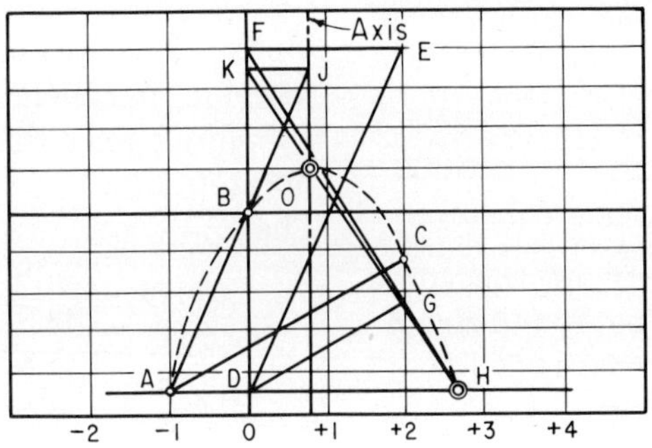

1. Draw horizontal base line through the left point, A, and draw vertical lines through points B and C.
2. Draw line DE parallel to AB, point E lies on the vertical line through point C; project E horizontally to establish point F on the vertical line through B.
3. Draw line DG parallel to AC, point G lies on the vertical line through point C.
4. Line FG intersects the base line at H. Axis of parabola bisects AH; draw the axis.
5. Intersection of line AB with the axis is at J; project J horizontally to K, which lies on the vertical line through point B.
6. Line KH intersects the axis at O, the maximum density point (vertex).

NOTE: If points A, B, and C are equally spaced horizontally (this is true when 2 points are obtained by adding water or when soil is dried exactly 2 percent) steps 2 and 3 above are eliminated. Point F coincides with point B and point G is halfway between the base line and point C. Hence, point H is obtained by drawing BG and point O is obtained by steps 5 and 6 as usual. See graph below.

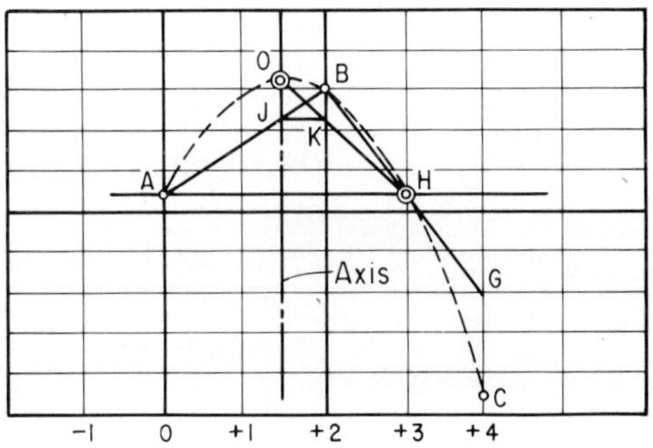

Figure E–11. **Graphical method of finding peak point of converted wet density curve. 101–D–326(B).**

Two examples will illustrate the method: In example 1, the fill water content is less than optimum. The test data are shown in figure E-8.

Fill wet density (minus No. 4 basis) $=127.5$ pounds per cubic foot.

Point	Wet density, pounds per cubic foot	Abscissa, percent	Converted wet density, pounds per cubic foot
1	123.4	0	123.4
2	128.6	2	126.1
3	124.6	4	119.8

The point of maximum density of the converted wet density curve may be found from the coordinates shown in figure E–10 as follows:

$Y_2=B-A=126.1-123.4=+2.7$

$Y_4=C-A=119.8-123.4=-3.6$

From figure E–10: $X_m=1.6$

$Y_m=2.9$

Maximum converted wet density $=123.4+2.9$ $=126.3$

Then:

$$D=\frac{127.5}{126.3}=101.0 \text{ percent}$$

$$C=\frac{127.5}{123.4}=103.3 \text{ percent}$$

$$w_o-w_f=+1.6+0.2=+1.8 \text{ percent (dry)}.$$

After the fill water content has been determined by drying a sample to constant weight at 110° C., the field density test is completed for record purposes as follows:

$w_f=15.0$ percent (determined by drying at 110° C.)

$$\gamma_{D_f}=\frac{127.5}{1.15}=110.9 \text{ lb. per cu. ft.}$$

$$\gamma_{D_m}=\frac{126.3}{1.15}=109.8 \text{ lb. per cu. ft.}$$

$$\gamma_{D_c}=\frac{123.4}{1.15}=107.3 \text{ lb. per cu. ft.}$$

$$w_o=0.15+(1.15)(0.016)=16.8 \text{ percent.}$$

In example 2, the fill water content is greater than optimum. The test data are as follows:

Fill wet density $=125.8$ pounds per cubic foot.

Point	Wet density, pounds per cubic foot	Abscissa, percent	Converted wet density, pounds per cubic foot
1	128.4	0	128.4
2	124.2	+2	121.8
3	123.7	−2.3	126.6
By parabola method:			
0		−0.7	128.9

Note: This problem must be solved by the graphic parabola method (fig. E–11) or by sketching since the points are not equally spaced horizontally.

Then:

$$D=\frac{125.8}{128.9}=97.6 \text{ percent}$$

$$C=\frac{125.8}{128.4}=98.0 \text{ percent}$$

$$w_o-w_f=-0.7-0.1=-0.8 \text{ percent (wet)}.$$

Completion of test for record purposes:

$w_f=18.0$ percent

$$\gamma_{D_f}=\frac{125.8}{1.18}=106.6 \text{ lb. per cu. ft.}$$

$$\gamma_{D_m}=\frac{128.9}{1.18}=109.2 \text{ lb. per cu. ft.}$$

$$\gamma_{D_c}=\frac{128.4}{1.18}=108.8 \text{ lb. per cu. ft.}$$

$$w_o=0.18+(1.18)(-0.007)=17.2 \text{ percent.}$$

Penetration resistance needle readings required for purposes of maintaining moisture control by means of the needle-moisture relation shown in figure 108 may be obtained during the rapid method test. The needle readings in the compaction cylinder at fill water content may be obtained directly after the weighing required to determine point 1 of the rapid method. To obtain the Proctor needle value at optimum water content, needle readings corresponding to points 2, 3, etc., should be obtained after weighing. All needle readings may then be plotted on figure E–8 to any convenient ordinate scale. The needle value at optimum is the intersection of a vertical line through the peak point of the converted wet density curve and the Proctor needle curve.

The following alternative method may be used to eliminate the drying requirement for soils that are close to optimum in water content.

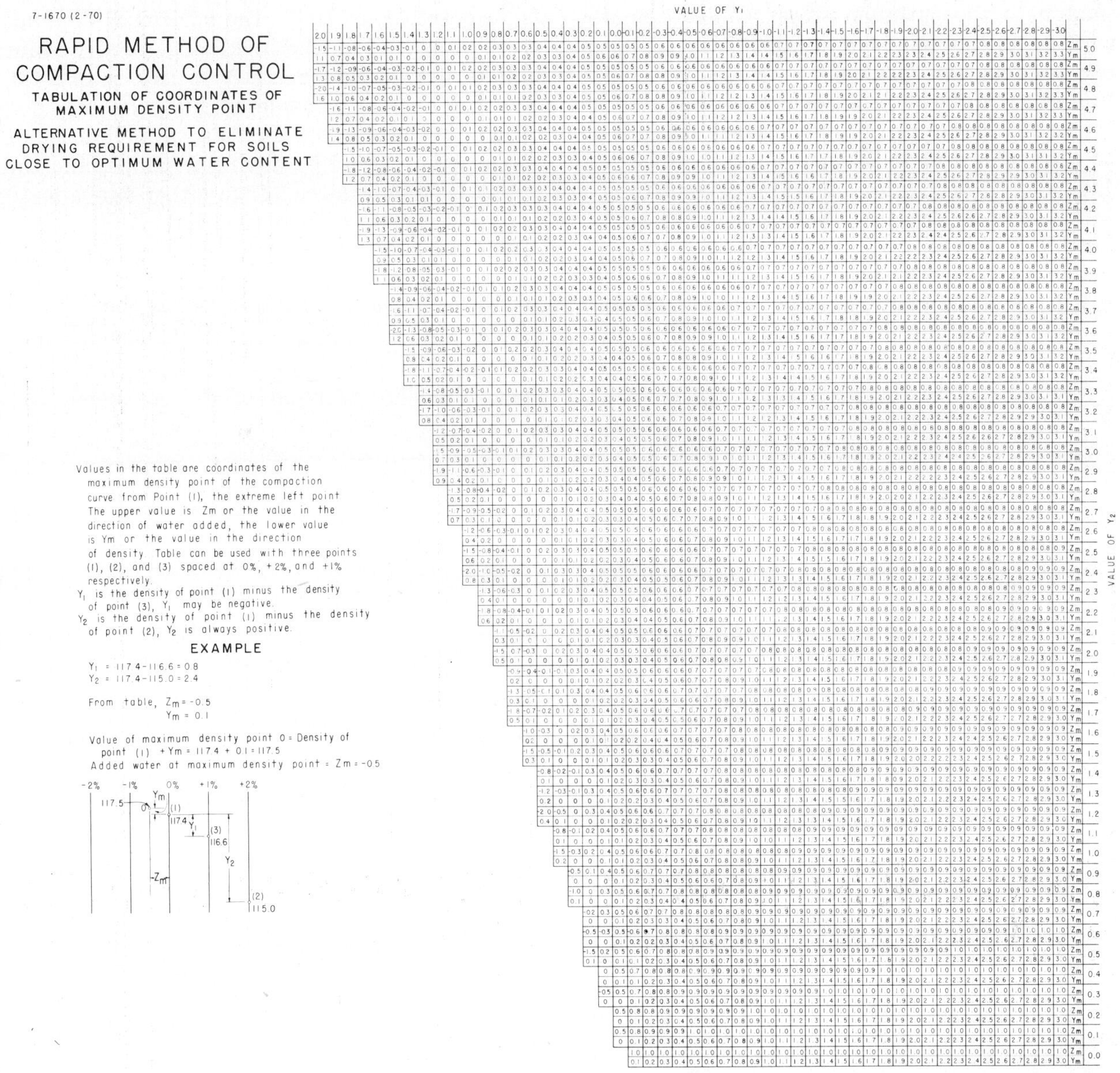

Figure E–12. Tabulation of coordinates of maximum density point; alternate method to eliminate drying requirement for soils close to optimum water content. 101–D–519.

Points 1 and 2 of the converted wet density curve are obtained in the usual manner. To obtain point 3 when the converted wet density of point 2 is smaller than point 1 but within 3 pounds per cubic foot of that point, proceed as follows: in lieu of drying 7.50 pounds of soil at fill water content, as required in the procedure for the rapid compaction control method, add 34 cubic centimeters (1 percent) of water to this quantity of soil; mix, and compact into a cylinder. Convert the wet density to wet density at fill water content by dividing by 1.01 and plot the converted wet density on the +1 percent vertical line. This division may be done graphically as follows: find the point on the +1 percent diagonal line corresponding to the wet density, project vertically to the 0 percent diagonal line, thence horizontally to plot point 3 on the +1 percent vertical line. The use of this method requires that the soil be close enough to the optimum water content so the portion of the compaction curve developed is represented by a parabola. This is generally true if the difference between the ordinates of points (1) and (2) is 3 pounds per cubic foot or less. This method should not be used if differences are greater unless prior experience with the soil indicates that it is still in the parabolic section.

The maximum density point of the converted wet density may now be easily found from the coordinates shown in figure E–12. This figure gives the horizontal distance (Z_m) and vertical distance (Y_m) to the maximum density point from point (1). If Z_m, the horizontal distance from point (1) to the maximum density point obtained from figure E–12, is less than –1.0, this method is considered inapplicable and should not be used. Instead, a fourth point should be obtained by drying a sample and the coordinates determined from either figure E–10 or the parabola method (fig. E–11).

The reasons for elimination of values of Z_m below –1.0 are discussed below.

The maximum density point may be found graphically as follows: if the converted wet density of point 3 is greater than point 1, the maximum density point of the converted wet

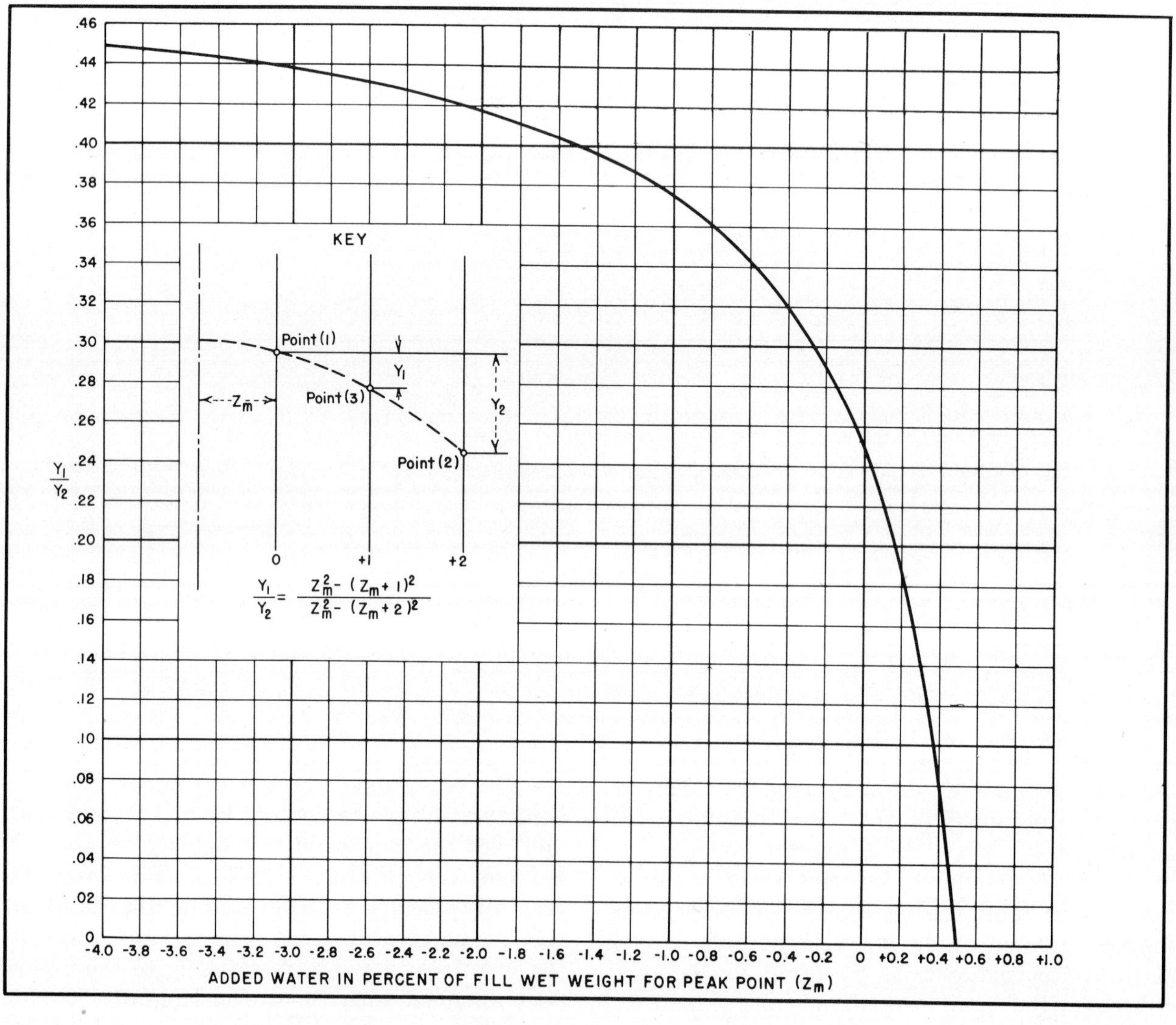

Figure E–13. Alternative procedure to eliminate requirement for drying soils that are close to optimum water content. It is recommended that this graph not be used for values of Y_1/Y_2 greater than 0.38. 288–D–2916.

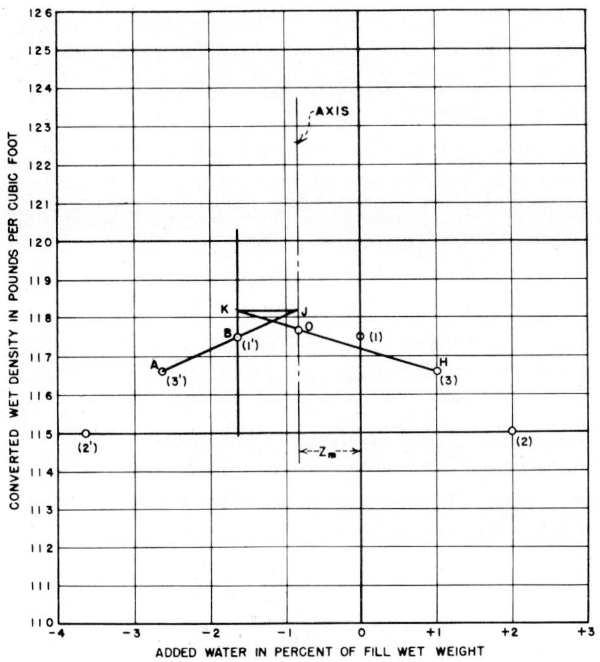

Figure E–14. Example of parabola construction for alternative method which eliminates drying requirement for a soil that is close to optimum water content. 288–D–2917.

density curve can be obtained graphically by the parabola method shown in figure E–11. If the converted wet density of point 3 is smaller than point 1, the graphical procedure for obtaining the maximum density point requires extrapolation, which reduces its accuracy. For this case, calculate Y_1/Y_2, where Y_1 is the difference in converted wet densities of points 1 and 3, and Y_2 is the difference in converted wet densities of points 1 and 2. Figure E–13 gives the location of the axis of the parabola (distance Z_m from the origin) for ratios Y_1/Y_2.

For example, if points 1, 2, and 3 are equal to 117.5, 115.0, and 116.6 pounds per cubic foot, respectively,

$$Y_1=117.5-116.6=0.9,$$

$$Y_2=117.5-115.0=2.5,$$

the coordinates of the maximum density point can be found from figure E–12 as follows:

$$Z_m=-0.8$$

$$Y_m=+0.2$$

Maximum converted wet density$=117.5+0.2$ $=117.7$ pcf.

The maximum density point may also be found by the graphical method from figure E–11 as follows:

Find the ratio Y_1/Y_2:

$$\frac{Y_1}{Y_2}=\frac{0.9}{2.5}=0.36$$

From figure E–13, $Z_m=-0.8$ percent.

With the axis of the parabola located 0.8 percent to the left of point 1 in figure E–14, the mirror images of points 1, 2, and 3 are plotted as shown by points 1′, 2′, and 3′. The maximum density point, point 0, is obtained by designating point 3′ as *A*, point 1′ as *B*, and point 3 as *H* and proceeding with the parabola construction as explained in figure E–11. The coordinates of the maximum density point of the converted wet density curve are (–0.8, 117.7).

It is important to note that for ratios of Y_1/Y_2 greater than 0.38, small changes in either Y_1 or Y_2 can cause large variations in Z_m values. For example, if the above value of Y_1 was 1.0 instead of 0.9, $Y_1/Y_2=0.40$ and $Z_m=-1.5$, which is much less than the –0.8 obtained previously. For this reason it is recommended that figure E–13 not be used for ratios of Y_1/Y_2 greater than 0.38 and that figure E-12 not be used for values of Z_m less than –1.0. Instead, a fourth point should be obtained by drying a sample, and the coordinates obtained from either figure E–10 or the parabola method of figure E–11.

E-6. *Pervious Fill.*—Permeable-type materials are used in rolled earthfill dams to provide an outer shell of high strength to support the impervious core, to secure favorable hydraulic conditions of drainage, and to act as filters between materials of wide variations in grain sizes. Control of construction of zones of sand and gravel is necessary to assure that (1) the material is formed into a homogeneous mass free from large voids, (2) the soil mass is free draining, (3) the material will not consolidate excessively under the weight of super-imposed fill, and (4) the soil has a high angle of internal friction.

The workability of a permeable-type soil is reduced considerably by the inclusion of even small amounts of silt or clay; hence, every effort should be made to insure that the contractor's operations in the borrow pits and on the fill are such that contamination of the pervious soil is held to a minimum. As the material is brought on the fill, it is directed to the proper zone. Within the pervious zone, individual loads should be placed so that the material will be graded in coarseness toward the outer slopes. When compacted thicknesses are specified, the thickness of loose layers should be determined by the inspector during the initial stages of construction. Since the field density will be checked by relatively few actual tests after satisfactory placing procedures have been established, the proper thickness of the loose layers must be maintained within close limits throughout the job. The thickness of specified compacted layers is usually made large enough to accommodate the size of rock encountered in the borrow area. In cases where cobbles or rock fragments greater in size than the specified thickness of layer occur, provisions are usually made for special embedding, removal to outer slopes of the pervious zone, or removal to other zones. In order to secure the best compaction, the inspector should see that the specified provisions for disposal of oversized rock are followed.

After the material has been placed and spread to the desired thickness of lift and oversized cobbles or rock fragments have been disposed of, the next important step is application of water. Thorough and uniform wetting of materials during or immediately prior to compaction is essential for best results. The most applicable method of adding and distributing water on the fill should be determined during the initial placement. It has been found that relaxation of the requirements of thorough wetting may result in densities far below the minimum requirements even with excessive compactive effort. Different pervious materials require different amounts of water for thorough wetting and best compaction. In general, it is desirable to add as much water to the material as it will readily absorb. Too much water cannot be added to an extremely pervious soil; however, permeable-type soils containing small amounts of silt or clay may become temporarily boggy if an excessive amount of water is used. For these soils care must be exercised in wetting the soil. The contractor's operations should be carefully controlled to avoid excessive wetting of the impervious zone adjacent to the pervious material being compacted.

When compacting permeable-type soils by the treads of crawler-type tractors, it is desirable to have the tractor operating at the highest practicable speed during the compaction operation. High speed is conducive to greater vibration which aids in the compaction. In inspection of compaction operations using tractor treads, it is important to require the tractor to make a complete coverage of the area to be compacted prior to making the second and subsequent passes. Different widths of areas to be compacted will require different numbers of tractor trips to obtain the same number of passes of the treads. The proper number of trips should be determined and enforced.

It is recommended that relative density tests and mechanical analyses be made during the initial placing operation at a frequency of about one test representing each 1,000 cubic yards placed. The procedure for making relative density tests is given in sections 119(a) and 120(f). After placement procedures have proved satisfactory, one relative density test for every 10,000 cubic yards of material placed will suffice unless significant changes in gradation occur. In the event of significant gradational changes in the borrow material, increased frequency of field tests may be needed to insure satisfactory compaction of the variable materials.

E-7. *Rockfill and Riprap.*—Rockfill zones are used in earthfill dams to provide stability for the embankment and to protect exposed surfaces of the fill. Riprap is a relatively thin layer of specially selected and graded rock fragments used for protecting earth slopes from erosion by water currents and waves. Rockfill and riprap are usually not compacted but are dumped or placed so as to obtain high shear strength by interlocking of the angular fragments. The most desirable riprap surface is well keyed but rough in order to resist wave action effectively. High permeability is desirable in rockfills; hence, the amount of fines permitted is limited. On the other hand, large unfilled voids are undesirable. The outer portion of a rockfill zone should contain the largest available rock in order to secure slope protection. Where very large rockfill sections are used, excessive settlement may be a problem, and sluicing may be required to compact the fill.

Inspection may be necessary both at the rock source and on the rockfill to insure that the material is selected to avoid an excessive amount of fines in the rockfill. Breakage in handling and transporting should be taken into account. Placing operations should be inspected to see that segregation is avoided and that no large voids are left in the rockfill. If sluicing is required, the contractor's operations should be carefully controlled to avoid excessive wetting of the impervious zone and to insure that a sufficient quantity of water is being used uniformly. Inspection of riprap placement consists of visual observation of the operation and of the completed product to insure that a dense, rough surface of well-keyed, graded rock fragments of the specified quality and sizes is obtained. Typical specifications provisions for placing rockfill and riprap are contained in appendix G.

E-8. *Miscellaneous Fills.*—Dam embankments on saturated fine-grain foundations may require toe support fills, the weight of which acts to improve stability. These are discussed in section 134. Excavation for foundations of the dam or for the appurtenant structures often produces material unsuitable for or in excess of the requirements for the structural zones of the dam. Such excavated material can be used for stabilizing fills at the toes of the dam. In localities where riprap of good quality is very expensive, fill materials from structural excavations have been used to flatten the upstream slope of the dam so as to permit use of poor quality rock or, in some cases, the omission of rock. In a few cases excess required excavation has been used in an isolated zone in the downstream portion of a dam merely to replace material which otherwise would have to be borrowed at greater expense.

The permeability of stabilizing fills is not important in the design, and such fills usually are not specially compacted. However, full use should be made of the compaction obtainable by routing of the hauling and placing equipment over layers of the material. Sometimes the nature of the available materials or the design requires that some compactive effort other than routing of equipment be used. For example, sheepsfoot rolling has been used to break up fairly large chunks of soft rocks to avoid excessive settlement. Compaction also may be required when the miscellaneous fill is to serve as an impervious blanket.

Inspection of miscellaneous fills is usually entirely visual; ordinarily no control tests are made. The main problem in inspection of miscellaneous fills is to assure that the specified thickness of lift is not exceeded and to see that the hauling equipment is not channelized by a roadway but is spread over the entire placed area so far as practicable.

E-9. *Records and Reports.*—Daily reports should be made by the inspector covering the activities for his shift. These reports should record the progress of construction, provide pertinent information for the inspector about to go on shift (including shutdowns and orders given to the contractor), and furnish data for use in compiling reports. The form of a daily report will vary to suit the requirements of each job, but all information called for on summary progress reports should be based on day-to-day records. It is desirable that a systematic method of identifying field density tests made on the embankment be used. A suggested scheme is to designate each test by the date, shift, number on that shift, and purpose. For example, "8–2–70–a–2–D" would define a field density test made August 2, 1970, on the first shift, the second test made on that shift, and for the purpose of checking an area of doubtful compaction. The legend is as follows: a, first shift; b, second shift; c, third shift; D, doubtful area; C, concentrated area; R, representative. The results of tests made daily on the embankment can be reported on appropriate forms.

E-10. *Control Criteria.*—Determination of the quality of the embankment being placed can be made by a simple statistical analysis of the test results as given by Davis [6]. Figures E–15 and E–16 show work sheets and curves for dry density control and moisture control, respectively, for compacted cohesive soils in an earthfill dam. By means of this analysis, the fre-

Example DAM ZONE I

$D = \frac{\text{FILL DRY DENSITY}}{\text{MAXIMUM LABORATORY DRY DENSITY}} \times 100$ (MINUS NO. 4 MATERIAL)

D	F (PREV.)	THIS PERIOD				TO DATE		
		FREQUENCY OF OCCURRENCE	F	CUM. F	CUM. %	F	CUM. F	CUM. %
93.0 - 93.9								
94.0 - 94.9								
95.0 - 95.9								
96.0 - 96.9		IIII	4	4	8			
97.0 - 97.9		IIII	4	8	16			
98.0 - 98.9		~~IIII~~ IIII	9	17	34			
99.0 - 99.9		~~IIII~~ II	7	24	48			
100.0 - 100.9		~~IIII~~ III	8	32	64			
101.0 - 101.9		~~IIII~~ ~~IIII~~	10	42	84			
102.0 - 102.9		~~IIII~~	5	47	94			
103.0 - 103.9		III	3	50	100			
104.0 - 104.9								
105.0 - 105.9								
TOTALS			50					

	PREV.	THIS PERIOD	TO DATE
Average max. lab. γ_D (P.C.F.)		117.1	
Average fill γ_D (P.C.F.)		117.1	
Mean variation from max. lab. γ_D (P.C.F.)		0.	
Average rock content (% of plus No. 4 by dry weight)		2.6	

PERIOD OF REPORT ______ TO ______

TESTS 9-26-70-d-IR TO 10-24-70-d-IR

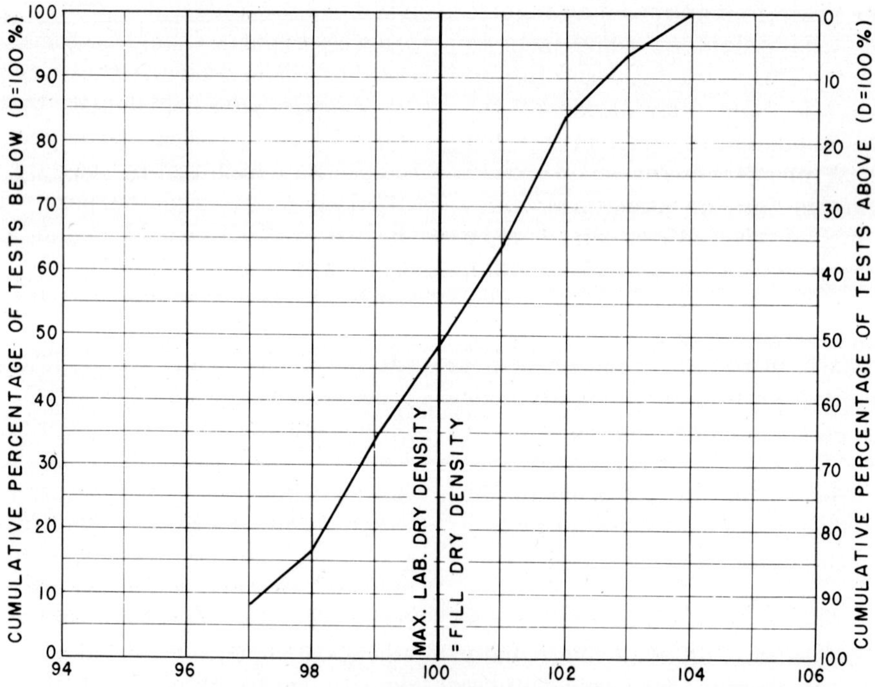

$D = \frac{\text{FILL DRY DENSITY}}{\text{MAXIMUM LABORATORY DRY DENSITY}} \times 100$

Figure E–15. Statistical analysis of field density tests for density control. 288-D-2567.

Example DAM ZONE I

OPTIMUM WATER CONTENT MINUS FILL WATER CONTENT ($W_o - W_f$) IN PERCENT OF DRY WEIGHT

		F (PREV.)	THIS PERIOD				TO DATE		
			FREQUENCY OF OCCURRENCE	F	CUM. F	CUM. %	F	CUM. F	CUM. %
W_f IS BELOW OPTIMUM	3.3-3.7								
	2.8-3.2								
	2.3-2.7		卌 I	6	6	12			
	1.8-2.2		卌	5	11	22			
	1.3-1.7		卌 I	6	17	34			
	0.8-1.2		卌 卌 I	11	28	56			
	0.3-0.7		卌 卌	10	38	76			
+0.2 TO -0.2			卌 II	7	45	90			
W_f IS ABOVE OPTIMUM	0.3-0.7		III	3	48	96			
	0.8-1.2		I	1	49	98			
	1.3-1.7								
	1.8-2.2								
	2.3-2.7		I	1	50	100			
	2.8-3.2								
	3.3-3.7								
TOTALS				50					

	PREV.	THIS PERIOD	TO DATE
Average optimum water content		13.7	
Average fill water content		12.8	
Mean variation from optimum water content		0.9	

PERIOD OF REPORT ______ TO ______
TESTS 9-26-70-a-IR TO 10-24-70-a-IR

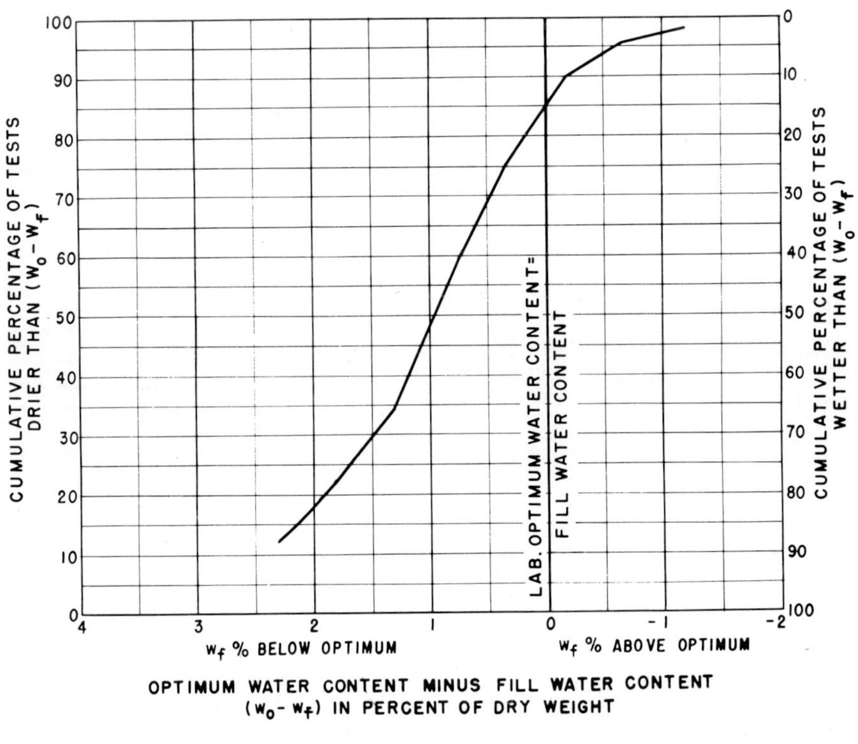

Figure E–16. Statistical analysis of field density tests for moisture control. 288–D–2568.

TABLE E-1.—*Criteria for control of compacted dam embankments*

Type of material	Percentage of +No. 4 fraction by weight of total material	Percentage based on minus No. 4 fraction		
		Minimum acceptable density	Desirable average density	Moisture limits, w_o-w_f
Cohesive soils controlled by Proctor test.	0–25	D=95	D=98	−2 to +2.
	26–50	D=92.5	D=95	
	More than 50 [1]	D=90	D=93	
Cohesionless soils controlled by relative density test.	Fine sands with 0–25.	D_d=75	D_d=90	Soils should be very wet.
	Medium sands with 0–25.	D_d=70	D_d=85	
	Coarse sands and gravels with 0–100.	D_d=65	D_d=80	

w_o-w_f is the difference between optimum water content and fill water content in percent of dry weight of soil.

D is fill dry density divided by Proctor maximum dry density, in percent.

D_d is relative density, as defined in appendix D, in percent.

[1] Cohesive soils containing more than 50 percent gravel sizes should be tested for permeability of the total material if used as a water barrier.

quency distribution of the test results is obtained from which statistical values such as the mean, standard deviation, and the percentage of tests falling outside specified limits can be determined.

Various criteria for quality control have been proposed. Table E–1 lists suggested limits of density and moisture control based on experience gained in compacting 44 cohesive soils and 18 cohesionless soils in Bureau of Reclamation earthfill dams. The soils were compacted by the equipment and methods specified in appendix G; hence, the values given in the table may not be possible of attainment with other methods of compaction or with lesser amounts of compactive effort. It is recognized that the normal frequency distribution curve for any desired average value permits a small percentage of very low tests. However, because of the relatively small number of samples recommended to be tested, the values listed in the table as "minimum acceptable" are suggested as a basis for requiring recompaction of the area represented by lower test values.

The effect of gravel content in cohesive soils is discussed in several papers [4, 7, 8]. Available data indicate that lesser percentages of density on the minus No. 4 basis are required for gravelly cohesive soils than for soils containing little or no gravel. This fact is reflected in table E–1.

E-11. *Bibliography.*

[1] Justin, J. D., "Earth Dam Projects," John Wiley & Sons, Inc., New York, N.Y., 1932, p. 188.

[2] Bassell, Burr, "Earth Dams," Engineering News Publishing Co., New York, N.Y., 1907.

[3] Proctor, R. R., "The Design and Construction of Rolled Earth Dams," Engineering News-Record, Aug. 31, Sept. 7, 21, and 28, 1933.

[4] Hilf, J. W., "Compacting Earth Dams With Heavy Tamping Rollers," ASCE Proceedings, vol. 83, April 1957, Journal of the Soil Mechanics and Foundations Division, No. SM2, Paper No. 1205.

[5] Hilf, J. W., "A Rapid Method of Construction Control for Embankments of Cohesive Soils," Engineering Monograph No. 26, Bureau of Reclamation, September 1961.

[6] Davis, F. J., "Quality Control of Earth Embankments," Third International Conference on Soil Mechanics and Foundation Engineering, Switzerland, 1953, vol. 1, p. 218.

[7] Walker, F. C., and Holtz, W. G., "Control of Embankment Material by Laboratory Testing," Trans. ASCE, vol. 118, 1953, p. 1.

[8] Holtz, W. G., and Lowitz, C. A., "Compaction Characteristics of Gravelly Soils," Conference on Soils for Engineering Purposes, ASTM Committee D–18 and Sociedad Mexicana de Mecanica de Suelos, Dec. 9–13, 1957, ASTM Special Technical Publication No. 232, American Society for Testing Materials, Philadelphia, Pa., p. 123.

Concrete in Construction

J. E. BACKSTROM,[1] L. C. PORTER,[1] E. L. ORE,[1] and G. B. WALLACE [2]

A. CONCRETE AND CONCRETING MATERIALS

F-1. ***Important Properties of Concrete.***—Concrete is one of the most durable and versatile of construction materials. It is composed of sand, gravel, crushed rock, or other aggregates held together by a hardened paste of hydraulic cement and water. The selection, testing, and evaluation of these materials, together with their processing and proportioning, form the subject of this appendix. Specifications for concrete are included in Appendix G. For complete coverage of concrete as a construction material the reader is referred to the Bureau of Reclamation's "Concrete Manual" [1].[3]

The characteristics of concrete discussed in the following sections should be considered on a relative basis and in terms of the degree of quality that is required for any given construction purpose. In addition to being adequately designed, a structure must be constructed properly of concrete which is strong enough to carry the design loads and which is economical, not only in first cost but also in terms of its ultimate service. In addition to strength, concrete must have the properties of workability and durability.

F-2. ***Workability.***—Workability has been defined as the ease with which a given set of materials can be mixed into concrete and subsequently handled, transported, and placed with a minimum loss of homogeneity. Workability is dependent on the proportions of the constituent materials as well as on their individual characteristics. The degree of workability required for proper placement and consolidation of concrete is governed by the dimensions and shape of the structure and by the spacing and size of the reinforcement. For example, concrete having suitable workability for a pavement slab would be difficult or impossible to economically place in a thin, heavily reinforced section. The slump test [2], coupled with judgment developed by experience, is a common means of evaluating concrete workability.

F-3. ***Durability.***—A durable concrete is one which will withstand, to a satisfactory degree, the effects of service conditions to which it will be subjected, such as weathering, chemical action, and wear.

(a) *Weathering Resistance.*—Disintegration of concrete by weathering is caused mainly by the disruptive action of freezing and thawing and by expansion and contraction, under restraint, resulting from temperature variations and alternate wetting and drying. Concrete can be made that will have excellent resistance to the effects of such exposures if careful attention is given to the selection of materials and to all other phases of job control. The purposeful entrainment of small bubbles of air helps greatly to improve concrete durability. It is also important that, where practicable, provision be made for adequate drainage of exposed concrete surfaces. In general, the more watertight the concrete, the more difficult it is for water to gain entrance and to fill the voids and the greater is the resistance to frost action.

(b) *Resistance to Chemical Deterioration.*—The common causes of chemical deterioration of

[1] Engineer, Concrete and Structural Branch, Bureau of Reclamation.

[2] Formerly Engineer, Concrete Laboratory Branch, now Chief, Polymer Concrete Group, Bureau of Reclamation.

[3] Numbers in brackets refer to items in the bibliography, section F-31.

concrete include: alkali-aggregate reactivity, in which alkalies in the cement react chemically with mineral constituents of concrete aggregates; deterioration resulting from contact with various chemical agents; and sulfate attack, in which salts (principally soluble sulfates) that are present in ground water or soil in contact with the concrete, attack the cement paste.

Alkali-aggregate reactivity is characterized by the following observable conditions: cracking, usually of random pattern on a fairly large scale (see fig. F-1); excessive internal and overall expansion; cracks that may be very large at the concrete surfaces (openings up to 1½ inches have been observed), but which extend into the concrete only a distance of from 6 to 18 inches; gelatinous exudations and whitish amorphous deposits, both on the surface and within the mass of the concrete, especially in voids and adjacent to some affected pieces of aggregate; peripheral zones of reactivity, alteration, or infiltration in the aggregate particles, particularly those particles containing opal and certain types of acid and intermediate volcanic rocks; and lifeless chalky appearance of the freshly fractured concrete.

Use of low-alkali cement, that is, cement having a total alkali content expressed as sodium oxide of not more than 0.6 percent as determined by summation of the percent of sodium oxide and 0.658 times the percentage of potassium oxide, provides an effective means of controlling expansive alkali-aggregate reaction, generally at no increase in cost. The use of a suitable pozzolan in concrete provides another effective method of obtaining such a control and provides added insurance when used in combination with low-alkali cement. However, the efficiency of different pozzolans in controlling this reaction varies widely, and it is therefore

Figure F–1. Typical pattern cracking on the exposed surface of concrete affected by alkali-aggregate reaction. TC–8–16.

necessary to test pozzolan sources individually to evaluate their effectiveness in controlling this expansive reaction.

Tests to evaluate reactive combinations of aggregate and cement are involved and expensive. Therefore, for jobs with a limited budget, inspection of existing concrete structures near the job site and determination of the source of the aggregate and cement used in the structures may provide valuable information regarding the quality of local materials for use in construction. Also, the need for protective measures frequently can be determined by examination of the prospective aggregate by an experienced petrographer.

Most prominent among aggressive substances which affect concrete structures are the sulfates of sodium, magnesium, and calcium. These salts, which are known as "white alkali," are frequently encountered in the "alkali" soils and ground waters of the western half of the United States. The sulfates react chemically with certain compounds in the cement to produce considerable expansion and disruption of the paste. The result of such action is shown in figure F–2. Sulfate attack is reduced by using the type of cement indicated in table F–1 for varying degrees of sulfate concentration. While use of the type of cement indicated in table F–1 is preferable, further increase in resistance of concrete to sulfate attack can be obtained through an increase in cement content and/or the use of fly ash or other effective pozzolan.

Where deposits of "white alkali" occur, it is advisable to examine existing concrete structures in the vicinity of the proposed work to determine whether protection against sulfate attack will be necessary. The presence of these white deposits of salts often indicates the need for testing the soil and ground water to determine if harmful sulfate concentrations are

Figure F–2. Disintegration of concrete caused by sulfate attack. PX–D–32050.

present. Testing is desirable because the white deposits may be *chloride* salts which, compared to sulfate salts, are relatively harmless to hardened concrete.

TABLE F-1.—*Attack on concrete by soils and waters containing various sulfate concentrations*

Relative degree of sulfate attack	Percent water-soluble sulfate (as SO_4) in soil samples	P.p.m. sulfate (as SO_4) in water samples
Negligible	0.00 to 0.10	0 to 150.
Positive [1]	0.10 to 0.20	150 to 1,000.
Considerable [2]	0.20 to 0.50	1,000 to 2,000.
Severe [2]	Over 0.50	Over 2,000.

[1] Use type II cement.
[2] Use type V cement.

F-4. *Effects of Curing on Strength* [7].—Experience has demonstrated that when the maximum permissible water-cement ratio has been established on the basis of durability requirements as shown in table F-2, concrete will usually develop adequate compressive strength if properly placed and cured. Figure F-3 shows the compressive strength development of concrete cured for various lengths of time and subsequently stored or dried. Concrete exposed to dry air from the time it is placed is only about 50 percent as strong at 6 months as concrete moist-cured 14 days before being exposed to dry air.

F-5. *Effects of Entrained Air on the Properties of Concrete.*—Except for compressive strength, all properties of concrete, including workability, durability, permeability, drying shrinkage, bleeding, etc., are materially improved by the purposeful entrainment of from 2 to 6 percent air, the optimum amount depending on the maximum size aggregate used. Supplementary benefits in the form of reduced water and cement requirements and an increase in ease of finishing may also be realized. Figure F-4 shows the effects of air content on the durability, compressive strength and required water content of concrete. Note that the durability increases rapidly to a maximum with an addition of air, and then decreases as the air content is further increased, while compressive strength and water content continue to decrease with the increase in air content. Figure F-5 shows

TABLE F-2.—*Allowable maximum net water-cement ratios for durability of concrete subjected to various degrees of exposure*

Type or location of concrete or structure, and degree of exposure	Water-cement ratio, *W/C*, by weight	
	Severe climate, wide range of temperature, long periods of freezing or frequent freezing and thawing	Mild climate, rainy or arid, rarely snow or frost
A. Concrete in portions of structures subject to exposure of extreme severity, such as the top 2 feet of walls, boxes, piers, and parapets; all of curbs, sills, ledges, copings, corners and cornices; and concrete in the range of fluctuating water levels or spray. These are parts of dams, spillways, wasteways, blowoff boxes, tunnel inlets and outlets, tailrace walls, valve houses, canal structures, and other concrete work.	0.45±0.02	0.55±0.02
B. Concrete in exposed structures and parts of structures where exposure is less severe than in A, such as portions of tunnel linings and siphons subject to freezing, the exterior of mass concrete, and the other exposed parts of structures not covered by A.	0.50±0.02	0.55±0.02
C. Concrete in structures or parts of structures to be covered with backfill, or to be continually submerged or otherwise protected from the weather, such as cutoff walls, foundations, and parts of substructures, dams, trashracks, gate chambers, outlet works, and control houses. (If severe exposure during construction appears likely to last several seasons, reduce *W/C* for parts most exposed by 0.05.)	0.58±0.02	0.58±0.02
D. Concrete that will be subject to attack by sulfate alkalies in soil and ground waters, and will be placed during moderate weather.	------------	0.50±0.02
E. Concrete that will be subject to attack by sulfate alkalies in soil and ground waters, but will be placed during freezing weather, when calcium chloride would normally be used in mix. Do not employ $CaCl_2$, but decrease *W/C* to the value shown.	0.45±0.02	------------
F. Concrete deposited by tremie in water	0.45±0.02	0.45±0.02
G. Canal lining	0.53±0.02	0.58±0.02
H. Concrete for the interior of dams	The properties of this concrete will be governed by the strength, thermal properties, and volume change requirements which will be established for each structure.	

the strength in relation to the water-cement ratio for both air-entrained and nonair-

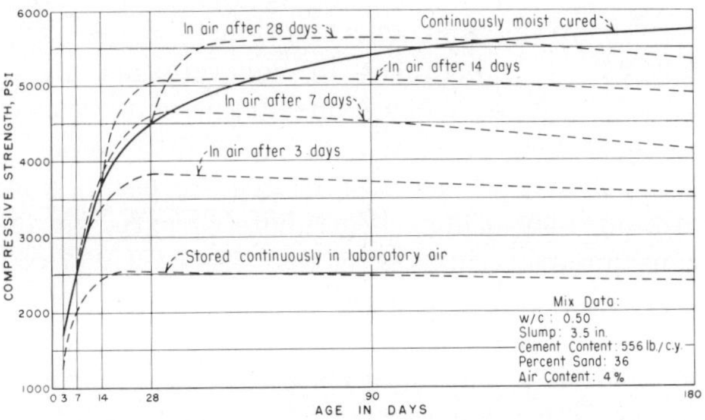

Figure F–3. **Compressive strength of concrete dried in laboratory air after preliminary moist curing. 288–D–2644.**

entrained concrete. Note that the strength decreases with an increase in water-cement ratio, and that the use of air entrainment also decreases the strength.

F-6. *Types of Portland Cement.*—Because of their size or physical location, structures often require the use of cements having special properties to assure adequate durability and economic life. At the present time there are five main types of cement, which will be briefly discussed. The differences in types are the result of changing the relative proportions of the four predominating chemical compounds.

Type I cement is for use in general concrete construction when the special properties of the other types of cement are not required. This type of cement is suitable for use when there is no exposure to sulfates in the soil or ground water. Usually, it is more economical than type II cement.

Type II cement is used where relatively low heat generation is desired or where moderate sulfate attack may occur. Concrete made with type II cement possesses all the good qualities inherent in that containing type I cement.

Type III cement is used where rapid strength development of concrete is essential, as in emergency construction and repairs, and in the construction of machine bases and gate installations. Where this type of cement is used, curing and protection of the concrete may be discontinued at an earlier age.

Type IV cement generates less heat than the other types and at a slower rate. It was developed to reduce the cracking resulting from high temperature rise and subsequent contraction with temperature drop that, in general, accompanies the use of type I or type II cements in massive concrete structures. In addition, concrete in which type IV is used has somewhat greater resistance to sulfate attack than for type I or type II, and has less rapid strength development but equal strength at advanced ages (particularly in the case of mass concrete).

Type V cement is especially beneficial where structures such as canal linings, culverts, and siphons will be in contact with soils and ground waters containing soluble sulfates in such concentrations as would cause serious deterioration of the concrete if other types of cement were used. Concrete containing type V cement is much more resistant to sulfate attack than concretes which contain any of the other types of cement. Compressive strength development, though generally not so rapid, ultimately is approximately equal to that developed by other types.

Any of the five types described above may be purchased to meet the low-alkali provisions of Federal Specifications SS-C-192. Air-entraining cement may also be purchased under these specifications.

F-7. *False Set in Cement.*—False set in cement (also called premature stiffening) is evidenced in concrete by a significant loss of workability during, or shortly after, mixing. This phenomenon is rare and results from the presence of unstable gypsum in the cement. False set is most likely to prove troublesome where concrete is mixed for a short time and then transported to the forms in nonagitating equipment. Workability can usually be restored by prolonged mixing or remixing of the concrete. The addition of more mixing water also restores workability, but it results in a higher water-cement ratio, which reduces the strength and should not be used as a corrective measure. False set is an extremely erratic condition and difficult to control. It can be severe enough to defeat efforts toward uniformity control, to delay construction, and to increase construction costs.

F-8. *Use of Pozzolans.*—Pozzolans are siliceous or siliceous and aluminous materials, which in themselves possess little or no cementitious value, but will, in finely divided form and in the presence of moisture, chemically react with calcium hydroxide (lime) at ordinary temperatures to form compounds possessing cementitious properties. The lime required for this reaction is provided by the portland cement. Materials having pozzolanic properties are some clays and shales, volcanic materials (including pumice, pumicite, etc.) and fly ash, a product of some coal-burning boilers.

Pozzolans usually should not be specified for concrete unless there are definite advantages in their use. Pozzolans may be used to improve the workability and quality of concrete, to effect economy, or to protect against disruptive expansion caused by alkali-aggregate reaction and/or sulfate attack. In addition to improved workability of concrete, most pozzolans will reduce heat generation and thermal volume change of the concrete when they are used to replace a portion of the cement. Concrete bleeding and permeability will also be reduced when pozzolans are used. Compressive strength development is at a slower rate than that of portland cement, but the ultimate strength developed is as great or possibly greater.

F-9. *Quality and Gradation of Aggregates.*— The procedures for quality and gradation tests for concrete aggregates are outlined in the appendix of the Bureau of Reclamation's "Concrete Manual" [1]. Concrete aggregate usually consists of natural sand and gravel, crushed rock, or mixtures of these materials. Natural sands and gravels are the most common and are used whenever they are of satisfactory quality and can be obtained economically in sufficient quantity. Crushed rock is widely used for coarse aggregate and occasionally is processed to produce sand when suitable materials from natural deposits are not economically available. Production of workable concrete using sharp, angular, crushed fragments usually requires more care and more cement and water than does concrete made with well-rounded sand and gravel. However, the difficulty of making workable concrete with crushed aggregate may be greatly reduced through the extra workability imparted by entrained air.

Aggregate is commonly contaminated by silt, clay, mica, coals, humus, wood fragments, or other organic matter, chemical salts, and surface coatings and encrustations. Some contaminating substances in concrete act in a variety of ways to cause unsoundness, decreased strength and durability, and unsightly appearance; their presence complicates processing and mixing operations. Fortunately, excesses of contaminating substances may frequently be removed by simple washing.

An aggregate is considered to be physically sound if it is adequately strong and is capable of resisting the agencies of weathering without disruption or decomposition. Mineral or rock particles that are physically weak, extremely absorptive, easily cleavable, or which swell when saturated are susceptible to breakdown through exposure to natural weathering processes. The use of such materials in concrete reduces strength or leads to premature deterioration by promoting weak bond between aggregate and cement paste, or by inducing cracking, spalling, or popouts. Shales, friable sandstones, some micaceous rocks, clayey rocks, some very coarsely crystalline rocks, and various cherts are examples of physically unsound aggregate materials.

Chemical soundness of an aggregate is also important. In many instances, excessive expansion causing premature deterioration of concrete has been associated with chemical reaction between reactive aggregate and the alkalies in cement. Known reactive substances are the silica minerals (opal, chalcedony, tridymite, cristobalite), zeolite, heulandite (and probably ptilolite), glassy to cryptocrystalline rhyolites, dacites and andesites and their tuffs, and certain phyllites.

An aggregate should and usually does have sufficient strength to develop the full strength of the cementing matrix. Generally, resistance of concrete to abrasion is directly related to its *compressive strength* regardless of the type of aggregate employed. Usually, quartz, quartzite, and many dense volcanic and siliceous rocks are well adapted for making strong and, therefore, wear-resistant concrete.

Volume change in aggregate resulting from wetting or drying is a common source of injury to concrete. Shales, clays, and some rock nodules are examples of materials that expand when they absorb water and shrink as they dry.

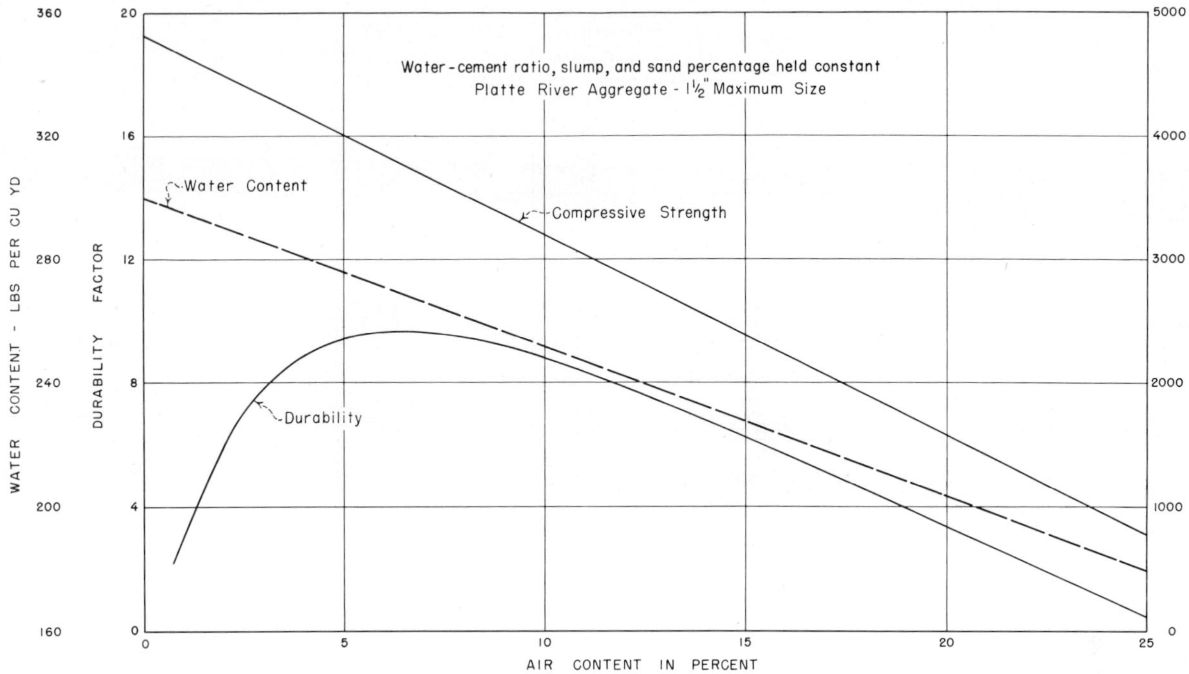

Figure F–4. Effects of air content on durability, compressive strength, and required water content of concrete. 288–D–1520.

Flat or elongated particles of aggregate have a detrimental effect on the workability of concrete and require more highly sanded mixes with consequent use of more cement and water. A moderate percentage of flat or elongated fragments in the larger sizes of coarse aggregate has little effect on the workability or cost of concrete.

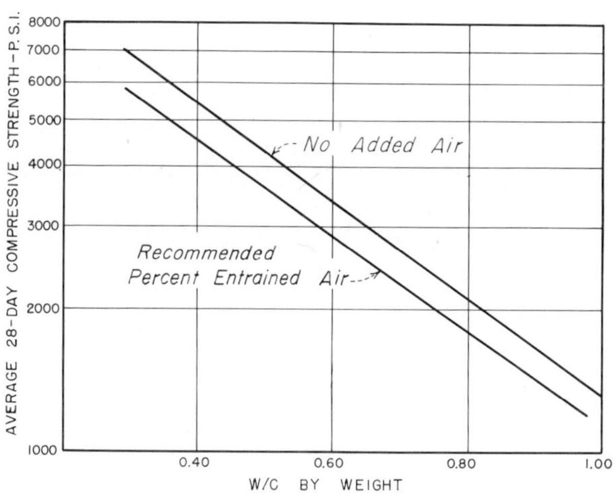

Figure F–5. Strength in relation to water-cement ratio for air-entrained and non-air-entrained concrete. 288–D–1524.

Specific gravity [3] is a useful, quick indicator of aggregate quality. Low specific gravity frequently indicates porous, weak, and absorptive material, and high specific gravity often indicates good quality. However, such indications are not infallible and should be confirmed by other tests. Specific gravity of aggregate in itself is of direct importance only in those cases where design or structural considerations require that the concrete have minimum or maximum weight. When lightness is desired, artificially prepared aggregates of low unit weight are frequently used in place of natural rock.

The particle-size distribution of aggregate as determined by separation with standard screens is known as its gradation [4]. Aggregate grading is important principally because of its effect on water-cement ratio and paste-aggregate ratio, which affect economy and placeability of concrete. A grading chart, similar to that shown in figure F–6, is useful for depicting the size distribution of the aggregate particles.

F-10. *Quality of Mixing and Curing Water.*—Mixing and curing water for concrete should be reasonably clean and free from objectionable

SCREEN SIZE	% RETAINED		COMB % RET	
	INDIVIDUAL	CUMULATIVE	INDIVIDUAL	CUMULATIVE
6 INCH	0	0	0	0
3 INCH	28	28	21	21
1½ INCH	26	54	20	41
¾ INCH	22	76	16	57
⅜ INCH	16	92	12	69
No. 4	8	100	6	75
No. 4	0	0	0	
No. 8	12	12	3	78
No. 16	20	32	5	83
No. 30	24	56	6	89
No. 50	24	80	6	95
No. 100	16	96	4	99
PAN	4	100	1	100
F M		2.76		
PERCENT SAND (clean separation) 25				
(Screen sizes are based on square openings)				

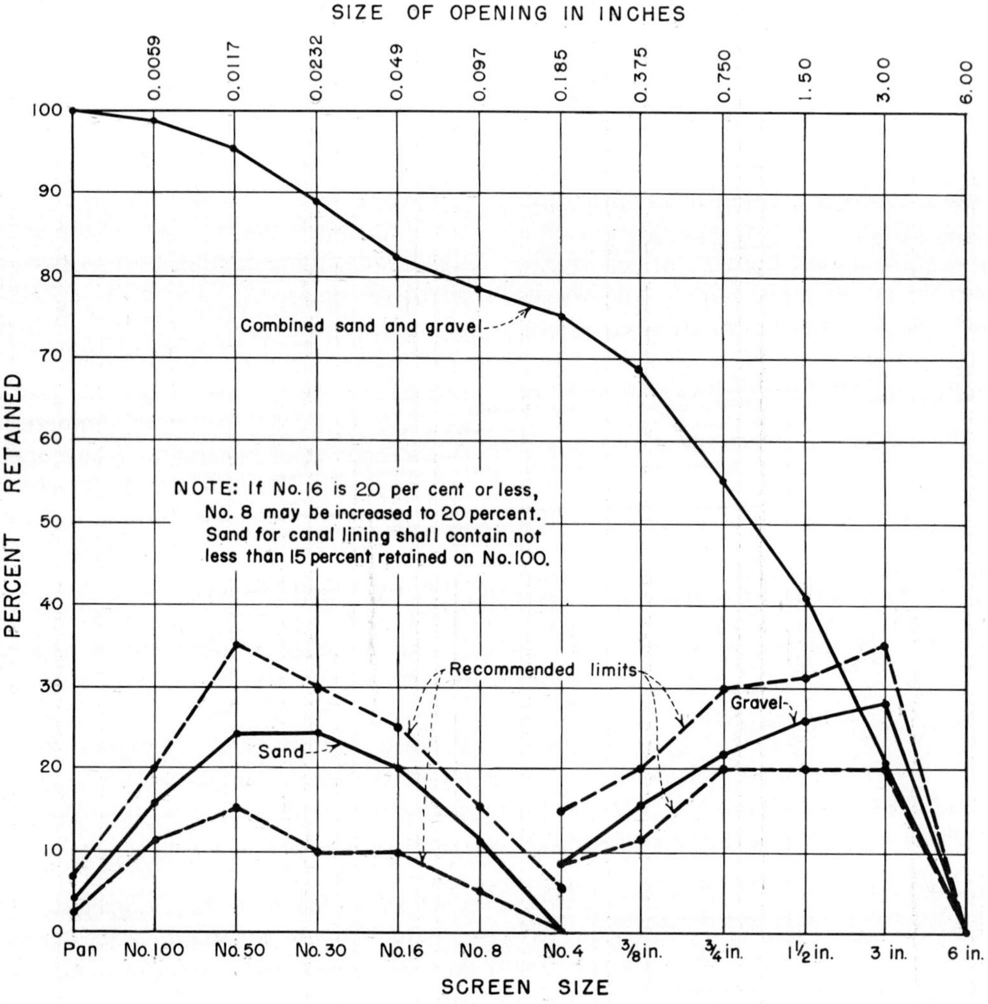

Figure F–6. Typical size distribution of suitably graded natural aggregate. 288–D–803.

quantities of silt, organic matter, alkali, salts, and other impurities. Preparatory to its use in concrete, water from a stream carrying an excessive quantity of suspended solids should be allowed to stand in settling basins or should be clarified by other means. Except for possible discoloration, moderate amounts of salts in water do not appear to have any harmful effects, and water containing not more than 3,000 parts per million of soluble sulfates may be used for mixing and curing concrete.

F-11. ***Use of Admixtures.***—The early strength of concrete can be materially increased by inclusion of an accelerator such as calcium chloride in the concrete mix. Increased early strength during cold weather affords better protection against damage of concrete from freezing at the end of the specified protection period. Also, high early strengths may be desirable for expediting form removal or to permit early loading of anchor devices.

However, calcium chloride should not be used in concrete in which aluminum or galvanized metalwork is to be embedded, nor in prestressed concrete because of the possibility of corrosion. Use of air-entrained concrete is a requirement for Bureau of Reclamation construction. The most important benefit of purposefully entraining air in concrete is that it greatly increases the resistance to the disintegrating action of freezing and thawing. Additionally, entrainment of air reduces bleeding and segregation, greatly facilitates the handling and placing of concrete, and permits the use of a wider range in aggregate gradations. It reduces sand and water requirements, and the curtailed bleeding permits finishing of concrete surfaces earlier and with less work. Also, resistance to chemical attack and permeability is improved by the reduction in capillarity produced by air entrainment. Among the factors that influence the amount of air entrained in concrete, for a given amount of air-entraining agent, are grading and particle shape of aggregate, richness of mix, mixing time, slump, and concrete temperature. Recommended percentages of entrained air are shown in table F–4. In mild climates these values may be reduced about one-fourth if strength development is critical and satisfactory workability can be maintained.

Water-reducing, set-controlling admixtures (WRA) are commonly required in Bureau concrete to further reduce water requirements and extend the length of time concrete can be consolidated by vibration [8]. Cement savings may result from water reductions and the risk of obtaining cold joints is reduced by extending the vibration period. However, the WRA dosage used should be determined by laboratory tests utilizing the WRA and project materials at job site temperatures to avoid excessive retardation.

F-12. ***Field Control.***—After concrete materials have been selected and the relative proportions determined, their use should be controlled closely. This field control governs the quality, uniformity, and ultimate economy of the concrete structure. Much of the potential value of first-class materials and optimum proportioning may be lost through ineffective control in batching, mixing, handling, placing, and curing. The poorer the quality of the ingredients, the greater the need for rigid control to attain satisfactory durability and strength and, therefore, maximum serviceable life in the structure.

The degree of uniformity of concrete strength is a measure of success in attaining adequate field control. Without adequate control of concrete manufacturing operations, wide variations in strength will occur and extra cement will be needed to insure that the quality of the concrete will meet minimum requirements.

B. DESIGN OF CONCRETE MIXES

F-13. ***Introduction.***—Concrete is composed essentially of water, cement, aggregate, and purposefully entrained air. Proportions of ingredients should be selected to make the most economical use of available materials that will produce concrete of the required workability, durability and strength. Mix proportions should be selected to produce concrete with:

(1) The stiffest consistency (lowest slump) that can be efficiently placed and consolidated

by vibration into a homogeneous mass,

(2) The lowest sand-aggregate ratio that is reasonably possible,

(3) The largest maximum size of aggregate economically available which is consistent with placement and strength requirements,

(4) Adequate durability to satisfactorily withstand weathering and other destructive agencies to which it may be exposed, and

(5) Sufficient strength to withstand the loads to be imposed without danger of failure.

F-14. *Estimate of Water Requirement.*—Overwet concrete should always be avoided; it is difficult to place without segregation and it is certain to be weak and lacking in durability. The proper consistency, as determined by the slump test [2], for placing and consolidating concrete in various types of structures is shown in table F–3.

TABLE F–3.—*Recommended maximum slumps for various types of concrete construction*

Type of construction	Maximum slump in inches[1]
Heavy mass construction	2
Canal lining	3
Slabs and tunnel inverts	2
Tops of walls, piers, parapets, and curbs	2
Sidewalls and arch in tunnel lining	4
Other structures	3

[1] These maximum slumps are for concrete after it has been deposited, but before it has been consolidated, and are for mixes having air contents as indicated in table F–4.

The quantity of water per unit volume of concrete required to produce a mix of desired consistency is influenced by the maximum size, particle shape, and grading of the aggregate and by the amount of entrained air. Within the normal range of mixes, the water requirement is relatively unaffected by the quantity of cement. The quantities of water given in table F–4 are of sufficient accuracy for preliminary estimates of proportions. They are the averages that may be expected for various maximum sizes of fairly well shaped and well graded aggregate. Flat-shaped aggregates with excess fines will require more water, and perfectly round shaped, well graded aggregates will not require as much water as shown in table F–4. The weight of water throughout the normal range of placing temperatures may be assumed to be 62.3 pounds per cubic foot.

F-15. *Estimate of Cement Requirement.*—A fundamental rule for designing plastic concrete mixes is that the strength and durability of hardened concretes, with the same air content, vary inversely with the ratio of the weight of water to the weight of cement. Table F–2 will serve as a guide in selecting maximum permissible water-cement ratios for different severities of exposure when proper use is made of air entrainment.

Table F–5 shows an approximation of the minimum strengths to be expected for air-entrained concrete with different water-cement ratios. This table is conservative and can be used in estimating the strength of concrete until verified by tests of compressive strength specimens.

The cement content is calculated, using the lowest water-cement ratio selected from table F–2 or table F–5 and the water requirement from table F–4. The calculation is accomplished by dividing the water requirement by the water-cement ratio. If a minimum cement content is specified, the corresponding water-cement ratio for estimating strength can be computed by dividing the water content by the cement content.

The term "cement" refers to portland cement or a combination of portland cement and pozzolan fully meeting the requirements of applicable Bureau of Reclamation specifications.

F-16. *Estimate of Admixture Requirements.*—When calcium chloride is used as an accelerator, it should not be used in excess of 2 percent by weight of cement. Calcium chloride should be added to the batch in solution as a part of the mixing water. If proprietary accelerators are used, the manufacturer's recommendations should be followed.

The amount of air-entraining agent required to produce a desired percentage of entrained air varies with the materials used, temperature of the concrete, richness of mix, and consistency of the fresh concrete. Decreasing the slump, or increasing the temperature or cement content of concrete, will usually require larger

TABLE F–4.—*Air and water contents for concrete containing natural sand and average coarse aggregate*

FOR FINENESS MODULUS (F.M.)[1] EQUAL TO 2.75 AND A SLUMP OF 2 TO 4 INCHES AT THE MIXER

	Maximum size of coarse aggregate, inches							
	3/8	1/2	3/4	1	1½	2	3	6
Percent dry-rodded unit weight of coarse aggregate per unit volume of concrete	41	52	62	67	73	76	81	87
Recommended air content, percent	8	7	6	5	4.5	4	3.5	3
Average water content, pounds per cubic yard	322	306	283	267	245	229	204	164
Sand, percent of total aggregate by solid volume	59	50	42	37	33	30	28	24

ADJUSTMENT OF VALUES FOR OTHER CONDITIONS

Changes in conditions stipulated	Effect on values		
	Unit water content	Percent sand	Percent of dry-rodded coarse aggregate
Each 0.1 increase or decrease in F.M. of sand	----------	±0.5	∓1
Each 1-inch increase or decrease in slump	±3%	----------	----------
Each 1 percent increase or decrease in air content	∓3%	(∓)0.5 to 1.0	----------
For less workable concrete, as in pavements	−8 lb.	−3	+6

[1] The fineness modulus of sand is computed by adding the cumulative percentages retained on the 6 standard screens, from the No. 4 to the No. 100, inclusive, and dividing the sum by 100.

TABLE F–5.—*Probable minimum compressive strength of concrete for various water-cement ratios, pounds per square inch*

Water-cement ratio by weight	Compressive strength of air-entrained concrete at 28 days
0.40	4,300
.45	3,900
.50	3,500
.55	3,100
.60	2,700
.65	2,400
.70	2,200

amounts of agent to maintain the desired air content. The manufacturer's recommendations should be used for the initial mix. More or less agent may be added to subsequent mixes based on fresh concrete tests. Tests to determine air content are described in the Bureau of Reclamation's "Concrete Manual" [5]. Agents should be added as a solution to the mix water, either before the mixer is charged or during charging. Some cements are manufactured with an air-entraining agent integrally blended with cement. However, if a uniform air content is to be maintained under varying conditions, the air-entraining agent should be added at the batch plant so that the amount added may be readily adjusted. Control of the amount of air is necessary for adequately uniform strength of concrete because a high air content will decrease the compressive strength.

Batch quantities of water-reducing, set-retarding admixtures (WRA) are based on the weight of cementitious material; that is, fluid ounces or pounds of WRA per 100 pounds of cement plus pozzolan. Again, manufacturer's dosage recommendations should be used for initial tests and then the dosage may be adjusted to obtain desired results. Actual results are affected by the materials used and the ambient temperature. The amount of set-retardation increases as the WRA dosage increases. The water reduction, however, only increases to a maximum with increasing WRA dosage and further additions will only extend the retardation. Consequently, in most cases the dosage used is dependent upon tolerable retardation.

F-17. *Estimate of Aggregate Requirement.*—Larger maximum sizes and quantities of well-graded coarse aggregate require less mortar to fill voids and provide workability. A good estimate of the optimum quantity of a given maximum-size coarse aggregate is shown in table F–4 as the percent of the dry-rodded unit weight per unit volume of concrete. Table F–6 gives the recommended maximum size of aggregate for various types of construction. The largest permissible aggregate size consistent with the conformation of the structure and the compressive strength requirement should be used because it permits minimum water and cement quantities. Determination of dry-rodded unit weight of coarse aggregate is described in the Bureau of Reclamation's "Concrete Manual" [6]. Basing the amount of coarse aggregate on a fixed percentage of the dry-rodded unit weight automatically makes allowance for differences in mortar requirements as reflected by void content. For example, angular aggregates have a higher void content and therefore require more mortar than rounded aggregate. The higher void content results in a lower dry-rodded unit weight and therefore decreases the amount of coarse aggregate obtained from the fixed percentage, which automatically produces a greater amount of mortar per unit volume of concrete.

When the coarse aggregate is estimated as described above, the amount of sand is calculated as follows:

Determine the volume of the known ingredients in cubic feet per cubic yard of concrete. The volume occupied by a material is equal to its weight divided by its density. (Density= specific gravity x 62.3.) The specific gravity of sand and coarse aggregate may be determined in accordance with the Bureau of Reclamation's "Concrete Manual" [3]. Cement density may be assumed to be 196.2 pounds per cubic foot. The volume of air in cubic feet per cubic yard of concrete equals 27 times the percent air divided by 100. Subtracting the total volume of all ingredients except sand from 27 gives the volume of sand in cubic feet per cubic yard. Multiplying the volume of sand by its density gives its weight in pounds per cubic yard.

F-18. *Computations of Proportions.*—The computations of proportions for concrete mixes can best be explained by means of specific examples. For these examples, the following design criteria and mix materials will be assumed:

(1) Type II cement with a specific gravity of 3.15.

(2) Coarse aggregate with a specific gravity of 2.68.

(3) Sand with a specific gravity of 2.63 and a fineness modulus of 2.75.

(4) Dry-rodded unit weight of coarse aggregate of 105 pounds per cubic foot.

(5) Sufficient air-entraining agent to entrain the percentages of air recommended in table F–4.

Example 1.—The first example involves a reinforced retaining wall, having a minimum thickness of 11 inches. Tables F–3 and F–6

TABLE F–6.—*Maximum sizes of aggregate recommended for various types of construction*

Minimum dimension of section, inches	Maximum size of aggregate,[1] in inches, for:		
	Reinforced walls, beams, and columns	Heavily reinforced slabs	Lightly reinforced or unreinforced slabs
5 or less		¾ to 1½	¾ to 1½.
6 to 11	¾ to 1½	1½	1½ to 3.
12 to 29	1½ to 3	3	3 to 6.
30 or more	1½ to 3	3	6.

[1] Based on square screen openings.

indicate that a 3-inch slump (under "Other structures") and 1½-inch-maximum-size aggregate will be satisfactory. The concrete will be subjected to rather severe climatic exposure and will fall in class B of table F–2. The design strength of the wall is 3,000 pounds per square inch; however, the concrete must be designed for a higher strength to compensate for material and testing variances. Experience has shown that satisfactory results can be obtained if 80 percent of the test cylinders have strengths greater than the design strength. Based on this figure and assuming a coefficient of variation of 15 percent, the mix will have to be designed for an average strength of 3,460 pounds per square inch. The computations for example 1 are shown in table F–7.

F-19. *Alternate Method of Computations of Proportions.*—If it is desirable to estimate the mix proportions without determining the dry-rodded unit weight of coarse aggregate, the percentages of sand shown in table F–4 may be utilized. The total volume of aggregate in cubic feet per cubic yard of concrete (both sand and coarse aggregate) is computed by subtracting from 27 the cubic feet of water, cement, and air per cubic yard of concrete. The volume of sand in cubic feet per cubic yard of concrete is obtained by multiplying the cubic feet of total aggregate in a cubic yard of concrete by the percentage of sand extracted from table F–4.

Computations utilizing this alternate method are illustrated in the following example.

Example 2.—The second example (table F–8) requires a concrete mix for a powerhouse

TABLE F–7.—*Computations of concrete mix proportions—Example 1*

Mix ingredients	Weight, pounds per cubic yard	Conversion of weight to volume	Solid volume, cubic feet per cubic yard
Water: Estimated value from table F–4	245	$\frac{245}{62.3}$	3.93
Cement:			
W/C for durability, class B concrete (table F–2)=0.50.			
W/C for strength (table F–5)=0.51 (durability controls, therefore use 0.50).			
$\text{Cement}=\frac{\text{water content}}{W/C}=\frac{245}{0.50}=$	490	$\frac{490}{3.15\times62.3}$	2.50
Air: From table F–4, 4.5 percent: 0.045×27=			1.21
Coarse aggregate:			
Percent of dry-rodded unit weight from table F–4=73 percent.			
Dry-rodded unit weight=105 pounds×0.73=76.6 pounds per cubic foot × 27=	2,070	$\frac{2,070}{2.68\times62.3}$	12.40
All ingredients except sand	2,805		20.04
Sand:			
Volume 27−20.04=			6.96
Weight 6.96×2.63×62.3=	1,140		
Total	3,945		27.00

foundation which will not be exposed to freezing and thawing. This condition will permit the use of class C concrete from table F–2. A design strength of 2,500 pounds per square inch at 28 days is specified for the foundation. An average strength of 2,880 pounds per square inch will be required to satisfy the same statistical requirements used in example 1. Tables F–3 and F–6 indicate that a 3-inch slump (again under "Other structures") and 3-inch-maximum-size aggregate will be satisfactory.

F-20. *Batch Weight Computations.*—The preceding trial-mix computations provide the batch quantities for a cubic yard of concrete. In cases where it is not possible or feasible to mix concrete in exact 1-cubic-yard batches, the design quantities must be decreased or increased to the desired volume. This adjustment can be accomplished by multiplying the 1-cubic-yard quantity of each ingredient by the desired batch volume, which in most cases is equivalent to the mixer capacity. Assuming a mixer capacity of 15.5 cubic feet, the batch quantities for example 1 will be as follows:

Ingredient	*Weight, pounds*	*Volume, cubic feet*
Water	0.575 × 245 = 141	0.575 × 3.93 = 2.26.
Cement (3 sacks)	0.575 × 490 = 282	0.575 × 2.50 = 1.44.
Sand	0.575 × 1,140 = 655	0.575 × 6.96 = 4.00.
Coarse aggregate	0.575 × 2,070 = 1,190	0.575 × 12.40 = 7.13.

The aggregates have been assumed to be in a saturated surface-dry state. Under field conditions, they will generally be moist, and the quantities to be batched in the mixer must be adjusted accordingly. Assume that tests show the sand to contain 5 percent and the coarse aggregate 1 percent free moisture. Since the quantity of saturated surface-dry sand required was 655 pounds, the amount of moist sand to be weighed is 655×1.05=688 pounds. Similarly, the weight of moist coarse aggregate is 1,190×1.01=1,202 pounds. Coarse aggregate is sometimes drier than saturated surface-dry. Assuming an absorption of 1 percent, the amount of dry aggregate required would be 1,190×0.99=1,178 pounds.

TABLE F–8.—*Alternate method of computations for concrete mix proportions—Example 2*

Mix ingredients	Weight, pounds per cubic yard	Conversion of weight to volume	Solid volume, cubic feet per cubic yard
Water: Estimated value from table F–4	204	$\frac{204}{62.3}$	3.27
Cement:			
W/C for durability, class C concrete (table F–2)=0.58.			
W/C for strength (table F–5)=0.58.			
Cementing materials= $\frac{\text{water content}}{W/C}=\frac{204}{0.58}=$ 352 pounds	352	$\frac{352}{3.15\times62.3}$	1.79
Air: From table F–4, 3.5 percent: 0.035×27=			0.95
All ingredients except aggregate	556		6.01
Aggregate:			
Volume=27−6.01=			20.99
Percent sand (table F–4)=28 percent.			
Volume of sand=0.28×20.99=			5.88
Volume of coarse aggregate=20.99−5.88=			15.11
Weight of sand=5.88×2.63×62.3=	963		
Weight of coarse aggregate=15.11×2.68×62.3=	2,523		
Total	4,042		27.00

The free water in the aggregate must be considered as part of the mixing water, whereas water must be added to allow for absorption in the case of dry aggregate. In our example, free water (mixing water) in the sand is 688—655=33 pounds, and in the coarse aggregate 1,202—1,190=12 pounds. If the coarse aggregate were dry, 1,190—1,178=12 pounds of water must be added to the mixing water to allow for absorption.

F-21. ***Determination of Actual Quantities per Cubic Yard.***—Because of the large number of variables introduced into a concrete mix through air entrainment, it is necessary to determine the actual quantities per cubic yard, or yield, of the concrete mix. The actual weight of each ingredient in pounds per cubic yard is equal to the batch weight of the ingredient divided by the batch volume in cubic yards. The actual batch volume is calculated from the unit weight test of fresh concrete. Assuming a unit weight of 147.4 pounds per cubic foot for example 1, the yield calculation is as follows:

$$\text{Batch volume}=\frac{\text{summation of batch weights}}{\text{measured unit weight} \times 27}=$$

$$\frac{2{,}268}{147.4 \times 27}=0.57 \text{ cubic yard}$$

Water content=

$$\frac{141}{0.570}=247 \text{ pounds per cubic yard}$$

Cement content=

$$\frac{282}{0.570}=495 \text{ pounds per cubic yard}$$

Sand content=

$$\frac{655}{0.570}=1{,}149 \text{ pounds per cubic yard}$$

Coarse aggregate content=

$$\frac{1{,}190}{0.570}=2{,}088 \text{ pounds per cubic yard.}$$

The percentage of entrained air in the concrete can be determined from both the unit weight and pressure air meter [5]. It is advantageous to utilize both methods because any appreciable difference in the results indicates an error which may lead to the discovery of mistakes in mix design computations, or in test methods. A variation in the air contents of as much as 0.3 percent is considered normal. Using the unit weight, the percentage of entrained air is calculated by dividing the volume of air by the total volume (27 cubic feet) and multiplying by 100. The volume of air is determined by subtracting the sum of the solid volumes from 27. Solid volumes of materials are calculated by dividing the weights of ingredients by their respective densities. The calculations for example 1 are shown below.

$$\text{Water volume}=\frac{247}{62.3}=3.96 \text{ cubic feet}$$

$$\text{Cement volume}=\frac{495}{3.15 \times 62.3}=2.52 \text{ cubic feet}$$

$$\text{Sand volume}=\frac{1{,}149}{2.63 \times 62.3}=7.01 \text{ cubic feet}$$

Coarse aggregate volume=

$$\frac{2{,}088}{2.68 \times 62.3}=12.51 \text{ cubic feet}$$

Total 26.00 cubic feet

The percent of air then equals:

$$\frac{27.00-26.00}{27.00} \times 100=3.7$$

Other yield items such as the water-cement ratio are calculated as they are during mix design computations, except yield quantities are used. Figure F–7 illustrates a convenient form for recording mix data. The data presented conform to the mix in example 1.

F-22. ***Adjustment of Trial Mix.***—The actual quantities per cubic yard determine whether a trial mix complies with the desired results and if adjustments in design quantities are necessary. In example 1, the slump was correct but the air content was low. To find the amount of agent required to produce 4.5 percent air, a linear adjustment is made. Since 7.9 fluid ounces produced 3.7 percent air and 4.5 percent is desired, then 4.5/3.7×7.9 or 9.6 fluid ounces should be used in the second trial mix. Also, table F–4 states that an air content increase of one percent will permit a decrease in water requirement of about 3.0 percent to maintain a 3-inch slump. The corrected water content then equals 247—(247×0.03)=240 pounds per cubic yard, which indicates that our original estimate for water was high. When these adjustments are made, the resulting concrete mix should yield the desired 3-inch slump and 4½ percent air.

Table F–4 shows other adjustments that may be necessary during trial mixing. The job

CONCRETE MIX COMPUTATION SHEET

SERIES DOSD DATE CAST 9-13-68

SAND GRADING %: PAN 3, #100 15, #50 29, #30 24, #16 15, #8 14; F.M. 2.75

C. AGG. GRADING %: #4-3/8" 20, 3/8-3/4" 30, 3/4-1½" 50, 1½-3", 3-6"

AIR METER NO. 4 FACTOR 4.028

UNIT WT. CAN NO. FACTOR

DRY RODDED UNIT WEIGHT LB./CU. FT. 105.0

INGREDIENTS	NAME	LAB NO.	SP. G.	DENSITY LBS./CU. FT.	% MOIST.
WATER	Tap		1.0	62.3	
CEMENT	Ideal	M-6800	3.15	196.2	
POZZOLAN	--	--	--	--	
SAND	Clear Creek	M-3864	2.63	163.8	5.0
C. AGG.	Clear Creek	M-3864	2.68	167.0	1.0

AEA NVX % SOLIDS 12 ADMIXTURE None

TRIAL MIX COMPUTATIONS

MIX NO.	$\frac{W}{C+P}$ BY WT.	% SAND OR % DRUW	AGENT		AIR	WATER	CEMENT	POZZ.	SAND	C. AGG.	TOTAL
					ESTIMATED QUANTITIES PER CUBIC YARD						
Example				LB.	4.5%	245	490	--	1140	2070	3945
1	0.50	73	7.8	CU. FT.	1.21	3.93	2.50	--	6.96	12.40	27

BATCH COMPUTATIONS

MIX NO.	VOL. OF BATCH IN CU. YD.	AGENT	WATER	CEMENT	POZZ.	SAT. SUR. DRY SAND	SAT. SUR. DRY C. AGG.	TOTAL	ADJUSTED WEIGHT SAND	ADJUSTED WEIGHT C. AGG.	MIXING WATER IN SAND	MIXING WATER IN C. AGG.	MIXING WATER ADDED	MIXING WATER TOTAL
		QUANTITIES PER BATCH							ADJUSTMENT FOR MOISTURE					
1	0.575	4.5	141	282	--	655	1190	2268	688	1202	33	12	96	141

RESULTS OF TESTS ON FRESH CONCRETE

MIX NO.	TEMP °F	SLUMP IN.	WEIGHT OF CONCRETE LBS.	UNIT WEIGHT LBS./CU. FT.	VOL. BATCH CU. FT./CU. YD*	AGENT		WATER	CEMENT	POZZ.	SAND	C. AGG.	TOTAL	% AIR GRAV.	% AIR PRESS.	$\frac{W}{C+P}$ BY WT.	% SAND OR % DRUW	SPECIMENS CAST	SPECIMEN NUMBERS
								ACTUAL QUANTITIES PER CUBIC YARD											
							LB.	247	495	--	1149	2088	3979						
1	70	3.0	36.59	147.4	0.570	7.9	CU. FT.	3.96	2.52	--	7.01	12.51	26.00	3.7	3.1	0.50	74		

MIX NOTES

* OR BATCHES PER CU. YD.

Figure F–7. Typical trial computations for concrete mix. 288–D–2865.

mix should not be adjusted for minor fluctuations in water-cement ratio. A variation in water-cement ratio of plus or minus 0.02, by weight, resulting from maintenance of constant slump, is considered normal.

F-23. *Mixes for Small Jobs.*—For small jobs, where time and personnel are not available to determine the proportions in accordance with the recommended procedure, mixes in table F–9 will provide concrete that is amply strong and durable if the amount of water added at the mixer is never large enough to make the concrete overwet. These mixes have been predetermined in conformity with recommended procedure by assuming conditions applicable to the average small job, and for aggregate of average specific gravity. Three mixes are given for each maximum size of coarse aggregate. Table F–6 may be used as a guide in selecting an appropriate maximum size of aggregate. Mix B for each size of coarse aggregate is intended for use as a starting mix in table F–9. If this mix is undersanded, change to mix A, or, if it is oversanded, change to mix C. Note that the mixes listed in the table apply where the sand is dry. If the sand is moist or very wet, make the corrections in batch weight prescribed in the note.

The approximate cement content in bags per cubic yard of concrete listed in the table will be helpful in estimating cement requirements for the job. These requirements are based on concrete that has just enough water in it to permit ready working into the forms without objectionable separation. Concrete should slide, not run, off a shovel.

C. MANUFACTURE, PLACEMENT, CURING, AND INSPECTION OF CONCRETE

F-24. *Aggregate Production and Control.*—The control of production and handling of concrete aggregates is often complicated by lack of uniformity in sources of supply and difficulty in maintaining uniformity in the finished production. It is a problem that requires the constant vigilance of the construction engineer. Deleterious materials are ordinarily removed by washing. Unsatisfactory grading requires corrections by wasting of surplus sizes or by supplying deficiencies, or both. Breakage must be minimized and the moisture content of the aggregate should be kept as uniform as practicable.

The gradation of sand as it comes from the pit often does not conform to the specifications and some form of processing is required. Defects in grading may be corrected by adding suitable blending sand, by crushing a portion of the excess of larger sizes, by removing portions of sizes present in excessive amounts, or by a combination of methods. Wet processing is more prevalent than dry processing for this purpose, because sand is commonly damp as it is excavated from the deposit.

Use of sand manufactured by crushing or grinding rock or gravel may result in a harsh mix and should be resorted to only when it is not practicable to obtain suitable natural sand at reasonable cost. Since the angular shape of crushed sand is its only disadvantage, it is important that crushing machines and equipment be used which will produce the best shape of particles from the material to be crushed. Sand produced by crushing in rollers is generally unsatisfactory because of the high percentage of thin and elongated particles. The product of a rod mill is much better in this respect. If the material is not too hard, as in the case of limestone, good results may be obtained with equipment of the impact type, more commonly known as the hammer mill, which excels in producing particles that approach a cubicle shape.

Natural river gravels are usually well shaped by stream action, and satisfactory coarse aggregate with the desired grading can be produced with a minimum of plant equipment.

TABLE F–9.—*Concrete mixes for small jobs* [1]

Maximum size of aggregate, inches	Mix designation	Approximate bags cement per cubic yard of concrete	Pounds of aggregate per 1-bag batch		
			Sand [2]		Gravel or crushed stone
			Air-entrained concrete [3]	Concrete without air	
½	A	7.0	235	245	170
	B	6.9	225	235	190
	C	6.8	225	235	205
¾	A	6.6	225	235	225
	B	6.4	225	235	245
	C	6.3	215	225	265
1	A	6.4	225	235	245
	B	6.2	215	225	275
	C	6.1	205	215	290
1½	A	6.0	225	235	290
	B	5.8	215	225	320
	C	5.7	205	215	345
2	A	5.7	225	235	330
	B	5.6	215	225	360
	C	5.4	205	215	380

[1] Procedure: Select the proper maximum size of aggregate. Then, using mix B, add just enough water to produce a sufficiently workable consistency. If the concrete appears to be undersanded, use mix A; and if it appears to be oversanded, use mix C.

[2] Weights are for dry sand. If damp sand is used, increase the weight of sand 10 pounds for a 1-bag batch; and if very wet sand is used, add 20 pounds for a 1-bag batch.

[3] Air-entrained concrete is specified for all Bureau of Reclamation work. In general, air-entrained concrete should be used in all structures that will be exposed to alternate cycles of freezing and thawing.

However, in some cases where natural coarse aggregate is not economically available, crushed aggregate is used. Although the shape of the individual particles is important, it is not so critical for coarse aggregate as it is for sand. Use of corrugated roll crushers to produce smaller sizes of coarse aggregate and of gyratory crushers to produce the large sizes generally results in the least amount of flat and elongated pieces. Because of the segragation and breakage which results, handling should be kept to a minimum during stockpiling operations. Figure F–8 shows correct and incorrect methods of stockpiling.

Since some breakdown of materials will occur regardless of the care exercised in stockpiling, it is desirable to finish screen coarse aggregates at the batch plant to assure production of uniform concrete. It is recommended that this requirement be specified when the quantity of concrete exceeds 10,000 cubic yards.

Periodic analysis of aggregate materials should be made to determine the specific gravity and moisture content of the aggregate, and to determine the relative percentages of the various size fractions. The frequency of these tests should be sufficient to assure that the aggregates are in accordance with specifications requirements.

F-25. *Batching Methods and Facilities at Concrete Mixing Plants.*—In order that full advantage of accurate weight batching may be realized, the weighted materials must be properly and carefully handled to the end that batches reaching the mixer will be uniform and complete when released by the measuring equipment.

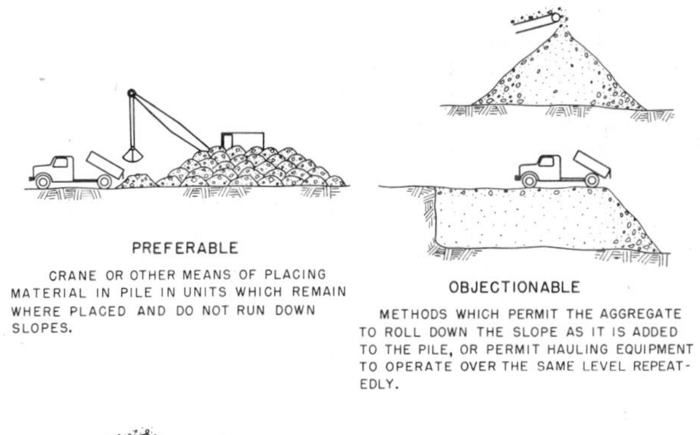

Figure F–8. Methods of stockpiling aggregates. 288–D–2655.

Tilting mixers are generally more efficient than other types because they can be discharged quickly with a minimum of segregation. Regardless of the type of mixer, to maintain efficiency the mixing blades should be properly spaced, inspected frequently, and repaired when worn, and the interior of the drum should be kept clean and free of deposits of hardened concrete or mortar.

More attention and effort are usually required to obtain uniform slump and mix proportions at minimum water content from truck mixers than from stationary mixers. There is often considerable slump loss in truck-mixed concrete, especially in warm weather. Such loss can be kept to a minimum by stopping initial mixing at about 30 revolutions and by avoiding overmixing. Other precautions which can be taken in warm weather are as follows:

(1) Mixer drums should be painted white and kept white.

(2) Materials should be kept as cool as practicable by shading and by light spraying to promote evaporative cooling.

(3) Water should be as cold as practicable and kept cold by shading and by painting tanks and surface lines white.

(4) Delays prior to the discharge and placement of the concrete should be avoided by organizing the work for prompt handling.

F-26. *Production of Quality Concrete.*—The assurance of uniform and economical concrete is largely dependent on inspection at the batching and mixing plants. Mix adjustments are made using results of aggregate gradation and moisture tests, and fresh concrete tests for consistency, temperature, air content, and unit weight. Concrete cylinders are made for the compressive strength tests necessary for quality control. The frequency of sampling and testing the concrete will vary with the type and size of job. In general, it will be sufficient to sample and test each class of concrete once during each shift. The samples should be representative of materials used and concrete placed during that shift.

F-27. *Preparations Preliminary to Placing.*—Before concrete is ordered for placing, adequate inspection should be performed to insure that: (1) foundations are properly prepared and ready to receive the concrete, (2) construction joints are free from defective concrete and clean, (3) forms are grout-tight, amply strong, and set to line and grade, (4) all reinforcement steel and embedded parts are clean, in their correct position, and securely held in place, and (5) adequate concreting equipment and facilities are on the job, ready to go, and capable of completing the placement without additional unplanned construction joints. Detailed requirements for these items are given in the "Concrete Specifications" portion of appendix G.

F-28. ***Transporting.***—Even though concrete is carefully designed and properly mixed, its quality may be seriously impaired by use of improper or careless methods in transporting and placing. Buckets, when designed for the job conditions and properly operated, are a satisfactory means for handling and placing concrete. They are not, however, to be used where they have to be hauled so far by truck or railroad that there will be noticeable separation or bleeding due to settlement, or there will be a loss of slump greater than 1 inch.

Dumpcrete trucks are convenient for the distribution of concrete from a central mixer to small and medium size structures. Care must be taken to avoid segregation during the filling and discharging of these units. No free water should be on the surface of the concrete as delivered, nor should there be an objectionable amount of settlement of coarse aggregate or caking at the bottom of the load. Such stratification or settlement can be reduced considerably by use of agitator bodies mounted on trucks, or preferably by mixing the concrete near the point of placement in portable mixers which are supplied by dry-batch trucks.

As ordinarily used, chutes are unsatisfactory devices for transporting concrete because they result in objectionable segregation and slump loss. To avoid these conditions, the following requirements must be fulfilled:

(1) The chute must be on a slope sufficiently steep to handle concrete of the least slump that can be worked and vibrated, and must be supported so that the slope will be constant for varying loads.

(2) If more than a few feet long, the chute must be protected from wind and sun in order to prevent slump loss.

(3) Effective end control that will produce a vertical drop and prevent separation of the concrete ingredients must be provided, preferably in the form of two sections of metal drop chutes, as shown in figure F–9.

(4) With pneumatic methods, separation of coarse aggregate will result from the impact of violently discharged concrete unless the end of the discharge line is always buried in fresh concrete. Specifications should therefore require that pneu-

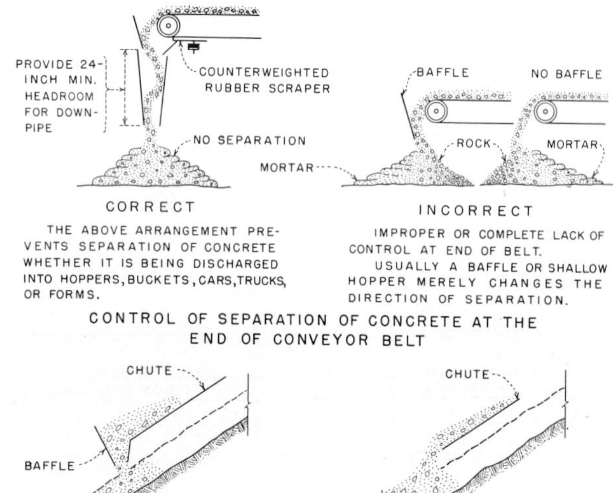

Figure F–9. **Methods of handling concrete at ends of conveyors and chutes. 288–D–854.**

matic equipment used in placing concrete be such as to permit introduction of the concrete into the forms without high-velocity discharge. A further objection to the pneumatic method is a loss of slump which occurs in the shooting process. Slump losses as great as 3½ inches between mixer and forms have been observed and a loss of 2 to 3 inches is not uncommon.

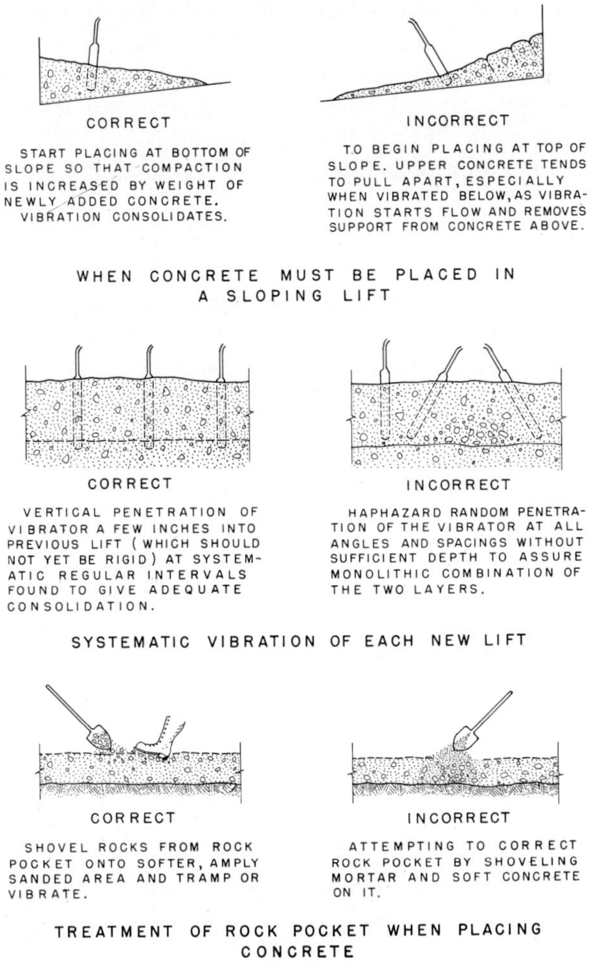

Figure F–10. Methods of vibrating and working of concrete. 288–D–856.

There is no objection to the use of belt conveyors if segregation and objectionable slump losses are prevented, and there is no loss of mortar on the return belt. Segregation, which occurs chiefly at transfer points and at the end of the conveyor, may be avoided by using suitable hoppers and drop chutes, as shown in figure F–9. Slump loss is largely preventable by protecting the belt from the sun and wind. A rubber or other suitable scraper should be placed on the return belt to prevent the loss of mortar and feed it into the concrete-receiving hopper.

Pumping through steel pipelines is one of the most satisfactory methods of transporting concrete where space is limited, such as in tunnels, bridge decks, powerhouses, and buildings. Although pump lines longer than 1,000 feet are not recommended, concrete has been pumped through straight, horizontal pipe under the most favorable conditions as far as 1,300 feet.

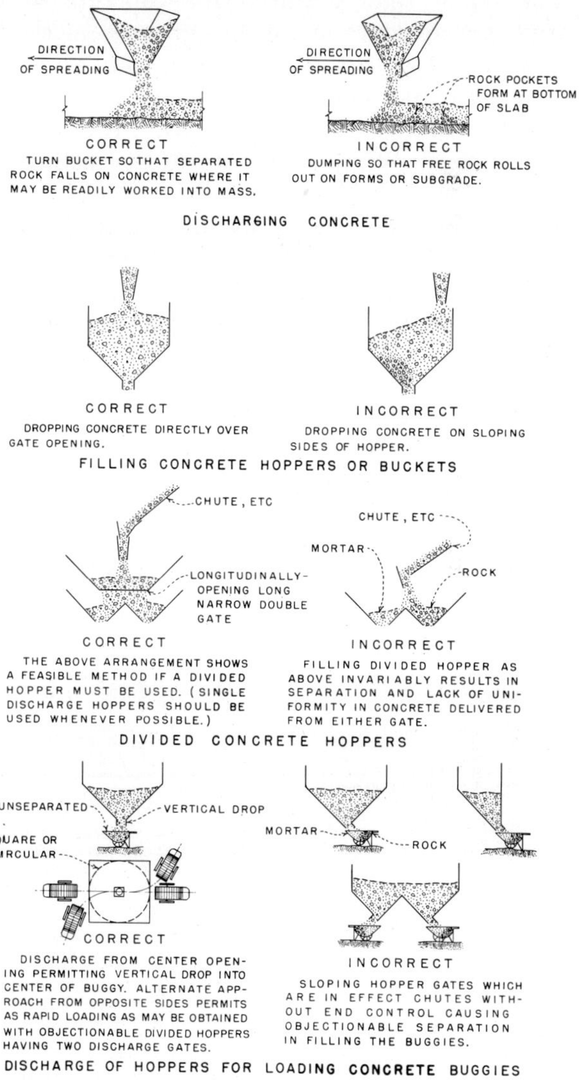

Figure F–11. Methods of handling concrete with buckets, hoppers, and buggies. PX–D–25316.

Curves, lifts, and harsh concrete material reduce the maximum pumping distance. For example, a 90° bend is equivalent to about 40 feet of straight, horizontal line, and each foot of head is equivalent to about 8 feet of line. Although manufacturers rate their largest equipment as capable of handling concrete containing aggregate up to 3 inches in size, experience indicates that operating difficulties will be materially lessened if the maximum size of aggregate pumped through such equipment is limited to 2½ or 2¾ inches. A pump will make good progress handling concrete with a slump of 3 to 4 inches and containing 2 to 3 percent more sand than required for concrete to be transported and placed by gravity methods. Normal rated capacities range from 15 to 65 cubic yards per hour.

F-29. *Placing.*—Properly placed concrete is free of segregation and the mortar is intimately in contact with the coarse aggregate, the reinforcement and other embedded parts. If any one detail of the placing inspector's many duties deserves special emphasis, it is that of guarding against objectionable segregation during concrete placement. Separation of coarse aggregate from the mortar may be minimized by avoiding or controlling the lateral movement of concrete during handling and placing operations as illustrated in figures F–9, F–10, and F–11. The concrete should be deposited as nearly as practicable in its final position. Placing methods which cause the concrete to flow in the forms should be avoided. These methods result in concentrations of less durable mortar in the ends of walls and corners where durability is most important and encourage the use of a mix that is wetter than necessary. The concrete should be placed in horizontal layers and each layer thoroughly vibrated. Practicable depths of layers for concrete range from 12 to 20 inches.

Hoppers for drop chutes should have throat openings of sufficient area to readily pass concrete of the lowest slump that is practicable to work and vibrate. If drop chutes are discharged directly through form ports, considerable separation results and rock pockets and honeycombs will probably be formed. Provision for an outside pocket below each port as

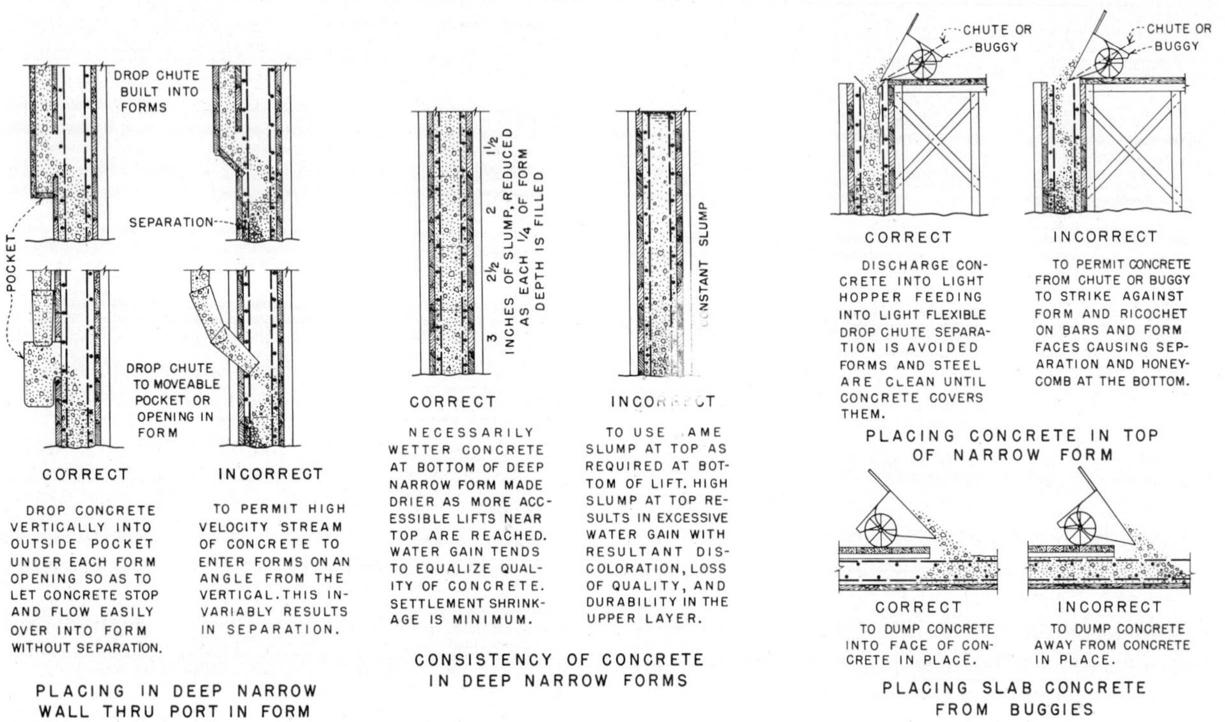

Figure F–12. Placing concrete in forms. 288–D–2657.

shown in figure F–12 will check the fall of the concrete and permit it to flow in the form with a minimum of separation.

Concrete for the top 2 feet of walls, piers, and columns should be of the lowest slump that can be adequately vibrated. After initial vibration, the concrete should be left for 1 or 2 hours to settle and complete the bleeding process. The surfaces should then be topped off with additional concrete as required and the top 2 feet revibrated to close bleeding channels.

When placing an unformed slab on a slope, there is a tendency to place the concrete using a stiff mix that will not slough. Drill cores have shown that the placement of such low-slump concrete without thorough vibration usually results in considerable honeycombing on the underside. To avoid such results, the consistency for this purpose should not be stiffer than a 2½-inch slump. Concrete with this consistency will barely stay on the slope, but it should not be drier. After spreading, the concrete should be thoroughly and systematically vibrated, preferably just ahead of a weighted steel-faced slipform screed working up the slope as shown in figure F–13.

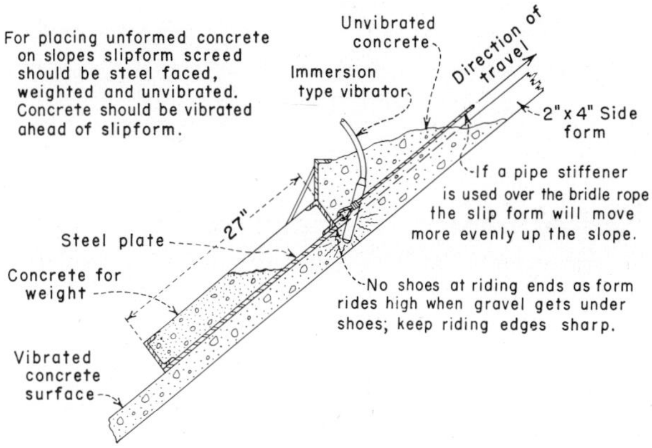

Figure F–13. Placing unformed concrete on slopes. PX–D–33031, PX–D–25252.

F-30. *Curing.*—Early drying must be prevented or concrete will not reach its full potential quality. In warm, dry, windy weather, corners, edges, and surfaces become dry very quickly. If these portions are prevented from drying so as to fully develop their hardness, it is certain that interior portions of the concrete will be adequately cured. Wet burlap in contact with the concrete is excellent for curing purposes. It not only shades the concrete, but it also holds the moisture needed for good curing. Wood forms left in place furnish good protection from the sun, but will not keep the concrete sufficiently moist for good curing of outdoor concrete. There is no better curing than that provided by well-moistened backfill. Ponding of floors, pavement, and other slabs also provides excellent curing and reduces crazing, cracking, and wear.

Where curing water is not economically available, or on vertical or sloping surfaces, it is often necessary to cure concrete by applying a sealing compound to the exposed surfaces immediately after form removal. The sealing compound is designed to restrict evaporation of the mixing water and when properly applied will retain enough moisture for adequate curing.

D. BIBLIOGRAPHY

F-31. *Bibliography.*

[1] "Concrete Manual," seventh edition, revised 1966, Bureau of Reclamation (for sale by the Superintendent of Documents, U.S. Government Printing Office, Washington, D.C. 20402).

[2] ———, appendix, designation 22.

[3] ———, appendix, designations 9 and 10.

[4] ———, appendix, designations 4 and 5.

[5] ———, appendix, designations 23 and 24.

[6] ———, appendix, designation 12.

[7] Price, W. H., "Factors Influencing Concrete Strength," American Concrete Institute Proceedings, vol. 47, 1951, pp. 417–432.

[8] Wallace, G. B., and Ore, E. L., "Structural and Lean Mass Concrete as Affected by Water-reducing, Set-retarding Agents," American Society for Testing Materials Special Technical Publication No. 266, 1960, pp. 38–94.

[9] Higginson, E. C., Wallace, G. B., and Ore, E. L., "Effect of Maximum Size Aggregate on Compressive Strength of Mass Concrete," American Concrete Institute Special Publication SP–6, pp. 219–246.

[10] "Concrete Manual," table 21, p. 188.

Appendix G

Sample Specifications

R. E. FINK,[1] N. F. LARKINS,[2] and E. R. LEWANDOWSKI [3]

G-1. ***Introduction.***—Designs are based on assumptions regarding the quality of work which will be obtained during construction. It is through the means of specifications that the assumed quality is described, and it is important that conformance to the specifications be obtained for all work.

This appendix includes specifications for the various items of work and structure components whose designs are treated in this text. For the construction of a particular dam, these specifications would have to be supplemented by descriptions of items for payment, description and classification of concrete items, and by specifications for other work such as painting, installation of equipment and metalwork, etc. Not all of the specifications included herein are applicable for any one dam as they cover alternative methods and a wide variety of construction details, not all of which would be performed at any one site.

These specifications are abstracted, with slight modifications, from guide specifications normally used by the Bureau of Reclamation. The designation "contracting authority" as used in these specifications applies to the owner of the dam or his authorized representative, as appropriate.

A. ENVIRONMENTAL QUALITY PROTECTION

G-2. ***Landscape Preservation*** [4].—(a) *General.*—The contractor shall exercise care to preserve the natural landscape and shall conduct his construction operations so as to prevent any unnecessary destruction, scarring, or defacing of the natural surroundings in the vicinity of the work. Except where clearing is required for permanent works, for approved construction roads and for excavation operations, all trees, native shrubbery, and vegetation shall be preserved and shall be protected from damage which may be caused by the contractor's construction operations and equipment. Movement of crews and equipment within the right-of-way and over routes provided for access to the work shall be performed in a manner to prevent damage to grazing land, crops, or property.

*No special reseeding or replanting will be required under these specifications; however, on completion of the work and in addition to all other requirements of the specifications, all work areas should be smoothed and graded in a manner to conform to the natural appearance of landscape. Where unnecessary destruction, scarring, damage, or defacing may occur as a result of the contractor's operations the same

[1] Civil Engineer, Earth Dams Section, Bureau of Reclamation.

[2] Civil Engineer, Concrete and Structural Branch, Division of General Research, Bureau of Reclamation.

[3] Formerly Civil Engineer, Earth Dams Section; now Regional Engineer, Region 2, Bureau of Reclamation.

[4] This guide paragraph is to be modified as applicable and supplemented to minimize construction scars and to control unnecessary clearing and defacement of the landscape. Accordingly, provisions covering local conditions should be added to limit, locate, or specify requirements for such items as temporary construction roads, campsite locations, trenching and backfilling for pipelines in cultivated and residential areas, and operations in borrow areas.

*Revise, as appropriate, when special reseeding or planting is specified elsewhere in the specifications.

shall be repaired, replanted, reseeded, or otherwise corrected at the contractor's expense.

(b) *Construction Roads.*—The location, alinement, and grade of construction roads shall be subject to approval of the contracting authority. When no longer required by the contractor, construction roads shall be made impassable to vehicular traffic and the surfaces shall be scarified and left in a condition which will facilitate natural revegetation.

(c) *Contractor's Campsite.*—The contractor's camp, shop, office, and yard area shall be located and arranged in a manner to preserve trees and vegetation to the maximum practicable extent. On abandonment, all camp, storage, and construction buildings, including concrete footings and slabs, and all construction materials and debris shall be removed from the site. The camp area shall be left in a neat and natural-appearing condition.

(d) *Borrow Areas and Quarry Sites.*—Borrow pits and quarry sites shall be so excavated that water will not collect and stand therein. Before being abandoned, the sides of borrow pits and quarry sites shall be brought to stable slopes, with slope intersections rounded and shaped to provide a natural appearance. All rubbish, contractor's equipment, and structures shall be removed from the site. Waste piles shall be leveled and trimmed to regular lines and shaped to provide a neat appearance.

(e) *Blasting Precautions.*—In addition to the requirements of section G-99, the contractor shall adopt precautions when using explosives which will prevent scattering of rocks, stumps, or other debris outside the work area.

(f) *Costs.*—The cost of all work required by this section shall be included in the price bid in the schedule for other items of work.

G-3. *Prevention of Water Pollution* [5].—The contractor shall comply with applicable Federal and State laws, orders, and regulations concerning the control and abatement of water pollution.

The contractor's construction activities shall be performed by methods that will prevent entrance, or accidental spillage, of solid matter, contaminants, debris, and other objectionable pollutants and wastes into streams, watercourses, lakes, and underground water sources. Such pollutants and wastes include, but are not restricted to, refuse, garbage, cement, concrete, sewage effluent, industrial waste, radioactive substances, oil and other petroleum products, aggregate processing tailings, mineral salts, and thermal pollution. Sanitary wastes shall be disposed of on land by burial at approved sites or by other approved methods.

Unwatering work for structure foundations or earthwork operations adjacent to, or encroaching on, streams or watercourses shall be conducted in a manner to prevent muddy water and eroded materials from entering the streams or watercourses by construction of intercepting ditches, bypass channels, barriers, settling ponds, or by other approved means.

Turbidity increases in a stream that are caused by construction activities shall be limited to the increases above the natural turbidities permitted under the State water quality standards prescribed for that stream. When necessary to perform required construction work in the stream channel, the prescribed turbidity limits may be exceeded, as approved by the contracting officer, for the shortest practicable period required to complete such work. This required construction work may include such work as diversion of a stream, construction or removal of cofferdams, specified earthwork in or adjacent to a stream channel, and construction of turbidity control structures. Mechanized equipment shall not be operated in flowing water except as necessary to construct crossings or to perform the required construction.

Waste waters from aggregate processing or other construction operations shall not enter streams, watercourses, or other surface waters without the use of such turbidity control methods as settling ponds, gravel-filter entrapment dikes, approved flocculating processes which are not harmful to fish, recirculation systems for washing of aggregates, or other approved methods. Any such waste waters discharged into surface waters shall be essentially free of

[5] Modify or delete any inapplicable provisions and add any requirement necessary to cover local pollution controls. Applicable State water quality standards for the class of water involved may be quoted as a help and convenience to contractors.

settleable material. For the purpose of these specifications, settleable material is defined as that material which will settle from the water by gravity during a 1-hour quiescent detention period.

The costs of complying with this section shall be included in the prices bid in the schedule for the various items of work.

G-4. *Abatement of Air Pollution* [6].—The contractor shall comply with applicable Federal, State, and local laws and regulations concerning the prevention and control of air pollution.

In conduct of construction activities and operation of equipment, the contractor shall utilize such practicable methods and devices as are reasonably available to control, prevent, and otherwise minimize atmospheric emissions or discharges of air contaminants.

The contractor's methods of storing and handling cement and pozzolans shall include means of eliminating atmospheric discharges of dust. Equipment and vehicles that show excessive emissions of exhaust gases due to poor engine adjustments, or other inefficient operating conditions, shall not be operated unless corrective repairs or adjustments are made.

*(Burning of materials resulting from clearing of trees and brush, combustible construction materials, and rubbish will be permitted only when atmospheric conditions for burning are considered favorable by appropriate State or local air pollution or fire prevention authorities. In lieu of burning, such combustible materials may be removed from the site, shipped or shredded, or buried as provided in section G-6.)

*(Burning of cleared timber, brush, and rubbish will not be permitted. Cleared timber and brush shall be removed from the site, chipped or shredded, or buried as provided in section G-6. Rubbish, trash, and other combustible materials shall be *(buried in a manner and in locations approved by the contracting authority or) removed from the site and disposed of in an approved manner.)

Storage and handling of flammable and combustible liquids and provisions for fire prevention shall be in accordance with sections 12 and 13 of the Bureau of Reclamation publication "Construction Safety Standards."

Dust nuisance resulting from construction activities shall be prevented in accordance with section G-5.

The costs of complying with this section shall be included in the prices bid in the schedule for the various items of work.

G-5. *Dust Abatement* [7].—During the performance of the work required by these specifications or any operations appurtenant thereto, whether on right-of-way provided by the Government or elsewhere, the contractor shall furnish all the labor, equipment, materials, and means required, and shall carry out proper and efficient measures wherever and as often as necessary to reduce the dust nuisance, and to prevent dust which has originated from his operations from damaging crops, orchards, cultivated fields, and dwellings, or causing a nuisance to persons. The contractor will be held liable for any damage resulting from dust originating from his operations under these specifications.

The cost of sprinkling or of other methods of reducing formation of dust shall be included in the prices bid in the schedule for other items of work.

B. EXCAVATION

G-6. *Clearing Damsite.*—(a) *General.*—The areas to be occupied by the permanent construction required under these specifications and the surfaces of all borrow pits[7A] (and

*Use applicable provision as required by local conditions.

[6] Insert provisions covering any applicable air quality standards in force for the geographical area involved, including any standards which have been established pursuant to subsection 4(b) of Executive Order 11507.

[7] Use this paragraph in contracts where necessary for controlling dust from contractor's operations. Additional provisions should be included where appropriate.

[7A] Revise or delete as appropriate.

stockpile and wastepile sites) shall be cleared of all trees, stumps, (exposed) roots, brush, rubbish, and other objectionable matter as determined by the contracting authority.**

*(The reservoir area below Elevation ______ as shown on Drawing No.______ shall be cleared of all trees, stumps, and brush 5 feet or more in height, regardless of diameter, and 2 inches or more in diameter, regardless of height.** Trees and stumps in the reservoir area shall either be uprooted or cut off so that the maximum allowable stump height shall be 6 inches as measured on the uphill side of the stump.) *(Brush in the reservoir area shall be cut off approximately flush with the ground level.)

*(All down timber, branches, and other floatable and combustible material 5 feet or more in length, regardless of diameter, and 2 inches or more in diameter, regardless of length, shall be cleared.)

No trees shall be cut outside of areas mentioned above without specific approval, and all trees designated by the contracting authority shall be protected from damage by the contractor's construction operations.

Clearing will consist primarily of the removal of ____________(insert description of major items)________________________.

The bidders and the contractor shall, by their own investigation, determine the extent of clearing required in accordance with this section.

The contractor may, at no cost, retain any materials of value from clearing operations for his own use or disposal by sale. Such material shall be removed from the construction area before the date of completion of the work under the specifications. Every reasonable effort shall be made by the contractor to channel materials of value resulting from clearing operations into beneficial use. The contracting authority assumes no responsibility for the protection or safekeeping of any materials so retained by the contractor.

(b) *Disposal of Materials.*—Materials from clearing operations shall become the property of the contractor and shall be, at the contractor's option, buried, *(burned), *(chipped or trimmed and cut to lengths), removed from the site of the work before the date of completion, or otherwise disposed of as approved. *(Burning of materials required to be cleared will not be permitted.)

Materials disposed of by burying shall be buried at locations approved by the contracting authority and shall be covered with not less than 2 feet of earth material. Approved locations will be natural or excavated depressions in the reservoir area which are not subject to erosion from streamflow or wave action.

*((Burning will be permitted only at times when conditions are considered favorable for burning and at locations approved by proper State or local authorities.) *(The contractor will be required to obtain a burning permit from ________________________________.) Materials to be burned shall be piled neatly and, when in a suitable condition, shall be burned completely. Piling for burning shall be done in such a manner and in such locations as to cause the least fire risk. All burning shall be so thorough that the materials are reduced to ashes. No logs, branches, or charred pieces shall be permitted to remain. The contractor shall at all times take special precautions to prevent fire from spreading to areas beyond the limits of the cleared areas and shall have available at all times suitable equipment and supplies for use in preventing and suppressing fires.)

*(Cut timber, down timber, dead timber, branches, and brush ____ inches and less in diameter, to be disposed of by chipping, shall be reduced to chips of one-half-inch maximum thickness. The chips may be sold by the contractor as provided in subsection G-6(a), or shall be disposed of by distributing uniformly on the ground surface in approved areas and the chips shall be mixed with the underlying earth so that they will not float or support combustion.)

Cut timber, down timber, dead timber, branches, and other floatable and combustible material over ___ inches in diameter, to be disposed of by cutting to lengths, shall be trimmed and cut into approximate 5-foot lengths. Such cut material may be sold by the contractor

**Insert other items to be cleared such as buildings, fences, and pipes.

*Revise or delete as appropriate.

as provided in subsection (a), or shall be hauled to stockpiles above Elevation ______, and at locations as approved by the contracting authority. Cut material shall be neatly stacked in piles.

(c) *Payment.*—*(Payment for clearing the areas to be occupied by the permanent construction and the surfaces of borrow pits *(and stockpile and wastepile sites) *(and the reservoir area below Elevation ______) will be made at the lump-sum price bid in the schedule for clearing. Estimates for progress payments for clearing will be made by the contracting authority on the basis of percentages of work completed. For the purpose of progress payments, estimates will be prepared on the following basis:

--).

*(No separate payment will be made for clearing and the cost of clearing shall be included in the prices bid in the schedule for other items of work.)

G-7. Classification of Excavation [8].—Except as otherwise provided in these specifications, material excavated will be measured and classified in excavation, to the lines shown on the drawings or as provided in these specifications, and will be classified for payment as follows:

Rock excavation.—For purposes of classification of excavation, rock is defined as sound and solid masses, layers, or ledges of mineral matter in place and of such hardness and texture that it:

(1) Cannot be effectively loosened or broken down by ripping in a single pass with a late model tractor-mounted hydraulic ripper equipped with one digging point of standard manufacturer's design adequately sized for use with and propelled by a crawler-type tractor rated between 210- and 240-net flywheel horsepower, operating in low gear, or

(2) In areas where it is impracticable to classify by use of the ripper described above, rock excavation is defined as sound material of such hardness and texture that it cannot be loosened or broken down by a 6-pound drifting pick. The drifting pick shall be Class D, Federal Specification GGG-H-506d, with handle not less than 34 inches in length.

All boulders or detached pieces of solid rock more than 1 cubic yard in volume will be classified as rock excavation.

Common excavation.—Common excavation includes all material other than rock excavation. All boulders or detached pieces of solid rock less than 1 cubic yard in volume will be classified as common excavation.

No additional allowance above the unit prices bid in the schedule for excavation will be made on account of any of the material being wet or frozen.

The contracting authority's representative and the contractor or his representative shall be present during classification of material excavated. On written request of the contractor, made within 10 days after the receipt of any monthly estimate, a statement of the quantities and classifications of excavation between successive stations or in otherwise designated locations included in said estimate will be furnished to the contractor within 10 days after the receipt of such request. This statement will be considered as satisfactory to the contractor unless specific objections thereto, with reasons therefor, are filed with the contracting authority in writing within 10 days after receipt of said statement by the contractor or the contractor's representative on the work. Failure to file such written objections with reasons therefor within said 10 days shall be considered a waiver of all claims based on alleged erroneous estimates of quantities or incorrect classification of materials for the work covered by such statement.

G-8. Classification of Excavation (Alternate) [9].—Materials excavated will not be classified for payment. Except as otherwise provided in these specifications, material excavated will be measured in excavation, to the lines shown on

*Revise or delete as appropriate.

[8] When excavations are not classified for payment, use section G-8 or G-9.

[9] When excavated materials are classified for payment, use section G-7 or G-9.

the drawings or as provided in these specifications, and all materials so required to be excavated will be paid for at the unit prices per cubic yard bid in the schedule for excavation. No additional allowance above the unit price bid in the schedule will be made on account of any of the material being wet or frozen. Bidders and the contractor must assume all responsibility for deductions and conclusions as to the nature of the materials to be excavated and the difficulties of making and maintaining the required excavations.

Where the terms "rock" and "rock excavation" and "common" and "common excavation" are used in these specifications, the following definitions shall apply:

Rock excavation.—Rock is defined as sound and solid masses, layers, or ledges of mineral matter in place and of such hardness and texture that it:

(1) Cannot be effectively loosened or broken down by ripping into a single pass with a late model tractor-mounted hydraulic ripper equipped with one digging point of standard manufacturer's design adequately sized for use with and propelled by a crawler-type tractor rated between 210- and 240-net flywheel horsepower, operating in low gear, or

(2) In areas where it is impracticable to classify by use of the ripper described above, rock excavation is defined as sound material of such hardness and texture that it cannot be loosened or broken down by a 6-pound drifting pick. The drifting pick shall be Class D, Federal Specification GGG-H-506d, with handle not less than 34 inches in length.

Common excavation.—Common excavation includes all material other than rock excavation.

G-9. *Definitions of Materials.*—* (Materials excavated will not be classified for payment.) For purposes of these specifications, * (other than for payment) materials of earthwork and embankment construction are defined in detail as follows:

(a) *Rock.*—A sound and solid mass, layer, or ledge of mineral matter in place and of such hardness and texture that it:

(1) Cannot be effectively loosened or broken down by ripping in a single pass with a late model tractor-mounted hydraulic ripper equipped with one digging point of standard manufacturer's design adequately sized for use with and propelled by a crawler-type tractor rated between 210- and 240-net flywheel horsepower, operating in low gear, or

(2) In areas where the use of the ripper described above is impracticable, rock is defined as sound material of such hardness and texture that it cannot be loosened or broken down by a 6-pound drifting pick. The drifting pick shall be Class D, Federal Specification GGG-H-506d, with handle not less than 34 inches in length.

(b) *Common Material.*—All earth materials which do not meet the requirements of rock as defined in (a) above.

(c) *Formation.*—Any sedimentary, igneous, or metamorphic material represented as a unit in geology, generally called rock but not necessarily meeting the classification requirements for rock in (a) above.

(d) *Cobbles.*—Rounded pieces of rock which are not greater than 12 inches, but are larger than 3 inches in maximum dimension.

(e) *Boulders.*—Detached pieces of rock, generally rounded but may be subrounded to angular, which are larger than 12 inches in maximum dimension.

(f) *Rock Fragments.*—Pieces of rock which generally are not rounded.

(g) *Soil Components.*—NOTE.—Soils in nature usually consist of a number of soil components. They are identified by the predominance of one of the components and other criteria given in the Unified Soil Classification System (see chapter V).

(1) *Clay.*—Plastic soil which passes a United States Standard No. 200 sieve.

(2) *Silt.*—Nonplastic soil which passes a United States Standard No. 200 sieve.

*Revise or delete as required.

(3) *Sand.*—Mineral grains which pass a United States Standard No. 4 sieve and are retained on a United States Standard No. 200 sieve.

(4) *Gravel.*—Pieces of rock which are not greater than 3 inches in maximum dimension, and are retained on a United States Standard No. 4 sieve.

(h) *Geologic Definitions of Other Materials.*—***Chalk.*—A material of variable hardness consisting of a consolidated aggregation of very fine particles, mainly calcium carbonate, which is usually buff in color but may range from white to dark gray.

Shale.—A consolidated, partially laminated, fine-grained material having a tendency to split among lamination or bedding planes. It may range from a clay-like material which, when cut or scraped with a knife, produces a slick surface with shiny appearance, to a partially cemented material which, although it can be scratched with a knife, can be cut only with difficulty and produces a dull, fine-grained surface.

Tuff.—A material composed of the finer kinds of volcanic detritus, usually more or less stratified, and in various states of consolidation or induration.

G-10. *Opencut Excavation, General* [10].—(a) *General.*—All opencut excavation required for the dam and appurtenant works shall be performed in accordance with this section and sections G-11 to G-14, inclusive. Opencut excavation in borrow areas shall be in accordance with section G-16 (Borrow Areas). Excavation shall be made to the lines, grades, and dimensions shown on the drawings or established by the contracting authority.

During the progress of the work, it may be found necessary or desirable to vary the slopes, grades, or the dimensions of the excavations from those specified herein. Any increase or decrease of quantities excavated as a result of such variations will be included in the estimates. However, if the contracting authority determines that the contractor's costs of performing the work will be increased by reason of such variations, an equitable adjustment will be made to cover such increased costs. Otherwise, the work will be paid for at the unit price bid therefor in the schedule regardless of such variations.

All necessary precautions, including control of blasting, shall be taken to preserve the material below and beyond the established lines of all excavation in the soundest possible condition. Whenever, in the opinion of the contracting authority, further blasting might injure the material upon or against which concrete is to be placed, the use of explosives shall be limited to light loads, and the excavation shall be completed by light blasting, wedging, barring, channeling, line drilling and broaching, or other suitable methods approved by the contracting authority. Any damage to the work due to the contractor's operations, including shattering of the material beyond the required excavation lines, shall be repaired at the expense of and by the contractor. Slopes shattered or loosened by blasting shall be taken down at the expense of and by the contractor.

Any and all excess excavation for the convenience of the contractor or overexcavation performed by the contractor for any purpose or reason, except as may be ordered in writing by the contracting authority, and whether or not due to the fault of the contractor, shall be at the expense of the contractor. Where required to complete the work, all such excess excavation and overexcavation shall be refilled with materials furnished and placed at the expense of and by the contractor: *Provided,* That payment will be made for cement used in concrete placed to refill such excess excavation or overexcavation unless such excess excavation or overexcavation is caused by careless excavation or is intentionally performed for the convenience of the contractor to facilitate his operations, as determined by the contracting authority.

All excavation for embankment and structure foundations shall be performed in the dry. No excavation shall be made in frozen materials without written approval. No additional allowance above the unit prices per cubic yard

**Geologic definitions to be inserted as applicable. Some typical definitions are listed. Delete if not applicable.

[10] This section is written for an earthfill dam. Revise section as required if used for specifications for concrete dam construction.

bid in the schedule for excavation will be made on account of any of the materials being wet or frozen.

Excavations shall be made to the full dimensions required and shall be finished to the prescribed lines and grades except that individual sharp points of undisturbed formation material will be permitted to extend within the prescribed lines not more than 6 inches where the excavation surfaces are not to be covered with concrete *(or are not required as a foundation for pipe).

All *(shale) foundations shall be protected from freezing and air slaking by leaving a temporary cover of ---- feet of unexcavated material. The final ---- feet of excavation above the *(shale) foundation surfaces shall be performed by continuous operation during nonfreezing weather. Such excavation shall be followed without delay by placement of embankment material as required, or by the application of protective coatings as provided in section G-15. Exposed finished excavated *(shale) surfaces shall be kept moist at all times to prevent evaporation of the natural moisture in the material; however, surfaces of foundations upon which concrete is to be placed shall be maintained free of standing water. Such surfaces shall also be protected from freezing. No allowance above the unit prices bid in the schedule will be made because of the requirements set forth herein for the protection of foundations from freezing or air slaking.

(b) *Structure Foundations.*—The bottom and side slopes of excavation upon or against which concrete is to be placed shall be excavated to the required dimensions as shown on the drawings or established by the contracting authority. No material will be permitted to extend within the neatlines of the structure.

If, at any point in rock or formation materials, the natural foundation material is disturbed or loosened, it shall be removed and replaced with concrete.

The bottom and side slopes of excavation in common material upon or against which concrete is to be placed shall be prepared by moistening and tamping or rolling with suitable tools and equipment to form a firm foundation for the concrete structure. If, at any point in common material, the natural foundation material is disturbed or loosened, for any purpose or reason, it shall be consolidated by tamping or rolling, or where directed, it shall be removed and replaced with selected earthfill material moistened and thoroughly compacted in 6-inch layers by tamping or rolling.

The cost of all work required in the preparation of structure foundations shall be included in the applicable unit price bid in the schedule for excavation.

Where the material is unsuited to form a firm foundation upon which to place concrete, additional excavation and refill will be ordered in writing by the contracting authority, and payment therefor will be made as follows:

(1) Payment for additional excavation when ordered in writing to remove unsuitable foundation materials will be made at the unit price per cubic yard bid in the schedule for excavation for the structure for which the excavation is made.

(2) Payment for concrete placed for refilling additional excavation in rock or formation materials will be made at the applicable unit price per cubic yard bid in the schedule for concrete in the structure involved.

(3) Payment for excavation and transportation of selected earthfill materials for use in refilling additional excavation in common material will be made at the unit price bid in the schedule for excavation of the material used for such refill.

(4) In additional excavations where compaction with the roller specified for use on the dam embankment is practicable and desirable, the foundations shall be prepared, and the refill materials shall be selected, placed, moistened, and compacted as provided in section G-21. The selected earthfill materials shall be placed and compacted to an excess depth of 18 inches above the established elevation of the structure foundation, and such excess material shall be excavated

*Revise or delete as appropriate.

to the established elevation of the structure foundation. Payment for placing and compacting refill materials with the roller as described above will be made at the unit price per cubic yard bid in the schedule for earthfill in dam embankment, Zone 1, and payment for excavation of the excess roller-compacted earthfill to established grade of the structure will be made at the unit price per cubic yard bid in the schedule for excavation for the structure for which the excavation is made.

(5) In excavations where compaction with the roller specified for use on the dam embankment is impracticable or undesirable, the refilling and compacting of refill material shall be performed in accordance with section G-22, and payment for placing and compacting the refill material will be made only of material required to refill additional excavation as ordered by the contracting authority and will be made at the unit price per cubic yard bid in the schedule for specially compacted earthfill.

(c) *Pipe Trenches.*—The trenches in which pipe is to be laid shall be excavated carefully to the established lines and grades to provide a firm and uniform bearing for the entire length of the pipe. If the character of the material at any point in the bottom of a pipe trench is such as might cause unequal settlement or provide unequal bearing for the pipe, or is otherwise unsuitable for a foundation for the pipe, the unsatisfactory materials shall be removed to such depth as may be directed by the contracting authority. Where additional excavation is required on account of encountering unsuitable materials in the bottom of a pipe trench, measurement for payment for all such excavation will be made to the prescribed lines and dimensions.

(d) *Excavated Materials.*—So far as practicable, as determined by the contracting authority, all suitable materials from excavations for specified permanent construction shall be used in the permanent construction required under these specifications.

Materials shall be selected as follows: (Insert descriptions of materials applicable to each part of the work.)

Excavated materials which are unsuitable for * (or are in excess of) dam embankment or other earthwork requirements, as determined by the contracting authority, shall be wasted as provided in section G-19.

The contractor's blasting and other operations in excavations shall be such that the excavations will yield as much suitable material for such construction as practicable, and shall be subject to the approval of the contracting authority. Where practicable, as determined by the contracting authority, suitable materials shall be excavated separately from the materials to be wasted and the suitable materials shall be segregated by loads during the excavation operations. The suitable materials shall be placed in the designated final locations directly from the excavation, or shall be placed in temporary stockpiles and later placed in the designated locations as directed by the contracting authority. In excavating materials which are suitable for use in the dam embankment, the contracting authority will designate the depths of cut which will result in the best gradation of materials, and the cuts shall be made to such designated depths.

Excavated materials which, after drainage or drying, are suitable for the impervious rolled earthfill portion of the dam embankment but which, when excavated, are too wet for immediate compaction in the embankment shall be placed temporarily in stockpiles until the moisture content is reduced sufficiently to permit them to be placed in the embankment, or may be placed on the embankment subject to the provisions of section G-21 relative to materials in which the moisture content is greater than that required for proper compaction.

Should cobbles, boulders, or rock fragments having maximum dimensions of more than 5 inches be found in otherwise approved earthfill materials, they shall be removed by the contractor either at the site of the excavation or after being transported to the earthfill but before the materials are rolled and compacted. Such rock materials shall be placed in other

*Delete if not applicable.

portions of the dam embankment or wasted, as directed.

If, after excavation, sand, gravel, and cobble fill material has a moisture content greater than that required for placement and compaction in embankment, the material shall not be placed on the embankment, but shall be placed temporarily in stockpiles and allowed to drain or dry until the moisture content is reduced sufficiently to permit it to be placed in the embankment.

(e) *Measurement and Payment.*—Excavated material will be measured, for payment, in excavation to the lines shown on the drawings or described in these specifications, and will include only material that is actually removed within the prescribed pay lines.

Where concrete is to be placed directly upon or against the excavations, such excavations shall be sufficient at all points to provide for the minimum dimensions of concrete. Where dimensions of a concrete structure are shown on the drawings or if the elevation of the foundation is indicated, such dimensions shall be considered as the minimum dimensions and such elevation shall be considered as the elevation determining the minimum dimensions of the structure. Where a dimension or an elevation is not indicated on the drawings, minimum dimensions will be established by the contracting authority.

Where concrete is to be placed directly upon or against the excavations and the character of the material cut into is such that the material can be trimmed efficiently to accurate dimensions by ordinary excavation finishing methods to the required lines of the concrete structure, as determined by the contracting authority, measurement for payment will be made only of the excavation within the neatlines of the concrete structure.

Where concrete is to be placed directly upon or against the excavations and the character of the material cut into is such that the material cannot be trimmed efficiently to accurate dimensions by ordinary excavation finishing methods, as determined by the contracting authority, measurement for payment thereof will be made to the prescribed average dimension lines. The prescribed average dimension lines shall be considered as 3 inches outside the neatlines of the concrete for the purposes of measurement, for payment, of excavation.

Where concrete protective coating is required to be placed for protection of finished structure foundations, as specified in section G-15, the minimum excavation lines shown on the drawings or established by the contracting authority shall be increased by 2 inches and measurement, for payment, of excavation for the structure will be made to the prescribed average dimension as specified above.

Measurement, for payment, of excavations upon or against which concrete is not required to be placed and excavation for trenches for pipe drains, including excavation for bedding, will be limited to the neatlines shown on the drawings or, where not shown on the drawings, to the most practicable lines, grades, and dimensions as established by the contracting authority.

Except as otherwise provided in section G-79 for diversion and care of river during construction and removal of water from foundations, the unit prices bid in the schedule for excavation in opencut shall include the cost of all labor, equipment, and materials for cofferdams and other temporary construction and of all pumping, bailing, draining, and all other work necessary to maintain the excavations in good order during construction and of removing such temporary construction where required.

The unit prices bid in the schedule for excavation in opencut shall also include the entire cost of:

(1) Excavating to designated depths.
(2) Segregating materials by loads.
(3) Draining or drying otherwise suitable materials.
(4) Transportation of materials from the excavation to points of final use, to disposal areas, to temporary stockpiles, and from temporary stockpiles to points of final use.
(5) Rehandling excavated materials which have been deposited temporarily in stockpiles.

(6) Removal of oversize materials from otherwise suitable materials and disposal of the same.

(7) Disposal of excavated waste materials.

All excavated materials actually placed in completed earthwork and embankment construction will again be included for payment under appropriate items of the schedule covering such construction.

No payment will be made for excavation performed in previously placed embankment, refill, or backfill; *except as provided in subsection (b) above.

G-11. *Drilling Line Holes for Rock Excavation.*—Rock excavation, where directed by the contracting authority, shall be formed by line drilling and broaching. The diameter of the holes for line drilling shall be subject to approval. The spacing of the holes shall be as directed by the contracting authority, and shall be sufficiently close to insure that the rock will break along the desired lines. Light blasting will be permitted in the holes along the sides of the excavation: *Provided,* That whenever, in the opinion of the contracting authority, further blasting might injure the rock upon or against which concrete is to be placed, the use of explosives shall be discontinued and the excavation shall be completed by wedging, barring, or other suitable methods.

Measurement or payment for drilling line holes for rock excavation will include only the length of holes actually drilled into the rock along the sides of the excavation as directed by the contracting authority and will include only holes drilled at spacings of **15 inches or less from center to center of the holes.

Payment for drilling line holes for rock excavation will be made at the unit price per linear foot bid therefor in the schedule, which unit price shall include the costs of light blasting, broaching, wedging, barring, or other methods used in conjunction with line drilling to complete the excavation.

G-12. *Excavation for Grout Cap.*—Excavation for grout cap shall be performed by the use of hand tools and approved mechanical equipment, in such a manner as to prevent shattering of the sides and bottom of the excavation. At the option of the contractor and with the approval of the contracting authority, line drilling and light blasting or other methods may be employed. If line drilling and light blasting are employed, the diameter, spacing, and depth of the holes shall be subject to the approval of the contracting authority, and the spacing shall be such as to insure that the material will break along the desired lines. The blasting shall be limited to approved methods which provide for successive fracturing of the worked face as the work is advanced by use of power tools and handwork. Whenever, in the opinion of the contracting authority, further blasting might injure the surfaces upon or against which concrete is to be placed, the use of explosives shall be discontinued.

When an excavation for grout cap crosses a fault or seam, the excavation shall be carried to depths shown on the drawings, or as may be directed, and shall be keyed into the formation on the sides of the fault or seam as directed: *Provided,* That if excavation is required to a greater depth than 8 feet measured normal to the finished surface of excavation, such excavation will be ordered in writing. The contractor shall furnish all materials to support the sides of the excavation where necessary, and all supports shall be removed before or during the placing of concrete.

Measurement, for payment, of excavation for grout cap will be made to the prescribed average dimension in width, and to the designated depth measured normal to the finished surfaces of excavation, *(and will not include ___-foot-thick temporary cover of unexcavated material as required in section G-10 (Opencut Excavation, General).

Payment for excavation for grout cap will be made at the applicable unit price per cubic yard bid in the schedule for excavation for grout cap up to 5 feet in depth and for excavation for grout cap between depths of 5 and 8 feet, which unit price shall include the entire cost of all work described in this section (including line drilling and light blasting if employed (NOTE.—If section G-11 is used, modify

*Revise or delete as appropriate.

**Revise as appropriate.

this statement accordingly.)) and the cost of furnishing, installing, and removing supports. The requirement for excavation for grout cap between depths of 5 and 8 feet will be determined by the contracting authority, and the contractor will be entitled to no additional allowance above the unit price bid therefor in the schedule by reason of any amount or none of the work for this item being required.

G-13. *Excavation for Dam Embankment Foundation (Earthfill Dam).*—(a) *General.*—Excavation for dam embankment foundation shall be in accordance with this section and section G-10, and includes all:

(1) Stripping for foundation of dam embankment.
(2) Excavation below stripping for foundation of dam embankment, including cutoff trench(es) and other areas if required, *but not including excavation for grout cap.
(3) Excavation for dam embankment drainage system.
(4) Excavation for open drain ditches.
(5) ______________

The alinements and excavation lines shown on the drawings are subject to such changes as may be found necessary by the contracting authority to adapt the dam foundation excavation *(and other features listed above) to the conditions disclosed by the excavation.

Accurate trimming of the slopes of the excavation will not be required, but the excavation shall conform as closely as practicable to the established lines and grades. *Loose rock shall be removed from foundation contacts and rock cliffs, ledges, overhangs and sharp irregularities shall be reduced to provide satisfactory foundation contours.

*(All foundation surfaces subject to damage from freezing or air slaking shall be protected as provided in section G-10.)

The finished foundation surfaces against which the earthfill portions of the dam embankment are to be placed shall be cleaned of all loose, soft, and disintegrated materials, including removal of such materials from pockets, irregularities, fissures, and depressions in the foundation.

(b) *Stripping.*—The entire area to be occupied by the dam embankment, *except the area to be covered by ________, zone ____, portions of embankment, shall be stripped to a sufficient depth, as determined by the contracting authority, to remove all unsuitable materials. The unsuitable materials to be removed by stripping shall include all surface boulders and loose rock; debris; topsoil; vegetable matter, including stumps and roots; and all other perishable and objectionable materials that are unsuitable for use in permanent construction required under these specifications or that might interfere with the proper bonding of the embankment with the foundation, or the proper compaction of the materials in the embankment, or that may be otherwise objectionable as determined by the contracting authority. *Topsoil shall be stockpiled for use on downstream slope of dam embankment. *(other) Materials from stripping operations shall be wasted as provided in section G-19 (Disposal of Excavated Materials).

(c) *Excavation Below Stripping.*—The cutoff trench(es) shall be excavated in the dam embankment foundation to firm formation, as shown on the drawings.

*Excavation below stripping outside of cutoff trench(es), where shown on the drawings or where directed, shall be to firm formation or to an otherwise suitable foundation as determined by the contracting authority.

(d) *Measurement and Payment.*—Measurement, for payment, of excavation for dam embankment foundation will be made as provided in section G-10.

*The dividing surface(s) for measurement, for payment, between excavation for dam embankment foundation and:

(1) Excavation for roadway will be a vertical plane normal to the centerline of the roadway at roadway station ______.
(2) Excavation, in opencut, for structures will be those surfaces described in section G-14.
(3) Excavation for grout cap will be the bottom of excavation for dam embank-

*Revise or delete as appropriate.

ment foundation as shown on the drawings.

(4) ____________.

Payment for excavation for dam embankment foundation will be made at the unit price per cubic yard bid therefor in the schedule, and shall include all costs as provided in section G-10, and all costs of cleanup of foundation surfaces as provided in subsection (a) above.

G-14. ***Excavation, in Opencut, for Structures.***—The item of the schedule for excavation, in opencut, for structures includes all opencut excavation, including stripping, as shown on the drawings, for the concrete structures as follows:

(Here usually are listed descriptions and locations of the excavations involved under this excavation pay item, together with the dividing limits for measurement for payment if these excavations are contiguous with other excavations for which separate payment is made. This section is also used for specifications for construction of a concrete dam.)

Measurement, for payment, of excavation, in opencut, for structures will be made as provided in section G-10.

Payment for excavation in opencut for structures will be made at the unit price per cubic yard bid therefor in the schedule, which price shall include all costs as provided in section G-10.

G-15. ***Protective Coating for Structure Foundations.***—(a) *General.*—Where designated by the contracting authority, a protective coating shall be applied to finished excavated foundation surfaces upon or against which concrete will be placed. The protective coating shall be applied only to foundation surfaces which are designated to receive protective coating where, in the judgment of the contracting authority, it is deemed necessary to prevent air slaking.

The protective coating shall be concrete protective coating, except an approved sprayed protective coating may be used on foundation surfaces of all structure cutoff excavations and on slopes steeper than 1:1.

Excavation shall not be performed within the final ___ feet of finished surfaces in areas which will require protective coatings until all equipment and facilities required for immediate application of protective coatings are available and in working condition. Areas requiring protection shall be subdivided into units under which the final ______ feet of excavation and the application of protective coating can be completed within one working shift. The work within these units shall be performed by continuous operation, and the application of protective coatings shall follow without delay. The protective coating shall be applied to these unit areas so that finished excavated surfaces will be exposed for the shortest possible period of time, which time shall not exceed ____ hour. Exposed finished excavated (*describe the formation, such as shale*) surfaces shall be kept moist at all times to prevent evaporation of the natural moisture in the material and shall be protected from freezing.

Surfaces to be coated shall be cleaned of all loose material, dirt, dust, mud, standing water, and other foreign matter. Coatings of protective material shall not be applied when air temperature is below 35° F., nor during other adverse weather as determined by the contracting authority. Temporary protective coatings, as approved by the contracting authority, will be permitted if necessary, until the permanent protective coatings are applied. Protective coatings damaged by frost, heat, traffic, or other causes shall be removed and replaced or repaired at the expense of and by the contractor as directed, including the cost of removal of any frost or other damaged foundation material below such protective coatings and replacement with compacted earth material or concrete as applicable. Protective coatings shall be covered with structural concrete as soon as practicable and in no event shall the protective coatings be left exposed for a period exceeding ___ days.

(b) *Sprayed Protective Coating.*—Sprayed protective coating shall be Aerospray 52 Binder, a product of American Cyanamid Co., Explosives and Mining Chemicals Dept., Wayne, N. J. 07470, or approved equal. The sprayed protective coating shall be applied to finished excavated surfaces in accordance with the manufacturer's recommendations and by methods approved by the contracting authority. The coating shall be applied to a thickness and in such a manner that a uniform coating free from pinholes will be produced which will protect the finished excavated surface from the air.

(c) *Concrete Protective Coating.*—Concrete protective coating shall be applied so that the contact surface is covered to a minimum depth of 2 inches. Where concrete protective coating is applied, the minimum excavation lines shown on the drawings or established by the contracting authority shall be increased by 2 inches and the protective coating shall be applied within such increased excavation lines to provide for full dimensions of structural concrete in overlying structures.

The concrete shall conform to the applicable provisions specified in sections G-59 to G-76, inclusive. The method of placing the concrete shall be subject to approval. Concrete for protective coating shall not contain aggregate larger than 3/4-inch maximum size.

Either water or membrane curing may be used on concrete protective coating, *(but if membrane curing is used, the sealing compound shall be removed by approved methods at the contractor's expense, where directed, before placement of structural concrete).

*(The exposed surfaces of concrete protective coating shall be considered as a construction joint and as such shall receive the same treatment as required in section G-71 prior to placement of structural concrete on the concrete coating.)

(d) *Measurement and Payment.*—Measurement, for payment, of protective coating will be made of the number of square yards of the finished excavated foundation area actually coated at the direction of the contracting authority. Payment for protective coatings for structured foundations will be made at the unit price per square yard bid therefor in the schedule, which unit price shall include the cost of all work equipment, and of furnishing the materials required for application of protective coating, except excavation for concrete protective coating and except cement used in protective coating. Concrete placed for protective coating will not be included for payment in other items of the schedule for concrete in the structure involved. Payment for cement used in concrete protective coating will be made at the unit price per barrel bid in the schedule for furnishing and handling cement *(for concrete.)

Payment for the additional excavation required for application of concrete protective coating will be made at the unit price per cubic yard bid in the schedule for excavation, in opencut, for structures.

G-16. Borrow Areas.—(a) *General.*—All materials required for the following construction, which are not available from excavations required for permanent construction under these specifications, shall be obtained from borrow areas _____, _____, _____, and _____. The locations of the borrow areas are shown on Drawing No. __________.

(1) Construction of dam embankment Zones _____, _____, and _____.
(2) Pervious backfill.
*(3) Bedding for riprap.
*(4) Selected surfacing.
*(5) ________________.

(Explorations in the borrow areas indicate that the materials are variable in nature and texture and contain variable amounts of moisture and plus 5-inch material. Approximate percentages of plus 5-inch material encountered in the explorations within borrow area __ are shown on the logs. The absence of percentages of oversize on any log of explorations within the area does not, however, imply that oversize materials will not be encountered in the vicinity of such explorations.)

Groundwater level encountered in the explorations, as shown on the logs, is for the indicated dates. The absence of a groundwater level or moisture content on any log of explorations within the area does not, however, imply that groundwater or variable moisture content will not be encountered in the vicinity of such explorations.

Bidders are cautioned that wide variation from the nature, texture, moisture content, and the percentage of oversize material as indicat-

*Revise or delete as appropriate.

ed by the explorations is to be anticipated. Bidders and the contractor must assume all responsibility for deductions and conclusions concerning the nature, moisture content, and texture of material, the percentages of oversize materials, the total yield of suitable materials, the difficulties of making excavations, of breaking down or removing the oversize materials, of obtaining a satisfactory moisture content, and of obtaining a uniform mixture of materials.

*(Some exploratory test pits in the borrow areas will be open for inspection and bidders should inspect the borrow areas and examine the test pits. Bidders are urged to sample and test material from borrow areas prior to submitting bids.)

The type of equipment used and the contractor's operations in the excavation of materials in borrow pits shall be such as will produce the required uniformity of mixture of each of the types of the materials at the borrow pits.

The location and extent of all borrow pits within borrow areas shall be as directed, and the contracting authority reserves the right to change the limits or location of borrow pits within the limits of the borrow areas in order to obtain the most suitable material, to minimize stripping, or for other reasons.

To avoid the formation of pools in borrow pits during the excavation operations, and in borrow pits above elevation ---------- after the excavation operations are completed, drainage ditches from borrow pits to the nearest outlets shall be excavated by the contractor where, in the opinion of the contracting authority, such drainage ditches are necessary.

Should any borrow pit be opened near the dam embankment and below the elevation of normal water surface in the reservoir, berms not less than ----- feet wide shall be left between the toe of the dam embankment and the edge of the borrow pit, with a slope of four to one to the bottom of the borrow pit. Excavated surfaces of borrow pits adjacent to the reservoir above the normal water surface shall be graded to slopes not steeper than ----------. *In other areas the contractor will not be required to excavate surfaces of borrow pits to any specified lines and grades, but such surfaces shall be left in a reasonably smooth and even condition and may require trimming, as directed by the contracting authority, to facilitate measurement for payment. Borrow pits shall be operated and left in a condition so as not to impair the usefulness or mar the appearance of any part of the work or any other property of the contracting authority. The surfaces of wasted material shall be left in a reasonably smooth and even condition.

*(b) *Roads, Buildings, and Utility Lines in Borrow Areas.*—Highway -------- traverses borrow area -------, as shown on Drawing No. -------. This road will be relocated *by others. Prior to the relocation of the road, the contractor shall not excavate material within 50 feet of the centerline of the road. The contractor shall conduct his operations in such a manner as to permit continued use of the road and to provide safety to the public until such time as the road has been relocated. The contractor shall permit access as necessary to others for the purpose of relocation of this road.

*The buildings located in borrow area ------, as shown on Drawing No. -------, will be disposed of by others. Prior to disposal of the buildings, they shall be protected from damage from the contractor's operations. The contractor shall permit access as necessary to others for the purpose of disposal of these buildings.

*Power transmission lines, pipelines, and telephone lines traversing the borrow areas as shown on the drawings will be relocated by others. Prior to relocation of the utility lines the contractor shall not excavate material within 50 feet of the centerline of any pipe, power transmission, or telephone lines. The contractor shall conduct his operations in a manner to prevent any interference with or damage to the utility lines and to permit access as necessary to others for the purpose of relocation of these utility lines.

(c) *Moisture and Drainage.*—The moisture content of the earthfill material prior to and

*Revise or delete as appropriate.

during compaction shall be in accordance with subsection G-21(d). As far as practicable, the material shall be conditioned in the borrow pits before excavation. If required, moisture shall be introduced into the borrow pits for the earthfill material by irrigation at least ___ days in advance of excavation operations, *or, at the option of the contractor, moisture may be added at the separation plant. When moisture is introduced into the borrow pits for earthfill material prior to excavation, care shall be exercised to moisten the material uniformly to produce the required moisture content during compaction, avoiding both excessive runoff and accumulation of water in depressions. The contractor is cautioned to control carefully the application of water and check on the depth and amount of water penetration during application so as to avoid over-irrigation.

If at any location in the borrow pits for earthfill material, before or during excavation operations, there is excessive moisture, as determined by the contracting authority, steps shall be taken to reduce the moisture by selective excavation to secure the drier materials, by excavating and placing in temporary stockpiles material containing excessive moisture, by excavating drainage ditches, by allowing adequate additional time for curing or drying, or by any other approved means.

The moisture content of the sand, gravel, and cobble fill material prior to and during compaction shall be in accordance with section G-23. Borrow pits for sand, gravel, and cobble fill material will not require preconditioning by irrigation but may require preconditioning by draining and lowering the water table below the elevation of borrow excavation. (Preconditioning by draining may be accomplished by any approved method, including lowering the water table in the borrow area prior to excavation or stockpiling.) If, after excavation, sand, gravel, and cobble fill material has a moisture content greater than that required for placement and compaction in embankment, the material shall not be placed on the embankment, but shall be placed temporarily in stockpiles and allowed to drain or dry until the moisture content is reduced sufficiently to permit it to be placed in the embankment.

In any event, the contractor will be required to excavate sufficient suitable material in portions of borrow areas ______ to complete the work under these specifications, regardless of whether overly wet conditions encountered are due to ground water, precipitation, difficulty of draining, or any other reason. To minimize operations with overly wet material, the contractor will be permitted to utilize portions of the borrow areas which contain dry material and which have been designated as suitable borrow pits to the greatest extent practicable consistent with obtaining suitable material.

The contractor shall be entitled to no additional allowance above the unit prices bid in the schedule on account of the requirement for excavating drainage ditches; for allowing additional time for curing or drying; for stockpiling and rehandling excavated materials which have been deposited temporarily in stockpiles; delays or increased costs due to stockpiling; poor trafficability in the borrow area, the haul roads, or the embankment; reduced efficiency of the equipment the contractor elects to use; or on account of any other operations or difficulties caused by overly wet materials.

No additional allowance above the unit prices bid in the schedule will be made because of variation in the proportion between wet and dry material which is required to be excavated in order to obtain adequate suitable material.

(d) *Stripping and Waste.*—Borrow pit sites shall be cleared as provided in section G-6. Borrow pits will be designated by the contracting authority as the work progresses, and stripping operations shall be limited only to designated borrow pits. The contractor shall carefully strip the sites of designated borrow pits of boulders, topsoil, sod, loam, and other matter which is unsuited for the purposes for which the borrow pit is to be excavated. The contractor shall maintain the stripped surfaces free of vegetation until excavation operations in the borrow pit are completed and the contractor shall be entitled to no additional allowance above the unit prices bid in the schedule because of this requirement. *Materials from stripping which are suitable for topsoil shall

*Revise or delete as appropriate.

be selected during stripping operations, temporarily stockpiled adjacent to borrow pits if necessary, and spread over exhausted portions of the borrow pits as directed by the contracting authority. Materials from stripping *(which are not suited for topsoil) shall be disposed of in exhausted borrow pits, or in approved areas adjacent to borrow pits, or as provided in section G-19.

If materials unsuitable or not required for permanent construction purposes are found in any borrow pit, such materials shall be left in place or excavated and wasted, as directed. Where excavation of such materials is directed, payment for such excavation and disposal of unsuitable or excess materials will be made at the (1) unit price per cubic yard bid in the schedule for excavation, stripping borrow pits, (2) applicable unit price per cubic yard bid in the schedule for excavation in borrow areas and transportation to dam embankment.[11]

(e) *Excavation and Transportation.*—The contracting authority will designate the depths of cut in all parts of the borrow pits, and the cuts shall be made to such designated depths. The earthfill materials delivered on the dam embankment shall be equivalent to a mixture of materials obtained from an approximately uniform cutting from the full height of the designated face of the borrow pit excavation. Shallow cuts will be permitted in the borrow areas if unstratified materials with uniform moisture content are encountered.

The contractor shall transport the materials to the dam embankment location designated by the contracting authority.

The contractor shall be entitled to no additional allowance above the unit prices bid in the schedule on account of the designation by the contracting authority of the various portions of the borrow area areas from which materials are to be obtained, on account of the depths of cut which are required to be made, or on account of the zone or location on embankment where materials are hauled.

(f) **Separation Plant.*—Materials from borrow areas ________ shall be separated before placement in dam embankment, Zone *1 or *2________________. The contractor shall construct separation plant facilities which will separate cobbles, boulders, and rock fragments having maximum dimensions greater than ___ inches from all other material. Material ___ inches or less in maximum dimensions shall be placed in dam embankment Zone *1 or *2, and cobbles, boulders, and rock fragments having maximum dimensions greater than ______ shall be placed in ______________: *Provided,* That cobbles, boulders, and rock fragments larger than ____ inches in maximum dimensions shall be placed in the outer slopes of _______ or shall be embedded in that zone, so as not to interfere with the compaction operations.

(g) *Measurement and Payment*[12].—Measurement, for payment, of excavation, stripping borrow pits will be made in excavation and will include only the stripping in locations and to the depths as directed by the contracting authority. Payment for excavation, stripping borrow pits will be made at the (1) unit price per cubic yard bid therefor in the schedule, which unit price shall include the costs of selecting, stockpiling, and spreading topsoil over exhausted portions of the borrow pits, or otherwise disposing of materials from stripping, (2) applicable unit price per cubic yard bid in the schedule for excavation in borrow areas and transportation to dam embankment, which unit price shall include the costs of selecting, stockpiling, and spreading topsoil over exhausted portions of the borrow pits, or otherwise disposing of materials from stripping.

Measurement for payment of excavation in borrow areas will be made in excavation only and to the excavation lines prescribed by the contracting authority. Payment for excavation in borrow areas and transportation to dam embankment, *and for excavation in borrow areas, separation, and transportation to dam

*Revise or delete as appropriate.

[11] Use (1) when haul distance is in excess of 2 miles, or when separation plant is required. Use (2) when haul distance is less than 2 miles and separation plant is not required.

[12] Use (1) when haul distance is in excess of 2 miles or when separation plant is required. Use (2) when haul distance is less than 2 miles and separation plant is not required. Modify as necessary.

embankment, will be made at the applicable unit prices per cubic yard bid therefor in the schedule, which unit prices shall include all costs of irrigation and unwatering of borrow pits, of conditioning the material properly, and all work *(other than stripping) required by this section. All materials from borrow pits placed in dam embankment, Zones ______ in ______ ______________________________, and in backfill will again be included for payment under the applicable items of the schedule for placing such earthwork.

If the contractor elects to obtain concrete aggregates, gravel for drains, ______________________, or other materials for which the cost of furnishing or procuring is included in other items of work, no payment will be made for stripping or excavation of such materials obtained from borrow areas. The contractor shall keep his operations for the production of these materials separate and distinct from his other borrow area operations.

G-17. *Rock Deposits (Rock to be Furnished by Contractor)* [13].—Rock fragments of the quality and gradations specified herein shall be furnished by the contractor for use in bedding for riprap and riprap to be placed and stockpiled and for other permanent construction required under these specifications.

(a) *Quality.*—The rock fragments shall meet the following requirements as to quality:

(1) Individual rock fragments shall be dense, sound, and resistant to abrasion and shall be free of cracks, seams, and other defects that would tend to increase unduly their destruction by water and frost actions.

(2) Samples prepared in accordance with applicable designations of the Bureau of Reclamation's "Concrete Manual," seventh edition, shall meet the following requirements when tested by the procedures described in the respective designations:

Test	Designation	Requirements
Specific gravity (saturated surface-dry basis).	10	Greater than **.
Soundness (sodium sulfate method).	19	Less than ** percent loss of weight after 5 cycles.
Abrasion (using Los Angeles machine grading A).	21	Less than ** percent loss of weight after 500 revolutions.

**To be inserted for each job depending on nature of material available and design considerations.

Samples of ______ and ______ [14] from the following locations have been tested and found suitable: (List locations.)

Bidders and the contractor are cautioned that the abovementioned deposits may be variable in quality and the sizes and quantity of rock fragments that may be obtained from any source is unknown. The contractor will be responsible for furnishing suitable rock fragments, for making necessary arrangements with property owners for rights-of-way, and for payment of required royalties.

(b) *Sampling and Testing.*—The contractor shall furnish to the contracting authority, at the damsite, without cost, such samples of rock fragments for testing as may be required by the contracting authority from proposed quarry sites and from rock fragments delivered to the damsite. The contracting authority reserves the right to make inspections of quarry sites and quarries. The approval of some rock fragments from a particular quarry site shall not be construed as constituting the approval of all rock fragments taken from that quarry, and the contractor will be held responsible for the specified quality and gradation of rock fragments delivered to the damsite. All rock fragments not meeting the requirements of these specifications, as determined by tests and/or inspection at the quarries and damsite, will be rejected.

G-18. *Rock Source (Source Furnished by Contracting Authority)* [15].—All rock materials required for construction of:

*Revise or delete as appropriate.

[13] Use this section when rock is to be furnished by the contractor and the rock is not to be used exclusively in riprap. Modify section G-32 accordingly.

[14] Designate rock types, for example granite and limestone.

[15] Use this section when rock is to be obtained from a source furnished by the contracting authority; it is not to be used exclusively in riprap. Modify section G-32 accordingly.

(1) ...., and

(2) .... shall be secured from the rock source shown on Drawing No. .... All operations within the rock source shall be subject to approval. The contracting authority reserves the right to designate the locations of excavations within the limits of the rock source in order to obtain suitable rock materials for construction purposes. The portions of the rock source to be excavated shall be cleared as provided in section G-6 (Clearing) and shall be stripped of all overburden and loose, soft, disintegrated rock as directed.

The contractor shall produce, by excavation in rock source and selection or processing, sufficient suitable rock fragments reasonably well graded, as determined by the contracting authority, up to .... inches in maximum dimensions for construction of * (rock fines fill in dam embankment, pervious backfill, and bedding for riprap). The contractor shall also produce by excavation in rock source and selection or processing sufficient suitable rock fragments reasonably well graded, as determined by the contracting authority, up to ...... cubic yards in volume for construction of * (rockfill in dam embankment and riprap). The type of equipment used and the contractor's operations in the rock source shall be such as will produce the required gradations of rock fragments at the rock source.

All suitable rock fragments shall be transported to points of final use, and all excavated materials unsuitable or in excess of requirements for construction purposes shall be disposed of in excavations in rock source or as directed.

The cost of all work described in this section, including clearing and stripping rock source, shall be included in the unit prices per cubic yard bid in the schedule for items of construction in which the rock fragments are used.

G-19. ***Disposal of Excavated Material.***—So far as practicable, as determined by the contracting authority, all suitable materials from excavation required under these specifications shall be used in the permanent construction. Excavated materials that are unsuitable for or are in excess of permanent construction requirements shall be wasted.

The disposal of all excavated materials that are to be wasted shall be subject to the approval of the contracting authority, but the contractor will not be required to haul materials to be wasted more than .... feet along the most practicable routes to the designated disposal areas. *Areas designated for disposal of waste material from excavation are shown on Drawings No. .... and ....

Waste piles shall be located where they will not interfere harmfully with the natural flow of the stream and drainage channels, with construction operations in the borrow areas, with the operation of the reservoir, or with the flow of water to or from the spillway or outlet works, and where they will neither detract from the appearance of the completed project nor interfere with the accessibility of the structures for operation. Waste piles shall be leveled and trimmed to reasonably regular lines.

The cost of transporting excavated materials from excavations to disposal areas or to points of final use, including stockpiling and rehandling, if required, and of disposing of all excavated materials that are wasted, as provided in this section, shall be included in the applicable unit prices per cubic yard bid in the schedule for excavation.

C. EMBANKMENT

G-20. ***Embankment Construction, General.***—(a) *General.*—For the purpose of these specifications, the term "dam embankment" includes all portions of the dam embankment, as follows:*

(1) The earthfill, Zone 1 portions designated on the drawings by the figure 1 encircled, including specially compacted earthfill.

(2) The earthfill, Zone 2 portions designated on the drawings by the figure 2

*Revise or delete as appropriate.

encircled, including specially compacted earthfill.

(3) The sand and gravel fill, Zone ...... portions designated on the drawings by the figure ...... encircled.

(4) The sand, gravel, and cobble fill, Zone ...... portions designated on the drawing by the figure ...... encircled.

(5) The miscellaneous fill, Zone ...... portions designated on the drawings by the figure ...... encircled.

(6) The ...., Zone ...... portions designated on the drawings by the figure ...... encircled.

(7) The rockfill, Zone ...... portions designated on the drawings by the figure ...... encircled.

(8) The bedding for riprap and the riprap on upstream slope of Zones ...... and ...... of dam embankment.

(9) The seeded topsoil cover on downstream slope of Zones ...... and ...... of dam embankment.

(10) Selected surfacing on the crest of the dam embankment.

Other items of embankment, which are not necessarily a part of the dam embankment, include (*list items, such as pervious backfill, etc.*).

The embankment shall be constructed in accordance with this section and sections G-21 through G-32. The completed embankment shall be to the lines and grades shown on the drawings. (Note any possible exceptions.)

All openings through the dam embankment required for construction purposes shall be subject to approval.

(b) *Foundation Preparation.*—No material shall be placed in any portion of the dam embankment until the foundation for each section has been unwatered, stripped, suitably prepared, and has been approved by the contracting authority. Stripping shall be in accordance with section G-13 (Excavation for Dam Embankment Foundation).

All cavities, depressions, and irregularities, either existing or resulting from removal of rock fragments found within the area to be covered by embankment, and which extend below or beyond the established lines of excavation for dam embankment foundations, shall be filled with embankment materials and compacted as specified for the overlying embankment.

(c) *Placing Embankment Materials.*—The suitability of each part of the foundation for placing embankment materials thereon, and of all materials for use in embankment construction, will be determined by the contracting authority. No embankment material shall be placed in the embankment when either the material or the foundation or embankment on which it would be placed is frozen.

No brush, roots, sod, or other perishable or unsuitable materials shall be placed in the embankment.

Each load of the material placed in the embankment, whether from excavation for other parts of the work or from borrow pits, shall be placed in the location designated by the contracting authority, and the contractor shall be entitled to no additional allowance above the unit prices bid in the schedule on account of this requirement.

In any separate portion of dam embankment being constructed, each layer of each zone shall be constructed continuously and approximately horizontal for the width and length of such portion at the elevation of the layer.

The contractor shall maintain the embankment in an approved manner, including maintaining surfaces free of weeds or other vegetation, until final completion and acceptance of all the work under the contract.

The contractor will be permitted to construct separate portions of the embankment below original ground surface, subject to the approval of the contracting authority. Above original ground surface, construction of the dam embankment shall be subject to the following conditions:

(1) Longitudinal bonding surfaces (surfaces parallel to the centerline crest of embankment) will not be permitted in Zone 1. In Zone .... and .... portions of embankment, longitudinal surfaces between previously constructed embankment and embankment to be constructed shall not be steeper than ......

(2) *A temporary gap through the dam embankment, for diversion purposes as described in section G-79, will be permitted: *Provided*, That the slopes of transverse bonding surfaces (surfaces normal to the centerline crest of embankment) between previously completed portions of embankment and embankment to be placed shall be not steeper than 4:1. *No other transverse bonding surfaces will be permitted in the dam embankment.

During construction of earthfill, Zone 1, embankment in the openings at the gap, the contractor shall construct a keyway trench in each 4:1 transverse bonding surface in the previously placed earthfill, Zone 1. The keyway trenches shall be excavated in the bonding surfaces to a minimum depth of 5 feet, shall have 4:1 side slopes, and shall have a minimum bottom width of 20 feet. The centerline of the trenches shall be located approximately midway between the upstream and downstream slopes of Zone 1. The trenches shall be refilled with earthfill, Zone 1, material subject to the provisions of section G-21.

(3) At any cross section, the difference in elevation between adjacent Zones ---- and ---- embankment shall not be greater than 5 feet.

Placing shall be performed in a manner to prevent damage to structures.

(d) *Measurement and Payment.*—Measurement, for payment, of the various items of embankment construction will be made of the material in place in the completed embankment to the lines, grades, slopes, and thicknesses shown on the drawings including crest camber, or established by the contracting authority. The cross sections obtained by surveys made after completion of excavation for dam embankment foundation will be used in computing the quantity of embankment placed. No allowance will be made in measurement for payment for settlement, shrinkage, and consolidation of the foundation or of the material in the embankment. In measuring embankment for payment, the volume of structures, of specially compacted earthfill, and of other work for which items for payment are provided in the schedule, will be deducted.

*Revise or delete as appropriate.

The dividing surfaces for measurement, for payment, between embankment items and

----

(insert dividing surfaces for embankment items which are adjacent to each other and are not otherwise defined, such as dam embankment and road embankment. Dividing surfaces may be by station, elevation or other description.)

Payment under all items of embankment construction shall include the costs of preparing the embankment foundations; of placing; of supplementary wetting on the fill, if necessary, and any additional work required on the embankment to accomplish uniform moisture application; of compacting where compaction is required; of preparing bonding surfaces; and all other operations required to secure adequate bond between embankment in place and embankment to be placed.

Payment for the embankment items will be in addition to payment made for excavation and transportation of the materials. (Note any exceptions.)

It may be feasible to transport some of the materials which are excavated for other parts of the work and which are suitable for embankment construction directly to the embankments at the time of making the excavations, but the contractor shall be entitled to no additional compensation above the unit prices bid in the schedule by reason of it being necessary, or required by the contracting authority, that such excavated materials be deposited temporarily in stockpiles and rehandled prior to being placed in the embankment.

No measurement or payment will be made for excavating keyway trenches in embankment or for refilling the trenches, and the cost of preparing bonding surfaces, including excavating keyway trenches and refilling such trenches in transverse bonding slopes and all other operations required to secure adequate bond between embankment in place and embankment to be placed shall be included in unit prices bid for items of constructing embankments.

G-21. ***Earthfill in Dam Embankment, Zone(s) 1 (and 2)*.***—(a) *General.*—The earthfill, Zone(s)

1 *(and 2) portion(s) of the dam embankment and earthfill placed for structure foundations (*insert structure description*) shall be constructed in accordance with this section and section G-20.

(b) *Materials.*—Zone 1 of the earthfill portion(s) of the dam embankment shall consist of a mixture of the ______, ______, ______, and ______, available from excavations required for the dam and appurtenant works or from borrow pits in borrow areas ______, ______, and ______. *(Zone 2 of the earthfill portion of the dam embankment shall consist of a mixture of the ______, ______, ______, and ______ available from excavations required for the dam and appurtenant works or from borrow pits in borrow areas ______, ______, and ______.)

The contractor's operations in the excavation of the materials for the earthfill shall be in accordance with sections G-10 and G-16.

Cobbles and rock fragments having maximum dimensions of more than 5 inches shall not be placed in the earthfill. Should cobbles and rock fragments of such size be found in otherwise approved earthfill materials, they shall be removed by the contractor either at the site of excavation *(at the separation plant) or after being transported to the earthfill but before the materials in the earthfill are rolled and compacted. Such cobbles and rock fragments shall be placed in the ______ or ______ portion(s) of the dam embankment or wasted, as approved by the contracting authority.

(c) *Preparation of Foundations.*—The foundation for the earthfill, except surfaces of formation materials, shall be prepared by leveling, moistening, and rolling so that the surface materials of the foundation will be as compact and will provide as satisfactory a bonding surface with the first layer of the earthfill as specified for the subsequent layers of the earthfill.

*(All shale and ______ foundation surfaces shall be protected from air slaking and freezing by leaving ______ feet temporary cover of unexcavated material, and the finish excavation shall be made to remove such temporary cover.) Immediately prior to placing the first layer of earthfill, all surfaces upon or against which the earthfill portions of the dam embankment are to be placed shall be cleaned of all loose and objectionable materials in an approved manner by handwork, barring, picking, brooming, sluicing, or by other effective means. Such surfaces shall have all water removed from depressions and shall be properly moistened and sufficiently clean to obtain a suitable bond with the earthfill.

(d) *Moisture and Density Control.*—

(1) *General.*—Each layer of the material on the embankment shall be compacted by 12 passes of the tamping roller as provided in subsection (g) below, which shall be the minimum compacting effort to be performed by the contractor. During compaction, the placement moisture content and dry density of the earthfill shall be maintained within the control limits specified below.

To determine that the moisture content and dry density requirements of the compacted earthfill are being met, field and laboratory tests will be made at frequent intervals on samples taken at embankment locations determined by the contracting authority. Field and laboratory tests will be made by the contracting officer in accordance with Designations E-11, E-24, and E-25 of the Bureau of Reclamation's "Earth Manual". The results of all completed earthwork tests will be available to the contractor at the office of the contracting authority.

Materials not meeting the specified moisture content and dry density requirements, as determined by the tests, shall be reworked until approved results are obtained. Reworking may include removal, rehandling, reconditioning, rerolling, or combinations of these procedures. The contractor shall be entitled to no additional allowance above the prices bid in the schedule by reason of any work required to achieve the placement moisture content and density specified in this section.

(2) *Moisture control.*—The standard optimum moisture content is defined as, "That moisture content which will result in a maximum dry unit weight of the soil when subjected to the Bureau of Reclamation Proctor Compaction Test." The maximum

*Revise or delete as appropriate.

dry weight, in pounds per cubic foot, obtained by the above procedure is the Proctor maximum dry density. The Bureau of Reclamation Proctor Compaction Test (Designation E-11 of "Earth Manual") is the same as ASTM Designation: D-698-70, Method A, except that a 1/20-cubic foot compaction mold is used and the rammer is dropped from a height of 18 inches.

The moisture content of the earthfill material prior to and during compaction shall be distributed uniformly throughout each layer of the material. The allowable ranges of placement moisture content are based on design considerations. The moisture control shall be such that the moisture content of compacted earthfill, as determined by tests performed by the contracting officer, shall be within the following limits:

a. Material represented by the samples tested having a placement moisture content more than **____ percent dry of the standard optimum condition, or more than **____percent wet of the standard optimum condition will be rejected and shall be removed or reworked until the moisture content is between these limits.

b. Within the above limits, and based on a continuous record of tests made by the contracting authority on previously placed and accepted material, the uniformity of placement moisture content shall be such that:

(aa) No more than 20 percent of the samples of accepted embankment material will be drier than **____ percent dry of the standard optimum moisture content, and no more than 20 percent will be wetter than **____ percent wet of the standard optimum moisture content.

(bb) The average moisture content of all accepted embankment material shall be between **____ and **____ percent dry of the standard optimum moisture content.

The contracting authority will inform the contractor when the placement moisture content is near or exceeds the limits of uniformity specified above, and the contractor shall immediately make adjustments in procedures as necessary to maintain the moisture content within the specified limits.

As far as practicable, the material shall be brought to the proper moisture content in the borrow pit before excavation, as provided in section G-16. *(At the option of the contractor, additional moisture, if required, may be added at the separation plant.) Supplementary water, if required, shall be added to the material by sprinkling on the earthfill and each layer of earthfill shall be conditioned so that the moisture is uniform throughout the layer.

(3) *Density control.*—Density control of compacted earthfill shall be such that the dry density of the compacted material, as determined by tests performed by the contracting authority, shall conform to the following limits:

a. Material represented by samples having a dry density less than **____percent of its Proctor maximum dry density will be rejected. Such rejected material shall be rerolled until a dry density equal to or greater than **____ percent of its Proctor maximum dry density is obtained.

b. Within the above limit and based on a continuous record of tests made by the contracting authority on previously placed and accepted embankment, the uniformity of dry density shall be such that:

(aa) No more than 20 percent of the material represented by the samples tested shall be at dry densities less than **____ percent of Proctor maximum dry density.

(bb) The average dry density of all accepted embankment material shall be not less than **____ percent of the average Proctor maximum dry density.

The contracting authority will inform the contractor when the dry density is near or exceeds the limits of uniformity specified above, and the contractor shall immediately make adjustments in procedures as necessary to maintain the dry density within the specified limits.

**Limits to be inserted for each job depending on nature of available material.

*Revise or delete as appropriate.

(e) *Placing.*—The distribution and gradation of the materials throughout the earthfill shall be such that the fills will be free from lenses, pockets, streaks, or layers of material differing substantially in texture, gradation, or moisture from the surrounding material. The combined excavation, *(separation,) and operations shall be such that the materials when compacted in the earthfill will be blended sufficiently to secure the best practicable degree of compaction and stability. Successive loads of material shall be dumped on the earthfill so as to produce the best practicable distribution of the material, subject to the approval of the contracting authority and for this purpose the contracting authority may designate the locations in the earthfill where the individual loads shall be deposited, to the end that the most impervious materials shall be placed in the central portion of the earthfill and the more pervious materials shall be placed so that the permeability of the fill will be gradually increased toward the upstream and downstream slopes of the earthfill.

The material shall be placed in the earthfill in continuous, approximately horizontal layers not more than 6 inches in thickness after being compacted. If, in the opinion of the contracting authority, the surface of the prepared foundation on the compacted surface of any layer of earthfill is too dry or smooth to bond properly with the layer of material to be placed thereon, it shall be moistened and/or worked with harrow, scarifier, or other suitable equipment, in an approved manner to a sufficient depth to provide a satisfactory bonding surface before the next succeeding layer of earthfill material is placed. If, in the opinion of the contracting authority, the compacted surface of any layer of the earthfill in place is too wet for proper compaction of the layer of earthfill material to be placed thereon, it shall be removed; allowed to dry; or be worked with harrow, scarifier, or other suitable equipment to reduce the moisture content to the required amount; and then it shall be recompacted before the next succeeding layer of earthfill material is placed.

The earthfill on each side of the *(concrete cutoff walls, conduit, and gate structure) and at the abutments shall be kept at approximately the same level as the placing of the earthfill progresses, and the walls, *(conduit, and gate structure) shall be protected against displacement or other damage. *(The upstream slope of the earthfill portion of the dam embankment above elevation ________ shall be compacted thoroughly, shall be reasonably true to line and grade, and all projections of more than 6 inches outside of the neat lines of the earthfill shall be removed at the expense of and by the contractor before the overlying material is placed.)

(f) *Rollers.*—Tamping rollers shall be used for compacting the earthfill. The rollers shall meet the following requirements:

(1) *Roller drums.*—Tamping rollers shall consist of two or more roller drums mounted side by side in a suitable frame. Each drum of a roller shall have an outside diameter of not less than 5 feet and shall be not less than 5 feet nor more than 6 feet in length. The space between two adjacent drums, when on a level surface, shall be not less than 12 inches nor more than 15 inches. Each drum shall be free to pivot about an axis parallel to the direction of travel. Each drum ballasted with fluid shall be equipped with at least one pressure-relief valve and with at least one safety head. The safety head shall be equal to union type safety heads as manufactured by Black, Sivals and Bryson, Inc., Kansas City, Mo., with rupture discs suitable for between 50- and 75-psi rupturing pressure.

The pressure-relief valve shall be a manually operated valve and shall be opened periodically. Personnel responsible for opening pressure-relief valves shall be instructed to ascertain that valve openings are free from plugging to assure that any pressure developed in roller drums is released at each inspection.

(2) *Tamping feet.*—At least one tamping foot shall be provided for each 100 square inches of drum surface. The space measured on the surface of the drum, between the centers of any two adjacent tamping feet, shall be not less than 9 inches. The

*Revise or delete as appropriate.

length of each tamping foot from the outside surface of the drum shall be not more than 11 inches and shall be maintained at not less than 9 inches. The cross sectional area of each tamping foot shall be not more than 10 square inches at a plane normal to the axis of the shank 6 inches from the drum surface, and shall be maintained at not less than 7 square inches nor more than 10 square inches at a plane normal to the axis of the shank 8 inches from the drum surface.

(3) *Roller weight.*—The weight of a roller when fully loaded shall be not less than 4,000 pounds per foot of length of drum.

The loading used in the roller drums and operation of the rollers shall be as required to obtain the specified compaction. If more than one roller is used on any one layer of fill, all rollers so used shall be of the same type and essentially of the same dimensions. Rollers operated in tandem sets shall be towed in a manner such that the prints of the tamping feet produced by the tandem units do not overlap. The design and operation of the tamping roller shall be subject to the approval of the contracting authority, who shall have the right at any time during the prosecution of the work to direct such repairs to the tamping feet, minor alterations in the roller, and variations in the weight as may be found necessary to secure optimum compaction of the earthfill materials. Rollers shall be drawn by crawler-type or rubber-tired tractors. The use of rubber-tired tractors shall be discontinued if the tires leave ruts that prevent uniform compaction by the tamping roller. Tractors used for pulling rollers shall have sufficient power to pull the rollers satisfactorily when drums are fully loaded with sand and water.

At the option of the contractor, self-propelled tamping rollers conforming with the above requirements may be used in lieu of tractor-drawn tamping rollers. For self-propelled rollers, in which steering is accomplished through the use of rubber-tired wheels, the tire pressure shall not exceed 40 psi. During the operation of rolling, the spaces between the tamping feet shall be maintained clear of materials which would impair the effectiveness of the tamping rollers.

(g) *Compaction.*—When each layer of material has been conditioned to have the specified moisture, as provided in subsection (d), it shall be compacted by passing the tamping roller over it 12 times, and when compacted the density shall be essentially uniform throughout the layer. If the uncompacted earthfill material is too wet for proper compaction, the earthfill material shall be worked with harrow, scarifier, or other suitable equipment to reduce the moisture content to the amount specified; shall be allowed to dry until such time as the material contains only the specified moisture content; or the material shall be removed from the embankment. Compacted earth material having a moisture content or dry density that do not meet the criteria specified in subsection (d) above shall be reworked and rerolled, as directed by the contracting authority, to obtain the specified moisture content and dry density of embankment in place.

(h) *Measurement and Payment.*—Measurement, for payment, of earthfill in dam embankment will be made of all earthfill compacted in place by rollers specified in subsection (f) and as provided in section G-20.

Payment for earthfill in dam embankment, Zone(s) 1, *(and 2) will be made at the unit price(s) per cubic yard bid therefor in the schedule, which price(s) shall include all costs of work required under this section and as provided in section G-20.

Where portions of the earthfill in dam embankment require special compaction, payment therefor will be made as provided in section G-22.

G-22. Specially Compacted Earthfill.—Where compaction of earthfill material by means of rollers specified for use on the dam embankment is impracticable or undesirable, the earthfill shall be specially compacted as specified herein at the following locations:

(1) Portions of the earthfill in dam embankment, where designated by the contracting authority, at steep and irregular abutments and on rough and irregular embankment foundations.

*Revise or delete as appropriate.

(2) Earthfill material placed to refill additional excavation, ordered in writing by the contracting authority, in common excavation for structure foundations.

(3) Portions of the earthfill, in dam embankment adjacent to structures and structure foundations, shown on the drawings as specially compacted earthfill.

(4) Earthfill material at locations outside the limits of the dam embankment as shown on the drawings or where designated by the contracting authority.

Material used in specially compacted earthfill shall conform to materials required for earthfill in dam embankment: *Provided,* That gravel, cobbles, and rock fragments having maximum dimensions of more than ---- inches shall not be placed in the specially compacted earthfill. The material shall be obtained from excavation required for permanent construction, and from excavation in borrow areas -------, -------, and --------.

All specially compacted earthfill material shall be placed in accordance with the applicable provisions of subsection G-21(e): *Provided,* That earthfill material to be specially compacted may require placement in layers thinner than those specified for roller compaction of earthfill material to obtain the desired compaction with the equipment used.

Where the foundation or compacted surface of any layer is too smooth to bond properly with the succeeding layer, it shall be scarified or otherwise roughened to provide a satisfactory bonding surface before the next layer of earthfill material is placed.

When each layer of material has been conditioned to have the required moisture content, it shall be compacted by special rollers, mechanical tampers, or by other approved methods. All equipment and methods used shall be subject to approval. The moisture and density shall be equivalent to that obtained in the earthfill placed in the dam embankment in accordance with subsection G-21(d).

Measurement, for payment, of specially compacted earthfill will be made of the material specially compacted, as provided in this section, and as provided in section G-20. Under (1) above, measurement, for payment, of specially compacted earthfill at steep and irregular dam abutments will be limited to a thickness of 2 feet measured horizontally from the average contacts where practicable, or as otherwise determined by the contracting authority, and measurement, for payment, of specially compacted earthfill on rough and irregular embankment foundations will be made in the most practicable manner, as determined by the contracting authority.

Payment for specially compacted earthfill will be made at the unit price per cubic yard bid therefor in the schedule, which unit price shall include all costs as provided in section G-20.

G-23. *Sand, Gravel, and Cobble Fill in Dam Embankment, Zone* -------.—(a) *General.*—The sand, gravel, and cobble fill in dam embankment shall be constructed in accordance with this section and section G-20.

(b) *Materials.*—The materials shall consist of a reasonably well-graded pervious mixture of sand, gravel, and cobbles selected from borrow pits in borrow area -----, in accordance with section G-16. Cobbles and boulders larger than ---- inches in maximum dimensions may be embedded in the fill or may be removed from the fill material and placed in the (*insert location*).

(c) *Preparation of Foundations.*—The foundation for Zone -------- shall be prepared by leveling, moistening, and compacting so that the surface materials will be as compact as specified for subsequent layers of Zone ------.

(d) *Moisture Control.*—The moisture content of the sand, gravel, and cobble-fill material, prior to and during compaction, shall be distributed uniformly throughout each layer of the material. The moisture content shall be sufficient to attain the maximum relative density of the material in place, when compacted by the specified compaction procedure as provided in subsection (e). In general, the material shall be thoroughly wetted to obtain the maximum practicable compaction but shall not contain moisture to the extent which will interfere with trafficability of the contractor's hauling, placing, or compacting equipment.

Moisture, as required, may be applied by sprinkling on the fill, or by other approved methods.

(e) *Placing and Compacting.*—The contractor's operations shall be such and he shall handle and place the material in such a manner as to prevent segregation.

The sand, gravel, and cobble material shall be placed in the dam embankment in continuous, approximately horizontal layers not more than 12 inches in compacted thickness. When each layer of material has been conditioned to have the required moisture, as provided in subsection (d), it shall be compacted by four passes of the treads of a crawler-type tractor weighing approximately 40,000 pounds, or as provided below. One pass of the treads is defined as the required number of successive tractor trips which, by means of sufficient overlap, will insure complete coverage of the entire surface of the layer by the tractor treads. Second and subsequent passes of the treads shall not be made until each pass, as defined above, is completed.

If the contractor elects to use methods of compaction other than the one specified above, the weight of the compactor, the number of passes, inflation pressures of tires (if rubber-tired compactors are used), and thickness of lift not to exceed 12 inches compacted, shall be such as to result in sand, gravel, and cobble fill in dam embankment compacted within the following limits:

(1) Material represented by samples having a relative density less than **____ percent will be rejected. Such rejected material shall be recompacted until a relative density equal to or greater than **____ percent is obtained.

(2) Within the above limit and based on a continuous record of tests made by the contracting authority on previously placed and accepted embankment, the uniformity of relative density shall be such that:

a. No more than **20 percent of the material represented by the samples tested shall be at relative densities less than **____ percent.

The contracting authority will inform the contractor when the relative density is close to or outside the limits specified above and the contractor shall immediately make adjustments in procedures as necessary to maintain the relative density within the specified limits.

The relative density of the compacted sand, gravel, and cobble fill material will be determined by the contracting authority for the full depth of each compacted layer in accordance with Part B of Designation E-12 of the Bureau of Reclamation's "Earth Manual."

(f) *Measurement and Payment.*—Measurement, for payment, of sand, gravel, and cobble fill in dam embankment will be made as provided in section G-20.

Payment for sand, gravel, and cobble fill in dam embankment will be made at the unit price per cubic yard bid therefor in the schedule, which unit price shall include all costs of work required under this section and as provided in section G-20.

G-24. Miscellaneous Fill in Dam Embankment Zone - - - [16].—(a) *General.*—The miscellaneous fill, Zone ____ portion of the dam embankment shall be in accordance with this section and section G-20.

(b) *Materials.*—The miscellaneous fill, Zone ____ portion of the dam embankment shall consist of miscellaneous mixtures of clay, silt, sand, gravel, cobbles, and rock fragments to ____ inches in maximum dimensions. The materials shall be obtained from excavations for permanent construction required under these specifications * (and from borrow areas).

Boulders and rock fragments larger than ____ inches in maximum dimensions shall be removed from otherwise approved miscellaneous fill material, either at the site of excavation or after the material has been placed on the embankment, but before the Zone ____ material is compacted. Such oversize boulders and rock fragments shall be placed in________ (insert location)________.

(c) *Preparation of Foundations.*—The foundation for Zone ____ shall be prepared by level-

**Limits to be inserted for each job depending on nature of available material.

[16] This paragraph may be modified to delete all compaction requirements and to require routing of equipment, in accordance with design requirements.

*Revise or delete as appropriate.

ing, moistening, and compacting so that the surface materials will be as compact as specified for subsequent layers of Zone ____.

(d) *Moisture Control.*—Prior to and during compaction, the material in each layer of Zone ____ fill material shall have the most practicable moisture content required for compaction purposes as determined by the contracting authority. Additional moisture as required may be applied by sprinkling on the dam embankment.

(e) *Placing and Compacting.*—The material shall be placed in the miscellaneous fill in continuous and approximately horizontal layers, not more than ____ inches in thickness after being compacted as herein specified. The combined excavation and placing operations shall be such that the materials, when compacted in the miscellaneous fill, will be sufficiently blended to secure the best practicable degree of compaction and stability.

When each layer of the material has been conditioned to have the proper moisture content, it shall be compacted by six trips of a 50-ton pneumatic-tired roller over each 15-foot horizontal width of the layer, as herein provided. Each trip of the roller shall be offset from the path of the previous trip so that the total compactive effort shall be distributed evenly over the entire horizontal layer of Zone ____ embankment.

The pneumatic-tired roller used for compaction shall have a maximum total capacity of 50 tons and shall have a minimum of four wheels equipped with pneumatic tires. The tires shall be of such size and ply as can be maintained at tire pressures between 80 and 100 psi for a 25,000-pound wheel load during roller operations. The roller wheels shall be located abreast, and be so designed that each wheel will carry approximately equal loads in traversing uneven ground. The spacing of the wheels shall be such that the distance between the nearest edges of adjacent tires will not be greater than 50 percent of the tire width of a single tire at the operating pressure for a 25,000-pound wheel load. The roller shall have a rigid steel frame provided with a body suitable for balanced loading such that the load per wheel may be maintained at 25,000 pounds.

Tractors used for pulling pneumatic-tired rollers shall have sufficient power to pull the fully-loaded roller satisfactorily under normal conditions of compaction.

(f) *Measurement and Payment.*—Measurement, for payment, of miscellaneous fill in dam embankment, Zone ____, will be made as provided in section G-20.

Payment for miscellaneous fill in dam embankment, Zone ____, will be made at the unit price per cubic yard bid therefor in the schedule, which unit price shall include all costs as provided in section G-20.

G-25. *Rockfill in Dam Embankment, Zone ---- (Earthfill Dam).*—The rockfill, Zone ____, portion of the dam embankment shall be in accordance with this section and section G-20.

Materials for rockfill, Zone ____, shall consist of cobbles, boulders, and rock fragments reasonably well graded up to ____ *(cubic feet) *(cubic yards) in volume, selected from excavation for permanent construction required under these specifications *(or from borrow area ____.) *(or from ____.) Rock fragments larger than ____ *(cubic feet) *(cubic yards) in volume may be embedded in the rockfill, Zone ____, or may be placed on the outer slope of Zone ____.

The rockfill, Zone ____, portions of the dam embankment shall be placed in approximately horizontal layers not exceeding 3 feet in thickness. Equipment shall be routed over the layers already in place and traffic shall be distributed evenly over the entire width of fill so as to obtain the maximum amount of compaction possible. Successive loads of the material shall be dumped so as to secure the best practicable distribution of the material, as determined by the contracting officer, with the larger pieces placed toward the outer slopes. The material shall be dumped and roughly leveled off in an approved manner so as to maintain reasonably uniform surfaces on the fill and on the outer slopes of the embankment, and to insure that the completed fill will be stable and that there will be no large unfilled spaces within the fill.

Measurement, for payment, of rockfill in

*Revise or delete as appropriate.

dam embankment, Zone ......, will be made as provided in section G-20.

Payment for rockfill in dam embankment, Zone ......, will be made at the unit price per cubic yard bid therefor in the schedule, which unit price shall include all costs as provided in section G-20.

G-26. *Impervious Backfill.*—Impervious backfill shall be placed .... (insert location) .... and elsewhere as shown on the drawings or as directed. The materials to be used for impervious backfill shall be obtained from excavation for the dam and appurtenant works, or from borrow pits, as directed. The material used for impervious backfill, the amount thereof, and the manner of placing shall be subject to approval.

Measurement, for payment, of impervious backfill will be made of the material in place about the structure to the prescribed lines, grades, and dimensions. Payment for impervious backfill will be made at the unit price per cubic yard bid therefor in the schedule, which unit price shall include the cost of all work connected therewith, except the excavation, transportation, and compaction of the impervious backfill materials. Where compaction of impervious backfill is required, the compacting shall be performed as provided in section G-27. Payment for compacting impervious backfill will be made at the unit price per cubic yard bid therefor in the schedule, which payment will be in addition to the payment for impervious backfill.

G-27. *Compacting Impervious Backfill* [17].—(a) *General.*—Where compacting of impervious backfill is required the materials shall be deposited in uniform layers and compacted as specified in this section. The distribution of materials shall be such that the compacted material will be homogeneous and free from lenses, pockets, streaks, or other imperfections.

The material shall be deposited in horizontal layers not more than 6 inches thick after being compacted. The excavating and placing operations shall be such that the material when compacted will be blended sufficiently to secure the best practicable degree of compaction, impermeability, and stability. Prior to and during compaction operations, the material shall have the optimum moisture content required for the purpose of compaction, as determined by the contracting authority, and the moisture content shall be uniform throughout each layer.

Insofar as practicable, as determined by the contracting authority, moistening of the material shall be performed at the site of excavation but such moistening shall be supplemented by sprinkling at the site of compaction if necessary. If the water content is less than optimum for compaction, the compaction operations shall not proceed, except with the specific approval of the contracting authority, and if the water content is greater than optimum for compaction, the compaction operations shall be delayed until such time as the material has dried to the optimum water content, and no adjustment in price will be made on account of any operation of the contractor in drying the material or on account of delays occasioned thereby.

When the material has been conditioned as hereinbefore specified, it shall be compacted by tamping rollers having staggered and uniformly spaced knobs and of sufficient weight for proper compaction, by hand or power tampers, or by other means or equipment approved by the contracting authority. When tamping rollers are used, the tamping knobs and cleaner bars shall be properly maintained and the spaces between the tamping feet shall be kept clear of material which impairs the effectiveness of the tamping roller.

The dry density of the soil fraction in the compacted material shall not be less than 95 percent of the laboratory standard maximum soil density (dry) as determined by the Proctor compaction test for the materials being compacted.

The compaction tests will be made by the contracting authority. The standard laboratory maximum soil density is the dry weight per cubic foot of the soil compacted at optimum moisture content by laboratory procedure. The compaction test will be made by the Bureau of Reclamation procedure using a 1/20-cubic-foot compaction mold.

[17] This section may be used in combination with section G-26, or this type of work may be included in sections G-21 or G-22, as appropriate.

Measurement, for payment, of compacting impervious backfill will be made of the material in place about the structure to the prescribed lines, grades, and dimensions. Payment for compacting impervious backfill will be made at the unit price per cubic yard bid therefor in the schedule.

G-28. *Pervious Backfill.*—Pervious backfill shall be *(furnished and) placed to the lines and dimensions as shown on the drawings and as directed at the following locations:

(1) ____
(2) ____
(3) ____
(4) Elsewhere as shown on the drawings and as directed by the contracting authority.

The materials to be used for pervious backfill shall be selected pervious mixtures of sand, gravel, and cobbles, reasonably well graded to ___ inches maximum size. Fragments larger than ___ inches may be used if well distributed in the backfill. The pervious backfill shall not contain more than 5 percent, by weight, of material passing a United States Standard No. 200 sieve. The materials shall *(be furnished from any approved source) *(be selected from excavation for permanent construction required under these specifications, or from borrow pits in borrow areas ____, ____, ____, as directed) *(shall be washed or otherwise processed to remove excess fines).

The material shall be handled and placed in such a manner as to prevent segregation. The method of placing pervious backfill shall be subject to approval. *(The pervious backfill on either side of each structure shall be kept approximately at the same level as the placing of the backfill progresses.) Pervious backfill shall be placed and roughly leveled off in layers not more than ___ inches thick. *(Compaction of pervious backfill is not required.) *(Water shall be added as necessary so that the moisture content shall be uniformly distributed throughout each layer and shall be sufficient to attain the specified relative density of the material in place. The pervious backfill shall be compacted by ____ (list compaction method alternatives) ____ so that the relative density of the compacted material shall be not less than ___ percent as determined by the contracting authority in accordance with Part B of Designation E-12 of the Bureau of Reclamation's "Earth Manual".)

*Revise or delete as appropriate.

Measurement, for payment, of pervious backfill will be made of the material in place about the structures to the prescribed lines, grades, and dimensions. Payment for pervious backfill will be made at the unit price per cubic yard bid therefor in the schedule, which unit price shall include only the cost *(of furnishing, and) *(of washing or otherwise processing to remove fines, and) of placing *(moistening, and compacting) the pervious backfill material. *(Payment for excavation and transportation of the material will be made at the unit prices per cubic yard bid in the schedule for excavation *(in borrow area ____.) *(of the material used.))

G-29. *Rock Backfill.*—Rock backfill shall be *(furnished and) placed to the lines and dimensions as shown on the drawings and as directed at the following locations:

(1) ___ (insert locations) ____
(2) ____
(3) Elsewhere as shown on the drawings and as directed by the contracting authority.

The material used for rock backfill shall be selected angular rock fragments, reasonably well graded, from 3 to ___ inches in minimum and maximum dimensions; and shall be *(obtained from the same rock source, and be of the same quality, as required for riprap in section G-32.) *(selected from rock materials from required excavations.) *(obtained from borrow area ____.) The rock backfill shall contain not more than 5 percent, by weight, of material passing a United States Standard No. 200 sieve. Materials less than 3 inches in maximum size may be used in an amount only sufficient to fill the voids in the coarser material.

The method of placing shall be subject to approval. The rock fragments shall be placed and spread in layers not more than 24 inches thick. The rock backfill on each side of the structures shall be kept approximately at the same level as the placing of the backfill progresses.

Measurement, for payment, of rock backfill will be made of the rock backfill in place to the lines, grades, and dimensions shown on the drawings or as established by the contracting authority. Payment for rock backfill will be made at the unit price per cubic yard bid therefor in the schedule, which unit price shall include *(the cost of furnishing, and) *(the cost of all operations required to produce and complete the rock backfill as specified in this section.) *(only the cost of placing the rock backfill material.) Payment for excavation, selection, and transportation of material will be made at the unit prices per cubic yard bid in the schedule for excavation of the material used.

G-30. ***Filters.***—Graded sand and gravel filters shall be constructed under the *(apron, weir, spillway floor lining) as shown on the drawings or as directed. All materials for the filters shall be furnished by the contractor.

Trenches for the filters shall be excavated to lines, shapes, and dimensions shown on the drawings. Overexcavation in a manner to disturb the compacted foundations will not be permitted, and any material outside of the required lines which is disturbed shall be removed, and shall be replaced at the expense of the contractor in the manner described in section G-10. The sand and gravel shall be placed and tamped into place in such a manner that mixing of sand with gravel in the filter or with foundation or backfill materials will not occur. The graded sand and gravel shall be placed and tamped to the dimensions shown. *(After the graded sand and gravel in the filter have been shaped and compacted to the required depths, surfaces of the filter over which concrete is to be placed shall be covered with a layer of mortar 1 inch thick to provide a covering that will prevent the filter material from being displaced during the placing of the concrete. The mortar coating shall be applied carefully to the required thickness. The consistency of the mortar and methods of application shall be such as to avoid unnecessary filling of the voids in the filter material.)

Materials for filters shall be as follows:

(1) Gravel under the __________ shall be clean, well-graded gravel from 3/16 inch to 1½ inches in size.

(2) Sand shall conform to the requirements specified for concrete in section G-63.

Measurement, for payment, of graded sand and gravel in filters will be of the volume of sand and gravel in the completed filter. Payment for graded sand and gravel for filters will be made at the unit price per cubic yard bid therefor in the schedule, which price shall include the cost of furnishing, delivery, handling, placing, and compacting the graded sand and gravel *(and furnishing sand for and mixing and placing the mortar covering). Excavation for filters will be paid for in accordance with section G-10.

G-31. ***Bedding for Riprap.***—Bedding for riprap shall be placed___(insert location)__________

____

and elsewhere as directed, to the lines, grades, and thicknesses shown on the drawings.

The materials to be used for bedding for riprap shall be selected pervious mixtures of _____(insert material description)__________ reasonably well graded from sand size to_____ inches. Cobbles up to_____inches in maximum dimensions may be used if well distributed in the bedding. The bedding materials shall not contain more than_____percent of material passing a No. 200 United States Standard sieve. The materials shall be selected from _____(describe source)__________ as directed.

The contractor's operations shall be such and he shall handle and place the material in such a manner as to prevent segregation. The bedding need not be compacted in place, but shall be placed in such a manner as will result in uniform layers of bedding for riprap of the specified thickness.

Measurement, for payment, of bedding for riprap will be made of the bedding in place to the established lines and grades and on the basis of the specified thickness. Payment for bedding for riprap will be made at the unit price per cubic yard bid therefor in the schedule, which price shall include only the cost of

*Revise or delete as appropriate.

placing the bedding for riprap.

Payment for excavation, selection, and transportation of the material will be made at the unit price per cubic yard bid in the schedule for excavation of the material used.

G-32. *Riprap.*—(a) *General.*—Riprap shall be *(furnished and) placed to the prescribed outlines and thicknesses as shown on the drawings and as directed for the protection of slopes, channels, and structures at the following locations:

(1) ....

(2) ....

(3) ....

(4) ....

(5) ....

(6) Elsewhere as shown on the drawings and as directed.

*(The contractor shall also stockpile ...... (tons) (cubic yards) of riprap at the location shown on Drawing No. .... The dimensions and form of the stockpile shall be subject to approval.)

The contractor will be responsible for furnishing rock fragments meeting the requirements of these specifications, for making all necessary arrangements with property owners for right-of-way *(for obtaining use permits from Government agencies), and for payment of required royalties.

The contractor shall furnish to the contracting authority at the site of the work, without cost to the contracting authority, such samples of rock materials for testing as may be required by the contracting authority from proposed quarry sites and from rock materials delivered to the damsite.

(b) *Rock Source.*—Rock for riprap may be obtained from any approved source.

Samples of rock from..........

....

....

have been tested and the samples were found to be of suitable quality.

Tests have not been made to determine the relative difficulty of obtaining rock fragments of the specified sizes. Bidders and the contractor are cautioned that the above-mentioned deposits may be variable in quality, and the sizes and quality of rock fragments that may be obtained from any source are unknown.

*(The source of material located..........

....

....

is on lands controlled by the contracting authority and no charge will be made to the contractor for materials taken from this source and used in the work covered by these specifications.)

If riprap is to be obtained from a source not previously tested by the contracting authority, the contractor shall submit representative samples for testing at least 30 days before the riprap is required for use. The samples shall consist of three or four rock fragments totaling 200 pounds.

The contracting authority reserves the right to make inspections of quarry sites and quarries. The approval by the contracting authority of some rock fragments from a particular quarry site shall not be construed as constituting the approval of all rock fragments taken from the quarry, and the contractor will be held responsible for the specified quality and gradation of rock fragments delivered to the damsite. All rock fragments not meeting the requirements of these specifications, as determined by tests and/or inspection at the quarry and damsite, will be rejected.

(c) *Quarry Operations.*—The contractor shall clear, strip, develop, and operate the quarries; excavate and transport the rock materials; dispose of waste material; construct and maintain haul routes; and perform all other operations required to produce acceptable riprap materials. Quarry operations shall be in conformance with section G-2.

The location and extent of all quarries within the rock source and all quarrying operations shall be subject to approval. The contracting authority reserves the right to change the limits or location of quarries within the rock source in order to obtain the best available materials for riprap.

Blasting in the quarry shall be so controlled that, so far as joint pattern in the quarry will permit, the rock fragments shall meet gradation requirements specified herein. Unsatisfactory blasting procedures will not be

*Revise or delete as appropriate.

permitted.

Methods of sorting and loading blasted material in the quarry shall be such as to produce riprap of the highest practicable quality, and shall be subject to the approval of the contracting authority.

Waste materials shall be disposed of in the exhausted areas of the rock source or in approved areas adjacent to quarries.

(d) *Quality.*—The rock fragments shall meet the following requirements as to quality:

(1) Individual rock fragments shall be dense, sound, and resistant to abrasion and shall be free from cracks, seams, and other defects that would tend to increase unduly their destruction by water and frost actions.

(2) Samples prepared in accordance with applicable designations of the Bureau of Reclamation's "Concrete Manual," seventh edition, shall meet the following requirements when tested by the procedures described in the respective designations:

Test	Designation	Requirements
Specific gravity (saturated surface-dry basis).	10	Greater than **.
Soundness (sodium sulfate method).	19	Less than ** percent loss of weight after 5 cycles.
Abrasion (using Los Angeles machine grading A).	21	Less than ** percent loss of weight after 500 revolutions.

**To be inserted for each job depending on nature of available material and design considerations.

(e) *Gradation.*—Riprap shall be reasonably well graded within the following limits:

Nominal thickness of riprap, inches	Percentage of rock fragments of various weights [1] (pounds)			
	Maximum size	40 to 50 percent greater than	50 to 60 percent from—to	0 to 10 percent less than [2]
30	2,500	1,250	75—1,250	75
36	4,500	2,250	100—2,250	100

[1] Sand and rock dust shall be less than 5 percent, by weight, of the total riprap material.

[2] The percentage of this size material shall not exceed an amount which will fill the voids in larger rock.

*(Riprap material in stockpile shall meet the grading requirements of the ....-inch thickness of riprap.)

(f) *Placing.*—The riprap need not be compacted but shall be placed to grade in a manner to insure that the larger rock fragments are uniformly distributed and the smaller rock fragments serve to fill the spaces between the larger rock fragments in such a manner as will result in well-keyed, densely placed, uniform layers of riprap of the specified thickness. Hand placing will be required only to the extent necessary to secure the results specified above.

(g) *Measurement and Payment.*—*Rock materials for riprap shall be weighed on platform scales furnished at the expense of and by the contractor. The scales shall be approved by the contracting authority and shall be tested and sealed, at the expense of the contractor, as often as the contracting authority may deem necessary to insure their accuracy. A weigher, appointed and compensated by the contracting authority, will weigh all materials required to be weighed as herein provided. The scales shall be located in the vicinity of the work at the damsite.

Measurement, for payment, of riprap will be made of the number of tons of riprap placed to the lines, grades, and thicknesses shown on the drawings or as established by the contracting authority. Payment for riprap will be made at the unit price per ton bid therefor in the schedule, which unit price shall include the cost of all operations required to furnish, produce, and place the riprap as specified in this section.

Measurement, for payment, of riprap in the stockpile will be made of the number of tons of rock fragments placed in approved stockpile. Payment for stockpiling the rock fragments will be made at the unit price per ton bid in the schedule for riprap.

*(Measurement, for payment, of riprap will be made of riprap in place to the lines and grades on the basis of the thicknesses shown on the drawings or as established by the contracting authority. Payment for riprap will be made at the unit price per cubic yard bid therefor in the schedule, which unit price shall

*Revise or delete as appropriate.

include the cost of all operations required to furnish, produce, and place the riprap as specified in this section. Measurement, for payment, of riprap in the stockpile will be made of riprap in the stockpile to the approved dimensions and forms. Payment for stockpiling the rock fragments will be made at the unit price per cubic yard bid in the schedule for riprap.)

D. TUNNEL AND SHAFT CONSTRUCTION [18]

G-33. *Construction of Tunnel and Shaft, General.* —Construction of tunnel and shaft as used in these specifications includes the excavation by tunneling methods, placing of concrete lining, and other related work for the following:

(1) Tunnel. ____ (give description and limits such as Stations 1+39 and 11+06) ________

(2) Shaft. ____ (give description and limits)

(3) Gate chamber. ____ (give description and limits) ________________

(4) Other features as required. ________

The locations of the tunnel portals and elevation of the shaft portal may be moved, at the direction of the contracting authority, to accommodate the conditions encountered during excavation operations. If the portals are moved, all stations or elevations of portal structures and limits of tunnel and shaft construction will be changed accordingly.

The tunnel shall be constructed in accordance with the applicable sections in these specifications and payment for the various items of required work will be made at the applicable unit prices bid therefor in the schedule, except that for mortar protection of excavated surfaces, for structural-steel tunnel and shaft supports, for rock bolt supports, for intermediate expansion bolts, and for chain link fabric, payment will be made to the contractor at fixed unit prices stated in the bidding schedules and in accordance with sections G-36, G-37, G-39, and G-40.

Tunnel supports shall be installed to support the roof and sides of the tunnel, shaft, and gate chamber where conditions encountered are such as to require support. The type and amount of supports (structural-steel or rock bolts with or without chain link fabric) at any location shall be subject to the approval of the contracting authority.

Nothing contained in these specifications shall prevent the contractor, at his own expense, from erecting such amounts of temporary supports as he may consider necessary, or from using more rock-support bolts or heavier structural-steel supports than approved by the contracting authority if use of such additional rock bolts or heavier supports results in no increased cost to the contracting authority, including the cost of additional excavation. No statement in these specifications shall be constructed to relieve the contractor from sole responsibility for the safety of the tunnel, gate chamber, and shaft, or from liability for injuries to or deaths of persons or damage to property.

G-34. *Draining, Lighting, and Ventilating Tunnel.*—The contractor shall drain the tunnel of water by gravity flow or pumping as necessary to obtain satisfactory working conditions. If substantial flows of water should be encountered, the contractor may seal off water entering the tunnel in order to decrease the amount of water to be pumped or drained. The contractor shall adequately light and ventilate the tunnel during all construction operations. The contractor shall provide all labor and materials for cofferdams and other temporary construction as necessary to obtain satisfactory working conditions.

The cost of all work described in this section shall be included in the unit prices bid in the schedule for other items of work.

G-35. *Tunnel and Shaft Excavation.*—(a) *General.*—The tunnel and shaft shall be excavated to the lines, grades, and dimensions shown on the drawings or established by the contracting authority. Variations in the alinement of the tunnel shown on the drawings or established by the contracting authority will be allowed to the extent permitted in section G-70. No varia-

[18] Applicable to tunnels, shafts, gate chambers, and other underground construction for spillways, outlet works, and diversion works. The wording of the text should be revised depending on the actual construction involved.

tions in grade of the tunnel will be permitted except for the tolerances permitted in section G-70.

The general dimensions, arrangements, and details of typical sections are shown on the drawings. The contracting authority does not represent that excavation can be performed within the established "B" lines for tunnel excavation; however, payment for excavation will be made to the "B" lines in all cases and overbreak of excavation outside the "B" lines shall be at the contractor's expense irrespective of the cause of such overbreak or overexcavation.

All final excavated surfaces of the tunnel which are subject to deterioration by air slaking shall, within 1 hour after exposure, be protected by a protective coating of pneumatically applied cement-mortar as provided in section G-36.

Where supports are required to support the roof and sides of tunnel and shaft they shall be furnished and installed in accordance with sections G-37 through G-40.

(b) *Definitions of "A" and "B" Lines.*—The "A" lines shown on the typical sections of the drawings are lines within which no unexcavated material of any kind, no timbering, and no metallic or other supports shall be permitted to remain. The "B" lines shown on the typical sections are the outside limits to which measurement, for payment, of excavation will be made and measurement for payment will in all cases be made to the "B" lines regardless of whether the limits of the actual excavation fall inside or outside the "B" lines.

(c) *Locations of "A" and "B" Lines.*—The locations of the "A" and "B" lines for the tunnel sections are shown on Drawing No. ------.

(d) *Miscellaneous Requirements.*—The contractor shall use every precaution in his operations to avoid loosening material beyond the "B" lines. All drilling and blasting shall be carefully performed. Any damage to or displacement of tunnel supports and damage to any other part of the work caused by blasting or any other operations of the contractor shall be repaired at the expense of and by the contractor in an approved manner except as provided in section G-39.

The excavation in the crown of unsupported and rock bolt supported tunnel sections shall be sufficient to accommodate the concrete discharge pipe between the forms or reinforcement and the excavated surfaces. The excavation in the crown of steel rib supported tunnel sections shall be sufficient to permit the placing of peaked structural-steel supports to provide for the concrete discharge pipe as shown on the drawings. No payment will be made for excavation in the crown that is outside the "B" line and is performed to provide for the concrete discharge pipe and for the placing of peaked structural-steel supports.

Immediately following excavation in unsupported sections, or in supported sections as directed, all loosened material either inside or outside of the "B" lines, that in the opinion of the contracting authority is liable to fall, shall be removed.

Loose material on the bottom surface of the tunnel shall be removed in accordance with section G-41.

All material projecting inside the "A" lines shall be removed by the contractor in accordance with section G-41.

(e) *Measurement and Payment.*—Measurement, for payment, of excavation in tunnel, gate chamber, and shaft will be limited to the specified sectional dimensions shown on the drawings and will be made along the located centerlines of the tunnel and shaft only for such reaches as are excavated by tunneling and shaft driving methods.

Payment for excavation in tunnel, gate chamber, and shaft will be made at the unit price per cubic yard bid therefor in the schedule, which unit price shall include the entire cost of excavating, transporting, and disposing of excavated materials, and maintaining the excavation in satisfactory condition until the concrete lining is placed. No additional allowance above the unit price bid in the schedule will be made on account of the class, nature, or condition of any of the material encountered.

Overbreak of excavation outside the "B" lines will be at the contractor's expense irrespective of the cause of such overbreak or overexcavation.

G-36. *Mortar Protection of Excavated Surfaces.*

—(a) *General.*—The contractor shall apply a protective coating of pneumatically applied cement-mortar on finished excavated surfaces in the tunnel, shaft, and gate chamber as directed. The protective coating shall be applied to portions of the surfaces which are subject to deterioration by air slaking and which are designated to receive protective coating where in the judgment of the contracting authority it is deemed necessary. The final excavated surfaces shall be protected within 1 hour after exposure. *(Exposed finished excavated shale surfaces shall be kept moist at all times to prevent evaporation of the natural moisture in the material and shall be protected from freezing.) Surfaces to be coated shall be cleaned of all loose material, dirt, dust, mud, and other foreign matter. Temporary protective coverings, at the contractor's expense and as approved by the contracting authority, will be permitted, if necessary, until the permanent protective coating is applied. Protective coating damaged by frost, heat, traffic, or other causes shall be removed and replaced or repaired at the expense of and by the contractor as directed.

(b) *Pneumatically Applied Mortar.*—The pneumatically applied mortar coating shall be not less than one-half inch in thickness.

The mortar coating shall consist of a mixture of portland cement, sand, and water placed under pneumatic pressure. The mortar shall be mixed in the proportions of one part of cement to the equivalent of four parts, by weight, of surface-dry sand and shall be mixed thoroughly before being fed into the placing machine. The placing machine and its operation shall be subject at all times to the approval of the contracting authority. The amount of water used shall be that required to produce mortar of suitable consistency, quality, and uniformity with the minimum amount of rebound.

Cement for mortar shall be in accordance with section G-60. Sand for mortar shall be in accordance with section G-63.

The contractor shall take such precautions as the contracting authority may consider necessary to avoid or minimize accumulations of rebound.

The pneumatically applied mortar shall be cured, where required, by membrane curing or by water curing, at the option of the contractor. Curing shall be such as to prevent the development of an objectionable degree of cracking due to rapid loss of moisture following placement.

The curing shall be sufficient to assure adequate hydration of the cement. Membrane curing, if used, shall be in accordance with section G-76 and need not be removed.

(c) *Measurement and Payment.*—Measurement, for payment, of mortar protection of the excavated surfaces in tunnel, gate chamber, and shaft will be made at the number of square yards of finished surface coated as directed on the net area measured at the "B" line as shown on the drawings. Payment for mortar protection of excavated surfaces in tunnel, gate chamber, and shaft will be made at the fixed price of (enter established unit price here) ---------------- per square yard, which price shall include the cost of furnishing all materials and applying the coating. No separate payment will be made for the cement used in pneumatically applied mortar coating.

G-37. Structural-Steel Tunnel Supports.—(a) *General.*—Where structural-steel supports are used to support the roof and sides of the tunnel and gate chamber and sides of the shaft they shall be installed as shown on the drawings, as prescribed in this section, and as approved by the contracting authority.

The structural-steel supports, steel lagging, and other approved structural-steel members shall be furnished complete with bolts, nuts, washers, plates, tie rods, and other accessories required for installing the supports. The contractor may furnish structural steels having minimum yield point strengths under 45,000 psi or structural steels having minimum yield point strengths equal to or over 45,000 psi and payment therefor will be made as provided in subsection (c).

Acceptable types of steel rib supports are indicated on the typical tunnel support draw-

*Revise or delete as appropriate.

ings. These types may be modified by the contractor, subject to the approval of the contracting authority. The size, weight, miscellaneous details, and spacing of the supports shall be subject to the approval of the contracting authority. The clear distance between flanges of the installed supports shall not be less than 12 inches.

(b) *Installation.*—The steel supports shall be installed to the proper lines and grades and shall be maintained by the contractor in the proper condition and alinement until the concrete lining is placed about them.

Improper installations of supports shall be corrected by the contractor within 48 hours after the improperly installed supports are called to his attention. In supported sections of the tunnel, the contractor shall securely brace the supports with spreaders and the minimum practicable amount of blocking and wedges as provided in section G-38.

The contractor may, to facilitate his operations, place structural-steel supports a greater distance from the inside finished surface of concrete lining than shown on the drawings or prescribed by the contracting authority: *Provided,* That any increase in the quantity of steel supports and by any excavation and concrete lining outside the "B" lines shown on the drawings or described in section G-35 required thereby shall be at the contractor's expense.

(c) *Measurement and Payment.*—Measurement, for payment, of structural-steel tunnel supports will be made of the weights of supports and steel accessories as required by the drawings or as are approved by the contracting authority and only for the quantities which, in the judgment of the contracting authority, are necessary for satisfactory construction.

Payment for furnishing and installing structural-steel tunnel supports will be made at fixed unit prices as follows:

(1) Structural-steel tunnel supports of steels with minimum yield strengths under 45,000 psi ....(enter established unit price)........per pound. For the purpose of comparison of bids, the estimated quantity shown in the schedule is based on steels with minimum yield strengths under 45,000 psi.

(2) That portion of structural-steel tunnel supports consisting of steels with minimum yield strengths equal to or over 45,000 psi ....(enter established unit price).................... per pound.

Structural-steel tunnel supports shall include the steel ribs, butt plates, footplates, tie bolts, bearing plates, steel lagging where specified, and other steel members as are approved or directed. Payment will only be made for structural-steel supports actually approved for installation and installed in the tunnel, gate chamber, and shaft. For payment purposes, the contractor shall furnish the contracting authority certified copies of mill reports identifying the type of steel used in the supports.

G-38. *Foot Blocks, Lagging, Blocking, and Spreaders.*—(a) *General.*—The dimensions and amount of the foot blocks, lagging, spreaders, wedges, and blocking are not shown on the drawings, but shall in all cases be as necessary to serve their functions and for safety. The material used for foot blocks, blocking, and spreaders may, at the contractor's option, be timber, steel, or precast concrete. All timber shall be well-seasoned, sound timber of rectangular cross section. All lagging shall be steel or approved precast concrete.

Timber blocking, timber wedges, and spreaders of any material shall be removed in accordance with section G-41.

(b) *Placement of Lagging and Blocking.*—Except as authorized in specific instances, lagging and blocking shall be placed in the form of open cribbing and shall be arranged to permit the ready flow of concrete through and around the lagging and blocking, so that the concrete lining will be in contact with at least one-half of the excavated surface area bounded by the centerlines of adjacent steel supports, and so that at least one-half of the distance between structural-steel rib supports will be free of timber.

(c) *Costs.*—No separate payment will be made for furnishing, placing, and removing lagging, spreaders, blocking, and wedges, and the costs thereof shall be included in the unit prices bid in the schedule for other items of work. No separate payment will be made for furnishing and placing foot blocks and the cost thereof shall be included in the unit prices bid in the schedule for other items of work. Payment for furnishing and placing steel lagging placed permanently in the work will be made at the unit price per pound as stated in section G-37 for structural-steel tunnel supports of steels with minimum yield strengths under 45,000 psi. Where permanent precast concrete lagging is used, payment therefor will be made at a price not to exceed the contracting authority's cost if permanent steel lagging has been used.

G-39. *Rock Bolt Supports*.—(a) *General.*—Where rock bolt supports are used to support the roof and sides of the tunnel, gate chamber, and shaft they shall be of either the expansion-anchor type with both ends threaded, expansion-anchor type with fixed heads at one end and threaded at the other end, or slot-and-wedge type. The rock bolts shall be installed as shown on Drawing No. .... (typical tunnel supports), as prescribed in this section, and as approved by the contracting authority.

The rock bolts shall be furnished complete with all accessories, including bearing plates, anchor devices, bevel washers, machine washers, nuts, rust-preventive compound, lubricant, and other materials required for installation of the rock bolts. The minimum length of bolts shall be 6 feet.

(b) *Materials.*—

(1) *Bolts.*—

a. For expansion-anchor type.—The bolts shall be not less than 3/4-inch nominal diameter and shall be either bolts with both ends threaded or bolts with fixed heads at one end and threaded at the other end. The threaded ends of the bolts shall have rolled threads at least 8 inches in length. The bolts shall be made of steel conforming to ASTM Designation: A 306, Grade 80.

b. For slot-and-wedge-type anchor.—The bolts shall be not less than 1-inch nominal diameter with slot at one end with rolled threads at least 8 inches in length at the other end, and shall be made of steel conforming to ASTM Designation: A 306, Grade 60. The slot shall be approximately one-eighth inch wide and 6 inches long.

(2) *Anchors.*—

a. The shells and plugs for the expansion-type anchor shall be malleable iron castings conforming to ASTM Designation: A 47, Guide 32510, or forged steel with a maximum sulphur content of 0.23 percent.

b. The steel wedges for the slot-and-wedge-type anchor shall be approximately 5½ inches long, and shall be tapered from 1/16 by 3/4 inch to approximately 3/4 by 3/4 inch, and shall be made from steel conforming to ASTM Designation: A 306, or shall be made with cast malleable iron conforming to ASTM Designation: A 47, Grade 32510.

(3) *Bearing plates.*—The bearing plates shall be 3/8-inch flat steel plates providing not less than 36-square-inch area for each bolt or, if approved, a length of structural-steel channel or other structural shape may be used.

(4) *Washers and nuts.*—The bevel washers shall be steel or malleable iron. Machine washers shall be hardened steel. All nuts shall be the manufacturer's heavy hexagonal type.

(5) *Corrosion-preventive compound.*—The threads of the bolts and nuts shall be coated at the factory with a coat of corrosion-preventive compound. The compound shall conform to Military Specifications, MIL-C-16173D, Grade 1.

(6) *Lubricant.*—Lubricant for threads shall be of molybdenum disulfide base such as Molykote "G" as manufactured by the Alpha-Molykote Corp., 65 Harvard Ave., Stamford, Conn., or equal.

(c) *Installation.*—

(1) Rock bolt installations shall follow the advancement of the heading. The bolts or bolting pattern shall be advanced to approximately 5 feet from the face following each blasting operation. Such installation shall be completed within 6 hours following blasting.

(2) Except for applying lubricant to the bolt threads as specified in subsection (c) (5), the diameter of the holes and installation of the supports shall be in accord-

ance with the manufacturer's instructions unless otherwise directed. Two copies of the manufacturer's instructions shall be submitted to the contracting authority.

(3) The equipment used to drill the holes, to effectively seat and tighten the rock bolt anchorage in the hole, and to tighten the bolt to the required tension shall be subject to the approval of the contracting authority.

(4) Rock bolts shall be installed with flat steel bearing plates or, if approved, with a length of structural-steel channel or other structural shapes to which one or more bolts are connected. Beveled washers where needed shall be placed between the bearing plate and the nut to insure uniform bearing on the bearing plate. A machine washer shall be placed between the nut or bolthead and the bevel washer, and shall be coated on both sides with lubricant.

(5) Each hole shall be cleaned of all drill cuttings, sludge, and debris before the bolt is inserted. All threads of the rock bolts shall be free from rust, burrs, and other foreign matter. Immediately prior to the installation of the rock bolts any excess corrosion-preventive compound shall be wiped off the threads and the threads then shall be given a liberal coating of the lubricant specified in subsection (b) (6). Where slot-and-wedge-type rock bolts are used, the wedges shall also be lubricated with the lubricant specified in subsection (b) (6), just prior to insertion into the bolt slot.

(6) Each rock bolt shall be tightened with a controlled torque impact wrench or jack so that the bolt is stressed to not less than 60 percent and not more than 80 percent of the yield point stress based on the net area of the bolt, except that reduced stresses may be permitted in certain rock types as determined by the contracting authority.

The torque required to produce the required stress in the bolts will be determined by the contracting authority by means of tests performed at the site. Conformance with the required torque or bolt tension shall be periodically tested with an approved torque wrench or jack.

At periodic intervals, as determined by the contracting authority, the contractor shall check and where necessary retighten all rock bolts to the required torque or bolt tension until completion of the work under these specifications. If a bolt 5 feet or more from the heading has been damaged or made ineffective by blasting operations, the damage shall be repaired and, if necessary, additional bolts shall be installed to replace the damaged or ineffective bolts, all at the expense of the contractor. If installing of rock bolts closer than 5 feet from the heading is approved by the contracting authority, and such bolts are damaged during blasting operation, payment will be made for repair of the damaged bolts and for any additional bolts required to be installed to replace the damaged or ineffective bolts at the fixed price per linear foot for furnishing and installing rock bolts in tunnel .

(d) *Measurement and Payment.*—Measurement, for payment, of furnishing and installing rock bolts in tunnel, gate chamber, and shaft will be made of the length of the rock bolts approved for installation.

Payment for furnishing and installing rock bolts in tunnel, gate chamber, and shaft will be made at the fixed unit price of .... (enter established unit price) .... per linear foot. The unit price will include the cost of drilling the holes, of furnishing and installing the rock bolts complete with all accessories except bearing plates and approved structural-steel shapes, and of retightening the bolts as required.

Payment for furnishing and installing the flat steel bearing plates and other approved structural-steel shapes used in conjunction with rock bolts will be made at the fixed unit price of .... (enter established unit price) .... per pound.

G-40. Chain Link Fabric.—(a) *General.*—Chain link fabric shall be furnished and installed against the excavation rock surfaces as part of the support system for the roofs or sides of the tunnel, gate chamber, and shaft where and as approved by the contracting authority.

(b) *Materials.*—

(1) Chain link fabric.—Federal Specifi-

cation RR-F-00191e, Part 1, Class 1, Coating A, 2-inch mesh, No. 9 gage (0.148-inch diameter of coated wire).

(2) Intermediate expansion bolts.—Expansion bolts for fastening the fabric at intermediate points between rock bolts shall be 18 inches long and shall be ¾-inch expansion-anchor-type steel bolts in accordance with section G-39.

(3) Bearing plates and nuts.—Bearing plates for fastening chain link fabric to intermediate expansion bolts shall be flat steel bearing plates not less than 4 by 4 by ¼ inches in size. Nuts shall be the rock bolt manufacturer's heavy duty hexagonal nuts.

(c) *Installation.*—Where rock conditions require chain link fabric, the fabric shall be placed at the time rock bolts are installed. The fabric shall be placed between the rock surface and the bearing plates. Where conditions warrant, and if approved by the contracting authority, the fabric may be placed following the completed rock bolt installation. The chain link fabric placed over installed rock bolts may be held to the rock surface by means of intermediate expansion bolts, or by other methods approved by the contracting authority. The contractor shall lap sections of fabric a minimum of 6 inches where practicable: *Provided,* That at connections where it is impracticable to maintain 6-inch laps, as determined by the contracting authority, the contractor will be permitted to extend laps in lieu of cutting along regular lines. The final layout of the fabric and extent of lapping shall be subject to approval of the contracting authority.

Where approved, intermediate expansion bolts shall be installed at intermediate points for fastening the chain link fabric to the rock. The intermediate expansion bolts shall be installed in such manner that the chain link fabric is held securely between the rock surface and the bearing plate.

(d) *Measurement and Payment.*—Measurement, for payment, of furnishing and installing chain link fabric will be made of the actual number of square yards of chain link fabric installed, including overlaps at joints.

Payment for furnishing and installing chain link fabric in tunnel will be made at the fixed unit price of (enter established unit price) ______ per square yard, which unit price shall include the cost of furnishing and installing the chain link fabric complete with all accessories, except that payment for intermediate expansion bolts and bearing plates therefor will be made as provided below.

Payment for furnishing and installing intermediate expansion bolts for chain link fabric will be made at the fixed unit price of (enter established unit price) ______ per bolt, which unit price shall include the cost of drilling the holes and of furnishing and installing the bearing plates, nuts, and all accessories.

G-41. *Preparations for Placing Concrete Tunnel Lining.*—(a) *General.*—The contractor shall prepare tunnel, gate chamber, and shaft for placing concrete lining in accordance with section G-71 and the requirements of this section.

Where appreciable quantities of water flow from the material surrounding the tunnel, it shall be excluded from the space to be filled with concrete by grouting, by caulking, by diverting with pipe, pans, or other means, or by pumping from sumps until the concrete has hardened and gained sufficient strength to be unaffected by action of the water through percolation, hydrostatic pressure, or erosion.

All material projecting inside the "A" lines shall be removed by the contractor before concrete is placed in the lining. The removal of such projections within the "A" lines may be performed at any time during the progress of the work: *Provided,* That immediately before the concrete lining is placed, the contractor will be required to remove all materials then extending within the "A" lines.

Loose excavation at the base of the sidewalls and on the invert shall be removed to clean undisturbed surfaces before the concrete is placed.

Timber blocking and timber wedges shall be removed as completely as practicable before concrete lining is placed. Where steel or concrete is used for lagging, blocking, or wedges, such lagging, blocking, and wedges may be left in place and lining concrete placed about them as approved by the contracting authority.

All timber spreaders used to brace the structural-steel supports shall be removed before the concrete tunnel lining is placed.

(b) *Cost.*—No direct payment will be made for preparing the tunnel, gate chamber, and shaft for placing concrete lining and for excluding water from spaces to be filled with concrete and the costs thereof shall be included in the unit prices bid in the schedule for other items of work.

G-42. Concrete in Tunnel and Shaft Linings.—(a) *General.*—The item of the schedule for concrete lining in ____(insert schedule item)____ includes concrete shown on the drawing(s) as first-stage concrete in ____(give description and limits) ____.

*(Pipe and fittings shall be furnished and installed in the ____(enter name of structure(s)) lining for grouting the joints between first- and second-stage concrete.) Two-inch standard pipes to be used for air vents when placing second-stage concrete in the ____(enter name of structure) ____shall be furnished and installed in accordance with the drawings, and the costs thereof shall be included in the applicable item of concrete lining.

Pipes set in the concrete for backfill grouting shall be in accordance with section G-43.

Concrete in tunnel and shaft lining shall conform to the requirements of sections G-59 through G-76.

Before placing concrete lining in the tunnel and shaft, the tunnel and shaft shall be prepared therefor in accordance with section G-41.

The tunnel and shaft shall be lined with concrete having minimum thicknesses equal to the radial distance at any point in the perimeter of the tunnel between the finished inside surfaces of the concrete lining and the "A" lines as shown on the typical tunnel sections of the drawing. The finished interior surfaces of the tunnel shall conform to the shape, alinement, grades, and sections shown.

All spaces outside of the minimum required thickness of concrete lining shall be filled completely and solidly with concrete and special care shall be taken to force concrete into all irregularities in the contact surfaces and to completely fill the tunnel arch.

After the placing of concrete lining has been completed, the contractor shall fill all spaces between the rock surfaces and the concrete lining of the tunnel with backfill grout. Where metal lagging, liner plates, and other continuous supports are used, both the spaces between such supports and excavated surfaces, and spaces between such supports and concrete lining shall be filled with backfill grout. The backfill grouting shall be performed in accordance with section G-43.

The distance between transverse construction joints in the tunnel linings shall not exceed 50 feet and the distance between transverse construction joints in the shaft lining shall not exceed 30 feet. Rubber waterstops shall be placed in all transverse construction joints. No waterstops will be required in longitudinal construction joints. Plane panel forming will be permitted for horizontal curves of the tunnel linings provided that cord lengths shall not exceed 5 feet.

(b) *Measurement and Payment.*—Measurement, for payment, of concrete in tunnel, gate chamber, and shaft lining will be made to the "B" lines shown on the drawings. No deductions will be made for rock projections inside the "B" lines or for timber, concrete and metal supports, struts, lagging, and blocking remaining within the prescribed "B" lines. No measurement for payment will be made for concrete required to be placed outside of the "B" lines due to overbreakage or excess excavation to facilitate the contractor's operations, or for additional excavation to accommodate the concrete discharge pipe.

Payment for concrete in tunnel, gate chamber, and shaft linings will be made at the applicable unit price per cubic yard bid therefor in the schedule, which prices shall include the entire cost of furnishing all materials and of placing all concrete coarse aggregate or rock fragments, pipes and fittings, and backfill grout as described in section G-43, except that payment for cement used in the concrete lining and in backfill grout will be made at the applicable unit price per barrel bid in the schedule for furnishing and handling cement. Except as provided below for cement, no payment will be made for furnishing and placing materials re-

*Revise or delete as appropriate.

quired for filling spaces outside the "B" lines: *Provided,* That no payment will be made for cement used in filling spaces outside the 'B" lines caused by careless excavation or excavation intentionally performed by the contractor to facilitate his operations, as determined by the contracting authority. Payment for furnishing and placing rubber waterstops in the transverse construction joints in the linings will be made at the unit price per linear foot bid in the schedule for waterstops.

No direct payment will be made for excluding water from the spaces to be filled with concrete and the cost thereof shall be included in the unit prices bid in the schedule for other items of work.

G-43. Backfill Grouting.—(a) *General.*—The contractor shall backfill with grout spaces between the rock surfaces and concrete lining resulting from any cause at the locations described in section G-42. The grout shall be injected through pipes set in the concrete lining for this purpose or through holes drilled through the concrete lining. The placing of the grout shall be done at low pressures, not exceeding 30 psi, as directed by the contracting authority. Approximately four grout holes will be required along each 20 linear feet of tunnel and they shall be located and staggered as directed.

(b) *Pipes for Backfill Grouting.*—Metal pipes to be embedded in the concrete lining for backfill grout connections and air vents shall be furnished and placed to extend through the linings at such points as may be designated by the contracting authority. The grout pipes shall be set so that grout can flow freely therefrom to the voids between the rock surfaces and concrete lining. The pipes shall end not less than 1 inch back from the finished inside surface of the concrete lining, and recesses shall be provided therefrom to the surface of the lining. The recesses shall be filled with concrete or mortar after grouting operations have been completed, in accordance with section G-74. The size of the grout pipe and vent pipe for each hole will be determined by the contracting authority to meet the requirements of the drilling and grouting equipment used.

Pipes shall be held firmly in position and protected from damage while concrete lining is being placed. Caps or other devices shall be attached to the end of the pipe to prevent entry of concrete or other foreign materials prior to grouting and to facilitate location of the grout pipe after forms are removed. After embedment in concrete and prior to grouting, a hole shall be extended through each grout and air-vent pipe to the rock surface. Care shall be taken to avoid clogging or obstructing the pipes before grout hookups are made thereto, and any pipe that becomes clogged or obstructed shall be thoroughly cleaned by and at the expense of the contractor.

Pipe shall be Weight A, Class 1, in accordance with Federal Specification WW-P-406c, and shall be cut and fabricated as required. Pipe fittings shall be malleable iron or steel, Type 1, in accordance with Federal Specification WW-P-521f.

(c) *Drilling Grout Holes for Backfill Grouting.*—Where grout holes for backfill grouting are required and grout pipes have not been embedded in the concrete lining in accordance with subsection (b), grout holes shall be drilled through the concrete lining at such points as may be designated by the contracting authority and through lagging or liner plates if encountered. The depth of the holes shall be only to the surface of undisturbed material around the lined section unless otherwise directed. The diameter at the bottom of each hole shall be not less than 1½ inches. Care shall be taken to avoid clogging or obstructing the holes before grout hookups are made thereto and any hole that becomes clogged or obstructed shall be thoroughly cleaned by and at the expense of the contractor.

The contractor will have the option of setting regular pipe nipples in the drilled holes for grout connections, or using an expansion-type packer for such grout hole connections.

(d) *Composition.*—The grout shall be composed of cement and water, or cement, sand, bentonite, and water. The proportions of cement, sand, and water will be determined by the contracting authority. The contractor shall use 2 percent bentonite, by weight of the cement, in grout containing sand.

The bentonite shall be a suitable, commercially processed powdered bentonite, such as Volclay produced by the American Colloid Co., 5100 Suffield Court, Skokie, Ill.; National Bentonite produced at Osage, Wyo. plant of the National Lead Co. (Baroid Sales Division), 111 Broadway, New York, N. Y.; Big Horn Brand produced by the Wyo-Ben Products Co., P. O. Box 1079, Billings, Mont.; Wyo.-Gel produced by the Archer-Daniels-Midland Co., P. O. Box 532, Minneapolis, Minn.; or equal.

Sand shall be clean and of such fineness that 100 percent will pass a No. 8 standard sieve; not more than 5 percent shall be retained on a No. 16 sieve; not less than 10 percent, nor more than 30 percent, will pass a No. 100 standard sieve; and not more than 5 percent shall pass the No. 200 sieve.

Cement for grout shall be in accordance with section G-60.

(e) *Grouting Operations.*—The equipment for mixing and placing backfill grout shall be in accordance with applicable provision of section G-48 and shall be capable of effectively mixing and stirring the grout and injecting it into the holes at the specified pressure. No grouting operations shall be performed until all concrete within a distance of 50 feet has been in place for at least 7 days. Grout holes adjacent to a grout hookup shall be left open during grouting operations to facilitate the escape of air and water from pockets in the space surrounding the tunnel lining. Where grout is found to flow from adjacent grout connections in sufficient quantity to interfere seriously with the grouting operations or to cause appreciable loss of grout, such connections may be temporarily capped.

(f) *Filling Spaces between Liner Plates or Lagging and Excavated Surfaces.*—In sections of tunnel and gate chamber which are supported by structural-steel ribs with liner plates or continuous metal lagging, or which are supported by liner plates only, all spaces between the liner plates or lagging and the excavated surface and between concrete lining and liner plates or lagging shall be filled completely with backfill grout.

In sections of the shaft in which liner plates, continuous metal lagging, or other continuous supports are used, all spaces between such supports and the excavated surface of the shaft shall be filled as completely and compactly as practicable with clean coarse aggregate. Clean gravel or crushed rock from the approved sources of concrete aggregates or from other approved sources will be satisfactory for filling spaces outside of continuous supports. Not more than 5 percent of this material, by weight, shall be smaller than 5/8-inch size. After placing the concrete shaft lining, the voids in the coarse aggregate or rock fragments placed between the liner plates or lagging and the excavated surfaces shall be completely filled with backfill grout.

Such filling as described above and placed spaces outside the liner plates or lagging and lying within the "B" lines will be considered as concrete and paid for as such. No payment will be made for furnishing and placing coarse aggregate and backfill grout in spaces lying outside the "B" lines.

(g) *Filling Spaces in Arches.*—All spaces in the arches of tunnel and gate chamber remaining unfilled after the placing of concrete lining has been otherwise completed shall be filled with backfill grout. In tunnel and gate chamber arches which are supported by liner plates and/or continuous metal lagging, care shall be taken to fill both the spaces between the liner plates and lagging, and excavated surfaces, and spaces between liner plates or lagging and concrete lining. The backfill grout for filling spaces in the arches shall be placed through pipes or through holes drilled through the concrete lining as described in subsections (b) and (c). Vent holes shall be provided through the lining at the highest breaks in excavation.

(h) *Costs.*—No separate payment will be made for backfill grouting, metal grout pipes, or other materials, except cement used in backfill grouting, or for drilling grout holes for backfill grouting. Backfill grouting placed within the "B" lines will be included in the quantities of concrete lining paid for in accordance with section G-42. Payment for cement used in backfill grouting, including cement used outside of the "B" line in accordance with section G-42, will be made at the unit price per barrel bid in the schedule for furnishing and handling cement for pressure grouting.

E. PRESSURE GROUTING

G-44. *Requirements for Grouting, General.*—The contractor shall drill and grout under pressure:

(1) The rock foundation of the dam;
(2) The rock surrounding the portion(s) of the tunnel;
(3) The rock formation and structures at other locations as shown on the drawings, or as directed; and
(4) Any faults, joints, shear zones, springs, and other foundation defects that may require grouting as determined by the contracting authority.

The contractor shall grout under pressure:

*(1) The joint between the first-stage and second-stage concrete of the diversion inlet plug in the outlet-works intake structure; and
*(2) The joints between the first-stage and second-stage concrete in the outlet-works gate structure.

The amount of drilling and pressure grouting that will be required is uncertain and the contractor shall be entitled to no additional compensation above the unit prices bid in the schedule by reason of increase or decrease of the schedule quantities except as provided herein.

Drilling and grouting shall be accomplished through standard pipe placed into the concrete or formation the minimum distances shown on the drawings. The pipe and the installation of the pipe shall be in accordance with the provisions of section G-46.

The location, direction, and depth of each grout hole shall be as shown on the drawings or as directed by the contracting authority. The order in which the holes are drilled and the manner in which each hole is drilled and grouted, the proportions of cement and water used in grout, the time of grouting, the pressures used in grouting, and all other details of the grouting operations shall be as directed by the contracting authority.

Unless otherwise directed, the first grout holes shall be spaced widely and shall be drilled and grouted before the intermediate holes are drilled and grouted. In this manner the drilling and grouting of holes shall be completed with such final spacing as the grouting results show to be necessary.

Wherever required, due to water loss or hole caving, the drilling and grouting shall be performed in successive stages from the collar of the hole downward, such operations consisting in each case of: drilling the hole to a limited depth; washing intersected rock crevices, seams, or faults; seating a packer just above the section to be grouted; water testing and grouting that section; cleaning out the grout hole by washing or other suitable means before the grout in the hole has set sufficiently to require redrilling; allowing the grout surrounding the grout hole to attain its initial set; drilling the hole to an additional depth; and thus successively drilling and grouting the hole at various depths within the stages until the required depth of hole is completely drilled and grouted, all as determined by the contracting authority. Redrilling required because of the contractor's failure to clean out a hole before the grout has set shall be performed at the contractor's expense. When grout has been allowed to set in a hole by direction of the contracting authority, the required redrilling will be paid for at the rate of 50 percent of the unit price per linear foot bid in the schedule for drilling grout holes in stage between depths of 0 foot and 30 feet regardless of depth. No additional allowance above the unit prices bid in the schedule for drilling grout holes in stage will be made on account of the requirement for cleaning out holes before further drilling, or on account of moving of equipment that may be necessary due to the requirement for such successive stage drilling and grouting.

After holes in an area have been drilled and grouted, it may be found necessary to drill and grout additional holes. As the construction work progresses, the development of leakage or the condition of the surrounding foundation may indicate that parts of the foundation already covered with concrete require additional grouting. In this event, holes shall be

*Revise or delete as appropriate.

drilled through concrete into the underlying formation in such locations and to such depths as may be directed. No allowance above the unit prices bid in the schedule will be made for the drilling and grouting of such holes nor for the expense of moving equipment to other locations and returning to a previously drilled area.

All foundation grouting, including the formation around tunnels, shall be completed prior to constructing drains and prior to placing concrete structures, except placing concrete in cutoffs and tunnels, within ____ feet of the drains or concrete structures.

Drilling and pressure grouting the formation surrounding tunnels, gate chambers, adits, and shafts shall be done after placing of concrete linings within ____ feet of the pressure grouting operations. Furthermore, placement of backfill grout shall be completed for all lined sections within a distance of ____ feet of foundation pressure grouting operations.

*(The joints between first- and second-stage concrete in the outlet-works intake structure and in the outlet-works gate structure shall be grouted after the second-stage concrete has set and cooled to the satisfaction of the contracting authority.)

G-45. ***Drilling Grout Holes.***—Grout holes shall be drilled into the rock at the locations as described in section G-44.

The holes shall be drilled with rotary-type drills and the use of rod dope, grease, or other lubricants on the drill rods or in the grout holes will not be permitted, except that an approved neutral liquid soap may be added to the drill water. Drilling equipment and techniques shall be such as to minimize causing a hole to cave or become oversize. Drilling grout holes with percussion-type drills will not be permitted, except for installing pipe nipples forming the collars of the grout holes when needed. The requirements as to location, depth, spacing, and direction of the holes are approximate and subject to revision during the drilling, testing, and grouting. It is expected that the required depth of holes will not exceed ____ feet.

The minimum diameter of each grout hole shall be not less than that produced by the Commercial Standard EX size bit, approximately $1\frac{1}{2}$ inches.

When the drilling of each hole or stage of hole has been completed, clean water shall be circulated through the hole until it is flushed free of all drill cuttings. The hole shall then be temporarily capped or otherwise suitably protected to prevent the hole from becoming clogged or obstructed until it is grouted. Any hole that becomes obstructed before it is grouted shall be opened by and at the expense of the contractor.

Measurement, for payment, of drilling grout holes will be made only of the linear feet of holes actually drilled in formation and concrete as directed. Measurement for determining stage depths of drilling grout holes will be made from the collar of the hole at the exposed surface of the formation or concrete.

Except as otherwise provided for redrilling and for drilling holes for backfill grout, payment for drilling grout holes in formation or in concrete will be made at the applicable unit price per linear foot bid in the schedule for drilling grout holes in stage between the depths specified in the schedule, which unit price shall include the cost of furnishing all labor, materials, tools, and equipment required for drilling the holes, maintaining the holes free from obstructions until grouted, and all incidental work connected therewith.

The cost of drilling holes for placing backfill grout shall be included in the unit price per cubic yard bid in the schedule for concrete involved.

G-46. ***Pipe for Foundation and Joint Grouting.***—Standard black pipe for foundation and construction joint grouting and for grouting rock surrounding *tunnels, *gate chambers, *adits, and *shafts shall be set in the concrete or rock in the locations described in section G-44. Pipes for grouting shall be set over springs, crevices in the rock, faults, or other foundation defects as directed. Grout pipes set in concrete, except those in grout caps which may be left projecting above the footings, shall end not less than 3 inches inside the finished surface of the concrete. A standard coupling and wrapped nipple to facilitate removal after

*Revise or delete as appropriate.

grouting shall be attached to the grout pipe and shall extend beyond the finished surface of the concrete as shown on the drawings. The holes left after removal of wrapped nipples and other grout holes shall be filled immediately and completely with drypack in accordance with section G-74 (Repair of Concrete).

The size of the grout pipe for each hole and the depth of the holes for setting pipe for grouting shall be as shown on the drawings or as directed. The grout pipes shall be anchored into the rock or concrete into which they are inserted, and the spaces around the pipes shall be carefully cleaned and sealed with oakum, grout, or other suitable caulking material to prevent entry of concrete or other foreign materials prior to grouting. All pipe and fittings to be embedded in rock or concrete shall be cleaned thoroughly of all dirt, grease, grout, and mortar immediately before being embedded in the rock or concrete. The pipe and fittings shall be carefully assembled and placed and shall be held firmly in position and protected from damage while the concrete is being deposited. Care shall be taken to avoid clogging or obstructing the pipes before being grouted, and any pipe that becomes clogged or obstructed from any cause shall be cleaned satisfactorily or replaced at the expense of and by the contractor.

*(Systems of standard pipe, fittings, and special grouting outlets shall be placed in the joints between the first- and second-stage concrete of the outlet-works gate structure and the diversion inlet in the outlet-works intake structure, as shown on the drawings or as directed by the contracting authority. The special grouting outlets consist of modified electrical conduit boxes as shown on the drawings. The pipe and fittings shall be carefully assembled and placed, and shall be held firmly in position and protected from damage while the concrete is being deposited. Care shall be exercised to insure that the two companion members of each conduit-box-cover grouting outlet are maintained in accurate alinement and position with respect to each other and that each member becomes an integral part of and moves with the concrete mass to which it is anchored. The method of attaching the first member of each grouting outlet to the forms and, in turn, the second member to the first shall be as shown on the drawings or as directed. Care shall also be taken to insure that all parts of the system are maintained free from dirt and any other foreign substance.)

*(After the grouting system is placed and before any concrete is placed around it, and at such other times as the contracting authority may direct, the pipe shall be tested by forcing air or water under pressure through it, after which it shall be immediately capped temporarily or otherwise closed to avoid the possibility of any foreign substance entering it until it is pressure grouted. Any pipe that becomes clogged due to any cause before final acceptance of the work shall, if practicable, be cleaned or opened. For any plugged pipe which the contractor fails to open or replace, the contractor shall pay to the contracting authority as fixed, agreed, and liquidated damages, the sum of ________ dollars ($________) per linear foot of the total length of pipe which is thereby made ineffective, as determined by the contracting authority.)

All pipe and fittings, special grouting outlets, oakum, asphalt emulsion for sealing purposes, nails, tie wire, and temporary supports required for installation of the foundation grout pipe and the joint grouting systems shall be furnished by the contractor. The pipe shall be Weight A, Class 1, standard black pipe in accordance with Federal Specification WW-P-406c. The pipe fittings shall be malleable iron or steel, Type 1, in accordance with Federal Specification WW-P-521f. The pipe shall be cut, threaded if necessary, fabricated as required, and placed by the contractor.

Measurement for payment will be made only of the weight of pipe and fittings actually installed and left in place as directed by the contracting authority. Payment for furnishing and placing standard black pipe, fittings, and special grout outlets for grouting will be made at the unit price per pound bid in the schedule for furnishing and placing pipe and fittings for grouting, which unit price shall include the

*Revise or delete as appropriate.

cost of furnishing caulking materials and supports; of cleaning and sealing joints and cracks for anchoring grout pipe; of protecting the pipe from injury and clogging; and of removing nipples in exposed faces and drypacking the holes left by removal of nipples from grouting operations: *Provided,* That the cost of furnishing and placing pipe and fittings to place backfill grout shall be included in the unit prices per cubic yard bid in the schedule for the concrete involved. No additional allowance above the unit price bid in the schedule will be made on account of the varying size, lengths, or number of pipes required.

G-47. *Hookups to Grout Holes and Connections.* —Each drilled hole for foundation grouting will be hooked onto for pressure grouting, and payment therefor will be made as provided in this section. Connections to cracks, crevices, or seams in the foundations, *(and connections to grout headers for grouting the joint between the first- and second-stage concrete of the outlet-works gate structure and the diversion inlet plug) when required, will be considered as hookups and will be paid for as such.

The number of grout holes and connections for foundation grouting requiring hookups as shown in the schedule is only approximate, and the contractor shall be entitled to no additional compensation above the unit price bid in the schedule by reason of increase or decrease of the number of hookups actually required to complete the grouting.

Payment for hookups to grout holes and connections will be made only once for each grout hole or connection actually hooked onto at the direction of the contracting authority, regardless of the additional number of times packer settings are made or the same hole is hooked onto, and regardless of the volume of water or grout actually injected into the grout hole or grout connection.

Payment for hookups to foundation grout holes and connections will be made at the unit price per hookup bid therefor in the schedule.

The cost of hooking to connections, for placing backfill grout, shall be included in the unit prices per cubic yard bid in the schedule for concrete involved.

*Revise or delete as appropriate.

G-48. *Pressure Grouting.*—(a) *General.*—The item of the schedule for pressure grouting includes all operations required for cleaning, water testing, and pressure grouting the grout holes in all stages of depth.

All pressure grouting operations shall be performed in the presence of an authorized representative of the contracting authority.

Each drilled grout hole and grout connection for pressure grouting shall have grout composed of cement and water forced into it under pressure. In areas where cavities or fissures are encountered, or where the quantity of grout injected becomes excessive, sand, bentonite, and other suitable admixtures may be required as ingredients of the grout. Bentonite, where required, shall be furnished by the contractor and all costs incidental to its use shall be included in the unit price per cubic yard bid in the schedule for furnishing and handling sand for pressure grouting.

The proportions of cement and water, or of cement, sand, and water used in mixing the grout, the time of grouting, the pressures used for grouting, and all other details of the grouting operations shall be as determined by the contracting authority. It is anticipated that the water:cement ratio of the grout mixture may vary from 10:1 to 0.8:1 by volume. The contractor shall use 2 percent bentonite, by weight of the cement, in grout containing sand. The requirement for sand will be determined by conditions encountered during grouting operations and the contractor shall be entitled to no compensation above the unit price per cubic yard bid in the schedule for furnishing and handling sand for pressure grouting by reason of any amount or no sand being required.

(b) *Materials.*—Bentonite shall be a suitable commercially processed powdered bentonite and shall be subject to the approval of the contracting authority.

Sand shall be clean and well graded and when tested by means of standard screens (Designation 4 of the seventh edition of the Bureau of Reclamation's "Concrete Manual") shall conform to the following limits:

Screen Number	Cumulative percent, by weight retained on screen
8	0
16	0 to 5
30	15 to 40
50	50 to 80
100	70 to 90
200	95 to 100

Cement shall be type ____ in accordance with Federal Specification SS-C-192g, and shall be in accordance with section G-60. Water shall meet the requirements of section G-62.

(c) *Plant and Equipment.*—Communication facilities between the grout plant and the holes being grouted shall be furnished by the contractor where necessary or desirable, as determined by the contracting authority.

The grout supply pipes and packers shall be furnished by the contractor. The packers shall consist of expansible tubes or rings of rubber, leather, or other suitable material attached to the end of the grout supply pipe. The packers shall be designed so that they can be expanded to seal the drill hole at the specified elevations and when expanded shall be capable of withstanding without leakage, for a period of 5 minutes, water pressure equal to the maximum grout pressures to be used.

The apparatus for mixing and placing grout shall be of a type and size approved by the contracting authority and shall be capable of mixing and stirring the grout effectively and of forcing it into the grout holes or grout connections in a continuous, uninterrupted flow at any specified pressure up to a maximum of ____ psi. The mixer shall be operated mechanically and provided with an accurate meter, reading in cubic feet to tenths of a cubic foot, for controlling the amount of mixing water used in the grout. In addition to the grout mixer, a holdover mechanical agitator tank shall be provided. All grout shall be pumped with a duplex piston-type pump, helical-screw rotor-type pump, or other approved type of pumping equipment. A standby grout pump shall be included as part of the grout plant. The grouting equipment shall be maintained in a manner approved by the contracting authority and so as to insure continuous and efficient performance during any grouting operation. The arrangement of the grouting equipment shall be such as to provide a supply line and return line from the grout pump to the grout hole. Provisions shall be made to permit continuous circulation and accurate control of grouting pressures and grout flows into the grout holes and connections.

The elevation of the grout pump shall not exceed more than 20 feet above the collar of the hole being grouted, where practicable. For each stage being water-tested or grouted, two accurate pressure gages shall be installed for pressure measurements, one gage at the grout pump and one gage in the manifold at the collar of the hole.

*(Expansion-type plugs, if used for grout hole connections in grouting of tunnels, gate chambers, shafts, and adits shall be furnished by the contractor. The plugs shall consist of expansible tubes or rings of rubber, leather, or other suitable material, pipe, and fittings. The plugs shall be designed so that they can be expanded to seal the drill holes, and when expanded shall be capable of withstanding, without leakage, water pressure equal to the maximum grout pressures to be used.)

(d) *Stages and Sequence of Operations.*—Each grout hole shall be water tested and grouted in sections or stages located between depths in the hole best suited to treat the geology and defects of the foundation as determined by the contracting authority: *(*Provided,* That the maximum length of each stage shall be 20 feet between depths of 0 to 60 feet, 25 feet between depths of 60 and 160 feet, and 50 feet between depths of 160 to 310 feet).

Where directed, the grouting shall be done in successive stages from the collar of the hole downward as described in section G-44.

Where grouting in successive stages downward as described in section G-44 is not required, grouting shall be performed by: drilling the hole to the full depth; attaching a packer to the end of the grout supply pipe; lowering the grout supply pipe into the hole to the top of the deepest section that is required to

*Revise or delete as appropriate.

be grouted; water-testing the stage; grouting at the required pressure; allowing the packer to remain in place until there is no back pressure; withdrawing the grout supply pipe and packer to the top of the next deepest section that is required to be grouted, and thus successively water testing and grouting the hole in sections at the specified grouting pressures until the entire hole is completely grouted.

(e) *Water Testing Grout Holes.*—Prior to grouting, each stage of each grout hole shall be thoroughly tested with clean water under continuous pressure up to the required grouting pressure, as determined by the contracting authority, in order to clean and lubricate intersected cracks and seams and to determine rate of take and extent of leakage. Sufficient water shall be made available to develop the desired pressure, if practicable, and pressure gages and water-meters shall be provided to measure the amount of water pumped and the pressure during the test. Pumps, piping, gages, and meters shall be of a type and capacity approved by the contracting authority.

(f) *Grouting Procedures.*—Different grouting pressures will be required for grouting different sections of the grout holes. Pressures as high as necessary but which, as determined by trial, are safe against formation, embankment, and concrete displacement shall be used in the grouting. Except as provided below, the maximum pressures measured at the collar shall not exceed .... psi per foot of depth, the depth being the minimum from the packer to the ground surface measured normal to the ground surface slope: *Provided,* That a maximum of 10 psi, at the collar, may be used for grouting the top stage. Higher pressures may be required for grouting of formation around tunnels, gate chambers, adits, and shafts.

If, during the grouting of any hole, grout is found to flow from adjacent grout holes or connections in sufficient quantity to interfere seriously with the grouting operation or to cause appreciable loss of grout, such grout holes and connections shall be capped temporarily. When grouting is being done with packers and grout returns from adjacent holes, the pressure of the returning grout shall be measured by seating a packer in the adjacent hole and such pressures shall be kept below the allowable pressures for that stage of that hole. Where such capping is not essential, ungrouted holes shall be left open to facilitate the escape of air and water as the grout is forced into other holes. Before the grout has set, the grout pump shall be connected to adjacent capped holes and to other holes from which grout flow was observed, and grouting of all holes shall be completed at the pressure specified for grouting. If during the grouting of any hole, grout is found to flow from points in the foundations, abutments, any parts of the concrete structure, or other locations, such flows or leaks shall be plugged or caulked by the contractor as directed.

The grouting of any stage of any grout hole or connection shall be continued until that stage of the hole or connection takes grout at the rate of less than 1 cubic foot of the grout mixture in 20 minutes if pressures of 50 psi or less are being used, in 15 minutes if pressures between 50 and 100 psi are being used, in 10 minutes if pressures between 100 and 200 psi are being used, and in 5 minutes if plus 200 psi are being used, or when less than two sacks of cement per hour are being injected when pumping continuously. So far as practicable, the full grouting pressures shall be maintained constantly during grout injections. However, as a safeguard against formation or concrete displacement, excessive grout travel, or while grout leaks are being caulked, the contracting authority may require the reduction of the pumping pressure, intermittent pumping, or the discontinuance of pumping.

The consistency of the grout mix shall be varied, as directed by the contracting authority, depending on the conditions encountered. Where the grout hole or connection continues to take a large amount of grout after the mix has been thickened, the contracting authority may require that pumping be done intermittently, waiting up to 8 hours between pumping periods to allow grout in the foundation or formation to set. After the grouting is complete, the pressure shall be maintained by means of stopcocks, or other suitable valve devices, until the grout has set sufficiently so that it will be retained in the holes or connections being grouted.

After completing the pressure grouting in an area, all grout holes shall be filled with grout by gravity.

(g) *(*Joint Grouting.*—As soon as possible after the second-stage concrete in the outlet-works gate structure and the diversion inlet plug of the outlet-works intake structure has cooled the desired amount, as determined by the contracting authority, the joints between the first-stage concrete and the second-stage concrete shall be pressure grouted with cement grout. The grouting of each joint shall be completed before the grout takes its set in the grout pipe systems, but shall not be grouted so rapidly that the grout will not settle in the joint, and in no case shall the time consumed in filling each joint be less than 30 minutes. Before each joint is grouted, it shall be washed and tested thoroughly with air and water under pressure. Where the air or water tests indicate a probable loss of grout from any joint or crack, such joint or crack shall be tightly caulked with lead wool or other suitable material.

The grout shall be pumped into the bottom headers of the piping systems for each joint and the vent or highest header shall be left open until each joint is filled with grout of proper consistency for retention in the joint, whereupon the header shall be closed and the required pressure applied. Pressures as high as practicable, but not to exceed ____ psi, shall be used in the grouting. All other details shall be in accordance with these specifications and as directed by the contracting authority.)

(h) *Measurement and Payment.*—Measurement, for payment, of furnishing and handling cement for pressure grouting will be made of all cement actually forced into the holes or grout connections or required to fill permanent pipes at the direction of the contracting authority, and will include a reasonable amount for line waste.

Measurement, for payment, of furnishing and handling sand for pressure grouting will be made of the cubic yards of sand actually forced into the holes or grout connections at the direction of the contracting authority.

Measurement, for payment, of pressure grouting will be made on the basis of the number of cubic feet of cement, and cubic feet of sand, counted separately, actually forced into the holes or grouting connections at the direction of the contracting authority. In measuring grout for payment, the volume of 1 sack of cement will be considered as 1 cubic foot.

The quantity of pressure grouting which will be required is uncertain and cannot be accurately estimated. Accordingly, the quantity stated in the schedule is solely for the purpose of comparison of bids, and the actual quantity may vary widely therefrom. Since the quantity of pressure grouting is uncertain, payment for pressure grouting will be made in accordance with the following formulae:**

(1) If the final pressure grouting quantity averages less than 1.00 cubic foot of cement and sand per linear foot of drilled grout hole, payment will be made at the bid price stated in the schedule for pressure grouting for the grout actually placed, plus an allowance of 75 percent of the bid price for pressure grouting for the additional cubic feet required to average 1.00 cubic foot per linear foot of drilled grout hole.

(2) If the final pressure grouting quantity averages between 1.00 cubic foot and 2.00 cubic feet of cement and sand per linear foot of drilled grout hole, payment will be made at the bid price stated in the schedule for pressure grouting.

(3) If the final pressure grouting quantity exceeds an average of 2.00 cubic feet of cement and sand per linear foot of drilled grout hole, payment will be made at the bid price stated in the schedule for pressure grouting for the amount required to average 2.00 cubic feet per linear foot of drilled hole, and all pressure grouting in excess of the 2.00 average will be paid for at 50 percent of the bid price stated in the schedule for pressure grouting.

*Revise or delete as appropriate.

**Revise or delete average grout takes as appropriate, based on anticipated grout take. The averages shown, 1.00 and 2.00, are based on an anticipated take of 1.50 cubic feet per foot of hole.

Progress payments will be made for the quantity actually placed at the unit price bid in the schedule for pressure grouting. Upon completion of the grouting operation, the average amount per linear foot of hole will be determined by dividing the total volume of cement and sand in cubic feet for which payment for pressure grouting is made by the total linear feet of drilled holes in formation or rock *(excluding the portions of holes above the elevations at which pressure grouting is required) and excluding any redrill. Any adjustment under subsections (1) and (3) above will then be made. Except as provided in this section, the contractor shall be entitled to no extra compensation above the unit price bid in the schedule by reason of increased or decreased quantity of pressure grouting.

The unit price per cubic foot bid in the schedule for pressure grouting shall include the cost of furnishing all labor, materials, and equipment, including communication facilities required for the grouting, plugging and caulking leaks, using packers, and testing grout holes with water, except that payment for furnishing and handling cement for pressure grouting will be made at the unit price per barrel bid therefor in the schedule, payment for furnishing and handling sand for pressure grouting will be made at the unit price per cubic yard bid therefor in the schedule, and payment for hooking onto each grout hole or grout connection will be made as described in section G-47. No payment will be made for grout or for cement, or for sand used in grout, lost due to improper anchorage of grout pipes or connections, or rejected by the contracting authority on account of improper mixing, or lost by leakage due to the failure of the contractor to caulk surface leaks when directed by the contracting authority.

F. CONCRETE SPECIFICATIONS

G-49. *Introduction*.—(a) *General.*—Two different concrete specifications are included herein, one for small quantities of concrete where the structures are relatively simple, and one for more complex work requiring a larger quantity of concrete in which the structures are such that more detailed coverage of the work and the manner in which the work is to be done is desirable. The specification sections, included herein for work requiring detailed control, do not include all the requirements that would be needed for a mass concrete dam. Normally, additional investigations and specification requirements pertaining to the design of concrete mixes and concrete construction procedures are required for a mass concrete dam.

When a mass concrete dam is to be constructed, it may be desirable to require, prior to installation, review and advance approval of the contractor's plants, equipment, and construction procedures. This would particularly apply to the plants and equipment for processing, handling, transporting, storing, and proportioning concrete ingredients, and for mixing, transporting, and placing concrete. In cases where such is considered desirable, the contractor should be required to submit drawings to and for approval by the contracting authority, showing proposed plant arrangement, and furnish a description of the equipment he proposes to use in sufficient detail that an adequate review can be accomplished. The drawings and descriptions should be submitted in sufficient time before plant erection to permit a reasonable time for review, and for any alterations to the plant if inadequacies are noted. Where specific types of equipment or specific procedures are provided in the specifications, the contractor should be permitted to use alternative procedures or types of equipment if such are demonstrated to the satisfaction of the contracting authority that equivalent results would be obtained. Approval by the contracting authority of the con-

*Revise or delete as appropriate.

tractor's plants and equipment or their operation, or of any construction procedure, should not be cause to waive or modify the requirements for quality of materials of the finished work.

The specifications are written on the basis that the concrete mixes to be used in the work will be designed and controlled by the purchaser (referred to in the specifications as the contracting authority) within the maximum water-cement ratio and slump limitations specified, the limitations for quality and grading of aggregates, and the limitations for the other materials as specified, except that for the small job specifications, cement content of the concrete may be specified in lieu of a maximum water-cement ratio requirement. The percentages of sand and each size coarse aggregate to be used in the mixes are also determined by the purchaser. The quality limitations shown in the specifications for sand and coarse aggregate are considered standard limits. These limits may be reduced when only substandard materials are available within reasonable hauling distance and provided it has been determined by tests of concrete made with such aggregate that durable and good strength concrete can be produced. Under these specifications as written, the quantity of cement used in small jobs can be included in the cost of the concrete or paid for separately; for the large job specifications, the quantity of cement used in the work must be paid for separately. In applying these specifications, the purchaser may have his own engineering personnel or arrange with a separate engineering organization to accomplish testing of proposed aggregates, the design of mixes, and the handling of the control thereof throughout the life of the construction contract. However, if the purchaser desires to have the mix design and control the responsibility of the contractor, the specifications should be revised to provide for such. In this case, the specifications should include specific design compressive strength requirements (such as 3,750 psi) for the concrete. The mixes should be designed to take into consideration the need for increased average compressive strengths at 28-days' age of standard 6- by 12-inch test cylinders to satisfy the 28-days' age design compressive strength requirements. Criteria generally accepted in Bureau of Reclamation work require that the compressive strength of 80 percent of the test cylinders be greater than the design compressive strength (see the Bureau of Reclamation's "Concrete Manual," seventh edition).

Unless stated otherwise, all references to "designations" in part D of this appendix refer to designations listed in the appendix of the Bureau of Reclamation's "Concrete Manual," seventh edition [1] [19].

(b) *Ready-Mixed Concrete.*—The specifications included herein permit the manufacture of concrete from a central plant at the jobsite, by use of concrete from ready-mix plants and by use of truck mixers. The more complete specifications contain specific provisions for a central batching and mixing plant for concrete manufactured at the jobsite or for supplying specification concrete from a ready-mix plant using truck mixers.

(c) *Reference Specifications.*—The materials which are specified by reference to Federal Specifications, ASTM Standard Specifications, or other standard specifications or codes should be in compliance with the latest editions or revisions thereof in effect on the date bids are received or award of contract is made, whichever is appropriate. Specifications for specific work should contain stipulations providing for such conditions.

1. Concrete Specifications for Small Jobs

G-50. *Source.*—The following sections have been prepared from guide specifications normally used by the Bureau of Reclamation for less than 200 cubic yards of concrete.

G-51. *Materials.*—Concrete shall be composed of cement, sand and coarse aggregate, water, air-entraining agent, and calcium chloride [20] as required, all well mixed and brought to the proper consistency. In general, cement, air-

[19] Numbers in brackets refer to references listed in the bibliography, section G-102.

[20] Determine type of cement to be used. If the aggregates to be used in concrete are reactive with alkalies in cement, use of low-alkali cement should be specified. If aggregates are not reactive, do not include this requirement. In areas where the hardened concrete will

entraining agent, and sealing compound will be accepted on manufacturer's certification of compliance with specifications requirements. Permission to ship on manufacturer's certification shall in no way relieve the contractor of the responsibility for furnishing materials meeting specifications requirements.

(a) *Cement.*—Cement shall be type ...., low alkali,[20] in accordance with Federal Specification SS-C-192 for portland cement or with the "Standard Specifications for Portland Cement" of the American Society for Testing Materials: Designation C 150, and shall meet the false-set limitation specified therein.

(NOTE.—One specification or the other should be indicated, not both. Federal specifications are usually required for Government work.) The cement shall be free from lumps and damaged cement when used in concrete.

(b) *Water.*—Water shall be free from objectionable quantities of silt, organic matter, alkalies, salts, or other impurities.

(c) *Sand and Coarse Aggregate.*—Sand and coarse aggregate shall be furnished by the contractor from any approved source. The sand particles shall be clean, hard, dense, durable, uncoated rock fragments that will pass a screen having 3/16- or 1/4-inch square openings. The sand shall be well graded from fine to coarse and shall be free from injurious amounts of dirt, organic matter, and other deleterious substances.

The coarse aggregate shall consist of natural gravel or crushed rock or a mixture of natural gravel and crushed rock and shall be clean, hard, dense, durable, uncoated rock fragments, free from injurious amounts of thin pieces, organic matter, or other deleterious substances. The coarse aggregate shall be reasonably well graded from 3/16 inch to 1½ inches, and shall be separated into two sizes by an intermediate screen having 3/4-inch square openings. Screens having openings of other sizes and shapes may be used, if equivalent results are obtained. The contracting authority reserves the right to test the sand and coarse aggregate and, if required, the contractor shall submit, for preliminary tests and approval, representative samples of the sand and coarse aggregate proposed for use in the concrete work. (NOTE.—In lieu of the above requirements for sand and coarse aggregate, the specifications could require the aggregates to conform to the "Standard Specifications for Concrete Aggregates" of the American Society for Testing Materials: Designation C 33.)

(d) *Air-Entraining Agent.*—The air-entraining agent shall conform to the provisions of ASTM Designation C 260.

(e) *Reinforcing Bars *(and Fabric).*—Reinforcing bars shall conform to Federal Specification QQ-S-632, Type II, Class B40, B50, B60, C40, C50, or R50. (NOTE.—Delete any of those classes which will not be permitted, when and if applicable.)

*(Fabric shall be electrically welded wire fabric conforming to ASTM Designation A 185.)

(f) *Sealing Compound.*—Sealing compound shall be white-pigmented compound conforming to "Standard Specifications for Liquid Membrane-Forming Compounds for Curing Concrete," ASTM Designation C 309.

G-52. Composition.—Sand and coarse aggregate shall be mixed in proportions as directed by the contracting authority. The concrete shall contain not less than six bags of cement per cubic yard for concrete containing 1½-inch-maximum-size aggregate and seven bags per cubic yard for concrete containing 3/4-inch-maximum-size aggregate.

Calcium chloride shall not be used in concrete in which aluminum or galvanized metalwork is to be embedded or in concrete where it may come in contact with prestressed steel. The contractor shall use 1 percent of calcium chloride, by weight of the cement, in all other concrete placed when the weather is cold enough to require protection from freezing. The use of calcium chloride in the concrete

be in contact with soil and/or water containing harmful quantities of soluble sulfates, Type II or V cement should be specified, the type to be specified being dependent on the severity of the sulfate conditions occurring in the soil and/or water. If only negligible sulfate conditions occur, Type I cement is suitable for use in concrete. When Type V cement is needed and specified, the use of calcium chloride in the concrete should be prohibited as the calcium chloride reduces the resistance of concrete containing such cement to attack by sulfates in soils and/or water that will be in contact with the concrete. In this case, the words "and calcium chloride as required," in the first sentence of section G-51 should be deleted. Refer to table F-1 (app. F).

*Revise or delete as appropriate.

shall in no way relieve the contractor of the responsibility for compliance with the requirements of these specifications governing protection and curing of the concrete. (NOTE.—Calcium chloride shall not be used in concrete containing Type V cement. When Type V cement is specified, the requirements for use of calcium chloride are to be deleted. See footnote 20.)

Air-entraining agent shall be used in such amount as will effect the entrainment of from 4 to 6 percent of air, by volume, of the concrete as discharged from the mixer. When calcium chloride is being used, the portion of mixing water containing the air-entraining agent shall be introduced separately into the mixer.

The slump of the concrete shall not exceed 2 inches for slabs, and 3 inches for all other concrete.

G-53. ***Batching and Mixing.***—The sand and coarse aggregate shall be weighed and shall be proportioned on the basis of integral bags of cement unless the cement is weighed. Weighing equipment of the beam type may be used. The contractor shall provide equipment and shall maintain and operate the equipment as required to accurately determine and control the amount of each separate ingredient entering the concrete. Batching shall be such that combined inaccuracies in feeding and measuring the materials will not exceed 1½ percent for water and weighed cement and 2 percent for sand and each size of coarse aggregate. The concrete shall be uniform in composition and consistency throughout the mixed batch, and from batch to batch, except where changes in composition or consistency are directed. The mixing time shall be at least 1½ minutes. Excessive overmixing requiring the addition of water to preserve the required consistency will not be permitted. The temperature of the concrete when it is being placed shall be not more than 90° F. and not less than 40° F. in moderate weather, or 50° F. when the mean daily temperature drops below 40° F.

Truck mixers will be permitted only when the mixers and their operation are such that the concrete throughout the mixed batch and from batch to batch is uniform with respect to consistency and grading. Any concrete retained in truck mixers so long as to require additional water to permit satisfactory placing shall be wasted.

G-54. ***Forms, Preparations for Placing, and Placing*** [21].—Forms shall be used to shape the concrete to the required lines. The surfaces of construction joints shall be clean, rough, and surface dry when covered with fresh concrete or mortar. Cleaning shall consist of the removal of all laitance, loose or defective concrete, coatings, sand, sealing compound if used, and other foreign material.

The methods and equipment used for transporting concrete, and the time that elapses during transportation, shall be such as will not cause appreciable segregation of coarse aggregate or slump loss in excess of 1 inch in the concrete as it is delivered to the work. Retempering of concrete will not be permitted. Any concrete which has become so stiff that proper placing cannot be assured shall be wasted. Formed concrete shall be placed in continuous approximately horizontal layers, the depth of which generally shall not exceed 20 inches.

Concrete shall be vibrated until it has been consolidated to the maximum practicable density, is free from rock pockets of coarse aggregate, and closes snugly against all surfaces of forms and embedded materials.

All porous and fractured concrete shall be removed by chipping openings into the concrete with dimensions as directed, and the chipped openings shall be filled with drypack, mortar, or concrete, as directed.

Exposed unformed surfaces of concrete shall be brought to uniform surfaces and worked with suitable tools to a reasonably smooth wood-float or steel-trowel finish as directed.

G-55. ***Protection and Curing.***—The contractor shall protect all concrete against injury until final acceptance by the contracting authority. (NOTE.—The requirements for protection of concrete against freezing, including protection at 50° F. for the first 72 hours after the concrete

[21] Additional requirements for forming, preparations for placing, and placing may be appropriate for some small jobs, depending on the type of structure involved. See requirements for these phases of construction work as contained in sections G-44 through G-62.

is placed, as contained in the specifications for large jobs in section G-61 (Protection), will also apply for small jobs.)

The concrete shall be cured by water curing or by membrane curing. If concrete is cured by water curing, the concrete shall be kept continuously moist for at least 14 days after being placed by sprinkling or spraying or by other methods approved by the contracting authority. Membrane curing of concrete shall be by application of sealing compound, and application of the sealing compound shall be in accordance with the procedures contained in the Bureau of Reclamation's "Concrete Manual" [1]. Any concrete found to be damaged or defective by reason of the contractor's operations, at any time before completion and acceptance of the work, shall be removed and replaced by the contractor with acceptable concrete at no additional cost.

G-56. ***Reinforcement.***—Steel reinforcing bars *(and fabric) shall be placed in the concrete where shown on the drawings. Before reinforcement is placed, the surfaces shall be cleaned of heavy flaky rust, loose mill scale, dirt, grease, or other foreign substances. Reinforcement shall be accurately placed and secured in position so that it will not be displaced during the placing of the concrete.

Reinforcement will be inspected for compliance with requirements as to size, shape, length, splicing, position, and amount after it has been placed.

G-57. ***Measurement and Payment.***—Measurement, for payment, of reinforcement will be made only of the weight of the reinforcement placed in the concrete, in accordance with the drawings or as directed. Payment for furnishing and placing reinforcing bars *(and fabric) will be made at the unit price per pound bid therefor in the schedule, which unit price shall include the cost of furnishing and attaching wire ties and metal or other approved supports, if used, and of cutting, bending, cleaning, and securing and maintaining in position all reinforcement as shown on the drawings or as directed.

Measurement, for payment, of concrete will be on the basis of concrete having the dimensions shown on the drawings or prescribed by the contracting authority. Payment for concrete will be made at the unit price per cubic yard bid therefor in the schedule, which unit price shall include the cost of furnishing all labor and materials, except that payment for furnishing and placing reinforcing bars *(and fabric) will be made at the unit price bid therefor in the schedule.

If the contracting authority requires the contractor to use cement in excess of the minimum quantities specified, an equitable adjustment in payment under the contract will be made for such additional quantities of cement required.

(NOTE.—Where the dimensions of the concrete, as shown on the drawings, are firm, and no increase or decrease in the quantity of concrete required for the work will be involved, the cost of the concrete could be included in the lump-sum price bid in the schedule for construction of the structure concerned. Also, in such cases, if the quantity of reinforcement required in the concrete is firm, the cost of the reinforcement could be included in the lump-sum price bid for construction of the structure concerned. When considered appropriate, the foregoing specifications requirements for including the cost of furnishing and handling cement in the unit price per cubic yard bid in the schedule for concrete could be changed to provide for separate measurement and payment for furnishing and handling cement (see sections G-77 and 78). In such cases the cement content requirements would be deleted and maximum water-cement ratio requirements would be substituted, the cement contents to be used to be determined by the contracting authority.)

2. Concrete Specifications for Large Jobs

G-58. ***Source.***—The following sections have been prepared from guide specifications normally used by the Bureau of Reclamation for work involving concrete quantities of 200 cubic yards or more, and when quality control is necessary.

G-59. ***Composition.***—(a) *General.*—Concrete shall be composed of cement, sand, coarse aggregate, water, and admixtures as spec-

*Revise or delete as appropriate.

ified, all well-mixed and brought to the proper consistency.

(b) *Maximum Size of Aggregate.*—The maximum size of coarse aggregate in concrete for any part of the work shall be the largest of the specified sizes, the use of which is practicable from the standpoint of satisfactory consolidation of the concrete by vibration. Three-inch-maximum-size aggregate shall be used in walls that are 15 inches or more in thickness and in slabs that are 8 inches or more in thickness, except that where reinforcement is unusually congested, an appropriately smaller maximum size of aggregate shall be used. Three-inch-maximum-size aggregate shall also be used in any location where concrete is placed against excavated surfaces and the thickness of the concrete will be sufficiently greater than that shown on the drawings to permit the use of the larger size aggregate. (NOTE.—Delete requirements for 3-inch aggregate when use of this size aggregate is not specified. Normally, use of 3-inch aggregate is appropriate only where the quantity of concrete is sufficiently large to warrant its use from an economic standpoint.)

(c) *Mix Proportions.*—The proportions in which the various ingredients are to be used for different parts of the work shall be as determined from time to time during the progress of the work and as tests are made of samples of the aggregates and the resulting concrete. The mix proportions and appropriate water-cement ratio will be determined on the basis of procuring concrete having suitable workability, density, impermeability, durability, and required strength, without the use of an excessive amount of cement. The net water-cement ratio of the concrete (exclusive of water absorbed by the aggregates) shall not exceed 0.60 by weight, and water-cement ratios shall be in accordance with those contained in table F-2 (app. F), except that the water-cement ratios shall be reduced, if necessary, to attain required strengths. Tests of the concrete will be made by the contracting authority, and the mix proportions shall be adjusted whenever necessary for the purpose of securing the required economy, workability, density, impermeability, durability, or strength. The contractor shall be entitled to no additional compensation because of such adjustments.

(d) *Consistency.*—The amount of water used in the concrete shall be regulated as required to secure concrete of the proper consistency and to adjust for any variation in the moisture content or grading of the aggregates as they enter the mixer. Addition of water to compensate for stiffening of the concrete before placing will not be permitted. Uniformity in concrete consistency from batch to batch will be required. The slump of the concrete, after the concrete has been deposited but before it has been consolidated, shall not exceed 2 inches for concrete in the tops of walls, piers, parapets, and curbs, and in slabs that are horizontal or nearly horizontal, 4 inches for concrete in sidewalls and arch of tunnel lining, and 3 inches for all other concrete. The contracting authority reserves the right to require a lesser slump whenever concrete of such lesser slump can be consolidated readily into place by means of the vibration specified in section G-58 (Placing). The use of buckets, chutes, hoppers, or other equipment which will not readily handle and place concrete of such lesser slump will not be permitted.

(e) *Tests.*—The compressive strength of the concrete will be determined by the contracting authority through the medium of tests of 6- by 12-inch cylinders made and tested in accordance with the Bureau of Reclamation's "Concrete Manual" [1], Designations 29 to 33, inclusive, except that, for all concrete samples from which cylinders are to be cast, the pieces of coarse aggregate larger than 1½ inches will be removed by screening or hand picking. Slump tests will be made by the contracting authority in accordance with Designation 22. The contractor shall provide such facilities as may be necessary for procuring and handling representative test samples.

G-60. Cement.—(a) *General.*—To prevent undue aging of sacked cement after delivery, the contractor shall use sacked cement in the chronological order in which it was delivered on the job. Each shipment of sacked cement shall be stored so that it may readily be distinguished from other shipments. Bins in which bulk cement is stored shall be weathertight.

The bins shall be emptied and cleaned by the contractor when so directed; however, the intervals between required cleanings will normally not be less than 4 months. The cement shall be free from lumps and shall be otherwise undamaged when used in concrete. The cement shall be type ...., low alkali,[22] in accordance with Federal Specification SS-C-192 for portland cement or with the "Standard Specifications for Portland Cement" of the American Society for Testing Materials: Designation C 150. (NOTE.—One specification or the other should be indicated, not both. Federal specifications are usually required for Government work.)

(b) *Inspection.*—If directed by the contracting authority, the cement will be sampled and tested by an authorized representative of the contracting authority in accordance with Federal Test Method Standard No. 158, or the applicable method of test cited in the "Standard Specifications for Portland Cement" of the American Society for Testing Materials: Designation C 150. (See note above.)

(c) *Measurement and Payment.*—Measurement, for payment, of bagged cement will be on the basis of the number of bags of cement used at the mixer. Measurement, for payment, of bulk cement will be on the basis of batch weights at the batching plant. Any cement, either bulk or bagged, used for grouting, finishing, or other miscellaneous work will be measured for payment in the most practicable manner. One barrel shall be considered as 4 bags of bagged cement or 376 pounds of bulk cement.

Payment will be made for cement used in concrete placed within the paylines for concrete; and for cement used in concrete placed outside the concrete paylines, unless the requirement for such concrete is determined by the contracting authority to be the result of careless excavation, or excavation intentionally performed by the contractor to facilitate his operations.

(NOTE.—When any precast items, such as concrete pipe, are involved in the work, no direct payment is made for the cement used therein but instead payment is made for the precast pipe on a per linear foot basis for each size and type of pipe. For other precast items, payment is usually made on a unit price per item bid therefor in the schedule, the cost of all materials, including cement, being included.)

No payment will be made for cement used as follows: cement used in wasted concrete, mortar, or grout; cement used in the replacement of damaged or defective concrete; cement used in extra concrete required as a result of careless excavation; and cement used in concrete placed by the contractor in excavation intentionally performed by the contractor to facilitate his operations. (NOTE.—Payment for cement for pressure grouting foundations and/or backfill grouting is normally paid for separately.)

Payment for furnishing and handling cement will be made at the unit price per barrel bid therefor in the schedule, which unit price shall include the cost of rail and truck transportation of the cement from the mill to the jobsite and the cost of storing the cement.

G-61. Admixtures.—(a) *Accelerator.*—Calcium chloride shall not be used in concrete in which aluminum or galvanized metalwork is to be embedded or in concrete where it may come in contact with prestressed steel.

The contractor shall use 1 percent of calcium chloride, by weight of the cement, in all other concrete placed when the mean daily temperature in the vicinity of the worksite is lower than 40° F. Calcium chloride shall not be used otherwise, except upon written approval. Requests for such approval shall state the reason for using calcium chloride and the percentage of calcium chloride to be used, and the location of the concrete in which the contractor desires to use the calcium chloride. Calcium chloride shall not be used in excess of 2 percent, by

[22] Determine type of cement to be used. If the aggregates to be used in concrete are reactive with alkalies in cement, use of low-alkali cement should be specified. If aggregates are not reactive, do not include this requirement. In areas where the hardened concrete will be in contact with soil and/or water containing harmful quantities of soluble sulfates, Type II or V cement should be specified, the type to be specified being dependent on the severity of the sulfate conditions occurring in the soil and/or water. If only negligible sulfate conditions occur, Type I cement is suitable for use in concrete. When Type V cement is needed and specified, the use of calcium chloride in the concrete should be prohibited as the calcium chloride reduces the resistance of concrete containing such cement to attack by sulfates in soils and/or water that will be in contact with the concrete. Refer to table F-1 (app. F).

weight, of the cement. Calcium chloride shall be measured accurately and shall be added to the batch in solution in a portion of the mixing water. Use of calcium chloride in the concrete shall in no way relieve the contractor of responsibility for compliance with the requirements of these specifications governing protection and curing of the concrete. (NOTE.—Calcium chloride should not be used in concrete containing Type V cement. When Type V cement is specified, the requirements for use of calcium chloride are to be deleted.)

(b) *Air-Entraining Agents.*—The contractor shall use an air-entraining agent in all concrete. The agent used shall conform to the requirements of ASTM Designation C 260, and shall be of uniform consistency and quality within each container and from shipment to shipment. Agents will be accepted on manufacturer's certification of conformance with specifications, but permission to ship on certification shall in no way relieve the contractor of responsibility for furnishing an agent meeting specifications requirements. Agents shall be subject to sampling and testing. The amount of air-entraining agent used in each concrete mix shall be such as will effect the entrainment of the percentage of air shown in the following tabulation in the concrete as discharged from the mixer:

Coarse aggregate, maximum size in inches:	*Total air, percent by volume of concrete*
¾	6 ±1
1½	4.5±1
3	3.5±1

The agent in solution shall be maintained at a uniform strength and shall be added to the batch in a portion of the mixing water. This solution shall be batched by means of a mechanical batcher capable of accurate measurement which shall be so constructed that the full measure of solution added to each batch of concrete can be observed in a sight gage by the plant operator prior to discharge of the solution into the mixer. When calcium chloride is being used in the concrete, the portion of the mixing water containing the air-entraining agent shall be introduced separately into the mixer.

(c) *Water-Reducing, Set-Controlling Admixture.*—The contractor shall, except as hereinafter provided, furnish and use an approved water-reducing, set-controlling admixture, referred to herein as WRA, in all concrete. The WRA used shall be either a suitable lignosulfonic-acid or hydroxylated-carboxylic-acid type conforming to Bureau of Reclamation "Specifications and Method of Test for Water-Reducing, Set-Controlling Admixtures for Structural and Mass Concrete," dated July 1, 1966.

The WRA shall be of uniform consistency and quality within each container and from shipment to shipment.

The quantity of WRA to be used in each concrete batch shall be determined by the contracting authority and for the lignosulfonic-acid type shall not exceed 0.40 percent, by weight of cement, of solid crystalline lignin, and for the hydroxylated-carboxylic-acid type shall not exceed 0.50 percent, by weight of cement, of liquid.

Since the quantity of WRA required will vary with changing atmospheric conditions, the quantity used shall be commensurate with the prevailing conditions. The contracting authority reserves the right to use lesser quantities or no WRA in concrete for any part of the work, depending on climatic or other job conditions, and the contractor shall be entitled to no additional compensation by reason of reduction in or elimination of WRA in any concrete to be placed under these specifications.

The WRA solution shall be measured for each batch by means of a reliable visual mechanical dispenser. The WRA, in a suitably dilute form, may be added to water containing air-entraining agent for the batch, providing the materials are compatible with each other, or shall be introduced separately to the batch in a portion of the mixing water if the two are incompatible.

The contracting authority reserves the right to test the WRA prior to its use and the contractor shall, if required, submit a sample of the WRA proposed for use together with 5 bags (94 pounds each) of the cement proposed to be used, at least ninety (90) calendar days before use may be expected. The size of sample to be submitted in the case of a lignosulfonic-acid type shall be 2 liquid gallons containing 30 to 50 percent solids, or the equivalent of solid lignin. The sample of hydroxylated-carboxylic-acid type shall be 1 liquid gallon, similar in concentration to that proposed for use on project.

When requested, the contractor shall submit to the contracting authority test data by the manufacturer showing effects of the WRA on mixing water requirements, setting time of concrete, and compressive strength at various ages up to 1 year.

The contractor shall submit a certificate with each delivery stating that the WRA is of the same quality and chemical composition as the WRA used in the acceptance tests.

The contractor shall be responsible for any difficulties arising or damages occurring as a result of the selection and use of a WRA, such as delay or difficulty in concrete placing or damage to the concrete during form removal. The contractor shall be entitled to no additional compensation by reason of such difficulties.

(d) *General.*—Air-entraining agent, accelerator, and WRA, as required, shall be furnished by the contractor, and the cost of the materials and all costs incidental to their use shall be included in the applicable prices bid in the schedule for concrete in which the materials are used.

(NOTE.—The foregoing requirements for use of WRA are normally specified by the Bureau of Reclamation for work involving at least 2,000 cubic yards of concrete except where the concrete work consists of small quantity placements which extend over a large area, such as in concrete footings for transmission line towers. When these WRA requirements are not specified, and the contractor requests permission to use such an admixture, the contracting authority could approve such use under the foregoing conditions provided the WRA is furnished and used by the contractor at no increase in cost to the contracting authority.)

G-62. Water.—The water used in concrete, mortar, and grout shall be free from objectionable quantities of silt, organic matter, alkali, salts, and other impurities.

G-63. Sand.—(a) *General.*—The term "sand" is used to designate aggregate in which the maximum size of particles is three-sixteenths of an inch. Sand for concrete, mortar, and grout shall be natural sand, except that crushed sand may be used to make up deficiencies in the natural sand grading. Sand, as delivered to the batching plant, shall have a uniform and stable moisture content.

(b) *Quality.*—The sand shall consist of clean, hard, dense, durable, uncoated rock fragments. The maximum percentages of deleterious substances in the sand, as delivered to the mixer, shall not exceed the following values:

	Percent, by weight
Material passing No. 200 screen (designation 16)	3
Lightweight material (designation 17)	2
Clay lumps (designation 13)	1
Total of other deleterious substances (such as alkali, mica, coated grains, soft flaky particles, and loam)	2

The sum of the percentages of all deleterious substances shall not exceed 5 percent, by weight. Sand producing a color darker than the standard in the colorimetric test for organic impurities (designation 14) may be rejected. Sand having a specific gravity (designation 9, saturated surface-dry basis) of less than 2.60 may be rejected. The sand may be rejected if the portion retained on a No. 50 screen, when subjected to five cycles of the sodium sulfate test for soundness (designation 19), shows a weighted average loss of more than 8 percent, by weight.

(c) *Grading.*—The sand as batched shall be well graded, and when tested by means of standard screens (designation 4), shall conform to the following limits:

Screen No.:	*Individual percent, by weight, retained on screen*
4	0 to 5
8	5 to 15 [1]
16	10 to 25 [1]
30	10 to 30
50	15 to 35
100	12 to 20
Pan	3 to 7

[1] If the individual percent retained on the No. 16 screen is 20 percent or less, the maximum limit for the individual percent retained on the No. 8 screen may be increased to 20 percent.

G-64. Coarse Aggregate.—(a) *General.*—The term "coarse aggregate" is used to designate aggregate graded from 3/16 inch to 3 inches, or any size or range of sizes within such limits. The coarse aggregate shall be reasonably well graded within the nominal size ranges hereinafter specified. Coarse aggregate for concrete shall consist of natural gravel or crushed rock, or a mixture of natural gravel and crushed rock. Coarse aggregate, as delivered to the batching plant, shall have a uniform and stable moisture content.

(b) *Quality.*—The coarse aggregate shall consist of clean, hard, dense, durable, uncoated rock fragments. The percentages of deleterious substances in any size of coarse aggregate, as delivered to the mixer, shall not exceed the following values:

	Percent, by weight
Material passing No. 200 screen (designation 16)	1
Lightweight material (designation 18)	2
Clay lumps (designation 13)	½
Other deleterious substances	1

The sum of the percentages of all deleterious substances in any size, as delivered to the mixer, shall not exceed 3 percent, by weight. Coarse aggregate may be rejected if it fails to meet the following test requirements:

Los Angeles rattler test (designation 21).—If the loss, using grading A, exceeds 10 percent, by weight, at 100 revolutions or 40 percent, by weight, at 500 revolutions.

Sodium sulfate test for soundness (designation 19).—If the weighted average loss after five cycles is more than 10 percent, by weight.

Specific gravity (designation 10).—If the specific gravity (saturated surface-dry basis) is less than 2.60.

(c) *Separation.*—The coarse aggregate shall be separated into nominal sizes and shall be graded as follows:

Designations of size, inches	Nominal size range, inches	Minimum percent retained on screens indicated
¾	3/16 to ¾	50 percent on ⅜ inch
1½	¾ to 1½	25 percent on 1¼ inch
3	1½ to 3	25 percent on 2½ inch

Coarse aggregate shall be screened at the batching bins over vibrating screens having openings three-sixteenths of an inch in the narrow dimension: *Provided,* That the contractor may use stationary sloping screens having slotted openings three-sixteenths of an inch in the narrow dimension if the crushed and broken portion of the total coarse aggregate, as determined by the contracting authority, is less than 30 percent, by weight. Material passing the 3/16-inch screens shall be wasted as directed. The method and rate of feed shall be such that the screens will not be overloaded and will operate properly in a manner that will result in a finished product which consistently meets the grading requirements of these specifications. The vibrating screens shall be mounted on the batching plant or, at the option of the contractor, the screens may be mounted on the ground adjacent to the batching plant. The vibrating screens shall be so mounted that the vibration of the screens will not be transmitted to or affect the accuracy of the batching scales. The finished products shall pass directly to the individual batching plant bins. Stationary slotted sloping screens, where permitted, shall meet the following requirements:

(1) The screens shall be at least 6 feet in length and shall be not less than 4 feet wide. The flow of the material shall be perpendicular to the narrow dimension of the slots.

(2) The slots shall be not more than 4 inches long.

(3) The slope of the screens shall be not less than 30° and not more than 50° from the horizontal, and shall be readily adjustable within that range.

(NOTE.—When large quantities of concrete are involved in the work, normally in excess of 10,000 cubic yards, the specifications should, in lieu of requiring screening of the coarse aggregate over stationary sloping screens, require the coarse aggregate to be finish screened over vibrating screens mounted over the batching plant or, at the option of the contractor, mounted on the ground adjacent to the batching plant. When finish screening is specified, the specifications should also provide for the finish screens, if installed over the batching plant, to be so mounted that the vibration of the screens will not be transmitted to, or affect the accuracy of the batching scales, that the method and rate of feed shall be such that the screens will not be overloaded and will operate properly in a manner that will result in a finished product which consistently meets the grading requirements of the specifications, and that the finished product, after finish screening, shall pass directly to the batching plant bins. In such cases, the percentage of material passing the undersize test screen (significant undersize), as specified below, should not exceed 2 percent, by weight.)

Separation of the coarse aggregate into the specified sizes shall be such that, when the aggregate as batched is tested by screening on the screens designated in the following tabulation, the material passing the undersize test screen (significant undersize) shall not exceed 3 percent by weight, and all material shall pass the oversize test screen:

Size of aggregate, inches	Size of square opening in screen	
	For undersize test	For oversize test
¾	No. 5 mesh (United States standard screen).	⅞ inch
1½	⅝ inch	1¾ inches
3	1¼ inches	3½ inches

Screens used in making the tests for undersize and oversize shall conform to the requirements of ASTM Designation E 11, with respect to permissible variations in average openings. (NOTE.—If the concrete work is such that use of 3-inch aggregate is not desirable, the requirements contained in this section for 3-inch aggregate should be deleted.)

G-65. Batching.—(a) *General.*—The contractor shall provide equipment and shall maintain and operate the equipment as required to accurately determine and control the amount of each separate ingredient entering the concrete. The amounts of bulk cement, sand, and each size of coarse aggregate entering each batch of concrete shall be determined by weighing, and the amount of water shall be determined by weighing or volumetric measurement. Where sacked cement is used, the concrete shall be proportioned on the basis of integral sacks of cement unless the cement is weighed.

When bulk cement and aggregates are hauled from a central batching plant to the mixers, the cement for each batch shall either be placed in an individual compartment which, during transit, will prevent the cement from intermingling with the aggregates and will prevent loss of cement, or be completely enfolded in and covered by the aggregates by loading the cement and aggregates for each batch simultaneously into the batch compartment. The bins of batch trucks shall be provided with suitable covers to protect the materials therein from wind or wet weather.

Each batch compartment shall be of sufficient capacity to prevent loss in transit and to prevent spilling and intermingling of batches as compartments are being emptied. If the cement is enfolded in aggregates containing moisture, and delays occur between filling and emptying the compartments, the contractor shall, at his own expense, add extra cement to each batch in accordance with the following schedule:

*Hours of contact between cement and wet aggregates:	*Additional cement required, percent*
0 to 2	0
2 to 3	5
3 to 4	10
4 to 5	15
5 to 6	20
Over 6	(1)

[1] Batch will be rejected.

*(The contracting authority reserves the right to require addition of cement for shorter periods of contact during periods of hot weather and the contractor shall be entitled to no additional compensation by reason of the shortened period of contact.)

(b) *Equipment.*—The weighing equipment shall conform to the applicable requirements of Federal Specification AAA-S-121 for such equipment, except that accuracy to within 0.4 percent over the entire range of equipment will be satisfactory, and the equipment shall be capable of ready adjustment for compensating for the varying weight of any moisture con-

*Revise or delete as appropriate.

tained in the aggregates and for effecting changes in concrete mix proportions. Batching equipment shall be constructed and operated so that the combined inaccuracies in feeding and measuring the materials will not exceed 1 percent for water, 1½ percent for weighed cement, 2 percent for each size of aggregate, and 3 percent for admixtures. Weighing equipment shall be sufficiently insulated against vibration or movement of other equipment about the plant so that the accuracy of the delivered quantities of materials will not be impaired.

Bulk cement shall be weighed in an individual hopper and shall be kept separate from the aggregates until the batch ingredients are discharged from the batching hopper. The cement hopper may be attached to a separate scale for individual weighing, or may be attached to the aggregate scale for cumulative weighing. If materials are weighed cumulatively by dial scales, the cement shall be weighed before the other ingredients. If the materials are weighed cumulatively by beam scales, a separate beam shall be provided for weighing each material. Weighing equipment and the water-measuring device shall be in full view of the operator.

The contractor shall provide standard test weights and any other auxiliary equipment required for checking the operating performance of each scale or other measuring device and shall make periodic tests over the ranges of measurements involved in the batching operations. The tests shall be made in the presence of a representative of the contracting authority, and shall be adequate to prove the accuracy of the measuring devices. Unless otherwise directed, tests of equipment in operation shall be made at least once every month. The contractor shall make such adjustments, repairs, or replacements as may be necessary to meet the specified requirements for accuracy of measurement.

The operating mechanism in the water-measuring device shall be such that leakage will not occur when the valves are closed. Water tanks on portable mixers shall be constructed so that the indicating device will register, within the specified limit of accuracy, the quantity of water discharged, regardless of the inclination of the mixer setting.

Dispensers for air-entraining agent and calcium chloride solutions shall have sufficient capacity to measure at one time the full quantity of the properly diluted solution required for each batch, and shall be maintained in a clean and freely operating condition. Equipment for measuring shall be designed for convenient confirmation of the accuracy of the measurement for each batch and shall be so constructed that the required quantity can be added only once to each batch.

G-66. *Mixing.*—(a) *General.*—The concrete mixers shall be designed and operated so that the concrete, as discharged from the mixer, shall be uniform in composition and consistency throughout the mixed batch, and from batch to batch, except where changes in composition or consistency are required. The adequacy of mixing shall be determined by the contracting authority by mixer performance tests in accordance with designation 26. Mixers, when tested, shall meet the following criteria:

(1) The unit weight of air-free mortar in samples taken from the first and last portions of the batch as discharged from the mixer shall not vary more than plus or minus 0.8 percent from the average of the two mortar weights, and the variation in unit weight of any six consecutive tests shall not exceed plus or minus 0.5 percent of the average of these tests.

(2) The weight of coarse aggregate per cubic foot in the samples taken from the first and last portions of the batch shall not vary more than plus or minus 5.0 percent from the average of the two weights of coarse aggregate.

The contracting authority reserves the right to either reduce the size of batch to be mixed or to increase the mixing time when the charging and mixing operations fail to produce a concrete batch which conforms throughout to the above-numbered criteria and in which the ingredients are uniformly distributed and the consistency is uniform. Water shall be added prior to, during, and following the mixer-charging operations. Overmixing, requiring addition of water to preserve the required consistency, will not be permitted. Any concrete retained in mixers so long as to require additional water in excess of 3 percent of the design mix water (net water-cement ratio water, not including water absorbed by aggregates) to permit satisfactory placing shall be wasted. Any mixer that at any time produces unsatisfactory results shall be repaired promptly and effectively or shall be replaced.

(b) *Central Mixers.*—Mixers in centralized mixing plants shall be arranged so that mixing action in the mixers can be observed by the inspector and/or plant operator from a location convenient to the mixing plant operator's station. Mixers shall not be loaded in excess of their rated capacity unless specifically authorized. The concrete ingredients shall be mixed in a batch mixer for not less than the period of time hereinafter indicated for various mixer capacities after all of the ingredients except the full amount of water are in the mixer, except that the mixing time may be reduced if, as determined by the contracting authority, thorough mixing conforming to subsections (a) (1) and (a) (2) can be obtained in less time.

Capacity of mixer:	*Time of mixing*
2 cu. yd. or less	1½ minutes
3 cu. yd.	2 minutes
4 cu. yd.	2½ minutes
Larger than 4 cu. yd.	To be determined by tests performed by the contracting authority.

Each mixer shall be equipped with a mechanically operated batch counter, and a timing and signaling device which will indicate completion of the required mixing period.

(c) *Truck Mixers.*—Each truck mixer shall be equipped with (1) an accurate watermeter between supply tank and mixer, the meter to have indicating dials and totalizer, (2) a reliable revolution counter, which can be readily reset to zero for indicating the total number of revolutions of the drum for each batch. Each mixer shall have affixed thereto a metal plate on which the drum capacities for both mixing and agitating are plainly marked in terms of volume of concrete in cubic yards and the maximum and minimum speeds of rotation of the drum in revolutions per minute.

Mixing shall be continued for not less than 50 or more than 100 revolutions of the drum at the manufacturer's rated mixing speed after all the ingredients, except approximately 5 percent of the water which may be withheld, are in the drum. The mixing speed shall be not less than 5 or more than 20 rpm. Thereafter, additional mixing, if any, shall be at the speed designated by the manufacturer of the equipment as agitating speed, except that after the addition of withheld water, mixing shall be continued at the specified mixing speed until the water is dispersed throughout the mix. After a period of agitation, a few revolutions of the drum at mixing speed will be required just prior to discharging. In no case shall the specified maximum net water-cement ratio be exceeded.

When a truck mixer or agitator is used for transporting concrete, the concrete shall be delivered to the site and the discharge completed within 1½ hours after the introduction of the cement into the mixer. Each batch of concrete, when delivered at the jobsite from commercial ready-mix plants, shall be accompanied by a written certificate of batch weights and time of batching.

Mixers shall be examined daily for changes in condition due to accumulation of hard concrete or mortar or to wear of blades. No mixer shall be charged in excess of its rated capacity for mixing or agitating; however, if any mixer cannot produce concrete meeting the requirements heretofore specified when mixing at rated capacity, within the specified limitation on the number of revolutions of the mixing drum at mixing speed, the size of batch mixed in that mixer may be reduced until, upon testing, a uniformly mixed batch, conforming to the mixer performance tests as provided in subsection (a), is obtained.

G-67. *Temperature of Concrete.*—The temperature of concrete, when it is being placed, shall be not more than 90° F., and not less than 40° F. in moderate weather or 50° F. in weather during which the mean daily temperature drops below 40° F. Concrete ingredients shall not be heated to a temperature higher than that necessary to keep the temperature of the mixed concrete, as placed, from falling below the specified ___° F. Methods of heating concrete ingredients shall be subject to approval.

If concrete is placed when the weather is such that the temperature of the concrete will exceed 90° F., as determined by the contracting authority, the contractor shall employ effective means, such as precooling of aggregates and mixing water and placing at night, as necessary to maintain the temperature of the concrete, as it is placed, below 90° F. The contractor shall be entitled to no additional compensation on account of the foregoing requirements.

G-68. Forms.—(a) *General.*—Forms shall be used, wherever necessary, to confine the concrete and shape it to the required lines. Forms shall have sufficient strength to withstand the pressure resulting from placement and vibration of the concrete, and shall be maintained rigidly in position. Forms shall be sufficiently tight to prevent loss of mortar from the concrete. Chamfer strips shall be placed in the corners of forms so as to produce beveled edges on permanently exposed concrete surfaces. Interior angles on such surfaces and edges at formed joints will not require beveling unless requirement for beveling is indicated on the drawings. (NOTE.—If any tunnel construction is included in the work and concrete linings are involved, additional specifications requirements concerning forming and openings therein for vibration and inspection behind the forms may be needed. Bureau of Reclamation requirements provide that inside forms for nearly horizontal circular tunnels having an inside diameter of 12 feet or more shall be constructed to cover the arch and sides, the bottom 60° (possibly more) of the inside circumference to be placed without forming. Also, the number of openings needed in the crown and side walls of the forms is dependent on the diameter of the tunnel.)

(b) *Form Sheathing and Lining.*—Wood sheathing or lining shall be of such kind and quality, or shall be so treated or coated, that there will be no chemical deterioration or discoloration of the formed concrete surfaces. Where pine is used for form sheathing, the lumber shall be *pinus ponderosa* in accordance with the Standard Grading Rules of the Western Wood Products Assn., or shall be other lumber of a grading equivalent to that specified for pine. Plywood used for form sheathing or lining shall be concrete form, Class I, Grade B-B, exterior, mill oiled and edge sealed, in accordance with Product Standard PSI-66 of the U.S. Department of Commerce. Form sheathing or lining shall be of wood or steel conforming to the following requirements, or may be of other materials producing equivalent results:

Required finish of formed surface	Wood sheathing or lining	Steel sheathing or lining [1]
F1	Any grade—S2E	Steel sheathing or steel lining permitted.
F2	No. 2 common or better, pine shiplap, or plywood sheathing or lining.	Steel sheathing permitted. Steel lining permitted if approved.

[1] Steel "sheathing" denotes steel sheets not supported by a backing of wood boards. Steel "lining" denotes thin steel sheets supported by a backing of wood boards.

The type and condition of form sheathing and lining, and the fabrication of forms for finish F2 shall be such that the form surfaces will be even and uniform. The ability of forms to withstand distortion caused by placement and vibration of concrete shall be such that formed surfaces will conform with applicable requirements of these specifications pertaining to finish of formed surfaces.

(c) *Form Ties.*—Embedded ties for holding forms shall remain embedded and, except where F1 finish is permitted, shall terminate not less than two diameters or twice the minimum dimension of the tie in the clear of the formed faces of the concrete. Where F1 finish is permitted, ties may be cut off flush with the formed surfaces. The ties shall be constructed so that removal of the ends or end fasteners can be accomplished without causing appreciable spalling at the faces of the concrete. Recesses resulting from removal of the ends of form ties shall be filled in accordance with the provisions for repair of concrete.

(d) *Cleaning and Oiling of Forms.*—At the time the concrete is placed in the forms, the surfaces of the forms shall be free from encrustations of mortar, grout, or other foreign material. Before concrete is placed, the surfaces of the forms shall be oiled with a commercial form oil that will effectively prevent sticking and will not soften or stain the concrete surfaces or cause the surfaces to become chalky or dust producing. For wood forms, form oil shall consist of straight, refined, pale, paraffin-base mineral oil. For steel forms, form oil shall consist of refined mineral oil suitably compounded with one or more ingredients which are appropriate for the purpose.

(e) *Removal of Forms.*—To facilitate satisfactory progress with the specified curing, and enable earliest practicable repair of surface imperfections, forms shall be removed as soon as the concrete has hardened sufficiently to prevent damage by careful form removal. Forms on upper sloping faces of concrete, such as forms on the water sides of warped transitions, shall be removed as soon as the concrete has attained sufficient stiffness to prevent sagging. Any needed repairs or treatment required on such sloping surfaces shall be performed at once, and be followed immediately by the specified curing.

To avoid excessive stresses in the concrete that might result from swelling of the forms, wood forms for wall openings shall be loosened as soon as this can be accomplished without damage to the concrete. Forms for the openings shall be constructed so as to facilitate such loosening. Forms for conduits and tunnel lining shall not be removed until the strength of the concrete is such that form removal will not result in perceptible cracking, spalling, or breaking of edges or surfaces, or other damage to the concrete. Forms shall be removed with care so as to avoid injury to the concrete, and any concrete so damaged shall be repaired in accordance with the provisions for repair of concrete.

(f) *Costs.*—The cost of all labor and mateials for forms and for any necessary treatment or coating of forms shall be included in the unit prices bid in the schedule for the concrete for which the forms are used.

G-69. *Reinforcing Bars *(and Fabric).* —(a) *Furnishing Reinforcement.*—The contractor shall furnish all the reinforcement required for completion of the work.

Reinforcing bars shall conform to Federal Specification QQ-S-632, Type II, Class B40, B50, B60, C40, C50, or R50: *Provided,* That all bars required to be bent to a radius of bend of 50-bar diameters or less shall be billet steel (Class B40, B50, or B60). (NOTE.—Delete any of those classes which will not be permitted, when and if applicable.)

*(Reinforcement fabric shall be a standard type of electrically welded wire fabric conforming to the requirements of ASTM Designation A185.)

(b) *Placing Reinforcement.*—Steel reinforcement shall be placed in the concrete wherever shown on the drawings, or where directed. Unless otherwise shown on the drawings, or directed, measurements made in placing the bars shall be to the centerlines of the bars. Reinforcement will be inspected for compliance with requirements as to size, shape, length, splicing, position, and amount after it has been placed.

Before reinforcement is embedded in concrete, the surfaces of the reinforcement and the surfaces of any bar supports shall be cleaned of heavy flaky rust, loose mill scale, dirt, grease, or other foreign substances which, in the opinion of the contracting authority, are objectionable. Heavy flaky rust that can be removed by firm rubbing with burlap or equivalent treatment is considered objectionable. After being placed, the reinforcement shall be maintained in a clean condition until it is completely embedded in the concrete.

Reinforcement shall be accurately placed and secured in position so that it will not be displaced during placing of the concrete, and special care shall be exercised to prevent any disturbance of the reinforcement in concrete that has already been placed. Chairs, hangars, spacers, and other supports for reinforcing bars may be of concrete, metal, or other approved material. Where portions of such sup-

*Revise or delete as appropriate.

ports will be exposed on concrete surfaces designated to receive F2 finish, the exposed portions of the supports shall be galvanized or noncorrodible metal.

The reinforcement in structures shall be so placed that there will be a clear distance of at least 1 inch between the reinforcement and any anchor bolts or other embedded metalwork.

G-70. *Tolerances for Concrete Construction.*—Permissible surface irregularities for the various classes of concrete surface finish are specified in section G-73 (Finishes and Finishings); they are defined as "finishes," and are to be distinguished from tolerances as described herein. The intent of this section is to establish tolerances that are consistent with modern construction practice, yet governed by the effect that permissible deviations will have upon the structural action or operational function of the structure. Deviations from the established lines, grades, and dimensions will be permitted to the extent set forth herein: *Provided,* That the contracting authority reserves the right to diminish the tolerances set forth herein if such tolerances impair the structural action or operational function of a structure.

Where tolerances are not stated in the specifications or drawings for any individual structure or feature thereof, permissible deviations will be interpreted conformably to the provisions of this section. Notations on the drawings of specific maximum or minimum tolerances in connection with any dimensions shall be considered as supplemental to the tolerances specified herein. The contractor shall be responsible for setting and maintaining concrete forms sufficiently within the tolerance limits so as to insure completed work within the tolerances specified herein. Concrete work that exceeds the tolerance limits specified in the following tabulation shall be remedied or removed and replaced at the expense of and by the contractor.

G-71. *Preparations for Placing.*—(a) *General.*—No concrete shall be placed until all formwork, installation of parts to be embedded, and preparation of surfaces involved in the placing have been approved. No concrete shall be placed in water except with the written permission of the contracting authority, and the method of depositing the concrete shall be subject to his approval. Concrete shall not be placed in running water and shall not be subjected to the action of running water until after the concrete has hardened. All surfaces of forms and embedded materials that have become encrusted with dried mortar or grout from concrete previously placed shall be cleaned of all such mortar or grout before the surrounding or adjacent concrete is placed.

(b) *Foundation Surfaces.*—Immediately before placing concrete, all surfaces of foundations upon or against which the concrete is to be placed shall be free from standing water, mud, and debris. All surfaces of rock upon or against which the concrete is to be placed shall, in addition to the foregoing requirements, be clean and free from oil, objectionable coatings, and loose, semi-detached, or unsound fragments. Earth foundations shall be free from frost or ice when concrete is placed upon or against them. The surfaces of absorptive foundations against which concrete is to be placed shall be moistened thoroughly so that moisture will not be drawn from the freshly placed concrete.

(c) *Surfaces of Construction and Contraction Joints.*—Concrete surfaces upon or against which concrete is to be placed, and to which new concrete is to adhere, that have become so rigid that the new concrete cannot be incorporated integrally with that previously placed, are defined as construction joints. The surfaces of construction joints shall be clean, rough, and surface dry when covered with fresh concrete or mortar. Cleaning shall consist of the removal of all laitance, loose or defective concrete, coatings, sand, sealing compound if used, and other foreign material.

The surfaces of construction joints shall be wet-sandblasted, washed thoroughly with air-water jets, and surface dried prior to placement of adjoining concrete. The sandblasting, washing, and surface drying shall be performed at the last opportunity prior to placing of concrete. Drying of the surface shall be complete and may be accomplished by air jet. The surfaces of all contraction joints shall be cleaned thoroughly of accretions of concrete or other foreign material by scraping, chipping,

Tolerances in Dams and Appurtenant Works

ALL STRUCTURES	
Variation of the constructed linear outline from established position in plan.	In 20 feet ½ inch. In 40 feet ¾ inch.
Variation of dimensions to individual structure features from established positions.	In 80 feet or more 1¼ inches. In buried construction, twice the above amounts.
Variation from the plumb, from the specified batter, or from the curved surfaces of all structures, including the lines and surfaces of columns, walls, piers, buttresses, arch sections, vertical joint grooves, and visible arrises.	In 10 feet ½ inch. In 20 feet ¾ inch. In 40 feet or more 1¼ inches. In buried construction, twice the above amounts.
Variation from the level or from the grades indicated on the drawings in slabs, beams, soffits, horizontal joint grooves, and visible arrises.	In 10 feet ¼ inch. In 30 feet or more ½ inch. In buried construction, twice the above amounts.
Variation in cross-sectional dimensions of columns, beams, buttresses, piers, and similar members.	Minus ¼ inch. Plus ½ inch.
Variation in the thickness of slabs, walls, arch sections, and similar members.	Minus ¼ inch. Plus ½ inch.
FOOTINGS FOR COLUMNS, PIERS, WALLS, BUTTRESSES, AND SIMILAR MEMBERS	
Variation of dimensions in plan	Minus ½ inch. Plus 2 inches.
Misplacement of eccentricity	2 percent of footing width in the direction of misplacement but not more than 2 inches.
Reduction in thickness	5 percent of specified thickness.
SILLS AND SIDE WALLS FOR RADIAL GATES AND SIMILAR WATERTIGHT JOINTS	
Variation from the plumb and level	Not greater than a rate of ⅛ inch in 10 feet.

Tolerances for Concrete Tunnel Lining and Monolithic Conduits

Departure from established alinement or from established grade	Free-flow tunnels and conduits 1 inch. High-velocity tunnels and conduits ½ inch. Railroad tunnels 1 inch.
Variation in thickness, at any point	Tunnel lining Minus 0. Conduits Minus 2½ percent or ¼ inch, whichever is greater. Conduits Plus 5 percent or ½ inch, whichever is greater.
Variation from inside dimensions	½ of 1 percent.

Tolerances for Placing Reinforcement Steel

Variation of protective covering	With 2-inch cover ¼ inch. With 3-inch cover ½ inch.
Variation from indicated spacing	1 inch.

or other means approved by the contracting authority. (NOTE.—Wet-sandblasting of surfaces of construction joints is usually required for those structures or portions of structures where watertight construction joints are essential. Delete references to this requirement when inapplicable, or apply the requirement to applicable features only.)

G-72. Placing.—(a) *Transporting.*—The methods and equipment used for transporting concrete and the time that elapses during transportation shall be such as will not cause appreciable segregation of coarse aggregate, or slump loss in excess of 1 inch, in the concrete as it is delivered into the work.

(b) *Placing.*—The surfaces of all rock against which concrete is to be placed shall be cleaned and, except in those cases where seepage or other water precludes drying of the rock face, shall be dampened and brought to a surface-dry condition. Except for tunnels, surfaces of highly porous or absorptive horizontal or nearly horizontal rock foundations to which concrete is to be bonded shall be covered with a layer of mortar approximately three-eighths of an inch thick prior to placement of the concrete. The mortar shall have the same proportions of water, air-entraining agent, cement, and sand as the regular concrete mixture, unless otherwise directed. The water-cement ratio of the mortar in place shall not exceed that of the concrete to be placed upon it, and the consistency of the mortar shall be suitable for placing and working in the manner hereinafter specified. The mortar shall be spread uniformly, and shall be worked thoroughly in all irregularities of the surface. Concrete shall be placed immediately upon the fresh mortar.

A mortar layer shall not be used on concrete construction joints. Unless otherwise directed in formed work, structural concrete placements shall be started with an oversanded mix containing 3/4-inch-maximum-size aggregate; a maximum net water-cement ratio of 0.47, by weight; 6 percent air, by volume of concrete; and having a maximum slump of 4 inches; placed several inches deep on the joint at the bottom of the placement.

Retempering of concrete will not be permitted. Any concrete which has become so stiff that proper placing cannot be assured shall be wasted. Concrete shall be deposited in all cases, as nearly as practicable, directly in its final position and shall not be caused to flow such that the lateral movement will permit or cause segregation of the coarse aggregate, mortar, or water from the concrete mass. Methods and equipment employed in depositing concrete in forms shall be such as will not result in clusters or groups of coarse aggregate particles being separated from the concrete mass, but if clusters do occur, they shall be scattered before the concrete is vibrated. A few scattered individual pieces of coarse aggregate that can be restored into the mass by vibration will not be objectionable.

Concrete in tunnel lining may be placed by pumping or any other approved method. Where the concrete in the invert is placed separately from the concrete in the arch and without inside forms, it shall not be placed by pneumatic placing equipment unless an approved type of discharge box which prevents segregation is provided and used. The equipment used in placing the concrete, and the method of its operation, shall be such as will permit introduction of the concrete into the forms without high-velocity discharge, and resultant separation. After the concrete has been built up over the arch at the start of a placement, the end of the discharge line shall be kept well buried in the concrete during placement of the arch and sidewalls to assure complete filling. The end of the discharge line shall be marked so as to indicate the depth of burial at any time. Special care shall be taken to force concrete into all irregularities in the rock surfaces and to completely fill the tunnel arch.

Placing equipment shall be operated by experienced operators only. Cold joints in tunnel lining shall be avoided where practicable. In the event of equipment breakdown, or if for any other reason continuous placing will be interrupted, the contractor shall thoroughly consolidate the concrete at such joints to a reasonably uniform and stable slope while the concrete is plastic. The concrete at the surface of such cold joints shall be cleaned and shall be surface dry as required for construction joints before being covered with fresh concrete.

Except as intercepted by joints, all formed concrete, except concrete in tunnel lining, shall be placed in continuous approximately horizontal layers, the depths of which generally shall not exceed 20 inches. Lesser depths of layers will be required where concrete in 20-inch layers cannot be placed in accordance with the requirements of these specifications. All intersections of construction joints with concrete surfaces which will be exposed to view, shall be made straight and level or plumb.

In placing unformed concrete on slopes so steep as to make internal vibration of the concrete impracticable without forming, the concrete shall be placed ahead of a nonvibrated slip-form screed extending approximately 2½ feet back from its leading edge. Concrete ahead of the slip-form screed shall be consolidated by internal vibrators so as to insure complete filling under the slip-form.

(c) *Consolidation.*—Concrete shall be consolidated to the maximum practicable density, so that it is free from pockets of coarse aggregate and entrapped air, and closes snugly against all surfaces of forms and embedded materials. Consolidation of concrete in structures and in tunnel lining invert shall be by electric- or pneumatic-driven, immersion-type vibrators. Consolidation of concrete in the side walls and arch of tunnel lining shall be by electric- or pneumatic-driven form vibrators, supplemented where practicable by immersion-type vibrators. Concrete vibrators shall be operated at speeds of at least 7,000 revolutions per minute when immersed in the concrete. Form vibrators shall be rigidly attached to the forms and shall operate at speeds of at least 8,000 revolutions per minute when vibrating concrete. In consolidating each layer of concrete, the vibrator shall be operated in a near vertical position and the vibrating head shall be allowed to penetrate and revibrate the concrete in the upper portion of the underlying layer. Layers of concrete shall not be placed until the layers previously placed have been worked thoroughly as specified. Care shall be exercised to avoid contact of the vibrating head with surfaces of the forms.

G-73. *Finishes and Finishing.*—(a) *General.*—Allowable deviations from plumb or level and from the alinement, profile grades, and dimensions shown on the drawings, as specified for tolerances for concrete construction, are defined as "tolerances," and are to be distinguished from irregularities in finish as described herein. The classes of finish and the requirements for finishing of concrete surfaces shall be as herein specified or as indicated on the drawings. Finishing of concrete surfaces shall be performed only by skilled workmen. Concrete surfaces will be tested by a representative of the contracting authority where necessary to determine whether surface irregularities are within the limits hereinafter specified.

Surface irregularities are classified as "abrupt" or "gradual." Offsets caused by displaced or misplaced form sheathing or lining of form sections, or by loose knots in forms or otherwise defective form lumber, will be considered as abrupt irregularities, and will be tested by direct measurements. All other irregularities will be considered as gradual irregularities, and will be tested by use of a template consisting of a straightedge or the equivalent thereof for curved surfaces. The length of the template will be 5 feet for testing of formed surfaces and 10 feet for testing of unformed surfaces.

(b) *Formed Surfaces.*—The classes of finish for formed concrete surfaces are designated by use of symbols F1 and F2. No sack rubbing or sandblasting will be required on formed surfaces. No grinding will be required on formed surfaces, other than that necessary for repair of surface imperfections. Unless otherwise specified or indicated on the drawings, the classes of finish shall apply as follows:

F1.—Finish F1 applies to formed surfaces upon or against which fill material or concrete is to be placed, and to that portion of the upstream face of a concrete dam that will be covered by water during the greater part of the life of the dam. The surfaces require no treatment after form removal except for repair of defective concrete and filling of holes left by the removal of fasteners from the ends of tie rods, as required in section G-74, (Repair of Concrete), and the specified curing. Correction of surface irregularities will be required for depressions only, and only for those which exceed 1 inch.

F2.—Finish F2 applies to all formed surfaces not permanently concealed by fill material or concrete, or not required to receive finish F1, such as inside surfaces of tunnel linings; many of the structures that are appurtenant to earthfill dams, including surfaces of outlet works and open spillways; galleries and tunnels in dams; and concrete dams. Surface irregularities shall not exceed one-fourth of an inch for abrupt irregularities and one-half of an inch for gradual irregularities. (NOTE.—If the work involved includes any surfaces of structures which are considered to be of special importance, such as those that will be prominently exposed to public inspection, or surfaces for which accurate alinement and evenness of surface are considered of paramount importance from the standpoint of eliminating destructive effects of water action, the allowable irregularities should be reduced [2]. Also, when flow velocities downstream from the gates of an outlet works will be in excess of 40 feet per second, abrupt irregularities on inside formed surfaces downstream from the gates should, for a predetermined distance downstream from the gates (depending on the flow velocities involved), be completely eliminated by grinding to a specified bevel, ratio of height to length; and for all other flow surfaces subject to such flow velocities, abrupt irregularities should be reduced as needed to prevent cavitation problems [2].)

(c) *Unformed Surfaces.*—The classes of finish for unformed concrete surfaces are designated by the symbols U1 and U2. Interior surfaces shall be sloped for drainage where shown on the drawings or directed. Surfaces which will be exposed to the weather and which would normally be level shall be sloped for drainage. Unless the use of other slopes or level surfaces is indicated on the drawings or directed, narrow surfaces, such as tops of walls, should be sloped approximately three-eighths of an inch per foot of width; broader surfaces, such as platforms and decks, shall be sloped approximately one-fourth of an inch per foot. Unless otherwise specified or indicated on the drawings, these classes of finish shall apply as follows:

U1.—Finish U1 (screeded finish) applies to unformed surfaces that will be covered by fill material or by concrete. Finish U1 is also used as the first stage of finish U2. Finishing operations shall consist of sufficient leveling and screeding to produce even, uniform surfaces. Surface irregularities shall not exceed three-eighths of an inch.

U2.—Finish U2 (floated finish) applies to unformed surfaces not permanently concealed by fill material or concrete, such as the inverts of tunnels; floors of spillways, outlet works, and stilling basins; floors of service tunnels, and temporary diversion conduits; and tops of walls. Floating may be performed by use of hand or power-driven equipment. Floating shall be started as soon as the screeded surface has stiffened sufficiently, and shall be the minimum necessary to produce a surface that is free from screed marks and is uniform in texture. Surface irregularities shall not exceed one-fourth of an inch. Joints and edges shall be tooled where shown on the drawings, or directed. (NOTE.—If steel troweling of a surface is necessary, finishes U1 and U2 will be performed as the first stages for the steel-troweled finish, the steel troweling to be performed immediately after the floated surface has hardened sufficiently to prevent excess of fine material from being drawn to the surface; the troweling to be performed with firm pressure such as will flatten the sandy texture of the floated surface and produce a dense, uniform surface free from blemishes and trowel marks [2]. When flow velocities downstream from the gates of an outlet works will be in excess of 40 feet per second, irregularities on inside unformed surfaces downstream from the gates, such as may occur at construction joints or elsewhere, should, for a predetermined distance downstream from the gates (depending on the flow velocities involved), be completely eliminated by grinding to a specified bevel, ratio of height to length;

and for all other flow surfaces subject to such flow velocities, the irregularities should be reduced as needed to prevent cavitation problems [2].)

G-74. *Repair of Concrete.*—(NOTE.—For further information concerning repair of concrete refer to reference [7] in the bibliography, section G-102.) Repair of concrete shall be performed by skilled workmen. The contractor shall correct all imperfections on the concrete surfaces as necessary to produce surfaces that conform to the requirements specified for finishes and finishing. Unless otherwise approved, repair of imperfections in formed concrete shall be completed within 24 hours after removal of forms. Fins and encrustations shall be neatly removed from surfaces for which finish F2 is specified, and encrustations shall be removed from surfaces for which finish U2 is specified.

Concrete that is damaged from any cause, concrete that is honeycombed, fractured, or otherwise defective, and concrete which, because of excessive surface depressions, must be excavated and built up to bring the surface to the prescribed lines, shall be removed and replaced with dry pack, mortar, or concrete, as hereinafter specified. If removal of the ends of form ties results in recesses larger than one-fourth of an inch in diameter or in the minimum dimension, the recesses shall be filled with dry pack: *Provided,* That filling of recesses in surfaces designated to receive finish F1 will be required only where the surfaces are required to be coated with dampproofing, and where the recesses are deeper than 1 inch in walls less than 12 inches thick.

Where bulges and abrupt irregularities protrude outside the limits specified for finishes and finishing on formed surfaces for which finish F2 is required, the protrusions shall be reduced by bushhammering and grinding so that the surfaces are within the specified limits. Dry pack shall be used for filling holes having at least one surface dimension little, if any, greater than the hole depth, for narrow slots cut for repair of cracks for grout pipe recesses, and for tie-rod fastener recesses as specified. Dry pack shall not be used for filling behind reinforcement or for filling holes that extend completely through a concrete section. Mortar filling, placed under impact by use of a mortar gun, may be used for repairing defects on surfaces designated to receive F1 and F2 finishes where the defects are too wide for dry pack filling, too shallow for concrete filling, and no deeper than the far side of the reinforcement that is nearest the surface. Concrete filling shall be used for holes extending entirely through concrete sections; for holes in which no reinforcement is encountered and which are greater in area than 1 square foot, and deeper than 4 inches; and for holes in reinforced concrete which are greater in area than one-half square foot and which extend beyond reinforcement.

All materials used in the repair of concrete shall conform to the requirements of these specifications and the repairs shall be made in accordance with the procedures of the Bureau of Reclamation's "Concrete Manual" [3]. All fillings shall be bonded tightly to the surfaces of the holes and shall be sound and free from shrinkage cracks and drummy areas after the fillings have been cured and have dried.

The cost of all materials, labor, and equipment required in the repair of concrete shall be borne by the contractor.

G-75. *Protection.*—The contractor shall protect all concrete against injury until final acceptance by the contracting authority. Immediately following the first frost in the fall, the contractor shall be prepared to protect all concrete against freezing. After the first frost, and until the mean daily temperature in the vicinity of the work site falls below 40° F. for more than 1 day, the concrete shall be protected against freezing temperatures for not less than 48 hours after it is placed. After the mean daily temperature in the vicinity of the worksite falls below 40° F. for more than 1 day, the concrete shall be maintained at a temperature not lower than 50° F. for at least 72 hours after it is placed.

Concrete cured by membrane curing will require no additional protection from freezing if the protection at 50° F. for 72 hours is obtained by means of approved insulation in contact with the forms or concrete surfaces; otherwise, the concrete shall be protected against freezing temperatures for 72 hours immediately following the 72 hours of protection at 50° F. Concrete cured by water curing shall be protected against freezing temperatures for 3 days immediately following the 72 hours of protection at 50° F.

Discontinuance of protection against freezing temperatures shall be such that the drop in temperature of any portion of the concrete will be gradual and will not exceed 40° F. in 24 hours. After March 15, when the mean daily temperature rises above 40° F. for more than three successive days, the specified 72-hour protection at a temperature not lower than 50° F. may be discontinued for as long as the mean daily temperature remains above 40° F.: *Provided,* That the concrete shall be protected against freezing temperatures for not less than 48 hours after placement. Where artificial heat is employed, special care shall be taken to prevent the concrete from drying.

Use of unvented heaters will be permitted only when unformed surfaces of concrete adjacent to the heaters are protected for the first 24 hours from an excessive carbon-dioxide atmosphere by application of sealing compound: *Provided,* That the use of sealing compound on such surfaces for curing of the concrete is permitted by and the compound is applied in accordance with section G-76.

G-76. Curing.—(a) *General.*—Concrete shall be cured either by water curing in accordance with subsection (b), or by membrane curing in accordance with subsection (c), except as otherwise hereinafter provided. The unformed top surfaces of walls and piers shall be moistened by covering with water-saturated material or by other effective means as soon as the concrete has hardened sufficiently to prevent damage by water. These surfaces and steeply sloping and vertical-formed surfaces shall be kept completely and continually moist, prior to and during form removal, by water applied on the unformed top surfaces and allowed to pass down between the forms and the formed concrete faces. This procedure shall be followed by the specified water curing or membrane curing.

(b) *Water Curing.*—Concrete cured with water shall be kept wet for at least 14 days immediately following placement of the concrete, or until covered with fresh concrete: *Provided,* That water curing of concrete may be reduced to 6 days during periods when the mean daily temperature in the vicinity of the work site is less than 40° F.: *Provided further,* That during the prescribed period of water curing when temperatures are such that concrete surfaces may freeze, water curing shall be temporarily discontinued. The concrete shall be kept wet by covering with water-saturated material or by a system of perforated pipes, mechanical sprinklers, porous hose, or by any other approved method which will keep all surfaces to be cured continuously (not periodically) wet. Water used for curing shall meet the requirements of these specifications for water used for mixing concrete.

(c) *Membrane Curing.*—Membrane curing shall be by application of a sealing compound which forms a water-retaining membrane on the surfaces of the concrete. The sealing compound shall be white-pigmented and shall conform to "Standard Specifications for Liquid Membrane Forming Compounds for Curing Concrete," ASTM Designation C 309. The compound shall be of uniform consistency and quality within each container and from shipment to shipment.

Sealing compounds shall be applied to the concrete surfaces by spraying in one coat to provide a continuous, uniform membrane over all areas. Coverage shall not exceed 150 square feet per gallon, and on rough surfaces coverage shall be decreased as necessary to obtain the required continuous membrane. The repair of surface imperfections shall not be made until after application of sealing compound.

When sealing compound is used on unformed concrete surfaces, application of the compound shall commence immediately after finishing operations are completed. When sealing compound is to be used on formed concrete surfaces, the surfaces shall be moistened with a light spray of water immediately after the forms are removed and shall be kept wet until the surfaces will not absorb more moisture. As soon as the surface film of moisture disappears, but while the surface still has a damp appearance, the sealing compound shall be applied. Special care shall be taken to insure ample coverage with the compound at edges, corners, and rough spots of formed surfaces. After application of the sealing compound has been completed, and the coating is dry to touch, any required repair of concrete surfaces shall be performed. Each repair, after being finished, shall be moistened and coated with sealing compound in accordance with the foregoing requirements.

Equipment for applying sealing compound and the method of application shall be in accordance with the provisions of the Bureau of Reclamation's "Concrete Manual" [2]. Traffic and other operations by the contractor shall be such as to avoid damage to coatings of sealing compound for a period of not less than 28 days. Where it is impossible because of construction operations to avoid traffic over surfaces coated with sealing compound, the membrane shall be protected by a covering of sand or earth not less than 1 inch in thickness or by other effective means. The protective covering shall not be placed until the sealing membrane is completely dry. Before final acceptance of the work, the contractor shall remove all sand or earth covering in an approved manner. Any sealing membrane that is damaged or that peels from concrete surfaces within 28 days after application shall be repaired without delay and in an approved manner.

Sealing compounds will be accepted on manufacturer's certification of compliance with specifications, but permission to ship on certification shall in no way relieve the contractor of the responsibility for furnishing compound meeting specifications requirements.

(d) *Costs.*—The costs of furnishing and applying all materials used for curing concrete shall be included in the price bid in the schedule for the concrete in which the curing materials are used.

G-77. *Measurement of Concrete.*—Measurement, for payment, of concrete required to be placed directly upon or against surfaces of excavation will be made to the lines for which payment for excavation is made.

Measurement, for payment, of concrete in tunnel lining will be made to the "B" lines shown on the drawings, with no deductions being made for rock projections inside the "B" lines or for timber, concrete and metal supports, struts, lagging, and blocking remaining within the prescribed "B" lines. No measurement for payment will be made for concrete required to be placed outside of the "B" lines due to overbreakage or excess excavation to facilitate the contractor's operations or for additional excavation to accommodate the concrete discharge line. (NOTE.—See section G-35 for definition and typical location, respectively, of "B" lines.)

Measurement, for payment, of all other concrete will be made to the neat lines of the structures, unless otherwise specifically shown on the drawings or prescribed in these specifications. In measuring concrete for payment, the volume of all openings, recesses, ducts, embedded pipes, woodwork, and metalwork, each of which is larger than 100 square inches in cross section, will be deducted.

G-78. *Payment for Concrete.*—Payment for concrete in the various parts of the work will be made at the applicable unit prices per cubic yard bid in the schedule, which unit prices shall include the cost of all labor and materials required in the construction, except that payment for furnishing and handling cement, and payment for furnishing and placing reinforcing bars *(and fabric) will be made at the applicable prices bid in the schedule.

G. MISCELLANEOUS

G-79. *Diversion and Care of River During Construction and Removal of Water from Foundations.* —(a) *General.*—The contractor shall construct and maintain all necessary cofferdams, channels, flumes, drains, sumps, and/or other temporary diversion and protective works; shall

*Revise or delete as appropriate.

furnish all materials required therefor; and shall furnish, install, maintain, and operate all necessary pumping and other equipment for removal of water from the various parts of the work and for maintaining the foundations and other parts of the work free from water. After having served their purpose, all cofferdams or other protective works downstream from the dam shall be removed from the river channel, or leveled to give a sightly appearance, so as not to interfere in any way with the operation or usefulness of the reservoir, and in a manner approved by the contracting authority. All cofferdams or other protective works constructed upstream from the dam and not a part of the permanent dam embankment shall be removed or leveled and graded to the extent required to prevent obstruction in any degree whatever of the flow of water to the spillway or outlet works. The contractor shall be responsible for and shall repair at his expense any damage to the foundations, structures, or any other part of the work caused by floods, water, or failure of any part of the diversion or protective works.

(b) *Plan.*—Prior to beginning any work on diversion and care of river and removal of water from foundations, the contractor shall submit for approval a water control plan showing his proposed method for the diversion and care of the river during construction and removal of water from foundations. The plan may be placed in operation upon approval, but nothing in this section shall relieve the contractor from full responsibility for the adequacy of the diversion and protective works.

The hydrographs of .... and discharge curves for the spillway and outlet works *(and for diversion) are shown on the drawings solely for the information of the contractor in timing his construction operations to prepare for such flood storage and/or to bypass such flow as may be necessary. The contracting authority does not guarantee the reliability or accuracy of any of these curves and assumes no responsibility for any deductions, interpretations, or conclusions which may be made from the curves.

*Revise or delete as appropriate.

(c) *Diversion and Care of River.*—Except as otherwise provided below, the contractor shall not interrupt nor interfere with the natural flow of .... through the damsite for any purpose without the written approval of the contracting authority. *(The contractor shall at all times, except during the closure period, pass the full natural flow of the stream through the damsite to meet prior rights for .... purposes.) *(Final closure shall be made between the following dates: ....)

(d) *Removal of Water from Foundations.*—The contractor's method of removal of water from foundation excavations shall be subject to the approval of the contracting authority. Where excavation for cutoff trenches in embankment foundations extends below the water table in common material, the portion below the water table shall be dewatered in advance of excavation. The dewatering shall be accomplished in a manner that will prevent loss of fines from the foundation, will maintain stability of the excavated slopes and bottom of the cutoff trench, and will result in all construction operations being performed in the dry. The use of a sufficient number of properly screened wells or other equivalent methods will be approved for dewatering. The contractor will also be required to control seepage along the bottom of the cutoff trench, which may require supplementing the approved dewatering systems by pipe drains leading to sumps from which the water shall be pumped. Such pipe drains shall be of uniform diameter for each run, shall be provided with grout connections and returns at 50-foot intervals, and shall be embedded in reasonably well-graded gravel or like material.

During the placing and compacting of the embankment material in a cutoff trench, the water level at every point in the cutoff trench shall be maintained below the bottom of the embankment until the compacted embankment

*Delete or revise as appropriate and add pertinent information on the following:

(1) Downstream requirements.

(2) What contractor will be permitted and will not be permitted to do.

(3) Work to be accomplished before permanent construction may be used for diversion purposes.

(4) Work to be accomplished before final closure can be made.

in the cutoff trench at that point has reached a depth of 10 feet, after which the water level shall be maintained at least 5 feet below the top of the compacted embankment. When the embankment has been constructed to an elevation which will permit the dewatering systems to maintain the water level at or below the designated elevations, as determined by the contracting authority, the pipe drains, including surrounding gravel, shall be filled with grout composed of water and cement or clay.

(e) *Cleanup and Reservoir Regulation.*—After completion of all *(second-stage) concrete in the spillway and outlet works, the stilling basins shall be unwatered and all sediment, rocks, and debris shall be removed from the tunnel, chutes, and stilling basins, and all drainage outlets shall be inspected and cleaned as necessary.

After placement of all *(second-stage) concrete and installation and testing of gates and other metalwork and equipment in the outlet works structures, the contracting authority will operate the gates and regulate the flow in the river and reserves the right to commence storage in the reservoir.

**(f) *Cost.*—The cost of furnishing all labor, equipment, and materials for constructing cofferdams, dikes, channels, flumes, and other diversion and protective works; removing or leveling such works, and disposing of materials, where required; diverting the river; making required closures; maintaining the work free from water as required; grouting drains and sumps; inspecting and cleaning the tunnel, chutes, stilling basins, and drainage outlets; and all other work required by this section shall be included in the prices bid in the schedule for items of construction work.

G-80. *Concrete or Cement-Bound Curtain.*—The contractor shall construct a mixed-in-place cement-bound curtain, or a concrete or cement-bound curtain constructed by other methods that will be equivalent to a mixed-in-place cement-bound curtain. At least 30 days before beginning any work on the concrete or cement-bound curtain, the contractor shall submit, for approval, plans and specifications for the work. The contractor shall furnish all materials and equipment required to construct the mixed-in-place or other type concrete or cement-bound curtain. The curtain shall have a minimum effective thickness of 12 inches and shall be constructed to the depths and elevations shown on the drawings or established by the contracting authority and as close as practicable to the established lines. The cement used in the curtain shall be portland cement conforming to the requirements of section G-60. Particular efforts should be directed toward maintaining a continuous, unbroken cutoff by bonding around the end of the previously placed "pile" whenever shutdown periods occur and construction is again resumed. The curtain, when completed, shall provide a reasonably impervious barrier to the passage of subsurface flows. Any royalties due because of the mixed-in-place cement-bound curtain or other type concrete or cement-bound curtain being placed by patented methods, or use of patented materials, shall be paid by the contractor.

Measurement, for payment, of mixed-in-place cement-bound curtain or other type concrete or cement-bound curtain will be made of the area of curtain placed, based on the depth and length of the curtain along the centerline of the curtain as shown on the drawings or established by the contracting authority. Payment for constructing concrete or cement-bound curtain will be made at the unit price per square yard bid therefor in the schedule, which unit price shall include the cost of all labor, materials and equipment required to complete the cutoff curtain, regardless of the type of curtain constructed: *Provided,* That payment for furnishing and handling cement will be made at the unit price per barrel bid therefor in the schedule.

G-81. *Steel Sheet Piling.*—Steel sheet piling shall be driven in locations as shown on the drawings or as directed by the contracting authority. The steel sheet piling, including taper piles and special intersections, shall be furnished by the contractor. The piling shall be driven to the depths shown on the drawings

*Revise or delete as appropriate.

**Modify to provide a lump sum item if this is a substantial part of work.

or prescribed by the contracting authority, with the top of each pile at the elevation shown on the drawings or established by the contracting authority. Where required, the tops of piling shall be cut off to the required elevation. The piling shall be driven as close as practicable to the lines shown on the drawings or established by the contracting authority, so as to ensure perfect interlocking throughout the entire length of each pile. The method of driving shall be subject to the approval of the contracting authority.

Assembling frames of capped wooden piles or other suitable temporary guide structures furnished by the contractor shall be used, and the steel piling shall be assembled against the guides so that each pile is plumb at both edge and side. Such temporary guide structures shall be removed by the contractor when they have served their purpose. A cast-steel driving head shall be used for driving steel sheet piling. Piles ruptured in the interlock or otherwise injured in driving shall be pulled and new piles driven. Should boulders be encountered, the contractor shall make every effort to drive the piling to the required depth, either by moving or shattering the boulder or by deviations in the line of the piling. If at any time the forward edge of the steel piling wall is found to be out of plumb, the piling already assembled and partly driven shall be driven to the required depth, and taper piles shall then be used to bring the forward edge plumb before additional piling is assembled or driven. The maximum permissible taper in a single pile shall be one-fourth inch per foot of length. Where welding of piles is required for field connections, the welding shall be by approved methods.

The steel sheet piling furnished by the contractor shall be in accordance with the following requirements:

(1) *Inspections and tests.*—The steel piling shall be manufactured from steel conforming to ASTM Designation A 328 and shall be equal to United States Steel Standard Section MP-115 (nominal weight 22 pounds per square foot). All material furnished and all work done shall be subject to rigid inspection and no material shall be shipped until all tests, analyses, and final inspection have been made, or certified copies of reports of tests and analyses or manufacturer's guarantees shall have been accepted. If test specimens are required for analyses they shall be properly boxed and prepared for shipment.

(2) *Type and general description of piling.*—The piling shall have a continuous interlock rolled integral with the pile throughout its entire length. The interlock shall permit an angular movement between adjoining piles of not less than 10° in either direction from the centerline. In order to withstand difficult driving requirements, the steel piling shall be hard and tough to resist battering and shock, and shall also be free from objectionable brittleness.

Measurement, for payment, of steel sheet piling will be made of the area remaining in place and on the basis of 22 pounds per square foot. Payment for furnishing and driving steel sheet piling will be made at the unit price per pound bid therefor in the schedule, which unit price shall include the cost of driving the piling, welding of field connections, cutting the tops, and pulling and replacing unsatisfactory piling as described in this section.

G-82. *Consolidation of Dam Foundation by Flooding.*—Consolidation of the dam foundation by flooding will be required in the embankment area as outlined on the drawings. The consolidation of the dam foundation will be accomplished by progressively flooding separate sections of the foundation. The flooding shall be such that complete saturation of the foundation will be accomplished and a continuous water cover to a minimum depth of 12 inches shall be maintained. Before flooding is started, the excavation for the dam foundation shall be completed as provided in section G-13. At such time as is approved by the contracting authority, the contractor shall begin consolidation of the foundation by flooding with water. The flooding operations shall be continued until the desired penetration and saturation has been attained: *Provided,* That the contractor will not be required to provide plant capacity sufficient to supply water in excess of ____ gallons per day: *Provided further,* That the contractor will not be required to maintain continuous flooding of any individual section to a minimum depth of 12 inches for longer than 60 days. In the event of plant failure, the maximum time limit of 60 days for any section will be extended by the same number of days as pumping is interrupted by plant failure, if the desired penetration and saturation has not been attained prior to that time. Additional pumping, if required in any section beyond the 60-day period, except where extension of the 60-day period is due to failure of plant, will be ordered in writing by the contracting authority and payment thereof will be made as extra work.

The quantity stated in the schedule for water for consolidation of dam foundation is an estimate only of the amount to be required, and the contractor shall be entitled to no additional compensation above the unit prices bid in the schedule by reason of any amount or none being required. The contractor shall be responsible for, and shall repair at his own expense, any damage to the dam foundation or any part of the work caused by excessive pumping or failure of pipelines or temporary dikes. No payment will be made for water lost due to failure of any part of the contractor's water-supply system or dikes. The consolidation of the dam foundation shall be scheduled so that as short a time as practicable will elapse between the time the consolidation is completed and the earth embankment is placed. Measurement, for payment, of water for consolidation of dam foundation will be made by metering the water near the point or points of discharge into the dam foundation. Meters furnished and installed by the contractor shall be tested for accuracy prior to use. Payment for water for consolidation of dam foundation will be made at the unit prices per million gallons bid therefor in the schedule, which unit prices shall include the cost of all water and the cost of all labor, materials, and operations required for continuously flooding the dam foundation as described in this section, including the cost of constructing, maintaining, and removing the necessary temporary dikes: *Provided,* That the cost of all plants shall be included in the unit price bid in the schedule for water for consolidation of dam foundation, less than ____ gallons.

G-83. *Topsoil for Seeding.*—The item of the schedule for topsoil for seeding consists of loading, hauling, placing, spreading, and rolling selected topsoil material. The topsoil shall be placed on the downstream slope of the dam embankment at locations shown on the drawings or designated by the contracting authority. All operations involved in the placing, spreading, and rolling of the topsoil shall be subject to the approval of the contracting authority. Selected topsoil shall be obtained from approved stockpiles of materials from excavation for dam embankment foundation or from stripping from borrow areas, or from other approved sources. The material shall contain the most fertile loam available from approved sources and shall be free from excessive quantities of grass, roots, weeds, sticks, stones, or other objectionable materials. Areas to receive topsoil shall be brought to within 1 foot of the prescribed final cross section at all points and finished smooth and uniform before the topsoil is applied. Topsoil shall be evenly placed and spread over the graded area and compacted in two layers, each by one pass of a roller weighing not less than 50 pounds per linear inch of length of drum. Topsoil shall not be placed when the subgrade is frozen or in a condition otherwise detrimental to proper grading and seeding as determined by the contracting authority.

Measurement, for payment, for topsoil will be by volume, compacted in place within the lines shown on the drawings or as established by the contracting authority. Payment for topsoil for seeding will be made at the unit price per cubic yard bid therefor in the schedule, which unit price shall include the cost of loading, hauling, placing, spreading, and rolling the topsoil, and shall also include the cost of excavating additional suitable topsoil material if sufficient quantity of such material is not obtained from approved stockpiles.

G-84. *Water for Seeded Areas.*—It is expected that it will be necessary to irrigate the seeded areas in preparing the seedbed and to promote germination and growth of the plants. The frequency of application and quantities of irrigation water used will be determined by the contracting authority. The contractor shall provide a temporary sprinkler irrigation system of pipelines or mobile water tanks to provide for uniform application of water over the entire seeded areas and complete control of the amount of water at all times to eliminate erosion. The contractor shall repair at his own expense any damage to the slopes or any part of the work caused by excessive or irregular application of irrigation water. The quantity stated in the schedule for water for seeded areas is an estimate only of the amount to be required, and the contractor shall be entitled to no additional compensation above the unit price bid in the schedule by reason of any amount or none being required.

Measurement, for payment, of water for seeded areas will be made by metering the delivery pipeline, using meters furnished and installed by the contractor, or by tank gallonage delivered through the sprinkler heads and applied to the seeded areas. Payment for water for seeded areas will be made at the unit price per thousand gallons bid therefor in the schedule, which unit price shall include the cost of all labor, materials, plant, and operations required for sprinkler irrigation of the seeded areas as described in this section, including the cost of removing the pipelines, if used.

G-85. Seeding.—(a) *General.*—Seeding shall consist of ground preparation, furnishing and planting approved seed, furnishing and placing mulch, and furnishing and spreading approved commercial fertilizer. Except for placing, spreading, and rolling topsoil, all seeding operations shall be performed in accordance with provisions of this section.

The areas to be seeded are as follows:

(list specific areas)

All excavation surfaces composed of rock, and excavated slopes with a slope of 1:1 or steeper, shall be excluded from the seeded areas.

The limits to which the above described surfaces are to be seeded shall be as prescribed by the contracting authority.

If directed, areas within the right-of-way lines shall be seeded where construction operations have removed the existing grass cover, where such areas have otherwise been denuded of grass cover, or where due to other reasons grass seeding is determined to be necessary.

The contractor shall maintain the seeded areas until final acceptance thereof and any damage caused to the seeded areas by the contractor's operations shall be repaired by and at the expense of the contractor.

(b) *Seedbed Preparation.*—

(1) *Where topsoil for seeding is required.*—The contractor shall place, spread, and roll topsoil over the areas to be seeded in accordance with section G-83. The surfaces of the topsoil shall be maintained smooth and even until they are seeded. If these surfaces are disturbed and loosened by the contractor's operations, he shall smooth and reroll them before starting the seeding operations, and the costs thereof shall be at the contractor's expense.

(2) *Where topsoil for seeding is not required.*—The contractor shall prepare the areas to be seeded to provide a firm, well-packed condition suitable for establishing grass stands just prior to sowing the seed.

Where required, the contractor shall scarify the ground to prepare a proper seedbed. No separate payment will be made for either compacting or scarifying areas to be seeded, and all cost of seedbed preparation shall be included in the unit price per acre bid in the schedule for seeding.

(c) *Seeds.*—Seed and seeding mixtures shall be free of all prohibited noxious weed seeds and shall not contain more than 0.5 percent by weight of restricted noxious weed seeds. Prohibited and restricted noxious weeds shall be those as classified by the State seed department.

All seed containers must be sealed and labeled to comply with existing__________ (insert State) seed laws and regulations or in accordance with U.S. Department of Agriculture rules and regulations under the Federal Seed Act, if shipped in interstate commerce. All different grass species specified shall be separately packaged and labeled and shall be uniformly and thoroughly mixed after they are received on the job.

The contractor shall furnish and sow a uniform seed mixture composed of the seeds listed in the following table:

Seed required for 1 acre [1]

Kind of seed	Pounds pure live seed/acre
Creeping red fescue	11
Chewing fescue	8
Newport blue grass	5
Perennial rye grass	3
Dutch white clover	3

[1] Sample table, revise as appropriate.

(d) *Sowing Periods.*—Seeding shall be done at such times of the year when climatic conditions of temperature and moisture are most adaptable for growth. Unless otherwise approved, seeding shall not be performed between ____________ and ____________ of each year.

(e) *Mulching.*—The contractor shall furnish and uniformly place, after seeding, a minimum of 2 tons per acre of hay or straw mulch on all seeded areas or, as an alternate, the contractor may hydromulch in accordance with subsection (h). Mulch material shall be reasonably free of mold or other evidence of decomposition and weed seed. The major portion of the mulch material in place shall exceed 12 inches in length. The mulch material shall be firmly anchored with an "IMCO" treader or equivalent. The treader shall be operated at a depth of 3 to 4 inches and crosswise to all slopes. Mulch will be anchored at 6- to 12-inch intervals across the slope. The cost of mulching shall be included in the unit price bid in the schedule for seeding.

(f) *Sowing Seed.*—

(1) *Method.*—The mixture specified herein shall be sown by drilling with either an approved disc or shoe-type grass drill, an approved hydroseeder, or by mechanical or hand broadcasting.

(2) *Drill seeding.*—If the drill seed method is used, the drill shall be regulated to uniformly distribute the seed at the rate specified herein on the areas to be seeded. Where possible to safely operate equipment, as determined by the contracting authority, drilling shall be done crosswise to the general slope. The drill shall be regulated so that the seed is properly placed in the soil and covered with soil to a depth of ___ to ___ inches. In the event the drill is equipped with an approved fertilizer attachment which will uniformly distribute fertilizer, up to ___ pounds of the nitrogen and ___ pounds of the phosphorous may be applied simultaneously with the drilling of the seed. The balance of the fertilizer shall be broadcast or drilled in prior to or after seeding of grass.

(3) *Hydroseeding.*—Seeding with an approved hydroseeder will be acceptable, provided wind velocities permit uniform distribution of the seed and fertilizer slurry on the areas to be seeded. In hydroseeding operations, the mixture of seed and the fertilizer specified herein shall be properly mixed with water to form a slurry. The slurry mixture shall be prepared immediately prior to application, and shall be promptly applied on the areas to be seeded and fertilized. Slurry mixtures prepared more than 1 hour prior to application are not acceptable.

The hydroseeder shall be designed so as to insure seed and fertilizer being uniformly applied at the rates per acre specified herein. The hydroseeder shall be equipped with a paddle-type agitator and recirculation pump that will continually stir and mix the slurry to prevent settling of solids in corners and at the bottom of the tank, and maintain a uniform mixture of seed, fertilizer, and water at all times during the entire seeding operation.

Immediately after the slurry mixture is applied to the soil surface, the seed shall be properly covered with soil to the depths prescribed above by means of rotary hoe, Dunham packer, spike tooth harrow, or other acceptable implements as approved by the contracting authority. Covering seed by dragging a log chain or similar device will not be permitted. Seed applied around structures shall be properly covered with soil with a hand rake or float.

(4) *Broadcast seeding.*—Broadcast seeding with either a mechanical broadcaster or by hand is acceptable only on areas inaccessible to the basic method employed, which may be either drill seeding or hydroseeding.

a. *Mechanical broadcasting.*—A mechanical broadcaster of either the centrifugal type or pull type similar to fertilizer spreaders is acceptable. Any equipment of this type used for broadcast seeding shall be designed and regulated to insure that the proper seeding rate per acre specified herein is uniformly applied on areas to be seeded. When this method is used, seed and fertilizer may not be applied in the same mixture simultaneously, but each shall be broadcast separately.

b. *Hand broadcasting.*—Hand broadcasting may be performed only on small, inaccessible areas as approved by the contracting authority. Seed application may be performed by using an approved hand broadcaster or by broadcasting the seed by hand from a sack or other suitable container. Whichever means is used, the seed shall be uniformly applied at the rates specified herein. In the employment of this method, both the seed and fertilizer shall be broadcast separately.

Immediately after broadcasting the seed by either mechanical or hand methods, it shall be properly covered with soil to the depths prescribed above by means of a spike tooth harrow, rotary hoe, Dunham packer, or other acceptable implements. Covering broadcast seed by dragging a log chain or similar device will not be permitted. Seed broadcast around structures shall be covered with soil with a hand rake or float.

(g) *(*Fertilizer.*—The contractor shall furnish and apply commercial fertilizer at the minimum rate per acre of ____ pounds of actual nitrogen and ____ pounds of actual phosphorus. The fertilizer shall be packaged and labeled showing the guaranteed analysis and net weight per bag when received on the job.

Fertilizer may be applied prior to seeding by suitable mechanical spreaders, blowers, or hydraulic equipment. Fertilizer may also be applied when seeding as specified in subsections (f) (2) and (3) above.)

(h) *Hydromulching.*—As an alternate to other methods of seeding, fertilizing, and mulching herein specified, application by an approved hydromulcher will be acceptable. Wood cellulose fiber mulch or other approved material to be used in hydromulching shall contain no germination or growth-inhibiting factors. It shall be dyed an appropriate color to allow visual metering of its application. The mulching material shall have the property of becoming evenly dispersed and suspended when agitated in water. When sprayed uniformly on the surface of the soil, the fibers shall form a blotter-like ground cover which readily absorbs water and allows infiltration to the underlying soil. Weight specifications from the suppliers, and for all applications, shall refer only to air dry weight of the fiber, a standard equivalent to 10 percent moisture. The mulch material shall be supplied in packages marked by the manufacturer to show the air dry weight content. Suppliers shall be prepared to certify that laboratory and field testing of their product has been accomplished and that it meets all the foregoing requirements.

The application by hydromulcher shall be done in the following manner: the mulching material shall be added to the water slurry in the hydraulic seeder after the proportionate quantities of grass seed and fertilizer materials. All slurry ingredients shall be mixed to form a homogeneous slurry. Slurry mixtures prepared more than 1 hour prior to application are not acceptable. Using the color of the mulch as a metering agent, the operator of the hydraulic seeder shall spray-apply the slurry mixture uniformly to the prepared seedbed to correspond with per acre requirements of all materials. Wood cellulose fiber mulch shall be applied at a minimum rate of 1,000 pounds per acre. Other approved mulching material shall be applied at an equivalent effective rate per acre. The contracting authority will verify, by inspections of tank loading and spray ap-

*Revise or delete as appropriate.

plication, that materials applied correspond with per acre requirements.

(i) *Measurement and Payment.*—Measurement, for payment, of seeding will be made along the surface of the areas actually seeded as directed. Payment for seeding will be made at the unit price per acre bid therefor in the schedule, which price shall include the cost of all labor, materials, and equipment required for preparation of seedbed (except for placing, spreading, and rolling topsoil), the cost of furnishing and spreading fertilizer, the cost of furnishing, sowing, and covering the seed, and the cost of furnishing and placing the mulch.

G-86. *Joints, Grooves, and Edges in Concrete.*— (a) *Construction Joints.*—Construction joints are joints which are either purposely placed in structures to facilitate construction or which occur in structures as a result of inadvertent delays in concrete placing operations. Construction joints are located to facilitate the contractor's operations, to reduce initial shrinkage stresses and cracks, to allow time for the installation of embedded metalwork, or to allow for the subsequent placing of other concrete, blockout concrete, or second-stage concrete. Bond is required at construction joints regardless of whether or not reinforcement is continuous across the joint.

The location of all construction joints in concrete work shall be subject to approval, and the joints shall be constructed in accordance with sections G-71 and G-72.

(b) *Contraction Joints.*—Contraction joints are joints placed in structures or slabs to provide for volumetric shrinkage of a monolithic unit or movement between monolithic units. The joints shall be so constructed that there will be no bond between the concrete surfaces forming the joint.

Contraction joints of the types shown on the drawings shall be constructed at the locations shown. The joints shall be made by forming the concrete on one side of the joint and allowing it to set before concrete is placed on the other side of the joint. The surface of the concrete first placed at a contraction joint shall be coated with sealing compound before the concrete on the other side of the joint is placed. The sealing compound shall conform to the requirements of section G-76.

(c) *Grooves and Edges.*—Formed grooves shall be constructed as shown on the drawing. The formed grooves shall be formed neatly and accurately at the required locations.

Concrete edges shall be tooled as shown on the drawings or as required.

(d) *Cost.*—Except as otherwise provided for furnishing and placing joint materials, the entire cost of constructing joints, grooves, and edges in concrete shall be included in the unit prices bid in the schedule for items for concrete in which the joints, grooves, or edges are required.

G-87. *Rubber Waterstop.*— (a) *General.*—Rubber waterstop shall be furnished and placed (plastic waterstop in accordance with section G-88 may be furnished and placed at the contractor's option) in the joints at the locations shown on the drawings or where directed. Rubber waterstop shall be furnished in accordance with the details shown on Drawing No. .... The contractor shall also furnish all labor and materials for making field splices in rubber waterstop. The contractor shall take suitable precautions to support and protect the waterstop during the progress of the work and shall repair or replace any damaged waterstop.

Rubber waterstop may be shipped in rolls to facilitate handling, but if any roll of waterstop is not to be installed in a structure within 6 months after receipt of the material, the roll shall be loosened. All waterstop shall be stored in as cool a place as practicable, preferably at 70° F. or less. Water stop shall not be stored in the open or where it will be exposed to the direct rays of the sun. All waterstop shall be protected from oil or grease.

(b) *Materials.*—

(1) *Rubber waterstop.*—The rubber waterstop shall be fabricated from a high-grade, tread-type compound. The basic polymer shall be natural rubber or a synthetic rubber. The material shall be compounded and cured to have the following physical characteristics shown on page 748.

	Natural rubber	Synthetic rubber
Tensile strength, pounds per square inch, minimum	3,500	3,000
Tensile strength at 300-percent modulus, pounds per square inch, minimum	1,450	1,150
Elongation at break, percent, minimum	500	450
Shore durometer (type A)	60 to 70	60 to 70
Specific gravity	1.15 ±0.03	1.15 ±0.03
Absorption of water, by weight, percent maximum (2 days at 70° C.)	5	5
Compression set (constant deflection) percent of original deflection, maximum	30	30
Tensile strength after oxygen pressure test (48 hours, 70° C., 300 pounds per square inch), percent of tensile strength before aging, minimum	80	80
Elongation after oxygen pressure test (48 hours, 70° C., 300 pounds per square inch), percent of elongation before aging, minimum	80	80

(2) *Gum rubber and rubber cement.*—Gum rubber and rubber cement shall be suitable for making field connections in rubber waterstop as described in subsection (e) (2) below.

(c) *Fabrication.*—The rubber waterstop may be molded or extruded. All material shall be molded or extruded and cured in such a manner that any cross section will be dense, homogeneous, and free from porosity and other imperfections. The following minor surface defects will be acceptable:

(1) Lumps and depressions not exceeding ¼ inch in longest lateral dimensions and 1/16 inch deep with no limit to the frequency of occurrence.

(2) Lumps and depressions between ¼ and ½ inch in longest lateral dimension and 3/32 inch deep as long as the frequency of occurrence does not exceed six in a 50-foot length, and there are at least 2 inches between any two such defects.

(3) Marks resulting from the tubing operation or handling during manufacture, with no limit to width or frequency of occurrence as long as the thickness of material below the mark is not less than the minimum thickness.

(4) Coarse or grainy surface texture.

(5) Suck-back along flash lines of molded goods if not more than 1/16 inch wide, 1/16 inch deep, and not more than 2 feet long.

The tolerances shown on the drawing shall govern all cross-sectional dimensions. Any defects which are not within the above limitations either shall be repaired as approved by the contracting authority or shall be removed from the finished product by cutting out a length of waterstop containing such defects and splicing the waterstop at that point. All factory splices shall be molded splices. Molded splices shall be made by vulcanizing the splices in a steel mold for a time sufficient to produce maximum strength in the splice. All molded splices shall withstand being bent 180° around a 2-inch-diameter pin without any separation at the splice.

*(Rubber waterstop for joints in the barrel or box portions of siphons, culverts, or other pressure conduits shall be furnished in continuous circular hoops. The hoops shall be fabricated from straight strip waterstop with the ends spliced together with molded splices to form a closed ring of the specified circumferential length. A tolerance of plus or minus 2 inches will be permitted in the circumferential length of each hoop.)

(d) *Tests.*—

(1) *General.*—Rubber waterstop shall be subject to laboratory tests before shipment. Material for tests shall be furnished by the manufacturer and all tests shall be made at the place of the manufacturer of the rubber waterstop.

Except as otherwise provided below, general sampling procedures shall be in accordance with section 6 of Federal Test Method Standard No. 601.

(2) *Sampling for tests.*—Samples for laboratory tests to determine physical properties of the compound shall be taken at random to obtain the following number of test units from each separate purchase order:

Size of purchase order, linear feet:	*Number of test units*
500 or less	1
501 to 1,000	2
1,001 to 5,000	4
5,001 to 10,000	8
Over 10,000	15

*Revise or delete as appropriate.

At the option of the manufacturer, laboratory tests to determine physical properties of the rubber waterstop required to be furnished under these specifications shall be performed on test specimens cut from (a) test units taken from the finished rubber product, or (b) substitute samples furnished in accordance with paragraph 3.5 of section 6, Federal Test Method Standard No. 601.

(3) *Methods of tests.*—Tests shall be made in accordance with the following methods described in Federal Test Method Standard No. 601:

Tensile strength	Method 4111
Elongation	Method 4121
Durometer	Method 3021
Specific gravity	Method 14011
Change in volume	Method 6211
Compression set	Method 3311
Oxygen pressure test	Method 7111, except the time interval of test shall be 48 ± ½ hours

(4) *Data to be furnished by manufacturer.*—One certified copy of all laboratory test reports representing each shipment of waterstop shall be mailed to ....

(e) *Installation.*—The waterstop shall be installed with approximately one-half of the width of the material embedded in the concrete on each side of the joint. Care shall be exercised in placing and vibrating the concrete about the waterstop to insure complete filling of the concrete forms under and about the waterstop, and to obtain a continuous bond between the concrete and the waterstop at all points around the periphery of the waterstop. In the event the waterstop is installed in the concrete on one side of a joint more than 1 month prior to the scheduled date of placing the concrete on the other side of the joint, the exposed waterstop shall be covered or shaded to protect it from the direct rays of the sun during the exposure.

(1) *Field splices.*—All field splices in continuous circular hoops shall be molded splices. All molded splices shall be made by vulcanizing the splices in a steel mold, as follows: The adjoining ends at splices shall be beveled at an angle of 45° or flatter by the use of a saw and miter box, so that the ends to be spliced will be pressed together when the mold is closed. The beveled ends and the sides for at least one-fourth inch back from the ends shall be buffed thoroughly to provide clean, rough surfaces. All buffed surfaces shall be given two thin coats of rubber cement and each coat shall be permitted to dry thoroughly. A piece of gum rubber cut to the same dimensions as the beveled face shall then be applied to the end of one strip after removing the cloth backing from the gum rubber. The adjoining strip shall then be placed accurately in position, and all edges shall be stitched thoroughly together with a suitable hand stitcher. The mold shall be heated to a temperature of 290° F. before the splice is placed in the mold. The prepared splice shall be placed in the mold with the splice in the center of the mold, and the mold shall be closed tightly to prevent slipping during the vulcanizing process. The splice shall remain in the mold 25 minutes after the mold is closed completely, during which time the mold shall be maintained at a temperature of 290° F.

The contractor shall furnish all materials for molded splices, all field splicing molds, and electrical energy for heating the molds.

Each finished splice shall withstand a bend test by bending the waterstop 180° around a 2-inch-diameter pin without showing any separation at the splice.

(f) *Measurement and Payment.*—Measurement, for payment, of waterstop will be made of the number of linear feet of waterstop in place, measured along the centerline of the waterstop, and no allowance will be made for lap at splices and intersections. Payment for furnishing and placing waterstop will be made at the unit price per linear foot bid therefor in the schedule, which unit price shall include the cost of furnishing all material, making the splices, and installing the waterstop.

G-88. *Plastic Waterstop.*—(a) *General.*—If plastic waterstop is furnished in lieu of rubber waterstop, it shall be in accordance with the details shown on Drawing No. .... The contractor shall furnish all labor, equipment, and materials for making field splices in plastic waterstop. The contractor shall take suitable precautions to support and protect the waterstop during the progress of the work and shall repair or replace any damaged waterstop.

(b) *Material.*—Plastic waterstop shall be fabricated from an elastomeric plastic compound, the basic resin of which shall be polyvinyl-chloride (PVC). The compound shall contain any additional resins, plasticizers, stabilizers, or other materials needed to insure that, when the material is compounded, it will have the physical characteristics specified herein. No reclaimed polyvinyl-chloride shall be used.

Tensile strength, pounds per square inch, minimum	3,000
Ultimate elongation, precast minimum	300
Durometer	65 to 80
Specific gravity	1.40 ± 0.2
Water absorption, percent, maximum	0.15
Tensile strength after accelerated extraction test, percent of tensile strength before extraction test, minimum	80
Ultimate elongation after accelerated extraction test, percent of ultimate elongation before extraction test, minimum	80

(1) *Effect of alkali test.*—The material, when tested in accordance with the effect of alkali test as described in subsection (d) (3) below, shall not show an increase in weight of more than 0.25 percent or a loss in weight of more than 0.10 percent after 7 days, or more than 0.40-percent increase or more than 0.30-percent loss in weight after 28 days. After 28 days' immersicn, the dimensions of the sample shall not differ from those of the original sample by more than 1 percent. After 7 days' immersion, the durometer hardness reading of the sample shall not differ by more than plus or minus 5 from the reading on the original sample.

(2) *Cold bend.*—When tested as described in subsection (d) (3) below, the material shall show no signs of cracking at the bend.

(3) *Impact resistance.*—When subjected to the impact test described in subsection (d) (3) below, the material shall show no signs of cracking or chipping.

(c) *Fabrication.*—All water stops shall be molded or extruded in such a manner that any cross section will be dense, homogeneous and free from porosity and other imperfections. The waterstop shall be fabricated in accordance with detail dimensions and tolerances shown on Drawing No. ....: *Provided,* That longitudinal ribs on the webs between bulbs may be incorporated in the cross section at the option of the manufacturer.

(d) *Inspection and Tests.*—

(1) *General.*—All waterstops shall be subject to laboratory tests before shipment. Material for tests shall be furnished by the manufacturer and all tests shall be made at the place of manufacture of the waterstops. Except as otherwise provided below, general sampling procedures and definition of terms used herein shall be in accordance with section 6 of Federal Test Method Standard No. 601.

(2) *Sampling for tests.*—Samples for laboratory tests to determine physical properties of the compound shall be taken in accordance with a random process to obtain the following number of test units from each separate purchase order:

Size of purchase order, linear feet:	*Number of test units*
500 or less	1
501 to 1,000	2
1,001 to 5,000	4
5,001 to 10,000	8
Over 10,000	15

At the option of the manufacturer, laboratory tests to determine physical properties of the waterstop required to be furnished under these specifications shall be performed on test specimens cut from (a) test units taken from the finished product, or (b) substitute samples furnished in accordance with paragraph 3.5 of section 6, Federal Test Method Standard No. 601.

(3) *Methods of tests.*—Tests shall be made in accordance with the following methods:

Tensile strength.—Federal Test Method Standard No. 601, Method 4111.

Elongation.—Federal Test Method Standard No. 601, Method 4121.

Durometer.—Federal Test Method Standard No. 601, Method 3021.

Specific gravity.—Federal Test Method

Standard No. 601, Method 14011.

Water absorption.—Federal Test Method Standard No. 601, Method 6251, except that time of immersion shall be 24 hours.

Accelerated extraction test.—Five tensile test specimens, conforming to the shape and dimensions given to Federal Test Method 4111, Die III, shall be prepared and the group weighed to the nearest milligram. The specimens shall be placed in a 1-liter tall-form beaker with spout. The beaker shall be filled within 2 inches of the top with a solution made by dissolving 5.0 grams of chemically pure sodium hydroxide and 5.0 grams of chemically pure potassium hydroxide in 1 liter of distilled water. The specimens shall be completely immersed and the top of the beaker covered with a watchglass. The beaker shall then be placed in a constant temperature bath which shall maintain the temperature of the solution between 140° and 150° F. A ¼-inch-diameter glass tube shall be inserted into the spout of the beaker to within ½ inch of the bottom of the beaker. Air shall then be gently bubbled through the solution at the rate of not less than three bubbles per second. The solution shall be changed every 24 hours, the new solution being warmed to 150° F. before replacing the old.

Once daily, each of the five specimens shall be removed from the beaker (preferably at the time of renewing the solution) and shall be rinsed lightly with distilled water. Each specimen shall then be superficially dried by means of a clean cloth. Ten minutes after the specimens have been thus dried, the group of five specimens shall be weighed and the weight recorded. The sequence testing shall be continued until three consecutive weighings, calculated as a percentage change from the original weight, do not differ from one another by more than 0.05 percent. In any case, extraction shall proceed for a minimum of 14 days. After completion of extraction immersions as provided above, the specimens shall be tested for tensile strength and ultimate elongation, as provided in Federal Test Methods 4111 and 4121. The tensile strength and elongation after extraction shall be the median of the values obtained from the five specimens tested.

Effect of alkali.—A specimen, not more than ¼-inch thick weighing about 150 grams, shall be prepared for this test. Its dimensions to the nearest 0.001 inch and weight to the nearest milligram shall be recorded. The durometer reading shall be taken in accordance with Federal Test Method 3021.

The specimen shall be completely immersed in a freshly made solution containing 0.5 gram of chemically pure sodium hydroxide and 0.5 gram of chemically pure potassium hydroxide in 1 liter of distilled water, kept at 70° to 75° F. The solution shall be replaced by a similar fresh solution at a similar temperature every 7 days. At the end of 7 days, the specimen shall be removed, rinsed with distilled water, and the surfaces wiped with a clean cloth and allowed to dry in laboratory air for approximately 1 hour. The weight and durometer shall be measured and recorded.

The specimen shall be reimmersed for a further period of 21 days, at the end of which time the dimensions and weight shall again be measured. At the age of 7 and 28 days any change in weight, and after 28 days any change in dimensions, shall be recorded as a percentage of the weight and dimensions of the original sample.

Cold bend test.—A specimen of the material 1 inch in width, $\frac{1}{16}$- to ⅛-inch thick, and about 6 inches long shall be subjected to a temperature of minus 20° F. for a period of 2 hours. Upon removal from the cooling chamber, the specimen shall be immediately bent 180° around a rod of ¼-inch diameter with enough force to hold the specimen in intimate contact with the rod. The tests shall be carried out on at least three specimens, and the results reported on each individual specimen.

Impact resistance.—A 4- to 6-inch length of the finished waterstop shall be fixed in a horizontal position in the jaws of a vise or by other suitable means. One-half of the width of the material shall project free of the jaws. The vise shall be fixed on a base in such a manner as to provide a vertical clearance of 5 inches between the specimen and the top surface of the base. The assembly shall be placed in a cooling chamber having forced circulation and subject to a temperature of minus 20° F. for a period of 3 hours.

Without removing the assembly from the cooling chamber, an 8-pound ball or weight shall then be dropped freely from a height of 5 feet so as to strike the projecting portion of the specimen squarely. The height of the drop shall be measured from the top surface of the specimen to the bottom of the ball or weight. The tests shall be carried out on at least three specimens and the results reported on each individual specimen.

(4) *Data to be furnished by the manufacturer.*—Certified copies of the laboratory test reports of the physical properties of the plastic compound used in all plastic waterstops shall be furnished to the contracting authority.

(e) *Installation.*—The waterstop shall be installed with approximately one-half of the width of the material embedded in the concrete on each side of the joint. Care shall be exercised in placing and vibrating the concrete about the waterstop to insure complete filling of the concrete forms under and about the waterstop and to obtain a continuous bond between the concrete and the waterstop at all points around the periphery of the waterstop.

All field splices shall be made with fusion-welded butt splices made in accordance with instructions furnished by the manufacturer of the waterstop. All splices will be inspected for continuity of the fusion weld and for watertightness, and each splice shall withstand a bend test by bending the waterstop 180° around a 2-inch-diameter pin without showing any separation at the splice.

(f) *Measurement and Payment.*—Measurement, for payment, of furnishing and placing waterstop will be made of the number of linear feet of waterstop in place, measured along the centerline of the waterstop, with no allowance for lap at splices and intersections.

Payment for furnishing and placing waterstop will be made at the unit price per linear foot bid therefor in the schedule, which unit price shall include the cost of furnishing all material, making field splices and intersections, and installing the waterstop.

G-89. Sponge Rubber Joint Filler.—(a) *General.*—Sponge rubber joint filler of the sizes shown shall be furnished and placed in joints in concrete where shown on the drawings.

(b) *Materials.*—

(1) *Sponge rubber.*—Sponge rubber shall conform to Federal Specification HH-F-341e, Type II, Class A, sponge rubber: *Provided,* That the load required to compress the test specimen to 50 percent of its thickness before test shall be not less than 50 nor greater than 150 psi. Sponge rubber shall be stored in as cool a place as practicable, preferably at 70° F. or less and in no case shall the rubber be stored in the open, exposed to the direct rays of the sun. Sponge rubber shall be gray in color.

(2) *Copper nails.*—Copper nails shall conform to Federal Specification FF-N-105a, common copper wire nails.

(3) *Adhesive.*—Adhesive for fastening the sponge rubber in place shall be a non-bituminous adhesive as recommended by the manufacturer of the filler material.

(c) *Installation.*—The sponge rubber filler shall be cut to size and shape of the joint surface to receive joint filler. The filler shall be secured to the concrete in an approved manner, with copper nails embedded in the first-placed concrete in such a manner that the nail points will protrude from the joint surface to be covered at approximately 12-inch centers, or by adhesive applied between the filler and the first-placed concrete.

Joints between adjoining portions of the filler material shall be sufficiently tight to prevent mortar from seeping through such joints. Unless otherwise shown or directed, the edges of the sponge rubber filler shall be placed flush with the finished surface of the concrete or to the bottom edge of chamfers.

*((d) *Measurement and Payment.*—Measurement, for payment, of sponge rubber joint filler will be made of the area of material in place. Payment for furnishing and placing the various sizes of sponge rubber joint filler

*Revise or delete as appropriate.

will be made at the applicable unit price per square foot bid therefor in the schedule.)

*((e) *Costs.*—The entire cost of furnishing and placing sponge rubber joint filler shall be included in the unit price per cubic yard bid in the schedule for concrete in which the filler is placed.)

G-90. *Metal Seals.*—(a) *General.*—Metal seals of corrosion-resisting metal shall be placed in joints in the structures where shown on the drawings. Metal seal connections shall be as shown on the drawings. The seals shall be joined carefully together by brazing so as to form continuous watertight diaphragms in the joints. Provisions shall be made to support and protect the seals during the progress of the work. Before any seal is embedded in the concrete, the seal shall be tested with soapy water to assure that all connections are tight. Additional vibration, over and above that used for mass concrete placement, shall be employed near the seal to assure complete embedment of the seal in concrete. Large pieces of aggregate near the seal shall be removed by hand during embedment to assure complete contact between the seal and the surrounding concrete. The contractor shall replace or repair any metal seals punctured or damaged.

(b) *Materials.*—The contractor shall furnish all materials for metal seals, including metal strip, materials for brazing, and washers and nails for fastening the seals to the forms. Materials for the metal seals shall conform to the following specifications:

(1) *Metal seals.*—Metal seals shall be made from one of the following materials:

Copper.—Federal Specification QQ-C-576b, 0.032-inch thick, soft annealed.

Corrosion-resisting steel.—Federal Specification QQ-S-766c, Class 6, 18 percent CR, 8 percent Ni (0.10 percent C maximum). (Ti or Cb stabilized), Condition A (annealed), hot- or cold-rolled finish, No. 20 gage, United States Standard (0.0375 inch thick).

(2) *Nails.*—FF-N-101.

(3) *Materials for joining* together the individual lengths of seals shall be of a type and composition approved by the contracting authority for brazing the seals specified above.

(c) *Measurement and Payment.*—Measurement, for payment, of metal seals will be made of the linear feet of seals in place, and no allowance will be made for lap at joints. Payment for furnishing and placing metal seals will be made at the unit price per linear foot bid therefor in the schedule, which unit price shall include the cost of bending the seals to the prescribed shape; furnishing all materials for brazing and brazing; installing the seals; and maintaining the seals free from damage during the progress of the work.

G-91. *Metal Waterstops.*—(a) *General.*—Metal waterstops shall consist of plates of 3/16- by 8-inch wrought iron bent and welded to the size and shape of the structure at the joint and coated with plastic compound. The metal plates, welding rods, and plastic compound shall be furnished by the contractor.

(b) *Placing.*—The waterstops shall be located in the concrete of the structures as shown on the drawings. The contractor shall cut, shape, and weld the plates and place them in the construction joints as herein provided. The exposed half of the metal waterstops shall be coated completely to a thickness of approximately 1/16 inch with plastic compound applied cold. Adequate provisions shall be made to support and protect the metal stops during the progress of the work. The contractor shall replace or repair any waterstops which are damaged before final acceptance of the work.

(c) *Materials.*—

(1) *Wrought-iron plates.*—The wrought-iron plates shall conform to ASTM Designation: A 42.

(2) *Plastic compound.*—The plastic compound shall consist of asphalt dispersed in water by means of a mineral emulsifying agent. The compound shall be heavy-bodied, of smooth plastic consistency, and free from lumps and sediment. When applied at room temperature in one coat to a smooth metal surface at a thickness of 1/16 inch and the specimen is then immediately placed in a vertical position, the compound shall be tightly adherent and shall not run or sag

*Revise or delete as appropriate.

while the coating is drying and after it has dried. The asphalt compound shall comprise at least 50 percent, by weight, of the compound, shall have a softening point (ASTM Designation: D 36) of 100° F. to 130° F., a penetration at 77° F. (ASTM Designation: D 5) of 60 to 120, and a ductility at 77° F. (ASTM Designation: D 113) of not less than 100 centimeters. The ash content shall be not more than 18 percent, as determined by ASTM Designation: D 128.

Welding rods.—Welding rods shall be of a type and composition approved by the contracting authority.

(d) *Payment.*—Payment for furnishing and placing metal waterstops will be made at the unit price per linear foot bid therefor in the schedule, which unit price shall include the cost of cutting, welding, placing, and coating the metal waterstops.

G-92. *Drilling Holes for Anchor Bars and Grouting Bars in Place.*—Anchor bars shall be reinforcing bars, grouted into holes drilled into formation, for anchoring concrete in the ___ (name structure and location) ___.

The dimensions of the anchor bars and the locations, diameters, and depths of the anchor-bar holes shall be as shown on the drawings or as directed. The diameter of each anchor-bar hole shall not be less than 1½ times the diameter or greatest transverse dimension of the anchor bar specified for that hole. The depth of holes shown on the drawings for receiving the bars are minimum depths to be drilled into the formation. In overbreak areas, longer bars will be required to maintain the required depth of hole and extend the top of the anchor-bar hook to the position shown on the drawings.

The contractor shall furnish anchor bars and all materials for grout. Anchor bars shall be in accordance with section G-69. Bars shall be bent to the shapes and dimensions shown on the drawings prior to being grouted in place.

Cement shall be in accordance with section G-60.

Anchor bars shall be cleaned thoroughly before being placed. The holes shall be cleaned thoroughly and shall be completely and compactly filled with grout mix in the proportions and to the consistency prescribed by the contracting authority. The anchor bars shall be forced into place before the grout takes its initial set and, where practicable, shall be vibrated or rapped until the entire surface of the embedded portions of the bars is in intimate contact with the grout. Special care shall be taken to insure against any movement of the bars which have been placed.

Measurement, for payment, of drilling holes for anchor bars and grouting bars in place will be based upon the length of holes required to be drilled beyond the surface of the excavation. Payment for drilling holes for anchor bars and grouting bars in place will be made at the unit price per linear foot bid therefor in the schedule, which unit price shall include the cost of furnishing materials for grout, except cement, drilling the holes, and grouting the bars in place. Payment for furnishing and placing the anchor bars will be made at the unit price per pound bid in the schedule for furnishing and placing reinforcing bars. Payment for furnishing and handling cement will be made at the unit price per barrel bid therefor in the schedule.

G-93. *Drainage, General.*—All drains shall be constructed at the locations shown on the drawings or as directed. Care shall be taken to avoid clogging the drains during the progress of the work, and should any drain become clogged or obstructed from any cause before final acceptance of the work, it shall be cleaned out in a manner approved by the contracting authority or replaced by and at the expense of the contractor. No pipe which has been damaged shall be used in the work if, in the opinion of the contracting authority, the pipe is unfit for use. Travel over drainpipe will not be permitted until the pipe has been covered to a depth sufficient to prevent damage to or breakage of the pipe.

G-94. *Dam Embankment Drainage System.*—(a) *General.*—The dam embankment drainage system shall consist of ___- and ___-inch-diameter perforated *(and nonperforated) (describe type of pipe) ___ drains as shown on Drawing No. ___.

---

*Revise or delete as appropriate.

The pipe shall be laid to lines and grades as shown on the drawing or as directed by the contracting authority. No portion of the drain system shall be placed with adverse slopes.

All pipe in the dam embankment drainage system shall be perforated, except the pipe shall be nonperforated where it is placed beyond the limits of the dam embankment as shown on Drawing No. ....

(b) *Materials.*—

*(1) *Pipe.*—The pipe shall be:

a. Perforated clay sewer pipe, Class A (standard strength, bell-and-spigot type), in accordance with Federal Specification SS-P-359b, and nonperforated clay sewer pipe, Class 1 (standard strength), Type I (bell and spigot), in accordance with Federal Specification SS-P-361b.

b. Perforated concrete sewer pipe, C'ass 1 (standard strength), Type 1 (bell-and-spigot), manufactured in accordance with Federal Specification SS-P-371e, except that perforated concrete sewer pipe shall be perforated in accordance with the provisions for perforating clay pipe as set forth in Federal Specification SS-P-359b. Cement used in concrete sewer pipe shall conform to cement specified in section G-60.

c. Asbestos-cement perforated underdrain pipe, Type II, in accordance with ASTM Designation: C508-63T, and asbestos-cement nonpressure sewer pipe, Class 2400, Type II, in accordance with ASTM Designation: C428-67T. The asbestos-cement pipe shall not contain more than 1.0 percent uncombined calcium hydroxide when tested in accordance with the specified test method.

d. Perforated and nonperforated corrugated-metal pipe and connecting bands, conforming to the requirements of Federal Specification WW-P-405a, Class ......, Shape ......, Coating ......, and No. ...... gage.

(2) *End caps, screen, couplings, and fittings.*—The end caps, screen, couplings, and fittings shall be suitable for use with the specified pipe. The screen shall be of approximately ½-inch mesh.

(3) *Bedding materials.*—The coarse bedding material adjacent to the perforated drain pipe shall consist of graded gravel or crushed rock, or a mixture of gravel and crushed rock, clean and well graded from 3/16 to 1½ inches in size, as approved by the contracting authority, and may contain up to 10 percent, by weight of the total material, of particles smaller than 3/16-inch size: *Provided,* That no more than 5 percent, by weight of the total material, shall pass a United States No. 200 sieve. The coarse bedding material may be obtained from approved sources of coarse aggregate for concrete, or may be produced by screening the desired sizes from selected material from approved deposits.

The graded sand bedding material shall consist of material meeting the requirements of sand for concrete. (See section G-63.)

(4) *Backfill material.*—Backfill material shall be the material excavated from the trenches for the pipe drains, or material from borrow area ......, as approved by the contracting authority.

(c) *Constructing Drains.*—The pipe trench shall be excavated in accordance with section G-13, and shall be kept free of water which might impair construction of the drainage system.

The perforated pipe shall be embedded in graded sand and graded gravel or crushed rock bedding and the nonperforated pipe shall be embedded in backfill, both as shown on Drawing No. ....

The methods of lowering the pipe into the trench and placing pipe in position shall be such as to prevent getting dirt inside of the pipe and coupling, and to prevent damage to the pipe. *(For bell-and-spigot pipe, holes shall be dug in the pipe subgrade to accommodate the bells. The pipe shall be laid carefully on the bedding with the bell end upgrade and with the spigot end placed concentrically in the bell.) Before and during assembly of a joint, all parts shall be clean and shall be free of mud, ice, oil, or grease. *(The joints between pipe of different diameter shall be closed joints equal in quality to the joints between pipe of

*Revise or delete as appropriate.

the same diameter.) All joints shall be made in accordance with the manufacturer's recommendations.

End caps shall be fastened in place on the upper ends of the drains, and a screen shall be fastened to the discharge end of the pipe.

(1) *Perforated pipe.*—A minimum 6-inch layer of graded sand bedding material shall be placed and tamped in the bottom excavated surfaces of the trench. Graded gravel or crushed rock shall then be placed over the sand bedding material and shall be tamped to provide equal bearing under the lower half of the pipe. The perforated pipe shall be laid carefully in the coarse bedding material with the perforations symmetrical about the vertical centerline and below the horizontal centerline. The pipe shall then be covered with the additional minimum thickness of bedding materials as shown on the drawing. The bedding materials shall be placed and tamped about the pipe so as not to disturb the pipe and to hold it securely in position while the overlying material is being placed.

(2) *Nonperforated pipe.*—When the surface in which a pipe drain is to be placed has reached an elevation approximately one-fourth the diameter of the pipe above the prescribed elevation of the invert of the pipe, the material shall be excavated carefully to the established lines and grades to provide a firm and uniform bearing for the entire length of pipe.

As each unit of pipe is laid, sufficient backfill material shall be tamped under and about the pipe to hold it rigidly in place until the joints are completed. After the joints are completed, the backfilling shall be completed to original ground.

*(The backfill shall be compacted in accordance with section G-27.)

(d) *Measurement and Payment.*—Measurement, for payment, of the dam embankment drainage system will be made along the centerline at midheight of the pipe, from end to end of the pipe in place, and no allowance will be made for lap at joints.

*Revise or delete as appropriate.

Payment for the dam embankment drainage system will be made at the applicable unit price per linear foot bid in the schedule for furnishing (pipe description) .... and constructing dam embankment drains, which price shall include the entire cost of furnishing and installing the pipe, furnishing all materials for and preparing and placing bedding material under and about the pipe, furnishing all materials for and making joints in the pipe, furnishing and installing end caps and screens, and placing backfill in the trenches outside the limits of the dam embankment.

Payment for excavation for trenches for the dam embankment drainage system will be made at the unit price per cubic yard bid in the schedule for excavation for dam embankment foundation.

*(Payment for compacting backfill will be made at the unit price per cubic yard bid in the schedule for compacting impervious backfill.)

G-95. *Constructing Perforated and Nonperforated Sewer-Pipe Drains.*—(a) *General.*—Perforated and nonperforated sewer-pipe drains ... inches in diameter shall be constructed (describe locations) .... as shown on the drawings, or as directed. The sewer pipe shall be either clay or concrete at the contractor's option. All sewer pipe, including stoppers, strainers, fittings, burlap, crushed-rock or gravel for bedding and sand for filter, concrete for lean concrete pads, and caulking material shall be furnished by the contractor.

All sewer pipe shall be perforated, except the pipe shall be nonperforated where it is embedded in concrete and where shown on the drawings as nonperforated.

(b) *Perforated Sewer Pipe.*—The pipe for the perforated sewer pipe drains shall be perforated clay sewer pipe, Class A (standard strength), (bell-and-spigot type), in accordance with Federal Specification SS-P-359b; or shall be perforated concrete sewer pipe, Class 1 (standard strength), Type 1 (bell-and-spigot pattern), manufactured in accordance with Federal Specification SS-P-371e, except that the pipe shall be perforated in accordance with the provisions for perforating clay pipe, as set forth in Federal Specification SS-P-359b. Cement used in concrete sewer pipe shall conform to cement specified in section G-60. Stoppers or strainers at ends of drains shall be standard clay or concrete to fit the pipe used and shall be cemented in place. Burlap shall be 40 inches in width, 10 ounces per linear yard, in accordance with Federal Specification CCC-C-467b.

Perforated sewer-pipe drains shall be placed on a lean concrete pad and shall be covered with a crushed-rock or gravel bedding and sand filter material as described below, and as shown on the drawings.

Concrete for lean concrete pads shall contain ......-inch maximum aggregate size, and shall be in accordance with sections G-59 through G-76.

The coarse bedding shall consist of crushed-rock or gravel material placed over the top and at the sides of the pipe and the sand filter shall be placed between the coarse bedding and the foundation or embankment material. The coarse bedding material adjacent to the pipe shall be natural gravel or crushed rock, or a mixture of natural gravel and crushed rock, clean and well graded, as approved by the contracting authority, and shall be obtained from approved sources of aggregate for concrete. Sand for filter shall be obtained from approved sources of sand for concrete. The gradation of the crushed-rock or gravel bedding placed adjacent to the drains shall bear a definite relation to the gradation of the sand in the outer layer, depending upon the type of material used. The grain size ratios shall be within the limits shown below:

	Gravel	*Crushed rock*
$\frac{\text{15 percent grain size coarse aggregate}}{\text{15 percent grain size sand}}$	=12 to 40	6 to 18
$\frac{\text{50 percent grain size coarse aggregate}}{\text{50 percent grain size sand}}$	=12 to 58	9 to 30

where 15 percent grain size indicates the size of screen opening that will pass 15 percent, by weight, of the total material, and 50 percent grain size indicates the size of screen opening that will pass 50 percent, by weight, of the total material.

The pipe shall be placed immediately after placement of the lean concrete pad and worked into the fresh, lean concrete to insure continuous support under the total length of the pipe. The bell end of the pipe shall be laid upgrade with spigot end concentrically in the bell. The pipe shall be laid so that perforations are symmetrical about the vertical centerline, with the 90° nonperforated segment on bottom. Special care shall be taken to avoid plugging the perforations. The coarse bedding material shall then be placed to provide minimum thicknesses, as shown on the drawings, over the top and at the sides of the pipe. The coarse bedding shall be placed and tamped carefully about the pipe so as not to disturb the pipe after being laid, and to hold it securely in position while the overlying material is being placed. Sand filter shall be placed as shown on the drawings.

Where concrete is to be placed over or against the bedding and filter of the drains, the bedding and filter shall be covered with a layer of burlap before the concrete is placed.

(c) *Nonperforated Sewer Pipe.*—Clay or concrete sewer pipe shall be constructed with caulked joints at the locations shown on the drawings or as directed.

The pipe shall be clay sewer pipe, Class 1 (standard strength), Type 1 (bell and spigot), in accordance with Federal Specification SS-P-361d; or concrete sewer pipe, Class 1 (standard strength), Type 1 (bell and spigot), in accordance with Federal Specification SS-P-371e. Cement used in concrete sewer pipe shall conform to cement used in structures as provided in section G-60.

The pipe shall be laid to the prescribed lines and grades. The ends of the pipe shall fit closely with the bell end upgrade and the spigot ends shall be placed concentrically in the bells and the joints shall be packed with oakum and caulked. The pipe shall be placed accurately in position and held securely in place during the placing and hardening of the surrounding concrete.

At the contractor's option, the joints may be of factory-fabricated types using resilient materials in accordance with ASTM Designation: C 425.

(d) *Measurement and Payment.*—Measurement, for payment, of furnishing perforated and nonperforated sewer pipe and constructing drains will be made along the centerlines of the pipe from end to end of the pipe in place, and no allowance will be made for lap at joints.

Payment for furnishing ______-inch-diameter perforated and nonperforated sewer pipe and constructing drains will be made at the unit price per linear foot bid therefor in the schedule, which unit price shall include the costs of furnishing materials, except cement, for and constructing lean concrete pads, furnishing materials for and preparing and placing the coarse bedding and sand filter about the pipe, furnishing and placing the burlap covering for the drains, furnishing materials for and caulking joints of nonperforated pipe, and all other materials and work necessary to complete the drain installation.

Payment for the cement used in lean concrete pads will be made at the unit price per barrel bid in the schedule for furnishing and handling cement.

Measurement, for payment, of excavation for the sewer-pipe drains, including excavation for bedding and filter materials and for lean concrete pads, will be made only to the neatlines as shown on the drawings or as directed, and payment for such excavation will be made at the unit price per cubic yard bid in the schedule for excavation, in opencut, for structures.

G-96. *Drilling Drainage Holes.*—Drainage holes shall be drilled through the concrete lining of the *spillway and *outlet works tunnel(s), **through the steel-liner plates or steel lagging, if encountered, and into the surrounding formation as shown on the drawings or as directed. The contractor will be permitted to furnish and install pipe inserts to form the holes in the concrete lining. The drainage holes shall be drilled at the angles of inclination shown on the drawings or as directed. All drainage holes shall be drilled to a minimum depth of ______ feet but not deeper than ______ feet as directed, from the inside of the concrete lining with approved rotary drills, and the holes shall be not less than 1½ inches in diameter. Drainage holes shall not be drilled until all adjacent grout holes within a minimum distance of 150 feet have been drilled and grouted.

If, after a given area has been grouted and drilled for drainage, it is found desirable to drill and grout additional grout holes, the contractor, after completion of grouting, shall open previously drilled drain holes to their original diameter and depth to provide satisfactory drainage. Opening of previously drilled drain holes by redrilling, if required, will be ordered in writing by the contracting authority, and payment therefor will be made at one-half the unit price per linear foot bid in the schedule for drilling drainage holes.

Measurement, for payment, of drilling drainage holes will be made of the depth of holes from the inside face of the concrete lining, including portions of holes formed by pipe inserts, if used. Payment for drilling drainage holes will be made at the unit price per linear foot bid therefor in the schedule, which unit price shall include the entire cost of furnishing all materials and labor to complete the work as described in this section.

G-97. *Cast Iron Pipe Drains.*—A ______-inch cast iron soil pipe drain shall be installed under and through the spillway floor to drain the spillway ______ upstream from spillway station ______ as shown on the drawings. The cast iron soil pipe, fittings, and joint material shall be furnished and installed by the contractor as shown on the drawings or as directed by the contracting authority.

Materials shall conform to the following specifications:

(1) Cast iron soil pipe and fittings, hub-and-spigot pattern, Class XH, Federal Specification WW-P-401.
(2) Jute caulking, Type II (tarred), Federal Specification HH-P-117.
(3) Caulking lead, Type I, Federal Specification QQ-C-40.

The spigot ends of cast iron soil pipe and fittings shall be placed concentrically in the hubs, and all joints shall be packed with tarred jute or similar material thoroughly caulked with suitable caulking tools so as to leave 2 inches in the bell for lead. Joints shall be poured full of molten lead in one operation. The lead shall be retained in the joints by

*Revise or delete as appropriate.
**List other locations as applicable.

suitable joint runners, and after the lead has cooled sufficiently, it shall be caulked tightly. Pipe to be embedded in concrete shall be held firmly in position while the concrete is being placed.

Measurement, for payment, for furnishing and installing ......-inch-diameter cast iron soil pipe spillway drain will be made from end to end of the pipe in place, and no allowance will be made for lap at joints. Payment for furnishing and installing ......-inch-diameter cast iron soil pipe spillway drain will be made at the unit price per linear foot bid therefor in the schedule.

G-98. *Dry-Rock Paving for Open Drains.*—Dry-rock paving shall be placed on a bedding of sand and gravel or crushed rock as lining for open drains, as shown on the drawings. The quality of the bedding material and rock used for dry-rock paving shall be equivalent to materials for bedding for riprap and riprap. The overall thickness of the finished bedding and paving at any point shall be not less than 15 inches. The dimensions of the paving stones normal to the face of the paving, for not less than two-thirds of the surface area of the paving, shall be not less than 12 inches, and the dimensions normal to the face of the paving of any stone forming the surface of the paving shall be not less than 8 inches. The paving stones shall have an average volume of not less than one-sixth cubic foot, and not more than 25 percent of the pieces shall be less than one-ninth cubic foot in volume. Stones of the same dimensions normal to the face of the paving shall be distributed uniformly throughout the paving. Rock materials may be selected from materials for bedding for riprap and riprap furnished by the contractor. The stones shall have roughly squared and reasonably flat upper faces; the stones shall be hand-placed, with close joints, to the established lines and grades; and the spaces between the stones shall be filled with spalls and gravel or crushed rock.

Measurement, for payment, of bedding and dry-rock paving will be made on the basis of the 15-inch thickness and on the basis of the area of finished surface of rock paving in place to the lines shown on the drawings or established by the contracting authority. Payment for dry-rock paving for open drains will be made at the unit price per square yard bid therefor in the schedule, which unit price shall include the cost of procuring, handling, hauling, and placing bedding and paving materials. Payment for excavation for open drains, to the lines and grades established by the contracting authority, will be made at the applicable unit price per cubic yard bid in the schedule for excavation for the structure involved.

G-99. *Explosives, Drilling and Blasting.*—(a) *Explosives.*—The transportation, handling, storage, and use of explosives shall be subject to the provisions of section 15, "Blasting Operations," of "Construction Safety Standards." The contractor shall maintain an inventory record of storage and withdrawal of all explosives. This record shall be available to the contracting authority and the authority shall be promptly notified of any loss or theft of explosives. The contractor shall provide such reasonable and adequate protective facilities as are necessary to prevent loss or theft of explosives. Overnight storage of explosives and detonators outside of the magazines will not be permitted.

(b) *Drilling Dust Control.*—When drilling in rock or other dust-producing material, the dust shall be controlled within the threshold limit values set forth in table 9 of "Construction Safety Standards."

(c) *Percussion Drilling.*—All percussion-type drilling shall be performed with drilling apparatus equipped with water or chemical dust-control systems or other equivalent means of controlling the dust. Pressure tanks used in the suppression equipment shall conform to ASME Boiler and Pressure Vessel Code, section VIII, for Unfired Pressure Vessels. Equipment and solutions used shall be suitable for operation in freezing weather. Except for underground excavation, dust-control devices are not required on jackhammers provided the operators wear approved-type dust respirators when dust concentrations exceed the threshold limit values set forth in table 9, "Construction Safety Standards."

(d) *Blasting.*—Blasting will be permitted only after adequate provision has been made for the protection of persons, the work, and public and private property. Approval by the contracting authority or its representatives of any of the contractor's blasting operations shall not relieve the contractor of his responsibility for the safety of persons and property. Damage shall be repaired by and at the expense of the contractor.

(e) *Vibration and Damage Control.*—Blasting in or adjacent to cofferdams, buildings, dwellings, structures, or other facilities shall be carefully planned and controlled to eliminate possibility of damage to such facilities and structures. When appropriate, the contracting authority may require the contractor to submit a controlled blasting plan which incorporates provision for monitoring blasts with an approved-type blasting seismograph.

G-100. ***Information as to Subsurface Investigations.***—All available samples and cores recovered in subsurface investigations may be inspected by bidders at the office of the contracting authority in (insert address) .... *(Arrangements for inspection of available samples and cores can be made through the office of the contracting authority, attention: (insert name) ....)

The contracting authority does not represent that the available cores and samples show the conditions that will be encountered in performing the work, and represents only that any such cores or samples show conditions encountered at the particular point for which such cores or samples were obtained. Bidders must assume all responsibility for deductions and conclusions which may be made as to the nature of the materials to be excavated, the difficulties of making and maintaining the required excavation, and of doing other work affected by the geology at the site of the work.

*(Solely for the convenience of bidders, logs, as abstracted by the contracting authority, of the core drilling and subsurface explorations are included in the drawings. The complete logs of all holes, including any available percolation tests and core recovery data, are available for examination by bidders at the (insert name and address) ....)

*Revise or delete as appropriate.

*(Bidders must make their own determinations as to whether the abstracted data properly and adequately reflect the information shown on the complete drill logs, as well as whether the complete logs and the abstracted logs properly and adequately reflect the information shown by the available cores and samples.) Bidders may obtain their own samples and perform tests on the soils and rock materials to determine unit weights and evaluate shrinkage and swell factors and other properties which the bidder believes to be significant in arriving at a proper bid.

*(Bidders are cautioned that the contracting authority disclaims responsibility for any opinions, conclusions, interpretations, or deductions that may be expressed or implied in any of the information presented or made available to bidders; it being expressly understood that the making of deductions, interpretations, and conclusions from all of the accessible factual information is the bidder's sole responsibility.)

G-101. ***Soil-Cement Slope Protection.***—(a) *General.*—Soil-cement slope protection on the upstream slope of the dam embankment shall consist of a combination of soil, portland cement, and water and shall be uniformly mixed, placed, compacted, finished, and cured in accordance with these specifications. The soil-cement slope protection shall be constructed to the lines, grades, and dimensions shown on the drawings or established by the contracting authority.

(b) *Materials.*—

(1) *Cement.*—Cement for the soil-cement shall be furnished by the contractor and shall be in accordance with the applicable provisions of section G-60.

(2) *Water.*—The water used in soil-cement shall be free from objectionable quantities of organic matter, alkali, salts, and other impurities, and shall contain no more than 3,000 parts per million soluble sulfate.

(3) *Soil.*—The soil for soil-cement shall be composed of selected nonplastic material of such gradation that 100 percent by dry weight shall pass 1-inch screen size, a minimum of 85 percent by dry weight shall pass a No. 4 sieve, and not less than 10 percent

nor more than 25 percent by dry weight shall pass a No. 200 United States Standard series sieve.

Soil material introduced into the mixing unit shall contain no clay balls larger than 1 inch in size.

The soil material shall be obtained from (insert location) .... The contractor's excavation operations and selection of material shall be such as will result in an acceptable gradation of the materials when incorporated in the soil-cement slope protection as approved by the contracting authority. *(It is expected that it will be necessary to adjust the depth of cut in borrow pits to obtain a mixture meeting the requirements specified above.)

*(All operations in borrow area .... shall be in accordance with section G-16.)

(c) *Plant and Equipment.*—

(1) *Mixing plant.*—Mixing of the soil, cement, and water to be used in the soil-cement mixture shall be accomplished in a stationary mixing plant. The mixer shall be an approved twin pugmill-type mixer unit. The plant may be either a batch type or a continuous-mixing type designed for either weight or volume proportioning. The plant shall have a rated capacity of at least 100 cubic yards per hour and shall be designed, coordinated, and operated so as to produce a uniform mixture within the limits required by these specifications. Facilities for efficiently storing, handling, and proportioning unmixed materials shall be provided at the plant.

Satisfactory means shall be provided to obtain the proper amount of cement, soil, and water. All measuring devices shall be sensitive to a 2 percent variation above or below the actual weight in pounds required. Proportioning may be on a volume basis: *Provided,* That the sensitivity specified for the weight basis is maintained.

The plant shall be equipped with a positive, adjustable governor for controlling the mixing time of each batch. The mixing time shall be considered as the interval between the time the cement contacts the soil and water and the time the mixture leaves the mixing unit.

Batching plants designed for weight proportioning shall include means for accurately weighing soil and cement in a weight box or hopper suspended on scales, ample in size to hold a full batch without hand raking or running over. The weight box or hopper shall be supported on fulcrums and knife edges so constructed that they will not be easily thrown out of alinement or adjustment. Scales may be either of the beam type with over-and-under indicator or springless-dial type and shall be of a standard make and design, sensitive to one-half of 1 percent of the maximum load that may be required. If the beam-type scale is used, there shall be included a separate beam for soil and for cement, each beam being connected so as to actuate the over-and-under indicator, and a tare beam for balancing the hopper.

Plants designed for continuous mixing shall include a means for accurately proportioning soil and cement and shall be equipped to insure positive interlocking control of the flow of soil and cement from bins.

(2) *Transportation equipment.*—Trucks for transporting the soil-cement mixture shall have tight, clean, smooth beds and protective covers.

(3) *Spreader.*—The equipment for spreading the soil-cement mixture shall be suitable for the purpose and shall be capable of discharging the mixture in layers to produce reasonably smooth uniform surfaces. The equipment shall be controllable so as to produce layers, which, when compacted, will be approximately of the specified thickness and will meet all of the requirements of these specifications.

(4) *Compaction equipment.*—The compaction equipment used for soil-cement shall consist of tamping rollers and pneumatic-tired rollers.

a. *Tamping rollers.*—Tamping rollers used for compaction of soil-cement shall conform to the following requirements:

*Revise or delete as appropriate.

1. *Roller drums.*—Each drum of a roller shall have an outside diameter of not more than 4 feet and shall be not more than 5 feet in length. Each drum shall be free to pivot about an axis parallel to the direction of travel. Each drum of ballasted rollers shall be equipped with at least one pressure-relief valve and with at least one approved 2-inch safety head.

The safety head shall be of the Union type with ruptured discs suitable for 60-psi rupturing pressure. The pressure-relief valve shall be a manually operated valve and shall be opened periodically. Personnel responsible for opening pressure-relief valves shall be instructed to ascertain that valve openings are free from plugging to assure that any pressure developed in roller drums is released at each inspection.

2. *Tamping feet.*—At least one tamping foot shall be provided for each 100 square inches of drum surface. The spaces measured on the surface of the drum, between the centers of any two adjacent tamping feet, shall be not less than 8 inches. The length of each tamping foot from the outside surface of the drum shall be not less than 6 inches nor more than 8 inches. The cross-sectional end area of each tamping foot shall be maintained at not less than 5 square inches nor more than 7 square inches.

3. *Roller weight.*—The weight of a roller when fully loaded shall be not less than 2,000 pounds per foot of length of drum.

The loading used in the roller drums and operation of the rollers shall be as required to obtain the desired compaction. If more than one roller is used on any one layer of fill, all rollers so used shall be of the same type and essentially of the same dimensions. Tractors used for pulling rollers shall have sufficient power to pull the rollers satisfactorily when drums are fully loaded with sand and water. Self-propelled tamping rollers which meet all the foregoing requirements will be permitted. During the operation of rolling, the tamping feet and the spaces between the tamping feet shall be maintained free of materials which would impair the effectiveness of the tamping rollers.

b. *Pneumatic-tired rollers.*—Pneumatic-tired rollers may be either of the towed or self-propelled type. Roller wheels shall be located abreast in not more than two rows and be so designed that each wheel will carry approximately equal loads during compaction. The tires shall be of such size and ply as can be maintained at tire pressures between 80 and 100 psi for a 10,000-pound wheel load.

The roller shall have a rigid steel frame provided with a body suitable for ballast loading such that the load per wheel may be varied from 4,000 to 10,000 pounds. The character and efficiency of this equipment shall be subject to the approval of the contracting authority. Tractors used for pulling towed rollers and power systems used for self-propelled rollers shall be sufficient to pull fully loaded towed rollers or to propel fully loaded self-propelled rollers satisfactorily under normal conditions of compaction.

(5) *Watering equipment.*—Watering equipment consisting of pumps, water tanks, sprinklers, or other approved equipment, shall be furnished by the contractor. Adequate equipment, subject to the approval of the contracting authority, shall be available at all times to provide water as required for curing and protection of permanently exposed slopes and for moistening compacted surfaces that are to receive an overlying layer of soil-cement, all as provided in this section. Water trucks or sprinklers for adding water to in-place layers of soil-cement shall be equipped with fog-type sprayers.

(6) *Power broom.*—The equipment for power brooming shall be a power-driven steel broom suitable for thoroughly cleaning and striating the bonding surfaces.

(7) *Inspection of plant and equipment.*—The contracting authority shall have access at all times for checking the adequacy of the equipment in use for inspecting the operation of the plant and for verification of

weights or proportions and character of material.

(8) *Safety requirements.*—Adequate and safe stairways to the mixer platform and guarded ladders to other plant units shall be placed where required for accessibility to all plant operations. All gears, pulleys, chains, sprockets, and other dangerous moving parts shall be thoroughly guarded and protected. Ample and unobstructed space shall be provided on the mixing platform. A clean and unobstructed passage shall be maintained at all times in and around the truck loading space. This space shall be kept free from droppings from the mixing platform.

(d) *Preparation of Foundation.*—The upstream surface of the dam embankment upon which soil-cement slope protection is to be placed shall be firm and compact, and shall be true to the line and grades shown on the drawings. If loose material is placed on the slope to bring the slope to line and grade, such material shall be compacted by special rollers, mechanical tampers, or other approved methods, to a degree equivalent to that obtained in the dam embankment. The embankment surface shall also be moistened and shall be kept moist until overlying soil-cement is placed.

(e) *Proportioning and Mixing.*—

(1) *General.*—The soil, cement, and water shall be accurately measured and conveyed into the mixer in the proportionate amounts necessary to meet the specified requirements. All ingredients shall be mixed for at least 30 seconds or longer as may be necessary to insure a thorough, uniform, and intimate mix of the soil, cement, and water, and until the resulting mixture is homogeneous and uniform in appearance.

(2) *Cement content.*—The approximate amount of cement shall be ___ percent by dry weight of soil. The exact amount of cement required will be determined by the contracting authority.

(3) *Water content.*—Water shall be added and mixing shall be continued until a thorough, uniform, and intimate mixture of soil, cement, and water is obtained as determined by the contracting authority. When mixing is completed, the percentage of moisture in the mixture on a basis of ovendry weight shall be such that at time of compaction it will be within 1 percentage point dry to 1 percentage point wet of the Bureau of Reclamation laboratory standard optimum condition, as defined in section G-21.

(f) *Transporting Soil-Cement Mixture.*—The soil-cement mixture shall be transported from mixing plant to the site of the work in trucks having tight, clean, and smooth beds. Haul time shall not exceed 30 minutes. Haul time shall be considered to mean the time elapsed from the time the material leaves the mixer until the same material is spread on the fill to specified thickness. The contractor shall protect the soil-cement mixture if transported during unfavorable weather. Any loads wet excessively by rain will be subject to rejection. Equipment shall not be operated on a finished compacted layer of the soil-cement, except where specifically permitted, and any damage resulting to the soil-cement from such operation shall be repaired at the expense of and by the contractor.

(g) *Placing Soil-Cement.*—The soil-cement mixture shall be placed and distributed in such a manner as to produce a reasonably smooth uniform surface in layers of such uncompacted thickness that when compacted each layer will be approximately 6 inches in thickness. In general, soil-cement shall be placed in horizontal layers: *Provided,* That the contractor will be permitted to place the material in sloping layers to accommodate compacting equipment if such sloping layers are not steeper than 8:1. Each successive layer in a section shall be placed and compacted as soon as practicable after the preceding layer is completed.

The contractor shall striate the entire bonding surface of each layer of soil-cement by power brooming with the power-driven steel broom. A minimum of one-half hour shall be allowed to elapse after completion of compaction of each layer and initiation of power brooming, and the timing of brooming shall be as directed by the contracting authority.

Compacted surfaces of soil-cement that are to receive an overlying or adjacent layer of soil-cement shall be clean and shall be kept moist continuously until placement of the overlying or adjacent layer of soil-cement: *Provided,* That the contractor will not be required to keep the surfaces moistened for a longer period than 7 days. Immediately prior to the placement of soil-cement for the overlying layer, the bonding surface shall be cleaned thoroughly including brooming again with the power-driven steel broom.

Equipment operating on soil-cement layers shall be routed so as not to disrupt or damage the layers and no equipment shall be operated over the finished edges of the soil-cement on the dam slope unless proper protection is provided. Transverse joints at stoppages of work shall be trimmed to form straight joints and with as steep a taper as possible. Soil-cement materials removed in brooming, smoothing, or trimming the layers after the time limits specified for transporting and compacting the layers shall be wasted.

During seasons of probable freezing temperature, soil-cement shall not be mixed or placed when the air temperature is below 45° F.: *Provided,* That if the temperature is 40° F. or above, soil-cement slope protection may be placed if the temperature is rising. Soil-cement shall not be placed if the soil to be processed is frozen or if, in the judgment of the contracting authority, weather conditions are such that the material being processed cannot be completely compacted and protected before the advent of freezing temperatures.

(h) *Compaction.*—The soil-cement mixture shall be compacted by six passes of a tamping roller followed by four passes of the fully loaded pneumatic-tired roller. One pass of the roller is defined as the required number of successive roller trips which, by means of sufficient overlap, will insure complete coverage of the entire surface by the roller. Second and subsequent passes of the roller shall not be made until each previous pass, as defined above, has been completed.

The specified moisture content shall be maintained uniformly throughout the layer of material being compacted. The compaction operations shall proceed in such a manner that the length of time between completion of spreading operations of the soil-cement mixture and completion of compaction by both tamping and pneumatic-tired rollers shall not exceed 1 hour for each layer. If necessary, water shall be added and incorporated during the compaction operation to maintain the uncompacted mixture at its optimum moisture content. If, in the opinion of the contracting authority, the surface of a layer of soil-cement has been rutted or compacted unduly by hauling equipment to reduce the effectiveness of compaction by the tamping roller, the contractor will be required to scarify such surfaces, as directed, prior to compacting with the tamping roller.

When any of the compaction operations are interrupted prior to completion of compaction by both types of rollers so that the mixture which has not been completely compacted is left unworked for more than 30 minutes for any reason, or when the soil-cement mixture, before completion of compaction, is wetted by rain so that the average moisture content exceeds the tolerance given in section G-101(e)(3) at the time of final compaction, the entire layer affected, as determined by the contracting authority, shall be removed and shall be replaced in accordance with these specifications at the expense of the contractor and no payment will be made for cement used in the removed material.

Trimming and removal of excess soil-cement outside of the line of the upstream slope of the dam will not be required.

(i) *Protecting and Curing Surfaces of Soil-Cement.*—All surfaces of soil-cement which will remain exposed permanently shall be water cured or, at the option of the contractor, may be cured by providing a moist earth covering: *Provided,* That during the prescribed period of water curing when temperatures are such that soil-cement surfaces may freeze, water curing shall be discontinued and such surfaces shall be protected against freezing by application of a moist earth covering. Soil-cement surfaces cured with water shall be kept wet for a minimum period of 7 days immediately following placement of the soil-cement.

If a protective covering of moist earth is used for curing or protecting soil-cement other than at locations to be used for haul ramps, it shall have a minimum thickness of 6 inches.

Portions of the soil-cement facing over which haul ramps will be constructed shall be protected by a layer of earth of sufficient thickness, as determined by the contracting authority (at least 2 feet are required), to preclude any damage to the soil-cement facing by the hauling equipment. All surfaces to be cured and/or protected by moist earth covering shall be kept wet until the earth cover is applied, after which the earth covering shall be kept moist until the soil-cement surface has been subjected to a total curing time of 7 days following placement of the soil-cement.

The earth material used for curing or for protective covering, including that used for haul ramps, may be obtained from any approved source but no measurement for payment will be made for excavation, transportation and placement of the materials. The contractor will not be required to remove the earth after the curing period: *Provided,* That all haul ramps above elevation ______ shall be removed.

(j) *Measurement and Payment.*—Measurement, for payment, of soil-cement slope protection, will be made of the soil-cement in place after compaction to the lines, grades, and dimensions shown on the drawings or established by the contracting authority and on the basis of the prescribed thickness. No measurement, for payment, of soil-cement slope protection will be made of soil-cement which, after compaction, lies ouside of the line of the upstream slope of the dam as shown on the drawings: *Provided,* That measurement, for payment, for cement in soil-cement will include all cement used in soil-cement, except cement used in soil-cement which is intentionally used by the contractor to facilitate his operations or which is carelessly wasted, as determined by the contracting authority.

Payment for soil-cement slope protection will be made at the unit price per cubic yard bid therefor in the schedule, which unit price shall include the cost of all labor, materials, and equipment required for the completion of the soil-cement construction as described above, including excavation and transportation of soil material for soil-cement, mixing, placing and spreading soil-cement; compaction by both tamping and pneumatic-tired rollers; water curing or furnishing and placing moist earth cover for curing or for protection and removing haul ramps above elevation ______: *Provided,* That payment for furnishing and handling cement for soil-cement will be made at the unit price per barrel bid therefor in the schedule.

H. BIBLIOGRAPHY

G-102. *Bibliography.*

[1] "Concrete Manual," seventh edition, revised 1963, Bureau of Reclamation (for sale by the Superintendent of Documents, U.S. Government Printing Office, Washington, D.C. 20402).

[2] ———, ch. VI.

[3] ———, ch. VII.

[4] "Earth Manual," first edition, revised, second printing, 1968, Bureau of Reclamation (for sale by the Superintendent of Documents, U.S. Government Printing Office, Washington, D.C. 20402).

[5] Abbett, R. W., "Engineering Contracts and Specifications," fourth edition, John Wiley & Sons, New York, 1963.

[6] Seeyle, E. E., "Specifications and Costs," third edition, John Wiley & Sons, New York, 1957.

[7] "Standard Specifications for Repair of Concrete," Bureau of Reclamation, November 1970.

Appendix H

Sedimentation

ROBERT I. STRAND [1]

A. RESERVOIR SEDIMENTATION

H-1. *General.*—All reservoirs formed by dams on natural water courses are subject to some degree of sedimentation. The problem confronting the project planner is to estimate the rate of sedimentation and the period of time before the sediment will interfere with the useful functions of the reservoir. Provisions should be made for sufficient sediment storage in the reservoir at the time of design so as not to impair the reservoir functions during the useful life of the project or during the period of economic analysis. The replacement cost of storage lost to sediment accumulation in American reservoirs amounts to millions of dollars annually [1][2].

A dam on a stream upsets the natural equilibrium of the stream by changing the characteristics of discharge and the sediment transport capability. Additional problems of aggradation and degradation may be created upstream and downstream from the structure.

H-2. *Sediment Inflow.*—Sediment is the end product of erosion or wearing away of the land surface by the action of water, wind, ice, and gravity. Water resource development projects are most affected by sediment which is transported by water. The total amount of onsite sheet and gully erosion in a watershed is known as the *gross erosion.* However, all the eroded material does not enter the stream system; some of the material is deposited at natural or man-made barriers within the watershed and some may be deposited within the channels and their flood plains. The portion of eroded material which does travel through the drainage network to a downstream measuring or control point is referred to as the *sediment yield.* The sediment yield per unit of drainage area is the *sediment yield rate.*

(a) *Sediment Yield Rate Factors.*—The factors which determine the sediment yield of a watershed can be summarized as follows:

(1) Rainfall—amount and intensity.
(2) Soil type and geologic formation.
(3) Ground cover.
(4) Land use.
(5) Topography.
(6) Nature of drainage network—density, slope, shape, size, and alinement of channels.

Some researchers have deemed it necessary to include some additional factors; however, even the six above are interrelated. As an example, a heavy vegetative cover is dependent upon at least a moderate amount of rainfall; however, the ground cover conditions could be upset by tillage practices, overgrazing, or fire.

Systems of weighting the individual sediment influencing factors have been devised [2] to arrive at a sediment yield rate for an individual drainage basin. This type of analysis is best applied to preliminary planning studies and has its greatest reliability when the yield rates can be correlated with a measured sediment yield from an adjacent basin or subbasin.

(b) *Reservoir Resurvey Data.*—A resurvey of an existing reservoir in the same general area as the proposed development will provide

[1] Hydraulic Engineer, Sedimentation Section, Bureau of Reclamation.

[2] Numbers in brackets refer to items in the bibliography, section H-5.

valuable data in determining sediment yield rates. The Agricultural Research Service of the U.S. Department of Agriculture, in behalf of member agencies of the Water Resources Council, periodically publishes a summary of these reservoir resurveys [3]. Adjustments in the sediment yield rate will usually be necessary to account for variation in drainage area characteristics. One of the most important variations is the size of the drainage basin. Some investigators have found that the sediment yield varies with the 0.8 power of the drainage area size [1]. Figure H–1 is a plot of sediment yield rate versus drainage area, which was developed from selected reservoir resurvey data in southwestern United States.

(c) *Sediment Sampling Data.*—Sampling is the surest method of obtaining an accurate determination of the sediment load being carried by a stream at a particular location and is the preferred method for determining the sediment inflow to a reservoir. A thorough discussion of sediment samplers and the techniques of sample collection is beyond the scope of this appendix. The reader is referred to two series of reports prepared by the Federal Inter-Agency Sedimentation Project [4].

There are two basic types of suspended sediment data collection programs, periodic and daily. Collection and analysis of sediment samples is an expensive process and daily sampling yields a good deal of duplication at base flows. For these reasons, periodic or miscellaneous sampling programs are more common. Results from either type of collection program are used to develop a correlation between sediment load and water discharge. This correlation is commonly referred to as a sediment-rating curve. The data are normally plotted on logarithmic paper, with the sediment discharge as the abscissa and the water discharge as the ordinate. A line of best fit is then drawn through the plotted points, usually by visual examination of the plotted data, which allows the use of a curve expressing more than one mathematical relationship between water

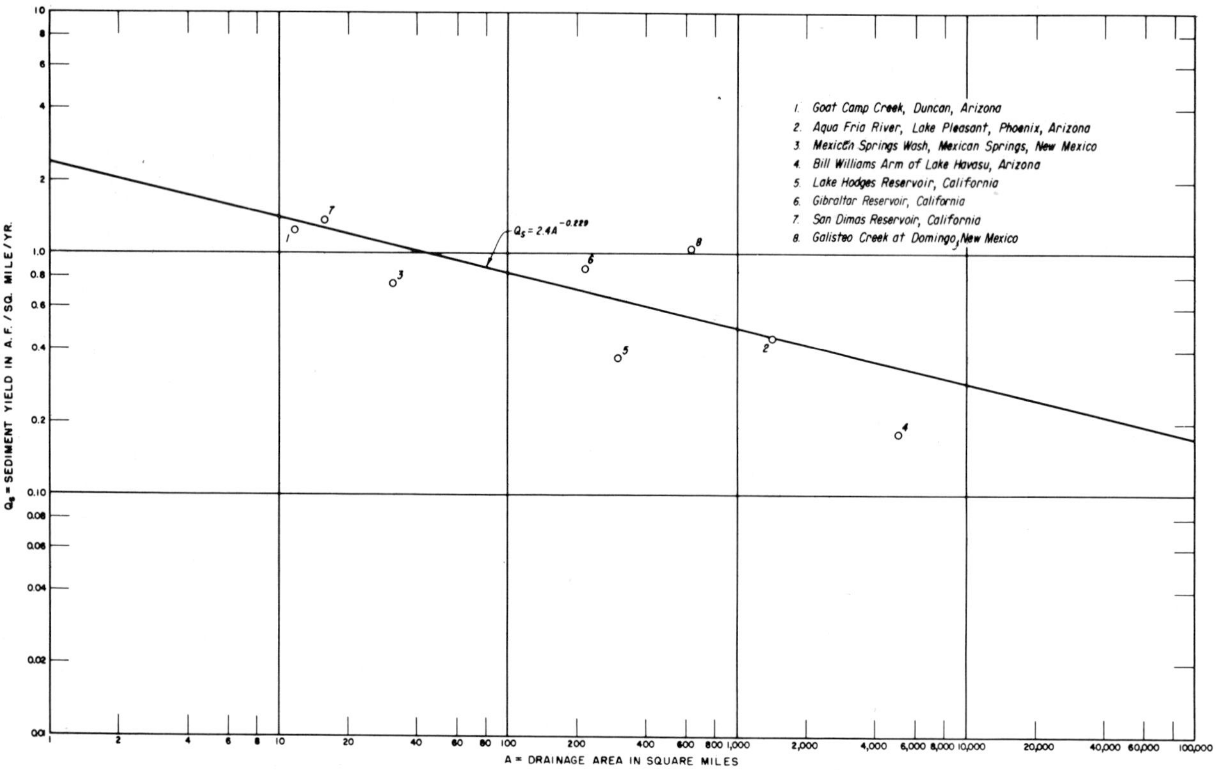

Figure H–1. Average annual sediment yield rate versus drainage area size. 288–D–2842.

and sediment discharge. Periodic sampling data often do not provide adequate definition for the upper portion or flood discharge region of the sediment-rating curve which may significantly affect the average rate of sediment transport because of the extremely large loads carried during flood periods.

When the source of runoff is either snowmelt or rainstorm, it may be necessary to develop individual sediment-rating curves for each of the seasons. Runoff from thunderstorms will carry larger sediment concentrations than runoff from other sources. Figure H–2 is a sample of seasonal rating curves for 19 years of sediment sampling data from the San Juan River at Bluff, Utah.

The water discharge records at a gaging station are usually available for a longer period and are more complete than the sediment data.

TABLE H–1.—*Class limits for discharges on flow-duration table*

Range in daily discharge, c.f.s.-days				
1 log cycle	2 log cycles	3 log cycles	4 log cycles	5 log cycles
10	10	10	10	10
11	12	15	15	15
12	14	20	20	20
13	17	25	30	30
14	20	30	40	50
15	25	40	50	70
16	30	50	70	100
18	35	60	100	150
20	40	80	150	etc.
22	45	100	etc.	
24	50	150		
26	60	etc.		
28	70			
30	80			
33	100			
36	120			
40	etc.			
45				
50				
55				
60				
65				
70				
75				
80				
90				
100				

NOTE.—Table shows sequence of numbers for five ranges in discharge. Locate decimal point and starting discharge to suit conditions. In general, use cycle closest to observed range. Where considerable data have been tabulated at a station by using other rules for class intervals, additional data would normally be tabulated by using the class intervals previously used.

These data are normally used to construct a flow-duration curve, which is really a cumulative frequency that shows the percent of time that specific discharges were equaled or exceeded in a given period. A flow-duration curve based upon a time unit greater than a day has little value in preparing sediment load estimates. To prepare a flow-duration curve, the recorded daily flows covering the period of record are arranged according to magnitude and the percent of time that the flow equaled or exceeded the specified values is computed. In a strict sense, the flow-duration curve applies only to the period for which data were used to develop the curve; however, if streamflow data represent the long-term flow of the stream, the curve may be considered a probability curve and used for estimation of future flows.

The data should be divided into class intervals, which will provide 20 to 30 well-distributed points on the flow-duration curve [5]. Table H–1 shows suggested class intervals in cubic feet per second for ranges in discharge from 1 to 5 log cycles.

Table H–2 is an example of how the data for a one-year period are tabulated using the records of the San Juan River near Bluff, Utah, gaging station, which represents a stream having a 3 log cycle range in discharge. A similar table is used to define the flow-duration curve for the total period of record, except that the "Total" column will contain the number of events in each class interval for the total period of record. The events recorded in each class interval are equal to or greater than the figure in the "Discharge" column, but less than the discharge shown on the next line. It is usually advisable to plot the maximum discharge as an individual point to aid in defining the upper portion of the curve.

The curves shown in figure H–3 represent the 19-year period of discharge record (1930-48) for the San Juan River near Bluff, Utah. The flow duration has been divided to correspond with the seasonal sediment-rating curves. The curves have been plotted on logarithmic probability paper, which is recommended for general use because it straightens the curve and expands the end portion where flood discharge occurrences must be read.

TABLE H-2.—*Tabulation of flow-duration data for San Juan River at Bluff, Utah, for water year 1941.* 288-D-2860.

Daily Discharge Data - San Juan River near Bluff, Utah

Water Year 1941

Date 1-27-70 Sheet 1 of 1

Discharge (c.f.s.)	Oct	Nov	Dec	Jan	Feb	Mar	April	May	June	July	Aug	Sept	Year	Total	Cum Total	% Time
600		卌	|	||||								卌|||	18	18	365	100.0
800	|||	卌卌卌|	卌卌|||	卌|||	||						|	卌	48	48	347	95.1
1,000	卌卌|	卌	卌卌	卌卌|||	卌|||						卌|		53	53	299	81.9
1,500	卌	||	|||	||||	卌|||	卌||					|||		33	33	246	67.4
2,000	|||		|			卌|					卌|		16	16	213	58.4
2,500	|	|	||	|	|||						卌		13	13	197	54.0
3,000	||||	|	|	|	|	卌				|	卌	卌||	26	26	184	50.4
4,000	|				卌|	卌||||	卌卌||||			|||	|	|||	37	37	158	43.3
5,000	|					||||	||||			|	|	|||	14	14	121	33.2
6,000	|						卌|||			卌卌	|	|	21	21	107	29.3
8,000	|									|||	|	|	6	6	86	23.6
10,000							|||	||||	卌卌卌|	卌卌|||		||	38	38	80	21.9
15,000							|	卌|||	卌卌|				20	20	42	11.5
20,000								卌卌|	|||				14	14	22	6.0
25,000								卌|					6	6	8	2.19
30,000								|					1	1	2	0.55
33,800								|					1	1	1	0.27
	31	30	31	31	28	31	30	31	30	31	31	30	365	365		

Short-term records at a gaging site may be adjusted to long-term representations by several methods. The preferred method now used by the U.S. Geological Survey [5] consists of first developing a flow-duration curve for the period of concurrent record at both the station at which records are to be expanded and at a station in the same area having similar climatic and physical drainage basin characteristics, but a longer period of recorded discharge. A relationship is then established between the two stations for the shorter period of concurrent record by plotting the discharges for specific duration points against the corresponding discharges at the other. Then, using the discharges from the same duration points on the long-term flow-duration curve to enter the correlation curve, new long-term discharge points can be read for the shorter term station.

Tables H-3 and H-4 are the computation forms used to compute the mean annual sediment load for the San Juan River near Bluff, Utah, gaging station. Figures from column 3 of the form are used to enter the flow-duration curve and the discharges so obtained are entered in column 4. The sediment-rating curve is then entered with discharge figures from column 4 and the corresponding sediment discharges are entered in column 5. Figures in columns 6 and 7 are then the products of the decimal equivalent of column 2 and columns 4 and 5, respectively. The sum of column 6 should equal the mean daily discharge of the station for the duration period covered—in this case, 1930-48. If the deviation is greater than one percent, the flow-duration curve should be adjusted.

(d) *Unmeasured Sediment Load.*—To complete the total load computation as shown in tables H-3 and H-4, some method of estimating the unmeasured portion of the total sediment load is required. The following terms will recur in the following discussion and are defined as follows [4]:

Bed material—the sediment mixture of which the streambed is composed.

Bedload—sediment that moves by saltation, rolling, or sliding on or near the streambed.

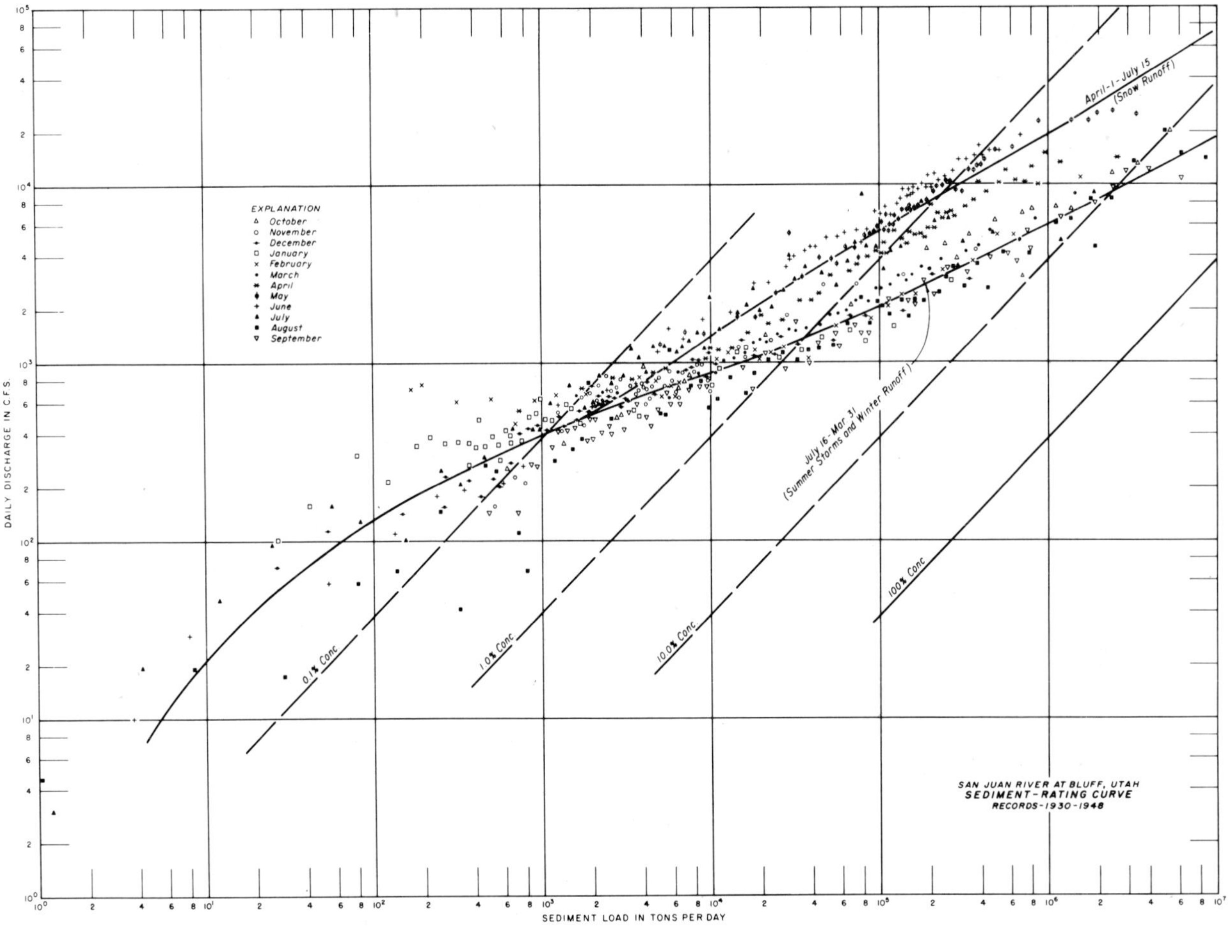

Figure H–2. Sediment-rating curves. 288–D–2843.

Bed material load—that part of the sediment load which consists of grain sizes represented in the bed.

Wash load—that part of the sediment load which consists of grain sizes finer than those of the bed.

Suspended load—particles moving outside the bed layer.

Unsampled zone—the 3 or 4 inches from the streambed up to the lowest point of the sampling vertical. Most suspended sediment samplers cannot sample within this zone.

Computations of sediment load based upon suspended sediment sampling alone assume that the measured concentrations are also applicable to the unsampled zone. Total load measurements have shown that this assumption is not valid and that some method of adjustment for greater concentrations in the unsampled zone is advisable. Therefore, the adjustment to the measured sediment loads is referred to as an unmeasured load adjustment because it should include additions for bedload movement and proportionally larger suspended loads within the unsampled zone.

Several methods or formulae for computing the quantity of bedload transport have been advanced by various investigators over the years. These bedload formulae are based on the principle that the capacity of the stream to transport sediment along its bed varies directly with the difference between the shear stress acting on the bed particles and the critical shear stress required for initiation of particle motion [6]. Some of the earlier formulae, such as those of Schoklitsch (1934) [7] and Meyer-Peter and Muller (1948) [8, 21], were

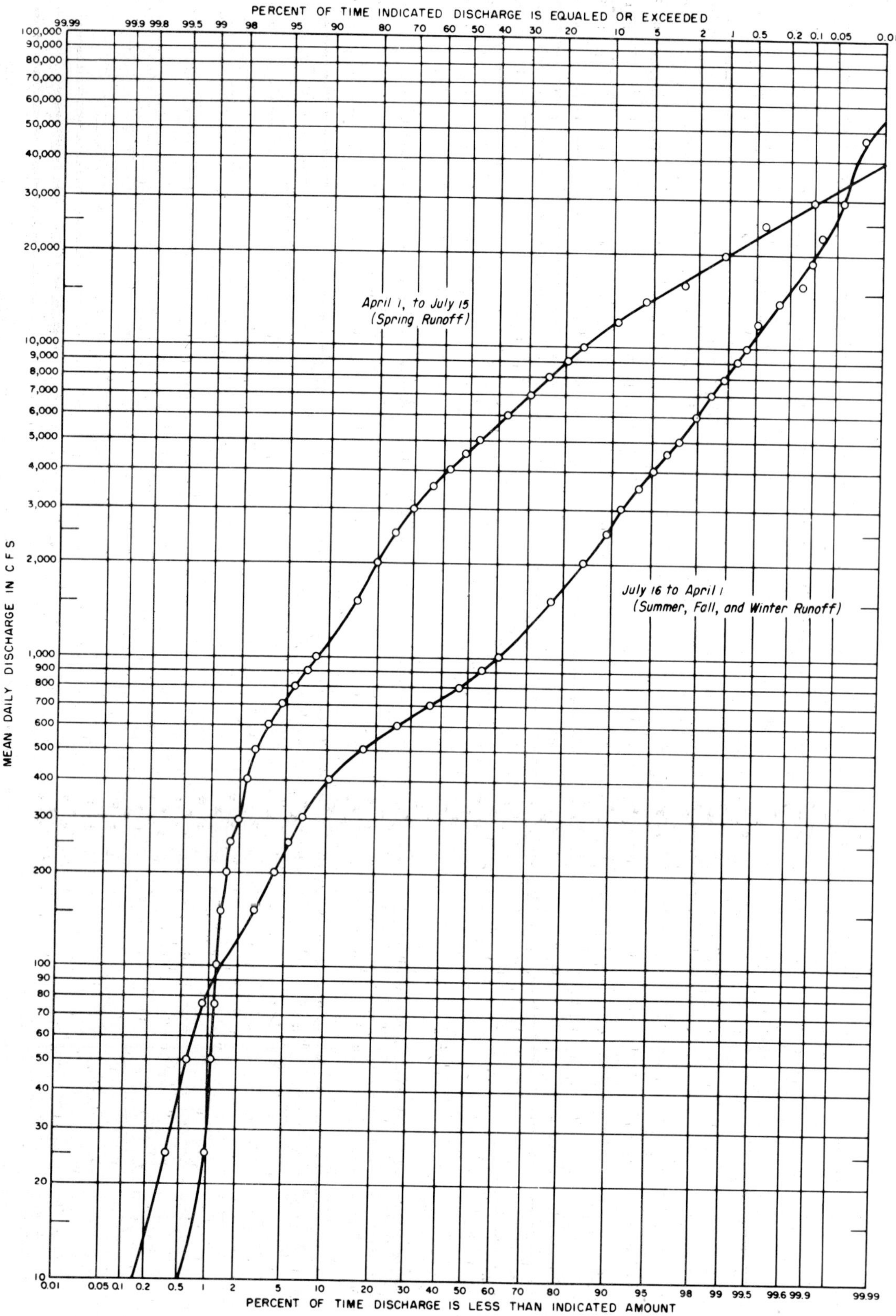

Figure H–3. Flow-duration curves. 288–D–2844.

based purely on experimental results. More recent formulae, such as Einstein (1950) [9], have a semi-theoretical background. Statistical and probability theories are used as a basis for the formulae and experimental results are used for evaluation of various constants and indices.

The U.S. Geological Survey has developed a Modified Einstein procedure for computing total sediment discharge that is dependent upon a suspended sediment concentration and size analysis in addition to the data required by the other formulae [10, 11, 12].

A description of the theory and development of the bedload formulae is much beyond the scope of this narrative and the reader is directed to the listed references for this information. The remainder of the discussion will be limited to examples of applications of the various formulae.

The Schoklitsch formula may be expressed:

$$G=43.2\ B\sum_{i=1}^{i=n}\left[\frac{437.9}{\sqrt{D_i}}S^{3/2}\left(q-q_{o_i}\right)\right] \tag{1}$$

in which

G=bedload transport in tons/day
B=stream width in feet
D_i=geometric mean diameter of each individual size fraction in millimeters
S=hydraulic gradient
q=unit discharge in cfs/foot of width
$q_{o_i}=\frac{0.00021\ D_i}{S^{4/3}}$ unit discharge necessary to initiate movement in c.f.s per foot of width
n=number of individual size fractions into which the bed material is divided.

Application of this formula will yield an estimate of the bedload discharge rate for one discharge. At some discharges the formula may also yield a part of the sand load in the unmeasured zone.

The Meyer-Peter, Muller formula may be used in the form:

$$G=1.606\ B\left[3.306\left(\frac{Q_B}{Q}\right)\left(\frac{D_{90}{}^{1/6}}{n_s}\right)^{3/2}dS-0.627\ D_m\right]^{3/2} \tag{2}$$

in which

G=bedload in tons/day
B=stream width in feet
Q_B=water discharge quantity directly over the area of bedload transport in c.f.s.
Q=total water discharge quantity in c.f.s.
D_{90}=particle size in millimeters at which 90 percent of the bed material is finer
n_S=Manning's n value for the bed of the stream
D_m=effective size of bed material in millimeters usually determined as the weighted mean diameter
d=mean channel depth in feet
S=hydraulic gradient.

Reference [8] provides nomographs to be used in solving for values of n_S and Q_B; however, the following equations may be used to solve for these values if the nomographs are not available:

$$n_s=n_m\left[1+\frac{2d}{B}\left\{1-\left(\frac{n_w}{n_m}\right)^{3/2}\right\}\right]^{2/3} \tag{3}$$

where

n_m=Manning's n for the total channel section
n_w=Manning's n for the channel side walls and

$$Q_B=\frac{Q}{1+\frac{2d}{B}\left(\frac{n_w}{n_s}\right)^{3/2}} \tag{4}$$

The Meyer-Peter, Muller formula will also give an estimate of the bedload movement for a single discharge.

Tables H–5 and II–6 are examples of results of Modified Einstein and Einstein Bedload Function computations, as obtained from a digital computer program developed by the Bureau of Reclamation. Computers have eliminated the laborious computation procedures previously required for these methods and have greatly enhanced their applicability. The total of the column "Computed Total Load" in tables H–5 and H–6 represents the combination of bed material moving in suspension and as bedload; however, table H–6 does not include the quantity of the wash load.

TABLE H-3.—*Computation of sediment load for San Juan River at Bluff, Utah—spring runoff.* 288-D-2861.

HD-32b
(Nov. 52)
Bureau of Reclamation

SEDIMENT LOAD

SECTION	PERIOD 1930-1948 April 1 to July 15	RIVER San Juan at Bluff
COMPUTED BY	CHECKED BY	DATE 2-19-51

1 % LIMITS	2 % INTERVAL	3 % MID. ORD.	4 Qw	5 Qs	6 2x4 Qw. DISCH.	7 2x5 Qs. DISCH.
0.00-0.02	0.02	0.01	40,000	3,500,000	8.0	700
0.02-0.1	0.08	0.06	32,000	2,400,000	25.6	1,920
0.1 -0.5	0.4	0.3	25,000	1,550,000	101.2	6,200
0.5 -1.5	1.0	1.0	20,700	1,130,000	207.0	11,300
1.5 -5.0	3.5	3.25	16,300	740,000	570.5	20,450
5-15	10	10	11,700	400,000	1,170.0	40,000
15-25	10	20	8,900	245,000	890.0	24,500
25-35	10	30	7,000	157,000	700.0	15,700
35-45	10	40	5,650	107,000	565.0	10,700
45-55	10	50	4,720	79,000	472.0	7,900
55-65	10	60	3,750	52,000	375.0	5,200
65-75	10	70	2,940	33,500	294.0	3,350
75-85	10	80	1,950	17,000	195.0	1,700
85-95	10	90	1,100	7,000	110.0	700
95-98.5	3.5	96.75	540	2,100	18.9	74
98.5-99.5	1.0	99.0	25	12	0.25	0.12
99.5-99.9	0.4	99.7	7.5	4.5	0.03	0.02
99.9-99.98	0.08	99.94	2.5	2.0	0.02	0.0
99.98-100	0.02	99.99	0	0	0	0
				TOTAL	5,702.5	150,394

Qw.A.D. = 5,702.5 D.D.x106x1.9835 = 1,200,000 (AF)/yr.

Qs.A.D. = 150,394 D.D.x106 = 15,950,000 Tons/yr.

20 Percent Correction for Bedload = 3,200,000 Tons/yr.

Total Sediment Discharge = 19,150,000 Tons/yr.

Sediment

A.D. = $\frac{19,150,000}{1361}$ $\frac{\text{Tons/yr.}}{\text{Tons/(AF)}}$ = 14,070 (AF)/yr.

Yield = $\frac{19,150,000}{1361 \times 23,000}$ $\frac{\text{Tons/yr.}}{\text{Tons/(AF)xD.A.}}$ = 0.612 (AF)/sq.mi.

Concentration = $\frac{15,950,000 \times 100}{1,200,000 \times 1361}$ $\frac{\text{Qs A.D.x100}}{\text{Qw A.D.x1361}}$ = 0.978 Percent

Runoff

Rate = $\frac{1,200,000}{23,000}$ $\frac{\text{Qw A.D.}}{\text{D.A.}}$ = 52.17 (AF)/sq.mi.

D.D.= Daily Discharge
A.D.= Annual Discharge
D.A.= Drainage Area
= lbs./cubic foot
= Tons/acre foot

TABLE H-4.—*Computation of sediment load for San Juan River at Bluff, Utah—summer and winter runoff.* 288–D–2862.

HD-32b
(Nov. 52)
Bureau of Reclamation

SEDIMENT LOAD

SECTION	PERIOD 1930-1948 July 16 to April 1	RIVER San Juan at Bluff
COMPUTED BY	CHECKED BY	DATE 2-19-51

1	2	3	4	5	6	7
% LIMITS	% INTERVAL	% MID. ORD.	Qw	Qs	2x4 Qw. DISCH.	2x5 Qs. DISCH.
0.00–0.02	0.02	0.01	56,000	78,000,000	11.2	15,600
0.02–0.1	0.08	0.06	24,800	16,000,000	19.84	13,000
0.1 –0.5	0.4	0.3	13,200	5,100,000	52.8	20,400
0.5 –1.5	1.0	1.0	8,300	2,000,000	83.0	20,000
1.5 –5.0	3.5	3.25	4,800	630,000	168.0	22,000
5–15	10	10	2,650	117,500	265.0	11,750
15–25	10	20	1,660	60,000	166.0	6,000
25–35	10	30	1,230	28,000	123.0	2,800
35–45	10	40	985	14,700	98.5	1,470
45–55	10	50	830	8,700	83.0	870
55–65	10	60	730	5,600	73.0	560
65–75	10	70	640	3,900	64.0	390
75–85	10	80	530	2,250	53.0	225
85–95	10	90	378	960	37.8	96
95–98.5	3.5	96.75	165	160	5.8	5.6
98.5–99.5	1.0	99.0	79	47	0.8	0.5
99.5–99.9	0.4	99.7	19.5	9.4	0.08	0.04
99.9–99.98	0.08	99.94	4.0	2.8	0.01	0.0
99.98–100	0.02	99.99	0	0	0	0
				TOTAL	1,304.8	93,167

Qw.A.D. = 1,304.8 D.D.x259x1.9835 = 670,000 (AF)/yr.

Qs.A.D. = 93,167 D.D.x 259 = 24,100,000 Tons/yr.

10 Percent Correction for Bedload = 2,410,000 Tons/yr.

Total Sediment Discharge = 26,510,000 Tons/yr.

Sediment

A.D. = $\frac{26,510,000}{1361}$ $\frac{\text{Tons/yr.}}{\text{Tons/(AF)}}$ = 19,478 (AF)/yr.

Yield = $\frac{26,510,000}{1361 \times 23,000}$ $\frac{\text{Tons/yr.}}{\text{Tons/(AF)xD.A.}}$ = 0.847 (AF)/sq.mi.

Concentration = $\frac{24,100 \times 100}{670,000 \times 1361}$ $\frac{\text{Qs A.D.x100}}{\text{Qw A.D.x1361}}$ = 2.64 Percent

Runoff

Rate = $\frac{670,000}{23,000}$ $\frac{\text{Qw A.D.}}{\text{D.A.}}$ = 29.13 (AF)/sq.mi.

D.D.= Daily Discharge
A.D.= Annual Discharge
D.A.= Drainage Area
= lbs./cubic foot
= Tons/acre foot

TABLE H–5.—*Computer output of sediment load computations using the Modified Einstein Procedure*

Output
Determination of total sediment load in a stream
Method of computation—Modified Einstein Date of computation—February 1968
Jobident—Middle Loup River at Dunning, Nebraska—Section C2
Date of sample—May 22, 1951 Time of sample—12:15 Temperature—61.0° F Slope of energy gradient—.00127 ft./ft.
Discharge—499 CFS Concentration in ppm—726 Sampled sediment—978 tons/day
D65 = .920 millimeter D35 = .430 millimeter
Area—172 ft. sq. Top width—0. ft. Equiv. depth—2.05 ft. Equiv. slope = 0. ft./ft.
Velocity—2.90 ft./sec. Equiv. width—84.0 ft. Average bottom depth—0. ft. Hydr. radius = 2.05 ft.
Distance between sampler and bed (DSUBN)—.40 ft. Average depth from sample verticals (DSUBS)—2.05 ft.

Size fraction		Percent of material		IBQB ton/day	Qprime Subs (t/d)	Z—Values		Computational factors		Computed total load
Inch	Millimeters	Suspended	Bed			Computed	Fitted	F (J)	F (I)+1	
.0020	.0625	27.00	0.	0.	221.7	0.	.00	1.20	0.	265.2
.0625	.1250	27.00	1.00	.08	221.7	.08	.09	0.	3534.25	276.7
.1250	.2500	36.00	9.00	1.99	295.6	.38	.32	0.	235.78	469.9
.2500	.5000	9.00	35.00	21.92	73.9	.89	.81	0.	13.25	290.6
.5000	1.0000	1.00	23.00	40.74	8.2	1.38	1.60	0.	3.74	152.3
1.0000	2.0000	0.	17.00	43.41	0.	0.	2.63	0.	1.69	73.4
2.0000	4.0000	0.	9.00	6.61	0.	0.	3.95	0.	1.41	9.3
4.0000	8.0000	0.	6.00	.10	0.	0.	5.76	0.	1.28	.1
Totals......	....	100.00	100.00	....	821.2	....	....	....	....	1537.6

Once the rate of unmeasured sediment movement has been determined for several discharges, a bedload rating curve can be drawn and computations similar to those in tables H–3 and H–4 can be made for bedload discharge.

For preliminary planning purposes, the information in table H–7 [13] may be used to estimate the quantity of bedload correction to be used to complete the total load computations.

H-3. *Reservoir Sediment Deposition.*—Once the estimated sediment inflow to a reservoir has been established, attention must be given to the effect the deposition of this sediment will have upon the life and daily operation of the reservoir. The mean annual sediment inflow, the trap efficiency of the reservoir, the ultimate specific weight of the deposited sediment, and the distribution of the sediment within the reservoir all must be considered in the design of the dam. Usually to prevent premature loss of usable storage capacity, an additional volume of storage equal to the anticipated sedi-

TABLE H–6.—*Computer output of sediment load computations using the Einstein Bedload Function*

Output
Determination of total sediment load in a stream
Method of computation—Original Einstein Date of computation—February 1968
Jobident—Middle Loup River at Dunning, Nebraska—Section C2
Date of sample—May 22, 1951 Time of sample—12:15 Temperature—61.0° F Slope of energy gradient—.00127 ft./ft.
Discharge—499 CFS Concentration in ppm—726 Sampled sediment—978 tons/day
D65 = .920 millimeter D35 = .430 millimeter
Area—172 ft. sq. Top width—0. ft. Equiv. depth—2.05 ft. Equiv. slope = 0. ft./ft.
Velocity—4.92 ft./sec. Equiv. width—84.0 ft. Average bottom depth—0. ft. Hydr. radius = 1.28 ft.
Distance between sampler and bed (DSUBN)—.40 ft. Average depth from sample verticals (DSUBS)—2.05 ft.

Size fraction		Percent of material		IBQB ton/day	Qprime Subs (t/d)	Z—Values		Computational factors		Computed total load
Inch	Millimeters	Suspended	Bed			Computed	Fitted	F (J)	F (I)+1	
.0020	.0625	0.	0.	0.	0.	0.	0.	0.	0.	0.
.0625	.1250	0.	1.00	.07	0.	.21	0.	0.	1243.50	85.7
.1250	.2500	0.	9.00	14.23	0.	.65	0.	0.	45.58	648.5
.2500	.5000	0.	35.00	122.40	0.	1.54	0.	0.	2.78	340.8
.5000	1.0000	0.	23.00	60.60	0.	2.86	0.	0.	1.51	91.5
1.0000	2.0000	0.	17.00	15.61	0.	4.50	0.	0.	1.30	20.3
2.0000	4.0000	0.	9.00	.55	0.	6.52	0.	0.	1.21	.7
4.0000	8.0000	0.	6.00	0.	0.	9.19	0.	0.	1.16	0.
Totals......	....	0.	100.00	....	0.	....	....	....	....	1187.5

TABLE H-7.—*Bedload correction table*

Concentration of suspended load, ppm	Type of material forming the channel of the stream	Texture of suspended material	Percent bed load in terms of measured suspended load
Less than 1,000	Sand	Similar to bed material.	25 to 150
Less than 1,000	Gravel, rock, or consolidated clay.	Small amount of sand.	5 to 12
1,000 to 7,500	Sand	Similar to bed material.	10 to 35
1,000 to 7,500	Gravel, rock, or consolidated clay.	25 per cent sand or less.	5 to 12
Over 7,500	Sand	Similar to bed material.	5 to 15
Over 7,500	Gravel, rock, or consolidated clay.	25 per cent sand or less.	2 to 8

ment deposition during the economic life of the reservoir is included in the original design. The Bureau of Reclamation requires that provisions be made for sediment storage space whenever the anticipated sediment accumulation during the period of project economic analysis exceeds 5 percent of the total reservoir capacity.

(a) *Trap Efficiency.*—The trap efficiency of a reservoir is defined as the ratio of the quantity of deposited sediment to the total sediment inflow and is dependent primarily upon the sediment particle fall velocity and the rate of flow through the reservoir. Particle fall velocity may be influenced by size and shape of the particle, viscosity of the water, and chemical composition of the water. The rate of flow through the reservoir is determined by the volume of inflow with respect to available storage and the rate of outflow.

Methods for estimating reservoir trap efficiency are empirically based upon measured sediment deposits in a large number of reservoirs. Gunnar Brune [14] has presented a set of envelope curves for use with normal ponded reservoirs using the capacity-inflow relationship of the reservoirs. These curves are reproduced in figure H-4. The reservoirs used to develop this relationship are typical storage reservoirs and the curves are not recommended for use in computing trap efficiencies of desilting basins, flood retarding structures, or semidry reservoirs.

Using data from Tennessee Valley Authority reservoirs, M. A. Churchill [15] developed a relationship between the percent of incoming sediment passing through a reservoir and the *sedimentation index* of the reservoir. The sedimentation index is defined as the ratio of the period of retention to the mean velocity through the reservoir. The Churchill curve is shown in figure H-5 with some additional data points which were added by the Bureau of Reclamation. Some of these data points represent desilting basins and semidry reservoirs and it appears the Churchill curve is more applicable in defining trap efficiencies of these types of reservoirs than the Brune relationship.

The following description of terms will be helpful in using the Churchill curve:

Capacity—capacity of the reservoir in the mean operating pool for the period to be analyzed.

Inflow—average daily inflow rate during the study period.

Period of retention—capacity divided by inflow rate. Capacity is in cubic feet and inflow in cubic feet per second.

Length—reservoir length in feet at mean operating pool level.

Velocity—mean velocity in feet per second, which is arrived at by dividing the inflow by the average cross sectional area in square feet. The average cross sectional area can be determined from capacity divided by the length.

Sedimentation index—period of retention divided by velocity.

When the anticipated sediment accumulation is a large percentage of the reservoir capacity, it is necessary that the trap efficiency be analyzed for incremental periods of the reservoir life. Theoretically, the reservoir trap efficiency will decrease continuously once storage is begun; however, it is not practical to analyze the trap efficiency in intervals of less than 10 years. The variability of the annual sediment inflow is sufficient reason not to use shorter periods of analysis.

(b) *Unit Weight of Deposited Sediment.*—Generally the estimated sediment inflow to a reservoir is estimated in terms of weight per time, such as tons per day, and must be converted to a volume equivalent by means of an estimated unit weight [16]. The classification of sediment according to size as proposed by the

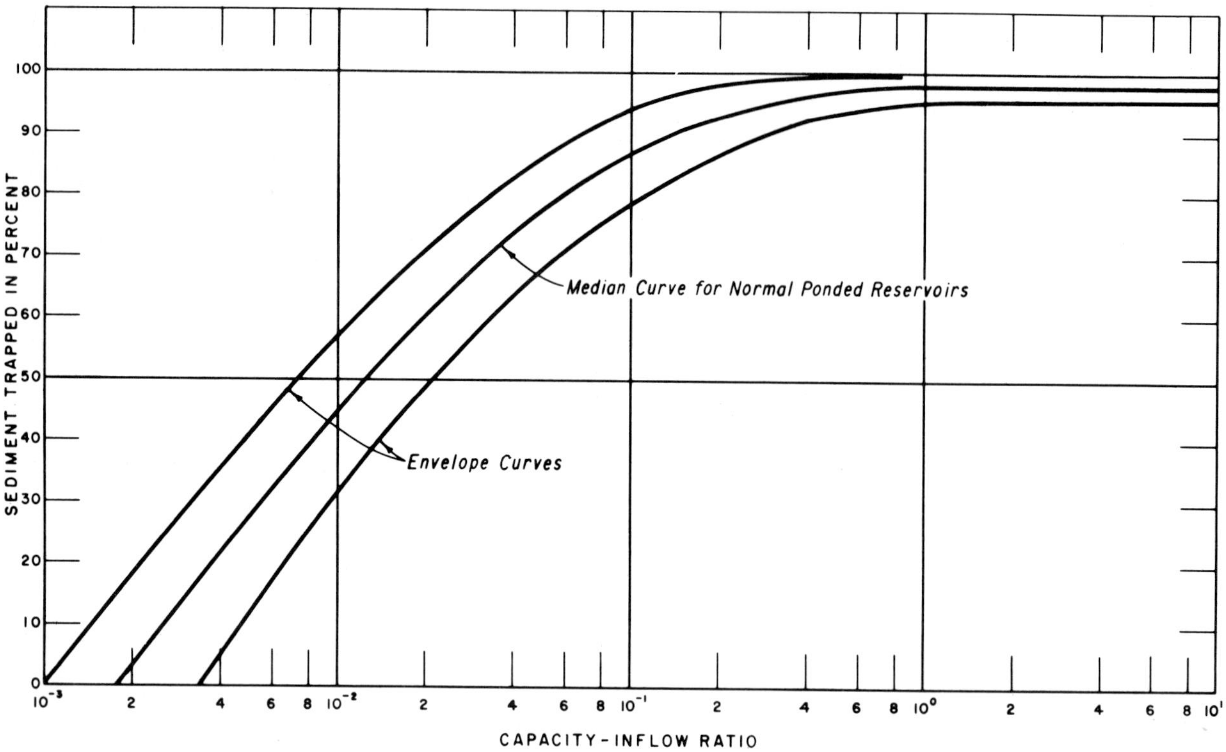

Figure H–4. Reservoir trap efficiency. Adapted from Brune [14]. 288–D–2845.

American Geophysical Union will be used herein:

Sediment type:	*Size range, millimeters*
Clay	less than 0.004
Silt	0.004 to 0.0625
Sand	0.0625 to 2.0

There are several factors influencing the unit weight of sediment that deposits in a reservoir. Some of those having the most pronounced effect are: (1) the manner in which the reservoir is to be operated, (2) the texture and size of the sediment particles, (3) the compaction or consolidation rate, and (4) other less influential factors such as action of density currents, thalweg slope of the incoming stream, and effect of vegetation in the reservoir headwater area.

The reservoir operation is probably the most influential of these factors. Sediments that have deposited in reservoirs subjected to considerable drawdown are exposed for long periods and undergo a greater amount of consolidation. Reservoirs operating with a fairly stable pool do not allow the sediment deposits to dry out and consolidate to the same degree.

The size of the incoming sediment particles has a significant effect upon the unit weight. Sediment deposits composed of silt and sand will have higher unit weights than those in which clay predominates.

Based on the results of unit weight and size distribution analyses of 1,316 samples, Lara and Pemberton [16] developed a method for estimating the initial unit weight of sediment deposits when the size analysis of the incoming sediment and the proposed reservoir operation scheme are known. Reservoir operations were classified according to type as follows:

Type	*Reservoir operation*
1	Sediment always submerged or nearly submerged.
2	Normally moderate to considerable reservoir drawdown.
3	Reservoir normally empty.
4	Riverbed sediments.

Selection of the proper reservoir type can usually be made from the operation study prepared for the reservoir.

Once the reservoir type has been selected, the initial unit weight of the sediment deposits

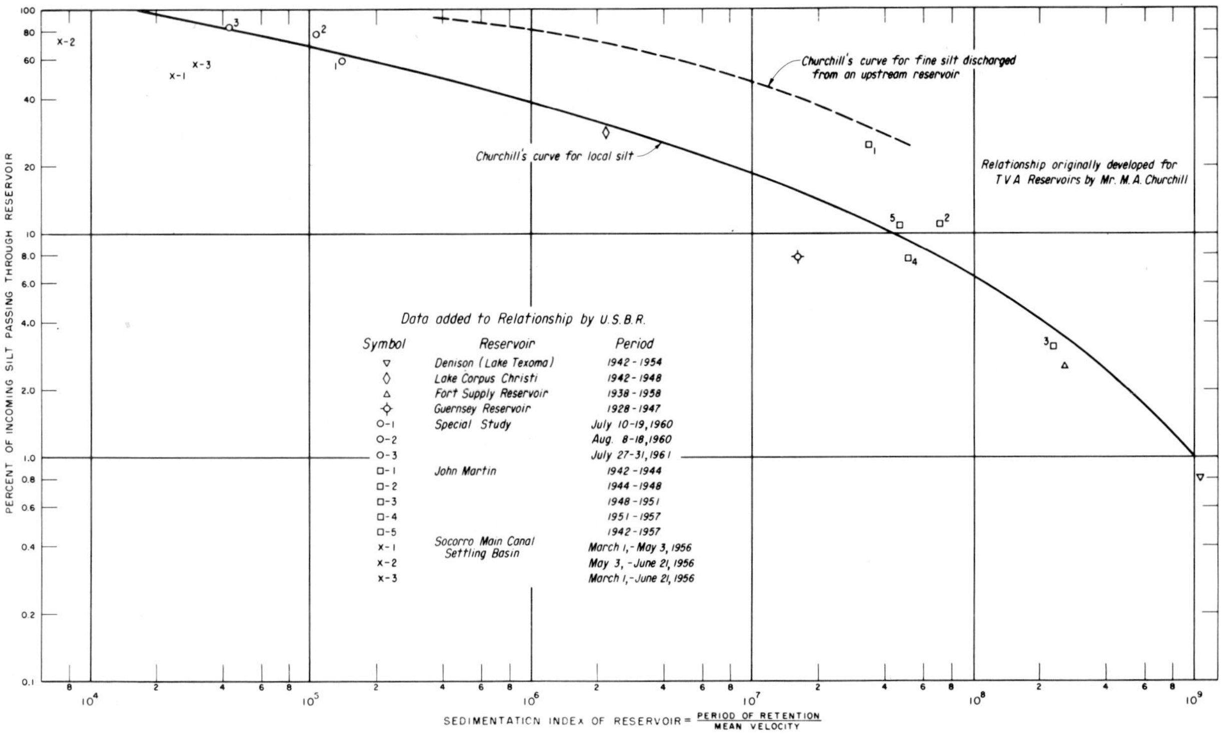

Figure H–5. Reservoir trap efficiency. Adapted from Churchill [15]. 288–D–2846.

can be estimated using the following equation:

$$\gamma = W_c p_c + W_m p_m + W_s p_s \qquad (5)$$

where:

γ=unit weight in pounds per cubic foot.

p_c, p_m, p_s=percentages of clay, silt, and sand, respectively, of the incoming sediment.

W_c, W_m, W_s=coefficients of clay, silt, and sand, respectively, which can be obtained from the following tabulation:

Reservoir type	W_c	W_m	W_s
1	26	70	97
2	35	71	97
3	40	72	97
4	60	73	97

As an example, the following data are known for a proposed reservoir:

Reservoir type: Type 1

Size analysis: 23 percent clay, 40 percent silt, and 37 percent sand

Then:

$$\gamma = 26\ (0.23) + 70\ (0.40) + 97\ (0.37) = 5.98 + 28.00 + 35.89 = 69.87 \text{ or } 70 \text{ lb./ft.}^3$$

The unit weight of the deposited sediment will increase each year it remains in the reservoir. The equation,

$$W = W_1 + K \log_{10} T \qquad (6)$$

in which

W=unit weight after T years of compaction

W_1=the initial unit weight

K=constant dependent upon the size analysis of the sediment, has been presented for use in determining the unit weight of sediment deposits in reservoirs after a period of reservoir operation [17]. However, part of the sediment will deposit in the reservoir in each of the T years of operation and each year's deposits will have a different compaction time. Miller [18] developed an approximation of the integral for determining the average unit weight of all sediment deposited in T years of operation as follows:

$$W_T = W_1 + 0.4343K\left[\frac{T}{T-1}(\log_e T) - 1\right] \qquad (7)$$

where

W_T=average unit weight after T years of reservoir operation

W_1=initial unit weight as derived from equation 5

K=constant based on type of reservoir operation and sediment size analysis as obtained from the following table:

Reservoir type	K		
	Sand	Silt	Clay
1	0	5.7	16.0
2	0	1.8	8.4
3	0	0.	0.

Using the same example as was used for the initial unit weight computation, the unit weight after 100 years can be computed as follows:

$$K = 16.0\ (0.23) + 5.7\ (0.40) + 0.0\ (.37)$$
$$= 3.68 + 2.28 + 0.0 = 5.96$$

$$W_{100} = 69.87 + 0.4343\ (5.96)\left[\frac{100}{99}(4.60517) - 1\right]$$
$$= 69.87 + 2.588\ (3.652) = 69.87 + 9.45$$
$$= 79.3 \text{ or } 79 \text{ lb./ft.}^3$$

This value may then be used to convert the weight of incoming sediment to the volume it will occupy in the reservoir after 100 years.

(c) *Sediment Distribution within a Reservoir.*—The forces acting upon a sediment particle entering a reservoir include a horizontal component in the direction of flow due to the force of the water movement and vertical components due to gravity and the water turbulence. A sediment particle will remain in suspension and be transported into the reservoir as long as the turbulence force equals or exceeds the force of gravity. When a flow enters the reservoir the increased cross sectional area results in a decrease in velocity and turbulence until they become ineffective in transporting the sediment and the particles are deposited.

The eventual distribution of sediment within the reservoir is dependent upon several interrelated factors, including texture of the sediment, inflow-outflow relations, size and shape of the reservoir, and reservoir operating procedure [1]. The sediment deposition is not necessarily confined to the lower storage increments of the reservoir.

Sediment accumulations in a reservoir are usually distributed below the top of the conservation pool or normal water surface. However, if the reservoir has a flood control pool and it is anticipated that the water surface will be held within this pool for significant periods of time, a portion of the sediment accumulation may be deposited within this pool. Figure H–6 is a plot of data from 11 Great Plains reservoirs which may be used as a guide in estimating the portion of the total sediment accumulation which will deposit above the normal water surface. This plot should be regarded as a rough guide only and the estimate obtained from it should be tempered with some judgment based upon the proposed reservoir operation and the nature of the incoming sediment. This curve is based on a limited amount of data and may be revised as more information becomes available.

The term *flood pool index* is computed as the ratio of the flood control pool depth to the depth below the pool, multiplied by the percent of time the reservoir water surface will be

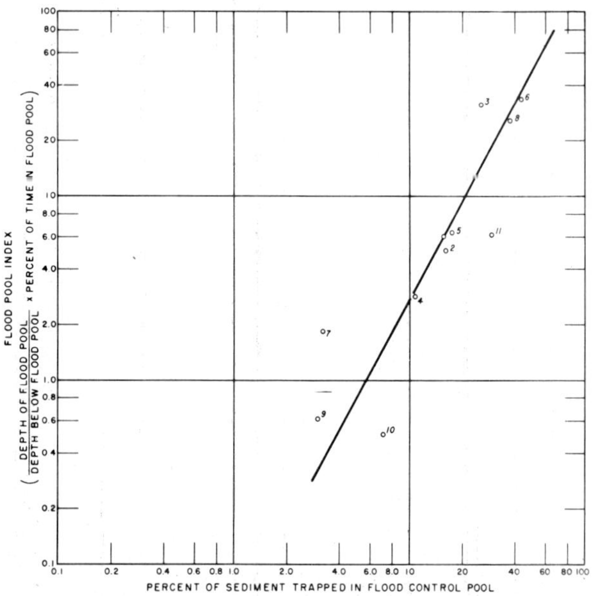

Figure H–6. Sediment deposited in flood control pool. 288–D–2847.

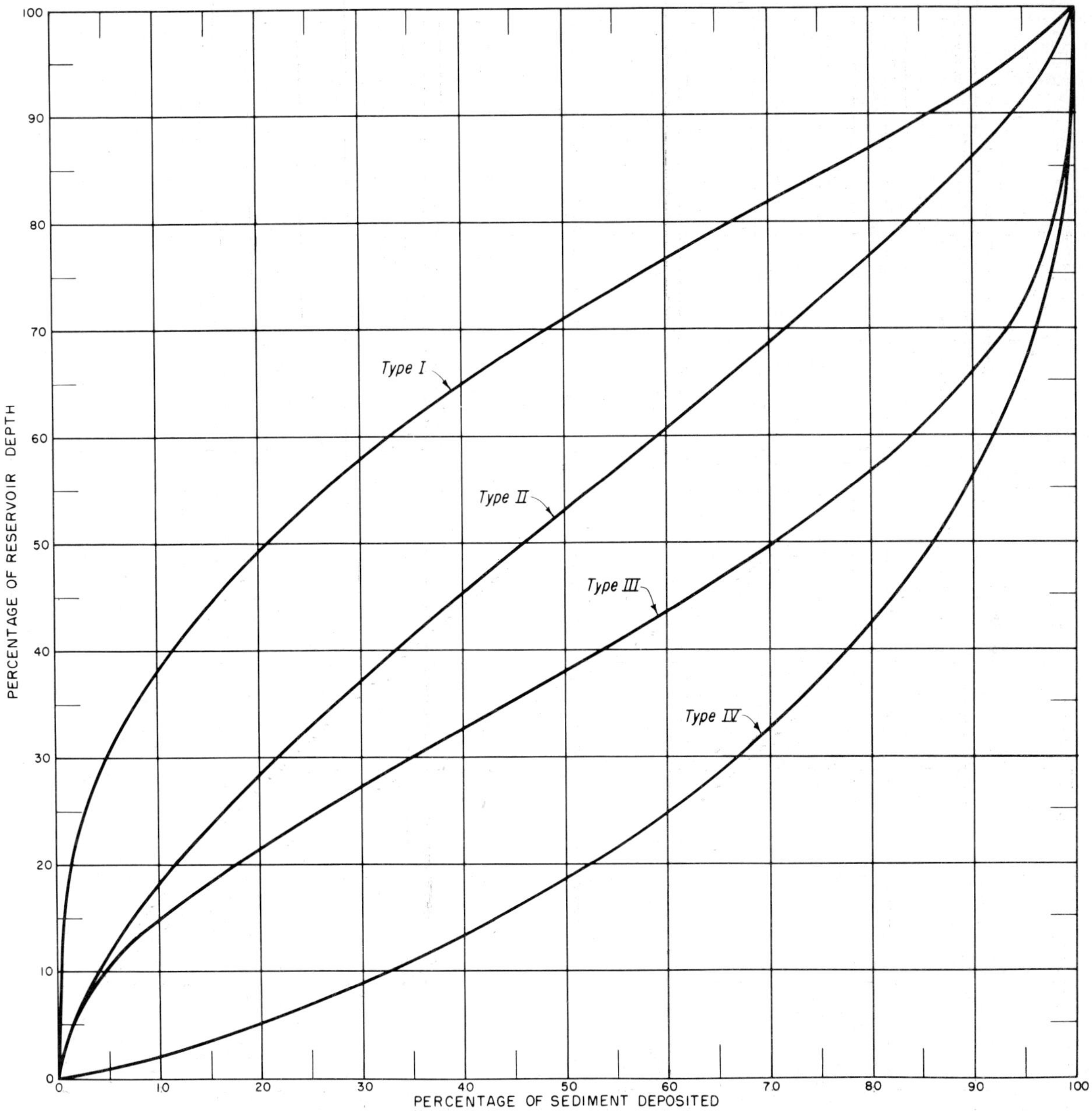

Figure H–7. Reservoir storage design curves. 288–D–2848.

within the flood control pool. This information for a proposed reservoir must be obtained from the reservoir operation study.

Once the quantity of sediment which will deposit below the normal water surface has been established, the Empirical Area-Reduction Method may be used to estimate the distribution at any given time [19] [20]. This method was developed from data gathered in the resurvey of 30 reservoirs. The data indicated that a definite relationship exists between the reservoir shape and the percentage of sediment deposited at various depths throughout the reservoir. The shape of the reservoir is defined by the depth to capacity relationship and the classification of reservoirs on this basis is as follows:

Reservoir type	Classification	m
I	Lake	3.5—4.5
II	Floodplain-foothill	2.5—3.5
III	Hill	1.5—2.5
IV	Gorge	1.0—1.5

where m is the reciprocal of the slope of the depth versus capacity plot on logarithmic paper. It should be noted that a lake type does not have to be on the plains nor does a gorge type have to be in the mountains. Occasionally, the type of reservoir operation or sediment size may override the classification due to shape. If a type III shaped reservoir is to be drawn down severely at frequent intervals or the sediment is predominantly clay, it should be classified as a type IV because a greater portion of the sediment is deposited near the bottom of a type IV reservoir. The same rationalization should be used when a reservoir falls on a boundary line between types.

The basic equation used to develop the procedure is:

$$S=\int_{o}^{y_o} A\,dy+\int_{y_o}^{H} K\,a\,dy \qquad (8)$$

where:

S=total sediment to be deposited in the reservoir
o=the original zero elevation at the dam
y_o=the zero elevation at the dam after the sediment inflow period
A=reservoir surface area
dy=incremental depth
H=total depth of the reservoir at normal water surface
K=a constant of proportionality for converting relative sediment areas to actual areas for a given reservoir
a=relative sediment area.

By integration and simplification of this equation, the following relationship can be developed:

$$\frac{1-v_o}{a_o}=\frac{S-V_o}{H\,A_o} \qquad (9)$$

where:

v_o=relative reservoir volume at new zero depth
a_o=relative reservoir area at new zero depth
V_o=total reservoir volume at new zero depth
H=original depth of reservoir
A_o=total reservoir area at zero depth.

Then by defining two new terms:

$$h_p=\frac{1-v_p}{a_p} \qquad (10)$$

and

$$h'_p=\frac{S-V_{pH}}{H\,A_{pH}} \qquad (11)$$

where:

p=relative depth, i.e., any fractional part of the reservoir depth measured from the streambed
V_{pH}=total reservoir volume at depth $_{pH}$
A_{pH}=total reservoir area at depth $_{pH}$.

it can be seen from equation 9 that h_p is equal to h'_p at the new zero elevation, y_o.

Using observed data from the reservoir surveys, dimensionless storage design curves were drawn for each of the four reservoir types and area design curves were derived from them. The storage design curves are shown in figure H–7 and the area design curves are plotted in figure H–8. Figure H–9 contains plotted curves of the full range of h_p values for all four reservoir types as developed from the storage and area design curves. A plot of h'_p can be superimposed on figure H–9 and the p value at which it crosses the proper type curve will give the relative depth, p_o, for the new zero elevation. Elevation y_o can then be computed by adding the product p_oH to the original streambed elevation.

The following example is an application of the Empirical Area-Reduction Method for Webster Reservoir, an existing Bureau of Reclamation reservoir located on the South Fork Solomon River in northcentral Kansas. The reservoir has a total capacity of 67,100 acre-feet at elevation 1889.6, the top of the active conservation pool. Table H–8 gives the pertinent area-capacity data necessary for completion of the sediment distribution study. A 50-year sediment accumulation of 9,300 acre-feet is estimated for this distribution computaton.

Figure H–10 is a logarithmic plot of the depth-capacity relationship for Webster Reservoir, from which the reservoir was assigned a type classification. Although the upper portion of the reservoir falls slightly above the type II slope limit, the storage design curve indicates approximately 70 percent of the sedi-

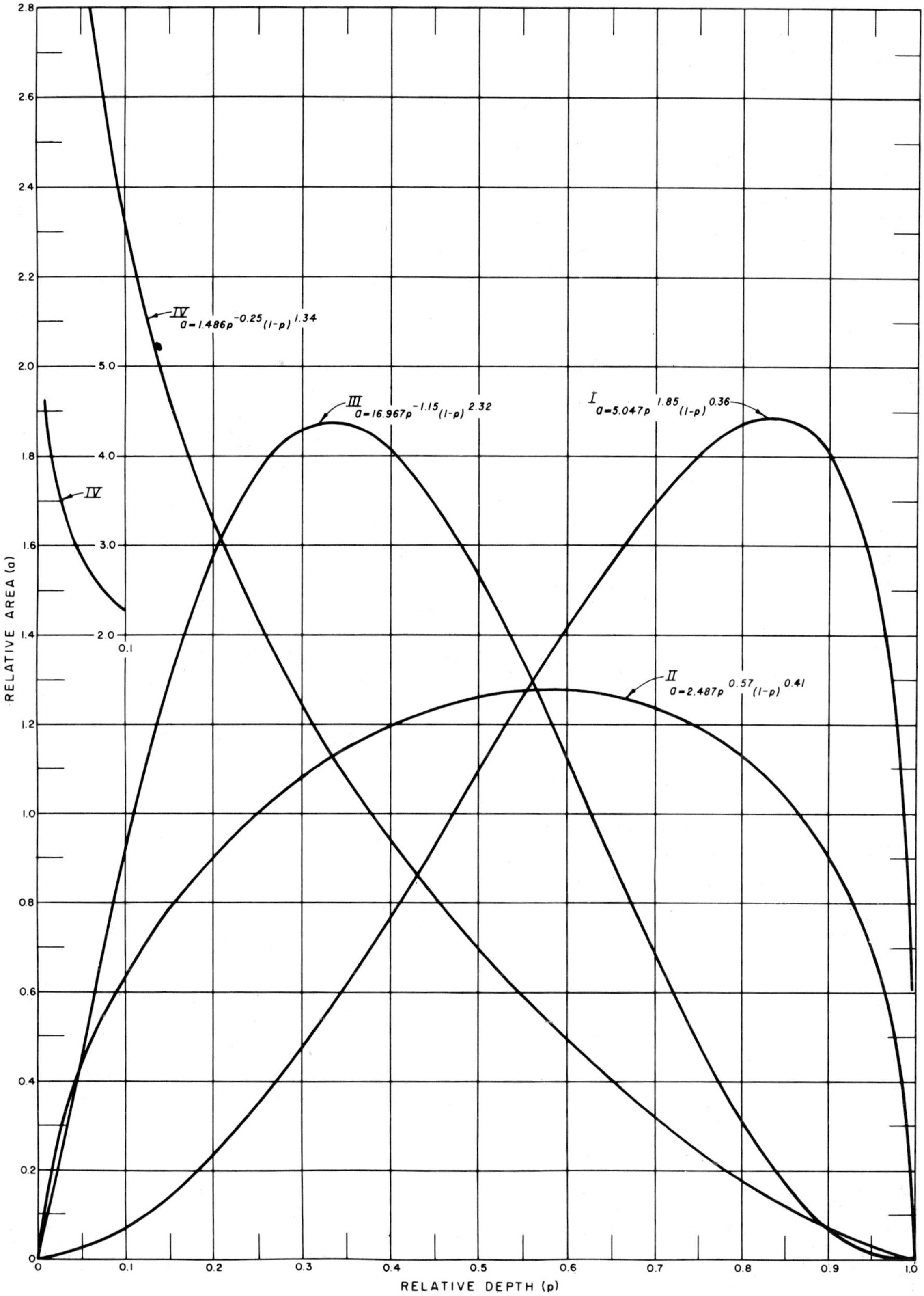

Figure H–8. Reservoir area design curves. 288–D–2849.

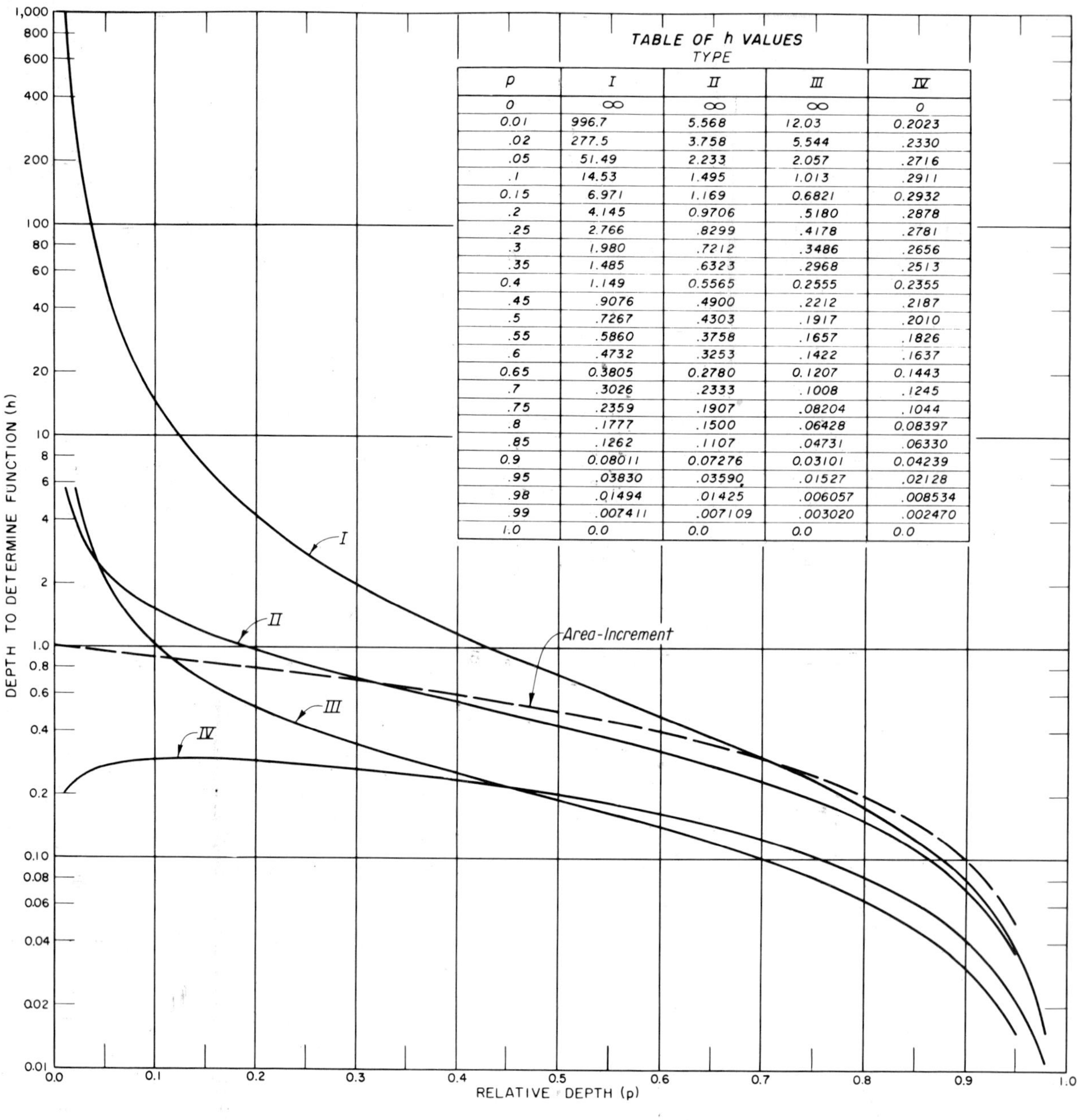

TABLE OF h VALUES

p	TYPE I	TYPE II	TYPE III	TYPE IV
0	∞	∞	∞	0
0.01	996.7	5.568	12.03	0.2023
.02	277.5	3.758	5.544	.2330
.05	51.49	2.233	2.057	.2716
.1	14.53	1.495	1.013	.2911
0.15	6.971	1.169	0.6821	0.2932
.2	4.145	0.9706	.5180	.2878
.25	2.766	.8299	.4178	.2781
.3	1.980	.7212	.3486	.2656
.35	1.485	.6323	.2968	.2513
0.4	1.149	0.5565	0.2555	0.2355
.45	.9076	.4900	.2212	.2187
.5	.7267	.4303	.1917	.2010
.55	.5860	.3758	.1657	.1826
.6	.4732	.3253	.1422	.1637
0.65	0.3805	0.2780	0.1207	0.1443
.7	.3026	.2333	.1008	.1245
.75	.2359	.1907	.08204	.1044
.8	.1777	.1500	.06428	0.08397
.85	.1262	.1107	.04731	.06330
0.9	0.08011	0.07276	0.03101	0.04239
.95	.03830	.03590	.01527	.02128
.98	.01494	.01425	.006057	.008534
.99	.007411	.007109	.003020	.002470
1.0	0.0	0.0	0.0	0.0

Figure H–9. **Curves of h_p for determining depth of sediment deposits at a dam. 288–D–2850.**

ment would deposit in the portion of the reservoir that is classified as type II. Therefore, a type II distribution was assumed to apply to the entire reservoir.

First, the 50-year zero elevation at the dam was computed, as shown in table H–9 and figure H–11. Values of h'_p were computed for four elevations and plotted in figure H–11. A line drawn through these points intersects the type II curve at p_o=0.237. Then $y_o = p_o H =$ 0.237 (50.6) = 12.0 feet and the 50-year zero elevation is 1839.0+12.0=1851.0.

TABLE H-8.—*Area and capacity data for Webster Reservoir*

Elevation	Area, acres	Capacity, acre-feet
1839	0	0
1845	48	80
1848	118	331
1850	154	603
1851	197	779
1852	239	996
1854	340	1,575
1855.5*	473	2,184
1860	906	5,300
1865	1,270	10,700
1870	1,678	18,100
1875	2,060	27,420
1880	2,487	38,790
1885	2,960	52,380
1889.6**	3,445	67,100

*Top of dead storage pool.
**Top of active storage pool.

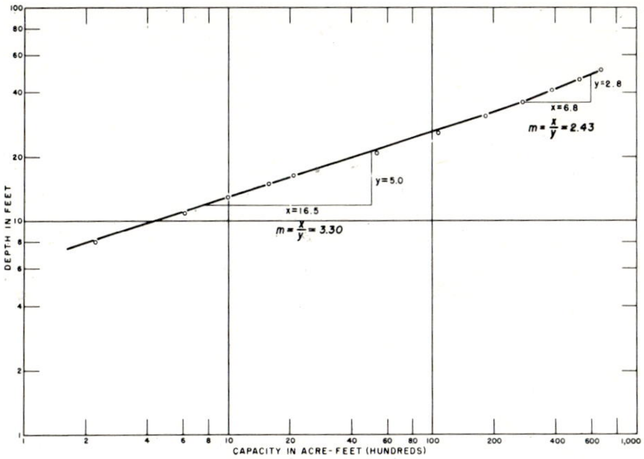

Figure H–10. **Depth versus capacity—Webster Reservoir. 288–D–2851.**

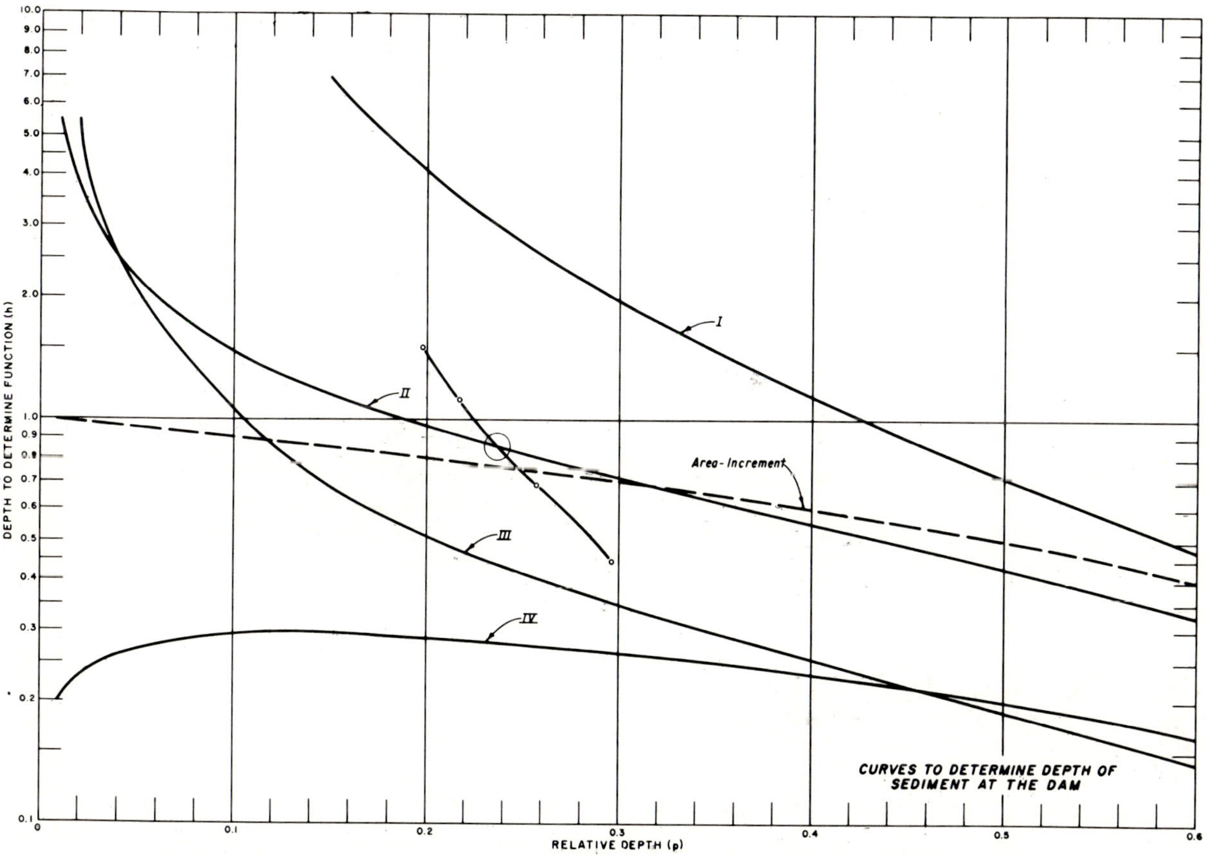

Figure H–11. **Depth of sediment—Webster Dam. 288–D–2852.**

TABLE H-9.—*Determination of elevation of sediment at Webster Dam.* 288-D-2863.

DIRECT DETERMINATION OF ELEVATION OF SEDIMENT DEPOSITED AT THE DAM

(Empirical Area-Reduction Method)

Reservoir *Webster* Project *Solomon Div. - M.R.B.P.*

S= *9,300 A.F.* H = *50.6 FT.*

① ELEV. (ft.)	② p	③ V (pH)	④ S-V(pH)	⑤ HA(pH)	⑥ h′(p)
1848	0.178	331	8,969	5,971	1.502
1850	0.217	603	8,697	7,792	1.116
1852	0.257	996	8,304	12,093	0.687
1854	0.296	1,575	7,725	17,204	0.449

p_o = *0.237*

$p_o H$ = *12.0*

Bottom elevation = *1839.0*

Elevation of sediment deposited at dam = *1851.0*

NOTATION OF SYMBOLS

p = relative depth of reservoir.

V (pH) = reservoir capacity in acre-feet at a given elevation.

S = total sediment inflow in acre-feet.

H = height of dam in feet.

A (pH) = reservoir area in acres at a given elevation.

h′(p) = a function of the reservoir and its anticipated sediment storage expressed as follows:

$$h'(p) = \frac{S - V(pH)}{HA(pH)}$$

The following steps were then used to complete the sediment distribution shown in table H–10.

1. Complete columns 1, 2, and 3 with area-capacity data from table H–8.
2. Compute p values in column 4 by dividing the original depth represented by elevation in column 1 by H.
3. Read relative area values from the Type II curve of figure H–8 and place them in column 5.
4. Compute K as shown in the supplement by dividing the area at the new zero elevation by the relative area at that elevation.
5. Complete column 6 by multiplying each of the values in column 5 by the K value.
6. Compute sediment volumes in column 7 with the average end area method using areas from column 6; i.e.,
$$\frac{(A_1+A_2)\ \Delta \text{ elevation}}{2}.$$
7. Accumulate values in column 7 to complete column 8. If the accumulated total sediment volume does not agree with the value of S, compute a new K as follows:
$$K_2=K_1\left(\frac{S}{S_1}\right)$$
and repeat steps 5 through 7.
8. Compute column 9 as the difference between column 2 and column 6.
9. Complete column 10 by finding the difference between column 3 and column 8.

New area and capacity curves can be drawn from the data in columns 9 and 10 of table H–10.

Occasionally a reservoir will be found in which the h'_p curve does not intersect the h_p curve for the proper type reservoir. The Area-Increment Method is then sometimes employed to estimate the new zero depth at the dam. The basic equation for this method is

$$S=A_o(H-y_o)+V_o \tag{12}$$

in which all terms are as previously defined for the Empirical Area-Reduction Method. An examination of this equation shows the basic assumption to be that the area of sediment deposition remains constant throughout the reservoir depth. A curve of h_p values (dashed curve) for the area-increment assumption is also plotted on figure H–9.

Regardless of the method used for distributing sediment within a reservoir, the Bureau of Reclamation seldom recommends placing the sill of the outlet works less than 10 feet above the streambed elevation. There is always a danger of plugging the inlet with trash and debris at the time the reservoir is initially filled if the sill is placed any closer to the streambed.

(d) *Delta Deposits.*—Another phenomenon of reservoir sediment deposition is the formation of delta deposits in the headwater areas of the reservoir. The major consequence of these delta deposits is the raising of the backwater elevations in the channel upstream from the reservoir. Therefore, the delta may cause a flood potential that would not be anticipated from pre-project channel conditions and proposed reservoir operating water surfaces. Figure H–12 is a definition sketch of a reservoir and delta formation.

The prediction of delta formation is as yet an empirical procedure based upon observed data deposits in reservoirs that have been resurveyed. The topset slope may be computed using either the Meyer-Peter, Muller formula for beginning transport:

$$S=\frac{0.19\ \frac{Q}{Q_B}\left(\frac{n_s}{D_{90}{}^{1/6}}\right)^{3/2}D}{d} \tag{13}$$

where all terms are as defined for equation 2, or by the Schoklitsch formula for no bedload transport:

$$S=\left(\frac{0.00021DB}{Q}\right)^{3/4} \tag{14}$$

where:

D=mean bed material particle size in mm., D_{50}

Q=discharge in c.f.s.

The discharge to be used in these equations is referred to as the *dominant discharge* and is usually determined by either the channel bankfull flow or as the mean annual flood. These equations will yield a slope at which the bed material will no longer be transported, which must necessarily be true for the delta to form.

TABLE H-10.—*Sediment distribution in Webster Reservoir.* 288-D-2864

HD-67
(6-62)

SEDIMENT DISPOSITION COMPUTATIONS

(Empirical Area-Reduction Method)

Reservoir *Webster* Project *Solomon Div.-M.R.B.P.*

Total Sediment Inflow *9,300* acre feet Computed by *R.I.S.* Date *9-24-68*

① Elevation (feet)	② Original Area (acres)	③ Original Capacity (acre-feet)	④ Relative Depth	⑤ A_p Type *II*	⑥ Sediment Area (acres)	⑦ Sediment Volume (acre-feet)	⑧ Accumulated Sediment Volume (acre-feet)	⑨ Revised Area (acres)	⑩ Revised Capacity (acre-feet)
1889.6	3,445	67,100	1.0	0.0	0		9,300	3,445	57,800
						440			
1885.0	2,960	52,380	0.909	0.875	179		8,860	2,781	43,520
						1,015			
1880.0	2,487	38,790	0.810	1.113	227		7,845	2,260	30,945
						1,195			
1875.0	2,060	27,420	0.711	1.23	251		6,650	1,809	20,770
						1,280			
1870.0	1,678	18,100	0.613	1.277	261		5,370	1,417	12,730
						1,300			
1865.0	1,270	10,700	0.514	1.267	258		4,070	1,012	6,630
						1,265			
1860.0	906	5,300	0.415	1.21	247		2,805	659	2,495
						1,070			
1855.5	473	2,184	0.326	1.117	228		1,735	245	449
						956			
1851.0	197	779	0.237	0.975	197		779	0	0
						699			
1845.0	48	80	0.119		48		80	0	
						80			
1839.0	0	0	0		0		0	0	

SUPPLEMENT

1851.0	197		0.237	0.975	$K_1 = 197 \div 0.975 = 202$

$K_2 = 202\ (9{,}300 / 9210)$

GPO 856918

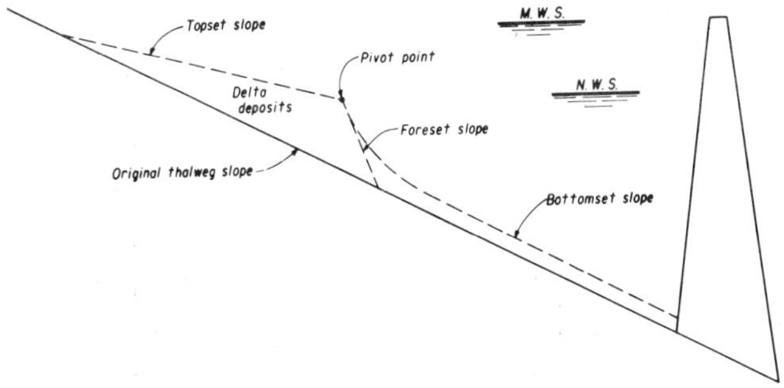

Figure H–12. Delta definition sketch. 288–D–2853.

It has also been found in most reservoirs that the topset slope will closely approximate one-half the original channel slope. This value may serve as a verification of the slope computed by equations 13 and 14.

The location of the pivot point between the topset and foreset slopes depends primarily on the operation of the reservoir and the existing channel slope in the delta area. If the reservoir is operated near the top of the conservation pool a large portion of the time, the elevation of the top of the conservation pool will be the pivot point elevation. If the original channel is meandering and has frequent overbank flows, the top of the conservation pool also sets the pivot point elevation. Conversely, if the reservoir water surface has frequent fluctuations and a deeply entrenched inflow channel, a mean operating pool elevation should be used to establish the pivot point.

The upstream end of the delta is set at the intersection of the maximum water surface and the original streambed, and the topset slope is projected from that point to the anticipated pivot point elevation to begin the first trial.

The average of foreset slopes observed in Bureau of Reclamation reservoir resurveys is 6.5 times the topset slope. However, some reservoirs exhibit a foreset slope considerably greater than this; i.e., Lake Mead foreset slope is 100 times the topset. By adopting a foreset slope of 6.5 times the topset, the first trial delta fit can be completed.

The volume of sediment computed from the channel cross sections with the delta imposed on them should agree with the volume of sand size or larger material anticipated to come from the delta stream. The quantity of sediment in the delta above normal water surface elevation should also agree with that estimated to deposit above the normal operating level. If the adjustment necessary to attain agreement is minor, it can usually be accomplished by a small change in the foreset slope.

If a significant change in delta size is necessary, it will be necessary to move the pivot point forward or backward in the reservoir, while maintaining the previously determined elevation of the point. The topset slope is then projected backward from the new pivot point location and the delta volume is again computed. There should be no cause for alarm if the intersection of the delta topset and the original streambed falls above the maximum water surface elevation as this condition has often been observed in small reservoirs.

B. DOWNSTREAM CHANNEL EFFECTS

H-4. *Channel Degradation*.—A natural flowing stream is usually in a state of quasi-equilibrium; that is, there is no long-term trend toward aggradation or degradation. A stream in equilibrium is said to be a stream in regime [22]. The state of stream equilibrium may be expressed qualitatively by the following equation:

$$Q_s d = k Q_w S \qquad (15)$$

where:

Q_s=bed material discharge
d=sediment particle diameter
Q_w=water discharge
S=slope of the stream
k=constant of proportionality.

If any one of the four variables is altered, one or more of the other variables must undergo change to return the stream to a state of equilibrium. In the case of a storage reservoir, the sediment load is eliminated or at least greatly decreased and a decreased slope downstream of the dam can be anticipated and the sediment particles remaining in the streambed will be the coarser fractions of the original material. This process of removing sediment particles from the streambed and banks is referred to as degradation.

There are two distinct approaches for estimating the depth or amount of degradation that will occur downstream from a dam or similar structure, each dependent on the type of material forming the bed of the river channel [23].

In cases where the streambed is composed of transportable material and the material extends to depths greater than that to which the channel can be expected to degrade, the approach most useful is that of computing the stable channel slope, the volume of expected degradation, and then determining a 3-slope channel profile which fits these values. However, if large size or coarse material, which cannot be transported by normal river discharges, exists in sufficient quantities an armor layer will develop as the finer material is sorted out and transported downstream. Vertical degradation will proceed at a progressively slower rate until the armor is of sufficient depth to inhibit further degradation.

(a) *Dominant Discharge.*—The dominant discharge is defined as the discharge which, if allowed to flow constantly, would have the same overall channel shaping effect as the natural fluctuating discharges would. The dominant discharge used in channel stabilization work is usually considered to be either the bank-full discharge or that peak discharge having a recurrence interval of approximately 2 years on an uncontrolled stream. With regulation of the streamflow by an upstream dam, the problem becomes more complex because detailed data on future releases are usually not available. If the releases from the reservoir are fairly uniform and flood discharges are a relatively rare occurrence, the average daily discharge may be used as the dominant discharge. However, if the releases are subject to considerable fluctuation due to floods, the peak discharge which is equalled or exceeded on the average of once every two years would be considered the dominant discharge.

(b) *Channel Hydraulic Properties.*—The next step in degradation computations is determination of the average channel hydraulic properties for the dominant discharge. These data can usually be obtained from the tailwater study that has been prepared for the dam. The properties of all the tailwater cross sections when carrying the dominant discharge are averaged to arrive at a generalized cross section which will be representative of the degradation reach. The water surface slope may be assumed equal to the hydraulic gradient for the computations requiring a slope.

(c) *Degradation Limited by Armoring.*—In the case where sufficient coarse material exists in the channel bed downstream from a dam to inhibit degradation after some time period, the formation of an armoring layer will control the depth of degradation. The formation of an armoring layer can usually be anticipated to control degradation when approximately 10 percent or more of the bed material is of armoring size or larger.

There are several ways to compute the size of bed material required for armoring and each method is regarded as a check on the others. The armoring size computed by each method will be different and some judgment may be required in selecting the proper size.

Several laboratory investigations have shown that the size of a particle plucked from a streambed is proportional to the velocity of flow near the bed. The velocity at which the particle starts to move is referred to as the competent bottom velocity [24]. The competent bottom velocity has been found to be approximately equal to 0.7 times V_m, the mean channel velocity. Figure H–13 is a plot of competent bottom velocity, V_b, versus the size of a moveable sediment particle which has been used for determining the armoring size.

Tractive force is the drag or shear acting on the wetted area of the channel bed and can be expressed as:

$$T.F.=\gamma dS \qquad (16)$$

where

$T.F.$=tractive force in lb./ft.^{2}
γ=unit weight of water (62.4 lb./ft.^{3})
d=mean water depth in feet
S=stream gradient in ft./ft.

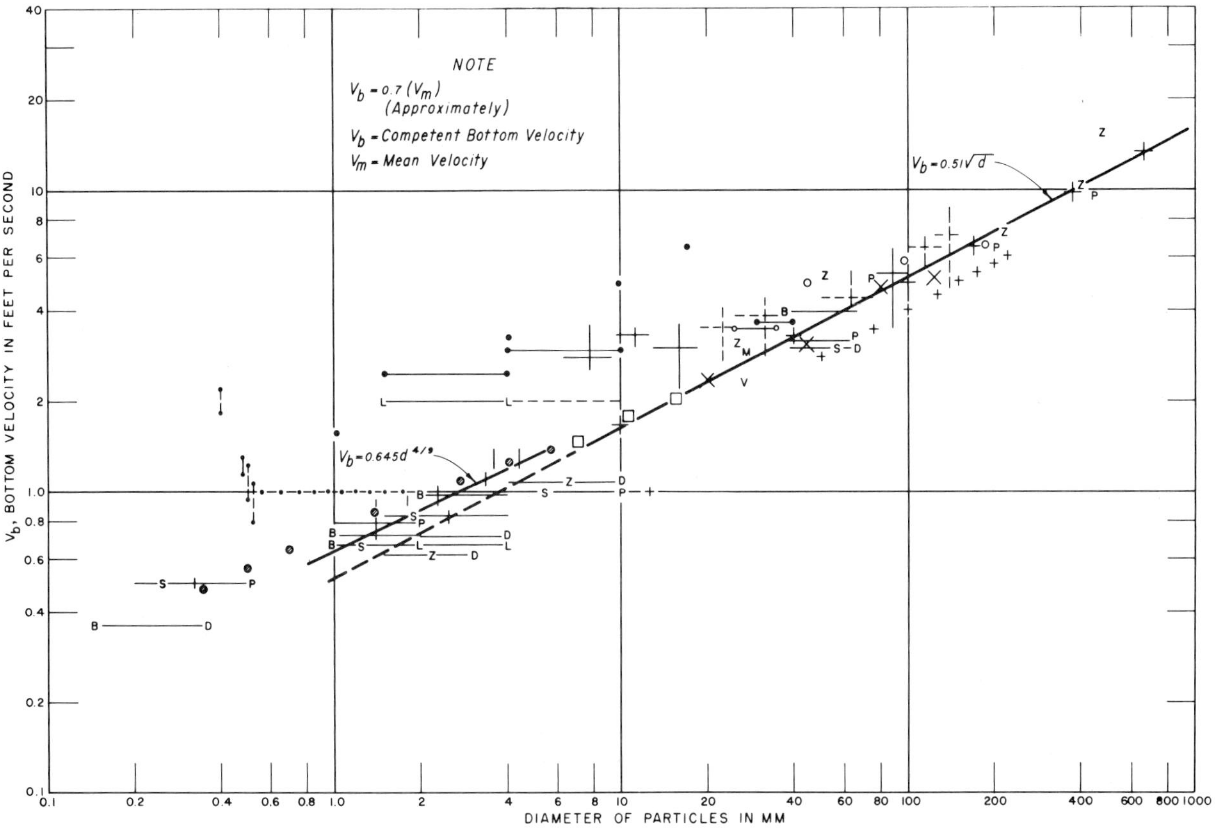

Figure H–13. Competent bottom velocity versus transportable sediment size. 288–D–2854.

When the tractive force has been computed for the dominant discharge, the curves of tractive force shown in figure H–14 may be used to determine the mean diameter of the armoring size of the bed material.

By rearranging the bedload equations (equations 13 and 14) to the following forms

$$D=\frac{5.26\ Sd}{(n_{s/D_{90}}{}^{1/6})^{3/2}} \qquad (17)$$

(Q/Q_B is assumed to be unity)

and

$$D=\frac{4762\ S^{4/3}\ Q}{B} \qquad (18)$$

the minimum transportable or armoring size material can be computed.

Figure H–15 is a definition sketch showing the way in which the degradation can be computed when an armoring layer is formed. From the sketch it can be seen that:

$$y_a = y - y_d \qquad (19)$$

where:

y_a=thickness of armoring layer

y=depth from original streambed to bottom of the armoring layer

y_d=depth from original streambed to top of armoring layer or the depth of degradation

and by definition:

$$y_a = (\Delta p)y \qquad (20)$$

where Δp=decimal percentage of material larger than the armoring size.

Then by combining the two equations, the following equation is developed:

$$y_d = y_a\left(\frac{1}{\Delta p}-1\right) \qquad (21)$$

The depth, y_a, necessary to armor the bed varies with the size particle needed to form the armoring layer. A rough guide of 3 armoring particle diameters or 0.5 foot, whichever is

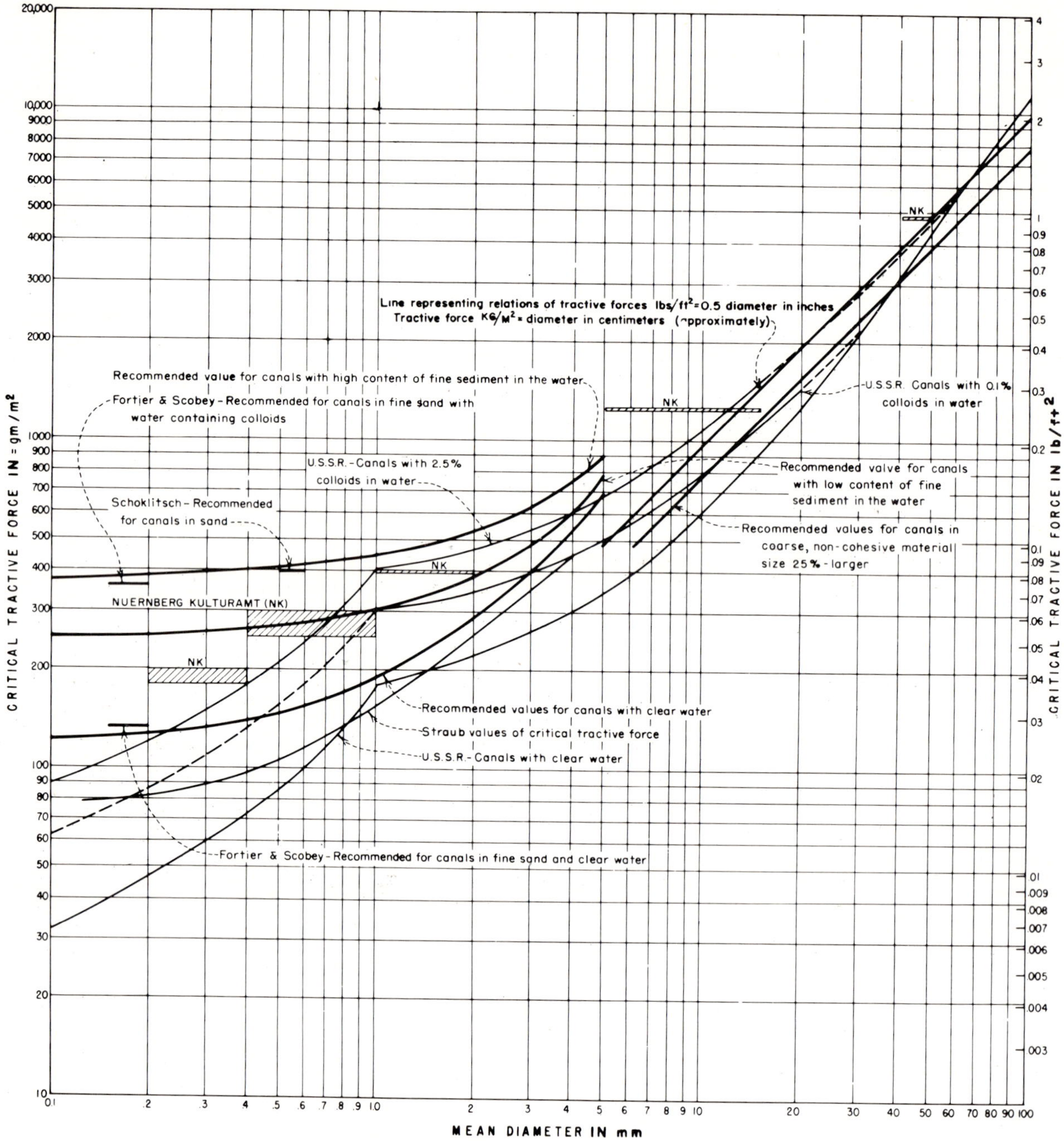

Figure H–14. Tractive force versus transportable sediment size. 288–D–2855.

smaller, has generally been used as the depth of y_a.

As an example of degradation limited by armoring, the following data are known for the channel downstream of a storage dam:

Dominant discharge	$=Q$	$=500$ cu. ft./sec.
Mean channel velocity	$=V_m$	$=3.4$ ft./sec.
Channel width	$=B$	$=60$ ft.
Mean channel depth	$=d$	$=4.0$ ft.
Stream gradient	$=S$	$=0.0021$ ft./ft.

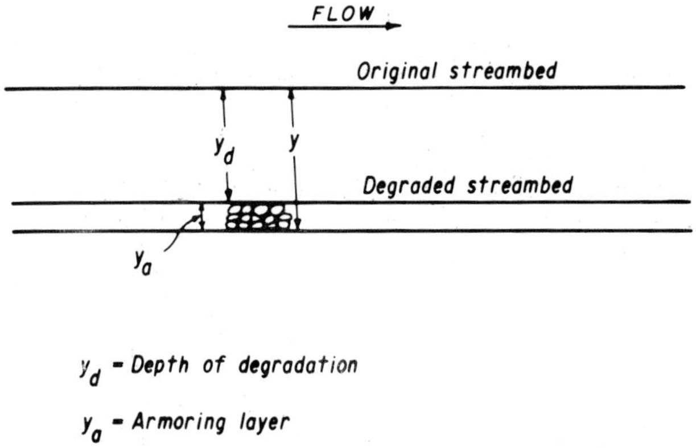

Figure H–15. Armoring definition sketch. 288–D–2856.

The size analysis of streambed material shown in figure H–16 is also available. Computed armoring sizes, D_a, are:

(1) $V_b = 0.7(3.4) = 2.4$
From figure H-13 $D_a = 22$ mm.

(2) $T.F. = 62.4(4.0)(0.0021) = 0.52$ lb./ft.^{2}
From figure H-14 $D_a = 31$ mm.

(3) $D_a = \dfrac{5.26(0.0021)(4.0)}{(0.03/(34)^{1/6})^{3/2}} = 20$ mm.

(4) $D_a = \dfrac{4762(0.0021)^{4/3}\,500}{60} = 11$ mm.

A mean of the above armoring sizes is 21 mm., which was adopted as a representative armoring size. By use of equation 21 and figure H–16, the depth of degradation is:

$$y_a = 3\,D_a = 3(21) = 63 \text{ mm. or } 0.206 \text{ ft.}$$

$$y_d = 0.206\left(\frac{1}{0.19} - 1\right) = 0.9 \text{ ft.}$$

(d) *Degradation Limited by a Stable Slope.*—This method is based on the limiting or stable slope which will result in no transport of the material forming the bed of the stream channel, and is used where there is not enough coarse material available to form an armor on the channel bed. The stable slope can be computed for a river reach using equations 13 and 14.

In addition to the limiting or stable slope of the degraded channel, it is also necessary to determine the volume of material that can be expected to be removed from the channel. If there are no downstream channel controls that could limit degradation, it is sometimes assumed that the river will pick up a load of coarse sediment ($>$0.0625 mm.) equal to that portion of the historic load in the same size range. This necessarily assumes little depletion in quantity of streamflow and minor regulation by the upstream reservoir. If the streamflow is significantly depleted or regulated, the sediment load to be picked up will be less than the historic load and appropriate adjustments in the load to be picked up will have to be made.

Having determined the stable slope and the volume of material that can be expected to be removed, the depth of degradation at the dam and the degraded channel profile can be estimated if the following assumptions can be reasonably well satisfied:

(1) The degradation reach is sufficiently uniform to permit the use of average cross sections and slope throughout the reach.
(2) The bed and bank material throughout the reach is similar enough that an average composition can be used and that there are no existing nonerodible barriers in the bed or banks to prevent the stream from attaining the average stable section at the stable slope.
(3) The degradation will be such that the vertical component will predominate and horizontal movement will be limited to bank sloughing resulting from the vertical degradation.

Experience has shown that a degraded stream profile can be represented by a typical three-slope arrangement, as shown in figure H–17 [23]. The derivation of the formulae for the longitudinal area, A_T, and the degradation reach length, L, are included in the figure. The previously determined volume may be expresed by:

$$V = \frac{A_T B}{43{,}560} \qquad (22)$$

where
V=volume of material to be degraded in acre-feet
A_T=longitudinal area of degradation in square feet
B=degraded channel width in feet.

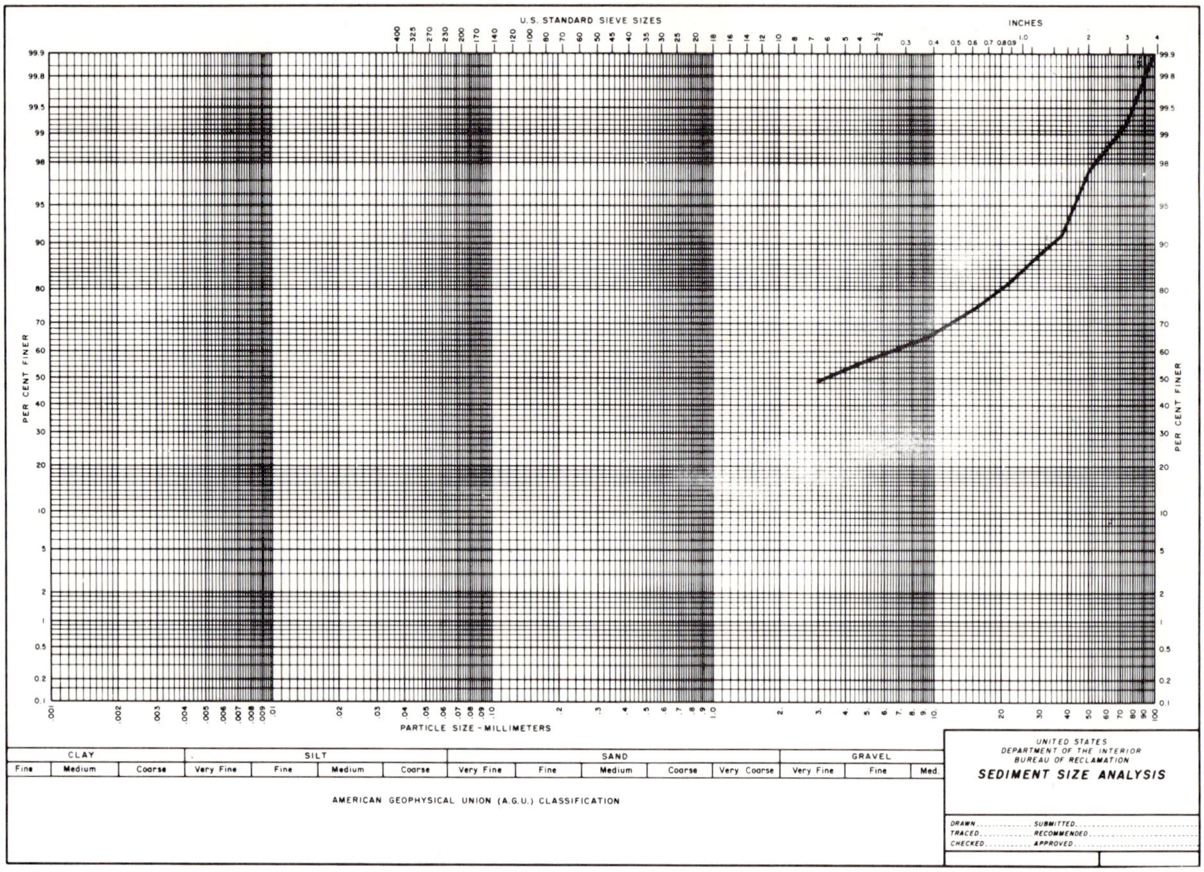

Figure H–16. Bed material size analysis. 288–D–2857.

By rearranging equation 22 the longitudinal area may be found:

$$A_T=\frac{43{,}560\ V}{B} \tag{23}$$

Once a value has been found for A_T, the depth of degradation may be computed using the following equation:

$$D=\left(\frac{64\ A_T \Delta S}{39}\right)^{1/2} \tag{24}$$

where ΔS=the difference between the existing slope and the stable slope and the length of the degraded reach can be computed by:

$$L=\frac{13\ D}{8\Delta S} \tag{25}$$

If it is anticipated that lateral degradation will be a significant factor, additional study will be necessary to determine the degraded channel width. Because part of the material will be coming from the stream banks, the extent of vertical degradation will ordinarily not be as great. Lateral movement should always be suspect when the banks are composed of the same material as the bed and there is not a great deal of vegetation to hold them. References [25] and [26] are recommended as guides when these conditions prevail.

If a permanent control exists at some point within the degradation reach, equation 25 may be used to solve for the depth of degradation directly.

As an example of degradation computations when the stable slope controls, the following

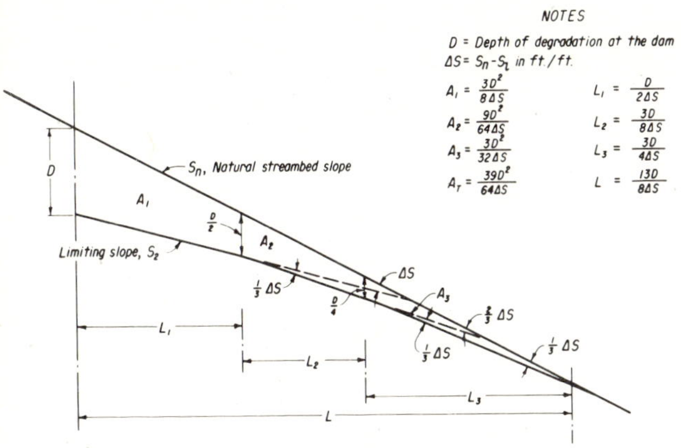

Figure H–17. Degraded channel by the three-slope method. 288–D–2858.

data are known about a river channel below a diversion dam:

Dominant discharge	$=Q=780$ c.f.s.
Channel width	$=B=350$ ft.
Mean channel depth	$=d=1.05$ ft.
Existing stream gradient	$=S_n=0.0014$
Bed material—$D_m=D_{50}$	$=0.3$ mm.
D_{90}	$=0.96$ mm.
Manning's n value	$=0.027$.

It was computed that 2,160 acre-feet of sand would deposit behind the diversion dam during the 100-year economic life of the structure, and that an equal volume of sand would be eroded from the downstream channel.

The stable slope computations are:

$$S_1=\left(\frac{0.00021\,(0.3)\,(350)}{780}\right)^{3/4}=0.000387$$

and

$$S_1=\frac{0.19[0.027/(0.96)^{1/6}]^{3/2}\,0.3}{1.05}=0.000244$$

(Q/Q_B is assumed equal to unity)

A mean of the two computed slopes is 0.000315 and it is assumed to be the stable slope. The difference between the existing and degraded slope, ΔS, is 0.001085.

The longitudinal degradation area is:

$$A_T=\frac{43{,}560\,(2{,}160)}{350}=268{,}830\text{ ft.}^2,$$

the depth of degradation at the dam is:

$$D=\left(\frac{64\,(268{,}830)\,(0.001085)}{39}\right)^{1/2}=21.9\text{ ft.},$$

and the length of the degradation reach is:

$$L=\frac{13\,(21.9)}{8\,(0.001085)}=32{,}800\text{ ft.}$$

C. BIBLIOGRAPHY

H-5. *Bibliography.*

[1] Chow, Ven Te, editor-in-chief, "Handbook of Applied Hydrology," McGraw-Hill, New York, N.Y., 1964.

[2] "Factors Affecting Sediment Yield and Measures for the Reduction of Erosion and Sediment Yield," Pacific Southwest Inter-Agency Committee, October 1968.

[3] "Summary of Reservoir Sediment Deposition Surveys Made in the United States through 1965," Miscellaneous Publication No. 1143, U.S. Department of Agriculture, Agriculture Research Service, May 1969.

[4] "A Study of Methods Used in Measurement and Analysis of Sediment Loads in Streams," Reports No. 1 through 14 and A through T, Subcommittee on Sedimentation—Inter-Agency Committee on Water Resources.

[5] Searcy, James K., "Flow-Duration Curves," U.S. Department of the Interior, Geological Survey Water Supply Paper 1542–A, reprinted 1963.

[6] Herbertson, John G., "A Critical Review of Conventional Bed Load Formulae," Journal of Hydrology, vol. VIII, No. 1, May 1969.

[7] Shulits, Samuel, "The Schoklitsch Bedload Formula," Engineering, vol. 139, 1935, pp. 644–646, 687.

[8] Sheppard, John R., "Investigation of Meyer-Peter, Muller Bedload Formulas," Bureau of Reclamation, June 1960.

[9] Einstein, Hans Albert, "The Bed-load Function for Sediment Transportation in Open Channel Flows," U.S. Department of Agriculture, Soil Conservation Service Technical Bulletin No. 1026, September 1950.

[10] Colby, B. R., and Hembree, C. H., "Computations of Total Sediment Discharge, Niobrara River near Cody, Nebraska," U.S. Department of the

Interior, Geological Survey Water Supply Paper 1357, 1955.

[11] "Step Method for Computing Total Sediment Load by the Modified Einstein Procedure," Bureau of Reclamation, Denver, Colo., July 1955.

[12] "Computation of 'Z's' for Use in the Modified Einstein Procedure," Bureau of Reclamation, Denver, Colo., June 1966.

[13] Lane, E. W., and Borland, W. M., "Estimating Bedload," American Geophysical Union Transactions, vol. 32, No. 1, 1951, pp. 121–123.

[14] Brune, Gunnar M., "Trap Efficiency of Reservoirs," Transactions of American Geophysical Union, vol. 34, No. 3, June 1953.

[15] Churchill, M. A., Discussion of "Analysis and Use of Reservoir Sedimentation Data," by L. C. Gottschalk, pp. 139–140, Proceedings of Federal Inter-Agency Sedimentation Conference, Denver, Colo., January 1948.

[16] Lara, J. M., and Pemberton, E. L., "Initial Unit Weight of Deposited Sediments," Proceedings of Federal Inter-Agency Sedimentation Conference–1963, Miscellaneous Publication No. 970, pp. 818–845, U.S. Department of Agriculture, Agriculture Research Service, June 1965.

[17] Lane, E. W., and Koelzer, V. A., "Density of Sediments Deposited in Reservoirs," A Study of Methods Used in Measurement and Analysis of Sediment Loads in Streams, Report No. 9, Inter-Agency Committee on Water Resources, 1943.

[18] Miller, Carl R., "Determination of the Unit Weight of Sediment for Use in Sediment Volume Computations," Bureau of Reclamation, Denver, Colo., 1953.

[19] Borland, W. M., and Miller, C. R., "Distribution of Sediment in Large Reservoirs," Transactions, American Society of Civil Engineers, vol. 125, pt. I, 1960, pp. 166–180.

[20] Lara, J. M., "Revision of the Procedure to Compute Sediment Distribution in Large Reservoirs," Bureau of Reclamation, Denver, Colo., May 1962.

[21] Meyer-Peter, E., and Muller, R., "Formulas for Bed-load Transport," The Second Meeting of the International Association for Hydraulic Structure Research, app. 2, June 1948, pp. 39–64.

[22] Lane, E. W., "The Importance of Fluvial Morphology in Hydraulic Engineering," Proceedings of American Society of Civil Engineers—Hydraulics Division, vol. 81, Paper No. 745, July 1955.

[23] "Guide for Computing Degradation," Bureau of Reclamation, Denver, Colo., June 1963.

[24] Mavis, F. T., and Laushey, L. M., "A Reappraisal of the Beginning of Bed-Movement-Competent Velocity," International Association for Hydraulic Research, Second Meeting, Stockholm, June 1948.

[25] "Stable Channel Profiles," Hydraulic Laboratory Report No. Hyd. 325, Bureau of Reclamation, Denver, Colo., September 1951.

[26] "Progress Report on Results of Studies on Design of Stable Channels," Hydraulic Laboratory Report Hyd-352, Bureau of Reclamation, Denver, Colo., June 1952.

Appendix I

Conversion Factors[1]

I-1. *English and Metric Conversion Factors.*—

LENGTH

To convert from	To	Multiply by
1. Inches	feet	0.083333
	angstrom units	2.54×10^{8}
	microns	25400
	millimeters	25.4
	centimeters	2.54
	meters	0.0254
2. Feet	inches	12.0
	angstrom units	3.048×10^{9}
	microns	304800
	millimeters	304.80
	centimeters	30.48
	meters	0.3048
3. Angstrom units	inches	3.9370079×10^{-9}
	feet	3.28084×10^{-10}
	microns	0.0001
	millimeters	1×10^{-7}
	centimeters	1×10^{-8}
	meters	1×10^{-10}
4. Microns	inches	3.9370079×10^{-5}
	feet	3.2808399×10^{-6}
	angstrom units	1×10^{4}
	millimeters	1×10^{-3}
	centimeters	1×10^{-4}
	meters	1×10^{-6}
5. Millimeters	inches	3.9370079×10^{-2}
	feet	3.2808399×10^{-3}
	angstrom units	1×10^{7}
	microns	1×10^{3}
	centimeters	1×10^{-1}
	meters	1×10^{-3}
6. Centimeters	inches	0.39370079
	feet	0.032808399
	angstrom units	1×10^{8}
	microns	1×10^{4}
	millimeters	10
	meters	1×10^{-2}
7. Meters	inches	39.370079
	feet	3.2808399
	angstrom units	1×10^{10}
	microns	1×10^{6}
	millimeters	1×10^{3}
	centimeters	1×10^{2}

AREA

To convert from	To	Multiply by
1. Square meters	square feet	10.763910
	square centimeters	1×10^{4}
	square inches	1550.0031
2. Square feet	square meters	9.290304×10^{-2}
	square centimeters	929.0304
	square inches	144
3. Square centimeters	square meters	1×10^{-4}
	square feet	1.0763910×10^{-3}
	square inches	0.15500031
4. Square inches	square meters	6.4516×10^{-4}
	square feet	6.9444×10^{-3}
	square centimeters	6.4516

VOLUME

To convert from	To	Multiply by
1. Cubic centimeters	cubic meters	1×10^{-6}
	cubic feet	3.5314667×10^{-5}
	cubic inches	0.061023744
	cubic yards	1.3079506×10^{-6}
2. Cubic meters	cubic feet	35.314667
	cubic centimeters	1×10^{6}
	cubic inches	61023.74
	cubic yards	1.3079506
3. Cubic inches	cubic meters	1.6387064×10^{-5}
	cubic feet	5.7870370×10^{-4}
	cubic centimeters	1.6387064×10^{-5}
	cubic yards	2.1433470×10^{-5}
4. Cubic feet	cubic meters	0.028316847
	cubic centimeters	28316.847
	cubic inches	1728

FORCE

To convert from	To	Multiply by
1. Pounds (avdp)	dynes	4.44822×10^{5}
	grams	453.59237
	kilograms	0.45359237
	tons (long)	4.464286×10^{-4}
	tons (short)	5×10^{-4}
	kips	1×10^{-3}
	tons (metric)	4.5359237×10^{-4}
2. Kips	pounds	1000
	tons (short)	0.500
	kilograms	453.59237
	tons (metric)	0.45359237
3. Tons (short)	kilograms	907.18474
	pounds	2000
	kips	2
	tons (metric)	0.907185
4. Kilograms	dynes	980665
	grams	1000
	pounds	2.2046226
	tons (long)	9.8420653×10^{-4}
	tons (short)	11.023113×10^{-4}
	kips	2.2046226×10^{-3}
	tons (metric)	0.001
5. Tons (metric)	grams	1×10^{6}
	kilograms	1000
	pounds	2204.6226
	kips	2.2046226
	tons (short)	1.1023113

[1] A more extensive list of conversion factors may be found in the "CRC Handbook of Chemistry and Physics," latest edition, Chemical Rubber Co.

I-1. *English and Metric Conversion Factors.*—

TIME

To convert from	To	Multiply by
1. Milliseconds	seconds	10^{-3}
	minutes	1.66666×10^{-5}
	hours	2.777777×10^{-7}
	days	1.1574074×10^{-8}
	months	3.8057×10^{-10}
	years	3.171416×10^{-11}
2. Seconds	milliseconds	1000
	minutes	1.66666×10^{-2}
	hours	2.777777×10^{-4}
	days	1.1574074×10^{-5}
	months	3.8057×10^{-7}
	years	3.171416×10^{-8}
3. Minutes	milliseconds	60000
	seconds	60
	hours	0.0166666
	days	6.944444×10^{-4}
	months	2.283104×10^{-5}
	years	1.902586×10^{-6}
4. Hours	milliseconds	3600000
	seconds	3600
	minutes	60
	days	0.0416666
	months	1.369860×10^{-3}
	years	1.14155×10^{-4}
5. Days	milliseconds	86400000
	seconds	86400
	minutes	1440
	hours	24
	months	3.28767×10^{-2}
	years	0.0027397260
6. Months	milliseconds	2.6283×10^{9}
	seconds	2.6283×10^{6}
	minutes	43800
	hours	730
	days	30.416666
	years	0.08333333
7. Years	milliseconds	3.1536×10^{10}
	seconds	3.1536×10^{7}
	minutes	525600
	hours (mean solar)	8760
	days (mean solar)	365
	months	12

STRESS

To convert from	To	Multiply by
1. Pounds/square foot	pounds/square inch	0.0069445
	feet of water	0.016018
	kips/square foot	1×10^{-3}
	kilograms/square centimeter	0.000488243
	tons/square meter	0.004882
	atmospheres	4.72541×10^{-4}
2. Pounds/square inch	pound/square foot	144
	feet of water	2.3066
	kips/square foot	0.144
	kilograms/square centimeter	0.070307
	tons/square meter	0.70307
	atmospheres	0.068046
3. Tons (short)/square foot	atmospheres	0.945082
	kilograms/square meter	9764.86
	tons (metric)/square meter	9.76487
	pounds/square inch	13.8888
	pounds/square foot	2000
	kips/square foot	2.0
4. Feet of water (at 39.2° F)	pounds/square inch	0.43352
	pounds/square foot	62.427
	kilograms/square centimeter	0.0304791
	tons/square meter	0.304791
	atmospheres	0.029499
	inches of Hg	0.88265
5. Kips/square foot	pounds/square inch	6.94445
	pounds/square foot	1000
	tons (short)/square foot	0.5000
	kilograms/square centimeter	0.488244
	tons (metric)/square meter	4.88244
6. Kilograms/square centimeter	pounds/square inch	14.223
	pounds/square foot	2048.1614
	feet of water (39.2°F.)	32.8093
	kips/square foot	2.0481614
	tons/square meter	10
	atmospheres	0.96784
7. Tons (metric)/square meter	kilograms/square centimeter	0.10
	pounds/square foot	204.81614
	kips/square foot	0.20481614
	tons (short)/square foot	0.102408

I-1. *English and Metric Conversion Factors.—*

To convert from	To	Multiply by	To convert from	To	Multiply by
UNIT WEIGHT					
1. Grams/cubic centimeter	tons (metric)/ cubic meter	1.00	4. Pounds/cubic inch	grams/cubic centimeter	27.679905
	kilograms/cubic meter	1000.00		tons (metric)/ cubic meter	27.679905
	pounds/cubic inch	0.036127292		kilograms/cubic meter	27679.905
	pounds/cubic foot	62.427961		pounds/cubic foot	1728
2. Tons (metric)/ cubic meter	grams/cubic centimeter	1.00	5. Pounds/cubic foot	grams/cubic centimeter	0.016018463
	kilograms/cubic meter	1000.00		tons (metric)/ cubic meter	0.016018463
	pounds/cubic inch	0.036127292		kilograms/cubic meter	16.018463
	pounds/cubic foot	62.427961		pounds/cubic inch	$5.78703704 \times 10^{-4}$
3. Kilograms/ cubic meter	grams/cubic centimeter	0.001			
	tons (metric)/ cubic meter	0.001			
	pounds/cubic inch	3.6127292×10^{-5}			
	pounds/cubic foot	0.062427961			
VELOCITY					
1. Centimeters/ second	microns/second	10000	3. Feet/minute	centimeters/ second	0.508001
	meters/minute	0.600		microns/second	5080.01
	feet/minute	1.9685		meters/minute	0.3048
	miles/hour	0.022369		miles/hour	0.01136363
	feet/year	1034643.6		feet/year	525600
2. Microns/ second	centimeters/second	0.0001	4. Feet/year	microns/second	0.009665164
	meters/minute	0.000060		centimeters/second	0.0000009665164
	feet/minute	0.00019685		meters/minute	5.79882×10^{-7}
	miles/hour	0.0000022369		feet/minute	1.9025×10^{-6}
	feet/year	103.46436		miles/hour	2.16203×10^{-8}

Index

☆ U.S. GOVERNMENT PRINTING OFFICE: 1973 O—444—004